MW00632922

Soil Mechanics

SERIES IN SOIL ENGINEERING

Edited by

T. William Lambe
Robert V. Whitman
Professors of Civil Engineering
Massachusetts Institute of Technology

BOOKS IN SERIES:

Soil Testing for Engineers by T. William Lambe, 1951
Soil Mechanics by T. William Lambe and Robert V. Whitman 1968
Soil Mechanics, SI Version by T. William Lambe and Robert V. Whitman, 1979
Fundamentals of Soil Behavior by James K. Mitchell

The aim of this series is to present the modern concepts of soil engineering, which is the science and technology of soils and their application to problems in civil engineering. The word "soil" is interpreted broadly to include all earth materials whose properties and behavior influence civil engineering construction.

Soil engineering is founded upon many basic disciplines: mechanics and dynamics; physical geology and engineering geology; clay mineralogy and colloidal chemistry; and mechanics of granular systems and fluid mechanics. Principles from these basis disciplines are backed by experimental evidence from laboratory and field investigations and from observations on actual structures. Judgment derived from experience and engineering economics are central to soil engineering.

The books in this series are intended primarily for use in university courses, at both the undergraduate and graduate levels. The editors also expect that all of the books will serve as valuable reference material for practicing engineers.

T. William Lambe and Robert V. Whitman

Soil Mechanics, SI Version

T. William Lambe • Robert V. Whitman

Massachusetts Institute of Technology

With the assistance of

H. G. Poulos

University of Sydney

John Wiley & Sons

New York Chichester Brisbane Toronto

Library of Congress Cataloging in Publication Data

Lambe, T. William
Soil mechanics, SI version.

(Series in soil engineering)
Bibliography: p. ii
Includes index.
1. Soil mechanics. I. Whitman, Robert V., 1928– joint author. II. Title.
TA710.L246 624'.1513 77-14210
ISBN 0-471-02491-0

Printed in the United States of America
10 9 8 7 6 5 4 3

Board of Advisors, Engineering

KARL TERZAGHI

Karl Terzaghi, born October 2, 1883 in Prague and died October 25, 1963 in Winchester, Massachusetts, is generally recognized as the Father of Soil Mechanics. His early professional life was spent in a search for a rational approach to earthwork engineering problems. His efforts were rewarded with the publication in 1925 of his famous book on soil mechanics: this publication is now credited as being the birth of soil mechanics.

Between 1925 and 1929, Terzaghi was at M.I.T. initiating the first U.S. program in soil mechanics and causing soil mechanics to be widely recognized as an important discipline in civil engineering. In 1938 he joined the faculty at Harvard University where he developed and gave his course in engineering geology.

Terzaghi's amazing career is well documented in the book *From Theory to Practice in Soil Mechanics* (Wiley, 1960). All of Terzaghi's publications through 1960 (256) are listed in this book. Terzaghi won many honors, including the Norman Medal of the American Society of Civil Engineers in 1930, 1943, 1946, and 1955. Terzaghi was given nine honorary doctorate degrees coming from universities in eight different countries. He served for many years as President of the International Society of Soil Mechanics and Foundation Engineering.

Not only did Terzaghi start soil mechanics, but he exerted a profound influence on it until his death. Two days before he died he was diligently working on a professional paper. Terzaghi's writings contain significant contributions on many topics, especially consolidation theory, foundation design and construction, cofferdam analysis, and landslide mechanisms. Probably Terzaghi's most important contribution to the profession was his approach to engineering problems, which he taught and demonstrated.

To commemorate Terzaghi's great work, the American Society of Civil Engineers created the Terzaghi Lecture and the Terzaghi Award.

PREFACE

Soil Mechanics is designed as a text for an introductory course in soil mechanics. An intensive effort was made to identify the truly fundamental and relevant principles of soil mechanics and to present them clearly and thoroughly. Many numerical examples and problems are included to illustrate these key principles. This text has been used successfully both in an introductory undergraduate course and in an introductory graduate course. Although *Soil Mechanics* has been written principally for the student, practicing engineers should find it valuable as a reference document.

The book is divided into five parts. Part I describes the nature of soil problems encountered in civil engineering and gives an overall preview of the behavior of soil. Part II describes the nature of soil, especially the transmission of stresses between soil particles. Part III is devoted primarily to dry soil since many aspects of soil behavior can best be understood by considering the interaction of soil particles without the presence of water. Part IV builds upon the principles given in Parts II and III to treat soils in which the pore water is either stationary or flowing under steady conditions. Part V considers the most complex situation in soil mechanics, that wherein pore pressures are influenced by applied loads and hence the pore water is flowing under transient conditions. This organization of the book permits the subject matter to be presented in sequential fashion, progressively building up to the more complex principles.

Parts III, IV, and V all have the same general format. First there are several chapters that set forth the basic principles of soil behavior. Then follow several chapters in which these principles are applied to the practical analysis and design of earth retaining structures, earth slopes, and shallow foundations. For example, chapters concerning shallow foundations appear in Part III, Part IV, and Part V. Special chapters on deep foundations and soil improvement appear at the end of Part V. These problem-oriented chapters illustrate the blending of theory, laboratory testing, and empirical evidence from past experience to provide practical but sound methods for analysis and design. *Soil Mechanics* does not attempt to cover all the details of these practical problems; numerous references are provided to guide the student in additional study.

Soil Mechanics deliberately includes far more material than can or should be covered in a single introductory course, thus making it possible for the instructor to choose the topics to be used to illustrate the basic principles. We have found that numerical examples of practical problems should be introduced very early in an introductory course in soil mechanics—preferably within the first eight periods. Thus we organize the early portion of our courses as follows:

1. Part I is covered in two lectures, giving students motivation for the study of soil mechanics and an understanding of the organization of the course.
2. Chapter 3 is covered in detail, but Chapters 4 to 7 are surveyed only hastily. As questions arise later in the course, reference is made to the material in these chapters.
3. Chapter 8 describes several basic methods for calculating and displaying the stresses; students must master these techniques. Chapters 9 to 12 contain certain key concepts concerning soil behavior plus descriptive matter and tables and charts of typical values. These chapters may be covered rapidly, stressing only the key concepts. Then the student is ready for an intensive study of retaining structures (Chapter 13) and shallow foundations (Chapter 14). Three or four periods may profitably be spent on each chapter. Chapter 15 serves to introduce the increasingly important problem of soil dynamics, and serves as supplementary reading for an introductory course.
4. In Chapter 16 the definition and manipulation of effective stress is emphasized and the mechanistic interpretation of effective stress serves as supplementary reading. The depth of study of

Chapters 17 to 19 will vary with the treatment given this topic in other courses such as fluid mechanics. Chapters 20 to 22 are largely descriptive and may be studied quickly with emphasis on the main features of soil behavior. This again leaves time for a detailed study of topics from Chapters 23 to 25. The choice of topics will depend on the interests of the instructor and the material to be covered in later courses. We attempt to cover only portions of two of the three chapters.

5. Chapters 26 and 27 contain key concepts and computational procedures and detailed coverage is needed for comprehension. Similarly, detailed coverage of Chapter 28 is required to give understanding of the key connection between drained and undrained behavior. Chapters 29 and 30 are largely descriptive and can be covered quickly leaving time for detailed study of selected topics from the remaining chapters.

The material not covered in an introductory course will serve to introduce students to more advanced courses and can be used as reference material for those courses.

As the reader will see, our book presents photographs of and biographical data on six pioneers in soil mechanics. These men have made significant contributions to soil mechanics knowledge and have had a major impact on soil mechanics students. There is a second generation of leaders whose works are having and will have an impact on soil mechanics. The extensive references in our book to the works of these people attest to this fact.

We thank the many authors and publishers for permission to reproduce tables and figures. The Council of the Institution of Civil Engineers granted permission to reproduce material from *Géotechnique*.

The early stages of the preparation of this text was supported in part by a grant made to the Massachusetts Institute of Technology by the Ford Foundation for the purpose of aiding in the improvement of engineering education. This support is gratefully acknowledged. We thank Professor Charles L. Miller, Head of the Department of Civil Engineering, for his encouragement in the undertaking of this book. We also acknowledge the contributions of our many colleagues at M.I.T. and the important role of our students who subjected several drafts to such careful scrutiny and criticism. Special recognition is due Professors Charles C. Ladd and Leslie G. Bromwell who offered comments on most of the text. Professor Bromwell contributed revisions to Part II and to Chapter 34. Professor John T. Christian helped considerably with theoretical portions of the text, as did Dr. Robert T. Martin on Part II and Dr. David D'Appolonia on Chapters 15 and 33. Professor James K. Mitchell of the University of California (Berkeley) contributed valuable comments regarding Part II, as did Professor Robert L. Schiffman of the University of Illinois (Chicago) regarding Chapter 27. Finally, we thank Miss Evelyn Perez and especially Mrs. Alice K. Viano for their indefatigable and meticulous typing of our many drafts.

T. William Lambe
Robert V. Whitman

MIT
1969

PREFACE TO SI VERSION

The original version of *Soil Mechanics* employed a variety of units, resulting from the many worldwide sources from which we drew the material for the book. In this new version, the aim has been to convert—insofar as practical—all quantities into the internationally standardized system of SI units. Otherwise the book remains unchanged.

SI units are now used as a matter of course for geotechnical engineering work in much of the world—including many English-speaking countries—and their use in the remaining areas is growing. It is quite appropriate that students commencing their study of soil mechanics should learn to think of the subject in terms of SI units, and that engineers accustomed to other units should begin to gain familiarity with this relatively new system. Actually, most geotechnical engineers have long been as familiar with the meter and the kilogram as with the foot and the pound. Hence, for most, the "new" system will prove different only in the careful distinction between force and mass and in the unit—the newton—used for force. Those using this book no doubt will continue to encounter mixed units for some time to come, and extensive conversion tables may be found in Appendix B.

When converting numbers appearing in the original text, rounded figures have been chosen wherever possible to preserve readability and ease of understanding. Thus quantities appearing in the original and new verson are not, in general, exactly equivalent. In some cases, where actual data are involved and accuracy must be maintained, much more exact conversions have been employed.

The International System of Units—SI for short—is an outgrowth of the metric system originated by French scholars in the early 1800s. In time, the need to standardize even within this one system became apparent, and several worldwide bodies were established for this purpose. The SI system arose from an extensive revision and simplification adopted in 1960. The practical use of this system has been made possible by the International Organization for Standardization (ISO)—a federation of national standards institutes—which has carried through the task of standardizing on symbols, preparing conversion tables, and so on. In the United States, conversion to SI units is supported by most professional organizations. The U.S. representative to ISO—the American National Standards Institute (ANSI)—voted for adoption of the system. In 1970, the American Society for Civil Engineers adopted the policy that all publications of the society should list measurements in SI units as well as in the customary units of the English system. A complete guide to the SI system has been published by the American Society for Testing and Materials (ASTM E-380).

In the SI system, the basic units of measure are *length*, *time*, and *mass* (plus four other units of measure required in electricity, thermodynamics, chemistry, and lighting). The basic unit of *length* is the *meter* (m), although it is also considered appropriate to express distances in millimeters (mm) as well. The basic unit of *time* is the *second* (s), although the use of hours and days and years is also permissible. The basic unit of *mass* (M) is the *kilogram* (kg). One thousand kilograms (1000 kg) is known as a *tonne* (t).

Mass expresses the quantity of matter in a body. The mass of a body is independent of the gravitational force. On the other hand, *weight* (W) is a measure of the gravitational force acting upon a mass at a particular location. Based upon Newton's law,

$$W = Mg$$

where g is the acceleration of gravity. Since g varies somewhat over the surface of the earth, the weight of an object will also vary slightly from place to place. However, this small variation is usually neglected, and the standard acceleration of gravity (approximately 9.806 m/s^2) is used to convert between mass and weight.

Weight is expressed in units of force, and the basic unit of force is the *newton* (N). One newton will accelerate a mass of 1 kg at 1 m/s^2. Thus, an object with a mass of 1000 kg has a weight (on earth) of 9810 N, or 9.81 kilonewtons (kN). However, is this object is placed upon a conventional scale, the scale will read 1000 kg, and we will say (somewhat sloppily) that the object *weighs* 1000 kg. In effect, such a scale is rigged to weigh mass—at least in a location where the acceleration of gravity is at its standard value.

The same distinction applies between mass density (ρ) and unit weight (γ). Mass density (units of M/L^3) is an inherent measure of the denseness of material, while unit weight is a function of the acceleration of gravity. Thus

$$\gamma = \rho g$$

The basic SI units for mass density are kilograms per cubic meters or tonnes per cubic meters. The mass density of water at standard conditions is 1000 kg/m^3 or 1 t/m^3. The unit weight of water at standard conditions is approximately 9806 N/m^3 or 9.806 kN/m^3. For most engineering purposes, the unit weight of water may be taken as 10 kN/m^3.

Stress and *pressure* have units of force per unit area. Therefore, for the basic unit of stress and pressure, we have newtons per square meter (*not* kg/m^2!). It is appropriate to use kilonewtons per square meter (kN/m^2) and meganewtons per square meter (MN/m^2). Thus, the pressure 1 m beneath the surface of a body of water is approximately 10 kN/m^2.

In an older form of the metric system, force was expressed in units of "kilogram-force" (kgf) or "ton-force" (tF), and many engineers will be familiar with these quantities—though usually without recognizing the distinction between kilograms and kilogram-force or between ton or ton-force. Where it is reasonable to accept the approximation that $9.806 \approx 10$, then

$$1\ \text{kgf/cm}^2 = 10\ \text{tf/m}^2 \approx 100\ \text{kN/m}^2$$

Thus engineers will readily learn to convert values to which they are accustomed, such as shear strength, into SI units.

T. William Lambe
Robert V. Whitman

MIT
1978

PART I *Introduction*

PART II *The Nature of Soil*

PART III *Dry Soil*

PART IV *Soil with Water—No Flow or Steady Flow*

PART V *Soil with Water—Transient Flow*

Soil Mechanics

DONALD WOOD TAYLOR

Donald Wood Taylor was born in Worcester, Massachusetts in 1900 and died in Arlington, Massachusetts on December 24, 1955. After graduating from Worcester Polytechnic Institute in 1922, Professor Taylor worked nine years with the United States Coast and Geodetic Survey and with the New England Power Association. In 1932 he joined the staff of the Civil Engineering Department at M.I.T. where he remained until his death.

Professor Taylor was active in both the Boston Society of Civil Engineers and the American Society of Civil Engineers. Just prior to his death he had been nominated for the Presidency of the Boston Society. From 1948 to 1953 he was International Secretary for the International Society of Soil Mechanics and Foundation Engineering.

Professor Taylor, a quiet and unassuming man, was highly respected among his peers for his very careful and accurate research work. He made major contributions to the fundamentals of soil mechanics, especially on the topics of consolidation, shear strength of cohesive soils, and the stability of earth slopes. His paper "Stability of Earth Slopes" was awarded the Desmond Fitzgerald Medal, the highest award of the Boston Society of Civil Engineers. His textbook, *Fundamentals of Soil Mechanics*, has been widely used for many years.

PART I

Introduction

Part I attempts to motivate the beginning student and alert him to the few really fundamental concepts in soil mechanics. Chapter 1 gives a general picture of civil engineering problems that can be successfully attacked by using the principles of soil mechanics. Chapter 2 describes, in terms familiar to the budding engineer, the essential principles that are covered in detail in the main portion of the book.

CHAPTER 1

Soil Problems in Civil Engineering

In his practice the civil engineer has many diverse and important encounters with soil. He uses soil as a foundation to support structures and embankments; he uses soil as a construction material; he must design structures to retain soils from excavations and underground openings; and he encounters soil in a number of special problems. This chapter deals with the nature and scope of these engineering problems, and with some of the terms the engineer uses to describe and solve these problems. Several actual jobs are described in order to illustrate the types of questions that a soil engineer must answer.

1.1 FOUNDATIONS

Nearly every civil engineering structure—building, bridge, highway, tunnel, wall, tower, canal, or dam—must be founded in or on the surface of the earth. To perform satisfactorily each structure must have a proper foundation.

When firm soil is near the ground surface, a feasible means of transferring the concentrated loads from the walls or columns of a building to the soil is through *spread footings*, as illustrated in Fig. 1.1. An arrangement of spread footings is called a *spread foundation*. In the past, timber or metal grillages, cobble pads, etc., were used to form spread footings, but today footings almost invariably are of reinforced concrete.

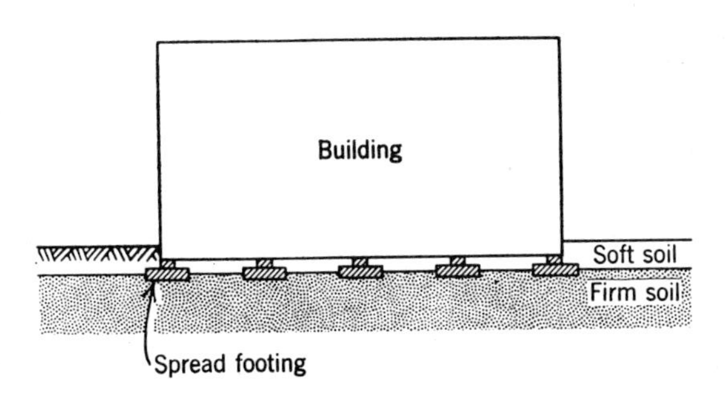

Fig. 1.1 Building with spread foundation.

When firm soil is not near the ground surface, a common means of transferring the weight of a structure to the ground is through vertical members such as *piles* (Fig. 1.2), *caissons*, or *piers*. These terms do not have sharp definitions that distinguish one from another. Generally, caissons and piers are larger in diameter than piles and are installed by an excavation technique, whereas piles are installed by driving. The weight of the building is carried through the soft soil to firm material below with essentially no part of the building load being applied to the soft soil.

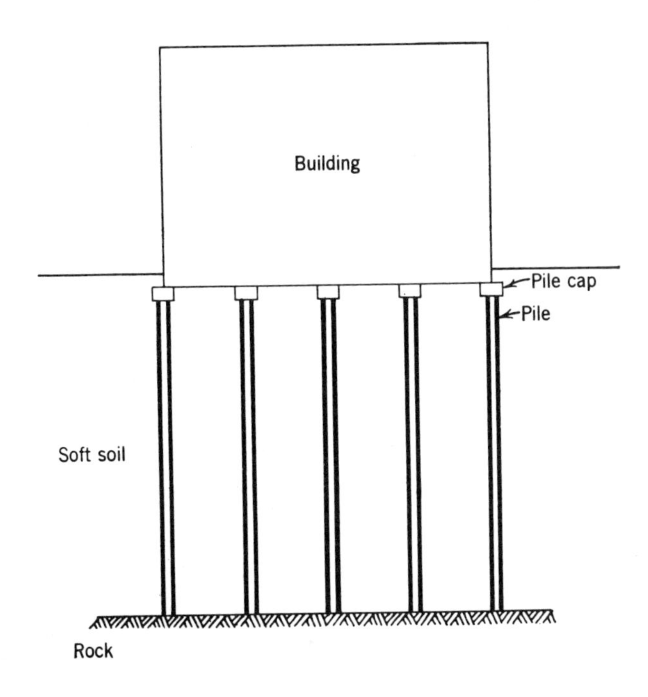

Fig. 1.2 Building with pile foundation.

There is much more to successful foundation engineering than merely selecting sizes of footings or choosing the right number and sizes of piles. In many cases, the cost of supporting a building can be significantly reduced by applying certain treatments to the soil. Further, some structures, such as steel storage tanks, can be supported directly on a specially prepared pad of soil without the benefit of intervening structural members. Thus the word *foundation* refers to the soil under the structure as well as any intervening load carrying member, i.e., *foundation* refers to the material whose behavior the civil engineer has analyzed in order to provide satisfactory and economical support for the structure. Indeed, the word foundation is used to describe the material that supports any type of engineering structure such as building, dam, highway embankment, or airfield runway. In modern usage, the term *shallow foundation* is used to describe an arrangement where structural loads are carried by the soil directly under the structure, and *deep foundation* is used for the case where piles, caissons, or piers are used to carry the loads to firm soil at some depth.

In the design of any foundation system, the central problem is to prevent settlements large enough to damage the structure or impair its functions. Just how much settlement is permissible depends on the size, type, and use of structure, type of foundation, source in the subsoil of the settlement, and location of the structure. In most cases, the critical settlement is not the *total settlement* but rather the *differential settlement*, which is the relative movement of two parts of the structure.

In most metropolitan areas of the United States and Western Europe, owners of buildings usually are unwilling to accept settlements greater than a few centimeters, since unattractive cracks are likely to occur if the settlements are larger. For example, experience has shown that settlements in excess of approximately 125 mm will cause the brick and masonry walls of buildings on the M.I.T. campus to crack.

However, where soil conditions are very bad, owners sometimes willingly tolerate large settlements with resulting cracking in order to avoid very significant additional costs of deep foundations over shallow ones. For example, along the waterfront in Santos, Brazil, 15-story apartment buildings are founded directly upon soft soil. Settlements as large as 0.3 m are common. Cracks in these buildings are apparent, but most of them have remained in continuous use.

Perhaps the classic case of bad foundation conditions exists in Mexico City. Here, for example, one building, the Palace of Fine Arts, shown in Fig. 1.3, is in continuous use even though it has sunk 3.6 m into the surrounding soil. Where a visitor used to walk up steps to the first floor, he must now walk down steps to this floor because of the large settlement.

With structures other than buildings, large settlements often are tolerated. Settlements as large as several meters are quite common in the case of flexible structures such as storage tanks and earth embankments. On the other hand, foundation movements as small as 0.2 mm may be intolerable in the case of foundations for precision tracking radars and nuclear accelerators.

Example of Shallow Foundation

Figure 1.4 shows the M.I.T. Student Center, which has a shallow foundation consisting of a slab under the entire building. Such a slab is called a *mat*. The subsoils at the site consist of the following strata starting from ground surface and working downward: a 4.5 m layer of soft fill

Fig. 1.3 Palacio de las Bellas Artes, Mexico City. The 2 m differential settlement between the street and the building on the right necessitated the steps which were added as the settlement occurred. The general subsidence of this part of the city is 7 m (photograph compliments of Raul Marsal).

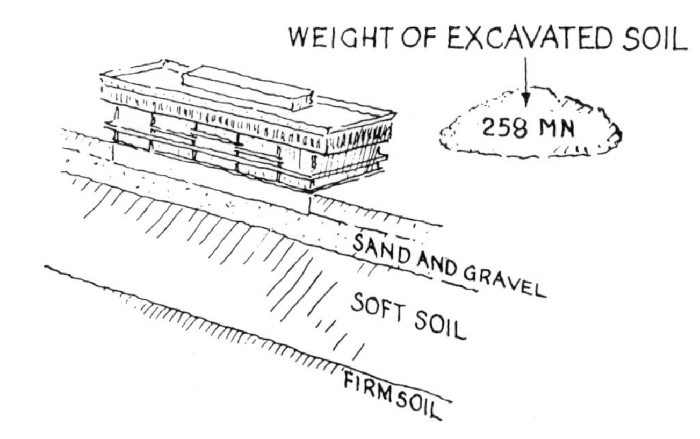

Fig. 1.4 Building on shallow mat foundation.

Weight of building	=	285 MN
Weight of furniture, people, etc. (time average)	=	45 MN
		330 MN
Weight of excavated soil	=	258 MN
Net load to clay	=	72 MN

and organic silt; a 6 m layer of sand and gravel; 22.5 m of soft clay; and, finally, firm soil and rock. The weight of the empty building (called the *dead load*) is 285 MN. The weight of furniture, people, books, etc. (called *live load*) is 45 MN. Had this building with its total load of 330 MN been placed on the ground surface, a settlement of approximately 300 mm would have occurred due to the compression of the soft underlying soil. A settlement of such large magnitude would damage the structure. The solution of this foundation problem was to place the building in an excavation. The weight of excavated soil was 258 MN, so that the net building load applied to the underlying soil was only 72 MN. For this arrangement the estimated settlement of the building is 50–75 mm, a value which can be tolerated.

This technique of reducing the net load by removing soil is called *flotation*. When the building is partly compensated by relief of load through excavation the technique is called *partial flotation*; when entirely compensated, it is called *full flotation*. Full flotation of a structure in soil is based on the same principle as the flotation of a boat. The boat displaces a weight of water equal to the weight of the boat so that the stress at a given depth in the water below the boat is the same, independent of the presence of the boat. Since the building in Fig. 1.4 has an average unit weight equal to about one-half that of water, and the unit weight of the excavated soil is about twice that of water, the building should be placed with about one-fourth of its total height under the ground surface in order to get full flotation.

On this particular project, the soil engineer was called upon to study the relative economy of this special shallow foundation versus a deep foundation of piles or caissons. After having concluded that the shallow foundation was desirable, he had to answer questions such as the following.

1. Just how deep into the soil should the building be placed?
2. Would the excavation have to be enclosed by a wall during construction to prevent cave-ins of soil?
3. Would it be necessary to lower the water table in order to excavate and construct the foundation and, if so, what means should be used to accomplish this lowering of the ground water (*dewatering*)?
4. Was there a danger of damage to adjacent buildings? (In later chapters it will be demonstrated that lowering the water table under a building can cause serious settlements. The question of just how and for what duration the water table is lowered can thus be very important.)
5. How much would the completed building settle and would it settle uniformly?
6. For what stresses and what stress distribution should the mat of the building be designed?

Example of Pile Foundation

Figure 1.5 shows the M.I.T. Materials Center, which has a deep pile foundation. The subsoils at the site of the Materials Building are similar to those at the Student Center with the important exception that there is little or no sand and gravel at the Materials Center site. The total load of the Materials Building is 249 MN made up of a dead load of 142 MN and a live load of 107 MN. The dead load of the Materials Center is less than that of the Student Center primarily because the Materials Building is made of lighter materials than the Student

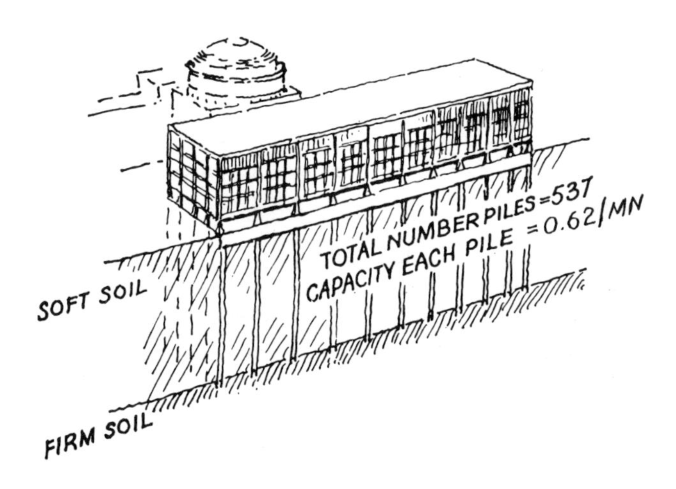

Fig. 1.5 Building on deep pile foundation.

Weight of building	=	142 MN
Weight of equipment, books, people, etc.	=	107 MN
Maximum total weight	=	249 MN

Center, and the live load is greater because of the much heavier equipment going into the Materials Building. The three major reasons that the Materials Building was placed on piles to firm soil rather than floated on a mat were:

1. The intended use of the Materials Building was such that floor space below ground surface, i.e., basement space, was not desirable.
2. There was little or no sand and gravel at the site on which to place the mat.
3. There were many underground utilities, especially a large steam tunnel crossing the site, which would have made the construction of a deep basement difficult and expensive.

The foundation selected consisted of 537 piles, each with a capacity of 0.62 MN. The piles were constructed by boring a hole about three-quarters of the way from ground surface to the firm soil, placing a hollow steel shell of 325 mm diameter in the bored hole and driving it to the firm soil, and then filling the hollow shell with concrete. (The hollow shell was covered with a steel plate at the tip to prevent soil from entering.) Such a pile is called a *point-bearing pile* (it receives its support at the point, which rests in the firm soil, as opposed to a *friction pile*, which receives its support along a large part of its length from the soil through which it goes) and a *cast-in-place concrete pile* (as opposed to a pile which is *precast* and then driven). Soil was removed by augering for three-quarters of the length of the pile in order to reduce the net volume increase below ground surface due to the introduction of the piles. Had *preaugering* not been employed, the surface of the ground at the building site would have risen almost 300 mm because of the volume of the 537 piles. The rise of the ground surface would have been objectionable because it would have raised piles that had already been driven, and it would have been dangerous because of possible disturbance of the nearby dome shown in the background in Fig. 1.5.

Among the questions faced by the soil engineer in the design and construction of the pile foundation were:

1. What type of pile should be used?
2. What was the maximum allowable load for a pile?
3. At what spacing should the piles be driven?
4. How should the piles be driven?
5. How much variation from the vertical should be permitted in a pile?
6. What was the optimum sequence for driving piles?
7. Would the driving of piles have an influence on adjacent structures?

Example of Embankment on Soft Soil

Figure 1.6 shows a 10.7 m embankment of earth placed on a 9.8 m layer of soft soil. The original plan was to place a tank, shown by dashed lines in Figure 1.6, 15.3 m in diameter and 17 m high, at the site. Had the tank been placed on the soft foundation soil with no special foundation, a settlement in excess of 1.5 m would have occurred. Even though a steel storage tank is a flexible structure, a settlement of 1.5 m is too large to be tolerated.

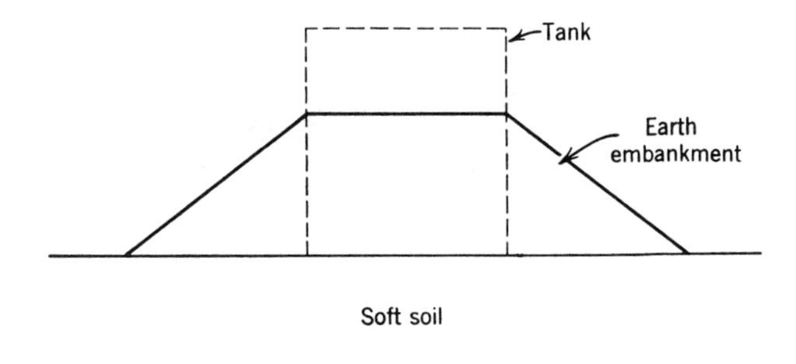

Fig. 1.6 Embankment on soft soil.

Soil engineering studies showed that a very economical solution to the tank foundation problem consisted of building an earth embankment at the site to compress the soft soil, removing the embankment, and finally placing the tank on the prepared foundation soil. Such a technique is termed *preloading*.

Since the preload was to be removed just prior to the construction of the tank, and the tank pad brought to the correct elevation, the magnitude which the preload settled was of no particular importance. The only concern was that the fill not be so high that a shear rupture of the soil would occur. If the placed fill caused *shear stresses* in the foundation which exceeded the *shear strength*, a rupture of the foundation could occur. Such a rupture would be accompanied by large movements of earth with probably serious disturbance to the soft foundation soil and possible damage to nearby tanks. Among the questions that had to be answered on this project were the following ones.

1. How high a fill could be placed?
2. How fast could the fill be placed?
3. What were the maximum slopes for the fill?
4. Could the fill be placed without employing special techniques to contain or drain the soft foundation soil?
5. How much would the fill settle?
6. How long should the fill be left in place in order that the foundation be compressed enough to permit construction and use of the tank?

Example of Foundation Heave

The foundation engineer faces not only problems involving settlement but also problems involving the upward movement (heave) of structures. Heave problems arise when the foundation soil is one that expands

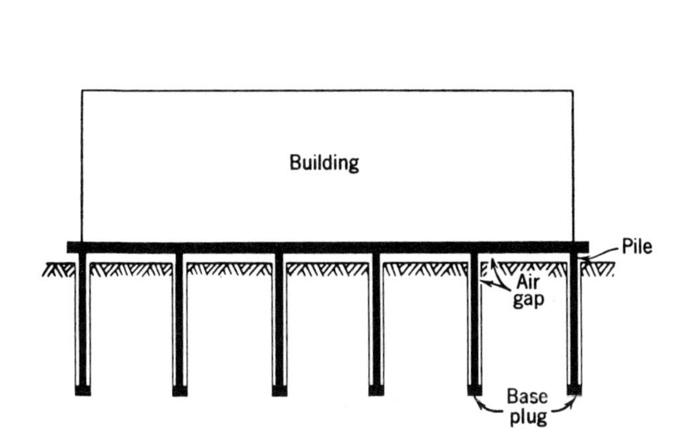

Fig. 1.7 Building on expansive soil.

when the confining pressure is reduced and/or the water content of the soil is increased. Certain soils, termed *expansive soils*, display heaving characteristics to a relatively high degree.

Heave problems are particularly common and economically important in those parts of the world with arid regions, e.g., Egypt, Israel, South Africa, Spain, Southwestern United States, and Venezuela. In such areas the soils dry and shrink during the arid weather and then expand when moisture becomes available. Water can become available from rainfall, or drainage, or from capillarity when an impervious surface is placed on the surface of the soil, thereby preventing evaporation. Obviously, the lighter a structure, the more the expanding soil will raise it. Thus heave problems are commonly associated with light structures such as small buildings (especially dwellings), dam spillways, and road pavements.

Figure 1.7 shows a light structure built in Coro, Venezuela. In the Coro area the soil is very expansive, containing the mineral *montmorillonite*. A number of buildings in Coro have been damaged by heave. For example, the floor slab and entry slab of a local hotel, resting on the ground surface, have heaved extensively, thus cracking badly and becoming very irregular. The building in Fig. 1.7 employs a scheme that avoids heave troubles but clearly is much more expensive than a simple shallow mat. First, holes were augered into the soil, steel shells were placed, and then concrete base plugs and piles were poured. Under the building and around the piles was left an air gap, which served both to reduce the amount of heave of the soil (by permitting evaporation) and to allow room for such heave without disturbing the building.

The main question for the soil engineer was in selecting the size, capacity, length, and spacing of the piles. The piles were made long enough to extend below the depth of soil that would expand if given access to moisture. The depth selected was such that the confining pressure from the soil overburden plus minimum load was sufficient to prevent expansion.

1.2 SOIL AS A CONSTRUCTION MATERIAL

Soil is the most plentiful construction material in the world and in many regions it is essentially the only locally available construction material. From the days of neolithic man, earth has been used for the construction of monuments, tombs, dwellings, transportation facilities, and water retention structures. This section describes three structures built of earth.

When the civil engineer uses soil as a construction material, he must select the proper type of soil and the method of placement, and then control the actual placement. Man-placed soil is called *fill*, and the process of placing it is termed *filling*. One of the most common problems of earth construction is the wide variability of the source soil, termed *borrow*. An essential part of the engineer's task is to see that the properties of the placed material correspond to those assumed in the design, or to change the design during construction to allow for any difference between the properties of the constructed fill and those employed in the design.

Example of an Earth Dam

Figure 1.8 is a vertical cross section of an earth dam built to retain a reservoir of water. The two main zones of the dam are the *clay core* and the *rock toe*: the core with its impermeable clay keeps leakage low; and the heavy, highly permeable rock toe adds considerable stability to the dam. Between these zones is placed a

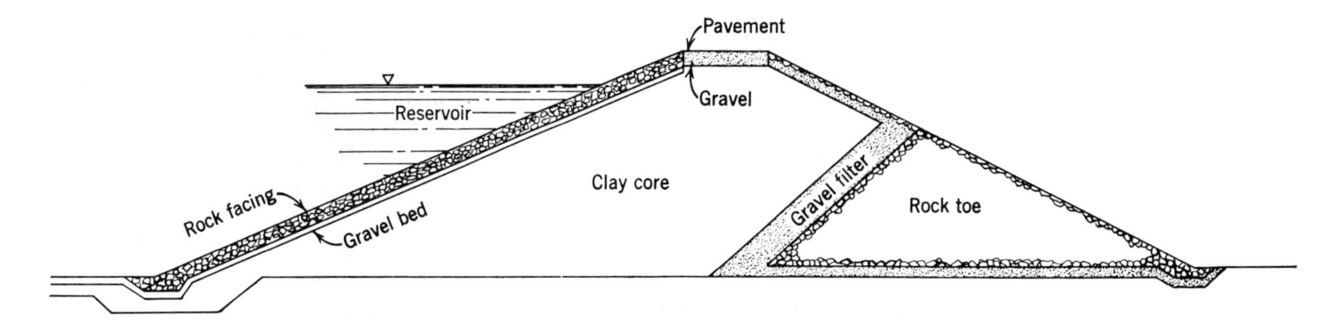

Fig. 1.8 Earth dam.

gravel filter to prevent washing of soil particles from the core into the voids of the rock toe. Between the core and the reservoir is a rock facing placed on a gravel bed. The rock facing prevents erosion of the core by rain or water in the reservoir. The gravel bed prevents large rocks on the face from sinking into the clay. This type of dam is called a *zoned earth dam* to differentiate it from a *homogeneous earth dam* in which the same type of material is used throughout the cross section.

The popularity of earth dams compared to concrete dams is increasing steadily for two major reasons. First, the earth dam can withstand foundation and abutment movements better than can the more rigid concrete structure. Second, the cost of earth construction per unit volume has remained approximately constant for the last 50 years (the increased cost of labor has been offset by the improvements in earth-handling equipment), whereas the cost of concrete per unit volume has steadily increased. One would thus expect earth dams to become increasingly popular.

The relative sizes of the zones in an earth dam and the types of material in each zone depend very much on the earth materials available at the site of the dam. At the site of the dam shown in Fig. 1.8, excavation for the reservoir yielded clay and rock in about the proportions used in the dam. Thus none of the excavated material was wasted. The only scarce material was the gravel used for the filter and the bed. This material was obtained from stream beds some distance from the site and transported to the dam by trucks.

Dam construction was carried out for the full length and the full width of the dam at the same time; i.e., an attempt was made to keep the surface of the dam approximately horizontal at any stage of construction. The toe, consisting of rock varying in size from 150 to 900 mm, was *end-dumped* from trucks and washed as dumped with water under high pressure. The clay and gravel were placed in horizontal lifts of 150 to 300 mm in thickness, then brought to a selected moisture content, and finally *compacted* by rolling compaction equipment over the surface.

The following questions were faced by the civil engineer during the design and construction of the earth dam.

1. What should be the dimensions of the dam to give the most economical, safe structure?
2. What is the minimum safe thickness for the gravel layers?
3. How thick a layer of gravel and rock facing is necessary to keep any swelling of the clay core to a tolerable amount?
4. What moisture content and compaction technique should be employed to place the gravel and clay materials?
5. What are the strength and permeability characteristics of the constructed dam?
6. How would the strength and permeability of the dam vary with time and depth of water in the reservoir?
7. How much leakage would occur under and through the dam?
8. What, if any, special restrictions on the operation of the reservoir are necessary?

Example of a Reclamation Structure

There are many parts of the world where good building sites are no longer available. This is particularly true of harbor and terminal facilities, which obviously need to be on the waterfront. To overcome this shortage, there is an increasingly large number of reclamation projects wherein large sites are built by filling. The soil for such projects is usually obtained by dredging it from the bottom of the adjacent river, lake, or ocean and placing it at the location desired. This process is *hydraulic filling*.

Figures 1.9 and 1.10 show a successful reclamation project built in Lake Maracaibo, Venezuela. The island was constructed by driving a wall of concrete piles enclosing an area 850 m long by 600 m wide. Soil was then dredged from the bottom of Lake Maracaibo and pumped into the sheet pile enclosure until the level of the hydraulic fill reached the desired elevation. Three factors—the lack of available land onshore, the deep water

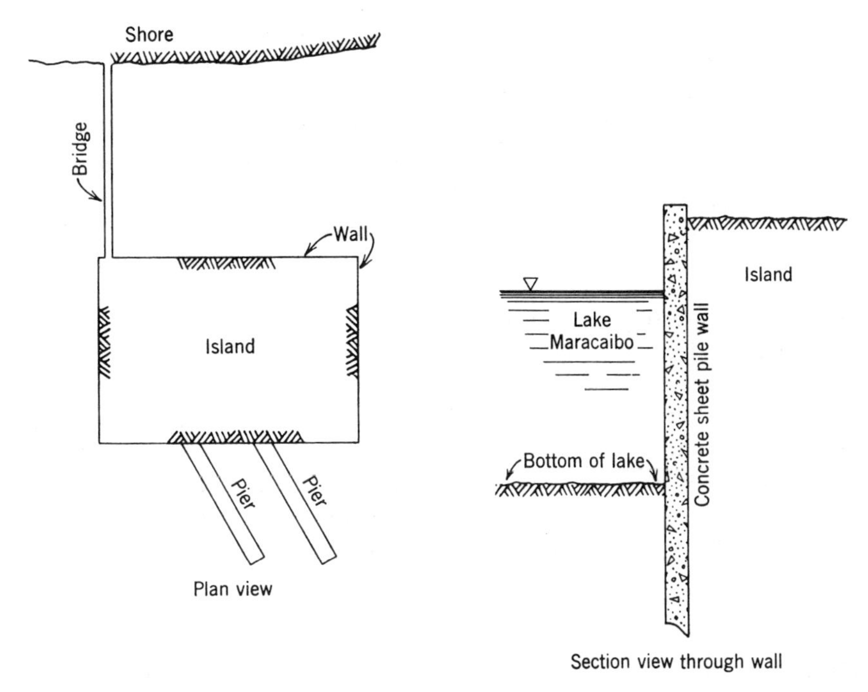

Fig. 1.9 Marine terminal built of hydraulic fill.

needed for large ships to come near the terminal, and the need to dredge a channel in the lake—combined to make the construction of this man-made island an excellent solution to the need of terminal facilities at this location.

On the island, storage tanks for various petroleum products were constructed. The products are brought by pipelines from shore to the tanks on the island and then pumped from the tanks to tankers docked at the two piers shown in Fig. 1.9.

Many exploratory borings were made in the area to be dredged in order to permit the soil engineer to estimate the type of fill that would be pumped onto the island. This fill consisted primarily of clay in the form of hard chunks varying in size from 25 to 150 mm plus a thin slurry of water with silt and clay particles in suspension. Upon coming out of the dredge pipe on the island, the large particles settled first, and the finer particles were transported a considerable distance from the pipe exit. At one corner of the island was a spillway to permit the excess water from the dredging operation to re-enter the lake.

Following are some questions faced by the civil engineer on this terminal project.

1. How deep should the sheet pile wall penetrate the foundation soil?
2. How should these piles be braced laterally?
3. What is the most desirable pattern of fill placement—i.e., how should the exit of the dredge pipe be located in order to get the firmer part of the fill at the locations where the maximum foundation loads would be placed?
4. What design strength and compressibility of the hydraulic fill should be used for selecting foundations for the tanks, buildings, and pumping facilities to be placed on the island?
5. Where did the soil fines in the dirty effluent which went out of the island over the spillway ultimately settle?

Example of Highway Pavement

One of the most common and widespread uses of soil as a construction material is in the pavements of roads and airfields. Pavements are either flexible or rigid. The primary function of a flexible pavement is to spread the

Fig. 1.10 La Salina Marine Terminal (compliments of the Creole Petroleum Corporation).

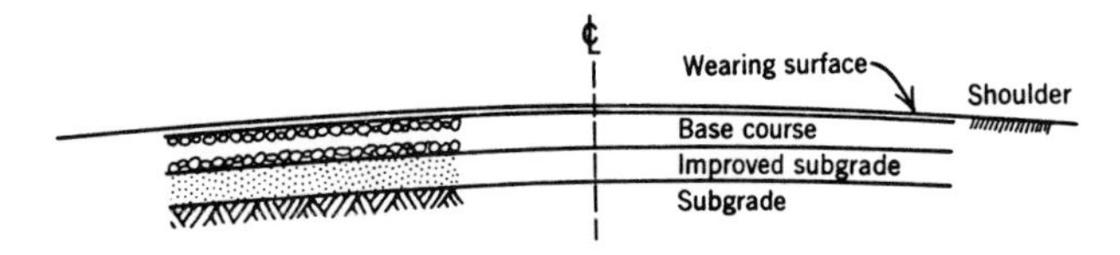

Fig. 1.11 Highway pavement.

concentrated wheel loads over a sufficiently large area of the foundation soil to prevent overstressing it. The rigid pavement constructed of reinforced concrete possesses sufficient flexural strength to bridge soft spots in the foundation. Which pavement is better for a given problem depends on the nature of the foundation, availability of construction materials, and the use to be made of the pavement.

Figure 1.11 shows a highway flexible pavement designed for 100 passes per lane per day of a vehicle having a maximum wheel load of 66.8 kN. The selected pavement consisted of an *improved subgrade* made by compacting the top 150 mm of the *in situ* soil; a *base course* consisting of a 150 mm layer of soil from the site mixed with 7% by soil weight of portland cement, brought to proper moisture content, mixed and then compacted; and a *wearing surface* consisting of a 50 mm-thick layer of hot-mixed sand-asphalt.

Commonly, the base course of a pavement consists of gravel or crushed stone. In the desert, where the pavement shown in Fig. 1.11 was built, there was a shortage of gravel but an abundance of desert sand. Under these circumstances, it was more economical to improve the properties of the local sand (*soil stabilization*) than to haul gravel or crushed stone over large distances. The most economical soil stabilizer and the method of preparing the stabilized base were chosen on the basis of a program of laboratory tests involving various possible stabilizers and construction techniques.

Among the questions faced by the engineer on the design and construction of this road were the following ones.

1. How thick should the various components of the pavement be to carry the expected loads?
2. What is the optimum mixture of additives for stabilizing the desert sand?
3. Is the desert sand acceptable for the construction of the wearing surface?
4. What grade and weight of available asphalt make the most economical, satisfactory wearing surface?
5. What type and how much compaction should be used?

1.3 SLOPES AND EXCAVATIONS

When a soil surface is not horizontal there is a component of gravity tending to move the soil downward, as illustrated by the force diagram in Fig. 1.12*a*. If along a potential slip surface in the soil the shear stress from gravity or any other source (such as moving water, weight of an overlying structure, or an earthquake) exceeds the strength of the soil along the surface, a shear rupture and movement can occur. There are many circumstances in natural slopes, compacted embankments, and excavations where the civil engineer must investigate the stability of a slope by comparing the shear stress with the shear strength along a potential slip surface—i.e., he must make a *stability analysis*.

Figure 1.12*a* shows a natural slope on which a building has been constructed. The increased shear stress from the building and the possible decrease of soil shear strength from water wasted from the building can cause a failure of the slope, which may have been stable for many years before construction. Such slides are common in the Los Angeles area.

The earth dam shown in Fig. 1.8 has a compacted earth slope which had to be investigated for stability. During the design of this dam, the civil engineer compared the shear stress with the shear strength for a number of potential slip surfaces running through the clay core.

Figures 1.12*b* and *c* illustrate excavations for a building and a pipe. The building excavation is a *braced excavation* and that for the pipe is an *unbraced excavation*. A designing engineer must be sure that the shear strength of the slope is not exceeded, for this would result in a cave-in.

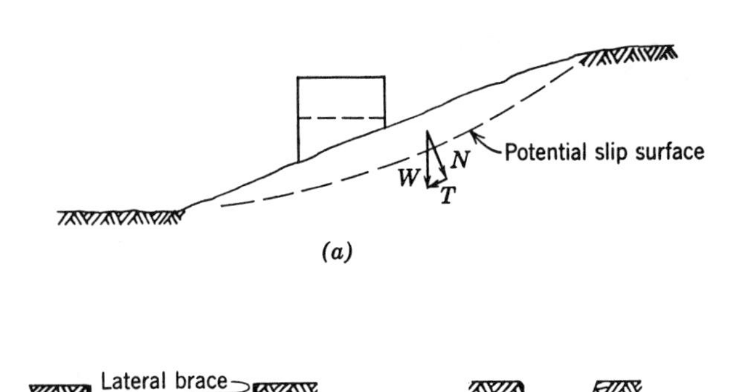

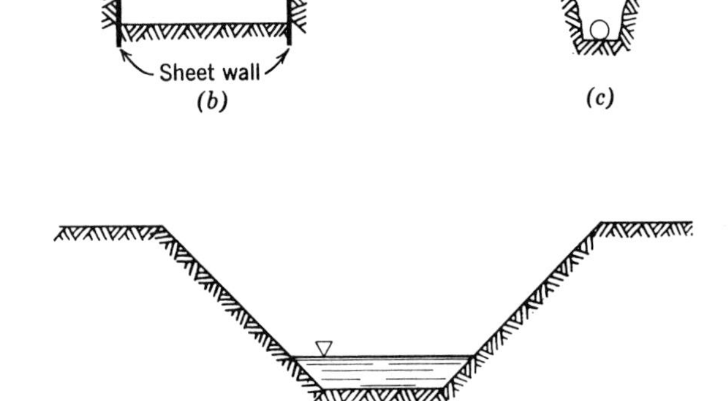

(d)

Fig. 1.12 Slopes and excavations. (*a*) Natural slope. (*b*) Excavation for building. (*c*) Excavation for pipe. (*d*) Canal.

Fig. 1.13 Landslide in a quick clay (compliments of Laurits Bjerrum).

Figure 1.12*d* shows a sketch of a canal. Canals usually are built by excavating through natural materials, but sometimes they are built by compacting fill. The slopes of the canal must be safe both against a shear failure, as described previously, and against the effects of moving water. If protection against moving water is not furnished the canal banks may erode, requiring continuous removal of eroded earth from the canal and possibly triggering a general shear failure of the canal sides.

Figure 1.13 shows a dramatic landslide of a natural slope of *quick clay*. Quick clay is a very sensitive clay deposited in marine water and later leached by ground water. The removal of the salt in the soil pores results in a soil that loses much of its strength when disturbed. The soil in the landslide zone in Fig. 1.13 had been leached for thousands of years until it became too weak to support the natural slope. Some excavation at the toe of the slope or added load may have triggered the slide. Landslides of this type are common in Scandinavia and Canada.

Panama Canal

Figure 1.14 shows one of the world's most famous canals—the Panama Canal. Excavation on the Panama Canal was started in February 1883 by a French company that intended to construct a sea level canal across the Isthmus of Panama joining the Atlantic and Pacific oceans. Excavation proceeded slowly until the end of 1899 when work ceased because of a number of engineering problems and the unhealthy working conditions.

In 1903, the United States signed a treaty with Colombia granting the United States rights for the construction, operation, and control of the Panama Canal. This treaty was later rejected by the Colombian Government. Following a revolt and secession of Panama from Colombia, the United States signed a treaty with Panama in 1903 for control of the Canal Zone in perpetuity.

Engineers studying the canal project developed two schemes: (*a*) a canal with locks estimated to cost $147,000,000 and requiring 8 years to build; and (*b*) a sea level canal costing $250,000,000 and requiring 12–15 years to build. Congress chose the high level lock canal, and construction was started in 1907 and finished in 1914. The actual construction cost was $380,000,000.

The canal is 82.4 km from deep water to deep water and required a total excavation of 316 600 000 cubic meters of which 128 700 000 cubic meters came from the Gaillard cut shown in Fig. 1.14. The minimum width of the canal was originally 91 m (through the Gaillard cut) and it was later widened to approximately 150 m. The minimum depth of the canal is 11.3 m (in Balboa Harbor at low tide).

Many shear slides occurred during construction, especially in the Cucaracha formation, a notoriously soft shale. (The slides contributed to the high construction cost.) The canal was opened to traffic in August 1914; however, landslides closed the canal on several occasions for periods of a few days to 7 months. The last closure was in 1931, although constrictions have occurred on several occasions since then. Removal of soil from slides and erosion now requires continuous maintenance by

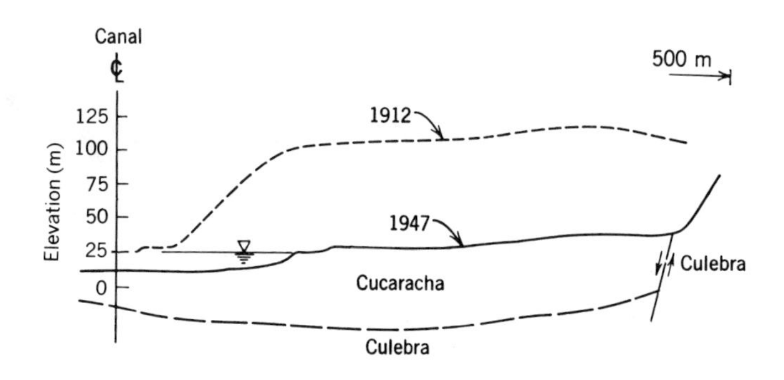

Fig. 1.14 The Panama Canal. (*a*) Cross section through East Culebra Slide. (*b*) Cucaracha Slide of August 1913. (*c*) Ship in Canal in 1965.

canal dredges. At one location the side slopes are even today moving into the canal at the rate of 8.5 m per year.

The long-term strength characteristics of soft shales, such as those lining parts of the Panama Canal, are a perplexing problem to the soil engineer. Since the slides along the canal appear to be related to cracks in the rocks and special geologic features, the analysis of slopes in these materials cannot be done solely on the basis of theoretical considerations and laboratory test results. The solution of this type of problem depends very much on an understanding of geology, and illustrates the importance geology can have to the successful practice of civil engineering.

1.4 UNDERGROUND AND EARTH RETAINING STRUCTURES

Any structure built below ground surface has forces applied to it by the soil in contact with the structure. The design and construction of underground (subterranean) and earth retaining structures constitute an important phase of civil engineering. The preceding pages have already presented examples of such structures; they include the pipe shells that were driven for the foundation shown in Fig. 1.2, the basement walls of the buildings shown in Figs. 1.4 and 1.5, the concrete sheet pile wall encircling the island shown in Fig. 1.9, and the bracing for the excavation shown in Fig. 1.12*b*. Other common examples of underground structures include tunnels for railways or vehicles, underground buildings like powerhouses, drainage structures, earth retaining structures, and pipelines.

The determination of forces exerted on an underground structure by the surrounding soil cannot be correctly made either from a consideration of the structure alone or from a consideration of the surrounding soil alone, since the behavior of one depends on the behavior of the other. The civil engineer therefore must be knowledgeable in *soil-structure interaction* to design properly structures subjected to soil loadings.

Example of Earth Retaining Structure

A common type of earth retaining structure is the *anchored bulkhead*, as illustrated in Fig. 1.15. Unlike a *gravity retaining* wall, which has a large base in contact with the foundation soil and enough mass for friction between the soil and the wall base to prevent excessive lateral movement of the wall, the anchored bulkhead receives its lateral support from penetration into the foundation soil and from an anchoring system near the top of the wall.

The bulkhead shown in Fig. 1.15 was built as part of a

ship-loading dock. Ships are brought alongside the bulkhead and are loaded with cargo stored on the land side of the wall. The loading is done by a crane moving on rails parallel to the bulkhead.

To determine the proper cross section and length of the bulkhead wall, the engineer must compute the stresses exerted by the soil against the wall (*lateral soil stresses*). The distribution of these stresses along the wall depends very much on the lateral movements that occur in the soil adjacent to the wall, and these strains in turn depend on the rigidity of the wall—a problem in soil-structure interaction.

The selection of the length and section of the bulkhead and the design of the anchoring system was only part of the problem. Consideration had to be given to the stability of the entire system against a shear rupture in which the slip surface passed through the backfill and through the soil below the tip of the bulkhead. This type of overall stability can be a much more serious problem with anchored bulkheads than is the actual design of the bulkhead itself.

The following questions had to be answered in planning the design of the anchored bulkhead.

1. What type of wall (material and cross section) should be used?
2. How deep must the wall penetrate the foundation soil in order to prevent the wall from kicking out to the left at its base?

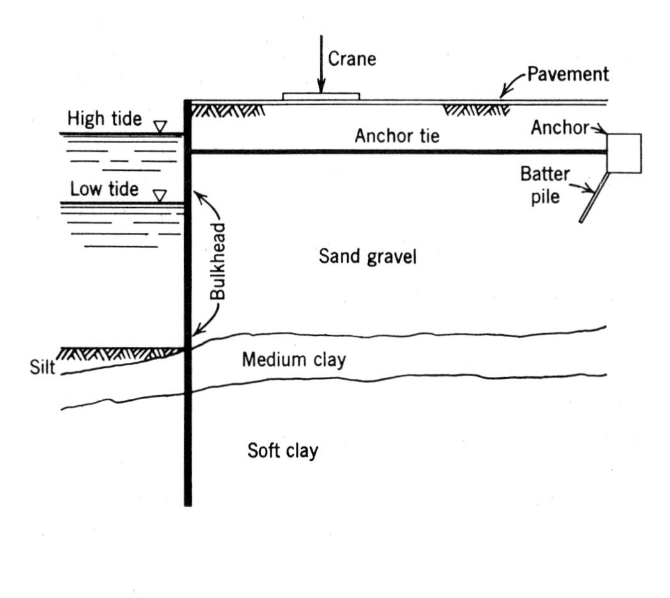

Fig. 1.15 Anchored bulkhead.

3. At what height on the wall should the anchor tie be located?
4. How far from the wall should the anchor tie extend?
5. What type of anchoring system should be employed at the onshore end of the anchor tie? (One way to anchor the wall is to use a large mass of concrete, i.e., *deadman*. Another way is to use a system of piles including some driven at a slope with the vertical; such a sloping pile is termed a *batter pile*.)
6. What was the distribution of stresses acting on the wall?
7. What type of drainage system should be installed to prevent a large differential water pressure from developing on the inside of the wall?
8. How close to the wall should the loaded crane (578 kN when fully loaded) be permitted?
9. What restrictions, if any, are necessary on the storage of cargo on the area back of the wall?

Example of Buried Pipeline

Frequently a pipe must be buried under a high embankment, railway, or roadway. The rapid growth of the pipeline industry and the construction of superhighways have greatly increased the frequency of buried pipe installations. Buried pipes are usually thin-wall metal or plastic pipes, called *flexible pipes*, or thick-wall reinforced concrete pipes, called *rigid pipes*.

There have been very few recorded cases of buried pipes being crushed by externally applied loads. Most of the failures that have occurred have been associated with: (*a*) faulty construction; (*b*) construction loads in excess of design loads; and (*c*) pipe sag due to foundation settlement or failure. Faced with the impressive performance record of the many thousands of buried pipes, we are forced to conclude that the design and construction procedures commonly used result in safe installations. There is little published information, however, indicating just how safe these installations are and whether or not they are far overdesigned, thus resulting in a great waste of money.

Figure 1.16 shows an installation of two steel pipes, each 750 mm in diameter with a wall thickness of 9.5 mm, buried under an embankment 24.4 m high at its center line. Use of the commonly employed analytical method yielded a value for the maximum pipe deflection of 190 mm. Current practice suggests a value of 5% of the pipe diameter, 37.5 mm for the 150 mm diameter pipe, as the maximum allowable safe deflection.

At this stage in the job, laboratory soil tests and field experimentation with installations were carried out. Use of the soil data obtained from these tests resulted in a computed pipe deflection of 8.1 mm, a safe value. The maximum value of pipe deflection actually measured in the installation was only 4.3 mm. These stated deflections indicate the merit of a controlled installation (and also

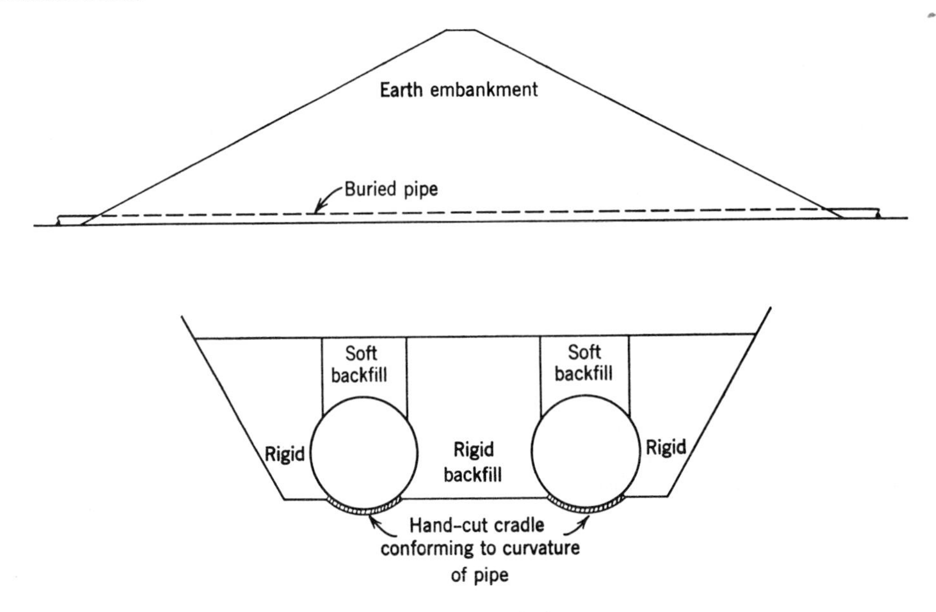

Fig. 1.16 Buried pipes.

the inaccuracy of commonly employed techniques to estimate the deflection of buried pipes).

The method of pipe installation employed is indicated in Fig. 1.16 and consisted of the following: carrying the fill elevation above the elevation of the tops of the pipes; cutting a trench for the pipes; shaping by hand a cradle for each pipe conforming to the curvature of the pipe; backfilling under carefully controlled conditions to get a rigid backfill on the sides of the pipes and a soft spot on top of each pipe.

The rigid side fills were to give strong lateral support to the pipes and thus reduce their lateral expansion. The soft spots were to encourage the fill directly over the pipes to tend to settle more than the rest of the fill, thus throwing some of the vertical load to the soil outside the zone of the pipes; this phenomenon is called *arching*.

Since the vertical load on the two pipes is related to the height of the fill, one would expect the settlement of the pipes to be the maximum at the midpoint of the embankment. Such was the case with 170 mm of settlement occurring at the center of the embankment and about 10 mm at the two toes of the embankment. The flexible steel pipes, more than 100 m in length, could easily withstand the 160 mm sag.

The civil engineer on this project had to select the thickness of the pipe wall, and work out and supervise the installation of the pipes.

1.5 SPECIAL SOIL ENGINEERING PROBLEMS

The preceding sections have discussed and illustrated some common civil engineering problems that involve soil mechanics. There are, in addition, many other types of soil problems that are less common but still important. Some of these are noted in this section in order to give a more complete picture of the range of problems in which soil mechanics is useful.

Vibrations

Certain granular soils can be readily densified by vibrations. Buildings resting on such soils may undergo significant settlement due to the vibration of their equipment, such as large compressors and turbines. The effects of a vibration can be particularly severe when the frequency of the vibration coincides with the natural frequency of the soil foundation. Upon deciding that vibrations can cause deleterious settlement in a particular structure, the engineer has the choice of several means of preventing it. He can increase the mass of the foundation, thus changing its frequency, or densify or inject the soil, thereby altering its natural frequency and/or compressibility.

Explosions and Earthquakes

Civil engineers have long been concerned with the effects on buildings of earth waves caused by quarry blasting and other blasting for construction purposes. The ground through which such waves pass has been found to influence greatly the vibrations that reach nearby buildings.

This problem has received an entirely new dimension as the result of the advent of nuclear explosives. The military has become increasingly interested in the design of underground facilities that can survive a very nearby nuclear explosion. The Atomic Energy Commission has established the Plowshare program to consider the peaceful uses of nuclear explosions, such as the excavation of canals or highway cuts. The possibility of excavating a sea level Panama Canal by such means has received

Fig. 1.17 Oil storage reservoir (compliments of the Creole Petroleum Corporation).

special attention and raises a whole new set of questions, such as the stability of slopes formed by a cratering process.

Similar problems arise as the result of earthquakes. The type of soil on which a building rests and the type of foundation used for the building influence the damage to a building by an earthquake. The possible effects of earthquakes on large dams have recently received much attention. The 1964 earthquake in Alaska caused one of the largest earth slides ever recorded.

The Storage of Industrial Fluids in Earth Reservoirs

Section 1.2 describes an earth structure for the retention of water. Because earth is such a common and cheap construction material, it has considerable utility for the construction of reservoirs and containers to store industrial fluids. One of the most successful applications of this technique is the earth reservoir for the storage of fuel oil shown in Fig. 1.17. This structure, with a capacity of 1 750 000 m^3, was built at one-tenth the cost of conventional steel tankage and resulted in a saving of approximately $20,000,000. Because of the interfacial tension between water and certain industrial fluids, compacted, fine-grained, wet soil can be used to store such fluids with no leakage.

Another example of this special application is the use of reservoirs for the storage of refrigerated liquefied gas. Earth reservoirs have been built for the retention of liquefied propane at −42°C and for liquefied natural gas at −162°C. Introducing a liquid at such low temperatures into a water-wet soil freezes the pore water in the soil. If the soil has enough water so that there are not continuous air channels in the soil, it will become impervious to both liquid and gas upon freezing of the pore water.

Frost

Because certain soils under certain conditions expand on freezing, the engineer may be faced with *frost heave* problems. When *frost susceptible* soils are in contact with moisture and subjected to freezing temperatures, they can

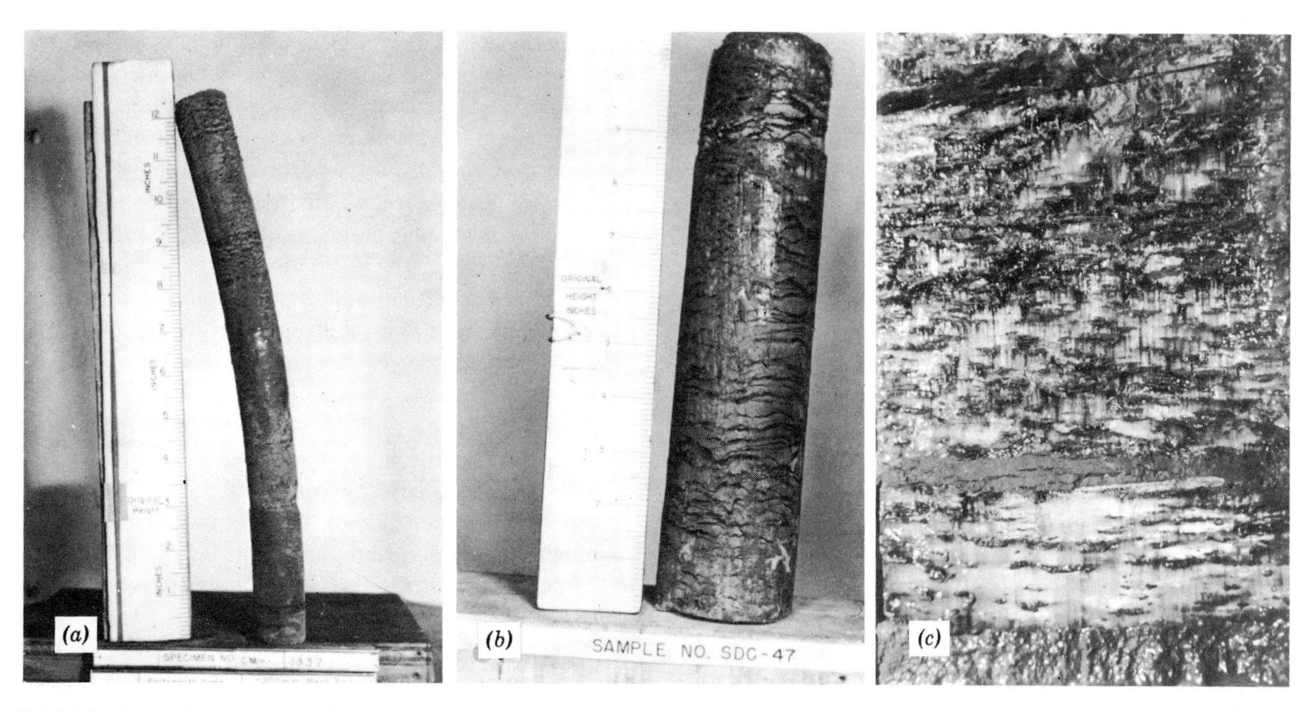

Fig. 1.18 Frost heave. (*a*) Soil sample which heaved from 79 to 320 mm on freezing. (*b*) Soil sample which heaved from 150 to 300 mm on freezing. (*c*) A close view of frozen soil. (Photographs compliments of C. W. Kaplar of U.S. Army CRREL.)

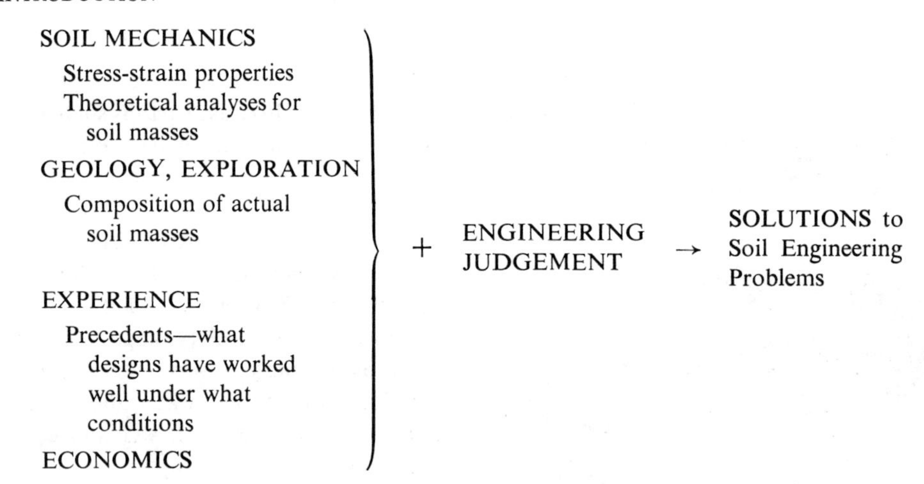

Fig. 1.19 The solution of soil engineering problems.

imbibe water and undergo a very large expansion. Figure 1.18 dramatically illustrates the magnitude a soil can heave under ideal conditions. Such heave exerts forces large enough to move and crack adjacent structures, and can cause serious problems on thawing because of the excess moisture. The thawing of frozen soil usually proceeds from the top downward. The melt water cannot drain into the frozen subsoil, thus becomes trapped, greatly weakening the soil. The movement of icehouses and ice-skating rinks is an interesting example of this phenomenon, but nowhere nearly as important and widespread as the damage to highway pavements in those areas of the world that have freezing temperatures. Frost heaves and potholes, which develop when the frost thaws, are sources of great inconvenience and cost to many northern U.S. areas, such as New England.

The civil engineer designing highway and airfield pavements in frost areas must either select a combination of base soil and drainage that precludes frost heave, or design his pavement to withstand the weak soil that occurs in the spring when the frost melts.

Regional Subsidence

Large-scale pumping of oil and water from the ground can cause major settlements over a large area. For example, a 41 km^2 area of Long Beach, California, has settled as the result of oil pumping, with a maximum settlement to date of 7.5 m. As a result, the Naval Shipyard adjacent to the settled area has had to construct special sea walls to keep out the ocean, and has had to reconstruct dry docks. Mexico City has settled as much as 9 m since the beginning of the twentieth century as the result of pumping water for domestic and industrial use. The first step in minimizing such regional subsidence is to locate the earth materials that are compressing as the fluid is removed, and then consider methods of replacing the lost fluid.

1.6 THE SOLUTION OF SOIL ENGINEERING PROBLEMS

Thus far this chapter has described some of the problems the civil engineer encounters with construction on soil, in soil, and of soil. The successful solution of each problem nearly always involves a combination of soil mechanics and one or more of the components noted in Fig. 1.19.

Geology aids the soil engineer because the method of forming a mass influences its size, shape, and behavior. Exploration helps establish the boundaries of a deposit and enables the engineer to select samples for laboratory testing.

Experience, as the term is used here, does not mean merely *doing* but the doing coupled with an *evaluation* of results of the act. Thus, when the civil engineer makes a design or solves a soil problem and then evaluates the outcome on the basis of measured field performance, he is gaining experience. Too much emphasis is usually placed on the *doing* component of experience and too little emphasis on the *evaluation* of the outcome of the act. The competent soil engineer must continue to improve his reservoir of experience by comparing the predicted behavior of a structure with its measured performance.

Economics is an important ingredient in the selection of the *best* solution from among the possible ones. Although a detailed economic evaluation of a particular earth structure depends on the unit costs at the site of a planned project, certain economic advantages of one scheme over another may be obvious from the characteristics of the schemes.

This book is limited to one component of the solution of soil engineering problems—*soil mechanics:* the science underlying the solution of the problem. The reader must remember that science alone cannot solve these problems.

Nearly all soil problems are statically indeterminate to a high degree. Even more important is the fact that natural soil deposits possess five complicating characteristics:

1. Soil does not possess a linear or unique stress-strain relationship.
2. Soil behavior depends on pressure, time, and environment.
3. The soil at essentially every location is different.
4. In nearly all cases the mass of soil involved is underground and cannot be seen in its entirety but must be evaluated on the basis of small samples obtained from isolated locations.
5. Most soils are very sensitive to disturbance from sampling, and thus the behavior measured by a laboratory test may be unlike that of the *in situ* soil.

These factors combine to make nearly every soil problem unique and, for all practical purposes, impossible of an exact solution.

Soil mechanics can provide a solution to a mathematical model. Because of the nature and the variability of soil and because of unknown boundary conditions, the mathematical model may not represent closely the actual problem. As construction proceeds and more information becomes available, soil properties and boundary conditions must often be re-evaluated and the problem solution modified accordingly.

The interpretation of insufficient and conflicting data, the selection of soil parameters, the modification of a solution, etc., require experience and a high degree of intuition—i.e., *engineering judgment*. While a sound knowledge of soil mechanics is essential for the successful soil engineer, *engineering judgment* is usually the characteristic that distinguishes the outstanding soil engineer.

PROBLEMS

1.1 List three events of national and/or international importance that involve soil mechanics (e.g., the extensive damage from the 1964 Alaskan earthquake).

1.2 Note the type of foundation employed in a building constructed recently in your area. List obvious reasons why this type of foundation was selected.

1.3 On the basis of your personal experience, describe briefly an engineering project that was significantly influenced by the nature of the soil encountered at the site of the project.

1.4 Note several subsoil and building characteristics that would make a pile foundation preferable to a spread foundation.

1.5 List difficulties you would expect to result from the large settlement of the Palacio de las Bellas Artes shown in Fig. 1.3.

1.6 Note desirable and undesirable features of building flotation.

CHAPTER 2

A Preview of Soil Behavior

This chapter presents a preliminary and intuitive glimpse of the behavior of homogeneous soil. This preview is intended to give the reader a general picture of the way in which the behavior of soil differs from the behavior of other materials which he has already studied in solid and fluid mechanics, and also to indicate the basis for the organization of this book. To present clearly the broad picture of soil behavior, this chapter leaves to later chapters a consideration of exceptions to, and details of, this picture.

2.1 THE PARTICULATE NATURE OF SOIL

If we examine a handful of beach sand the naked eye notices that the sand is composed of discrete particles. The same can be said of all soils, although many soil particles are so small that the most refined microscopic techniques are needed to discern the particles. The discrete particles that make up soil are not strongly bonded together in the way that the crystals of a metal are, and hence the soil particles are relatively free to move with respect to one another. The soil particles are solid and cannot move relative to each other as easily as the elements in a fluid. Thus soil is inherently a *particulate system*.[1] It is this basic fact that distinguishes soil mechanics from solid mechanics and fluid mechanics. Indeed, the science that treats the stress-strain behavior of soil may well be thought of as *particulate mechanics.*

The next sections examine the consequences of the particulate nature of soil.

2.2 NATURE OF SOIL DEFORMATION

Figure 2.1 shows a cross section through a box filled with dry soil, together with a piston through which a vertical load can be applied to the soil. By enlarging a portion of this cross section to see the individual particles, we are able to envision the manner in which the applied force is transmitted through the soil: contact forces develop between adjacent particles. For convenience, these contact forces have been resolved into components normal N and tangential T to the contact surfaces.

The individual particles, of course, deform as the result of these contact forces. The most usual type of deformation is an elastic or plastic strain in the immediate vicinity of the contact points. Particle crushing can be important in certain situations (as later chapters discuss). These deformations lead to an enlargement of the contact area between the particles, as shown in Fig. 2.2*a*, and thus permit the centers of the particles to come closer together. If platelike particles are present, these particles will bend,

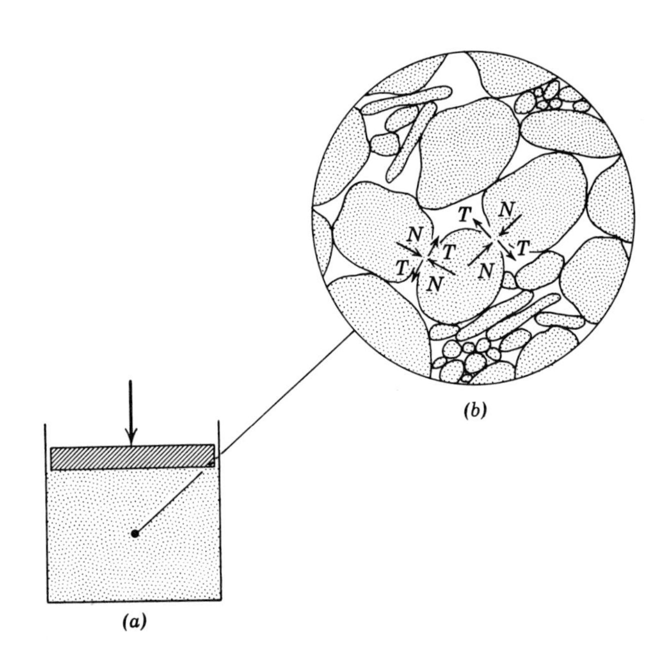

Fig. 2.1 Schematic representation of force transmission through soil. (*a*) Cross section through box filled with soil. (*b*) Enlargement through portion of cross section showing forces at two of the contact points.

[1] The word "particulate" means "of or pertaining to a system of particles."

as in Fig. 2.2*b*, thus allowing relative movements between the adjacent particles. In addition, once the shear force at a contact becomes larger than the shear resistance at the contact, there will be relative sliding between the particles (Fig. 2.2*c*). The overall strain of a soil mass will be partly the result of deformation of individual particles and partly the result of relative sliding between particles. However, experience has shown that interparticle sliding, with the resultant rearrangement of the particles, generally makes by far the most important contribution to overall strain. The *mineral skeleton* of soil usually is quite deformable, due to interparticle sliding and rearrangement, even though the individual particles are very rigid.

Thus we see the first consequence of the particulate nature of soil: *the deformation of a mass of soil is controlled by interactions between individual particles, especially by sliding between particles.* Because sliding is a nonlinear and irreversible deformation, we must expect that the stress-strain behavior of soil will be strongly nonlinear and irreversible.[2] Moreover, a study of phenomena at the contact points will be fundamental to the study of soils, and we shall inevitably be concerned with concepts such as friction and adhesion between particles.

There are, of course, a fantastically large number of individual contact points within a soil mass. For example, there will be on the order of 5 million contacts within just 1 cm³ of a fine sand. Hence it is impossible to build up a stress-strain law for soil by considering the behavior at each contact in turn even if we could describe exactly what happens at each contact. Rather it is necessary to rely upon the direct experimental measurement of the properties of a system involving a large number of particles. Nonetheless, study of the behavior at typical contact points still plays an important role; it serves as a guide to the understanding and interpretation of the direct experimental measurements. This situation may be likened to the study of metals: knowledge of the behavior of a single crystal, and the interactions between crystals, guides understanding the behavior of the overall metal and how the properties of the metal may be improved.

If the box in Fig. 2.1 has rigid side walls, the soil will normally decrease in volume as the load is increased. This volume decrease comes about because individual particles nestle closer and closer together. There are shear failures (sliding) at the many individual contact points, but there is no overall shear failure of the soil mass. The vertical load can be increased without limit. Such a process is volumetric *compression*. If the applied load is removed, the soil mass will increase in volume through a reverse process again involving rearrangement of the particles. This process of volume increase is called *expansion*, or in some contexts, *swell*.

[2] This statement means that a plot of stress-strain is not a straight line and is not unique for load-unload cycles.

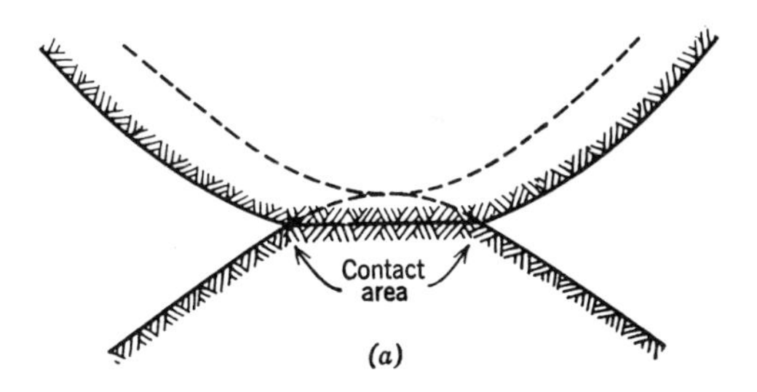

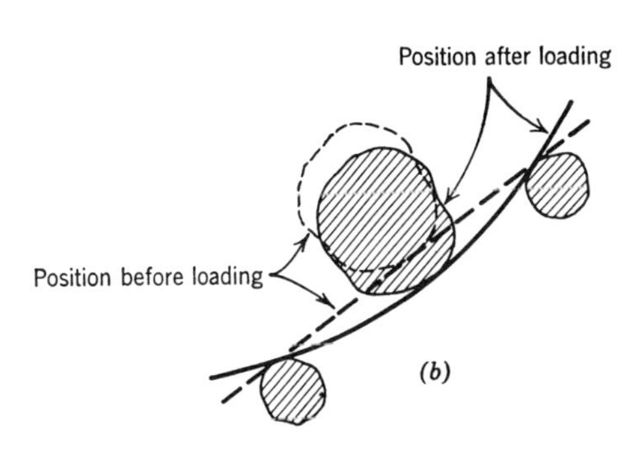

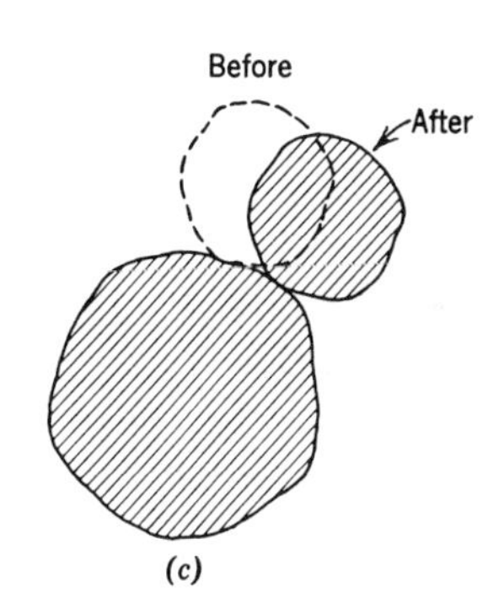

Fig. 2.2 Causes of relative motions among soil particles. (*a*) Motion of particles due to deformation of contacts. Solid lines show surfaces of particles after loading (the lower particle was assumed not to move); dashed lines show surfaces before loading. (*b*) Relative motion of particles due to bending of platelike particles. (*c*) Relative motion of particles due to interparticle sliding.

If, on the other hand, the box has flexible side walls, an overall shear failure can take place. The vertical load at which this failure occurs is related to the *shear strength* of the soil. This shear strength is determined by the resistance to sliding between particles that are trying to move laterally past each other.

The properties of compressibility, expansibility, and shear strength will be studied in detail in later chapters.

2.3 ROLE OF PORE PHASE: CHEMICAL INTERACTION

The spaces among the soil particles are called *pore spaces*. These pore spaces are usually filled with air

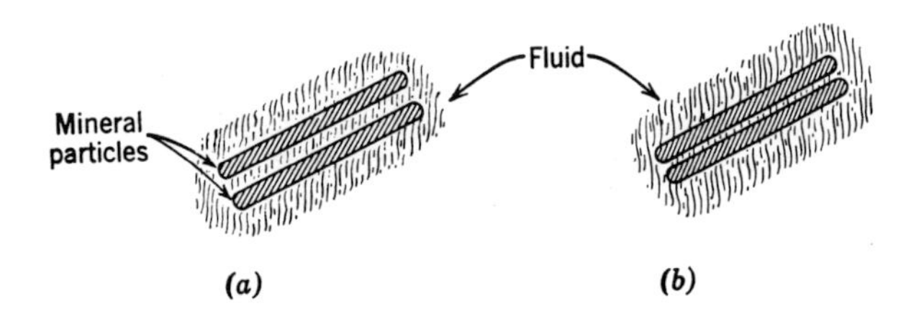

Fig. 2.3 Fluid films surrounding very small soil particles. (a) Before load. (b) Particles squeezed close together by load.

and/or water (with or without dissolved materials). Thus soil is inherently a *multiphase system* consisting of a mineral phase, called the *mineral skeleton*, plus a fluid phase, called the *pore fluid.*

The nature of the pore fluid will influence the magnitude of the shear resistance existing between two particles by introducing chemical matter to the surface of contact. Indeed, in the case of very tiny soil particles, the pore fluid may completely intrude between the particles (see Fig. 2.3). Although these particles are no longer in contact in the usual sense, they still remain in close proximity and can transmit normal and possibly also tangential forces. The spacing of these particles will increase or decrease as the transmitted compressive forces decrease or increase. Hence a new source of overall strain in the soil mass is introduced.

Thus we have a second consequence of the particulate nature of soil: *soil is inherently multiphase, and the constituents of the pore phase will influence the nature of the mineral surfaces and hence affect the processes of force transmission at the particle contacts.* This interaction between the phases is called *chemical interaction.*

2.4 ROLE OF PORE PHASE: PHYSICAL INTERACTION

Let us now return to our box of soil, but now consider a soil whose pore spaces are completely filled with water—a *saturated soil.*

First we assume that the water pressure is hydrostatic; i.e., the pressure in the pore water at any point equals the unit weight of water times the depth of the point below the water surface. For such a condition there will be no flow of water (see Fig. 2.4*a*).

Next we suppose that the water pressure at the base of the box is increased while the overflows hold the level of the water surface constant (Fig. 2.4*b*). Now there must be an upward flow of water. The amount of water that flows will be related to the amount of excess pressure added to the bottom and to a soil property called *permeability*. The more *permeable* a soil, the more water will flow for a given excess of pressure. Later parts of this book consider the factors that determine the permeability of a soil.

If the excess water pressure at the base is increased, a pressure will be reached where the sand is made to boil by the upward flowing water (Fig. 2.4*c*). We say that a *quick* condition is created. Obviously, there has been a *physical interaction* between the mineral skeleton and the pore fluid.

At this stage, the soil will occupy a somewhat greater volume than initially, and clearly the soil has less shear strength in the quick condition than in the normal condition. These changes have occurred even though the total weights of sand and water pressing down have remained unchanged. But we have seen that changes in volume and shear strength come about through changes in the forces at contacts between particles. Hence these contact forces must have been altered by the changes in pressure in the pore phase; that is, these contact forces must be related

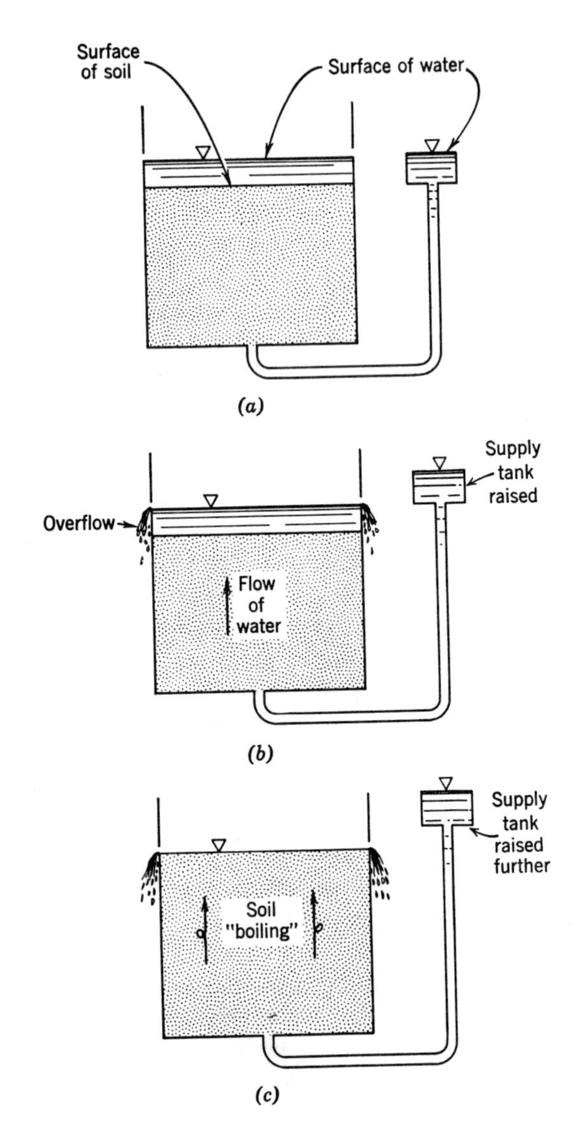

Fig. 2.4 Physical interaction between mineral and pore phases. (a) Hydrostatic condition: no flow. (b) Small flow of water. (c) Quick condition.

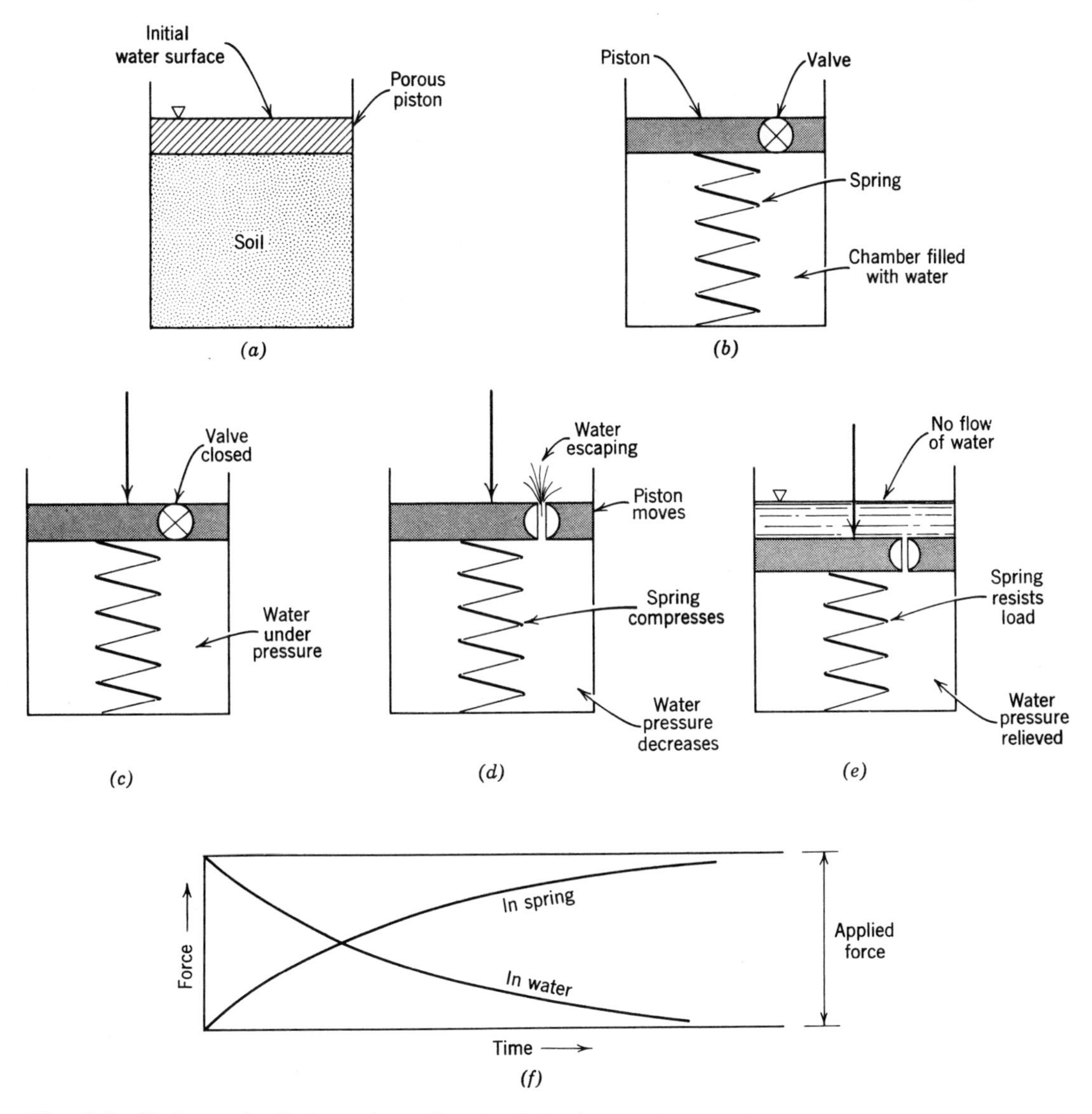

Fig. 2.5 Hydromechanical analogy for load-sharing and consolidation. (*a*) Physical example. (*b*) Hydromechanical analog; initial condition. (*c*) Load applied with valve closed. (*d*) Piston moves as water escapes. (*e*) Equilibrium with no further flow. (*f*) Gradual transfer of load.

to the difference between the stress pressing downward (the *total stress*) and the *pore pressure*. These observations form the basis for the very important concept of *effective stress*.

We have now seen the third consequence of the particulate nature of soil: *water can flow through soil and thus interact with the mineral skeleton, altering the magnitude of the forces at the contacts between particles and influencing the compression and shear resistance of the soil.*

2.5 ROLE OF PORE PHASE: SHARING THE LOAD

Finally, because soil is a multiphase system, it may be expected that a load applied to a soil mass would be carried in part by the mineral skeleton and in part by the pore fluid. This "sharing of the load" is analogous to partial pressures in gases.

The sketches in Fig. 2.5 help us to understand load sharing. Figure 2.5*a* shows a cylinder of saturated soil. The porous piston permits load to be applied to the soil and yet permits escape of the fluid from the pores of the soil. Part (*b*) shows a hydromechanical analog in which the properties of the soil have been "lumped": the resistance of the mineral skeleton to compression is represented by a spring; the resistance to the flow of water through the soil is represented by a valve in an otherwise impermeable piston.

Now suppose a load is applied to the piston of the hydromechanical analog but that the valve is kept closed. The piston load is apportioned by the water and the spring in relation to the stiffness of each. The piston in our

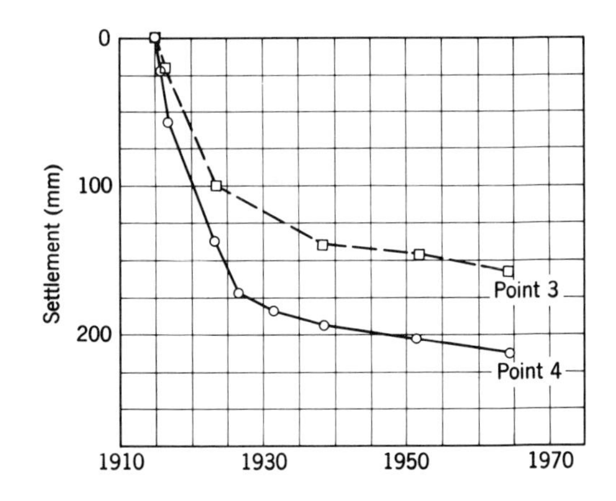

Fig. 2.6 Settlement of Building 10 at M.I.T.

hydromechanical analogy will move very little when the load is applied because the water is relatively incompressible. Since the spring only shortens slightly it will carry very little of the load. Essentially all of the applied load is resisted by an increase in the fluid pressure within the chamber. The conditions at this stage are represented in Fig. 2.5*c*.

Next we open the valve. The fluid pressure within the chamber will force water through this valve (Fig. 2.5*d*). As water escapes the spring shortens and begins to carry a significant fraction of the applied load. There must be a corresponding decrease in the pressure within the chamber fluid. Eventually a condition is reached (Fig. 2.5*e*) in which all of the applied load is carried by the spring and the pressure in the water has returned to the original hydrostatic condition. Once this stage is reached there is no further flow of water.

Only a limited amount of water can flow out through the valve during any interval of time. Hence the process of transferring load from the water to the spring must take place gradually. This gradual change in the way that the load is shared is illustrated in Fig. 2.5*f*.

Sharing the load between the mineral and pore phases also occurs in the physical example and in actual soil problems, although the pore fluid will not always carry all of the applied load initially. We shall return to this subject in detail in Chapter 26. Moreover, in actual problems there will be the same process of a gradual change in the way that the load is shared. This process of gradual squeezing out of water is called *consolidation*, and the time interval involved is the *hydrodynamic time lag*. The amount of compression that has occurred at any time is related not only to the applied load but also to the amount of stress transmitted at the particle contacts, i.e., to the difference between the applied stress and the pore pressure. This difference is called *effective stress*. Consolidation and the reverse process of swelling (which occurs when water is sucked into a soil following load removal) are treated in several chapters.

Here then is the fourth consequence of the particulate nature of soil: *when the load applied to a soil is suddenly changed, this change is carried jointly by the pore fluid and by the mineral skeleton. The change in pore pressure will cause water to move through the soil, hence the properties of the soil will change with time.*

This last consequence was discovered by Karl Terzaghi around 1920. This discovery marked the beginning of modern soil engineering. It was the first of many contributions by Terzaghi, who is truly the "father of soil mechanics."

The most important effect of the hydrodynamic time lag is to cause delayed settlement of structures. That is, the settlement continues for many years after the structure has been completed. Figure 2.6 shows the time-settlement record of two points on Building 10 on the campus of the Massachusetts Institute of Technology. The settlement of this building during the first decade after its completion was the cause of considerable alarm. Terzaghi examined the building when he first came to the United States in 1925, and he correctly predicted that the rate of settlement would decrease with time.

A further look at consolidation. At this stage it is essential that the student have a general appreciation of the duration of the hydrodynamic time lag in various typical soil deposits. For this purpose it is useful to make an intuitive analysis of the consolidation process to learn which soil properties affect the time lag and how they affect it. (Chapter 27 presents a precise derivation and solution for the consolidation process.)

The time required for the consolidation process should be related to two factors:

1. The time should be directly proportional to the volume of water which must be squeezed out of the soil. This volume of water must in turn be related to the product of the stress change, the compressibility of the mineral skeleton, and the volume of the soil.
2. The time should be inversely proportional to how

fast the water can flow through the soil. From fluid mechanics we know that velocity of flow is related to the product of the permeability and the hydraulic gradient, and that the gradient is proportional to the fluid pressure lost within the soil divided by the distance through which the pore fluid must flow.

These considerations can be expressed by the relation

$$t \sim \frac{(\Delta\sigma)(m)(H)}{(k)(\Delta\sigma/H)} \tag{2.1}$$

where

t = the time required to complete some percentage of the consolidation process
$\Delta\sigma$ = the change in the applied stress
m = the compressibility of the mineral skeleton
H = the thickness of the soil mass (per drainage surface)
k = the permeability of the soil

Hence the time required to reach a specified stage in the consolidation process is

$$t \sim \frac{mH^2}{k} \tag{2.2}$$

This relation tells us that the consolidation time:

1. Increases with increasing compressibility.
2. Decreases with increasing permeability.
3. Increases rapidly with increasing size of soil mass.
4. Is independent of the magnitude of the stress change.

The application of this relation is illustrated by Examples 2.1 and 2.2.

▶ Example 2.1

A stratum of sand and a stratum of clay are each 3 m thick. The compressibility of the sand is $\frac{1}{5}$ the compressibility of the clay and the permeability of the sand is 10,000 times that of the clay. What is the ratio of the consolidation time for the clay to the consolidation time of the sand?

Solution.

$$\frac{t_{clay}}{t_{sand}} = \frac{1}{1/5}\,\frac{10\,000}{1} = 50\,000$$ ◀

▶ Example 2.2

A stratum of clay 3 m thick will be 90% consolidated in 10 years. How much time will be required to achieve 90% consolidation in a 12 m-thick stratum of this same clay?

Solution.

$$t \text{ for 12 m stratum} = 10 \text{ years} \times \frac{12^2}{3^2} = 160 \text{ years}$$ ◀

Soils with a significant clay content will require long times for consolidation—from one year to many hundreds of years. Coarse granular soils, on the other hand, will consolidate very quickly, usually in a matter of minutes. As we shall see, this difference in consolidation time is one of the main distinctions among different soils.

2.6 ORGANIZATION OF BOOK

This chapter has described the important consequences of soil being particulate and hence multiphase. As shown in Table 2.1, these consequences are used as the basis for organizing this book.

Part II will study individual particles, the way that they are put together, and the chemical interaction between

Table 2.1 Soil Is a Different Type of Material because It Is Particulate, and Hence Multiphase

Consequence	Example of Importance	Discussed in Part	Concepts Needed from Previous Studies
Stress-strain behavior of mineral skeleton determined by interactions between discrete particles	Great compressibility of soil Strength is frictional in nature, and related to density	II, III	Stress and strain; continuity; limiting equilibrium; Mohr circle
There are chemical interactions between pore fluid and mineral particles	Affects density (and hence strength) which soil will attain under given stress Quick clays	III	Principles of chemical bonding
There are physical interactions between pore fluid and mineral skeleton	Quicksands Slope instabilities affected by ground water	IV	Fluid mechanics: potential flow, laminar flow
Loads applied to soil masses are "shared" by mineral skeleton and pore phase	Consolidation time-lag Delayed failure of slopes	V	

these particles and the pore phase. Part III will study the processes of volume change and shear strength in situations where there is no physical interaction between the phases, i.e., in dry soils. Part IV will then analyze the consequences of the physical interaction between the phases in the cases where the flow of water is governed by natural ground water conditions. Part V will study the transient phenomena that occur after a change in load is applied to a soil.

PROBLEMS

2.1 Cite at least three passages from Chapter 1 that refer to physical interaction between the mineral skeleton and pore phase.

2.2 Cite at least one passage from Chapter 1 that refers to the hydrodynamic time-lag or consolidation effect.

2.3 The time for a clay layer to achieve 99% consolidation is 10 years. What time would be required to achieve 99% consolidation if the layer were twice as thick, five times more permeable, and three times more compressible?

2.4 List the possible components of soil deformation.

2.5 Which component listed in the answer to Problem 2.4 would be most important in each of the following situations:

a. loading a loose array of steel balls;
b. loading an array of parallel plates;
c. unloading a dense sample of mica plates and quartz sand.

ARTHUR CASAGRANDE

Arthur Casagrande was born (August 28, 1902) and educated in Austria. He immigrated to the United States in 1926. There he accepted a research assistantship with the Bureau of Public Roads to work under Terzaghi at M.I.T. While at M.I.T., Professor Casagrande worked on soil classification, shear testing, and frost action in soils. In 1932 he initiated a program in soil mechanics at Harvard University.

Professor Casagrande's work on soil classification, seepage through earth structures, and shear strength has had major influence on soil mechanics. Professor Casagrande has been a very active consultant and has participated in many important jobs throughout the world. His most important influence on soil mechanics, however, has been through his teaching at Harvard. Many of the leaders in soil mechanics were inspired while students of his at Harvard.

Professor Casagrande served as President of the International Society of Soil Mechanics and Foundation Engineering during the period 1961 through 1965. He has been the Rankine Lecturer of the Institution of Civil Engineers and the Terzaghi Lecturer of the American Society of Civil Engineers. He was the first recipient of the Karl Terzaghi Award from the ASCE.

PART II

The Nature of Soil

Part II, consisting of Chapters 3 to 7, examines the nature of soil. Chapter 3 considers an assemblage of soil particles. This chapter is placed at the start of Part II because the student must master the definitions and terms relating the phases in a soil mass before he proceeds with his study of soil mechanics. Chapter 4 looks closely at the individual particles that make up a soil mass. Chapters 5 and 6 consider stress transmission between soil particles on a microscopic scale and the influence of water on these stresses. Part II closes with a treatment (Chapter 7) of the natural soil profile.

CHAPTER 3

Description of an Assemblage of Particles

This chapter considers the description of an assemblage of particles. It presents relationships among the different phases in the assemblage, and discusses particle size distribution and degree of plasticity of the assemblage. The phase relationships are used considerably in soil mechanics to compute stresses. The phase relationships, particle size characteristics, and Atterberg limits are employed to group soils and thus facilitate their study.

3.1 PHASE RELATIONSHIPS

By being a particulate system, an element of soil is inherently "multiphase." Figure 3.1 shows a typical element of soil containing three distinct phases: *solid* (mineral particles), *gas*, and *liquid* (usually water). Figure 3.1*a* represents the three phases as they would typically exist in an element of natural soil. In Part (*b*) the phases have been separated one from the others in order to facilitate the development of the phase relationships. The phases are dimensioned with volumes on the left and weights on the right side of the sketch.

Below the soil elements in Fig. 3.1 are given expressions that relate the various phases. There are three important relationships of volume: *porosity*, *void ratio*, and *degree of saturation*. *Porosity* is the ratio of void volume to total volume and *void ratio* is the ratio of void volume to solid volume. Porosity is usually multiplied by 100 and thus the values are given in percent. Void ratio is expressed in a decimal value, such as a void ratio of 0.55, and can run to values greater than unity. Both porosity and void ratio indicate the relative portion of void volume in a soil sample. This void volume is filled with fluid, either gas or liquid, usually water. Although both terms are employed in soil mechanics, void ratio is the more useful[1] of the two. Two relationships between porosity n and void ratio e are

$$n = \frac{e}{1 + e} \qquad \text{and} \qquad e = \frac{n}{1 - n}$$

The degree of saturation indicates the percentage of the void volume which is filled with water. Thus a value of $S = 0$ indicates a dry soil, $S = 100\%$ indicates a saturated soil, and a value between 0 and 100% indicates a partially saturated soil.

The most useful relationship between phase weights is *water content*, which is the weight of water divided by the weight of solid in a soil element. The water content of a soil sample is readily obtained by: weighing the natural soil; drying it in an oven; weighing the dry soil; and, finally, computing the water content as the difference in initial and dry weights divided by the dry weight. This procedure assumes that all of the volatiles are water, an acceptable assumption except when working with organic soils or soils containing additives such as asphalt. For a saturated soil the water content and void ratio are uniquely related, as one can see by examining the expressions for the two terms. Since it is much easier to obtain weights than to obtain volumes, the soil engineer makes considerable use of changes in water content of a saturated soil to measure volumetric strain.

The lower part of Fig. 3.1 gives expressions for various unit weights, i.e., the weight of a given volume. The *total unit weight* γ_t is, for example, the weight of the entire soil element divided by the volume of the entire element.[2] The *dry unit weight*, often called *dry density*,[3] is the weight of mineral matter divided by the volume of the entire element. Unit weights appear in units of force per volume such as kilonewtons per cubic meter.

[1] During a typical compression of a soil element, both the numerator and the denominator of the porosity decrease, whereas only the numerator of the void ratio decreases. This fact results in void ratio being more useful than porosity for studying soil compression.

[2] The symbol γ is also used for total unit weight.

[3] The two terms are not strictly identical; the units of density are mass per unit volume so that density is $1/g$ times the unit weight, where g = acceleration due to gravity.

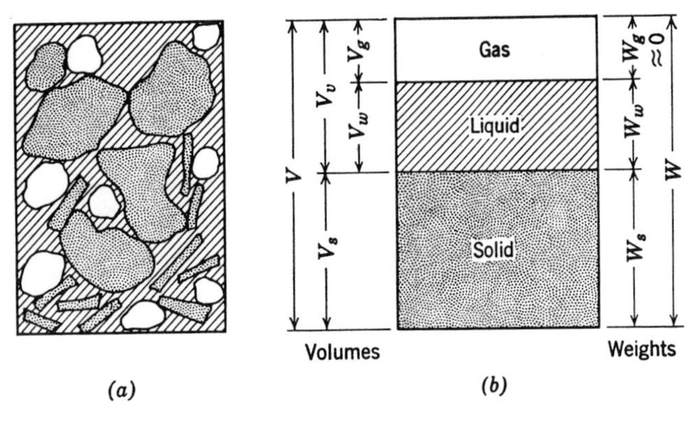

Fig. 3.1 Relationships among soil phases. (*a*) Element of natural soil. (*b*) Element separated into phases.

Volume

Porosity:

$$n = \frac{V_v}{V}$$

Void ratio:

$$e = \frac{V_v}{V_s}$$

Degree of saturation:

$$S = \frac{V_w}{V_v}$$

$$n = \frac{e}{1+e}; \qquad e = \frac{n}{1-n}$$

Weight

Water content:

$$w = \frac{W_w}{W_s}$$

Specific Gravity

Mass:

$$G_m = \frac{\gamma_t}{\gamma_0}$$

Water:

$$G_w = \frac{\gamma_w}{\gamma_0}$$

Solids:

$$G = \frac{\gamma_s}{\gamma_0}$$

γ_0 = Unit weight of water at 4°C $\approx \gamma_w$

Note that $Gw = Se$

Unit Weight

Total:

$$\gamma_t = \frac{W}{V} = \frac{G + Se}{1+e}\gamma_w = \frac{1+w}{1+e}G\gamma_w$$

Solids:

$$\gamma_s = \frac{W_s}{V_s}$$

Water:

$$\gamma_w = \frac{W_w}{V_w}$$

Dry:

$$\gamma_d = \frac{W_s}{V} = \frac{G}{1+e}\gamma_w = \frac{G\gamma_w}{1 + wG/S} = \frac{\gamma_t}{1+w}$$

Submerged (buoyant):

$$\gamma_b = \gamma_t - \gamma_w = \frac{G - 1 - e(1-S)}{1+e}\gamma_w$$

Submerged (saturated soil):

$$\gamma_b = \gamma_t - \gamma_w = \frac{G-1}{1+e}\gamma_w$$

Specific gravity is the unit weight divided by the unit weight of water. Values of specific gravity of solids G for a selected group of minerals[4] are given in Table 3.1.

Table 3.1 Specific Gravities of Minerals

Quartz	2.65
K-Feldspars	2.54–2.57
Na–Ca-Feldspars	2.62–2.76
Calcite	2.72
Dolomite	2.85
Muscovite	2.7–3.1
Biotite	2.8–3.2
Chlorite	2.6–2.9
Pyrophyllite	2.84
Serpentine	2.2–2.7
Kaolinite	2.61[a]
	2.64 ± 0.02
Halloysite (2 H_2O)	2.55
Illite	2.84[a]
	2.60–2.86
Montmorillonite	2.74[a]
	2.75–2.78
Attapulgite	2.30

[a] Calculated from crystal structure.

The expression $Gw = Se$ is useful to check computations of the various relationships.

The student in soil mechanics must understand the meanings of the relationships in Fig. 3.1, convince himself once and for all that they are correct, and add these terms to his active vocabulary. These relationships are basic to most computations in soil mechanics and thus are an essential part of soil mechanics.

Typical Values of Phase Relationships for Granular Soils

Figure 3.2 shows two of the many possible ways that a system of equal-sized spheres can be packed. The dense packings represent the densest possible state for such a system. Looser systems than the simple cubic packing can be obtained by carefully constructing arches within the packing, but the simple cubic packing is the loosest of the stable arrangements. The void ratio and porosity of

[4] Chapter 4 discusses the common soil minerals.

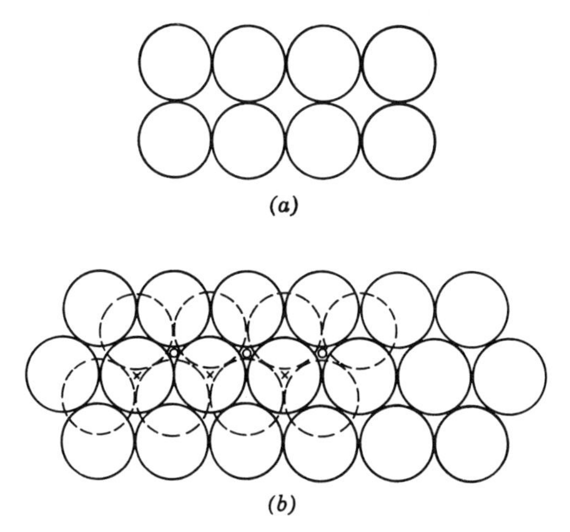

Fig. 3.2 Arrangements of uniform spheres. (*a*) Plan and elevation view: simple cubic packing. (*b*) Plan view: dense packing. Solid circles, first layer; dashed circles, second layer; ∘, location of sphere centers in third layer: face-centered cubic array; ×, location of sphere centers in third layer: close-packed hexagonal array. (From Deresiewicz, 1958.)

these simple packings can be computed from the geometry of the packings, and the results are given in Table 3.2.

This table also gives densities for some typical granular soils in both the "dense" and "loose" states. A variety of tests have been proposed to measure the maximum and

Table 3.2 Maximum and Minimum Densities for Granular Soils

Description	Void Ratio		Porosity (%)		Dry Unit Weight (kN/m³)	
	e_{max}	e_{min}	n_{max}	n_{min}	$\gamma_{d\,min}$	$\gamma_{d\,max}$
Uniform spheres	0.92	0.35	47.6	26.0	—	—
Standard Ottawa sand	0.80	0.50	44	33	14.5	17.3
Clean uniform sand	1.0	0.40	50	29	13.0	18.5
Uniform inorganic silt	1.1	0.40	52	29	12.6	18.5
Silty sand	0.90	0.30	47	23	13.7	20.0
Fine to coarse sand	0.95	0.20	49	17	13.4	21.7
Micaceous sand	1.2	0.40	55	29	11.9	18.9
Silty sand and gravel	0.85	0.14	46	12	14.0	22.9

After B. K. Hough, *Basic Soils Engineering*. Copyright © 1957, The Ronald Press Company, New York.

minimum void ratios (Kolbuszewski, 1948). The test to determine the maximum density usually involves some form of vibration. The test to determine minimum density usually involves pouring oven-dried soil into a container. Unfortunately, the details of these tests have not been entirely standardized, and values of the maximum density and minimum density for a given granular soil depend on the procedure used to determine them. By using special measures, one can obtain densities greater than the so-called maximum density. Densities considerably less than the so-called minimum density can be obtained, especially with very fine sands and silts, by slowly sedimenting the soil into water or by fluffing the soil with just a little moisture present.

The smaller the range of particle sizes present (i.e., the more nearly uniform the soil), the smaller the particles, and the more angular the particles, the smaller the minimum density (i.e., the greater the opportunity for building a loose arrangement of particles). The greater the range of particle sizes present, the greater the maximum density (i.e., the voids among the larger particles can be filled with smaller particles).

A useful way to characterize the density of a natural granular soil is with *relative density* D_r, defined as

$$D_r = \frac{e_{max} - e}{e_{max} - e_{min}} \times 100\%$$

$$= \frac{\gamma_{d\,max}}{\gamma_d} \times \frac{\gamma_d - \gamma_{d\,min}}{\gamma_{d\,max} - \gamma_{d\,min}} \times 100\% \qquad (3.1)$$

where

e_{min} = void ratio of soil in densest condition
e_{max} = void ratio of soil in loosest condition
e = in-place void ratio
$\gamma_{d\,max}$ = dry unit weight of soil in densest condition
$\gamma_{d\,min}$ = dry unit weight of soil in loosest condition
γ_d = in-place dry unit weight

Table 3.3 characterizes the density of granular soils on the basis of relative density.

Table 3.3 Density Description

Relative Density (%)	Descriptive Term
0–15	Very loose
15–35	Loose
35–65	Medium
65–85	Dense
85–100	Very dense

Values of water content for natural granular soils vary from less than 0.1% for air-dry sands to more than 40% for saturated, loose sand.

Typical Values of Phase Relationships for Cohesive Soils

The range of values of phase relationships for cohesive soils is much larger than for granular soils. Saturated sodium montmorillonite at low confining pressure can exist at a void ratio of more than 25; saturated clays

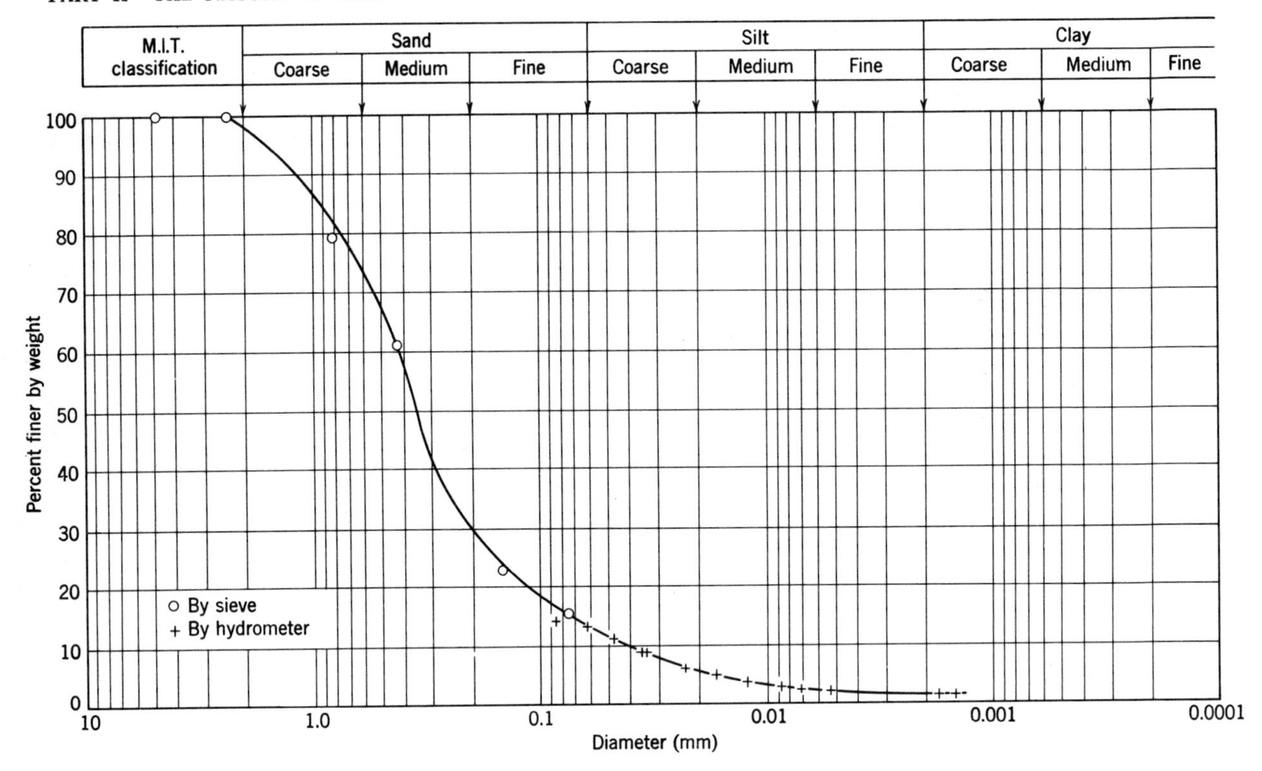

Fig. 3.3 Particle size distribution curve (From Lambe, 1951).

compressed under the high stresses (e.g., 70 MN/m²) that exist at great depths in the ground can have void ratios less than 0.2.

Using the expression $Gw = Se$ (Fig. 3.1), we can compute the water contents corresponding to these quoted values of void ratio:

Sodium montmorillonite	900%
Clay under high pressure	7%

If a sample of oven-dry Mexico City clay sits in the laboratory (temperature = 21°C, relative humidity = 50%), it will absorb enough moisture from the atmosphere for its water content to rise to $2\frac{1}{2}$% or more. Under similar conditions, montmorillonite can get to a water content of 20%.

3.2 PARTICLE SIZE CHARACTERISTICS

The particle size distribution of an assemblage of soil particles is expressed by a plot of percent finer by weight versus diameter in millimeters, as shown in Fig. 3.3. Using the definition for sand, silt, and clay noted at the top of this figure[5] we can estimate the make-up of the soil sample as:

Gravel	2%
Sand	85%
Silt	12%
Clay	1%

[5] This set of particle size definitions is convenient and widely used. A slightly different set is given in Tables 3.5 and 3.6.

The uniformity of a soil can be expressed by the *uniformity coefficient*, which is the ratio of D_{60} to D_{10}, where D_{60} is the soil diameter at which 60% of the soil weight is finer and D_{10} is the corresponding value at 10% finer. A soil having a uniformity coefficient smaller than about 2 is considered "uniform." The uniformity of the soil whose distribution curve is shown in Fig. 3.3 is 10. This soil would be termed a "well-graded silty sand."

There are many reasons, both practical and theoretical, why the particle size distribution curve of a soil is only approximate. As discussed in Chapter 4, the definition of particle size is different for the coarse particles and the fine particles.

The accuracy of the distribution curves for fine-grained soils is more questionable than the accuracy of the curves for coarse soils. The chemical and mechanical treatments given natural soils prior to the performance of a particle size analysis—especially for a hydrometer analysis—usually result in effective particle sizes that are quite different from those existing in the natural soil. Even if an exact particle size curve were obtained, it would be of only limited value. Although the behavior of a cohesionless soil can often be related to particle size distribution, the behavior of a cohesive soil usually depends much more on geological history and structure than on particle size.

In spite of their serious limitations, particle size curves, particularly those of sands and silts, do have practical value. Both theory and laboratory experiments show

Table 3.4 Atterberg Limits of Clay Minerals

Mineral	Exchange-able Ion	Liquid Limit (%)	Plastic Limit (%)	Plasticity Index (%)	Shrinkage Limit (%)
Montmorillonite	Na	710	54	656	9.9
	K	660	98	562	9.3
	Ca	510	81	429	10.5
	Mg	410	60	350	14.7
	Fe	290	75	215	10.3
	Fe[a]	140	73	67	—
Illite	Na	120	53	67	15.4
	K	120	60	60	17.5
	Ca	100	45	55	16.8
	Mg	95	46	49	14.7
	Fe	110	49	61	15.3
	Fe[a]	79	46	33	—
Kaolinite	Na	53	32	21	26.8
	K	49	29	20	—
	Ca	38	27	11	24.5
	Mg	54	31	23	28.7
	Fe	59	37	22	29.2
	Fe[a]	56	35	21	—
Attapulgite	H	270	150	120	7.6

Data from Cornell, 1951.
[a] After five cycles of wetting and drying.

that soil permeability and capillarity are related to some effective particle diameter. These relationships are discussed later in the book.

The method of designing inverted filters for dams, levees, etc., uses the particle size distribution curves of the soils involved. This method is based on the relationship of particle size to permeability, along with experimental data on the particle size distribution required to prevent the migration of particles when water flows through the soil. Also, the most common criterion for establishing the susceptibility of soils to frost damage is based on particle size.

3.3 ATTERBERG LIMITS

Largely through the work of A. Atterberg and A. Casagrande (1948), the Atterberg limits and related indices have become very useful characteristics of assemblages of soil particles. The limits are based on the concept that a fine-grained soil can exist in any of four states depending on its water content. Thus a soil is *solid* when dry, and upon the addition of water proceeds through the *semisolid*, *plastic*, and finally *liquid* states, as shown in Fig. 3.4. The water contents at the boundaries between adjacent states are termed *shrinkage limit*, *plastic limit*, and *liquid limit*. The four indices noted in Fig. 3.4 are computed from these limits.

The liquid limit is determined by measuring the water content and the number of blows required to close a specific width groove for a specified length in a standard liquid limit device. The plastic limit is determined by measuring the water content of the soil when threads of the soil 3 mm in diameter begin to crumble. The shrinkage limit is determined as the water content after just enough water is added to fill all the voids of a dry pat of soil. The detailed procedures for determining these three limits are given in Lambe (1951). Table 3.4 gives Atterberg limits for some common clay minerals.

Physical Significance of the Atterberg Limits

The concept of a soil existing in various states, depending on its moisture content, is a sound one. The greater the amount of water a soil contains, the less interaction there will be between adjacent particles and the more the soil should behave like a liquid.

In a very general way, we may expect that the water that is attracted to surfaces of the soil particles should not behave as a liquid. Thus, if we compare two soils A and B, and if soil A has a greater tendency to attract water to the particle surfaces, then we should expect that the water content at which the two soils begin to behave as a liquid will be greater for soil A than for soil B. That is, soil A should have a larger liquid limit than soil B. We might expect that the same reasoning would apply to the plastic limit, and hence to the plasticity index.

However, the limits between the various states have

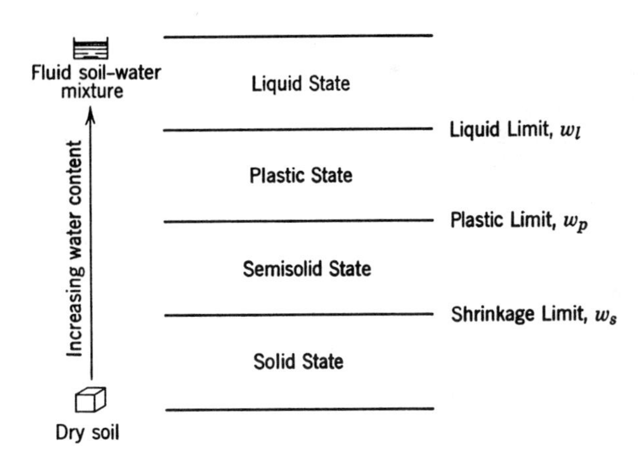

Fig. 3.4 Atterberg limits and related indices.

Plasticity Index:

$$I_p \text{ or } PI = w_l - w_p$$

Flow Index:

I_f = Slope of flow curve (flow curve is plot of water content vs number of blows, log scale)

Toughness Index:

$$I_t = \frac{I_p}{I_f}$$

$$\left.\begin{array}{r}\text{Water-plasticity Ratio } B\\ \text{Liquidity Index } LI \text{ or } I_L\end{array}\right\} = \frac{w_n - w_p}{w_l - w_p}$$

w_n = natural water content

been set arbitrarily, and thus it is unlikely that the limits per se can be completely interpreted fundamentally. That is, it is unlikely that the magnitude of the liquid limit for a particular soil can be tied quantitatively to the thickness of the adsorbed water layer.

The difficulty of interpreting fundamentally the Atterberg limits should not deter their use. The student should think of them as approximate boundaries between the states in which fine-grained soils can exist, and not worry too much about attaching significance to the exact value of the arbitrarily determined limits.

Relationship of Atterberg Limits to Composition of Soil

Let us make further use of the concept that the Atterberg limits for a soil are related to the amount of water that is attracted to the surfaces of the soil particles. Because of the great increase in surface area per mass with decreasing particle size (as discussed in Chapter 5), it may be expected that the amount of attracted water will be largely influenced by the amount of clay that is present in the soil. On the basis of this reasoning, Skempton (1953) defined a quantity he called activity:

$$\text{Activity of a clay} = \frac{\text{plasticity index}}{\%\text{ by weight finer than } 2\ \mu\text{m}} \tag{3.2}$$

Figure 3.5 shows some results obtained on prepared mixtures of various percentages of particles less than and greater than 2 μm. In Part (*a*) several natural soils were separated into fractions greater and less than 2 μm and then the two fractions were recombined as desired. The results in the right diagram were obtained on clay minerals mixed with quartz sand.

Engineering Use of Atterberg Limits

The Atterberg limits and related indices have proved to be very useful for soil identification and classification, as shown in the next section. The limits are often used directly in specifications for controlling soil for use in fill and in semiempirical methods of design.

The plasticity index, indicating the magnitude of water content range over which the soil remains plastic, and the liquidity index, indicating the nearness of a natural soil to the liquid limit, are particularly useful characteristics of soil. One must realize, however, that all of the limits and indices with the exception of the shrinkage limit are determined on soils that have been thoroughly worked into a uniform soil-water mixture. The limits therefore give no indication of particle fabric or residual bonds between particles which may have been developed in the natural soil but are destroyed in preparing the specimen for the determinations of the limits.

3.4 SOIL CLASSIFICATION

The direct approach to the solution of a soil engineering problem consists of first measuring the soil property needed and then employing this measured value in some rational expression to determine the answer to the problem. Examples of this approach are:

1. To determine the rate of water flowing through a sample of soil, measure the permeability of the soil, and use this value together with a flow net and Darcy's law to determine the flow.
2. To determine the settlement of a building, measure the compressibility of the soil, and use this value in

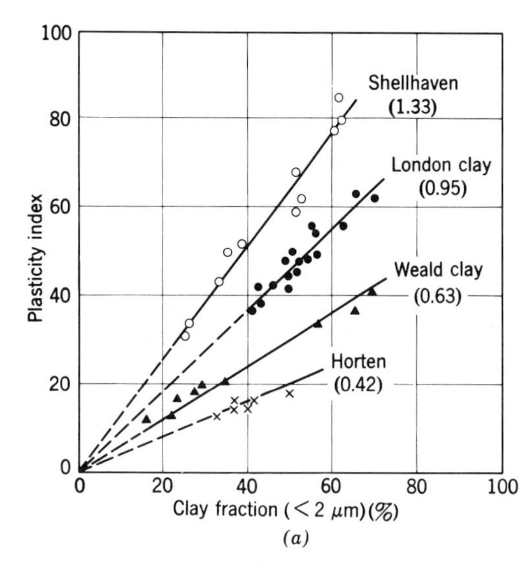

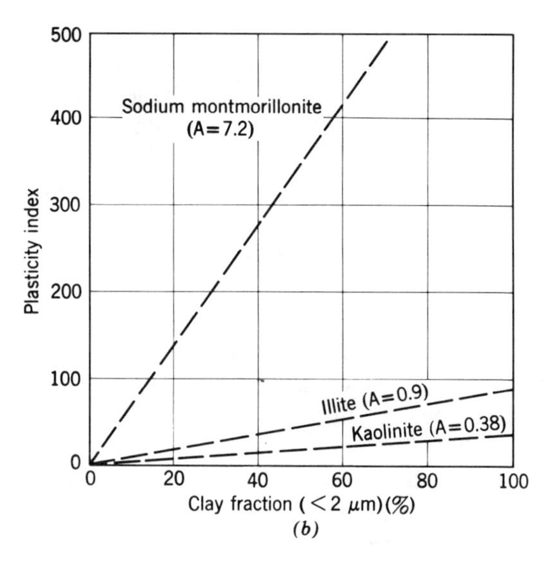

Fig. 3.5 Relation between plasticity index and clay fraction. Figures in parentheses are the "activities" of the clays. (From Skempton, 1953.)

Table 3.5 Unified Soil Classification

			Field Identification Procedures (Excluding particles larger than 75 μm and basing fractions on estimated weights)	Group Symbols [a]	Typical Names	Information Required for Describing Soils	Laboratory Classification Criteria	
Coarse-grained soils. More than half of material is *larger* than 75 μm sieve size[b] (The 75 μm sieve size is about the smallest particle visible to naked eye)	Gravels. More than half of coarse fraction is larger than 4 mm sieve size	Clean gravels (little or no fines)	Wide range in grain size and substantial amounts of all intermediate particle sizes	*GW*	Well graded gravels, gravel-sand mixtures, little or no fines	Give typical name; indicate approximate percentages of sand and gravel; maximum size; angularity, surface condition, and hardness of the coarse grains; local or geologic name and other pertinent descriptive information; and symbols in parentheses	$C_U = \frac{D_{60}}{D_{10}}$ Greater than 4; $C_C = \frac{(D_{30})^2}{D_{10} \times D_{60}}$ Between 1 and 3	
			Predominantly one size or a range of sizes with some intermediate sizes missing	*GP*	Poorly graded gravels, gravel-sand mixtures, little or no fines		Not meeting all gradation requirements for *GW*	
		Gravels with fines (appreciable amount of fines)	Nonplastic fines (for identification procedures see *ML* below)	*GM*	Silty gravels, poorly graded gravel-sand-silt mixtures	For undisturbed soils add information on stratification, degree of compactness, cementation, moisture conditions and drainage characteristics	Atterberg limits below "A" line, or *PI* less than 4	Above "A" line with *PI* between 4 and 7 are *borderline* cases requiring use of dual symbols
			Plastic fines (for identification procedures, see *CL* below)	*GC*	Clayey gravels, poorly graded gravel-sand-clay mixtures		Atterberg limits above "A" line, with *PI* greater than 7	
	Sands. More than half of coarse fraction is smaller than 4 mm sieve size	Clean sands (little or no fines)	Wide range in grain sizes and substantial amounts of all intermediate particle sizes	*SW*	Well graded sands, gravelly sands, little or no fines	Example: *Silty sand*, gravelly; about 20% hard, angular gravel particles 12 mm maximum size; rounded and subangular sand grains coarse to fine, about 15% non-plastic fines with low dry strength; well compacted and moist in place; alluvial sand; (*SM*)	$C_U = \frac{D_{60}}{D_{10}}$ Greater than 6; $C_C = \frac{(D_{30})^2}{D_{10} \times D_{60}}$ Between 1 and 3	
			Predominantly one size or a range of sizes with some intermediate sizes missing	*SP*	Poorly graded sands, gravelly sands, little or no fines		Not meeting all gradation requirements for *SW*	
		Sands with fines (appreciable amount of fines)	Nonplastic fines (for identification procedures, see *ML* below)	*SM*	Silty sands, poorly graded sand-silt mixtures		Atterberg limits below "A" line or *PI* less than 5	Above "A" line with *PI* between 4 and 7 are *borderline* cases requiring use of dual symbols
			Plastic fines (for identification procedures, see *CL* below)	*SC*	Clayey sands, poorly graded sand-clay mixtures		Atterberg limits below "A" line with *PI* greater than 7	

Use grain size curve in identifying the fractions as given under field identification.

Determine percentages of gravel and sand from grain size curve. Depending on percentage of fines (fraction smaller than 75 μm sieve size) coarse grained soils are classified as follows: Less than 5% *GW, GP, SW, SP*; More than 12% *GM, GC, SM, SC*; 5% to 12% *Borderline* cases requiring use of dual symbols.

Identification Procedures on Fraction Smaller than 380 μm Sieve Size

		Dry Strength (crushing characteristics)	Dilatancy (reaction to shaking)	Toughness (consistency near plastic limit)	Group Symbols [a]	Typical Names	Information Required for Describing Soils
Fine-grained soils. More than half of material is *smaller* than 75 μm sieve size (The 75 μm sieve size is about the smallest particle visible to naked eye)	Silts and clays liquid limit less than 50	None to slight	Quick to slow	None	*ML*	Inorganic silts and very fine sands, rock flour, silty or clayey fine sands with slight plasticity	Give typical name; indicate degree and character of plasticity, amount and maximum size of coarse grains; colour in wet condition, odour if any, local or geologic name, and other pertinent descriptive information, and symbol in parentheses
		Medium to high	None to very slow	Medium	*CL*	Inorganic clays of low to medium plasticity, gravelly clays, sandy clays, silty clays, lean clays	
		Slight to medium	Slow	Slight	*OL*	Organic silts and organic silt-clays of low plasticity	For undisturbed soils add information on structure, stratification, consistency in undisturbed and remoulded states, moisture and drainage conditions
	Silts and clays liquid limit greater than 50	Slight to medium	Slow to none	Slight to medium	*MH*	Inorganic silts, micaceous or diatomaceous fine sandy or silty soils, elastic silts	
		High to very high	None	High	*CH*	Inorganic clays of high plasticity, fat clays	Example: *Clayey silt*, brown; slightly plastic; small percentage of fine sand; numerous vertical root holes; firm and dry in place; loess; (*ML*)
		Medium to high	None to very slow	Slight to medium	*OH*	Organic clays of medium to high plasticity	
Highly Organic Soils		Readily identified by colour, odour, spongy feel and frequently by fibrous texture			*Pt*	Peat and other highly organic soils	

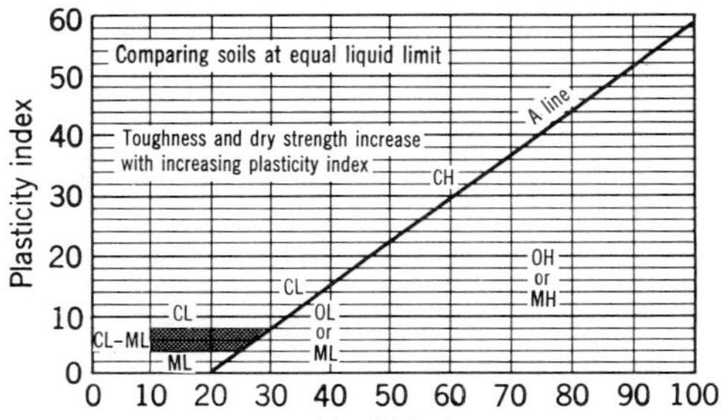

Plasticity index (0–60) vs. Liquid limit (0–100): Comparing soils at equal liquid limit; Toughness and dry strength increase with increasing plasticity index; A line; CH; CL; CL-ML; ML; OL or ML; OH or MH

Plasticity chart
for laboratory classification of fine grained soils

From Wagner, 1957.
[a] *Boundary classifications*. Soils possessing characteristics of two groups are designated by combinations of group symbols. For example *GW–GC*, well graded gravel-sand mixture with clay binder.
[b] All sieve sizes on this chart are U.S. standard.

Field Identification Procedure for Fine Grained Soils or Fractions

These procedures are to be performed on the minus 380 μm sieve size particles. For field classification purposes, screening is not intended, simply remove by hand the coarse particles that interfere with the tests.

Dilatancy (Reaction to shaking):
After removing particles larger than 380 μm sieve size, prepare a pat of moist soil with a volume of about 8000 mm³. Add enough water if necessary to make the soil soft but not sticky.
Place the pat in the open palm of one hand and shake horizontally, striking vigorously against the other hand several times. A positive reaction consists of the appearance of water on the surface of the pat which changes to a livery consistency and becomes glossy. When the sample is squeezed between the fingers, the water and gloss disappear from the surface, the pat stiffens and finally it cracks or crumbles. The rapidity of appearance of water during shaking and of its disappearance during squeezing assist in identifying the character of the fines in a soil.
Very fine clean sands give the quickest and most distinct reaction whereas a plastic clay has no reaction. Inorganic silts, such as a typical rock flour, show a moderately quick reaction.

Dry Strength (Crushing characteristics):
After removing particles larger than 380 μm sieve size, mould a pat of soil to the consistency of putty, adding water if necessary. Allow the pat to dry completely by oven, sun or air drying, and then test its strength by breaking and crumbling between the fingers. This strength is a measure of the character and quantity of the colloidal fraction contained in the soil. The dry strength increases with increasing plasticity.
High dry strength is characteristic for clays of the CH group. A typical inorganic silt possesses only very slight dry strength. Silty fine sands and silts have about the same slight dry strength, but can be distinguished by the feel when powdering the dried specimen. Fine sand feels gritty whereas a typical silt has the smooth feel of flour.

Toughness (Consistency near plastic limit):
After removing particles larger than the 380 μm sieve size, a specimen of soil about 12 mm cube in size, is moulded to the consistency of putty. If too dry, water must be added and if sticky, the specimen should be spread out in a thin layer and allowed to lose some moisture by evaporation. Then the specimen is rolled out by hand on a smooth surface or between the palms into a thread about 3 mm in diameter. The thread is then folded and re-rolled repeatedly. During this manipulation the moisture content is gradually reduced and the specimen stiffens, finally loses its plasticity, and crumbles when the plastic limit is reached.
After the thread crumbles, the pieces should be lumped together and a slight kneading action continued until the lump crumbles.
The tougher the thread near the plastic limit and the stiffer the lump when it finally crumbles, the more potent is the colloidal clay fraction in the soil. Weakness of the thread at the plastic limit and quick loss of coherence of the lump below the plastic limit indicate either inorganic clay of low plasticity, or materials such as kaolin-type clays and organic clays which occur below the A-line.
Highly organic clays have a very weak and spongy feel at the plastic limit.

Table 3.6 Soil Components and Fractions

Soil	Soil Component	Symbol	Grain Size Range and Description	Significant Properties
Coarse-grained components	Boulder	None	Rounded to angular, bulky, hard, rock particle, average diameter more than 300 mm	Boulders and cobbles are very stable components, used for fills, ballast, and to stabilize slopes (riprap). Because of size and weight, their occurrence in natural deposits tends to improve the stability of foundations. Angularity of particles increases stability.
	Cobble	None	Rounded to angular, bulky, hard, rock particle, average diameter smaller than 300 mm but larger than 150 mm	
	Gravel	*G*	Rounded to angular bulky, hard, rock particle, passing 75 μm sieve retained on 4 mm sieve	Gravel and sand have essentially same engineering properties differing mainly in degree. The 4 mm sieve is arbitrary division, and does not correspond to significant change in properties. They are easy to compact, little affected by moisture, not subject to frost action. Gravels are generally more perviously stable, resistant to erosion and piping than are sands. The well-graded sands and gravels are generally less pervious and more stable than those which are poorly graded (uniform gradation). Irregularity of particles increases the stability slightly. Finer, uniform sand approaches the characteristics of silt: i.e., decrease in permeability and reduction in stability with increase in moisture.
	Coarse		75 to 19 mm	
	Fine		19 to 4 mm	
	Sand	*S*	Rounded to angular, bulky, hard, rock particle, passing 4 mm sieve retained on 75 μm sieve	
	Coarse		4 mm to 1.7 mm sieves	
	Medium		1.7 mm to 380 μm sieves	
	Fine		380 μm to 75 μm sieves	
Fine-grained components	Silt	*M*	Particles smaller than 75 μm sieve identified by behavior: that is, slightly or non-plastic regardless of moisture and exhibits little or no strength when air dried	Silt is inherently unstable, particularly when moisture is increased, with a tendency to become quick when saturated. It is relatively impervious, difficult to compact, highly susceptible to frost heave, easily erodible and subject to piping and boiling. Bulky grains reduce compressibility; flaky grains, i.e., mica, diatoms, increase compressibility, produce an "elastic" silt.
	Clay	*C*	Particles smaller than 75 μm sieve identified by behavior; that is, it can be made to exhibit plastic properties within a certain range of moisture and exhibits considerable strength when air dried	The distinguishing characteristic of clay is cohesion or cohesive strength, which increases with decrease in moisture. The permeability of clay is very low, it is difficult to compact when wet and impossible to drain by ordinary means, when compacted is resistant to erosion and piping, is not susceptible to frost heave, is subject to expansion and shrinkage with changes in moisture. The properties are influenced not only by the size and shape (flat, plate-like particles) but also by their mineral composition; i.e., the type of clay-mineral, and chemical environment or base exchange capacity. In general, the montmorillonite clay mineral has greatest, illite and kaolinite the least, adverse effect on the properties.
	Organic matter	*O*	Organic matter in various sizes and stages of decomposition	Organic matter present even in moderate amounts increases the compressibility and reduces the stability of the fine-grained components. It may decay causing voids or by chemical alteration change the properties of a soil, hence organic soils are not desirable for engineering uses.

From Wagner, 1957.
Note. The symbols and fractions were developed for the Unified Classification System. The sand fractions are not equal divisions on a logarithmic plot; the 1.7 mm was selected because of the significance attached to that size by some investigators. The 380 μm size was chosen because the "Atterberg limits" tests are performed on the fraction of soil finer than the 380 μm.

Table 3.7 Engineering Use Chart

Typical Names of Soil Groups	Group Symbols	Important Properties				Relative Desirability for Various Uses									
						Rolled Earth Dams			Canal Sections		Foundations		Roadways		
													Fills		
		Permeability when Compacted	Shearing Strength when Compacted and Saturated	Compressibility when Compacted and Saturated	Workability as a Construction Material	Homogeneous Embankment	Core	Shell	Erosion Resistance	Compacted Earth Lining	Seepage Important	Seepage not Important	Frost Heave not Possible	Frost Heave Possible	Surfacing
Well-graded gravels, gravel-sand mixtures, little or no fines	*GW*	pervious	excellent	negligible	excellent	—	—	1	1	—	—	1	1	1	3
Poorly graded gravels, gravel-sand mixtures, little or no fines	*GP*	very pervious	good	negligible	good	—	—	2	2	—	—	3	3	3	—
Silty gravels, poorly graded gravel-sand-silt mixtures	*GM*	semipervious to impervious	good	negligible	good	2	4	—	4	4	1	4	4	9	5
Clayey gravels, poorly graded gravel-sand-clay mixtures	*GC*	impervious	good to fair	very low	good	1	1	—	3	1	2	6	5	5	1
Well-graded sands, gravelly sands, little or no fines	*SW*	pervious	excellent	negligible	excellent	—	—	3 if gravelly	6	—	—	2	2	2	4
Poorly graded sands, gravelly sands, little or no fines	*SP*	pervious	good	very low	fair	—	—	4 if gravelly	7 if gravelly	—	—	5	6	4	—
Silty sands, poorly graded sand-silt mixtures	*SM*	semipervious to impervious	good	low	fair	4	5	—	8 if gravelly	5 erosion critical	3	7	8	10	6
Clayey sands, poorly graded sand-clay mixtures	*SC*	impervious	good to fair	low	good	3	2	—	5	2	4	8	7	6	2
Inorganic silts and very fine sands, rock flour, silty or clayey fine sands with slight plasticity	*ML*	semipervious to impervious	fair	medium	fair	6	6	—	—	6 erosion critical	6	9	10	11	—
Inorganic clays of low to medium plasticity, gravelly clays, sandy clays, silty clays, lean clays	*CL*	impervious	fair	medium	good to fair	5	3	—	9	3	5	10	9	7	7
Organic silts and organic silt-clays of low plasticity	*OL*	semipervious to impervious	poor	medium	fair	8	8	—	—	7 erosion critical	7	11	11	12	—
Inorganic silts, micaceous or diatomaceous fine sandy or silty soils, elastic silts	*MH*	semipervious to impervious	fair to poor	high	poor	9	9	—	—	—	8	12	12	13	—
Inorganic clays of high plasticity, fat clays	*CH*	impervious	poor	high	poor	7	7	—	10	8 volume change critical	9	13	13	8	—
Organic clays of medium to high plasticity	*OH*	impervious	poor	high	poor	10	10	—	—	—	10	14	14	14	—
Peat and other highly organic soils	*Pt*	—	—	—	—	—	—	—	—	—	—	—	—	—	—

From Wagner, 1957.

settlement equations based on Terzaghi's theory of consolidation.

3. To evaluate the stability of a slope, measure the shear strength of the soil and substitute this value in an expression based on the laws of statics.

To measure fundamental soil properties like permeability, compressibility, and strength can be difficult, time consuming, and expensive. In many soil engineering problems, such as pavement design, there are no rational expressions available for the analysis for the solution. For these reasons, sorting soils into groups showing similar behavior may be very helpful. Such sorting is *soil classification.*

Soil classification is thus the placing of a soil into a group of soils all of which exhibit similar behavior. The correlation of behavior with a group in a soil classification system is usually an empirical one developed through considerable experience. Soil classification permits us to solve many types of simple soil problems and guide the test program if the difficulty and importance of the problem dictate further investigation.

Most soil classifications employ very simple index-type tests to obtain the characteristics of the soil needed to place it in a given group. Clearly a soil classification loses its value if the index tests become more complicated than the test to measure directly the fundamental property needed. The most commonly used characteristics are particle size and plasticity.

Since soil classifications are developed as an attempt to aid in the solution of problems, classifications for many types of problems have grown. Thus, for use in flow problems, soils are described as having degrees of permeability such as high, medium, low, very low, practically impermeable. The Corps of Engineers has developed a frost susceptibility classification in which, on the basis of particle size, we can classify soil in categories of similar frost behavior. The Bureau of Public Roads developed a classification for soils in highway construction. The Corps of Engineers and FAA each developed a classification for airfield construction. In 1952 the Bureau of Reclamation and the Corps of Engineers developed a "unified system" intended for use in all engineering problems involving soils. This classification is presented in Tables 3.5 and 3.6. Table 3.7 gives a general indication of the permeability, strength, and compressibility of the various soil groups along with an indication of the relative desirability of each group for use in earth dams, canal sections, foundations, and runways.

Soil classification has proved to be a valuable tool to the soil engineer. It helps the engineer by giving him general guidance through making available in an empirical manner the results of field experience. The soil engineer must be cautious, however, in his use of soil classification. Solving flow, compression, and stability problems merely on the basis of soil classification can lead to disastrous results. As this book will show in subsequent chapters, empirical correlations between index properties and fundamental soil behavior have many large deviations.

3.5 SUMMARY OF MAIN POINTS

1. There are a number of terms (given in Fig. 3.1) used to express the phase relationships in an element of soil. These terms are an essential component of soil mechanics.
2. The looseness of a sand is expressed by its relative density, which is a most reliable indicator of the behavior of the sand.
3. The particle size distribution and the Atterberg limits are useful index tests for classifying soils. Since the conduct of these tests inherently involves disturbance of the soil, they may not give a good indication of the behavior of the *in situ,* undisturbed soil.

PROBLEMS

3.1 Four soil samples, each having a void ratio of 0.76 and a specific gravity of 2.74, have degrees of saturation of 85, 90, 95, and 100%. Determine the unit weight for each of the four samples.

3.2 A cubic meter of soil in its natural state weighs 17.75 kN; after being dried it weighs 15.08. The specific gravity of the soil is 2.70. Determine the degree of saturation, void ratio, porosity, and water content for the soil as it existed in its natural state.

3.3 The mass of a container of saturated soil was 113.27 g before it was placed in an oven and 100.06 g after it remained in the oven overnight. The mass of the container alone is 49.31 g; the specific gravity of the soil is 2.80. Determine the void ratio, porosity, and water content of the original soil sample.

3.4 A saturated soil has a unit weight of 18.85 kN/m^3 and a water content of 32.5%. Determine the void ratio and specific gravity of the soil.

3.5 A sample of dry sand having a unit weight of 16.50 kN/m^3 and a specific gravity of 2.70 is placed in the rain. During the rain the volume of the sample remains constant but the degree of saturation increases to 40%. Determine the unit weight and water content of the soil after being in the rain.

3.6 Determine the submerged unit weight of each of the following chunks of saturated soil:

a. A silty sand, total unit weight = 20.58 kN/m^3;

b. A lean clay, total unit weight = 19.17 kN/m^3;

c. A very plastic clay, total unit weight = 16.65 kN/m^3.

Assume reasonable values for any additional data which you need.

3.7 For a soil with a specific gravity of 2.70 prepare a chart in which total unit weight (units of kN/m^3, range of 10–25) is plotted as ordinate and void ratio (range of 0.2–1.8) is plotted as abscissa. Plot for the three percentages of saturation of 0, 50, and 100%.

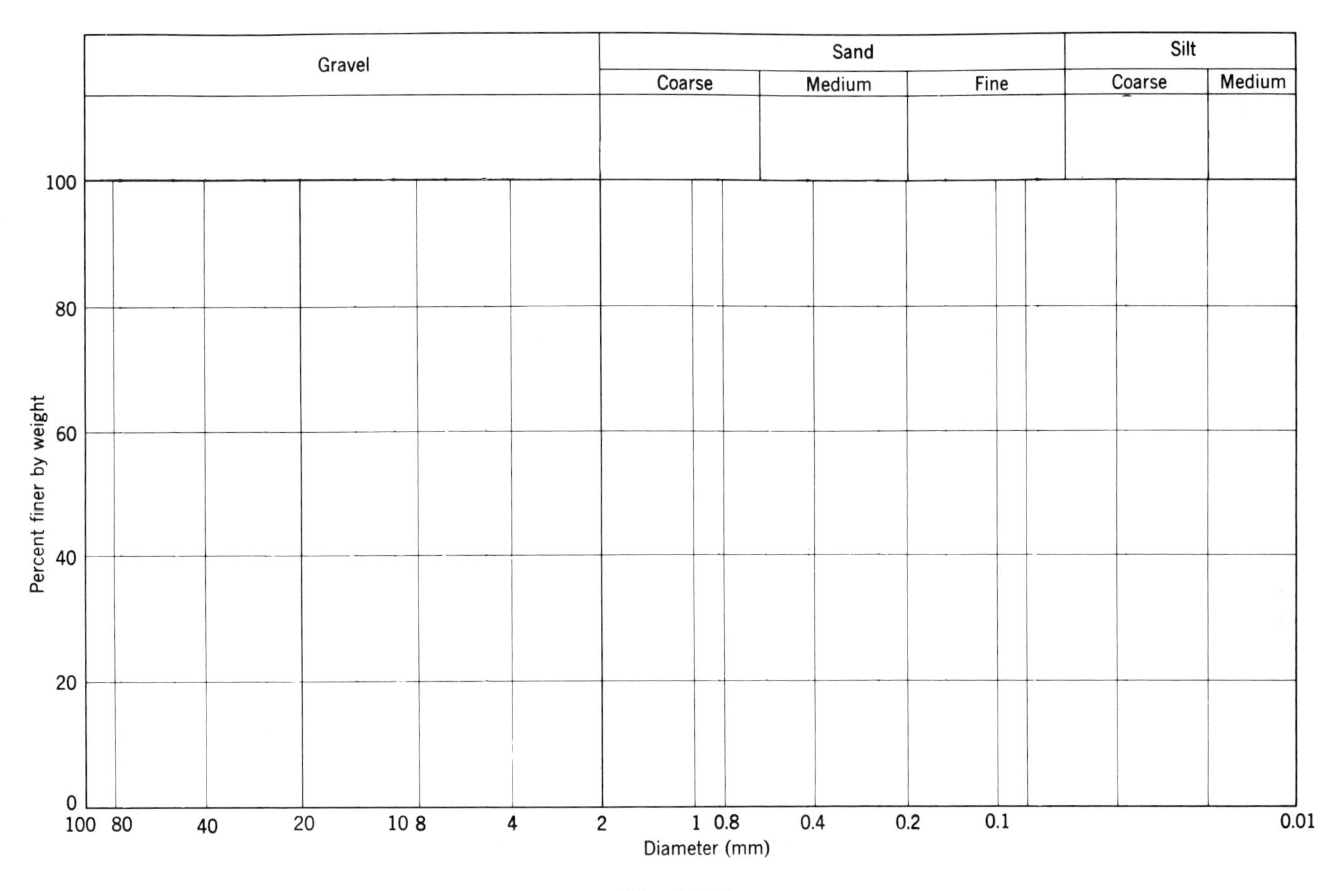

Fig. P3.10

3.8 Prove $Gw = Se$

3.9 A sample of parallel kaolinite particles (all have the size shown in Fig. 5.6) is saturated. The water content is 30%. What is the average particle spacing?

3.10 A sieve analysis on a soil yields the following results:

Sieve	75 mm	50 mm	25 mm	12 mm	4 mm	1.7 mm
Percentage passing	100	95	84	74	62	55

Sieve	840 μm	380 μm	250 μm	150 μm	75 μm
Percentage passing	44	32	24	16	9

a. Plot the particle size distribution of this soil on Fig. P3.10 and classify the soil on the basis of the scale shown in the figure.

b. Comment on the suitability of this soil as drainage material behind a concrete retaining wall.

Hints. (*a*) Use Tables 3.5–3.7 to predict whether or not soil will be pervious, easy to work as construction material, etc. (*b*) A common guide for frost susceptibility is percentage finer than 0.02 mm must be less than 3% for material to be nonfrost susceptible.

3.11 Prove that the identity given by Eq. 3.1 is correct.

CHAPTER 4

Description of an Individual Soil Particle

A sample of soil consists of an assemblage of many individual soil particles with air and/or liquid filling the voids among the particles. This chapter examines the individual soil particle.

4.1 APPEARANCE OF A SOIL PARTICLE

Particle Size

The size of a particle, other than a sphere or cube, cannot be uniquely defined by a single linear dimension. The meaning of "particle size" therefore depends on the dimension that was recorded and how it was obtained. Two common ways of determining particle size are a *sieve analysis*[1] for particles larger than approximately 0.06 mm and a *hydrometer analysis*[1] for smaller particles. In the sieve analysis, the soil particles are shaken on a sieve with square openings of specified size. Thus the "size" of a particle larger than 0.06 mm is based on the side dimension of a square hole in a screen. In the hydrometer analysis, the "size" of a particle is the diameter of a sphere which settles in water at the same velocity as the particle.

Soil particles vary in size from 1×10^{-6} mm (10 Å), up to large rocks several meters in thickness, a range of one to more than a billion. The tremendous magnitude of this range can be better grasped by noting that the size range between a moth ball or child's marble and our earth is also a one to a billion.

In describing the size of a soil particle, we can cite either a dimension or a name that has been arbitrarily assigned to a certain size range. Table 4.1 gives such a set of names with their corresponding particle size ranges. (Noted in Table 4.1 in parentheses are other numerical values which are also used.) The word "clay" is also used to describe a fine-grained soil having plasticity, as was discussed in Chapter 3. We can avoid confusion by employing "clay size" rather than merely "clay" to denote a particle smaller than 2 μm.

Table 4.1 Particle Size

Boulder	>300 mm
Cobble	150 to 300 mm
Gravel	2.0 mm (or 4.76 mm) to 150 mm
Sand	0.06 (or 0.076 mm) to 2.0 mm (or 4.76 mm)
Silt	0.002 to 0.06 mm (or 0.074 mm)
Clay	< 0.002 mm

In Fig. 4.1 are plotted the sizes of various particles and the ranges of some methods of detecting particle size. A widely used soil particle size classification is shown at the top of Fig. 4.1. A study of this figure will give perspective to particle size and its determination.

Particle Shape

The preceding discussion noted that the size of a particle could be given by a single number only when the particle was equidimensional, as a cube or sphere. This situation is not too far from true for soil particles in the silt range and coarser, but it is far from true for particles in the clay size range. This is illustrated in Figs. 4.2 and 4.3, which show sand particles, and in Fig. 4.4, which shows clay particles. Sheetlike particles, such as mica, do occur in the silt and larger size portions of soil, and particles having shapes such as those shown in Figs. 4.2 and 4.3 do occur in the clay size range. In general, however, most of the particles in the silt range and coarser are approximately equidimensional and most of those in the clay size are far from equidimensional. The most common shape for clay size particles is platey, as are the kaolinite particle and illite particle shown in Fig. 4.4. Rods and laths, however, are found in soils, generally in the clay size fraction.

Geologists working with rocks describe particle shapes using such terms as disk, sphere, blade, and rod on the basis of dimension ratios. The civil engineer generally

[1] The detailed procedures for these analyses are given in Lambe (1951).

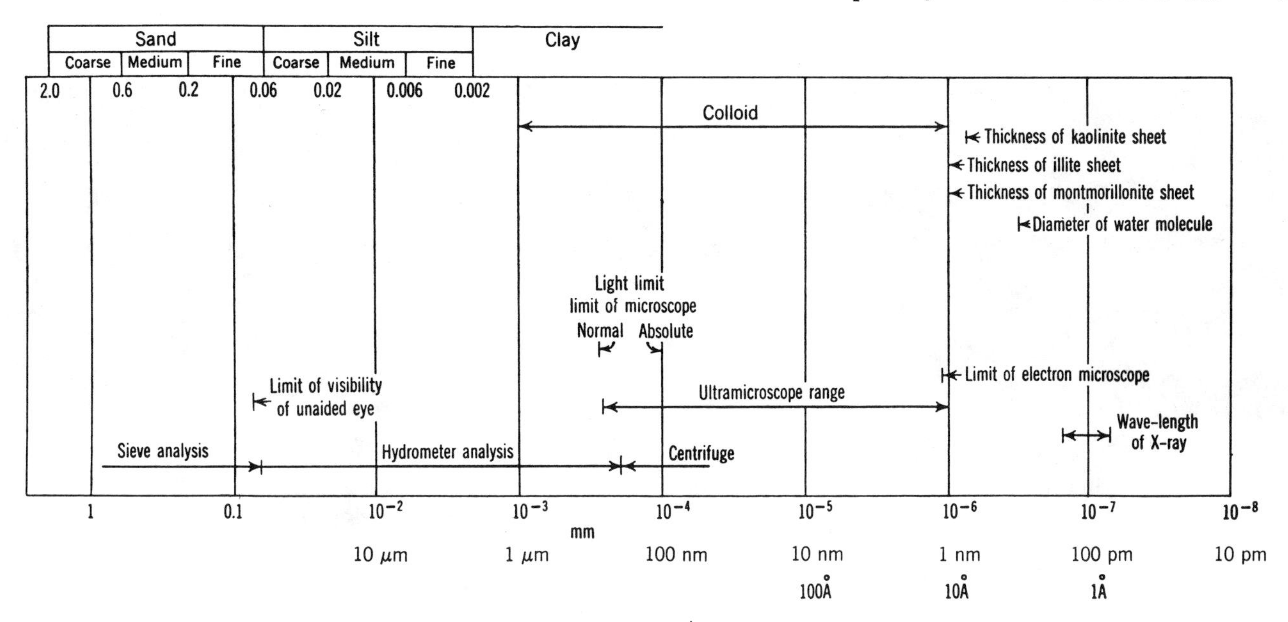

Fig. 4.1 Size.

finds it impractical to characterize numerically particle shape because of the small-sized particles with which he normally works.

Degree of Roundness, Surface Texture, and Color

The degree of roundness refers to the sharpness of the edges and corners of a particle. Figure 4.5 shows five levels of degree of roundness.

Minor features of a surface of a particle, independent of size, shape, or degree of roundness, are termed "surface texture" of the particle. Some terms used to describe surface texture are dull or polished, smooth or rough, striated, frosted, etched, or pitted.

Color is a useful particle characteristic to the geologist working in mining, but it is of little value to the soil engineer. The soil engineer does, however, frequently use color to describe an assemblage of particles, e.g., Boston *blue* clay. He must use color descriptions with caution, since the color of a soil mass can change with a change in moisture content or chemical composition.

The soil particles in Figs. 4.2, 4.3, and 4.4 illustrate several features of particle appearance. The Ottawa sand and Raguba particles are well rounded and frosted. The particles of sand formed by crushing large mineral chunks (Fig. 4.2*d*, *e*, and *f*) have sharp edges and corners, and their surfaces are not striated, frosted, or etched. The photographs of the Venezuelan sand show that compression to high pressures may cause considerable crushing of particles. The natural Venezuelan sand (Fig. 4.2*h*) had 4% of its particles finer than 0.074 mm, whereas after compression (Fig. 4.2*i*) it had 20% of its particles finer than 0.074 mm.

All of the Libyan sands shown except the Raguba sand are from near the Mediterranean Sea and are 70–90% carbonate minerals. The Raguba sand came from the desert 160 km away from the sea and is 98% quartz. The carbonate sands, especially those in Fig. 4.3*a*, have a high degree of aggregation (i.e., joining together of particles by cementation), as can be seen. This aggregation inevitably affects the behavior of the soil. For example, tests on undisturbed specimens of the aggregated sand displayed a significant time dependency of the stress-strain behavior. However, tests on reconstituted specimens in which the aggregation has been disintegrated show less time dependency.

The kaolinite particle in Fig. 4.4 is about 1 μm across and 0.08 μm thick. Two smaller kaolinite particles can be seen on top of the large one. The surface of the kaolinite particle appears to be smooth to a scale of probably 10 nm. The smallest clay particles, montmorillonite, can and do commonly exist in platelets as thin as 1 nm and are smooth to within 0.1 nm.

4.2 COMPOSITION OF A SOIL PARTICLE

The beginning student in soil mechanics usually reasons with apparent logic that the composition of the individual particles in an element of soil is an important characteristic of soil. This belief is false since there are few useful, general relationships between soil composition and soil behavior. On the other hand, this belief is true as far as a fundamental understanding of soil behavior is concerned. The nature and arrangement of the atoms in a soil particle—i.e., composition—have a significant influence on permeability, compressibility, strength, and stress transmission in soils, especially in fine-grained soils. There

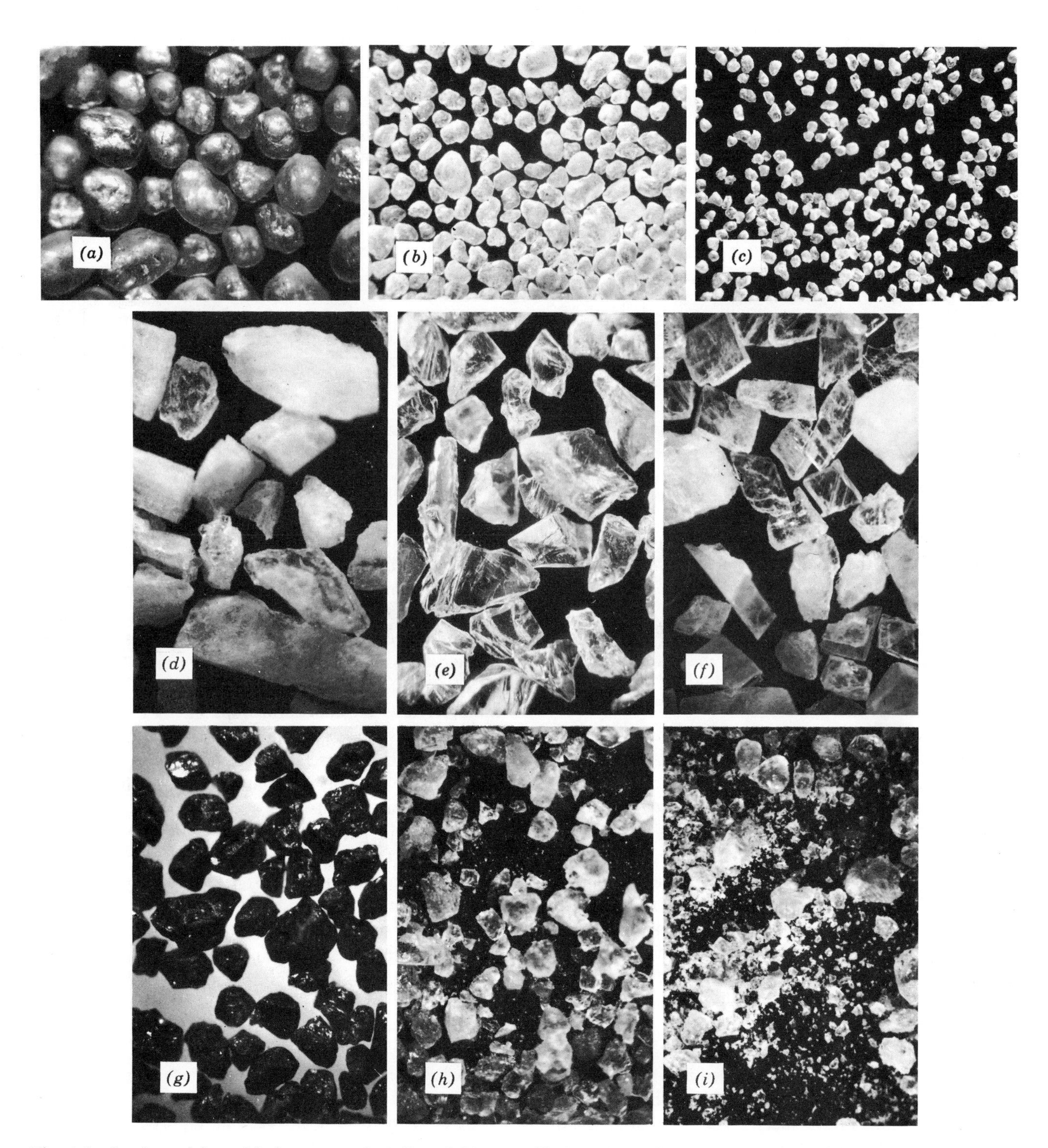

Fig. 4.2 Sand particles. (*a*) Ottawa sand, 0.42 to 0.84 mm. (*b*) Ottawa sand, 0.19 to 0.42 mm. (*c*) Ottawa sand, 0.11 to 0.19 mm. (*d*) Ground feldspar, 0.19 to 0.42 mm. (*e*) Ground quartz, 0.19 to 0.42 mm. (*f*) Ground dolomite, 0.19 to 0.42 mm. (*g*) Hawaiian beach sand. (*h*) Venezuelan sand. (*i*) Venezuelan sand (sand *h* after compression to 140 MN/m^2). (From Roberts 1964.)

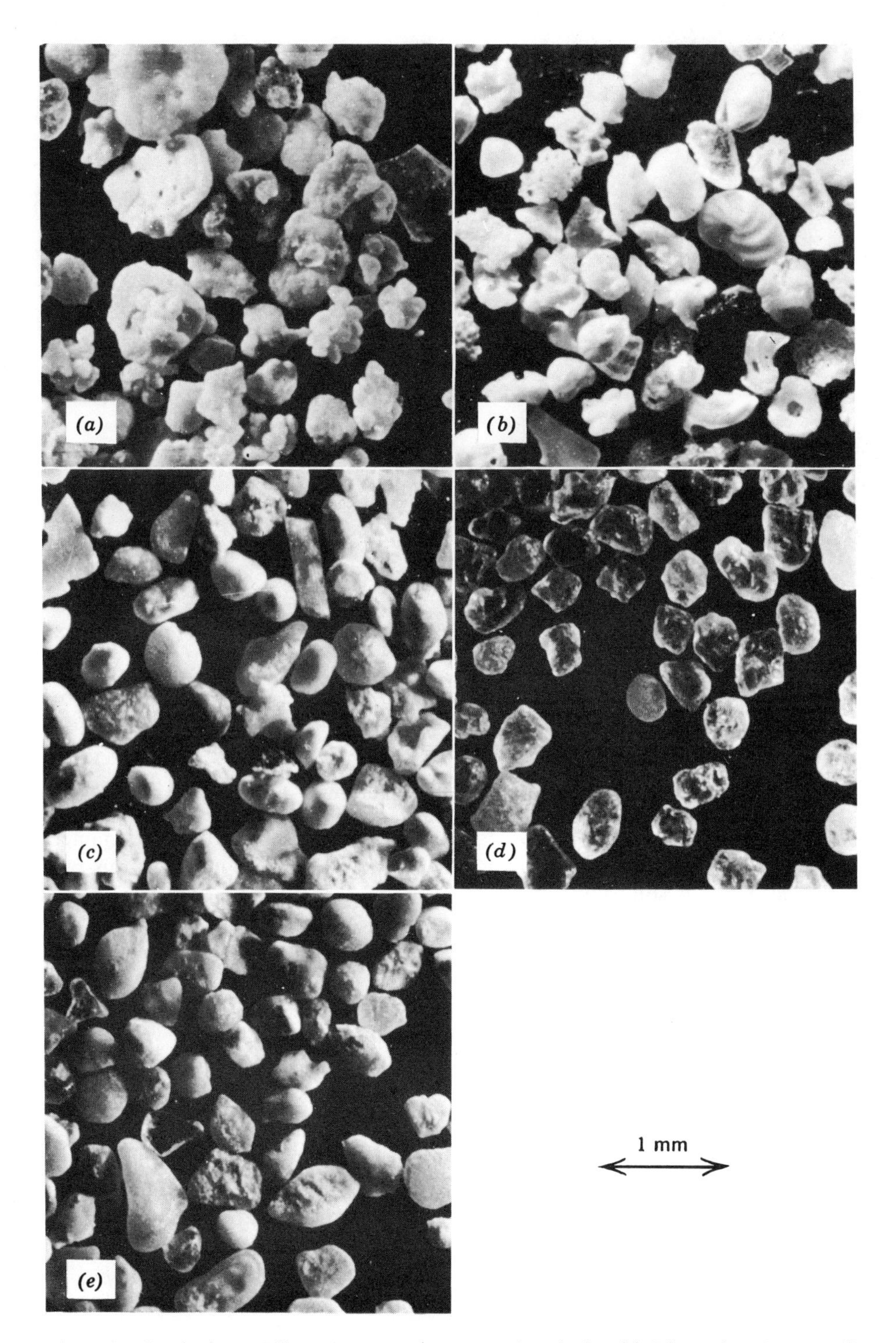

Fig. 4.3 Sands from Libya (0.15 to 0.25 mm fraction). (*a*) Plant site, Brega. (*b*) Harbor bottom, Brega. (*c*) Gas plant, Brega. (*d*) Raguba. (*e*) Crude tank farm, Brega. (Sands supplied by ESSO Libya; Photos by R. T. Martin, M.I.T.)

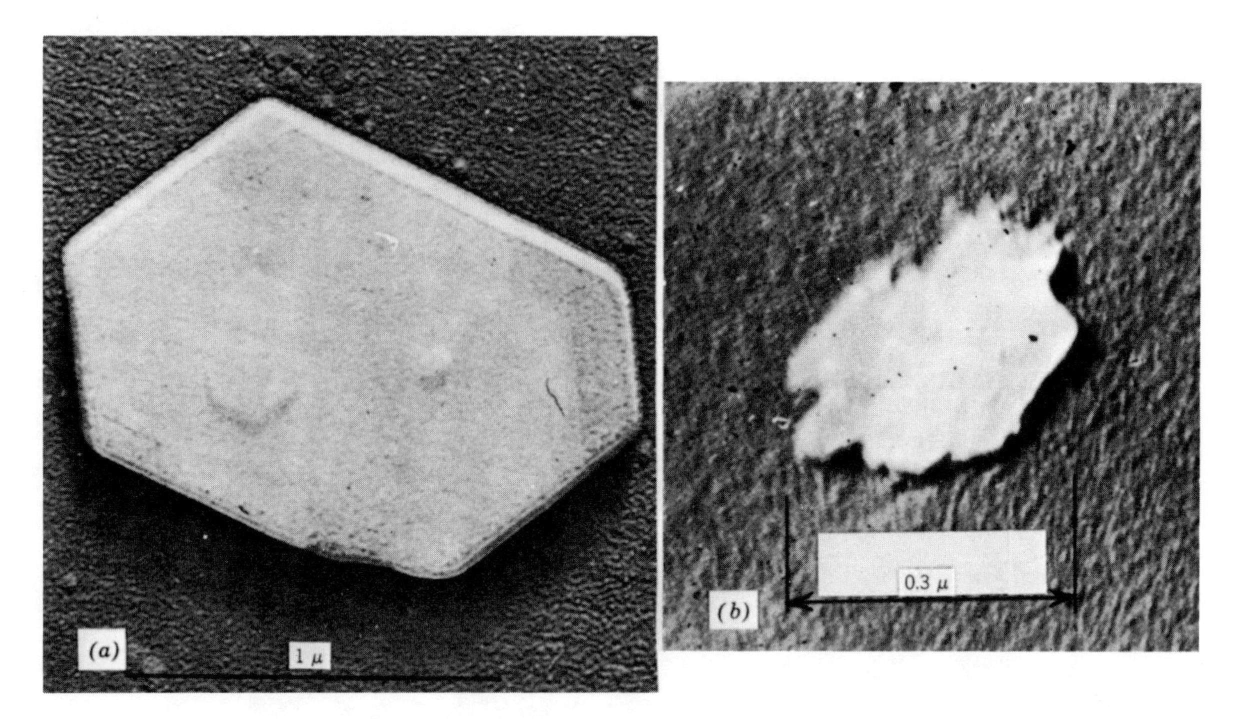

Fig. 4.4 Clay particles. (*a*) Kaolinite (From Lambe, 1951). (*b*) Illite (by R. T. Martin, M.I.T.).

are certain minerals that can give a soil containing them unusual properties. Two examples are montmorillonite and halloysite. Montmorillonite can cause a soil to be highly expansive and halloysite can cause soil to have a very low unit weight. These and other relationships between composition and behavior are noted in later chapters. Thus the student needs to study soil composition if he is to understand the fundamentals of clay behavior and particularly the dependence of this behavior on time, pressure, and environment. In explaining soil behavior, later chapters in this book will make use of the material presented in the remaining part of this chapter on soil composition.

A soil particle may be either *organic* or *inorganic*. Little is known about the composition of organic soil; in fact, at the present state of knowledge, the engineer makes little attempt to identify the actual organic compounds in soil. There are soils that are composed entirely of organic particles, such as peat or muskeg,[2] and there are soils that contain some organic particles and some inorganic particles, such as "organic silt."

An inorganic soil particle may be either a mineral or a rock. A *mineral* is a naturally occurring chemical element or compound (i.e., has a chemical composition expressible by a formula) formed by a geologic process. A *rock* is the solid material which comprises the outer shell of the earth and is an aggregate of one or more minerals or glasses.

The rest of this chapter presents certain principles of mineralogy and describes a few minerals of interest to the soil engineer. The intent of this presentation is to give the reader some understanding of the nature and arrangement of atoms in soil particles so that he can then grasp why certain particles are small plates that are chemically active and other particles are large, approximately equidimensional chunks that are relatively inactive. For a detailed consideration of mineralogy the reader should see books devoted entirely to this subject, such as Grim (1953), Dana (1949), and *Proceedings of National Conference on Clays and Clay Minerals*.[3]

Minerals have been classified on the basis of both the *nature* of the atoms and the *arrangement* of the atoms. The first classification has headings such as carbonates, phosphates, oxides, and silicates. This classification is of limited value to the civil engineer since the most abundant and important minerals are silicates. In fact, if all of the soil in the world were placed in one pile, over 90% of the weight of the pile would be silicate minerals.

Table 4.2 (p. 50) is a classification of silicates based on the arrangement of atoms in the mineral. This classification has merit for several reasons. First, it is a

[2] The National Research Council of Canada has had a group studying muskeg for a number of years. The various proceedings of Muskeg Research Conferences sponsored by NRC are an excellent source of information on muskeg.

[3] Available from the Publications Office of the National Academy of Sciences, National Research Council, 2101 Constitution Avenue, Washington, D.C. 20525.

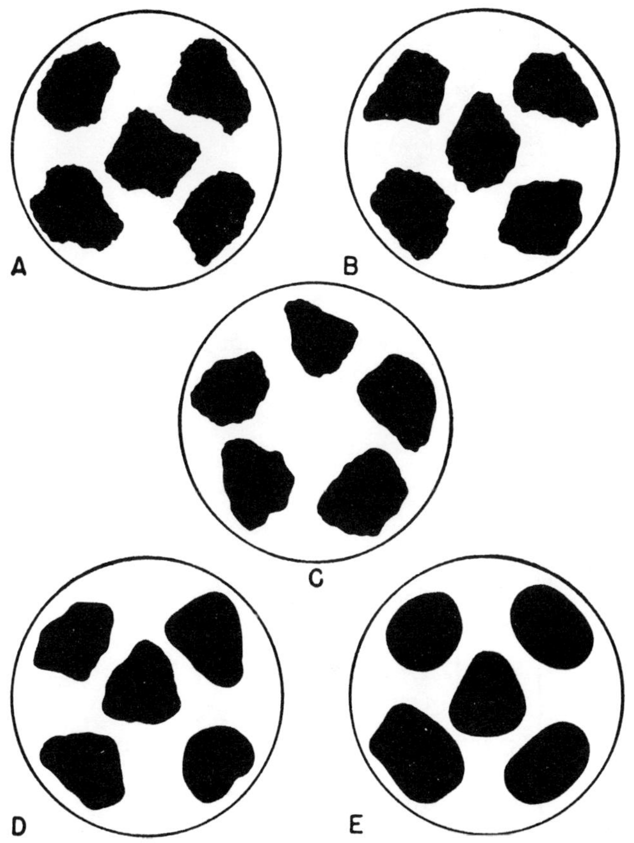

Chart to show roundness classes. *A*, angular; *B*, subangular; *C*, subrounded; *D*, rounded; *E*, well rounded.

Fig. 4.5 Degree of particle roundness [Fig. 24 *Sedimentary Rocks* (1949) by F. J. Pettijohn, by permission of Harper & Row, Publishers.]

definite grouping since there is only one known silicate (vesuvianite) that falls into more than one group. Second, there is a relationship between the atomic arrangement in a mineral and its physical, optical, and chemical properties.

Soils are usually the products of rock weathering and thus the most abundant soil minerals are common rock-forming minerals and those that are most resistant to chemical and physical weathering. The sheet and framework silicate minerals are therefore the most abundant and common soil minerals.

Basic Structural Units

The study of the silicate mineral structures may be facilitated by "building" a mineral out of basic structural units. This approach is an educational technique and not necessarily the method whereby the mineral is actually formed by nature. The structures presented in this chapter are idealized. The typical crystal in a clay is a complex structure similar to the idealized arrangement but usually having irregular substitutions and interlayering. Figure 4.6 shows a group of basic silicate units. The silicon-oxygen tetrahedron consists of four oxygens nestled around a silicon atom to form the unit shown in Fig. 4.6*a*. The atoms are drawn to scale on the basis of the radii in units of angstroms (0.1 nm) given in Fig. 4.6*h*. The table at the right of each unit gives the valences.

Figure 4.6*c* shows the aluminum-oxygen octahedron and Fig. 4.6*d* the magnesium-oxygen octahedron. Combining the silicon-oxygen tetrahedrons gives the silica sheet shown in Fig. 4.6*e*. Combining the aluminum-oxygen octahedrons gives gibbsite (Fig. 4.6*f*), and combining the magnesium-oxygen octahedrons gives brucite (Fig. 4.6*g*). A study of the valences in Fig. 4.6 shows that the tetrahedron and two octahedrons are not electrically neutral and therefore do not exist as isolated units. Gibbsite and brucite are, however, electrically neutral and exist in nature as such.

Two-Layer Sheet

If we stack a brucite unit on top of a silicate unit we get serpentine, as shown in Fig. 4.7. This figure shows both the atomic structure and the symbolic structure. Combining in a similar way gibbsite and silica gives the mineral kaolinite shown in Fig. 4.8.

The actual mineral particle does not usually consist of only a few basic layers as suggested by the symbolic structures in Figs. 4.7 and 4.8. Instead, a number of sheets are stacked one on top of another to form an actual crystal—the kaolinite particle shown in Fig. 4.4 contains about 115 of the two-layer units. The linkage between the basic two-layer units consists of hydrogen bonding and secondary valence forces.

In the actual formation of the sheet silicate minerals the phenomenon of *isomorphous substitution* frequently occurs. Isomorphous (meaning "same form") substitution consists of the substitution of one kind of atom for another. For example, one of the sites filled with a silicon atom in the structure in Fig. 4.8 could be occupied by an aluminum. Such an example of isomorphous substitution could occur if an aluminum atom were more readily available at the site than a silicon atom during the formation of the mineral; furthermore, aluminum has coordinating characteristics somewhat similar to silicon, thus it can fit in the silicon position in the crystal lattice. Substituting the aluminum with its +3 valence for silicon with its +4 valence has two important effects:

1. A net unit charge deficiency results per substitution.
2. A slight distortion of the crystal lattice occurs since the ions are not of identical size.

The significance of the charge deficiency is discussed in Chapter 5. The distortion tends to restrict crystal growth and thus limits the size of the crystals.

In the kaolinite mineral there is a very small amount of isomorphous substitution, one possibility being one aluminum replacing one silicon in the silica sheet of the

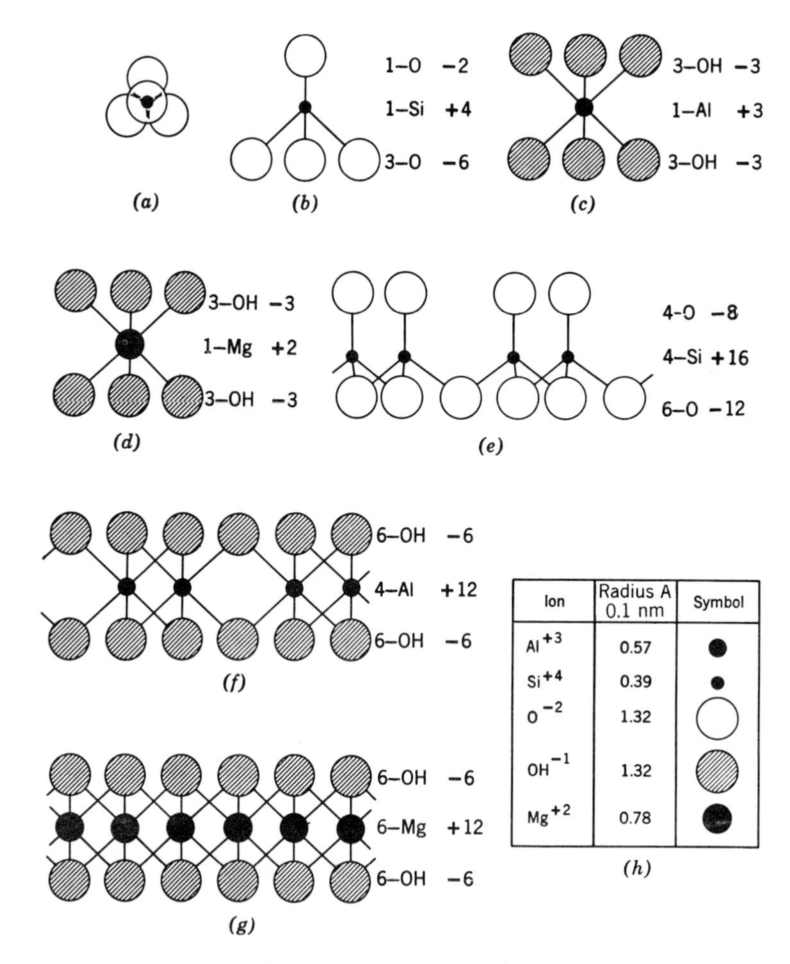

Fig. 4.6 Basic silicate units. (*a*) and (*b*) Silicon tetrahedron. (*c*) Aluminum octahedron. (*d*) Magnesium octahedron. (*e*) Silica. (*f*) Gibbsite. (*g*) Brucite.

mineral. The amount of this substitution needed to explain the charge on kaolinite is one substitution for every four hundredth silicon ion.

Since the basic structure of kaolinite consists of a layer of gibbsite on top of a layer of silicate, this mineral is called a "two-layer" mineral. Kaolinite is the most important and most common two-layer mineral encountered by the engineer. Halloysite, having essentially the same composition and structure as kaolinite, is an interesting and not uncommon member of the two-layer silicate group. The main difference between halloysite and kaolinite is the presence of water between the basic sheets of halloysite, which results in halloysite existing in tubular particles.

Three-Layer Sheets

The three-layer sheet minerals are formed by placing one silica on the top and one on the bottom of either a gibbsite or brucite sheet. Figure 4.9 shows the mineral pyrophyllite made of a gibbsite sandwiched between two silica sheets. Figure 4.10 shows the structure of the mineral muscovite, which is similar to pyrophyllite except that there has been isomorphous substitution of aluminum for silicon in muscovite. The net charge created by this substitution is balanced by potassium ions, which serve to link the three-layer sandwiches together, as indicated in the symbolic structure in Fig. 4.10.

The two most common three-layer structures in soil are montmorillonite and illite type minerals. Montmorillonite has a structure similar to pyrophyllite with the exception that there has been isomorphous substitution of magnesium for aluminum in the gibbsite sheet.

Figure 4.11 gives a summary of the sheet silicate minerals of importance to the civil engineer.

Frameworks

Quartz, a framework silicate structure, has a very low ratio of oxygen to silicon (2:1), as noted in Table 4.2. It is thus one of the most weather resistant minerals. The feldspars have higher oxygen to silicon ratios (2.7 to 4.0)

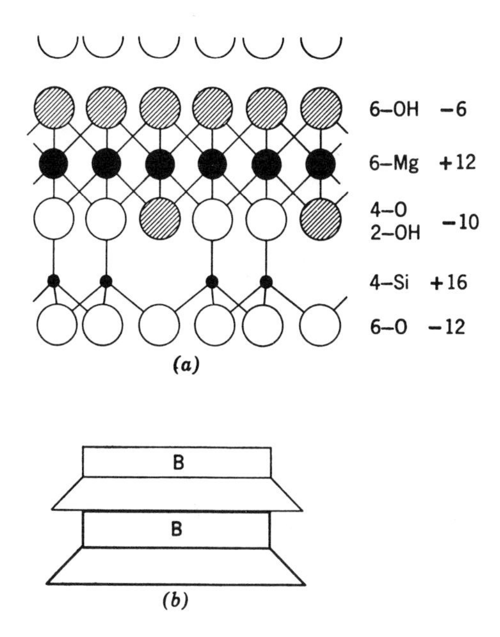

Fig. 4.7 The structure of serpentine. (*a*) Atomic structure. (*b*) Symbolic structure.

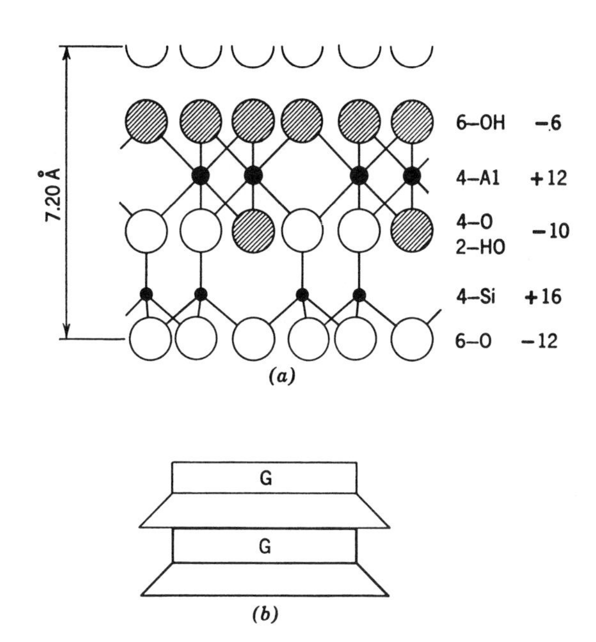

Fig. 4.8 The structure of kaolinite. (*a*) Atomic structure. (*b*) Symbolic structure.

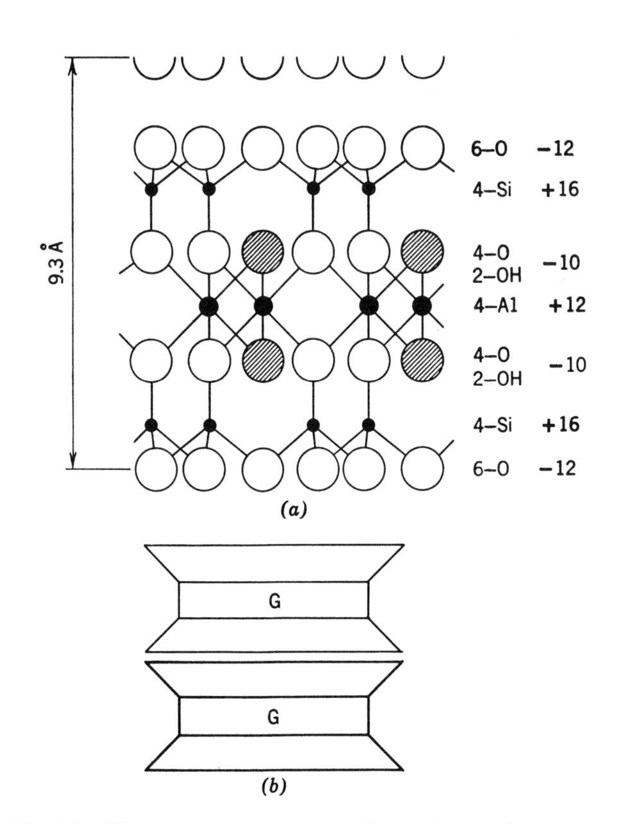

Fig. 4.9 The structure of pyrophyllite. (*a*) Atomic structure. (*b*) Symbolic structure.

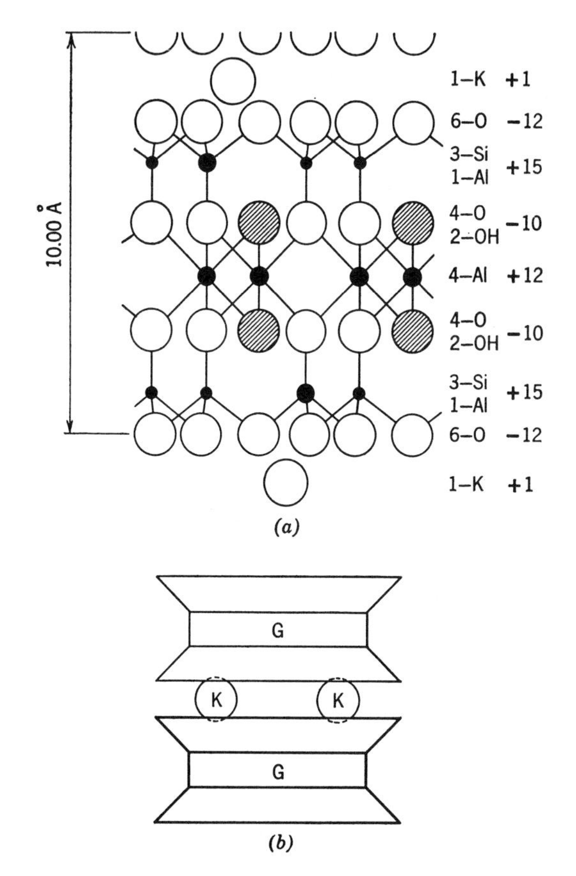

Fig. 4.10 The structure of muscovite. (*a*) Atomic structure. (*b*) Symbolic structure.

Mineral	Structure Symbol	Isomorphous Substitution (nature and amount)	Linkage between Sheets (type and strength)	Specific Surface (m^2/g)	$\frac{1}{\text{Charge Density}}$ ($Å^2$/ion)	Potential Exchange Capacity (me/100 g)	Actual Exchange Capacity (me/100 g)	Particle Shape	Particle Size
Serpentine	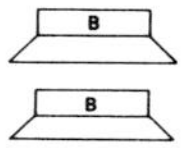 B / B	none	H-bonding + secondary valence			1	1	Platy or fibrous	
Kaolinite	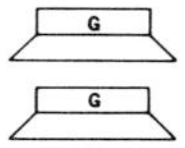 G / G	Al for Si 1 in 400	H-bonding + secondary valence	10–20	83	3	3	Platy	$d = 0.3$ to $3\ \mu m$ thickness $= \frac{1}{3}$ to $\frac{1}{10}d$
Halloysite ($4H_2O$)	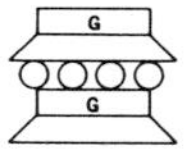 G / G	Al for Si 1 in 100	Secondary valence	40	55	12	12	Hollow rod	OD $= 0.07\ \mu m$ ID $= 0.04\ \mu m$ $L = 0.5\ \mu m$
Halloysite ($2H_2O$)	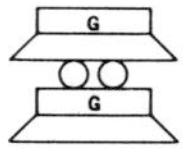 G / G	Al for Si 1 in 100	Secondary valence	40	55	12	12	Hollow rod	OD $= 0.07\ \mu m$ ID $= 0.04\ \mu m$ $L = 0.5\ \mu m$
Talc	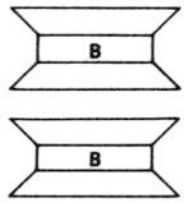 B / B	none	Secondary valence			1	1	Platy	
Pyrophyllite	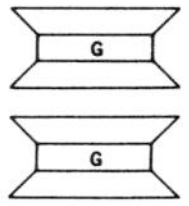 G / G	none	Secondary valence			1	1	Platy	

Fig. 4.11 Sheet silicate minerals.

Muscovite	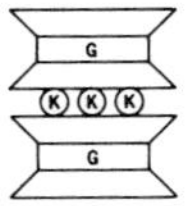 G / K K K / G	Al for Si 1 in 4	Secondary valence + K linkage			250	5–20	Platy	
Vermiculite	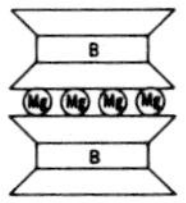 B / Mg Mg Mg Mg / B	Al, Fe, for Mg Al for Si	Secondary valence + Mg linkage	5–400	45	150	150	Platy	$t = \frac{1}{10}d$ to $\frac{1}{30}d$
Illite	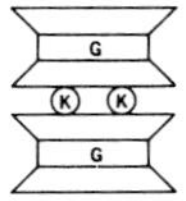 G / K K / G	Al for Si, 1 in 7 Mg, Fe for Al Fe, Al for Mg	Secondary valence + K linkage	80–100	67	150	25	Platy	$d = 0.1$ to $2\ \mu m$ $t = \frac{1}{10}d$
Montmorillonite	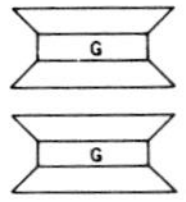 G / G	Mg for Al, 1 in 6	Secondary valence + exchangeable ion linkage	800	133	100	100	Platy	$d = 0.1$ to $1\ \mu m$ $t = \frac{1}{100}d$
Nontronite	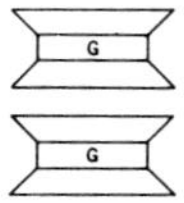 G / G	Al for Si, 1 in 6	Secondary valence + exchangeable ion linkage	800	133	100	100	Lath	$l = 0.4$ to $2\ \mu m$ $t = \frac{1}{100}l$
Chlorite	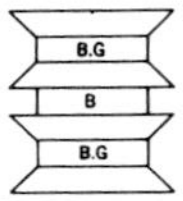 B.G / B / B.G	Al for Si, Fe, Al for Mg	Secondary valence + brucite linkage	5–50	700	20	20	Platy	

Fig. 4.11 Sheet silicate minerals.

Table 4.2 The Silicate Structures

Structural Group	Diagrammatic Representation	No. Shared Oxygens per Silicon	Oxygen to Silicon Ratio	Si-O Unit and Charge	Example
Independent tetrahedrons	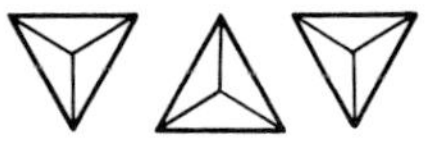	0	4:1	$(SiO_4)^{-4}$	Zircon $ZrSiO_4$
Double tetrahedrons	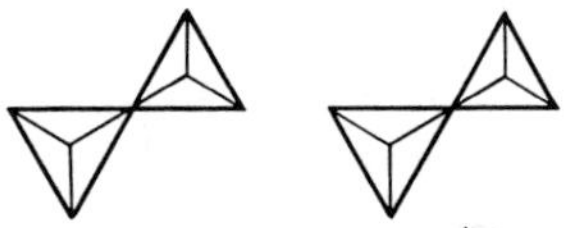	1	7:2	$(SiO_{3\frac{1}{2}})^{-3}$	Akermanite $Ca_2Mg(Si_2O_7)$
Rings	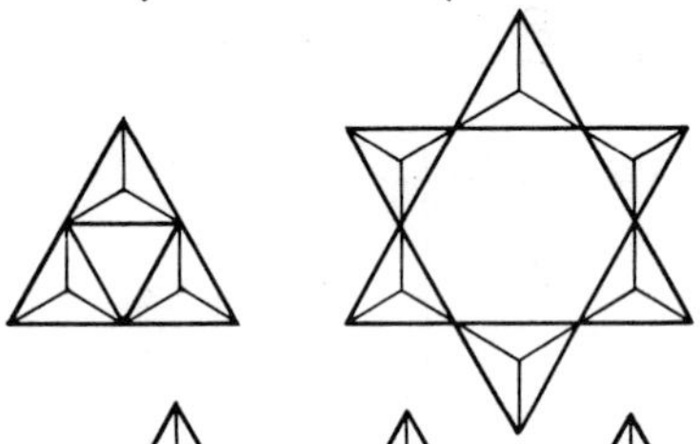	2	3:1	$(SiO_3)^{-2}$	Beryl $Be_3Al_2Si_6O_{18}$
Chains Single	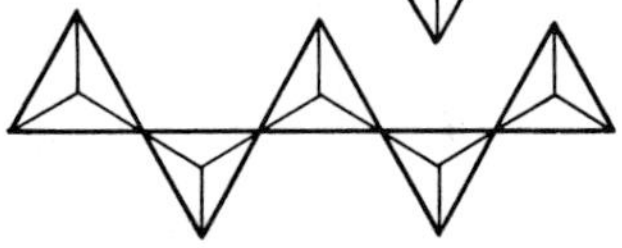	2	3:1	$(SiO_3)^{-2}$	Enstatite $MgSiO_3$
Double	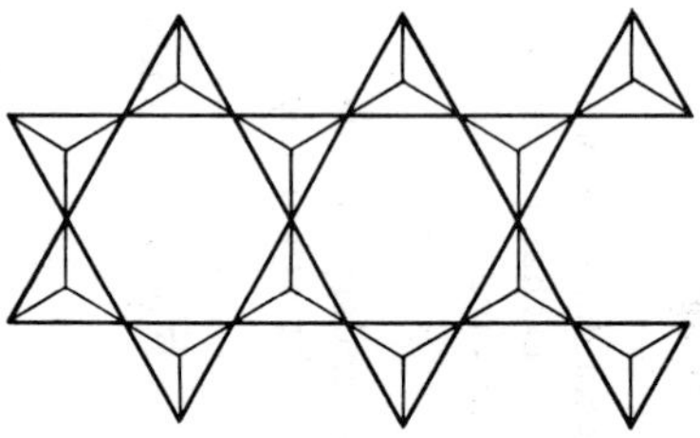	$2\frac{1}{2}$	11:4	$(SiO_3)^{-2}$ and $(SiO_{2\frac{3}{4}})^{-1\frac{1}{2}}$	Tremolite $Ca_2Mg_5Si_8O_{22}(OH)_2$
Sheets	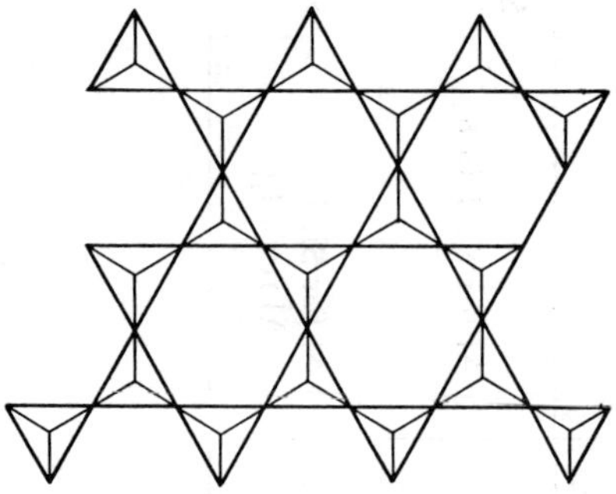	3	5:2	$(SiO_{2\frac{1}{2}})^{-1}$	Pyrophyllite $Al_2Si_4O_{10}(OH)_2$
Frameworks	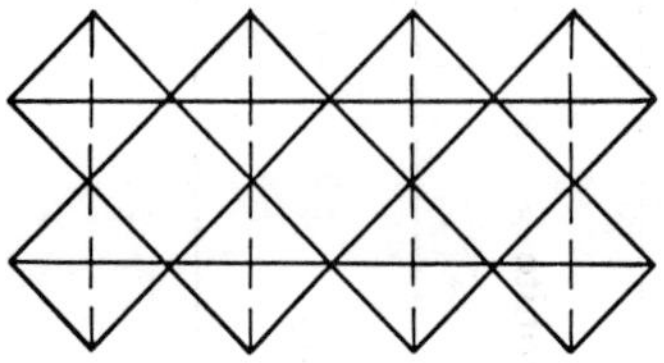	4	2:1	$(SiO_2)^0$	Quartz SiO_2

and can change through weathering into clay minerals. Because they are very common rock-forming minerals, the frameworks, especially quartz and feldspars, are very abundant in soils. While these minerals sometimes occur in clay size particles, they are most common in silt size and larger. Because of the nature of their structure, particles of the framework minerals tend to exist in shapes that are approximately equidimensional.

4.3 SUMMARY OF MAIN POINTS

This chapter gives a very condensed and selected treatment of an extensive and complex body of knowledge. The significant points in this treatment are the following:

1. Soil particles range in size from very small to very large.
2. Generally, sand and silt particles are approximately equidimensional, but clay particles are plate-shaped or, less commonly, lath-shaped or rod-shaped.
3. Particle size, shape, and activity can be explained in terms of crystal chemistry of the particle.
4. Isomorphous substitution, common in the sheet minerals, tends to retard crystal growth and gives the crystal a net electrical charge.

PROBLEMS

4.1 Would you expect the sand shown in Fig. 4.3*a* to exhibit a particle size distribution dependent on the treatment given the sand prior to sieving? Why?

4.2 Would you disaggregate the sand (Fig. 4.3*a*) prior to sieving if the purpose of the particle size testing was an attempt to relate particle size and permeability? Why?

4.3 Draw the atomic structure of montmorillonite. [*Hint*. Alter the pyrophyllite structure (Fig. 4.9) in accordance with the isomorphous substitution noted in Fig. 4.11.]

4.4 Using Fig. 4.5 as a guide, classify the sands shown in Fig. 4.2 as to roundness.

CHAPTER 5

Normal Stress between Soil Particles

Chapter 4 considered soil particles as individual, isolated units. This chapter examines the interaction of adjacent soil particles; i.e., the stresses that develop between adjacent soil particles and the way in which these stresses affect the way that adjacent particles fit together. This presentation deals primarily with the normal stresses acting between small particles which are not in contact. Chapter 6 treats shear stresses and normal stresses between particles in contact with each other.

In a highly schematic way, the types of forces that exist between two adjacent soil particles are (see Fig. 5.1):

F_m = the force where the contact is mineral-mineral
F_a = the force where the contact is air-mineral or air-air
F_w = the force where the contact is water-mineral or water-water
R' = the electrical repulsion between the particles
A' = the electrical attraction between the particles

5.1 THE ELECTRICAL CHARGE ON A SOIL PARTICLE

Every soil particle carries an electrical charge. This fact can readily be demonstrated by mixing a fine-grained soil with water in a beaker and then inserting at different locations in the beaker two electrodes which are components of an electrical circuit containing a battery and an ammeter. The ammeter will indicate that the electrical charge in the circuit is transmitted through the soil-water mixture. Although theoretically a soil particle can carry either a net negative or a net positive charge, only negative charges have been measured. This net electrical charge may arise from any one or a combination of the five following factors:

1. Isomorphous substitution.
2. Surface disassociation of hydroxyl ions.
3. Absence of cations in the crystal lattice.
4. Adsorption of anions.
5. Presence of organic matter.

Of these five possible causes the first—isomorphous substitution—is the most important.

In addition to a net charge, a soil particle can carry a distribution charge because the seat of the positive charge and the seat of the negative charge do not coincide. Similarly, the crystal bonding in a soil particle results in local charges.

Since the magnitude of the electrical charge is directly related to the particle surface area, the influence of this charge on the behavior of the particle relative to the influence of mass forces (i.e., the weight of the particle) will be directly related to the surface area per mass of

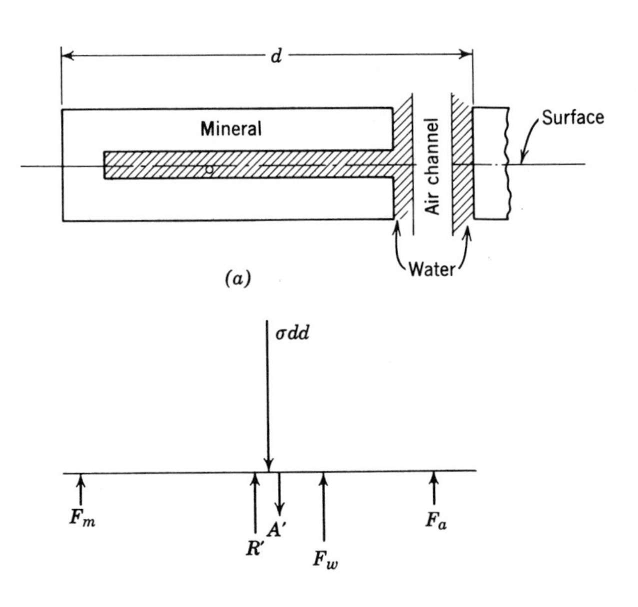

Fig. 5.1 Forces between particles. (*a*) Adjacent soil particles. (*b*) Forces across surface.

particles. The magnitude of the surface area per mass, *specific surface*,[1] is therefore a good indication of the relative influence of electrical forces on the behavior of the particle. The term *colloid* is used to describe a particle whose behavior is controlled by the surface-derived forces rather than mass-derived forces.

A clay particle is a colloid because of its small size and irregular shape. The smaller a particle size the larger its specific surface, as can be seen in Table 5.1. From this

Table 5.1

Length of Cube Side (cm)	Number of Particles	Total Volume (cm^3)	Total Surface Area (cm^2)	Surface Area ÷ Volume $\left(\frac{1}{cm}\right)$
1	1	1	6	6
1 μm = 10^{-4}	10^{12}	1	60,000	60,000
1 nm = 10^{-7}	10^{21}	1	60,000,000	60,000,000

table we can see that the specific surface goes up directly as the particle size goes down. As a matter of fact, the surface area per volume of a cube is $6/L$ and of a sphere is $6/D$.

The size range of colloids has been more or less arbitrarily set as 1 nm to 1 μm, as noted in Fig. 4.1. Below 1 nm lie the diameters of atoms and molecules. Most particles larger than approximately 1 μm are predominantly influenced by forces of mass. A specific surface of 25 m^2/g has also been suggested as the lower limit of the colloidal range. The principles of colloidal chemistry are helpful in the understanding of clay behavior.

Particles silt size and larger have specific surface values of less than 1 m^2/g, i.e., considerably smaller than the lower limit of the colloidal range. The "specific surface" column in Fig. 4.11 gives typical values of specific surface for clay particles. Note particularly the large difference in specific surface between kaolinite (10 to 20 m^2/g) and montmorillonite (800 m^2/g). The tremendous surface area of montmorillonite can be visualized when one realizes that 6 g of montmorillonite has approximately the same surface area as an entire football field—or only 12 g of montmorillonite would be needed to cover an entire football field (to cover the field requires 2 × 6 g since the area on both faces of the clay particles contribute to the specific surface).

$$\begin{array}{c} Na_{0.33} \\ \uparrow \\ (Al_{1.67}Mg_{0.33})Si_4O_{10}(OH)_2 \end{array}$$

Fig. 5.2 Montmorillonite formula.

[1] Specific surface is sometimes defined as surface area per volume.

Unit Mass:

$$\begin{aligned} Al&: 1.67 \times 26.97 = 45.0 \\ Mg&: 0.33 \times 24.32 = 8.0 \\ Na&: 0.33 \times 23.0 = 7.6 \\ Si&: 4 \times 28.06 = 112.4 \\ O&: 12 \times 16.00 = 192.0 \\ H&: 2 \times 1.00 = 2.0 \\ &\text{Total } 367.0 \end{aligned}$$

$$\begin{aligned} \text{Charge Deficiency} &= \tfrac{1}{3} \text{ electrical equivalent per 367 gram-mol-weights} \\ &= \frac{0.333}{367} = 0.091 \frac{\text{equivalents}}{100 \text{ grams}} \\ &= 91 \text{ me}/100 \text{ g} \end{aligned}$$

Fig. 5.3 Computation of net charge.

A soil particle in nature attracts ions to neutralize its net charge. Since these attracted ions are usually weakly held on the particle surface and can be readily replaced by other ions, they are termed *exchangeable ions*. The soil particle with its exchangeable ions is neutral.

As an illustration of the net charge on a soil particle, let us consider a montmorillonite crystal approximately 1000 Å (100 nm) in lateral dimension and one basic three-layer-unit thick. Figure 5.2 gives the structural formula for montmorillonite. The net negative charge of one-third of a unit charge is shown as being balanced by an exchangeable sodium. It is a common convention to represent the exchangeable ion as sodium in a structural formula, although any one or a combination of a large number of cations can exist in the exchangeable positions. Calcium is a very common exchangeable ion in soil.

Figure 5.3 shows the computation for the formula mass of montmorillonite. The formula mass of 367 g and the charge deficiency of one-third per formula can be expressed in terms of milliequivalents per 100 g of clay, abbreviated as me/100 g. The computation yields 91 me/100 g as the theoretical charge deficiency, or *ion exchange capacity*, of montmorillonite. The measured exchange capacity for montmorillonite is close to the theoretical value of 91.

We can also make theoretical calculations of the specific surface and the surface area per charge of a soil crystal. Figures 5.4 and 5.5 show these computations for a

$$\begin{aligned} \text{Surface area per unit} &= 92.6 \text{ Å}^2 \\ \text{Volume per unit} &= \frac{92.6}{2} \text{ Å}^2 \times 10 \text{ Å} = 463 \text{ Å}^3 \\ \text{Mass per unit} &= 463 \text{ Å}^3 \times 10^{-24} \text{ cm}^3/\text{Å}^3 \times 2.76 \text{ g/cm}^3 \\ &= 1278 \times 10^{-24} \text{ g} \\ \text{Surface area per gram} &= \frac{92.6 \text{ Å}^2 \times 10^{-20} \text{ m}^2/\text{Å}^2}{1278 \times 10^{-24} \text{ g}} \\ &= 725 \text{ m}^2/\text{g} \end{aligned}$$

Fig. 5.4 Computation of specific surface.

Charge deficiency $= \frac{1}{3}$ per 2 cations in gibbsite
$\therefore$ Charge deficiency $= \frac{2}{3}$ per 4 cations—i.e., unit in Fig. 4.9
Surface area for unit is:

2(top and bottom surfaces) $\times$ 8.9 Å $\times$ 5.2 Å $=$ 92.6 Å^2
(8.9 and 5.2 are dimensions of unit as determined from atomic structure)

Surface area per unit charge deficiency:

$$92.6 \text{ Å}^2 \times \tfrac{3}{2} = 139 \text{ Å}^2$$

Thus one net charge per surface area of 139 Å^2

Fig. 5.5 Computation of surface area per charge.

montmorillonite unit with four cations in the gibbsite sheet. The computed value of 725 m^2/g is close to the experimental value given in Fig. 4.11 of 800 m^2/g. (The value of 800 m^2/g was obtained from a laboratory test in which the mineral was permitted to adsorb a monomolecular thickness of adsorbate.) The value of net charge per surface area, given in units of 1/Å^2, is the "charge density" of the mineral. The theoretical value of 139 from Fig. 5.5 compares well with the measured value (133) given in Fig. 4.11.

5.2 PARTICLE WITH WATER AND IONS

Let us now consider the nature of a soil particle in water since this is the state in which the civil engineer is nearly always interested. To give perspective to this consideration two typical clay particles will be employed. Figure 5.6*a* shows a typical particle of montmorillonite, which is one of the smallest and most water sensitive minerals encountered in clay; Fig. 5.6*b* shows a typical kaolinite particle, one of the larger and less water sensitive minerals encountered in clay. Figure 5.7 shows a portion of the lateral surface of each of these clay particles with the sites of the exchangeable ions.

The two typical clay particles contain about 14,000 exchangeable monovalent ions on the montmorillonite and 4,000,000 monovalent ions per kaolinite particle. Figure 5.8 shows the computation of the number of

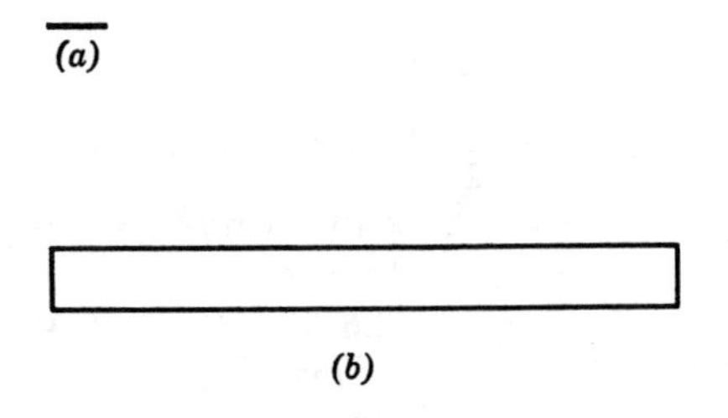

Fig. 5.6 Typical clay particles. (*a*) Montmorillonite, 1000 Å by 10 Å thick. (*b*) Kaolinite, 10,000 Å by 1000 Å thick.

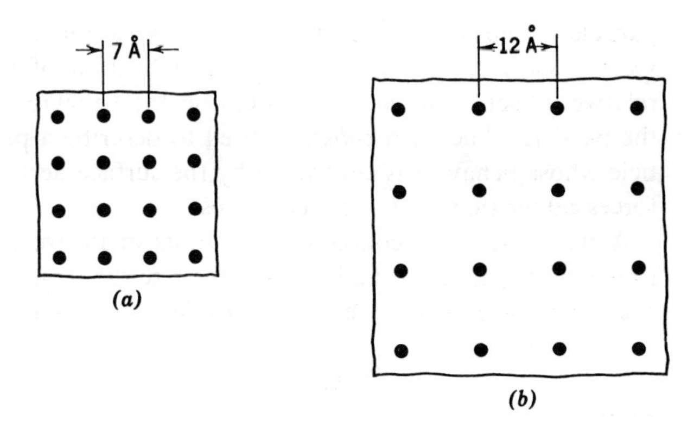

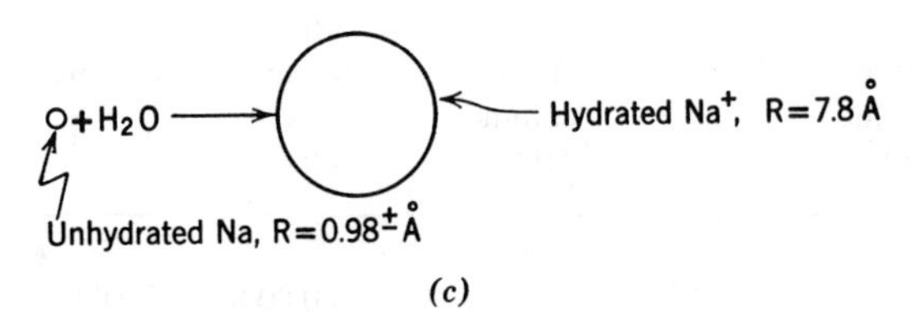

Fig. 5.7 Soil surface with exchangeable ions. (*a*) Surface of dry kaolinite + sodium ions. (*b*) Surface of dry montmorillonite + sodium ions. (*c*) Hydration of sodium ion.

charge deficiencies or sites for monovalent exchangeable ions on the montmorillonite particle. In this discussion, sodium has been chosen as the exchangeable ion for illustrative purposes. Thus the montmorillonite particle in Fig. 5.6 would carry 14,000 sodium ions and the kaolinite particle 4,000,000 sodium ions.

If the individual clay particles are now dropped into water, both the mineral surfaces and the exchangeable ions pick up water, i.e., hydrate. Upon hydration, the sodium ion grows about sevenfold, as is illustrated in Fig. 5.7. As the scaled drawings indicate, the hydrated sodium ions are too large to fit into a monoionic layer on the mineral particles even if they wanted to. Actually, the exchangeable ions with their shells of water move away from the mineral surfaces to positions of equilibrium. The ions are attracted to the mineral surface to satisfy the negative charge existing within the surface; they also desire to move away from each other because of their thermal energies; the actual positions they occupy are compromises between these two types of forces. Thus,

For montmorillonite particle 0.1 μ $\times$ 0.1 μ $\times$ 10 Å
Surface area of particle:

$$1000 \text{ Å} \times 1000 \text{ Å} \times 2 = 2 \times 10^6 \text{ Å}^2$$

Number of charge deficiencies:

$$2 \times 10^6 \text{ Å}^2 \times \frac{1 \text{ charge}}{139 \text{ Å}^2} = 14{,}400$$

Fig. 5.8 Number of charges on montmorillonite particle.

when the individual particles are dropped into water the ions move away from the surface to form what is termed the *double layer*.[2] In Fig. 5.9 the clay particles are shown with the fully developed double layers they would have in pure water. Figure 5.10 shows in three dimensions the same surface sections presented in Fig. 5.7. From Fig. 5.10, we can get some idea of the approximate spacings of the hydrated ions in the double layer. These spacings represent a maximum since the pore fluid in this case is distilled water. Figure 5.11 shows the double layers of the sodium kaolinite particle and the sodium montmorillonite particle to the same scale as in Fig. 5.10. In Fig. 5.11*a* the ions around our selected surface sections are shown as point charges. Figure 5.11*b* shows plots of the concentration of sodium ions versus distance from the particle surface. At a distance of approximately 400 Å, which is the thickness of the double layer, the concentration of sodium ions has become equal to that in the "pore" or "free" water. In Fig. 5.11*c* plots of electrical potential versus distance from the surface are shown. Electrical potential is the work required to move a unit charge from infinity to the point in question and is negative for clay surfaces. The double-layer thickness is thus the distance from the surface required to neutralize the net charge on the particle, i.e., the distance over which there is an electrical potential.

The water in the double layer is under an attractive force to the soil particle since this water is attached to the exchangeable ions which are in turn attracted to the soil surface. Water is also attracted to the mineral surface by other forces (the force between the polar water and the stray electrical charges on the mineral surface,

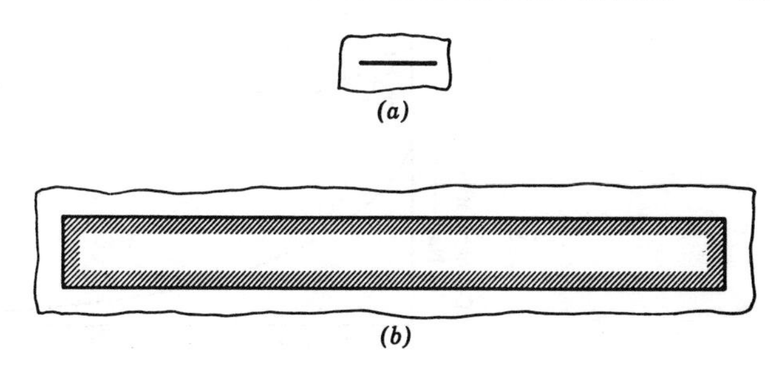

Fig. 5.9 Soil particles with water and ions. (*a*) Sodium montmorillonite. (*b*) Sodium kaolinite.

hydrogen bonding, and van der Waals forces). Although there is controversy as to the exact nature of the water immediately next to the mineral surface, it is generally recognized that at least the first few molecular layers of water around the soil particle are strongly attracted to the particle.

In order to illustrate the importance of this adsorbed or attracted water, let us calculate for typical soil particles the water content corresponding to a 5 Å layer (about two water molecules thick).

Table 5.2, which is highly approximate and intended only to show trends, illustrates the great importance of particle size on the amount of water which can be bound to a particle. To illustrate the significance of these results

Table 5.2

Particle	Specific Surface (m^2/g)	Water Content[a] (for 5 Å layer) (%)
0.1 mm sand	0.03	1.5×10^{-4}
Kaolinite	10	0.5
Illite	100	5
Montmorillonite	1000	50

[a] The water content was calculated as:

Water content = (specific surface) × (thickness of layer of water) × (unit mass of water)

For kaolinite,

$$\text{Water content} = (10\ \text{m}^2/\text{g}) \times (5 \times 10^{-10}\ \text{m}) \times (10^6\ \text{g/m}^3) = 5 \times 10^{-3} \text{ or } 0.5\%$$

[2] The Gouy Chapman double layer theory can be used to calculate the distribution of ions in the double layer (Verwey and Overbeek, 1948).

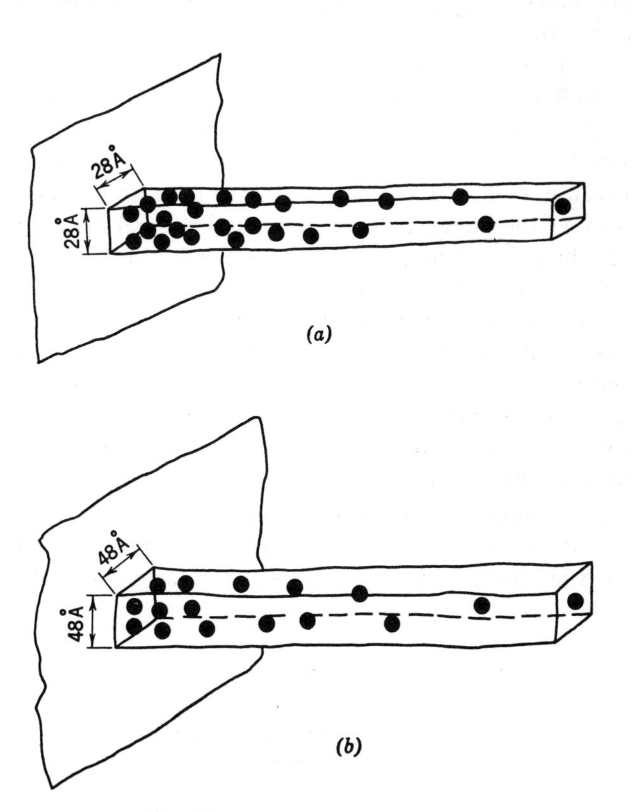

Fig. 5.10 Particle surfaces with water and ions. (*a*) Sodium kaolinite. (*b*) Sodium montmorillonite.

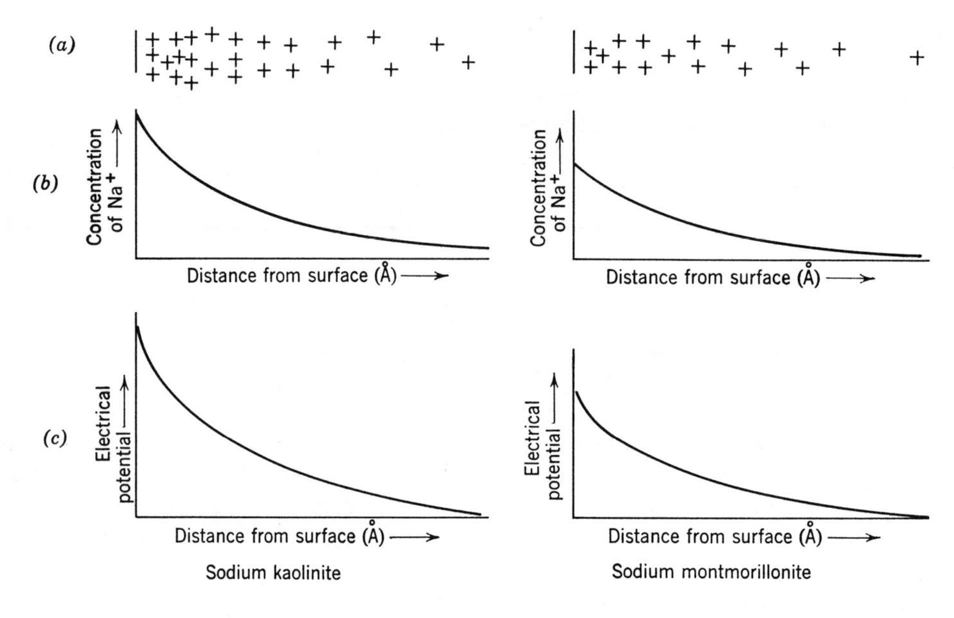

Fig. 5.11 Particles with double layers.

a typical illitic clay in nature may have a water content of 50%. From our calculation, we see that almost all of this water is "free"; i.e., not strongly attracted to the mineral skeleton and thus really constituting a separate phase. On the other hand, in many highly montmorillonitic clays, it may be quite difficult to separate the mineral and pore phases.

There are certain soil minerals that have the ability to immobilize a relatively large amount of water. The most common such mineral is halloysite, which has a crystal structure similar to kaolinite, as indicated in Fig. 4.11. Because of the ability of halloysite to adsorb water between the sheets, soils containing this mineral can exist at a high water content and a low density. Clays containing halloysite have successfully been used for dam cores even though they showed compacted dry unit weights of 7.8–9.4 kN/m³ (half of that for normal clays) and water contents of 30–50% (two or more times the usual compaction water content for clay). Several examples of this unusual behavior are given by Lambe and Martin (1953–1957).

In the preceding discussion, sodium has been selected as the exchangeable ion. The ions adsorbed on soil particles can readily be exchanged, as illustrated in the symbolic reaction shown in Fig. 5.12. The addition of calcium chloride to a soil-water system results in the replacement of sodium by calcium. The nature of the exchangeable ion on the soil particle has an important influence on the behavior of the soil system. Note from Table 3.4, for example, how much the Atterberg limits of clay can depend on the nature of the exchangeable ion.

A reaction, such as that shown in Fig. 5.12, results in a depression of the double layer around the soil particle, i.e., the thickness of the ion-water layer around the soil particle reduces. This reduction of particle double layer results in a change of the properties of a mass of particles. There are general principles that control the rate and direction of exchange reactions. These principles involve the valence of the exchanged cations, concentration of cations, etc.

5.3 THE FORCES *R'* AND *A'*

If we take two clay particles in water which are far apart and then bring them toward each other, they will reach an interparticle spacing at which they begin to exert forces on each other. Since each particle carries a net negative charge, the two particles repel each other

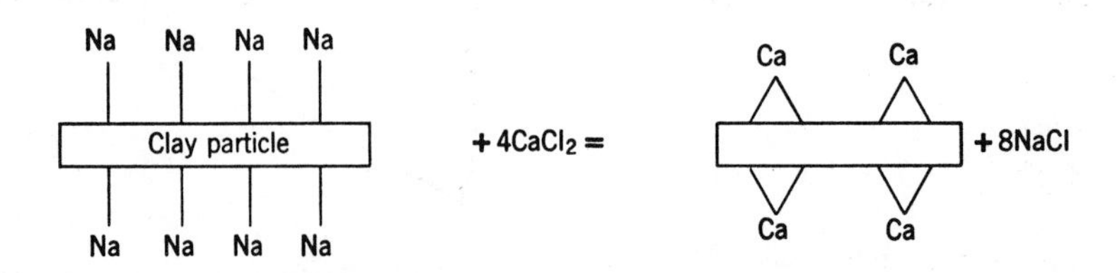

Fig. 5.12 Ion exchange reaction.

because of the Coulombic electrical force between like charges. This is the force R'. This repulsion of the approaching clay particles is like that between two bar magnets when the negative poles are pushed toward each other (or the positive poles are pushed toward each other).

Since the negative charge on a clay particle is balanced by the cations in the double layer, the two advancing particles begin to repel each other when their double layers come into contact with one another. The repulsive force between the adjacent particles for any given spacing is therefore directly related to the sizes of the double layers on the two particles, and any change in the characteristics of the soil-water system that reduces the thickness of the double layers reduces this repulsive force for the same interparticle spacing. Figure 5.13 illustrates the influences of various characteristics of the system on the electrical potential ψ, and therefore R', at any given distance x from the particle surface.

In addition to a repulsive force between the approaching clay particles, there is also a component of attractive force A' between the two particles. This attractive force is the van der Waals' force, or secondary bonding force, which acts between all adjacent pieces of matter. This attractive force between the clay particles is essentially independent of the characteristics of the fluid between the particles.

At this stage, it is convenient to distinguish two cases: (*a*) the case where the total force between particles is very small, i.e., equivalent to the weight of soil contained in an ordinary beaker; and (*b*) the case where the total force is equivalent to the weight of a building or to the weight of three or more meters of overburden.

The first case is encountered as a sedimentary soil is first formed, and a study of this case leads to an understanding of how particles may be arranged within a sedimentary soil. This case is studied in Section 5.4. It suffices to consider R' and A' only in qualitative terms.

The second case is typical of that encountered in engineering practice, and a study of this case (see Section 5.5) leads to an understanding of just how forces are transmitted between particles. In this study it will be necessary to treat R' and A' in quantitative terms.

5.4 FLOCCULATION AND DISPERSION

If the net effect of the attractive and repulsive forces between the two clay particles is attractive, the two particles will tend to move toward each other and become attached—*flocculate*. If the net influence is repulsive they tend to move away—*disperse*. Since the repulsive force component is highly dependent on the characteristics of the system and the attractive component of force is not strongly influenced by the characteristics of the system, a tendency toward flocculation or dispersion may be caused by an alteration in the system characteristics,

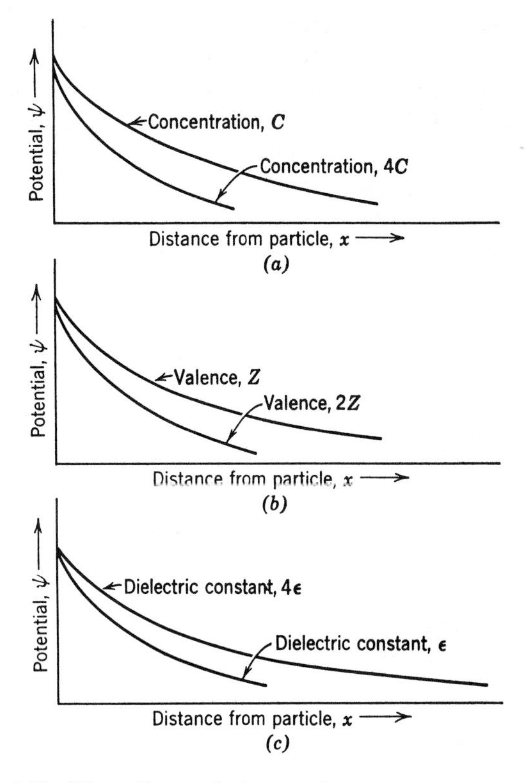

Fig. 5.13 The effects of changes in system properties on double layers. (*a*) Concentration only variable. (*b*) Valence only variable. (*c*) Dielectric constant only variable.

which alters the double-layer thickness. A tendency toward flocculation is usually caused by increasing one or more of the following:

Electrolyte concentration
Ion valence
Temperature

or decreasing one or more of the following:

Dielectric constant
Size of hydrated ion
pH
Anion adsorption

Most of the effects of varying the characteristics of the soil-water system on the tendency toward flocculation or dispersion can be demonstrated with a soil-water suspension in a test tube. In each demonstration the same weight of soil particles is employed. These are illustrated in Fig. 5.14.

The two types of interparticle forces discussed so far have two important characteristics:

1. They originate from within the mineral crystal.
2. They can act over relatively large distances, i.e., several hundred angstroms.

These are the only two types of forces considered by colloidal theories. There is convincing evidence that there are other electrical forces which can become very important when the spacing between clay particles decreases to very small distances such as are typical in the deposit with which the civil engineer usually works. The most important force not considered by the colloidal theories is that arising from the net positive charge at the edges of the soil particles. This net charge is small relative to the net negative charge on the particle arising from isomorphous substitution and thus plays a minor role when adjacent particles are hundreds of angstroms apart. When, however, the particles are very close together, this edge charge can participate in an edge-to-face linkage between particles of an electrostatic type.

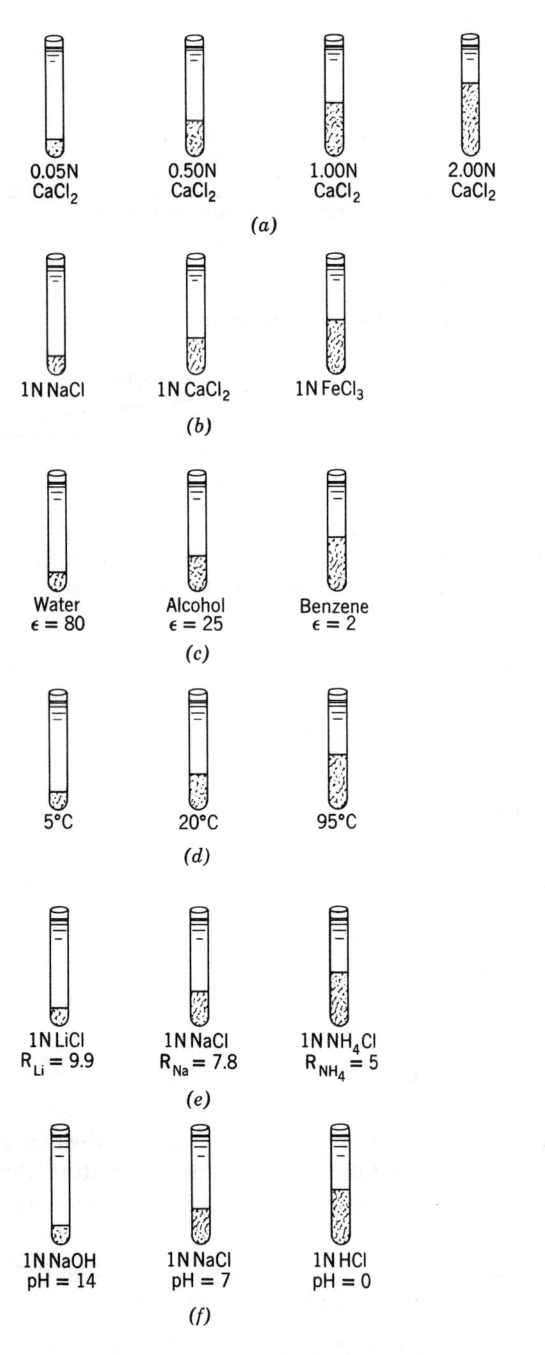

Fig. 5.14 Effects of system characteristics on soil sediment. (*a*) Effect of electrolyte concentration. (*b*) Effect of ion valence. (*c*) Effect of dielectric constant. (*d*) Effect of temperature. (*e*) Effect of size of hydrated ion. (*f*) Effect of pH. Each tube has same concentration of soil in liquid.

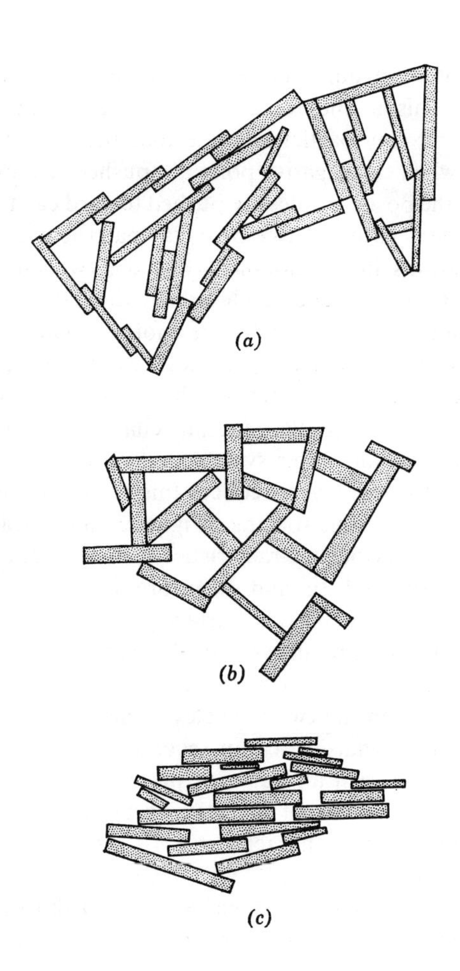

Fig. 5.15 Sediment structures. (*a*) Salt flocculation. (*b*) Nonsalt flocculation. (*c*) Dispersion.

In the demonstrations illustrated in Fig. 5.14, the flocculated sediments consisted of particles attracted to one another to form loose arrays. The particles in the sediment which repelled each other could stack in efficient arrays much like playing cards in a deck. Figure 5.15 illustrates particle arrangements in the soil sediments. Where there is salt-type flocculation (that treated by the colloidal theories), there is a measurable degree of parallelism between adjacent particles since the attraction between the particles is of the secondary valence type. In the edge-to-face or nonsalt-type flocculation, the particles tend to be perpendicular to each other since the attraction is an electrostatic one between the edge of one particle and the face of another. As shown in Fig. 5.15*c*, the dispersed sediment tends to have particles that are in a parallel array.

5.5 THE TRANSMISSION OF FORCE THROUGH SOIL

Figure 5.16 shows two parallel platens to which a normal force of 63.1 N is applied. Each platen is square, 25 mm on a side, thus having an area of 625 mm². The normal stress between the two platens is the total force 63.1 divided by the area of 625 mm² and thus equals 0.101MN/m².

Each platen is now coated with a layer of wet sodium montmorillonite particles oriented parallel to the platens. For a system of parallel sodium montmorillonite particles Bolt (1956) obtained experimentally the curve of spacing versus stress shown in Fig. 5.16*b*. Since the parallel clay particles cover essentially all of the area of the platens facing each other, the stress between particles is 0.101 MN/m² and, from Bolt's data, are at a spacing of about 115 Å. In other words, the stress carried by the clay particles is essentially the same as that carried by the platens. Further, the particle spacing and interparticle stress are related such that the greater the stress between the particles the smaller the spacing. A stress of about 550 MN/m² is required to force the two montmorillonite particles into mineral-mineral contact, thereby squeezing out the adsorbed water between them.

Next the platens are coated with sand particles, as shown in Fig. 5.16*c*. Each particle has a diameter of approximately 0.06 mm. For such a parallel array of particles between the platens, the stress at points of contact between the sand particles is equal to the force divided by the actual contact area. Measurements of this contact area show that typically it is approximately 0.03% of the total area. Thus contact stress is equal to 63.1 divided by 0.188 mm², which equals approximately 335 MN/m² This contact stress is high enough to extrude the adsorbed water.

The example in Fig. 5.16 illustrates the fact that normal stress can be transmitted through a highly dispersed clay system by long-range electrical stresses, and there is no direct mineral-mineral contact between the particles. On the other hand, in a flocculated soil, such as that shown in Fig. 5.15*a* or 5.15*b*, the particles are effectively in mineral-mineral contact and the normal stresses are transmitted in a fashion similar to that for the sand system shown in Fig. 5.16*c*.

The particles in a natural soil are not the same size and shape as the colloidal theories assume. Nearly all natural soils contain particles of many shapes and many sizes, and almost all soils contain particles of different compositions and impurities. Silt-sized particles occur in most natural clays and these nonplate-shaped particles affect the arrangements of the plate-shaped particles. Moreover, even the plate-shaped clay particles do not in general have perfectly smooth surfaces. For example, irregularities can be seen on the surface of the kaolinite particle shown in Fig. 4.4*a*. Such irregularities may be 100 Å high, which is equal to the distance over which significant long-range electrical forces can act.

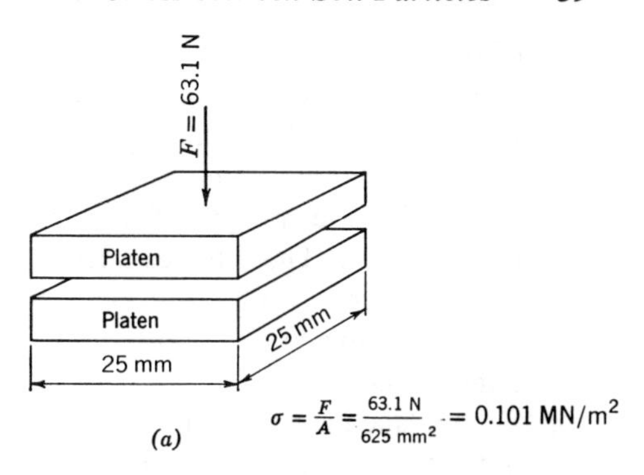

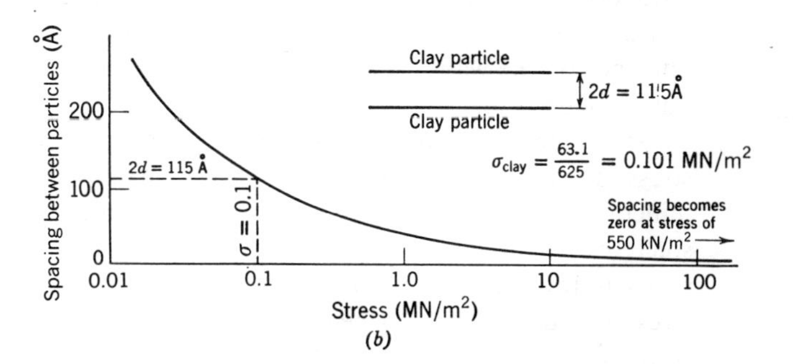

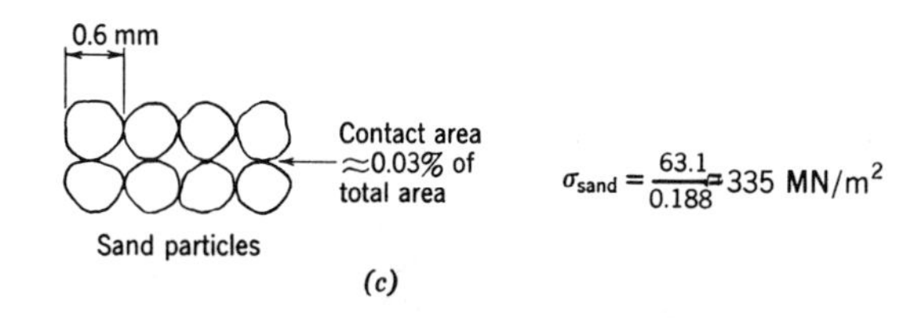

Fig. 5.16 Stress transmission through soil.

Thus the mechanism of stress transmission between soil particles in natural clays must lie between the extreme situations of equidimensional particles and of parallel clay particles. The behavior in general is nearer to that of soils with equidimensional particles.

Because of these difficulties, and because the theories neglect certain forces which probably are important when the particle spacing is less than 100 Å, the principles of colloidal chemistry have been of very little quantitative help in the study of clay behavior. The colloidal principles are, however, very useful to the civil engineer in helping to give an understanding of the fundamental behavior of fine-grained soils.

5.6 SUMMARY OF MAIN POINTS

1. Every soil particle carries an electrical charge on its surface and therefore attracts ions to this surface in order to achieve overall electrical neutrality.
2. These ions in turn attract molecules of water and, in addition, water is attracted directly to the surfaces

of soil particles. Hence all soil particles tend to be surrounded by a layer of water.

3. Forces of attraction and repulsion act between all soil particles, but again are more important (with respect to the weight of the particles) in fine-grained soils. These forces affect the way in which particles are arranged during sedimentation and can cause fine-grained soils to have a very open mineral skeleton of low unit weight.
4. Factors such as temperature and ion concentration in the pore water have an influence on the forces of attraction and repulsion between particles, and hence the environment of deposition can have an influence on the way that the particles are arranged during deposition.
5. In soils composed of equidimensional particles, stress is transmitted through the soil by means of mineral-mineral contact forces. In soils composed solely of small clay platelets oriented face-to-face, stress is transmitted through long-range electrical stresses, and particles may be separated by distances of 100 Å or even more. Stress transmission through natural clay soils lies between these extreme situations.

PROBLEMS

5.1 Estimate the specific surface in units of square meters per gram for the sand in Fig. 4.2*a*. Take the specific gravity equal to 2.65.

5.2 Compute the ion exchange capacity in units of me/100 g for kaolinite having isomorphous substitution of one Al for every four hundredth Si.

5.3 For the kaolinite particle shown in Fig. 4.4*a* compute:

a. The total surface area.

b. The specific surface in m^2/g.

c. The surface area (in $Å^2$) per unit charge from isomorphous substitution.

d. The number of sodium ions required to satisfy the ion exchange computed in Problem 5.2.

The specific gravity of kaolinite is 2.62.

5.4 If calcium chloride were added to the wet clay in Fig. 5.16*b*, would the platens move together or apart? Why?

5.5 If the contact stress required to crush quartz is 6900 MN/m^2, what force would have to be applied to the platens in Fig. 5.16*c* to crush quartz sand?

CHAPTER 6

Shear Resistance between Soil Particles

This chapter considers the fundamental nature of shear resistance between soil particles. Section 6.1 discusses the mechanism of shear resistance in general terms and indicates its typical magnitude. Sections 6.2 to 6.5 present a more detailed treatment for those who wish to delve more deeply into the subject.

6.1 GENERAL DISCUSSION OF SHEAR RESISTANCE BETWEEN PARTICLES

Chapter 2 stated that relative sliding between particles is the most important mechanism of deformation within a soil mass. Hence the resistance of soil to deformation is influenced strongly by the shear resistance at contacts between particles. A knowledge of the possible magnitude of this shear resistance, and of the factors that influence this resistance, is basic to the mastery of soil mechanics.

It must be emphasized that the shear resistance between mineral surfaces is only part of the resistance of a soil mass to shear or compression. Also very important is the interlocking of particles, which is largely a function of packing density. Interlocking is treated in Part III. The fundamental considerations of this chapter, however, apply no matter how the particles are packed.

Mechanism of Shear Resistance

The shear resistance between two particles is the force that must be applied to cause a relative movement between the particles. The source of the shear resistance is the attractive forces that act among the surface atoms of the particles. These attractive forces lead to chemical bond formation at points of contact of the surfaces. Thus the frictional resistance between two particles is fundamentally of the same nature as the shear resistance of a block of solid, intact material such as steel.

The strength and the number of bonds that form at the interface between two particles are influenced very much by the physical and chemical nature of the surfaces of the particles. Hence an understanding of the magnitude of the shear resistance between particles involves an understanding of the factors that influence the interaction between the two surfaces at their points of contact. A detailed explanation of this interaction effect is presented in the later sections of this chapter. However, we can summarize the interaction effect by saying that the total shear resistance (the product of the strength of each bond and the total number of bonds) is proportional to the normal force that is pushing the two particles together. If this normal force decreases, either the strength or the number of bonds decreases, and thus the total shear resistance decreases. Hence we say that interparticle shear resistance is *frictional* in nature.

There are some situations in which part of the total shear resistance between particles is independent of the normal force pushing the particles together; i.e., even if the normal force is decreased to zero there is still a measurable shear resistance. In such cases, we say that there is *true cohesion* between particles. True cohesion can develop between soil particles that have remained in stationary contact over a long period of time. In some cases this true cohesion can be very important, as when cementation turns sand into sandstone. Generally, however, the magnitude of true cohesion between particles is very small, and its contribution to the strength of a soil is also very small. Later chapters of this book will discuss a few of the situations in which true cohesion between particles is important. The reader should regard frictional behavior as the normal situation for soils and cohesive behavior as the exception.

Two alternative ways of expressing frictional resistance are in common use. The first is to use the coefficient of friction f. Thus, if N is the normal force across a surface, the maximum shear force on this surface is $T_{\max} = Nf$. The second is to use a friction angle ϕ_μ defined such that

$$\tan \phi_\mu = f$$

The geometric interpretation of ϕ_μ is shown in Fig. 6.1.

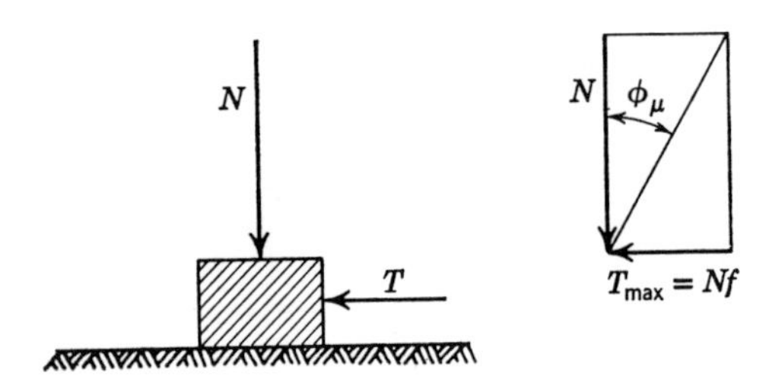

Fig. 6.1 Definition of friction angle ϕ_μ.

6.2 FUNDAMENTALS OF FRICTIONAL BEHAVIOR

Basic Laws of Friction

There are two basic laws of frictional behavior:

1. The shear resistance between two bodies is proportional to the normal force between the bodies.
2. The shear resistance between two bodies is independent of the dimensions of the two bodies.

The second law can be illustrated by pulling a brick over a flat surface. The pulling force will be the same whether the brick rests on a face or on an edge.

As is the case with many of the "laws" of science, the "laws of friction" merely state the result of many empirical observations. Exceptions to these rules are easy to find. Nevertheless, they remain a good starting point for understanding frictional behavior. These laws were first stated by Leonardo da Vinci in the late 1400s, largely forgotten, and then rediscovered by the French engineer Amontons in 1699. They are often called Amontons' laws.

Mechanism of Friction

The basic explanation of the friction process is embodied in the following statements:

1. On a submicroscopic scale most surfaces (even carefully polished ones) are actually rough, hence two solids will be in contact only where the high points (termed *asperities*) touch one another; i.e., actual contact is a very small fraction of the apparent contact area (see Fig. 6.2).
2. Because contact occurs at discrete sites, the normal stresses across these contacts will be extremely high and even under light loading will reach the yield strength of the material at these sites. Thus the actual area of contact A_c will be

$$A_c = \frac{N}{q_u} \tag{6.1}$$

where N is the normal load and q_u is the normal stress required to cause yielding (i.e., plastic flow). Since q_u is fixed in magnitude, an increase in total normal load between the bodies must mean a proportional increase in the area of actual contact. This increase is a result of plastic flow of the asperities.
3. The high contact stresses cause the two surfaces to adhere at the points of actual contact; i.e., the two bodies are joined by chemical bonds. Shear resistance is provided by the adhesive strength of these points. Thus the maximum possible shear force T_{max} is

$$T_{max} = sA_c \tag{6.2}$$

where s is the shear strength of the adhered junctions and A_c is the actual area of contact.[1] For the moment, we will not say whether or not s is equal to the shear strength s_m of the material composing the particles.

Combining these ideas leads to the relation

$$T_{max} = N\frac{s}{q_u} \tag{6.3}$$

Since s and q_u are material properties, T_{max} is proportional to N. The friction factor f should equal the ratio s/q_u.

Terzaghi (1925) stated this hypothesis in his pioneering book on soil mechanics,[2] but his ideas on this subject were overlooked for many years. The hypothesis was independently stated and shown to describe the frictional behavior of a wide variety of materials by Bowden, Tabor, and their colleagues starting in the late 1930s [see, for example, Bowden and Tabor (1950) and (1964)]. It is called the Adhesion Theory of Friction and now serves as the starting point for essentially all friction studies. The following paragraphs will discuss its applicability to the friction of soil minerals.

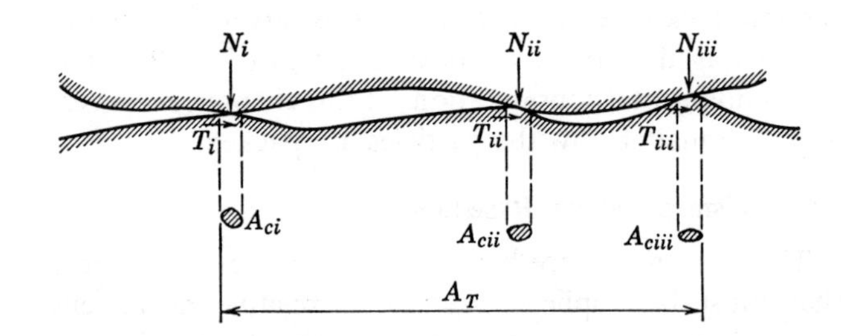

Fig. 6.2 Microscopic view of frictional resistance.

$$N = \sum N_i = \sum A_{ci} q_u$$
$$T = \sum T_i = \sum A_{ci} \tau_m$$
$$\mu = \frac{T}{N} = \frac{\tau_m}{q_u}$$

[1] A_c is the area of mineral-mineral contact. The term A_m is also used for this area.

[2] An English version of this section of *Erdbaumechanik* is in Terzaghi (1960).

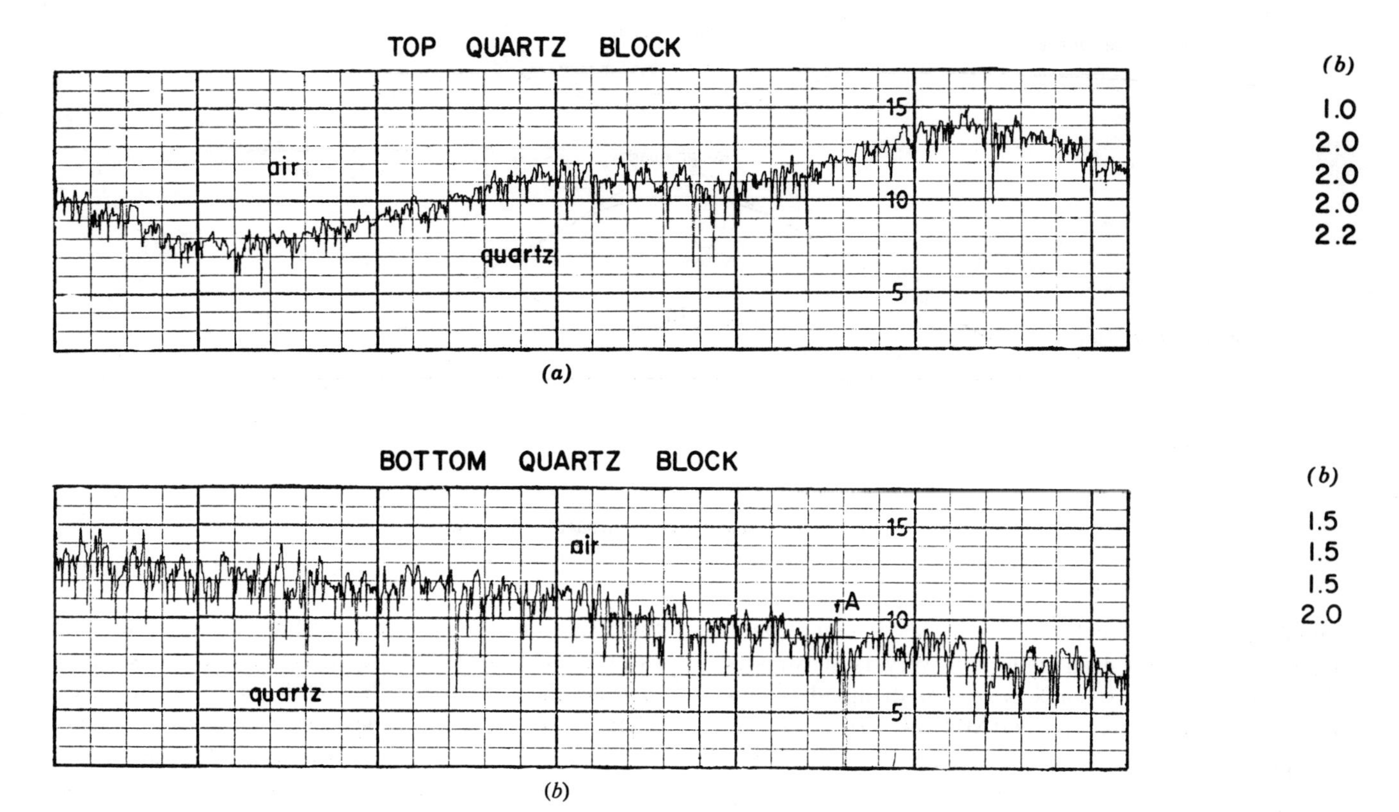

Fig. 6.3 Profile of "smooth" quartz surfaces. (*a*) Surface traces. Scale: vertical, 1 division = 50×10^{-6} mm; horizontal, 1 division = 50×10^{-3} mm. (*b*) Center line average roughness in 25×10^{-6} mm. (From Dickey, 1966.)

Surface Roughness

Various techniques can be used to measure the roughness of a surface. For example, a sharp diamond stylus, called a "profilometer", may be slid across the surface and measurements made of the up and down movement of the stylus. Figure 6.3 shows the trace made by such an instrument as it moved across two "smooth" quartz surfaces (Dickey, 1966). The surfaces had been ground with a very fine diamond wheel (600 grit) and appeared to be mirror smooth. However, the profilometer shows the surfaces to be composed of peaks and valleys having an average height of 50 nm (500 Å).

It should be noted that the horizontal sensitivity of the profilometer is 1000 times less than the vertical and therefore the asperities were not really jagged, as the trace seems to indicate. Rather, they were very flat, as shown in Fig. 6.4, which is a true-scale drawing of one of the asperities. Figure 6.4 also shows a typical asperity from a "rough" quartz surface, prepared by grinding with a No. 220 grit diamond wheel. The average roughness of this surface was about 500 nm. The asperities were sharper than for the smooth surface, but they were still quite gentle, having an average included angle of about 120°. Both of these surfaces are probably smoother than the surfaces of most granular soil particles.

Magnitude of Contact Stress

When two surfaces are brought into contact they will be supported initially on the summits of the highest asperities. As the normal load is increased, the "bearing capacity" of the contacting asperities will be exceeded and they will deform plastically. The magnitude of stress required to cause plastic flow is termed q_u. It can be determined by indentation hardness measurements.[3]

For quartz, with a hardness of about 10 800 MN/m^2 (Brace, 1963), the stress on an asperity must exceed 10 300 MN/m^2 to produce plastic deformation. Whether or not this stress is reached for a significant number of asperities in a granular soil mass is not known, but it seems likely that it is. If q_u is not reached, the asperities deform *elastically*, and then the behavior becomes quite different. According to the Hertz contact theory (see Bowden and Tabor, 1964) the contact area A_c increases as $N^{2/3}$. Thus the coefficient of friction will probably *decrease* with increasing load. Such behavior has been

[3] Indentation hardness is measured by pushing a suitably shaped indenter into a flat test surface. The high confining stresses that develop around the tip of the indenter inhibit brittle fracture in a material such as quartz. The analogy between this test and the asperity contacts between two surfaces is evident. The indentation hardness is defined as the normal load on the indenter divided by the residual deformed area after the indenter is removed.

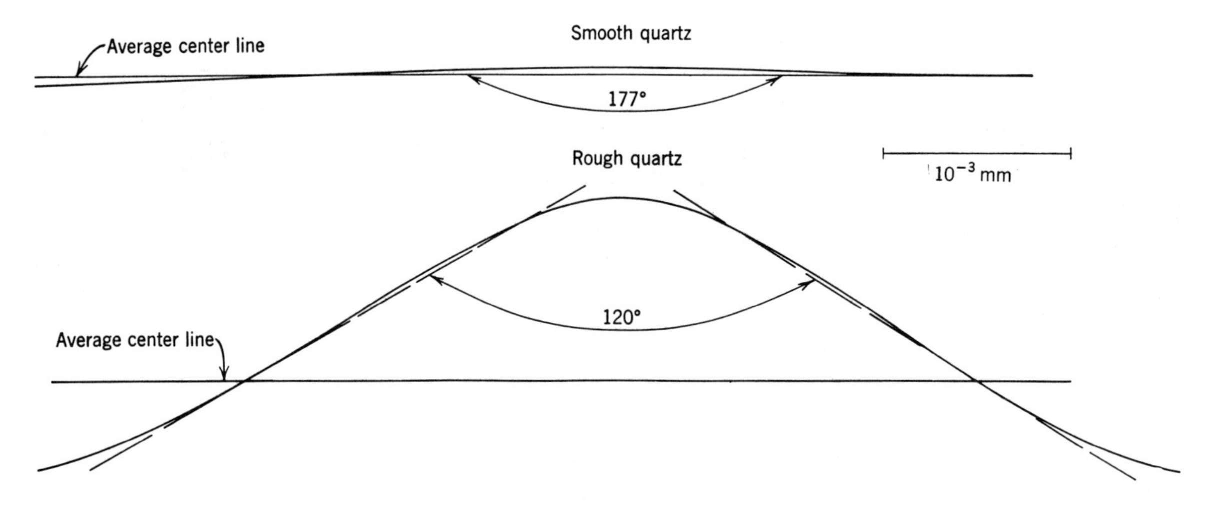

Fig. 6.4 Typical asperities on quartz surfaces.

observed for a diamond stylus on a flat diamond plate. Diamond is elastic even under the very large stresses developed at asperity contacts (q_u for diamond is estimated to be greater than 100 000 MN/m²). However, the complex contact conditions between two surfaces that have a very large number of asperities *can* lead to a nearly constant value of f even though the individual asperities are deforming elastically (e.g., see Archard, 1957).

Shear Resistance at Points of Contact

In Chapter 5 it was pointed out that water and other materials are attracted to the surfaces of minerals, where they are adsorbed and act as a contaminating layer. When two such contaminated surfaces are put together, the amount of actual solid-to-solid contact will be influenced by the type and amount of adsorbed material, as shown in Fig. 6.5.

The most important influence of the surface contaminants is to make the junctions weaker in shear than is the bulk solid. If the surface contaminants are removed, e.g., by heating the surfaces in a high vacuum chamber, the shear strength of the junctions approaches that of the bulk solid. This causes the coefficient of friction to increase. Under these conditions, ductile metals undergo a process known as junction growth, whereby the contacting asperities undergo large-scale plastic deformations. This leads to the phenomenon known as cold welding, which produces extremely high coefficients of friction ($f \gg 1$). Minerals and other brittle materials do not exhibit the large-scale plastic flow under shear stresses that is required for junction growth and hence do not cold-weld.

Effect of Surface Roughness

The Adhesion Theory implies that friction is independent of surface roughness. For metals this is found to be the case over a wide range of surface finishes. However, as surfaces become very rough, asperity interlocking may increase the value of f. It is difficult to define exactly what is "very rough". The frosted appearance of most granular soil particles indicates that they are rough. Electron photomicrographs indicate that many sheet

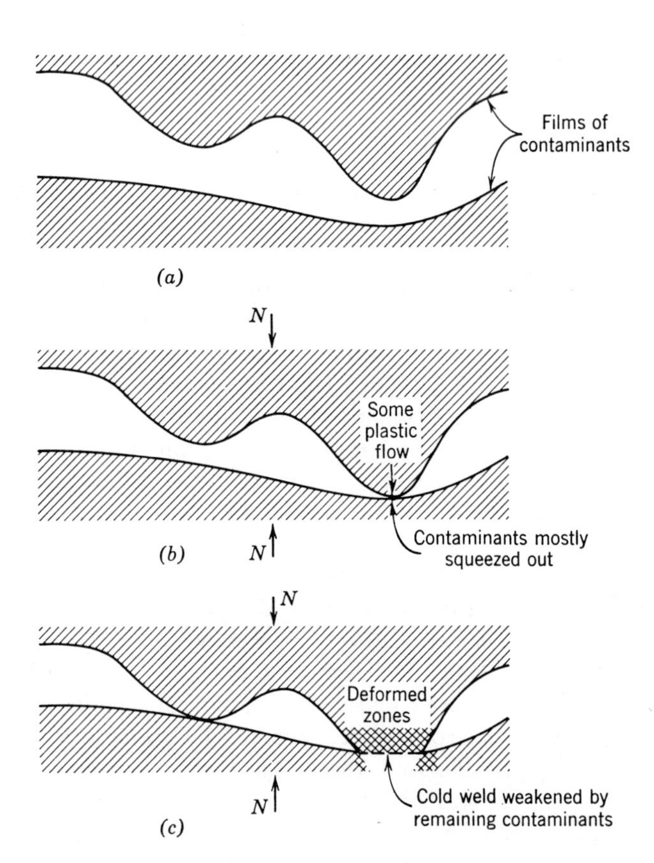

Fig. 6.5 The development of junctions. (*a*) Unloaded condition. (*b*) Light normal load. (*c*) Increased normal load; plastic flow at contact with constant normal stress.

minerals, on the other hand, have "supersmooth" surfaces. By assuming an average slope of angle θ for the asperities, the effect of surface roughness on the value of f can be estimated (see Problem 6.4).

Because the contact situation between two real surfaces is so complex, generally it is not possible to determine a value of θ to use for predicting f. Hence the relationship between friction and surface roughness must be determined experimentally.

Static Versus Kinetic Friction

The shear force required to initiate sliding between two surfaces is often greater than the force required to maintain motion (see Fig. 6.6*a*). That is, the static friction exceeds the kinetic (sliding) friction. This behavior is usually explained by assuming that bond formation at the junctions is time-dependent, either because creep causes gradual increases in the contact area or because the surface contaminants are gradually squeezed out from within the junction.

The difference between the static and the kinetic friction often leads to the phenomenon known as *stick-slip* (Fig. 6.6*b*). When sliding begins, part of the stored elastic energy in the loading mechanism is released, accelerating the slider and causing the measured shear force to drop below that required to maintain motion. The slider then stops and the shear force must be increased to the value associated with the static friction to induce sliding again. When sliding begins the whole procedure of intermittent motion is repeated. Under these conditions, one cannot determine accurately the value of the kinetic coefficient of friction.

Rolling Friction

When one body is rolled over another, junctions form at the contact points just as when two bodies are pressed together. As the rolling body moves on, these junctions are broken in tension, not in shear. Due to elastic rebound as the normal force goes to zero, the strength of the junctions in tension is usually almost zero. This explains why the adhesion between two surfaces that are pressed together is not generally observed—it only acts when the surfaces are under a compressive load. Hence rolling friction is generally quite small ($f \ll 0.1$) compared to static and sliding friction, and is essentially independent of surface cleanliness.

Summary

Section 6.2 has presented the fundamentals of frictional behavior: emphasis has been given to the following concepts, which are necessary in order to develop a quantitative explanation of observed frictional behavior:

1. The roughness and irregularity of apparently smooth surfaces.

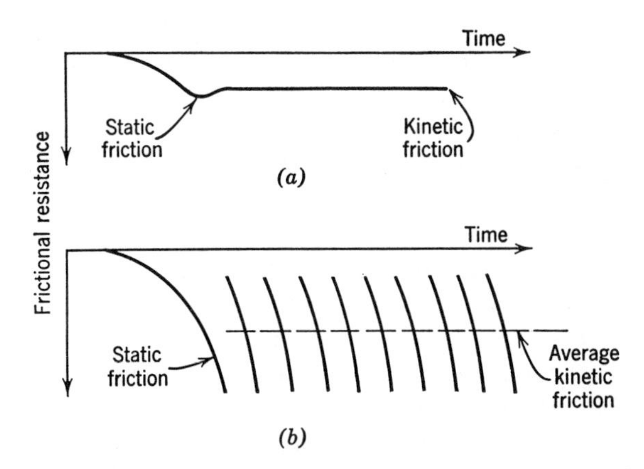

Fig. 6.6 Starting and kinetic friction. (*a*) Smooth sliding. (*b*) Slip-stick. *Note.* Shear displacement applied at constant rate through load cell with some flexibility; surfaces move at constant rate only during smooth sliding.

2. The very low ratio of actual contact area to apparent contact area.
3. The plastic flow that occurs at actual contact points.
4. The adhesion that occurs between two surfaces at the actual contact points.
5. The weakening influence of surface contaminants on the strength of this welded junction.

These concepts will now be used to explain the observed frictional behavior of soil minerals.

6.3 FRICTION BETWEEN MINERALS IN GRANULAR FORM

This section will treat friction between nonsheet minerals such as quartz, the feldspars, and calcite—those minerals that make up granular particles of silt size and larger. The following section will treat the behavior of sheet minerals. The friction of minerals has not been studied nearly as intensively as has the friction of metals. Therefore much of what follows is based on limited data and must be considered speculative.

General Nature of Contact

Particles of coarse silt have a minimum diameter of 0.02 mm (20 μm or 200,000 Å). The diameters of these and larger particles are clearly larger than the height of the asperities (about 1000–10,000 Å) that may be expected on the surfaces of these particles. Consequently, we would expect that each apparent point of contact between particles actually involves many minute contacts.

The surfaces of these soil particles are, of course, contaminated with water molecules and various ions and possibly other materials. These contaminants are largely squeezed out from between the actual points of contact, although some small quantity of contaminants remains to influence the shear strength of the junctions.

The minimum diameter of fine silt particles is 2 μm or 20,000 Å. Such small particles have dimensions of the same order as the height of the asperities on larger particles. For these small particles it makes more sense to talk about "corners" rather than asperities. Although the general nature of the frictional resistance is the same for either large or small granular particles, an apparent contact between very small granular particles may, in fact, consist of only one actual contact point.

The testing systems shown in Fig. 6.7 have been used to determine the frictional resistance for minerals. When fixed buttons or sliding blocks are used (Fig. 6.7*a*) the results give the static (and perhaps kinetic) coefficient of friction. When many sand particles are pulled over a flat surface (Fig. 6.7*b*), the results generally reflect some combination of sliding and rolling friction. Hence the friction factor as measured by the second type of test involving many particles may be different from the value measured by the first type of test.

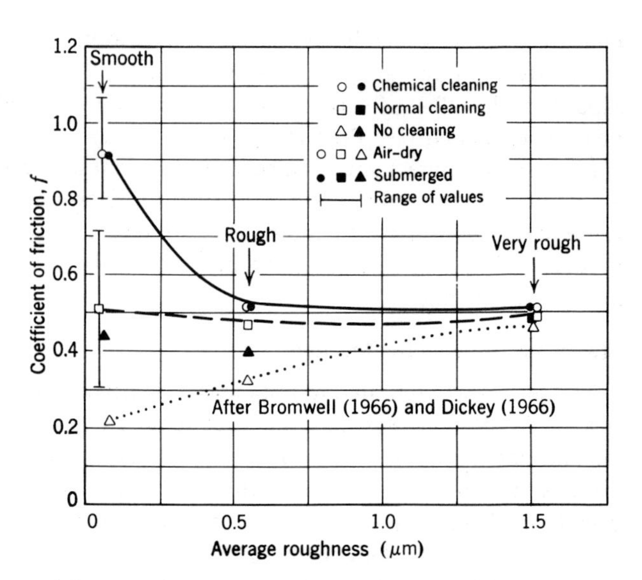

Fig. 6.8 Friction of quartz [after Bromwell (1966) and Dickey (1966)].

Effect of Surface Water and Surface Roughness

Figure 6.8 summarizes the friction factors observed for quartz under varying conditions of surface cleanliness, humidity, and surface roughness. The results show that the friction of smooth quartz varies from about $f = 0.2$ to $f = 1.0$ depending on the surface cleanliness.[4] For the more contaminated surfaces, water increases the friction; i.e., it acts as an antilubricant. However, for carefully cleaned surfaces, water has no effect. This indicates that water is intrinsically neutral on quartz. But if there is a contaminating layer (probably a thin film of organic material) the water disrupts this layer, reduces its effectiveness as a lubricant, and thereby increases the friction.

As the surfaces get rougher the effects of cleaning procedure on friction decrease, so that a "very rough" surface of 1.5 μm (15,000 Å) gives essentially the same value of f independent of surface cleanliness. This indicates that the ability of the contaminating layer to lubricate the surfaces decreases as the surface roughness increases. This is what we would expect for a thin lubricating layer which acts as a boundary lubricant (Bowden and Tabor, 1964).

The fact that the rougher surfaces do not give higher values of friction when they are carefully cleaned is more difficult to explain. The evidence seems to indicate that the rougher surfaces cannot be cleaned as effectively as smooth surfaces, although the reason for this is not clear.

From a practical point of view, the essentially constant value of $f = 0.5$ ($\phi_\mu = 26°$) for very rough quartz surfaces is of great significance, since essentially all quartz particles in natural soils have rough surfaces.

Values of friction for other nonsheet minerals are summarized in Table 6.1. The low values of f for these minerals in the air-dry condition probably have no

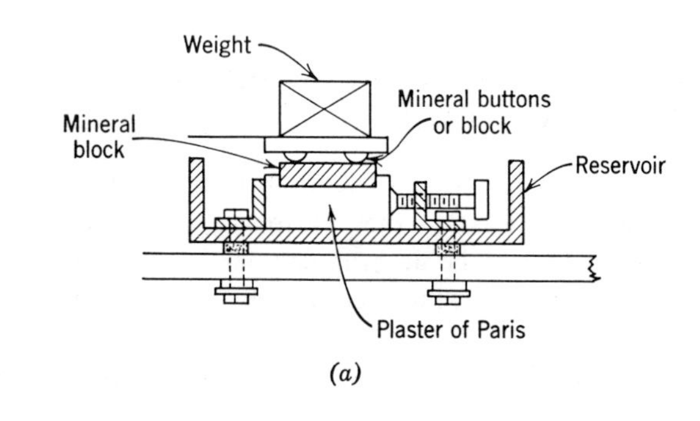

(*a*)

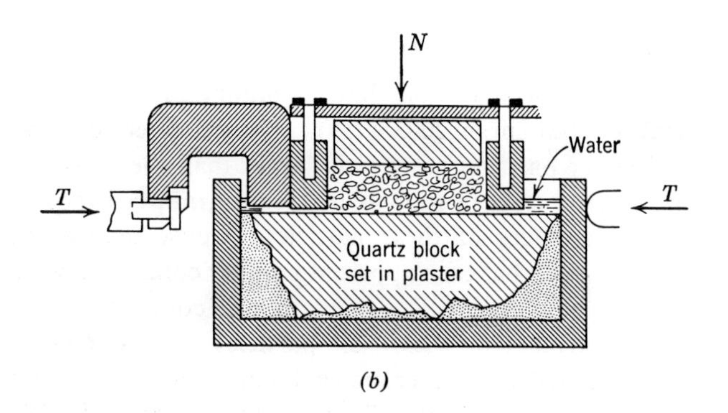

(*b*)

Fig. 6.7 Devices for measuring friction factor of mineral surfaces. (*a*) Sliding on buttons or on block. (*b*) Sliding of many particles.

[4] These results were obtained on ground surfaces [Bromwell (1966) and Dickey (1966)]. However, the trend of the results generally supports the data and conclusions of previous tests, which were usually run on polished surfaces [see, for example, Horn and Deere (1962)].

Table 6.1 Friction of Nonsheet Minerals

	Conditions of Surface Moisture		
Mineral	Oven-Dried	Oven-Dried; Air-Equilibrated	Saturated
Quartz[a]	0.13	0.13	0.45
Feldspar	0.12	0.12	0.77
Calcite	0.14	0.14	0.68

Notes. Tests run on highly polished surfaces. Data from Horn and Deere (1962).

[a] For effects of surface cleanliness and surface roughness on the friction of quartz, see Fig. 6.8.

practical significance, since they represent ineffective cleaning of smooth, polished surfaces. Much more data are needed for these other minerals before one can confidently choose f values.

Effect of Normal Load

The measured friction factors for nonsheet minerals have been found to be independent of normal load. Based on tests in which the normal load per contact varied by a factor of 50, Rowe (1962) reported that the friction angle ϕ_μ remained constant within $\pm 1°$.

On the other hand, Rowe's results show that the friction angle ϕ_μ is affected by the size of the particle involved in the test (Fig. 6.9). Rowe used the test procedure shown in Fig. 6.7*b*. For a given total normal load, the normal load per contact increases as the particle size increases. However, since the particle diameter in this case also increases, the average contact stress (N/A_c) did not change. Therefore arguments involving elastic deformation do not appear adequate to explain these data. One possible explanation is that the larger particles are able to roll more easily than the smaller particles, perhaps as a result of their center of gravity being further away from the plane of shear. Hence the measured friction angle, which involves both rolling and sliding components, is smaller for the larger particles.

6.4 FRICTION BETWEEN SHEET MINERALS

We are interested in minerals such as mica primarily because the frictional behavior of such minerals may be similar to the frictional behavior of clay-size particles.[5]

General Nature of Contact

Surfaces of mica do show irregularities, but in the form of mesas and plateaus rather than in the form of asperities. Moreover, the scale of these irregularities is quite different from that existing on the surfaces of granular particles. On fresh cleavage surfaces, the "steps" are only as high as the thickness of several repeating sheet units (about 10–100 Å). In the words of Bowden and Tabor (1964), the cleavage surfaces "· · · are molecularly smooth over large areas." Compared to the surfaces of smooth quartz particles, fresh cleavage surfaces are "supersmooth". There are reasons to believe that the surfaces of clay particles are similar. Unfortunately, the fundamentals of frictional resistance between supersmooth surfaces have received relatively little study, and hence the following explanations are still largely speculative.

[5] Data from M.I.T. and from the Norwegian Geotechnical Institute have been obtained on the friction angle ϕ_μ between clay particles. Values as low as 3° have been recorded.

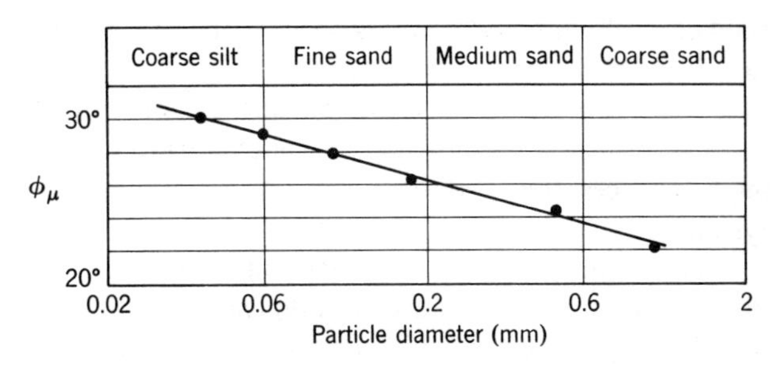

Fig. 6.9 Friction angle of quartz sands as function of grain size (after Rowe, 1962).

Two cleavage faces of mica give quite a different contact than do surfaces with asperities. Mica, and presumably clay, surfaces should come into close proximity over almost their entire area, but they may not actually come into direct contact. The contaminants on the surfaces, including adsorbed water, are not squeezed out from between the surfaces unless the normal stress exceeds about 550 MN/m². Rather, these contaminants participate in the transmission of the normal stress, as described in Chapter 5.

A more normal situation for clay particles is probably some sort of edge-to-face orientation. This type of contact is more nearly similar to the asperity contacts discussed for granular particles, except that in the case of clays, each contact probably consists of only one "asperity".

It still remains to discuss whether the shear resistance between very smooth surfaces is greater or less than the resistance between rough surfaces. To answer this, we must turn to experimental data.[6]

Effect of Surface Water

The data in Table 6.2 show that the water acts as a lubricant. A possible explanation for this behavior is as follows. In the oven-dried condition the surface ions are not completely hydrated. The actual mineral surfaces come close together, and the bonding is strong. As water is introduced, the ions hydrate and become less strongly

[6] These data come primarily from Horn and Deere (1962).

attached to the mineral surfaces. Hence the shear resistance drops as water is introduced.

It is important to contrast the role of the contaminants for the cases of very smooth and rough surfaces. With rough surfaces the contaminants serve to weaken the crystalline bond, and increasing the mobility of the contaminants with water helps get them out of the way and hence minimizes their adverse influence. With very smooth surfaces the contaminants are actually part of the mineral, and increasing their mobility decreases the shear resistance.

Table 6.2 Friction Factors for Several-Layer Lattice Materials Under Varying Conditions of Humidity

	Condition of Surface Moisture		
Mineral	Oven-Dried	Oven-Dried; Air-Equilibrated	Saturated
Muscovite mica	0.43	0.30	0.23
Phlogopite mica	0.31	0.25	0.15
Biotite mica	0.31	0.26	0.13
Chlorite	0.53	0.35	0.22

Notes. Starting and moving friction identical. Data from Horn and Deere (1962).

In the saturated condition the friction angle between sheet minerals can be low. Since clay minerals are always surrounded by water in practical situations, it is important to test these minerals in the saturated condition.

Static and Kinetic Friction

The kinetic friction of the sheet minerals is greater than 90% of the static friction and usually equals it. The slip-stick phenomenon has not been observed with these minerals. The friction factors of mica increase about 25% when the rate of sliding is increased from 0.3 to 2.5 mm/sec. Because the adhesive bonding is relatively weak in the case of these minerals, and the ions through which the bonding occurs are relatively free to move, this relatively small time effect is to be expected.[7]

Variation of Friction Angle with Normal Load

For the usual range of normal loads used, the friction angle of these minerals appears to be constant. However, nothing is known concerning the possible variation with very large changes in the normal load.

[7] Mitchell (1964), by invoking rate process theory, has provided an excellent description of the mechanisms underlying this time-dependent behavior of clay minerals.

6.5 MISCELLANEOUS CONCEPTS CONCERNING SHEAR RESISTANCE BETWEEN MINERAL SURFACES

At the present time, we are still not sure of the extent to which the theory in Section 6.4 may apply to shear resistance between clay particles. However, it is shown in Part IV that many natural clay soils, especially those in which montmorillonite and illite are important constituents, have shear resistances that are compatible with this theory.

The larger a particle, the greater the likelihood that there will be surface irregularities of any consequence. For example, cleavage steps can be seen on the surface of kaolinite platelets, as shown in Fig. 4.4*a*, which are on the order of 100 Å in height. Hence, when platelets of kaolinite are in a face-to-face configuration, it is certain that actual "contact" occurs only over part of the apparent contact surface, and unless the platelets are perfectly aligned it seems likely that contact is confined to relatively small zones right at the cleavage steps. As this situation develops, it is likely that the mechanism of shear resistance, and even the magnitude of shear resistance, becomes more and more analogous to the behavior of granular particles. The same would be true when particles come together in edge-to-face orientation.

Present knowledge regarding interparticle friction in soils can be summarized as follows:

1. The frictional behavior between granular particles is reasonably well understood.
2. The theory for sliding between ideal clay platelets probably applies to the smallest of clay particles in face-to-face array.
3. The mechanism of shear resistance in natural clay lies between the two extremes of granular particles and parallel clay platelets, often nearer to that of the granular particles.

True Cohesion between Clay Particles

Our study of the fundamentals of frictional behavior has helped us to understand the possibility of a true cohesion developing between clay particles. If clay platelets are in edge-to-face contact, there is a good chance that a true cohesion will develop, especially if a bonding has been developed over almost all of the area of the contact.

The discussion in Chapter 5 has already suggested that clay platelets in face-to-face contact may be pushed together so tightly that they will not move apart when the load is removed. Such an occurrence certainly represents true cohesion, and a new and thicker particle has in effect been created by this process. Time, weathering, and desiccation contribute to true cohesion.

6.6 SUMMARY OF MAIN POINTS

The foregoing discussion shows that it is very difficult to say just what the particle-to-particle friction factor will be for any particular case. Hence the main results of this chapter can be summarized in terms of some general principles and a range of possible shear resistances.

1. The shear resistance between particles is provided by adhesive bonding at the points of contact.
2. The shear resistance is determined primarily by the magnitude of the current normal load, so that the overall behavior is frictional in nature.
3. For quartz the friction angle ϕ_μ is generally in the range of 26° to 30°. Because the surfaces of such particles are rough, the presence or absence of water has little or no effect on the frictional resistance. The friction of other nonsheet minerals has received less study and typical values cannot be given.
4. For parallel clay particles whose faces are "supersmooth", the friction angle may be below 8° and may typically be about 13°. The bonding occurs over a rather large area but is relatively weak and may be somewhat time-dependent.
5. For most natural clays the frictional resistance is probably nearer that of granular particles than of parallel colloidal particles.
6. True cohesion can develop between particle surfaces. The true cohesion at any one contact is generally small, so that the overall effect is important only when there are many contacts—as in clay soils. Such bonds are most likely to develop at edge-to-face contacts and can be broken by very small strains.
7. The particle-to-particle friction angle ϕ_μ is but one of the factors which contribute to the strength of an actual soil. Other factors such as interlocking of particles will be discussed in later chapters.

PROBLEMS

6.1 Derive the equations that lead to the following results:

a. f is independent of N for materials that deform plastically.

b. f is a function of normal load for elastic materials. State the necessary assumptions for each equation.

6.2 The diameter of the contact area between two elastic solids (Fig. P6.2–1) is given by

$$d = \left[\frac{12(1-\nu^2)}{E} N \frac{R_1 R_2}{R_1 + R_2}\right]^{1/3}$$

ν = Poisson's ratio = 0.31 for quartz

E = Young's modulus = 75.8×10^3 MN/m² for quartz

N = normal load

R_1, R_2 = radii of curvature

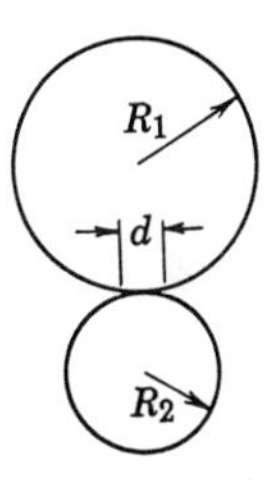

Fig. P6.2-1

a. Develop a relationship for the average contact stress (N/A) between two quartz spheres of equal radius.

b. Consider a system of dry silt-sized material. If the particles are perfect quartz spheres of radius 0.005 mm in a cubic array (see Fig. P6.2-2*a*), what all-around confining stress must be applied to cause plastic yielding? (Plastic yielding for quartz occurs at a normal stress of 10 300 MPa.)

Hint. Consider a horizontal plane through the system (see Fig. P6.2-2*b*) and calculate the contact area on this plane for various values of confining stress.

c. If each sphere in part (*b*) is not perfectly smooth but contacts its neighbors on one asperity which has a tip radius of 1000 Å (100 nm), what must the confining stress be to cause plastic deformation?

d. An actual silt may be expected to give plastic deformation at the contacts at some value of confining stress between the values given by parts (*b*) and (*c*). Why?

e. Would the stress required to cause some plastic yielding be greater for a loose or a dense silt? Explain.

6.3 Consider a flocculated clay made up of platelike particles 1 μm long × 0.1 μm wide × 0.01 μm thick. The void ratio is 2.0 and G_s = 2.70.

a. Calculate the number of particles per unit volume.

b. The number of particles contacting a unit area may be assumed equal to (no./unit volume)$^{2/3}$. If each particle makes two stress-carrying contacts on a horizontal plane, what is the number of contacts per unit area?

c. If each particle-particle contact is assumed to be an edge-to-face contact with the edge taken as a spherical indenter with radius 0.01 μm, what effective confining stress must be applied to the clay to cause plastic yielding? Use same elastic constants as for quartz, but assume plastic flow starts at N/A = 6 900 MN/m².

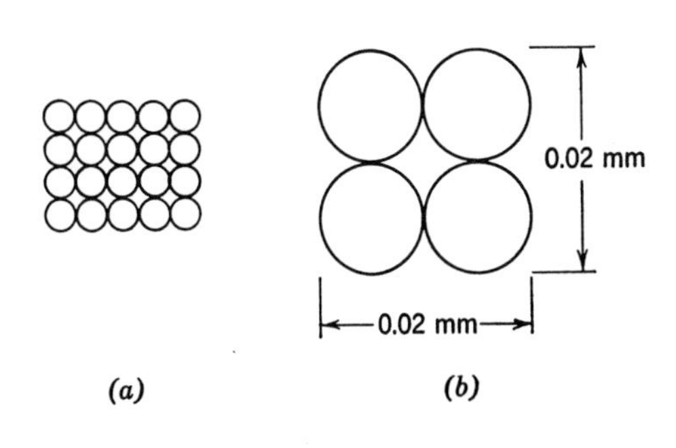

Fig. P6.2-2

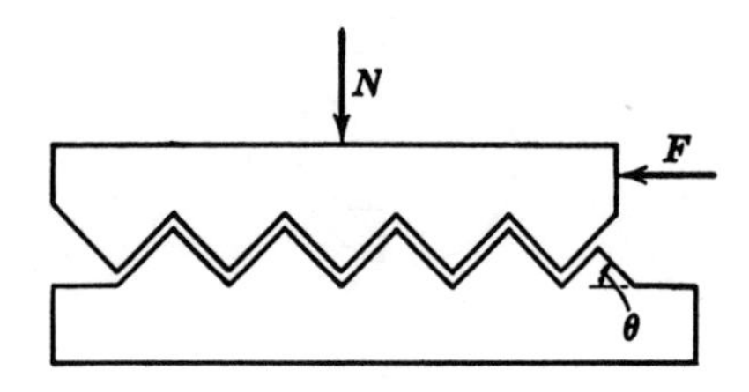

Fig. P6.4

6.4 The asperities on the idealized surface shown in Fig. P6-4 are inclined at an angle θ to the direction of sliding.

a. Derive a relationship for the horizontal force F required to initiate motion. Express F in terms of N, θ, and ϕ_μ (the friction angle for a surface with $\theta = 0$).

b. Calculate the coefficient of friction for the surfaces shown in Fig. P6-4 if $\phi_\mu = 15°$ for a quartz surface with $\theta = 0$. Discuss the validity of this calculation.

6.5 Three clay-water mixtures composed of flocculated platelike particles have the particle dimensions and void ratio given in following table:

Mixture	Particle Dimensions (μm)	Void Ratio
A	10 × 10 × 1	1
B	1 × 1 × 0.05	2
C	0.1 × 0.1 × 0.001	5

If the net attractive stress at particle contacts under zero normal load were equal to F where

$$F \text{ (newtons)} = 10^{-2} \text{ particle thickness} \quad \text{(nm)}$$

compute the cohesion (tensile stress) which must be overcome in order to pull apart the three mixtures. *Note:* Use the approach in Problem 6.3 to obtain the number of contacts per unit area.

Answer:

System	c (kN/m²)
A	5.9
B	35.3
C	1.29

CHAPTER 7

Soil Formation

Based on its method of formation, a soil is *sedimentary*, *residual*, or *fill*. In sedimentary soil the individual particles were created at one location, transported, and finally deposited at another location. A residual soil is one formed in place by the weathering of the rock at the location, with little or no movement of the individual soil particles. Fill is a man-made soil deposit. These three types of soil are discussed in order in this chapter. Preceding this discussion is a consideration of the concept of "structure", which will be used in the presentation of the soil types.

7.1 SOIL STRUCTURE

The term *soil structure* refers to the orientation and distribution of particles in a soil mass (also called "fabric" and "architecture") and the forces between adjacent soil particles. This discussion is limited primarily to small, plate-shaped particles and to the orientation of individual particles. The arrangement of larger particles is considered in later chapters. The force component of soil structure refers primarily to those forces that are generated within the particles themselves—electrochemical forces.

The two extremes in soil structure, as illustrated in Fig. 7-1, are *flocculated* structure and *dispersed* structure. In the flocculated structure the soil particles are edge to face and attract each other. A dispersed structure, on the other hand, has parallel particles which tend to repel each other. Between the two extremes there is an infinite number of intermediate stages. At the present development of knowledge and of techniques for measuring orientations and interparticle forces, there seems little justification in attempting to define structures between the two extremes. Thus the terms *flocculated* and *dispersed* are used in a general sense to describe soil elements which have the structures approaching those shown in Fig. 7.1.

Chapter 5 discussed the electrical forces between particles and introduced the concepts of flocculated and dispersed. No attempt is made to distinguish between the two types of flocculation suggested in Fig. 5.15.

A given soil structure can be significantly altered by introducing displacements between particles. Generally, displacements tend to break down the bonds between particles and to move particles toward a parallel array. If the flocculated structure in Fig. 7.1*a* were subjected to a horizontal shear displacement, the particles would tend to line up as shown in the dispersed structure on the right. A compression tends to cause adjacent particles to move toward a parallel array, probably resulting in small zones of approximately parallel particles, with an unlike orientation between zones. Physically working an element of soil until it becomes homogeneous (termed "remolding") tends to align adjacent particles and to destroy bonds between the particles.

The engineering behavior of a soil element depends very much on the existing structure. In general, an element of flocculated soil has a higher strength, lower compressibility, and higher permeability than the same element of soil at the same void ratio put in a dispersed state. The higher strength and lower compressibility in the flocculated state result from the interparticle attraction and the greater difficulty of displacing particles when they are in a disorderly array. The higher permeability in the flocculated soil results from the larger channels available for flow. Whereas a flocculated element and a dispersed element at the same void ratio have approximately the same total cross-sectional area available for

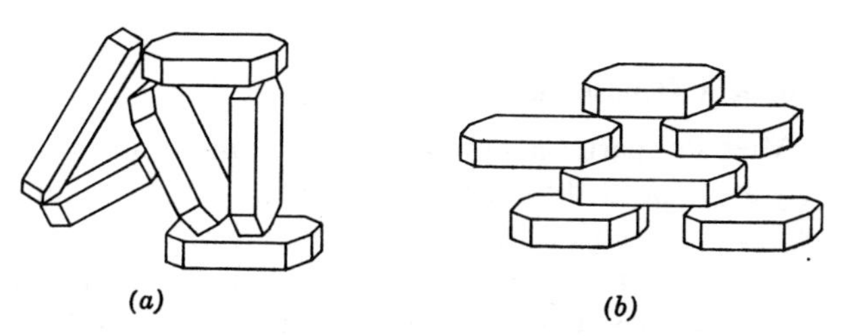

Fig. 7.1 Types of soil structure. (*a*) Flocculated. (*b*) Dispersed.

Table 7.1 Effects of Transportation on Sediments

	Water	Air	Ice	Gravity	Organisms
Size	Reduction through solution, little abrasion in suspended load, some abrasion and impact in traction load	Considerable reduction	Considerable grinding and impact	Considerable impact	Minor abrasion effects from direct organic transportation
Shape and roundness	Rounding of sand and gravel	High degree of rounding	Angular, soled particles (⏢)	Angular, non-spherical	
Surface texture	Sand: smooth, polished, shiny Silt: little effect	Impact produces frosted surfaces	Striated surfaces	Striated surfaces	
Sorting	Considerable sorting	Very considerable sorting (progressive)	Very little sorting	No sorting	Limited sorting

flow, in the flocculated soil the flow channels are fewer in number and larger in size. Thus there is less resistance to flow through a flocculated soil than through a dispersed soil.

7.2 SEDIMENTARY SOIL

The formation of sedimentary soils can best be presented by considering sediment formation, sediment transportation, and sediment deposition, respectively.

Sediment Formation

The most important manner of forming sediments is by the physical and chemical weathering of rocks on the surface of the earth. Generally silt-, sand-, and gravel-sized particles are formed by the physical weathering of rocks and clay-sized particles are formed by the chemical weathering of rocks. The formation of clay particles from rocks can take place either by the build-up of the mineral particles from components in solution, or by the chemical breakdown of other minerals.

Sediment Transportation

Sediments can be transported by any of five agents: water, air, ice, gravity, and organisms. Transportation affects sediments in two major ways: (*a*) it alters particle shape, size, and texture by abrasion, grinding, impact, and solution; (*b*) it sorts the particles. Table 7.1 summarizes some of the effects of the five transporting agents on sediments.

Sediment Deposition

After soil particles have been formed and transported they are deposited to form a sedimentary soil. The three main causes of deposition in water are velocity reduction, solubility decrease, and electrolyte increase. When a stream reaches a lake, ocean, or other large body of water it loses most of its velocity. The competency of the stream thus decreases and sedimentation results. A change in water temperature or chemical nature can result in a reduction in the solubility of the stream, with a resulting precipitation of some of the dissolved load. Figure 7.2 suggests the soil structure that might result from a sedimentary soil formed in salt water compared with one formed in fresh water. As was shown in Fig. 5.14, the soil deposited in salt water will be more highly flocculated and will thus have a much larger void ratio and water content.

7.3 RESIDUAL SOIL

Residual soil results when the products of rock weathering are not transported as sediments but accumulate in place. If the rate of rock decomposition exceeds the rate of removal of the products of decomposition, an accumulation of residual soil results. Among the factors influencing the rate of weathering and the nature of the products of weathering are climate (temperature and rainfall), time, type of source rock, vegetation, drainage, and bacterial activity.

The residual soil profile may be divided into three zones: (*a*) the *upper zone* where there is a high degree of weathering and removal of material; (*b*) the *intermediate zone* where there is some weathering at the top part of the zone, but also some deposition toward the bottom part of the zone; and (*c*) the *partially weathered zone* where there is the transition from the weathered material to the unweathered parent rock.

Temperature and other factors have been favorable to the development of significant thicknesses of residual soils in many parts of the world, particularly Southern Asia, Africa, Southeastern North America, Central America, the islands of the Caribbean, and South America. As we can infer from this distribution, residual soils tend to be more abundant in humid, warm regions that are favorable to chemical weathering of rock, and have sufficient vegetation to keep the weathering products from being easily transported as sediments. Even

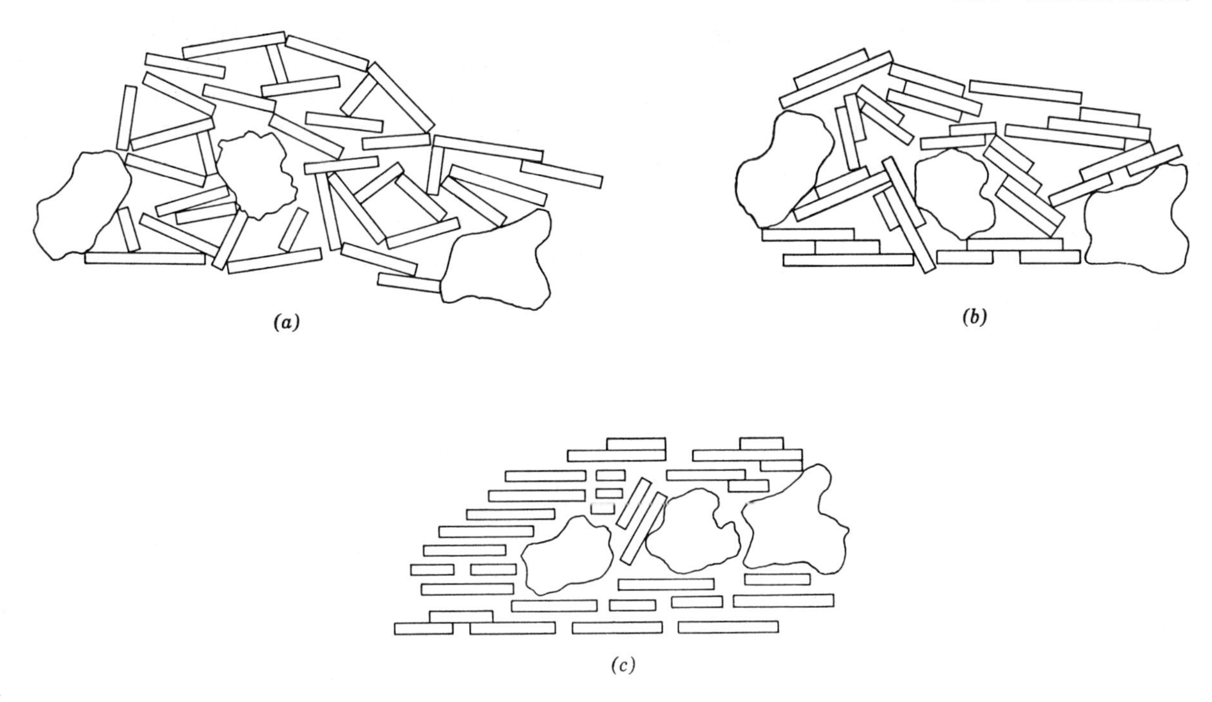

Fig. 7.2 Structure of natural soil (*a*) Undisturbed salt water deposit. (*b*) Undisturbed fresh water deposit. (c) Remoulded.

though residual soils are widely spread throughout the world, they have received relatively little study from soil mechanics experts because these soils are generally located in areas of undeveloped economics, as contrasted to the sedimentary soils which exist in most centers of population and industry.

Sowers (1963) gives the following typical depths of residual soils:

Southeastern United States	6 to 23 m
Angola	7.5 m
South India	7.5 to 15 m
South Africa	9 to 18 m
West Africa	10 to 20 m
Brazil	10 to 25 m

7.4 FILL

The two preceding sections considered the formation of soil deposits by nature. A man-made soil deposit is called a *fill* and the process of forming the deposit is called *filling*. A fill is actually a "sedimentary" deposit for which man carried out all of the formation processes. Soil, termed *borrow*, is obtained from a source or made by blasting, transported by land vehicle (such as truck, scraper, pan, or bulldozer) or water vehicle (barge) or pipe, and deposited by dumping. The fill can be left as dumped, such as the rock toe in the earth dam shown in Fig. 1.8 and the hydraulic fill shown in the terminal in Fig. 1.9, or can be processed and densified—*compacted*—as for the core in the dam shown in Fig. 1.8 or the highway pavement shown in Fig. 1.11. The principles of compaction and the properties of compacted soils are treated in Chapter 34.

7.5 ALTERATIONS OF SOIL AFTER FORMATION

The civil engineer working with soil must design his structure not only for the properties of the soil as it exists at the start of the project but also for the entire design-life of the structure. He thus needs to know both the properties of the soil at the start of the project and how these properties will vary during the design-life. Both the size and shape of a given deposit and the engineering properties of the soil in the deposit may change very significantly. Many of these changes occur independent of man's activity, whereas others are brought about by the construction activity itself.

The significant changes in engineering behavior that can and do occur during the life of a soil make soil engineering both difficult and interesting. The engineer soon learns that soil is not inert but rather very much alive and sensitive to its environment. Table 7.2 lists the factors with the greatest influence on the behavior of soil.

Stress

In general, an increase in stress on a soil element causes an increase in shear strength, a decrease in compressibility,

Table 7.2 Factors Influencing Soil Behavior

Factors Influencing Behavior of As-Formed Deposit	Factors Contributing to Changes in Soil Behavior
Sedimentary soil	
Nature of sediments	Stress
Methods of transportation and deposition	Time
Nature of deposition environment	Water
Compacted soil	
Nature of soil	Environment
Amount of molding water	
Amount and type of compaction	Disturbance

and a decrease in permeability; a reduction in stress causes the reverse. The changes brought about by a stress reduction are usually less than those caused by a stress increase of equal magnitude.

During the formation of a sedimentary soil the total stress at any given elevation continues to build up as the height of soil over the point increases. Thus the properties at any given elevation in a sedimentary soil are continuously changing as the deposit is formed. The removal of soil overburden, e.g., by erosion, results in a reduction of stress. A soil element that is at equilibrium under the maximum stress it has ever experienced is *normally consolidated*, whereas a soil at equilibrium under a stress less than that to which it was once consolidated is *overconsolidated.*

There are construction activities that result in an increase of the confining stress on soil and there are activities that result in a reduction of stress. For example, the embankment shown in Fig. 1.6 caused a very great increase of vertical stress on the soils underneath the embankment. When equilibrium was reached under this embankment load, the soil underneath the embankment had become much stronger. On the other hand, the excavation for the Panama Canal (Fig. 1.14) resulted in considerable unloading of the soil in the canal and immediately adjacent to it. This unloading resulted in a decrease in the strength of the shale immediately adjacent to the canal and contributed to the slides that occurred along the canal.

Time

Time is a dependent variable for the other factors contributing to change in soil behavior (especially stress, water, and environment). As noted in Chapter 2, for the full effects of a stress change to be felt, water usually must be extruded or imbibed in the soil element. Because of the relatively low permeability of fine-grained soil, time is required for this water to flow into or out of this type of soil. Time is an obvious factor in chemical reactions such as those occurring during weathering.

Water

As was discussed in Chapter 2, water can have two deleterious effects on soil. First, the mere presence of water causes the attractive forces between clay particles to decrease. Second, pore water can carry applied stress and thus influence soil behavior. A sample of clay which may have a strength approaching that of a weak concrete when it has been dried can become mud when immersed in water. Thus increasing the water content of a soil generally reduces the strength of the soil.

The activities of both nature and man serve to alter pore water conditions. In many parts of the world, there is a very marked variation in water conditions during the year. During the hot, dry season, there is a lack of rain and the ground water level drops; during the wet season, there is surface water and a general rising of the ground water. This seasonal variation in water conditions causes a marked change in soil properties through the year.

There are many construction operations that alter ground water conditions. For example, the dam shown in Fig. 1.8 impounded a reservoir, which subjected the foundation soils to a great increase in pore water pressure. Not only were the foundation soils given an increase in pore pressure, but many dry soils which had never been inundated were submerged with water from the reservoir. Construction for the two buildings shown in Figs. 1.4 and 1.5 required a lowering of the water table. This lowering caused a change in the properties of the subsoils.

Environment

There are several characteristics of the environment of a soil which may influence its behavior. The two considered here are the nature of the pore fluid and temperature. A sedimentary clay or compacted clay can be formed with a pore fluid of a certain composition, and at a certain temperature both of these can change during the life of the deposit. One example is a marine clay laid down in water high in salt, 35 g of salt/liter of water for a typical marine environment. A marine clay is frequently uplifted so that it is above the level of the sea, and the ground waters percolating through the clay are of much lower salt content than the sea water. Thus during the life of the sedimentary clay there can occur a gradual removal of the salt in the pore fluid so that, after many thousands of years of leaching, the pore fluid can be quite different from that at time of sediment formation. As discussed in Chapter 5, reducing the electrolyte content of the water around soil particles can reduce the net attraction between them. In other words, leaching of the salt in the pore fluid can cause a reduction in shear strength.

The most dramatic example of a reduction in shear strength brought about from pore water leaching is exhibited in the "quick clays". These marine clays were deposited in a highly flocculated condition. Despite the resulting high water content, these clays developed a moderately large strength because of the bonds that formed at the edge-to-face contacts. These clays then had most of the electrolyte in their pore fluid removed by years of leaching. For this new environmental condition, the clay would tend to be in a dispersed condition (see Fig. 7.2*c*), and for the same water content it would have very little strength. However, this change does not show up fully until the clay is subjected to enough disturbance to break the bonds built up by years of confining stress. Upon disturbance, the clay may lose all of its strength and become a soil-water slurry with zero shear strength. These quick clays have caused many engineering problems in the Scandinavian countries and in Canada where they are widespread. The landslide shown in Fig. 1.13 occurred in a quick clay.

The change in temperature from time of deposit formation to a given time under consideration can result in a change of soil behavior. Thus a clay deposited in a glacial lake undergoes a general warming during its life. Further, a soil existing at great depth in the ground, sampled and brought into the laboratory for testing, may undergo property changes due to the difference in temperature between the ground and the laboratory. Increasing the temperature of a cohesive soil normally causes an expansion of the soil as well as causing some of the air dissolved in the pore fluid to come out of solution.

The engineer can see from the discussion in this section that he must give thought to how the properties of the soil might change during the life of his structure, and not expect to make a proper design solely on the basis of the properties of the soil as it exists prior to construction. He could be faced with a disastrous failure if he designed his earth dam on the basis of the strengths of the soil which exist prior to the construction of the dam. Later chapters in this book will treat the principles needed to select the proper values of strength, permeability, and compressibility to be used in a given soil problem.

7.6 SOIL INVESTIGATION

Table 7.3 lists some of the methods of soil investigation[1] in general use. The proper program of soil investigation for a given project depends on the type of project, the importance of the project, and the nature of the subsoils involved. For example, a large dam project would usually require a more thorough subsoil investigation than would a highway project. A further example is soft clays, which usually require more investigation than do gravels.

The first four methods of soil investigation listed in Table 7.3 normally cover a large area and are intended to give the engineer a general picture of the entire site. Geophysical techniques make possible the detection of markedly different strata in the subsoils. These techniques permit a relatively large volume of subsoil to be explored in a given period of time.

Table 7.3 Methods of Soil Investigation

Methods of Soil Investigation
Reconnaissance
Visual inspection
Airphotos
Geologic reports and maps
Records of past construction
Exploration
Geophysical
Electrical
Pits—sampling and testing
Borings—sampling and testing
Field tests
Penetration tests
Vane tests
Water table—pore pressure tests
Pumping tests
Load tests
Compaction tests

Sampling either from pits or from borings followed by laboratory testing is widely used in soil investigation, especially for important structures and relatively uniform subsoils. The investigator can obtain high-quality undisturbed samples from open pits, but obviously pits can be advanced only to relatively shallow depths. Pits or trenches can be dug by hand or by power equipment such as a back hoe or dozer. Borings can be made by augers either with or without a casing.

There are difficulties in obtaining high-quality undisturbed soil samples, especially from considerable depths. The sampling operation, sample transportation, and specimen preparation require that the soil be subjected to stresses which are quite different from those that existed in the ground. This inherent change of stress system alters the behavior of the soil. Further, the sampling, transportation and preparation operation usually subjects the specimen to strains that alter the soil structure. For these reasons the determination of *in situ* properties by laboratory tests can be most difficult. Later chapters in this book discuss laboratory testing and

[1] The reader is referred to Terzaghi and Peck (1967) for a more detailed treatment of soil investigation.

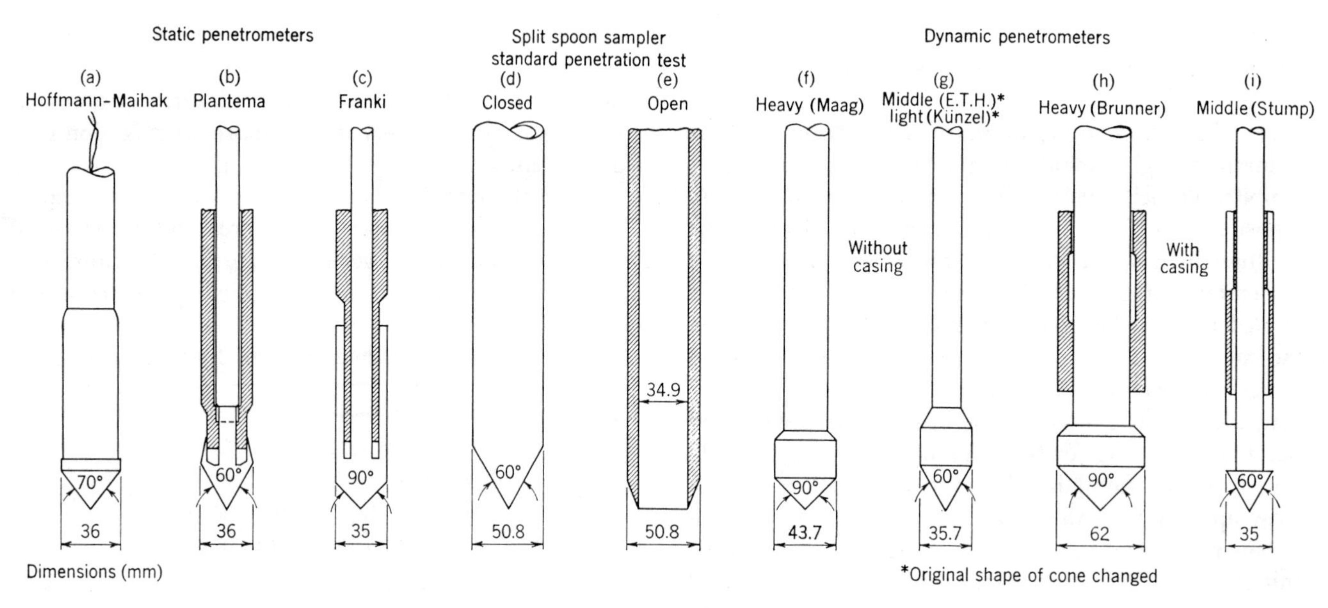

Fig. 7.3 Penetrometers (From Schultze and Knausenberger, 1957).

point out some of the significant influences of sample disturbance.

Field tests take on an increased importance in soils which are sensitive to disturbance and in subsoil conditions where the soils vary laterally and/or vertically. The most widely used field test method is penetration testing. Figure 7.3 shows some of the penetrometers that have been used for soil investigation. These penetrometers are driven or pushed into the ground and the resistance to penetration is recorded. The most widely used penetration test is the "standard penetration test", which consists of driving the spoon, shown in Fig. 7.4, into the ground by dropping a 63 kg mass from a height of 760 mm. The penetration resistance is reported in number of blows of the weight to drive the spoon 305 mm.

Table 7.4 presents a correlation of standard penetration resistance with relative density for sand and a correlation of penetration resistance with unconfined compressive strength for clay. The standard penetration test is a very valuable method of soil investigation. It should, however, be used only as a guide, because there are many reasons why the results are only approximate.

Figure 7.5 presents the results of some penetration tests run in a large tank in the laboratory. These test data show that the penetration resistance depends on factors other than relative density. As can be seen, the penetration resistance depends on the confining stress and on the type of sand. Further, the figures show that the test data scatter considerably. The influence of sand type on penetration resistance is particularly large at low densities—those of most interest. Another factor that may have a marked influence on the penetration resistance of a sand is the pore pressure conditions during the measuring operation. If the level of water in the drill hole is lowered prior to penetration measurement, a lowered resistance can result.

Experience has shown that the determination of the shear strength of a clay from the penetration test can be very unreliable.

The standard penetration test should be used only as an approximation or in conjunction with other methods of exploration.

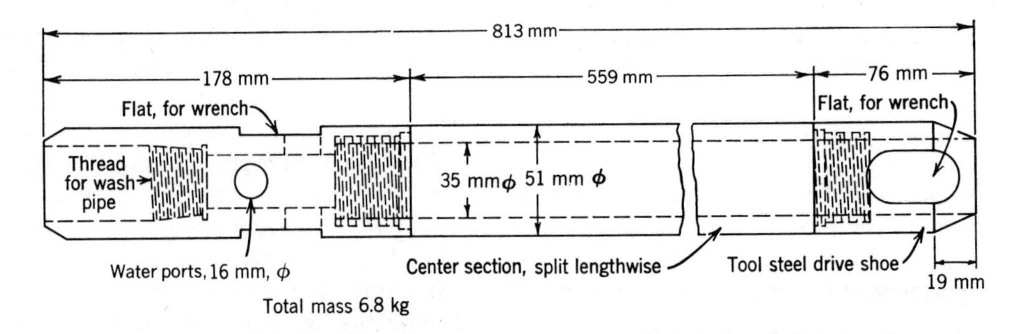

Fig. 7.4 Spoon for standard penetration test (From Terzaghi and Peck, 1967). The metric dimensions are approximate.

Table 7.4 Standard Penetration Test

Relative Density of Sand		Strength of Clay		
Penetration Resistance N (blows/305 mm)	Relative Density	Penetration Resistance N (blows/305 mm)	Unconfined Compressive Strength (kN/m²)	Consistency
0–4	Very loose	<2	<24	Very soft
4–10	Loose	2–4	24–48	Soft
10–30	Medium	4–8	48–96	Medium
30–50	Dense	8–15	96–192	Stiff
>50	Very dense	15–30	192–388	Very stiff
		>30	>388	Hard

From Terzaghi and Peck, 1948.

In certain countries, such as Holland, subsoil conditions are such that penetration testing has proved to be a relatively reliable technique. More sophisticated techniques [such as the friction jacket cone (Begemann, 1953)] have been widely used.

The vane test has proved to be a very useful method of determining the shear strength of soft clays and silts. Figure 7.6 shows various sizes and shapes of vanes which have been used for field testing. The vane is forced into the ground and then the torque required to rotate the vane is measured. The shear strength is determined from the torque required to shear the soil along the vertical and horizontal edges of the vane.

As later chapters in this book will show, a proper subsoil investigation should include the determination of water pressure at various depths within the subsoil. Methods of determining pore water pressure are discussed in Part IV. Part IV also notes how the permeability of a subsoil can be estimated from pumping tests.

Various load tests and field compaction tests may be highly desirable in important soil projects. In this type of test, a small portion of the subsoil to be loaded by the prototype is subjected to a stress condition in the field which approximates that under the completed structure. The engineer extrapolates the results of the field tests to predict the behavior of the prototype.

7.7 SUBSOIL PROFILES

Figures 7.7 to 7.17 present a group of subsoil profiles and Table 7.5 gives some information on the geological history of the various profiles. The purposes of presenting these profiles are to:

1. Indicate how geological history influences soil characteristics.
2. Give typical values of soil properties.
3. Show dramatically the large variability in soil behavior with depth.
4. Illustrate how engineers have presented subsoil data.

Three considerations were used in the selection of the profiles: first, examples were chosen with different types of geological history; second, most of the profiles are ones for which there are excellent references giving considerably more detail on the characteristics of the soil and engineering problems involved with the particular profile; and finally, most of the profiles selected have been involved in interesting and/or important soil engineering projects.

Some of the soil characteristics shown in the profiles have already been described in this book. These characteristics include water content, unit weight, void ratio, porosity, Atterberg limits, and particle size. Other characteristics, particularly those referring to strength and compressibility, will be discussed in detail in later portions of this book. Reference will then be made back to these profiles.

The profiles illustrate many concepts presented in the preceding parts of this book; some of them are discussed in the remaining part of this section.

Stress History

In a normally consolidated sedimentary soil both the void ratio and water content decrease with depth in the profile, and the strength therefore increases. This characteristic is illustrated in several of the profiles, e.g., the Norwegian marine clay (Fig. 7.7), the Thames Estuary clay (Fig. 7.10), and the Canadian clay (Fig. 7.11). The London clay is overconsolidated since it was compressed by a greater overburden than now exists. Erosion removed some of the original overburden. As would be expected, the overconsolidated London clay does not

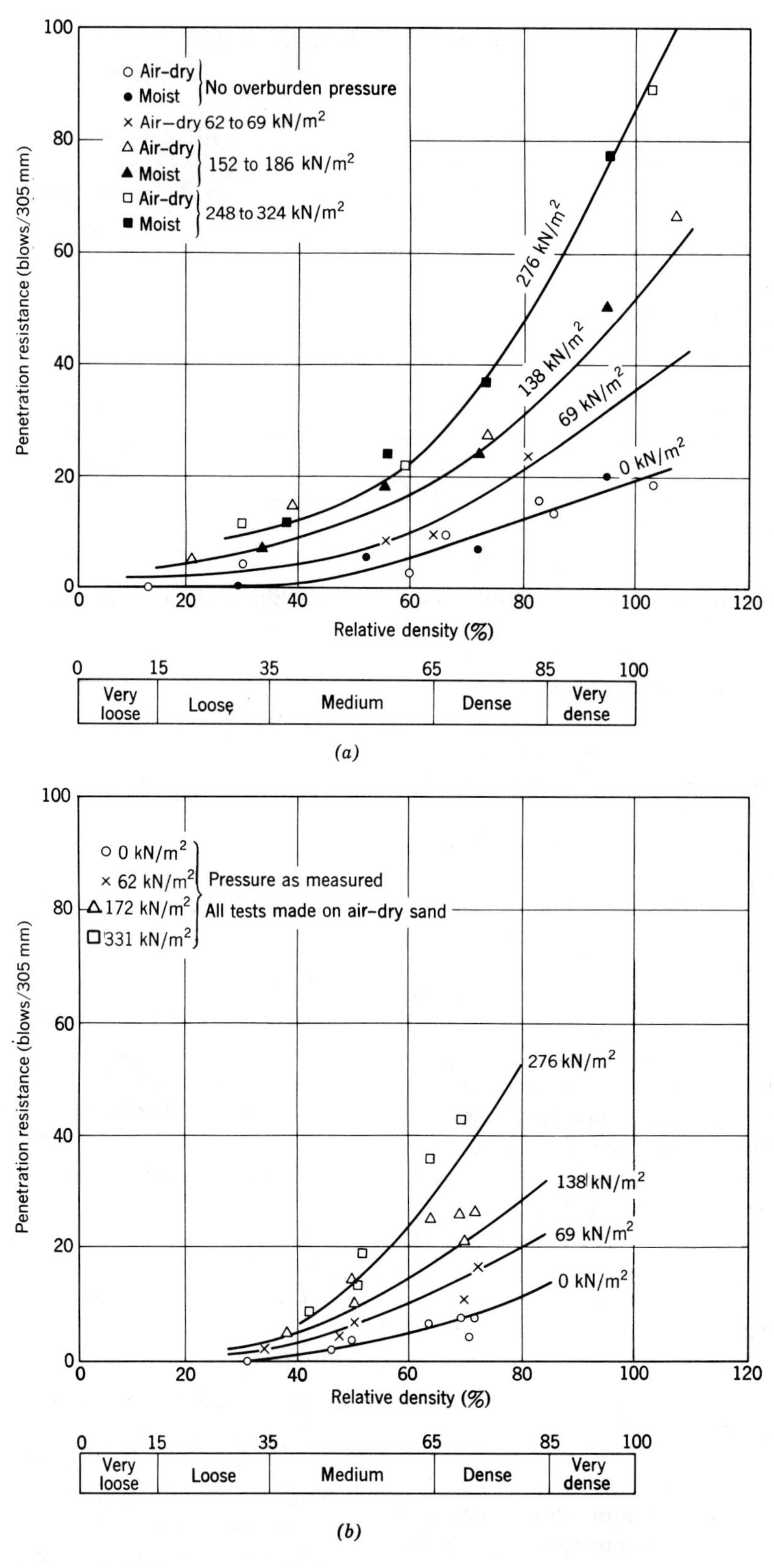

Fig. 7.5 Results of standard penetration tests. (*a*) Coarse sand. (*b*) Fine sand. (From Gibbs and Holtz, 1957.)

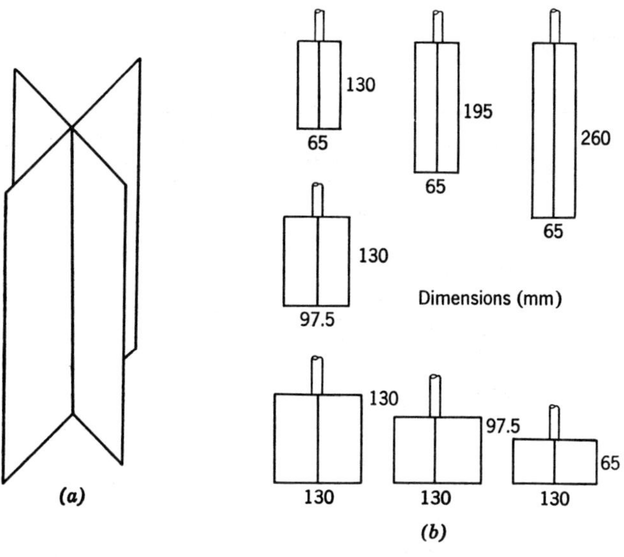

Fig. 7.6 Vanes. (*a*) Vane. (*b*) Vanes studied by Aas (1965).

Table 7.5 Subsoil Profiles

Number	Profile	Formation	Post Deposition	Comments	References
1	Norwegian marine clay	Sediments transported by rivers of melted glaciers of Pleistocene and deposited in sea	Sediments uplifted and leached. Surface dried and weathered	Normally consolidated below surface crust	Bjerrum, 1954
2	London clay	Deposited under marine conditions during the Eocene, roughly 30 million years ago	Uplift and erosion removed the overlying deposits and $\frac{1}{2}$ to $\frac{2}{3}$ of London clay	Overconsolidated to maximum past pressure $\approx$ 1900 kN/m^2	Skempton and Henkel, 1957 Ward, Samuels and Butler, 1959
3	Boston blue clay	Sediments transported by streams of melted glaciers of Pleistocene and deposited in the quiet marine waters of Boston Basin	Clay uplifted, submerged, and reuplifted	Overconsolidated at top, normally consolidated at bottom	Horn and Lambe, 1964 Skempton, 1948
4	Thames Estuary clay	Sediments transported by streams and deposited in an estuary during postglacial time		Normally consolidated below surface crust	Skempton and Henkel, 1953
5	Canadian varved clay	Sediments transported by streams from melting glaciers and deposited in cold lakes		Light, silt layer laid down in spring, summer; dark "clay" layer in winter	Milligan, Soderman and Rutka, 1962 Eden and Bozozuk, 1962
6	Mexico City clay	Sediments of volcanic origin deposited in lake in valley of Mexico during late Pleistocene	Pumping from water wells has lowered water table	In some parts of city, clay is normally consolidated, other parts overconsolidated	Marsal, 1957 Lo, 1962 Zeevaert, 1953
7	Chicago clay and sand	Clay deposited as till sheets by glaciers, during both advancing and receding stages, and deposited in glacial lakes	Clay surface desiccated, giving crust usually 1 to 2 m		Peck and Reed, 1954
8	South African clay				Jennings, 1953
9	Brazilian residual clay	Formed in place by weathering of rock			Vargas, 1953
10	Volga River sand	Alluvial sand of the Volga river		Data in Fig. 7.17 obtained on frozen samples from pits	Durante, Kozan, Ferronsky, and Nosal, 1957
11	Kawasaki subsoils	Alluvial deposits of the Holocene Epoch. Top 4 m hydraulic fill		Profile shown in Part IV	

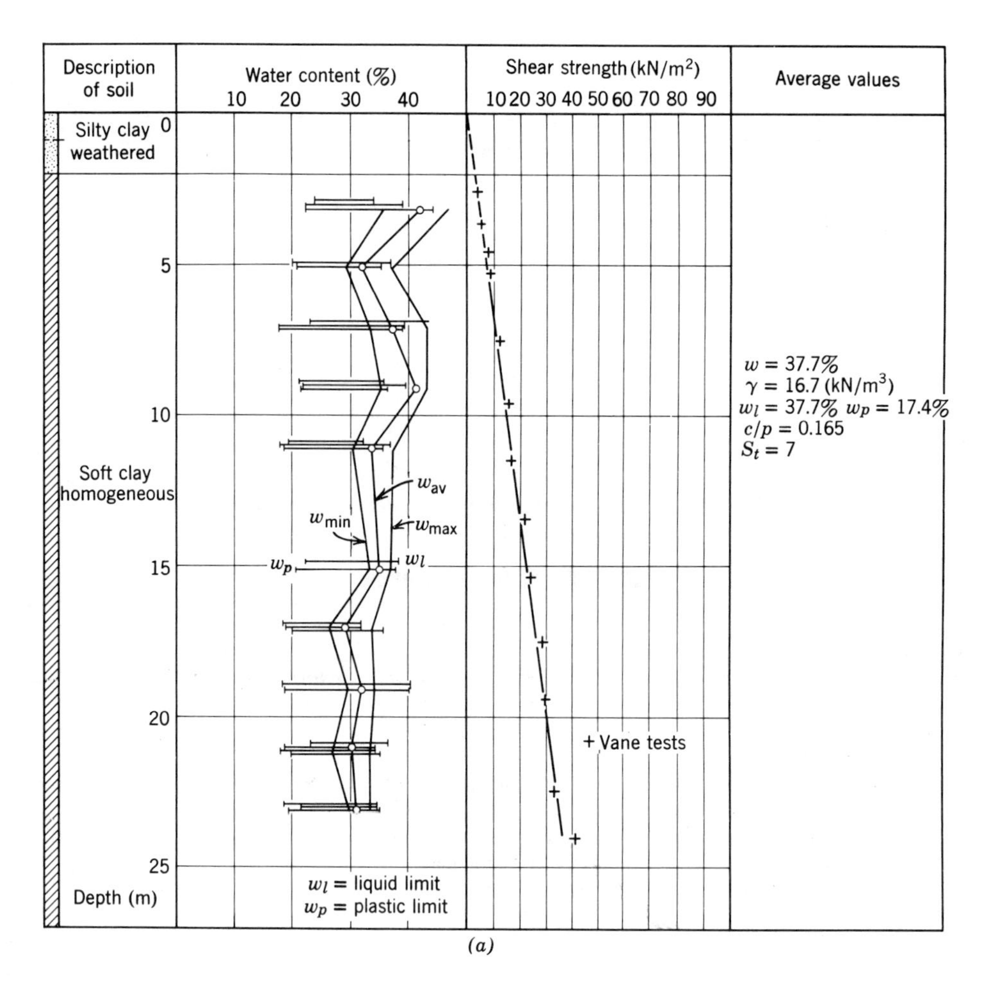

Fig. 7.7 Norwegian marine clay. (*a*) Results of a boring in Drammen. (*b*) Results of a boring at Manglerud in Oslo. (From Bjerrum, 1954.)

show a marked reduction in water content or increase in strength with depth.

The surfaces of most of the soil profiles contain crusts resulting from desiccation and weathering. Drying causes pore water tensions, which increase the stress between soil particles and overconsolidate the clay. Desiccation also encourages chemical weathering, especially oxidation, which gives the soil an apparent overconsolidation.

In both the Mexico City clay and London clay the pore water pressure in the soil is less than the static pressure. The importance of this reduced water pressure is discussed in detail in Parts IV and V of this book.

The Brazilian residual clay (Fig. 7.16) shows evidence of overconsolidation in the upper half of the stratum and normal consolidation in the lower half. It is doubtful that one should use the terms "overconsolidated" and "normally consolidated" in reference to residual soils, however.

Sensitivity

Time and changes in stress and environment since the time of formation may result in a soil having a higher strength in the undisturbed state than it does in the remolded state (after the soil has been thoroughly worked as was done prior to the liquid limit test described in Chapter 3). The term *sensitivity* is used to describe this difference in strength and is determined by the ratio of the strength in the undisturbed state to that in the remolded state. Sensitivity is related to liquidity index, since the greatest loss of strength should occur in a highly flocculated soil whose water content is very large compared to its liquid limit as measured using remolded soil. As discussed in the preceding section, sedimentary soils laid down in a marine environment and then leached after deposition tend to have high sensitivity. Any soil with a sensitivity equal to or greater than eight is termed "quick". The Manglerud clay (Fig. 7.7) is an extreme example of a quick clay, having a sensitivity ranging to above 500. The Canadian clay (Fig. 7.11) is also quick.

Variability in Soil

The profiles offer many examples of variability in soils, over both large and small distances. In the Manglerud clay and Thames Esturay clay, distinct strata of many feet of thickness can be detected. In the sedimentary clays there often exists a large variation in soil properties over distances of less than an inch. Such variations over small distances are dramatically shown in the Canadian varved clay (Fig. 7.12). Figure 7.12 shows the great differences in water content and plasticity between the dark ("clay") and light ("silt") layers.

Plasticity

The plasticity of the clay shown in the various profiles varies tremendously. The glacial clays, generally containing a significant amount of the clay mineral illite, tend to have relatively low plasticity. Values of plasticity index of 15 to 20 are shown for the glacial clays, e.g., the Norwegian marine clay—however, they can have much higher values, as indicated by the data on the Canadian varved clay, especially the dark layers.

The Mexico City clay, containing the clay mineral montmorillonite and volcanic ash, is one of the most plastic clays encountered by the soil engineer. As can be seen in Fig. 7.13, this clay can have values of PI in excess of 400. The South African soils (Fig. 7.15) can have high values of PI and fall above the *A* line on the plasticity chart. This characteristic is common of soils which present heave problems as do some of the South African clays.

7.8 SUMMARY OF MAIN POINTS

1. The determination of the soil profile is an essential step in almost all soil mechanics problems.
2. The properties of the soils in a profile depend on (*a*) the nature of soil components, (*b*) the method of profile formation, and (*c*) the alteration of the profile after formation.

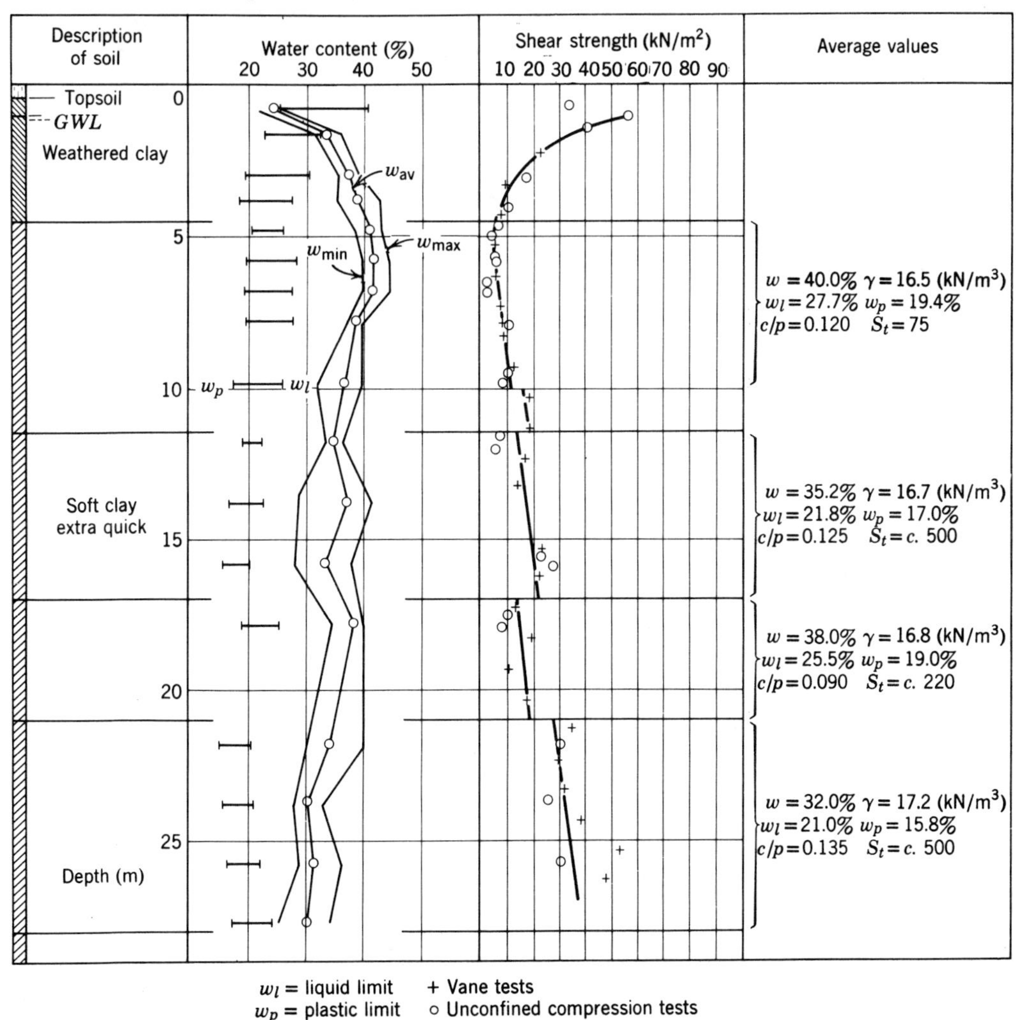

(*b*)

Fig. 7.7 (continued)

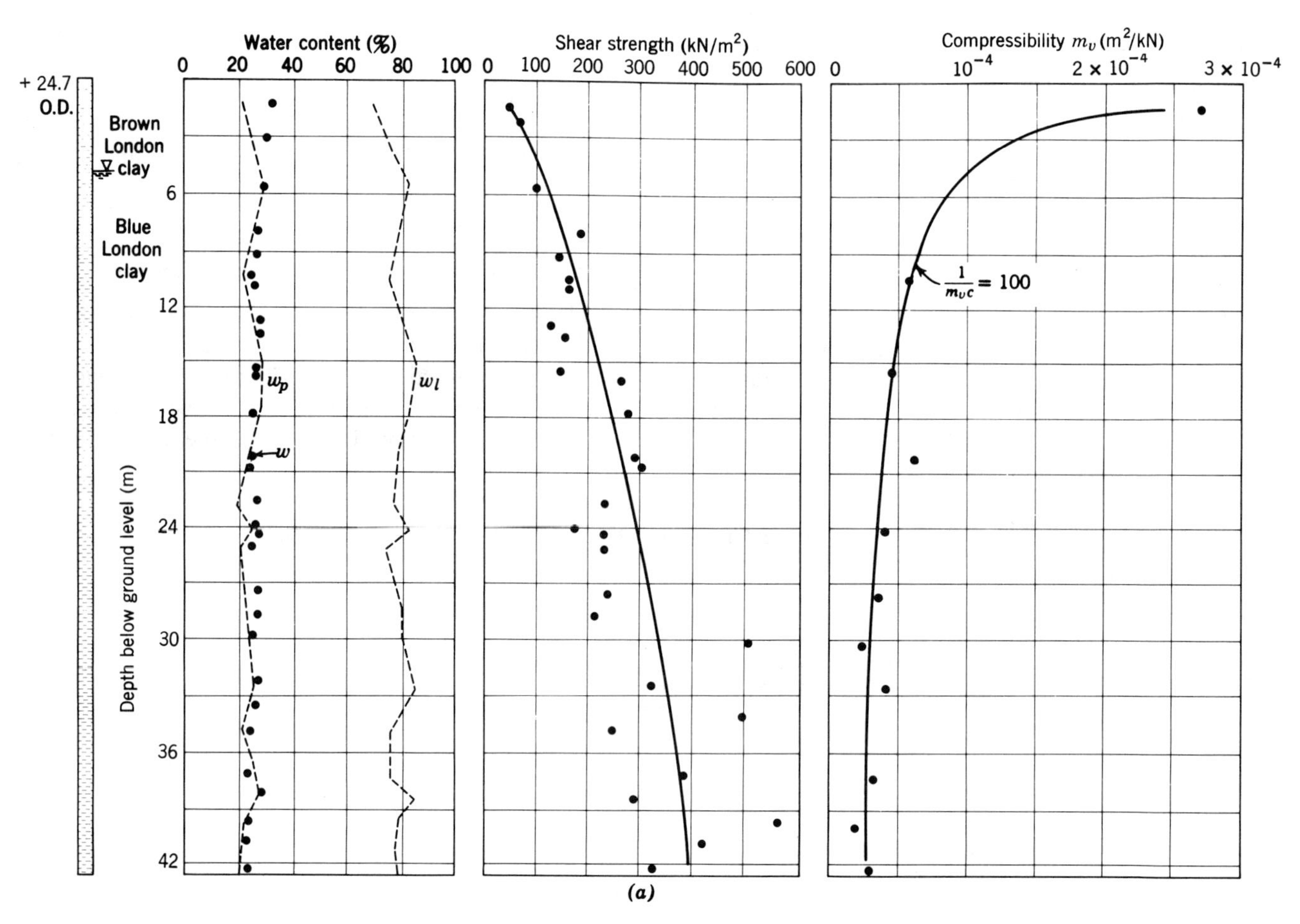

Fig. 7.8 London clay. (*a*) Test results at Paddington. (*b*) Test results at Victoria and South Bank. (From Skempton and Henkel, 1957.)

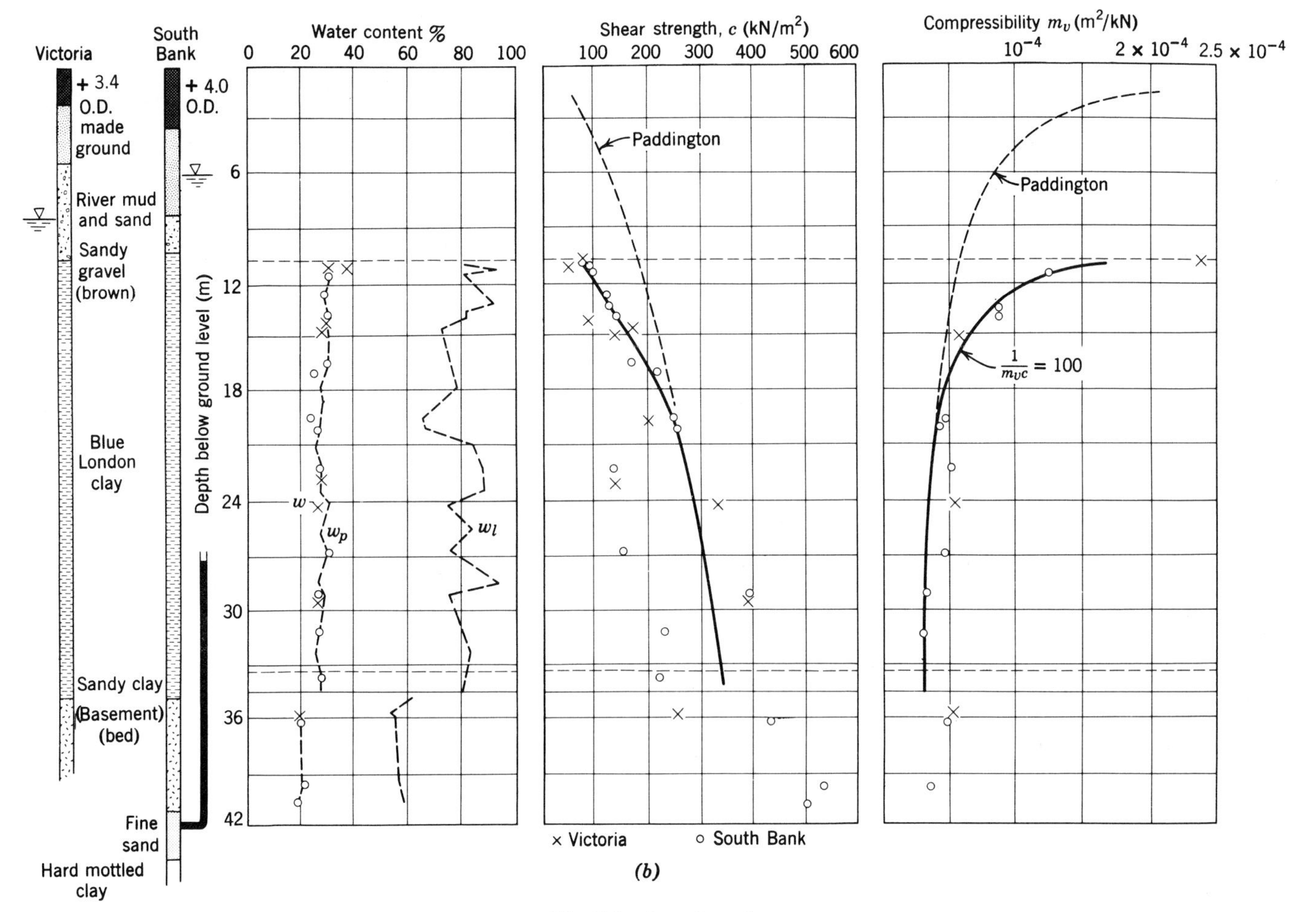

(b)

Fig. 7.8 (continued)

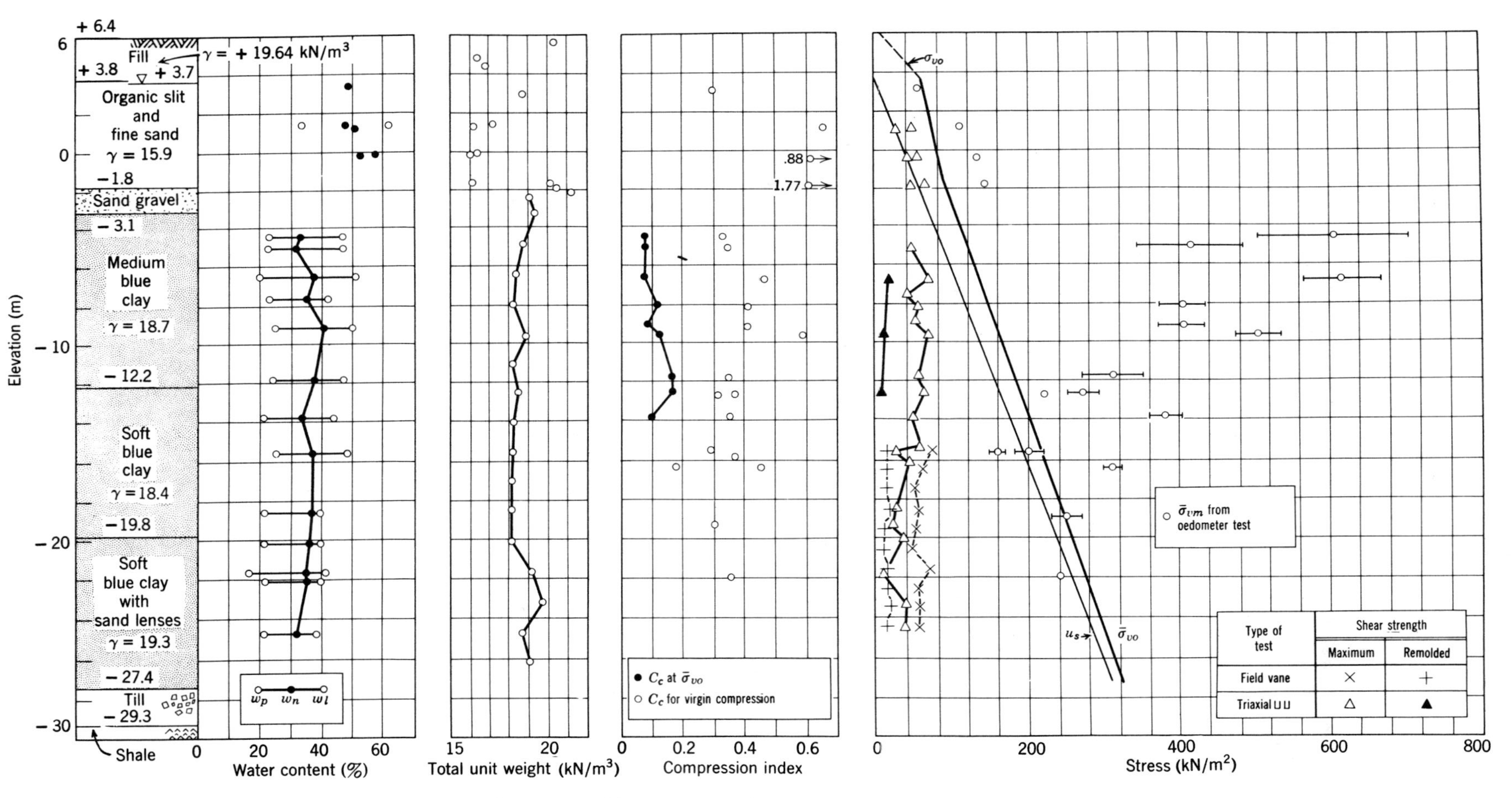

Fig. 7.9 Profile at M.I.T. Advanced Engineering Center.

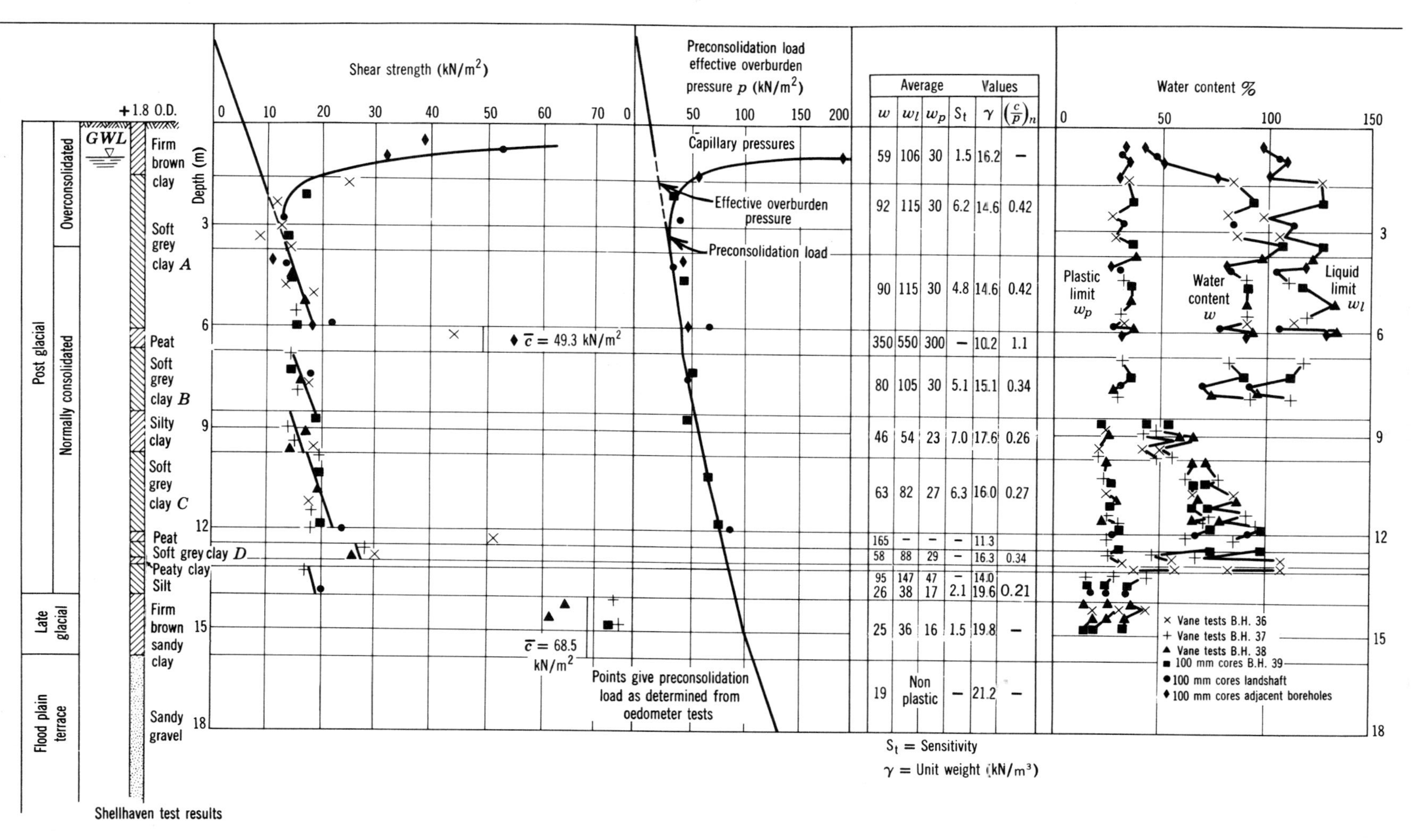

Fig. 7.10 Thames Estuary clay (From Skempton and Henkel, 1953).

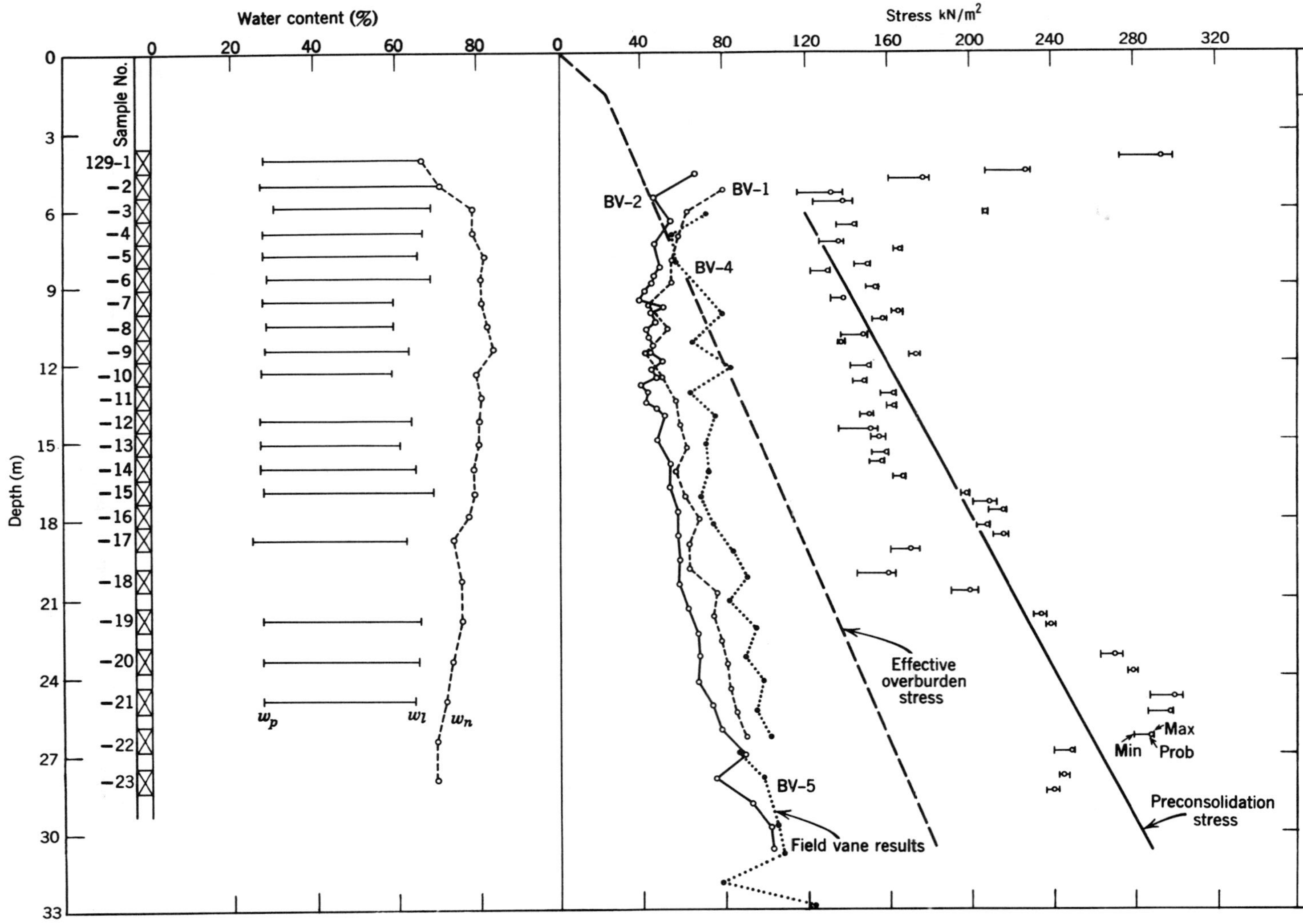

γ kg/m³	% clay	S.C. g/l	S_t
15.82	77	0.2	12
15.43	82	0.4	21
15.10	86		
15.11	78		48
15.19	80	0.6	37
15.02	82	0.8	54
15.00			150
14.91	80	0.5	127
14.83	85	0.9	100
15.10	83	0.8	128
15.02			74
14.97	88		
15.05	85	1.2	72
15.11	86		
15.03		1.5	76
			118
15.24	80	1.9	53
15.22			53
15.25		2.2	76
15.41	83	2.0	37
15.47	73	3.4	34
15.52			
15.54		2.5	

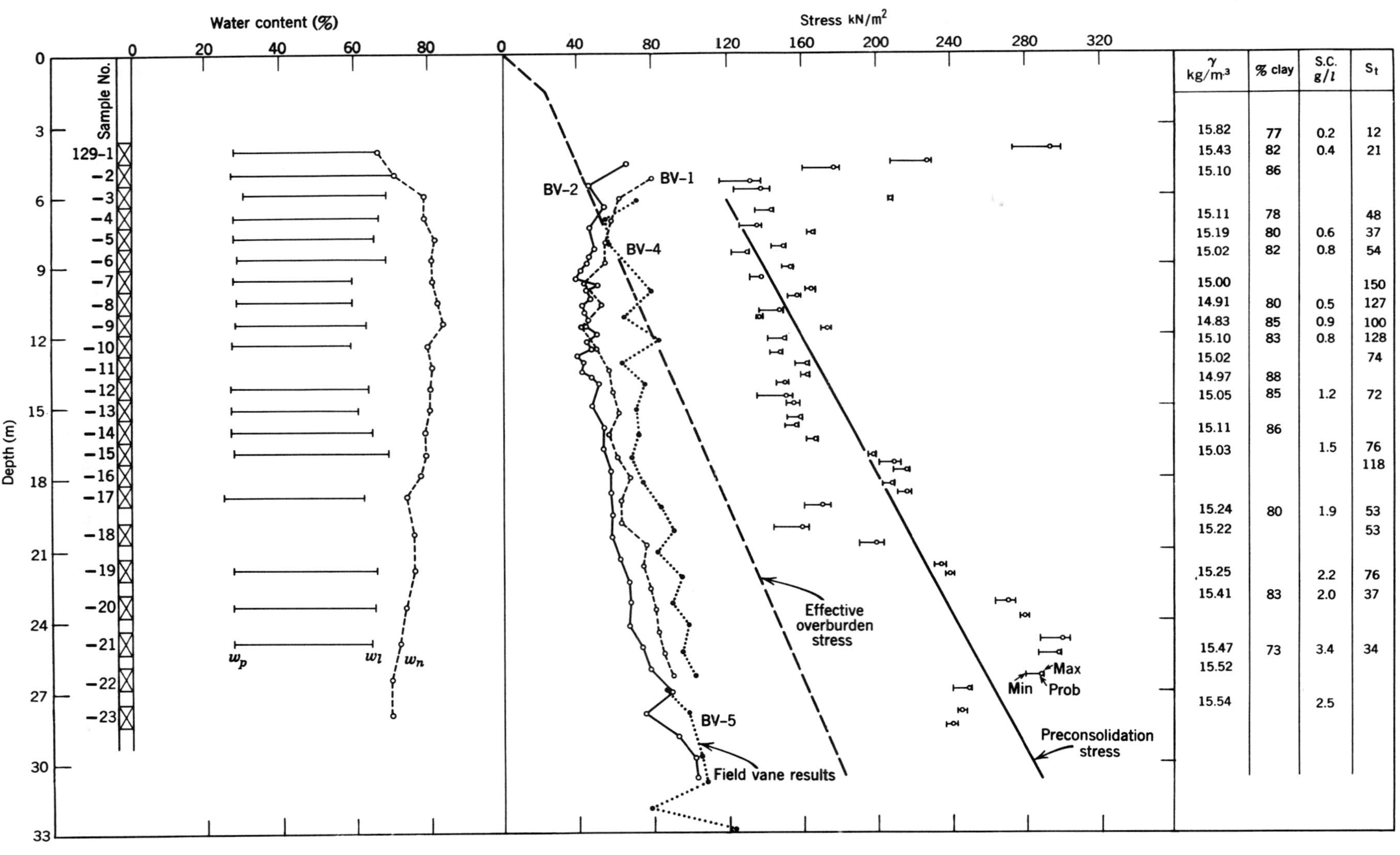

Fig. 7.11 Canadian clay (private communication, Div. of Building Research, National Research Council of Canada, Ottawa).

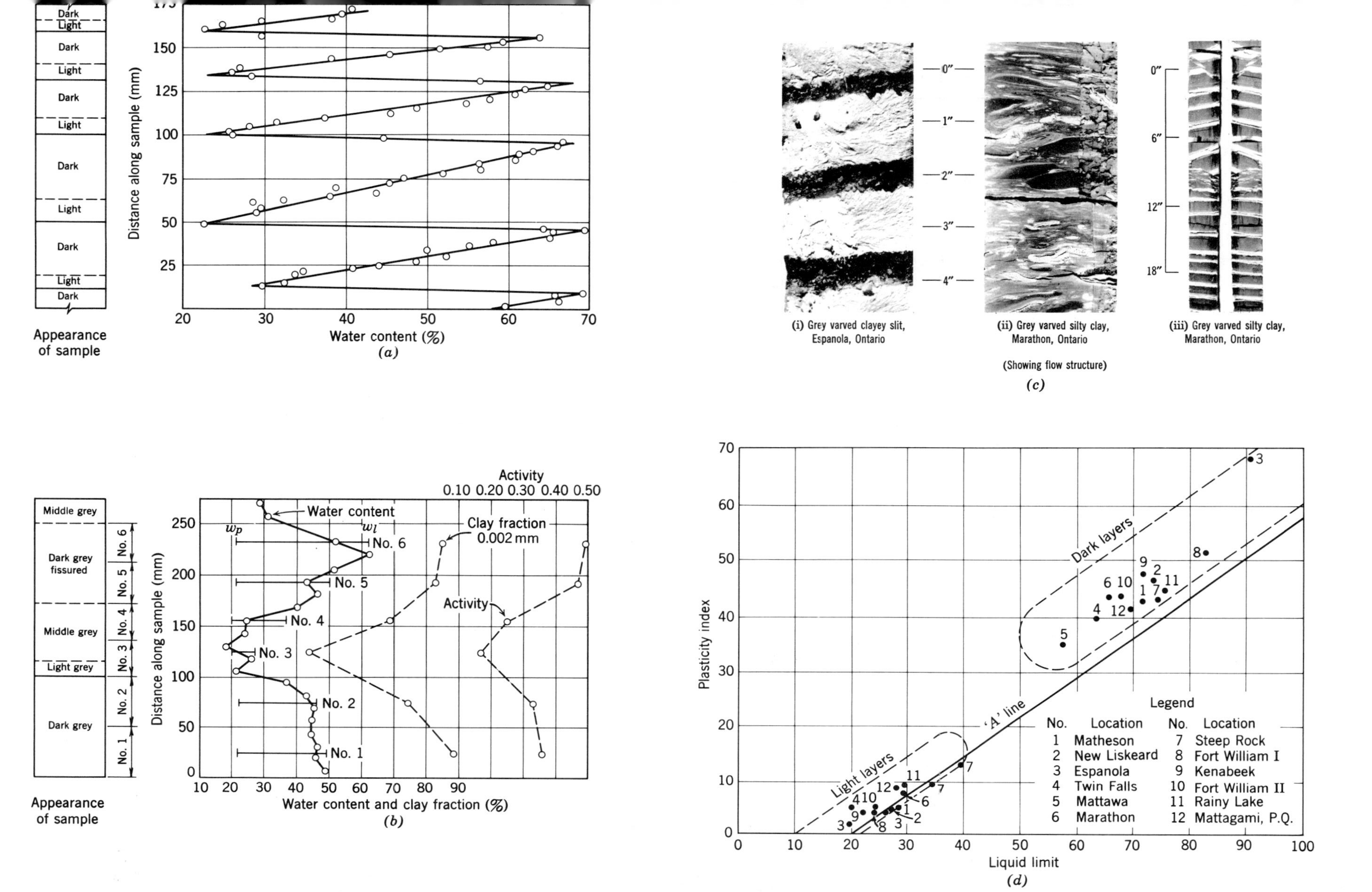

Fig. 7.12 Canadian varved clays. (*a*) Water content variation in a sample of varved clay (after De Lory, 1960). (*b*) Water contents, limits and clay fractions in a sample of varved clay (after De Lory, 1960). (*c*) Varved clay samples. (*d*) Casagrande classification chart for some varved clays. (From Milligan, Soderman, and Rutka, 1962.)

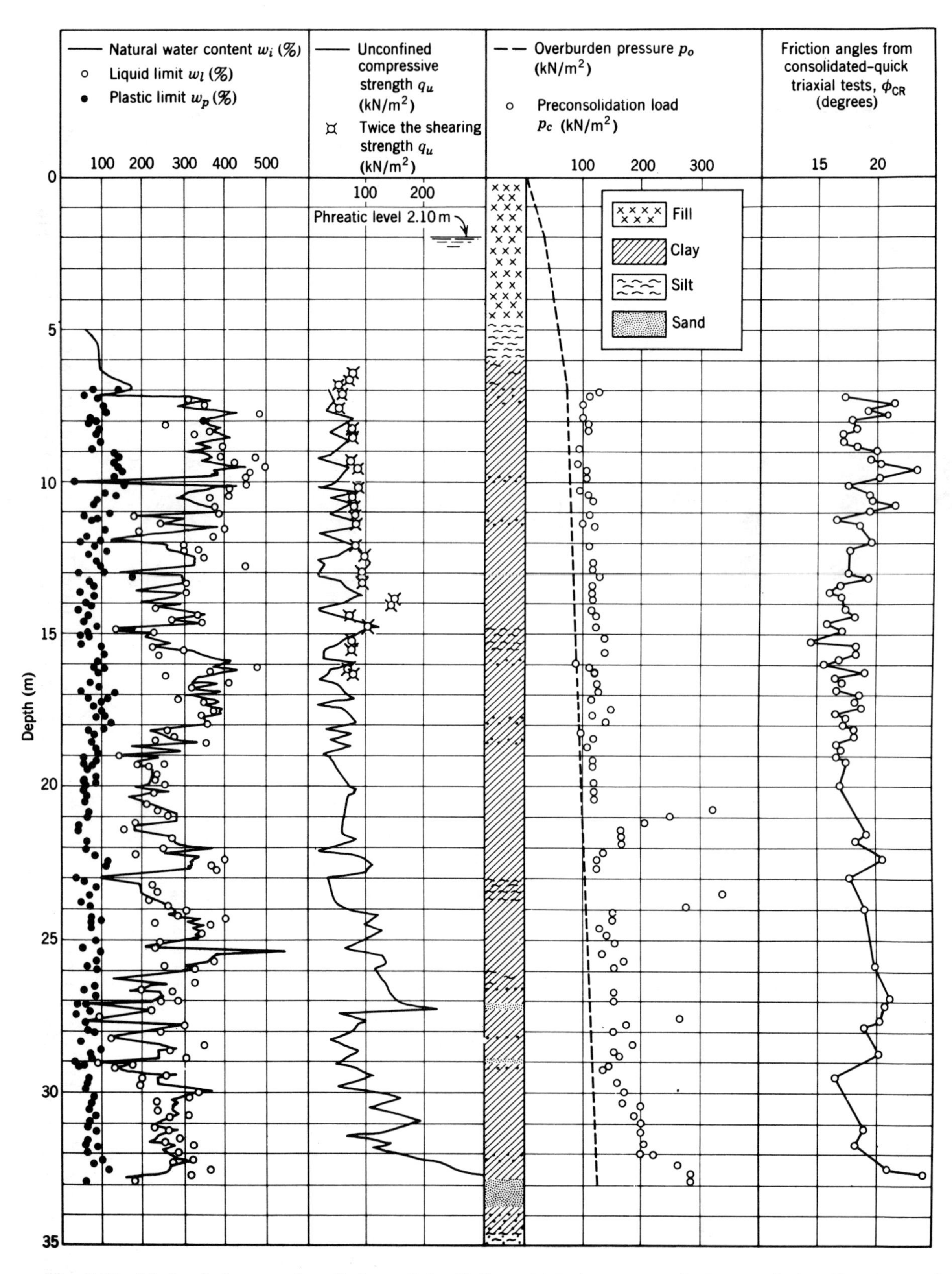

Fig. 7.13 Mechanical properties of clays of the Valley of Mexico at a typical spot of the City (From Marsal, 1957).

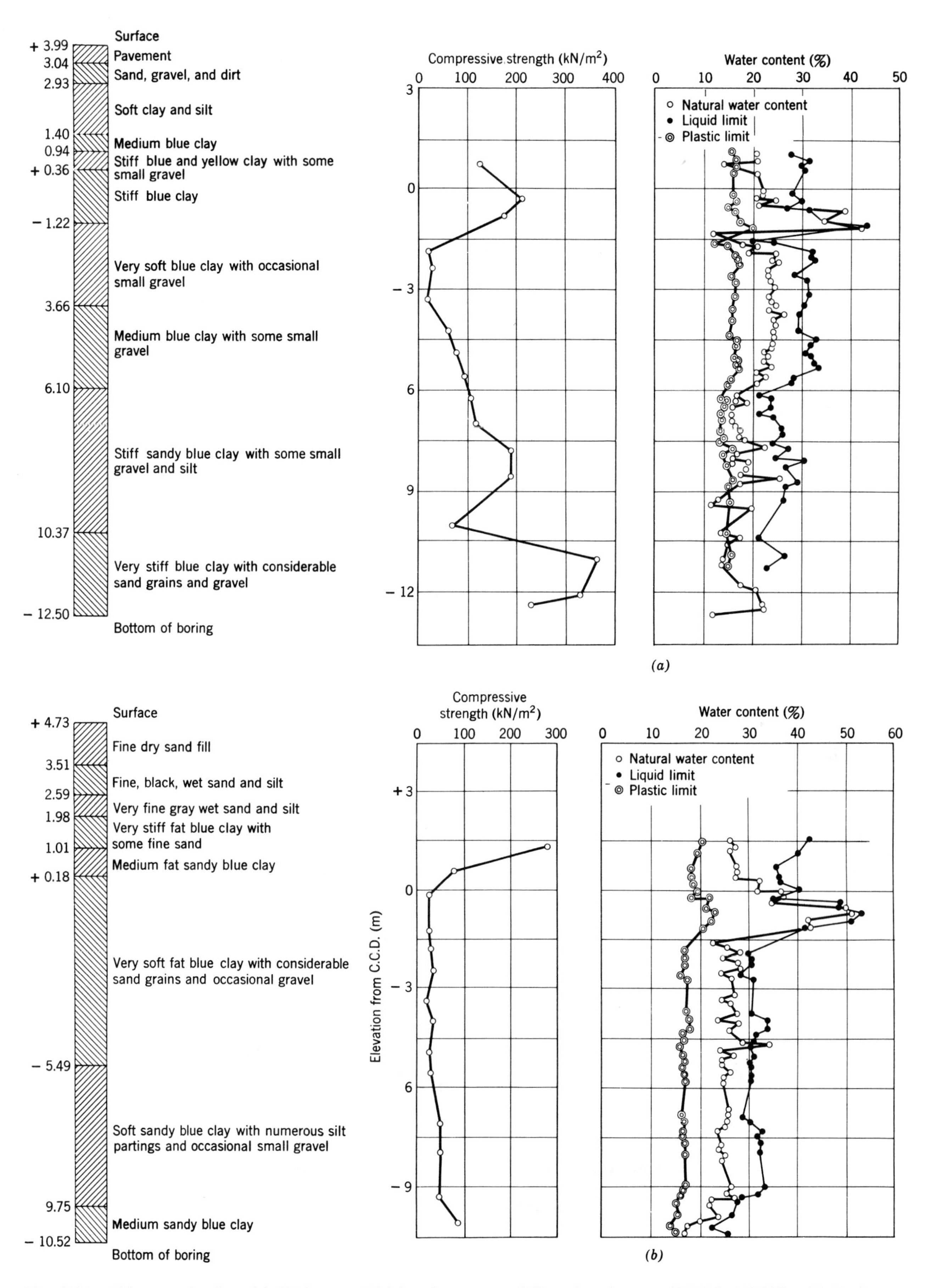

Fig. 7.14 Chicago subsoils. (*a*) Boring on Division Street near Milwaukee Avenue (1200N, 1600W). (*b*) Boring at Congress Street and Racine Avenue (500S, 1200W). (From Peck and Reed, 1954.)

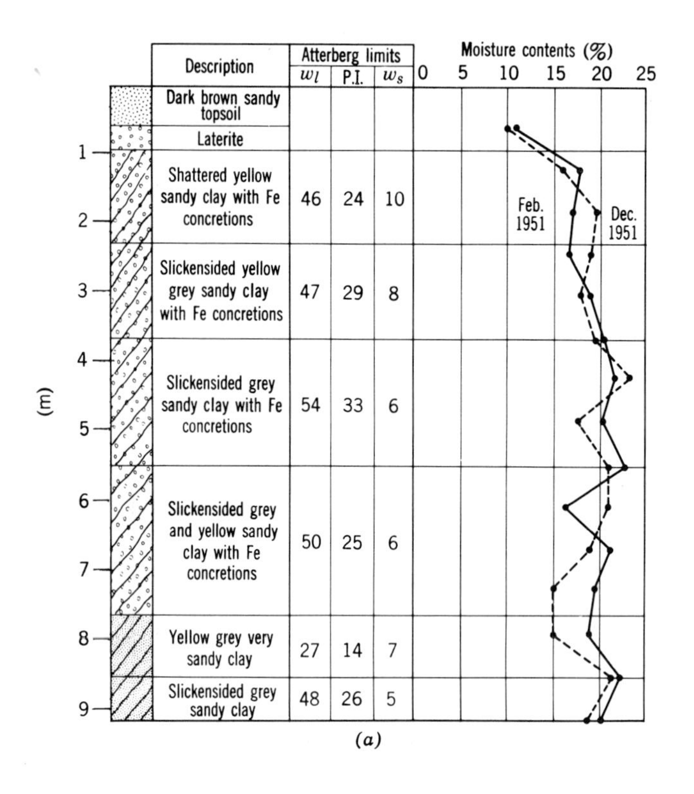

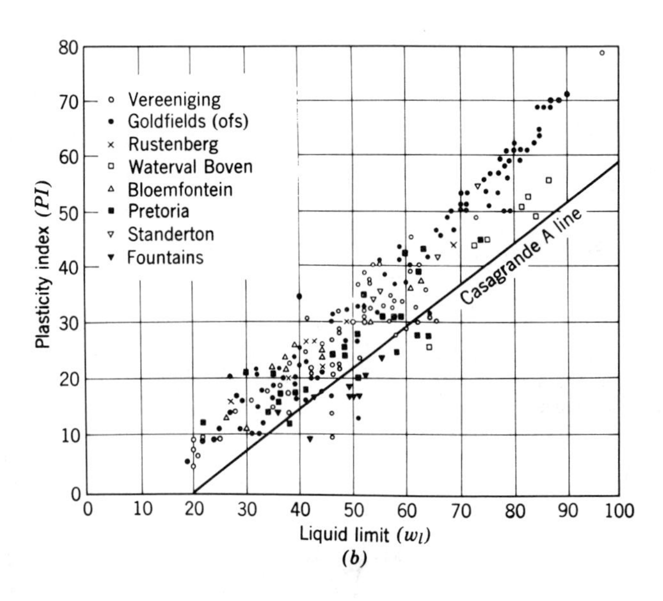

Fig. 7.15 South African clays. (*a*) Variations of direct measurements of moisture content in the soil under an impermeable slab. (*b*) Results of indicator tests on South African soils where heaving conditions have been observed. (From Jennings, 1953.)

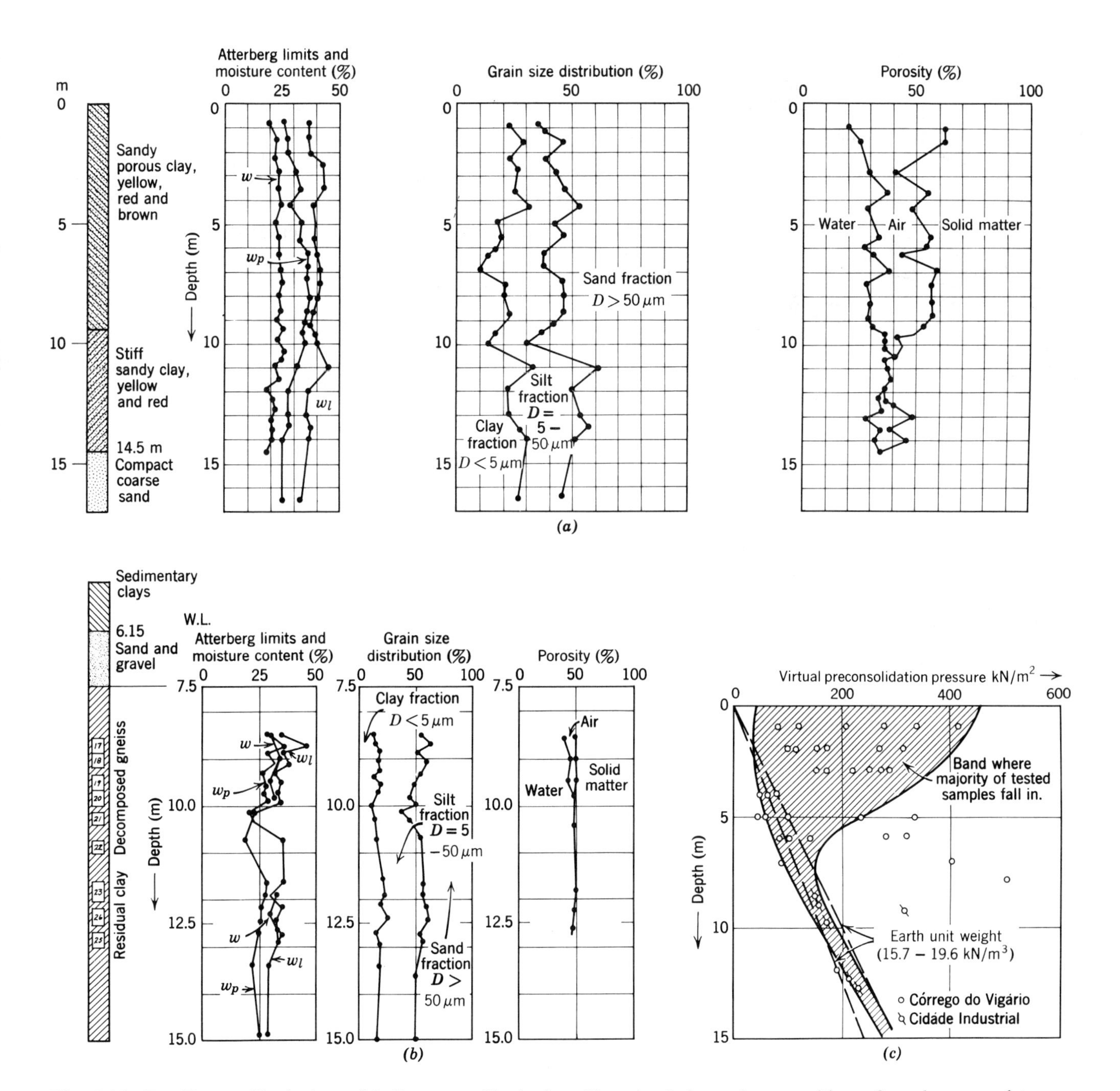

Fig. 7.16 Brazilian residual clay. (*a*) Porous residual clay (Campinas) from decomposition of a clayey sandstone. Variation of consistency, grain-size distribution, and porosity with depth. (*b*) Residual clay (Belo Horizonte) from decomposition of gneiss. Variation of consistency, grain-size distribution, and porosity with depth. (*c*) Virtual preconsolidation pressures against depth of samples. (From Vargas, 1953.)

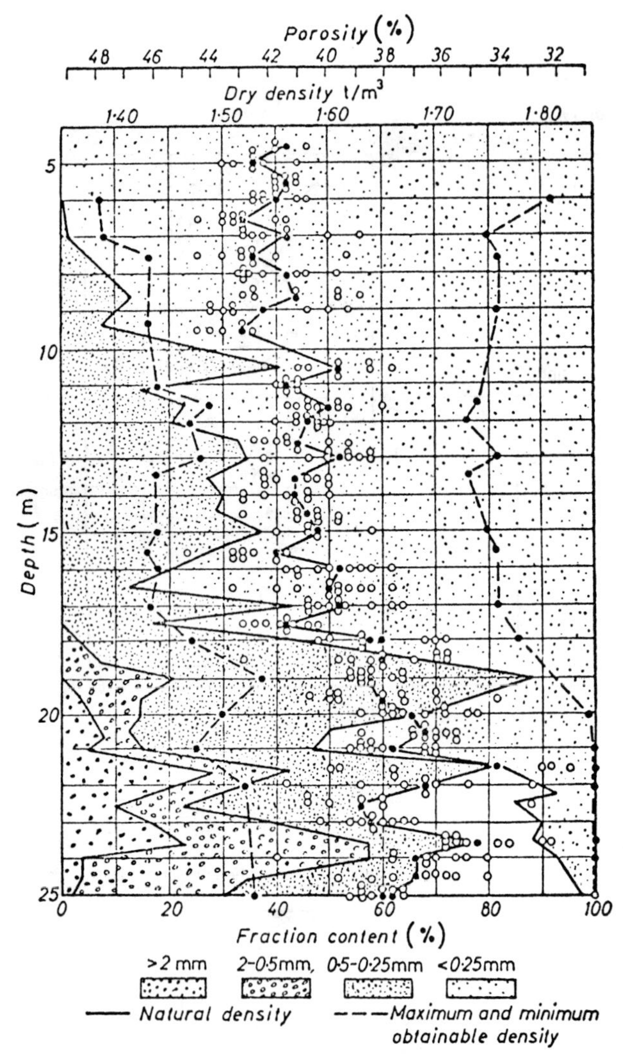

Natural density and grading of alluvial sands (vertical section of pit)

Fig. 7.17 Volga River sand (From Durante et al., 1957).

3. There are many methods available for investigating subsoils. The standard penetration test is very useful for giving an approximate, general portrayal of the subsoil profile. Sampling and laboratory and field testing are usually necessary to obtain design data.
4. Experience—as illustrated by the actual profiles shown in Figs. 7.7 to 7.17—emphasizes the importance of stress history and the great variability of soil properties in a given profile.

PROBLEMS

7.1 Suggest soil investigation methods for each of the following situations:

a. Dwelling house on sand.

b. Highway section on rock.

c. 30 m high embankment on 6 m deep deposit of soft clay.

d. Large compressor foundation on 3 m hydraulic sand fill.

7.2 Plot depth (ordinate) versus liquidity index (abscissa) for:

a. Manglerud clay (Fig. 7.7).

b. Paddington clay (Fig. 7.8).

c. Chicago clay (Fig. 7.14).

Comment on any relationship between liquidity index and geological history for each clay.

7.3 Plot sensitivity (ordinate) versus liquidity index (abscissa) using data from the following soils:

a. Drammen clay and Mangerud clay (Fig. 7.7).

b. Boston blue clay (Fig. 7.9).

Comment on any relationship between sensitivity and liquidity index.

7.4 Compute the activity of Boston blue clay, a Canadian varved clay, and a Brazilian residual clay.

7.5 For the Volga River sand (Fig. 7.17) plot relative density (abscissa) versus depth (ordinate).

RALPH B. PECK

Dr. Ralph Peck was born in Winnipeg, Canada, on June 23, 1912 and received his education at Rensselaer Polytechnic Institute and Harvard University. At Harvard in 1939, Dr. Peck began a long association with Dr. Terzaghi. As Dr. Terzaghi's representative during the initial construction of the Chicago subways, Dr. Peck was in charge of the soil laboratory and of the program of field measurements on one of the earliest large construction projects in which modern soil mechanics played a major role.

Dr. Peck has devoted his efforts primarily to the application of soil mechanics to design and construction, and to the evaluation and the presentation of the results of research in form suitable for ready use by the practicing engineer. Dr. Peck is one of the most respected consulting soil engineers in the world. An excellent lecturer, Dr. Peck has left his imprint on the many students he has had at the University of Illinois. His textbook, jointly authored with Dr. Terzaghi, *Soil Mechanics in Engineering Practice*, is a widely used book, both by students and by practicing engineers.

Dr. Peck has been the recipient of the Norman Medal and Wellington Prize of the ASCE, and has been a Terzaghi Lecturer.

PART III

Dry Soil

Part III establishes certain basic principles concerning the stress-strain behavior of the skeleton of the soil by considering cases (i.e., dry soils) in which there are no important interactions between the skeleton and the pore fluid. The principles concerning the properties of dry soils will be of the greatest value to the study of soils containing water in Parts IV and V.

When we talk about dry soil in Part III, we mean an air-dried soil. Even an air-dried sand actually contains a small amount of water (perhaps a water content as much as 1%). However, as long as the particle size is that of a coarse silt or larger, this small amount of moisture has little or no effect upon the mechanical properties of the soil. The principles established in Part III apply to a wide variety of dry soils including coarse silts, sands, and gravels.

CHAPTER 8

Stresses within a Soil Mass

Part II dealt with the forces that act between individual soil particles. In an actual soil it obviously is impossible to keep track of the forces at each individual contact point. Rather, it is necessary to use the concept of *stress*.

This chapter introduces the concept of stress as it applies to soils, discusses the stresses that exist within a soil as a result of its own weight and as a result of applied forces, and, finally, presents some useful geometric representations for the state of stress at a point within a soil.

8.1 CONCEPT OF STRESS FOR A PARTICULATE SYSTEM

Figure 8.1*a* shows a hypothetical small measuring device (Element *A*) buried within a mass of soil. We imagine that this measuring device has been installed in such a way that no soil particles have been moved. The sketches in Fig. 8.1*b*, *c* depict the horizontal and vertical faces of Element *A*, with soil particles pushing against these faces. These particles generally exert both a normal force and a shear force on these faces. If each face is square, with dimension a on each side, then we can define the stresses acting upon the device as

$$\sigma_v = \frac{N_v}{a^2}, \qquad \sigma_h = \frac{N_h}{a^2}, \qquad \tau_h = \frac{T_h}{a^2} \qquad \tau_v = \frac{T_v}{a^2} \tag{8.1}$$

where N_v and N_h, respectively, represent the normal forces in the vertical and horizontal directions; T_v and T_h, respectively, represent the shear forces in the vertical and horizontal directions; and σ_v, σ_h, τ_v, and τ_h represent the corresponding stresses. Thus we have defined four stresses which can, at least theoretically, be readily visualized and measured.

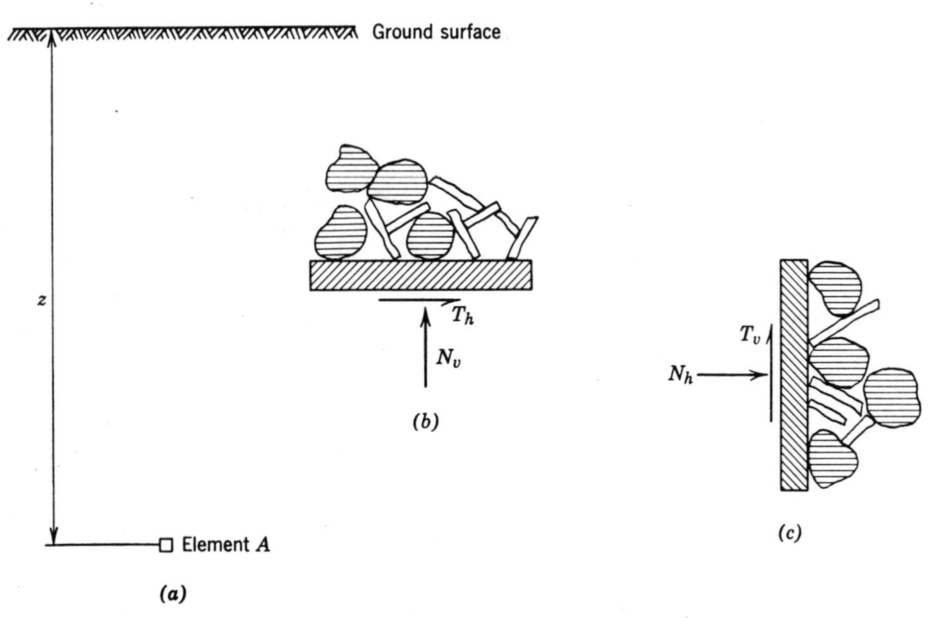

Fig. 8.1 Sketches for definition of stress. (*a*) Soil profile. (*b*) and (*c*) Forces at element *A*.

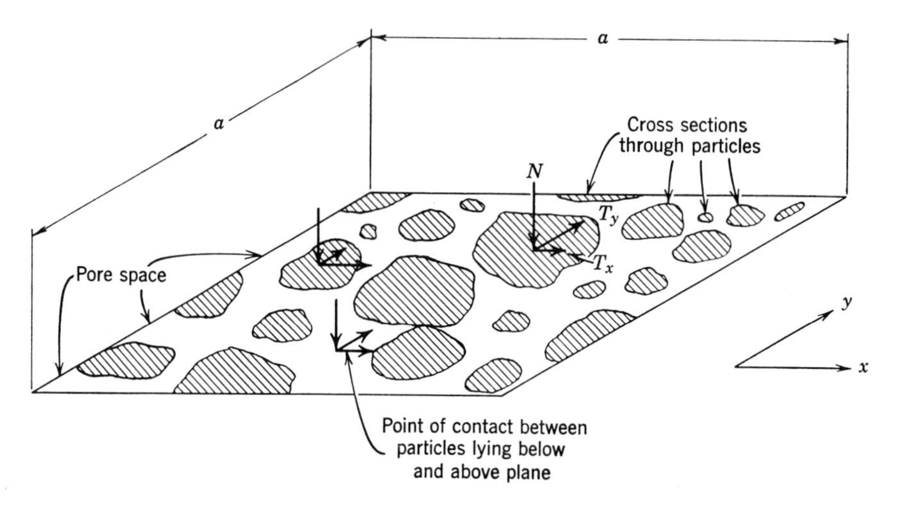

Fig. 8.2 Definition of stress in a particulate system.

$$\sigma = \frac{\Sigma N}{a \times a} \qquad \tau_x = \frac{\Sigma T_x}{a \times a} \qquad \tau_y = \frac{\Sigma T_y}{a \times a}$$

In Part III, except as noted, it will be assumed that the pressure within the pore phase of the soil is zero; i.e., equal to atmospheric pressure. Hence the forces N_v, N_h, T_v, and T_h arise entirely from force that is being transmitted through the mineral skeleton. In dry soil, stress may be thought of as the *force in the mineral skeleton per unit area of soil.*

Actually, it is quite difficult to measure accurately the stresses within a soil mass, primarily because the presence of a stress gage disrupts the stress field that would otherwise exist if the gage were not present. Hamilton (1960) discusses soil stress gages and the problems associated with them. In order to make our definition of stress apply independently of a stress gage, we pass an imaginary plane through soil, as shown in Fig. 8.2. This plane will pass in part through mineral matter and in part through pore space. It can even happen that this plane passes through one or more contact points between particles. At each point where this plane passes through mineral matter, the force transmitted through the mineral skeleton can be broken up into components normal and tangential to the plane. The tangential components can further be resolved into components lying along a pair of coordinate axes. These various components are depicted in Fig. 8.2. The summation over the plane of the normal components of all forces, divided by the area of the plane, is the *normal stress* σ acting upon the plane. Similarly, the summation over the plane of the tangential components in, say, the x-direction, divided by the area of this plane, is the *shear stress* τ_x in the x-direction.

There is still another picture which is often used when defining stress. One imagines a "wavy" plane which is warped just enough so that it passes through mineral matter only at points of contact between particles. Stress is then the sum of the contact forces divided by the area of the wavy plane. The summation of all the contact areas will be a very small portion of the total area of the plane, certainly less than 1%. Thus stress as defined in this section is numerically much different from the stress at the points of contact.

In this book, when we use the word "stress" we mean the macroscopic stress; i.e., force/total area, the stress that we have just defined with the aid of Figs. 8.1 and 8.2. When we have occasion to talk about the stresses at the contacts between particles, we shall use a qualifying phrase such as "contact stresses". As was discussed in Chapter 5, the contact stresses between soil particles will be very large (on the order of 700 MN/m²). The macroscopic stress as defined in this chapter will typically range from 10 to 10 000 kN/m² for most actual problems.

The concept of stress is closely associated with the concept of a continuum. Thus when we speak of the stress acting at a point, we envision the forces against the sides of an infinitesimally small cube which is composed of some homogeneous material. At first sight we may therefore wonder whether it makes sense to apply the concept of stress to a particulate system such as soil. However, the concept of stress as applied to soil is no more abstract than the same concept applied to metals. A metal is actually composed of many small crystals, and on the submicroscopic scale the magnitude of the forces between crystals varies randomly from crystal to crystal. For any material, the inside of the "infinitesimally small cube" is thus only statistically homogeneous. In a sense all matter is particulate, and it is meaningful to talk about macroscopic stress only if this stress varies little over distances which are of the order of magnitude of the size of the largest particle. When we talk about the stresses at a "point" within a soil, we often must envision a rather large "point".

Returning to Fig. 8.1, we note that the forces N_v, etc., are the sums of the normal and tangential components of the forces at each contact point between the soil particles and the faces. The smaller the size of the particle, the greater the number of contact points with a face of dimension a. Thus, for a given value of macroscopic stress, a decreasing particle size means a smaller force per contact. For example, Table 8.1 gives typical values for the force per contact for different values of stress and different particle sizes (see Marsal, 1963).

Table 8.1 Typical Values for Average Contact Forces within Granular Soils

Particle Description	Particle Diameter (mm)	Average Contact Force (N) for Macroscopic Stress (kN/m²)		
		7	70	700
Gravel	60	13	130	1300
	2.0	0.013	0.13	1.3
Sand				
	0.06	13×10^{-6}	13×10^{-5}	0.0013
Silt				
	0.002	13×10^{-9}	13×10^{-8}	13×10^{-7}

8.2 GEOSTATIC STRESSES

Stresses within soil are caused by the external loads applied to the soil and by the weight of the soil. The pattern of stresses caused by applied loads is usually quite complicated. The pattern of stresses caused by the soil's own weight can also be complicated. However, there is one common situation in which the weight of soil gives rise to a very simple pattern of stresses: when the ground surface is horizontal and when the nature of the soil varies but little in the horizontal directions. This situation frequently exists, especially in sedimentary soils. In such a situation, the stresses are called *geostatic stresses*.

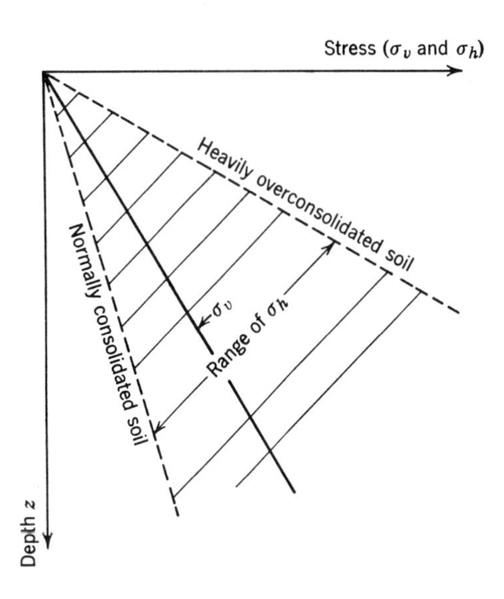

Fig. 8.3 Geostatic stresses in soil with horizontal surface.

Vertical Geostatic Stress

In the situation just described, there are no shear stresses upon vertical and horizontal planes within the soil. Hence the vertical geostatic stress at any depth can be computed simply by considering the weight of soil above that depth.

Thus, if the unit weight of the soil is constant with depth,

$$\sigma_v = z\gamma \tag{8.2}$$

where z is the depth and γ is the total unit weight of the soil. In this case, the vertical stress will vary linearly with depth, as shown in Fig. 8.3.

Of course, the unit weight is seldom constant with depth. Usually a soil will become denser with depth because of the compression caused by the geostatic stress. If the unit weight of the soil varies continuously with depth, the vertical stress can be evaluated by means of the integral

$$\sigma_v = \int_0^z \gamma \, dz \tag{8.3}$$

If the soil is stratified and the unit weight is different for each stratum, then the vertical stress can conveniently be computed by means of the summation

$$\sigma_v = \sum \gamma \, \Delta z \tag{8.4}$$

Example 8.1 illustrates the computation of vertical geostatic stress for a case in which the unit weight is a function of the geostatic stress.

▶ Example 8.1

Given. The relationship between vertical stress and unit weight

$$\gamma = 14.92 + 0.0023\sigma_v$$

where γ is in kN/m^3 and σ_v is in kN/m^2.

Find. The vertical stress at a depth of 30 meters for a geostatic stress condition.

Solution Using Calculus. From Eq. 8.3:

$$\sigma_v = \int_0^z (14.92 + 0.0023\sigma_v)dz \quad (z \text{ in meters})$$

or

$$\frac{d\sigma_v}{dz} = 14.92 + 0.0023\sigma_v$$

The solution of this differential equation is:

$$\sigma_v = 6\,487(e^{0.0023z} - 1)$$

For $z = 30$ m:

$$\sigma_v = 6\,487(1.0714 - 1) = 463 \text{ kN/m}^2$$

Alternative Approximate Solution by Trial and Error.

First trial: assume average unit weight from $z = 0$ to $z = 30$ m is 15.7 kN/m^3. Then σ_v at $z = 30$ m would be 471 kN/m^2. Actual unit weight at that depth would be 16.00 kN/m^3, and average unit weight (assuming linear variation of γ with depth) would be 15.46 kN/m^3.

Second trial: assume average unit weight of 15.45 kN/m^3. Then at $z = 30$ m, $\sigma_v = 464$ kN/m^2 and $\gamma = 15.99$ kN/m^3. Average unit weight is 455 kN/m^3, which is practically the same as for the previous trial.

The slight discrepancy between the two answers occurs because the unit weight actually does not quite vary linearly with depth as assumed in the second solution. The discrepancy can be larger when γ is more sensitive to σ_v. The solution using calculus is more accurate, but the user easily can make mistakes regarding units. The accuracy of the trial solution can be improved by breaking the 30 m depth into layers and assuming a uniform variation of unit weight through each layer. ◀

Horizontal Geostatic Stress

The ratio of horizontal to vertical stress is expressed by a factor called the *coefficient of lateral stress* or *lateral stress ratio*, and is denoted by the symbol K:

$$K = \frac{\sigma_h}{\sigma_v} \tag{8.5}$$

This definition of K is used whether or not the stresses are geostatic.

Even when the stresses are geostatic, the value of K can vary over a rather wide range depending on whether the ground has been stretched or compressed in the horizontal direction by either the forces of nature or the works of man. The possible range of the value of K will be discussed in some detail in Chapter 11.

Often we are interested in the magnitude of the horizontal geostatic stress in the special case where there has been no lateral strain within the ground. In the special case, we speak of the *coefficient of lateral stress at rest*[2] (or *lateral stress ratio at rest*) and use the symbol K_0.

As discussed in Chapter 7, a sedimentary soil is built up by an accumulation of sediments from above. As this build-up of overburden continues, there is vertical compression of soil at any given elevation because of the increase in vertical stress. As the sedimentation takes place, generally over a large lateral area, there is no reason why there should be significant horizontal compression during sedimentation. From this, one could logically reason that in such sedimentary soil the horizontal total stress should be less than the vertical stress. For a sand deposit formed in this way, K_0 will typically have a value between 0.4 and 0.5.

On the other hand, there is evidence that the horizontal stress can exceed the vertical stress if a soil deposit has been heavily preloaded in the past. In effect, the horizontal stresses were "locked-in" when the soil was previously loaded by additional overburden, and did not disappear when this loading was removed. For this case, K_0 may well reach a value of 3.

The range of horizontal stresses for the at rest condition have been depicted in Fig. 8.3.

8.3 STRESSES INDUCED BY APPLIED LOADS

Results from the theory of elasticity are often used to compute the stresses induced within soil masses by externally applied loads. The assumption of this theory is that stress is proportional to strain. Most of the useful solutions from this theory also assume that soil is *homogeneous* (its properties are constant from point to point) and *isotropic* (its properties are the same in each direction through a point). Soil seldom if ever exactly fulfills, and often seriously violates, these assumptions. Yet the soil engineer has little choice but to use the results of this theory together with engineering judgment.

[2] The phrase *coefficient of lateral pressure* is also used, but in classical mechanics the word pressure is used in connection with a fluid that cannot transmit shear.

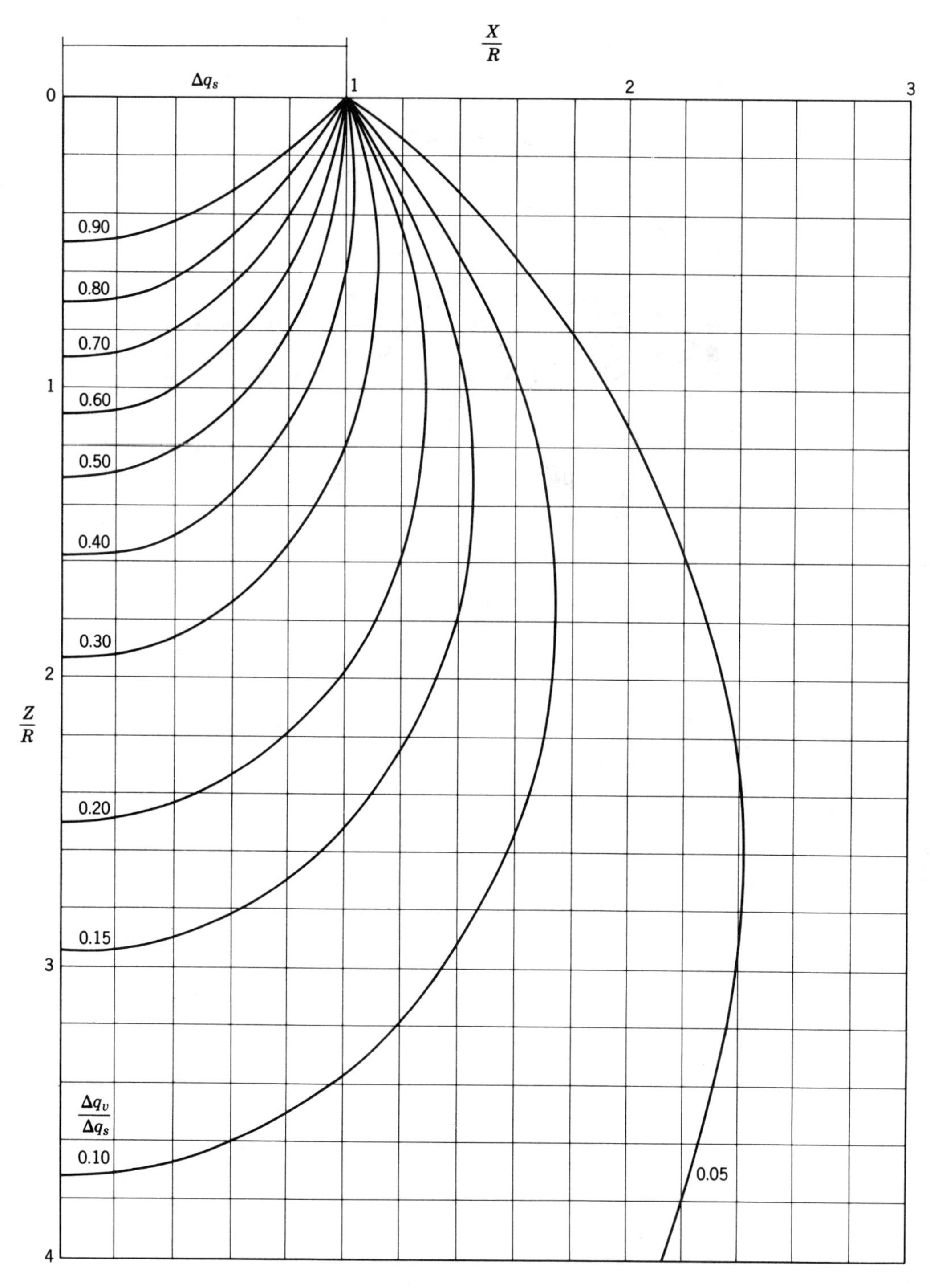

Fig. 8.4 Vertical stresses induced by uniform load on circular area.

It is a very tedious matter to obtain the elastic solution for a given loading and set of boundary conditions. In this book, we are concerned not with how to obtain solutions but rather with how to use these solutions. This section presents several solutions in graphical form.

Uniform load over a circular area. Figures 8.4 and 8.5 give the stresses caused by a uniformly distributed normal stress Δq_s acting over a circular area of radius R on the surface of an elastic half-space.[3] These stresses must be added to the initial geostatic stresses. Figure 8.4 gives

[3] In general, the stresses computed from the theory of elasticity are functions of Poisson's ratio μ. This quantity will be defined in Chapter 12. However, vertical stresses resulting from normal stresses applied to the surface are always independent of μ, and stresses caused by a strip load are also independent of μ. Thus of the charts presented in this chapter only those in Fig. 8.5 depend upon μ, and are for $\mu = 0.45$.

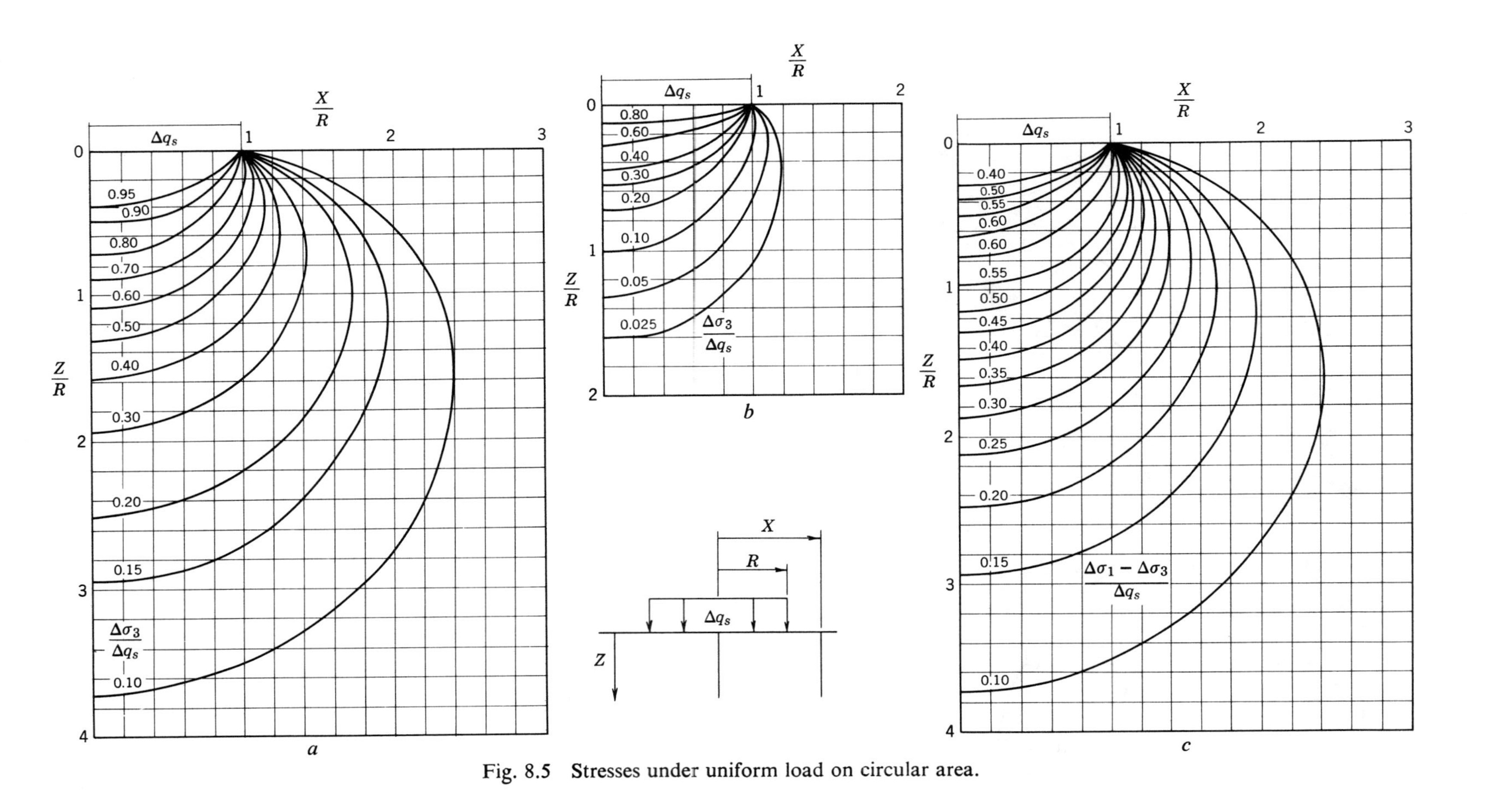

Fig. 8.5 Stresses under uniform load on circular area.

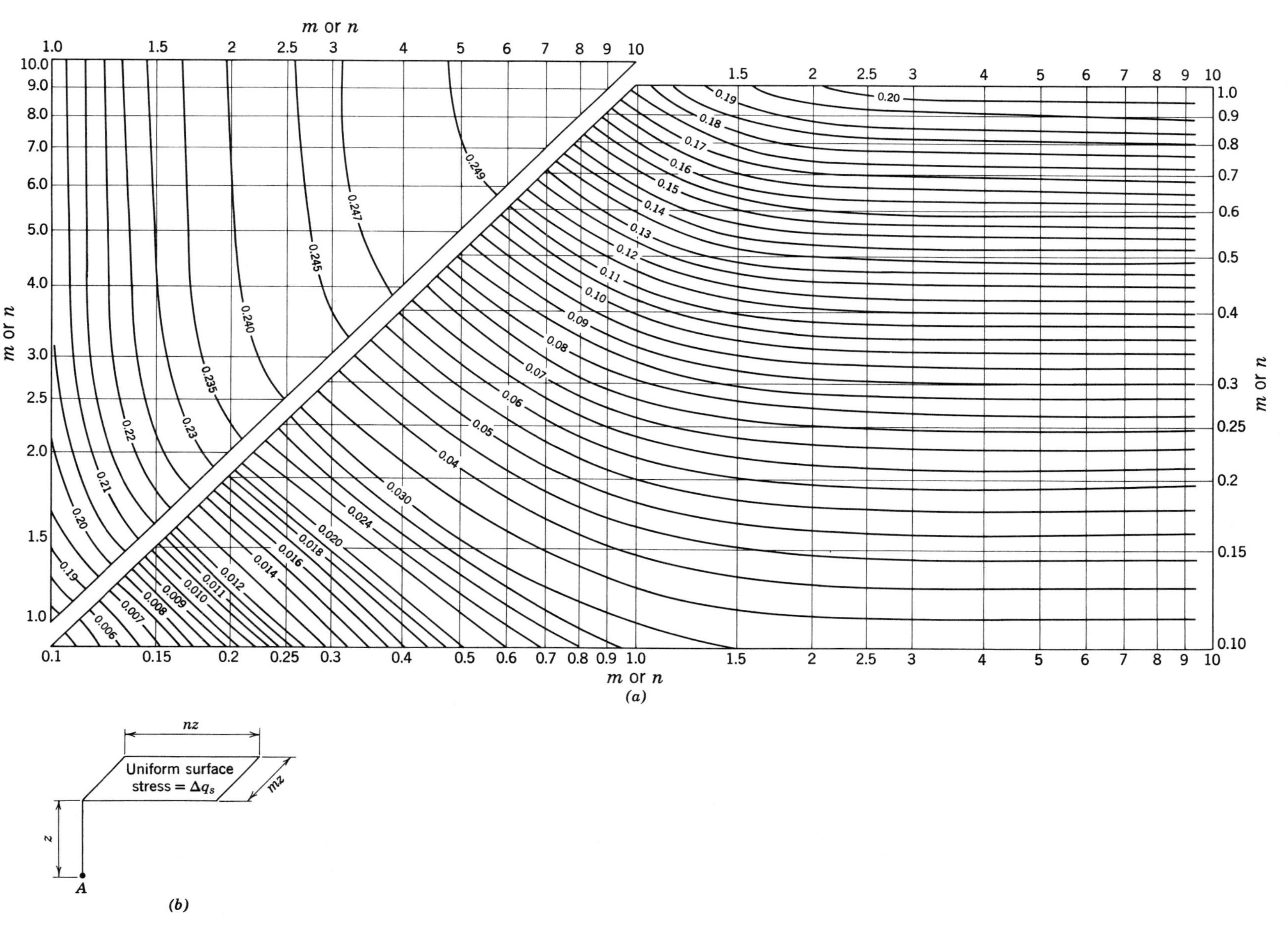

Fig. 8.6 (*a*) Chart for use in determining vertical stresses below corners of loaded rectangular surface areas on elastic, isotropic material. Chart gives $f(m, n)$. (*b*) At point A, $\Delta\sigma_v = \Delta q_s \times f(m, n)$. (From Newmark, 1942)

the vertical stresses. The significance of $\Delta\sigma_1$ and $\Delta\sigma_3$, given in Fig. 8.5, will be discussed in Section 8.4. For the moment it suffices to know that along the vertical center line

$$\Delta\sigma_1 = \Delta\sigma_v \quad \text{and} \quad \Delta\sigma_3 = \Delta\sigma_h$$

Example 8.2 illustrates the use of these charts. The stresses induced by a surface loading must be added to the geostatic stresses in order to obtain the final stresses after a loading.

Charts such as these give the user a feel for the spread of stresses through a soil mass. For example, the zone under a loaded area wherein the vertical stresses are significant is frequently termed the "bulb" of stresses. For a circular loaded area, the vertical stresses are less than $0.15\Delta q_s$ at a depth of $3R$ and less than $0.10\Delta q_s$ at a depth of $4R$. Usually the stress bulb is considered to be the volume within the contour for $0.1\Delta q_s$, but this choice is strictly arbitrary.

Uniform load over rectangular area. The chart in Fig. 8.6 may be used to find the vertical stresses beneath the corner of a rectangularly loaded area. Example 8.3 illustrates the way in which this chart may be used to find the stresses at points not lying below the corner of a load. Problems involving surface loads which are not uniformly distributed or which are distributed over an irregularly shaped area can be handled by breaking the load up into pieces involving uniformly distributed loads over rectangular areas.

Strip loads. Figures 8.7 and 8.8 give stresses caused by strip loadings; i.e., loadings which are infinitely long in the direction normal to the paper. Two cases are shown: load uniformly distributed over the strip and load distributed in a triangular pattern. Again, $\Delta\sigma_1 = \Delta\sigma_v$ and $\Delta\sigma_3 = \Delta\sigma_h$ along the center line.

Other solutions. Solutions are also available in chart form for other loading conditions, for layered elastic bodies, and for elastic bodies which are rigid in the horizontal directions but can strain in the vertical direction. With the digital computer, the engineer can readily

▶ **Example 8.2**

Given. Soil with $\gamma = 16.5$ kN/m³ and $K_0 = 0.5$, loaded by $\Delta q_s = 240$ kN/m² over a circular area 6 m in diameter.

Find. The vertical and horizontal stresses at a depth of 3 m under the center of the loaded area.

Solution.

	Vertical Stress (kN/m²)	Horizontal Stress (kN/m²)
Initial stresses	$\gamma z = 49.5$	$K_0\gamma z = 24.8$
Stress increments	Fig. 8.4 (0.64)(240) = 153.6	Fig. 8.5*b* (0.10)(240) = 24.0
Final stresses	203.1	48.8

◀

▶ **Example 8.3**

Given. The plan view of a loading shown in Fig. E8.3-1.

Find. The vertical stress at a depth of 3 m below point *A*.

Solution. The given loading is equivalent to the sum of the four loadings shown in Fig. E8.3-2.

Loading	*m*	*n*	Coefficient	$\Delta\sigma_v$ – kN/m²
I	1.5	2	0.223	22.3
II	0.5	2	0.135	−13.5
III	1.5	0.5	0.131	−13.1
IV	0.5	0.5	0.085	8.5
				4.2 kN/m²

◀

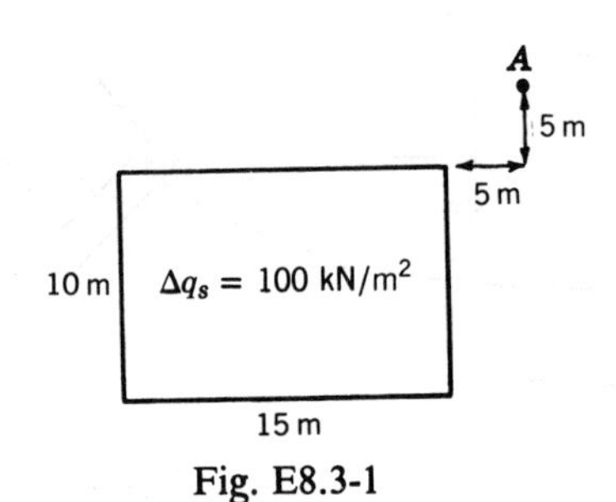

Fig. E8.3-1

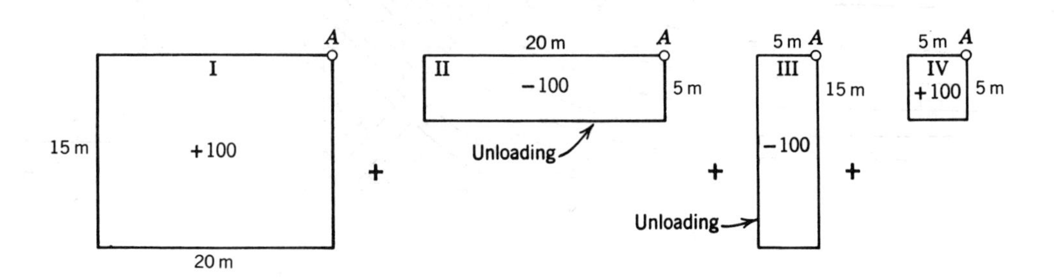

Fig. E8.3-2

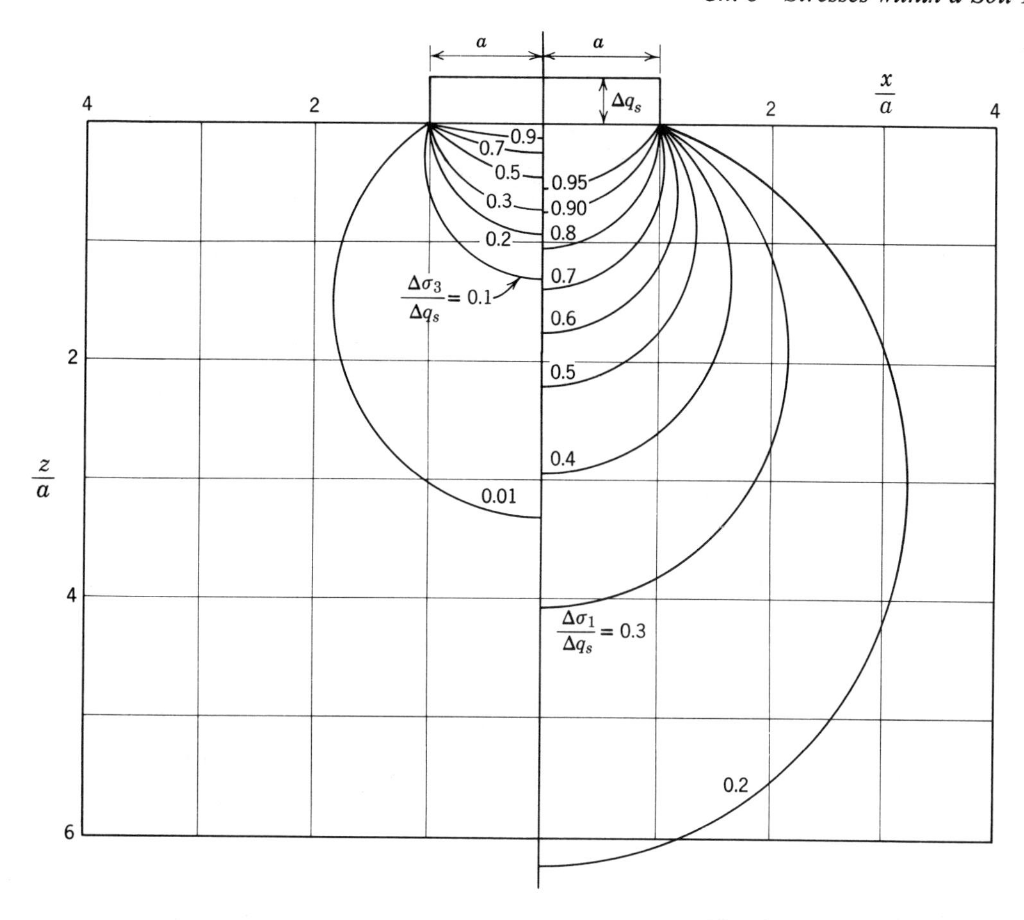

Fig. 8.7 Principal stresses under strip load.

obtain elastic stress distributions for almost any loading and boundary conditions. Charts such as those given here are useful for preliminary analysis of a problem or when the computer is not available.

Accuracy of calculated values of induced stresses. A critical question is: How accurate are the values of induced stresses as calculated from stress distribution theories? This question can be answered only by comparing calculated with actual stress increments for a number of field situations. Unfortunately, there are very few reliable sets of measured stress increments within soil masses (see Taylor, 1945 and Turnbull, Maxwell, and Ahlvin, 1961).

The relatively few good comparisons of calculated with measured stress increments indicate a surprisingly good agreement, especially in the case of vertical stresses. A great number of such comparisons are needed to establish the degree of reliability of calculated stress increments. At the present stage of knowledge, the soil engineer must continue to use stress distribution theories based on the theory of elasticity for lack of better techniques. He should realize, however, that his computed stress values may be in error by as much as $\pm 25\%$ or more.

8.4 PRINCIPAL STRESSES AND MOHR CIRCLE

As in any other material, the normal stress at a point within a soil mass is generally a function of the orientation of the plane chosen to define the stress. It is meaningless to talk of *the* normal stress or *the* shear stress at a point. Thus subscripts will usually be attached to the symbols σ and τ to qualify just how this stress is defined. More generally, of course, we should talk of the *stress tensor*, which provides a complete description for the state of stress at a point. This matter is discussed in textbooks on elementary mechanics, such as Crandall and Dahl (1959). The following paragraphs will state the essential concepts and definitions.

Principal Stresses

There exist at any stressed point three orthogonal (i.e., mutually perpendicular) planes on which there are zero shear stresses. These planes are called the *principal stress planes*. The normal stresses that act on these three planes are called the *principal stresses*. The largest of these three

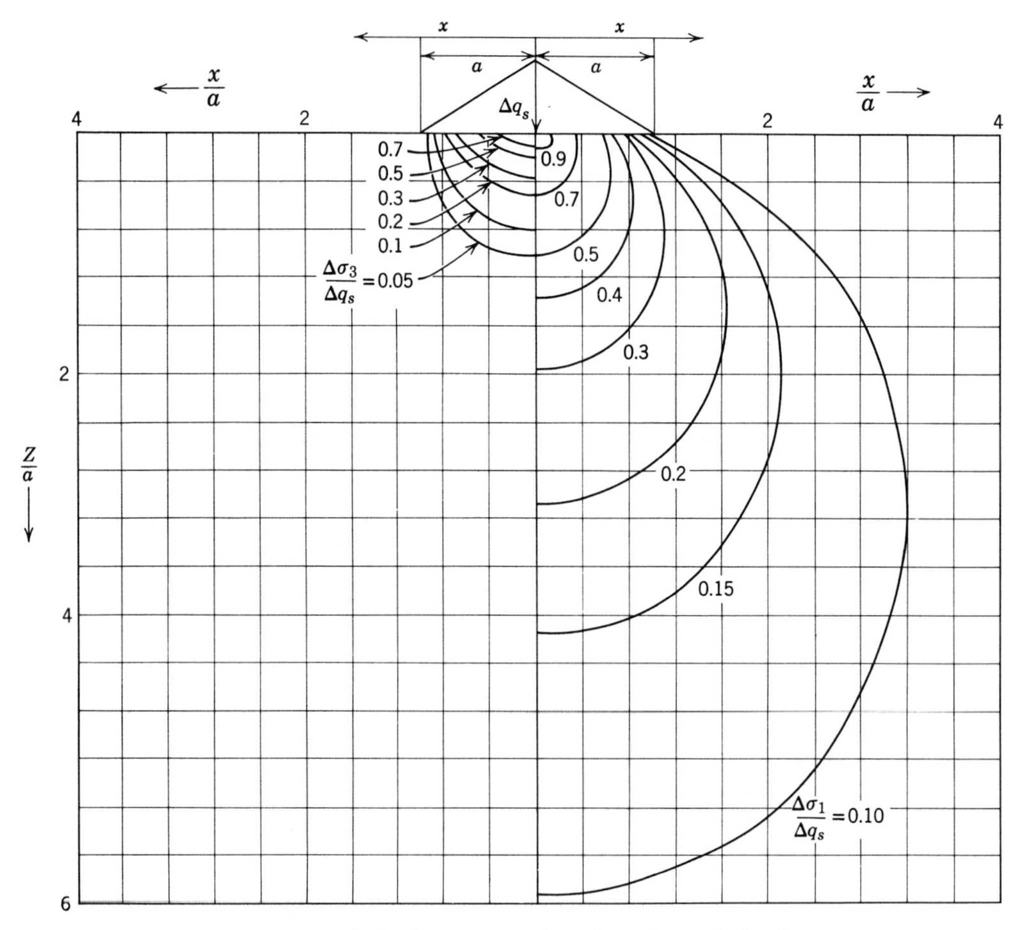

Fig. 8.8 Principal stresses under triangular strip load.

stresses is called the *major principal stress* σ_1, the smallest is called the *minor principal stress* σ_3, and the third is called the *intermediate principal stress* σ_2.

When the stresses in the ground are geostatic, the horizontal plane through a point is a principal plane and so too are all vertical planes through the point. When $K < 1$, $\sigma_v = \sigma_1$, $\sigma_h = \sigma_3$, and $\sigma_2 = \sigma_3 = \sigma_h$. When $K > 1$ the situation is reversed: $\sigma_h = \sigma_1$, $\sigma_v = \sigma_3$, and $\sigma_2 = \sigma_1 = \sigma_h$. When $K = 1$, $\sigma_v = \sigma_h = \sigma_1 = \sigma_2 = \sigma_3$ and the state of stress is *isotropic*.

We should also recall that the shear stresses on any two orthogonal planes (planes meeting at right angles) must be numerically equal. Returning to the definition of stresses given in Section 8.1, we must have $\tau_h = \tau_v$.

Mohr circle. Throughout most of this book, we shall be concerned only with the stresses existing in two dimensions rather than those in three dimensions.[4] More specifically, we shall be interested in the state of stress in the plane that contains the major and minor principal stresses, σ_1 and σ_3. Stresses will be considered positive when compressive. The remainder of the sign conventions are given in Fig. 8.9. The quantity $(\sigma_1 - \sigma_3)$ is called the *deviator stress* or *stress difference*.

Given the magnitude and direction of σ_1 and σ_3, it is possible to compute normal and shear stresses in any other direction using the equations developed from statics and shown in Fig. 8.9.[5] These equations, which provide a complete (in two dimensions) description for the state of stress, describe a circle. Any point on the circle, such as A, represents the stress on a plane whose normal is oriented at angle θ to the direction of the major principal stress. This graphical representation of the state of stress is known as the *Mohr circle* and is of the greatest importance in soil mechanics.

Given σ_1 and σ_3 and their directions, it is possible to find the stresses in any other direction by graphical

[4] The intermediate principal stress unquestionably has some influence upon the strength and stress-strain properties of soil. However, this influence is not well understood. Until this effect has been clarified, it seems best to work primarily in terms of σ_1 and σ_3.

[5] Equations 8.6 and 8.7 are derived in most mechanics texts; e.g., see Crandall and Dahl (1959), pp. 130–138.

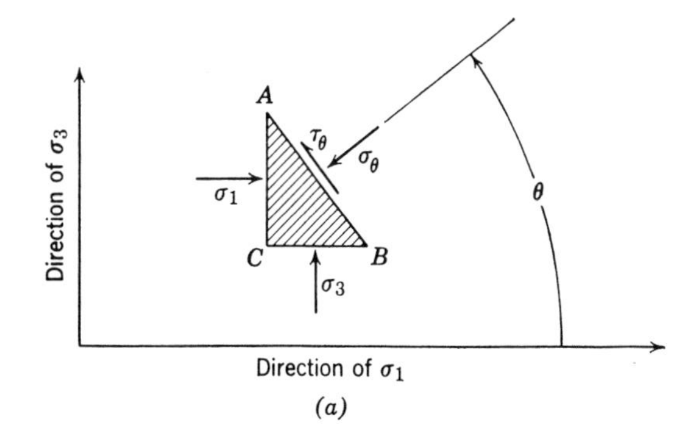

(*a*)

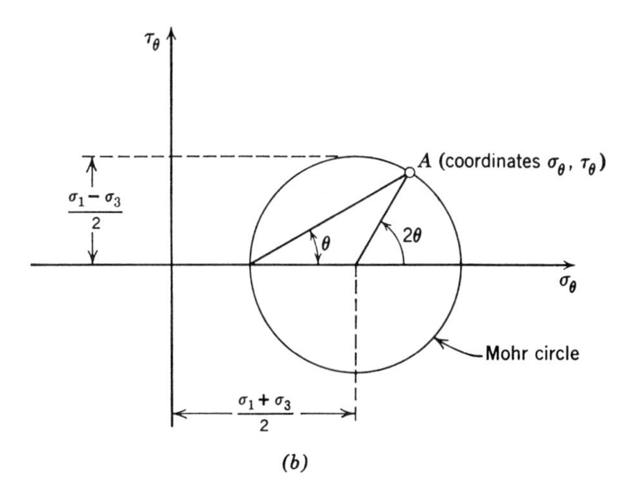

(*b*)

Fig. 8.9 Representation of stress by the Mohr circle. (*a*) Equations for state of stress at a point. (*b*) Mohr diagram for state of stress at a point. τ is positive when counterclockwise; θ is measured counter clockwise from the direction of σ_1.

$$\sigma_\theta = \sigma_1 \cos^2 \theta + \sigma_3 \sin^2 \theta = \frac{\sigma_1 + \sigma_3}{2} + \frac{\sigma_1 - \sigma_3}{2} \cos 2\theta \quad (8.6)$$

$$\tau_\theta = (\sigma_1 - \sigma_3) \sin \theta \cos \theta = \frac{\sigma_1 - \sigma_3}{2} \sin 2\theta \quad (8.7)$$

construction using the Mohr circle. Or, given the σ_θ and τ_θ that act on any *two* planes, the magnitude and direction of the principal stresses can be found. The notion of the *origin of planes* is especially useful in such constructions. The origin of planes is a point on the Mohr circle, denoted by O_P, with the following property: a line through O_P and any point A of the Mohr circle will be parallel to the plane on which the stresses given by point A act. Examples 8.4 to 8.7 illustrate the use of the Mohr circle and of the origin of planes. The reader should study these examples very carefully.

The maximum shear stress at a point, τ_{max}, is always equal to $(\sigma_1 - \sigma_3)/2$; i.e., the maximum shear stress equals the radius of the Mohr circle. This maximum shear stress occurs on planes lying at ±45° to the major principal stress direction.

If the stress condition is geostatic, the largest shear stress will be found upon planes lying at 45° to the horizontal. The magnitude of the maximum shear stress will be

$$\text{if } K < 1, \quad \tau_{max} = \frac{\sigma_v}{2}(1 - K)$$

$$\text{if } K > 1, \quad \tau_{max} = \frac{\sigma_v}{2}(K - 1)$$

$$\text{if } K = 1, \quad \tau_{max} = 0$$

8.5 *p-q* DIAGRAMS

In many problems it is desirable to represent, on a single diagram, many states of stress for a given specimen of soil. In other problems, states of stress for many different specimens are represented on one such diagram. In such cases it becomes cumbersome to plot Mohr circles, and even more difficult to see what is on the diagram once all circles are plotted.

An alternative scheme for plotting the state of stress is to plot a *stress point* whose coordinates are

$$p = \frac{\sigma_1 + \sigma_3}{2}$$

$$q = \pm \frac{\sigma_1 - \sigma_3}{2} \begin{cases} + \text{ if } \sigma_1 \text{ is inclined equal to or less than } \pm 45^\circ \text{ to the vertical} \\ - \text{ if } \sigma_1 \text{ is inclined less than } \pm 45^\circ \text{ to the horizontal} \end{cases} \quad (8.8)$$

In most cases for which the stress point representation is used, the principal stresses act on vertical and horizontal planes. Then Eq. (8.8) simplifies to

$$p = \frac{\sigma_v + \sigma_h}{2}, \quad q = \frac{\sigma_v - \sigma_h}{2} \quad (8.9)$$

Plotting a stress point is equivalent to plotting one single point of a Mohr circle: the uppermost point if q is positive or the bottom-most point if q is negative. Numerically, q equals one-half of the deviator stress.

Example 8.8 shows stress points corresponding to the state of stress worked out in Examples 8.4 to 8.6. Knowing the values of p and q for some state of stress, one has all of the information needed to plot the corresponding Mohr circle. However, the use of a *p-q* diagram is no substitute for the use of the Mohr circle construction to determine the magnitude of these principal stresses from a given state-of-stress.

▶ Example 8.4

Given. Figure E8.4-1.

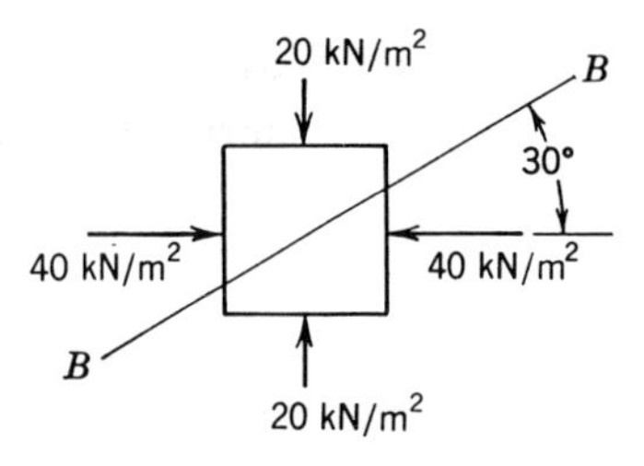

Fig. E8.4-1

Find. Stresses on plane *B–B*.
Solution. Use Fig. E8.4-2.

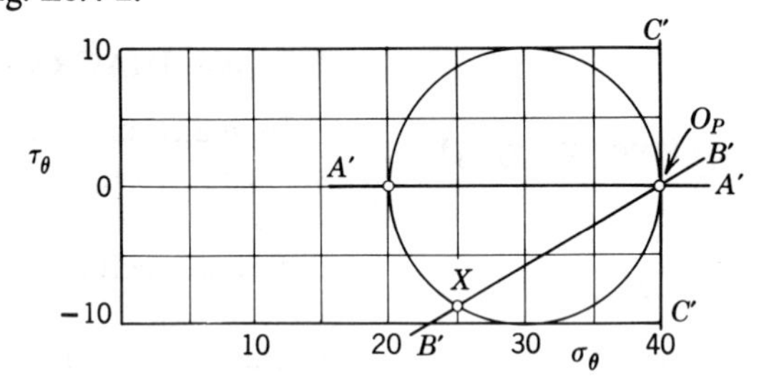

Fig. E8.4-2

1. Locate points with co-ordinates (40, 0) and (20, 0).
2. Draw circle, using these points to define diameter.
3. Draw line $A'A'$ through point (20, 0) and parallel to plane on which stress (20, 0) acts.
4. Intersection of $A'A'$ with Mohr circle at point (40, 0) is the origin of planes.
5. Draw line $B'B'$ through O_P parallel to BB.
6. Read coordinates of point X where $B'B'$ intersects Mohr circle.

Answer. See Fig. E8.4-3.

$$\text{on } BB \quad \begin{cases} \sigma = 25 \text{ kN/m}^2 \\ \tau = -8.7 \text{ kN/m}^2 \end{cases}$$

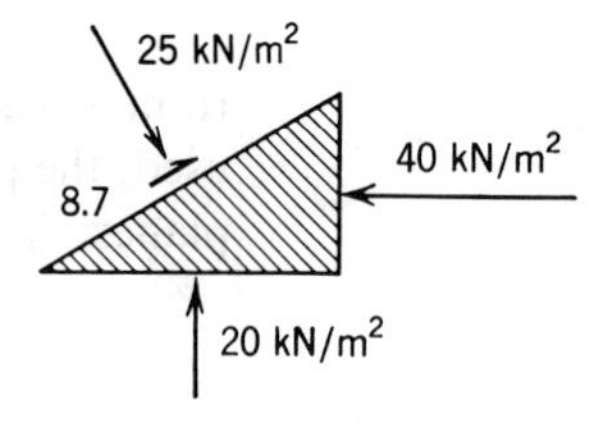

Fig. 8.4-3

Alternate Solution. Steps 1 and 2 same as above.

3. Draw line $C'C'$ through (40, 0) parallel to plane on which stress (40, 0) acts. $C'C'$ is vertical.

4. $C'C'$ intersects Mohr circle only at (40, 0), so this point is O_P. Steps 5 and 6 same as above.

Solution Using Eqs. 8.6 *and* 8.7.

$$\sigma_1 = 40 \text{ kN/m}^2 \quad \sigma_3 = 20 \text{ kN/m}^2 \quad \theta = 120°$$

$$\sigma_\theta = \frac{40 + 20}{2} + \frac{40 - 20}{2} \cos 240° = 30 - 10 \cos 60° = 25 \text{ kN/m}^2$$

$$\tau_\theta = \frac{40 - 20}{2} \sin 240° = -10 \sin 60° = -8.66 \text{ kN/m}^2$$ ◀

(*Questions for student.* Why is $\theta = 120°$? Would result be different if $\theta = 300°$?)

▶ Example 8.5

Given. Figure E8.5-1.

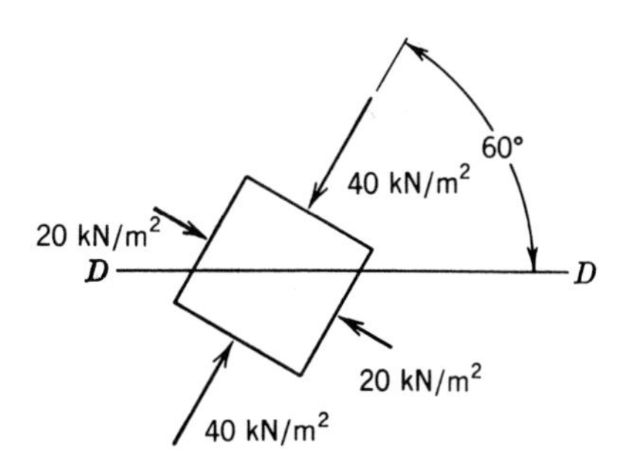

Fig. E8.5-1

Find. Stresses on horizontal plane *DD*.

Solution.

1. Locate points (40, 0) and (20, 0) on Mohr diagram (Fig. E8.5-2).

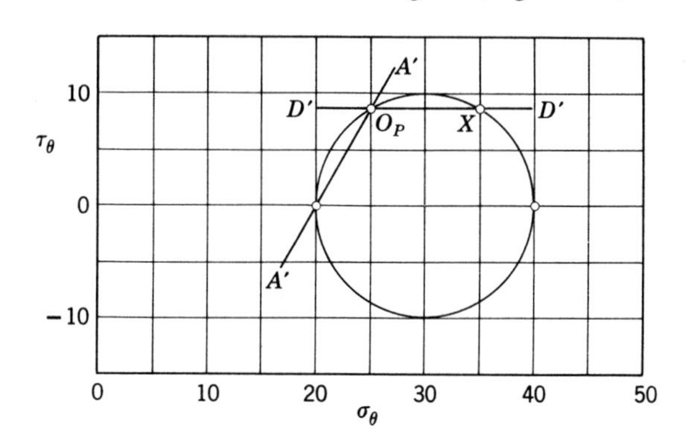

Fig. E8.5-2

2. Draw Mohr circle.
3. Draw line $A'A'$ through (20, 0) parallel to plane upon which stress (20, 0) acts.
4. Intersection of $A'A'$ with Mohr circle gives O_P.
5. Draw line $D'D'$ parallel to plane DD.
6. Intersection X gives desired stresses

Answer. See Fig. E8.5-3.

$$\text{on } DD \quad \begin{cases} \sigma = 35 \text{ kN/m}^2 \\ \tau = 8.7 \text{ kN/m}^2 \end{cases}$$

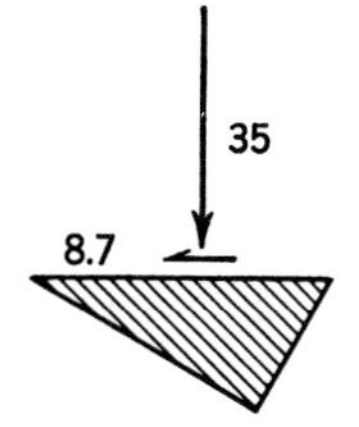

Fig. E8.5-3

◀

Example 8.6

Given. Figure E8.6-1.

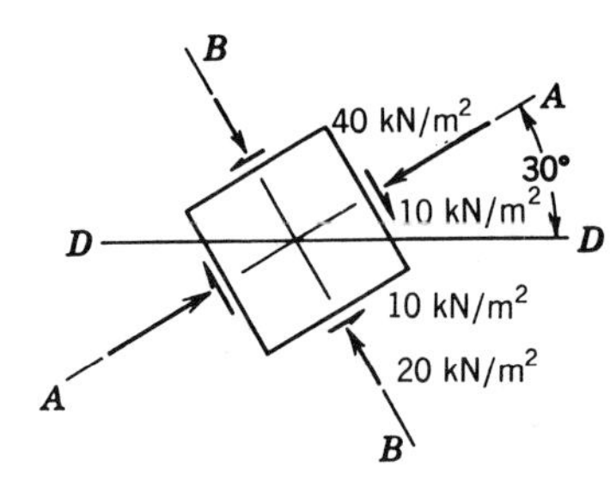

Fig. E8.6-1

Find. Magnitude and direction of the principal stresses.
Solution. Use Fig. E8.6-2.

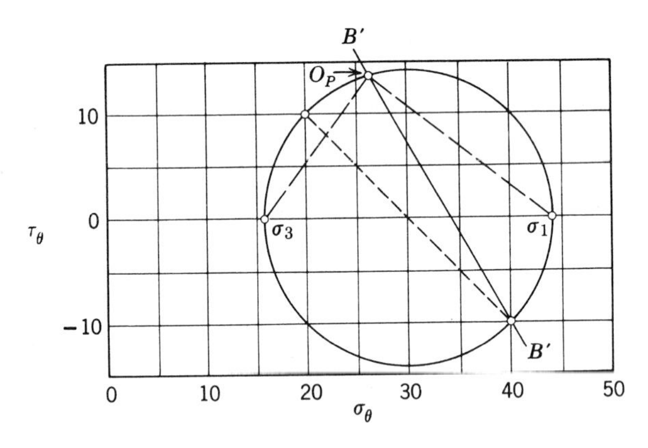

Fig. E8.6-2

1. Locate points (40, −10) and (20, 10).
2. Erect diameter and draw Mohr circle.
3. Draw $B'B'$ through (40,−10) parallel to BB.
4. Intersection of $B'B'$ with circle gives O_P.
5. Read σ_1 and σ_3 from graph.
6. Line through O_P and σ_1 gives plane on which σ_1 acts, etc. (see Fig. E8.6-3).

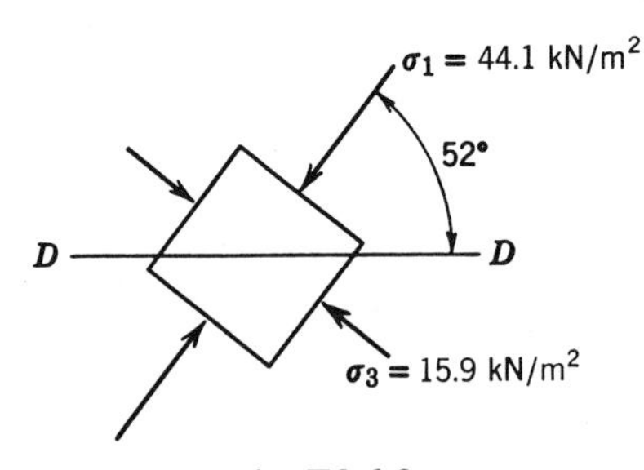

Fig. E8.6-3

Solution by Equations.
1. First make use of fact that sum of normal stresses is a constant:

$$\frac{\sigma_1 + \sigma_3}{2} = \frac{\Sigma\sigma_\theta}{2} = \frac{40 + 20}{2} = 30 \text{ kN/m}^2$$

2. Use relation

$$\left(\frac{\sigma_1 - \sigma_3}{2}\right) = \sqrt{\left[\sigma_\theta - \left(\frac{\sigma_1 + \sigma_3}{2}\right)\right]^2 + [\tau_\theta]^2}$$

with either pair of given stresses

$$\left(\frac{\sigma_1 - \sigma_3}{2}\right) = \sqrt{[20 - 30]^2 + [10]^2} = \sqrt{200} = 14.14 \text{ kN/m}^2$$

3.
$$\sigma_1 = \left(\frac{\sigma_1 + \sigma_3}{2}\right) + \left(\frac{\sigma_1 - \sigma_3}{2}\right) = 44.14 \text{ kN/m}^2$$

$$\sigma_3 = \left(\frac{\sigma_1 + \sigma_3}{2}\right) - \left(\frac{\sigma_1 - \sigma_3}{2}\right) = 15.86 \text{ kN/m}^2$$

4. Use stress pair in which σ_θ is largest; i.e. (40, −10)

$$\sin 2\theta = \frac{2\tau_\theta}{\sigma_1 - \sigma_3} = \frac{-20}{28.28} = -0.707$$

$$2\theta = -45°$$

$$\theta = -22\tfrac{1}{2}°$$

5. Angle from horizontal to major principal stress direction $= 30° - \theta = 52\frac{1}{2}°$. ◀

▶ **Example 8.7**

Given. A load of 50 kN/m² uniformly distributed over a circular area with a radius of 30 m.

Find. At 30 m depth under the edge of the loaded area, find the horizontal stress increment and the directions of the major and minor principal stress increments.

Solution. Figures 8.4 and 8.5 can be used to find $\Delta\sigma_v$, $\Delta\sigma_1$, and $\Delta\sigma_3$. These are plotted and the Mohr circle is constructed. The origin of planes is located by drawing a horizontal line through the point representing the vertical stress, and the problem is then completed.

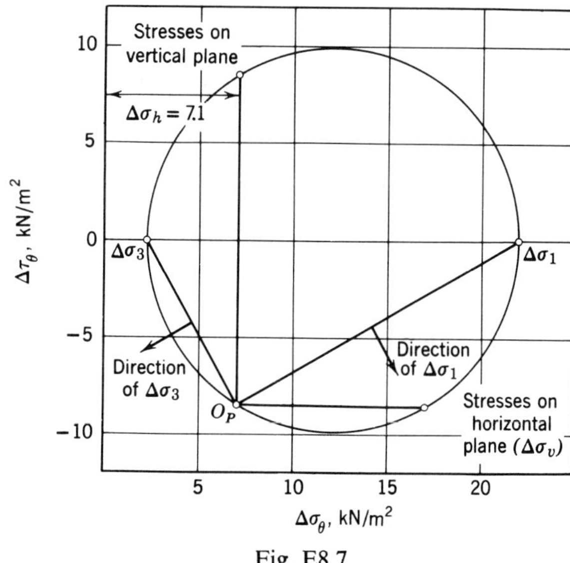

Fig. E8.7

Question for student. In order to construct the diagram, it was necessary to assume that the shear stress was negative on the horizontal plane. One way to test this assumption is to ask whether the directions of the principal stress increments are reasonable. Are they? ◀

▶ **Example 8.8**

On a p-q diagram, represent the states of stress given in Examples 8.4 to 8.6.
Solution. See Fig. E8.8. ◀

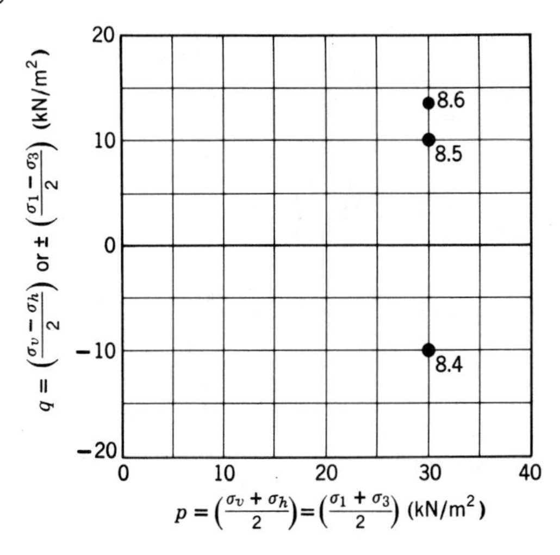

Fig. E8.8

8.6 STRESS PATHS

We shall often wish to depict the successive states of stress that exist in a specimen as the specimen is loaded. One way to do this is to draw a series of Mohr circles. For example, Fig. 8.10a shows successive states as σ_1 is increased with σ_3 constant. However, a diagram with many circles can become quite confusing, especially if the results of several tests are plotted on the same diagram. A more satisfactory arrangement is to plot a series of stress points, and to connect these points with a line or curve (Fig. 8.10b). Such a line or curve is called a *stress path*. Just as a Mohr circle or a stress point represents a state of stress, a stress path gives a continuous representation of successive states of stress.[6] Figure 8.11 shows a variety of stress paths that will be of interest to us in following chapters.

Figure 8.11a shows stress paths starting from a condition where $\sigma_v = \sigma_h$. This is a common initial condition in many types of laboratory tests. From this initial condition, we commonly either change σ_v and σ_h by the same amount ($\Delta\sigma_v = \Delta\sigma_h$), or else change one of the principal stresses while holding the other principal stress constant ($\Delta\sigma_v$ positive while $\Delta\sigma_h = 0$, or $\Delta\sigma_h$ negative while $\Delta\sigma_v = 0$). Of course many other stress paths are

[6] The terms *stress trajectory* and *vector curve* are also used to denote curves depicting successive states of stress, but the definitions of these other curves are somewhat different.

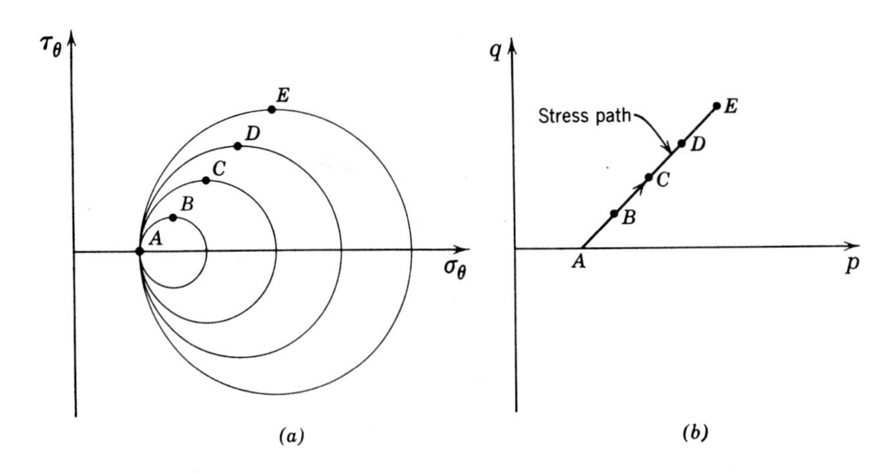

Fig. 8.10 Representation of successive states of stress as σ_1 increases with σ_3 constant. Points A, B, etc., represent the same stress conditions in both diagrams. (a) Mohr circles. (b) p-q diagram.

Major division		Field Identification Procedures (Excluding particles larger than 75 μm and basing fractions on estimated weights)	Group Symbols [a]	Typical Names	Information Required for Describing Soils	Laboratory Classification Criteria	
Coarse-grained soils: More than half of material is *larger* than 75 μm sieve size[b]. Gravels: More than half of coarse fraction is larger than 4 mm sieve size	Clean gravels (little or no fines)	Wide range in grain size and substantial amounts of all intermediate particle sizes	*GW*	Well graded gravels, gravel-sand mixtures, little or no fines	Give typical name; indicate approximate percentages of sand and gravel; maximum size; angularity, surface condition, and hardness of the coarse grains; local or geologic name and other pertinent descriptive information; and symbols in parentheses. For undisturbed soils add information on stratification, degree of compactness, cementation, moisture conditions and drainage characteristics. Example: *Silty sand*, gravelly; about 20% hard, angular gravel particles 12 mm maximum size; rounded and subangular sand grains coarse to fine, about 15% non-plastic fines with low dry strength; well compacted and moist in place; alluvial sand; (*SM*)	Use grain size curve in identifying the fractions as given under field identification. Determine percentages of gravel and sand from grain size curve. Depending on percentage of fines (fraction smaller than 75 μm sieve size) coarse grained soils are classified as follows: Less than 5% *GW, GP, SW, SP*; More than 12% *GM, GC, SM, SC*; 5% to 12% *Borderline* cases requiring use of dual symbols	$C_U = \frac{D_{60}}{D_{10}}$ Greater than 4; $C_C = \frac{(D_{30})^2}{D_{10} \times D_{60}}$ Between 1 and 3
		Predominantly one size or a range of sizes with some intermediate sizes missing	*GP*	Poorly graded gravels, gravel-sand mixtures, little or no fines			Not meeting all gradation requirements for *GW*
	Gravels with fines (appreciable amount of fines)	Nonplastic fines (for identification procedures see *ML* below)	*GM*	Silty gravels, poorly graded gravel-sand-silt mixtures			Atterberg limits below "A" line, or *PI* less than 4 — Above "A" line with *PI* between 4 and 7 are *borderline* cases requiring use of dual symbols
		Plastic fines (for identification procedures, see *CL* below)	*GC*	Clayey gravels, poorly graded gravel-sand-clay mixtures			Atterberg limits above "A" line, with *PI* greater than 7
Sands: More than half of coarse fraction is smaller than 4 mm sieve size	Clean sands (little or no fines)	Wide range in grain sizes and substantial amounts of all intermediate particle sizes	*SW*	Well graded sands, gravelly sands, little or no fines			$C_U = \frac{D_{60}}{D_{10}}$ Greater than 6; $C_C = \frac{(D_{30})^2}{D_{10} \times D_{60}}$ Between 1 and 3
		Predominantly one size or a range of sizes with some intermediate sizes missing	*SP*	Poorly graded sands, gravelly sands, little or no fines			Not meeting all gradation requirements for *SW*
	Sands with fines (appreciable amount of fines)	Nonplastic fines (for identification procedures, see *ML* below)	*SM*	Silty sands, poorly graded sand-silt mixtures			Atterberg limits below "A" line or *PI* less than 5 — Above "A" line with *PI* between 4 and 7 are *borderline* cases requiring use of dual symbols
		Plastic fines (for identification procedures, see *CL* below)	*SC*	Clayey sands, poorly graded sand-clay mixtures			Atterberg limits below "A" line with *PI* greater than 7

(The 75 μm sieve size is about the smallest particle visible to naked eye)

Identification Procedures on Fraction Smaller than 380 μm Sieve Size

Major division		Dry Strength (crushing characteristics)	Dilatancy (reaction to shaking)	Toughness (consistency near plastic limit)	Group Symbols [a]	Typical Names	Information Required for Describing Soils
Fine-grained soils: More than half of material is *smaller* than 75 μm sieve size	Silts and clays liquid limit less than 50	None to slight	Quick to slow	None	*ML*	Inorganic silts and very fine sands, rock flour, silty or clayey fine sands with slight plasticity	Give typical name; indicate degree and character of plasticity, amount and maximum size of coarse grains; colour in wet condition, odour if any, local or geologic name, and other pertinent descriptive information, and symbol in parentheses. For undisturbed soils add information on structure, stratification, consistency in undisturbed and remoulded states, moisture and drainage conditions. Example: *Clayey silt*, brown; slightly plastic; small percentage of fine sand; numerous vertical root holes; firm and dry in place; loess; (*ML*)
		Medium to high	None to very slow	Medium	*CL*	Inorganic clays of low to medium plasticity, gravelly clays, sandy clays, silty clays, lean clays	
		Slight to medium	Slow	Slight	*OL*	Organic silts and organic silt-clays of low plasticity	
	Silts and clays liquid limit greater than 50	Slight to medium	Slow to none	Slight to medium	*MH*	Inorganic silts, micaceous or diatomaceous fine sandy or silty soils, elastic silts	
		High to very high	None	High	*CH*	Inorganic clays of high plasticity, fat clays	
		Medium to high	None to very slow	Slight to medium	*OH*	Organic clays of medium to high plasticity	
Highly Organic Soils		Readily identified by colour, odour, spongy feel and frequently by fibrous texture			*Pt*	Peat and other highly organic soils	

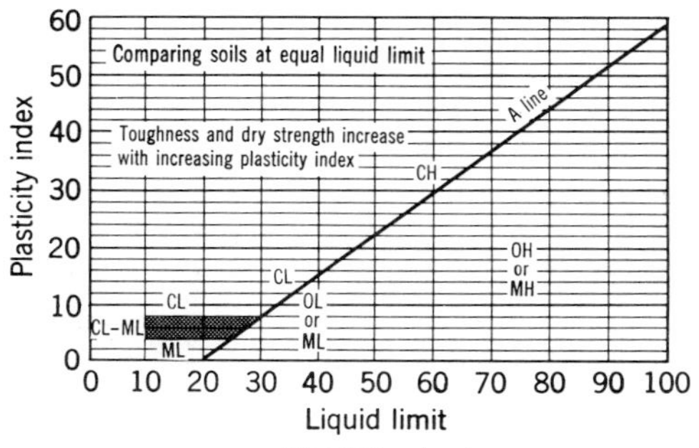

Comparing soils at equal liquid limit
Toughness and dry strength increase with increasing plasticity index
A line
CH
CL
CL–ML
ML
OL or ML
OH or MH
Plasticity index
Liquid limit
0 10 20 30 40 50 60 70 80 90 100
0 10 20 30 40 50 60

Plasticity chart for laboratory classification of fine grained soils

From Wagner, 1957.
[a] *Boundary classifications*. Soils possessing characteristics of two groups are designated by combinations of group symbols. For example *GW–GC*, well graded gravel-sand mixture with clay binder.
[b] All sieve sizes on this chart are U.S. standard.

Field Identification Procedure for Fine Grained Soils or Fractions

These procedures are to be performed on the minus 380 μm sieve size particles. For field classification purposes, screening is not intended, simply remove by hand the coarse particles that interfere with the tests.

Dilatancy (Reaction to shaking):
After removing particles larger than 380 μm sieve size, prepare a pat of moist soil with a volume of about 8000 mm³. Add enough water if necessary to make the soil soft but not sticky.
Place the pat in the open palm of one hand and shake horizontally, striking vigorously against the other hand several times. A positive reaction consists of the appearance of water on the surface of the pat which changes to a livery consistency and becomes glossy. When the sample is squeezed between the fingers, the water and gloss disappear from the surface, the pat stiffens and finally it cracks or crumbles. The rapidity of appearance of water during shaking and of its disappearance during squeezing assist in identifying the character of the fines in a soil.
Very fine clean sands give the quickest and most distinct reaction whereas a plastic clay has no reaction. Inorganic silts, such as a typical rock flour, show a moderately quick reaction.

Dry Strength (Crushing characteristics):
After removing particles larger than 380 μm sieve size, mould a pat of soil to the consistency of putty, adding water if necessary. Allow the pat to dry completely by oven, sun or air drying, and then test its strength by breaking and crumbling between the fingers. This strength is a measure of the character and quantity of the colloida fraction contained in the soil. The dry strength increases with increasing plasticity.
High dry strength is characteristic for clays of the CH group. A typical inorganic silt possesses only very slight dry strength. Silty fine sands and silts have about the same slight dry strength, but can be distinguished by the feel when powdering the dried specimen. Fine sand feels gritty whereas a typical silt has the smooth feel of flour.

Toughness (Consistency near plastic limit):
After removing particles larger than the 380 μm sieve size, a specimen of soil about 12 mm cube in size, is moulded to the consistency of putty. If too dry, water must be added and if sticky, the specimen should be spread out in a thin layer and allowed to lose some moisture by evaporation. Then the specimen is rolled out by hand on a smooth surface or between the palms into a thread about 3 mm in diameter. The thread is then folded and re-rolled repeatedly. During this manipulation the moisture content is gradually reduced and the specimen stiffens, finally loses its plasticity, and crumbles when the plastic limit is reached.
After the thread crumbles, the pieces should be lumped together and a slight kneading action continued until the lump crumbles.
The tougher the thread near the plastic limit and the stiffer the lump when it finally crumbles, the more potent is the colloidal clay fraction in the soil. Weakness of the thread at the plastic limit and quick loss of coherence of the lump below the plastic limit indicate either inorganic clay of low plasticity, or materials such as kaolin-type clays and organic clays which occur below the A-line.
Highly organic clays have a very weak and spongy feel at the plastic limit.

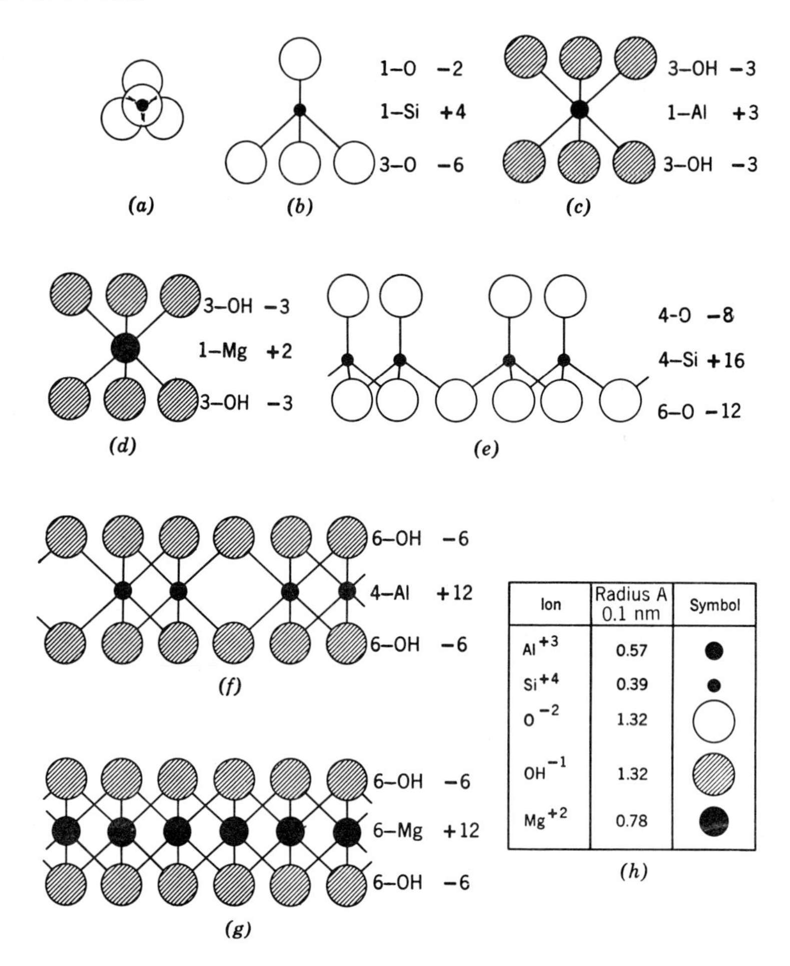

Ion	Radius A 0.1 nm	Symbol
Al^{+3}	0.57	
Si^{+4}	0.39	
O^{-2}	1.32	
OH^{-1}	1.32	
Mg^{+2}	0.78	

Fig. 4.6 Basic silicate units. (*a*) and (*b*) Silicon tetrahedron. (*c*) Aluminum octahedron. (*d*) Magnesium octahedron. (*e*) Silica. (*f*) Gibbsite. (*g*) Brucite.

mineral. The amount of this substitution needed to explain the charge on kaolinite is one substitution for every four hundredth silicon ion.

Since the basic structure of kaolinite consists of a layer of gibbsite on top of a layer of silicate, this mineral is called a "two-layer" mineral. Kaolinite is the most important and most common two-layer mineral encountered by the engineer. Halloysite, having essentially the same composition and structure as kaolinite, is an interesting and not uncommon member of the two-layer silicate group. The main difference between halloysite and kaolinite is the presence of water between the basic sheets of halloysite, which results in halloysite existing in tubular particles.

Three-Layer Sheets

The three-layer sheet minerals are formed by placing one silica on the top and one on the bottom of either a gibbsite or brucite sheet. Figure 4.9 shows the mineral pyrophyllite made of a gibbsite sandwiched between two silica sheets. Figure 4.10 shows the structure of the mineral muscovite, which is similar to pyrophyllite except that there has been isomorphous substitution of aluminum for silicon in muscovite. The net charge created by this substitution is balanced by potassium ions, which serve to link the three-layer sandwiches together, as indicated in the symbolic structure in Fig. 4.10.

The two most common three-layer structures in soil are montmorillonite and illite type minerals. Montmorillonite has a structure similar to pyrophyllite with the exception that there has been isomorphous substitution of magnesium for aluminum in the gibbsite sheet.

Figure 4.11 gives a summary of the sheet silicate minerals of importance to the civil engineer.

Frameworks

Quartz, a framework silicate structure, has a very low ratio of oxygen to silicon (2:1), as noted in Table 4.2. It is thus one of the most weather resistant minerals. The feldspars have higher oxygen to silicon ratios (2.7 to 4.0)

mass. Thus for most problems the value of ϕ based upon the peak of the stress-strain curve is properly used to represent the strength of the sand. There are some problems in which large strains occur, as when one must evaluate the resistance encountered by a tracked vehicle as it plows its way through a sand mass. For such problems, it would be appropriate to use ϕ_{cv} to represent the strength of the sand.

The foregoing comments apply to the internal strength of a sand. The engineer frequently needs to know the frictional resistance between sand and the surface of some structure, such as a retaining wall or pile. If this surface is very smooth, as in the case of sand sliding on unrusted steel, the friction angle is most likely equal to ϕ_μ for the sand. If the surface is at all rough, such as a typical concrete surface, then the friction between the surfaces probably approaches ϕ_{cv}.

Table 11.1 summarizes these recommendations concerning the type of friction angle that should be used for various situations. Values of ϕ_μ have already been presented in Chapter 6. Typical values for ϕ and ϕ_{cv} appear in this chapter. Subsequent chapters which treat engineering applications in detail will have still more to say about the choice of a friction angle for use in a particular situation.

11.3 EFFECT OF VARIOUS LOADING CONDITIONS

Intermediate Principal Stress

With the normal form of triaxial test (specimen failed by increasing axial stress while holding confining pressure constant), the intermediate principal stress is equal to the minor principal stress: $\sigma_2 = \sigma_3$ (stress path for vertical compression loading in Fig. 9.8). As shown in Fig. 9.8, a specimen can be failed in vertical extension, in which case $\sigma_2 = \sigma_1$.

Numerous investigators have compared the friction angle from compression with that from extension tests, with various results; see Roscoe (1963) et al. for a summary. Most investigators have concluded that the friction angle is the same for both cases, but a few have found that ϕ was greater, by several degrees, if $\sigma_2 = \sigma_1$ as compared with the results for $\sigma_2 = \sigma_3$ (as is the case in Figs. 10.20 and 10.22).

Figure 11.10 shows the results of a set of plane-strain tests; these are tests in which the sand can strain only in the axial direction and one lateral direction while its dimension remains fixed in the other lateral direction. The friction angle from these plane-strain tests exceeded the angle as obtained from conventional triaxial tests by as much as 4° for the densest specimens. Little or no difference in ϕ values was observed on loose specimens.

A plane-strain condition is often encountered in engineering practice problems, and for many problems

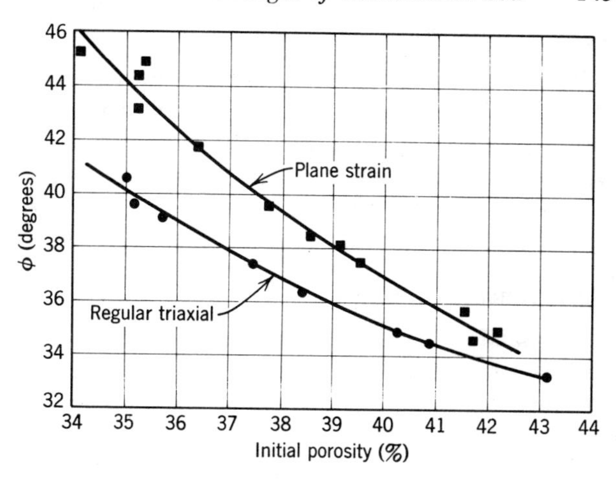

Fig. 11.10 Results of regular and plane-strain triaxial tests (from Cornforth, 1964).

a plane-strain test is more realistic than is the triaxial test. The authors feel that the plane-strain form of triaxial test will become increasingly popular with practicing engineers as well as with researchers.

The reason for the increased resistance in the plane-strain condition presumably comes about because the soil particles are given less freedom in the ways that they can move around their neighbors so as to overcome interlocking. A failure law in three dimensions is now needed. The possible form of such a law has been discussed numerous times (Kirkpatrick, 1957; Haythornthwaite, 1960), but the matter is still unresolved. Special testing devices, which permit greater flexibility in the types of loads that are applied, are needed for use in research to clarify the nature of the three-dimensional failure law.

Failure with Decreasing Stresses

In problems such as retaining walls (under active conditions), the soil fails as the result of decreasing stresses rather than increasing stresses; i.e., the stress path is more like that marked E in Fig. 8.11 or that for vertical compression unloading in Fig. 9.8, and those for tests 3, 4, and 5 in Fig. 10.20. As indicated in Fig. 10.22, the friction angle for unloading is virtually the same as for loading.

Rate of Loading

The friction angle of sand, as measured in triaxial compression, is substantially the same whether the sand is loaded to failure in 5 millisec or 5 min. The increase in tan ϕ from the slower to the faster loading rate is at most 10%, and probably is only 1–2% (see Whitman and Healy, 1963). It is possible that the effect might be somewhat greater if the sand is sheared in plane strain or if the confining pressure is in excess of 700 kN/m².

Vibrations and Repeated Loadings

Repeated loadings, whether changing slowly or quickly, can cause ϕ to change. A loose sand will densify, with resulting strength increase, and a dense sand can expand, with resulting strength decrease. A stress smaller than the static failure stress can cause very large strains if the load is applied repeatedly (see Seed and Chan, 1961).

Slight Amount of Moisture

Any sand, unless it has just been intentionally dried, possesses a small moisture content. The presence of this moisture can have some effect upon the mineral-to-mineral friction angle (see Chapter 6). However, since both shear tests and most practical situations really involve either air-dry or saturated sand, the presence of this small amount of moisture need seldom be taken into account.

Moisture can also introduce an apparent cohesion between particles by capillarity. In some situations, such as in model tests, this cohesion can be a significant component of strength. In practical problems, this small cohesion is of no consequence.

Testing Errors

Chapter 9 mentioned some of the errors which can develop in triaxial tests and in direct shear tests. The common tests can give rise to an error of as much as 2° in the measurement of the peak friction angle ϕ. Nonetheless, these tests suffice for most engineering purposes. For careful measurement of strength and volume change in research work, it is essential to use the improved devices.

Summary

This section has indicated that many factors have an influence on the friction angle of granular soils. Using the ordinary laboratory test, the measured value of ϕ may differ by several degrees from the friction angle actually available within the ground, even if the initial void ratio has been chosen accurately. If a more accurate evaluation of ϕ is needed, special care must be taken to establish the loading condition actually existing in the ground and to duplicate this condition in the laboratory by means of special tests.

11.4 EFFECT OF COMPOSITION

This section considers the effect of composition on (*a*) the ϕ versus e_0 relation for a small range of confining stresses, and (*b*) the change in ϕ over a wide range of confining stresses. Even when confining stresses are limited to conventional magnitudes (less than 700 kN/m²) the ϕ versus e_0 relation falls within a broad band as shown in Fig. 11.11. Since the value of ϕ_μ varies relatively little between various particle sizes or various

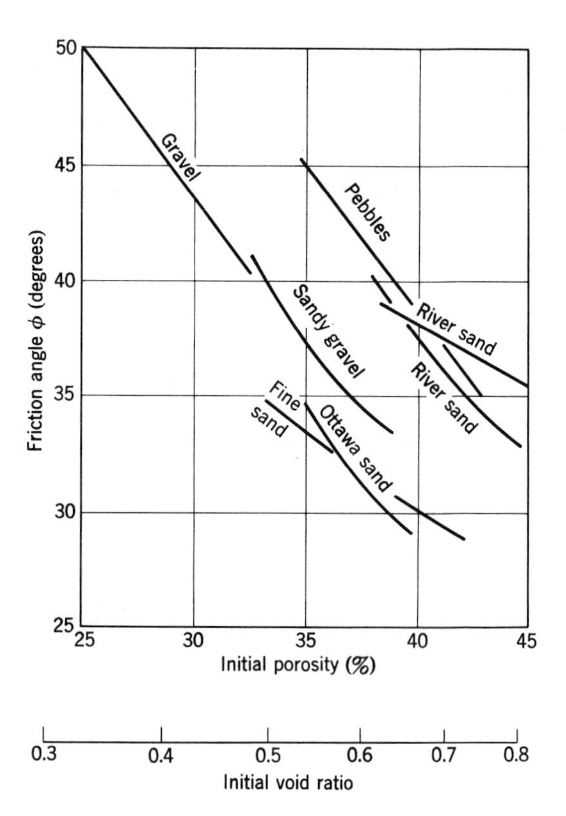

Fig. 11.11 Friction angle versus initial porosity for several granular soils.

minerals, these differences in ϕ for a given e_0 result primarily from different degrees of interlocking.

Composition affects the friction angle of a granular soil in two ways. First it affects the void ratio that is obtained with a given compactive effort, and second it affects the friction angle that is achieved for that void ratio. The effect of composition might be studied either by comparing friction angles at fixed e_0 or at fixed compactive effort. Because the role of composition is most important with regard to embankment construction the comparisons are often made at fixed compactive effort.

Average Particle Size

Figure 11.12*b* shows data for five soils all having a uniformity coefficient of 3.3, but having different average particle sizes. For a given compactive effort, these sands achieve different void ratios. However, the friction angle was much the same for each sand. The effect of the greater initial interlocking in the sand with the largest particles is compensated by the greater degree of grain crushing and fracturing that occurs with the larger particles because of the greater force per contact.

Crushing of particles, and the consequent curvature of the Mohr envelope, is most important with large particles, especially gravel-sized particles or rock fragments used for rockfills. This is because increasing the

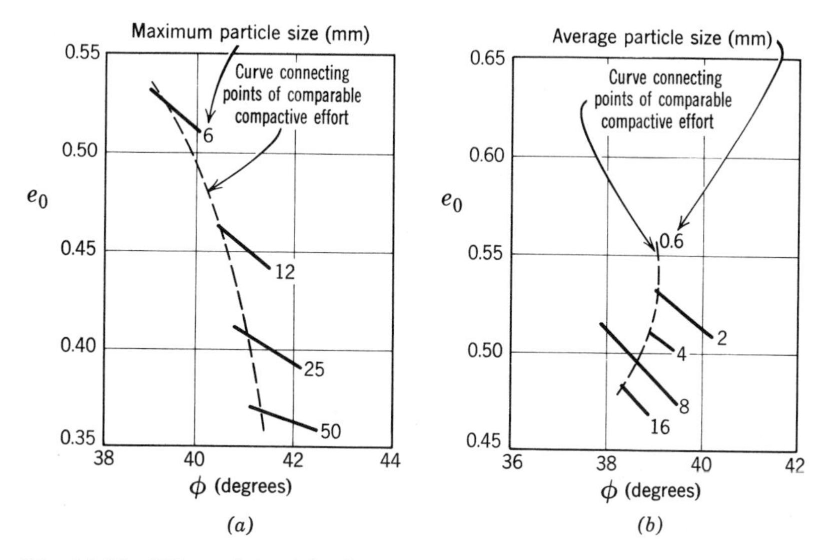

Fig. 11.12 Effect of particle size and gradation on friction angle. (*a*) Soils with same minimum particle size (0.5 mm). (*b*) Soils with same uniformity coefficient. Data from Leslie (1963).

particle size increases the load per particle, and hence crushing begins at a smaller confining stress. Spurred by the increasing popularity of rockfill dams, several laboratories have constructed triaxial testing systems which can accommodate specimens as large as 300 mm in diameter. An apparatus that can test specimens 1.13 m in diameter and 2.5 m long has been constructed in Mexico (Marsal, 1963).

Grading of the Sand

Figure 11.12*a* shows data for four soils all having the same minimum particle size but different maximum particle sizes. For comparable compactive efforts, the better graded sand has both a smaller initial void ratio and a larger friction angle. It is apparent that a better distribution of particle sizes produces a better interlocking. This trend is also shown by the data in Table 11.2, and is further confirmed by a series of tests reported by Holtz and Gibbs (1956).

In many soils, a few particles of relatively large size make up a large fraction of the total weight of the soil.

Table 11.2 Effect of Angularity and Grading on Peak Friction Angle

Shape and Grading	Loose	Dense
Rounded, uniform	30°	37°
Rounded, well graded	34°	40°
Angular, uniform	35°	43°
Angular, well graded	39°	45°

From Sowers and Sowers, 1951.

If these particles are numerous enough so that they interlock with each other, it is important that these large particles be present in the test specimen. However, if these larger particles are just embedded in a matrix of much smaller particles so that the shearing takes place through the matrix, then the large particles can safely be omitted from the specimen. Unfortunately, the profession still is lacking definitive guides as to what constitutes a satisfactory test upon a gravelly soil.

A well-graded soil experiences less breakdown than a uniform soil of the same particle size, since in a well-graded soil there are many interparticle contacts and the load per contact is thus less than in the uniform soil. Figure 11.13 illustrates that the better graded soil suffers less decrease in ϕ with increasing confining pressure.

Angularity of Particles

It would be expected that angular particles would interlock more thoroughly than rounded particles, and hence that sands composed of angular particles would have the larger friction angle. The data for peak friction angle presented in Table 11.2 confirm this prediction. Even when a sand is strained to its ultimate condition, so that no further volume change is taking place and the sand is in a loose condition, the sand with the angular particles has the greater friction angle. In gravels, the effect of angularity is less because of particle crushing.

Mineral Type

Unless a sand contains mica, it makes little difference whether the sand is composed primarily of quartz, one of the feldspars, etc. A micaceous sand will often have a large void ratio, and hence little interlocking and a low

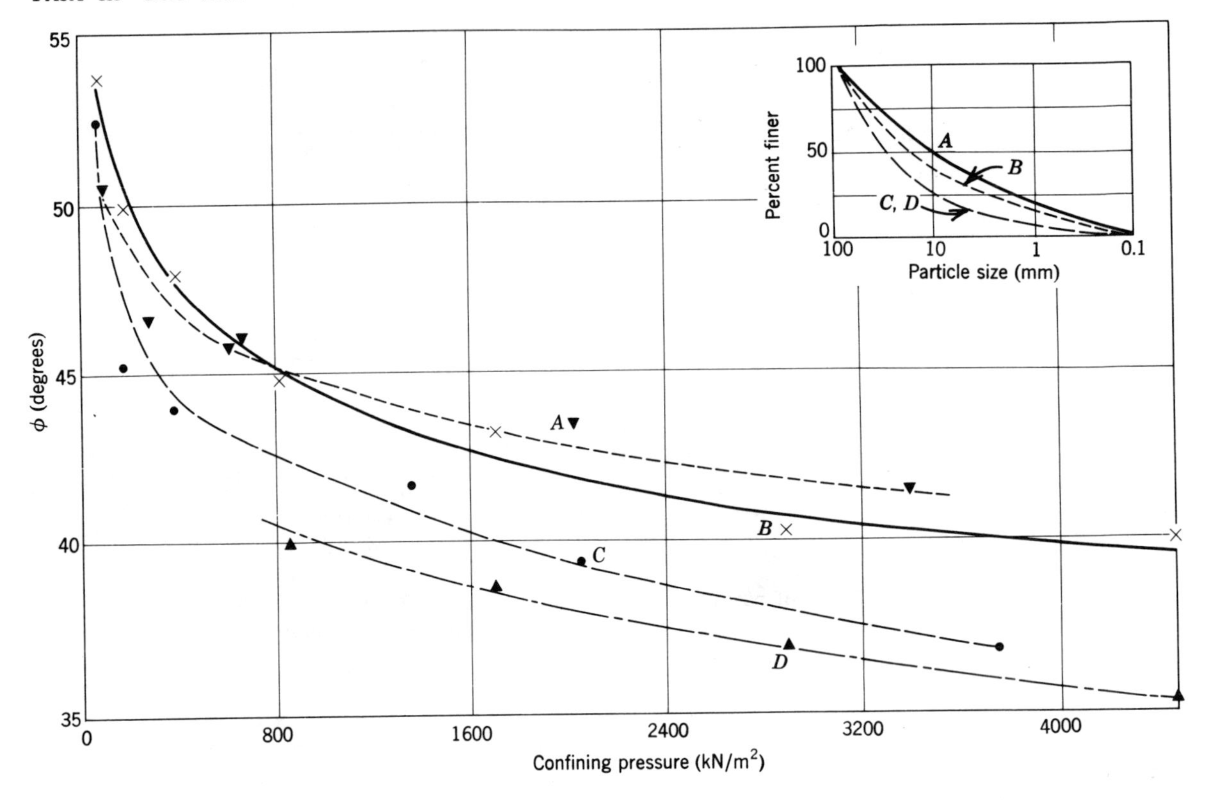

Fig. 11.13 Friction angle versus confining pressure (data from Leslie 1963).

friction angle. The smaller value of ϕ_μ for mica compared to that of quartz has relatively little to do with this result.

Tests (Horn and Deere, 1962) have been carried out using powdered mica with care taken to have the mica flakes oriented nearly parallel. The result was a friction angle (ϕ_{cv}) of 16°, compared to $\phi_\mu = 13\frac{1}{2}°$. There is some small amount of interlocking in such a case.

Where particles of gravel are an important constituent of soil, the origin of the gravel particles can have an important effect. If the gravel particles are relatively soft, crushing of these particles will minimize the interlocking effect and decrease the friction angle as compared to a comparable soil with hard gravel particles.

Summary

The composition of a granular soil can have an important influence upon its friction angle, indirectly by influencing e_0 and directly by influencing the amount of interlocking that occurs for a given e_0. Table 11.3 provides a summary of data that can be used for preliminary design. However, for final design of an embankment, the actual soil should be tested using the void ratio and stress system that will exist in the field.

11.5 DETERMINATION OF *IN SITU* FRICTION ANGLE

The results presented in the foregoing sections have emphasized the predominant role of the degree of interlocking upon magnitude of the friction angle. Thus, if we wish to determine the friction angle of a sand *in situ*, it is not enough to find the nature and shape of the particles composing the sand. It is essential to know how tightly together these particles are packed in their natural state.

It is extremely difficult to obtain samples of a sand without changing the porosity. Thus, except for

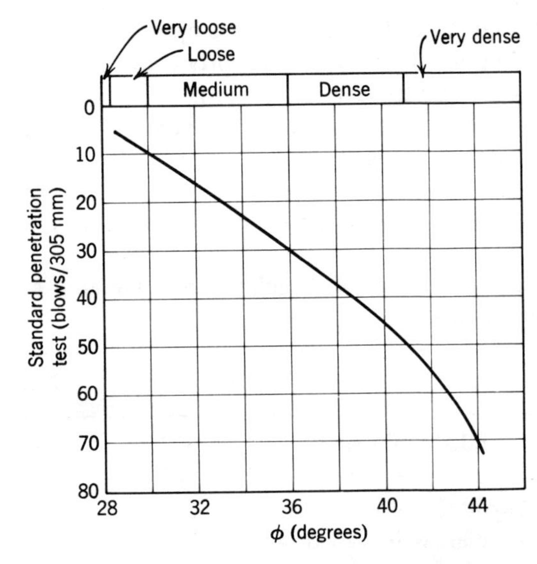

Fig. 11.14 Correlation between friction angle and penetration resistance (From Peck, Hanson, and Thornburn, 1953).

Table 11.3 Summary of Friction Angle Data for Use in Preliminary Design

	Friction Angles							
	Slope Angle of Repose		At Ultimate Strength		At Peak Strength			
					Medium Dense		Dense	
Classification	$i(^\circ)$	Slope (vert. to hor.)	$\phi_{cv}(^\circ)$	$\tan \phi_{cv}$	$\phi(^\circ)$	$\tan \phi$	$\phi(^\circ)$	$\tan \phi$
Silt (nonplastic)	26 to 30	1 on 2 1 on 1.75	26 to 30	0.488 0.577	28 to 32	0.532 0.625	30 to 34	0.577 0.675
Uniform fine to medium sand	26 to 30	1 on 2 1 on 1.75	26 to 30	0.488 0.577	30 to 34	0.577 0.675	32 to 36	0.675 0.726
Well-graded sand	30 to 34	1 on 1.75 1 on 1.50	30 to 34	0.577 0.675	34 to 40	0.675 0.839	38 to 46	0.839 1.030
Sand and gravel	32 to 36	1 on 1.60 1 on 1.40	32 to 36	0.625 0.726	36 to 42	0.726 0.900	40 to 48	0.900 1.110

From B. K. Hough, *Basic Soils Engineering*. Copyright © 1957, The Ronald Press Company, New York.
Note. Within each range, assign lower values if particles are well rounded or if there is significant soft shale or mica content, higher values for hard, angular particles. Use lower values for high normal pressures than for moderate normal pressure.

problems involving man-made fills, it is difficult to either measure or estimate the friction angle of a sand on the basis of laboratory tests alone. For these reasons, extensive use is made in practice of correlations between the friction angle of a sand and the resistance of the natural sand deposit to penetration.

Figure 11.14 shows an empirical correlation between the resistance offered to the standard penetration spoon (Chapter 7) and the friction angle. Inevitably, any such correlation is crude. The actual friction angle may deviate by $\pm 3^\circ$ or more from the value given by the curve. The given relation is intended to apply for depths of overburden up to 12 m, and is conservative for greater depths.

11.6 SUMMARY OF MAIN POINTS

1. The strength of soil can be represented by a Mohr envelope, which is a plot of τ_{ff} versus σ_{ff}. Generally the Mohr envelope of a granular soil is curved. For stresses less than 700 kN/m², the envelope usually is almost straight so that

$$\tau_{ff} = \sigma_{ff} \tan \phi$$

where ϕ is the friction angle corresponding to the peak point of the stress-strain curve.

2. The value of ϕ for any soil depends on ϕ_μ and upon the amount of interlocking; i.e., the initial void ratio and σ_{ff}.
3. Where sand is being subjected to very large strains, ϕ_{cv} should be used in the failure law. Unless the sand is very loose, ϕ_{cv} will be less than ϕ. Where the sand is sliding over the surface of a structure, the friction angle will vary from ϕ_μ to ϕ_{cv}, depending on the smoothness of the surface.
4. A knowledge of the effect of composition helps guide the selection of materials to be used in man-made fills
5. Materials to be used in man-made fills should be tested using the actual range of confining pressures which will be encountered in the fill.
6. For many practical problems, the friction angle of an *in situ* sand deposit can be determined by indirect means, such as the standard penetration test.

PROBLEMS

11.1 Given the following triaxial test data, plot the results (*a*) in a Mohr diagram and (*b*) in a p–q diagram, and determine ϕ by each method.

σ_3 (kN/m²)	Peak σ_1 (kN/m²)
68.9	190.3
137.9	375.8
206.8	579.2
275.8	758.4
344.7	958.4
413.7	1144.6

11.2 Suppose you had a sample of the sand used to obtain the test results shown in Fig. 10.18. This sand is at a void ratio of 0.7. For $\sigma_3 = 137.9$ kN/m², estimate:

a. The peak value of σ_1.
b. The ultimate value of σ_1.
c. The void ratio after considerable shearing.

11.3 Draw the stress path for the test on the loose specimen in Fig. 10.18.

11.4 A sand with $\phi = 30°$ is sheared in a triaxial test with $\sigma_1 = \sigma_3 = 138$ kN/m² initially. Both σ_1 and σ_3 are increased, with $\Delta\sigma_3 = \Delta\sigma_1/4$. What is the maximum value of σ_1 reached during the test?

11.5 A sand with a friction angle of 40° is tested in direct shear, using a normal stress of 345 kN/m². Making the simplest possible assumption concerning the stress condition within the shear box, determine how much shear stress must be applied before the sand will fail.

11.6 The blow count during a standard penetration test upon a sand at 6 m depth is 20 blows/305 mm. Estimate the friction angle of the sand. Suppose the blow count at 12 m depth is exactly the same. Is the sand at 12 m depth looser or denser or the same density as the sand at 6 m depth? Explain your answer.

11.7 Suppose two sandy soils are compacted with the same compactive effort. Sand *A* is uniform and has rounded particles. Sand *B* is well graded with angular particles

a. Which sand will have the larger void ratio?
b. Which sand will have the larger friction angle?

11.8 Estimate the value of ϕ for the following soils. Indicate which figures or tables you used to guide your estimate.

a. A well-graded sand to be densely compacted for a low embankment.
b. A gravel, with less than 20% sand sizes to be used for a rockfill dam 150 m high.
c. A natural deposit of fine sand, of medium density, which is to support a building.

11.9 Derive the relationships given in Fig. 11.6*b*.
Hint. Draw a Mohr circle and show both the Mohr envelope and K_f-line on this same diagram.

CHAPTER 12

Stress-Strain Relationships

Once an engineer has satisfied himself that a soil mass is not going to fail totally, he generally must then ascertain the amount of movement that will result from the application of loads and decide whether this movement is permissible. To do this, the engineer requires a stress-strain relationship for soil.

From our general study of stress-strain behavior in Chapter 10, we know that this behavior can be very complex. The amount of strain caused by a stress will depend on the composition, void ratio, past stress history of the soil, and manner in which the stress is applied. An equation giving the stress-strain relationship of one sand for any loading with constant direction of principal stresses has been developed by Hansen (1966). However, this expression is extremely complicated. Usually it is preferable to use formulas and data that are adapted to the particular problem at hand.

For many problems, the best approach often is to measure directly the strains produced in a laboratory test using stresses that will occur in the actual soil mass. This approach will be discussed in Chapter 14.

For other problems, it helps greatly to use concepts and formulas from the theory of elasticity. This means that the actual nonlinear stress-strain curves of a soil must be "linearized", i.e., replaced by straight lines. Then one speaks in terms of the *modulus* and *Poisson's ratio* of soil. Obviously, modulus and Poisson's ratio are not constants for a soil, but rather are quantities which approximately describe the behavior of a soil for a particular set of stresses. Different values of modulus and Poisson's ratio will apply for any other set of stresses. Especially when speaking of modulus, one must be very careful to specify what is meant.

The terms *tangent modulus* and *secant modulus* are used frequently. Tangent modulus is the slope of a straight line drawn tangent to a stress-strain curve at a particular point on the curve (see Fig. 12.1). The value of tangent modulus will vary with the point selected. The tangent modulus at the initial point of the curve is the *initial tangent modulus*. Secant modulus is the slope of a straight line connecting two separate points of the curve. The value of secant modulus will vary with the locations of both points. As the two points come closer together, the secant modulus becomes equal to the tangent modulus. For a truly linear material, all of these values of modulus are one and the same.

12.1 CONCEPTS FROM THE THEORY OF ELASTICITY

If we apply a uniaxial stress σ_z to an elastic[1] cylinder (Fig. 12.1), there will be a vertical compression and a lateral expansion such that

$$\epsilon_z = \frac{\sigma_z}{E} \tag{12.1}$$

$$\epsilon_x = \epsilon_y = -\mu\epsilon_z \tag{12.2}$$

where

$\epsilon_x, \epsilon_y, \epsilon_z$ = strains in the x, y, z directions, respectively (plus when compressive)
E = *Young's modulus* of elasticity
μ = *Poisson's ratio*

If shear stresses τ_{zx} are applied to an elastic cube, there will be a shear distortion such that

$$\gamma_{zx} = \frac{\tau_{zx}}{G} \tag{12.3}$$

where G = *shear modulus*. Equations 12.1 to 12.3 define the three basic constants of the theory of elasticity: E, G, and μ. Actually only two of these constants are needed, since

$$G = \frac{E}{2(1+\mu)} \tag{12.4}$$

[1] The word "elastic" actually denotes an ability of a material to recover its original size and shape after removal of stress. In this book, we use the word in a more restrictive sense to mean a material having a linear, reversible stress-strain curve.

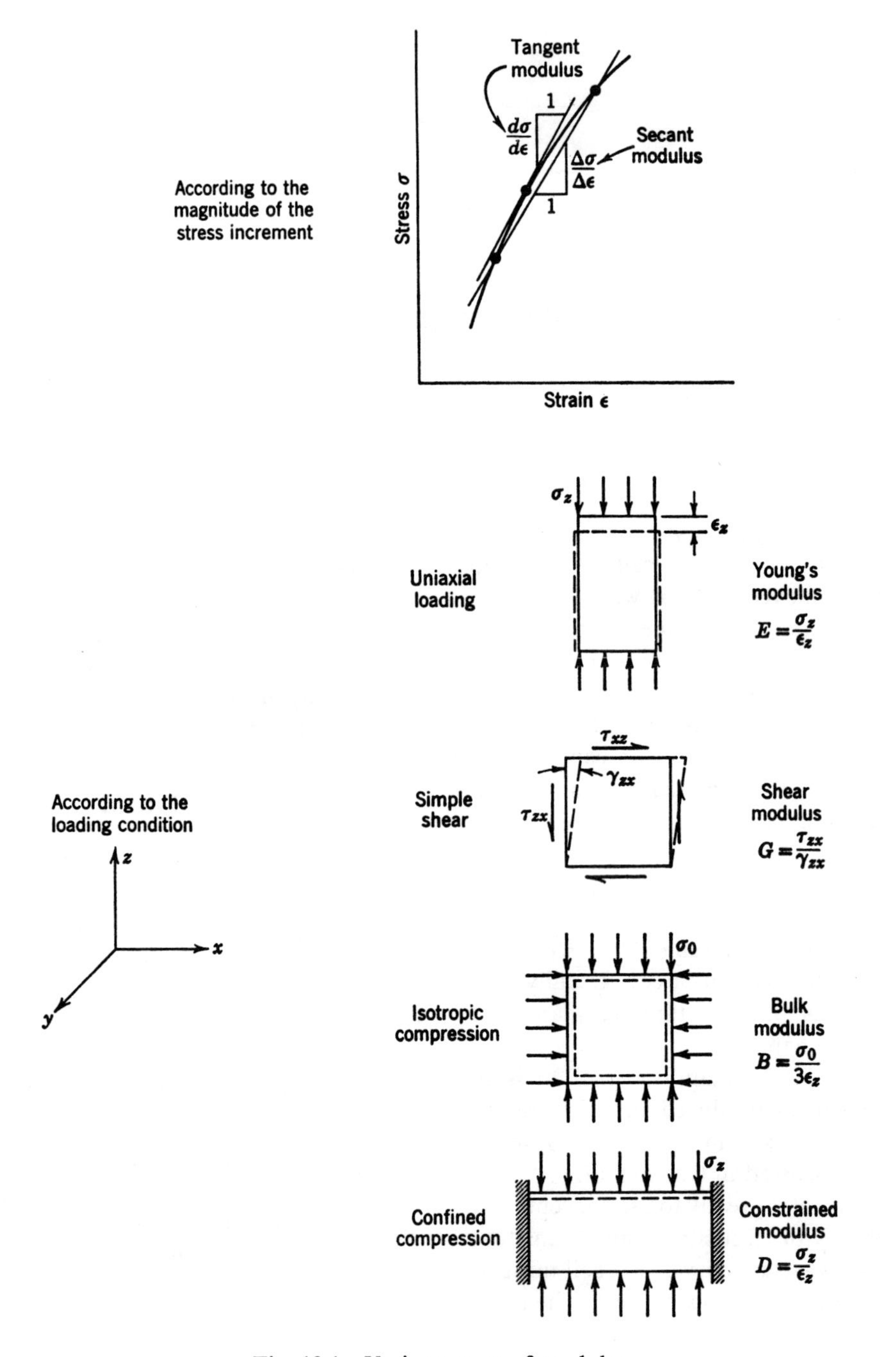

Fig. 12.1 Various types of modulus.

For an elastic material with all stress components acting, we can employ the principle of superposition to obtain

$$\epsilon_x = \frac{1}{E}[\sigma_x - \mu(\sigma_y + \sigma_z)] \tag{12.5a}$$

$$\epsilon_y = \frac{1}{E}[\sigma_y - \mu(\sigma_z + \sigma_x)] \tag{12.5b}$$

$$\epsilon_z = \frac{1}{E}[\sigma_z - \mu(\sigma_x + \sigma_y)] \tag{12.5c}$$

$$\gamma_{xy} = \frac{\tau_{xy}}{G} \tag{12.5d}$$

$$\gamma_{yz} = \frac{\tau_{yz}}{G} \tag{12.5e}$$

$$\gamma_{zx} = \frac{\tau_{zx}}{G} \tag{12.5f}$$

The volumetric strain is

$$\frac{\Delta V}{V} = \epsilon_x + \epsilon_y + \epsilon_z \tag{12.5g}$$

For the special case where $\sigma_x = \sigma_y = \sigma_z = \sigma_0$ and $\tau_{xy} = \tau_{yz} = \tau_{zx} = 0$, the volumetric strain equals

$$\frac{\Delta V}{V} = \frac{3\sigma_0}{E}(1 - 2\mu)$$

The *bulk modulus B* is defined as

$$B = \frac{\sigma_0}{\Delta V/V} = \frac{E}{3(1 - 2\mu)} \tag{12.6}$$

Still another special type of modulus is the *constrained modulus*, *D*, which is the ratio of axial stress to axial strain for confined compression (Fig. 12.1). This modulus can be computed from Eqs. 12.5 by setting $\epsilon_x = \epsilon_y = 0$. Thus

$$\sigma_x = \sigma_y = \frac{\mu}{1 - \mu}\sigma_z \tag{12.7}$$

$$D = \frac{E(1 - \mu)}{(1 + \mu)(1 - 2\mu)} \tag{12.8}$$

Uniaxial loading and confined compression involve both shear strain and volume change. This important fact is demonstrated in Example 12.1.

▸ Example 12.1

Find. Volumetric strain ($\Delta V/V$) and maximum shear strain during (*a*) uniaxial loading, (*b*) confined compression.

Solution.

Condition	Volumetric	Shear
Uniaxial loading	$\frac{\Delta V}{V} = \epsilon_x + \epsilon_y + \epsilon_z = (1 - 2\mu)\frac{\sigma_z}{E}$	$\tau_{\max} = \frac{\sigma_z}{2}$, $\gamma_{\max} = \frac{\sigma_z}{2G}$
Confined compression	$\frac{\Delta V}{V} = \epsilon_x + \epsilon_y + \epsilon_z = \frac{(1 + \mu)(1 - 2\mu)\sigma_z}{E(1 - \mu)}$	$\tau_{\max} = \frac{\sigma_z}{2}\frac{(1 - 2\mu)}{1 - \mu}$, $\gamma_{\max} = \frac{\sigma_z}{2G}\frac{(1 - 2\mu)}{1 - \mu}$

Note. The volumetric strain becomes zero for $\mu = \frac{1}{2}$. $\tau_{\max}$ occurs on planes inclined at 45° to the horizontal. $\gamma_{\max}$ occurs for an element whose sides are at 45° to the horizontal. ◂

For an elastic material, the foregoing equations apply for increments of stress starting from some initial stress, as well as for increments of stress starting from zero stress. Example 12.2 derives equations which may be used to find E and μ from measured strains.

▸ Example 12.2

Given. Strains $\Delta\epsilon_x = \Delta\epsilon_y$, $\Delta\epsilon_z$ caused by stresses $\Delta\sigma_x = \Delta\sigma_y$, $\Delta\sigma_z$ upon a cylinder of an elastic material.

Find. Expressions for Young's modulus and Poisson's ratio.

Solution. Eqs. 12.5a and 12.5c become

$$E\Delta\epsilon_x = \Delta\sigma_x - \mu(\Delta\sigma_x + \Delta\sigma_z)$$
$$E\Delta\epsilon_z = \Delta\sigma_z - 2\mu\Delta\sigma_x$$

These may be solved to give

$$E = \frac{(\Delta\sigma_z + 2\Delta\sigma_x)(\Delta\sigma_z - \Delta\sigma_x)}{\Delta\sigma_x(\Delta\epsilon_z - 2\Delta\epsilon_x) + \Delta\sigma_z\Delta\epsilon_z}$$

$$\mu = \frac{\Delta\sigma_x\Delta\epsilon_z - \Delta\epsilon_x\Delta\sigma_z}{\Delta\sigma_x(\Delta\epsilon_z - 2\Delta\epsilon_x) + \Delta\sigma_z\Delta\epsilon_z}$$ ◂

Wave Velocities

The velocity of wave propagation, or simply *wave velocity*, is defined as the distance moved by a wave in a unit of time (Fig. 12.2). There are several different wave velocities, each corresponding to a wave involving different types of strain:

Rod velocity $C_L = \sqrt{E/\rho}$ (12.9*a*)

Shear velocity $C_S = \sqrt{G/\rho}$ (12.9*b*)

Dilatational velocity $C_D = \sqrt{D/\rho}$ (12.9*c*)

where

ρ = mass density, equal to γ/g
g = acceleration of gravity
C_L and C_D = velocities of compressive waves for uniaxial loading and confined compression, respectively

Because of these simple relationships between modulus and velocity, velocity is often measured and used to evaluate modulus.

12.2 BEHAVIOR DURING CONFINED COMPRESSION

Figure 10.5 gives a typical stress-strain curve for a sand during confined compression. Since there is no lateral strain during this test, the axial strain is exactly equal to the volumetric strain. Example 12.3 gives values of

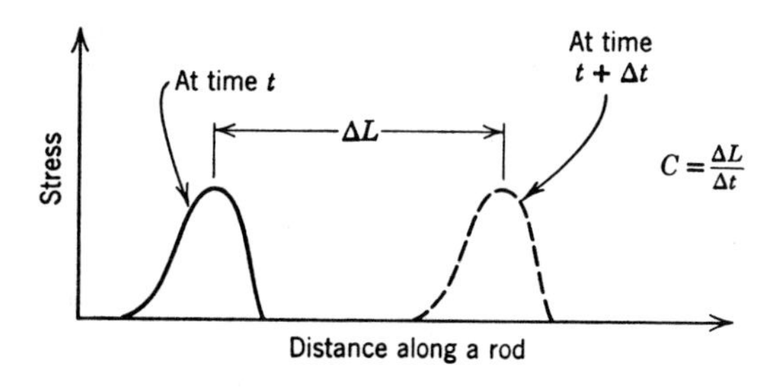

Fig. 12.2 Meaning of wave velocity.

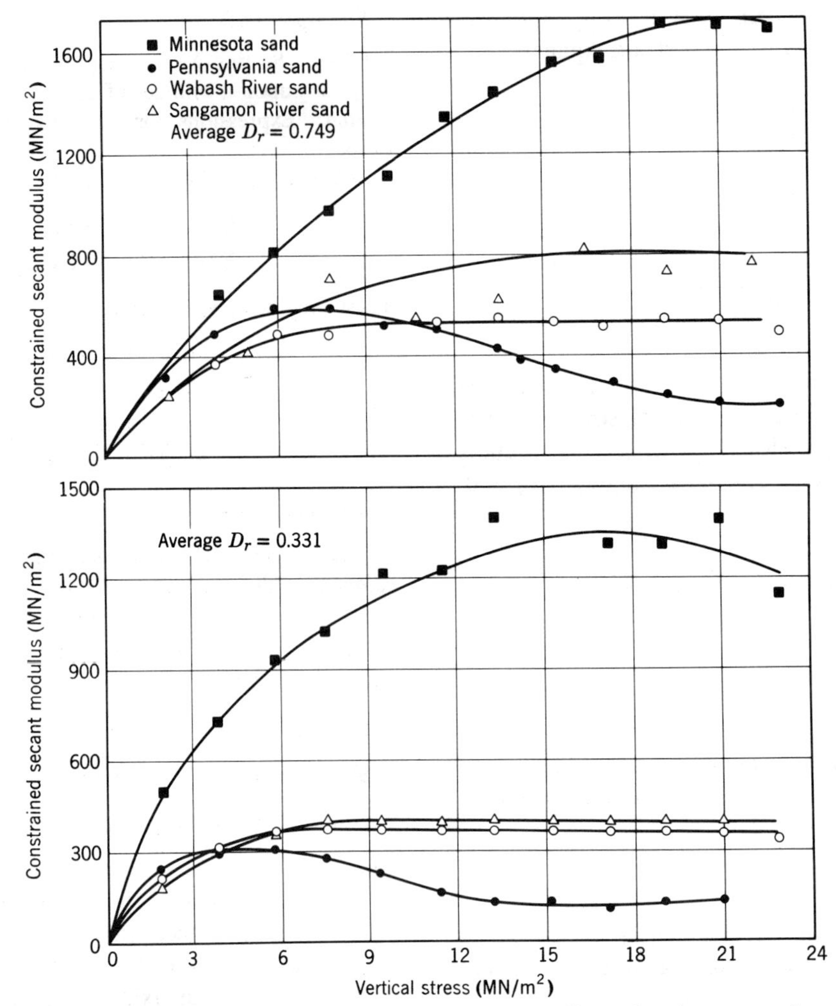

Fig. 12.3 Behavior of several sands during one-dimensional compression. Secant modulus from zero MN/m² to indicated stress. (From Hendron, 1963.)

▶ **Example 12.3**

Given. Stress-strain curve in Fig. 10.5.

Find.

a. Secant modulus from 0 to 100 kN/m², first loading.
b. Secant modulus from 1 to 800 kN/m², first loading.
c. Secant modulus from 1 to 800 kN/m², second loading.
d. Secant modulus from 1 to 800 kN/m², second unloading.
e. Tangent modulus at 100 kN/m², first loading.

Solution.

Case	$\Delta\sigma$ (kN/m²)	$\Delta\epsilon$	Modulus, (kN/m²)
a	100	0.0078	13 000
b	700	0.0120	58 000
c	700	0.0043	163 000
d	700	0.0031	230 000
e[a]	700	0.0298	23 000

[a] Measurements made along tangent line, from 100 to 800 kN/m².

◀

constrained modulus as measured from this curve. The general magnitude of the constrained modulus for a sand should be noted, together with the fact that the sand becomes stiffer as it is loaded and reloaded.

As was discussed in Chapter 10, crushing and breaking of particles become increasingly important for stresses greater than 3500 kN/m². Thus for large stresses the modulus tends to become constant, or may even decrease (Fig. 12.3). The Minnesota sand was composed of hard, rounded particles, whereas the Pennsylvania sand was made up of softer, angular particles. The other two curves illustrate the behavior of well-graded sands.

Initial Relative Density

As would be expected, the looser the soil the smaller the modulus for a given loading increment. This is illustrated by the results given in Table 12.1.

Repeated Loadings

Figure 12.4 illustrates the increase in modulus during successive cycles of loading. The modulus increases

Table 12.1 Secant Constrained Modulus for Several Granular Soils during Virgin Loading

Soil	Relative Density	Modulus (MN/m^2) $\Delta\sigma_1$ from 62 to 103 kN/m^2	Modulus (MN/m^2) $\Delta\sigma_1$ from 200 to 510 kN/m^2
Uniform gravel	0	30.3	60.0
1 mm < D < 5 mm	100	117.2	179.3
Well graded sand	0	13.8	25.5
0.02 mm < D < 1 mm	100	51.7	121.4
Uniform fine sand	0	14.5	35.2
0.07 mm < D < 0.3 mm	100	51.0	120.0
Uniform silt	0	2.8	17.2
0.02 mm < D < 0.07 mm	100	35.2	75.8

From Hassib, 1951.

markedly between the first and second loadings. The increase gradually becomes less and less during successive cycles, and after several hundred cycles the stress-strain curve stabilizes.

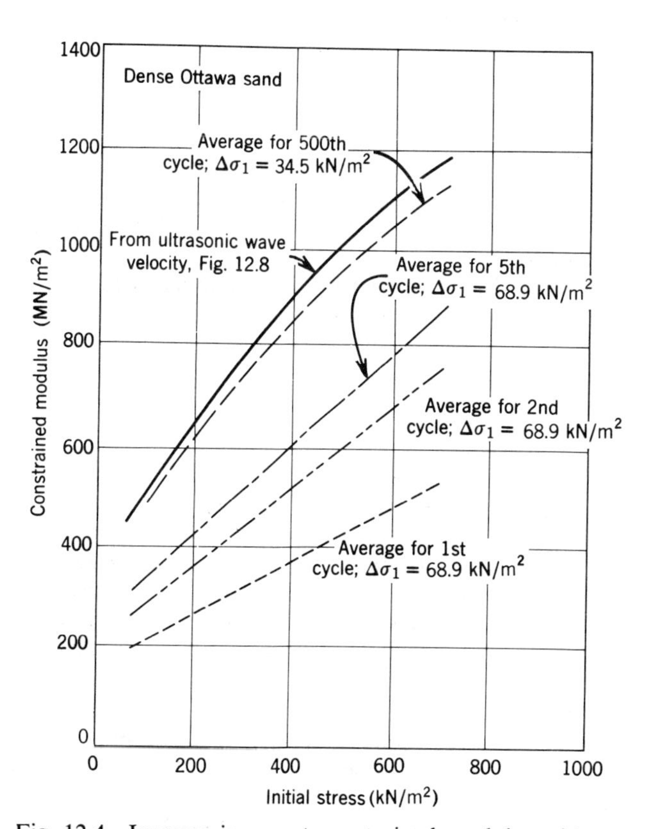

Fig. 12.4 Increase in secant constrained modulus with successive cycles of loading. *Note.* Average curves have been drawn through scattered data.

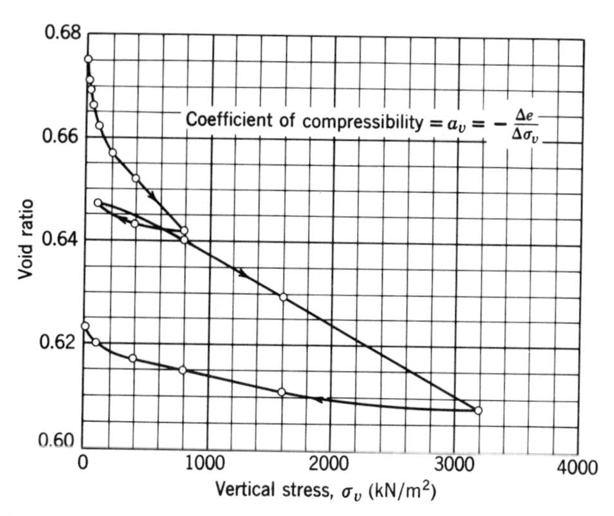

Fig. 12.5 Results of confined compression test plotted as void ratio versus stress on natural scale.

Rate of Compression

For an initial loading on a sand, the modulus is affected by the time required to achieve peak stress. The modulus may double if the loading time is 5 msec instead of the usual several seconds (see Whitman et al., 1964). The influence of the loading time is much less during subsequent cycles of a repeated loading.

Composition

As in the case of friction angle, modulus is affected in two ways by composition: composition affects the void ratio for a given relative density, and then it affects the modulus for that relative density. For a given relative density, the modulus of an angular sand will be less than that of a rounded sand. Table 12.1 indicates the influences of particle size and grading. In general, modulus decreases as the particle size leads to a larger void ratio for a given relative density. The effect of composition tends to disappear at very large stresses and during subsequent cycles of a repeated loading.

Alternate Methods of Protraying Data

In addition to the simple form of stress-strain curve in Fig. 10.5, two other methods of plotting stress-strain data are often used.

Figure 12.5 shows the results of Fig. 10.5 plotted as void ratio versus vertical stress σ_v. The slope of the resulting curve is defined as the *coefficient of compressibility* a_v:

$$a_v = -\frac{de}{d\sigma_v} \quad \text{or} \quad a_v = -\frac{\Delta e}{\Delta\sigma_v} \tag{12.10}$$

Figure 12.6 shows the same results plotted as void ratio versus the logarithm of vertical stress. This form of plot is useful for two reasons: (*a*) it is convenient for

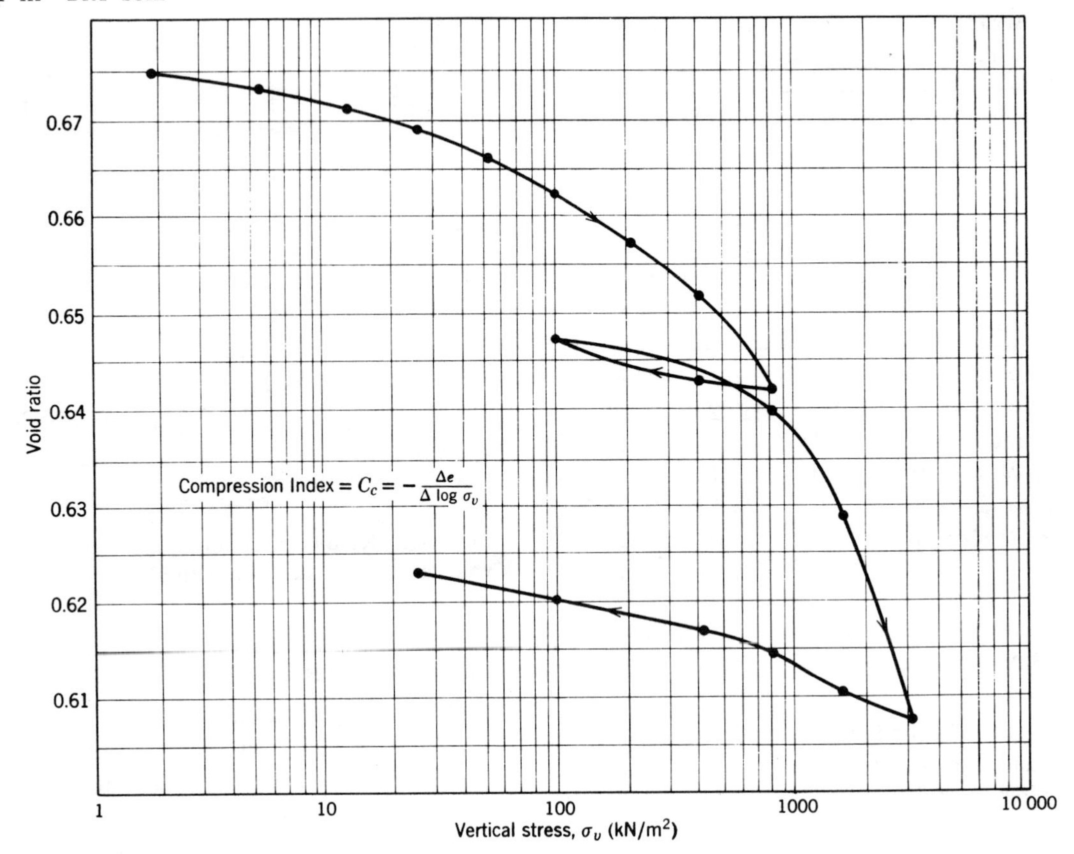

Fig. 12.6 Results of confined compression test plotted as void ratio versus stress on logarithmic scale.

showing stress-strain behavior over a wide range of stresses; and (*b*) such curves usually become more-or-less straight at large stresses. As will be seen in Part IV, this form of plot is especially useful for clays. Figure 12.7 shows the curves of Fig. 10.4 replotted in this way. At large stresses, the curves for the different sands tend to fall along a common path. The slope of this type of curve is the *compression index* C_c:

$$C_c = -\frac{de}{d(\log \sigma_v)} \quad \text{or} \quad C_c = -\frac{\Delta e}{\Delta(\log \sigma_v)} \tag{12.11}$$

C_c is thus the change in void ratio per logarithmic cycle of stress.

Still another term used to describe stress-strain behavior in confined compression is the *coefficient of volume change* m_v, which is simply the reciprocal of constrained modulus:

$$m_v = \frac{d\epsilon_v}{d\sigma_v} \quad \text{or} \quad m_v = \frac{\Delta\epsilon_v}{\Delta\sigma_v} \tag{12.12}$$

The relationships among D, m_v, a_v, and C_c are given in Table 12.2. The vertical strain during confined compression equals $\Delta e/(1 + e_0)$, where e_0 is the initial void ratio. Example 12.4 illustrates typical numerical values.

▶ **Example 12.4**

Given. Stress-strain curves in Figs. 10.5, 12.5, and 12.6.

Find. Values of m_v, a_v, and C_c for the same stresses used in Example 12.3.

Solution. The values may be scaled from the figures. They may be computed using the equations in Table 12.2, but this computation is inaccurate in the case of secant values of C_c, since the choice of the average stress σ_{va} greatly affects calculated values.

Case	m_v (m²/MN)	a_v (m²/MN)	C_c
a	0.078	0.130	0.0065
b	0.017	0.028	0.0225
c	0.006	0.010	0.0079
d	0.0045	0.0073	0.0066
e	0.045	0.065	0.0140

Note. C_c is dimensionless; a change per logarithmic cycle is the same for any set of units. ◀

Note that the compressibilities a_v and m_v decrease as the stress increases, but that C_c increases. The maximum value of C_c in Fig. 12.6 is 0.07.

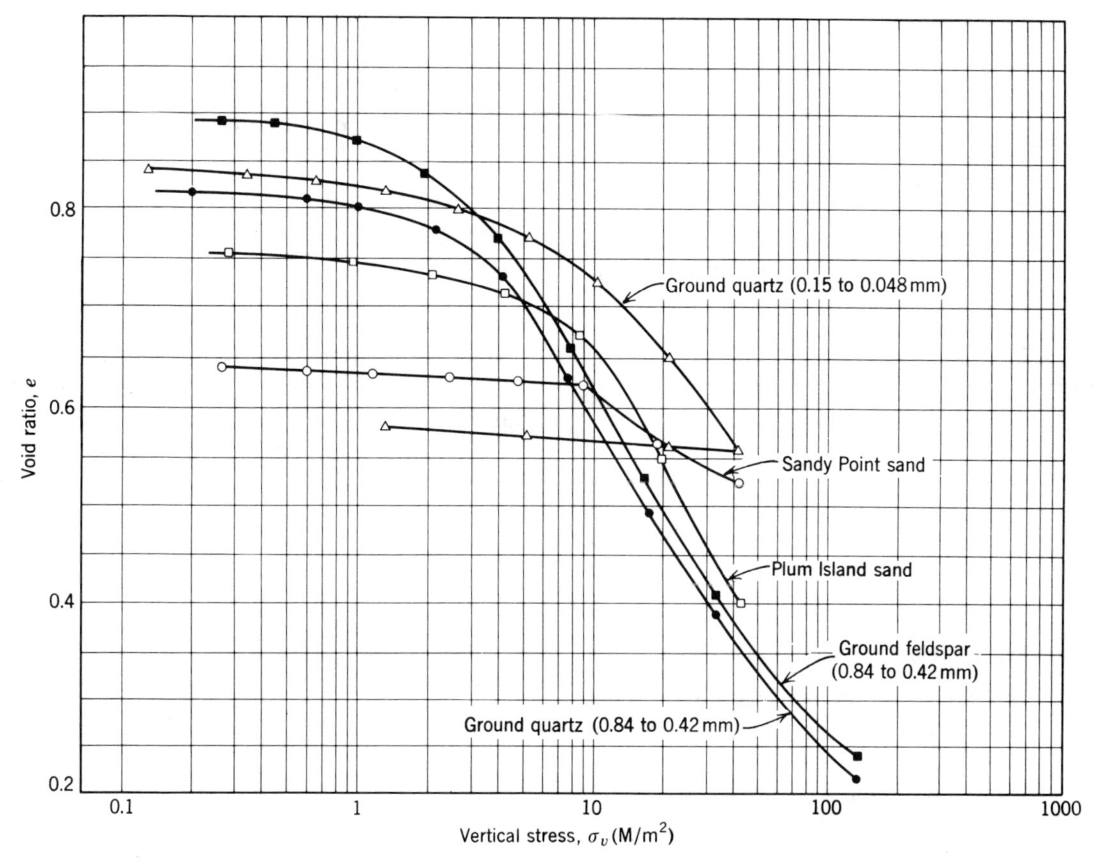

Fig. 12.7 Results of high-stress, confined compression tests on several sands (data from Roberts, 1964).

The stress-strain curve for an initial loading generally resembles a parabola. Hence the stress-strain relationship may be expressed as

$$\sigma_v = C(\epsilon_v)^n \tag{12.13}$$

The coefficient C varies with the type of soil and its initial void ratio. For a wide variety of soils, however, the exponent n has been found to be very close to 2. For a perfect packing of elastic spheres, this exponent would be 3. The difference between the theoretical and actual values for the exponent is the result of sliding among and rearrangement of the particles within an actual soil. Equation 12.13 implies that both secant modulus from zero stress and the tangent modulus should increase as $\sqrt{\sigma_v}$.

Table 12.2 Relations Between Various Stress-Strain Parameters for Confined Compression

	Constrained Modulus	Coefficient of Volume Change	Coefficient of Compressibility	Compression Index
Constrained modulus	$D = \dfrac{\Delta\sigma_v}{\Delta\epsilon_v}$	$D = \dfrac{1}{m_v}$	$D = \dfrac{1+e_0}{a_v}$	$D = \dfrac{(1+e_0)\sigma_{va}}{0.435C_c}$
Coefficient of volume change	$m_v = \dfrac{1}{D}$	$m_v = \dfrac{\Delta\epsilon_v}{\Delta\sigma_v}$	$m_v = \dfrac{a_v}{1+e_0}$	$m_v = \dfrac{0.435C_c}{(1+e_0)\sigma_{va}}$
Coefficient of compressibility	$a_v = \dfrac{1+e_0}{D}$	$a_v = (1+e_0)m_v$	$a_v = -\dfrac{\Delta e}{\Delta\sigma_v}$	$a_v = \dfrac{0.435C_c}{\sigma_{va}}$
Compression index	$C_c = \dfrac{(1+e_0)\sigma_{va}}{0.435D}$	$C_c = \dfrac{(1+e_0)\sigma_{va}m_v}{0.435}$	$C_c = \dfrac{a_v\sigma_{va}}{0.435}$	$C_c = -\dfrac{\Delta e}{\Delta \log \sigma_v}$

Note. e_0 denotes the initial void ratio. σ_{va} denotes the average of the initial and final stresses.

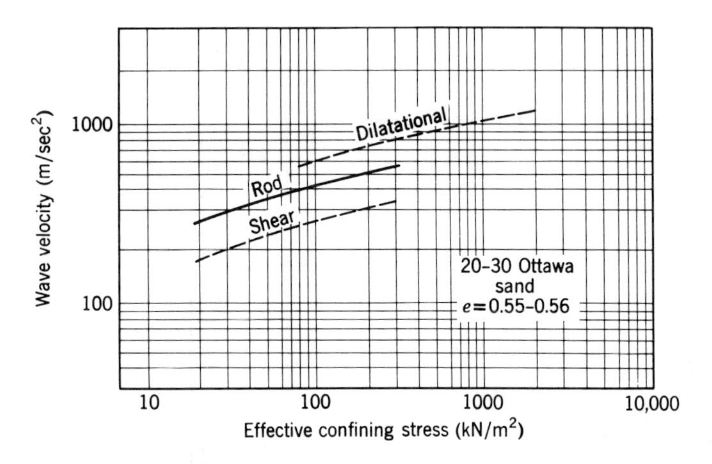

Fig. 12.8 Wave velocities through sand as function of confining stress. Dilatational and shear velocities from Whitman and Lawrence (1963); rod velocities from Hardin and Richart (1963).

Relationship to Wave Velocity

Figure 12.8 shows typical values for dilatational wave velocity through granular soils. The velocity typically increases as $\sigma_v^{0.25}$, which according to Eq. 12.9*c* means that constrained modulus should increase as $(\sigma_v)^{1/2}$. However, the modulus as computed from measured wave velocity using Eq. 12.9*c* generally is much larger than the constrained modulus as measured directly in an oedometer. This is illustrated by Example 12.5. The difference

▶ Example 12.5

Given. Wave velocity versus stress in Fig. 12.8 and modulus versus stress in Fig. 12.4.

Find. Constrained modulus for stress of 138 kN/m². Compare with modulus as measured directly.

Solution. C_D = 579 m/s. Typical value for γ = 16.5 kN/m², or ρ = 1.68 t/m².

$$D = \rho C_D^2 = 1.68 \times 579^2 \times \text{kN/m}^2 = 563\,205 \text{ kN/m}^2 = 563.2 \text{ MN/m}^2$$

versus 207 MN/m² as measured directly. ◀

arises because the small stresses associated with a seismic wave mainly cause elastic deformations of particles, whereas the large stresses applied in an oedometer test cause slippage between adjacent particles. This situation has been sketched in Fig. 10.10. If very small stress increments are used in the oedometer, then the modulus as measured directly becomes approximately equal to the modulus as calculated from wave velocity (Whitman et al., 1964). Furthermore, the modulus as measured after many cycles of loading, even using large stress increments, is also about equal to the modulus calculated from wave velocity (Fig. 12.4).

Hence wave velocity is not a useful direct measure of the compressibility of a soil during a single intense loading, but it does indicate the compressibility during repeated loadings. This appears to be true regardless of the frequency of the repeated loading.

For further discussion of wave velocity, see Hardin and Richart (1963), Whitman (1966).

12.3 BEHAVIOR DURING TRIAXIAL COMPRESSION TEST

The standard triaxial test (i.e., with constant confining stress and increasing axial stress) gives a direct measure of Young's modulus. Modulus decreases with increasing axial stress, and at the peak of the stress-strain curve the tangent modulus becomes zero.

When a value of Young's modulus is quoted for soil, it usually is the secant modulus from zero deviator stress to a deviator stress equal to $\frac{1}{2}$ or $\frac{1}{3}$ of the peak deviator stress. This is a common range of working stresses in actual foundation problems, since typically a safety factor of 2 or 3 is used in these problems. Example 12.6 illustrates

▶ Example 12.6

Given. Stress-strain curve for test in Fig. 10.13.

Find. Secant Young's modulus for deviator stress equal to $\frac{1}{2}$ of peak stress.

Solution.

$$\Delta\sigma_v \text{ at peak} = 380 \text{ kN/m}^2$$
$$\Delta\sigma_v \text{ at } \tfrac{1}{2} \text{ peak} = 190 \text{ kN/m}^2$$
$$\Delta\epsilon_v = 0.002$$
$$E = 95 \text{ MN/m}^2 = 95\,000 \text{ kN/m}^2$$ ◀

the computation of modulus from a typical stress-strain curve. For the scale to which this curve has been plotted, it is difficult to tell whether or not the curve is linear or curved up to $\frac{1}{2}$ the peak. However, the very precise data given in Fig. 12.9 show that the curve is nonlinear almost from the beginning of loading.

Kondner and Zelasko (1963) suggested that the stress-strain curves of sand in standard triaxial compression can be fitted by a hyperbolic equation of the form

$$\sigma_1 - \sigma_3 = \frac{\epsilon_1}{a + b\epsilon_1} \tag{12.14}$$

where a and b are constants.

Confining Stress

As the confining stress increases, the modulus increases. For the case where the initial stress σ_0 is isotropic, the modulus increases as $\sigma_0{}^n$ where n varies from 0.4 to 1.0. A reasonable average value is n = 0.5. The larger values of the exponent tend to apply to loose sands.

In most practical problems, the stresses before loading are not isotropic. The effect of the actual state of stress on modulus is not clear, but the best available rule is that modulus depends on the average of the initial principal

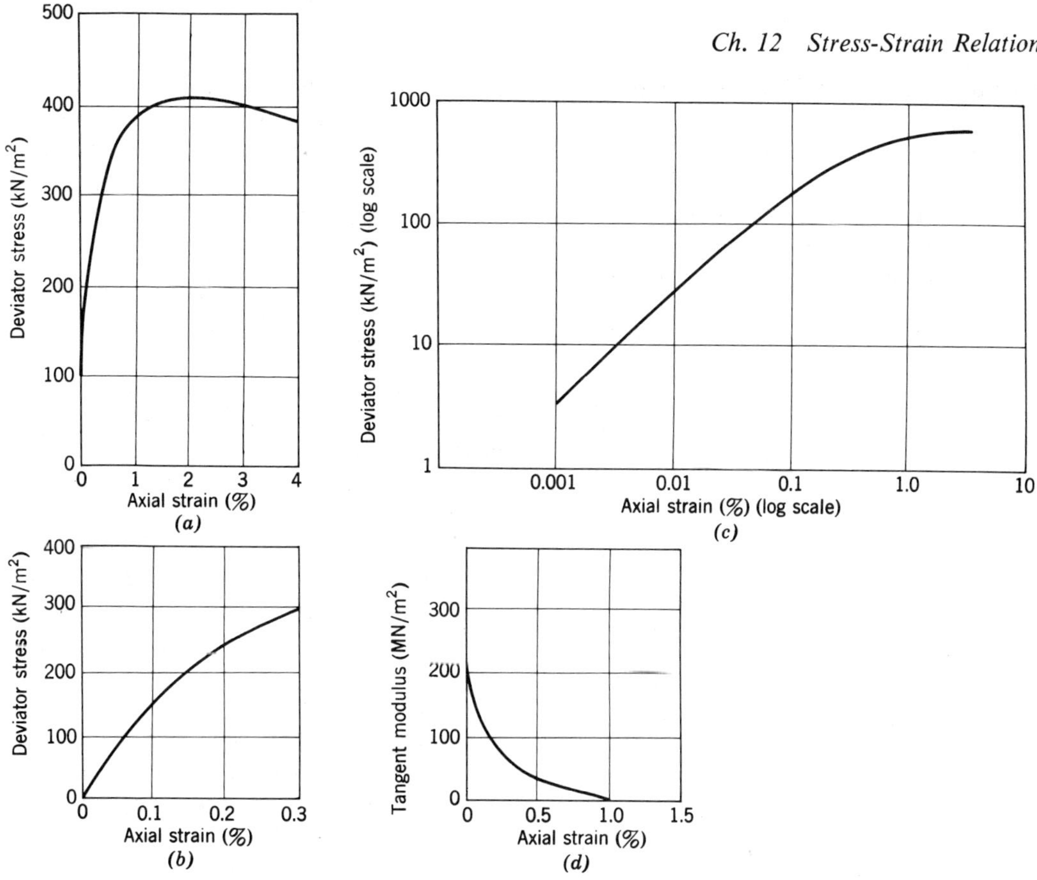

Fig. 12.9 Stress-strain data from a triaxial test. *Note.* Medium, subangular sand: porosity = 0.39; confining stress = 98.5 kN/m². (From Chen, 1948.)

stresses; thus

$$E \sim \sqrt{\sigma_v \frac{1 + 2K_0}{3}} \qquad (12.15)$$

where K_0 is the coefficient of lateral stress at rest. Equation 12.15 holds only when $\frac{1}{2} < K_0 < 2$ and when the factor of safety against failure is 2 or more.

Various Factors

The effect of void ratio, composition, stress history, and loading rate upon E is the same as their effect upon D. Table 12.3 indicates the general effect of void ratio and composition on E for a first loading to one-half the peak deviator stress. Table 12.4 gives values of E obtained after several cycles of loading. The values in Table 12.4 are also indicative of the initial tangent modulus and of the modulus which is computed from rod wave velocity.

Table 12.3 Young's Modulus for Initial Loading

	Loose (MN/m²)	Dense (MN/m²)
Angular, breakable particles	14	35
Hard, rounded particles	56	105

Note. Secant modulus to $\frac{1}{2}$ peak deviator stress, with 101.3 kN/m² (1 atm) confining stress.

Table 12.4 Young's Modulus for Repeated Loadings

Soil [101.3 kN/m² (1 atm) confining pressure]	Young's Modulus (MN/m²)	
	Loose	Dense
Screened crushed quartz, fine angular	117	207
Screened Ottawa sand, fine rounded	179	310
Ottawa Standard sand, medium, rounded	207	669
Screened sand, medium, subangular	138	241
Screened crushed quartz, medium, angular	124	186
Well graded sand, coarse, subangular	103	193

From Chen, 1948.

It is of interest to compare these values of E with those for the minerals of which the particles of a granular soil

Table 12.5 Poisson's Ratio and Young's Modulus for Various Materials

Material	Poisson's Ratio	Young's Modulus (MN/m^2)
Amphibolite	0.28–0.30	93.8–121.4 × 10^3
Anhydrite	0.30	68.3 × 10^3
Diabase	0.27–0.30	86.9–116.5 × 10^3
Diorite	0.26–0.29	75.2–107.6 × 10^3
Dolomite	0.30	110.3–121.3 × 10^3
Dunite	0.26–0.28	148.9–182.7 × 10^3
Feldspathic Gneiss	0.15–0.20	82.7–118.6 × 10^3
Gabbro	0.27–0.31	88.9–126.9 × 10^3
Granite	0.23–0.27	73.1– 86.2 × 10^3
Ice	0.36	7.10 × 10^3
Limestone	0.27–0.30	86.9–107.6 × 10^3
Marble	0.27–0.30	86.9–107.6 × 10^3
Mica Schist	0.15–0.20	79.3–101.4 × 10^3
Obsidian	0.12–0.18	64.8– 80.0 × 10^3
Oligoclasite	0.29	80.0– 84.8 × 10^3
Quartzite	0.12–0.15	82.1– 96.5 × 10^3
Rock salt	0.25	35.4 × 10^3
Slate	0.15–0.20	79.3–112.4 × 10^3
Aluminum	0.34–0.36	55.2– 75.8 × 10^3
Steel	0.28–0.29	200.0 × 10^3

Values for rock computed from compressibility measurements by Brace (1966) at confining stresses of 300–500 MN/m^2. Values for steel and aluminum from Lange (1956).

are composed, and with steel and aluminum (see Table 12.5). The great compressibility of soil, the result of its particulate nature, is evident from this comparison.

Poisson's Ratio

Poisson's ratio may be evaluated from the ratio of the lateral strain to axial strain during a triaxial compression test with axial loading. Figure 10.13 has shown values of this ratio at various stages during a typical test. During the early range of strains for which the concepts from theory of elasticity are of use, the Poisson's ratio is varying with strain. The Poisson's ratio for sand becomes constant only for large strains which imply failure, and then has a value greater than 0.5. Such a value of μ implies expansion of the material during a triaxial test (see Example 12.1). Poisson's ratio is less than 0.5 only during the early stages of such a test where the specimen decreases in volume.

Because of this behavior, it is very difficult to make an exact evaluation of the value of μ for use in any problem. Fortunately, the value of μ usually has a relatively small effect upon engineering predictions. For the early stages of a first loading of a sand, when particle rearrangements are important, μ typically has values of about 0.1 to 0.2. During cyclic loading, μ becomes more of a constant, with values from 0.3 to 0.4. The ratio of two different types of wave velocities is often used to estimate the value of μ applicable to a cyclic loading.

12.4 BEHAVIOR DURING OTHER TESTS

Simple Shear

The shear modulus of soil finds its widest use in connection with foundation vibration problems and is generally evaluated through a measurement of shear wave velocity. Figure 12.8 indicated the typical variation of shear wave velocity with confining stress. Figure 12.10 shows the effect of void ratio. Factors such as composition affect C_S by influencing void ratio. Figure 12.10 can be used for a wide variety of granular soils.

As is the case for constrained and rod modulus, the shear modulus from a static repeated loading is for practical purposes equal to the modulus calculated from the wave velocity for the same initial stress. This is true for stresses much less than those associated with failure. The confining stress may be taken equal to

$$\frac{\sigma_v}{3}(1 + 2K_0)$$

Special Triaxial Tests

In order to duplicate the type of loading expected within an actual mass of soil, both confining stress and axial stress are often varied during a triaxial test. Using the equations developed in Example 12.2, values of E and μ may still be evaluated from such a test. This is illustrated in Example 12.7.

▸ Example 12.7

Given. Data for Test B, Figs. 10.21 and 10.23.

Find. E and μ at end of first loading.

Solution. The first step is to find the values of $\Delta\sigma_z = \Delta\sigma_v$ and $\Delta\sigma_x = \Delta\sigma_h$.

$$\Delta\sigma_z = \Delta p + \Delta q = 152 + 81 = 233$$
$$\Delta\sigma_x = \Delta p - \Delta q = 152 - 81 = 71$$

The strains from this loading are

$$\Delta\epsilon_z = 0.00268$$
$$\Delta\epsilon_x = 0.00020$$

Then, from Example 12.2,

$$E = \frac{(233 + 2 \times 71)(162)}{71(0.00268 - 0.0004) + 233(0.00268)}$$

$$= \frac{375(162)}{0.162 + 0.625} = 77\ 200\ \text{kN/m}^2$$

$$\mu = \frac{71(0.00268) - 0.00020(233)}{0.787}$$

$$= \frac{0.189 - 0.047}{0.787} = 0.18$$

◂

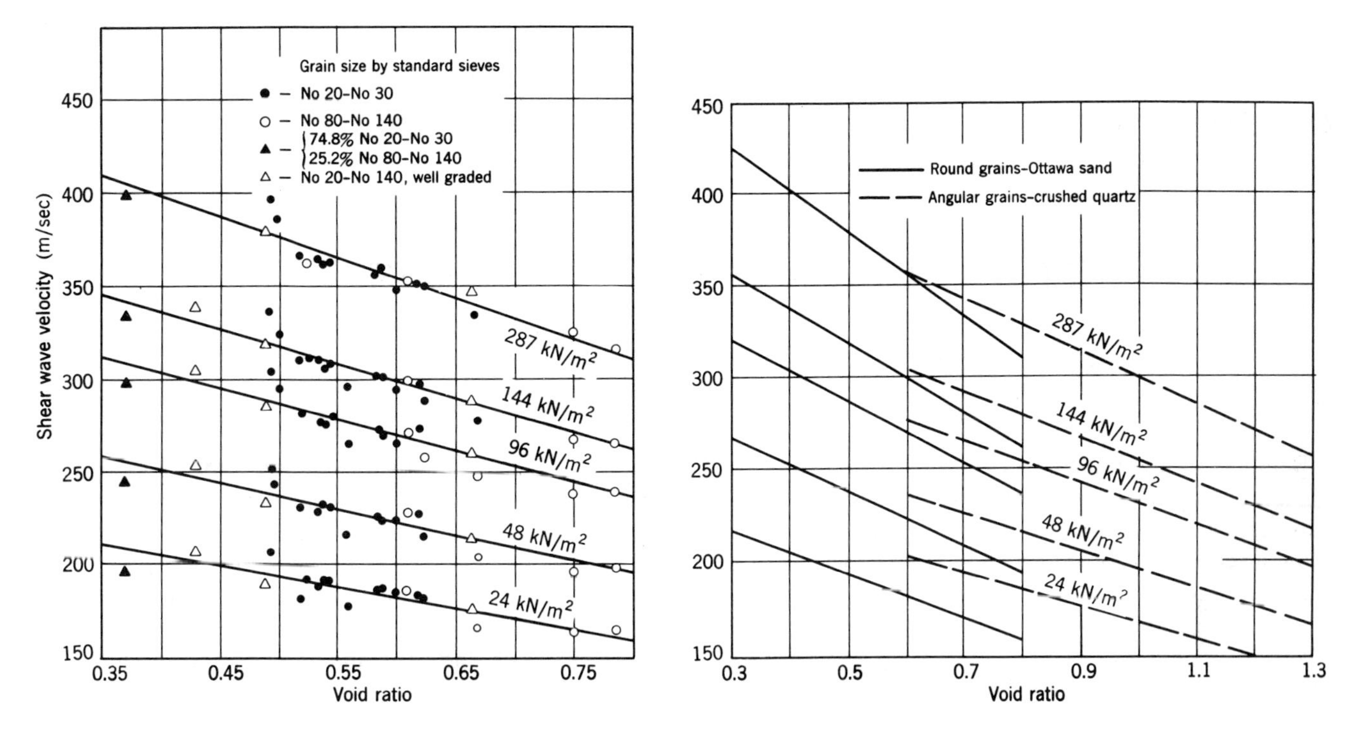

Fig. 12.10 Shear wave velocities through quartz sands (From Hardin and Richart, 1963).

12.5 SUMMARY OF MAIN POINTS

The concepts from the theory of elasticity apply to soil only in a very approximate way. Nonetheless, it is often useful to use these concepts and to use values of modulus and Poisson's ratio which apply approximately for a particular loading. Clearly, good judgment is needed when choosing values for these parameters.

The same factors that affect ϕ also affect modulus. However, the effect upon modulus is more marked. It is difficult to estimate values of modulus with much accuracy, and test data for the particular soil will be necessary whenever an accurate estimate is needed.

Since modulus depends on void ratio, and it is difficult to obtain undisturbed samples of granular soils, it is especially difficult to measure the modulus of granular soils reliably. From experience, it appears that the second cycle of loading during a laboratory test usually gives the best measure of *in situ* modulus. Apparently the effects of sample disturbance are compensated by the effects of the initial loading. There are no reliable correlations between modulus and blow count.

PROBLEMS

12.1 If $E = 110$ MN/m² and $\mu = 0.35$, evaluate the constrained modulus D and shear modulus G.

12.2 For the data given in Problem 12.1, compute the dilatational velocity C_D, rod velocity C_L, and shear velocity C_S. Assume a value of ρ which is reasonable for a dense sand.

12.3 K_0 for a sand is found to be 0.45. Assuming that sand is an elastic material, compute Poisson's ratio μ.

12.4 Refer to Figs. 10.21 and 10.23. For Test D, initial loading, compute E and μ for (*a*) the entire stress increment and (*b*) the increment to the first data point. First assume that E and μ can be computed as though this were an ordinary triaxial test using Eqs. 12.1 and 12.2. Then use the equations in Example 12.2.

12.5 Repeat Problem 12.4, using the results for Test A, second loading.

12.6 Estimate Young's modulus (secant modulus to $\frac{1}{3}$ of failure load for a first loading) for a well-graded, subangular, dense sand located at a depth of 60 m below ground surface. *Hint.* You will need to estimate several factors in order to arrive at a satisfactory estimate.

12.7 Using the data in Fig. 12.10, estimate the shear modulus at 6 m depth of a sand having $e = 0.6$, $G = 2.7$, $K_0 = 0.5$.

CHAPTER 13

Earth Retaining Structures and Slopes

Building on preceding chapters, this chapter considers earth retaining structures. Several examples of retaining structures were given in Chapter 1. Figures 1.9 and 1.15 show sheet pile bulkheads and Fig. 1.12*b* shows a braced excavation. Figure 13.1 illustrates an even more common retaining structure: a gravity retaining wall.

When designing retaining structures, an engineer often needs to ensure only that total collapse or failure does not occur. Movements of several centimeters and even several meters are often of no concern as long as there is assurance that even larger motions will not suddenly occur. Thus the approach to the design of retaining structures generally is to analyze the conditions that would exist at a collapse condition, and to apply suitable safety factors to prevent collapse. This approach is known as *limit design* and requires *limiting equilibrium mechanics*.

The early portions of this chapter present methods used to analyze the stability of structures that retain dry granular soils. There are many practical situations to which these methods can be applied directly. Generally, of course, water and the clay content of a soil are important to a practical problem, but the methods developed for dry granular soils form the basis for the methods (presented in Parts IV and V) used for these more complicated situations.

There are many situations in which the movements of retaining structures must be given serious consideration—situations where consideration of stability only is inadequate for a proper design. These situations arise especially with regard to clayey soils, but they can also arise with sandy soils. The later sections of this chapter consider such situations.

This chapter concludes with a brief discussion of the stability of slopes in dry granular soils.

13.1 APPROACH TO DESIGN OF GRAVITY RETAINING WALLS

A gravity retaining wall is typically used to form the permanent wall of an excavation whenever space requirements make it impractical to simply slope the side of the excavation. Such conditions arise, for example, when a roadway or storage area is needed immediately adjacent to an excavation. In order to construct the wall, a temporary slope is formed at the edge of the excavation, the wall is built, and then backfill is dumped into the space between the wall and the temporary slope. In earlier days masonry walls were often used. Today, most such walls are of unreinforced concrete although other special forms of construction are sometimes employed (see Huntington, 1957; Teng, 1962).

Figure 13.2 shows in a general way the forces that act upon a gravity retaining wall. The bearing force resists the weight of the wall plus the vertical components of other forces. The *active thrust*, which develops as the backfill is placed and as any surcharges are placed on the surface of the backfill, acts to push the wall outward. This outward motion is resisted by *sliding resistance* along the base of the wall and by the *passive resistance* of the soil lying above the toe of the wall. The active thrust also tends to overturn the wall around the toe. This overturning is resisted by the weight of the wall and the vertical component of the active thrust. The weight of the wall is thus important in two ways: it resists overturning and it causes frictional sliding resistance at the base of the wall. This is why such a wall is called a *gravity* retaining wall.

A gravity retaining wall, together with the backfill the wall retains and the soil that supports the wall, is a highly *indeterminate* system. The magnitudes of the forces that act upon a wall cannot be determined from statics alone, and these magnitudes will be affected by the sequence of construction and backfilling operations. Hence the design of such a wall is based not on an analysis to determine the expected forces but on analysis of the forces that would exist if the wall started to fail, i.e., to overturn or to slide outwards.

The first step in such an analysis is to envision the pattern of deformations that would accompany such a failure. These patterns have been studied by means of

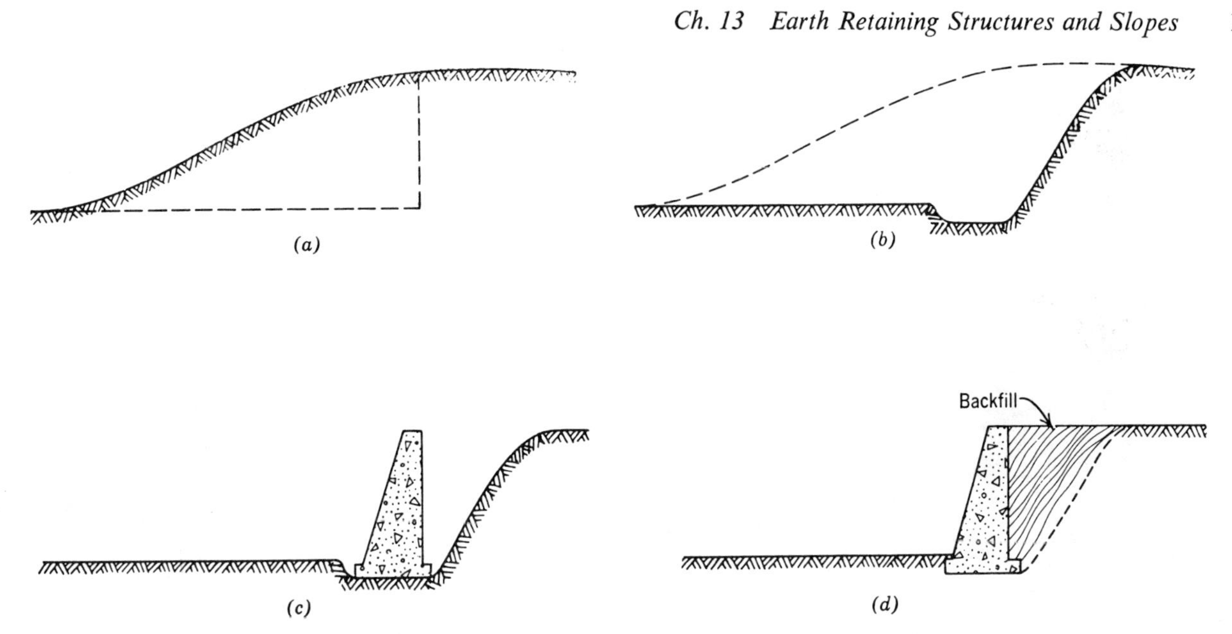

Fig. 13.1 Stages in construction of typical gravity retaining wall. (*a*) Proposed excavation. (*b*) Excavation completed. (*c*) Wall formed and poured. (*d*) Backfill placed.

small-scale tests which simulate actual retaining walls. Figure 13.3 shows the movements during such a test in which granular soil is simulated by rods.[1] These movements occurred as a support holding the wall was removed. Within the backfill the soil moved toward the wall and downward. These motions indicate that shear failure occurred throughout this *active zone;* i.e., the full frictional resistance was mobilized throughout this zone. A second zone of shear failure (the *passive zone*) developed at the toe of the wall where the wall was pushing against the soil.

Considering these patterns of deformations, an approach to the design of gravity retaining walls can then be stated. First, trial dimensions for the wall are selected. Next, the active thrust against the wall is determined, based on the assumption that shear failure occurs throughout the active zone. Then the resistance offered by the weight of the wall, the shear force at the base of the wall, and the passive zone at the toe of the wall are determined. Finally, the active thrust and total resistance are compared, and the resistance must exceed the active thrust by a suitable safety factor.

[1] There are many variations of this basic technique. There have been tests with sand contained between two glass plates. Use of horizontal metal rods, or even toothpicks, eliminates the need for glass side walls and the problem of friction between these walls and the sand. X-ray techniques have been used to observe the patterns of motion within soil masses (Roscoe et al., 1963).

The frame shown in Fig. 13.3 is used in the M.I.T. laboratories for student experiments and demonstrations (see also Fig. 13.30). The frame is 686 mm long by 737 mm high. The rods are 150 mm long and are of two shapes and sizes (round, 3 mm and 6 mm diameter; and hexagonal, 4.8 and 7.9 mm across flats) to simulate the interlocking which occurs in actual soils. Using this frame, students test their own designs for small-scale retaining structures, thereby gaining experience in the application of theoretical principles to design.

This approach to design will be illustrated in Section 13.6. First, however, we must consider methods for determining active thrust and passive resistance.

13.2 RANKINE ACTIVE AND PASSIVE STATES

As a first step in the evaluation of active thrust and passive resistance, we evaluate the conditions of limiting equilibrium for the geostatic state of stress, which occurs in a soil deposit with a horizontal surface and no shear stresses on horizontal and vertical surfaces.

Suppose that such a soil deposit is stretched in the horizontal direction. Any element of soil will then behave just like a specimen of a triaxial test in which the confining stress is decreased while the axial stress remains constant, as shown by the stress path in Fig. 13.4. When the horizontal stress is decreased to a certain magnitude, the full shear strength of the soil will be mobilized. No further decrease in the horizontal stress is possible. The horizontal stress for this condition is called the *active*

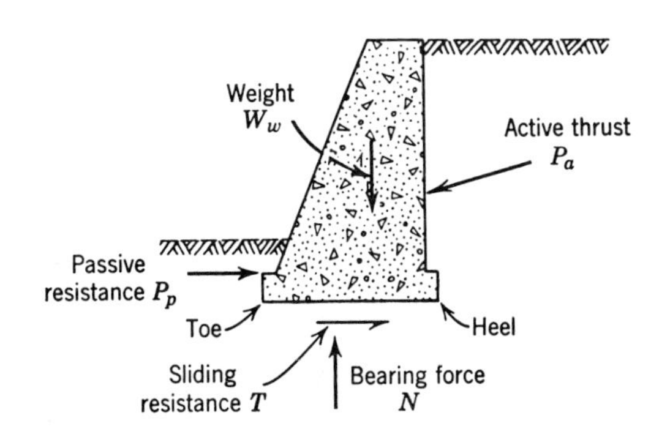

Fig. 13.2 Forces acting on gravity retaining wall.

Fig. 13.3 Double exposure showing movements of "soil" surrounding model retaining wall.

stress, and the ratio of horizontal to vertical stress is called the *coefficient of active stress* and is denoted by the symbol K_a.

Figure 13.5 shows the Mohr circle for the active state of stress.[2] From our analysis of the stresses at failure during a triaxial test (Section 11.1), we already know the ratio of the horizontal and vertical stresses for this case is

$$K_a = \frac{\sigma_{ha}}{\sigma_v} = \frac{\sigma_{3f}}{\sigma_{1f}} = \frac{1 - \sin \phi}{1 + \sin \phi}$$

$$= \tan^2 \left(45 - \frac{\phi}{2}\right) = \frac{1 - \tan \alpha}{1 + \tan \alpha} \tag{13.1}$$

Now let us suppose that the soil is compressed in the horizontal direction. Any element of soil is now in just the condition of a triaxial specimen being failed by increasing the confining pressure while holding the vertical stress constant [or, if we imagine that the triaxial specimen is placed on its side, increasing the axial stress while holding the confining pressure constant (see Fig. 13.4)]. The horizontal stress cannot be increased beyond a certain magnitude called the *passive stress*. The ratio of horizontal to vertical stress is called the *coefficient of passive stress* K_p. Figure 13.5 also shows the Mohr circle for this state of stress, and the magnitude of K_p is given by

$$K_p = \frac{\sigma_{hp}}{\sigma_v} = \frac{\sigma_{1f}}{\sigma_{3f}} = \frac{1 + \sin \phi}{1 - \sin \phi}$$

$$= \tan^2 \left(45 + \frac{\phi}{2}\right) = \frac{1 + \tan \alpha}{1 - \tan \alpha} \tag{13.2}$$

Ignoring any slight difference in ϕ for the two different stress paths (see Chapter 11), we see that $K_p = 1/K_a$.

[2] This chapter considers only cases where the failure law is $\tau_{ff} = \sigma_{ff} \tan \phi$. Methods for cases where it is appropriate to use $\tau_{ff} = c + \sigma_{ff} \tan \phi$ are discussed in Part IV.

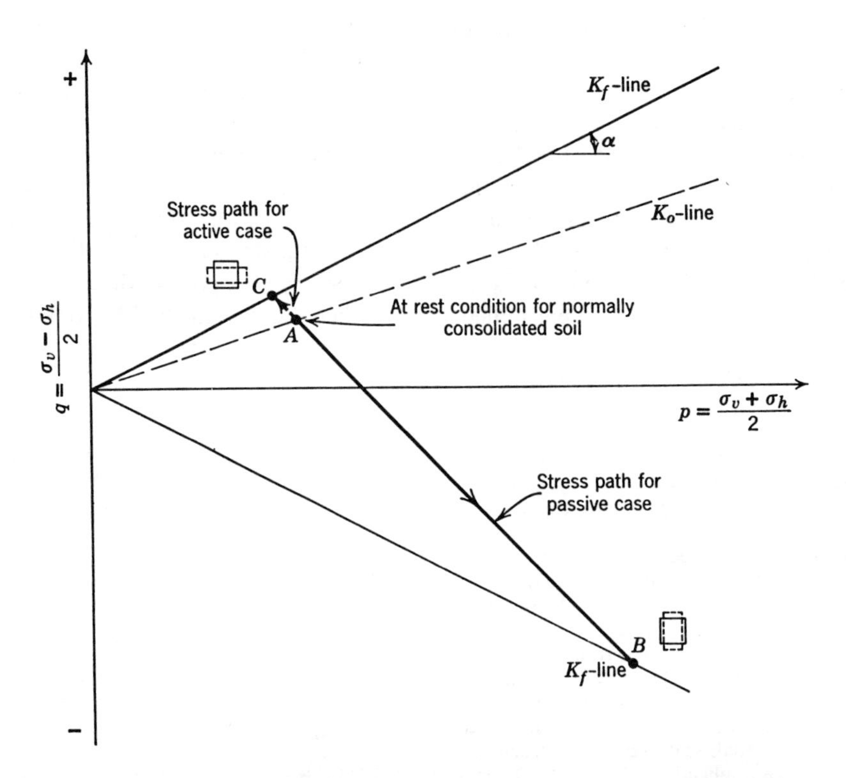

Fig. 13.4 Stress paths for Rankine active and passive conditions.

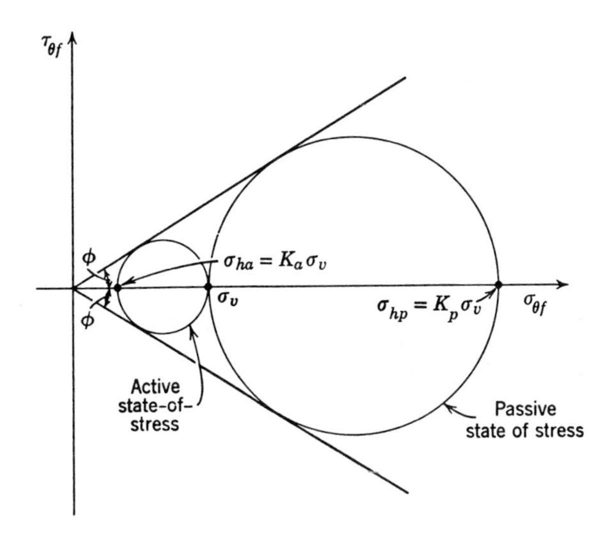

Fig. 13.5 Rankine states of stress for geostatic condition.

Thus for a given vertical geostatic stress σ_v, the horizontal stress can be only between the limits $K_a\sigma_v$ and $K_p\sigma_v$. These two limiting stresses are called *conjugate stresses.* The states of stress at the two extreme situations are called *Rankine states*, after the British engineer Rankine who in 1857 noted the relationship between the active and passive conditions. The inclinations of the slip lines for the two limiting cases are sketched in Fig. 13.6, which illustrates the use of the origin of planes to obtain these inclinations. In the active condition the shear stress opposes the effect of gravity. In the passive condition the shear stress acts together with gravity to oppose the large horizontal stress.

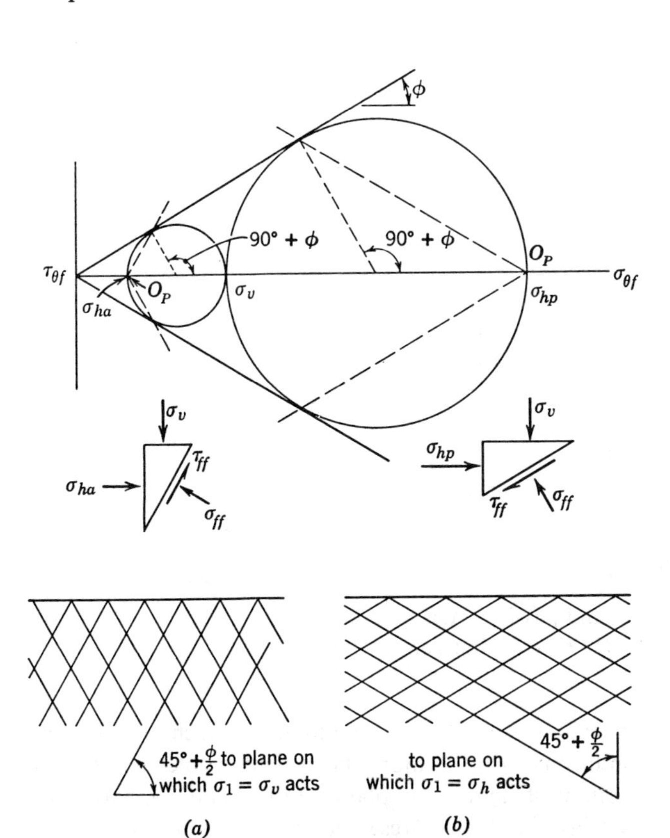

Fig. 13.6 Orientation of slip lines for Rankine states. (*a*) Active state. (*b*) Passive state.

Table 13.1 gives typical values for K_a and K_p. If the horizontal stretching or compressing of the soil causes very large strains, the friction angle ϕ_{cv} should be used to determine these coefficients. Generally, however, it is appropriate to use the peak friction angle ϕ. For $\phi =$ 30°, the theoretical failure lines will be at 60° to the horizontal for the active case and 30° to the horizontal for the passive case.

Table 13.1 Values of K_a and K_p for Rankine States of Geostatic Stress

ϕ	K_a	K_p
10°	0.703	1.42
15°	0.589	1.70
20°	0.490	2.04
25°	0.406	2.46
30°	0.333	3.00
35°	0.271	3.66
40°	0.217	4.60
45°	0.171	5.83

Strains Associated With Rankine States

The strains required to achieve active and passive conditions may be inferred from the results of triaxial tests such as those for tests 3 and 6 in Fig. 10.22. These results have been replotted in Fig. 13.7. Part (*a*) of this figure shows the stress paths and both the horizontal and vertical strains; part (*b*) shows the horizontal strain versus the stress ratio *K*. The important conclusions are:

1. Very little horizontal strain, less than −0.5%, is required to reach the active state.
2. Little horizontal compression, about 0.5%, is required to reach one-half of the maximum passive resistance.
3. Much more horizontal compression, about 2%, is required to reach the full maximum passive resistance.

These results are typical for most dense sands. For loose sands the first two conclusions remain valid, but the horizontal compression required to reach full passive resistance may be as large as 15%.

There are two reasons why less strain is required to reach the active condition than to reach the passive condition. First, an unloading (the active state) always

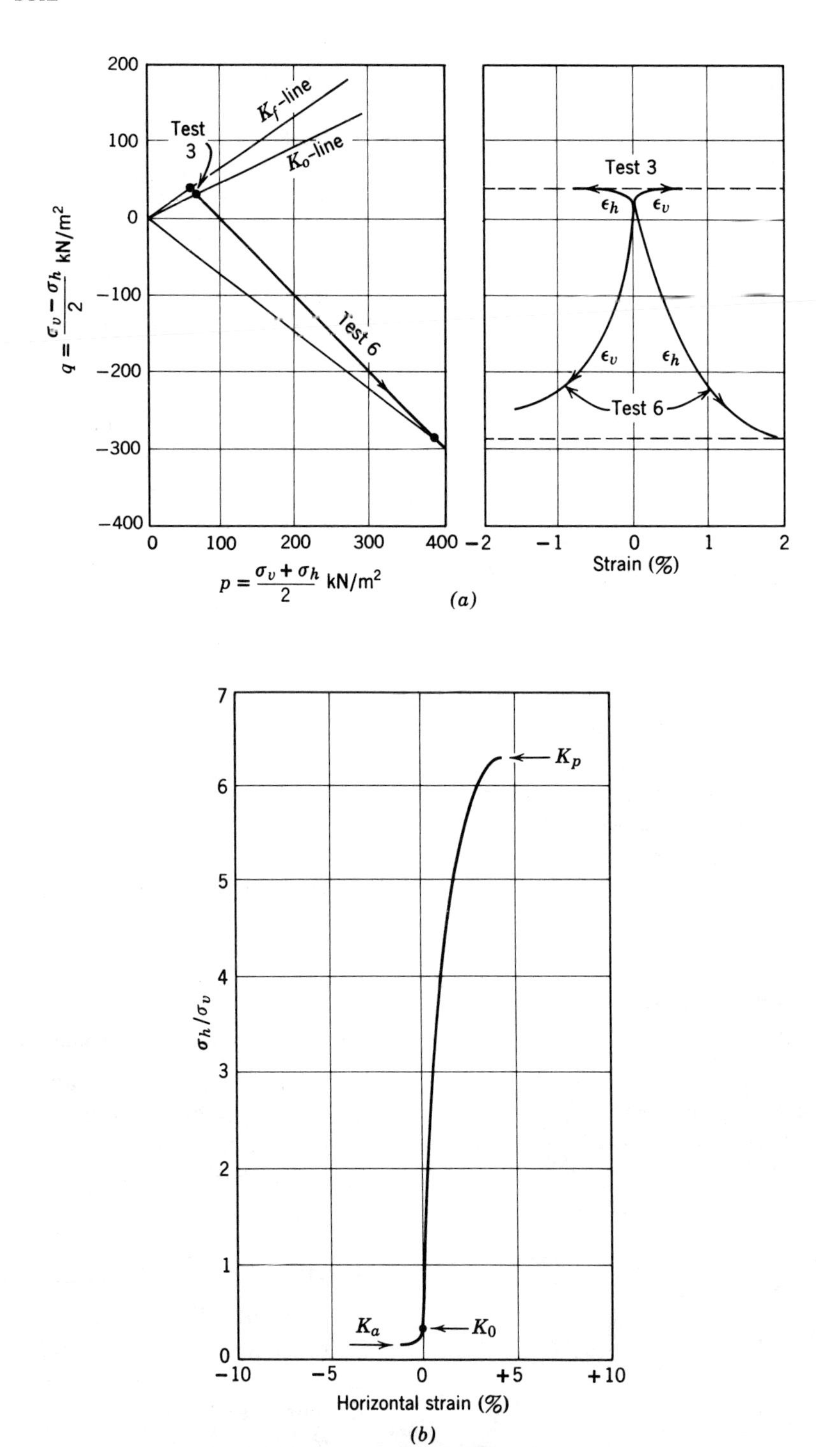

Fig. 13.7 Strains required to reach active and passive states in a dense sand. (*a*) Stress paths and q versus strain. (*b*) K versus horizontal strain.

involves less strain than a loading (the passive state). Second, the stress change in passing to the active state is much less than the stress change in passing to the passive state.

The foregoing results apply when the initial condition is a K_0 condition. If initially $\sigma_h/\sigma_v \neq K_0$, then somewhat different strains will be required to reach the limiting conditions. Furthermore, most field problems involving retaining structures are plane strain situations, and hence the triaxial data just presented are only indicative of those applicable to actual field problems. Data from plane strain tests are more appropriate.

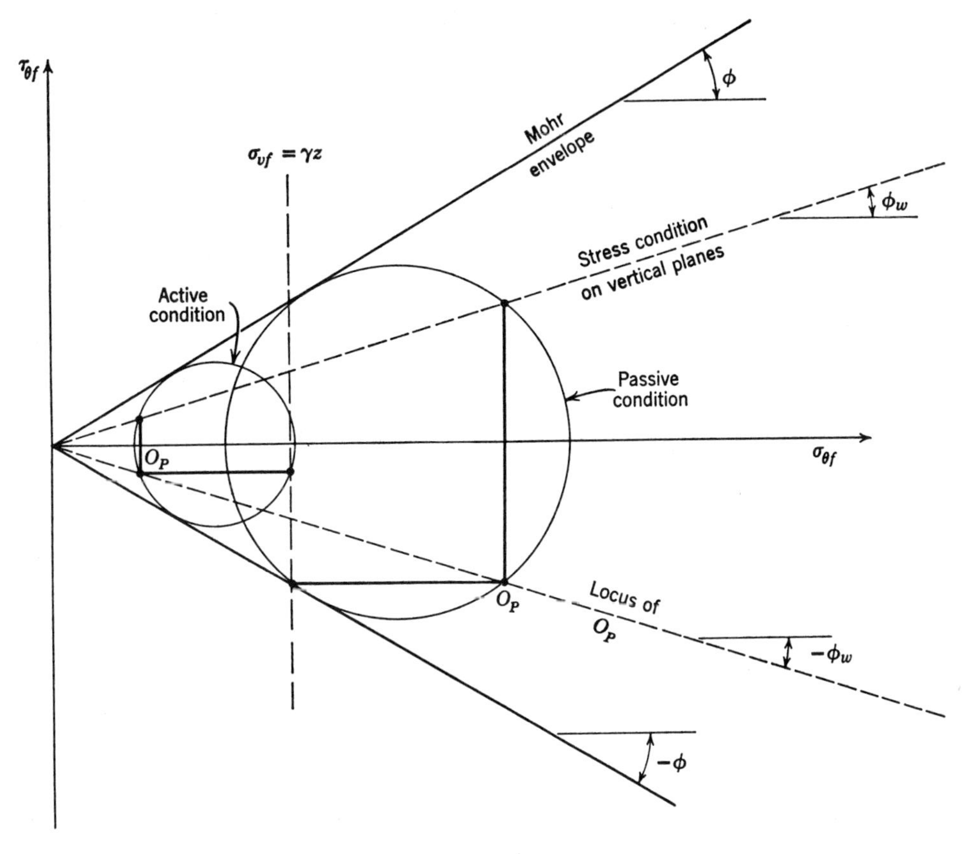

Fig. 13.8 Rankine states of stress for horizontal ground with shear stresses on vertical surfaces.

Other Than Geostatic Condition

The concepts of active and passive stress, and of conjugate stresses, apply to many problems in addition to the problem of horizontal geostatic stresses. For example, consider the case where the ground surface is level but there are equal shear stresses on all vertical planes. These stresses can be represented as $\tau_{vh} = \sigma_h \tan \phi_w$. Figure 13.8 shows the Mohr circle representation for the Rankine active and passive conditions for this case. The Mohr circles must satisfy the following conditions:

1. $\sigma_v = \gamma z$. The shear stresses on vertical planes do not alter this condition, since these shear stresses cancel on opposite sides of a column of soil.
2. The shear stress on a horizontal plane equals the given shear stress on a vertical plane but is of opposite sign.
3. The Mohr circle must pass through the point specified by conditions 1 and 2 and must be tangent to lines at $\pm\phi$.

Careful inspection of the figure will show that the origin of planes must lie along a line inclined at $-\phi_w$. Then a vertical line through the O_P will intersect the line at slope ϕ_w at a point giving the stresses on vertical planes. Example 13.1 illustrates the use of this construction. It is possible to derive equations giving the conjugate stresses for such situations (see Taylor, 1948). The form of the equations will differ from Eqs. 13.1 and 13.2, but the concepts remain the same.

13.3 SIMPLE RETAINING WALLS WITHOUT WALL FRICTION

Our next step is to consider the case of a simple retaining wall where (*a*) the backfill has a horizontal surface; (*b*) the face of the retaining wall in contact with the soil is vertical; and (*c*) there is no shear stress between the vertical face of the retaining wall and the soil. This simple case will serve to illustrate the concepts and methods needed for the solution of more complex problems. The active case will be considered first.

Active Thrust Using Rankine Zone

One way to evaluate the active thrust for this case is to assume that the active zone is a triangle and that everywhere within the triangle the soil is in the Rankine active condition. The slip lines for this assumed condition are shown in Fig. 13.9. Within the Rankine zone the horizontal stress at any depth z is

$$\sigma_h = K_a \gamma z \qquad (13.3)$$

where

γ = the unit weight of the soil
z = the depth below ground surface
K_a = the active stress coefficient, Eq. 13.1

The horizontal stress against the wall increases linearly with depth. Hence the total horizontal thrust against the wall will be

$$P_a = \tfrac{1}{2}\gamma H^2 K_a \tag{13.4}$$

where

H = height of the wall
P_a = active horizontal thrust

The resultant total thrust P_a will act at a point one-third of the distance from the bottom to the top of the wall.

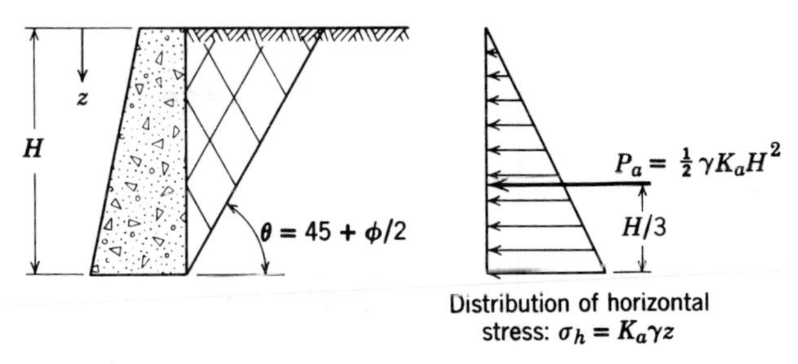

Fig. 13.9 Active thrust for simple Rankine case.

▸ Example 13.1

Given. Soil with horizontal surface, $\gamma = 17\ \text{kN/m}^3$, $\phi = 30°$. On vertical planes, $\tau_{vh} = -\sigma_h \tan 30°$.

Find. For active condition at depth of 6.47 m: horizontal stress, directions of principal stresses, orientation of slip lines.

Solution. A trial and error solution is necessary. First assume that the shear stress on the horizontal plane is given by point A' in Fig. E13.1. The Mohr circle corresponding to failure conditions is then as shown by the dashed circle. For this circle, the origin of planes is at $O_P{}'$ and the stresses on the vertical plane are given by point B'. This result does not satisfy the requirement that $\tau_{vh} = -\sigma_h \tan 30°$. Further trials show that the given conditions are satisfied only by the Mohr circle drawn as a solid line, with stresses on vertical planes at point A and stresses on horizontal planes at point B. ◂

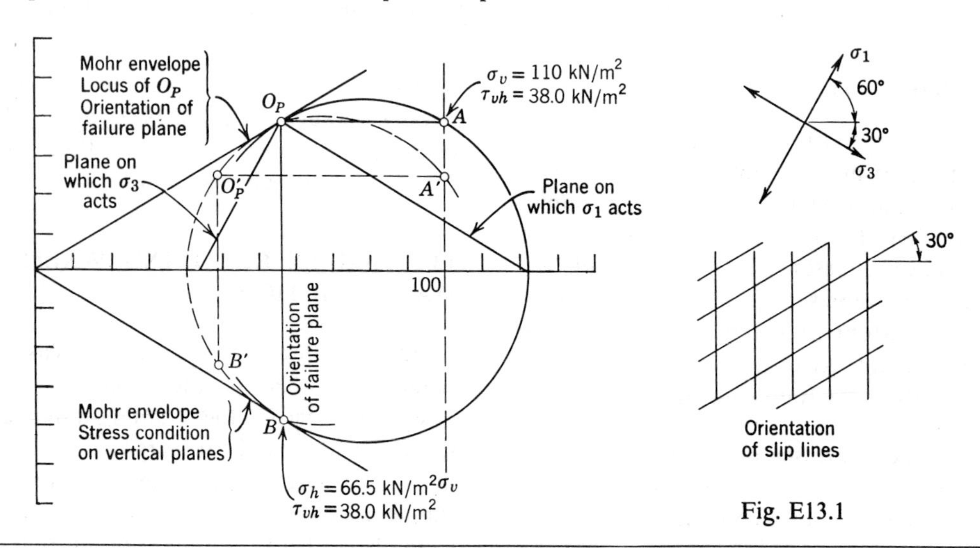

Fig. E13.1

▸ Example 13.2

Given. Retaining wall as shown in Fig. E13.2.

Find. For the active condition:

a. Horizontal stress at base of wall.
b. Total horizontal thrust.
c. Location of thrust.

Solution. From Table 13.1, find $K_a = 0.333$.

a. At base, $\sigma_h = (17.3)(6.1)(0.333) = 35.14\ \text{kN/m}^2$
b. $P_a = \tfrac{1}{2}(35.14)(6.1) = 107.18$ kN/m of wall
c. Thrust acts $6.1/3 = 2.03$ m above base of wall ◂

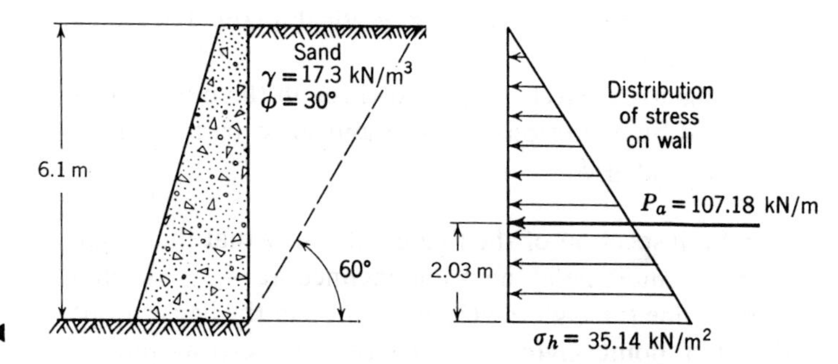

Fig. E13.2

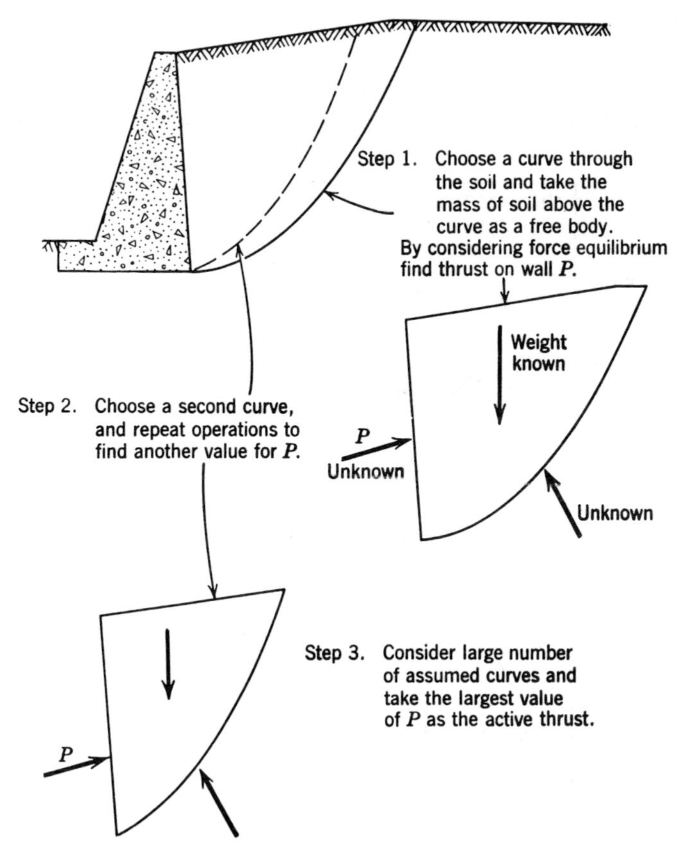

Step 1. Choose a curve through the soil and take the mass of soil above the curve as a free body.
By considering force equilibrium find thrust on wall P.
Weight known
P
Unknown
Unknown
Step 2. Choose a second curve, and repeat operations to find another value for P.
Step 3. Consider large number of assumed curves and take the largest value of P as the active thrust.
P

Fig. 13.10 Steps in trial wedge method of stability analysis.

Example 13.2 illustrates the computation of active thrust using these equations. The peak friction angle should be used to evaluate K_a. However, backfills are often in a rather loose condition; thus ϕ typically is about 30°.

This solution is intuitively satisfying. The requirement of equilibrium and the failure condition are fulfilled at each point within the Rankine zone, as are the boundary conditions along the surface of the backfill (no stress) and along the wall (no shear stress). However, this solution is not exact in the mathematical sense. This solution says nothing about the stresses outside of the failure zone; hence there is no complete assurance that the stresses outside the zone satisfy equilibrium without violating the failure law. There are other difficulties which will be discussed in Section 13.5.

Since the usefulness of Eqs. 13.1, 13.3, and 13.4 cannot be proved mathematically, this usefulness can only be demonstrated by comparing the predictions of these equations with actual measurements. Such comparisons have been made by Terzaghi (1934) and these equations have been found to give reasonable predictions for the conditions specified.

Active Thrust by Trial Wedges

The trial wedge method of analysis involves the following steps, which are illustrated in Fig. 13.10:

1. A mass of soil behind the wall is considered as a free body. The force P, which must exist between this free body and the wall, is found by writing the equations of equilibrium for the free body as a whole.
2. A different free body is considered, having a different boundary through the soil. Once again the required force P between the wall and the free body is found.
3. The actual force against the wall will be the *largest* value of P found as the result of considering all possible free bodies.

Even though the active thrust is the minimum possible thrust for which the backfill can be in equilibrium, we must seek the free body that gives the largest value of this thrust consistent with the assumption that the full shear strength of the soil is mobilized.

Figure 13.11 shows the application of the trial wedge method to the problem of a simple retaining wall without friction on the face. Example 13.3 illustrates the computations. Only those free bodies bounded by straight lines through the heel of the wall are considered. There are distributed normal stresses along IJ and JM and distributed shear stresses along JM, but the desired analysis can be carried out in terms of the resultants P and F of these distributed stresses.

▶ Example 13.3

Given. Retaining wall and backfill of Example 13.2.
Find. Active thrust by trial wedge method.
Solution. Figure E13.3-1 shows to scale the free bodies and the force polygons for $\theta = 45°$

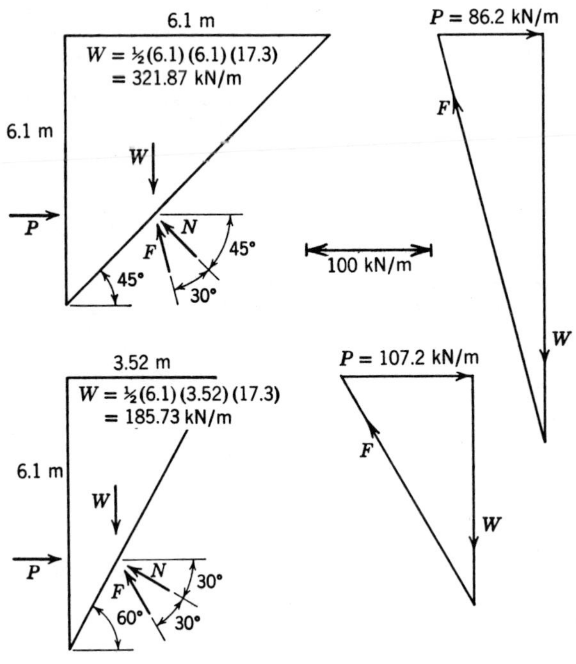

Fig. E13.3-1

and $\theta = 60°$. Equation 13.5 may be used to evaluate P for many values of θ.

θ	$\cot\theta$	$\tan(\theta - 30°)$	Product	P
55°	0.700	0.466	0.328	105.6
$57\frac{1}{2}°$	0.637	0.520	0.331	106.5
60°	0.577	0.577	0.333	107.2
$62\frac{1}{2}°$	0.521	0.637	0.331	106.5
65°	0.467	0.700	0.328	105.6

The plot in Fig. E13.3-2 shows graphically the manner in which P varies with θ.

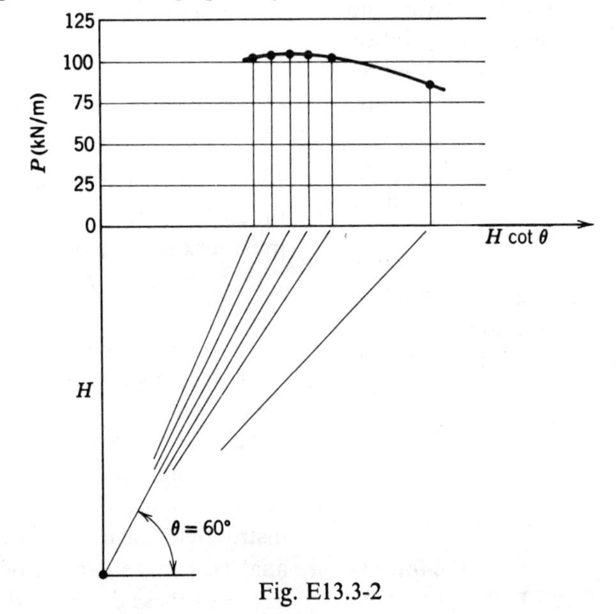

Fig. E13.3-2 ◀

Forces acting on the free body:
W = weight of soil $= \frac{1}{2}\gamma H^2 \cot\theta$
P = resultant of distributed stresses between soil and wall
N = resultant of normal stresses within soil along assumed plane
T = resultant of shear stresses within soil along assumed plane $= N\tan\phi$
F = resultant of N and T

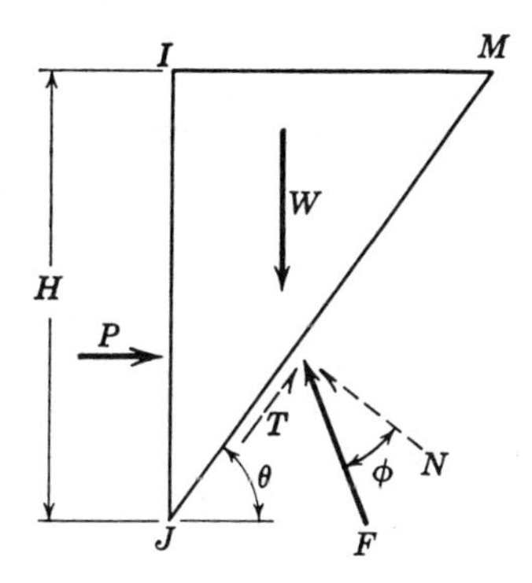

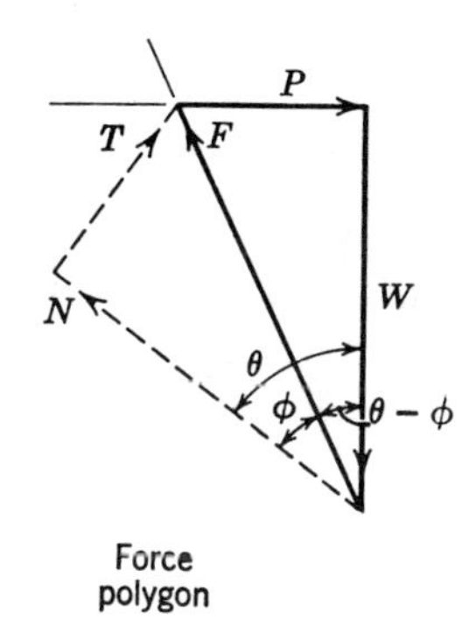

Equations of equilibrium

$$\Sigma V = O: \quad F = \frac{W}{\cos(\theta - \phi)}$$

$$\Sigma H = O: \quad P = W\tan(\theta - \phi)$$

$$\therefore \quad P = \tfrac{1}{2}\gamma H^2 \cot\theta \tan(\theta - \phi) \qquad (13.5)$$

Fig. 13.11 Equilibrium of trial wedge for simple retaining wall: active case.

Step 1. Place free body in equilibrium. The weight W is known in magnitude and direction. The resultant forces P and F are determined as to direction, but not as to magnitude. Hence there are two unknowns (the magnitudes of P and F) and two equations of force equilibrium. The problem is statically determinate and thus may be solved by statics alone.

In order to solve this equilibrium problem a *force polygon* is useful. The forces acting on the free body are plotted as vectors, with the tail of one vector connected to the head of another vector. In this problem the vector W is first plotted to some convenient scale. Then the directions of P and F are laid off and their intersection gives the closure of the force polygon. The magnitudes of P and F may be scaled from the diagram, or alternatively the force polygon may be used to guide the writing of a pair of equations which are then solved to give the magnitudes of P and F.

Steps 2 and 3. Search for critical free body. There are several ways in which the search for the most critical free body may be performed.

One way is to assume various inclinations of the failure line and determine the value of P corresponding to each inclination. Either Eq. 13.5 may simply be evaluated for several different values of θ, or a force polygon may be constructed for each θ and then P scaled graphically. Example 13.3 illustrates the variation of P with θ and shows a convenient way to plot the results. The thrust P is greatest when $\theta = 60°$. If P were to be less than the computed value, the backfill would fail along a slope with this inclination.

For this simple case it is possible to carry out the search mathematically (see Example 13.4). The

▶ Example 13.4

Given. Equation 13.5 (in Fig. 13.11) for P as a function of θ.

Find. Maximum value of P and θ for which this maximum occurs.

Solution.

$$\frac{\partial P}{\partial \theta} = \tfrac{1}{2}\gamma H^2\left[-\frac{\tan(\theta-\phi)}{\sin^2\theta} + \frac{\cot\theta}{\cos^2(\theta-\phi)}\right]$$

$$= \tfrac{1}{2}\gamma H^2 \frac{-\sin(\theta-\phi)\cos(\theta-\phi) + \sin\theta\cos\theta}{[\sin\theta\cos(\theta-\phi)]^2}$$

$$= \tfrac{1}{2}\gamma H^2 \frac{-\sin\theta\cos\theta(\cos^2\phi - \sin^2\phi - 1) - \sin\phi\cos\phi(\sin^2\theta - \cos^2\theta)}{[\sin\theta\cos(\theta-\phi)]^2}$$

$$= \tfrac{1}{2}\gamma H^2 \frac{\sin 2\theta\sin^2\phi + \sin\phi\cos\phi\cos 2\theta}{[\sin\theta\cos(\theta-\phi)]^2}$$

$$= \tfrac{1}{2}\gamma H^2 \frac{\sin\phi\cos(2\theta-\phi)}{[\sin\theta\cos(\theta-\phi)]^2}$$

This is zero when $\cos(2\theta - \phi) = 0$ or $2\theta_{cr} - \phi = 90°$ or $\theta_{cr} = 45 + \phi/2$

Substituting in Eq. 13.5,

$$P_a = \tfrac{1}{2}\gamma H^2 \cot\left(45 + \frac{\phi}{2}\right)\tan\left(45 - \frac{\phi}{2}\right)$$

$$= \tfrac{1}{2}\gamma H^2 \tan^2\left(45 - \frac{\phi}{2}\right) = \tfrac{1}{2}\gamma H^2 K_a$$ ◀

equilibrium equation contains the variable θ, which defines the boundary of the free body through the soil. By maximizing the expression for P with respect to θ, the actual thrust as well as the location of the critical plane of sliding can be found. A graphical procedure for finding the critical inclination is also available (see Taylor, 1948, p. 497).

The maximum thrust found by these procedures is the active thrust P_a.

Moment Equilibrium for Trial Wedge

The line of action of the vector W is through the centroid of the trial wedge. One possible location for the vectors P and F is shown in Fig. 13.12: P acts at the third-point of the wall and F acts at the third-point of the failure surface. These locations of P and F are consistent with a linear variation of stress with depth.

Critique of Trial Wedge Method

The trial wedge method does not consider stress conditions either within the trial wedge or outside of the wedge, and again there is no complete assurance that the stresses

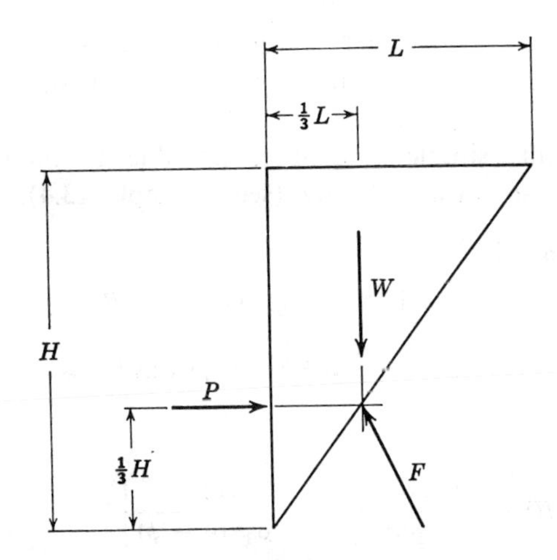

Fig. 13.12 Location of forces so that trial wedge is in moment equilibrium.

within and below the wedge satisfy equilibrium without violating the failure law. Thus, although the solution is intuitively satisfactory, it cannot be mathematically proven to be exact.

For the conditions considered in this section, the trial wedge method gives exactly the same result as does the solution using the Rankine zone. Indeed, for this case the trial wedge method only repeats the steps which led (in Chapter 8) to the equations for the Mohr circle. The difference between the methods becomes greater as we turn to more complex situations.

The trial wedge method was originated by the French engineer Coulomb in 1776, almost a century before Rankine (apparently without knowledge of Coulomb's work) published his analysis. Coulomb can thus be regarded as the founder of the theories for active earth thrust. The contribution of Rankine was to introduce the concept of passive stress and to tie together the two extreme cases of active and passive stress.

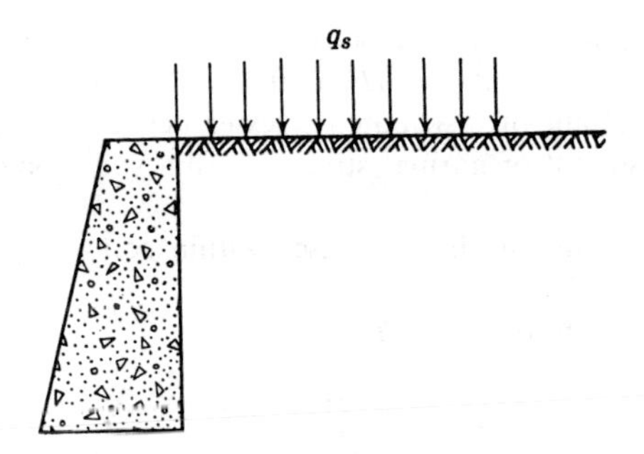

Fig. 13.13 Retaining wall with uniform surcharge.

Active Thrust with Uniform Surcharge

The methods of solution presented in the preceding paragraphs can readily be extended to cover situations in which there is a surcharge over the surface of the backfill behind the retaining wall. Such a surcharge might arise from stored material or parked vehicles.

With a uniform surcharge q_s (Fig. 13.13), the vertical stress at any depth is simply[3]

$$\sigma_v = q_s + \gamma z$$

The horizontal stress is $\sigma_h = K_a\sigma_v$ where K_a is still as given by Eq. 13.1. Hence the horizontal stress at any depth is

$$\sigma_h = (q_s + \gamma z)\frac{1 - \sin\phi}{1 + \sin\phi} = (q_s + \gamma z)K_a \qquad (13.6)$$

[3] Note that q_s denotes an entirely different quantity than does $q = (\sigma_v - \sigma_h)/2$.

▸ Example 13.5

Given. Retaining wall of Example 13.2, with surcharge of 50 kN/m²

Find. Active thrust against wall, and location of this thrust.

Solution. Additional thrust from surcharge (see Fig. E13.5) is

$$(50)(6.1)(\tfrac{1}{3}) = 101.67 \text{ kN/m}$$

$$P\bar{x} = 101.67(3.05) + 107.18(2.03) = 527.67$$

$$\bar{x} = \frac{527.67}{208.85} = 2.53 \text{ m}$$

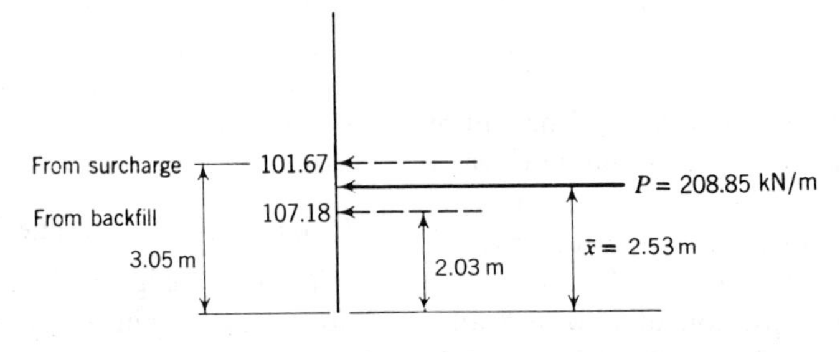

Fig. E13.5 ◂

▶ **Example 13.6**

Given. Retaining wall and backfill of Example 13.2.

Find. Passive thrust, passive stresses, location of slip line, location of resultant passive thrust.

Solution. See Fig. E13.6:

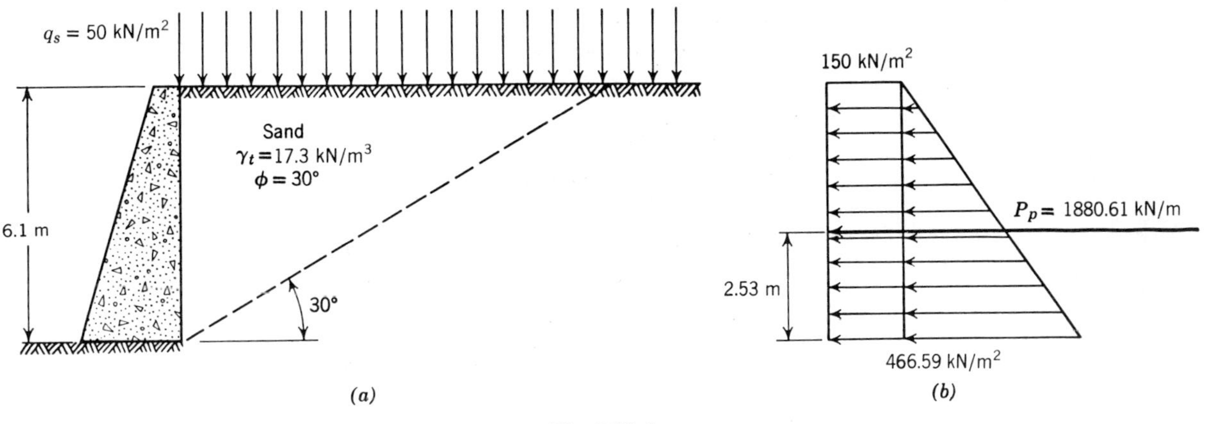

Fig. E13.6

At base: $\sigma_{hp} = [(6.1)(17.3) + 50](3) = [105.53 + 50](3) = 466.59 \text{ kN/m}^2$

From Fig. E13.6*b*:

$$P_p = [\tfrac{1}{2}(6.1)^2(17.3) + (6.1)(50)](3) = [321.87 + 305.0](3) = 1880.61 \text{ kN/m}$$

$$\bar{x} = \frac{965.61(6.1/3) + 915(3.05)}{1880.61} = 2.53 \text{ m}$$ ◀

The total active thrust against the wall is then given by

$$P_a = \tfrac{1}{2}\gamma H^2 K_a + q_s H K_a \tag{13.7}$$

Note that the horizontal stress resulting from the surcharge is distributed uniformly with depth, and hence the resultant force corresponding to the surcharge is located at midheight of the wall. Thus the resultant of the total thrust, reflecting the effects of surcharge and weight of soil, will lie between midheight and the third point. The location of the resultant of the total thrust is found by vectorial addition of the thrusts for each of the two components. This is illustrated in Example 13.5, in which Example 13.2 is extended to include the effects of a surcharge of 50 kN/m². The additional thrust of 101.67 kN/m² acts at midheight of the wall, or 3.05 m above the base. The resultant of this thrust plus that from the weight of the soil (see Example 13.5) acts 2.53 m above the base of the wall.

The trial wedge procedure can be used to obtain the same result. The surcharge causes another force on the free body, but this force simply adds to the weight vector W. The location of the critical surface is not affected.

Passive Resistance

Assuming that soil which offers passive resistance is in the passive Rankine condition, the passive stress and total passive resistance are given by

$$\sigma_h = \gamma z K_p + q_s K_p \tag{13.8}$$

$$P_p = \tfrac{1}{2}\gamma H^2 K_p + q_s H K_p \tag{13.9}$$

where K_p is as given by Eq. 13.2. Here H is the depth of the passive zone and q_s is the surcharge on the passive zone. The use of these equations is illustrated in Example 13.6.

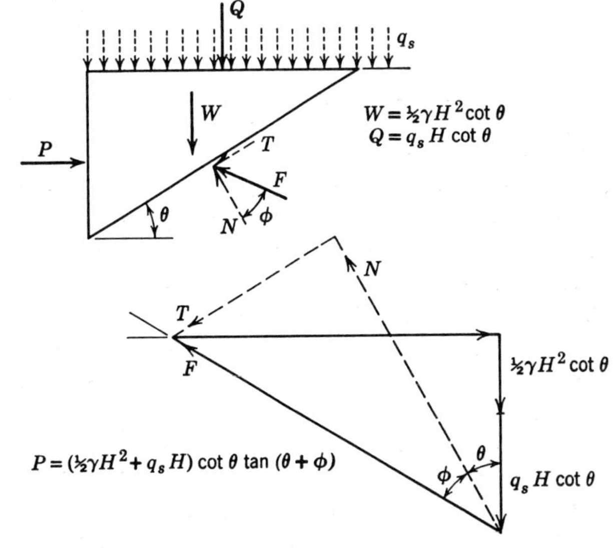

Fig. 13.14 Equilibrium of trial wedge for simple retaining wall: passive case.

▶ Example 13.7

Given. Retaining wall and backfill of Example 13.6.
Find. Passive thrust by trial wedge method.

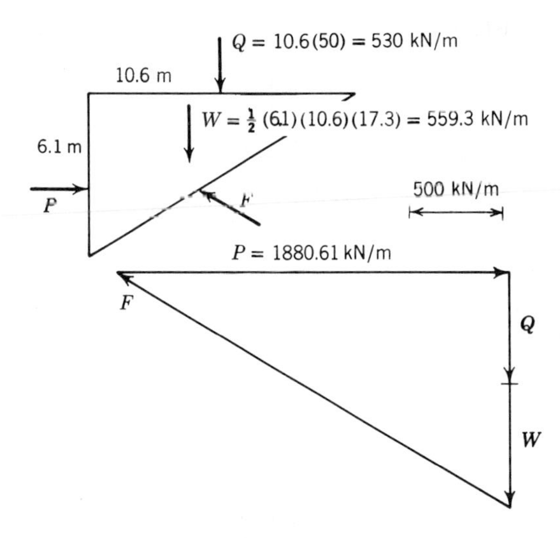

Fig. E13.7

Solution. The force polygon and free body for $\theta = 30°$ are shown in Fig. E13.7. Using equation in Fig. 13.14:

θ	$\cot\theta$	$\tan(\theta + 30°)$	$P/(\frac{1}{2}\gamma H^2 + q_s H)$
20°	2.75	1.192	3.28
25°	2.145	1.428	3.06
30°	1.732	1.732	3.00 ←
35°	1.428	2.145	3.06
40°	1.192	2.75	3.28

$$
\begin{aligned}
P_p &= 3[\tfrac{1}{2}(17.3)(6.1)^2 + 50(6.1)] \\
&= 3[321.87 + 305] \\
&= 1880.61 \text{ kN/m}
\end{aligned}
$$

As in Example 13.6, resultant thrust is located 2.53 m above base of wall. ◀

The trial wedge method for the passive case is basically similar to that for the active case, with but one significant difference: now the *shear stresses* on the failure surface act *together* with the *weight* of the soil to resist the horizontal thrust from the wall. Thus, even though the passive thrust is the maximum possible thrust for which the soil can be in equilibrium, we must seek the free body that leads to the smallest value for the thrust. If the wall applies a thrust greater than this smallest passive thrust, the soil will not be in equilibrium. Figure 13.14 shows the formulation of the passive problem using straight failure surfaces. Example 13.7 illustrates the method.

As was the case for the active thrust, both methods of solution give the same intuitively satisfying result for the case of a simple wall without wall friction. However, the only true justification for the use of Eqs. 13.2, 13.8, and 13.9 lies in the agreement between the predictions of those equations and actual observed results.

13.4 RETAINING WALLS WITH WALL FRICTION

Generally shear forces develop between the face of a retaining wall and the backfill because of relative motions between the wall and backfill. Figure 13.3 illustrated the typical patterns of motion. In the active zone, the outward stretching leads to downward motion of the soil relative to the wall. Because of friction between the soil and wall, this motion causes a downward shear force on the wall. Such a downward shear upon the wall is called positive wall friction for the active case (see Fig. 13.15). In the passive zone, the horizontal compression must be

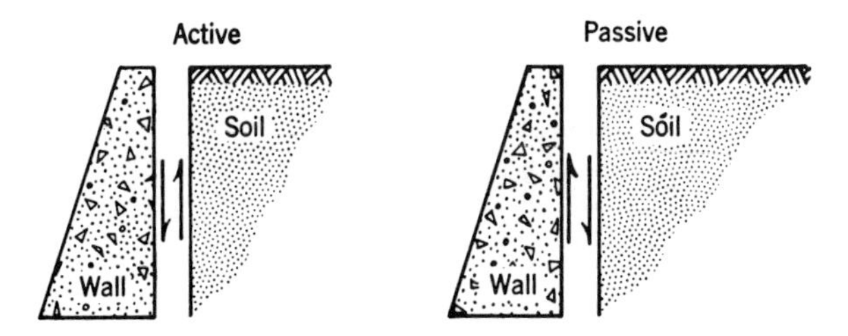

Fig. 13.15 Direction of positive wall friction.

accompanied by an upward bulging of the soil, and hence there tends to be an upward drag on the wall. Such an upward shear on the wall is called positive wall friction for the passive case. In the active case wall friction is almost always positive. Either positive or negative wall friction may develop in the passive case. Whether wall friction is present, and the sign of this friction, must be determined from a study of the motions expected for each problem.

The magnitude of this shear force is controlled by the friction angle ϕ_w between the soil and the wall. As noted in Chapter 11, ϕ_w usually is about equal to ϕ_{cv} and typically has a value of about 30°. For a loose backfill ϕ and ϕ_w will be numerically equal, whereas $\phi_w < \phi$ for a dense backfill.

Solution Using Failure Zone

The conditions that must be fulfilled along the boundaries of the failure zone are sketched in Fig. 13.16. Along the surface of the backfill there are no shear stresses on horizontal and vertical surfaces. Hence at this surface the slip lines must be inclined at $\pm(45 + \phi/2)$ to the horizontal. Along the wall, however, the ratio of shear to normal stresses must equal $\tan \phi_w$. Thus at the wall the stress conditions must be as sketched in Example 13.1, and the slip lines have the inclination shown in Fig. 13.16. Hence different Rankine states apply within different portions of the backfill.

The solution of this boundary value problem now becomes quite complicated. In order that equilibrium be satisfied within the failure zone, the stresses must satisfy the differential equations of equilibrium.[4]

$$\frac{\partial \sigma_v}{\partial z} - \frac{\partial \tau_{vh}}{\partial x} - \gamma = 0 \qquad (13.10a)$$

$$\frac{\partial \sigma_h}{\partial x} + \frac{\partial \tau_{vh}}{\partial z} = 0 \qquad (13.10b)$$

In addition, the failure condition must be fulfilled throughout the failure zone:

$$\tau_{ff} = \sigma_{ff} \tan \phi \qquad (13.11)$$

[4] See Crandall and Dahl (1959, p. 127) for a derivation of these equations. The special sign convention used in soil mechanics must be taken into account.

Combining Eqs. 13.10 and 13.11 leads to an equation called Kötter's equation. Solution of this equation, for the boundary conditions as shown in Fig. 13.16, gives the orientation of the slip lines together with the stresses at each point of the failure zone (see Sokolovski, 1965; Harr, 1966). A numerical integration technique is necessary in order to obtain this solution.

A complete derivation of Kötter's equation, and the numerical integration technique used for its solution, are beyond the scope of this text. Figure 13.16 illustrates the results by showing the slip-line field construction by this method for the case of $\phi = \phi_w = 30°$. The resulting coefficient of active stress is 0.31. Now K_a is no longer the ratio of vertical to horizontal stress but is the ratio

$$K_a = \frac{\sqrt{\tau_{vh}^2 + \sigma_h^2}}{\gamma z}$$

for stresses at the wall. Note that σ_v is *not* necessarily equal to γz, owing to the curvature of the slip-line field. The active thrust is

$$P_a = \tfrac{1}{2}\gamma H^2 K_a = 0.31(\tfrac{1}{2}\gamma H^2)$$

and is inclined to the horizontal at the angle of the wall friction. Along the wall all components of stress still increase linearly with depth, and so the resultant thrust still acts at the third-point of the wall.

A separate numerical integration must be made for each value of ϕ and ϕ_w. Sokolovski (1965) presents a table giving these results.

Active Thrust by Trial Wedges

Figure 13.17 shows the general formulation of this problem using straight failure surfaces and Example 13.8 illustrates a specific case. The force polygon is modified since P is now inclined instead of horizontal. Otherwise

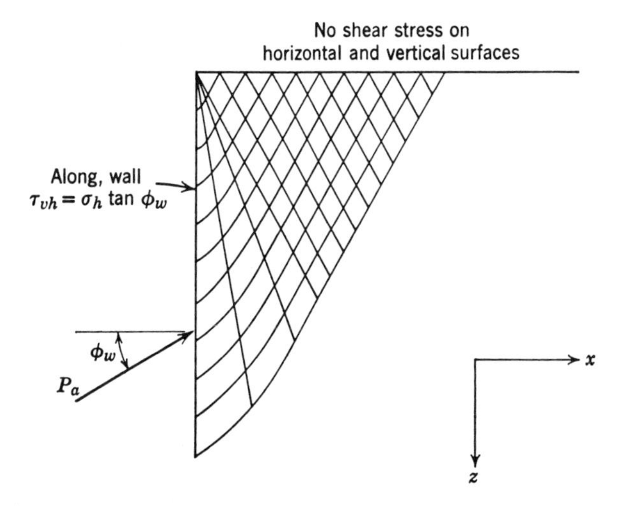

Fig. 13.16 Slip-line field and failure zone for case with wall friction. Slip-line field for $\phi = \phi_w = 30°$ by method of Sokolovski (1965).

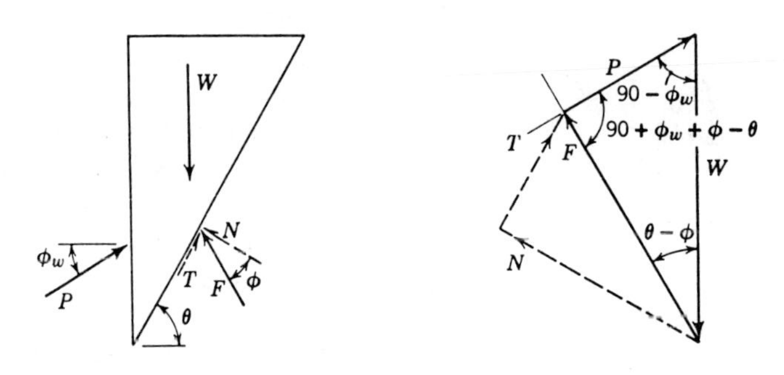

Fig. 13.17 Equilibrium of trial wedge for simple retaining wall with friction.

By law of sines:

$$P = W \frac{\sin(\theta - \phi)}{\sin(90 + \phi_w + \phi - \theta)}$$

$$= \tfrac{1}{2}\gamma H^2 \cot\theta \frac{\tan(\theta - \phi)}{\cos\phi_w + \sin\phi_w \tan(\theta - \phi)}$$

▶ Example 13.8

Given. Retaining wall and backfill of Example 13.2, except that now there is wall friction $\phi_w = 30°$.

Find. Active thrust by trial wedge method.

Solution. Figure E13.8 shows the free body and force polygon for $\theta = 60°$. The equation in Fig. 13.17 may be used to evaluate P for many values of θ.

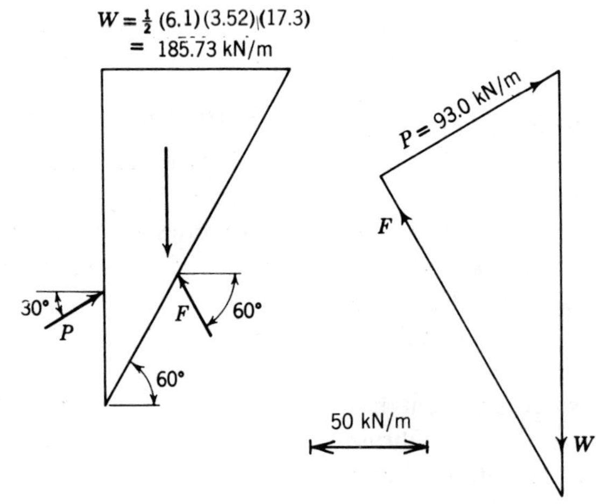

Figure E13.8

θ	$\cot\theta$	$\tan(\theta - 30°)$	$0.866 + \frac{1}{2}\tan(\theta - \phi)$	$\frac{P}{\frac{1}{2}\gamma H^2}$
50	0.839	0.364	1.048	0.292
$52\frac{1}{2}$	0.767	0.414	1.073	0.296
55	0.700	0.467	1.100	0.297 ←
$57\frac{1}{2}$	0.637	0.520	1.126	0.295
60	0.577	0.577	1.154	0.289

$$P_a = 0.297(\tfrac{1}{2}\gamma H^2) = 95.59 \text{ kN/m}$$ ◀

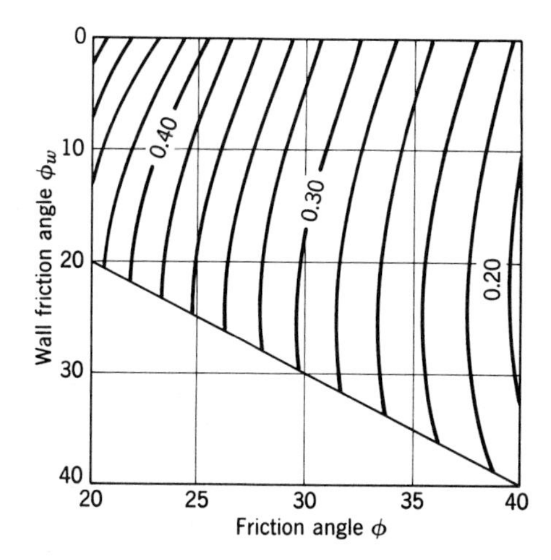

Fig. 13.18 Coefficient of active stress as function of wall friction.

the same general procedure is followed. Moment equilibrium is satisfied with P located at the third-point of the wall, but F is no longer located at the third-point of the failure surface.

The critical straight failure surface found by this method is an approximation to the more exact failure surface indicated in Fig. 13.16. The failure surface of the trial wedge satisfies the boundary conditions neither at the top surface of the backfill nor at the wall. Note that the inclination of this surface is no longer equal to $45 + \phi/2$.

Figure 13.18 gives values of K_a calculated using the trial wedge procedure with straight-line failure surfaces. These values of K_a may be used in Eqs. 13.3 or 13.4 to give the stress against the wall at any depth or the thrust against the wall. The stress thus calculated is $\sqrt{\tau_{vh}^2 + \sigma_h^2}$ rather than just the horizontal stress, and the thrust thus calculated is at the angle ϕ_w to the horizontal rather than the horizontal thrust. Thus wall friction has two effects upon active thrust: (*a*) on the magnitude of P_a and (*b*) on the direction of P_a. The second of these effects is usually the more important, as is shown by the comparison in Example 13.9. Wall friction changed the active thrust by only 7%, but decreased the horizontal component of this thrust by 24%.

▶ Example 13.9

Given. Retaining wall of height H with backfill having $\phi = 35°$ and unit weight γ.

Find. The effect of wall friction ($\phi_w = 35°$) upon (*a*) the active thrust and (*b*) the horizontal component of active thrust.

Solution. The difference is that between K_a and $K_a \cos \phi_w$. Fig. 13.18 was used for the following tabulation.

	$\phi_w = 0$	$\phi_w = 35°$	Percent difference
K_a	0.27	0.25	7%
$K_a \cos \phi_w$	0.27	0.204	24%

◀

Use of curved failure surfaces for trial wedges leads to slightly more critical free bodies and to slightly greater values of K_a. However, the differences in K_a are at most a few percent and generally so small they are undetectable in a plot such as Fig. 13.18. The critical curved failure surfaces, and the values of K_a, are almost exactly the same as those found by the method of Sokolovski. However, the important thing is that these results are in reasonable agreement with the few actual measurements which have been made in large-scale tests.

Passive Resistance

For the passive case the trial wedge method using straight failure surfaces significantly overestimates the resistance. That is to say, trial wedge solutions using curved failure surfaces (see Fig. 13.19) give a smaller passive resistance than the passive resistance computed using straight surfaces. The difference increases with increasing wall friction. The technique of solution using the trial wedge method with curved boundaries is described in detail in Terzaghi (1943) and Terzaghi and Peck (1967). Figure 13.20 gives passive stress coefficients obtained in this way. Alternatively, the method of Sokolovski (1965) may be used. Both approaches give essentially the same answer. The thrust computed using the coefficients in Fig. 13.20 is inclined to the horizontal at an angle corresponding to the wall friction.

The theoretical predictions regarding passive resistance with wall friction are not as well confirmed by experiment

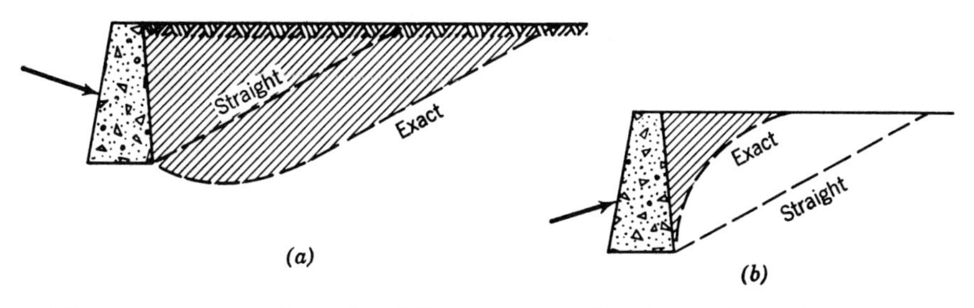

Fig. 13.19 Comparison of passive failure zone predicted by trial wedge method using straight and curved slip lines. (*a*) Positive wall friction. (*b*) Negative wall friction.

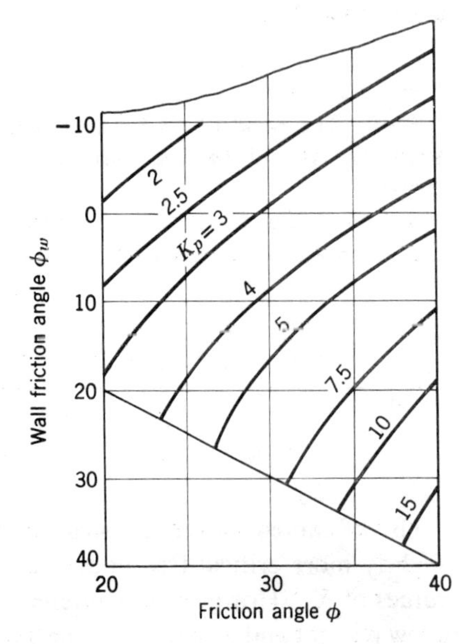

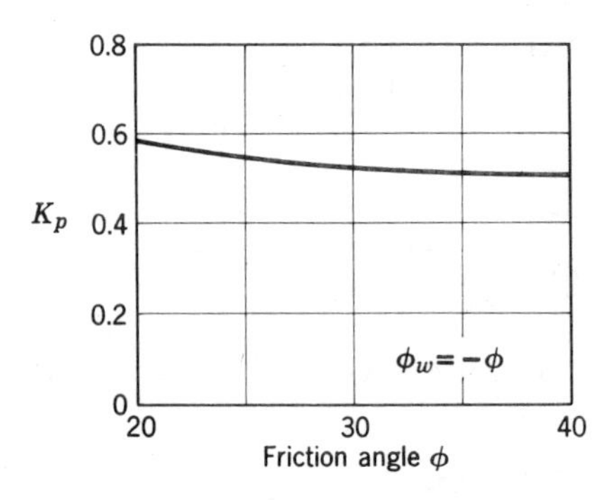

Fig. 13.20 Chart for passive stress coefficient (based on Caquot and Kérisel, 1949).

as the predictions regarding active thrust, and hence cannot be used with as much confidence. Tschebotarioff (1951) reports on the results of a few large-scale laboratory tests.

Surcharge

In general, addition of a surcharge changes somewhat the slip-line field as obtained by the method of Sokolovski or by trial wedges using curved failure surfaces. Hence Eqs. 13.7 and 13.9 do not apply exactly unless simple geostatic conditions exist; i.e., the thrust should be evaluated separately for each different combination of q_s and γ. However, within the accuracy needed for engineering computations (and keeping in mind the uncertainty as to just what is an "exact" solution) Eqs. 13.7 and 13.9 may still be used together with values of K_a or K_p computed for zero surcharge.

13.5 ACTIVE THRUST AND PASSIVE RESISTANCE FOR OTHER CONDITIONS

The foregoing sections have given results which can be applied to simple retaining walls, and, more important, they have illustrated the methods that can be used to handle more complicated situations.

Active thrust from a homogeneous backfill generally can be evaluated with reasonable accuracy using Eq. 13.7. For relatively simple boundary conditions, values of K_a can be obtained from tables, charts, and formulas so that it usually is not necessary to go through a series of trial wedge computations. Figure 13.21 gives a formula (Eq. 13.12) applicable for inclined retaining walls and backfills, including the effect of wall friction. The coefficient of $\frac{1}{2}\gamma H^2$ in this formula is K_a. The direction of P_a is as indicated in the figure. This formula was derived (by Coulomb in 1776!) by the trial wedge procedure using straight failure surfaces, but the general accuracy of the results has been confirmed by calculations using the method of Sokolovski.

Figure 13.22 gives values of K_a for the special case of zero wall friction. This table can be used to estimate the thrust for the case of wall friction, as is illustrated in Example 13.10. Example 13.11 shows the application of the active stress coefficients to a problem with surcharge. Note that q_s in Eq. 13.7 is the surcharge per horizontal area regardless of the slope of the backfill.

Similarly, Eq. 13.9 may be used to evaluate passive resistance in more complicated problems. The trial

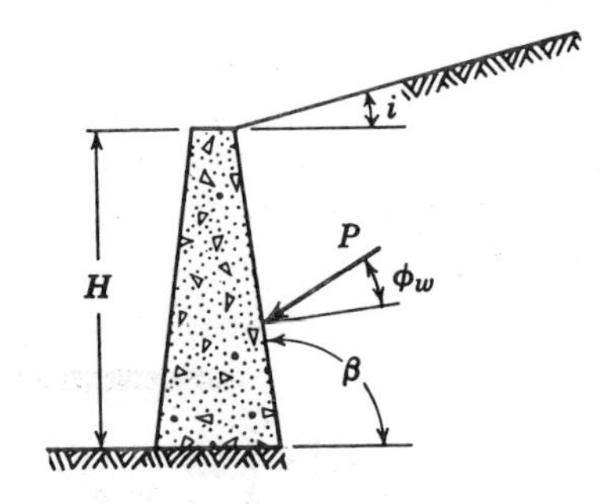

Fig. 13.21 Coulomb equation for sloping backfill and wall friction:

$$P_a = \frac{1}{2}\gamma H^2 \left\{ \frac{\csc\beta \sin(\beta - \phi)}{\sqrt{\sin(\beta + \phi_w)} + \sqrt{\dfrac{\sin(\phi + \phi_w)\sin(\phi - i)}{\sin(\beta - i)}}} \right\}^2 \tag{13.12}$$

	i =		−30°	−12°	±0	+12° 1:4.7	+30° 1:1.7
$\phi = 20°$	$\beta' = +20°$			0.57	0.65	0.81	
	$\beta' = +10°$			0.50	0.55	0.68	
	$\beta' = \pm 0°$			0.44	0.49	0.60	
	$\beta' = -10°$			0.38	0.42	0.50	
	$\beta' = -20°$			0.32	0.35	0.40	
$\phi = 30°$	$\beta' = +20°$		0.34	0.43	0.50	0.59	1.17
	$\beta' = +10°$		0.30	0.36	0.41	0.48	0.92
	$\beta' = \pm 0°$		0.26	0.30	0.33	0.38	0.75
	$\beta' = -10°$		0.22	0.25	0.27	0.31	0.61
	$\beta' = -20°$		0.18	0.20	0.21	0.24	0.50
$\phi = 40°$	$\beta' = +20°$		0.27	0.33	0.38	0.43	0.59
	$\beta' = +10°$		0.22	0.26	0.29	0.32	0.43
	$\beta' = \pm 0°$		0.18	0.20	0.22	0.24	0.32
	$\beta' = -10°$		0.13	0.15	0.16	0.17	0.24
	$\beta' = -20°$		0.10	0.10	0.11	0.12	0.16

for $\phi_w = 0$; $\beta' = \beta - 90°$

Fig. 13.22 Coefficient of active stress as function of inclination of wall and backfill.

wedge or Sokolovski method must usually be used to find K_p. These methods may also be employed to find either active thrust or passive resistance for more complicated situations such as stratified backfills, irregularly shaped backfills or walls, nonuniform surcharge, etc. These applications in connection with gravity retaining walls are discussed in Huntington (1957). Application of Kötter's equation to complex problems involving other types of retaining structures is given by Hansen (1953).

General Evaluation of Limiting Equilibrium Solutions

It has already been noted that the methods used to obtain the solutions given in Sections 13.3 and 13.4 are not exact in a mathematical sense. That is, it cannot be proved by mathematics alone that these solutions give a unique solution for the assumed boundary conditions.

A complete, exact solution for an active or passive condition of limiting equilibrium must meet the following five conditions:

1. Each point within the soil mass must be in equilibrium. Hence the pattern of stresses must satisfy the differential equations of equilibrium, Eqs. 13.10.
2. The Mohr-Coulomb failure condition must not be violated at any point; for any plane through any point,

$$\tau_\theta \leq c + \sigma_\theta \tan \phi \tag{13.13}$$

3. The strains that occur must be related to the stresses through a stress-strain relationship suitable for the soil.
4. The strains that occur at each point must be compatible with the strains at all surrounding points.
5. The stresses within the soil must be in equilibrium with the stresses applied to the soil.

The requirement of using a suitable stress-strain relationship imposes the greatest obstacle for obtaining an exact solution. It is necessary to consider the strains that occur once the failure condition is reached (such as the volume increase which accompanies shear distortion) as well as the strains for stresses less than failure. Progress has been made in developing methods for handling such complex stress-strain relationships, e.g., Christian (1966). Almost all limiting equilibrium solutions assume that soil is rigid-plastic, which means that there are no strains at any point until the failure condition is fulfilled. Haythornthwaite (1961) discusses the general theory of limiting equilibrium in rigid-plastic materials obeying the Mohr-Coulomb failure law. Upper and lower bound theorems have been developed. However, in view of the uncertainty as to a proper stress-strain relationship, the

▶ Example 13.10

Given. Retaining wall and backfill as shown in Fig. E13.10-1.

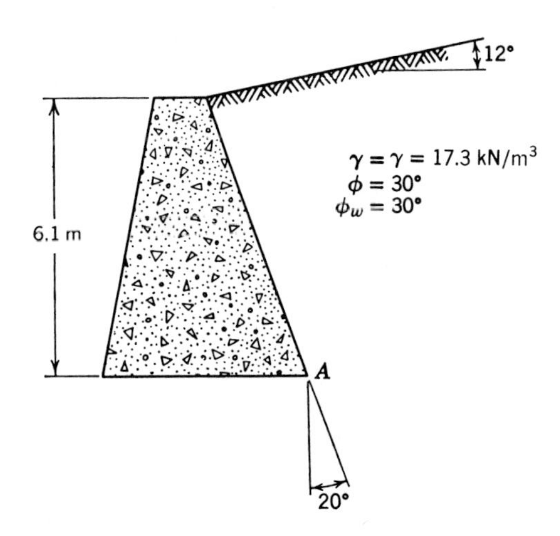

Fig. E13.10-1

Find. Moment of active thrust about point A.
Solution Using Eq. 13.12.

$$i = 12° \quad \beta = 110°$$

$$\csc 110° \sin 80° = \frac{\sin 80°}{\sin 70°} = 1.049$$

$$\sqrt{\sin 140°} = 0.803$$

$$\sqrt{\frac{\sin 60° \sin 28°}{\sin 98°}} = \sqrt{\frac{0.866 \times 0.470}{0.990}} = 0.641$$

$$P_a = \tfrac{1}{2}(17.3)(6.1)^2 \left[\frac{1.049}{0.803 + 0.641}\right]^2 = 321.87(0.528) = 169.95 \text{ kN/m}$$

Normal component of P_a:

$$P_a \cos 30° = 147.18 \text{ kN/m}$$

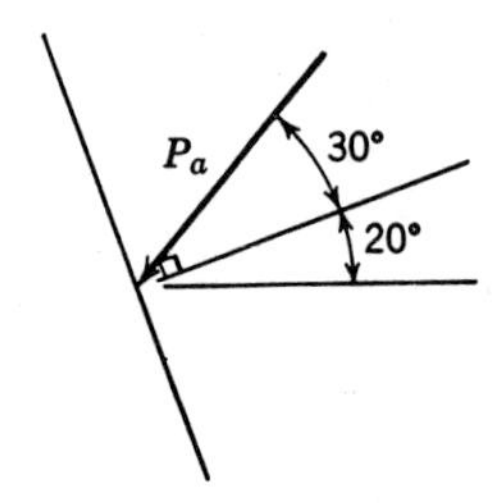

Fig. E13.10-2

P_a acts $\frac{1}{3}$ of way up wall, or at slant distance of 2.16 m above base (see Fig. E13.10-2).
Moment of P_a about point A = 147.18 × 2.16 = 317.90 kN m/m

Approximate Solution Using Fig. 13.22. Use K_a for $\phi_w = 0$, but incline P_a at $\phi_w = 30°$ to normal to wall.

$K_a = 0.59$ instead of 0.528 above, so that moment is overestimated by 12%. ◀

▶ Example 13.11

Given. Retaining wall and backfill as shown in Fig. E13.11-1. Surcharge = 24 kN/m² of slope.

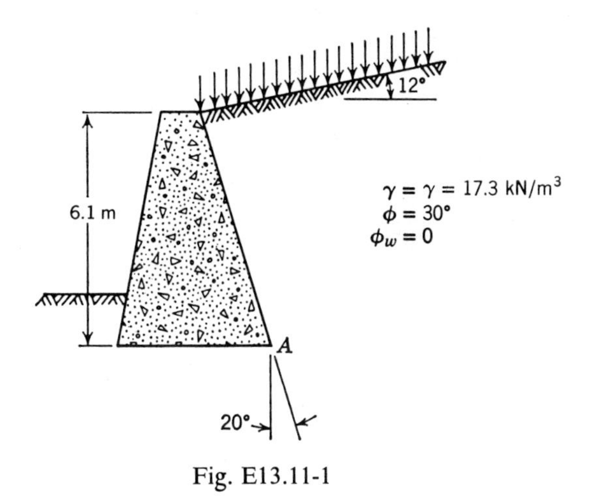

Fig. E13.11-1

Find. Moment of active thrust about point A.
Solution. See Fig. E13.11-2.

$$q_s = 24/\cos 12^\circ = 24.54 \text{ kN/horizontal m}^2$$

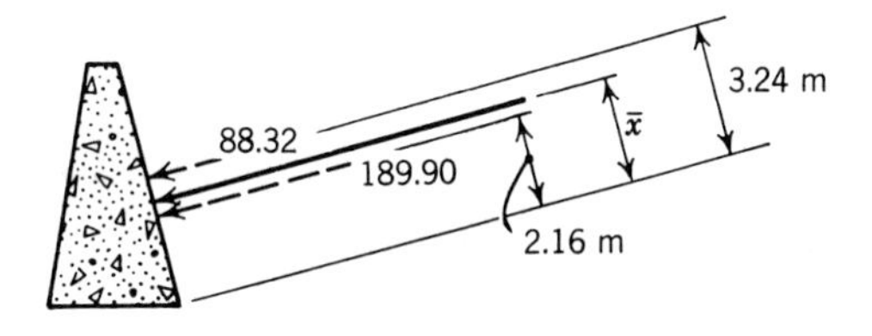

Fig. E13.11-2

From Fig. 13.22

$$K_a = 0.59$$

$$P_a = \tfrac{1}{2}(0.59)(17.3)(6.1)^2 + (0.59)(24.54)(6.1) = 189.90 + 88.32 = 278.22 \text{ kN/m of wall}$$

$$\text{Slant height of wall} = \frac{6.1}{\cos 20^\circ} = 6.49 \text{ m}$$

$$\bar{x} = \frac{(189.90)(2.16) + (88.32)(3.24)}{189.90 + 88.32} = 2.50 \text{ m}$$

Moment about point A = 278.22(2.50) = 695.55 kN m/m ◀

---

applicability of these theorems is uncertain. The development of methods for handling more realistic stress-strain relationships deserves much more attention.

Even when a rigid-plastic material is assumed there still are great difficulties. It is difficult to ensure that Eqs. 13.10 and 13.13 are fulfilled throughout the soil mass. Most solutions prove that these conditions are satisfied only in a limited portion of the mass within the failure zone. Even within these zones there is disagreement on the relationship between stress and strain because of the necessity of accounting for the volume changes that accompany shear strains, and hence there is uncertainty as to whether the strains associated with the stresses are compatible, or *kinematically admissible.*

In addition to these fundamental difficulties, the equations that must be solved (Kötter's equation) are complicated, and contact with physical reality is lost while carrying out the required numerical integrations. Whereas such solutions have received considerable attention in Europe, in the Americas the tendency has been to use the simpler trial wedge method. The solutions of Sokolovski and Hansen, which deserve more attention

▶ **Example 13.12**

Given. Retaining wall as shown in Fig. E13.12-1.

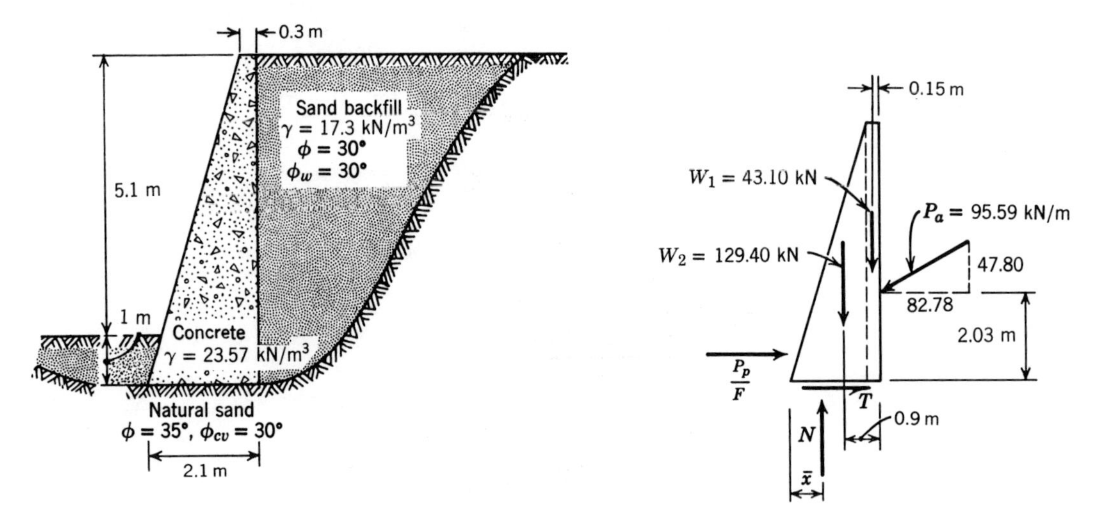

Fig. E13.12-1

Find. Adequacy of wall.

Solution. The first step is to determine the active thrust; see Example 13.8. The next step is to compute the weights:

$$W_1 = (0.3)(6.1)(23.57) = 43.10 \text{ kN/m}$$

$$W_2 = \tfrac{1}{2}(1.8)(6.1)(23.57) = 129.40 \text{ kN/m}$$

Next N and $\bar{x}$ are computed:

$$N = 129.40 + 43.10 + 47.80 = 220.30 \text{ kN/m}$$

$$\text{Overturning moment} = 82.78(2.03) - 47.80(2.1) = 168.04 - 100.38 = 67.66$$

$$\text{Moment of weight} = (1.95)(43.10) + (1.20)(129.40) = 84.05 + 155.28 = 239.33$$

$$\text{Ratio} = 3.54 \qquad \underline{\underline{\text{OK}}}$$

$$\bar{x} = \frac{239.33 - 67.66}{220.30} = \frac{171.67}{220.30} = 0.78 \text{ m} \qquad \underline{\underline{\text{OK}}}$$

The location of N shown in Fig. E13.12-2.

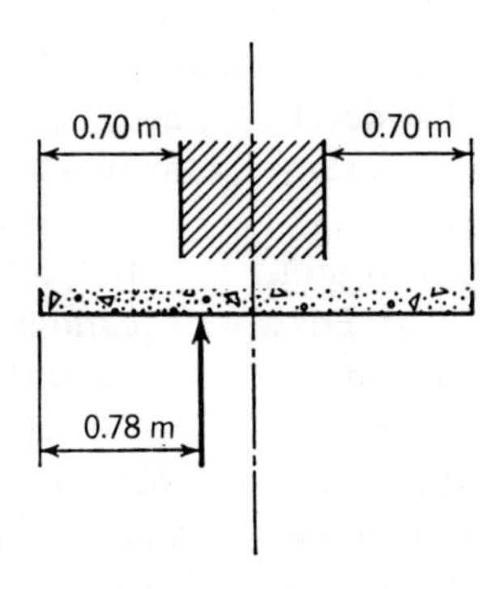

Fig. E13.12-2

Example 13.12 *(continued)*

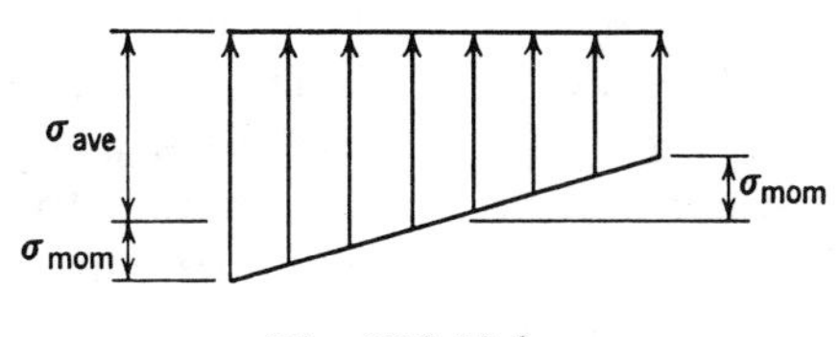

Fig. E13.12-3

Next the bearing stress is computed. The average bearing stress is 220.30/2.1 = 104.90 kN/m². Assuming that the bearing stress is distributed linearly, the maximum stress can be found (see Fig. E13.12-3), since

$$\sigma_{mom} = \frac{M}{S}$$

where

$$M = \text{moment about } \mathbb{C}\!\!\!\!L = 220.30(1.05 - 0.78) = 59.48 \text{ kN m/m}$$

$$S = \text{section modulus} = \tfrac{1}{6}B^2 = \tfrac{1}{6}(2.1)^2 = 0.74 \text{ m}^2$$

where B is width of base

$$\sigma_{mom} = \frac{59.48}{0.74} = 80.38 \text{ kN/m}^2$$

$$\text{Maximum stress} = 104.90 + 80.38 = 185.28 \text{ kN/m}^2$$

Finally, the resistance to horizontal sliding is checked. Assuming passive resistance without wall friction,

$$K_p = 3$$

$$P_p = \tfrac{1}{2}(17.3)(1^2)(3) = 25.95 \text{ kN/m}$$

With reduction factor of 2,

$$\frac{P_p}{F} = 12.98 \text{ kN/m}$$

$$T = 82.78 - 12.98 = 69.80 \text{ kN/m}$$

$$N \tan 30° = 127.19 \text{ kN/m}$$

$$\frac{N \tan \phi_{cv}}{T} = 1.82 < 2 \qquad \underline{\underline{\text{not OK}}}$$

Ignoring passive resistance

$$T = 82.78 \text{ kN/m}$$

$$\frac{N \tan \phi_{cv}}{T} = 1.54 > 1.5 \qquad \underline{\underline{\text{OK}}}$$

◀

than they have received, are still not exact in the sense that they do not fulfill all of the five conditions previously outlined.

Despite these many theoretical difficulties, the solutions presented in Sections 13.3 to 13.5 are useful in practical work. Their applicability has been verified in a limited number of situations by measurements of stresses and thrusts in large-scale model tests and in actual situations. After being verified by these observations, the results can be used with reason to predict stresses and thrusts in situations for which there are no actual data. For active conditions, the results presented in these sections will give the active thrust within ±10% provided that the friction angle is known accurately. For passive conditions, the uncertainty is greater—perhaps ±20%—especially if wall friction is present.

13.6 EXAMPLE OF DESIGN OF GRAVITY RETAINING WALL

In order to illustrate the design procedure for a gravity retaining wall, let us consider the problem shown in Example 13.12. The following steps should be noted:

1. The active thrust is computed using a value of K_a selected from Fig. 13.18 for the given ϕ and ϕ_w. This calculation is made in Example 13.8. This computation, of course, presumes that the failure surface passes entirely through the backfill rather than through the natural sand. It is convenient to break this thrust up into its vertical and horizontal components.
2. The weight of the wall is computed, breaking the

actual geometrical shape up into two simple shapes to facilitate the computation.

3. The bearing force N is computed. The location of the line of action of N is also computed. if $\bar{x}$ were less than zero, obviously the wall would not be stable. That is, such a result would mean that the overturning moment from the active thrust exceeds the resisting moment of the weight. Different engineers use different design rules to guard against such a possibility. One rule which also is used to limit the maximum bearing stress (discussed later) is that N should be within the middle third of the base. An alternate rule is to require that the ratio of resisting to overturning moment should be 1.5 or greater. This second rule is in effect a safety factor against a poor estimate of the active thrust. The wall of the example is adequate by either of these criteria.
4. The next question is: Can the natural sand safely support the vertical force N? A full answer to this question must wait until after we have studied bearing capacity in Chapter 14. An average bearing stress of 104 kN/m^2 will usually be tolerable. Because the resultant N does not act exactly in the center of the base, the maximum bearing stress at the toe will exceed the average bearing stress. As will be explained in Chapter 14, if N acts within the middle third of the base the maximum stress will be less than twice the average stress and will also be tolerable.
5. Different engineers also use different rules to check for sliding resistance. In one rule the passive resistance is considered, and the combined sliding resistance and passive resistance must exceed the horizontal component of thrust by a safety factor of two or greater. In a second and more common rule the passive resistance is ignored, and it is required that the resistance to sliding be at least 1.5 times the horizontal component of active thrust. The wall is adequate since it fulfills the second of these rules.

In this example, the stated safety factors represent the engineering judgment of engineers regarding the certainty with which the various forces and resistances can be estimated. By these standards the wall is barely adequate. Either smaller or greater safety factors may be required depending upon the circumstances of each individual problem.

The rest of this section discusses further several of the most important points.

Justification for Use of Active Thrust

In earlier sections, it has been pointed out that the active thrust is the minimum possible thrust that the soil may exert against a retaining wall. This question then arises: Should not such a wall be designed for the possibility that some larger thrust exists?

The first answer to this question is: As long as the backfill *is* a dry granular soil whose friction angle is known, the thrust against a gravity retaining wall generally does equal the theoretical active thrust. This was demonstrated by the very careful tests by Terzaghi in the 1920s. In these tests, the walls were held against horizontal movement as the backfill was placed and the thrust against the wall was measured. As expected, this thrust was greater than the active thrust. Then the walls were released and permitted to move horizontally or to rotate. After a movement of the top of the wall equal to only 0.001 times the height of the wall, the thrust had dropped to its theoretical active value.[5] This is a very small amount of movement (the angular rotation is only 0.06°), and it must be expected that a gravity retaining wall will rotate this much as the backfill is placed against it.

Even so, if, for some reason, the thrust against a retaining wall were greater than the active value, it would not mean that the wall potentially was in trouble. On the contrary, it would mean that the earth underlying the wall is much stronger than it need be. Long before a wall can fail, it must move enough to mobilize the shear strength of the soil and to drop the thrust to its active value. In other words, the strength of the backfill behind a retaining wall will be mobilized long before the shear strength of the soil that supports the wall is mobilized. Under such circumstances, it makes great sense to design the wall for the active thrust, and to use a safety factor on the quantity in the design about which the designer knows least: the bearing capacity of the soil supporting the wall.

Having emphasized how small the wall movements are, we now must turn around and emphasize how large they can be. If a retaining wall is 6 m high, a rotation of 1 in 1000 means a horizontal displacement of 6 mm at the top of the wall. In most situations where gravity retaining walls are used—highway or railway cuts, etc.—this amount of movement (or even several times this amount of movement) literally is of no consequence. However, there are problems where this amount of movement might cause trouble. A classic situation is a wall used for the abutment of a bridge. If the wall has been designed for the active thrust, and if the backfill is placed after the bridge is set in place, then there must be sufficient clearance between the wall and the girders to accommodate the outward movement of the wall.

There are numerous retaining structures that resemble gravity retaining walls, but often these should not be designed on the basis of active thrust. A braced

[5] In Section 13.2 it was stated that a horizontal strain of about 0.005 is required in passing from the at rest to the active condition. The horizontal width of the failure wedge is $H \cot (45 + \phi/2)$ or about $H/2$. Hence the horizontal displacement of the wall would be $0.0025H$. Thus behavior of sand during triaxial tests is in good agreement with Terzaghi's results.

excavation (Section 13.7) and sometimes anchored bulkheads (Section 13.8) are examples. The cantilever type of retaining wall shown in Fig. 13.23 is another example. Such walls, which have reinforcing steel, are sometimes used where space restrictions preclude the use of massive gravity walls. If a cantilever wall rests upon very firm soil so that the foundation experiences little or no sliding or rotation, active conditions within the backfill can develop only by bending of the cantilever. The amount of bending necessary to develop the active condition may cause severe cracking of the concrete and yielding of the steel. Cantilever retaining walls are often designed on the basis of K_0 rather than K_a.

The wall surrounding the basement of a building is an example of an unyielding wall. The magnitude of the stresses acting against a foundation wall will depend largely upon the degree of compaction given to the backfill. If a clean sand is dumped against the wall without compaction, the horizontal stresses may be almost as small as the active stresses. If light compaction is used, such as simply running a bulldozer over the several layers of the backfill, the horizontal stresses will likely equal the at-rest stress. With heavy compaction, stresses approaching the passive stresses might be developed. The usual practice is to design foundation walls for the at-rest stress; i.e., for a horizontal stress of approximately one-half the vertical stress. When a wall is designed on this basis, heavy compaction of the backfill must be avoided. Otherwise the foundation wall may be cracked.

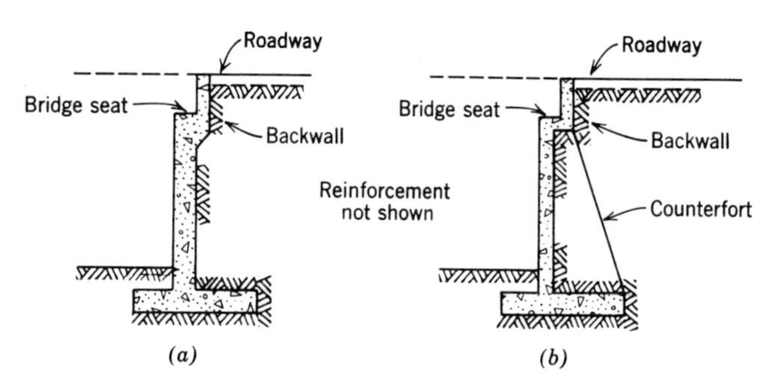

Fig. 13.23 Counterfort and cantilever retaining walls. (*a*) Cantilever. (*b*) Counterfort. (From Huntington, 1957.)

Choice of Friction Angle for Backfill

The peak friction angle of the backfill should be used for design computations. If the granular soil is simply dumped into place, this angle will be approximately ϕ_{cv}. Usually, however, backfill is given at least nominal compaction by a bulldozer, so that a medium dense state is generally achieved. The increase in friction angle achieved by moderate compaction will offset the disadvantageous increase in unit weight. However, intense compaction seldom is justified, since there is the danger that large outward wall movements will occur during compaction.

Role of Wall Friction

Wall friction greatly reduces the horizontal thrust and especially the overturning moment against a wall. The wall in Example 13.12 would not be adequate if it were not for wall friction (see Problem 13.8). Generally it is appropriate to take advantage of the beneficial effects of wall friction, since the downward drag will develop as the wall moves outward. However, an engineer must satisfy himself on this point in each case.

Evaluation of Passive Thrust

The horizontal width of the passive failure wedge is $H' \tan (45 + \phi/2)$, or about $2H'$, where H' is the depth to which the wall is embedded. For a loose sand, from 10 to 20% strain might be needed to mobilize full passive resistance. This strain would correspond to a displacement of 0.2H′ to 0.3H′. Thus for H′ = 1 m as in our retaining wall example, as much as 0.3 m of horizontal displacement might be necessary in order to mobilize full passive thrust at the toe. This is more base displacement than is desirable, and hence a relatively large safety factor is used whenever passive resistance is taken into consideration. As indicated by the curve in Fig. 13.7, not much displacement is needed to mobilize one-half of the full thrust. Usually wall friction in the passive zone is ignored, thus adding to the conservatism. If wall friction is included, the vertical component of passive thrust will cause a decrease in N, and this effect should be considered.

Some Design Suggestions

The foregoing details have been included to indicate the type of considerations that enter into design. Still other details may be found in Huntington (1957). Clearly the making of an adequate design requires much more "engineering" than simply the calculation of active thrust.

Use of cinders for backfill is sometimes considered as a means of reducing active thrust and economizing on design. Cinders have a small unit weight (7.8 kN/m^3) and yet have a friction angle as large as sand.

Sloping the wall in contact with the soil leads to a more favorable location of the resultant weight of the wall relative to the outside edge of the wall, and may thus make it possible to use a narrower base and yet keep the resultant N within the middle third of the base. This saving must be compared with the cost of added formwork.

13.7 BRACED EXCAVATIONS

A gravity retaining wall is a permanent structure, used when an excavation is permanent. In many cases, however, an excavation is only temporary. Examples are excavations for buildings or subways. Here the excavations are filled with a structure which then permanently retains the surrounding earth. If the temporary excavation is made in sand, the walls of the excavation must be supported during construction of the building by a system of bracing, as shown in Fig. 1.12*b*. The design of bracing for excavations will be discussed in some detail to illustrate one situation in which it may not be proper to design on the basis of active thrust.

Figure 13.24 shows two common systems for installing the bracing. In one system, *sheet piling* (a continuous line of piles) is driven in advance of excavation. As excavation proceeds, horizontal members known as *wales* are placed against the sheet piling, and additional horizontal members called *struts* are placed across the excavation and wedged against the wales. In the second general system, vertical members called *soldier beams* are driven at intervals along the line of excavation. As excavation proceeds, horizontal wooden planks called *lagging* are inserted against the earth and are supported by the soldier beams. Wales are again placed horizontally across several soldier beams, and struts are wedged in place between the opposite walls of the excavation.

There are, of course, many variations on these basic systems, depending on the size of the excavated area and the preferences of the individual contractor. Figure 13.25 shows struts braced against a block in the center of the foundation instead of against the opposite wall. There

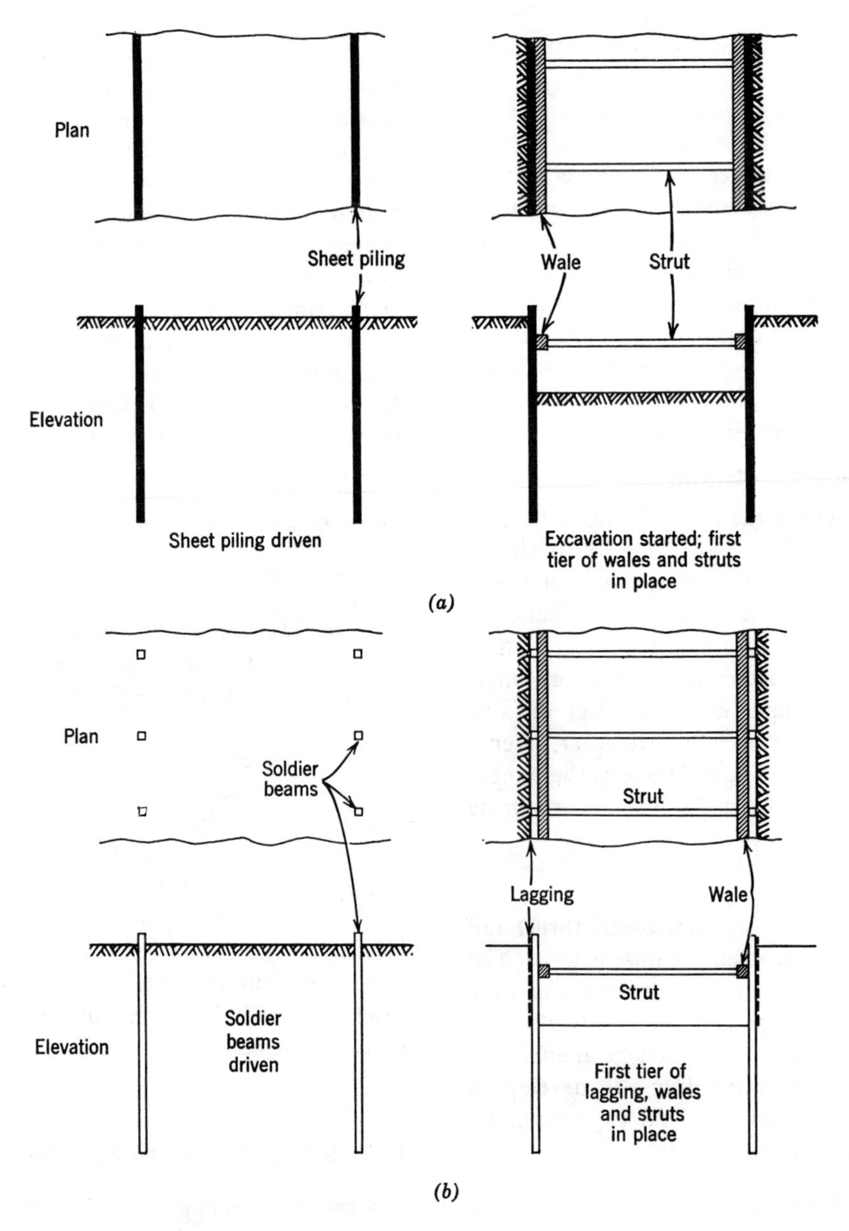

Fig. 13.24 Systems for installing bracing. (*a*) Braced excavation using sheet piling. (*b*) Braced excavation using soldier beams and lagging.

Fig. 13.25 Excavation for M.I.T. Center for Advanced Engineering Study.

is increasing use of *tiebacks*, anchors driven through the wall into the earth behind the wall. The use of tiebacks keeps the excavation free of obstacles.

Form and Magnitude of Stress against Bracing

Several field observations[6] have shown that the stress against the bracing (when the bracing is placed against a sandy soil) has the distribution shown in Fig. 13.26. Note that this distribution is quite different from the active stress distribution. Moreover, measurements have also indicated that the total thrust against the bracing may be somewhat larger than the thrust predicted for the active condition.

The observed pattern of stress may be understood if we examine the way in which the soil deforms as the excavation proceeds (see Fig. 13.27). The topmost strut, once installed and wedged tightly against the wale, will not permit any further appreciable horizontal displacement of the soil at that elevation. As soil is excavated at some lower elevation, the remaining soil at that lower elevation will move toward the excavation until it is in turn supported by a strut. Thus the overall pattern of soil movement is one of rotation about some point near the top of

[6] Terzaghi and Peck (1967) summarize the results of field measurements from braced excavations.

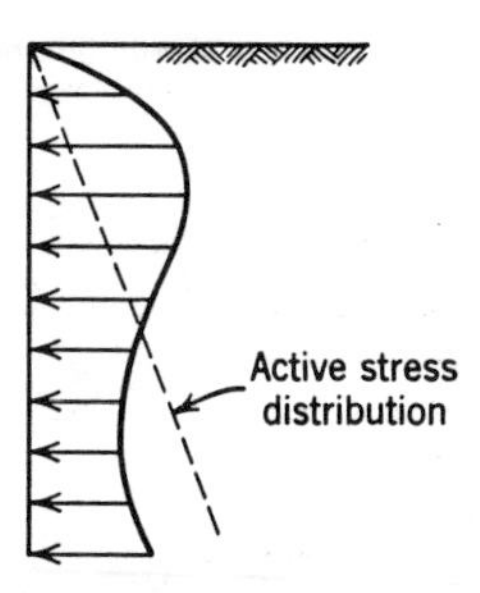

Fig. 13.26 Typical distribution of stress against bracing system.

the bracing. The soil near the top of the bracing is not allowed to move outward as is necessary to mobilize full shear resistance within the soil. Rather, the soil at a lower elevation exerts a drag type of shear force upon the overlying soil. Hence the soil near the top of the wall is more nearly in a passive state of stress than in an active state of stress.

Although the distribution of stress against the bracing is quite different than in the classical active stress situation, it is not necessarily true that the total thrust against the wall differs greatly from that predicted for the active condition. As long as full shear resistance is mobilized along the bottom boundary of the failure wedge, the total thrust exerted against the soil by the retaining structure is much the same whether the retaining structure is a gravity wall or a bracing system. However, the

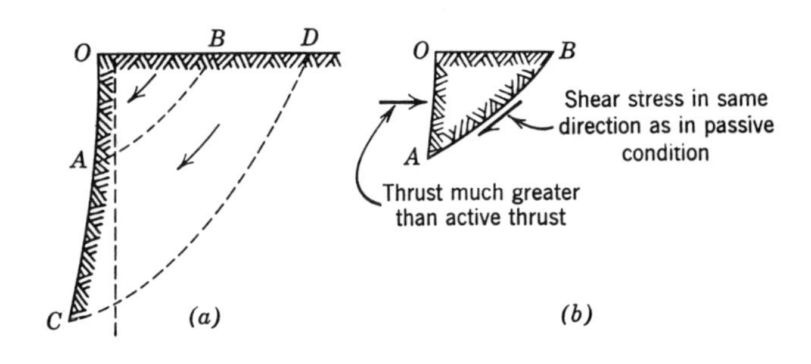

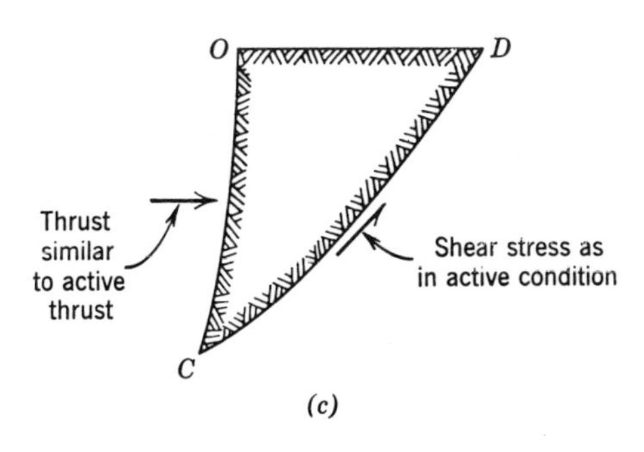

Fig. 13.27 Movements and stresses within soil. (*a*) Soil movement (greatly exaggerated). (*b*) Stresses on wedge *OAB*. (*c*) Stresses on wedge *OCD*.

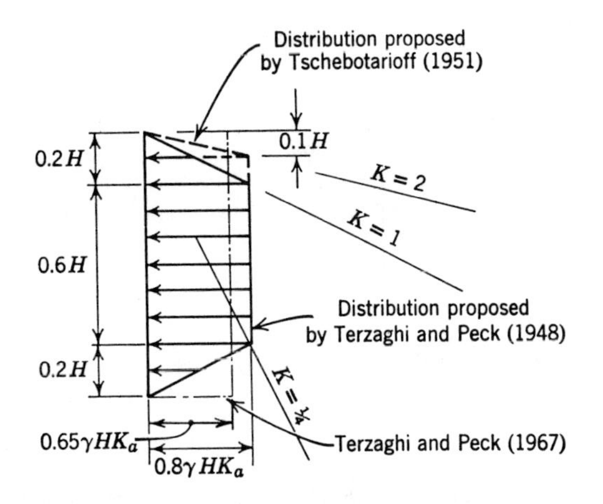

Fig. 13.28 Stress distribution used for design of bracing system.

pattern of deformation of the soil will influence somewhat the location of the critical theoretical failure line, and hence the thrust will change somewhat with the retaining system (Hansen, 1953). The total thrust against a braced wall may be 10–15% greater than that against a gravity wall.

The state of stress in the soil behind a braced cut has often been described as an *arching active* condition.

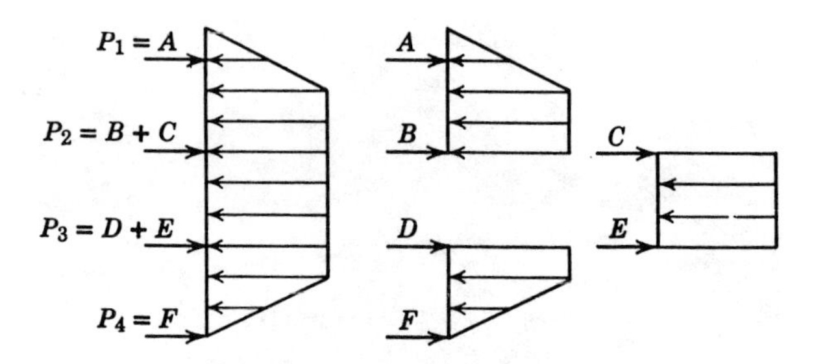

Fig. 13.29 Computation of strut loads.

Design Procedures

For purposes of designing a bracing system it usually is assumed that the distribution of stress against the sheeting or lagging is as shown in Fig. 13.28. The method of computing the strut loads from this distribution is indicated in Fig. 13.29 and is illustrated in Example 13.13.

According to the stress distribution proposed by Terzaghi and Peck (1948), the total thrust is $0.64\gamma H^2 K_a$, or 28% greater than the active thrust. Thus the proposed design stress distribution recognizes that the total thrust may exceed the active thrust. However, there is a second (and more important) reason why the total design thrust exceeds the active thrust. The actual stress distribution

---

▶ Example 13.13

Given. Excavation and bracing system as shown in Fig. E13.13-1.

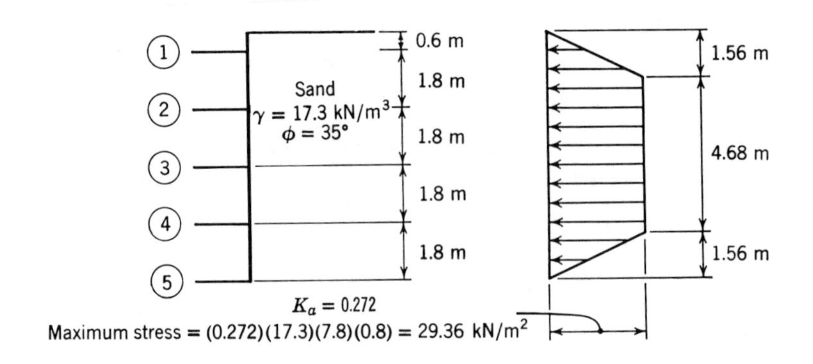

Fig. E13.13-1

Find. Design strut loads.

Solution. From Table 13.1, $K_a = 0.272$. Maximum stress is

$$(0.272)(17.3)(7.8)(0.8) = 29.36 \text{ kN/m}^2$$

From Fig. E13.13-2:

$$P_1(1.8) = 22.90(1.36) + (24.66)(0.42) = 31.14 + 10.36 = 41.50 \text{ kN}$$

$$P_1 = 23.06 \text{ kN/m}$$

$$B = 24.66 + 22.90 - 23.06 = 24.50 \text{ kN/m}$$

Example 13.13 (*continued*)

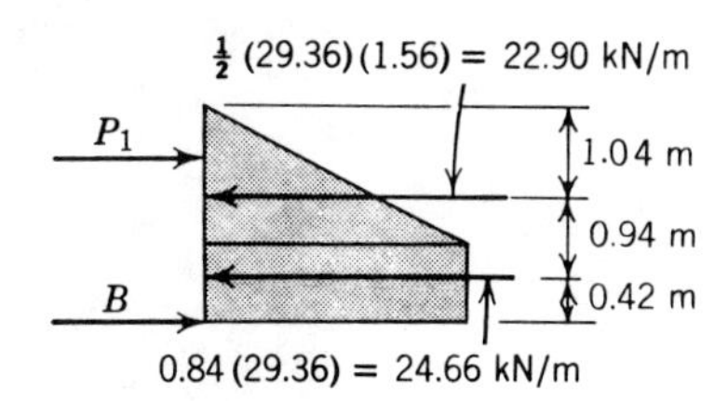

Fig. E13.13-2

From Fig. E13.13-3:

$$C = D = 52.85 \text{ kN/m}$$

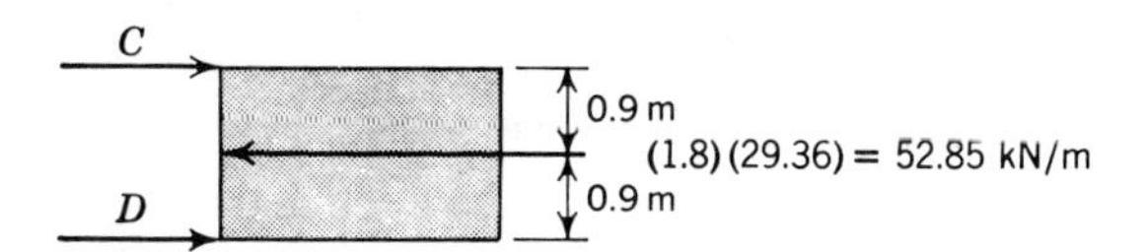

Fig. E13.13-3

From Fig. E13.13-4:

$$P_5(1.8) = 7.05(0.12) + (22.90)(0.76) = 0.85 + 17.40 = 18.25 \text{ kN}$$

$$P_5 = 10.14 \text{ kN/m}$$

$$E = 22.90 + 7.05 - 10.14 = 19.81 \text{ kN/m}$$

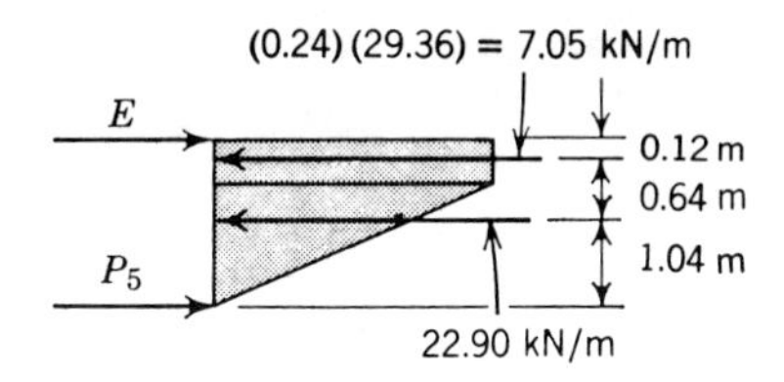

Fig. E13.13-4

Assembling these results:

$$P_1 = 23.06 \text{ kN/m}$$

$$P_2 = 24.50 + 52.85 = 77.35 \text{ kN/m}$$

$$P_3 = 2(52.85) = 105.70 \text{ kN/m}$$

$$P_4 = 52.85 + 19.81 = 72.66 \text{ kN/m}$$

$$P_5 = 10.14 \text{ kN/m}$$

If struts are located as 2 m intervals along wall, then design strut loads are

$$P_1 = 46.1 \text{ kN}$$

$$P_2 = 154.7 \text{ kN}$$

$$P_3 = 211.4 \text{ kN}$$

$$P_4 = 155.2 \text{ kN}$$

$$P_5 = 20.3 \text{ kN}$$

Struts should be designed for a safety factor appropriate for the material used for the strut.

◀

Fig. 13.30 Failure of model of braced excavation. (*a*) Stable. (*b*) About to fail. (*c*) Failing; note motions. (*d*) After failure.

will change from section to section depending on just how tightly the individual struts are wedged in place. The design stress distribution curve represents an envelope to the various possible actual distributions. Since struts fail in buckling, it is important that no single strut be overstressed. It is not permissible to say that if one strut is overstressed and starts to fail, the overstress will simply be transferred to an adjacent strut. If one strut even starts to buckle, its load-carrying capacity may drop to almost nothing and then the whole bracing system will be in jeopardy. Figure 13.30 illustrates the sudden rapid collapse of a braced excavation as one strut buckles. The use of an envelope to all possible stress distribution curves ensures that each strut will be designed for the largest load which might reach the strut. However, the sum total of the loads in all struts will undoubtedly be less than $0.64\gamma H^2 K_a$.

The two important points with regard to the design of a bracing system are: (*a*) the uppermost struts[7] will be subjected to loads much greater than would be predicted from the ordinary active stress distribution; and (*b*) struts in compression are a brittle system tending to collapse as soon as yielding begins. Limit design is not an appropriate procedure for a *brittle* system; in contrast, a gravity retaining wall is a *ductile* system where large foundation movements may occur without loss of foundation strength.

13.8 ANCHORED BULKHEADS

As described and illustrated (Fig. 1.15) in Chapter 1, an *anchored bulkhead* receives its lateral support from penetration into the foundation soil and from an anchoring system near the top of the wall. The sheet piling must be designed for the shears and bending moments which thus develop. The anchor system must be designed to take the lateral forces required to support the wall.

Anchored bulkheads are often used to form wharves or quays, since the soft soils that usually underlie such waterfront structures are unable to support the weight of massive gravity walls and since the use of anchored bulkheads is generally cheaper than supporting a gravity retaining wall on piles. The design and analysis of anchored bulkheads is a rather complicated subject.

The distribution of stresses from the backfill will depend strongly on the manner in which the wall is constructed. Tschebotarioff (1951) has suggested that we must distinguish among the three cases shown in Fig. 13.31.

1. If backfill is placed after the bulkhead is constructed, the stresses against the bulkhead down to the point of embedment will increase linearly with depth in accordance with the classical theories of active stress.

[7] The upper struts can receive greater loads at partial excavation than at full excavation.

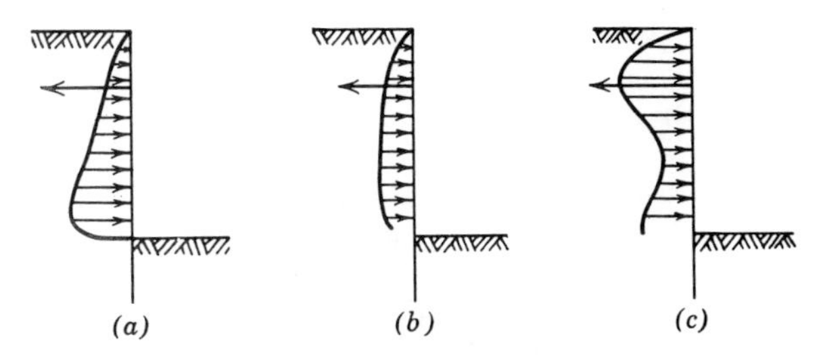

Fig. 13.31 Relation between construction procedure and stress distribution against anchored bulkhead. (*a*) Backfilling. (*b*) Dredging with normal yield of anchor. (*c*) Dredging with unyielding anchor.

2. If the bulkhead is driven into level ground, one side of which is then excavated, the stresses will be more or less uniform with depth—unless the anchor is unusually stiff.
3. If the anchor is unusually stiff, the stress distribution will be similar to that upon a bracing system. This situation might arise if a very heavy member is used for the anchor rod, or if a short rod is attached to a very massive anchor.

Furthermore, the magnitude of the maximum bending moment in the piling is influenced greatly by the distribution of stresses against that part of the piling which is embedded, and the stress conditions in this zone are quite complex. This effect cannot be predicted on the basis of simple theory, although the complex theories of Hansen are useful. Usually, experimental data plus field experience are used as a basis for design. Tschebotarioff (1951) and Rowe (1952) have presented such methods of design.

Often a deadman anchor (Fig. 1.15) is used to support the anchor rod. The design of such an anchor involves an interesting problem in the evaluation of passive resistance.

13.9 STABILITY OF SLOPES

There are many situations in which an earth mass need not be retained by a structure but left as an unretained slope. The inclination of the slope must be flat enough and/or the height low enough for the earth mass to be stable. The same principles of limiting equilibrium mechanics are used to evaluate the stability of an unretained earth mass as for a retained earth mass.

Parts (*a*) and (*b*) of Fig. 13.32 show two typical processes by which a slope is formed in a granular soil.[8] In (*a*) an embankment is being formed by end dumping from a truck; in (*b*) ore or sand or some other stockpiled material is dropped from a chute or from the end of a conveyor belt. In both of these situations, the material will tumble down the slope. From time to time during

[8] A slope can also be formed by excavation, as done for a canal (see, e.g., Fig. 1.14).

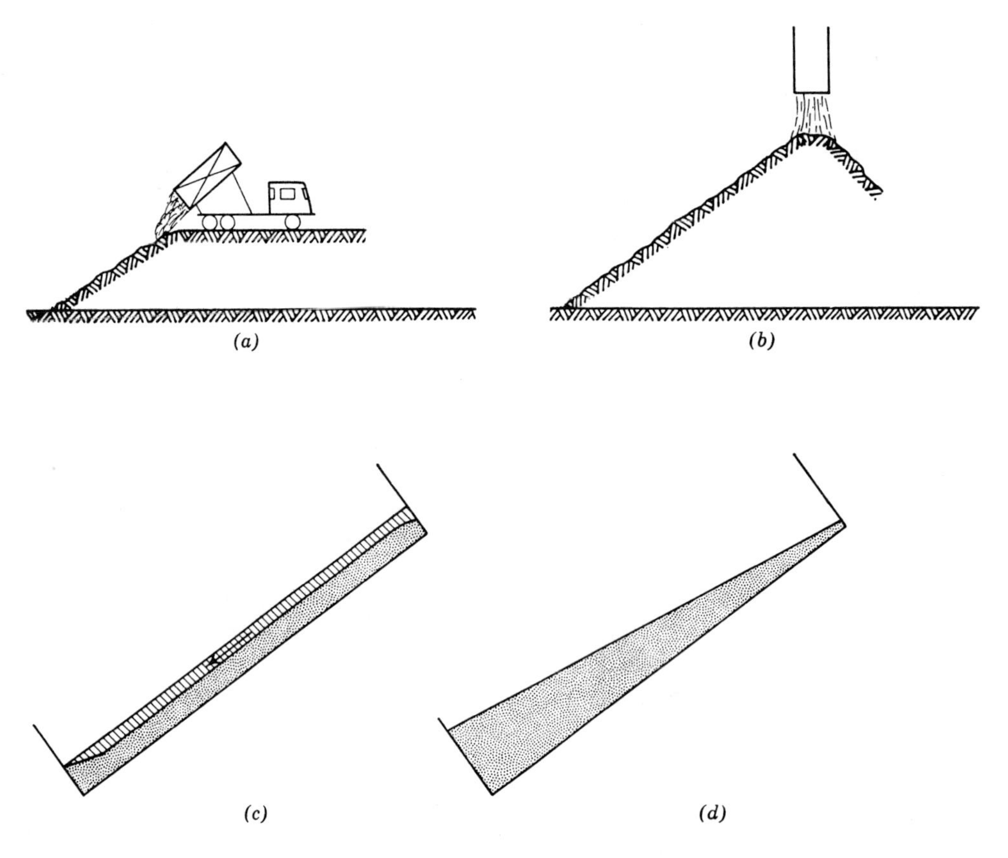

Fig. 13.32 Infinite slope problems. (*a*) Formation of slope by dumping. (*b*) Formation of ore pile. (*c*) Tipping experiment as sliding starts. (*d*) Tipping experiment after sliding.

dumping, material which has already come to rest on the slope will start moving again; i.e., a mass of material, with a thickness small compared to the height of the slope, will slide down the slope. The inclination of the slope once dumping has ceased—the maximum slope at which the material is stable—is called the *angle of repose*.

The behavior during the tipping experiment depicted in parts (*c*) and (*d*) of Fig. 13.32 is similar. As the angle of tipping is gradually increased, individual particles will start to tumble down the slope. Finally, as the angle is increased further, a mass of material will slide as a whole, as indicated in Fig. 13.32*c*. Once sliding ceases, the slope will have an average inclination roughly equal to the angle of repose for this same sand if dumped.

In all of these situations, the thickness of the unstable moving material is small compared to the height of the slope. In such situations, the slope is called an *infinite slope*. The failure surface is parallel to the slope.

Analysis of Free Body

In order to analyze the stability of this slope, we "cut" a free body element of soil from the slope, as shown in Fig. 13.33. We assume that the slope is very wide in the direction normal to the cross-section, and consider only the stresses that act in the plane of the cross-section.

In general, there will be stresses on three sides of this free body, as indicated in Fig. 13.33*a*. However, with an infinite slope it is reasonable to assume that the stresses on the two vertical faces are equal and exactly balance each other. If this were not true, the stresses on vertical faces would change depending on the location along the slope, and such a situation would be inconsistent with the observation that a thin veneer of the whole slope moves as a mass. Thus only the stresses on the face CD, together with the weight of the soil, enter into the equilibrium of the free body.

Part (*b*) of this figure analyzes the equilibrium of the free body in terms of the total forces T and N acting on the face CD. The result is: when full shear resistance is mobilized and sliding begins, the angle of inclination of the slope should equal the angle of internal friction. According to this analysis, sliding is equally likely to begin at any depth; i.e., the depth of the free body completely cancels out of the result.

Example 13.14 illustrates the computation of the stresses that exist beneath an infinite slope at the angle of repose.

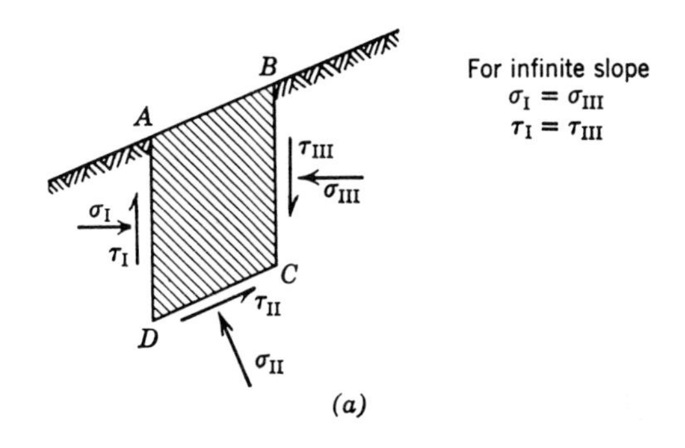

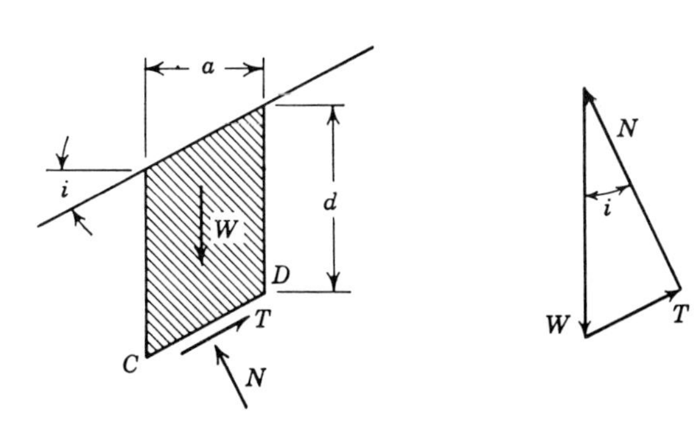

Fig. 13.33 Analysis of infinite slope. (*a*) Stresses upon element of soil (*b*) Analysis of equilibrium:

$$W = ad\gamma$$

$$\left.\begin{aligned} W - T\sin i - N\cos i &= 0 \\ T\cos i - N\sin i &= 0 \end{aligned}\right\} \qquad \begin{aligned} N &= W\cos i \\ T &= W\sin i \\ \therefore T &= N\tan i \end{aligned}$$

If the full shear resistance is mobilized so that $T = N\tan\phi$, then $i = \phi$.

Note that the vertical stress is not simply equal to the depth multiplied by the unit weight.

Choice of Friction Angle

The slope angle at which sliding commences in the tipping experiment is related to the peak friction angle ϕ.[9] Thus the maximum stable slope angle is fundamentally related to the peak friction angle. However, we know that ϕ is very much a function of the void ratio at which the sand exists.

Whenever sand or gravel is dumped the sand generally finds itself in a loose state. For the loose state, ϕ essentially equals ϕ_{cv}. Thus, the angle of repose for dumped sand or gravel is about equal to the angle of internal friction for the loose state, ϕ_{cv}. Typical angles of repose, together with the tangents of these angles, have already been listed in Table 11.3.

On the other hand, slopes steeper than the angle of repose can exist in a stable condition. In modern rock fill dams, the fill is carefully compacted as it is dumped in thin layers so as to bring the fill into a dense condition. Hence the friction angle available to resist sliding is greater than the angle of repose.

Safety Factor

The safety factor for an infinite slope usually is defined as

$$FS = \frac{\tan\phi}{\tan i}$$

The only unknown factor in the stability of an infinite slope is the appropriate value for the angle of internal friction. This quantity can be estimated with reasonable accuracy and, furthermore, the consequences of failure of such a slope are slight. Hence the safety factor does not need to be large. Usually an engineer will be conservative in his choice of $\phi = \phi_{cv}$, and will use $FS = 1$.

13.10 SUMMARY OF MAIN POINTS

The main objectives of this chapter have been to illustrate the methods used to calculate active thrust and passive resistance and to illustrate how these calculated forces are used in the design of typical retaining structures. The details of the methods are important, and the student should be competent to carry out an analysis of simple problems by the trial wedge method. In addition, the following concepts should be understood.

1. Limit design can be used for the design of most gravity retaining walls. The active thrust from the backfill is evaluated assuming that full shear resistance is mobilized within the backfill. Resistance is then provided (with an appropriate margin of safety) against the overturning and sliding caused by the active thrust.
2. Compression bracing for excavations should not in general be proportioned by limit design, since such bracing is a brittle system which will fail as soon as any portion becomes overstressed.
3. Other types of retaining structures must be studied carefully to learn how much and what types of movements may occur, and only then should the forces acting on the retaining structures be evaluated.
4. The maximum slope angle in a granular soil is equal to the friction angle of the soil.

PROBLEMS

13.1 A sand backfill has $\gamma = 17.3$ kN/m³, $\phi = 30°$, and $K_0 = 0.5$. Construct a p–q diagram showing the K_f- and K_0-

[9] This statement is based on an extrapolation of the results given by Seed and Goodman (1964). In model tests, the small cohesion intercept of a dry soil has some influence on slope stability and determines the actual depth of sliding.

▶ Example 13.14

Given. A 30° slope in a sand having $\phi = \phi_{cv} = 30°$. The unit weight is 15.7 knN/m³
Find. The stresses at a depth of 6.37 m
Solution. Referring to Fig. 13.33, the stresses on the failure plane *CD* are

$$\sigma_{ff} = \frac{N}{a/\cos i} = \gamma d \cos^2 i = 75.0 \text{ kN/m}^2$$

$$\tau_{ff} = \frac{T}{a/\cos i} = \gamma d \sin i \cos i = 43.3 \text{ kN/m}^2$$

The Mohr circle for this condition is shown in Fig. E13.14. ◀

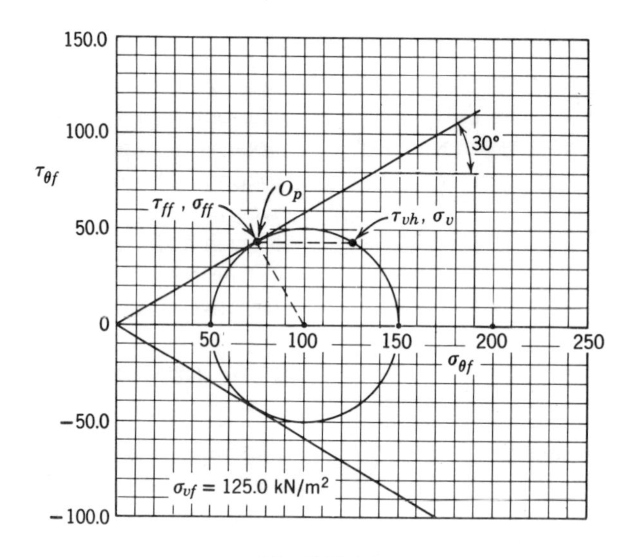

Fig. E13.14

lines, and the stress paths for an element of soil at 3 m depth for:

a. Stressing to the active condition.
b. Stressing to the passive condition.

Assume that there is zero shear stress on vertical planes.

13.2 From the results for Problem 13.1, what are the horizontal stresses for the active and passive conditions? How do these stresses compare with the average active (Example 13.2) and passive (Example 13.6, ignoring surcharge) stresses for a wall 6.1 m high?

13.3 Design a gravity retaining wall, 4.6 m high, to retain a backfill with γ = 16.5 kN/m³ and ϕ = 40° and having a horizontal surface. Assume that ϕ_w = 30° and that the coefficient of friction on the base of the wall is 0.5. Neglect passive resistance at the toe, and make the resultant force fall within the middle third of the base.

13.4 Draw the Mohr circles for the active and passive conditions of Problem 13.1.

13.5 Referring to Example 13.5, construct the force polygon for θ = 65°.

13.6 Referring to Example 13.7, construct the force polygons for θ = 25° and 35°.

13.7 Referring to Example 13.8, construct the force polygon for θ = 55°.

13.8 Evaluate the adequacy of the wall in Example 13.12, assuming ϕ_w = 0.

13.9 A wall, which supports a horizontal backfill with γ = 17.3 kN/m³ and ϕ = 35°, is to be used to provide a reaction for a *horizontal* load of 146 kN/m of wall. If the wall is to have a safety factor of 2 against failure, how high must the wall be? At what depth below the top of the wall should the load be placed?

13.10 A braced cut, holding back soil with γ = 16.5 kN/m³ and ϕ = 30°, is 6.1 m high. Struts, on 2 m centers horizontally, are located at depths of 0.6 m, 2.4 m, 4.2 m, and 6.1 m. Compute the strut loads.

13.11 Repeat Problem 13.3 for a case where the backfill is sloped at 1 vertical on 3 horizontal.

13.12 Repeat Problem 13.3 for a case where the backfill carries a surcharge of 20 kN/m².

13.13 A sand having ϕ = 35° is sloped at 35°. Find the normal and shear stresses on horizontal and vertical planes at a depth of 4.6 m (measured vertically) beneath the slope. Are either of these planes failure planes?

13.14 What should the design slope be for the sand whose friction angle behavior is given in Fig. 10.18, if the sand is to be poorly compacted?

CHAPTER 14

Shallow Foundations

14.1 GENERAL BEHAVIOR OF SHALLOW FOUNDATIONS

As described in Chapter 1, the term "shallow foundation" refers to a structure that is supported by the soil lying immediately beneath the structure. *Individual footings*, usually rectangular in plan view, are the most common shallow foundations for columns, whereas *strip footings* are used to support walls. In some instances structures are supported by *mats*.

The design of foundations is a trial-and-error procedure. A type of foundation and trial dimensions are selected. Analyses are then made to ascertain the adequacy of the proposed foundation. The foundation may be found to be adequate, in which case a check should be made to determine whether a cheaper foundation might also be adequate. If the proposed foundation is found to be inadequate, a larger foundation is considered. In some cases it may be impossible to design an adequate shallow foundation upon the given soil, in which case either deep foundations (Chapter 33) or improvement of soil (Chapter 34) must be considered.

The selection of a trial foundation and trial dimensions is often guided by tables of allowable bearing stresses. Most building codes contain such tables, based upon general experience with soils in the area to which the code applies. These allowable stresses usually lead to conservative designs for low buildings supported on spread footings, but they may lead to unconservative designs for unusual or large structures. In many cases a careful study will show that bearing stresses larger than those given by codes can be safely used.

This chapter discusses the "adequacy" of a foundation. The same general principles that apply to the analysis of settlement and stability of shallow foundations for structures also apply to embankments and dams on soft foundations. For a discussion of the many practical details and economic considerations involved in the design of a foundation, the reader is referred to Teng (1962) and U.S. Navy (1962), and for a discussion of foundation construction, to Carson (1965).

This chapter does not consider shallow foundations subjected to dynamic loads; they are treated in Chapter 15.

Behavior of Footing on Elastoplastic Material

To help understand the general behavior of shallow foundations, consider the situation shown in Fig. 14.1, where a stress increment Δq_s is applied to the surface of an idealized material.[1] This material is assumed to be elastic until the maximum shear stress τ_{max} reaches the value c. Once this condition is reached, further shear distortion can occur at constant shear stress. This material is assumed to be perfectly elastic with regard to volume change.

As Δq_s is increased the whole body first behaves elastically, and the stresses and settlements can be predicted

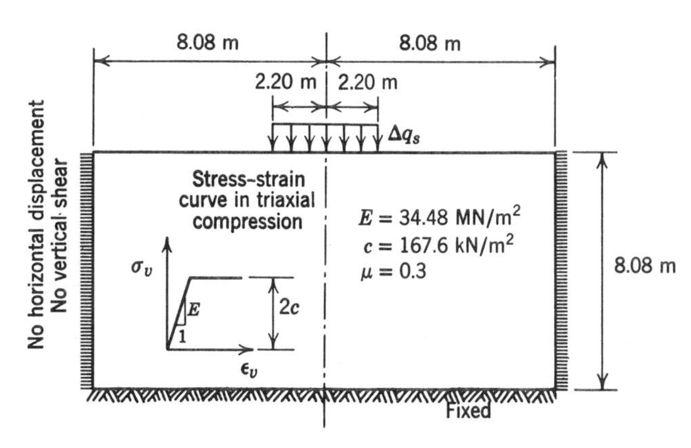

Fig. 14.1 Uniform strip load on hypothetical elastoplastic material.

[1] The results presented here were calculated with a digital computer using a finite difference technique (Whitman and Hoeg, 1966). The procedure has been extended to incorporate other stress-strain relations which are more similar to those of actual soils (Christian, 1966). Note that Δq_s or q_s as used in this chapter denotes applied surface stress and not $(\sigma_v - \sigma_h)/2$.

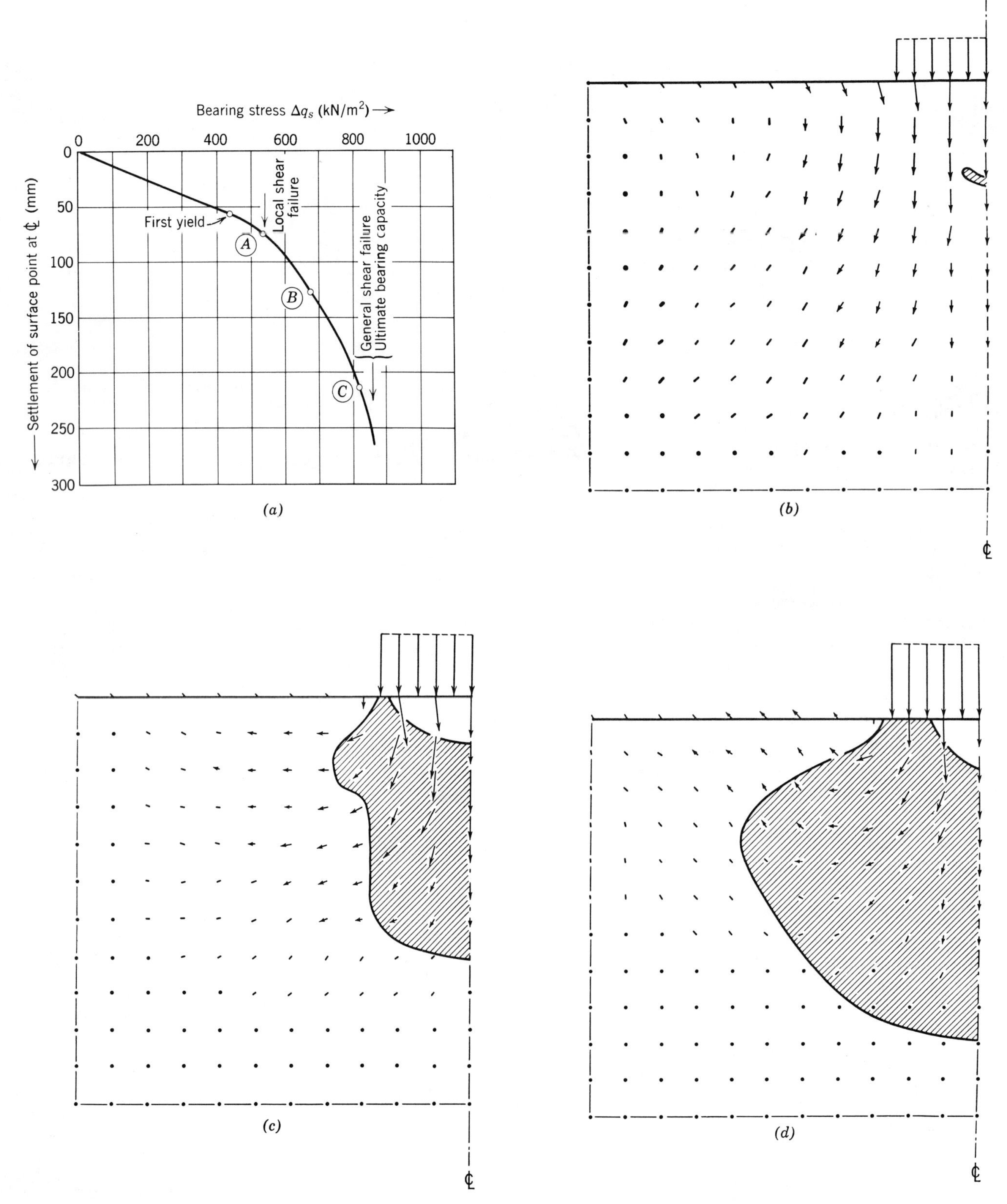

Fig. 14.2 (*a*) Load-settlement curve at centerline of hypothetical problem. (*b*) Displacements and first yielded point at load of 432.8 kN/m³. (*c*) Extent of yielded zone and motion field at load of 670.3 kN/m². (*d*) Extent of yielded zone and motion field at load of 814.0 kN/m².

Fig. 14.3 Failure zones under footing.

using elastic theory, as discussed in Chapter 8. As long as $\tau_{\max} < c$ at all points, the settlement is proportional to Δq_s. For any value of Δq_s, the largest value of $\tau_{\max}$ occurs along the center line at a depth roughly equal to one-half of the width of the loaded area. When $\Delta q_s = 432.8$ kN/m^2, $\tau_{\max} = c$ at this critical point, and this point yields. However, nothing catastrophic occurs at this stage because this yielded point is fully surrounded by material which can carry additional stress. A further increase in Δq_s causes *contained plastic flow* of the yielded point and additional elastic deformation of the surrounding points. Gradually the surrounding points also yield and the plastic zone grows.

Figure 14.2 shows *load-settlement curve* and the growth of the plastic zone. (Strictly speaking this should be called a "stress-settlement curve", but we shall use the common phrase here.) Shortly after yielding first begins, the load-settlement curve steepens (point *A*). This condition is called a *local shear failure*. The load-settlement curve steepens gradually until the plastic zone spreads beyond the loaded area (point *C*). Once this happens, the settlement increases rapidly and, finally, a condition is reached where it is not possible to increase Δq_s without very large settlement. This occurs at $\Delta q_s = 862$ kN/m^2. The condition at this stage is called a *general shear failure*, and the value of Δq_s at this condition is called the *ultimate bearing capacity*.

The arrows in Fig. 14.2*b* show the direction and relative magnitude of the motions of various points during the application of a small increment of load. During the elastic portion of the loading, points on the surface outside the loaded area move downward and toward the load. However, once yielding occurs these points begin to move upward and outward. The inset for the ultimate load shows the flow of soil from under the load, thence sideways and upwards. As would be expected, these motions are greatest within the zone that has yielded.

Behavior of Footings on Actual Soils

Figure 14.3 shows the pattern of motion at failure within a stack of rods loaded by a rigid punch. As discussed in Chapter 13, pictures such as this provide the basis for understanding the development of failure in granular soils. Note how the "soil" is pushed out from beneath the "footing" and the surface of the surrounding soil heaves. The pattern of motion is quite similar to that computed for the hypothetical material, as shown in Fig. 14.2*d*.

Figure 14.4 shows load-settlement curves observed during tests of circular plates from 50 to 200 mm in diameter resting on a dry sand. The curve for a medium dense sand (Fig. 14.4*b*) is very similar to that in Fig. 14.2 for the hypothetical material. There is a well-defined break-point or "knee" in the curve corresponding to a local shear failure. Beyond this point the curve becomes steeper and erratic until a general shear failure occurs. This actual load-settlement curve shows a gradually increasing resistance even after the general shear failure. As the footing penetrates, the soil above the base of the footing acts as a surcharge and increases the shear resistance of the soil.

For very loose sands the shear zones at the sides of the footing never become well-defined and little if any surface heave occurs. This behavior, which is simply an extreme case of the behavior described in the preceding paragraph, is termed a *punching failure*.

A footing on a very dense sand shows a somewhat different behavior. Here the load causing general shear

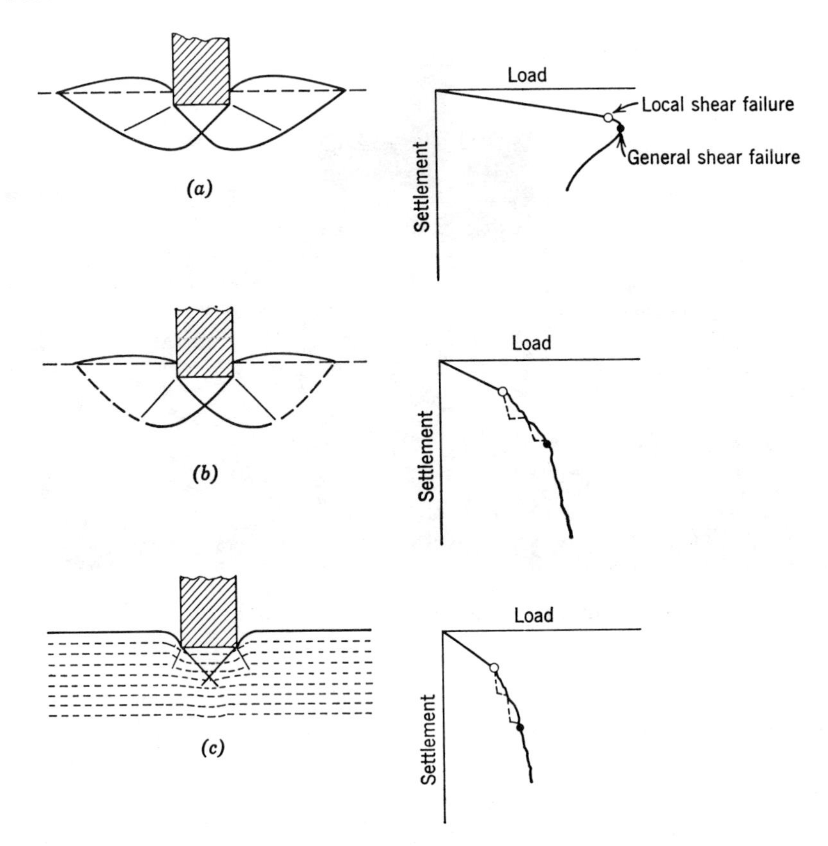

Fig. 14.4 Load-settlement curves and shear zones observed during model tests on sand. (*a*) Dense sand. (*b*) Medium dense sand. (*c*) Very loose sand. (After Vesic, 1963.)

failure is only slightly greater than the load causing local shear failure. Very sharply defined failure surfaces develop. Following the general shear failure, the resistance decreases because of the loss of interlocking resistance past the peak of the stress-strain curve for a dense sand. Although not shown, the resistance will eventually increase again because of the surcharge effect that develops once the footing has penetrated to considerable depth.

The behavior of actual foundations on natural soils appears to be similar to that observed in these small-scale tests, although there have been very few well-documented total failures of foundations resting on sand.

Design Criteria

The basic criterion governing the design of foundations is that the settlement must not exceed some permissible value. This value will vary from structure to structure, as discussed in Section 14.2. In order to ensure that this basic criterion is met, an engineer must make two considerations. First, for any foundation there is some value of the applied stress at which the settlements start to become very large and difficult to predict. This load is called the *bearing capacity*. The foundation must be designed so that the actual bearing stress is less than the bearing capacity, with an appropriate margin of safety to cover uncertainties in the estimate of both the bearing stress and the bearing capacity. The meaning of the terms "very large settlements" and "difficult to predict" involves judgment on the part of the engineer. Generally, the bearing capacity is taken as the bearing stress causing local shear failure; i.e., the stress corresponding to the "knee" of the stress-settlement curve. In a few problems, an engineer may feel that a larger load better fits the definition of bearing capacity. Clearly, however, the load that causes a general shear failure (i.e., the ultimate bearing capacity) is an upper limit for the bearing capacity.

Second, after determining the bearing capacity and ensuring that the bearing capacity exceeds the expected applied bearing stress with an adequate margin of safety, an engineer must estimate the settlement that will occur under the expected load and compare this estimated settlement with the permissible value.

Thus the three key steps in evaluating foundation design are:

1. Selection of the required factor of safety against a shear failure and the permissible settlement.

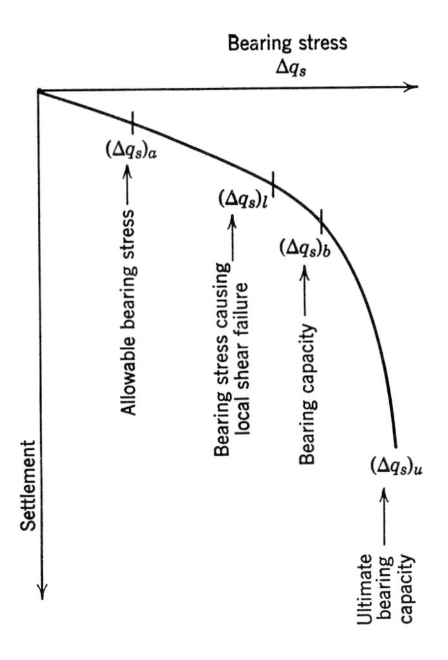

Fig. 14.5 Relationship between bearing stresses and bearing capacities.

2. Determination of the bearing capacity and the actual factor of safety under the expected load.
3. Estimation of the settlement and comparison with the permissible settlement.

In the foregoing discussion, the terms "bearing capacity" and "bearing stress" have been used in several different senses. The meaning of each of the various terms is summarized below and in Fig. 14.5.

Bearing stress Δq_s. This is the stress actually applied to the soil. In an actual foundation Δq_s must be no greater than the:

Allowable bearing stress $(\Delta q_s)_a$. The allowable bearing stress is selected after consideration of safety against instability, permissible settlement, and economy. Often $(\Delta q_s)_a$ is obtained by dividing a safety factor F into the:

Bearing capacity $(\Delta q_s)_b$. The bearing stress at which settlements begin to become very large and unpredictable because of a shear failure is the bearing capacity. Usually, $(\Delta q_s)_b$ is taken equal to the:

Bearing stress causing local shear failure $(\Delta q_s)_l$. This is the bearing stress at which the first major nonlinearity appears in the stress-settlement curve. In some carefully analyzed problems $(\Delta q_s)_b$ may exceed $(\Delta q_s)_l$. However, in any case $(\Delta q_s)_b$ must not exceed the:

Ultimate bearing capacity $(\Delta q_s)_u$. The ultimate bearing capacity is the bearing stress which causes a sudden catastrophic settlement of the foundation.

There are many problems in which $(\Delta q_s)_a$ must be less than $(\Delta q_s)_b$, owing to limitations upon settlement.

14.2 ALLOWABLE SETTLEMENT

Settlement can be important, even though no rupture is imminent, for three reasons: appearance of the structure; utility of the structure; and damage to the structure.

Settlement can detract from the appearance of a building by causing cracks in exterior masonry walls and/or the interior plaster walls. It can also cause a structure to tilt enough for the tilt to be detected by the human eye.

Settlement can interfere with the function of a structure in a number of ways, e.g., cranes and other such equipment may not operate correctly; pumps, compressors, etc., may get out of line; and tracking units such as radar become inaccurate.

Settlement can cause a structure to fail structurally and collapse even though the factor of safety against a shear failure in the foundation is high.

Some of the various types of settlement are illustrated in Fig. 14.6. Figure 14.6*a* shows *uniform settlement*. A building with a very rigid structural mat undergoes uniform settlement. Figure 14.6*b* shows a uniform *tilt*, where the entire structure rotates. Figure 14.6*c* shows a very common situation of *nonuniform settlement*,

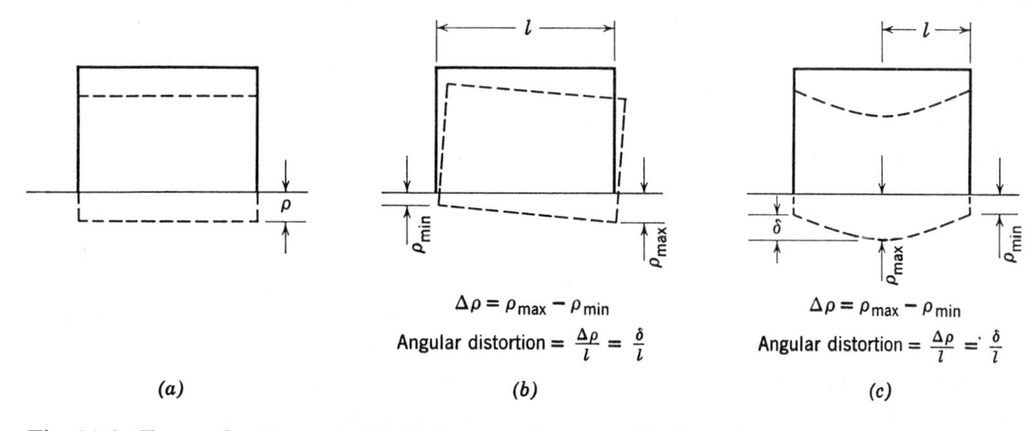

Fig. 14.6 Types of settlement. (*a*) Uniform settlement. (*b*) Tilt. (*c*) Nonuniform settlement.

Table 14.1 Allowable Settlement

Type of Movement	Limiting Factor	Maximum Settlement
Total settlement	Drainage	150–300 mm
	Access	300–600 mm
	Probability of nonuniform settlement:	
	Masonry walled structure	25–50 mm
	Framed structures	50–100 mm
	Smokestacks, silos, mats	75–300 mm
Tilting	Stability against overturning	Depends on height and width
	Tilting of smokestacks, towers	$0.004l$
	Rolling of trucks, etc.	$0.01l$
	Stacking of goods	$0.01l$
	Machine operation-cotton loom	$0.003l$
	Machine operation-turbogenerator	$0.0002l$
	Crane rails	$0.003l$
	Drainage of floors	$0.01–0.02l$
Differential movement	High continuous brick walls	$0.0005–0.001l$
	One-story brick mill building, wall cracking	$0.001–0.002l$
	Plaster cracking (gypsum)	$0.001l$
	Reinforced-concrete building frame	$0.0025–0.004l$
	Reinforced-concrete building curtain walls	$0.003l$
	Steel frame, continuous	$0.002l$
	Simple steel frame	$0.005l$

From Sowers, 1962.
Note. l = distance between adjacent columns that settle different amounts, or between any two points that settle differently. Higher values are for regular settlements and more tolerant structures. Lower values are for irregular settlements and critical structures.

"dishing." Nonuniform settlement can result from: (*a*) uniform stress acting upon a homogeneous soil; or (*b*) nonuniform bearing stress; or (*c*) nonhomogeneous subsoil conditions.

As shown in Fig. 14.6, ρ_{max} denotes the maximum settlement and ρ_{min} denotes the minimum settlement. The differential settlement $\Delta\rho$ between two points is the larger settlement minus the smaller. Differential settlement is also characterized by *angular distortion* δ/l, which is the differential settlement between two points divided by the horizontal distance between them.

The amount of settlement a structure can tolerate—the *allowable settlement* or *permissible settlement*—depends on many factors including the type, size, location, and intended use of the structure, and the pattern, rate, cause, and source of settlement. Table 14.1 gives one indication of allowable settlements. It might seem that the engineer designing a foundation would have the permissible settlement specified for him by the engineer who designed the structure. However, this is seldom the case and the foundation engineer frequently finds himself "in the middle" between the structural engineer who wants no settlement and the client who wants an economical foundation. Thus a foundation engineer must understand allowable settlements.

In the following paragraphs some of the salient aspects of allowable settlement are discussed and illustrated. The last portion of this section presents general guides for estimating the allowable settlement for a particular situation.

Total Settlement

Generally, the magnitude of total settlement is not a critical factor but primarily a question of convenience. If the total settlement of a structure exceeds 150 to 300 mm there can be trouble with pipes (for gas, water, or sewage) connected to the structure. Connections can, however, be designed for structure settlement. Figure 1.3 shows a classic example of a building that has undergone large settlements and yet remained in service. However,

(a)

(b)

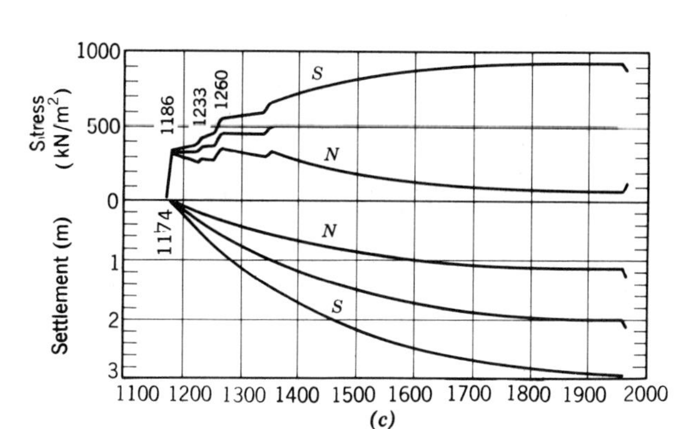

(c)

Fig. 14.7 The Leaning Tower of Pisa. (*a*) From 1964 ASCE Settlement Conference. (*b*) and (*c*) From Terracina, 1962.

there are situations where large total settlements can cause serious problems; e.g., a tank on soft clay near a waterfront can settle below water level.

Tilt

The classic case of tilt is the Leaning Tower of Pisa (Fig. 14.7). As can be seen from the time-settlement curve, the north side of the tower has settled a little over 1 m, whereas the south side has settled nearly 3 m, giving a differential settlement of 1.8 m. The tilt causes the bearing stress to increase on the south side of the tower, thus aggravating the situation. This much tilt in a tall building represents a potentially unstable, dangerous situation. Engineers are now studying methods to prevent further tilt (Terracina, 1962).

Nonuniform Settlement

The allowable angular distortion in buildings has been studied by theoretical analyses, by tests on large models of structural frames, and by field observations. Figure 14.8 gives a compilation of results from such studies. An extreme case is precision tracking radars where a tilt as small as $\delta/l = 1/50{,}000$ can destroy the usefulness of the radar system.

A steel tank for the storage of fluids is a particularly interesting structure. Most of the load is from the stored fluid, and owing to the flexibility of the tank's bottom the bearing stress has a uniform distribution. The flexibility also means that tanks can tolerate large differential settlements without damage, and owners of such tanks are seldom concerned by their appearance. Yet there is amazing disagreement among engineers, builders, and owners as to the allowable settlement of such tanks. A survey of this subject by Aldrich and Goldberg (unpublished) has revealed the following facts:

1. Tanks have settled more than 1.5 m and remained in service.
2. Tanks have failed structurally as the result of settlements as small as 175 mm.
3. Allowable settlements commonly used for the design of tank foundations vary from 25 to 450 mm.

The wide disparity of observed results and views as to allowable settlements illustrates vividly the difficulty faced by a soil engineer in establishing an allowable settlement. Although Table 14.1 and Fig. 14.8 give good

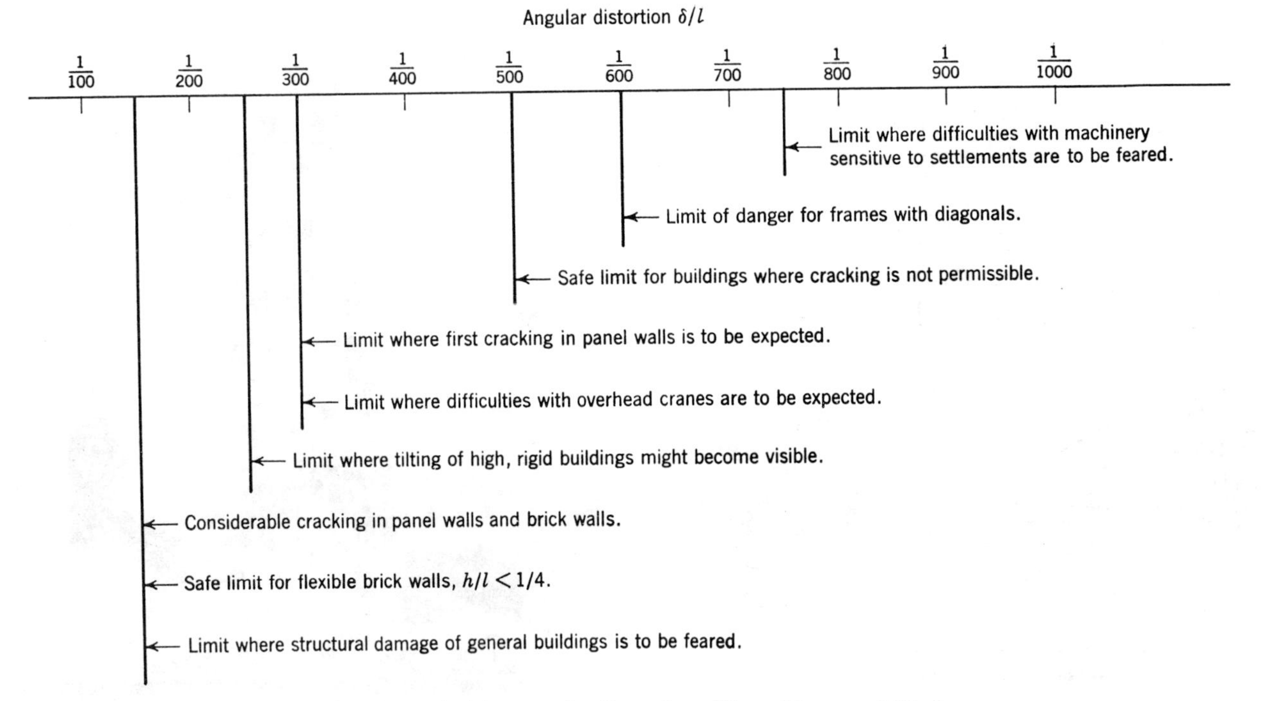

Fig. 14.8 Limiting angular distortions (From Bjerrum, 1963a).

general guidance that will suffice for routine jobs, each large project must receive additional careful study.

Relation of Total and Differential Settlement

As stated previously, it usually is the differential settlement (rather than the total settlement) that is of concern in the designing of a foundation. On the other hand, it is much more difficult to estimate differential settlement than it is to estimate the maximum settlement. This is because the magnitude of differential settlement is affected greatly by the nonhomogeneity of natural soil deposits, and also by the ability of structures to bridge over soft spots in the foundation. On a very important job, it usually is worthwhile to make a very detailed study of the subsoil to locate stronger and weaker zones, and to investigate comprehensively the relation between foundation movements and forces in the structures. On a less important job, it may suffice to use an empirical relationship between total settlement and differential settlement, and to state the design criterion in terms of an allowable total settlement.

Figure 14.9 presents results from actual buildings resting on granular soils. Part (*a*) gives observed values of angular distortion δ/l versus maximum differential settlement. Whereas δ/l is determined by the differential settlement between adjacent columns, the maximum differential settlement may well be between two columns which are far apart. The curve drawn on the figure gives the average for the observed points. Part (*b*) shows the relationship between maximum differential settlement and maximum settlement. The line drawn as an upper envelope indicates that the maximum differential settlement can be equal to the maximum settlement; i.e., there may well be one column which has almost no settlement. Generally, the maximum differential settlement is less than the maximum settlement.[2]

The use of these relationships is illustrated in Example 14.1. From the nature of the building a permissible δ/l is chosen. Then the curves are used to find first the maximum differential settlement and then the maximum permissible total settlement. The settlement as predicted by the methods discussed in Sections 14.8 through 14.10 should then be less than this allowable settlement. An allowable total settlement of 25 mm is a typical specification for commercial buildings.

▶ Example 14.1

Given. A one-story reinforced concrete building with brick curtain walls.

Find. Allowable total settlement which will ensure no cracking of the brick walls.

Solution. From Fig. 14.8, maximum $\delta/l = 1/500 = 0.002$. Table 14.1 would give 0.003. Use $\delta/l = 0.002$.

From Fig. 14.9*a*, maximum allowable differential settlement is 25 mm.

From Fig. 14.9*b*, using the upper bound, the allowable total settlement is also 25 mm.

[2] Maximum differential settlement greater than maximum total settlement can result when one portion of the structure heaves while another settles. This situation is not uncommon in tanks on sand.

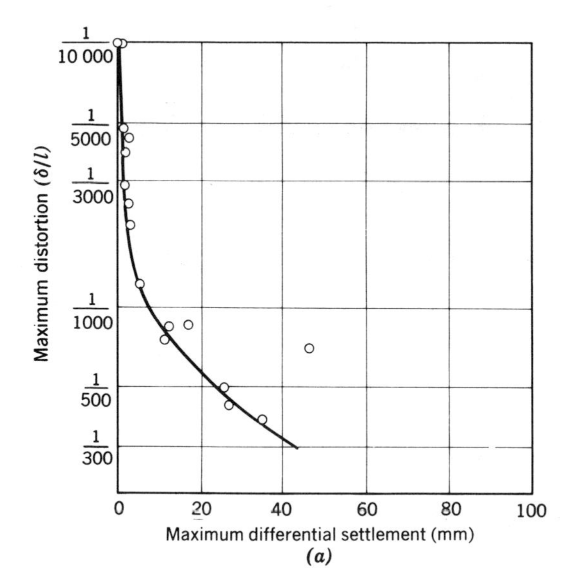

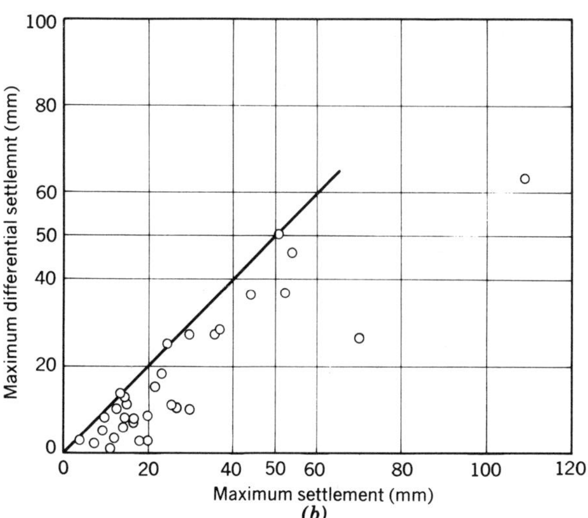

Fig. 14.9 Settlement of structures on sand (From Bjerrum, 1963a and 1963b).

14.3 ULTIMATE BEARING CAPACITY OF STRIP FOOTINGS

As a first step in our study of methods for establishing the bearing capacity of foundations, we shall study the ultimate bearing capacity $(\Delta q_s)_u$ of a footing which is very long compared to its width. This type of footing occurs under retaining walls and under building walls. Methods have been developed for predicting the ultimate bearing capacity of such footings. Subsequent sections will discuss how the theoretical results are modified by judgment and experience to account for the effects of local shear failure and for different shapes of footings.

A typical strip footing is depicted in Fig. 14.10. Because the footing is very long in comparison to its width, the problem is one of plane strain; i.e., the problem is two-dimensional. There are several reasons why the footing is generally located below ground surface rather than at the very surface: (*a*) to avoid having to raise the first-floor level well above ground surface; (*b*) to permit removal of the surface layer of organic soil; (*c*) to gain the additional bearing capacity that comes from partial embedment (see later portions of this section); and (*d*) to place the footing below the zone of soil which experiences volume changes because of frost action or other seasonal effects. In Boston, for example, the building code requires that exterior footings be 1.2 m or deeper below ground surface.

For purposes of analysis, the actual situation shown in Fig. 14.10*a* is usually replaced by the situation shown in Fig. 14.10*b*: the soil above the base of the footing is replaced by a uniform surcharge of intensity $q_s = \gamma d$, where

γ = the unit weight of the soil
d = the depth of the base of the footing below ground surface

The effect of the weight of the soil above the footing base is thus taken into consideration, but the shear resistance of this soil is neglected. The accuracy of this approximation will be discussed later in this section.

Solution Based on Rankine Wedges

We shall begin with an analysis which is much too approximate for practical use, but which illustrates in a simple way the factors that must be considered in a more accurate analysis. It is assumed that the failure zone is made up of two separate wedges, as shown in Fig. 14.11: a Rankine active wedge I, which is pushed downward and outward, and a Rankine passive wedge II, which is pushed outward and upward. There are corresponding patterns of motion on the other side of the center line.

The analysis begins with consideration of wedge II. Using Eq. 13.9, we can write an expression for the maximum thrust P (i.e., passive thrust) which can be applied to this wedge along the vertical face IJ (note $N_\phi = K_p$). Equation 14.1 includes the resistance resulting from friction and surcharge. This thrust P is also the maximum thrust available to hold the active wedge I in equilibrium under the application of the loading Q_{ult}/B. The value of this loading may therefore be found by using Eq. 13.7 for the active thrust.

Equation 14.3 may be written in the form[3]

$$\frac{Q_{\text{ult}}}{B} = (\Delta q_s)_u = \frac{\gamma B}{2} N_\gamma + q_s N_q \qquad (14.4)$$

where N_γ and N_q are dimensionless factors that depend only on the friction angle of the soil. Based on this

[3] The reason for writing $\gamma B/2$ is purely historical; i.e., this is the way it was first written.

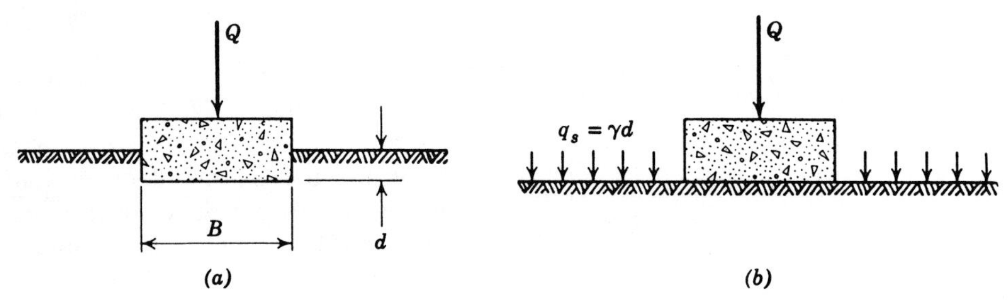

Fig. 14.10 Shallow strip footing under a vertical load. (*a*) Actual situation. (*b*) Assumed situation.

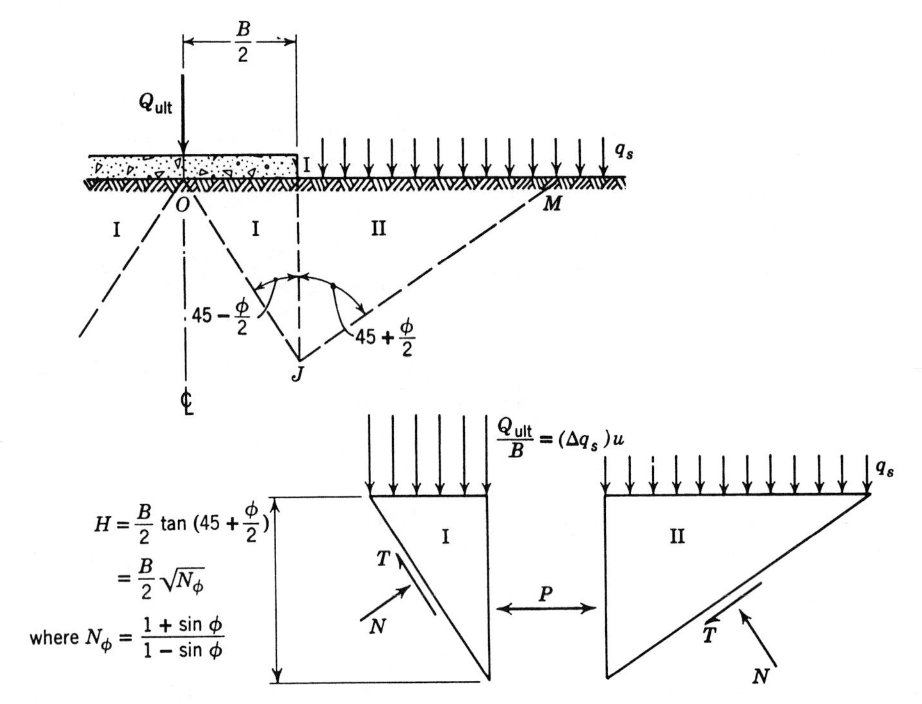

Fig. 14.11 Derivation of bearing capacity based on Rankine wedges.

Maximum force P that can be applied to passive wedge II

From Eq. 13.9:

$$P = q_s H N_\phi + \tfrac{1}{2}\gamma H^2 N_\phi$$

$$P = q_s \frac{B}{2} N_\phi^{3/2} + \tfrac{1}{8}\gamma B^2 N_\phi^{\,2} \qquad (14.1)$$

Maximum surcharge Q_{ult}/B that can be applied to active wedge I

From Eq. 13.7:

$$P = \frac{Q_{ult}}{B}\frac{H}{N_\phi} + \tfrac{1}{2}\gamma H^2 \frac{1}{N_\phi}$$

$$\frac{Q_{ult}}{B} = \frac{P}{H} N_\phi - \tfrac{1}{2}\gamma H$$

$$= \left(\frac{2P}{B} - \tfrac{1}{4}\gamma B\right)\sqrt{N_\phi}$$

$$\frac{Q_{ult}}{B} = q_s N_\phi^{\,2} + \tfrac{1}{4}\gamma B N_\phi^{5/2} - \tfrac{1}{4}\gamma B N_\phi^{1/2} \qquad (14.2)$$

$$\frac{Q_{ult}}{B} = \frac{\gamma B}{4}(N_\phi^{5/2} - N_\phi^{1/2}) + q_s N_\phi^{\,2} \qquad (14.3)$$

solution involving Rankine wedges, N_γ and N_q have the values

$$N_\gamma = \tfrac{1}{2}(N_\phi^{5/2} - N_\phi^{1/2})$$
$$N_q = N_\phi^{\,2} = K_p^{\,2} \tag{14.5}$$

where

$$N_\phi = K_p = \frac{1 + \sin\phi}{1 - \sin\phi}$$

Thus, according to Eq. 14.4, the ultimate bearing capacity of a strip footing can be written as the sum of the two terms. The first term depends on the unit weight of the soil and the width of the footing. The second term depends on the surcharge. By introducing the relation between the depth of embedment and the surcharge (Fig. 14.10), we have

$$\frac{Q_{\text{ult}}}{B} = (\Delta q_s)_u = \frac{\gamma B}{2} N_\gamma + \gamma d N_q \tag{14.6}$$

The dimensionless factors N_γ and N_q are called *bearing capacity factors* and depend only on ϕ.

The use of the foregoing results is illustrated by Examples 14.2 to 14.4. As mentioned earlier, the results obtained using the Rankine wedges are too approximate (too low) for use in practice, but the results do serve to illustrate the following important points which are also true of the more accurate solutions:

1. An important increase in ultimate bearing capacity comes about as the result of partial embedment.
2. There is a sharp increase in bearing capacity as the friction angle increases. The footing load, which of course causes the stresses that shear the soil, also causes normal stresses which act to increase the shear resistance. Figure 14.12 shows stress paths for points at mid-depth within the passive and active zones, assuming that initially the stresses are geostatic with $K_o = 1$.[4] The stress path for the point under the footing rises at a slope less than 45°. With increasing friction angle for the soil, a larger and larger footing load is required to make the stress path "catch up" with the failure line.

Note also that the bearing capacity of a footing on sand would be zero if the soil were weightless.

Other Solutions

There are two basic shortcomings to the foregoing solution based on simple Rankine wedges. First, the actual failure zone (see Fig. 14.4) is bounded by curves rather than by two straight surfaces. Second, the foregoing solution has neglected the shear stresses which

[4] In drawing the stress path for point R, we assume that the force P increases uniformly as the load is applied. The actual variation of P with load is discussed in Section 14.4, and actual stress paths for point R are curved rather than straight.

► **Example 14.2**

Given. Footing shown in Fig. E14.2.

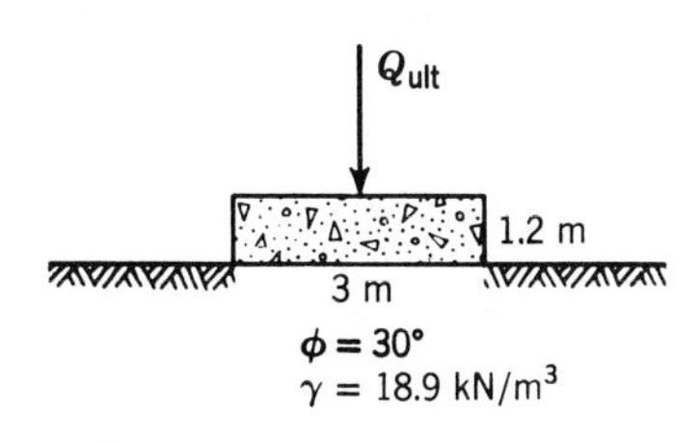

Fig. E14.2

Find. Q_{ult}.
Solution.

$$N_\phi = \frac{1 + \sin\phi}{1 - \sin\phi} = 3$$

$$N_\gamma = \tfrac{1}{2}(15.60 - 1.73) = 6.94$$

$$N_q = 3^2 = 9$$

$$\frac{Q_{\text{ult}}}{B} = (\Delta q_s)_u = (18.9)(3)\left(\frac{6.94}{2}\right) = 196.7 \text{ kN/m}^2$$

$$Q_{\text{ult}} = 590.1 \text{ kN/m of wall}$$ ◄

► **Example 14.3**

Given. Footing shown in Fig. E14.3.
Find. Q_{ult}.
Solution.

$$\frac{Q_{\text{ult}}}{B} = (\Delta q_s)_u = 196.7 + (18.9)(1.2)$$
$$= 196.7 + 204.1 = 400.8 \text{ kN/m}^2$$
$$Q_{\text{ult}} = 1202.4 \text{ kN/m of wall}$$ ◄

► **Example 14.4**

Given. Same as Example 14.3, but with $\phi = 40°$

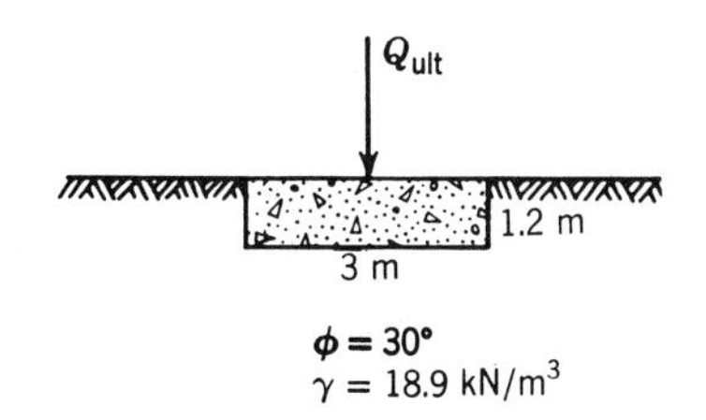

Fig. E14.3

Find. Q_{ult}
Solution.

$$N_\phi = 4.61$$
$$N_\gamma = \tfrac{1}{2}(45.8 - 2.15) = 21.6$$
$$N_q = 21.2$$

$$\frac{Q_{\text{ult}}}{B} = (\Delta q_s)_u = (18.9)(3)\left(\frac{21.6}{2}\right) + (18.9)(1.2)(21.2)$$
$$= 612.4 + 480.8$$
$$= 1093.2 \text{ kN/m}^2$$
$$Q_{\text{ult}} = 3279.6 \text{ kN/m of wall}$$ ◄

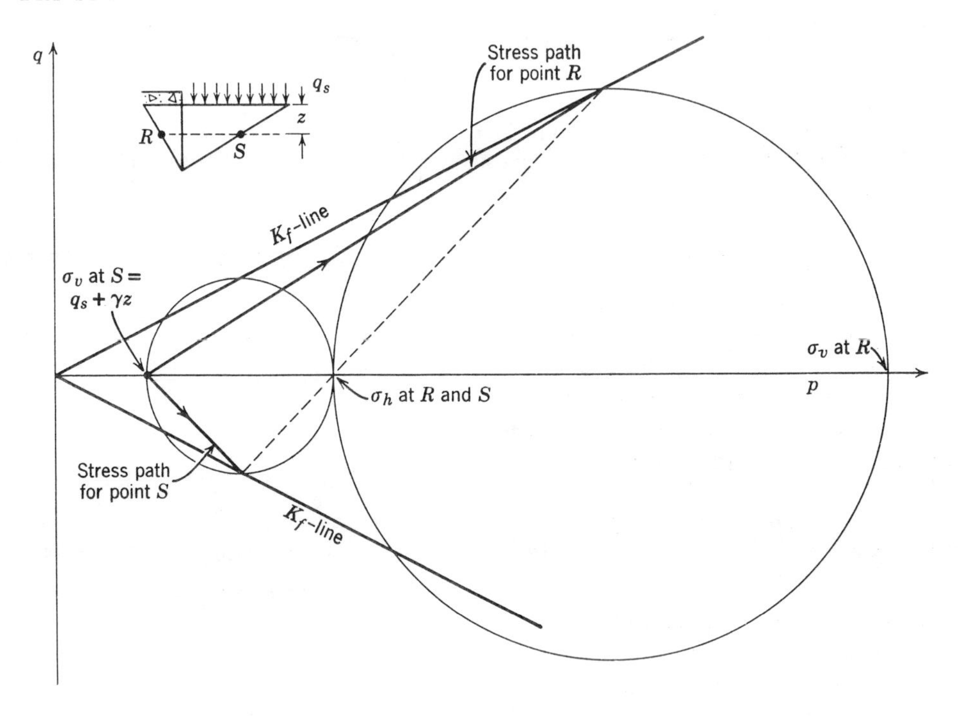

Fig. 14.12 Stress paths at points below foundation.

must act upon the line IJ in Fig. 14.11. Because of the second shortcoming the solution grossly underestimates the actual bearing capacity.

Many different types of solutions have been made in an attempt to overcome satisfactorily these shortcomings. Trial wedge solutions have been made using free bodies bounded by various combinations of straight lines, circles, and logarithmic spirals (Hansen, 1966). Solutions have been made by numerical integration of Kötter's equation (Sokolovski, 1965; Harr, 1966). Most of these solutions involve some degree of approximation and, as discussed in Chapter 13, it still is not clear just what is meant by an exact solution to a limiting equilibrium problem involving soil.

The most commonly used solution is that developed by Terzaghi (1943). This solution assumes that Eq. 14.6 is applicable; i.e., the resistance offered by the weight of the soil and by the surcharge can be evaluated independently of each other. This is not strictly true, since the location of the theoretical failure surface is somewhat different for each combination of ϕ, γ, and Δq_s. However, it has been shown that this assumption leads to a conservative result—to an underestimate of the bearing capacity. Having made this assumption, Terzaghi then evaluated N_γ and N_q by the trial wedge method using free bodies of the type shown in Fig. 14.13a. Values applicable to a rough footing, which is the typical case encountered in practice, are plotted versus ϕ in Fig. 14.13b. Values for smooth bases are also available. Examples 14.5 to 14.7 repeat the earlier examples, but use Terzaghi's values of N_q and N_γ and thereby obtain much larger estimates for the bearing capacity.

Table 14.2 compares values of N_q and N_γ as calculated by Terzaghi with average values deduced from small-scale footing tests. There was considerable scatter in the

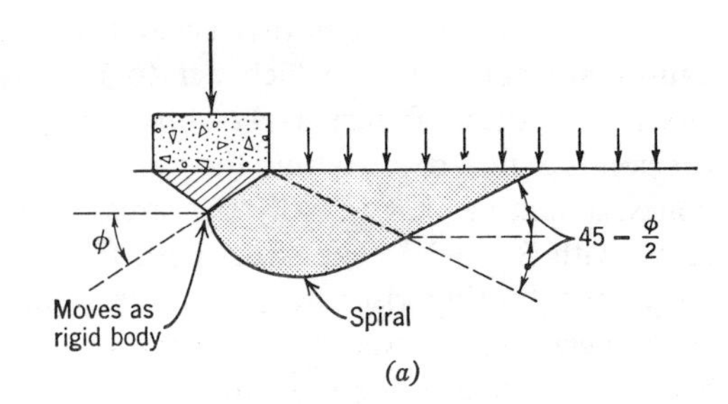

(a)

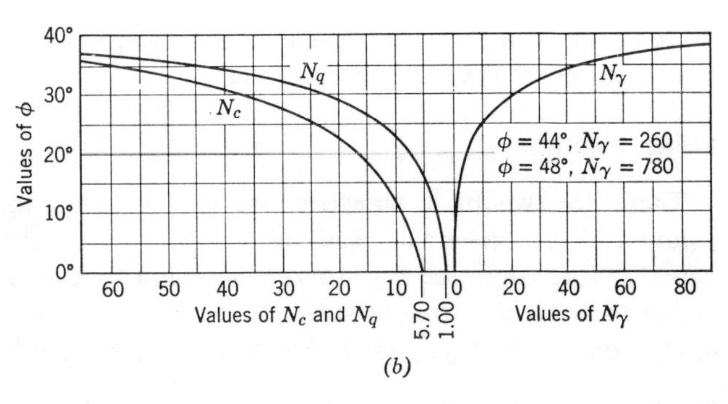

(b)

Fig. 14.13 (a) Shapes of failure surfaces for Terzaghi solution. (b) Bearing capacity factors according to Terzaghi (for footing with rough base).

▶ **Example 14.5**

Repeat Example 14.2, using Terzaghi's bearing capacity factors

$$N_\gamma = 20$$

$$(\Delta q_s)_u = (18.9)(3)(\tfrac{20}{2}) = 567.0 \text{ kN/m}^2$$ ◀

▶ **Example 14.6**

Repeat Example 14.3, using Terzaghi's bearing capacity factors

$$N_q = 22$$

$$(\Delta q_s)_u = 567.0 + (18.9)(1.2)(22)$$
$$= 567.0 + 499.0 = 1066 \text{ kN/m}^2$$ ◀

▶ **Example 14.7**

Repeat Example 14.4, using Terzaghi's bearing capacity factors

$$\left.\begin{aligned} N_\gamma &= 130 \\ N_q &= 80 \end{aligned}\right\} \text{ see Table 14.2}$$

$$(\Delta q_s)_u = (18.9)(3)(\tfrac{130}{2}) + (18.9)(1.2)(80)$$
$$= 3685.5 + 1814.4 = 5499.9 \text{ kN/m}^2$$ ◀

experimental data. These results indicate that Terzaghi's factors are conservative with regard to the average experimental results, especially for large friction angles. The value of ϕ as measured in conventional triaxial tests was used to deduce N_q and N_γ from the footing tests. Since a strip footing is a situation in plane strain, ϕ should have been assumed somewhat larger (see Section 1I.4.). Assuming a larger ϕ would cause one to deduce smaller values of N_q and N_γ inorder to give the observed bearing capacity, and thus would lead to better agreement between Terzaghi's values and the experimental values.

None of the other solutions has given appreciably better agreement between theoretical and measured bearing capacities, and hence the Terzaghi solution continues to be used.

A Further Look at Effect of Embedment

Equation 14.6 may be rearranged to read

$$\frac{Q_{\text{ult}}}{B} = (\Delta q_s)_u = \frac{\gamma B}{2} N_\gamma \left(1 + 2 \frac{d}{B} \frac{N_q}{N_\gamma}\right)$$

Examination of the results in Table 14.2 indicates that, for ϕ equal to 30°, the ratio N_q/N_γ is approximately equal to unity, although the value of the ratio may drop to 0.6 for denser sands. Several experimenters have reported values of from 0.7 to 1.0 for this ratio. As an approximation, we can assign a value of unity to this ratio and thus obtain the following approximate expression:

$$(\Delta q_s)_u = \frac{\gamma B}{2} N_\gamma \left(1 + 2 \frac{d}{B}\right) \tag{14.7}$$

Table 14.2 Comparison of Theoretical and Actual Bearing Capacity Factors

Factor	$\phi = 30°$	$\phi = 40°$
N_q—Terzaghi	22	80
experimental	23	400
N_γ—Terzaghi	20	130
experimental	33	170–210

Meyerhof (1951) has investigated the importance of the shear resistance of the soil lying above the base of the footing. For $d < B$, he found that the rules derived above (based on consideration only of the weight of this soil) were reasonably accurate. For deeper footings and for friction piles, it is necessary to take the resistance of this soil into account.

14.4 EFFECT OF LOCAL SHEAR FAILURE ON BEARING CAPACITY

There is no strictly theoretical method for estimating the load at which local shear failure occurs. In this section, we first examine the factors that make local shear failure more important in some soils than in others. Then we present semiempirical methods for estimating bearing capacity.

When the load equals the ultimate bearing capacity a general shear failure occurs: the full shear resistance of the soil is mobilized all along a failure surface which starts beneath the footing and extends to the surface of the soil beyond the footing. As explained in Section 14.1, at some smaller load there will be a local shear failure, at which time the shear resistance is reached along only a part of the ultimate failure surface. It was also noted that local shear failure increases in importance as a soil becomes looser. Figure 14.14 indicates the range of relative densities for which the several types of failure

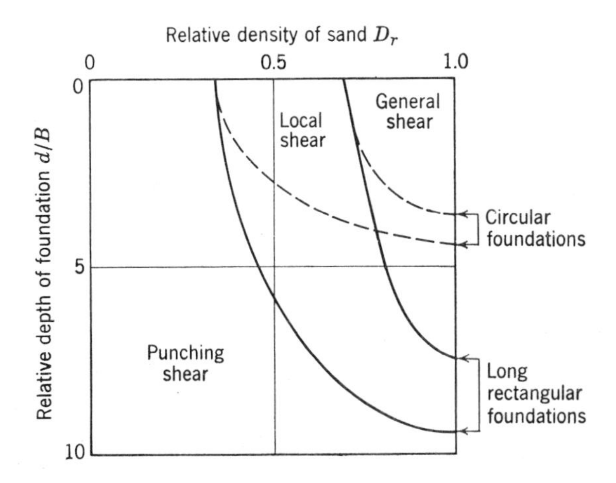

Fig. 14.14 Controlling type of failure as function of relative density and depth of embedment (From Vesic, 1963).

determine the load at which the load-settlement curve shows major yielding.

In order to understand the relationship between ultimate bearing capacity and the load causing local shear failure, it is necessary to consider: (*a*) the ratio of horizontal to vertical stress before loading, i.e., K_0; and (*b*) the way in which strains develop during loading. Representation of the failure zone by two Rankine wedges (Fig. 14.11) and of stress conditions by stress paths for two typical points (Fig. 14.12) provides a convenient basis for an approximate discussion of these two factors.

Loose Sand

Point O in Fig. 14.15*a* shows the stress conditions at the two typical points R and S before loading.

During the initial stage of loading, while the soil is still more-or-less elastic, there is relatively little change in σ_h at point R (see, for example, point C in Example 8.9). Thus during this early stage, the stress path for point R rises essentially as in an ordinary triaxial test (path OL in Fig. 14.15*a*) while the stresses at point S remain essentially unchanged. This situation continues until the stress path for point R reaches the failure line, at which time local shear failure occurs.

As the load is increased further, σ_h increases at both points R and S. The stress path for point S is ON in Fig. 14.15*a* and loading continues until this stress path reaches the failure line at N, at which time the ultimate bearing capacity is reached. Meanwhile, the stress path for point R runs along the failure line from L to M.

The load causing local shear failure may be computed using the derivation in Fig. 14.11. The assumption that σ_h remains constant during the early part of the loading means that the horizontal force P on surface IJ will be[5] $\frac{1}{2}\gamma H^2 K_0$. Using the expression immediately preceding Eq. 14.2, the load causing local shear failure is

$$(\Delta q_s)_l = \tfrac{1}{2}\gamma H(K_0 N_\phi - 1)$$

Dividing by the corresponding expression for $(\Delta q_s)_u$ gives

$$\frac{(\Delta q_s)_l}{(\Delta q_s)_u} = \frac{K_0 N_\phi - 1}{N_\phi{}^2 - 1} \tag{14.8}$$

Using typical values of $K_0 = 0.6$ and $N_\phi = 3$, the ratio is 0.1. While this analysis is too crude for practical use, it shows clearly that in a loose sand local shear failure will occur at a load much smaller than the ultimate bearing capacity.

During the early stage of loading, the soil immediately beneath the footing strains much as in an ordinary triaxial test starting from the K_0 condition. Since the sand is loose, there is relatively little horizontal strain when failure occurs in such a test. Hence there is very little outward push against the loose sand in zone II (Fig. 14.15) and thus σ_h stays essentially constant at points R and S. Once local failure occurs in zone I, then large horizontal strains occur in zone I as the load is increased farther and this outward push causes shear resistance to be developed in zone II.

[5] For this derivation the surcharge q_s is taken as zero.

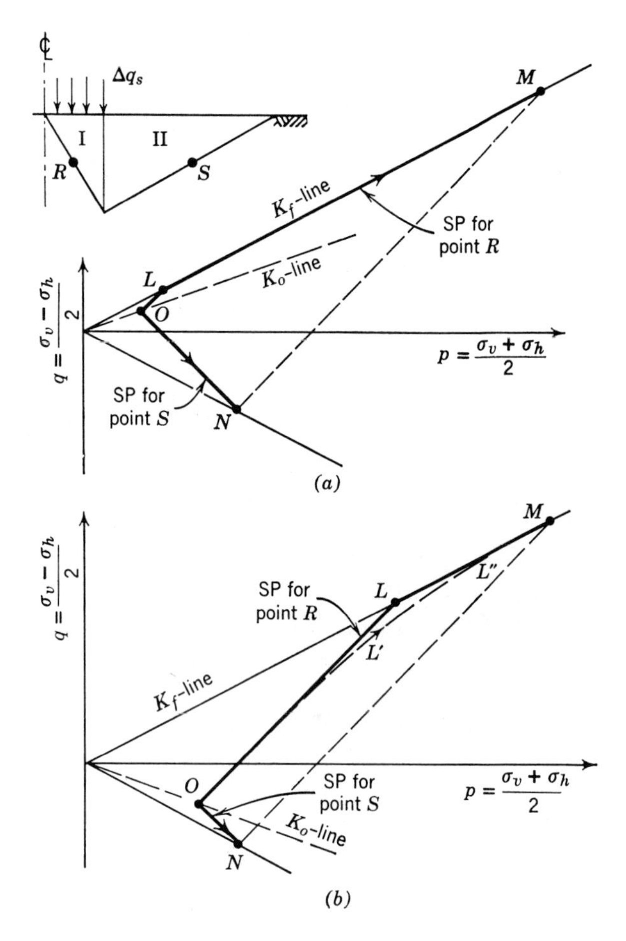

Fig. 14.15 Effect of density of sand on stress paths at two points under foundation. (*a*) Loose sand. (*b*) Dense sand.

Dense Sand

Figure 14.15*b* shows the corresponding stress paths for a dense sand, again assuming that the horizontal stresses remain constant until local shear failure occurs at point R. Using $K_0 = 2$ and $N_\phi = 4$, the ratio in Eq. 14.8 is 0.47; this is much greater than for a loose sand.

Actually the stress path for point R is more like $OL'L''M$. Since it is dense, the sand within zone I will begin to dilate before local shear failure can occur. The resulting horizontal strains cause an outward push against zone II, and since the sand is dense, relatively little push is necessary to develop significant shear resistance in zone II. Thus the ratio of $(\Delta q_s)_l$ to $(\Delta q_s)_u$ is greater than given by Eq. 14.8.

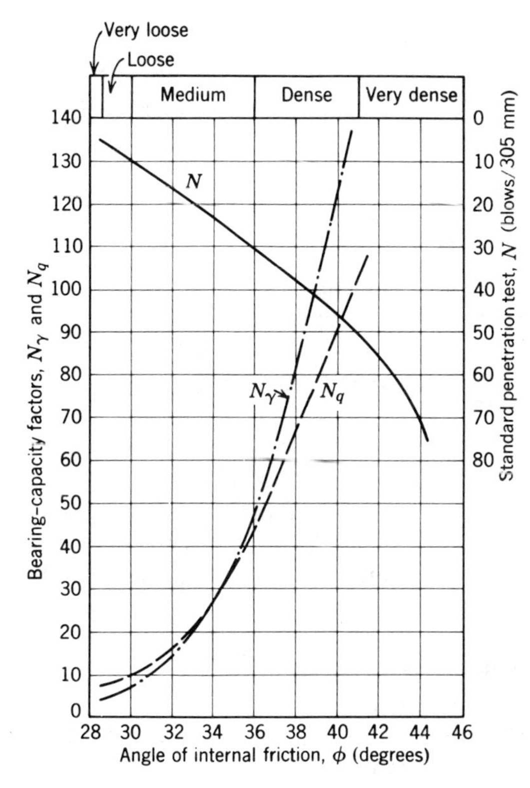

Fig. 14.16 Bearing capacity factors which automatically incorporate allowance for local shear failure (From Peck, Hansen, and Thornburn, 1953).

Combined Effect

These findings can be restated as follows. The soil which supports a footing begins to yield $(\Delta q_s)_l$ when the full shear resistance has been mobilized directly underneath the footing, but does not reach the ultimate bearing capacity $(\Delta q_s)_u$ until full resistance is reached all along the boundary of the failure wedge. For a dense sand, full resistance is mobilized almost simultaneously along all parts of the boundary. However, for a loose sand considerable footing movement is necessary before full resistance is reached along the outermost portions of the boundary. This difference occurs because of the differences in the initial stress conditions and compressibility of loose and dense sands. An accurate analysis of the development of local and general shear failures in loose and dense sands is given by Christian (1966) using finite difference techniques.

Empirical Solution for Bearing Capacity

Figure 14.16 gives factors N_γ and N_q, which may be used to estimate bearing capacity $(\Delta q_s)_b$ according to the equation

$$(\Delta q_s)_b = \frac{\gamma B}{2} N_\gamma + \gamma d N_q \qquad (14.10)$$

In this figure, ϕ denotes the peak friction angle of the soil. These factors, which take into account local shear, were obtained as follows. For $\phi \geq 38°$, the curves are the same as for the ultimate bearing capacity (Fig. 14.13). For $\phi \leq 28°$, the N_γ and N_q are equal to the values in Fig. 14.13 at $\phi = \tan^{-1}(\frac{2}{3}\tan\phi)$. Thus N_γ in Fig. 14.16 for $\phi = 28°$ equals N_γ in Fig. 14.13 for $\phi = 19.5°$. This strictly empirical correction to account for local shear in loose soils was suggested by Terzaghi from an analysis of experimental data. For $28° < \phi < 38°$, smooth transition curves were drawn.

14.5 FOOTING DESIGN

The bearing capacity results given in Sections 14.3 and 14.4 may be applied directly to the design of foundations for walls, as illustrated in Example 14.8.

▶ Example 14.8

Given. A wall which is 2.1 m wide at the base, and which rests 1 m below the surface of a sand with $\phi = 35°$ and $\gamma = 17.3$ kN/m³.

Find. Bearing capacity.

Solution. From Fig. 14.16 we find

$$N_\gamma = 35$$
$$N_q = 34$$

Hence.

$$\begin{aligned}(\Delta q_s)_b B &= \tfrac{1}{2}(17.3)(2.1)^2(35) + 1.0(17.3)(2.1)(34) \\ &= 1335.1 + 1235.2 \\ &= 2570.3 \text{ kN/m of wall}\end{aligned}$$

This wall and its supporting soil have the same properties as the wall and supporting soil in Example 13.12. In that example, the vertical component of the force on the supporting soil was 220.3 kN/m of wall, less than one-tenth of the bearing capacity just computed. Strictly speaking, of course, one should check for the effects of inclination and eccentricity of the actual loading upon the base of a retaining wall (see Section 14.7). However, with such a large factor of safety against bearing capacity failure, and considering that the resultant lies within the middle third of the base and that the wall checks for resistance to sliding, most designers would consider the wall of Example 13.12 safe. ◀

In such foundation design problems, it usually is necessary to rely upon the results of penetration tests to provide an estimate of the friction angle (see Section 11.5). Figure 14.16 can be used to relate blow count directly to bearing capacity factors. The fact that the proper value of ϕ is usually uncertain whenever blow count must be used is one reason why a rather liberal factor of safety (at least 3) should be used when checking the bearing capacity of foundations. A small uncertainty in ϕ causes a large uncertainty in the values for the bearing capacity factors. For example, the bearing capacity (2570.3 kN/m of wall) in Example 14.8 would only be 1121 kN/m if ϕ were reduced from 35° to 32°.

At this point the reader may well ask: Why did we study all the theory and then revert to crude empirical equations with a large safety factor? The reasons are simple. The theory has served an indispensable function. It has indicated how the bearing capacity should vary with such factors as the unit weight of soil and the width of the foundation. Moreover, the theory has provided numerical results for the ultimate bearing capacity. However, the theory is inadequate to provide accurate numerical values for the bearing capacity, taking into account the effects of local shear. Data from model tests and field experience must be used to fill this gap. Such experience has been incorporated in Fig. 14.16. Used together with a liberal safety factor, this approach will provide a conservative answer for any practical problem. If conservatism must be avoided, then alternative methods, such as a loading test on the site, must be used to evaluate the bearing capacity. Since such a load test can seldom be performed using a full-scale foundation, theory must be used to extrapolate from the actual loading test to the full-scale foundation (see Example 14.9).

▶ Example 14.9

Given. A plate bearing test shows a bearing capacity failure at a bearing stress of 345 kN/m². The plate is 0.3 m square and bears 0.9 m below the ground surface. The unit weight of the soil is estimated at 15.7 kN/m³.

Find. Bearing capacity for a footing 1.8 m square, to be founded 0.9 m below ground surface.

Solution. The first step is to find a value of ϕ which will satisfy Eq. 14.12:

$$345 \text{ kN/m}^2 = \tfrac{1}{2}(15.7)(0.3)(0.7)N_\gamma + 0.9(15.7)(1.2)N_q$$

After several trials, it is found that $\phi=32.5°$, giving $N_\gamma=16\frac{1}{2}$ and $N_q=18\frac{1}{2}$, satisfies the equation. Now these values of N_γ and N_q can be applied to the actual footing:

$$(\Delta q_s)_b = \tfrac{1}{2}(15.7)(1.8)(0.7)(16\tfrac{1}{2}) + 0.9(15.7)(1.2)(18\tfrac{1}{2})$$

$$= 477 \text{ kN/m}^2$$ ◀

14.6 ROUND AND RECTANGULAR FOOTINGS

Several very approximate theoretical analyses have been made for the bearing capacity of round footings. However, there are no theoretical analyses that give the ultimate bearing capacity of square or rectangular footings. There have been numerous model studies aimed at evaluating the ultimate bearing capacity of round, square, or rectangular footings but, unfortunately, the data from these tests are often conflicting. Data for surface footings by Vesic (1963) are shown in Fig. 14.17.

Many equations have been proposed for use in estimating the bearing capacity of round and rectangular footings. All are based on theoretical considerations plus experimental data, and from the practical standpoint the differences in the predictions are slight. The following are recommended:

Round footings:

$$\frac{Q_b}{(\pi/4)D^2} = (\Delta q_s)_b = (0.6)\tfrac{1}{2}\gamma D N_\gamma + \gamma\, d N_q \quad (14.11)$$

where D is the diameter (Terzaghi, 1943).

Rectangular and square footings:

$$\frac{Q_b}{BL} = (\Delta q_s)_b = \tfrac{1}{2}\gamma B N_\gamma\left(1 - 0.3\frac{B}{L}\right) + \gamma d N_q\left(1 + 0.2\frac{B}{L}\right) \quad (14.12)$$

where L is the length of the footing (Hansen, 1966). The values of N_γ and N_q are taken from Figs. 14.13 or 14.16 as appropriate. Example 14.10 illustrates the use of such equations.

▶ Example 14.10

Given. A footing 1.8 m by 3.6 m is to be founded 1.2 m below the surface of a sand with $\phi = 40°$ and $\gamma = 18.0$ kN/m³.

Find. Bearing capacity.

Solution. Equation 14.12 becomes

$$(\Delta q_s)_b = (0.85)\tfrac{1}{2}\gamma B N_\gamma + (1.1)\gamma\, d N_q$$

Using either Fig. 14.13 or Fig. 14.16 we find

$$N_\gamma = 120$$

$$N_q = 90$$

Consequently,

$$(BL)(\Delta q_s)_b = (1.8)(3.6)\,[\tfrac{1}{2}(0.85)(18.0)(1.8)(120) + (1.1)(18.0)(0.2)(90)]$$

$$= (6.48)[1652.4 + 2138.4]$$

$$= 24\,564 \text{ kN}$$ ◀

Figure 14.17 compares results predicted using these equations with failure loads observed in model tests. Note that there is considerable scatter in the experimental results. Except for dense sand with very large friction angles the equations adequately predict the general shear failures. Use of Fig. 14.16 overestimates the loads at which local shear failure occurs, primarily because this sand has a very high friction angle for a given relative density. The need for a large safety factor when using these equations is clear.

14.7 BEARING CAPACITY UNDER INCLINED AND ECCENTRIC LOADS

Meyerhof (1953) has suggested the following relation be used whenever strip footing loads are inclined and/or

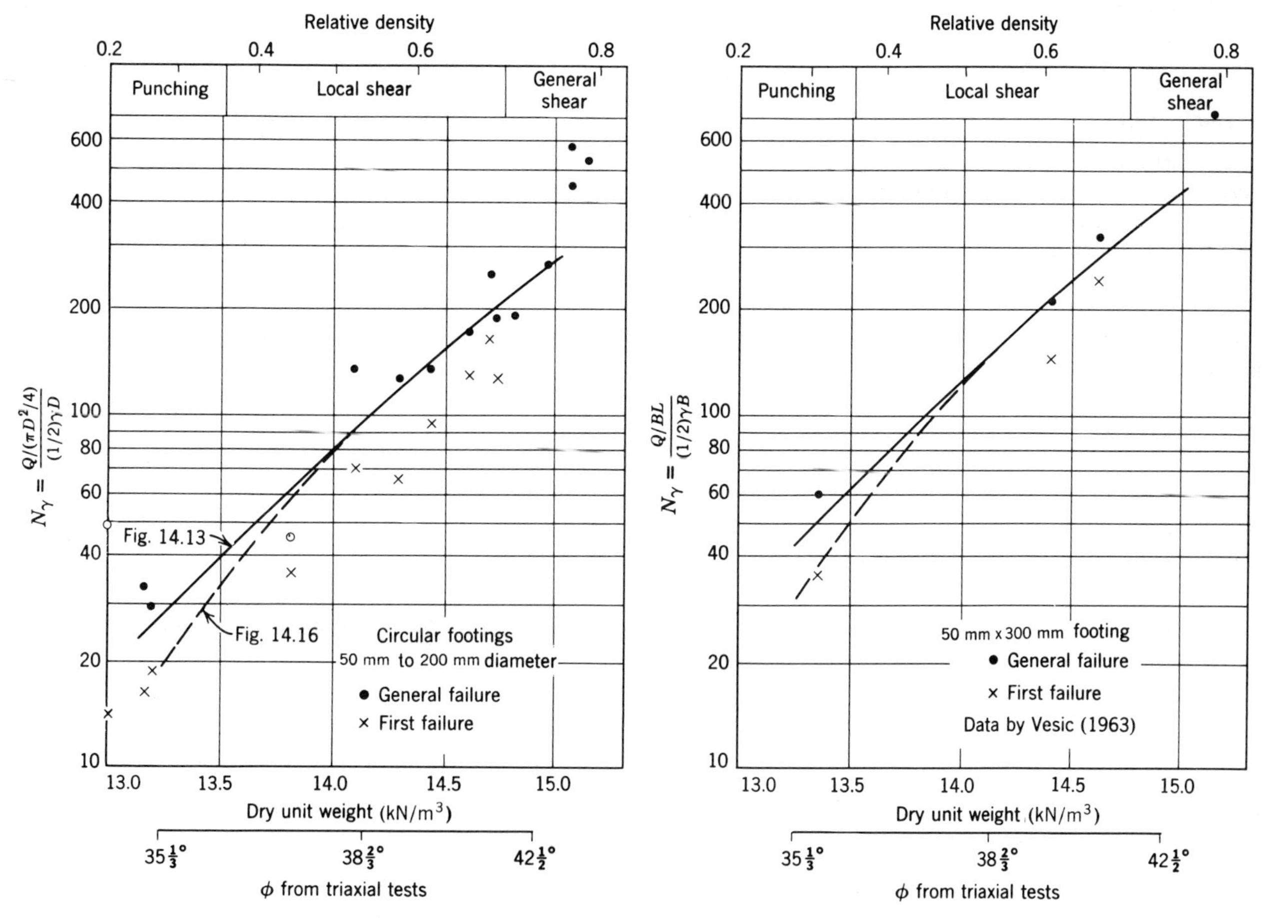

Fig. 14.17 Comparison of predicted and measured $N\gamma$.

eccentric to the centroid of the footing (see Fig. 14.18):

$$(\Delta q_s)_b = \frac{Q_b}{B} = \left(1 - \frac{2e}{B}\right)\left(1 - \frac{\alpha}{90°}\right)^2 \gamma d N_q + \left(1 - \frac{2e}{B}\right)^2\left(1 - \frac{\alpha}{\phi}\right)^2 \tfrac{1}{2}\gamma B N_\gamma \quad (14.13)$$

where

Q_b = limiting value for the vertical component of the load

N_γ and N_q = bearing capacity factors for vertical loading

e = the distance between the centroid of the base and the point of action of the resultant force on the base

α = the angle of inclination of the resultant force with respect to the vertical

Meyerhof developed Eq. 14.13 partly on rough theoretical grounds and partly on the basis of fitting a conservative envelope to experimental results. The equation no doubt is quite conservative. Note that the friction angle of the soil supporting the footing, rather than the friction angle between the soil and the footing, is used in this equation.

Example 14.11 illustrates the use of this equation. Comparing the results in Examples 14.9 and 14.11, we see that consideration of inclination and eccentricity leads to a large reduction in bearing capacity. Equation 14.13 should be used with a safety factor of 3 or more. On this basis the wall in Example 13.12 is still safe, since the safety factor is 775.8/220.3 = 3.5.

For a rectangular footing Eq. 14.13 may be used by including the correction factors appearing in Eq. 14.12,

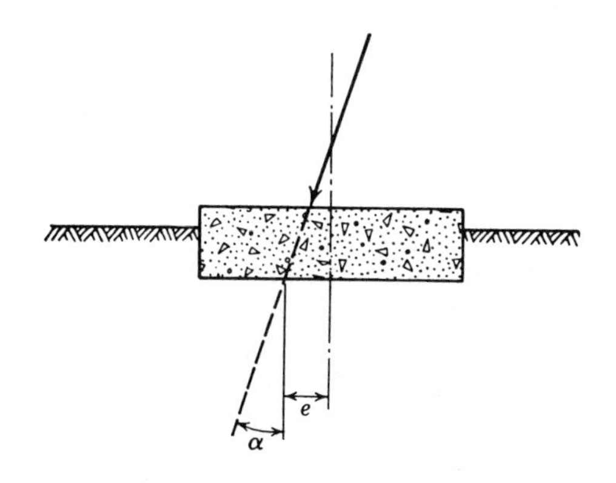

Fig. 14.18 Footing with eccentric and inclined load.

provided that the eccentricity is in the narrow direction of the footing. For the more general case of eccentric loads on rectangular footings see Harr (1966).

▶ **Example 14.11**

Given. Retaining wall in Example 13.12.

Find. Bearing capacity, considering eccentricity and inclination of force on base.

Solution. Eccentricity:

$$e = 0.27 \text{ m}$$

Inclination: horizontal component of active thrust − passive resistance = 69.8 kN/m

$$\tan \alpha = \frac{69.8}{220.3} = 0.32; \qquad \alpha = 17.6°$$

Eq. 14.13:

$$Q_b = \left(1 - \frac{0.54}{2.1}\right)\left(1 - \frac{17.6}{90}\right)^2 \gamma B d N_q + \left(1 - \frac{0.54}{2.1}\right)^2 \times \left(1 - \frac{17.6}{35}\right)^2 \frac{1}{2} \gamma B^2 N_\gamma$$

$$= (0.743)(0.647)(1235.2) + (0.552)(0.247)(1335.1)$$

$$= (593.8 + 182.0) = 775.8 \text{ kN/m of wall}$$ ◀

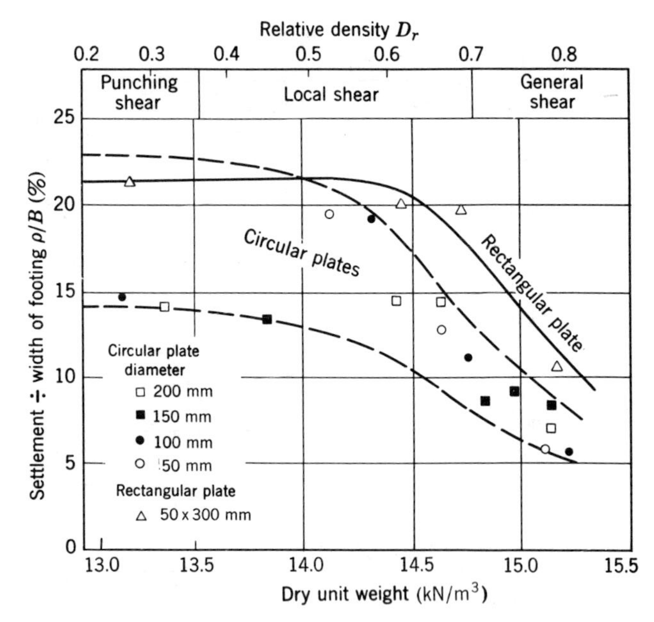

Fig. 14.19 Settlements of model footings at ultimate failure (From Vesic, 1963).

14.8 SETTLEMENTS AS PREDICTED BY ELASTIC THEORIES

Figure 14.19 shows the magnitude of the settlement, in ratio to footing width, at which ultimate bearing capacity was recorded in small-scale footing tests. For example, for a relative density of 0.7, the average settlement at failure for a round footing is 10% of the diameter. For a diameter of 3 m, the settlement would be 0.3 m. If the working load is one-third of the ultimate bearing capacity, i.e., a factor of safety of 3, the settlement at the working load would be about 75 to 100 mm. This amount of settlement would generally be unacceptable. Hence in foundation design it usually is not sufficient merely to determine the bearing capacity and apply a safety factor. The settlement under the working load must be determined and the foundation designed to make this settlement less than the permissible value.

If soil were elastic, homogeneous, and isotropic there would be no difficulty in predicting the settlement that would take place as a result of a surface loading. For such a simple situation there are formulas from the theory of elasticity giving the relationship between load and settlement. In actuality, however, it is very difficult to predict the magnitudes of settlements of footings on real soils. Not only are actual soils nonhomogeneous and nonisotropic, with the modulus generally increasing with depth, but there is the added difficulty of evaluating the *in situ* stress-strain properties.

Despite these complications, however, elastic theory plays a key role in settlement predictions. With judgment, results from the theory of elasticity can be used to give useful estimates of settlement. More important, results from the theory provide an understanding of the settlement phenomenon, which then provides the basis for establishing approximate methods for predicting settlements for practical work.

Hence our discussion of predicting settlements begins with the study of elastic theory. In this section we are concerned with concepts and principles. The problem of using these results in practice, and the all important question of selecting a modulus for use in these results, will be considered in Section 14.9.

Elastic Theory for Settlement under a Uniform Circular Load

Chapter 8 discussed the use of elastic theory to compute the stress increments developed within an elastic body as a result of a uniform stress applied over a circular area on the surface of an elastic material. An example of the calculation of these stresses was given in Example 8.9. Knowing these stresses, and using the equations presented in Chapter 12, we can compute the strains. For example, the strains corresponding to Example 8.9 are shown in Example 14.12, based upon assumed values of E and μ.[6]

By adding up the strains along any vertical line the settlement of the surface can be computed. In the case of an elastic body with a simple surface loading, this

[6] The choice of E is discussed in Section 14.9. μ is taken as 0.45 to be consistent with the stress distribution charts given in Chapter 8.

▶ **Example 14.12**

Given. The tank loading and subsoil of Example 8.9.

$$E = 95.8 \text{ MN/m}^2$$
$$\mu = 0.45$$

Find. The vertical and horizontal strains as a function of depth in the subsoil.
Solution. From Equation 12.5 we get

$$\epsilon_v = \frac{1}{E}(\Delta\sigma_v - 2\mu\Delta\sigma_h)$$

and

$$\epsilon_h = \frac{1}{E}[(1 - \mu)\Delta\sigma_h - \mu\Delta\sigma_v]$$

ϵ_v and ϵ_h versus depth are given in Example 8.9.

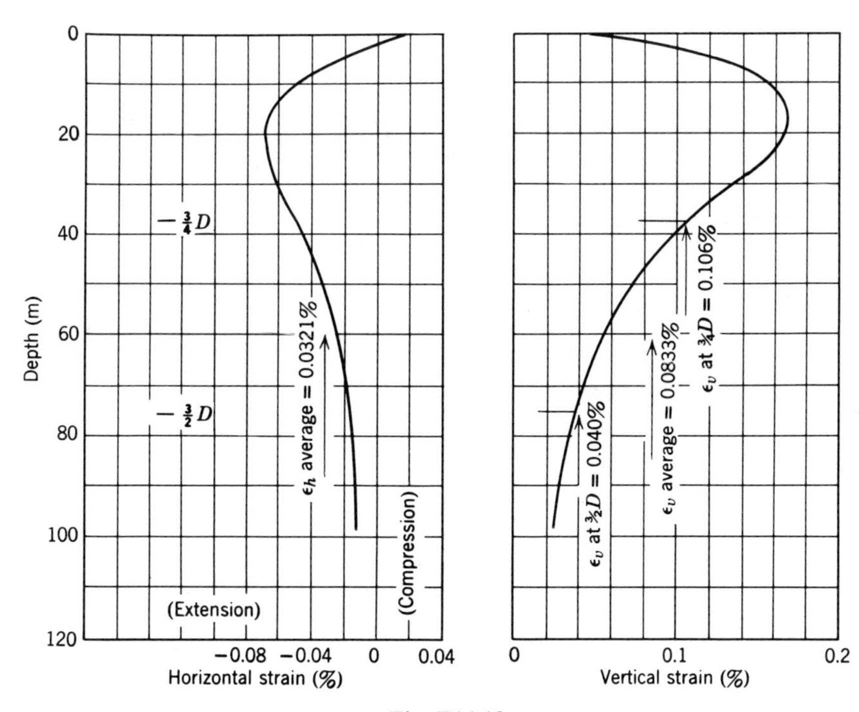

Fig. E14.12

ϵ_v and ϵ_h have been computed for every 3.8 m from depth zero to 91.2 m and plotted as shown in Fig. E14.12. The significance of the average strains will be discussed in Example 14.13. ◀

---

result can be obtained by direct integration of the equations for strain at a point:

$$\rho = \int_0^Z \epsilon_v \, dz$$

where

ρ = settlement
ϵ_v = vertical strain
z = depth measured from surface
Z = depth over which strains are to be summed

If the elastic body is of infinite depth, $Z = \infty$, the surface settlement may be expressed as

$$\rho = \Delta q_s \frac{R}{E} I_\rho \tag{14.14}$$

where

R = radius of the loaded area
I_ρ = an *influence coefficient*, which depends on Poisson's ratio μ and the radius to the point at which the settlement is being evaluated

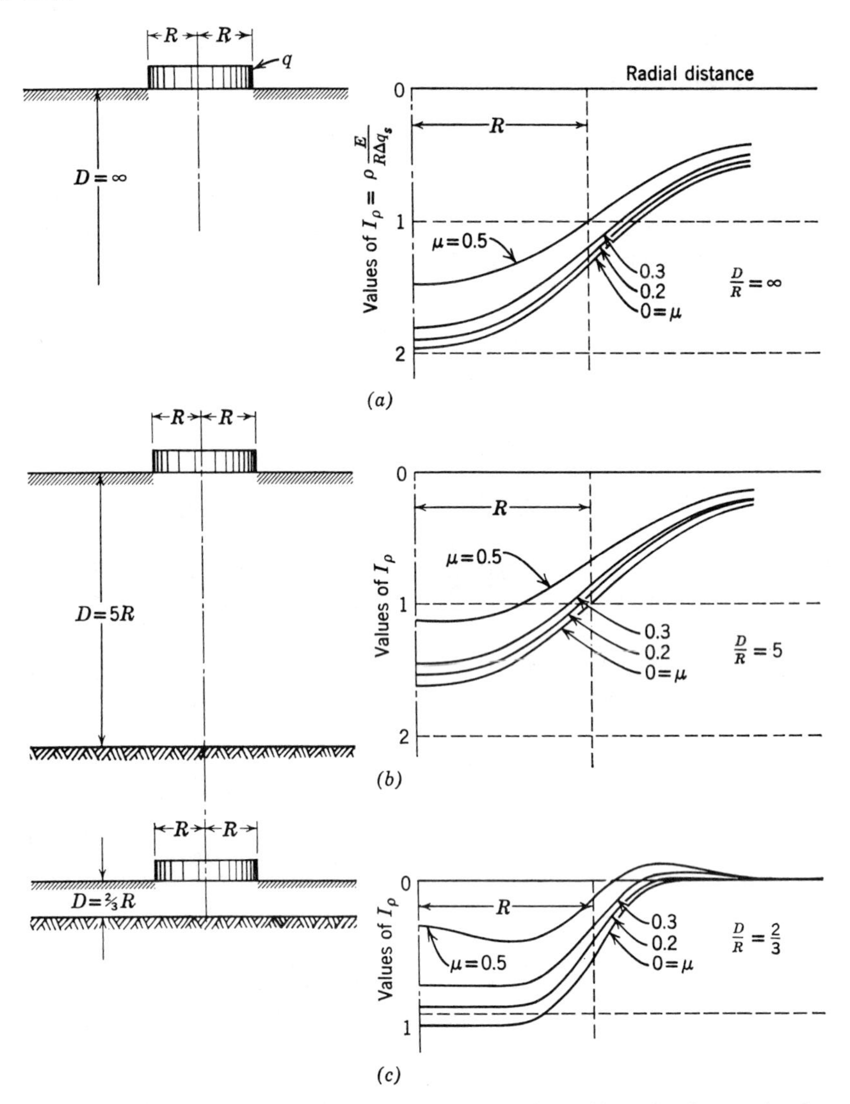

Fig. 14.20 Influence coefficients for settlement under uniform load over circular area (From Terzaghi, 1943).

Figure 14.20*a* gives values of this influence coefficient. Not only does the loaded area itself settle downward, but points on the surface outside of the loaded area also settle. The settlement at the edge of the loaded area is approximately 70% of that at the center line. A simple expression can be written for the settlement at the center line:

$$\rho = \Delta q_s \frac{R}{E} 2(1 - \mu^2) \tag{14.15}$$

Strains at considerable depth, although small, still contribute to the settlement of the surface. This is shown in Fig. 14.21, which indicates the error in the calculated settlement if strains below any depth are ignored. For example, strains within a depth of $4R$ account for only about 75% of the total settlement.

Example 14.13 illustrates the application of Eq. 14.15 to the computation of settlement. The example further shows that a reasonable estimate for the settlement can be obtained by (*a*) defining the bulb of stresses as being $3R$ deep, (*b*) finding the vertical strain at mid-depth of the bulb, $3R/2$, and (*c*) multiplying this "average" strain times the depth of the bulb. This procedure is useful for making approximate settlement estimates.

As may be seen in Example 14.12, the relative importance of horizontal and vertical strains changes markedly with depth. At most depths, the change in horizontal stress is small compared to the change in vertical stress, as is true in the standard triaxial compression test. Thus at most depths the horizontal strain is tensile and points move outward (see Fig. 14.2). On the other hand, at the surface under the loaded area the change in horizontal stress approximately equals the change in vertical stress, as in an isotropic compression test. Here the horizontal strain is compressive, and points on the surface must move *toward* the center line of the load. Outside of the

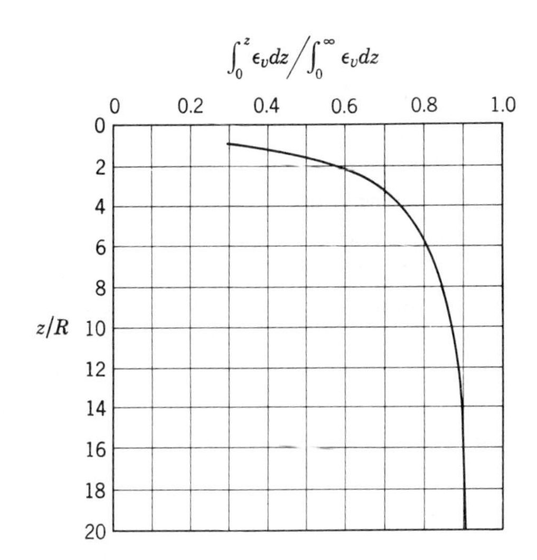

Fig. 14.21 Effect upon I_ρ of considering only a limited depth of strains.

▶ **Example 14.13**

Given. The tank loading and subsoil shown in Example 8.9.

$$E = 95.8\ \text{MN/m}^2$$

$$\mu = 0.45$$

Find. The settlement at the center of the tank for the condition of homogeneous, isotropic soil of infinite depth.

Solution.

$$\rho_{\text{℄}} = \Delta q_s \frac{R}{E} 2(1 - \mu^2) \qquad \text{Eq. 14.15}$$

$$\left.\begin{aligned} \Delta q_s &= 263.3\ \text{kN/m}^2 \\ R &= \frac{D}{2} = \frac{46.72}{2} \end{aligned}\right\} \text{given in Example 8.9}$$

$$\rho_{\text{℄}} = \frac{263.3\ \text{kN/m}^2 \times \dfrac{46.72\ \text{m}}{2} \times 2(1 - 0.45^2)}{95.8 \times 1000\ \text{kN/m}^2}$$

$$= \underline{\underline{0.102}}\ \text{m} = \underline{\underline{102}}\ \text{mm}$$

Settlement may be estimated by multiplying an average strain times the depth of the bulb of stresses. The following tabulation shows several ways in which this might be done.

Assumed Depth of Bulb	Average Strain	Settlement (mm)
$3R = 70.1$ m	Use strain at depth of $3R/2$: $\epsilon_v = 0.00106$	74
$4R = 93.4$ m	Use strain at depth of $2R$: $\epsilon_v = 0.00076$	71

The first method, using a bulb of depth $3R$, gives a closer estimate to the actual result of 102 mm. ◀

loaded area the horizontal strains at the surface must be tensile, and this can happen only if the horizontal stress increments are tensile. Circular tension cracks are often observed around heavy loads resting on the surface of grounds. This general pattern of horizontal strains is somewhat similar to that in a fixed-end beam carrying a concentrated load at midspan.

Equation 14.14 may also be used when the elastic body is of limited depth. However, a different value of I_ρ must be used. Figure 14.20 gives values of I_ρ for two cases of an elastic stratum of limited depth. As would be expected, decreasing the depth of the elastic body decreases the settlement. When the elastic body is thin in comparison to the dimensions of the load, points outside of the loaded area may heave instead of settling. Burmister (1956) has compiled charts and tables that are especially useful when dealing with settlements of strata of limited thickness.

Elastic Theory for Settlement under Other Uniform Loads

The settlement at the corner of a rectangular area carrying a uniform stress Δq_s may be calculated from

$$\rho = \Delta q_s \frac{B(1 - \mu^2)}{E} I_\rho \qquad (14.16)$$

where

B = the width (least dimension) of the rectangle
L = the length (greatest dimension) of the rectangle
I_ρ = an influence coefficient given by Fig. 14.22

Settlements for points other than the corner of a rectangular area, and for any shape of loaded area that can be divided into rectangles, can be obtained using the method of superposition, as explained in Chapter 8 in connection with computing stresses (see Example 8.3). In particular, the settlement at the center of a square loaded area is

$$\rho = \Delta q_s \frac{B}{E} 1.12(1 - \mu^2) \qquad (14.17)$$

As L/B becomes very great (i.e., for a strip footing), I_ρ gradually increases beyond all bounds. Thus a strip footing resting on an elastic body of infinite depth would experience an infinite settlement. In real problems, of course, soil strata are not infinite in depth and strip footings are not infinite in length. For a rectangular loaded area on an elastic stratum of thickness D overlying a rigid body, the approximate settlement at the corner of the area may be computed using Eq. 14.16 and

$$I_\rho = (1 - \mu^2)F_1 + (1 - \mu - 2\mu^2)F_2 \qquad (14.18)$$

where the functions F_1 and F_2 may be read from Fig. 14.23. Burmister (1956) also presents useful charts for such problems.

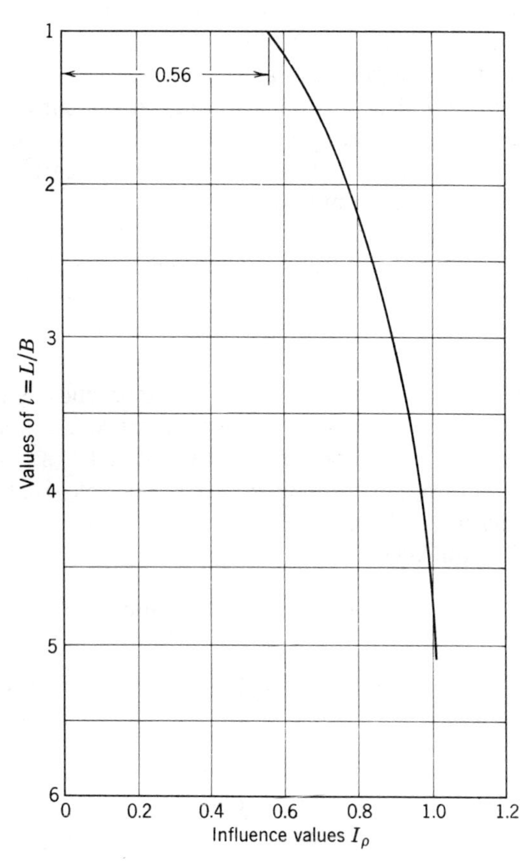

Fig. 14.22 Influence coefficient for settlement of rectangular loaded area (From Terzaghi, 1943).

Solutions have also been obtained for many other types of loading conditions, including loading by shear stresses. Scott (1963) gives a useful summary. With computer techniques, numerical values applicable to specific cases can be obtained.

Elastic Theory for Settlement under Rigid Footings

The condition of a uniformly distributed load occurs in practical problems, such as steel tanks for the storage of fluids. In many practical problems, however, the structural member (such as a footing) in contact with the soil will be quite rigid, and the settlement is more or less uniform over the area of contact between the footing and the soil. Since a uniform stress causes a dish-shaped pattern of settlement, in order to produce a uniform settlement the contact stresses must increase on the outside of the loaded area and decrease near the center line. The curves in Fig. 14.24 marked $K_r = \infty$ show the theoretical distribution of contact stress for the case of a truly rigid foundation. At the edge of the loaded area the contact stress theoretically is infinite.

A change in the distribution of stress over the contact area means a change in the relationship between load and settlement. For a circular rigid loaded area this becomes

$$\rho = \Delta q_s \frac{R}{E} \frac{\pi}{2} (1 - \mu^2) \tag{14.19}$$

where Δq_s = average stress over the loaded area. Comparing Eq. 14.19 with Eq. 14.15, we see that the settlement of a rigid footing is 21% less than the center line settlement under a uniform load. Whitman and Richart (1967) present load-settlement relationships for rigid rectangular footings with various types of loading.

In some problems the structural member in contact with the soil cannot be considered perfectly flexible or perfectly rigid. Figure 14.24 can be used to estimate the contact stresses for intermediate conditions.

14.9 THEORETICAL PROCEDURES FOR USE WITH SOILS

As was discussed in Chapters 10 and 12, a mass of soil does not behave as an elastic, homogeneous, and isotropic material. Nonelasticity influences (*a*) the distribution of stress increments caused by these loads, and (*b*) the strains resulting from these stress increments. At present there are no theoretical procedures that consider both these difficulties, although such procedures are under development. Fortunately, experience has shown that useful predictions of settlement can be made by using the distribution of stress increments predicted by elastic theory, but employing special procedures to determine the resulting strains.

Stress Path Method

As applied to estimating the settlements, the stress path method consists of the following four steps:

1. Select one or more points within the soil under the proposed structure.
2. Estimate for each point the stress path for the loading to be imposed by the structure.
3. Perform laboratory tests which follow the estimated stress paths.
4. Use the strains measured in these tests to estimate the settlement of the proposed structure.

This same general approach, which is a powerful aid to understanding and solving deformation and stability problems, has already been used in Chapter 13.

Example 14.14 illustrates the application of this approach to the tank foundation of Example 8.9. Stress paths for selected points have already been given in Example 8.9. Figure 10.23 presents stress-strain results from triaxial tests following the stress paths for points *A*, *B*, *D*, and *G*. The vertical and horizontal strains as measured in these tests have been plotted in Example 14.14. Integration of these strains over a depth of 91.2

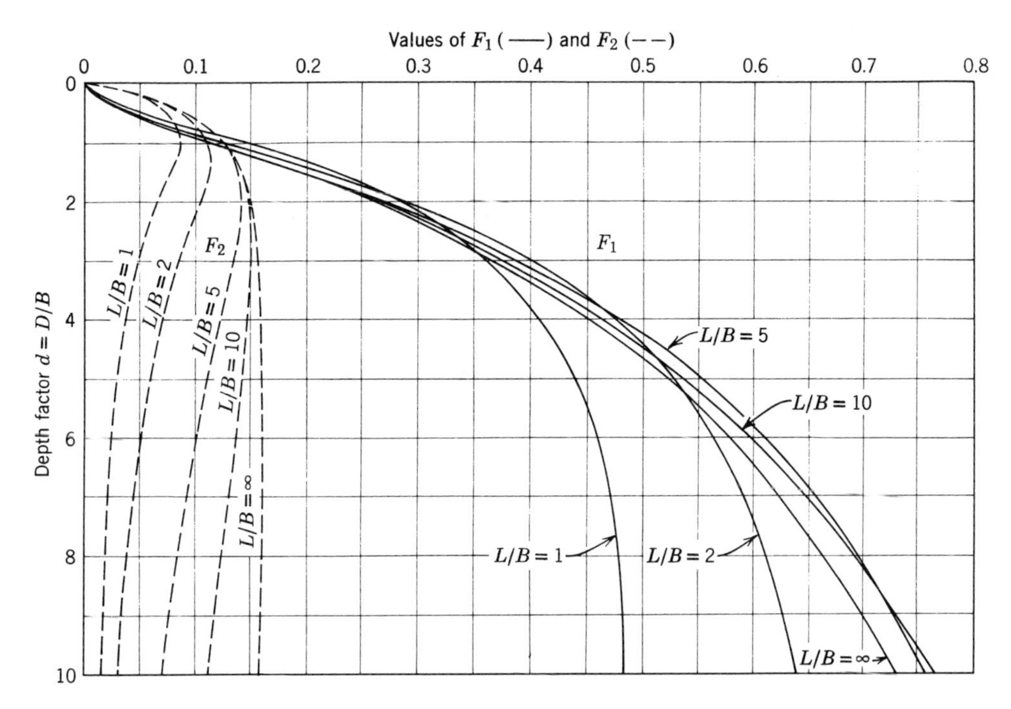

Fig. 14.23 Chart for factors in Eq. 14.18 (After Steinbrenner, 1934).

m gives a center line settlement of 114 m for the initial loading and 19 mm for the second loading. There are strains below a depth of 91.2 m. An estimate for the additional contribution of these deep-seated strains can be obtained from Fig. 14.21.

Stress Path Method Based on Average Point

A simple, and usually adequate, form of the stress path method involves use of a single "average point" together with the concept of a bulb of stresses. According to the discussion in Section 14.8, the bulb may be taken as $3R$ deep with the average point at a depth of $3R/2$. As can be seen in Example 8.9, the laboratory test run for point D closely represents the conditions at the average point under the tank. The vertical strains in this test were 0.14% for the first loading and 0.027% for the second loading. Multiplying these strains by $3R = 70.1$ m gives settlements of 98 and 19 mm, respectively.

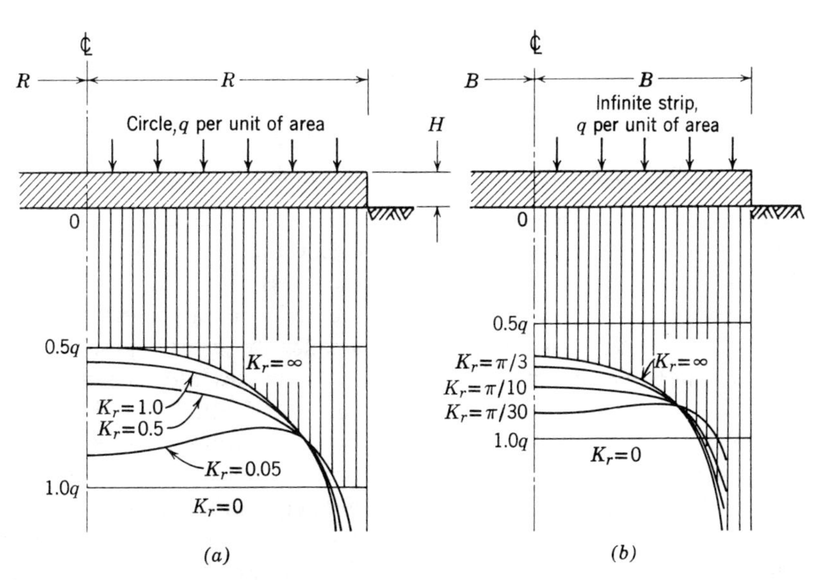

Fig. 14.24 Stress distributions under circular footings of varying rigidity (After Borowicka, 1936 and 1938).

▶ **Example 14.14**

Given. The same tank loading and subsoil as in Examples 8.9, 14.12, and 14.13.

Find. The settlement and distribution of subsoil strains by the stress path method.

Solution. A series of points (A to H) are selected and stress paths for them are drawn (Example 8.9).

Triaxial tests are run along paths A, B, D, and G; test results are plotted in Fig. 10.23.

The vertical and horizontal strains measured in the laboratory tests are plotted as shown in Fig. E14.14.

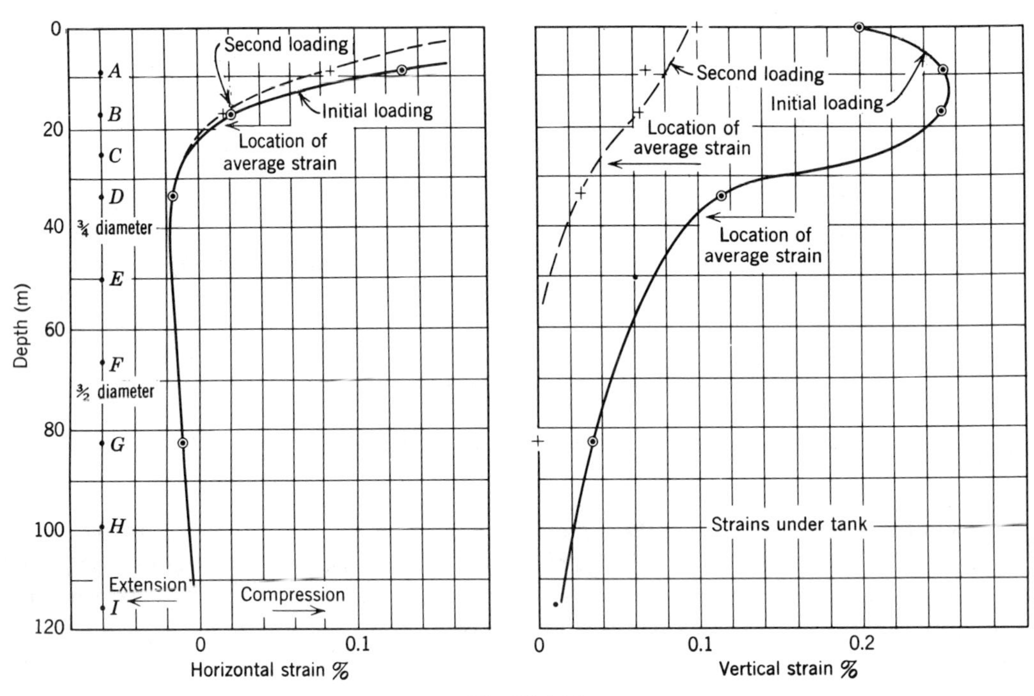

Fig. E14.14

The settlement under the center of the tank, found by the mechanical integration of strain-depth plot is:

Initial loading: $\rho_{℄} = 114$ mm

Second loading: $\rho_{℄} = 19$ mm ◀

Use of Stress Path Method to Determine Modulus

An alternate procedure is to determine a value of E from the stress path test for the average point, and then to compute settlement from an equation such as Eq. 14.15. The procedure illustrated in Example 12.7 may be used to determine the modulus E from the stress path test. In the case of test D, the change in horizontal stress is so small (i.e., this test is so much like a standard triaxial compression test) that it suffices to obtain E by dividing change in axial stress by change in axial strain. This gives $E = 95.8$ MN/m² for the first loading and $E = 359.1$ MN/m² for the second loading. The settlement for the first loading has already been computed in Example 14.13 as 102 mm; the settlement corresponding to the second loading is 27 mm.

Discussion of Methods

Figure 14.25 compares the strains as predicted by elastic theory (Example 14.12) with those predicted by the stress path method (Example 14.14). The stress path method gives larger strains near ground surface but gives smaller strains at depth. This is because the stress path method takes into account the increase in stiffness of the soil with depth. At shallow depths, the initial stress and hence the stiffness are small and relatively large strains occur. Conversely, at greater depth the stiffness is greater than the average stiffness at point D, and hence the strains at depth are smaller than those computed from an average modulus. Figure 14.25 shows that the predicted pattern of strain under the tanks agrees generally with those measured under a model footing.

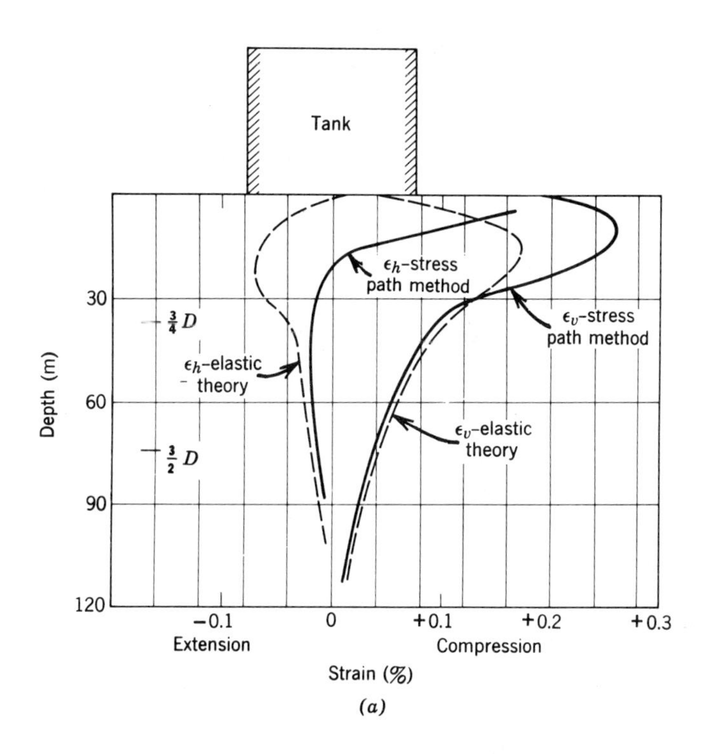

(a)

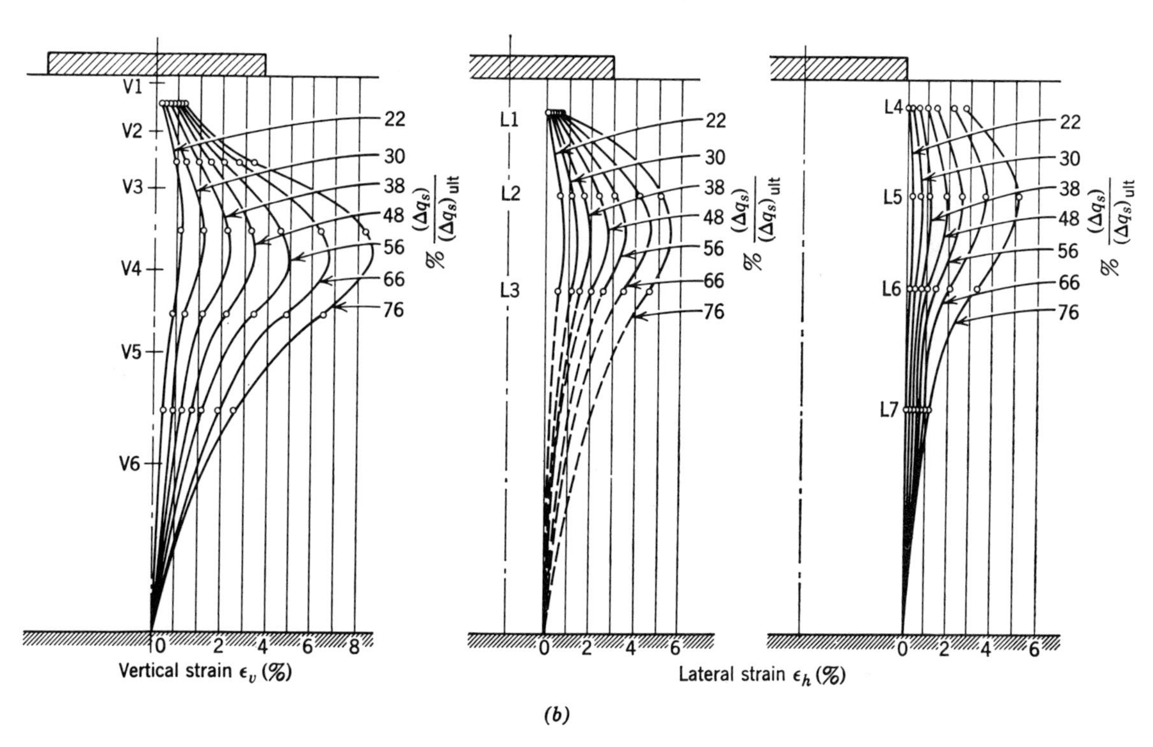

(b)

Fig. 14.25 Strains in subsoil [(*b*) From Eggestad, 1963.]

Each of the three methods involves approximations and each has its advantages. The stress path method involving several points best accounts for the many factors that affect the stiffness of soil, but must neglect the strains below some depth. The stress path method with a single average point is very simple, but involves several assumptions. The elastic method, using an average modulus from an average point, also involves doubtful assumptions, but is especially useful when the settlement must be known at many points other than just at the center line. The choice among the methods will depend upon the circumstances of each problem.

A major difficulty in making theoretical estimates of settlements is obtaining representative samples of the soil. Usually the process of sampling tends to decrease the stiffness of the sample compared to the *in situ* stiffness. Settlements estimated in the foregoing examples from the second loading (19 to 25 mm) are reasonable in view of the settlement actually measured under similar tanks in the same area, but the estimates based on the first loading are unreasonably large. Considerable evidence of this type suggests that stress-strain data from a second loading should be used when estimating the settlement of structures to be founded on sands.

In summary, any theoretical estimate of settlement is an approximation. At the present time, the best estimates can be obtained by the stress path method which (*a*) uses elastic theories to estimate stresses, (*b*) obtains strains or moduli from tests that duplicate the initial stresses and expected stress increments, and (*c*) relies upon experience to indicate how best to compensate for the effects of sample disturbance.

14.10 EMPIRICAL METHODS FOR PREDICTING SETTLEMENT

Because of the difficulties with a strictly theoretical approach, an engineer should always study the settlement experience of existing structures nearby. Empirical approaches, based on a large number of case studies, may also be used to supplement theoretical analyses or for crude preliminary estimates. The two most widely used empirical, or semiempirical, methods are the *load test* and the *penetration test*.

Load Test

In the load test the soil is subjected to a load increased in stages with the settlement under each increment of load being measured. The measured load-settlement data are then used to predict the behavior of the actual footing. Although a full-size footing can be used for the loading test, the normal practice is to employ a small plate of the order of 0.3 to 1 m in diameter. The use of larger test footings is usually impractical because of the expense and difficulty in obtaining large loads. The results from the load test are then extrapolated from the test footing to the prototype footing.

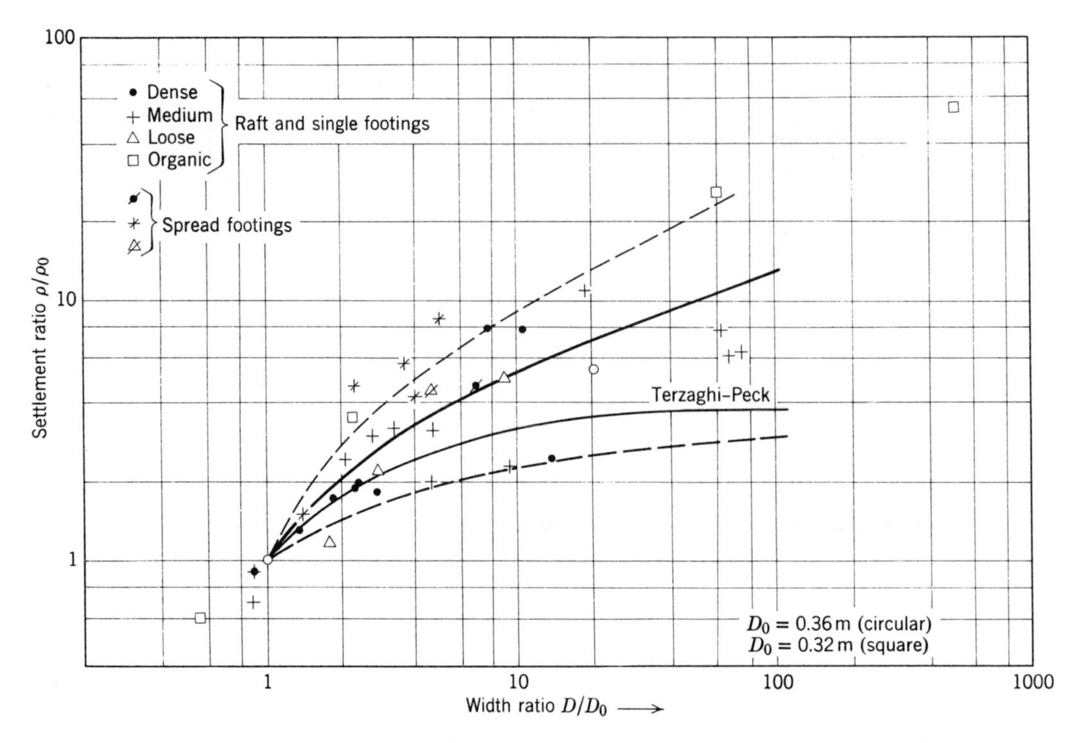

Fig. 14.26 Comparison between settlement and dimension of loaded area as derived from collected case records (From Bjerrum and Eggestad, 1963).

A widely used relation between settlement on sands and footing size is the empirical one developed by Terzaghi and Peck (1967):

$$\frac{\rho}{\rho_0} = \frac{4}{(1 + D_0/D)^2} \tag{14.20}$$

where

ρ = the settlement of the prototype
ρ_0 = the settlement of the test footing
D = the smallest dimension of the prototype
D_0 = the smallest dimension of the test footing

Figure 14.26 shows a plot prepared by Bjerrum and Eggestad (1963) from 14 sets of load-settlement data along with a plot of Eq. 14.20. This figure shows that the settlement of Eq. 14.20 is approximately correct, but that there is considerable scatter.

To get dependable results from a load test we must be sure that the soil under the test plate is not disturbed, and that the soil at the site is homogeneous for a depth which is large relative to the size of the actual footing. Figure 14.27, for example, shows a subsoil situation in which the results of the load test may be very misleading. The settlements under the test plate are due primarily to strains occurring within soil A, whereas under the actual footing the settlements are due primarily to strains occurring in soil B. If soils A and B have different stress-strain properties, the settlement predicted from the load test can bear little resemblance to what will actually occur under the prototype footing.

Penetration Tests

Various penetration tests—standard penetration tests, Dutch deep sounding tests, and a radio-isotropic probe (see Meigh and Nixon, 1961, for a comparison of these penetration tests)—have been used to predict the settlement of foundations on sand. The one most widely used, especially in the United States, is the standard penetration test described in Chapter 7.

Figure 14.28 gives the surface stress Δq_s required to cause a settlement of 25 mm for a footing resting on sand

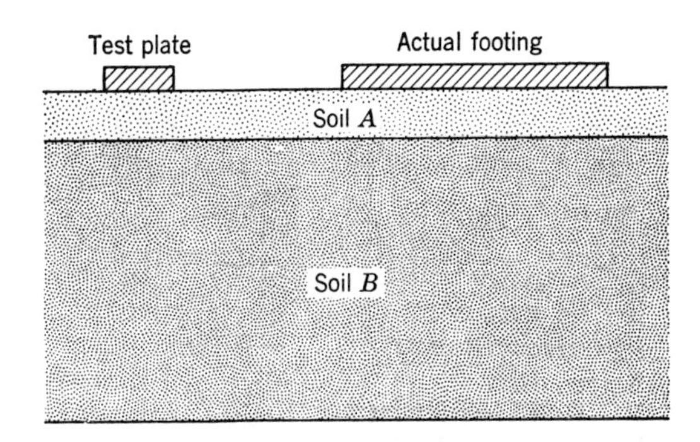

Fig. 14.27 Situation where load test results can be misleading.

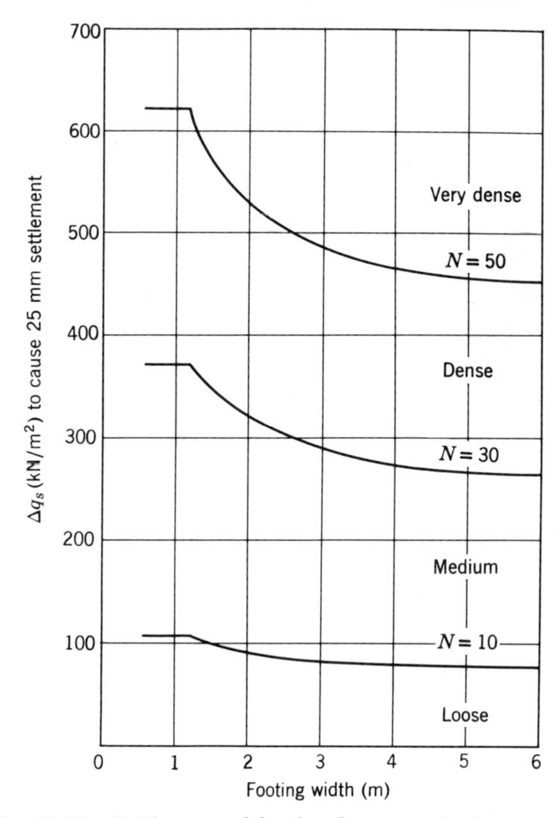

Fig. 14.28 Settlement of footing from standard penetration resistance N. (From Terzaghi and Peck, 1948).

as a function of the standard penetration resistance N and the footing width B. Another relation, proposed by Meyerhof (1965), gives,

$$\Delta q_s = 0.47\, N\rho \qquad B \leq 1.2 \text{ m}$$

$$\Delta q_s = 0.31\, N\rho \left(\frac{B + 0.3}{B}\right)^2 \qquad B > 1.2 \text{ m} \tag{14.21}$$

where Δq_s is the kN/m², B in m, and ρ in mm. Figure 14.29 (Meyerhof, 1965) shows a comparison of predicted and observed settlements for footings on sand and on gravel. It shows that the predicted settlements for the actual structure studied by Meyerhof are greater than the observed ones, and that there does not appear to be any significantly superior performance of either the standard penetration or the static cone penetration test.

14.11 THE INFLUENCE OF FOOTING SIZE ON BEARING CAPACITY AND SETTLEMENT

The preceding sections have shown that the bearing capacity and settlement of a footing resting on sand depend on the properties of the sand and on the size, shape, and embedment of the footing. Bearing capacity

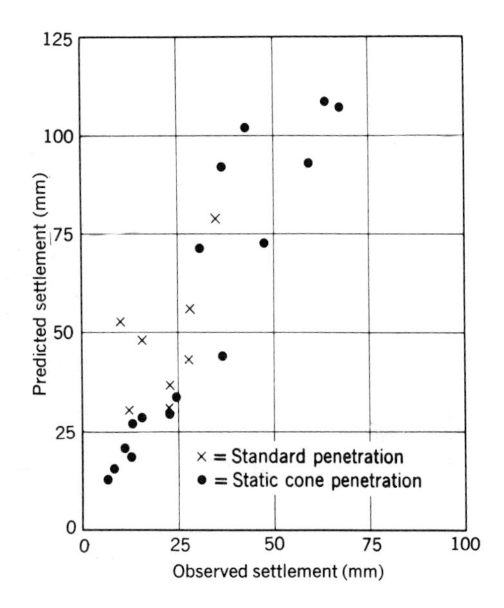

Fig. 14.29 Comparison of predicted and observed settlements of footing on sand and gravel (From Meyerhof, 1965).

depends significantly on the friction angle ϕ and on the relative density of the sand. The soil property which most significantly influences the settlement of a footing is the stress-strain modulus E. The rest of this section considers primarily the influence of footing size on bearing capacity and settlement, in an attempt to pull together some of the many concepts already presented and to leave the student with a simple, general picture of the behavior of footings on sand.

The bearing capacity equations show that, for footings on the surface of sand, the bearing capacity is directly proportional to the size of the footing. Further, the bearing capacity goes up significantly as the depth of the footing below the surface increases. The importance of these two variables (footing size and depth below the surface) is illustrated in Example 14.15.

▶ Example 14.15

Given. A round footing resting on sand with; $\phi = 34\frac{1}{2}°$ and $\gamma = 15.7$ kN/m³

Find. The bearing capacity for:

a. $D = 0.9$ m, $d = 0$.
b. $D = 0.9$ m, $d = 0.6$ m
c. $D = 1.8$ m, $d = 0$.
d. $D = 1.8$ m, $d = 0.6$ m

Solution.

$$(\Delta q_s)_b = (0.6)\tfrac{1}{2}\gamma D N_\gamma + \gamma d N_q \qquad (14.11)$$

From Fig. 14.16, $N_\gamma = N_q = 30$.

a. $(\Delta q_s)_b = (0.6)(\frac{1}{2})(15.7)(0.9)(30) = 127.2$ kN/m²
b. $(\Delta q_s)_b = (0.6)(\frac{1}{2})(15.7)(0.9)(30) + (15.7)(0.6)(30) = 409.8$ kN/m²
c. $(\Delta q_s)_b = (0.6)(\frac{1}{2})(15.7)(1.8)(30) = 254.4$ kN/m²
d. $(\Delta q_s)_b = (0.6)(\frac{1}{2})(15.7)(1.8)(30) + (15.7)(0.6)(30) = 537.0$ kN/m² ◀

▶ Example 14.16

Given. A 14.6 m-high tank is built on an infinite deposit of sand with:

$$\gamma = 20.3 \text{ kN/m}^3$$

$$\mu = 0.45$$

Find. The settlement of the center of the tank when filled with water for the following conditions:

a. $D = 30.5$ m, E constant and equals 191.5 MN/m²
b. $D = 61.0$ m, E constant and equals 191.5 MN/m².
c. $D = 30.5$ m; E varies as σ_{v0} and equals to 191.5 MN/m² at $d = 22.9$ m.
d. $D = 61.0$ m; E varies as σ_{v0} and equals 191.5 MN/m² at $d = 22.9$ m.
e. $D = 30.5$ m; E varies as $\sqrt{\sigma_{v0}}$ and equals 191.5 MN/m² at $d = 22.9$ m.
f. $D = 61.0$ m; E varies as $\sqrt{\sigma_{v0}}$ and equals 191.5 MN/m² at $d = 22.9$ m.

Solution.

$$\rho = \Delta q_s \frac{R}{E} 2(1 - \mu^2) \qquad (14.15)$$

$$\Delta q_s = 14.6 \text{ m} \times 9.8 \text{ kN/m}^3 = 143.1 \text{ kN/m}^2$$

$$2(1 - \mu^2) = 2(1 - 0.45^2) = 1.60$$

a. $$\rho = 143.1 \text{ kN/m}^2 \times \frac{15.25 \text{ m} \times 1.60}{191.5 \times 1000 \text{ kN/m}^2} = 0.018 \text{ m}$$

b. $$\rho = \frac{143.1 \times 30.5 \times 1.60}{191.5 \times 1000} = 0.036 \text{ m}$$

c. Since E varies as σ_{v0} and σ_{v0} varies as depth, E varies as depth. Take "average point" at depth $= \frac{3}{4}D$

$$E_{3D/4} = E_{75} = 191.5 \text{ MN/m}^2$$

ρ for case *c* same as for case *a*, i.e., $\rho \doteq 0.018$ m

d. Now $E_{3D/4} = E_{150}$, and hence the modulus is twice as large as in *c*. The radius R is also twice as large.

$$\rho = \frac{(143.1)(30.5)(1.60)}{2 \times 191.5 \times 1000} = 0.018 \text{ m}$$

e. ρ same as ρ for case *a*, i.e., $\rho = 0.018$ m

f. $$\rho = \frac{(143.1)(30.5)(1.60)}{\sqrt{\frac{45.8}{22.9} \times \frac{\gamma}{\gamma}} \times E \text{ at } 22.9} = \frac{(143.1)(30.5)(1.60)}{\sqrt{2} \times 191.5 \times 1000} = 0.026$$ ◀

Example 14.16 illustrates the influence of foundation size and the nature of variation of the modulus E with depth on settlement. The modulus at a depth of 22.9 m is given. This is the "average point" for the 30.5 m foundation. Different rules are used to extrapolate the given modulus to the average point for the larger foundation. Assuming modulus E is constant with depth, settlement is directly proportional to foundation size. If the modulus E varies directly with the vertical confining stress, the settlement is independent of foundation size.

▶ Example 14.17

Given. A round, rigid footing resting on sand with

$$\phi = 34\tfrac{1}{2}^\circ$$
$$\gamma = 15.7 \text{ kN/m}^3$$
$$\mu = 0.45$$

Find. Relationship among D (varying from 0.3 to 3 m), ρ, and $(\Delta q_s)_b$ for

a. $E = 9580$ kN/m^2
b. $E = 9580$ kN/m^2 at depth 3 m and varying as σ_{v0}.
c. $E = 9580$ kN/m^2 at depth 3 m and varying as $\sqrt{\sigma_{v0}}$.

Solution. Bearing capacity

$$(\Delta q_s)_b = (0.6)\tfrac{1}{2}\gamma D N_\gamma \tag{14.11}$$

From Fig. 14.16,

$$N_\gamma = 30$$

$$\therefore \quad (\Delta q_s)_b = (0.6)(\tfrac{1}{2})(15.7)D(30) = 141.3\ D \text{ in kN/m}^2$$

Settlement:

$$\rho = \Delta q_s \frac{R}{E}\frac{\pi}{2}(1 - \mu^2) \tag{14.19}$$

$$\frac{\pi}{2}(1 - 0.45)^2 = \left(\frac{\pi}{2}\right)(0.797) = 1.25$$

a.
$$\rho = \Delta q_s R \frac{1.25}{9580} = \Delta q_s R(1.30 \times 10^{-4})$$

b.
$$\rho = \Delta q_s R \frac{1.25}{(9580/3)(3R/2)} = \Delta q_s (2.60 \times 10^{-4})$$

c.
$$\rho = \Delta q_s R \frac{1.25}{(9580/\sqrt{3})\sqrt{\tfrac{3}{2}R}} = \Delta q_s \sqrt{R}(1.85 \times 10^{-4})$$

These equations are used to compute the following results

	$D = 1.5$ m	$D = 3$ m
$(\Delta q_s)_b$	212.0 kN/m^2	424 kN/m^2
a. ρ for $\Delta q_s = 141$	—	0.0275 m
ρ for $\Delta q_s = 70.5$	0.0069 m	0.0137 m
b. ρ for $\Delta q_s = 141$	—	0.0367 m
ρ for $\Delta q_s = 70.5$	0.0183 m	0.0183 m
c. ρ for $\Delta q_s = 141$	—	0.0319 m
ρ for $\Delta q_s = 70.5$	0.0113 m	0.0160 m

These results are plotted in Fig. E14.7. ◀

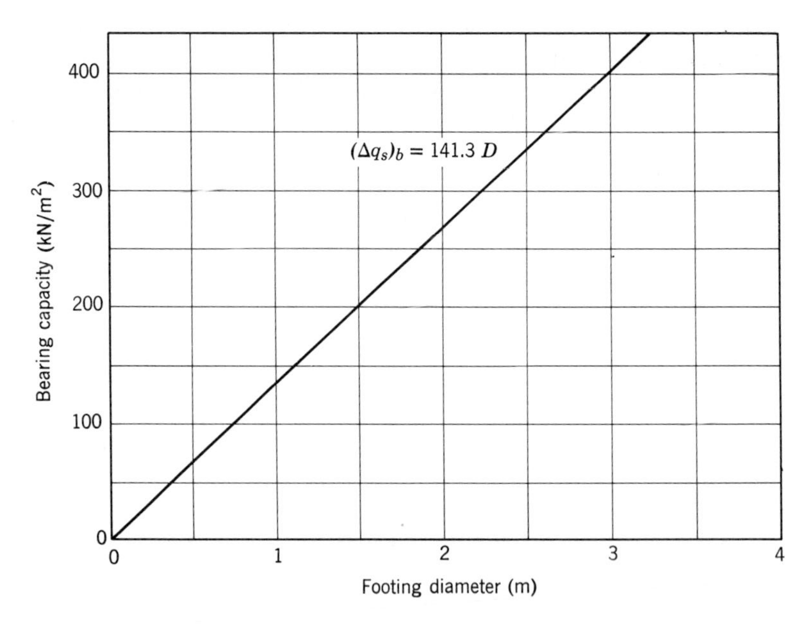

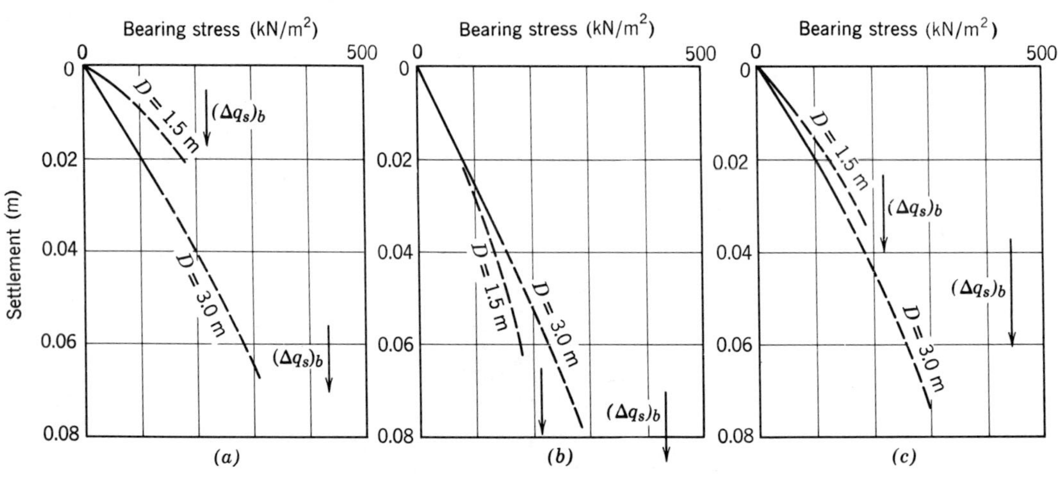

Fig. E14.17 (*a*) Constant E. (*b*) E varies as σ_{v0}. (*c*) E varies as $\sqrt{\sigma_{v0}}$.

If the modulus E varies as the square root of the confining stress—probably the best general relationship between E and confining stress—the result is intermediate.

Example 14.17 combines most of the variables of Examples 14.15 and 14.16 to show the relationship among footing size, settlement, and bearing stress. As illustrated in Fig. E14.17, the bearing capacity is directly related to the footing diameter and is equal to 0.9 of the diameter of the footing. The lower part of Fig. E14.17 shows the settlements versus bearing stress plots for footings of diameters of 1.5 and 3 m for the three conditions of modulus E. The situation shown in c is that which best represents the general relationship between stress and settlement for footings on sand.

It should be emphasized that the settlement equations, such as Eq. 14.14, hold only for bearing stresses which are small relative to the bearing capacity, e.g., factors of safety of 3 or greater. As the bearing stress approaches the bearing capacity, the settlements increase in an unpredictable fashion. This important fact is accounted for in Fig. E14.17 by showing the early portion of the stress-settlement curve as a solid line and that portion past the factor of safety of 3 as a dashed line.

14.12 SUMMARY OF MAIN POINTS

1. For a footing to be properly designed it must meet two conditions:

   a. The *bearing stress* Δq_s must be less than the *bearing capacity* $(\Delta q_s)_b$, which is the bearing stress that causes a shear failure within the foundation soil.

   b. The settlement must be less than the *allowable settlement*.

2. As a footing is loaded to failure the foundation soil first reaches *local shear* and then *general shear*.
3. Local shear occurs when the strength of the soil in a zone is reached and the zone becomes plastic. General shear occurs when all the soil along a slip surface is at failure.
4. In a loose sand, local shear occurs at a much lower bearing stress than does general shear. In a dense sand, local shear occurs at a bearing stress only slightly less than that which causes general shear.
5. Bearing capacity is seldom a controlling factor in the design of footings on sand other than small footings—less than 1 m—because the bearing capacity is usually far in excess of the bearing stress which causes the settlement to exceed the allowable settlement.
6. The allowable settlement is the maximum settlement a structure can tolerate and still perform properly.
7. It is usually the *differential settlement* or *angular distortion* between two points which is more serious to the structure than the total settlement. The allowable settlement is expressed as a function of total settlement rather than differential settlement or distortion because:

   a. The differential settlement is much more difficult to predict than the total settlement.

   b. There generally exists an empirical relationship between differential settlement and maximum settlement.

8. Available for predicting settlement are two theoretical methods—*elastic formulas* and *summation of strains*—and two empirical or semiempirical methods—*load test* and *penetration test*. Theoretical methods should be used in conjunction with empirical methods, since empirical methods reflect field experience.
9. Recommended for predicting settlement is the *stress path method*, either to help pick the modulus E for an elastic solution or to get measured strain for a direct summation of strains.
10. The inadequacy of methods for predicting settlement are due to:

    a. Difficulty in obtaining correct stresses in soil.

    b. Difficulty in obtaining appropriate *in situ* stress-strain data from laboratory tests (trouble caused primarily by sample disturbance).

    c. Soil is not a linearly-elastic, homogeneous, isotropic material.

    d. Soil varies considerably both in horizontal and vertical directions.

11. Bearing capacity and settlement of a footing on sand are related both to soil properties and to footing size and depth of embedment. Bearing capacity increases significantly with increased footing size and depth of embedment. Settlement increases somewhat with increasing footing size.

PROBLEMS

14.1 A footing 2.44 m square bears at 0.9 m depth in a sand with a friction angle of 36°. Find the bearing capacity and the ultimate bearing capacity. The sand weighs 18.07 kN/m^3.

14.2 The soil profile at a given site is as follows:

0–1.2 m cinders, with $\phi = 30°$ and $\gamma = 8.64\ kN/m^3$

1.2–15.2 m sandy gravel, with $\phi = 38°$ and $\gamma = 18.85\ kN/m^3$

Find the bearing capacity for a 3.05 m square footing bearing on top of the sandy gravel.

14.3 A load test was made on a square plate 0.3 m by 0.3 m on a dense sand having a unit weight of 18.07 kN/m^3. The bearing plate was enclosed in a box surrounded by a surcharge of the same soil 0.3 m deep. Failure occurred at a load of 31.1 kN. What would be the failure load per unit of area of the base of a footing 1.5 m square located at the same depth in the same material.

14.4 Assume that the footing in Problem 14.3 supports a light building frame which exerts not only a vertical load V but also a horizontal component $H = 0.15V$ and a moment $M = 0.15V$ (i.e., eccentricity 0.15 m). What is the allowable load V if a safety factor of 3 is used.

14.5 A foundation 15.2 m by 30.4 m rests upon a soil with an average E of 69 MN/m^2. The average bearing stress is 570 kN/m^2. Calculate the settlement at the corners and center of the foundation. Assume $\mu = 0.3$.

14.6 Repeat Problem 14.5, assuming that the sand is only 7.6 m thick and is underlain by rock.

14.7 A standard load test (0.3 m square plate) on a dense dry sand ($\gamma = 18.85\ kN/m^3$) gives the following data:

Load $(kN/m)^2$	Settlement (m)
71.8	0.003
143.6	0.006
215.5	0.012
287.3	0.024
359.0	0.076 (failure)

Another load test is run on the same soil but with the following differences:

Width = 1.5 m

Length = 15.0 m

Predict:

a. The ultimate bearing capacity.

b. The settlement at a load of 190 kN/m^2

14.8 Using the data of Problem 14.7, determine the allowable bearing stress for a 2.4 m square footing if the permissible settlement is 25 mm.

14.9 A sand has an average blow count of 20 blows/305 mm. Design a footing to carry a load of 1780 kN with a maximum settlement of 50 mm and a minimum safety factor of 3 against a shear failure.

14.10 The soil at the site of the tank in Example 14.12 has a standard penetration resistance varying between 15 and 25 blows/305 mm. Predict the tank settlement on the basis of (*a*) Eq. 14.21; and (*b*) Fig. 14.28.

14.11 On the basis of Figs. 14.8 and 14.9 select the maximum allowable settlement for a factory building to house equipment very sensitive to differential settlement.

14.12 A 1.8 m wide strip footing rests 0.9 m below the surface of sand having $\phi = 32°$ and $\gamma = 20.4$ kN/m^3. Compute the ultimate bearing capacity from: (*a*) Eqs. 14.4 and 14.5; and (*b*) Eq. 14.6 and Fig. 14.16. Which value is more nearly correct? Explain.

CHAPTER 15

Dynamic Loading of Soil

If the loads applied to a mass of soil change rapidly enough so that inertia forces become significant in comparison to static forces, special calculations become necessary in order to estimate the deformation of the soil. Typical problems of this type include machine foundations, slope stability during earthquakes, pile driving, and vibratory compaction. This chapter introduces some of the basic concepts from the important field of soil dynamics.

The rate of loading at which a problem "becomes dynamic" depends very much upon the size of the mass of soil involved. With the typical specimen used for laboratory tests, inertial forces generally do not become significant until the frequency of loading exceeds 25 Hz (cycles per second). On the other hand, a large earth dam may experience significant inertial forces with frequencies as low as 0.5 Hz.

15.1 FOUNDATIONS SUBJECTED TO DYNAMIC LOADS

The most common problem involving dynamic loading is that of foundations for machinery. Reciprocating machines and poorly balanced rotating equipment cause periodic dynamic forces Q:

$$Q = Q_0 \sin 2\pi f t \tag{15.1}$$

where

Q_0 = maximum amplitude of dynamic force
f = operating frequency
t = time

Typical operating frequencies range from 3 Hz for large reciprocating air compressors to about 200 Hz for turbines and high-speed rotary compressors. Punch presses and forging hammers also apply intermittent dynamic loads to foundations. A recent problem is that of providing foundations for precision tracking radars. The principles used to analyze the response of foundations to such applied loads may also be used to analyze the response of foundations to ground motions, such as those imposed by earthquakes, blasting, and nearby machinery.

As in the case of foundations subjected to static loadings, the basic criterion governing the design of machine foundations is permissible motion. In general, there is a prescribed limit for the dynamic motions to be permitted during operation, and also a prescribed limit upon the settlement that may develop during an extended period of operation.

Usually it is necessary to perform a dynamic analysis in order to assure that these criteria are met. In order to make such an analysis, the machine-foundation-soil system can be represented by an equivalent lumped mass-spring-dashpot system, which will vary from problem to problem (see Fig. 15.1) depending on the number of modes of motion which the actual system can experience. This section, which is based upon Whitman and Richart (1967), discusses systems having a single degree of freedom, usually vertical motion. For a fuller discussion of the problem, together with methods for handling more complicated types of problems, the reader may consult Barkan (1962). Field tests demonstrating the validity of these methods are described by Richart and Whitman (1967) and Whitman (1966).

Permissible Dynamic Motions

A foundation subjected to a periodic dynamic load will experience a dynamic motion ρ_d at the same frequency as the applied force. The peak velocities and accelerations of the foundation may be expressed in terms of the maximum motion and frequency as follows:

$$\dot{\rho}_d = 2\pi f \rho_d \tag{15.2a}$$

$$\ddot{\rho}_d = (2\pi f)^2 \rho_d \tag{15.2b}$$

where dots indicate differentiation with respect to time. To avoid damage to machines or machine foundations,

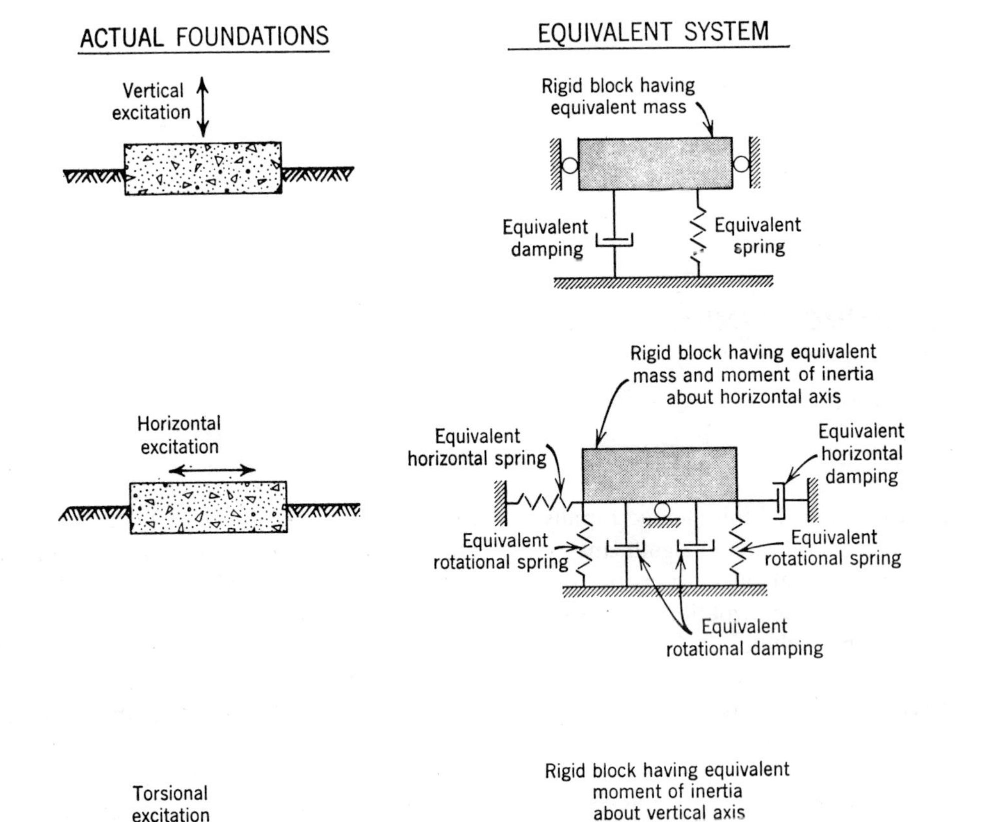

Fig. 15.1 Typical equivalent lumped systems.

the maximum velocity of vibration should not exceed 25 mm/sec. However, if people are to work near the equipment, even stricter requirements may be necessary. Vibrations begin to be troublesome to persons when the maximum velocity exceeds 2.5 mm/sec, and they are noticeable to persons if the velocity exceeds 0.25 mm/sec. At a frequency of 16.7 Hz, these velocities correspond to amplitudes of motion of 0.25, 0.025, and 0.0025 mm, respectively. At other frequencies of operation the permissible amplitude of motion will be different. Note that the motion which may be noticed by persons is approximately $\frac{1}{100}$ of that which is likely to cause damage to machines. Usually it is also necessary to impose a limit on the maximum acceleration that the foundations may experience. In some problems, such as when a stable base must be provided for precision machinery or calibration equipment, it may be necessary to restrict the acceleration to less than $10^{-4}\,g$.

The foundation engineer will find it necessary in all problems to work closely with the client to establish design criteria suitable to the particular problem at hand.

Concepts from Basic Dynamics

The response of a single degree of freedom mass-spring-dashpot system to a periodic applied load is given by the response curves in Fig. 15.2. The key characteristic determining the response of such a system is the undamped natural frequency f_n:

$$f_n = \frac{1}{2\pi}\left(\frac{k}{M}\right)^{1/2} \tag{15.3}$$

where k is the spring constant and M is the mass.

If the operating frequency f is much less than the undamped natural frequency f_n, then the applied force is resisted primarily by the spring, and damping and inertia are of little importance. The amplitude of motion in this case is simply the static response:

$$f \ll f_n \qquad \rho_d = \frac{Q_0}{k} \tag{15.4}$$

If $f \gg f_n$, then the applied force is resisted primarily by inertia and the spring and damping are of little

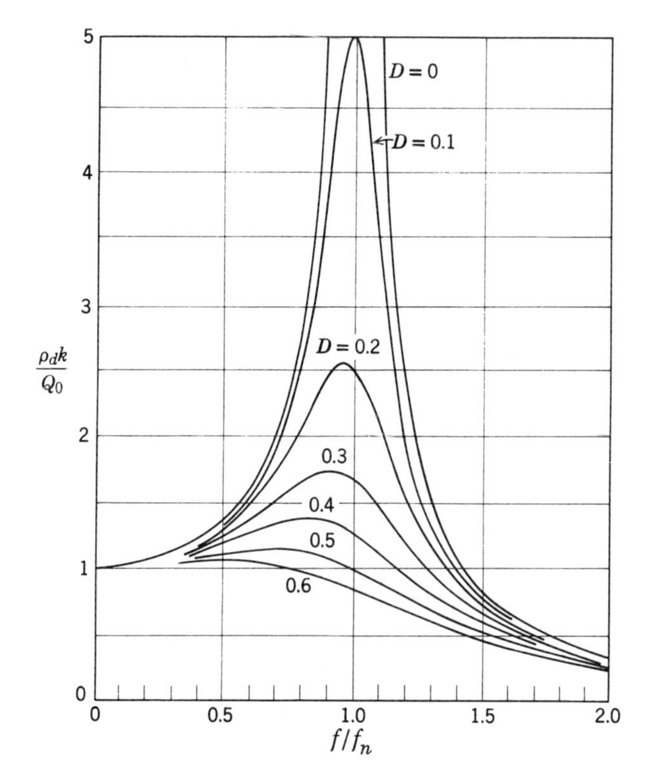

Fig. 15.2 Response of mass-spring-dashpot system.

importance. In this case the amplitude of motion is given by

$$f \gg f_n \qquad \rho_d = \frac{Q_0}{(2\pi f)^2 M} \tag{15.5}$$

If $f \simeq f_n$ then the system is said to be in resonance. The motions at resonance are determined by the damping ratio D, the ratio of the actual damping to the critical damping.

Design criteria for a dynamically loaded foundation are often written in such a way as to avoid resonance. If the damping present in the system is small, the motions at resonance may be very large, and it is indeed prudent to avoid the resonant condition in order to meet the specifications upon permissible dynamic motions. However, if moderate to large damping is present in the system, it may be possible to operate near the resonant condition and still keep the dynamic motions within permissible limits.

Analysis		Factors Required
Approximate estimate for resonant frequency		k and M
Approximate estimate for motions at frequencies well away from resonance	$\ll f_n$	k
	$\gg f_n$	M
Upper limit for motion at frequencies near resonant frequency		D and k or M

Fig. 15.3 Summary of parameters required for dynamic analysis.

Figure 15.3 summarizes the way in which the various parameters of a lumped system influence the response of the system. The approach to dynamic analyses and to design differs depending on the amount of damping present in the system. Hence the magnitude of damping that may exist in actual foundations is considered first in the following subsections. When damping is so small that resonance must be avoided, it becomes necessary to estimate the natural frequency, which requires that the spring constant and the mass be known. Since it is easier to make a reasonable estimate for the mass, it is considered next. Finally the spring constant, which is at the same time the most important and the most difficult parameter for the engineer to evaluate, is discussed.

Choice of Damping for Equivalent Lumped Systems

The dashpots of a lumped system represent the damping of the soil. There are two types of damping: loss of energy through propagation of waves away from the immediate vicinity of the footing, and the internal energy loss within the soil owing to hysteretic and viscous effects. The use of dashpots in the lumped system does not necessarily imply that the engineer believes that soil has viscous properties. Rather, dashpots are used in order to derive simple and useful mathematical expressions for the response of the lumped system. Damping ratios are chosen to represent an equivalent amount of damping, and not to represent a particular type of damping.

The damping due to wave propagation is often termed *radiation damping*. Each time that the foundation moves downward against the soil, a stress wave is originated (see Fig. 15.4). As this wave moves away from the foundation it carries with it some of the energy put into

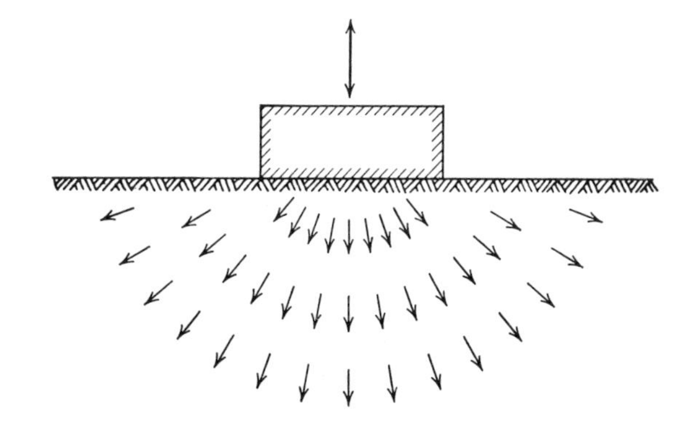

Fig. 15.4 Waves radiating away from vibrating foundation.

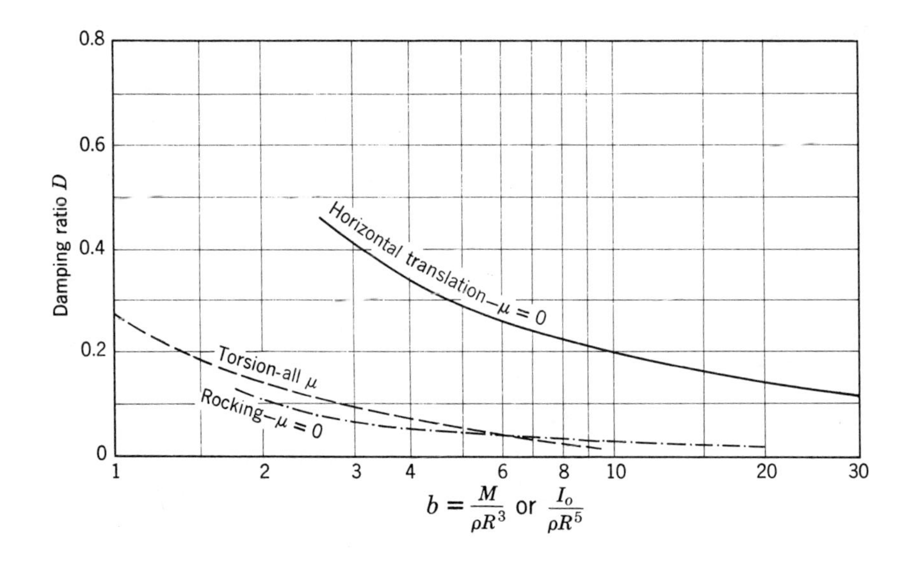

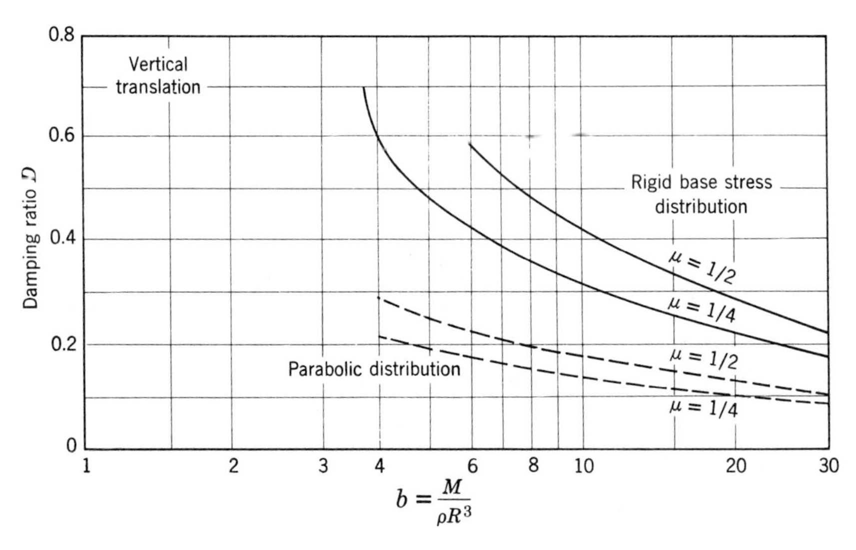

Fig. 15.5 Equivalent damping ratio for circular bases.

the soil. Since this energy is then not available to participate in a resonance phenomenon, a damping effect is introduced. The existence of radiation damping has been revealed by the theory for a rigid disk resting on an elastic half-space (Richart, 1960). This theory may also be used to evaluate an equivalent value of damping ratio. Figure 15.5 gives equivalent damping ratios for the case of circular foundations. The key parameter is the *mass ratio b*, defined as

$$b = \frac{M}{\rho R^3} \quad \text{for translation} \tag{15.6a}$$

and

$$b = \frac{I_0}{\rho R^5} \quad \text{for rotation} \tag{15.6b}$$

where

M = the mass of the foundation block plus machinery

I_0 = the mass moment of inertia of the foundation block plus machinery, evaluated about the vertical axis through the center of gravity for torsional motion, or about a horizontal axis through the centroid of the bottom of the foundation in the case of rocking

ρ = mass density of the soil

R = radius of the soil contact area at the foundation base

Note that the damping ratios are different for each mode of motion. Damping is most important for relatively light foundations, and is much greater for translations than for rotations. Values of D for rectangular foundations may be estimated by entering Fig. 15.5 with an equivalent radius given by

$$R = \begin{cases} \left(\dfrac{BL}{\pi}\right)^{1/2} & \text{for translation} \\ \left(\dfrac{BL^3}{3\pi}\right)^{1/4} & \text{for rocking} \\ \left[\dfrac{BL(B^2 + L^2)}{6\pi}\right]^{1/4} & \text{for twisting} \end{cases} \qquad (15.7)$$

where

B = width of foundation (along axis of rotation for case of rocking)
L = length of foundation (in plane of rotation for case of rocking)

The internal loss of energy due to hysteresis has already been discussed in Chapter 10. The magnitude of this energy loss is a function of the magnitude of the strains experienced by the soil. For the level of strains usually permitted under machine foundations, this hysteretic energy loss is equivalent to a damping ratio D of about 0.05.

Approximate values for the combined effects of radiation and internal damping can be obtained by adding $D = 0.05$ to the values of D given in Fig. 15.5. For horizontal translation and especially for vertical translation, internal damping appears to be relatively unimportant in comparison to radiation damping. For rotational motions, however, the radiation damping is small and the internal damping then becomes a significant part of the total damping.

Choice of Mass for Equivalent Lumped System

Clearly, the mass for an equivalent lumped system should at least include the mass of the foundation block plus the mass of the machinery. At first glance it might appear that an additional mass term should also be used to represent the inertia of the soil underlying the foundation block.

Actually, there is no such thing as an identifiable mass of soil which moves with the same amplitude and in phase with the foundation block. At any instant of time various points within the soil are moving in different directions with different magnitudes of acceleration. The use of an "effective mass" is justified only to the extent that a mass larger than that of the foundation block plus machinery is needed to make the response curve of the lumped mass fit the response curve of the actual system. If an "effective mass" is used, it must be regarded as a totally fictitious quantity which cannot be meaningfully related to any actual mass of soil.

The simplest assumption that can be made when choosing the mass of the lumped system is simply to take this mass equal to that of foundation and machinery and to ignore any "effective mass" of the soil. Moreover, for most foundation problems this simple assumption will give the resonant frequency and dynamic motion within 30% accuracy. Whitman and Richart (1967) provide estimates for the "effective mass" which may be used in those few cases where greater accuracy is justified.

Evaluation of Spring Constant

The determination of a spring constant for use with a dynamically loaded foundation involves essentially the same steps as determination of the load-settlement relationship for a statically loaded foundation. In each case the key is to subject a small mass of soil to the same initial stresses and stress changes as will be experienced under the actual foundation. In the case of dynamically loaded foundations, this means that the soil should be subjected to an initial static stress equal to the stress expected under the actual foundation as a result of the dead load of the foundation plus geostatic stresses, and to stress changes approximately equal to those expected as the result of the dynamic loading. The frequency with which the stress change is applied to the specimen is relatively unimportant.

The various methods described in Sections 14.8 to 14.10 may all be used for estimating a spring constant. The most commonly used approach is to employ formulas from the theory of elasticity. Formulas applicable to rectangular foundations are given in Table 15.1 and values of the coefficients appearing in these formulas are given in Fig. 15.6. The shear modulus G appearing in these equations can be evaluated by the methods described in Chapter 12. This is most often done by measuring the shear wave velocity, either *in situ* or upon laboratory samples using a resonant technique. The

Table 15.1 Spring Constants for Rigid Rectangular Base Resting on Elastic Half-Space

Motion	Spring Constant	Reference
Vertical	$k_z = \dfrac{G}{1-\mu}\beta_z(BL)^{1/2}$	Barkan (1962)
Horizontal	$k_x = 2(1+\mu)G\beta_x(BL)^{1/2}$	Gorbunov-Possadov (1961)
Rocking	$k_\phi = \dfrac{G}{1-\mu}\beta_\phi BL^2$	Gorbunov-Possadov (1961)

Note. Values for β_z, β_x, and β_ϕ are given in Fig. 15.6 for various values of L/B.

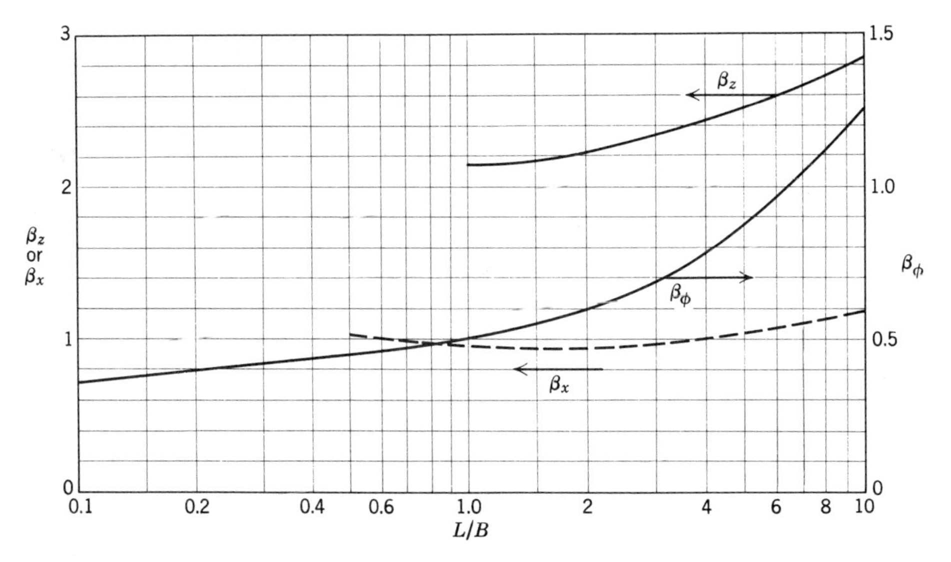

Fig. 15.6 Spring constant coefficients for rectangular foundations.

Poisson's ratio μ to be used in the equations can usually be estimated with satisfactory accuracy as 0.35 for soils of low saturation and 0.5 for fully saturated soils. Another satisfactory approach is to perform a small plate bearing test, using an initial dead load stress equal to that expected under the actual foundation plus a small repeated live load stress. The force-settlement ratio after about 10 cycles of this loading gives the spring constant for the small loaded area. The methods described in Section 14.10 must then be used to extrapolate the spring constant to the actual size of the foundation.

It should be apparent that the engineer must exercise considerable judgment in the selection of a spring constant, so as to take into account the effect of partial embedment of the foundation, stratification in the soil, etc.

Settlement Caused by Vibrations

The dynamic stresses within the soil beneath a machine foundation will cause settlement of the foundation, and excessive settlements must be avoided by proper design. As in the case of settlements resulting from a single static load, vibratory induced settlements of foundations on sand result partly from volume decrease but primarily from shear strains.

The best approach for predicting the magnitude of vibratory-induced settlement in a given case is to subject a sample of the soil to the initial stresses and dynamic stress changes expected below the foundation. Permissible settlements as a result of vibrations are essentially the same as permissible static settlements as discussed in Section 14.2.

In the absence of a detailed testing program, several design principles may be used to minimize the likelihood of excessive settlements. The sum of the static and dynamic bearing stresses is often held to less than one-half of the usual permissible static bearing stress. Another approach is to subject the soil to vibrations more intense than those to be expected under the actual foundation. Such vibratory compaction may be accomplished by vibratory surface rollers (see Section 15.2). Vibroflotation may also be used to compact soil (D'Appolonia, 1953). A typical requirement is that the soil should be densified to greater than 70% relative density.

15.2 DENSIFICATION BY DYNAMIC LOADS

In many problems, such as the design of machine foundations, the engineer must ensure that vibrations do not cause significant densification of soil. On the other hand, vibrations are often used deliberately to densify soil, as in vibratory compaction. Figure 15.7 shows a vibratory roller compacting a sand fill. Vibratory

Fig. 15.7 Vibratory compactor.

compaction, which has long been used to densify granular soils, is now often used for compacting clayey soils as well. This section will be concerned primarily with densification of sands.

Laboratory Studies

Figure 15.8 shows a form of test which has frequently been used in the laboratory to study densification of sand by vibration. A container, open at the top, is filled with sand in a loose condition. Sometimes a weight is placed upon the surface of the sand. The container is subjected to vibrations for several minutes, and then the vibrations are stopped while the depth of the soil is measured and a new unit weight computed. Then an increased level of vibration is applied, and so on.

Figure 15.9, obtained when such a container was vibrated by a shaking table causing periodic vertical motion, shows typical results of such tests. The sand initially was at about zero relative density. Very little densification occurred until the accelerations increased nearly to $1g$, and most of the densification occurred when the acceleration was at or about $1g$. A peak density was attained when the acceleration reached $2g$, but further increase in acceleration caused the sand to become less dense. In this particular sand, several different combinations of displacement and acceleration were used to achieve each acceleration, and, as shown in the figure, the results were substantially independent of the combinations employed.

Results of tests such as these have often been taken to imply that peak acceleration is the primary variable controlling densification (e.g., see Barkan, 1962). However, large accelerations alone, in the absence of significant changes in stress, may not cause densification (Ortigosa and Whitman, 1968). On the other hand, as discussed in Chapter 10, increase in stress produces relatively little densification of sand until the stresses become large enough to crush grains. Clearly some combination of events is producing the large observed densification of sand during vibration. The events during the vibratory test may be established by considering the dynamic equilibrium between forces at various times during a cycle of motion (see Fig. 15.10).

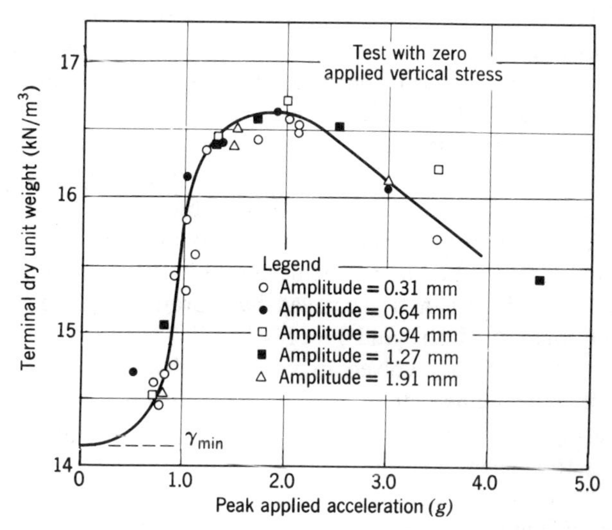

Fig. 15.9 Typical results obtained during laboratory study of densification by vibrations. (From D'Appolonia and D'Appolonia, 1967).

When the container accelerates upward, the inertia force acts to increase the stress above the static value. When the table accelerates downward, the inertia force is opposed to the weight of the soil. Thus, if the peak acceleration of the container is $0.5g$, the vertical stress at any depth within the soil fluctuates between 1.5 and 0.5 times the geostatic stress.

However, if the peak acceleration of the container exceeds $1g$, then events during the test are much more complicated (see Fig. 15.11). At the point in each cycle where the downward acceleration of the container reaches $1g$, the vertical stress within the soil drops to zero. Since sand cannot sustain tension, the sand is unable to follow the subsequent motion of the container and experiences *free fall* until it impacts against the container later in the cycle. Then the sand and container move upward together until separation once more occurs and the cycle is repeated.

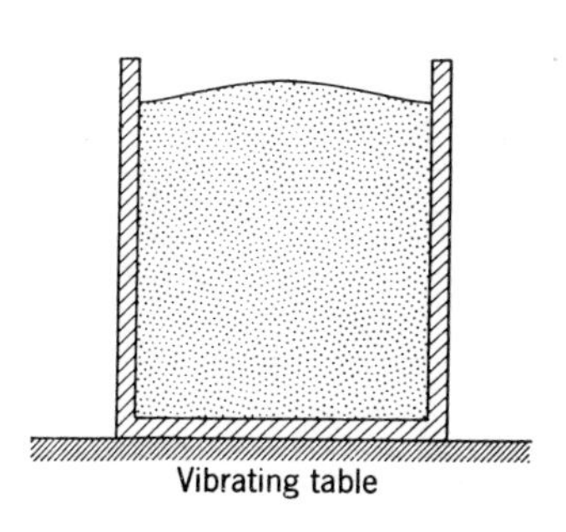

Fig. 15.8 Common laboratory test for study of densification of sand during vibration.

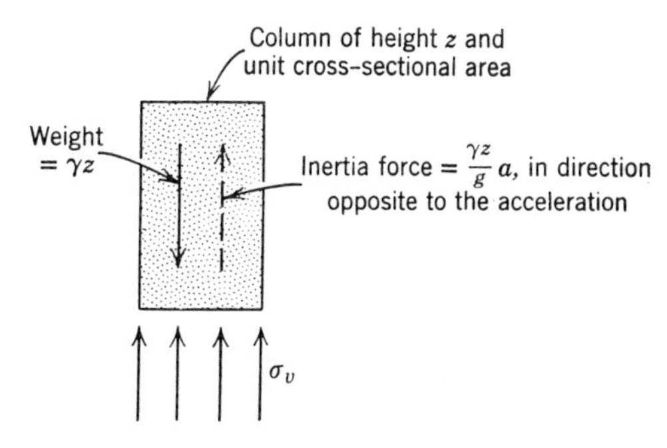

Fig. 15.10 Forces acting upon element of soil during vertical vibrations.

The occurrence of free fall is what distinguishes tests with accelerations of 1g or greater (where considerable densification occurs) from those tests with accelerations of less than 1g (where little densification occurs). During free fall, the particles become separated from one another and hence are free to seek positions of optimum packing when they fall back against a fixed surface. In a similar vein, it has been found that by sprinkling sand into a container it is possible to achieve a unit weight as great as that which can be achieved by vibration (see, e.g., Whitman, Getzler, and Hoeg, 1963). Thus, although the phenomena involved in vibratory densification are still poorly understood, the *absence of effective stress* during a part of each cycle of motion appears to be the key to efficient densification.

Vibrations are often used in tests to establish the *maximum unit weight* of a sand for purposes of studying relative density (see Section 3.1). From the foregoing discussion, it is evident that test conditions have a major influence upon the maximum unit weight obtained in a test. In relative density determinations, it is essential that standardized procedures be used to determine both maximum and minimum unit weight (see *ASTM*, 1967).

Vibratory Compaction

It is generally acknowledged that granular soils can be effectively compacted in the field using vibratory rollers,

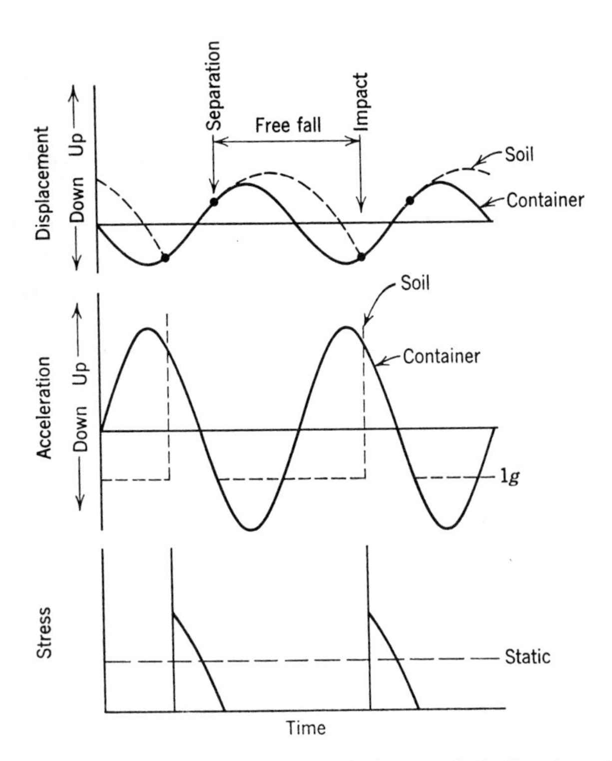

Fig. 15.11 Motions and stress during vertical vibration with peak acceleration greater than 1g. *Note.* Drawn for peak acceleration of container equal to 2g at 25 Hz.

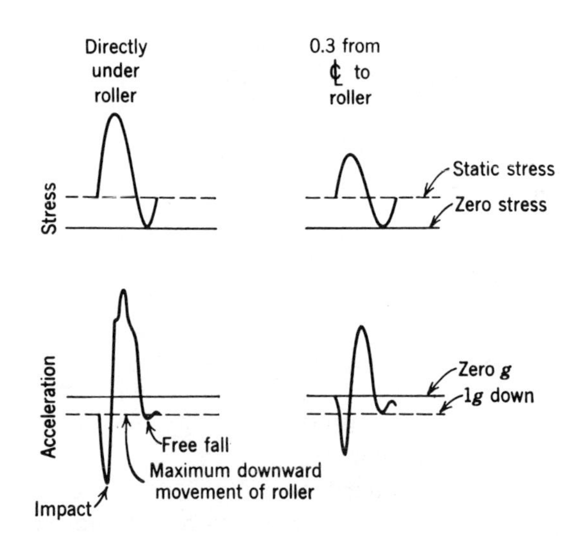

Fig. 15.12 Stress-time and acceleration-time at depth of 0.46 m in sand beneath vibratory roller (From D'Appolonia et al., 1968).

but there is a lack of facts on the possibilities and limitations of such compaction. One study has been made by Forssblad (1965). The results given in this subsection are from D'Appolonia et al. (1968).

The typical vibratory roller, such as that shown in Fig. 15.7, consists of a drum supported by heavy springs from a frame. Inside the drum an eccentric weight rotates rapidly about the axle of the drum, producing a periodic force against the drum. The drum itself typically weighs about 18 kN, but the periodic force is several times larger so that the drum is raised free of the ground during each cycle and then slammed down against the ground producing large impact stresses. Figure 15.12 shows stresses and accelerations measured within sand beneath a roller during 1 cycle of motion. The impact of the roller against the ground and the subsequent rebound of

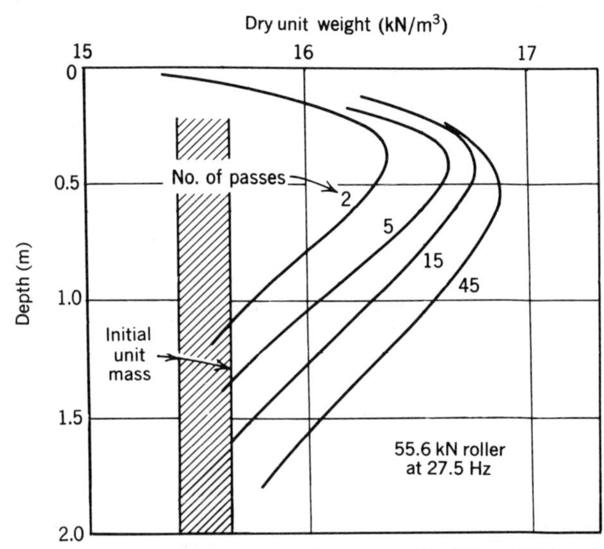

Fig. 15.13 Densification by vibratory roller (From D'Appolonia et al., 1968).

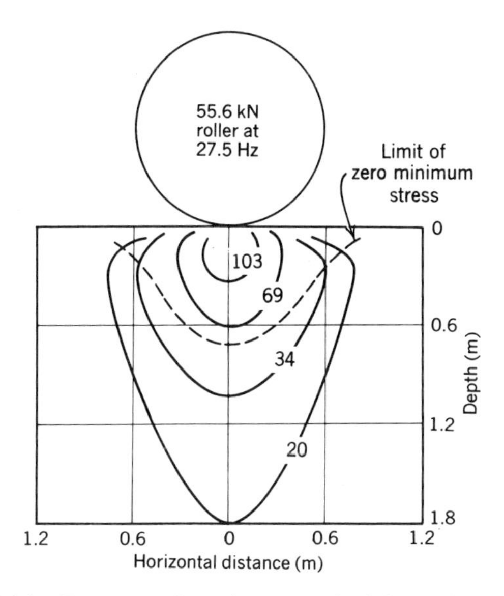

Fig. 15.14 Contours of maximum vertical dynamic stress beneath vibratory roller, kN/m² (From D'Appolonia et al., 1968).

the soil into a state of free fall may be identified in these records. Jumping of particles from the surface of the sand is visible in Fig. 15.7.

The increase in unit weight with depth for various numbers of passes of the roller is shown in Fig. 15.13. The most efficient compaction occurs at a depth of about 0.6 m, which, as is shown in Fig. 15.14, is the greatest depth at which zero effective stress occurs during rebound of the soil. By a great number of passes, some densification at a depth of 1.5 m can be obtained, and this relatively inefficient compaction presumably results from many cycles of dynamic stress. The topmost 0.15 m receives little compaction, probably because of the violent agitation (accelerations of more than $3g$ were observed) in this zone just after the center-line of the roller passed (compare with Fig. 15.9).

An interesting feature of these field observations was the large horizontal stresses built up as the result of several successive passes of the roller (Fig. 15.15). The resulting horizontal stresses exceeded the vertical geostatic stress.

It appears that the action of vibratory rollers with clays is quite different than with sands. Compaction of clay probably is accomplished by successive cycles of impact-caused stress.

Densification During Earthquakes

Earthquakes cause vertical acceleration of the surface of the ground, but these accelerations are too small (at most about $0.3g$) to cause densification. Earthquakes also cause horizontal accelerations which, as indicated in Fig. 15.16, give rise to shear stresses. The direction of these shear stresses reverses many times during a strong earthquake as the direction of the acceleration reverses.

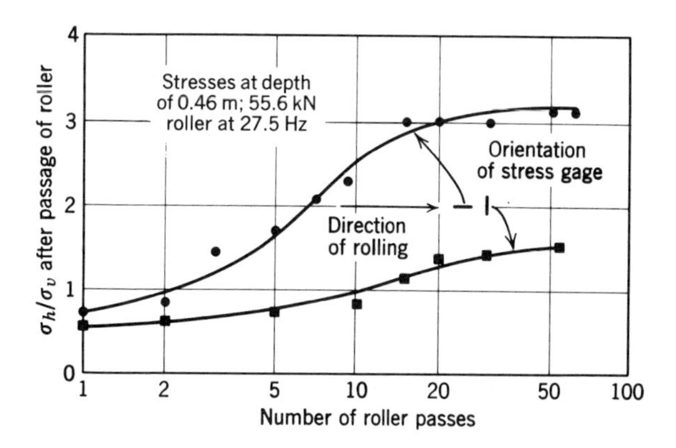

Fig. 15.15 Increase in horizontal stress beneath vibratory roller (From D'Appolonia et al., 1968).

Thus conditions in the ground during an earthquake are similar to those in a direct shear test with several reversals of the direction of shearing.

Subsidence of the ground has occurred during large earthquakes. Part of this subsidence is the result of tectonic movement of the underlying rock, but part results from densification of soil. In Valdivia, Chile, the subsidence due to densification during the 1960 earthquake amounted to more than 1 m. Some, but probably not all, of the subsidence during earthquakes is associated with the phenomenon of liquefaction (see Section 32.10).

15.3 DYNAMIC STABILITY OF SLOPES

When a slope is subjected to an earthquake, the shear stresses associated with ground acceleration (Fig. 15.16) will add to the shear stresses required for static equilibrium and may lead to temporary instability of the slope.

The key features of this problem may be studied by examining the problem of a block resting on an inclined plane (Fig. 15.17). If the block is to accelerate in a direction parallel to the slope, the shear force between the block and the slope must differ from T, the shear force required for static equilibrium. Since the shear

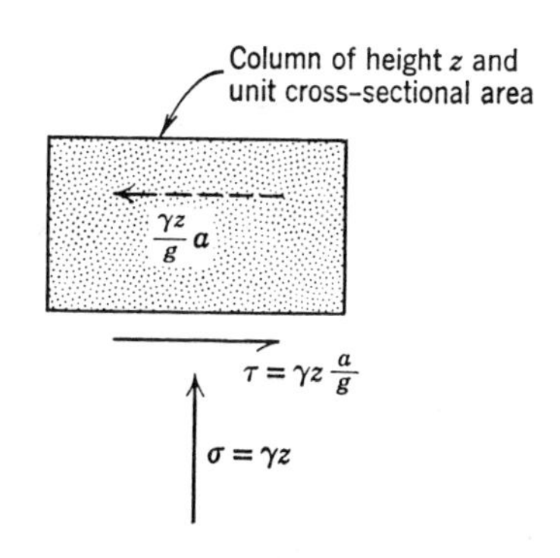

Fig. 15.16 Forces acting on element of soil during horizontal vibrations.

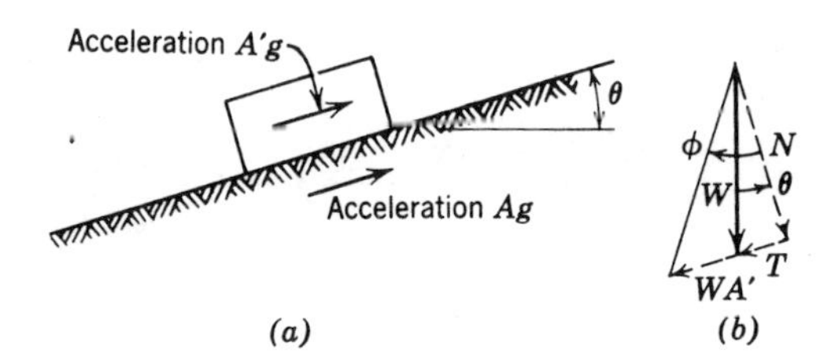

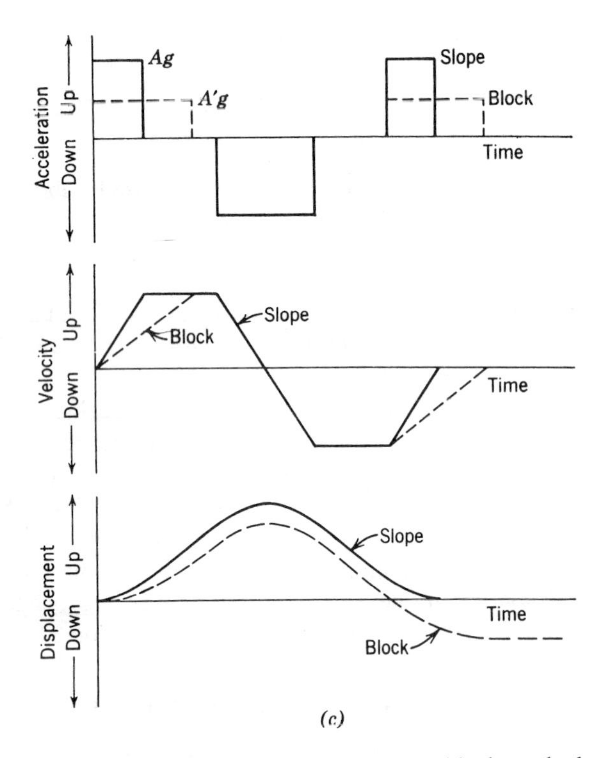

Fig. 15.17 Relative movement between block and slope during dynamic loading.

force is limited (it cannot exceed $N \tan \phi$, where N is the normal force and ϕ the friction angle), the acceleration the block can experience is limited:

maximum upslope acceleration:

$$\frac{W}{g} A'g = W \cos \theta \tan \phi - W \sin \theta$$

or

$$A' = \cos \theta \tan \phi - \sin \theta$$

maximum downslope acceleration:

$$A' = \cos \theta \tan \phi + \sin \theta$$

If the maximum acceleration coefficient A of the slope is less than A', then the block and the slope will move together without relative displacement. However, if $A' < A$, then relative displacement will develop as shown in Fig. 15.17*c*. The block cannot keep up with the slope as the slope accelerates uphill, and hence relative downhill displacement occurs. During downhill acceleration the block and slope are able to move together since A' is greater in this case.

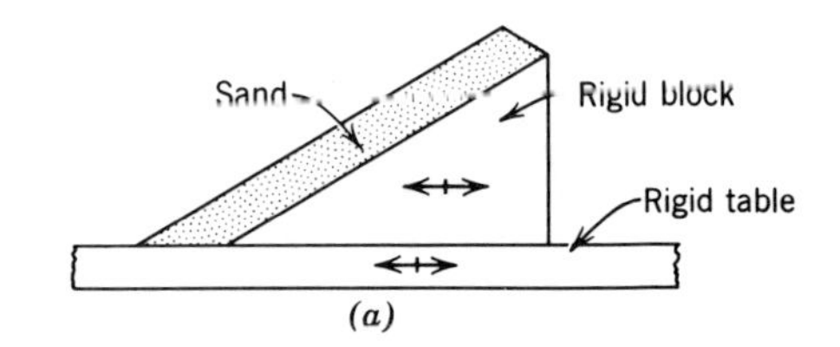

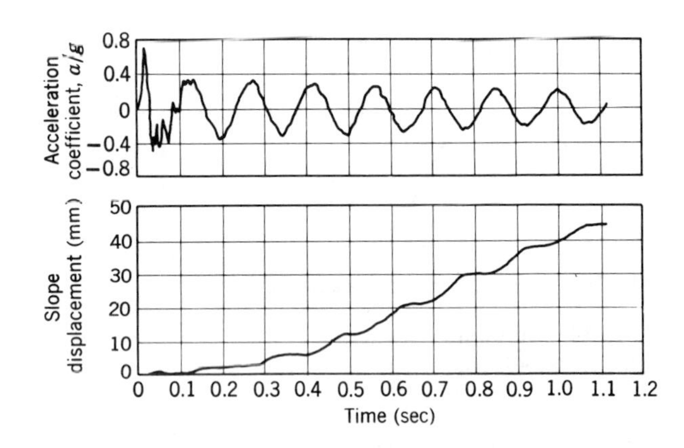

Fig. 15.18 Movement of sand slope during dynamic loading. (*a*) Schematic arrangement of sand bank for shaking test. (*b*) Acceleration and displacement of slope during shaking test. Monterey no. 20 sand, 31° slope (From Goodman and Seed, 1966).

A slope in sand, which behaves as an infinite slope (Section 13.9), will experience movements very similar to those of a block on a plane. Figure 15.18 shows downhill relative displacement each time that the applied uphill acceleration exceeds the acceleration corresponding to maximum shear strength. Test results such as these confirm the correctness of the theory for a material of constant strength in which very little strain is required to mobilize this strength.

This method of analysis has been developed in detail by Newmark (1965), who outlines methods for predicting maximum downhill movement during typical earthquake ground motions. The method can be applied in approximate fashion to slopes in materials other than sand. Application of the method will be discussed further in Section 31.8.

15.4 SUMMARY OF MAIN POINTS

The main point to be understood from this chapter is the role of inertia in modifying the stresses and displacements during dynamic loadings. This role has been illustrated for several relatively simple problems. Methods useful in certain practical problems have been presented, but the chapter has only introduced the complex and increasingly important subject of soil dynamics.

A. W. SKEMPTON

Dr. Skempton was born in Northampton, England, in 1914 and educated at Northampton School and Imperial College in the University of London, from which he graduated with the degree B.Sc. (Eng.) with first-class honors in 1935, and the degree of M.Sc. in 1936. In 1949 he was awarded the degree D.Sc. from Imperial College.

From 1936 until 1946, Dr. Skempton worked at the Building Research Station. In 1946 he established soil mechanics at Imperial College. From 1957 to 1961, Dr. Skempton was President of the International Society of Soil Mechanics and Foundation Engineering. In 1961 he was elected a Fellow of the Royal Society. He has been a Rankine Lecturer.

Dr. Skempton's interests have covered a wide range of problems in soil mechanics, rock mechanics, and geology, and in addition he has done extensive research into the history of civil engineering. Professor Skempton has made major contributions to soil mechanics on the fundamentals of the effective stress, pore pressures in clays, bearing capacity, and slope stability. Dr. Skempton has repeatedly shown his remarkable ability to identify the important components of a complex problem and to present a clear explanation of them. Under Dr. Skempton's leadership, Imperial College has become one of the top centers for soil mechanics in the world.

PART IV

Soil with Water-No Flow or Steady Flow

Pore water influences the behavior of soil in two ways: by affecting the way in which soil particles join together to form the mineral skeleton (*chemical interaction*) and by affecting the magnitude of the forces transmitted through the mineral skeleton (*physical interaction*). Chemical interaction has been discussed in Part II. Part IV introduces the concepts needed to understand physical interaction.

Part IV specifically considers situations in which the pore pressures within a soil mass are determined by the pore pressures at the boundaries of the mass and are independent of applied loads. Such situations exist whenever loads are applied slowly compared to the rate of consolidation and at some time (long compared to consolidation time) after a rapid loading. Situations in which pore pressures are influenced by loadings will be covered in Part V.

CHAPTER 16

The Effective Stress Concept

Our intuitive glimpse of soil behavior in Chapter 2 alerted us to this fact: the behavior of a chunk of soil is related to the difference between total stress and pore pressure. The present chapter examines this concept, one of the most essential to soil mechanics.

16.1 STRESSES IN THE SUBSOIL

Chapter 8 presented equations for determining the vertical geostatic stresses in dry soil. The same equations can also be used to determine the *total* vertical geostatic stress for wet soil. The soil unit weight contributing to these total stresses is, of course, the total unit weight and the equations corresponding to Eqs. 8.2 to 8.4 become, respectively,

$$\sigma_v = z\gamma_t \tag{16.1}$$

$$\sigma_v = \int_0^z \gamma_t \, dz \tag{16.2}$$

$$\sigma_v = \sum \gamma_t \Delta z \tag{16.3}$$

Figure 16.1 shows the same situation as Fig. 8.1 except that the soil is now saturated with static water. The location at which the pressure in the pore water is atmospheric (i.e., zero gage pressure), termed *water table* or *phreatic surface*, is noted by $\underline{\nabla}$, and the depth from this location to the element A is z_w. For the situation shown in Fig. 16.1 we have in addition to the total stresses σ_v and σ_h both vertical u_v and horizontal u_h water pressures. Since u_v and u_h are measured at the same elevation in our infinitesimally small element and since water cannot take a static shear stress,

$$u_v = u_h = u$$

and

$$u = z_w\gamma_w \tag{16.4}$$

16.2 EFFECTIVE STRESS PRINCIPLE

Perpendicular to any plane at element A in the soil profile, Fig. 16.1, there are acting a total stress σ and a pore water pressure u. Let us now define[1] the *effective stress* as equal to the total stress σ minus the pore pressure u:

$$\bar{\sigma} = \sigma - u \tag{16.5}$$

and

$$\bar{\sigma}_v = \sigma_v - u \tag{16.5a}$$

$$\bar{\sigma}_h = \sigma_h - u \tag{16.5b}$$

Equation 16.5 is the *effective stress equation.*

The coefficient of lateral stress is actually based on *effective* stress rather than total stress; thus

$$K = \frac{\bar{\sigma}_h}{\bar{\sigma}_v} \tag{16.6}$$

As indicated in Fig. 16.1, the total stress σ acts over the entire area under consideration, e.g., a^2 for the element A, and the pore water pressure acts over the area where there is water in contact with the total area under consideration, i.e., the total area minus the mineral contact area. The effective stress is approximately the force carried by the soil skeleton divided by the total area of the surface.

One can intuitively reason that the effective stress will more closely correlate with soil behavior than either total stress or pore water pressure. For example, increasing the effective stress should cause the soil particles to shift into a denser packing; however, equal increases in the total stress and the pore pressure, which would keep the effective stress constant, would logically have little or no effect on the particle packing. This expectation is supported by a considerable amount of experimental data.

The definition for effective stress and the fact that it correlates with soil behavior combine to give the *effective stress principle*, which can be stated as follows:

1. The effective stress is equal to the total stress minus the pore pressure.

[1] In this book the effective stress will be denoted by a horizontal line over the stress, e.g., $\bar{\sigma}$. A prime is also used to indicate an effective stress, e.g., σ'.

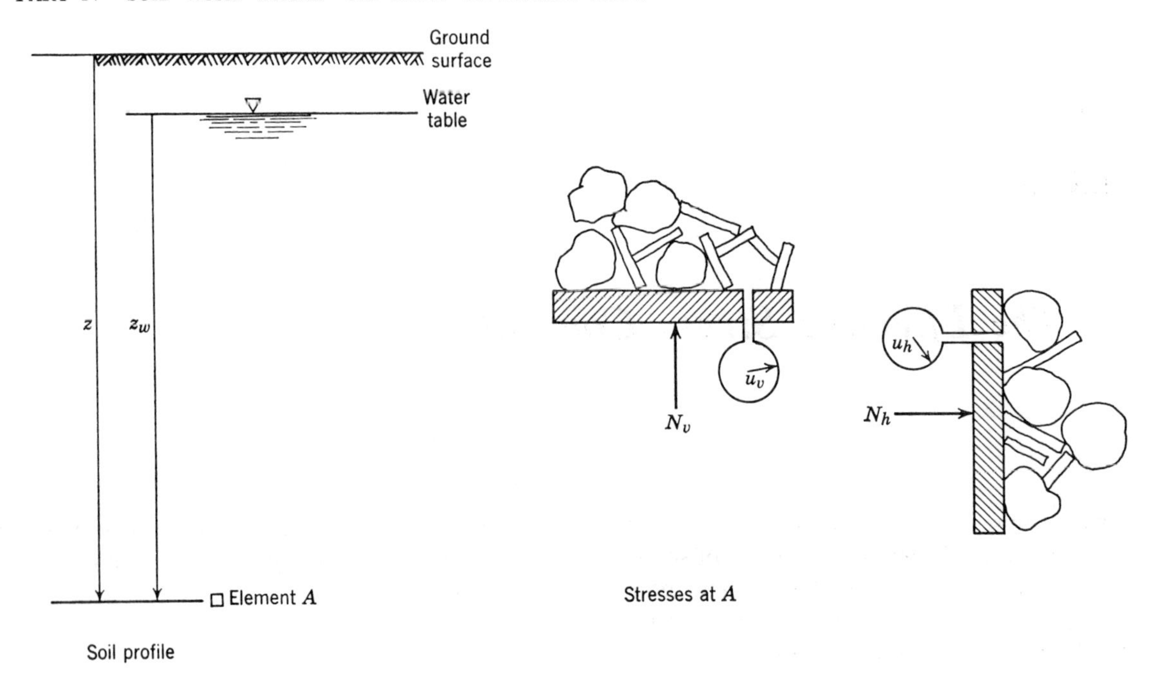

Fig. 16.1 Stresses in the subsoil.

2. The effective stress controls certain aspects of soil behavior, notably compression and strength.

This chapter is concerned with the first statement of the effective stress principle, and other portions of this book treat the second statement extensively.

The computation and portrayal of effective stress in a soil profile are illustrated in Example 16.1. Following are several features of the stress computation and portrayal that should be emphasized:

1. The plot of static water pressure with elevation is a straight line.
2. $p = \bar{p} + u$.
3. q can be computed from either total stresses or effective stresses:

$$q = \frac{\sigma_v - \sigma_h}{2} = \frac{(\bar{\sigma}_v + u) - (\bar{\sigma}_h + u)}{2} = \frac{\bar{\sigma}_v - \bar{\sigma}_h}{2}$$

4. The coefficient of lateral stress (Eqs. 8.5 and 16.6) refers to *effective* stresses not total stresses. (The stress paths shown in Fig. 8.11c are, in fact, for effective stresses. For the special case of zero pore pressure—the situation of Fig. 8.11—the stress paths are also for total stresses.)

16.3 PHYSICAL INTERPRETATION OF THE EFFECTIVE STRESS EQUATION

Section 16.2 defined the effective stress (Eq. 16.5) as the total stress minus the pore pressure. While this statement is all that is needed to solve most soil engineering problems, a physical interpretation of the effective stress equation helps in understanding soil behavior. This section presents a physical development of the effective stress equation.

Figure 16.2 considers on submicroscopic scale a horizontal surface through soil at any given depth. A truly horizontal surface cuts through many mineral particles as suggested by Fig. 16.2b, which is similar to the view shown in Fig. 8.2. As has been discussed at length in the preceding parts of this book, stress conditions at the contacts of particles, rather than within particles, are of primary concern to a consideration of strains and shear resistance within a soil mass. We are therefore really interested in a "horizontal" surface which goes through points of contact. Such a surface is termed a "wavy surface." As indicated in Fig. 16.2c, the mineral–mineral contact on the wavy surface is a small fraction of the total surface area.

Figure 5.1 showed an average interparticle condition for a soil mass and thus represents the whole of the wavy surface. An element $d \cdot d$ corresponds to a unit cell in a crystal; it is, in effect, the repeating unit. The condition is general as it includes all the possible interfaces, namely: mineral–mineral, air–mineral, water–mineral, water–water, air–air, and air–water. The forces acting between the particles are shown in Fig. 5.1 and defined in Chapter 5. The forces acting across the surface must be equal to the stress multiplied by the area:

$$\sigma dd = F_m + F_a + F_w + R' - A'$$

or

$$\sigma dd = \bar{\bar{\sigma}} A_m + u_a A_a + u_w A_w + R' - A'$$

where

$$F_m = \bar{\bar{\sigma}} A_m, \qquad F_a = u_a A_a, \qquad F_w = u_w A_w$$

or

$$\sigma = \bar{\bar{\sigma}} a_m + u_a a_a + u_w a_w + R - A \tag{16.7}$$

where

$$a_m = \frac{A_m}{d \cdot d}, \qquad a_a = \frac{A_a}{d \cdot d}, \qquad a_w = \frac{A_w}{d \cdot d},$$

$$R = \frac{R'}{d \cdot d}, \qquad A = \frac{A'}{d \cdot d}$$

Equation 16.7 is a statically correct relationship of the stresses acting normal to any given plane. The limitations to the use of this equation lie in the evaluation of the terms. Let us examine the terms.

In granular soils the contact stress $\bar{\bar{\sigma}}$ is usually very high and the contact area ratio very small. Figure 5.16 suggested a $\bar{\bar{\sigma}}$ value of 338 MN/m² for σ equal to 101 kN/m². The contact stresses in heavily loaded granular soils can exceed the particle crushing strength—which may be as high as 7930 kN/m². Experimental work suggests values of a_m for granular soils generally less than 0.03 and probably less than 0.01 (see Bishop and Eldin, 1950). On the other hand, in montmorillonite $\bar{\bar{\sigma}}$ and a_m can both be zero (see Chapter 5).

In saturated soil the terms u_a and a_a are zero and $a_m + a_w = 1$. The value of a_w is thus usually 0.97 to 1.00. The value of u_w which should go in Eq. 16.7 is the pore water pressure as measured by a standpipe or pressure gauge. Specifically, u_w is the pressure in the fluid at point 2 in Fig. 16.3 as measured by a standpipe inserted at point 2 and containing a fluid of the same composition and at the same temperature as that at point 2. At equilibrium the pressure in the fluid of this composition is the same throughout the system; if it were not, water would flow to equalize any difference in pressure. As noted in Chapter 5, however, there is a difference in cation concentration between the points 2 and 1. Thus in addition to the pore water pressure, as measured at point 2, there is at point 1 a pressure (which can be considered a cation partial pressure) resulting from the higher concentration of cations at point 1. In other words, one can consider the total fluid pressures at points 1 and 2 being different by an amount equal to the partial pressure of the excess cations at 1. This difference in total fluid pressure, the "osmotic pressure," between points 2 and 1 is numerically equal to the electrical repulsion between particles. In other words,

▸ **Example 16.1**

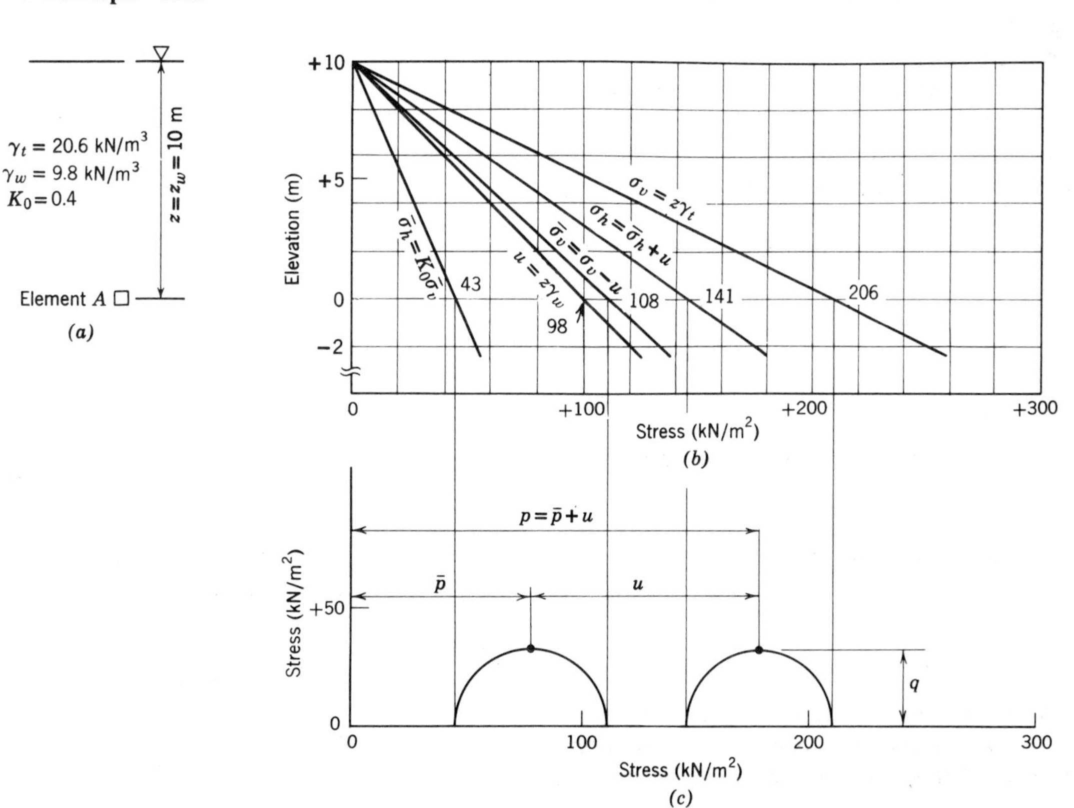

Fig. E16.1 Stresses in the subsoil. (*a*) Soil profile. (*b*) Stress versus elevation. (*c*) Stresses at *A*. ◂

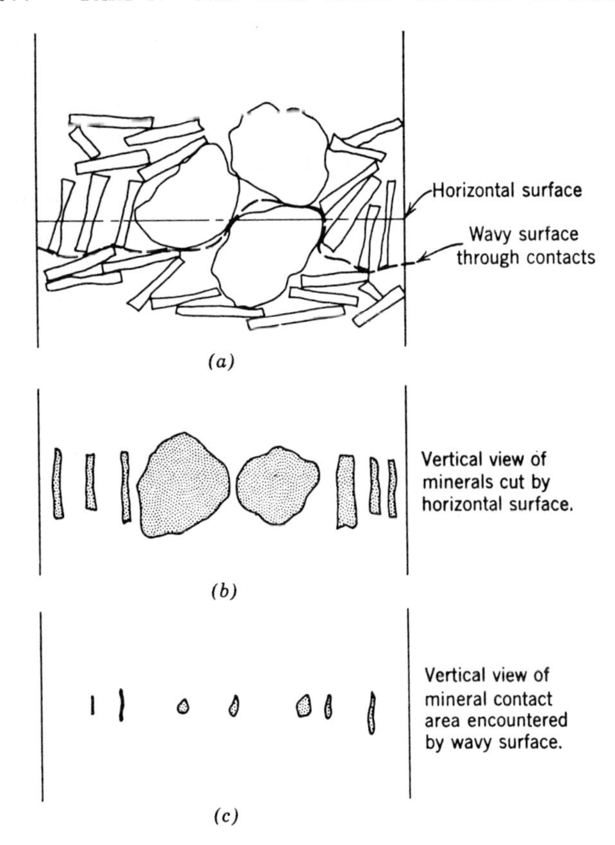

Fig. 16.2 Contact surface.

the electrical repulsive pressure R plus the pore water pressure u_w is the total fluid pressure existing at the mid-plane between the adjacent particles. The nature of the electrical forces between soil particles has been discussed in preceding chapters.

As has been pointed out, Eq. 16.7 is a general and correct expression of equilibrium of stresses acting normal to a given plane. For certain soil systems, or for systems where we can justifiably make approximations, there are special cases of Eq. 16.7. Some of these special cases follow.

1. Saturated soil:

$$\sigma = \bar{\bar{\sigma}} a_m + u_w a_w + R - A \tag{16.8}$$

2. Saturated soil with no mineral–mineral contact:

$$\sigma = u_w + R - A \tag{16.9}$$

3. Saturated soil with no net $R - A$:

$$\sigma = \bar{\bar{\sigma}} a_m + u_w a_w \tag{16.10}$$

or

$$\sigma = \bar{\bar{\sigma}} a_m + u_w(1 - a_m) \tag{16.10a}$$

Equation 16.8 holds for all saturated soil; Eq. 16.9 holds for highly plastic, dispersed systems such as montmorillonite (Fig. 5.16*b*); Eq. 16.10a holds for granular soils.

On the basis of the preceding discussion, we can see at least two circumstances where we can physically visualize the effective stress in saturated soil:

1. For the condition of no mineral-mineral contact:

$$\bar{\sigma} = R - A \tag{16.11}$$

2. For the condition of no net $R - A$:

$$\bar{\sigma} = (\bar{\bar{\sigma}} - u_w)a_m \tag{16.12}$$

and since $\bar{\bar{\sigma}}$ is so large,

$$\bar{\sigma} \approx \bar{\bar{\sigma}} a_m \tag{16.12a}$$

In other words, in a highly plastic, saturated, dispersed clay the effective stress is the net electrical stress transmitted between particles; and in a granular soil at a high degree of saturation, the effective stress is approximately equal to the contact stress multiplied by contact area ratio.

The foregoing discussion helps us to understand that effective stress is *closely related* to the stress transmitted through the mineral skeleton. For this reason, $\bar{\sigma}$ is often called *intergranular stress*. These detailed physical arguments lead to some disagreement as to the *exact* relationship between the stress in the mineral skeleton and effective stress (e.g., see Skempton, 1961). Be this as it may, $\bar{\sigma}$ as defined by Eq. 16.5 has proved to be the key to the interpretation of soil behavior.

Let us now return to Eq. 16.7 and consider partially saturated soil. If the air in a partially saturated soil exists within bubbles, the preceding discussion on saturated soil applies since we can take our wavy surface around the bubbles thereby avoiding air on our contact surface. If, however, there is an air channel, as shown

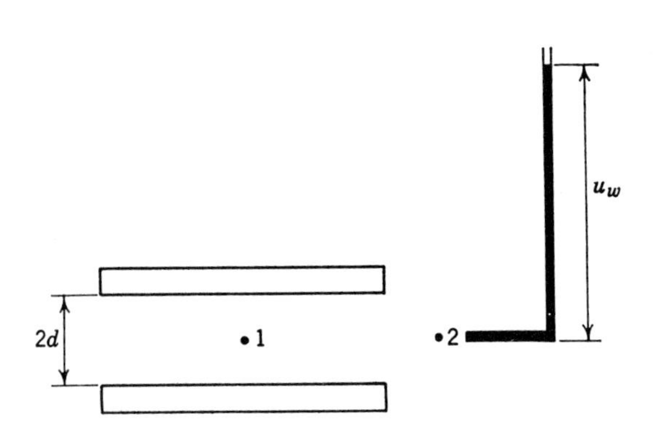

$$u_1 - u_2 = \text{osmotic pressure} = R$$
$$R = R_g T(n_1 - n_2)$$

where

R_g = gas constant
T = absolute temperature
n_1 = concentration of ions at 1
n_2 = concentration of ions at 2

Fig. 16.3 Pore pressure.

in Fig. 5.1, the air must be included in the equation of forces.

Let us define the effective stress in partially saturated soil as

$$\bar{\sigma} = \sigma - u^* \tag{16.13}$$

where u^* is an equivalent pore pressure and

$$u^* = u_a a_a + u_w a_w \tag{16.14}$$

Since

$$a_m \approx 0, \qquad a_a + a_w \approx 1$$

and

$$u^* \approx u_a + a_w(u_w - u_a) \tag{16.15}$$

and

$$\bar{\sigma} = \sigma - u_a + a_w(u_a - u_w) \tag{16.16}$$

Equation 16.16 is the form of effective stress equation for partially saturated soil proposed by Bishop et al. (1961).

In a general way, the principle of effective stress clearly applies to saturated soils. More research will be needed to learn whether Eq. 16.16 is really a useful quantity for the detailed interpretation of the behavior of partially saturated soils.

16.4 CAPILLARITY IN SOIL

There is much evidence that a liquid surface resists tensile forces because of the attraction between adjacent molecules in the surface. This attraction is measured by *surface tension*, a constant property of any pure liquid in contact with another given liquid or with a gas at a given temperature. An example of this evidence is the fact that water will rise and remain above the line of atmospheric pressure in a very fine bore, or capillary, tube. This phenomenon is commonly referred to as *capillarity*.

Capillarity enables a dry soil to draw water to elevations above the phreatic line; it also enables a draining soil mass to retain water above the phreatic line. The height of water column a soil can thus support is called *capillary head* and is inversely proportional to the size of soil void at the air-water interface.[2] Since any soil has an almost infinite number of void sizes, it can have an almost infinite number of capillary heads. In other words, the height of water column which can be supported is dependent on the size of void that is effective. There is therefore no such thing as *the* capillary head of a soil; there are limiting values of capillary head which can best be explained by the setup in Fig. 16.4.

[2] The height of rise h_c in a capillary tube is

$$h_c = \frac{2T_s}{R\gamma} \cos \alpha$$

where γ = unit weight of the liquid
T_s = surface tension of liquid
α = contact angle made between the liquid and the tube
R = radius of the tube

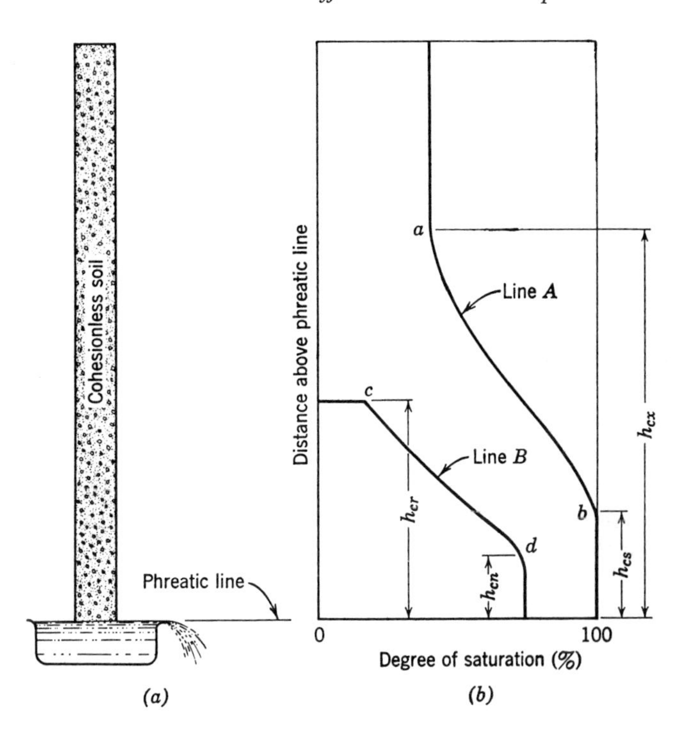

Fig. 16.4 Capillary heads in soil.

Figure 16.4*a* shows a tube of cohesionless soil; part (*b*) shows a plot of degree of saturation against distance above the phreatic line. If the tube of soil were initially saturated and allowed to drain until a static condition were reached, the distribution of moisture could be represented by line *A*. If, on the other hand, the tube of dry soil were placed in the container of water, line *B* would represent the distribution of moisture equilibrium.[3] The lines *A* and *B* represent the two limiting conditions of capillary moisture distribution for the tube of soil shown.

It would seem logical that point *a* on the drainage curve (Fig. 16.4) is the highest elevation to which any continuous channel of water exists above the free water surface. This distance therefore is taken as the *maximum capillary head* h_{cx}. Another critical point on the degree of saturation curve for a draining soil is the highest elevation at which complete saturation exists (point *b*, Fig. 16.4). The distance from the free water surface to this point is called the *saturation capillary head* h_{cs}.

On the distribution curve for capillary rise there are also two critical points. The distance from the free water surface to the highest elevation to which capillary water would rise (point *c*, Fig. 16.4) is called the *capillary rise* h_{cr}. The distance from the free water surface to the highest elevation at which the maximum degree of

[3] The times required for equilibrium to be reached depend greatly on the soil. Extremely long times are generally required to obtain line *B*.

saturation exists (point d, Fig. 16.4) is named the *minimum capillary head* h_{cn}.

The four capillary heads just described are limits of the possible range of capillary heads a soil may have. Any capillary head associated with drainage would lie between h_{cx} and h_{cs}, and any associated with capillary rise would lie between h_{cr} and h_{cn}. Since the size of void at the air-water interface determines the capillary head, it is reasonable, in the case of a dropping water column, for a small void to develop a meniscus which can support the water in larger voids below its surface, although it could not raise the water past these larger voids. That h_{cx} should be greater than h_{cr}, and h_{cs} be greater than h_{cn}, therefore, is to be expected.

Between the two extremes h_{cx} and h_{cn} there exist many capillary heads. The effective capillary head for any soil problem involving capillarity would depend on the particular problem, but would lie within the range of limiting heads described above. For comparing various soils and for certain drainage problems, the saturation capillary head h_{cs} is of much value. Also, h_{cs} is easily measured.

Table 16.1, presenting test data obtained by Lane and Washburn (1946), indicates the range of capillary heads for cohesionless soils.

Table 16.1 Capillary Heads

Soil	Particle Size D_{10} (mm)	Void Ratio	Capillary Head (cm) h_{cr}	h_{cs}
Coarse Gravel	0.82	0.27	5.4	6.0
Sandy Gravel	0.20	0.45	28.4	20.0
Fine Gravel	0.30	0.29	19.5	20.0
Silty Gravel	0.06	0.45	106.0	68.0
Coarse Sand	0.11	0.27	82.0	60.0
Medium Sand	0.02	0.48–0.66	239.6	120.0
Fine Sand	0.03	0.36	165.5	112.0
Silt	0.006	0.95–0.93	359.2	180.0

From Lane and Washburn, 1946.

The fact that water held by capillarity exists at an absolute pressure less than atmospheric pressure, i.e., negative gage pressure, is an essential concept. To illustrate this concept, let us consider the pressure in the pore water—pore pressure—at two points in the soil column shown in Fig. 16.5. At these two points, point 1 and point 2, are installed standpipes, as shown. The water in the open ends of both standpipes is at the level of the phreatic line; therefore the pore pressure at point 1 is negative and numerically equal to the vertical distance from the phreatic line to point 1 multiplied by the unit weight of water, and the pore pressure at point 2 is also minus and equal to the vertical distance from point 2 to the phreatic line multiplied by the unit weight of water.

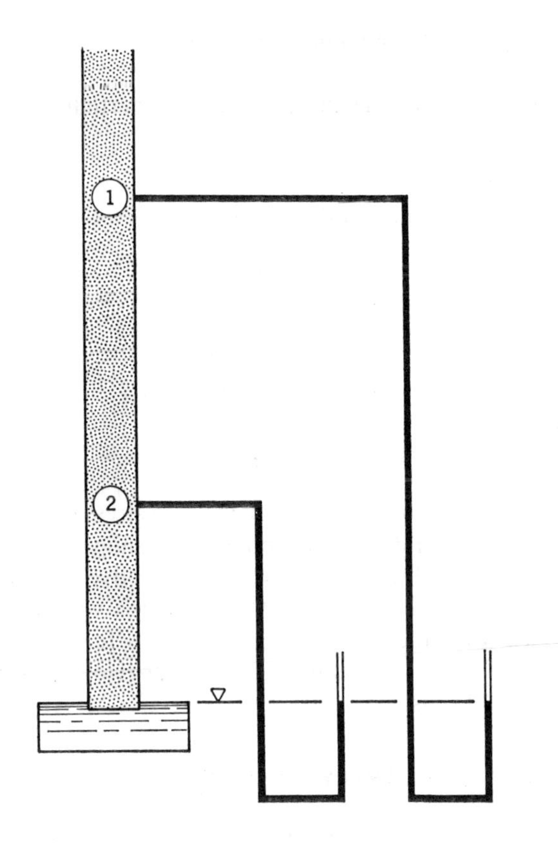

Fig. 16.5 Pressure in capillary water.

Whether the column of soil shown in Fig. 16.5 was initially dry and imbibed water or was initially saturated and drained has no influence on the pore pressure at a given point. The pore water pressure at any elevation in the soil column is equal to the height of that point above the phreatic line multiplied by the unit weight of water as long as there is continuous static water.

Figure 16.4 shows that there is no unique relationship between degree of saturation and height above the phreatic line; the relationship depends on the history of the sample. Figure 16.5 shows that for static equilibrium (and continuous channels of water) the pore water pressure at any point is exactly equal to the height of the point above the phreatic line multiplied by the unit weight of water, regardless of the degree of saturation.

Combining these two facts, we can conclude that pore water pressure is not a unique function of degree of saturation but depends also on the history of the sample.

16.5 THE COMPUTATION OF EFFECTIVE STRESS FOR STATIC GROUND WATER CONDITIONS

Example 16.1 illustrated the computation of stress within a highly idealized subsoil. Example 16.2 illustrates the computation for an actual soil profile at an industrial site in Kawasaki, Japan. Underlying 15 m of sand and silt are three strata of recently deposited sedimentary clays. The three clays are composed of the same minerals but were deposited under slightly different geological conditions and thus have different engineering properties. Underlying the three clay strata is a stratum of dense, clean sand.

A wall has been constructed around the property and a canal has been dredged in order that tankers may dock at a pier which runs perpendicular to the wall. The site development plan calls for a number of product storage tanks to be constructed near the canal. The sketch shows one of these tanks; it is 30 m in diameter, 13.8 m in height, and has a nominal storage capacity of 10,000 kl. This planned construction presents both a stability problem and a settlement problem which must be investigated. The factor of safety against a shear rupture through the soil beneath the tank must be adequate; the settlement of the tank must be predicted in order that proper allowance be made for piping connections to the tank. As we shall see later in this book, an early and essential step in both a stability and a settlement analysis is the determination of the effective stress existing in the subsoil prior to planned construction.

Shown in the sketch are plots of vertical total stress σ_v, pore pressure u, and vertical effective stress $\bar{\sigma}_v$ against elevation. Also shown is a table of computations for the stresses. The stresses between elevations +4.0 and +3.3 are based on full saturation and full capillarity.

The pore water pressure plotted is for the conditions of no ground water movement and the phreatic surface at elevation +3.3. The phreatic surface was taken as the level to which the water rose into holes excavated into the sand-silt or into the boreholes.[4] The assumption of no ground water movement is very common although not always sound, especially for a site such as Kawasaki where there has been recent placement of fill. Further, in the Kawasaki area wells have been installed into the deep sand stratum in order to secure water for industrial uses. Because of the recent site filling and because of pumping from wells in the area, the pore pressure plot and dependent effective stress plot are suspect. Pore water pressure measuring devices ($P1$ to $P6$) were therefore installed at the elevations shown.

If the devices had shown that the pore pressure was indeed equal to the distance from the device to the phreatic line multiplied by the unit weight of water, the plots of pore pressure and effective stress shown in Example 16.2 would be correct. In fact, however, there is considerable deviation in pore pressure from the static pressure plot and the assumption of a static condition is very poor for the Kawasaki site. The actual pore pressure measurements are presented in Chapter 17, and the details of computing the correct pore pressures and effective stresses are discussed there. Example 16.2 illustrates how the pore pressure plot for a static case is determined.

The increment of total stress $\Delta\sigma_v$ was obtained by multiplying the thickness of the stratum by the total unit weight of the soil. Thus the total weight used for this computation is the average for the stratum. The fact that only one void ratio or unit weight value is given for each stratum should not lead one to infer that these values are constant for the entire stratum. Fig. E16.2-2 presents the results of a compression test on a sample[5] of Clay II. This plot, in the form of effective stress $\bar{\sigma}_v$ on a log scale against void ratio e, is a straight line. As noted on the sketch, the total unit weight of Clay II actually varies from 14.62 kN/m^3 at the top of the clay stratum to 14.81 kN/m^2 at the bottom, a very small variation. The use of 14.72 kN/m^2 as a total unit weight for the entire stratum is thus acceptable. If, however, the actual variation in effective stress of 73.6 kN/m^2 had occurred in going from 19.6 to 93.2 rather than 192.3 to 266.8, the variation in void ratio would have been 0.67 rather than 0.15. In other words, we must remember that generally variations in void ratio are much larger for a given stress increment at a low stress level than for a high stress level. Where there is a marked decrease in void ratio with depth in a particular stratum, the plot of total stress with depth should not be plotted as a straight line, but rather as a curved one with a larger increment of total stress for a given change in elevation as one proceeds deeper. In such a situation, the clay stratum can be divided into several layers with the contribution of each layer to total stress determined.

Example 16.3 illustrates the determination of stresses in a nonsaturated soil. A column of fine, uniform sand at constant void ratio will be used. The soil was initially saturated and then permitted to drain with the phreatic line maintained at elevation zero. After static equilibrium had been reached, measurements of degree of saturation were made at a number of points in the sand and a plot of percent saturation versus elevation was prepared. To the right in the example are plotted various stresses as a function of elevation.

[4] One must wait until the water level becomes stationary for this technique to give the correct phreatic surface.

[5] The e versus log $\bar{\sigma}_v$ plot has been extended back to small stresses.

▶ **Example 16.2**

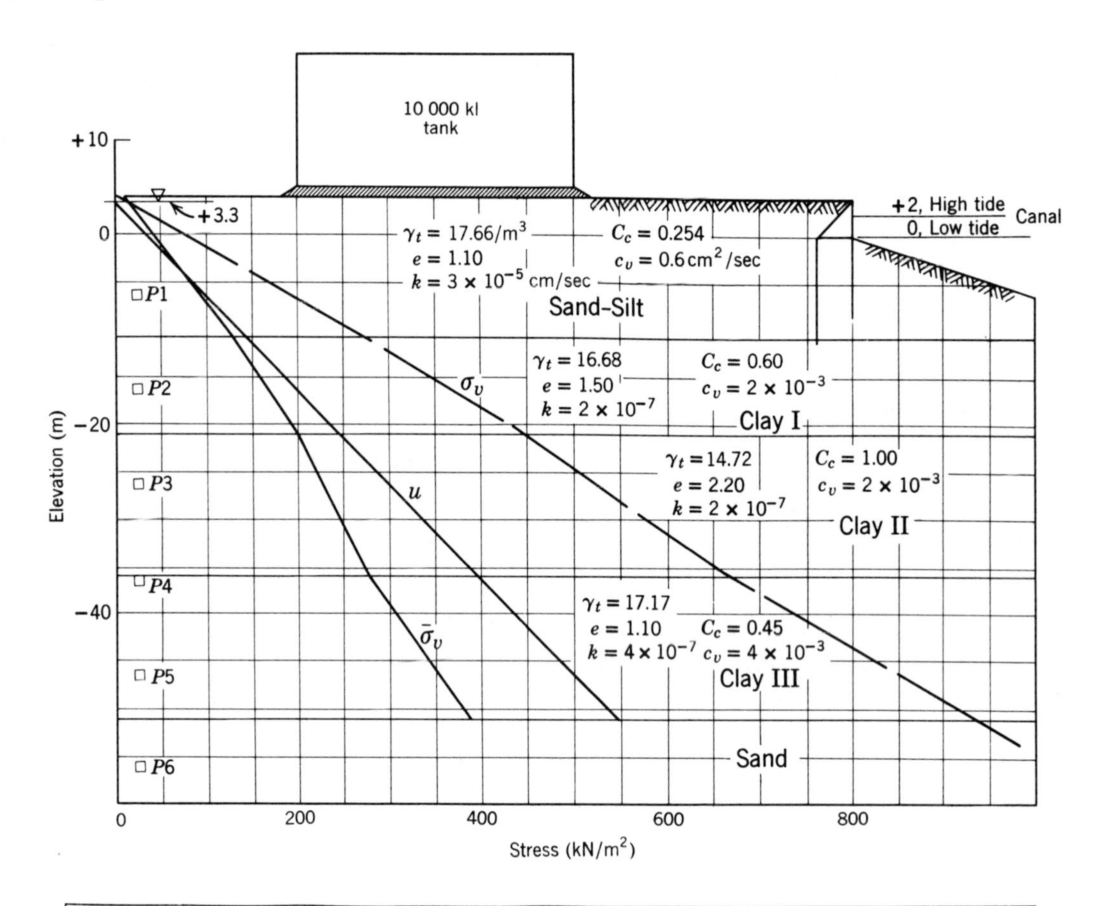

Elevation	Soil	γ_t	Δz	$\Delta\sigma_v$	σ_v	u	$\bar{\sigma}_v$
+4.0					0	−6.9	6.9
	Sand-Silt	17.66	0.7	12.4			
+3.3					12.4	0	12.4
	Sand-Silt	17.66	14.3	252.5			
−11					264.9	140.3	124.6
	Clay I	16.68	10	166.8			
−21					431.7	238.4	193.3
	Clay II	14.72	15	220.8			
−36					652.5	385.5	267.0
	Clay III	17.17	15	257.6			
−51					910.1	532.7	377.4

Fig. E16.2-1 Kawasaki subsoil profile.

▶ **Example 16.2** *continued*

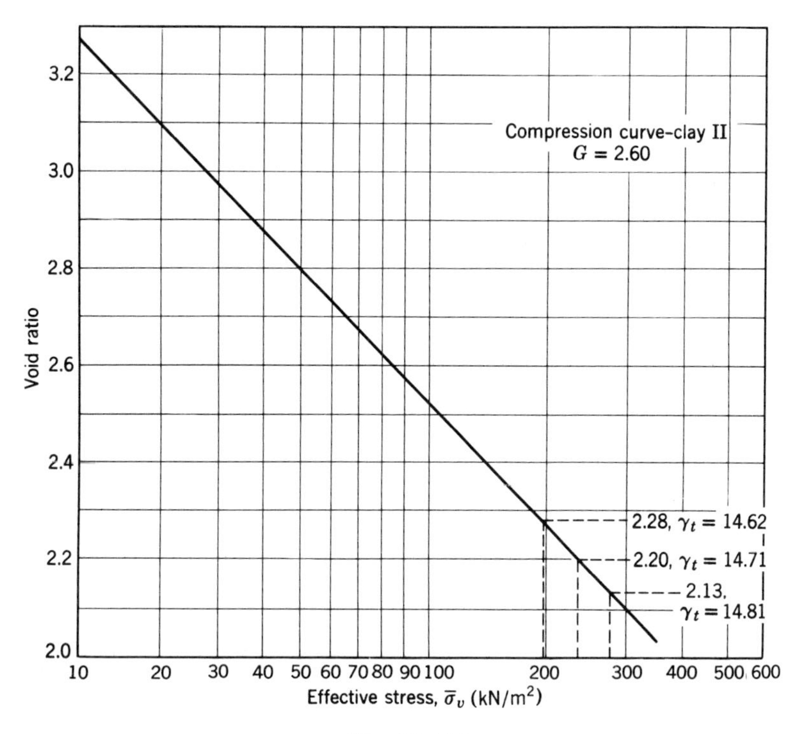

Fig. E16.2-2 ◀

▶ **Example 16.3**

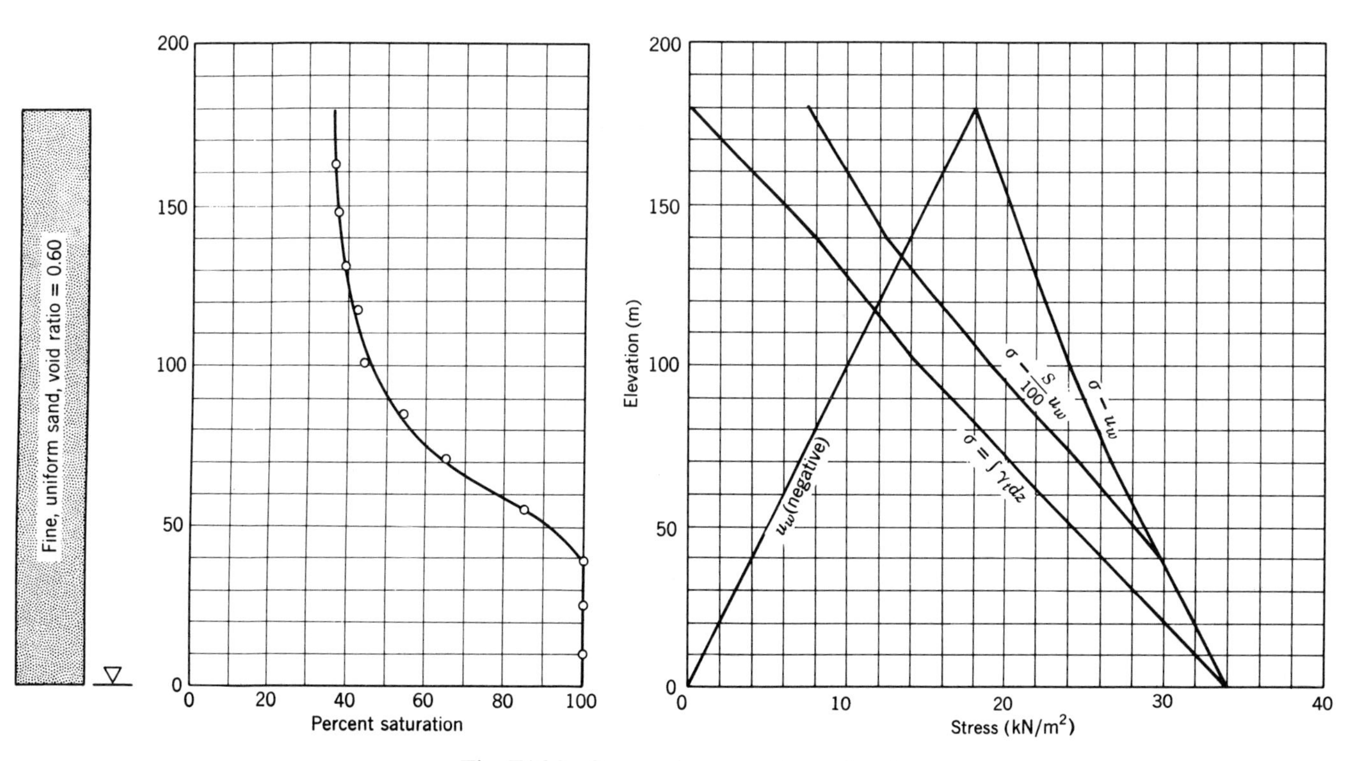

Fig. E16.3 Stresses in a nonsaturated soil. ◀

The plot of elevation versus pore water pressure is a straight line going from zero at the phreatic line, elevation 0, to minus 17.6 kN/m^2 at elevation 180 cm. The total vertical stress σ_v is the total weight of soil and water per unit area above the point in question. Since the degree of saturation varies with elevation, the total unit weight of soil varies with elevation, and the plot of total stress with elevation is not a straight line.

At the far right is plotted the difference between total stress and pore water pressure. It was obtained merely by subtracting the pore water pressure from the total stress; since the pore water pressure at all points is negative, the numerical value of the pore pressure was added to that of the total stress. Also shown is the plot of $\sigma - (S/100)u_w$, i.e., the effective stress as defined by Eq. 16.16, with $u_a = 0$ and $a_w = S/100$.

At some height in a column of fine sands, such as that in Example 16.3, the pore water ceases to be continuous. When the pore water is not continuous, the pore water pressure is no longer a unique function of height above the phreatic line. Water could be trapped in voids far above the phreatic line and could still exist at positive pressures.

As can be seen from Example 16.3, there is considerable difference between the values of $\sigma - u_w$ and $\sigma - (S/100)u_w$ above elevation 40. The stress $\sigma - (S/100)u_w$ is probably closer to the stress that best correlates with soil behavior than is $\sigma - u_w$.

16.6 SUMMARY OF MAIN POINTS

1. The effective stress is:
   a. Saturated soil:
   $$\bar{\sigma} = \sigma - u$$
   b. Partially saturated soil:
   $$\bar{\sigma} \approx \sigma - u_a + a_w(u_a - u_w)$$
2. For geostatic stresses and static pore water:
   a. $\sigma_v = \sum \gamma_t \, \Delta z$.
   b. $u = z_w \gamma_w$.
   c. $\bar{\sigma}_h = K\bar{\sigma}_v$.
3. In granular soils, the effective stress is approximately the contact stress multiplied by the ratio of contact area to total area.
4. In general, the effective stress is essentially the force carried by the soil skeleton divided by the total sectional area.
5. Within the capillary zone of continuous moisture the pressure is negative, and at any point it is numerically equal the height of the point above the phreatic line multiplied by the unit weight of water.

PROBLEMS

16.1 For the Thames Estuary clay in Fig. 7.10 make a plot of σ_v, u, and $\bar{\sigma}_v$ versus elevation for a depth of 12.2 m (Compare your plot of $\bar{\sigma}_v$ with that given in Fig. 7.10.)

16.2 Refer to the subsoil profile for the South Bank in Fig. 7.8. Which would be the larger value of $\bar{\sigma}_v$ at a depth of 42.7 m:

a. $\bar{\sigma}_v$ computed for static pore pressure.

b. $\bar{\sigma}_v$ computed for the pore pressure indicated by the standpipe in Fig. 7.8.

What is the magnitude of the difference between the two values of $\bar{\sigma}_v$?

16.3 The water table in Example 16.1 rises 2 m (the tide comes in) while the soil surface elevation remains constant. Compute the values of σ_v, $\bar{\sigma}_v$, u, σ_h, $\bar{\sigma}_h$ at element *A*.

16.4 The contact area ratio in the sand under the center of the tank at elevation -5 m in Example 16.2 is 0.1%. Discuss the likelihood of sand particle crushing following the filling of the tank with water.

16.5 The height of soil column in Fig. 16.4*a* is 2.0 m. The soil (fine sand) has the following properties:

$$e = 0.473 \text{ (constant with depth)}$$
$$G = 2.69$$
$$a_w = \frac{S}{100\%}$$
$$a_m = 0.05\%.$$

For the fully drained condition (line *A*) compute for point *a*:

$$u_w, \quad u^*, \quad \sigma_v, \quad \bar{\sigma}_v, \quad \bar{\bar{\sigma}}_v$$

Hint. The air pressure is atmospheric. Assume geostatic stresses.

16.6 Refer to Fig. 5.16*b*. Compute the effective stress for an interparticle spacing of 40 Å. The spacing is held constant while salt is added to water around the particles. Does the effective stress increase or decrease? Explain.

CHAPTER 17

One-Dimensional Fluid Flow

17.1 THE NATURE OF FLUID FLOW IN SOILS

Chapters 17, 18, and 19 deal with the flow of fluids—primarily water—through soil. The engineer must understand the principles of fluid flow in order to solve problems (*a*) involving the rate at which water flows through soil (e.g., the determination of rate of leakage through an earth dam); (*b*) involving compression (e.g., the determination of the rate of settlement of a foundation); and (*c*) involving strength (e.g., the evaluation of factors of safety of an embankment). The emphasis in these three chapters is on the influence of the fluid on the soil through which it is flowing; in particular, on the effective stress.

In general, all voids in soils are connected to neighboring voids. Voids that are isolated from neighboring voids are impossible in an assemblage of spheres, regardless of the type of packing. In the coarse soils—gravels, sands, and even silts—it is hard to imagine isolated voids. In the clays, consisting as they usually do of plate-shaped particles, a small percentage of isolated voids would seem possible. Electron photomicrographs of natural clays, however, suggest that even in the finest grained soils, all of the voids are interconnected.

Since the pores in soil are apparently interconnected, water can flow through the densest of natural soils. Thus in a tube of soil, such as that illustrated in Fig. 17.1, water can flow from point A to point B. Actually, the water does not flow from point A to point B in a straight line at a constant velocity, but rather in a winding path from pore to pore, as illustrated by the heavy line in Fig. 17.1. The velocity of a drop of water at any point along the winding path depends on the size of the pore and its position in the pore, especially on its distance from the surface of the nearest soil particle. However, in soil engineering problems, the water can be considered to flow from point A to point B along a straight line at an effective velocity.

17.2 DARCY'S LAW

In the 1850s, H. Darcy, working in Paris, performed a classical experiment. He used a setup similar to that shown in Fig. 17.2 to study the flow properties of water through a sand filter bed. He varied the length of sample L and the water pressure at the top and bottom of the sample; he measured the rate of flow Q that passed through the sand. Darcy experimentally found that Q was proportional to $(h_3 - h_4)/L$, and that

$$Q = k\,\frac{h_3 - h_4}{L}\,A = kiA \tag{17.1}$$

where

Q = the rate of flow
k = a constant, now known as Darcy's coefficient of permeability[1]
h_3 = the height above datum which the water rose in a standpipe inserted at the entrance end of the filter bed
h_4 = the height above datum which the water rose in a standpipe inserted at the exit end of the filter bed
L = the length of sample
A = the total inside cross-sectional area of the sample container

$$i = \frac{h_3 - h_4}{L}, \text{ the gradient}$$

Equation 17.1, now known as Darcy's law, is one of the cornerstones of soil mechanics. During the century since Darcy performed his monumental work, Eq. 17.1 has been subjected to numerous examinations by many experimenters. These tests have shown that Darcy's law is valid for most types of fluid flow in soils. For liquid

[1] Chapter 19 treats this important soil property, *permeability*. Values given in Chapter 19 show a very wide range, from greater than 1 cm/sec for gravel to less than 1×10^{-7} cm/sec for clay.

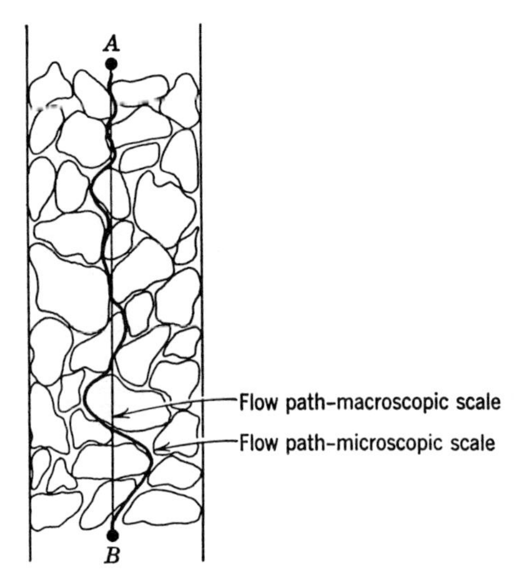

Fig. 17.1 Flow path in soil.

flow at very high velocity and for gas flow at very low or at very high velocity, Darcy's law becomes invalid. The validity of Darcy's law is treated later in this chapter.

17.3 FLOW VELOCITY

A further consideration of the velocity at which a drop of water moves as it flows through soil helps to understand fluid flow. Equation 17.1 can be rewritten as

$$\frac{Q}{A} = ki = v \tag{17.2}$$

Since A is the total open area of the tube above the soil (Fig. 17.2), v is the velocity of downward movement of a drop of water from position 1 to position 2. This velocity is numerically equal to ki; therefore k can be interpreted as the *approach velocity* or *superficial velocity* for a gradient of unity, i.e., $k = v/i$ or $k = v$ for a gradient equal to 1.

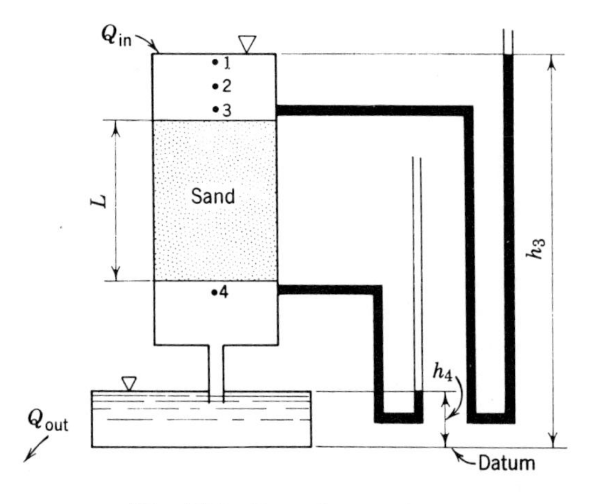

Fig. 17.2 Darcy's experiment.

From position 3 to position 4 in the sample shown in Fig. 17.2, a drop of water flows at a faster rate than it does from position 1 to 2 because the average area of channel available for flow is smaller. This reduced area of flow channel is represented in Fig. 17.3, which is the test setup in Fig. 17.2 with the mineral and void volumes separated. Using the principle of continuity, we can relate the velocity of approach v to the average effective velocity of flow through the soil v_s as follows:

$$Q = vA = v_s A_v$$

$$\therefore\ v_s = v\frac{A}{A_v} = v\frac{AL}{A_v L} = v\frac{V}{V_v} = \frac{v}{n}$$

The average velocity of flow through the soil v_s, termed the *seepage velocity*, is therefore equal to the velocity of approach divided by the porosity:

$$v_s = \frac{v}{n} = \frac{ki}{n} \tag{17.3}$$

Equation 17.3 gives the average velocity of a drop of water as it moves from position 3 to position 4; this is the straight line distance from 3 to 4 divided by the time required for the drop to flow from 3 to 4. As noted earlier in this section, a drop of water flowing through the soil actually follows a winding path with varying velocity; therefore v_s is a fictitious velocity for an assumed drop of water that moves in a straight line at a constant velocity from position 3 to position 4.

Even though approach velocity and seepage velocity are fictitious quantities, they both can be used to compute the time required for water to move through a given distance in soil, such as between positions 3 and 4.

17.4 HEADS

In the study of fluid flow it is convenient to express energy, both potential and kinetic in terms of *head*, which

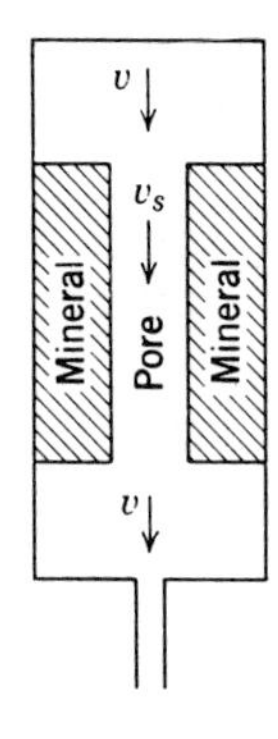

Fig. 17.3 Flow channel.

is energy per unit of mass.[2] The following three heads must be considered in problems involving fluid flow in soil:

1. *Pressure head*, h_p = the pressure divided by the unit weight of fluid.
2. *Elevation head*, h_e = the distance from the datum.
3. *Total head*, $h = h_p + h_e$ = the sum of pressure head and elevation head.

When dealing with flow through pipes and open channels, we must also consider velocity head. The velocity head in soils, however, is much too small to be of any consequence and thus can be neglected. (For example, a high velocity for fluid flow in soil is 0.6 m/min, a value which gives a velocity head of 0.000 006 m. This head is far less than the accuracy with which the engineer can normally measure pressure head or elevation head.) Those engineers dealing with pipe and channel flow define total head as velocity head + pressure head + elevation head; and they define piezometric head as pressure head + elevation head. For flow through soil with its negligible velocity head, the total head and piezometric head are equal.

Since both pressure and elevation heads can contribute to the movement of fluid through soils, it is the total head that determines flow, and the gradient to be used in Darcy's law is computed from the difference in total head. The importance and truth of this statement can be seen from the two situations shown in Fig. 17.4.

Figure 17.4*a* shows a common bucket full of water in a static condition. In Fig. 17.4*a* the heads for the two points, number 1 near the top and number 2 near the bottom of the bucket, are labeled and plotted. Between points 1 and 2 there exists both a pressure gradient and an elevation gradient; there is, however, no gradient of total head since the total head of the two points is identical and equal to h. Even though the pressure at point 2 is considerably greater than that at point 1, water does not flow from point 2 to point 1; further, flow does not occur from point 1 toward point 2 even though the elevation head at point 1 is three times that at point 2.

Figure 17.4*b* shows a capillary tube in which there is water standing to a height of h_c. As with the water in the bucket there is no total head gradient.

These very simple examples illustrate two important principles:

1. Flow between any two points depends only on the difference in total head.
2. Any elevation can be selected for datum as a base of elevation heads. The absolute magnitude of elevation head has little meaning; it is the difference in elevation head that is of interest and the difference of elevation head between any two points is the same regardless of where datum is taken.

17.5 PIEZOMETERS

In soil mechanics we are especially interested in pressure head, since the pore pressure needed to compute effective stress can be obtained from the pressure head. The pressure head at a point can be either measured directly or computed using principles of fluid mechanics.

Pore Pressure Measurements in the Laboratory

The pressure head or water pressure at a point in a soil mass is determined by a "piezometer," a word literally meaning "pressure meter." Figure 17.5 shows two simple piezometers. On the right side of the soil system is a manometer or standpipe connected to a porous tip located at midheight in the soil system. On the left of the tube is a regular Bourdon gauge connected to the porous tip.

As illustrated in Fig. 17.5, as the water flows in the soil sample a pressure head at its midheight of 0.15 m is indicated by the manometer and a pressure of 1.47kN/m² by the gauge. While the two piezometers shown in Fig. 17.5 are simple in principle, they both have a serious drawback: a flow of water from the soil into the measuring system is required to actuate each device. The drawback can be illustrated by considering what would happen if the reservoir height were increased 0.3 m. This would increase the pressure head at midheight by 0.15 m, meaning that a volume of water equal to 0.15 m times the inside area of the manometer tube would flow from the soil into the manometer. Even though considerably less flow would be required to actuate the gauge, a measurable amount of water would flow from the soil into the gauge. If the soil in the permeameter were pervious, the time required for re-establishment of an equilibrium flow condition would be minor. If, however, the soil were a silt or clay, an appreciable amount of time would be required for the water to flow from the soil into the manometer or

[2] Kinetic energy $= \dfrac{Mv^2}{2} = ML^2T^{-2}$

Elevation energy $= MLg = ML^2T^{-2}$

Pressure energy $= \dfrac{pM}{\rho} = \dfrac{ML^{-1}T^{-2}M}{ML^{-3}} = ML^2T^{-2}$

$$\text{Head} = \frac{\text{energy}}{g} = \frac{ML^2T^{-2}}{MLT^{-1}} = L$$

where

M = mass
v = velocity
g = acceleration of gravity
L = length
T = time
p = pressure
ρ = density

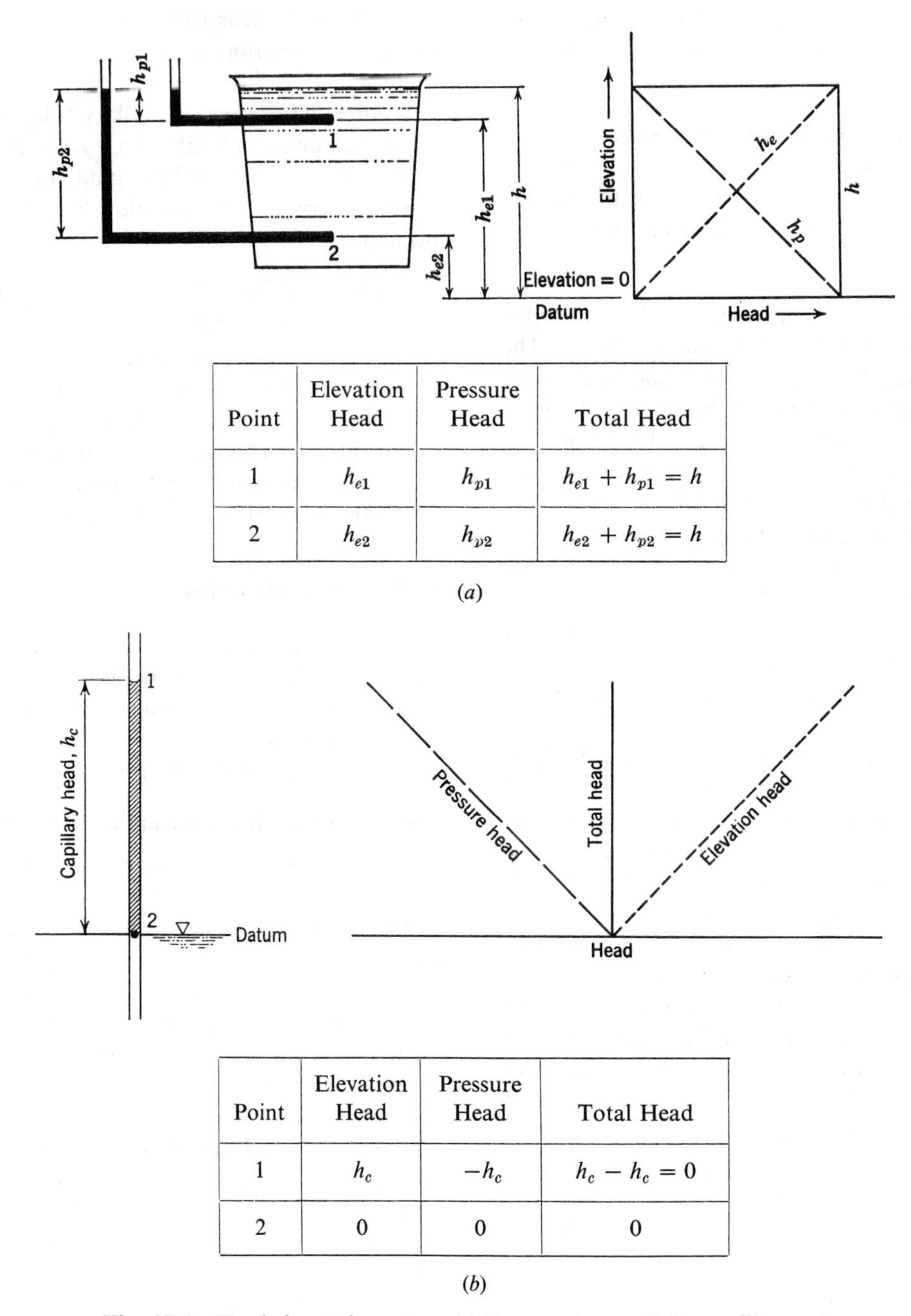

Point	Elevation Head	Pressure Head	Total Head
1	h_{e1}	h_{p1}	$h_{e1} + h_{p1} = h$
2	h_{e2}	h_{p2}	$h_{e2} + h_{p2} = h$

(*a*)

Point	Elevation Head	Pressure Head	Total Head
1	h_c	$-h_c$	$h_c - h_c = 0$
2	0	0	0

(*b*)

Fig. 17.4 Heads in static water. (*a*) In container. (*b*) In capillary tube.

pressure gauge. The two systems shown in Fig. 17.5 would therefore involve large time lags in measuring pore pressures in relatively impermeable soils. To measure pore pressures under "no-flow" conditions various types of piezometers have been developed (see, for example, Lambe, 1948; Bishop, 1961; Whitman et al. 1961; and Penman, 1961).

Figure 17.6 shows a most interesting and important test result. A stress of 138 kN/m² was applied to a saturated clay specimen and held until the pore water pressure was zero. The stress was instantaneously removed and the resulting negative pore pressure was recorded. The record shown in Fig. 17.6 indicates an almost immediate pore pressure response equal to the magnitude of the relieved stress, 138 kN/m². This measured pore pressure was 36.5 kN/m² *below absolute zero.* The pore water sustained this pure tension for a short period of time, but cavitation (probably in the measuring system) resulted in a recorded pressure of -101 kN/m²; i.e., absolute zero. From other measurements we can

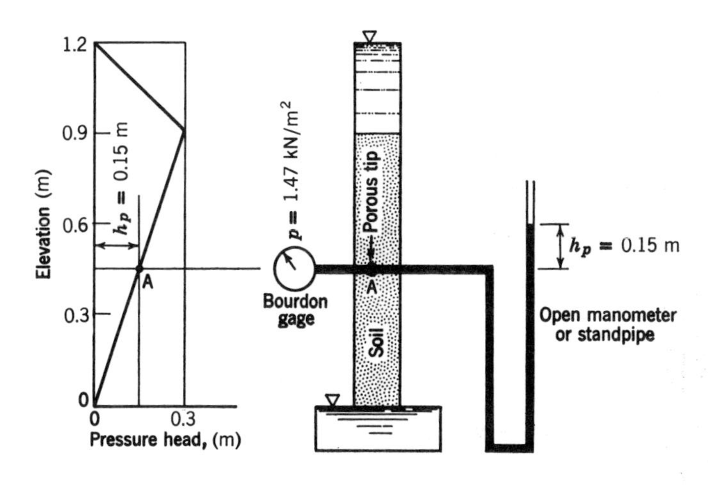

Fig. 17.5 Flow-type piezometer.

infer that pore pressures below absolute zero can exist over long periods of time, and that these negative pore pressures may be as large or larger than 414 kN/m².

The measurement of pore water pressure, especially values below atmospheric pressure, in partially saturated soil requires special precautions. From the nature of the measuring systems described, it is apparent that air must be kept out of the measuring devices. The easiest way to avoid difficulty with entrapped air is to use a sensing element which tends to prevent air from entering. If, for example, the sensing element is a fine porous stone initially saturated with water, capillarity will keep air from entering the stone until the air pressure exceeds the capillary pressure, or breakthrough pressure, of the stone. Porous stones with breakthrough pressures as high as 414 kN/m² are commercially available. Such high breakthrough pressure stones are desirable for measuring pore water pressures in partially saturated soils.

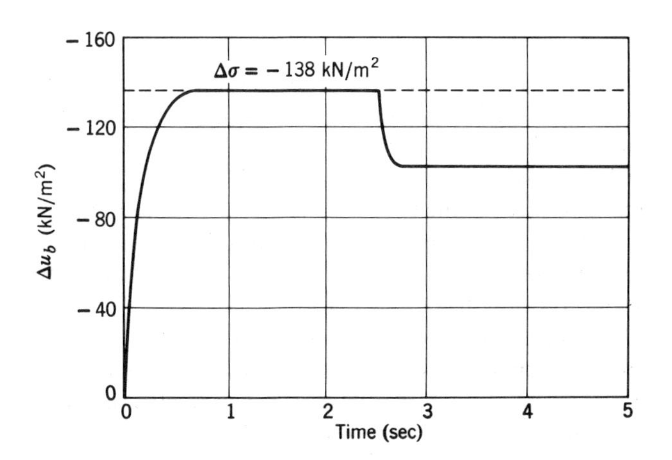

Fig. 17.6 Pore pressures measured by transducer piezometer.

Pore Pressure Measurements in the Field

The preceding subsection discussed the measurements of pore water pressure during laboratory tests. Such measurements are usually helpful (and often essential) in interpreting the test data. The basic principles involved in laboratory measuring systems can be used in field piezometers. Figure 17.7 shows a modified Casagrande piezometer. This piezometer was developed by A. Casagrande. It consists essentially of a porous ceramic stone inserted in a plug of clean sand. Since water must flow from the sand into the plug and up the tube to register an increase in pore water pressure, it is a "flow" piezometer. By having a relatively large zone (the sand surrounding the porous point) from which to draw water, this piezometer requires relatively short time lags. The Casagrande piezometer has been widely and successfully used in civil engineering field installations.

A crucial feature of any field piezometer is the necessity of sealing the sensing unit into the zone where pore pressure is desired.

Since flow of water from the soil at the point where pore pressures are being measured is required to actuate the Casagrande piezometer, it would not be a satisfactory device to measure rapidly changing pore water pressures. We can and should determine the response time of a Casagrande piezometer by filling the tube with water and determining the time required for attaining a steady reading (see Hvorslev, 1949).

The reader is referred to the Proceedings of the Conference on Pore Pressure and Suction in Soils (1961) for more information on piezometers.

17.6 CALCULATION OF PRESSURE HEAD

The principles of flow through porous media, which have been described in the preceding sections, can be made clearer by considering Figs. 17.8 to 17.11.

Figure 17.8 shows a tube of soil having a porosity of 0.33 and a permeability of 0.5 cm/sec in which water is flowing vertically downward. Atmospheric pressure is maintained at the top of the reservoir water (elevation 3.6) and at the bottom of the tail water (elevation 0). Datum is taken at the tail water; this selection is merely one of convenience, since any location can be selected for datum. To the right of the tube are plots of heads and velocity versus elevation.

In general, is it more convenient to first determine the elevation and total heads and then compute the desired pressure head by subtracting the elevation head from the total head. The elevation head is merely the elevation of the water at any point under consideration. Since the horizontal scale for head has been taken the same as the vertical scale for elevation, the slope of the elevation head versus elevation plot is one to one, i.e., 45°. The

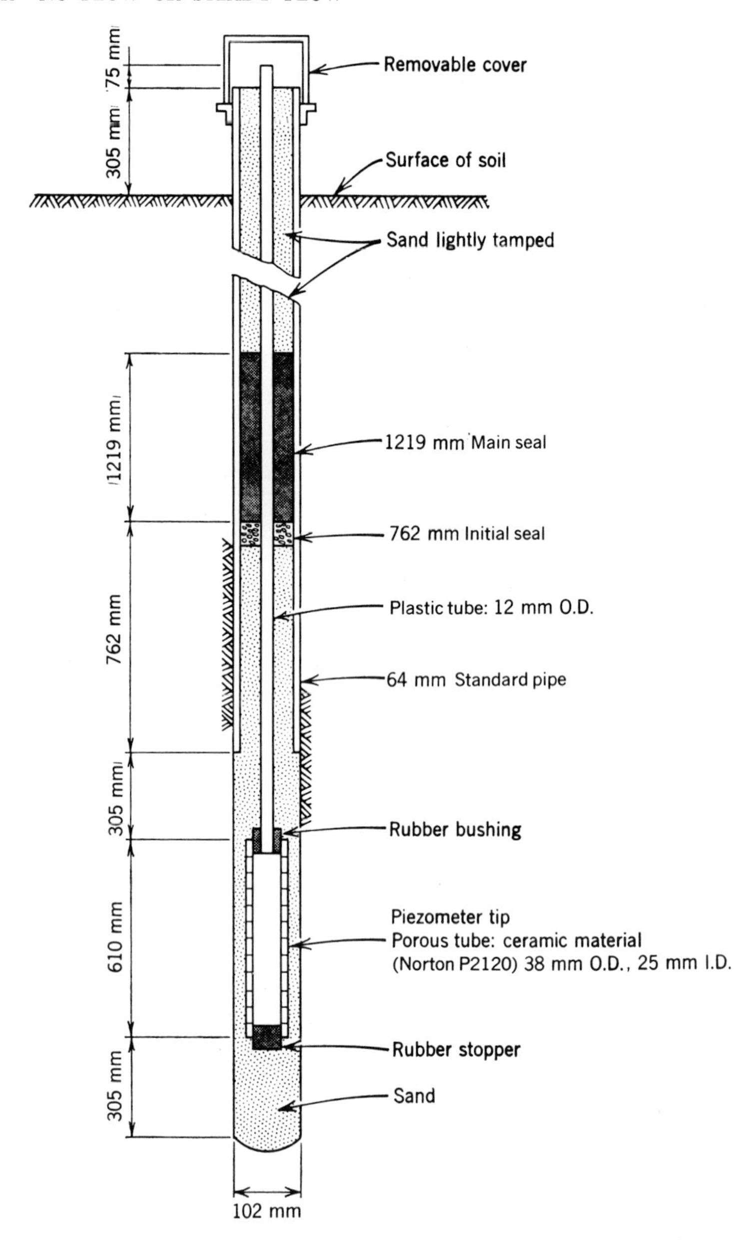

Fig. 17.7 Casagrande piezometer (not to scale).

total head at elevation 3.6 is the elevation head since the pressure head is 0. In flowing from elevation 3.6 to elevation 2.4 "no" total head is lost so that the total head at elevation 2.4 is also 3.6 m. Similarly, we note that the total head at both elevation 0 and elevation 0.6 is 0. Since the soil has uniform permeability and porosity the dissipation of total head in flowing through the soil must be uniform; the total head plot is therefore a straight line running from a value of 3.6 at elevation 2.4 to a value of 0 at elevation 0.6. The pressure head is obtained by subtracting the elevation head from the total head for any point under consideration.

To draw the total head line vertically between elevation 3.6 and 2.4, and between 0.6 and 0, assumes that the friction loss of flow in the entrance and exit parts of the tube is negligible compared to the head lost in flow through the soil. The validity of this assumption is easily checked by computing the head lost in the entrance and exit parts of the tube. Using a reasonable frictional coefficient and the principles of hydraulics, we compute the head lost in this 1.8 m of tube as 0.9×10^{-6} m. This is indeed negligible.

The velocity of flow in the entrance and exit parts of the tube was computed from Darcy's law, $v = k \times i = 0.5$ cm/sec $\times \frac{3.6}{1.8} = 1$ cm/sec. The seepage velocity is equal to the approach velocity/porosity, i.e., $1/0.33 = 3$ cm/sec.

Figure 17.9 presents no new principle except that the water is flowing upward through the soil. Figure 17.10

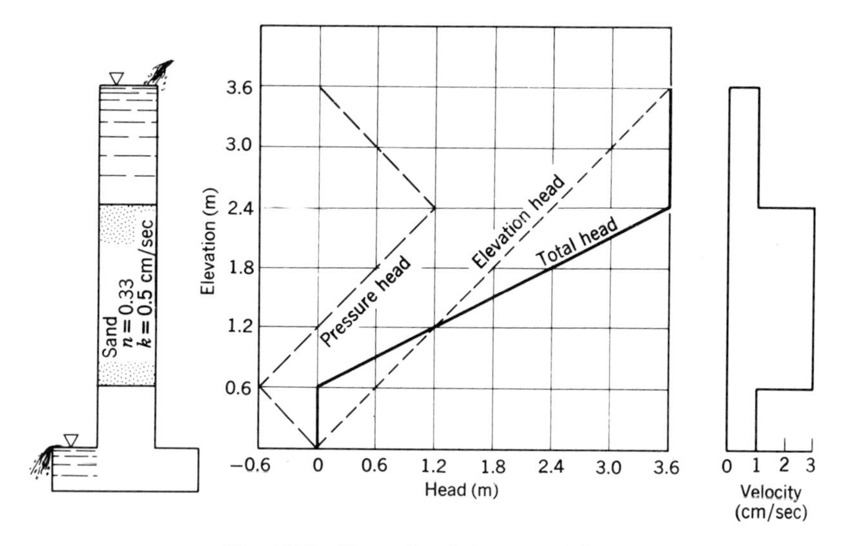

Fig. 17.8 Example of downward flow.

involves horizontal flow in which the elevation head is constant. In Fig. 17.11 the area of the permeameter and the properties of the soil change at elevation 1.2. Since no water is added to or removed from the system, the rate of flow of water in soil I must be equal to that in soil II. Thus

$$k_{I} i_{I} A_{I} = k_{II} i_{II} A_{II}$$

From this equation we find that the total head lost in soil I is half of that lost in soil II, or that a total head of 1.2 m is lost in soil I and 2.4 m in soil II. The total head line in Fig. 17.11 can be drawn on this basis, and the pressure head can be obtained by subtracting the elevation head from the total head.

The four preceding examples of flow through porous media illustrate the following five interesting points.

1. *The velocity head in soils is negligible.* The maximum velocity in the four examples was 6 cm/sec. The velocity head for this maximum case is 1.8×10^{-4} m, a value that is indeed negligible.
2. *All head is lost in soil.* Our computation for Fig. 17.8 clearly indicates that in flow through tubes of soils we can neglect completely the head lost in the portions of tube with no soil.
3. *Negative pore pressure can exist.* In the examples given in Figs. 17.8, 17.10, and 17.11 we have pore pressures which are less than atmospheric, so-called

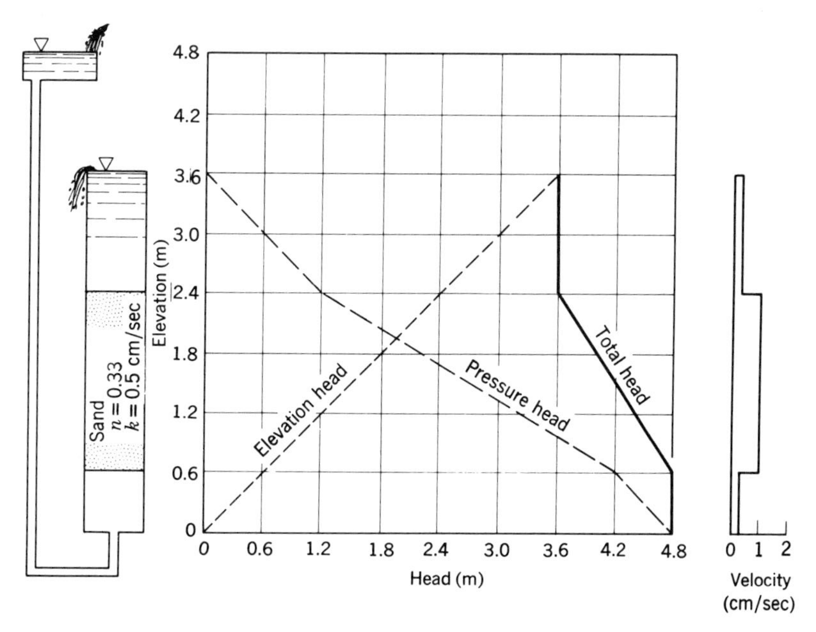

Fig. 17.9 Example of upward flow.

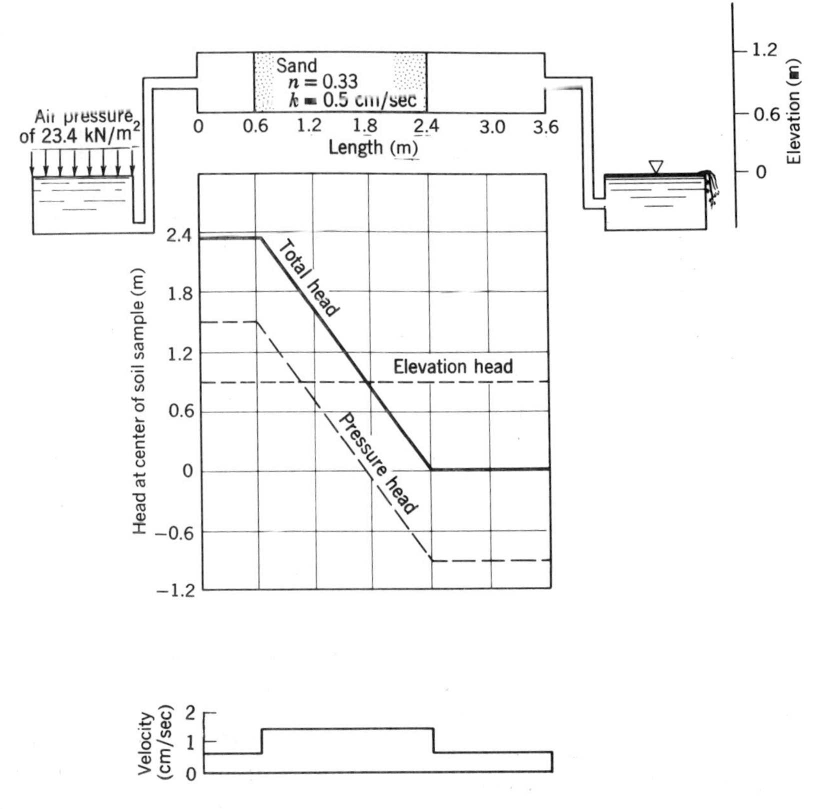

Fig. 17.10 Example of horizontal flow.

negative pore pressures. We already have noted in the preceding section that pore pressures are possible even below absolute zero.

4. *Direction of flow determined by total head difference.* The four examples present cases where flow was toward increasing elevation head, toward decreasing elevation head, toward increasing pressure head, and toward decreasing pressure head. These facts illustrate a point we have already made: it is the total head that determines flow and not pressure head or elevation head alone.
5. *Method of head determination.* Our four examples illustrate the general method of determining heads: first determine elevation and total head, and then

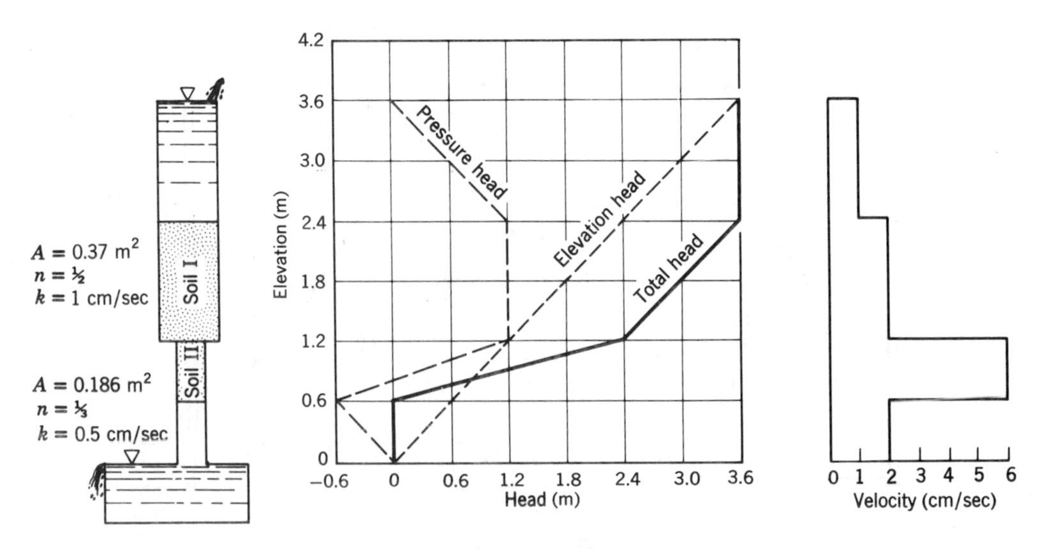

Fig. 17.11 Example of flow through two soils in series.

compute the desired pressure head from these two. In a simple situation, such as Fig. 17.8, we can readily determine the pressure head at any point without knowing the elevation or total head; in a more complex situation such a procedure may not apply. In many practical examples, however, the engineer will have a measure of the pressure head and elevation head, and thus may compute total head from these two.

Let us summarize our discussion of heads by noting that there are three of interest to the soil engineer:

1. *Elevation head.* Its absolute magnitude depends on the location of datum.
2. *Pressure head.* The pressure head magnitude is of considerable importance since it indicates the actual pressure in the water. The pressure head is the height to which the water rises in the piezometer above the point under consideration.
3. *Total head.* The total head is the sum of the elevation and pressure head and is the only head that determines flow. The total head is used in Darcy's law to compute gradient.

17.7 EFFECTIVE STRESS IN SOIL WITH FLUID FLOW

Chapter 16 treated the computation of effective stress for static ground water conditions. Thus far this chapter has considered the computation of heads for simple situations of fluid flow. Let us now combine these two

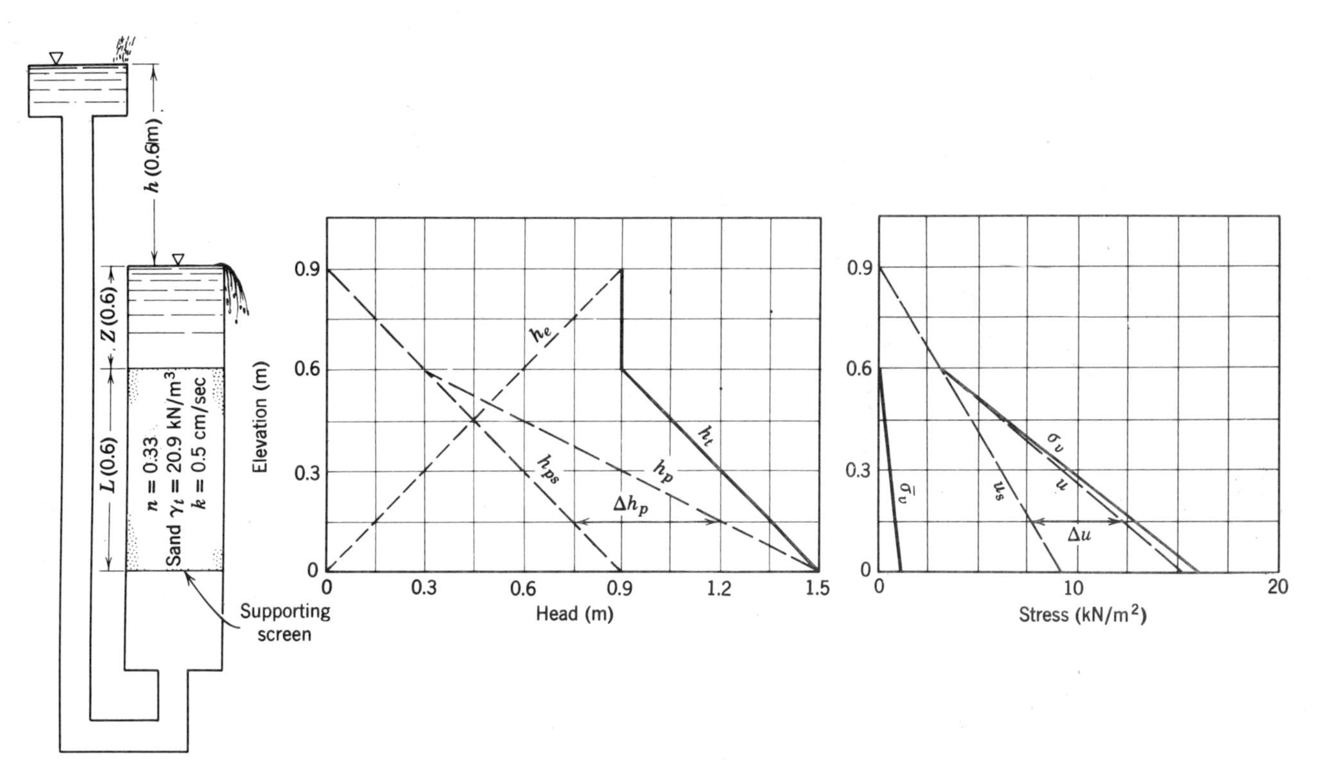

Elevation (m)	$\Delta\sigma_v$ (kN/m²)	σ_v (kN/m²)	u (kN/m²)	$\bar{\sigma}_v$ (kN/m²)
0.9		0	0	0
	$0.3\times\gamma_w = 2.94$			
0.6		2.94	$0.3\times\gamma_w = 2.94$	0
	$0.6\times\gamma_t = 12.54$			
0		15.48	$1.5\gamma_w = 14.72$	0.76

Fig. 17.12 Heads and stresses for fluid flow.

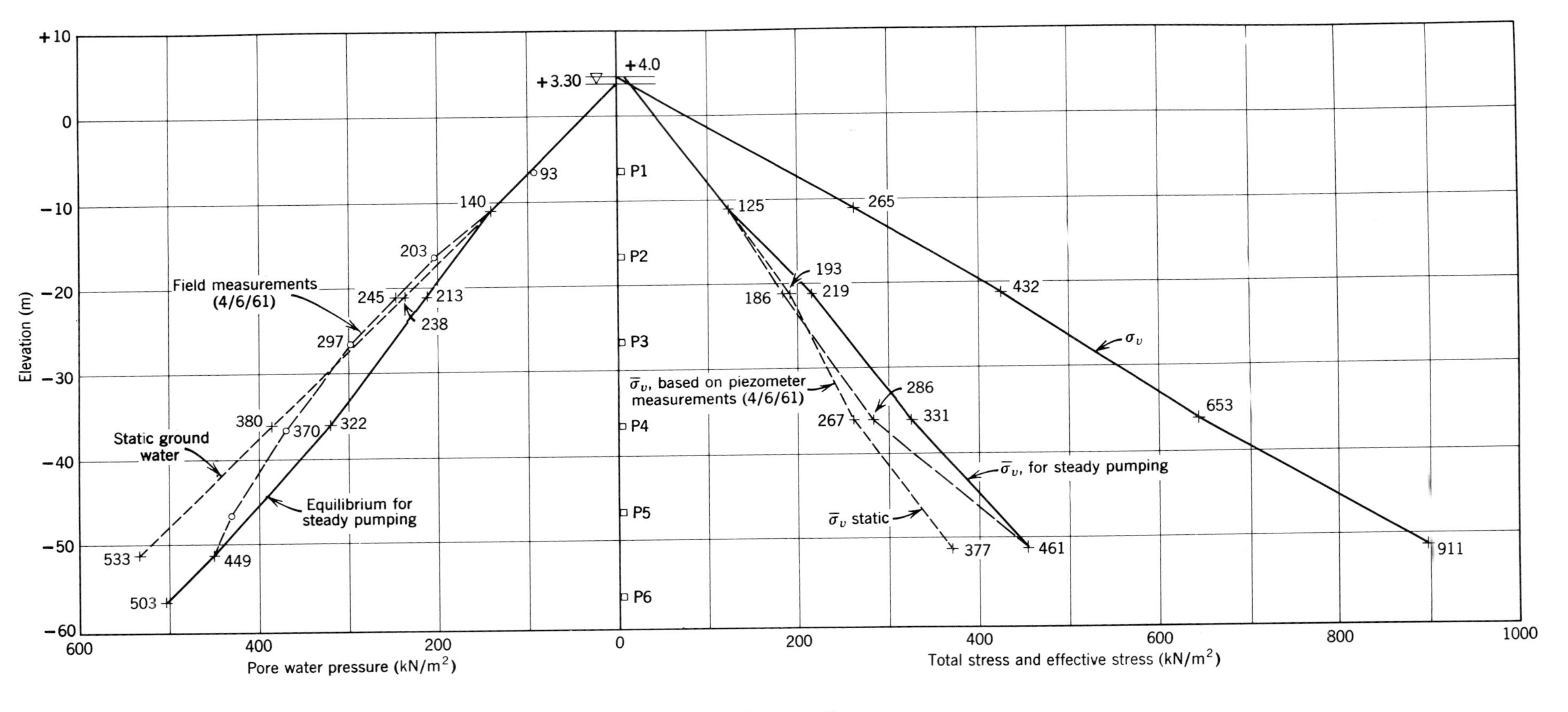

Stress Computation

Elevation	$\Delta\sigma_v$	σ_v	Static		Steady Seepage		Measured	
			u	$\bar{\sigma}_v$	u	$\bar{\sigma}_v$	u	$\bar{\sigma}_v$
+4.0		0	−7	7	−7	7	−7	7
	0.7 × 17.66 = 12							
+3.3		12	0	12	0	12	0	12
	14.3 × 17.66 = 253							
−11.0		265	140	125	140	125	140	125
	10 × 16.68 = 167							
−21.0		432	238	194	213	219	245	187
	15 × 14.72 = 221							
−36.0		653	386	267	322	331	367	286
	15 × 17.17 = 258							
−51.0		911	533	378	449	461	449	461

Fig. 17.13 Refinery subsoil condition (Kawasaki, Japan).

operations to compute effective stresses in soil through which fluid flow is occurring.

Figure 17.12 shows a setup (very similar to that in Fig. 17.9) in which water is flowing vertically upward through a sand. Plotted in this figure are the heads versus elevation and stresses versus elevation.

Pore pressure has to be computed from known flow conditions or be measured. In Fig. 17.12 we have a measurement of pore pressure at elevations 0.6 and 0. Since essentially all total head is lost in the flow through the soil, the tube from elevation 0 back to the head water and the tube from elevation 0.6 to elevation 0.9 serve as piezometers, i.e., the pore pressure at elevation 0 is 1.5 m × 9.81 kN/m³ and at elevation 0.6 is 0.3 m × 9.81 kN/m³. Also plotted are the static pressure head h_{ps} and the static pore pressure u_s.

Let us now return to the consideration of stresses existing in the Kawasaki subsoil used in Example 16.2. Figure 17.13 shows this section with plots of total stress, pore pressure, and effective stress versus elevation. Six piezometers, P_1 to P_6, were installed in the Kawasaki subsoil at the elevations shown and the pore pressures measured by these piezometers are plotted in Fig. 17.13. An investigation showed that a number of wells had been installed in the sand below elevation −51 m. The withdrawal of water from these wells had reduced the piezometric head at elevation −56.5 to a value of 51.3 m. On the basis of this measured pore pressure and the observed phreatic line at elevation +3.3, let us now construct the pore pressure line that would exist for steady seepage of water from the top sand layer down to the bottom sand. For this condition, the head lost in each of the strata can be computed from

$$Q_{\text{sand-silt}} = Q_{\text{clay I}} = Q_{\text{clay II}} = Q_{\text{clay III}}$$

as was done in Fig. 17.11.

On the basis of the computed values of head lost in the clay strata, the plot of pore pressure for steady-state seepage shown in Fig. 17.13 was made. From this plot and the total stress plot, the effective stress plot for steady seepage was made. The details of the computations are shown in the figure.

As shown, the actual pore pressure in the clay is greater than that computed for the steady-state seepage case. This difference in pore pressure was caused by the recent placement of fill at the surface of the site. In Part V of this book, excess pore pressures resulting from the application of a stress to an element of soil will be considered.

17.8 SEEPAGE FORCE

The vertical water pressures acting on the block of soil in Fig. 17.12 are shown in Fig. 17.14. (The horizontal pressures on the vertical faces of the sample cancel.) The

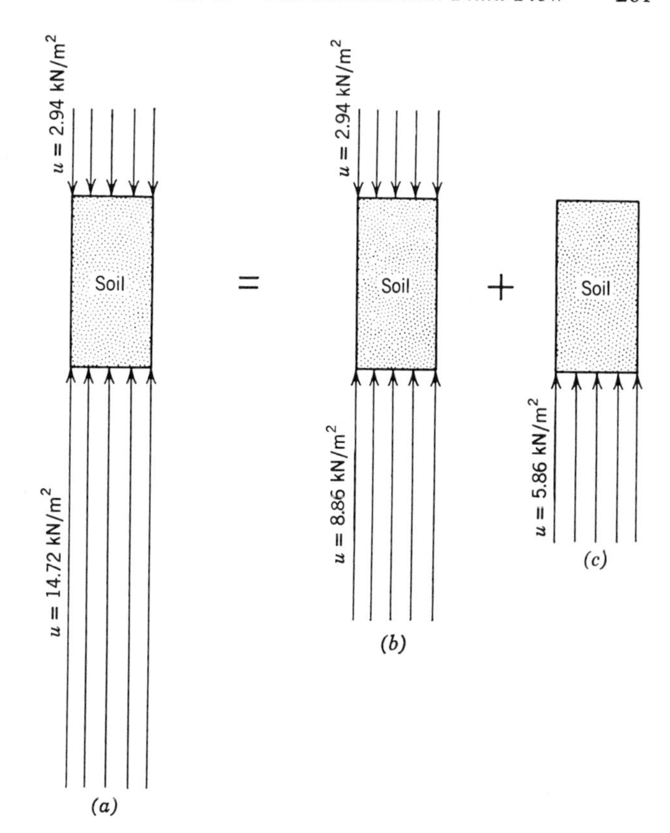

Fig. 17.14 Water pressures on soil sample. (*a*) Boundary water pressures. (*b*) Bouyancy water pressures (static). (*c*) Pressure lost in seepage.

water pressure on the top of the sample is the product of the pressure head and unit weight of water, 0.3 × 9.81 kN/m³ = 2.94 kN/m². Similarly, the upward water pressure on the bottom of the sample is 1.5 m × 9.81 kN/m³ = 14.72 kN/m². These vertical water pressures are those acting on the boundaries of the soil sample and are thus termed *boundary water pressures*.

Figure 17.14*b* shows the water pressures resulting from buoyancy, i.e., the pressures that would exist if there were no flow. These are exactly the same pressures that would act on a similar volume of any material submerged to a depth of 0.3 m in water. These two water pressures give the *static* water effect on our soil sample.

The difference between the total boundary water pressures (Fig. 17.14*a*) and the buoyancy pressures (Fig. 17.14*b*) gives the seepage pressure (Fig. 17.14*c*). The seepage pressure, exerted by the flowing water, is uniformly and completely spent in upward flow through the soil.

The water pressures in Fig. 17.14 have been converted to forces, through multiplying them by the total cross-sectional area of the soil sample, and are shown in Fig. 17.15. The magnitudes of the two buoyancy forces depend on the height Z, but the difference between them does not. The net buoyancy force always acts upward

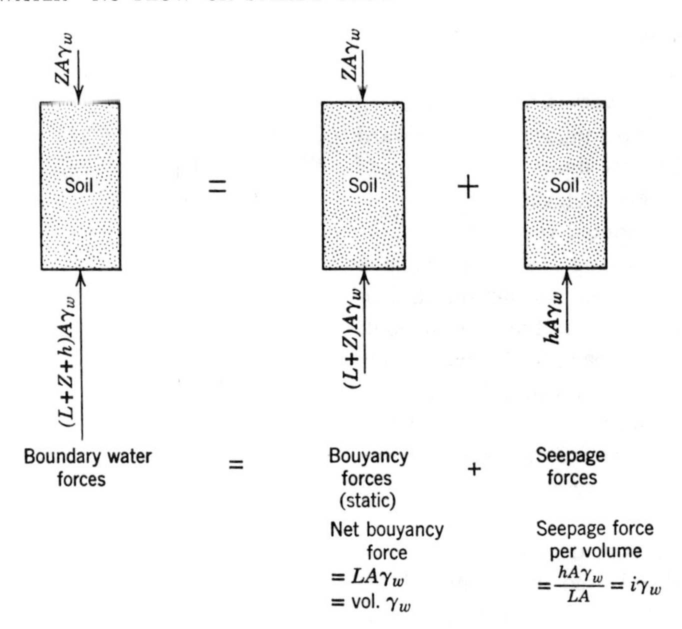

Fig. 17.15 Water forces on soil sample.

and is equal to the total volume of the soil sample times the unit weight of water. This is, of course, Archimedes' principle. Archimedes made his discovery some 2000 years ago while checking the gold content in a crown made for King Hiero II.

The seepage force is applied by the moving water to the soil skeleton through frictional drag. In other words, a pressure related to the loss in total head is transferred from pore pressure to effective stress. In an isotropic soil seepage force always acts in the direction of flow. A convenient expression for the seepage force is the force per unit of volume of soil, thus

$$j = \frac{\text{Seepage force}}{\text{Volume of soil}} = \frac{hA\gamma_w}{LA} = i\gamma_w \tag{17.4}$$

In solving a soil problem we can work either with the total boundary forces or with the buoyancy force. plus seepage force. This fact is illustrated in Fig. 17.16, where

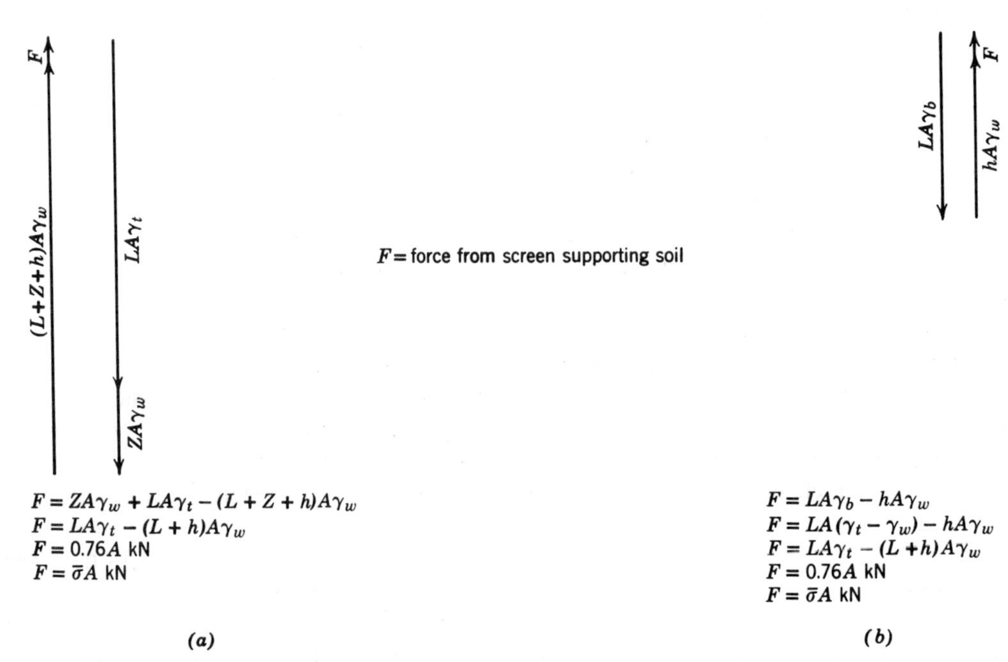

F
$(L+Z+h)A\gamma_w$
$LA\gamma_t$
$ZA\gamma_w$
F = force from screen supporting soil
$LA\gamma_b$
$hA\gamma_w$
F
$F = ZA\gamma_w + LA\gamma_t - (L+Z+h)A\gamma_w$
$F = LA\gamma_t - (L+h)A\gamma_w$
$F = 0.76A$ kN
$F = \bar{\sigma}A$ kN
$F = LA\gamma_b - hA\gamma_w$
$F = LA(\gamma_t - \gamma_w) - hA\gamma_w$
$F = LA\gamma_t - (L+h)A\gamma_w$
$F = 0.76A$ kN
$F = \bar{\sigma}A$ kN
(a)
(b)

Fig. 17.16 Force equilibrium. (*a*) Total weight plus boundary water forces. (*b*) Submerged weight plus seepage force.

the force applied to the supporting screen by the soil sample in Fig. 17.12 is computed first from total soil weight plus boundary water forces, and then by submerged soil weight plus seepage force (it is assumed no force is transferred to the tube by friction between the soil and tube). As can be seen, the two methods give exactly the same answer as, in fact, they must. If boundary water forces are used to compute the equilibrium forces on an element of soil, then the seepage force *must not* be added. To do so means that the effect of flowing water is included twice.

When using boundary water forces plus total weight, we are in effect considering the equilibrium of the entire soil. The seepage force is an internal drag force by water against skeleton and reaction by skeleton against water. This force between phases has no effect on the equilibrium of the overall soil. When using seepage force plus submerged unit weight, we are in effect working with the equilibrium of the mineral skeleton.

In this section we have analyzed a simple flow situation in considerable detail. A number of important and useful principles have been presented for this case; these principles hold for the most complex flow situations. A restatement of these important principles follows.

1. The *boundary water* forces acting on a soil element equal the *buoyancy* force plus the *seepage* force.
2. To analyze the forces acting on an element we can use either *boundary water* forces with *total* weights or *seepage* forces with *submerged* weights. (Although the two methods give exactly the same answer, the use of boundary forces and total weights is nearly always the more convenient method.)
3. The seepage force per unit soil volume j is equal to the total head gradient i times the unit weight of permeant γ. In an isotropic soil the seepage force always acts in the direction of flow.

17.9 QUICK CONDITION

As discussed in Chapter 11, the shear strength of cohesionless soil is directly proportional to the effective stress. When a cohesionless soil is subjected to a water condition that results in zero effective stress, the strength of the soil becomes zero; a *quick condition* then exists. A quick condition is, in other words, one where the shear strength of a soil is zero due to the absence of effective stress. Because cohesive soils can have strength even at zero effective stress, they do not necessarily become quick at zero effective stress.

The effective stress obviously is zero when the pore pressure equals the total stress. There are two common situations in soil mechanics where this equality arises:

1. An upward fluid flow of such magnitude that the total upward water force equals the total soil weight (for an unloaded soil element), i.e., the seepage force equals the submerged soil weight.
2. A shock on certain loose soils which causes a volume decrease in the soil skeleton with the result that the effective stress is transferred to pore pressure (see Part V).

As we saw in Fig. 17.12, the effective stress at the bottom of the soil element was almost zero. As a matter of fact, from the information in Fig. 17.16, we can determine the upward gradient necessary to make the effective stress at the bottom of the soil sample equal to zero. For $\bar{\sigma}$ to be zero,

$$LA\gamma_b - hA\gamma_w = 0$$

or

$$\frac{h}{L} = i = \frac{\gamma_b}{\gamma_w}$$

or

$$i_{\text{critical}} = i_c = \frac{\gamma_b}{\gamma_w} \tag{17.5}$$

The gradient required to cause a quick condition, termed critical gradient i_c, is equal to buoyant weight/unit weight of water. Since the ratio γ_b/γ_w is usually close to unity, the critical gradient is approximately equal to 1. Note that the flow must be vertically upward—opposite in action to the soil unit weight—for Eq. 17.5 to hold; also, Eq. 17.5 requires that the soil element be unloaded—the vertical effective stress in the element with no flow must depend only on the buoyant unit weight. In any soil where strength is proportional to effective stress, an upward gradient of γ_b/γ_w will cause zero strength, or a quick condition.

A more general way to determine the likelihood of a quick condition is to work with the effective stress equation. Consider, for example, the situation shown in Fig. 17.17. The effective stress at point A is

$$\bar{\sigma}_v = \sigma_v - u = (a\gamma_w + b\gamma_t + \Delta q_s) - h\gamma_w$$

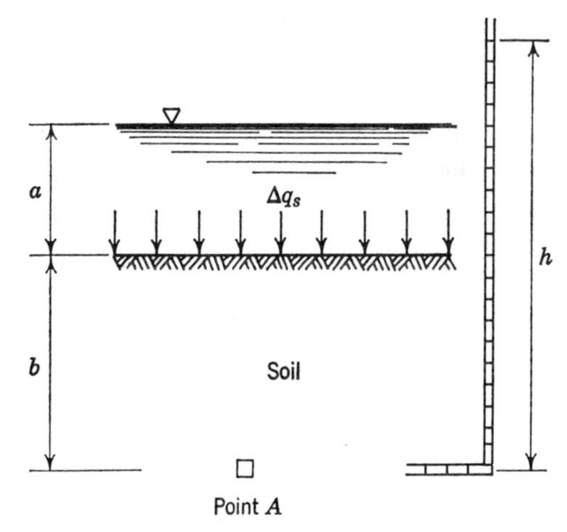

Fig. 17.17 Upward flow in soil.

If the soil is cohesionless, the strength becomes zero when $\bar{\sigma}_v$ becomes zero. For the special case of $\Delta q_s = 0$ and $a = 0$, our equation reduces to

$$\bar{\sigma}_v = b\gamma_t - h\gamma_w$$

and

$$i = \frac{\Delta h_t}{b} = \frac{h - b}{b} \qquad \text{or} \qquad h = b(i + 1)$$

and

$$\bar{\sigma}_v = b\gamma_t - b(i + 1)\gamma_w$$

For quick condition, $\bar{\sigma}_v = 0$ and $i = i_c$, so that

$$\gamma_t = \gamma_w + i_c\gamma_w$$

and

$$i_c = \frac{\gamma_t - \gamma_w}{\gamma_w} = \frac{\gamma_b}{\gamma_w}$$

This expression is the same as Eq. 17.5.

The nontechnical literature[3] abounds with stories of "quicksand" sucking victims beneath the surface of the soil and drowning them. In fact, quicksand is a liquid of unit weight twice that of water. There is therefore no suction and a person would float with about half of his body out of quicksand.

In summary, we can note:

1. "Quick" refers to a condition and not a material.
2. Two factors are required for a soil to become quick: strength must be proportional to effective stress and the effective stress must be equal to zero.
3. The upward gradient needed to cause a quick condition in unloaded cohesionless soil is equal to γ_b/γ_w and is approximately equal to 1.
4. The amount of flow required to maintain a quick condition increases as the permeability of the soil increases.

17.10 THE VALIDITY OF DARCY'S LAW

The material in this chapter is based on Darcy's law. Let us now consider the conditions wherein this expression is valid.

Studying flow through pipes, Reynolds found a critical velocity, v_c, which he expressed in terms of the Reynolds number R, where

$$R = \frac{v_c D\gamma_w}{\mu g}$$

where

D = pipe diameter

γ_w = unit weight of water

μ = viscosity of water

g = acceleration of gravity

[3] See, for example, "Quicksand—Nature's Terrifying Death Trap," *Reader's Digest*, p. 140 (Dec. 1964).

Many experimenters have attempted to use Reynolds' concept to determine the upper limit of the validity of Darcy's law. This work is described and discussed by Muskat (1946) and Scheidegger (1957). In soils, D is taken as the average particle diameter or the average pore diameter. The values of R for which the flow in porous media become turbulent have been measured as low as 0.1 and as high as 75 (see Scheidegger, 1957). Scheidegger feels the main reason that porous media do not exhibit a definite critical Reynolds number is probably because soil cannot accurately be represented as a bundle of straight tubes.

The value of D corresponding to a Reynolds number of 1 is approximately $\frac{1}{2}$ mm, which is in the coarse sand range. At any rate, coarse sand appears to be the most pervious soil through which laminar flows occurs.

There is some experimental evidence to suggest that in soils of very low permeability a threshold gradient of as much as 20 to 30 may be required to initiate flow. However, much of these data have been questioned. Scheidegger (1957) discusses several reasons that flow through very small openings may not follow Darcy's law.

There is an overwhelming mass of undisputed evidence which shows that Darcy's law holds in silts through medium sands. The same can be said for steady-state flow through clays. For soils more pervious than a medium sand, the actual relationship between gradient and velocity should be determined experimentally for the particular soil and void ratio under study.

17.11 SUMMARY OF MAIN POINTS

1. In soils finer than coarse sands $v = ki$.
2. There are three heads of importance to flow through porous media: elevation head, pressure head, and total head.
3. Flow depends on differences in total head.
4. The seepage force per volume of soil is $i\gamma_w$ and (for isotropic soil) acts in the direction of flow.
5. The equilibrium of a soil element can be evaluated on the basis of boundary water forces and total weights *or* of seepage forces and submerged weights.
6. "Quick" refers to a condition wherein a cohesionless soil loses its strength because the upward flow of water makes the effective stress become zero.

PROBLEMS

17.1 For the flow situation shown in Fig. 17.9, compute the vertical effective stress in the sand at elevation +1.2 m $G = 2.60$ and $S = 100\%$.

17.2 In a certain sand deposit the water table is at the ground surface. Compute the total stress, pore pressure, and effective stress on a horizontal plane at a depth of 4.5 m for each of the following cases:

a. Static ground water.
b. Upward flow under a gradient of $\frac{1}{2}$.
Make and list reasonable assumptions for any data needed.

17.3 A jar 100 cm high and 10 cm² in cross-sectional area is filled with soil and water having an overall average unit weight of 10.57 kN/m³. The specific gravity of the soil is 2.80. For each of the following three cases compute σ_v, u, and $\bar{\sigma}_v$ at the bottom of the jar:
a. Uniform slurry.
b. Sediment of soil 5 cm thick and sea water, $\gamma_w = 10.07$ kN/m³.
c. Sediment of soil 6 cm thick and pure water, $\gamma_w = 9.81$ kN/m³.
For cases (*b*) and (*c*) compute the void ratio of the sediments.

17.4 For the setup shown in Fig. P17.4, plot to scale elevation head, pressure head, total head, and seepage velocity versus distance along the sample axis.

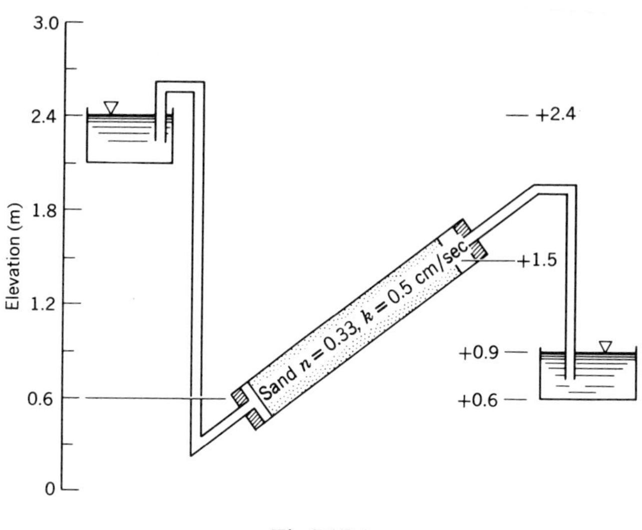

Fig P17.4

17.5 For the setup shown in Fig. P17.5, compute the vertical force exerted by the soil on screen *A* and that on screen *B*. Neglect friction between the soil and tube. $G = 2.75$.

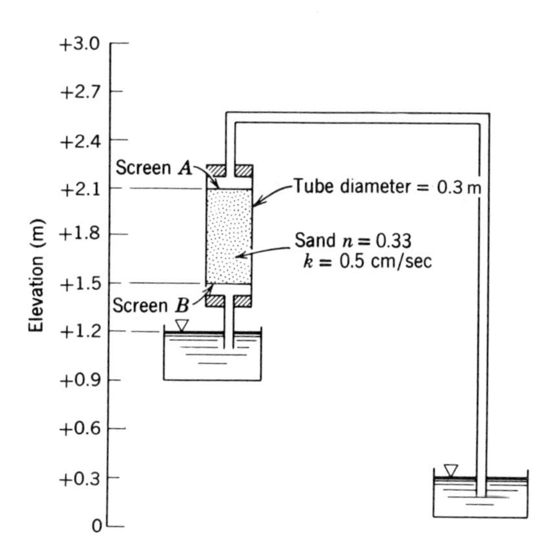

Fig. P17.5

17.6 In the profile shown in Fig. P17.6, steady vertical seepage is occurring. Make a scaled plot of elevation versus pressure head, pore pressure, seepage velocity, and vertical effective stress. Determine the seepage force on a 0.3 m cube whose center is at elevation −4.5 m. *G* for all soils = 2.75.

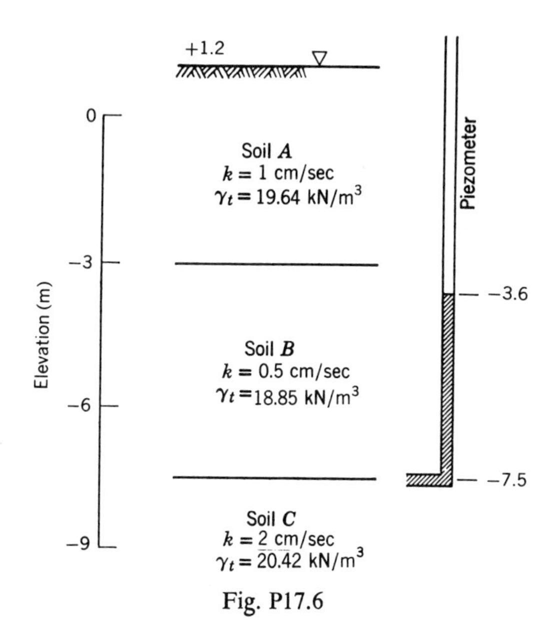

Fig. P17.6

CHAPTER 18

Two-Dimensional Fluid Flow

Chapter 17, which introduced fluid flow in soils, was confined to one-dimensional flow. This chapter considers two-dimensional flow, including the cases of non-homogeneous and anisotropic soil. The following approach is used: (*a*) the *flow net* is introduced in an intuitive manner using a one-dimensional flow situation such as was treated in Chapter 17; (*b*) the flow net solution of several two-dimensional problems is given; (*c*) the basic equation for flow through soil is derived (this equation shows the theoretical basis for the flow net); and (*d*) the basic flow equation is employed to study the seepage of fluid through anisotropic soil.

18.1 FLOW NET FOR ONE-DIMENSIONAL FLOW

Figure 18.1*a* shows a tube 1.2 m × 1.2 m in cross section by 4.8 m high through which steady-state vertical flow is occurring. This flow situation is similar to those studied in Chapter 17 (e.g., Fig. 17.8). The values of total head, elevation head and pressure head are plotted in Fig. 18.1*b*. The rate of seepage through the tube, as computed by Darcy's law, is equal to

$$Q = kiA = 0.05 \times 10^{-2} \text{ m/sec} \times \tfrac{4.8}{3.0} \times 1.2 \text{ m} \times 1.2 \text{ m}$$
$$= 0.00\,115 \text{ m}^3/\text{sec}$$

If we placed a dye on the top of the soil (elevation 3.6) and traced on a macroscopic scale the movement of the dye through the soil we would get a vertical *flow line* or *flow path* or *streamline*, just as was shown in Fig. 17.1. That is, each drop of water that goes through the soil follows a flow line. In the 1.2 m × 1.2 m tube we have an infinite number of flow lines. For convenience, five flow lines are shown on one vertical cross section in Fig. 18.1*c*; three flow lines (indicated by solid lines) are at the quarter points and one flow line is at each vertical boundary of the soil. These five flow lines divide the area into four *flow channels* of equal dimension, 0.3 m wide. Since all flow is vertical and thus all flow lines are parallel, there is no flow from one channel into another. We can thus determine the total quantity of flow through the tube by multiplying the quantity of flow in one channel by the number of channels. In the tube in Fig. 18.1*c*, there are four flow channels per 0.3 m perpendicular to the page; thus there is a total of 16 channels.

In the sketch in Fig. 18.1*c* are also shown lines along which the total head in the flowing water is a constant. The values of total head h in Fig. 18.1*b* have been recorded at the right of each horizontal line in Fig. 18.1*c*. These horizontal lines are called *equipotential lines* since they are lines drawn through points of equal total head. Just as there was an infinite number of flow lines there is an infinite number of equipotential lines. Dividing the length of the tube with equipotential lines at equal spaces means that the total head lost between any two pair of adjacent equipotential lines is the same.

A system of flow lines and equipotential lines, as shown in Fig. 18.1*c*, constitutes a *flow net*. In isotropic soil the flow lines and equipotential lines intersect at right angles, meaning that the direction of flow is perpendicular to the equipotential lines. The intersecting flow lines and equipotential lines form an orthogonal net. The simplest pattern of orthogonal lines is one of squares. Whereas any orthogonal pattern can be used in flow nets, the simple system of squares is the one commonly employed. From a flow net the soil engineer can determine three very useful items of information: rate of flow; head; and gradient.

First, let us see how the flow net can be used to determine the rate of flow through the soil. Consider square A in our flow net as noted in Fig. 18.1*c*. The rate of flow q_A through this square is

$$q_A = ki_A a_A$$

The total head lost in square A is equal to H/n_d where H is the total head lost in flow and n_d is the number of head drops in the net, and i_A is equal to $H/n_d l$ where l is the

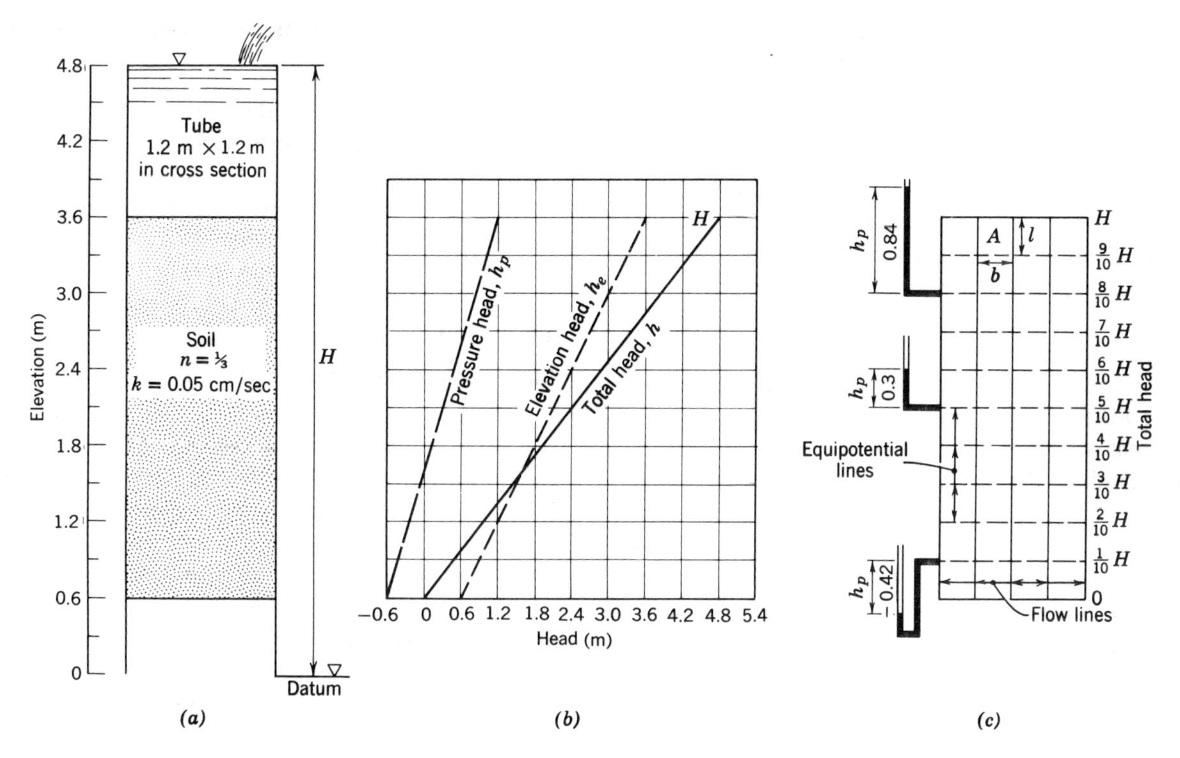

Fig. 18.1 One-dimensional flow.

vertical dimension of A. The cross-sectional area a_A of square A, as seen in plan view, is b as dimensioned in the figure times L where $L = 0.3$ m is the dimension perpendicular to the page. Therefore

$$q_A = \frac{kHb}{n_d l}$$

Since the shapes in the net are square, $b = l$ and $q_A = k(H/n_d)$. Since all of the flow which goes through the flow channel containing square A must pass through square A, the flow through square A is that for the entire flow channel. To obtain the flow per unit length L perpendicular to the page, we must multiply q_A by the number of flow channels, n_f:

$$\frac{Q}{L} = q_A n_f = k\frac{H}{n_d} n_f$$

or

$$\frac{Q}{L} = kH\frac{n_f}{n_d} = kH\$ \tag{18.1}$$

The ratio $\$ = n_f/n_d$ is a characteristic of the flow net[1] and is independent of the permeability k and the total head loss H; it is termed the *shape factor* of the net and is represented by the symbol $\$$. The value of $\$$ for the net in Fig. 18.1c is

$$\$ = \frac{n_f}{n_d} = \frac{4}{10} = 0.4$$

and

$$\frac{Q}{L} = k \times H \times \$ = 0.05 \times 10^{-2}\ \text{m/sec} \times 4.8\ \text{m} \times 0.4$$

$$= .00096\ \text{m}^3/\text{sec/m}$$

and

$$Q = \frac{Q}{L} \times 1.2 = .00096\ \text{m}^3/\text{sec/m} \times 1.2\ \text{m}$$

$$= 0.00115\ \text{m}^3/\text{sec}$$

This value of total seepage is, of course, the same that we got by our initial computation using Darcy's law directly.

Next let us see how we can use the flow net to determine head at any point. Since H is the total head lost in flow and since there are ten equal head drops, $H/10$ is lost in flow from one equipotential to the next. At the right of the flow net are shown the values of total head. It is essential to realize that the equipotentials in the flow net are drawn through points of equal *total head*, since it is the total head that controls flow. Having the total head and the elevation head for any given point, we can readily determine the pressure head, as was done in Chapter 17. For example, consider a point in the soil at elevation 3 where the total head $h = \frac{8}{10} \times H = 3.84$ m and the h_e equals the elevation of the point, 3 m. The

[1] It is not necessary that n_f and n_d be integers. Example 18.4 and Problem 18.7 show nets in which the lowest flow path involves rectangles rather than squares; i.e., these are not full flow paths. In Example 18.4, $n_f \approx 2.6$.

pressure head is therefore equal to $h - h_e = 0.84$ m. The pressure head at any point is the height above the point to which water will rise in a piezometer installed at the point. Thus the water stands 0.84 m above elevation 3 in the piezometer sketched at the left of the flow net. The pore water pressure at elevation 3 is 0.84 m $\times$ 9.81 kN/m^3 = 8.24 kN/m^3. In similar fashion, the pressure heads at elevations 2.1 and 1 are 0.3 m and -0.42 m, respectively, and are shown at the left of the flow net.

Finally, let us use our flow net to determine the gradient at any point in the net. The value of gradient i for any square is equal to the total head lost in the square divided by the length of the square, $i = \Delta h/l$. Since for the flow net in Fig. 18.1c all of the squares are the same size, the gradient for any square is equal to $\Delta h/l$ equal to $H/(10 \times 0.3) = 1.6$.

Thus by the techniques described in the preceding paragraphs, a flow net can be used to determine the rate of flow, the head at any point, and gradient at any point. The example selected for this demonstration is so simple that these quantities could have been obtained easily without the flow net. The technique employed with the flow net to obtain these values holds for the most complex net, whereas the simple techniques described in Chapter 17 are not practical for complex two-dimensional flow systems. Hence the purpose of the example in Fig. 18.1 is to show what a flow net is and how it is used, and, further, to show that the values of flow, head, and gradient are exactly correct when obtained from the flow net if the net used for the determination is itself exactly correct.

18.2 FLOW NET FOR TWO-DIMENSIONAL FLOW

This section discusses the use of flow nets for three situations involving two-dimensional fluid flow. The first, flow under a sheet pile wall, and the second, flow under a concrete dam, are cases of confined flow since all the boundary conditions are completely defined. The third, flow through an earth dam, is unconfined flow since the top flow line is not definitely defined (in advance of constructing the flow net).

Flow under Sheet Pile Wall

Chapter 1 noted several actual civil engineering problems involving a sheet pile wall: a wall to retain a building excavation, the wall around the marine terminal, the anchored bulkhead for the ship dock, etc. Methods of analyzing the stability of such a wall, ignoring the effects of water, were discussed in Chapter 13. Chapter 23 will discuss stability computations which include the effects of water. Example 18.1 shows a sheet pile wall driven into a silty soil having a permeability of 0.5×10^{-6} cm/sec. The sheet pile wall runs for a considerable length in a direction perpendicular to the page and thus the flow underneath the sheet pile wall is two-dimensional.

The boundary conditions for the flow under the sheet pile wall are: kb, upstream equipotential; hl, downstream equipotential; beh, flow line; mn, flow line. Within these boundaries the flow net shown has been drawn. Having the flow net, we can compute the seepage under the wall, the pore pressure at any point in the subsoil, and the gradient at any location in the

---

▶ Example 18.1

Given. Flow net in Fig. E18.1.
Find. Pore pressures at points a to i; quantity of seepage; exit gradient.
Solution.

Point	Elevation Head (m)	Total Head (m)	Pressure Head (m)	Water Pressure (kN/m^2)
a	27.0	27.0	0	0
b	18.0	27.0	9.00	88.3
c	14.7	26.06	11.36	111.4
d	11.7	25.13	13.43	131.7
e	9.0	23.25	14.25	139.8
f	11.7	21.38	9.68	95.0
g	14.7	20.44	5.74	56.3
h	18.0	19.5	1.50	14.7
i	19.5	19.5	0	0

Seepage under wall:

$$\frac{Q}{L} = kH \text{ [illegible] } = 5 \times 10^{-9} \text{ m/sec} \times 7.5 \times \tfrac{4}{8}$$

$$= 18.75 \times 10^{-9} \text{ m}^3/\text{sec(m)}$$

Exit gradient:

$$i = \frac{\Delta h}{l} = \frac{7.5/8}{3.45} = 0.27$$

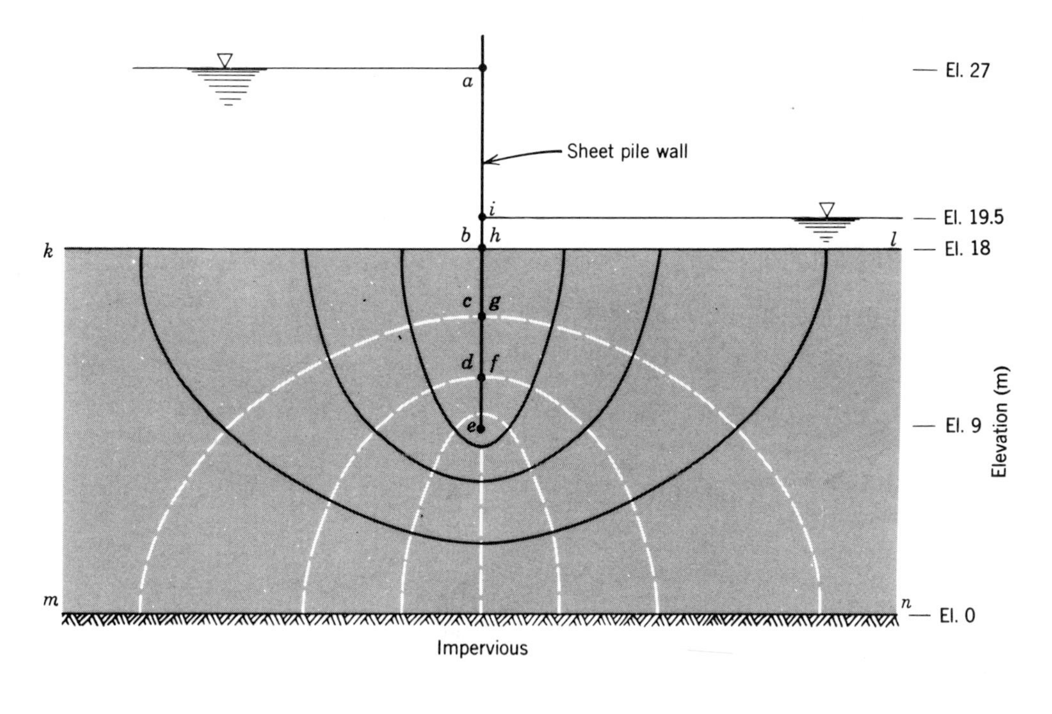

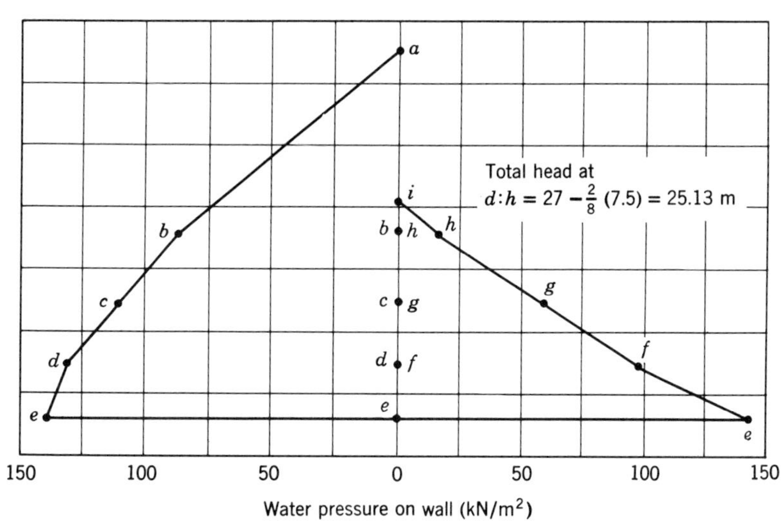

Fig. E18.1 Flow under sheet pile wall.

---

subsoil. These computations are performed in Example 18.1.

A water pressure plot, such as that in Example 18.1, is useful in the structural design of the wall and in a study of the water pressure differential tending to cause leakage through the wall.

The critical part of the section for possible liquefaction is where the flow near the ground surface is upward and the gradient the maximum. For the sheet pile wall this occurs at point h—the flow is upward, as indicated by the vertical flow lines, and the vertical gradient is a maximum here, as indicated by the fact that the square at h is the smallest for any square through which the flow is vertically upward. The gradient in the square next to the wall is equal to 0.27, as the computation in the example shows. Even though this gradient is considerably below that necessary to cause a quick condition (approximately 1), it is relatively high. A large factor of safety against a quick condition on the downstream portion of a structure can economically be obtained and is usually desirable because of the seriousness of a quick condition, and the fact that minor variations in soil might cause relatively large errors in the computation for exit gradient.

Flow under Concrete Dam

Example 18.2 shows a concrete dam resting on an isotropic soil having a permeability of 0.05 cm/sec. The section shown is actually a spillway, since water flows over the dam at certain periods. At the present time, the upstream water is at elevation 28.2 and the tail water at 20.4. The lines *AB* and *GH* represent impervious cutoff walls, usually formed by driving sheet piling into the soil.[2] Using the same principles as were used for the example involving flow under the sheet pile wall, the seepage under the dam, the pore pressure head along the base of the dam, and the gradient in the figure *X* have been computed as shown in Example 18.2.

Example 18.3 shows three concrete dams resting on pervious soil. The three cases shown are identical except in case I there is no underground cutoff, case II has a sheet pile cutoff at the upstream face, and case III has a cutoff at the downstream face. The flow nets for the three cases are shown. The flow net for case I is symmetrical about the center line of the dam and the flow net for cases II and III are identical but reversed. For each of the three cases, the table gives the shape factor of the flow net, the quantity of underseepage, the exit gradient, and the uplift pressure head at point *A*. From the flow nets and the results in the table, we can readily make a comparison among the three cases. Dams II and III have the same amount of underseepage, which is less than that for dam I. Dam III has the lowest exit gradient, indicating that the downstream toe of the dam is the most effective cutoff location for reducing exit gradient. Dam III, however, has the greatest uplift pressure.

Example 18.3 well illustrates how powerful a tool a flow net is in developing a design. The engineer can readily evaluate various schemes.

Flow through Earth Dam

Example 18.4 shows a cross section of an earth dam resting on an impervious foundation. Also shown is the flow net for the steady-state seepage through the dam. The line *AB* is the upstream equipotential, and *AD* is a flow line. These two boundary conditions are definitely determined. The line *BC* is a boundary flow line and has the special characteristic that at all points on the line the pressure head is zero; thus it is a phreatic line. Hence the difference in total head between two equipotentials must equal the change in elevation between the points where these equipotentials intersect the boundary flow line. In other words, *BC* is a flow line along which the total head is equal to the elevation head. The location of this top flow line is not known until the flow net is constructed. The line CD is neither equipotential nor flow line, but total head equals elevation head everywhere on CD.

[2] Although such sheet piles are generally assumed to be impervious, actually they often are far from being so.

If there were no rock toe in the dam in Example 18.4, the top flow line would exit on the downstream slope of the dam, as shown in Fig. 18.2*a*. The face *AB* would gradually erode away—the water flowing out of the face will carry soil particles with it. This process will eventually cause the entire dam to fail. In order to prevent such a failure, it is necessary to provide drains that lower the position of the top flow line. The rock toe of Example 18.4 is one possible form of drain; other common schemes are shown in Fig. 18.2. Design of a satisfactory drainage system is one of the most important problems involved in the design of an earth dam.

Even if drainage has been provided, it still is necessary to consider the stability of the entire downstream slope against a shear rupture. A stability analysis is made by comparing the forces tending to cause movement of a mass of earth—actuating forces—with those tending to resist the movement—resisting forces—as discussed in Chapter 13. One of the actuating forces on the downstream slope is the force of water. This water force can be obtained from a flow net, as is illustrated in Example 18.4. The curved line *DE* is any trial failure surface along which the forces are evaluated. The water pressure head diagram along the curved line *DE* was obtained by the same procedure as was used to plot the water pressure

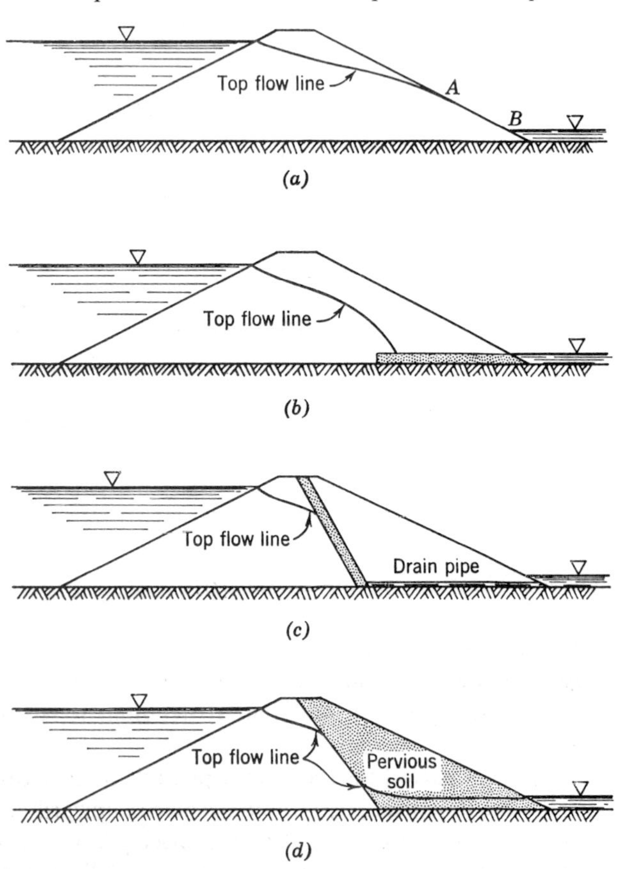

Fig. 18.2 Types of internal drainage for earth dam. (*a*) Homogeneous dam without internal drain. (*b*) Homogeneous dam with underdrain. (*c*) Homogeneous dam with chimney drain. (*d*) Zoned dam.

diagram on the sheet pile wall in Example 18.1 and the uplift diagram on the concrete dam in Example 18.2. The computation for water pressure along the curved line can be facilitated by using the fact that the pressure head on each equipotential where it intersects the boundary flow line is zero. In other words, the water pressure at any point on an equipotential is merely the difference in elevation between the point under consideration and the point where the equipotential intersects the top flow line. This characteristic was employed to draw the pressure diagram in Example 18.4. Proper design of internal drains will reduce the pore pressures within the downstream portion of the dam, and hence will help prevent a shear rupture.

▶ Example 18.2

Given: Flow net in Fig. E18.2

Find: Pressure heads at points A to H; quantity of seepage; gradient in X

Solution: The pressure heads are shown in Fig. E18.2.

Seepage:

$$n_f = 4 \qquad n_d = 12.6 \qquad k = 0.05 \text{ cm/sec} \qquad \$ = \frac{n_f}{n_d} = 0.317$$

$$\frac{Q}{L} = kH\$ = 0.05 \times 10^{-2} \text{ m/sec} \times 7.8 \text{ m} \times 0.317 = 0.00124 \text{ (m}^3\text{/sec)(m)}$$

Gradient in X:

$$i_X = \frac{\Delta h}{l} = \frac{0.62}{3.3} = 0.19$$

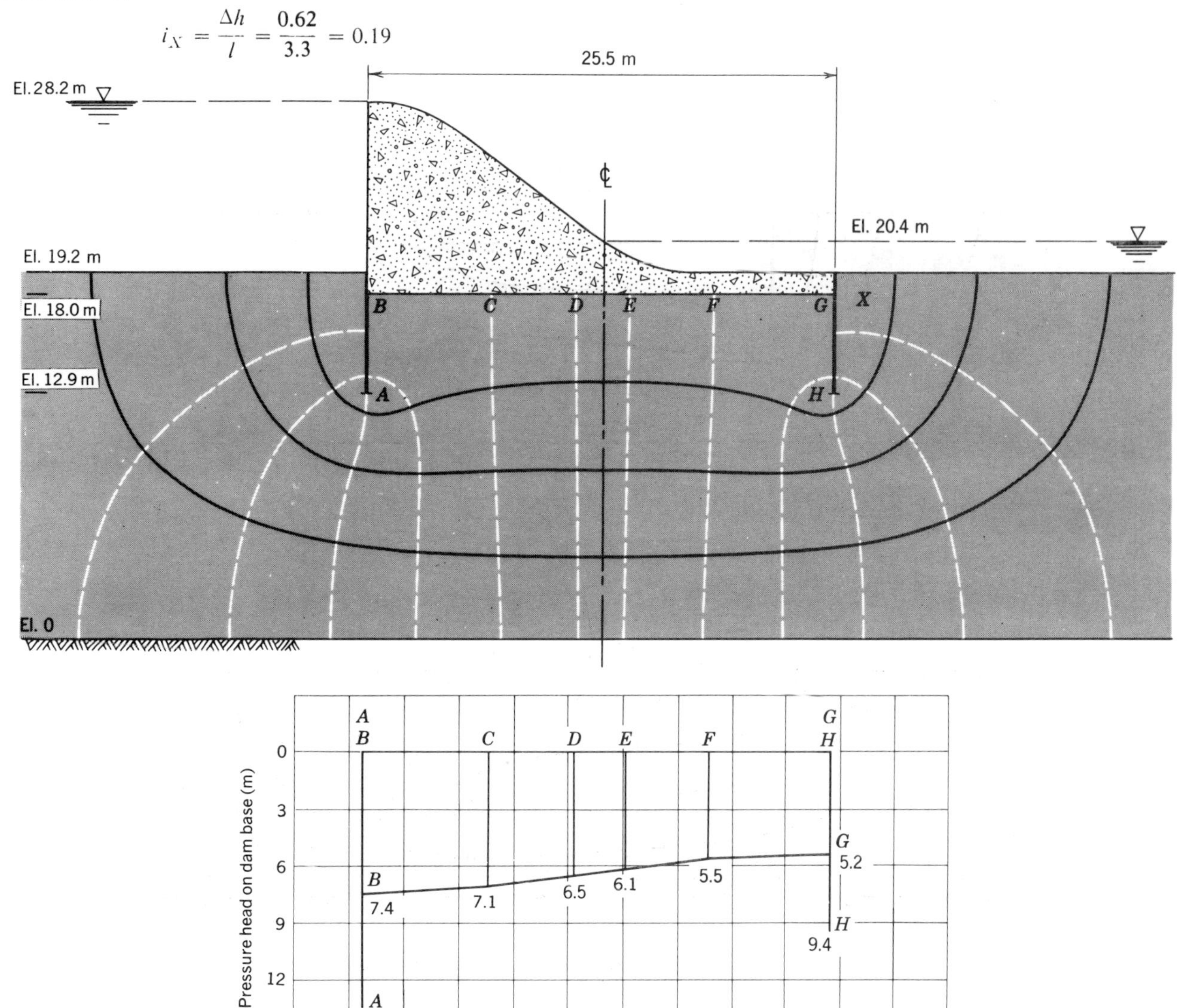

Fig. E18.2 Flow under dam. ◀

▶ **Example 18.3**

For the dams shown in Fig. E18.3, determine the quantity of seepage, the uplift pressure at point A, and the maximum exit gradient.

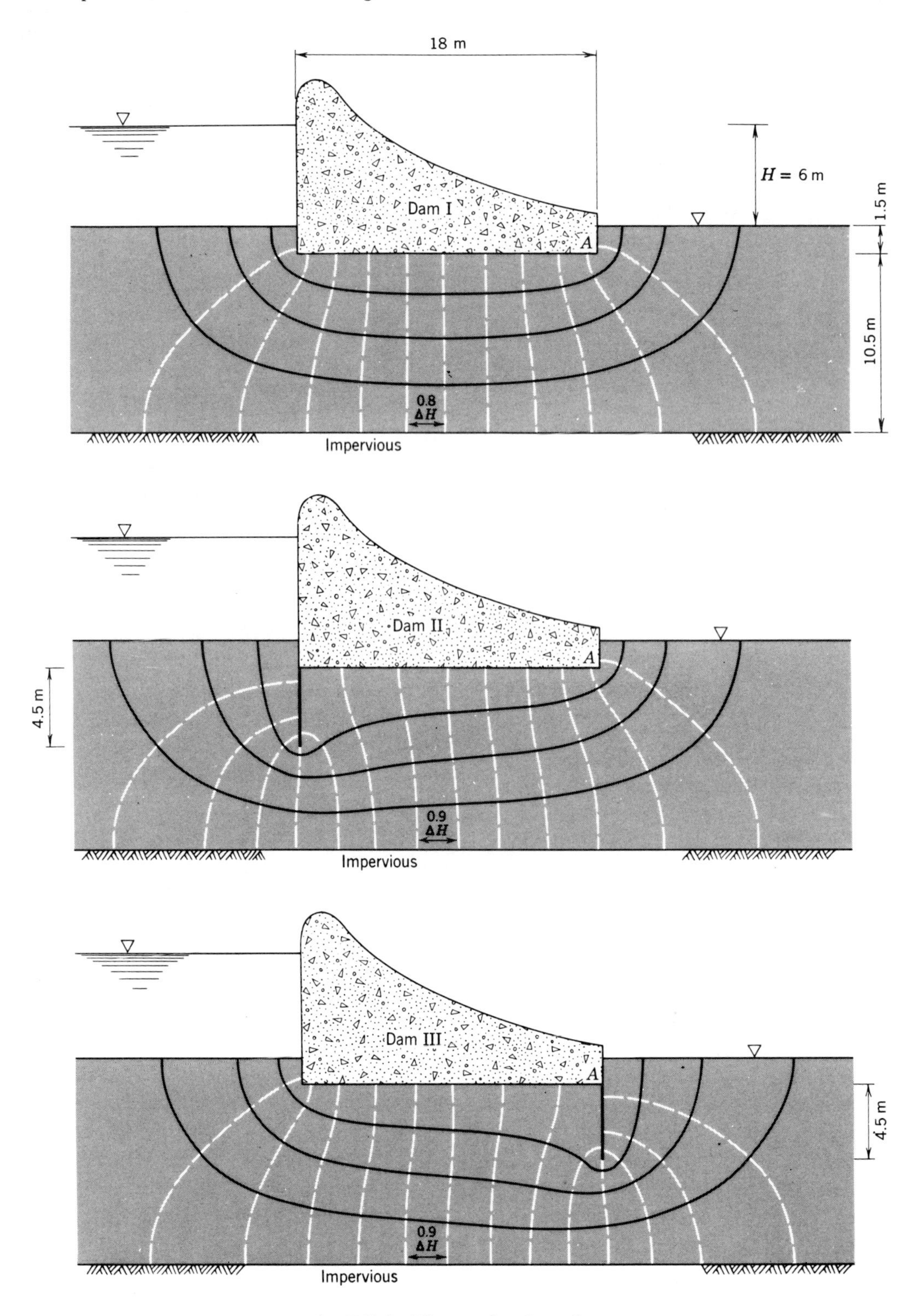

Fig. E18.3 Flow under three dams.

Example 18.3 *continued*

Dam	Shape Factor of Flow Net	Seepage Under Dam [m³/sec(m)]	Uplift Water Pressure Head at A (m)	Exit Gradient
I	4/12	10.29×10^{-6}	2.25	0.42
II	4/14	8.84×10^{-6}	2.13	0.34
III	4/14	8.84×10^{-6}	3.87	0.18

◀

▶ Example 18.4

Given: Flow net in Fig. E18.4

Find: Quantity of seepage, gradient in square I, pore pressures along trial failure surface ED.

Solution: The seepage under the dam is equal to

$$\frac{Q}{L} = kH\$$$

where k = permeability $= 1.52 \times 10^{-4}$ m/sec, $H = 12$ m, $\$ = \frac{n_f}{n_d} = \frac{2.65}{9} = 0.294$

Therefore, $\frac{Q}{L} = 1.52 \times 10^{-4} \times 12 \times 0.294 = 5.36 \times 10^{-4}$ m³/sec(m)

The gradient in the square I equals $i_{\mathrm{I}} = \frac{\Delta h}{l_{\mathrm{I}}} = \frac{12/9}{3.36} = 0.40$

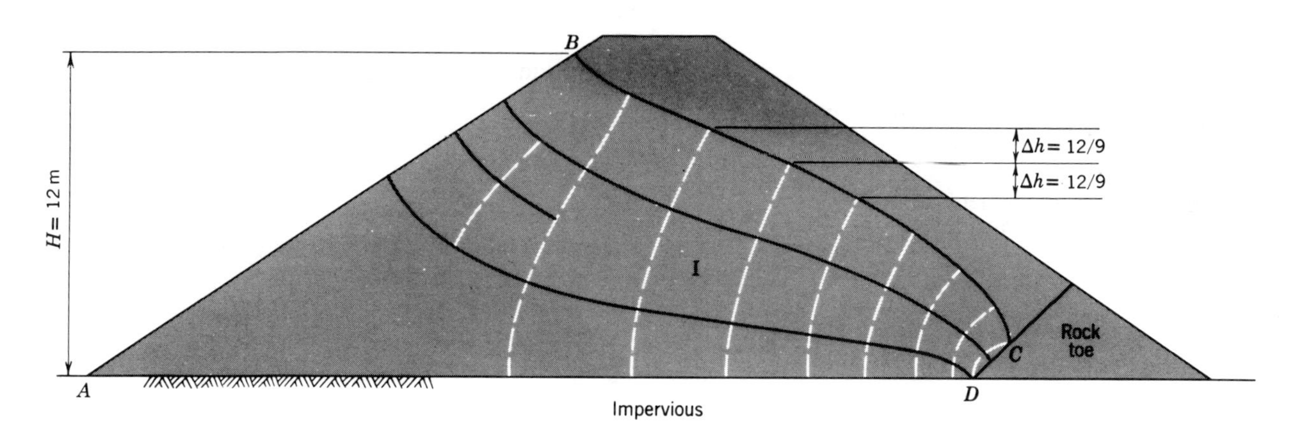

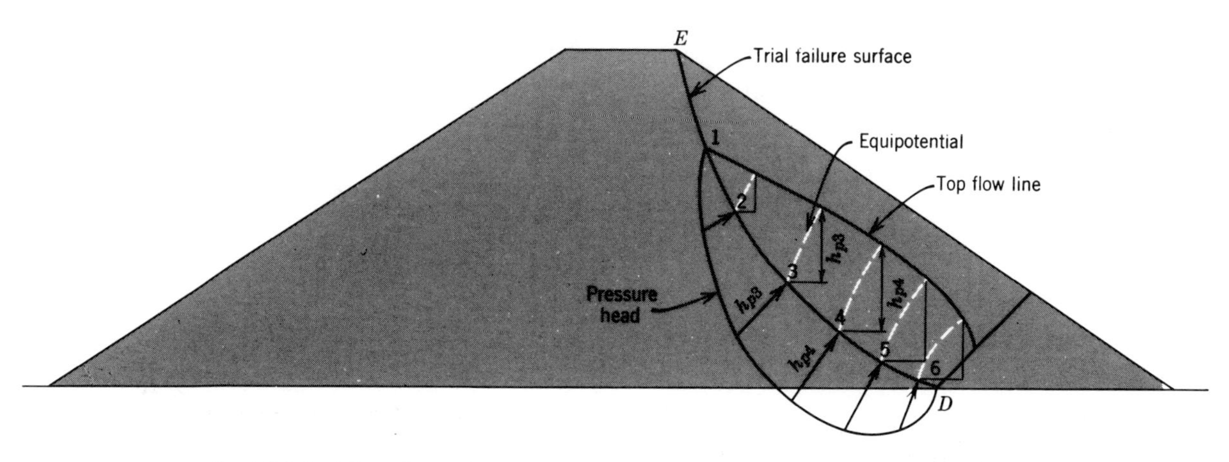

Fig. E18.4 Flow through earth dam (flow net from Corps of Engineers, 1952). ◀

18.3 BASIC EQUATION FOR FLOW IN SOIL

The flow net, which has been used in the preceding two sections, was introduced in an intuitive manner with little theoretical justification. This section derives the equation for flow in soils which forms the basis for the flow net as well as other methods of solving flow problems.

Let us consider an element of soil, Fig. 18.3, through which is occurring laminar flow q with components in the x, y, z directions:

$$q = q_x + q_y + q_z$$

Using Darcy's law we can write the following expressions for the vertical component of flow q_z.

Flow into the bottom of element $q_z = kia$, where a is the area of the bottom face:

$$q_z = k_z\left(-\frac{\partial h}{\partial z}\right) dy\, dx$$

Flow out of the top of element:

$$q_z = \left(k_z + \frac{\partial k_z}{\partial z} dz\right)\left(-\frac{\partial h}{\partial z} - \frac{\partial^2 h}{\partial z^2} dz\right) dy\, dx$$

where

k_z = permeability in z direction at point x, y, z
h = total head

The net flow into the element from vertical flow = Δq_z = flow into bottom − flow out of top

$$\Delta q_z = k_z\left(-\frac{\partial h}{\partial z}\right) dy\, dx - \left(k_z + \frac{\partial k_z}{\partial z} dz\right) \times \left(-\frac{\partial h}{\partial z} - \frac{\partial^2 h}{\partial z^2} dz\right) dy\, dx$$

$$\Delta q_z = \left(k_z \frac{\partial^2 h}{\partial z^2} + \frac{\partial k_z}{\partial z}\frac{\partial h}{\partial z} + \frac{\partial k_z}{\partial z} dz \frac{\partial^2 h}{\partial z^2}\right) dx\, dy\, dz$$

For the condition of constant permeability $\partial k_z/\partial z = 0$,

$$\Delta q_z = \left(k_z \frac{\partial^2 h}{\partial z^2}\right) dx\, dy\, dz$$

Similarly, the net flow in the x direction is

$$\Delta q_x = \left(k_x \frac{\partial^2 h}{\partial x^2}\right) dx\, dy\, dz$$

For the condition of two-dimensional flow $q_y = 0$,

$$\Delta q = \Delta q_x + \Delta q_z = \left(k_z \frac{\partial^2 h}{\partial z^2} + k_x \frac{\partial^2 h}{\partial x^2}\right) dx\, dy\, dz$$

The volume of water V_w in the element is

$$V_w = \frac{Se}{1+e} dx\, dy\, dz$$

and the rate of change of the water volume is equal to

$$\Delta q = \frac{\partial V_w}{\partial t} = \frac{\partial}{\partial t}\left(\frac{Se}{1+e} dx\, dy\, dz\right)$$

Since $dx\, dy\, dz/(1+e)$ = volume of solids in element and is a constant,

$$\Delta q = \frac{dx\, dy\, dz}{1+e}\frac{\partial(Se)}{\partial t}$$

Equating the two expressions for Δq gives

$$\left(k_z \frac{\partial^2 h}{\partial z^2} + k_x \frac{\partial^2 h}{\partial x^2}\right) dx\, dy\, dz = \frac{dx\, dy\, dz}{1+e}\frac{\partial(Se)}{\partial t}$$

which reduces to

$$k_z \frac{\partial^2 h}{\partial z^2} + k_x \frac{\partial^2 h}{\partial x^2} = \frac{1}{1+e}\left(e\frac{\partial S}{\partial t} + S\frac{\partial e}{\partial t}\right) \quad (18.2)$$

Equation 18.2 is the basic equation for two-dimensional laminar flow in soil. Looking at the e and S terms on the right of Eq. 18.2 we see four possible types of flow:

1. e and S both constant.
2. e varies and S constant.
3. e constant and S varies.
4. e and S both vary.

Type 1 is *steady flow* which has been treated in Chapter 17 and this chapter, and types 2, 3, and 4 are *nonsteady flow* situations. Type 2 is *consolidation* for e decrease and *expansion* for e increase, and is considered

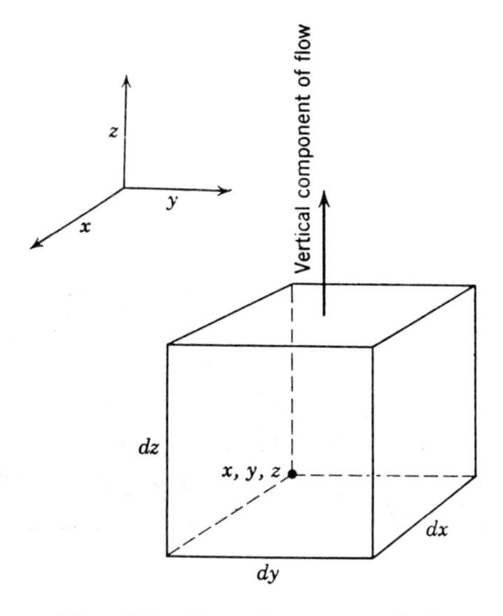

Fig. 18.3 Flow through a soil element.

in Part V. Type 3 is constant-volume *drainage* for S decrease and *imbibition* for S increase. Type 4 includes compression and expansion problems. Types 3 and 4 are complex flow conditions for which satisfactory solutions have not been found.[3]

For steady flow (e and S both constant) Eq. 18.2 reduces to

$$k_z \frac{\partial^2 h}{\partial z^2} + k_x \frac{\partial^2 h}{\partial x^2} = 0 \tag{18.3}$$

and when the condition of permeability is the same in all directions ($k_z = k_x$) Eq. 18.3 reduces to

$$\frac{\partial^2 h}{\partial z^2} + \frac{\partial^2 h}{\partial x^2} = 0 \tag{18.4}$$

Equation 18.4 is Laplace's equation. It says that the change of gradient in the z direction plus the change of gradient in the x direction is zero. The fact that the basic equation of steady flow in isotropic soil, Eq. 18.4, satisfies Laplace's equation means that the flow lines intersect at right angles with the equipotential lines in a flow net. In other words, the flow net as drawn in the preceding two sections is a theoretically sound solution to the flow situation.

18.4 FLOW IN NONHOMOGENEOUS AND ANISOTROPIC SOIL

Although Eq. 18.2 was derived for quite general conditions, the preceding numerical examples considered only soil that does not vary in properties from point-to-point vertically or horizontally—*homogeneous* soil—and soil that has similar properties at a given location on planes at all inclinations—*isotropic* soil. Unfortunately, soils are generally nonhomogeneous and anisotropic. As discussed in Chapter 7, sedimentary soils are built up over a period of many years. During this time, the nature of sediments and the environment of deposition change, with the result that the soil in a deposit varies vertically and, under certain conditions (such as deposition near a shore line), horizontally as well. The actual subsoil profiles shown in Chapter 7 and elsewhere in this book indicate marked variation in soil properties with depth; e.g., Example 16.2 shows a subsoil profile with four different soils within the top 55 m. As noted on this profile, the permeabilities vary for the four soils from 3×10^{-5} cm/sec to 2×10^{-7} cm/sec.

As was discussed in Chapter 8, the process of forming a sedimentary soil is such that the vertical compression is larger than the horizontal compression and thus the horizontal effective stress is about one-half of the vertical effective stress in a normally consolidated soil. Because of the higher vertical effective stress in a sedimentary soil, the platelike clay particles tend to have a horizontal alignment resulting in lower permeability for vertical flow than for horizontal flow.

Because of the variation in nature of sediments in a vertical direction and because of particle alignment, a ratio of horizontal to vertical permeability of two to ten is not unusual in normally consolidated sedimentary clay.

In man-made as well as natural soil, the horizontal permeability tends to be larger than the vertical. The method of placement and compaction in earth fills is such that stratifications tend to be built into the embankments. Ratios of horizontal to vertical permeability in compacted fills tend to be even larger than those in normally consolidated sedimentary clays.

Nonhomogeneous Soil

As a vehicle for considering flow through nonhomogeneous soil consider the two cases shown in Example 18.5. In part (*a*) a 1.2 m layer of soil B having a permeability of 0.005 cm/sec is overlain by a 1.2 m layer of soil A having a permeability of 0.05 cm/sec. In part (*b*), soil A and soil B are placed next to each other with a vertical interface between the two. In each case, steady-state flow is occurring through the soil with a total head of 3.6 m of water being lost.

Example 18.5 shows the seepage and total head drops for each of the two setups. When the flow is perpendicular to the two soil strata, the amount of flow is, of course, the same through both, and the majority of head is lost in the soil with the lower permeability. When the flow is parallel to the two soil strata, the plot of total head lost is the same for both soils and the majority of the flow is through the soil with the higher permeability. From these two simple examples we see that (*a*) for flow perpendicular to strata, the head loss and rate of flow is influenced primarily by the less pervious soils and (*b*) for flow parallel to the strata, the rate of flow is essentially controlled by the more pervious soil.

Figure 18.4 shows a flow channel (part of a two-dimensional flow net) going from soil A to soil B. The permeability of soil A is twice that of soil B. Based on the principle of continuity, i.e., the same rate of flow exists in the flow channel in soil A as in soil B, we can derive the relationship between the angles of incident of the flow paths with the boundary for the two flow channels. This is shown in Fig. 18.4. Not only does the direction of flow change at a boundary between soils with different permeabilities, but also the geometry of the figures in the flow net changes. As can be seen in Fig. 18.4, the figures in soil B are not squares as is the case in soil A, but rather rectangles in which the width of the flow path is twice the distance between equipotentials.

[3] Equation 18.2 is strictly applicable only for small strains.

▸ Example 18.5

Given the two soil-filled tubes in Fig. E18.5, find the quantity of flow and total head vs. elevation.

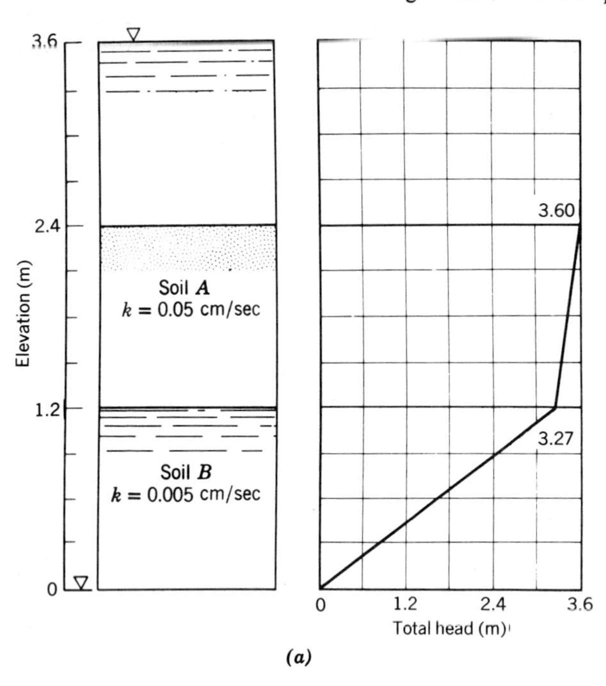

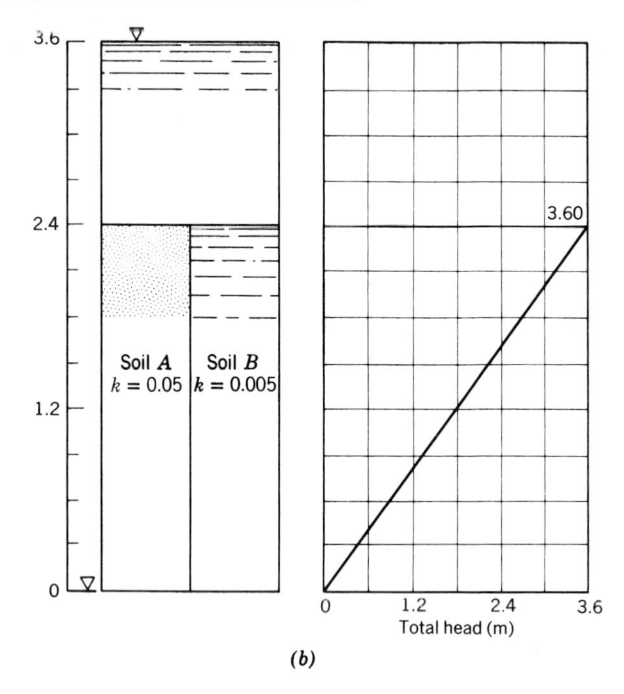

Soil A:

$$\frac{Q}{L} = 0.05 \times 10^{-2}\ \text{m/sec} \times \frac{0.33}{1.2} \times 1.2\ \text{m} = 1.65 \times 10^{-3}\ (\text{m}^3/\text{sec})(\text{m})$$

Soil B:

$$\frac{Q}{L} = 0.005 \times 10^{-2}\ \text{m/sec} \times \frac{3.27}{1.2} \times 1.2\ \text{m} = 1.65 \times 10^{-3}\ (\text{m}^3/\text{sec})(\text{m})$$

Soil A:

$$\frac{Q}{L} = 0.05 \times 10^{-2}\ \text{m/sec} \times \frac{3.6}{2.4} \times 0.6 = 0.45 \times 10^{-3}\ (\text{m}^3/\text{sec})(\text{m})$$

Soil B:

$$\frac{Q}{L} = 0.005 \times 10^{-2}\ \text{m/sec} \times \frac{3.6}{2.4} \times 0.6 = 0.045 \times 10^{-3}\ (\text{m}^3/\text{sec})(\text{m})$$

Fig. E18.5 Flow through nonhomogeneous soil. ◂

Anisotropic Soil

The Laplace equation for flow, Eq. 18.4, was based on the permeability being the same in all directions. Before stipulating an isotropic soil in the derivation of the Laplace equation, we had

$$k_z \frac{\partial^2 h}{\partial z^2} + k_x \frac{\partial^2 h}{\partial x^2} = 0 \tag{18.3}$$

We can reduce Eq. 18.3 to the form

$$\frac{\partial^2 h}{\partial z^2} + \frac{\partial^2 h}{(k_z/k_x)\,\partial x^2} = 0$$

and we can further reduce this equation to

$$\frac{\partial^2 h}{\partial z^2} + \frac{\partial^2 h}{\partial {x_T}^2} = 0 \tag{18.5}$$

where

$$x_T = \left(\frac{k_z}{k_x}\right)^{1/2} x \tag{18.6}$$

In other words, if we transform all x dimensions in our cross section by using Eq. 18.6, we get Eq. 18.5, which is a Laplace equation. We can therefore prepare for any anisotropic subsoil profile a flow net by first performing a transformation and then sketching the net on the transformed section. The permeability to be used with the transformed section is

$$k_e = \sqrt{k_x k_z} \tag{18.7}$$

where k_e is the effective permeability.

From a transformed section, we can determine directly the rate of seepage using Eq. 18.1 with the substitution of effective permeability k_e for k (see Fig. 18.5). Further, the transformed section can be used to determine the pressure head at any point. When determining gradient, however, it is important to remember that the dimensions on the transformed section must be corrected in computing the distance over which a given total head is lost. This can be seen from Fig. 18.6, which shows a portion of a flow net in anisotropic soil. At the left is the flow net in transformed section where the figures are squares. In the natural section, the figures are no longer squares but parallelograms. To

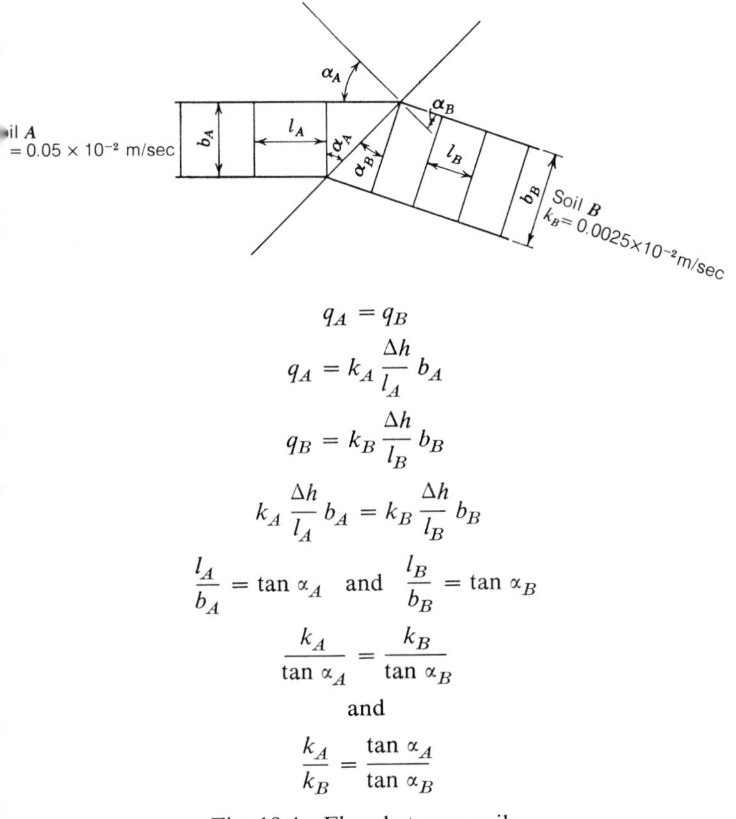

Fig. 18.4 Flow between soils.

compute the gradient occurring in the net, we divide the head loss between equipotentials by the distance l_N, which is the perpendicular distance between equipotentials on a natural scale, not by l_T, which is the distance between equipotentials on a transformed scale.

Figure 18.6 also illustrates the important concept that flow is perpendicular to the equipotentials only in isotropic soil. As can be seen in the natural section of the flow pattern, flow is not perpendicular to the equipotential in Fig. 18.6.

18.5 METHODS OF SOLVING FLOW PROBLEMS

This chapter on steady-state flow has treated the use of the flow net to obtain the amount of seepage through and/or under a structure, and the pore pressure and the gradient at any point in the flow net. Although the fundamental principles on which the flow net is based have been presented, little has been said about the method of obtaining the flow net. Let us now briefly consider methods for obtaining the flow net, and other methods of solving seepage problems.[4]

In the simple, one-dimensional, steady-state flow situations treated in Chapter 17 and at the start of this chapter, the seepage, gradient, and head at any point could be obtained merely by using Darcy's law and the expression (total head = elevation head + pressure head). In two-dimensional flow problems, these two principles alone are insufficient to make solutions. The derivation of the basic equation for flow in soil resulted in Eq. 18.4, Laplace's equation. This derivation involved the original two expressions—Darcy's law and $h = h_e + h_p$. The four methods of solving flow problems described next all are based on Laplace's equation. The aim of all of these methods is to obtain the flow net for the given problem.

Flow Net Sketching

A flow net for a given cross section is obtained by first transforming the cross section (if the subsoil is anisotropic), noting the boundary conditions, and then sketching the net by trial and error. The flow lines and equipotential lines must intersect one another at right angles and the various rules concerning boundary conditions and interfaces between zones of different permeability must be observed.

Flow net sketching was first suggested by Forchheimer and further developed by A. Casagrande (1937). This method has the desirable feature of helping the sketcher develop a feel for the problem. The sketcher can readily

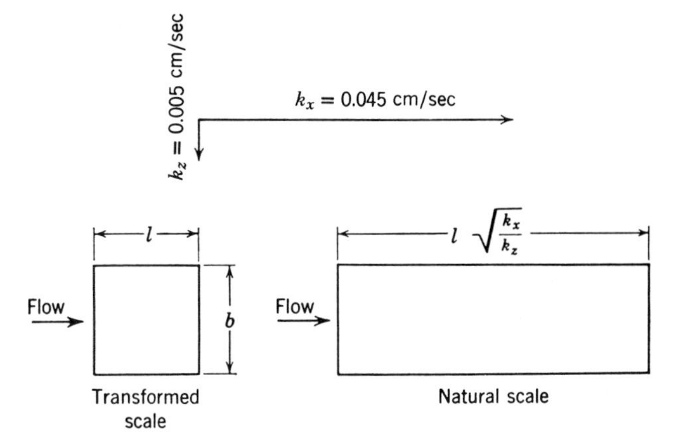

Transformed section:

$$q_T = k_e i a = k_e \frac{\Delta h}{l} b = k_e \, \Delta h$$

Natural section:

$$q_N = k_x i a = k_x \frac{\Delta h}{l\sqrt{k_x/k_z}} b = k_x \frac{\Delta h}{\sqrt{k_x/k_z}}$$

Since $q_T = q_N$,

$$k_e \, \Delta h = k_x i a = k_x \frac{\Delta h}{\sqrt{k_x/k_z}}$$

$$k_e = \sqrt{k_x k_z}$$

Fig. 18.5 Flow in anisotropic soil.

[4] A detailed treatment of methods of solving flow problems is beyond the scope of this book. The reader interested in such information is referred to Harr (1962) or Scott (1963). Cedergren (1967) gives a good treatment of flow nets.

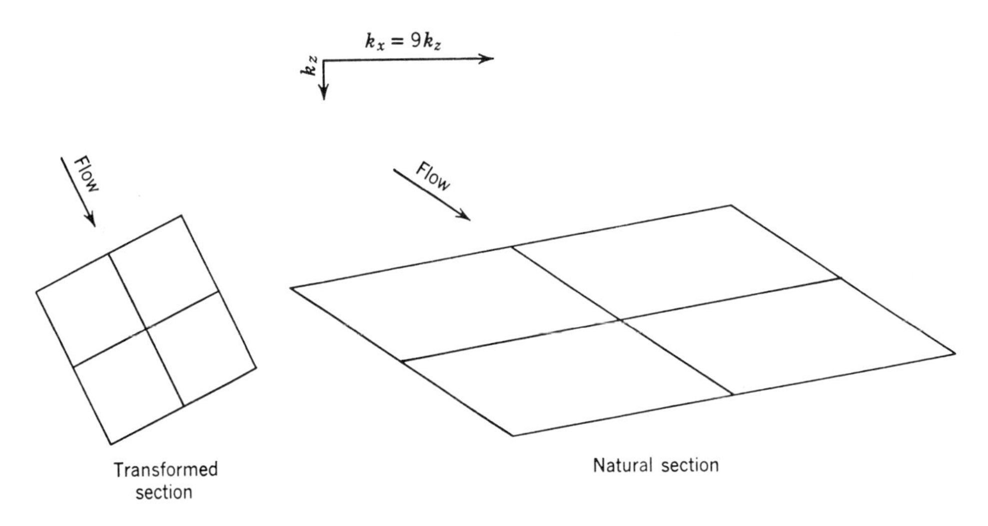

Fig. 18.6 Portion of flow net in anisotropic soil.

see how various alterations in the design affect the solution to the problem.

The undesirable feature of the flow net sketching technique lies in the difficulty of sketching the net. A polished net for even a simple two-dimensional flow situation can require many hours of sketching. Unfortunately, many people are not inherently talented in sketching and find it difficult to draw good nets. This feature is partially offset by the happy fact that the solution of a two-dimensional problem is relatively insensitive to the quality of the flow net. Even a crude flow net generally permits an accurate determination of seepage, pore pressure, and gradient. Further, the literature on soil mechanics contains good flow nets for many common situations.

Analytical Methods

There are certain flow problems for which a theoretical solution has been made. The best known theoretical solution is one for flow through an earth dam somewhat similar to that shown in Example 18.4. If the upstream equipotential is a parabola and the toe drain is a horizontal one, the flow net consists of a system of confocal parabolas. This solution was made by Kozeny in 1933. A. Casagrande has developed approximations to the Kozeny parabola to account for the upstream face of the structure being a straight line rather than a parabola. He also worked out modifications to the Kozeny equation to account for flow that does not end in a horizontal drain.

The sheet pile wall in Example 18.1 is another problem for which a theoretical solution is available (see Harr,

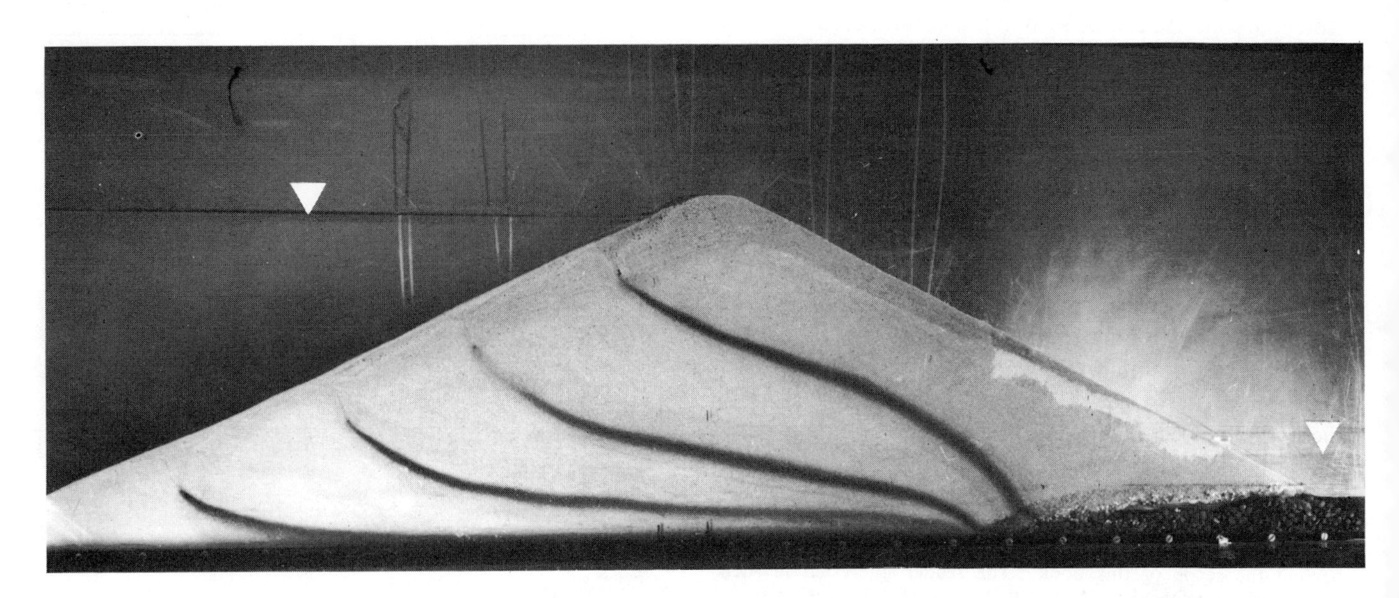

Fig. 18.7 Flow through a model dam.

1962). For the geometry of the flow situation shown, the maximum exit gradient is

$$i_{\text{exit}} = \frac{1}{\pi} \times \frac{\text{total head loss}}{\text{depth of wall penetration}}$$
$$= 0.32 \times \tfrac{7.5}{9.0} = 0.267$$

The theoretical gradient (0.267) is essentially the same as that obtained by the flow net (0.27).

Models

A flow problem can be solved by constructing a scaled model and analyzing flow in the model. For example, Fig. 18.7 shows flow through a model of an earth dam similar to that in Example 18.4. The model consists of sand placed between parallel lucite plates 100 mm apart. The dam is 375 mm high and has side slopes of two horizontal to one vertical. Steady flow is occurring into the toe drain, as traced by the dye lines. Piezometer tubes can be seen in the photograph.

Models are especially useful to illustrate the fundamentals of fluid flow. The model in Fig. 18.7 was used in a student laboratory. The students predicted the rate of flow and the pore pressure at various locations in the dam, and then compared the predictions with values measured in the model. By first constructing and testing a dam without a filter, the students got a dramatic illustration of what can happen to a dam when seepage breaks out on the downstream slope—failure.

Soil models, however, are of limited use in the general solution of flow problems because of the time and effort required to construct such a model and because of the difficulties caused by capillarity. The engineer can sketch many flow nets and investigate the influence of various design features in a shorter period of time than he can construct one soil model. Although the flow that occurs above the top seepage line may be of little importance in the prototype, it can be of considerable importance in a soil flow model. If the engineer uses a fine sand in his model, he encounters water movement above the top flow line. This zone of capillarity can be a large fraction of the height of his model.

Models employing viscous fluids have also been used to study flow problems. A model with transparent plates (glass or plastic) spaced closely together and filled with a fluid such as glycerin can be used to solve a flow situation, since the viscous material will follow the same flow laws as water in a soil. Whereas such models have been successfully used to study in detail a certain type of flow problem, such as flow into a well, their use to solve practical problems is limited because of the difficulty in constructing them.

Analogy Methods

Laplace's equation for fluid flow also holds for electrical and heat flow. Although practical difficulties are encountered with trying to use heat flow models to solve fluid flow problems, considerable use has been made of electrical models. In the electrical model voltage corresponds to total head, conductivity to permeability, and current to velocity. Measuring voltage enables one to locate the equipotentials, which can then be used to sketch a flow pattern. Electrical flow models are valuable for instructional purposes, and since they are easier to construct than soil models and can be adapted to a wide variety of boundary conditions, they are valuable for solving problems too complex to handle by flow net sketching.

Numerical Analysis

Laplace's flow equation can be solved approximately by the techniques of numerical analysis. One can obtain by a series of approximations the total heads at various points in a network. Relaxation methods are based on this principle.

When the high-speed digital computers became readily accessible, the importance of numerical analysis to solve fluid flow problems greatly increased. After programs are written for flow problems, solutions can be obtained very rapidly.

Summary Concerning Techniques for Solving Flow Problems

As the reader can see, there are a number of methods of solving fluid flow problems. The techniques used in Chapter 17 are most valuable because they are based on fundamentals of fluid flow which are essential for the student to understand. The flow net is a valuable tool in that it gives insight into the flow problem. The future will probably see a greatly increased role of the digital computer in solving complex fluid flow problems. With the computer it will be possible to solve and plot up the results for many typical flow situations. The engineer can then get an approximate solution to practical problems by comparing his particular problem with one for which a solution has been obtained.

18.6 SUMMARY OF MAIN POINTS

1. A flow net is a system of squares or rectangles formed by flow lines intersecting equipotential lines.
2. From a flow net the engineer can obtain (*a*) rate of flow, (*b*) pore pressure, and (*c*) gradient.
3. The rate of flow is q, where $q = kH\mathfrak{S}$. A flow net provides the shape factor $\mathfrak{S}$.
4. In anisotropic soil, the soil section must be transformed before the flow net can be sketched.

PROBLEMS

18.1 The tube in Fig. 17.9 has a square cross section 0.3 m by 0.3 m. Draw the flow net for the seepage conditions shown.

From the flow net determine (*a*) rate of flow, (*b*) pore pressure at elevation +1.8, and (*c*) the exit gradient.

18.2 Draw the flow net on both natural and transformed sections for the refinery subsoil profile in Fig. 17.13 for the case of steady seepage under pumping. Refer to Example 16.2 for soil properties.

18.3 For the sheet pile wall in Example 18.1 make a scaled plot of vertical effective stress on the horizontal surface *mn*. The total unit weight of soil is 20.42 kN/m³ and geostatic stresses exist.

18.4 Isolate figure *X* in Example 18.2 and plot on it the boundary water pressures.

18.5 Draw the flow net in Example 18.1 to natural scale if $k_v = 1\ \mu m/sec$ and $k_h = 10\ \mu m/sec$.

18.6 Steady-state, two-dimensional flow is occurring into the double row of sheet piles shown in Fig. P18.6. Draw the flow net and compute the rate of flow per meter of wall length. Determine the maximum exit gradient and the factor of safety against liquefaction. Plot to scale the water pressure along the both sides of one of the sheet pile walls.

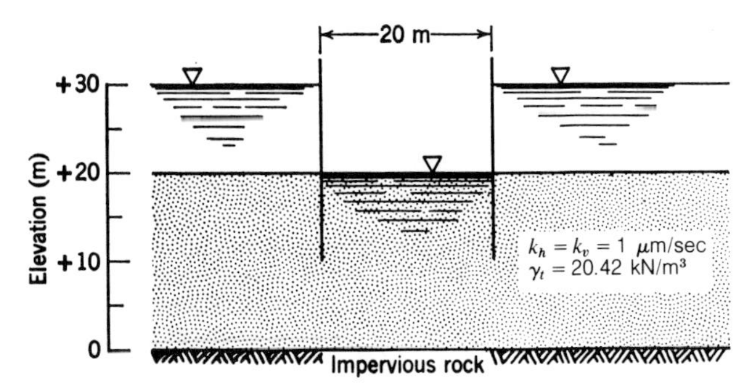

Fig. P18.6

18.7 Compute the seepage in cubic meters per day per meter of dam length through the dam of Fig. P18.7. For the Point *B* determine the pore pressure and gradient.

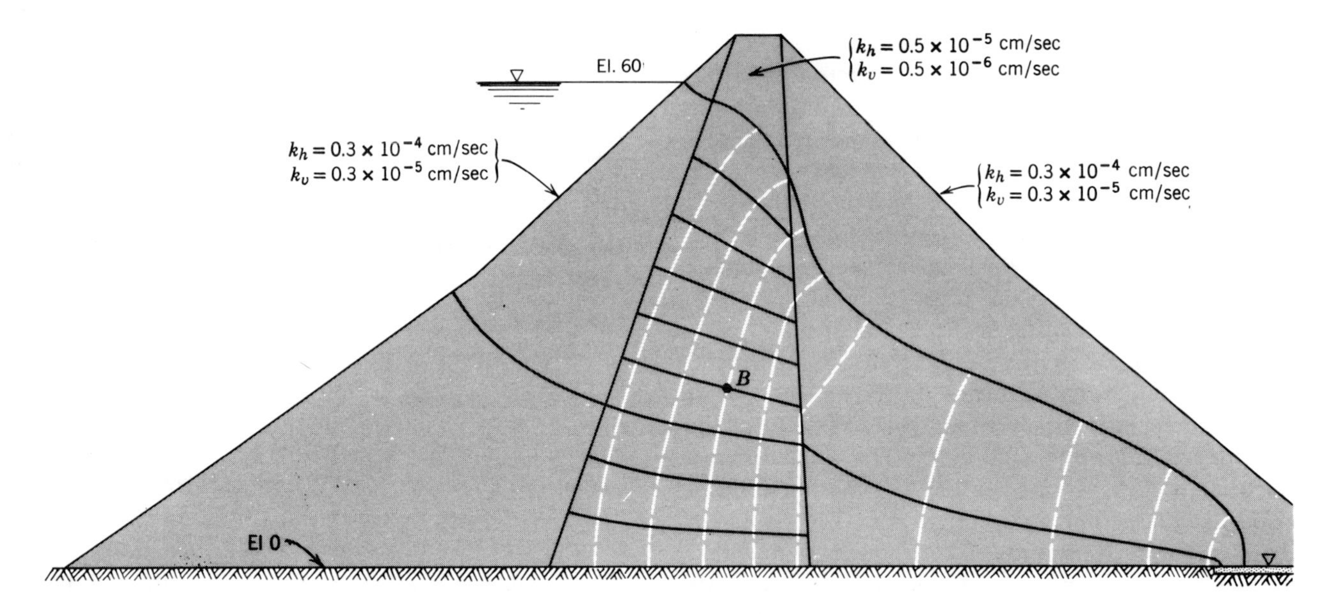

Fig. P18.7

CHAPTER 19

Soil Permeability and Filter Requirements

Chapters 17 and 18 have shown that the fundamental soil property involved in fluid flow is *permeability*. This chapter describes the determination of permeability, discusses the factors that affect it, and finally considers filters. Filters are included because their behavior is so closely related to permeability.

19.1 THE DETERMINATION OF PERMEABILITY

Soil permeability can be measured in either the laboratory or the field; laboratory determinations are much easier to make than field determinations. Because permeability depends very much on soil fabric (both microstructure—the arrangement of individual particles—and macrostructure—such as stratification) and because of the difficulty of getting representative soil samples, field determinations of permeability often are required to get a good indication of the average permeability. Laboratory tests, however, permit the relationship of permeability to void ratio to be studied and are thus usually run whether or not field measurements are made.

Among the methods used in the laboratory to determine permeability are:

1. Falling, or variable, head permeameter.
2. Constant head permeameter.
3. Direct or indirect measurement during an oedometer test.

Since a relatively large permeability is required to obtain good precision with the falling head test, it is limited to pervious soils. Further, the degree of saturation of an unsaturated soil changes during the variable head test, thus the variable head test should be used only on saturated soils. Since oedometer tests are usually run only on plastic soils, the determination of permeability from this test is normally done only with low permeability soils. The constant head permeability test is widely used on all types of soils.

Figure 19.1 shows a setup for the variable head permeability test. The coefficient of permeability can be computed from

$$k = 2.3 \frac{aL}{A(t_1 - t_0)} \log_{10}\left(\frac{h_0}{h_1}\right) \tag{19.1}$$

in which

a = the cross-sectional area of the standpipe
L = the length of soil sample in the permeameter
A = the cross-sectional area of the permeameter
t_0 = the time when the water level in the standpipe is at h_0
t_1 = the time when the water level in the standpipe is at h_1
h_0, h_1 = the heads between which the permeability is determined

Figure 19.2 shows two setups for the constant head permeability test. The coefficient of permeability can be computed from

$$k = \frac{QL}{thA} \tag{19.2}$$

in which

Q = the total quantity of water that flowed through the soil in elapsed time t
h = the total head lost

Both Eqs. 19.1 and 19.2 are derived by using Darcy's law, Eq. 17.1, for the flow situations in the permeameters.

The permeability at temperature T, k_T, can be reduced to that at 20°C, $k_{20°C}$, by using

$$k_{20°C} = \frac{\mu_T}{\mu_{20°C}} k_T \tag{19.3}$$

in which

$k_{20°C}$ = the permeability at temperature 20°C
k_T = the permeability at temperature T
$\mu_{20°C}$ = the viscosity of water at temperature 20°C
μ_T = the viscosity of water at temperature T

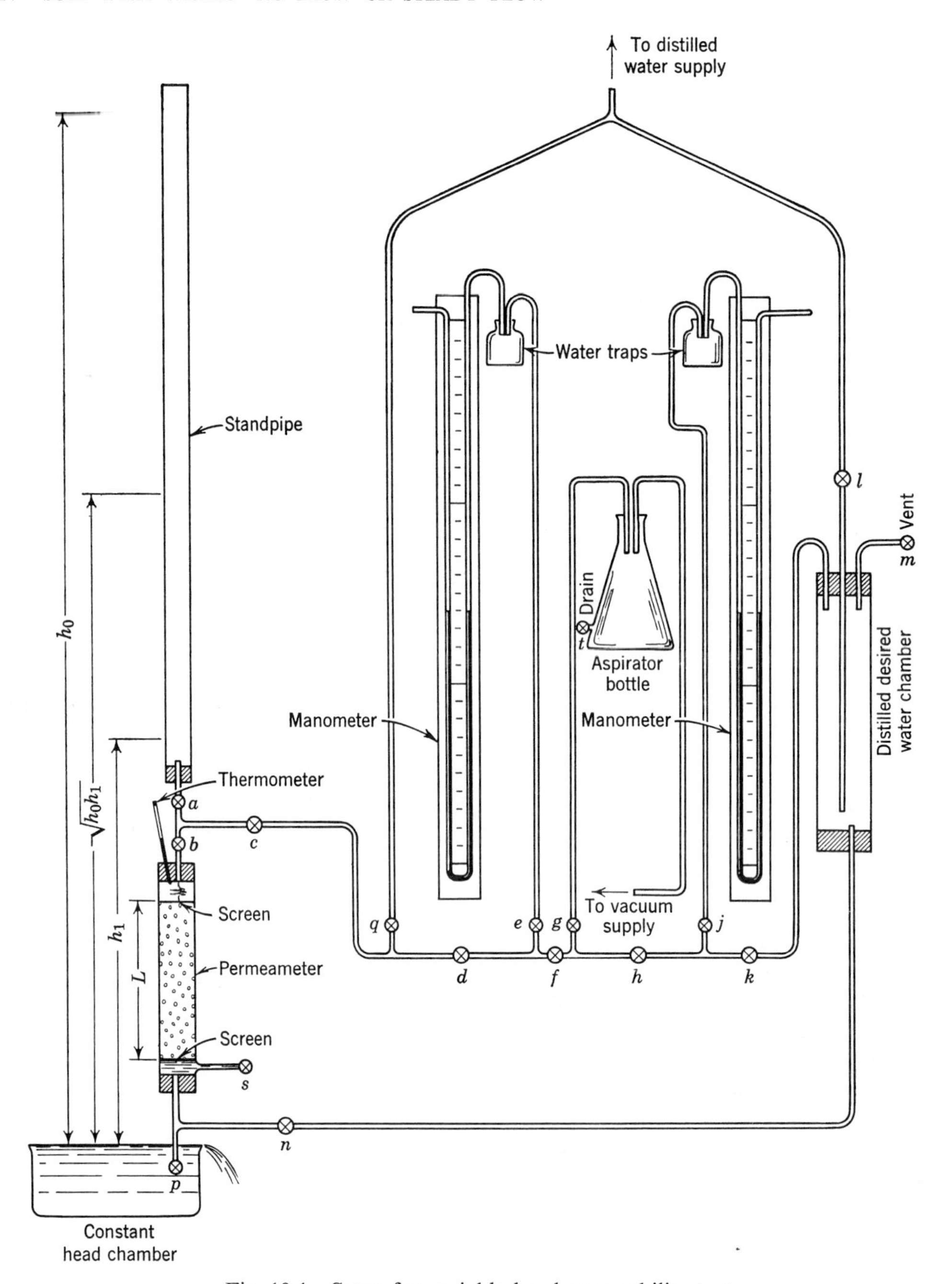

Fig. 19.1 Setup for variable-head permeability test.

As discussed in Part V of this book, the rate of consolidation of a soil depends directly on the permeability. Thus, using the appropriate relationships, we can compute the permeability from the measured rate of consolidation. This determination is far from precise because there are several terms in addition to permeability that enter into the rate of consolidation-permeability relationship. These other terms cannot easily be determined with precision. At the end of a compression increment, a constant head permeability test can be run on a sample of soil in the consolidation apparatus. This determination, being a direct measurement of permeability, is much more precise than a value computed from compression rate data.

Direct permeability measurement on soils of low permeability require certain modifications to the setups shown in Fig. 19.2 in order to get reasonable precision. Figure 19.3 shows a setup that has been successfully used to measure permeability, even of very plastic clays.

The laboratory measurement of soil permeability is fundamentally straightforward, but it requires careful technique to obtain reliable data. The reader is referred

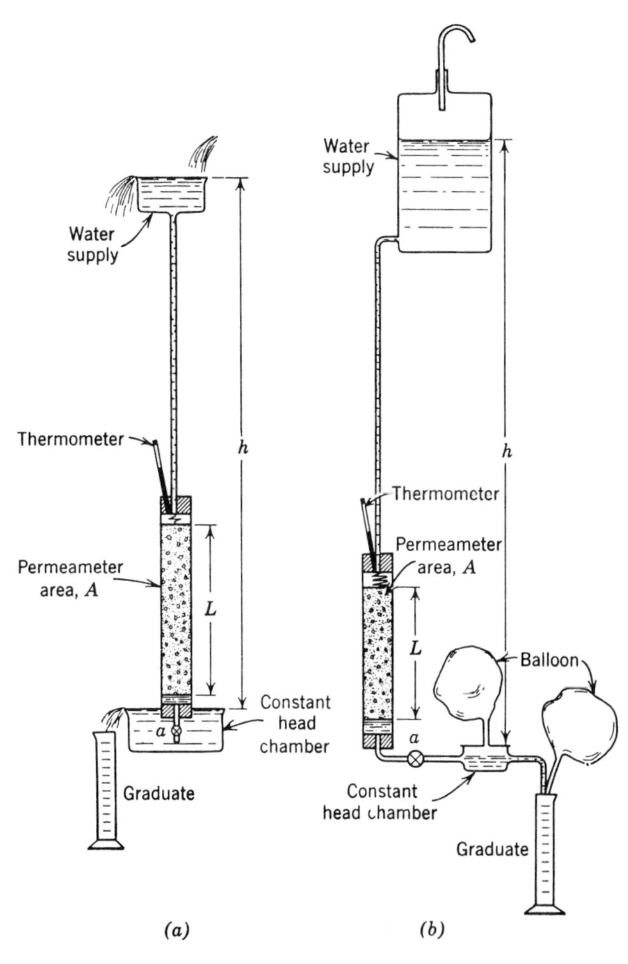

Fig. 19.2 Setup for constant-head permeability test.

to Lambe (1951) for a thorough treatment of permeability measurement.

Figure 19.4 from Hvorslev (1949) notes a number of setups which can be used to measure the permeability of soil in the field. Field measurements of permeability are not usually precise, because the soil and water conditions that exist at the location the permeability measurement is being made are not certain.

19.2 VALUES OF PERMEABILITY

Table 19.1 lists the permeability of a number of common types of soils. Table 19.2 gives a classification of soil on the basis of permeability. Figure 19.5 presents laboratory peremeability test data on a variety of soils. One can also get a perspective on the permeability of soil from the data given in the next section.

19.3 FACTORS AFFECTING PERMEABILITY

The coefficient of permeability as used by the soil engineer is the superficial or approach velocity of the permeant for flow through soil of unit area under a unit gradient. The permeability value thus depends on the characteristics of both the permeant and the soil.

An equation reflecting the influence of the permeant and the soil characteristics on permeability was developed by Taylor (1948) using Poiseuille's law. This equation is based on considering flow through a porous media similar to flow through a bundle of capillary tubes. The equation

$$k = D_s^2 \frac{\gamma}{\mu} \frac{e^3}{(1+e)} C \tag{19.4}$$

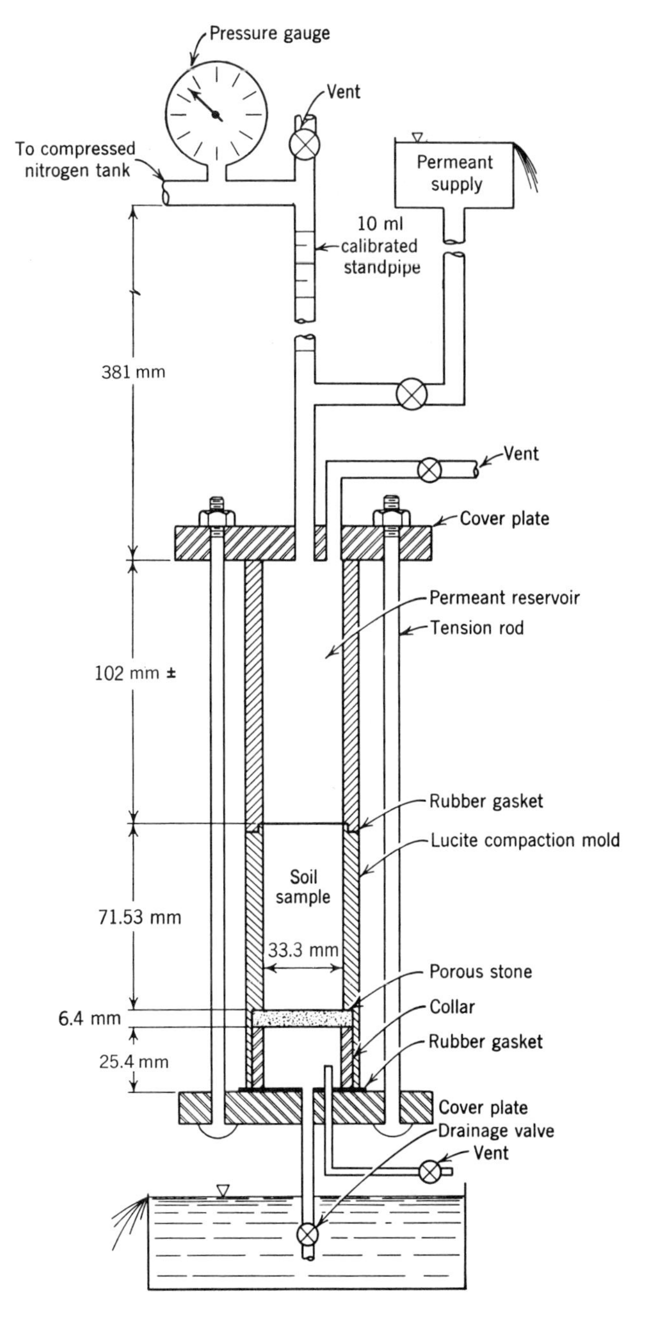

Fig. 19.3 Permeability test setup.

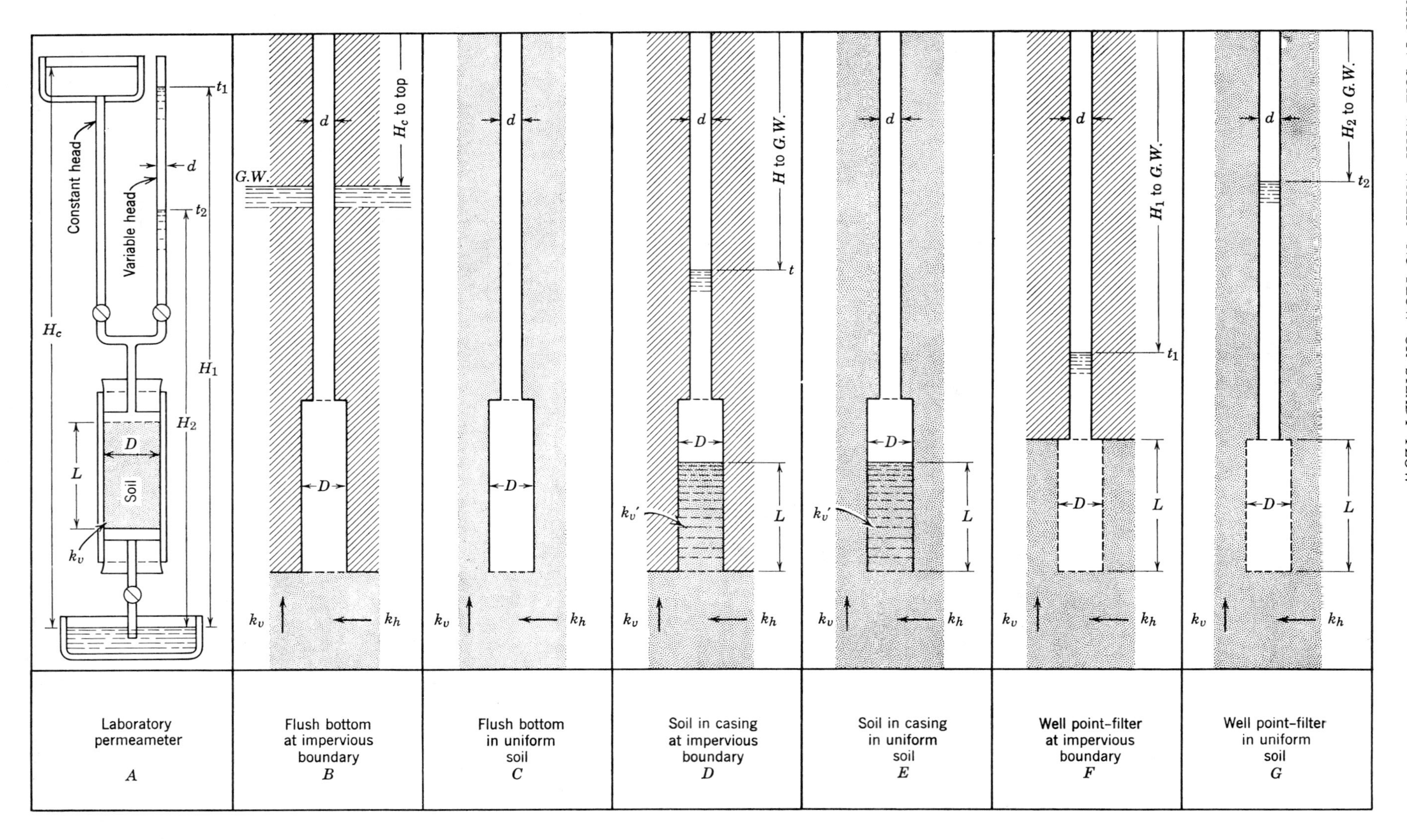
Constant head
Variable head
Soil
H_c
H_1
H_2
t_1
t_2
d
D
L
k_v
G.W.
H_c to top
H to G.W.
t
k_v'
H_1 to G.W.
H_2 to G.W.
k_h
Laboratory permeameter
A
Flush bottom at impervious boundary
B
Flush bottom in uniform soil
C
Soil in casing at impervious boundary
D
Soil in casing in uniform soil
E
Well point–filter at impervious boundary
F
Well point–filter in uniform soil
G

Case	Constant Head	Variable Head	Basic Time Lag	Notation
A	$k_v = \dfrac{4 \cdot q \cdot L}{\pi \cdot D^2 \cdot H_c}$	$k_v = \dfrac{d^2 \cdot L}{D^2 \cdot (t_2 - t_1)} \ln \dfrac{H_1}{H_2}$ $k_v = \dfrac{L}{t_2 - t_1} \ln \dfrac{H_1}{H_2}$ for $d = D$	$k_v = \dfrac{d^2 \cdot L}{D^2 \cdot T}$ $k_v = \dfrac{L}{T}$ for $d = D$	D = Diam, intake, sample (cm) d = Diameter, standpipe (cm) L = Length, intake, sample (cm) H_c = Constant piez. head (cm) H_1 = Piez. head for $t = t_1$ (cm) H_2 = Piez. head for $t = t_2$ (cm) q = Flow of water (cm^3/sec) t = Time (sec) T = Basic time lag (sec) k_v' = Vert. perm. casing (cm/sec)
B	$k_m = \dfrac{q}{2 \cdot D \cdot H_c}$	$k_m = \dfrac{\pi \cdot d^2}{8 \cdot D \cdot (t_2 - t_1)} \ln \dfrac{H_1}{H_2}$ $k_m = \dfrac{\pi \cdot D}{8 \cdot (t_2 - t_1)} \ln \dfrac{H_1}{H_2}$ for $d = D$	$k_m = \dfrac{\pi d^2}{8 \cdot D \cdot T}$ $k_m = \dfrac{\pi \cdot D}{8 \cdot T}$ for $d = D$	
C	$k_m = \dfrac{q}{2.75 \cdot D \cdot H_c}$	$k_m = \dfrac{\pi \cdot d^2}{11 \cdot D \cdot (t_2 - t_1)} \ln \dfrac{H_1}{H_2}$ $k_m = \dfrac{\pi \cdot D}{11 \cdot (t_2 - t_1)} \ln \dfrac{H_1}{H_2}$ for $d = D$	$k_m = \dfrac{\pi \cdot d^2}{11 \cdot D \cdot T}$ $k_m = \dfrac{\pi \cdot D}{11 \cdot T}$ for $d = D$	
D	$k_v' = \dfrac{4 \cdot q\left(\dfrac{\pi}{8} \cdot \dfrac{k_v'}{k_v} \cdot \dfrac{D}{m} + L\right)}{\pi \cdot D^2 \cdot H_c}$	$k_v' = \dfrac{d^2 \cdot \left(\dfrac{\pi}{8} \cdot \dfrac{k_v'}{k_v} \cdot \dfrac{D}{m} + L\right)}{D^2 \cdot (t_2 - t_1)} \ln \dfrac{H_1}{H_2}$ $k_v = \dfrac{\dfrac{\pi}{8} \cdot \dfrac{D}{m} + L}{t_2 - t_1} \ln \dfrac{H_1}{H_2}$ for $\begin{cases} k_v' = k_v \\ d = D \end{cases}$	$k_v' = \dfrac{d^2 \cdot \left(\dfrac{\pi}{8} \cdot \dfrac{k_v'}{k_v} \cdot \dfrac{D}{m}\right) + L}{D^2 \cdot T}$ $k_v = \dfrac{\dfrac{\pi}{8} \cdot \dfrac{D}{m} + L}{T}$ for $\begin{cases} k_v' = k_v \\ d = D \end{cases}$	k_v = Vert. perm. ground (cm/sec) k_h = Horz. perm. ground (cm/sec) k_m = Mean coeff. perm. (cm/sec) m = Transformation ratio $k_m = \sqrt{k_h \cdot k_v}$ $m = \sqrt{k_h/k_v}$ $\ln = \log_e = 2.3 \log_{10}$
E	$k_v' = \dfrac{4 \cdot q \cdot \left(\dfrac{\pi}{11} \cdot \dfrac{k_v'}{k_v} \cdot \dfrac{D}{m} + L\right)}{\pi \cdot D^2 \cdot H_c}$	$k_v' = \dfrac{d^2 \cdot \left(\dfrac{\pi}{11} \cdot \dfrac{k_v'}{k_v} \cdot \dfrac{D}{m} + L\right)}{D^2 \cdot (t_2 - t_1)} \ln \dfrac{H_1}{H_2}$ $k_v = \dfrac{\dfrac{\pi}{11} \cdot \dfrac{D}{m} + L}{t_2 - t_1} \ln \dfrac{H_1}{H_2}$ for $\begin{cases} k_v' = k_v \\ d = D \end{cases}$	$k_v' = \dfrac{d^2 \cdot \left(\dfrac{\pi}{11} \cdot \dfrac{k_v'}{k_v} \cdot \dfrac{D}{m} + L\right)}{D^2 \cdot T}$ $k_v' = \dfrac{\dfrac{\pi}{11} \cdot \dfrac{D}{m} + L}{T}$ for $\begin{cases} k_v' = k_v \\ d = D \end{cases}$	
F	$k_h = \dfrac{q \cdot \ln\left[\dfrac{2mL}{D} + \sqrt{1 + \left(\dfrac{2mL}{D}\right)^2}\right]}{2 \cdot \pi \cdot L \cdot H_c}$	$k_h = \dfrac{d^2 \cdot \ln\left[\dfrac{2mL}{D} + \sqrt{1 + \left(\dfrac{2mL}{D}\right)^2}\right]}{8 \cdot L \cdot (t_2 - t_1)} \ln \dfrac{H_1}{H_2}$ $k_h = \dfrac{d^2 \cdot \ln\left(\dfrac{4mL}{D}\right)}{8 \cdot L \cdot (t_2 - t_1)} \ln \dfrac{H_1}{H_2}$ for $\dfrac{2mL}{D} > 4$	$k_h = \dfrac{d^2 \cdot \ln\left[\dfrac{2mL}{D} + \sqrt{1 + \left(\dfrac{2mL}{D}\right)^2}\right]}{8 \cdot L \cdot T}$ $k_h = \dfrac{d^2 \cdot \ln\left(\dfrac{4mL}{D}\right)}{8 \cdot L \cdot T}$ for $\dfrac{2mL}{D} > 4$	H_0 (Log. scale) $0.37\ H_0$ $t = T$ Piezometric head H $0.10\ H_0$ $t = 2.3 \cdot T$ Time t (linear scale)
G	$k_h = \dfrac{q \cdot \ln\left[\dfrac{mL}{D} + \sqrt{1 + \left(\dfrac{mL}{D}\right)^2}\right]}{2 \cdot \pi \cdot L \cdot H_c}$	$k_h = \dfrac{d^2 \cdot \ln\left[\dfrac{mL}{D} + \sqrt{1 + \left(\dfrac{mL}{D}\right)^2}\right]}{8 \cdot L \cdot (t_2 - t_1)} \ln \dfrac{H_1}{H_2}$ $k_h = \dfrac{d^2 \cdot \ln\left(\dfrac{2mL}{D}\right)}{8 \cdot L \cdot (t_2 - t_1)} \ln \dfrac{H_1}{H_2}$ for $\dfrac{mL}{D} > 4$	$k_h = \dfrac{d^2 \cdot \ln\left[\dfrac{mL}{D} + \sqrt{1 + \left(\dfrac{mL}{D}\right)^2}\right]}{8 \cdot L \cdot T}$ $k_h = \dfrac{d^2 \cdot \ln\left(\dfrac{2mL}{D}\right)}{8 \cdot L \cdot T}$ for $\dfrac{mL}{D} > 4$	Determination basic time lag T

ASSUMPTIONS

Soil at intake, infinite depth, and directional isotropy (k_v and k_h constant). No disturbance, segregation, swelling, or consolidation of soil. No sedimentation or leakage. No air or gas in soil, well point, or pipe. Hydraulic losses in pipes, well point, or filter negligible.

Fig. 19.4 Formulas for determination of permeability (From Hvorslev, 1951).

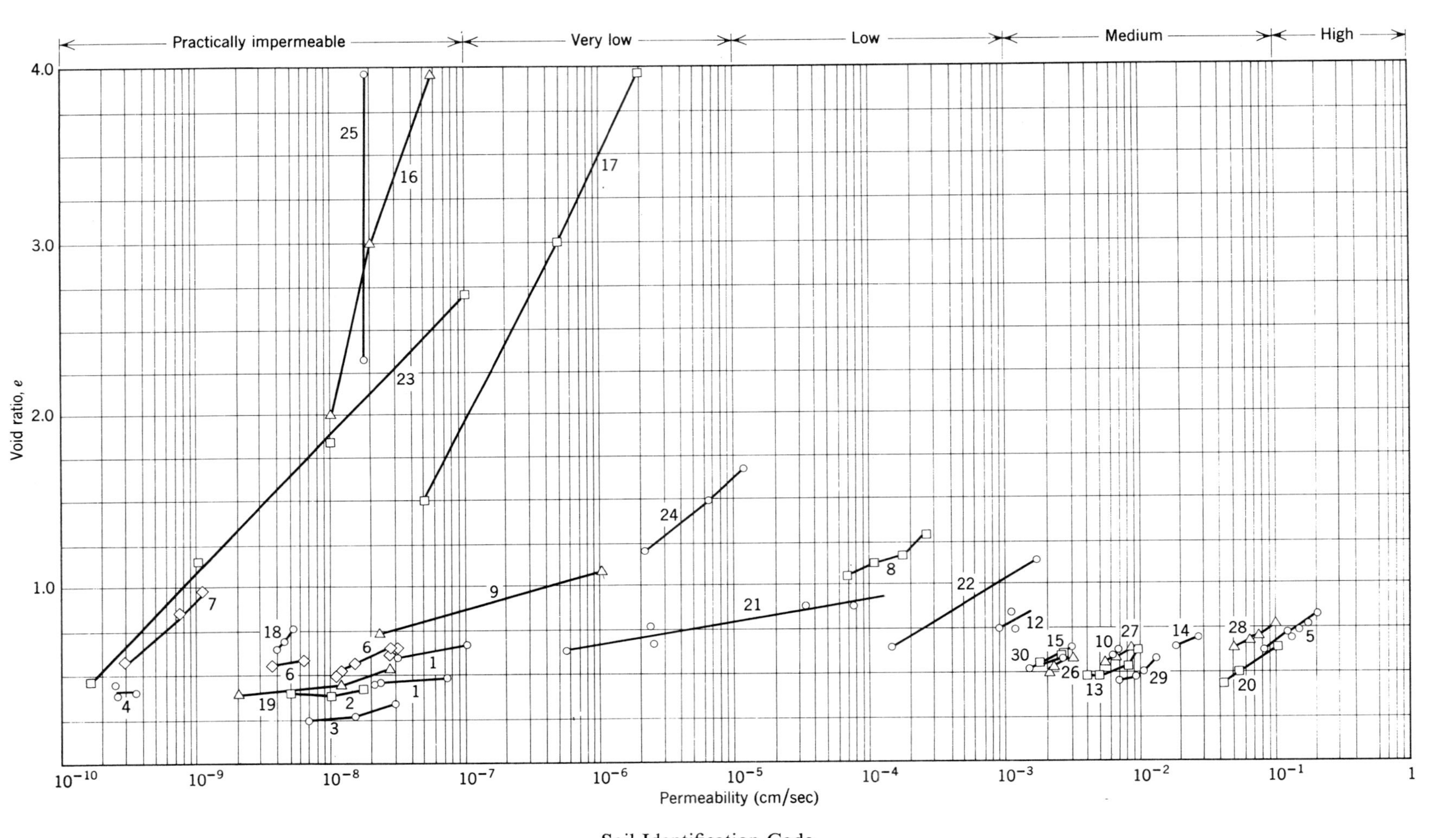

Soil Identification Code

1 Compacted caliche
2 Compacted caliche
3 Silty sand
4 Sandy clay
5 Beach sand
6 Compacted Boston blue clay
7 Vicksburg buckshot clay
8 Sandy clay
9 Silt—Boston
10 Ottawa sand
11 Sand—Gaspee Point
12 Sand—Franklin Falls
13 Sand—Scituate
14 Sand—Plum Island
15 Sand—Fort Peck
16 Silt—Boston
17 Silt—Boston
18 Loess
19 Lean clay
20 Sand—Union Falls
21 Silt—North Carolina
22 Sand from dike
23 Sodium—Boston blue clay
24 Calcium kaolinite
25 Sodium montmorillonite
26–30 Sand (dam filter)

Fig. 19.5 Permeability test data.

in which

k = the Darcy coefficient of permeability
D_s = some effective particle diameter
γ = unit weight of permeant
μ = viscosity of permeant
e = void ratio
C = shape factor

The following is an expression for the permeability of porous media, known as the *Kozeny-Carman equation* since it was proposed by Kozeny and improved by Carman:

$$k = \frac{1}{k_0 S^2} \frac{\gamma}{\mu} \frac{e^3}{(1 + e)} \tag{19.5}$$

in which

k_0 = factor depending on pore shape and ratio of length of actual flow path to soil bed thickness
S = specific surface area

Since D_s is defined as the diameter of particle having a specific surface of S, Eq. 19.4 can be considered a simplification of the Kozeny-Carman equation.

Table 19.1 Coefficient of Permeability of Common Natural Soil Formations

Formation	Value of k (cm/sec)
River deposits	
Rhone at Genissiat	Up to 0.40
Small streams, eastern Alps	0.02–0.16
Missouri	0.02–0.20
Mississippi	0.02–0.12
Glacial deposits	
Outwash plains	0.05–2.00
Esker, Westfield, Mass.	0.01–0.13
Delta, Chicopee, Mass.	0.0001–0.015
Till	Less than 0.0001
Wind deposits	
Dune sand	0.1–0.3
Loess	0.001 ±
Loess loam	0.0001 ±
Lacustrine and marine offshore deposits	
Very fine uniform sand, U^a = 5–2	0.0001–0.0064
Bull's liver, Sixth Ave., N.Y., U = 5–2	0.0001–0.0050
Bull's liver, Brooklyn, U = 5	0.00001–0.0001
Clay	Less than 0.0000001

[a] U = uniformity coefficient.
From Terzaghi and Peck, 1967.

Table 19.2 Classification of Soils According to Their Coefficients of Permeability

Degree of Permeability	Value of k (cm/sec)
High	Over 10^{-1}
Medium	10^{-1}–10^{-3}
Low	10^{-3}–10^{-5}
Very low	10^{-5}–10^{-7}
Practically impermeable	Less than 10^{-7}

From Terzaghi and Peck, 1967.

Equation 19.4 or 19.5 aids considerably in the following examination of the variables affecting permeability. In this examination those characteristics related to the permeant are considered first and then those related to the soil composition are treated.

Permeant

Equations 19.4 and 19.5 show that both the viscosity and the unit weight of the permeant influence the value of permeability. These two permeant characteristics can be eliminated as variables by defining another permeability, the *specific* or *absolute* permeability, as:

$$K = \frac{k\mu}{\gamma} \tag{19.6}$$

Since k is in units of velocity, K is in units of length^2; e.g., if k is in cm/sec, the corresponding unit for K is cm^2. K is also expressed in terms of darcys; 1 darcy = 0.987×10^{-8} cm^2. For water at 20°C, the following two equations permit one to convert k in cm/sec to K in cm^2 or in darcys:

$$K \text{ in cm}^2 = k \text{ in cm/sec} \times 1.02 \times 10^{-5} \tag{19.7}$$

$$K \text{ in darcys} = k \text{ in cm/sec} \times 1.035 \times 10^{3} \tag{19.8}$$

Figure 19.6 is a chart for the conversion of permeability values from one set of units to another. (Conversion factors are given in the appendix.)

While viscosity and unit weight are the only variables of the permeant that influence the permeability of pervious soils, other permeant characteristics can have a major influence on the permeability of relatively impervious soils. The magnitude of influence for characteristics other than viscosity and unit weight are illustrated in Fig. 19.7. In this figure values of permeability of saturated kaolinite are plotted for various permeants. The permeability is expressed in terms of the absolute permeability, thus the influences of viscosity and unit weight have been eliminated. The data in Fig. 19.7 show that the nature of the permeant can be very important, with variations of many hundred percent in absolute permeability depending on the actual permeant. The

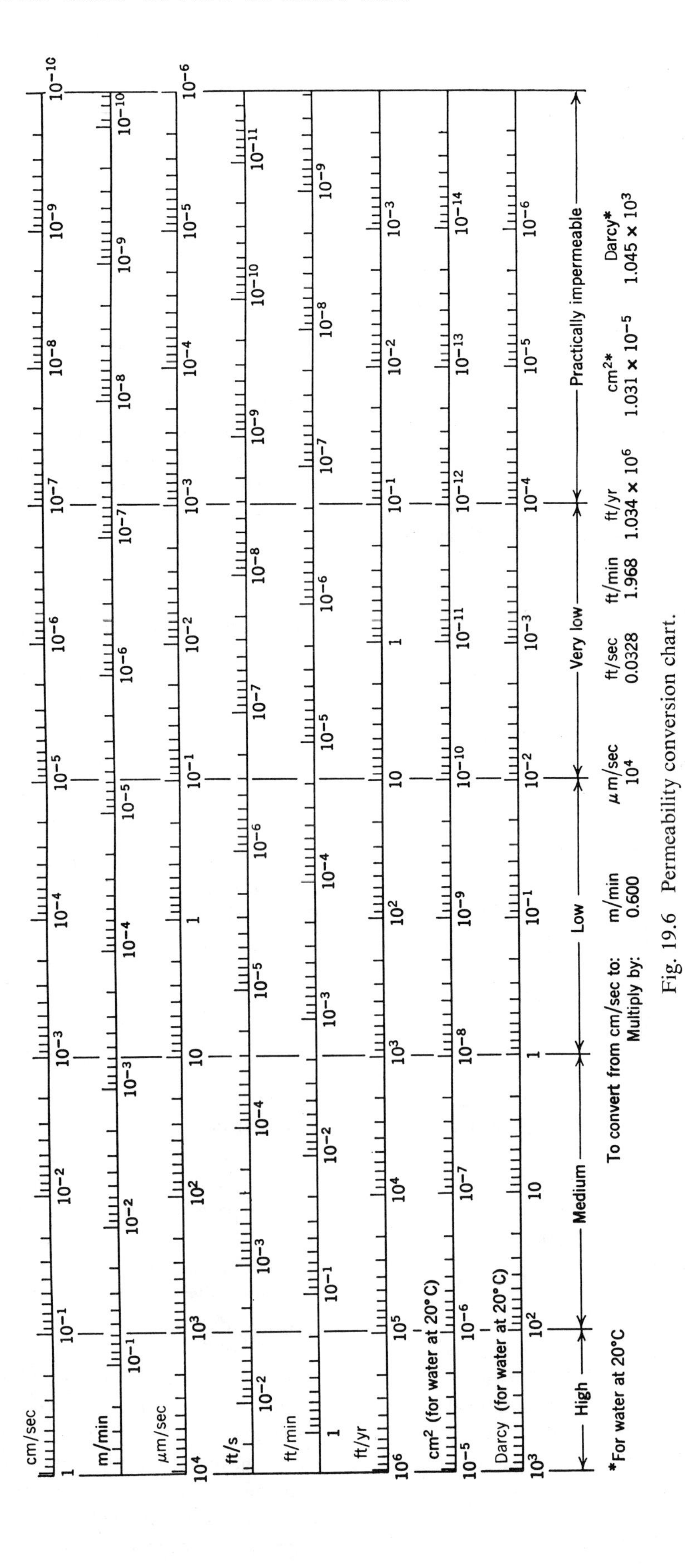

Fig. 19.6 Permeability conversion chart.

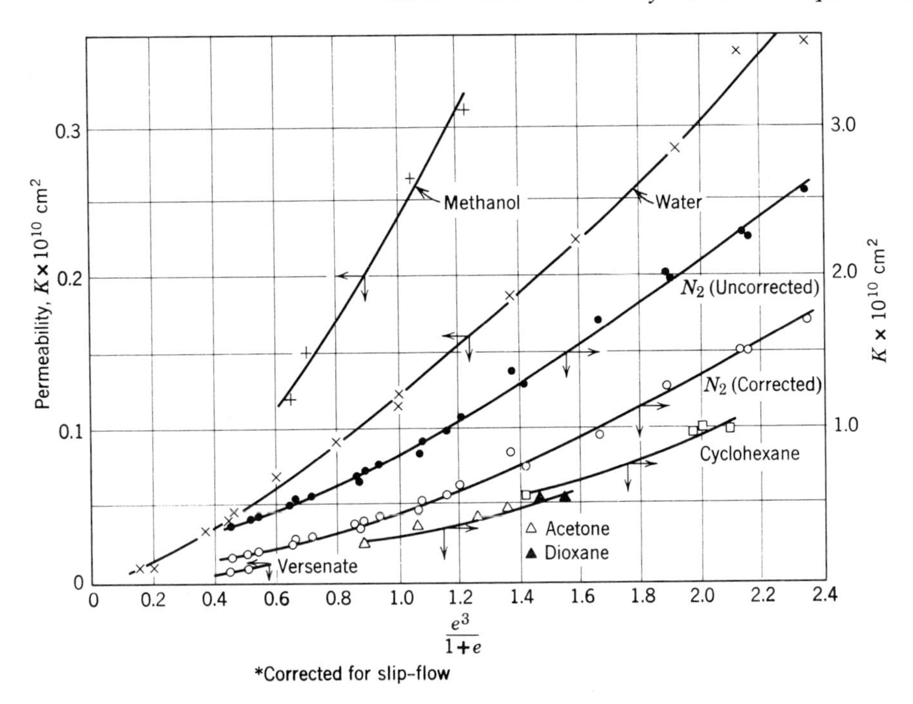

Fig. 19.7 Permeability of kaolinite to various fluids as a function of $e^3/(1 + e)$ where e = void ratio. (From Michaels and Lin, 1954.)

data in Fig. 19.7 were obtained from tests in which the kaolinite was molded in the fluid which was to be used as the permeant. In Fig. 19.8 are presented results of tests in which water was used as the molding fluid and the initial permeant; each succeeding permeant displaced the previous one. Figure 19.8 shows that, although different permeabilities were obtained for different permeants, the differences are much smaller than those shown in Fig. 19.7.

The large differences in permeability at a given void ratio, as shown in Fig. 19.7, can be explained by the

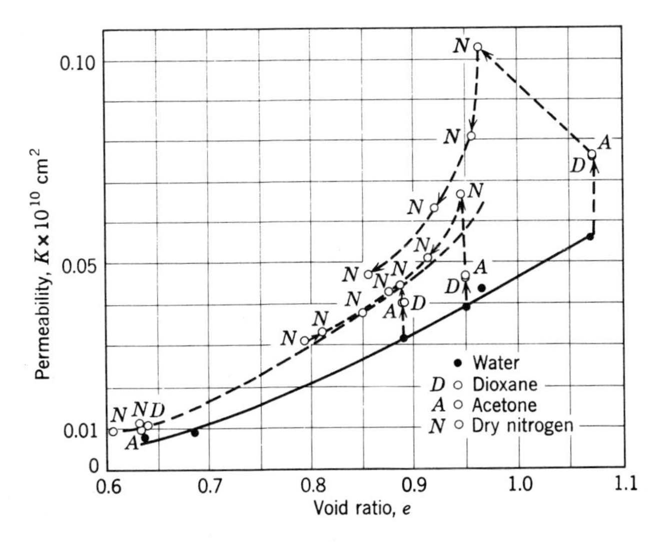

Fig. 19.8 Effect of desolvation on kaolinite permeability; initial permeant, water. (From Michaels and Lin, 1954.)

differences in soil fabric resulting from sample preparation in the different fluids. This large difference in initial fabric is eliminated in the tests shown in Fig. 19.8. The comparison of the data in the two figures illustrates the conclusion drawn from the Michaels-Lin work: the major influence of the different permeants was on the fabric of the soil. (Chapter 5 discussed the influence of pore fluid on the fabric of the soil sediment.)

This leads us to conclude that viscosity and density are not the only permeant characteristics, as suggested by the theoretical equations, that influence the permeability of fine-grained soils. Since the electro-osmotic backflow (the movement of permeant in the opposite direction to net fluid flow because of the electrical potential generated by the fluid flow) and the mobility of the fluid immediately adjacent to the soil particles depend on the polarity of the pore fluid, some measure of polarity should be included in the equations.

Soil

The following five characteristics influence permeability:

1. Particle size.
2. Void ratio.
3. Composition.
4. Fabric.
5. Degree of saturation.

Equations 19.4 and 19.5 consider directly only particle size and void ratio, while the other three characteristics

Table 19.3 Permeability Test Data

Soil	Particle Size, D_{10} (cm)	Permeability (μm/sec)	k/D_{10}^2 (1/sec cm)
Coarse gravel	0.082	1100	16
Sandy gravel	0.020	160	40
Fine gravel	0.030	71	8
Silty gravel	0.006	4.6	11
Coarse sand	0.011	1.1	1
Medium sand	0.002	0.29	7
Fine sand	0.003	0.096	1
Silt	0.0006	0.15	42
		Average =	16

Permeability and particle size data from "Capillarity Tests by Capillarimeter and by Soil Filled Tubes" by K. S. Lane and D. E. Washburn, *Proc. HRB*, 1946.

are treated indirectly or ignored. Unfortunately, the effects of one of the five are hard to isolate since these characteristics are closely interrelated—e.g., fabric usually depends on particle size, void ratio, and composition.

Equation 19.4 suggests that permeability varies with the square of some particle diameter. It is logical that the smaller the soil particles the smaller the voids, which are the flow channels, and thus the lower the permeability. A relationship between permeability and particle size is much more reasonable in silts and sands than in clays, since in silts and sands the particles are more nearly equidimensional and the extremes in fabric are closer together. From work on sands, Hazen proposed

$$k = 100 D_{10}^2 \tag{19.9}$$

where k is in cm/sec and D_{10} is in cm.

Listed in Table 19.3 are some permeability and particle size test data and the corresponding values of k/D_{10}^2. As shown, the values of K/D_{10}^2 vary from 1 to 42 with an average of 16.

Logic and experimental data suggest that the finer particles in a soil have the most influence on permeability. Hazen's equation, for example, uses D_{10} as "the" diameter for relating particle size and permeability. This relation assumes that the distribution of particle sizes is spread enough to prevent the smallest particles from moving under the seepage force of the flowing water, i.e., the soil must have "hydrodynamic stability." Uniform coarse soils containing fines frequently do not possess hydrodynamic stability. Flow in such soils can wash out the fines and thereby cause an increase in permeability with flow. Particle size requirements to prevent such migration of fines are given in the next section.

The permeability equations indicate that a plot of k versus $e^3/(1 + e)$ should be a straight line. Other theoretical equations have suggested that k versus $e^2/(1 + e)$ or k versus e^2 should be a straight line. There are considerable experimental data which indicate that e versus $\log k$ is frequently a straight line. Figure 19.9 presents experimental data in the form of k versus functions of e. The test data on this sand show that the plot of k versus $e^3/(1 + e)$ and $\log k$ versus e are both

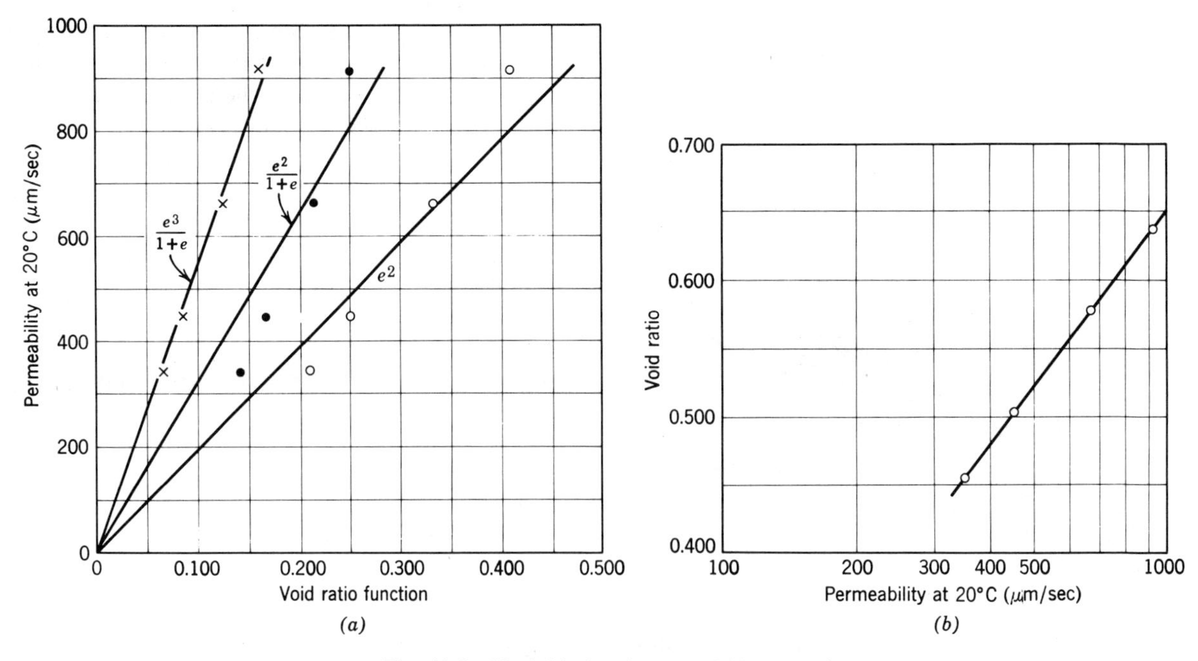

Fig. 19.9 Variable-head permeability test data.

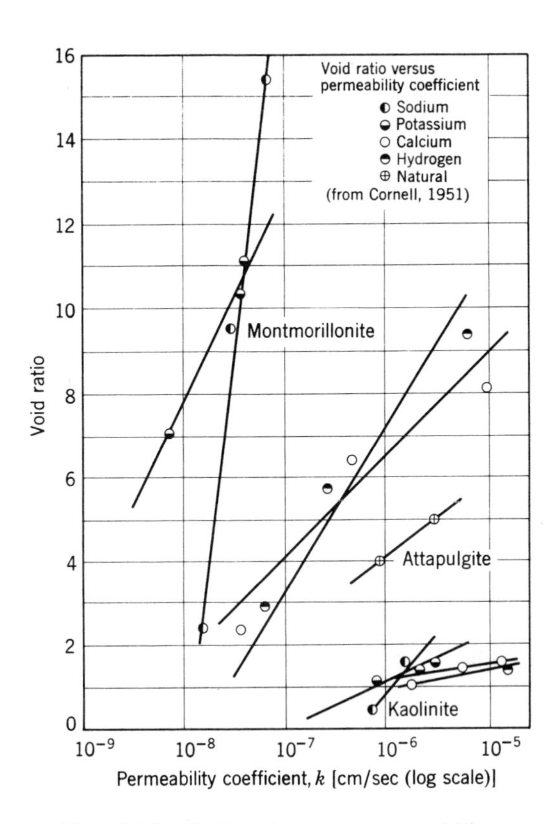

Fig. 19.10 Void ratio versus permeability.

relatively close to straight lines. The test data in Fig. 19.7 show that a plot of k versus $e^3/(1 + e)$ for kaolinite is not a straight line. In general, e versus $\log k$ is close to a straight line for nearly all soils, as suggested by Fig. 19.5.

The influence of soil composition on permeability is generally of little importance with silts, sands, and gravels (mica and organic matter are two exceptions); it is of major importance with clays. The very large influence composition can have on clay permeability is illustrated in Fig. 19.10. As indicated by the data in the figure, of the common exchangeable ions sodium is the one that gives the lowest permeability to a clay. Figure 19.10 shows that at a void ratio as high as 15, sodium montmorillonite has a permeability less than 10^{-7} cm/sec. Sodium montmorillonite is one of the least permeable soil minerals and is therefore widely used by the engineer as an impermeabilizing additive to other soils.

The magnitude of permeability variation with soil composition ranges widely. Figure 19.10 shows that the ratio of permeability of calcium montmorillonite to that of potassium montmorillonite at a void ratio of 7 is approximately 300. It further shows that the permeability of kaolinite is a hundred times that of montmorillonite. The lower the ion exchange capacity of a soil, the lower, of course, the effect of exchangeable ion on permeability.

The fabric component of structure is one of the most important soil characteristics influencing permeability, especially for fine-grained soils. Comparing soil specimens at the same void ratio, we find that the specimen which is in the most flocculated state will have the highest permeability, and the one in the most dispersed state will have the minimum permeability. The more dispersed the particles, i.e., the more nearly parallel they are, the more tortuous is the flow path if the flow is normal to the particles. This increased tortuosity helps explain some of the low permeability caused by improved structure. The main factor, however, is that in a flocculated soil there are some large channels available for flow. Since flow through one large channel will be much greater than flow through a number of small channels having the same size of total channel area as the one large channel, it is readily apparent that the larger a channel for a given void volume, the higher the permeability.

To suggest how large the effect of fabric on permeability can be, Table 19.4 gives test results obtained on a compacted clay.

The first comparison, between a sample compacted dry of optimum and one wet of optimum, shows two samples of essentially the same void ratio and degree of saturation having a permeability ratio of approximately 60. The second comparison, also between samples at the same void ratio and degree of saturation, shows a permeability ratio of greater than 3.

Test data which further illustrate the influence of structure on permeability are presented in Fig. 19.11. Figure 19.11*a* shows that physically mixing or blending of a soil can have a major effect on permeability. Figure 19.11*b* shows the large influence on permeability obtained by mixing in 0.1% (based on soil dry weight) of a polyphosphate dispersant. The dispersant, increasing the repulsion between fine particles, permits them to move to positions of greater hydraulic stability, resulting in a reduction of permeability.

The preceding discussion on structure has been

Table 19.4

Soil	Dry Unit Weight or Void Ratio	Degree of Saturation	Permeability (cm/sec)
Jamaican clay	18.07 kN/m^3	Approximately same	4×10^{-6}
	18.22 kN/m^3		7×10^{-8}
Virginia sandy clay	1.3	100%	1×10^{-3}
	1.3	100%	2.7×10^{-4}

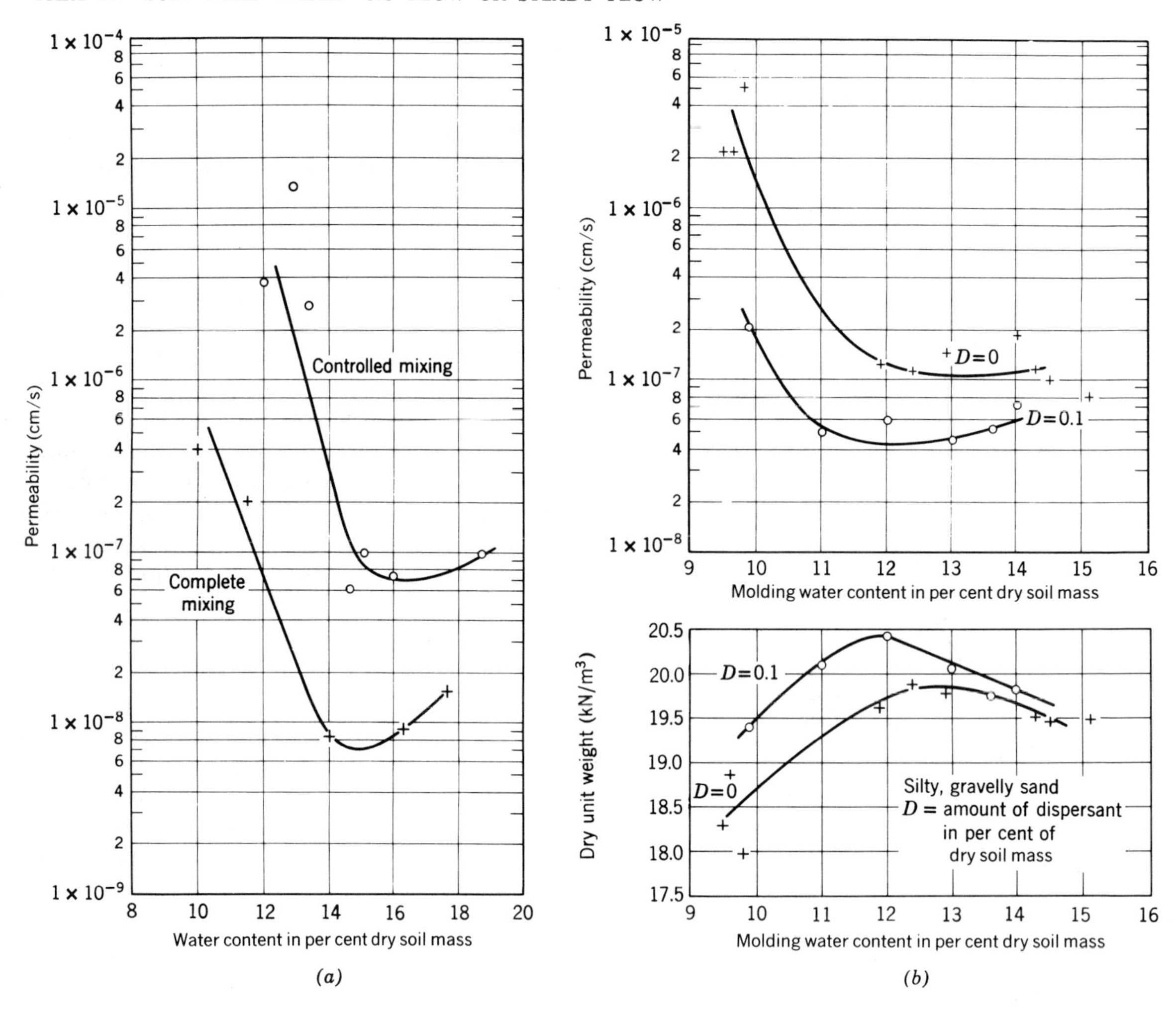

Fig. 19.11 Effect of structure on permeability. (*a*) Effect of mixing on permeability. Jamaica clay. (*b*) Effect of dispersion on permeability. (From Lambe, 1955.)

devoted primarily to "microstructure" or "microfabric." "Macrostructure" is also of considerable importance. A stratified soil consisting, for example, of a layer of silt and a layer of sand, has a much higher permeability for flow parallel to the stratifications than it does for flow perpendicular to the stratifications. This fact was illustrated by Example 18.5.

The degree of saturation of a soil has an important influence on its permeability. The higher the degree of saturation, the higher the permeability. Figure 19.12, presenting test data on four sands, shows that the influence on permeability is much more than would be explained merely by a reduction in the flow channels available for water flow. Although the plots in Fig. 19.12 suggest a unique relationship between degree of saturation and permeability, the development of a relationship between the two is not feasible because of the great influence of fabric.

The preceding discussion on the factors influencing permeability emphasizes the importance of duplicating field conditions when determining field permeability in the laboratory.

19.4 FILTER REQUIREMENTS

There are certain situations in earth structures that require filters. First, water cannot be permitted to exit on the slope of a dam, as was discussed in Chapter 18. Second, the movement of particles from one soil to another or from a soil into a drainage structure by flowing water cannot be permitted. If this were permitted, the resulting soil erosion could cause serious stability difficulties with the earth structure. Soil erosion is prevented by soil layers, called filters.

The design of a proper filter consists of choosing the dimensions of the filter and of choosing a material for the filter such that:

1. Sufficient head is lost in flow through the filters.
2. No significant invasion of soil is permitted into the filter.

The selection of a filter to meet the first requirement depends on both the type of soil and the flow pattern in the earth structure under consideration. Figure 19.13 presents a useful plot for the design of a filter for flow out of a slope. For a given slope and permeability in the structure, Fig. 19.13 enables one to select combinations of filter thickness and permeability. This figure was developed from flow nets, as illustrated by the two nets shown.

The requirements of a filter to keep soil particles from invading the filter significantly are based on particle size. These requirements were developed from tests by Terzaghi which were later extended by the Corps of Engineers at Vicksburg. The resulting filter specifications relate the grading of the protective filter to that of the soil being protected by the following:

$$\frac{D_{15}\,\text{Filter}}{D_{85}\,\text{Soil}} < 5 \tag{19.10}$$

$$4 < \frac{D_{15}\,\text{Filter}}{D_{15}\,\text{Soil}} < 20 \tag{19.11}$$

$$\frac{D_{50}\,\text{Filter}}{D_{50}\,\text{Soil}} < 25 \tag{19.12}$$

where D_{15}, D_{50}, and D_{85} are the particle sizes from a particle size distribution plot at 15, 50, and 85%,

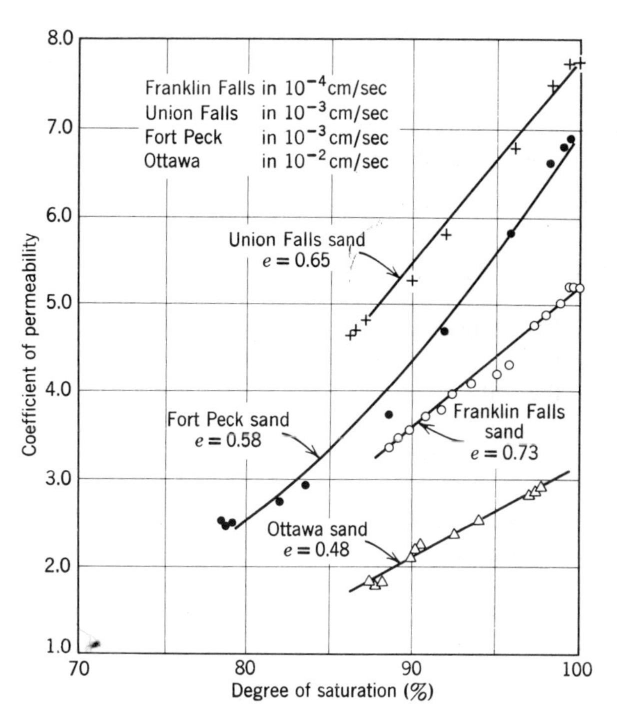

Fig. 19.12 Permeability versus degree of saturation for various sands (From Wallace, 1948).

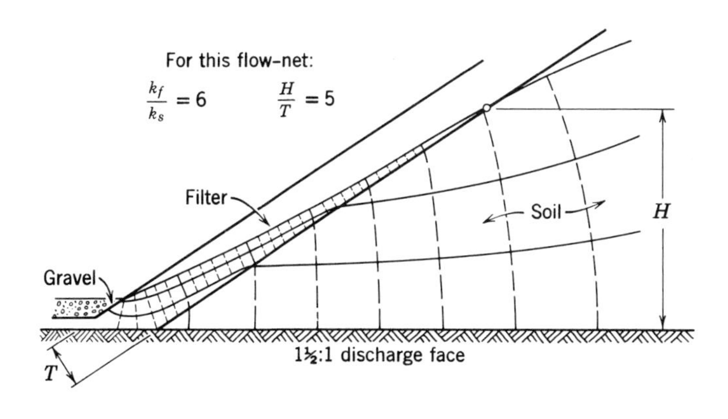

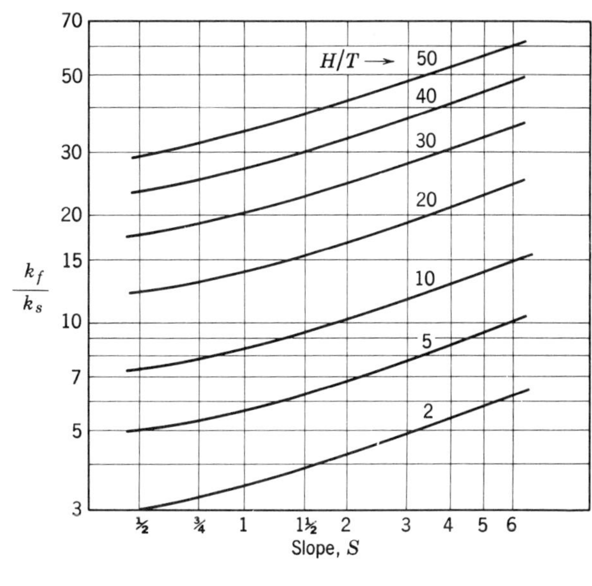

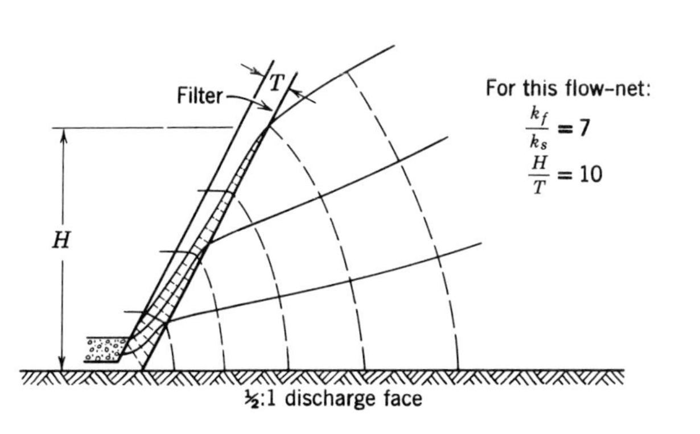

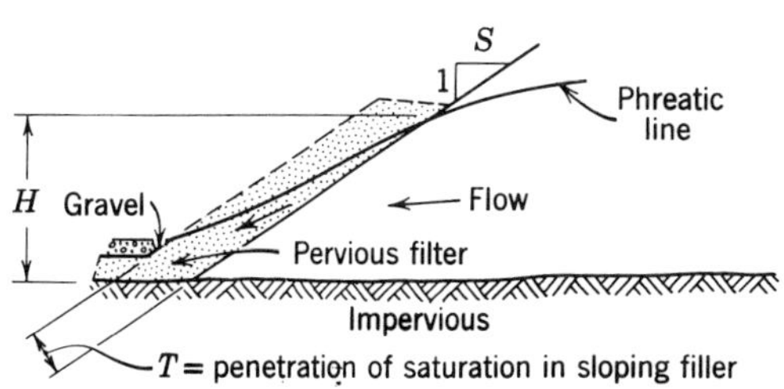

Fig. 19.13 Filter design (From Cedergren, 1960).

respectively, finer by weight. The expressions not only limit particle movement from the soil into the filter to a small zone at the interface between the soil and filter but also ensure that the permeability of the filter is considerably greater than that of the soil. A better method of selecting filter permeability is to use the pattern of flow for the actual problem at hand, such as Fig. 19.13 for a slope.

19.5 SUMMARY OF MAIN POINTS

1. Permeability is the soil property that indicates the relative ease with which a fluid will flow through the soil.
2. The range of permeability is extremely large, going from 1 cm/sec for gravel to below 10^{-8} cm/sec for clay.
3. Permeability depends on the characteristics of both the permeant and the soil. Viscosity, unit weight, and polarity are the major permeant characteristics. Particle size, void ratio, composition, fabric, and degree of saturation are the major soil characteristics.
4. Filters are essential features of most water retention structures of soil. They serve to give the desired flow pattern and to prevent internal erosion.

PROBLEMS

19.1 Derive Eq. 19.1.

19.2 Estimate the permeability for the soil whose particle size distribution curve is given in Fig. 3.3.

19.3 Estimate the "percent passing a 75 μm sieve" for each soil *A* and soil *B* in Example 18.5.

19.4 On the basis of the permeability data given for the zones in the dam in Problem 18.7, identify the type of soil in each zone.

19.5 Water is to flow from the soil whose particle size distribution is given in Fig. 3.3 into a gravel drain. The gravel consists of uniform particles .05 m in diameter. On a plot of "Percent Finer" versus "Particle Diameter (mm)" plot the curve in Fig. 3.3 and that for a filter material meeting the requirements stated by Eqs. 19.10, 19.11, and 19.12.

19.6 A soil ($k = 10^{-1}$ cm/sec) is to be used as a filter for a soil ($k = 5 \times 10^{-3}$ cm/sec) which exists in an embankment with a discharge face of 1 vertical to 2 horizontal. The flow breaks out of the embankment at a height of 7.5 m. Select the thickness for the filter on the basis of Fig. 19.13.

CHAPTER 20

General Aspects of Drained Stress-Strain Behavior

This chapter describes the general aspects of the stress-strain behavior of saturated soil. The water in the soil pores is allowed to flow into or out of the soil during compression, and thus the pore water pressure is either zero or some other static value at the moment the strain is recorded. The static pore pressure may be a pressure applied during a laboratory test (backpressure) or an equilibrium value of a field pore pressure. The word *drained* is used to describe the condition in which pore water is allowed to flow freely into or out of a soil and thus dissipate any excess pore pressure. *Undrained* and *partially drained* conditions are treated in Part V.

At the outset of our study of drained behavior, let us recognize a most important fact: the behavior of dry cohesionless soil, as presented in Part III, is virtually identical with the drained behavior of cohesionless saturated soil. In fact, many of the stress-strain data presented in Part III were actually obtained from drained tests on saturated sands. In dry sands under normal loading conditions the pore pressure is zero and thus all total stresses are also effective stresses.

Consider Fig. 20.1, which presents stress paths for isotropic compression, for confined compression, and for triaxial compression. The element at the left, for example, is in equilibrium under an isotropic effective stress $\bar{\sigma}_c$ and then subjected to an isotropic stress increment $\Delta\sigma$. Below the element is shown the effective stress path (ESP) for this loading. This effective stress path could be obtained from either a test on dry sand (see Fig. 9.1) or a test on saturated sand with zero water pressure. Thus the line AB is both the effective stress path and the total stress path. At the bottom of the figure is shown the stress paths for the isotropic compression in which a saturated sample of sand was compressed isotropically under drained conditions with the pore water pressure equal to a static value u_s. The effective stress path and the total stress path (TSP) are noted in the figure.

Stress paths for confined compression and triaxial compression are similarly presented in Fig. 20.1. Note that for the drained test on saturated soils the total stress path is displaced horizontally from the effective stress path by a value equal to the static pore pressure. If u_s is positive, the total stress path is to the right of the effective stress path, and if u_s is negative, the total stress path is to the left of the effective stress path. The total stress path and effective stress path are always parallel and displaced horizontally, since pore water cannot take a static shear stress.

This chapter extends Chapter 10 by including the effects of water on soil behavior. Chapter 10 was limited to granular soils; Chapter 20 will describe the behavior of clays and compare it with the behavior of sands.

20.1 MECHANISMS OF STRAIN

As stated in Chapter 10, the strains experienced by an element of soil result from strains within individual soil particles and relative motions among the many particles composing the element. As applied to soils containing clay particles, these mechanisms of strain take the forms shown in Fig. 20.2.

It is easy to imagine that the bending of a platey particle contributes to strain. If we load a mixture of mica and quartz, the situation suggested in Fig. 20.2*a* is likely to result. On applying the load F, the mica particle is bent to the position shown. This type of strain should be recoverable on the removal of the load F.

Earlier parts of this book, e.g., Figs. 7.1 and 7.2, indicate that clay particles can be rearranged into more effective packings. Figure 20.2*b* indicates a vertical strain resulting from the application of a vertical force. This type of strain mechanism is one of the most important contributions to the deformation of undisturbed natural clays. Fabric measurements have indicated that

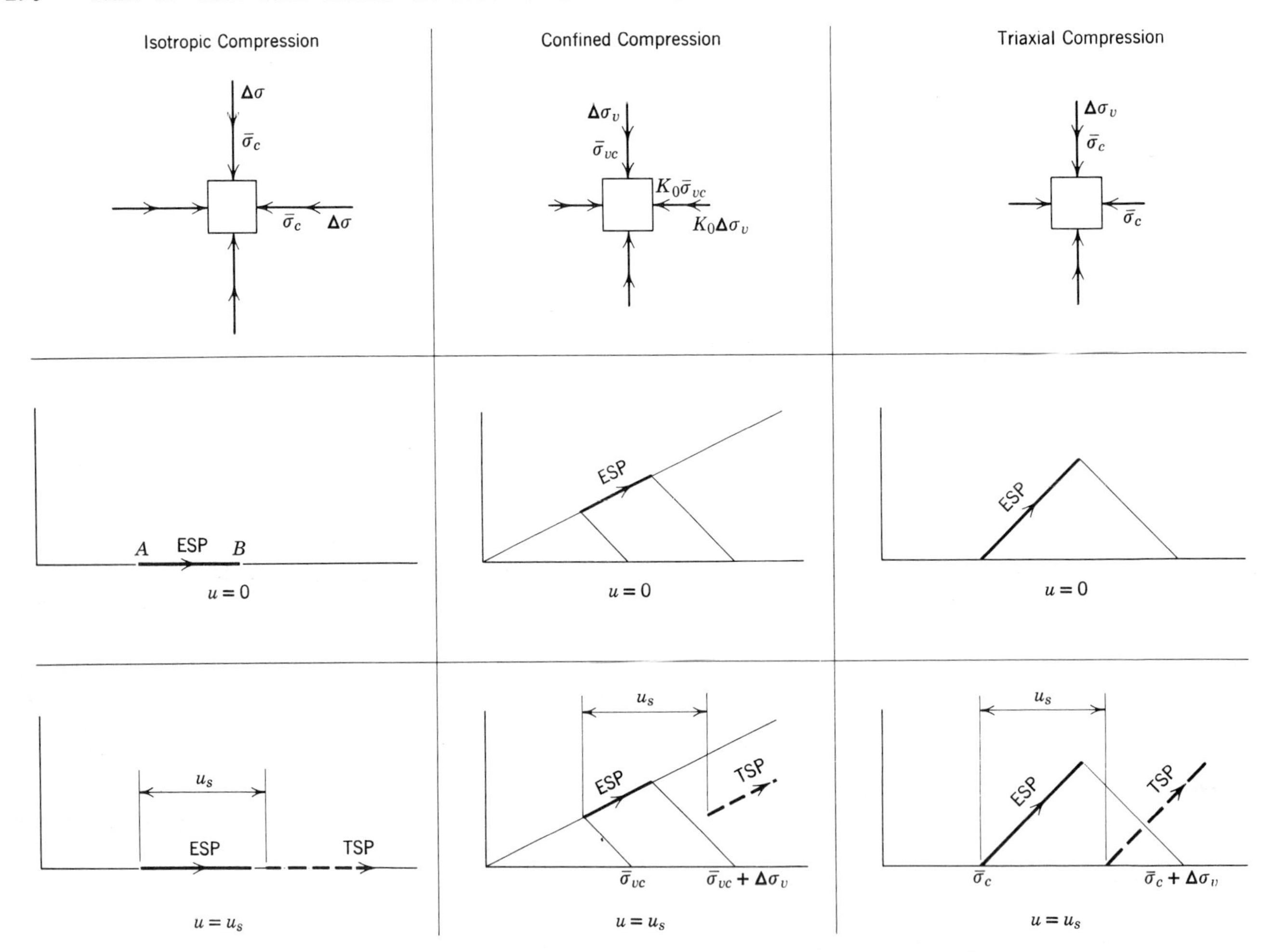

Fig. 20.1 Stress paths for compression tests.

strains, especially along the shear surface, tend to align particles into a parallel array. It seems logical that most of the strain resulting from the reorientation of clay particles is not recoverable upon the removal of the applied stress.

In pure clays a major component of strain is the alteration of the spacing between particles. This phenomenon was discussed in Chapter 5. Figure 5.16*b* showed a plot of spacing between particles versus applied stress. Spacing between pure clay particles may be changed not only by means of applied stress, but also by altering the environment. For example, increasing the salt in the pore fluid or reducing the pH of the pore fluid reduces the spacing between particles. Strains resulting from an alteration in particle spacing are recoverable.

In general, strains in natural soils subjected to loads under 14 MN/m² are caused primarily by the relative movement and rearrangement of individual particles. In remolded, highly plastic clays the alteration of particle spacing is a significant contributor to strain. The fracturing of soil particles at high stress is limited to granular particles.

20.2 STRESS-STRAIN BEHAVIOR DURING CONFINED COMPRESSION

As pointed out in Chapter 10, one-dimensional compression is important in soil mechanics because it approximates a situation commonly found in practice; moreover, it is a very convenient type of loading to obtain in the laboratory. In the one-dimensional laboratory compression test a soil sample is placed in the oedometer (Figs. 9.2 and 9.3) and subjected to an increment of total stress. With a saturated soil sample the increment in total stress is carried initially by the pore water and transferred gradually to the soil skeleton, as described and illustrated in Chapter 2. When compression ceases, a reading is taken to permit the determination of the vertical strain and the test is continued by applying another increment of vertical stress. The oedometer test on a saturated soil sample can be run with the boundary pore water pressure either kept at zero or maintained at some static value by backpressure. Backpressure is helpful because it prevents dissolved air from coming out of the pore water and also more closely represents the pore water pressure conditions in the field.

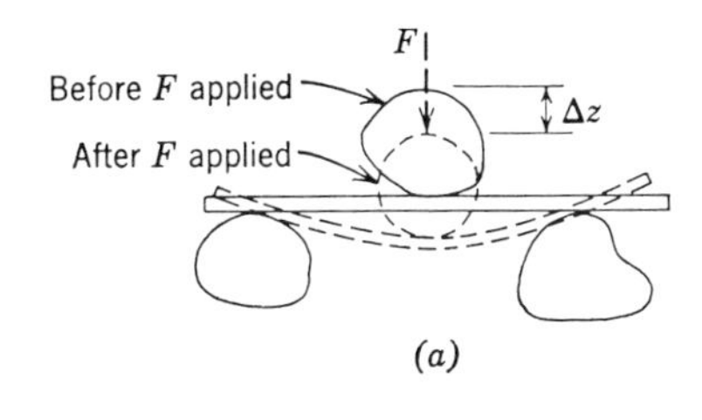

(a)

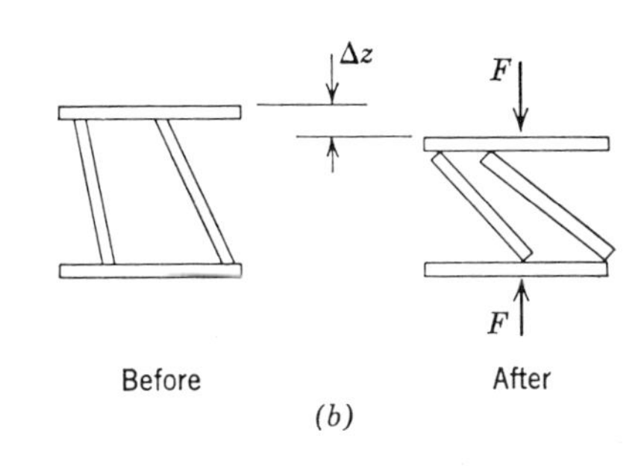

(b)

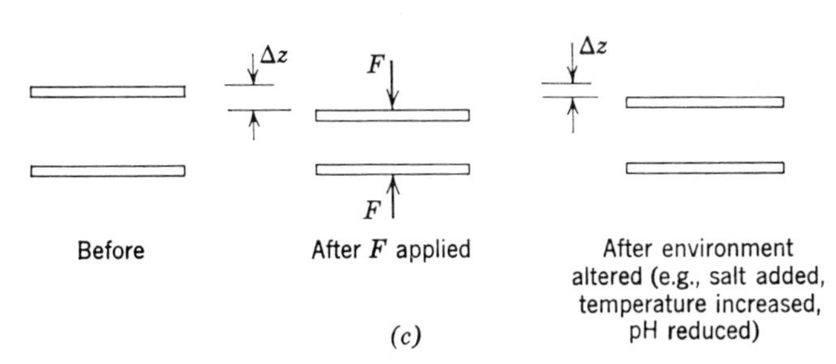

(c)

Fig. 20.2 Strain mechanisms. (a) Deformation of mica particle. (b) Reorientation of particles. (c) Reduction of particle spacing.

Behavior during Loading and Unloading

Figure 20.3 presents the results of an oedometer test on a sample of clay from the Cambridge profile shown in Fig. 7.9. The sample was obtained at a depth of 22.9 m where the *in situ* vertical stress was $\bar{\sigma}_{v0} = 216$ kN/m². In the laboratory test the sample was loaded in increments to 785 kN/m², unloaded back to 98, reloaded to 1570, and finally unloaded back to 49 kN/m². For comparison, the results of the oedometer test on sand, given in Fig. 10.5, are replotted in Fig. 20.3. From this figure we observe that the general stress-strain behavior of the sand and clay is similar, but that the clay is much more compressible than the sand.

In the stress range normally encountered in civil engineering problems, clays are usually much more compressible than the sands. At stresses high enough to cause particle crushing, however, a sand can become as compressible as clay. This fact is illustrated in Fig. 20.4, which presents the results of oedometer tests on montmorillonite, the most compressible clay mineral, kaolinite, one of the least compressive clay minerals, and a 380-840 μm mesh Ottawa sand. Up to a stress of about 21 MN/m², the clays are much more compressible than the sand. Above 21 MN/m², sand particles crush and the sand becomes very compressible. As would be expected, the sand demonstrates very little rebound on unloading.

Maximum Past Consolidation Stress

The stress history of a soil can best be studied from the results of a compression test plotted in the form of void ratio (or strain) versus log of effective stress, as in Fig. 20.5. The results of the oedometer test on the Cambridge clay presented in Fig. 20.5 show a sharper difference between recompression and virgin compression than the plot in Fig. 20.3. Also, $e - \log \bar{\sigma}_v$ for virgin compression tends to be a straight line.

Using the results of an oedometer test plotted in the form of $e - \log \bar{\sigma}_v$, *A.* Casagrande (1936) proposed the following technique for estimating the value of the maximum consolidation stress, termed *maximum past consolidation stress*, and represented by $\bar{\sigma}_{vm}$. His technique, shown in Fig. 20.6, consists of these steps:

1. Locate on the $e - \log \bar{\sigma}_v$ curve the point of minimum radius of curvature T.
2. Through T lay off a horizontal line h and a tangent to the curve t.
3. Bisect the angle formed by t and h, i.e., locate line c.
4. Extend the virgin part of the compression curve backward I, and where this line intersects line c, note D, which is the estimated value of the maximum past consolidation stress $\bar{\sigma}_{vm}$.

There are a number of reasons why the maximum past stress, as estimated by the Casagrande method, is not precise and, in fact, why the laboratory consolidation curve does not reproduce the field compression curve. The most important reason is the change of stress and the change of fabric inherent in the sampling, specimen preparation, and testing procedures. The difference in temperature between ground and laboratory as well as the test details can also be important.

The Casagrande technique is, however, a powerful tool for the soil engineer as long as he realizes that the value of the stress determined is only an estimate. There are other methods available for estimating the maximum past stress (such as the Burmister method and the Schmertmann method), and these are discussed by Leonards (1962). By using the Casagrande technique, the maximum past pressure for the Cambridge clay (Fig. 20.5) is estimated at 304 kN/m². The computed value of vertical effective stress on the sample in the field is 216 kN/m². Thus the specimen tested is a slightly overconsolidated clay having an *overconsolidation ratio* equal to $\bar{\sigma}_{vm}/\bar{\sigma}_{v0} = 1.4$. At vertical effective stresses greater than 304 the clay is normally consolidated.

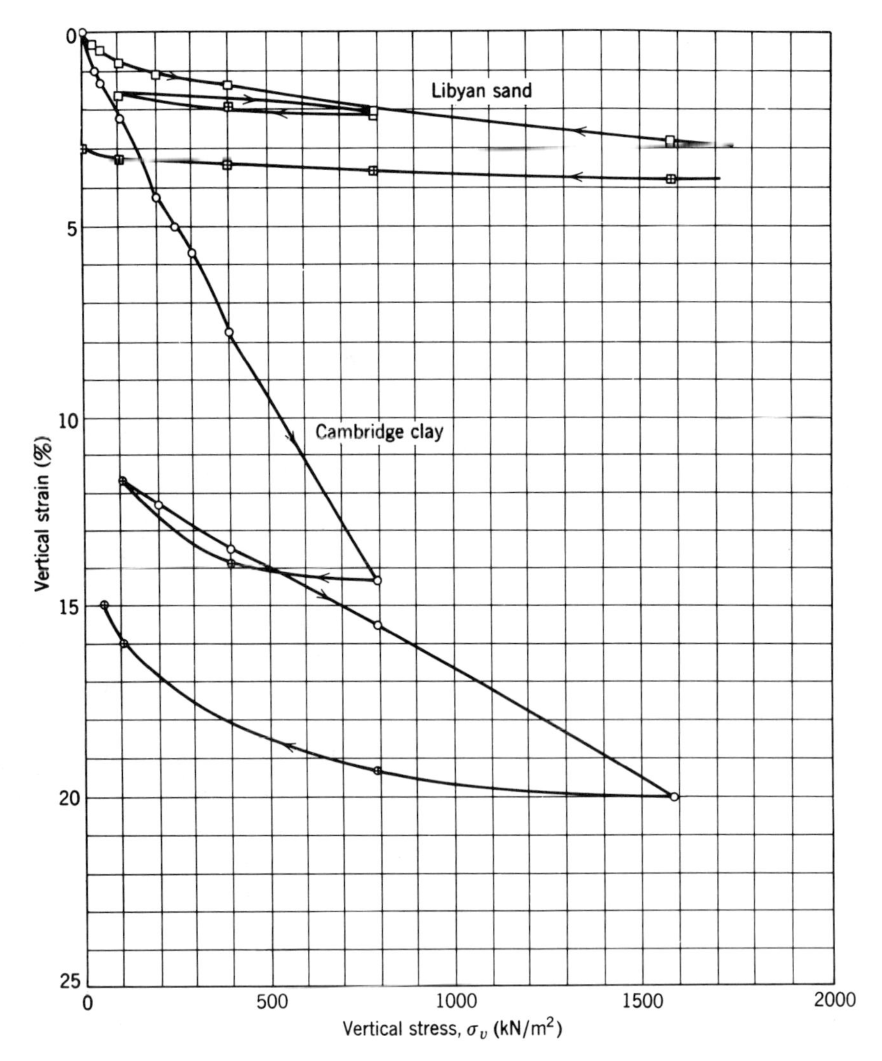

Fig. 20.3 Results of oedometer tests.

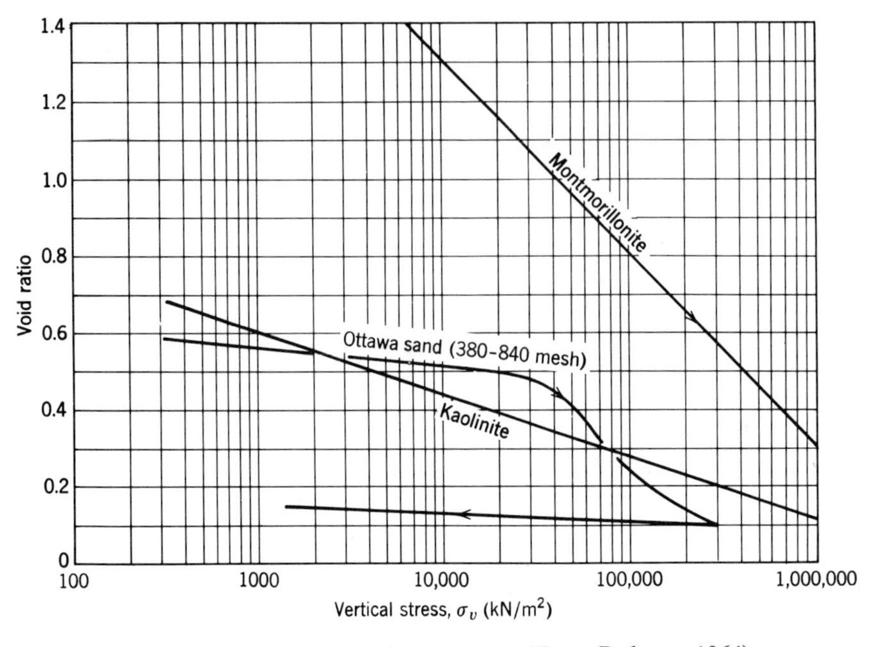

Fig. 20.4 Results of oedometer tests (From Roberts, 1964).

Time Effects

When a sample of saturated soil is subjected to an increment of vertical stress in the oedometer, the strain that occurs is time dependent. This time dependency results from two phenomena: *hydrodynamic time lag* and *secondary compression.* The hydrodynamic time lag was discussed and illustrated in Chapter 2. The increment of vertical stress is initially carried fully by the pore water and is transferred to the soil skeleton as water flows out under the induced excess pore pressure. The time required for the water to flow out of the soil depends on the nature of the stress applied, the soil compressibility, the soil thickness, the type of drainage, and the soil permeability. Chapter 27 presents a theory for estimating the time required for the excess pore pressure to become zero.

The standard oedometer test employs a specimen about 32 mm in thickness. For such a specimen and drainage at the top and bottom of the sample, the excess pore pressure induced in a saturated sand sample during an oedometer test is essentially dissipated in less than a minute. In a clay, from approximately 10 min to more than 1000 min can be required for the excess pore pressure to become zero.

Even after all of the vertical stress has been transferred from the pore water pressure to effective stress, compression continues. In a sand, at usual stresses, this action is completed so rapidly that it usually is not noticed (see Fig. 10.8). A sand exhibits considerable secondary compression at high stress because of particle fracturing (see Fig. 10.9). For a sand with solid mineral particles this phenomenon becomes important only at stresses 14 MN/m² and higher. Sands with softer particles or weakly cemented particles exhibit significant secondary compression at stresses typically encountered by the civil engineer.

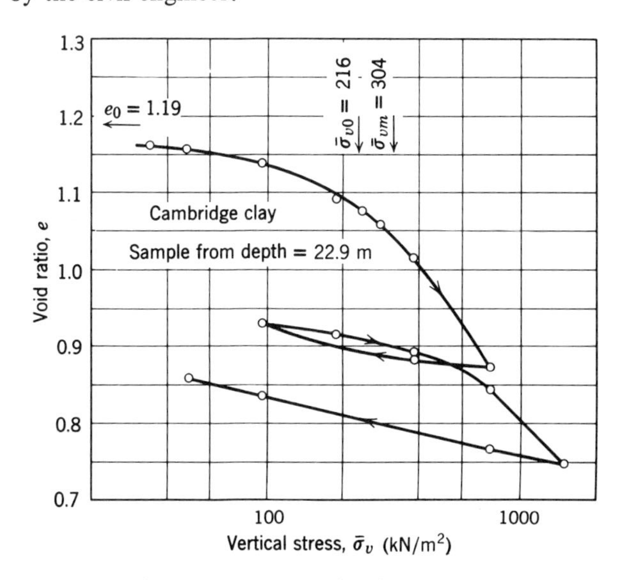

Fig. 20.5 Results of oedometer test.

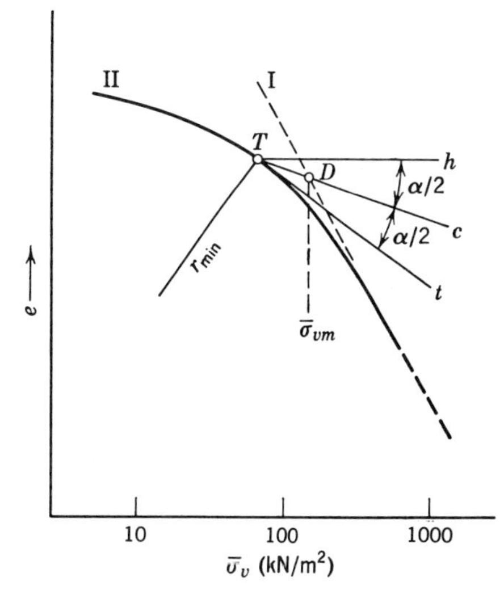

Fig. 20.6 Estimation of maximum past consolidation stress (After Casagrande, 1936).

In highly plastic soils, especially organic soils, secondary compression can be very large. In fact, in such soils the compression occurring after the excess pore pressures become zero can exceed the compression that occurs during the transfer of excess pore pressure to effective stress. The exact cause of secondary compression is not known. It is probably caused by continued reorientation of particles, possibly influenced by the extrusion of water which is held by attractive forces from the soil particles.

Lateral Stresses during Confined Compression

Figure 20.7 shows the effective stress path for an oedometer test in which the sample starts at A under a vertical stress of 400 kN/m². The vertical stress is increased to 800 kN/m² and the stress path for the loading is from A to B along the K_0-line. Experiments have shown that the K_0-line is essentially straight. Its slope is designated by $\bar{\beta}$, the angle whose tangent is $(1 - K_0)/(1 + K_0)$. At point B the vertical effective stress is reduced from 800 kN/m² back to 400, and the soil sample expands, as suggested by the shapes in Fig. 20.7. The stress path for this expansion is BC.

At all locations along the K_0-line for virgin compression of this soil, K_0 is equal to $\frac{1}{2}$. In proceeding from point B back to point C, both the overconsolidation ratio and K_0 are increasing. At point C the overconsolidation ratio is 2, and K_0 is equal to 0.8. In an oedometer test on clay line BC can continue until it reaches $q = 0$ i.e., $K_0 = 1$ and, in fact, can go to negative values of q, i.e., $K_0 > 1$.

By running a series of oedometer tests in which not only the vertical effective stress but the lateral effective stress was measured (see Fig. 9.3), Hendron, Brooker, and Ireland (Brooker and Ireland, 1965) developed the

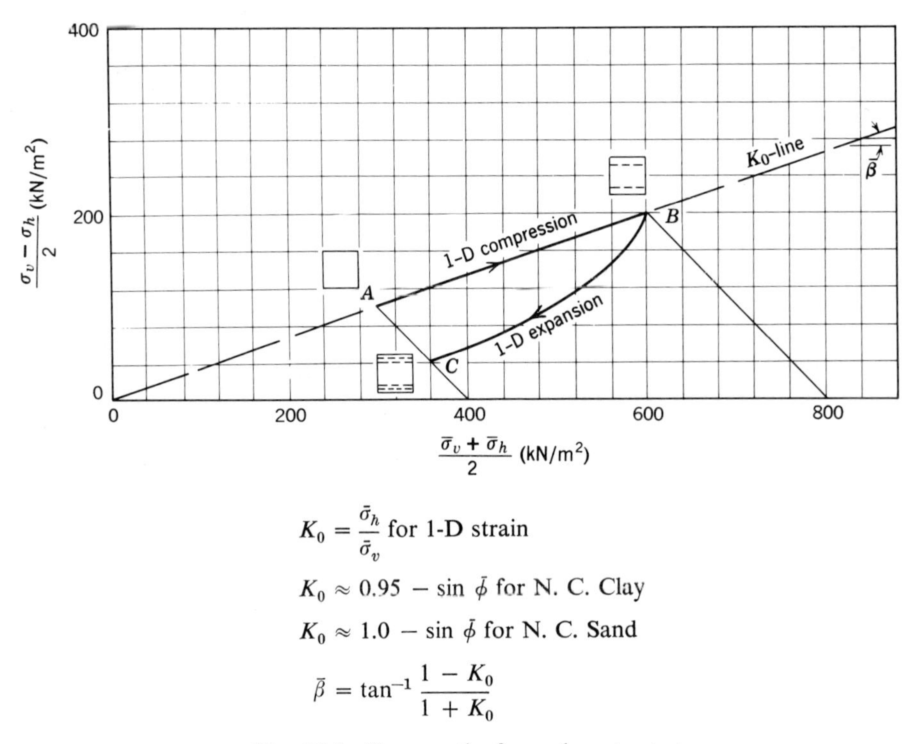

$$K_0 = \frac{\bar{\sigma}_h}{\bar{\sigma}_v} \text{ for 1-D strain}$$

$$K_0 \approx 0.95 - \sin \bar{\phi} \text{ for N. C. Clay}$$

$$K_0 \approx 1.0 - \sin \bar{\phi} \text{ for N. C. Sand}$$

$$\bar{\beta} = \tan^{-1} \frac{1 - K_0}{1 + K_0}$$

Fig. 20.7 Stress paths for oedometer test.

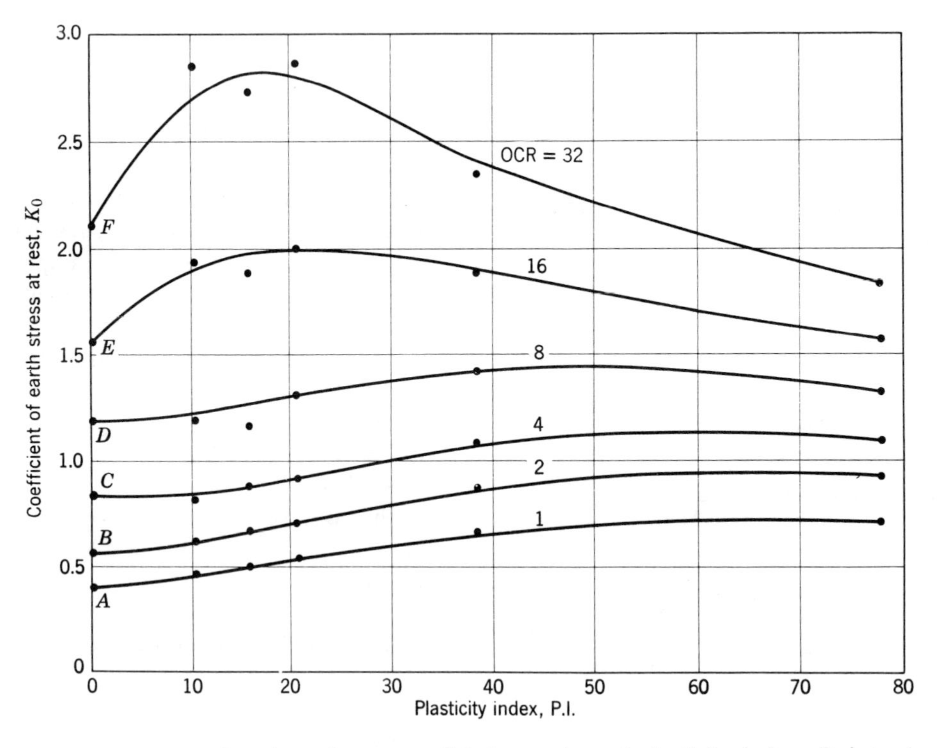

Fig. 20.8 K_0 as function of overconsolidation ratio and plasticity index. Points A to F interpolated from Hendron's data. Overconsolidation ratio—OCR (From Brooker and Ireland, 1965).

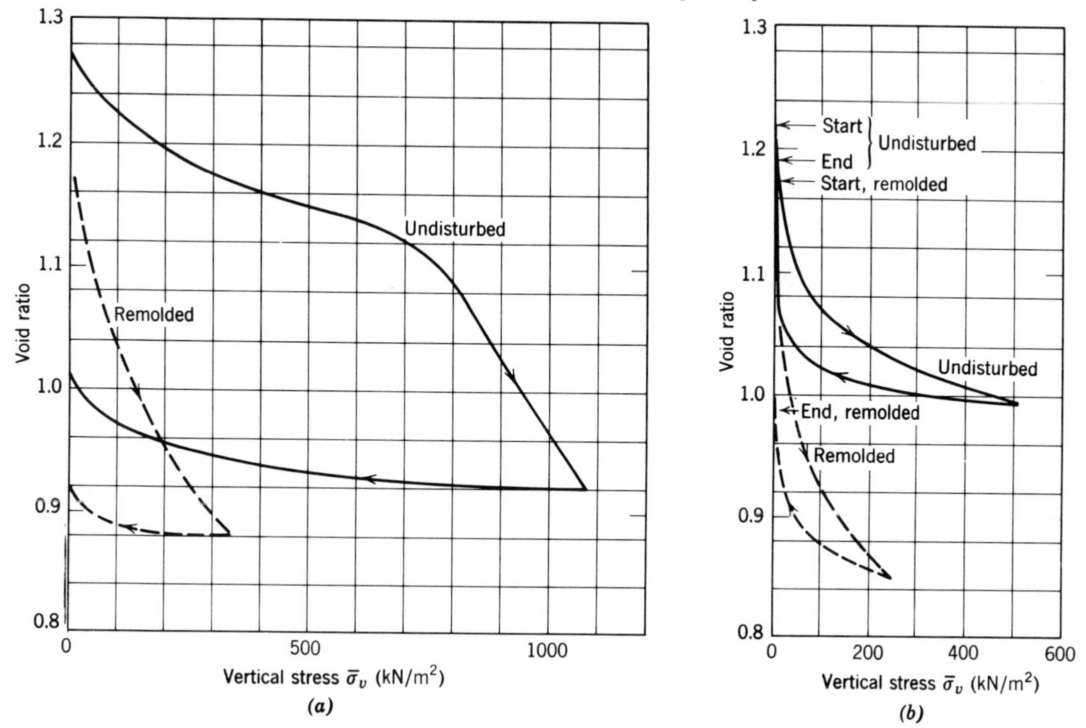

Fig. 20.9 Effect of disturbance on stress-strain behavior. (*a*) Laurentian clay. (*b*) Boston clay. (Data from Casagrande, 1932.)

relationships among K_0, overconsolidation ratio, and plasticity index shown in Fig. 20.8. As an average for all normally consolidated clays,

$$K_0 = 0.95 - \sin \bar{\phi} \tag{20.1}$$

Alpan (1967) recommends

$$K_0 = 0.19 + 0.233 \log \text{P.I.} \tag{20.2}$$

for normally consolidated clay, where P.I. = plasticity index in %.

Effect of Disturbance on Stress-Strain Behavior

As described in Chapter 7, a natural clay develops a structure that depends on conditions during formation of the clay and changes in environment occurring after formation. As already noted in this chapter, strains tend to disturb this structure, rearranging particles. An indication of the effect of extreme disturbance on the behavior of a clay can be obtained by testing the clay in an undisturbed state and then after it has been thoroughly worked—"remolded." Figure 20.9 presents the results of oedometer tests on two clays. The test data in Fig. 20.9 indicate that destroying the natural structure in a clay by remolding greatly increases the compressibility, especially at low stress levels.

20.3 STRESS-STRAIN BEHAVIOR IN TRIAXIAL COMPRESSION

As shown in Fig. 20.1, the loading in the standard triaxial test consists of consolidating the soil sample to an isotropic stress of $\bar{\sigma}_c$, and then shearing the sample by applying a vertical stress while holding the horizontal stress constant. The major principal stress is thus the vertical stress and the minor (and intermediate) principal stress is the horizontal stress.

Figure 20.10 shows the results of two standard drained triaxial tests on initially remolded Weald clay. The normally consolidated clay was consolidated to 207 kN/m² and then failed by increasing the vertical stress. The overconsolidated specimen was first consolidated isotropically to 827 kN/m² and then allowed to swell back to an isotropic stress of 34.5 kN/m². The specimen was then failed by increasing the vertical stress. In Fig. 20.10 are plotted strain versus stress expressed as $(\sigma_1 - \sigma_3)/2\bar{\sigma}_c = q/\bar{\sigma}_c$; i.e., the deviator stress has been normalized, as discussed in Chapter 10.

Comparing the behavior of the normally consolidated clay and the heavily overconsolidated clay, we observe two important characteristics:

1. The overconsolidated clay is stronger and stiffer. It has a peak shear strength which it loses with further strain, and thus the normally consolidated and over-consolidated clay strengths approach each other at large strains.
2. The overconsolidated clay first decreases and then increases in volume, whereas the normally consolidated clay decreases in volume throughout the test.

A comparison of the clay stress-strain behavior in Fig. 20.10 with that for sands in Fig. 10.18 illustrates a

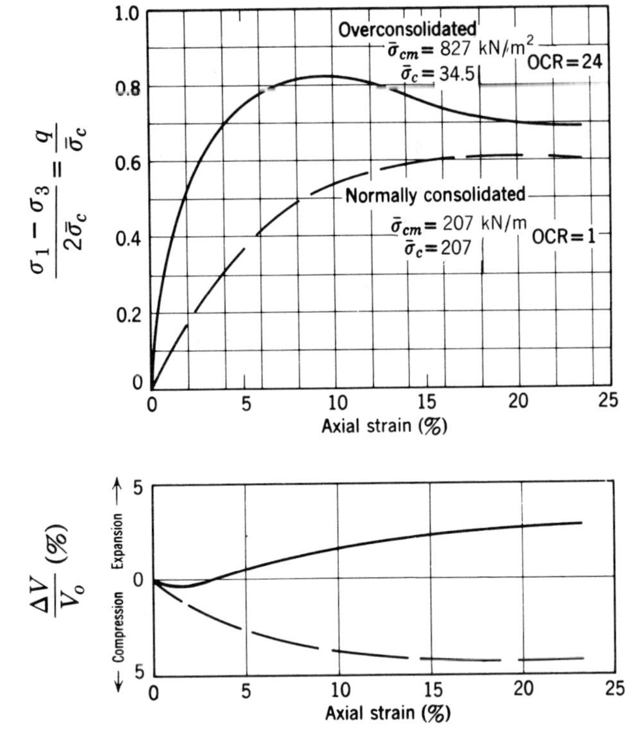

Fig. 20.10 Drained triaxial tests on Weald clay (Data from Henkel, 1956).

most important general fact: the stress-strain curve of an overconsolidated clay is similar to that of a dense sand, whereas the behavior of a normally consolidated clay is similar to that of a loose sand. In general, the strains required to reach peak strength are greater in clay than in sand. During loading, loose sand tends to reduce then increase in volume, whereas normally consolidated clay shows little tendency to expand after volume decrease. The ultimate strengths of overconsolidated clay and of normally consolidated clay approach each other at large strains just as occurred with dense and loose sands.

20.4 Other Loading Conditions

This chapter so far has considered the stress-strain behavior of soils for confined loading and standard triaxial loading. The oedometer and the triaxial machine are widely used devices for measuring the stress-strain behavior of soils. The triaxial machine is a particularly versatile device since it can run many types of loading other than the standard one of $\Delta\sigma_h = 0$. Figure 9.8 shows some of the other types of loadings which can be obtained in the triaxial machine. In the triaxial machine, however, σ_2 must be equal to either σ_1 or σ_3. There are certain field situations which present loadings that cannot be duplicated in the triaxial machine. For example, a long embankment imposes plane strain in the underlying soil. A plain strain device is thus needed to simulate this field condition.

The direct shear test, described in Fig. 9.1, is useful for running drained tests on clay. Since the direct shear machine employs a thin sample, we can run a drained direct shear test in a fifth to a tenth of the time required to run a drained standard triaxial test. Further, it is more convenient to subject a sample of clay to large strains and to cycles of strain in the direct shear machine than in the triaxial machine.

The direct shear test is widely used to study the strength of an overconsolidated clay at very large strains, i.e., the *ultimate strength* or *residual strength.* Figure 20.11 shows the results of a direct shear test on a specimen of Cucaracha clay-shale from the Panama Canal. As can be seen, the residual strength is a small fraction of the peak strength; this is one reason why slides occurred in the Panama Canal. This loss in strength resulted partially from destruction of bonds between particles and partially from a reorientation of particles.

20.5 SUMMARY OF MAIN POINTS

1. The effective stress-strain behavior of granular soil is virtually the same for dry and saturated conditions.
2. At normal stresses (up to 1400 kN/m²) the major components of strain are the relative movements between adjacent particles and the rearrangement of particles. At stresses above 14 MN/m² the crushing of granular particles can be a major cause of strain.
3. The stress-strain behavior of a clay is greatly dependent on the stress history of the sample—the higher the overconsolidation ratio, the stiffer the soil.
4. The stress-strain behavior of a loose sand is similar to that of a normally consolidated clay; whereas the behavior of a dense sand is similar to that of an overconsolidated clay.

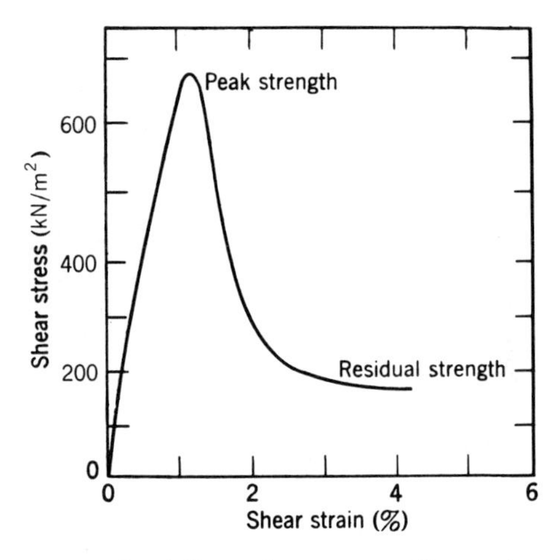

Fig. 20.11 Drained direct shear test on Cucaracha clay-shale (From Panama Canal, 1947).

5. K_0 for a normally consolidated clay is approximately equal to $0.95 - \sin \bar{\phi}$. K_0 increases with overconsolidation ratio.

PROBLEMS

20.1 Using Figs. 20.8 and 7.9 estimate K_0 for normally consolidated Cambridge clay.

20.2 A static pore pressure of 100 kN/m² was used in the oedometer test whose results are plotted in Fig. 20.5. Plot the total stress path and effective stress path for the second load-unload cycle of the test, i.e., 100 to 1600 to 50 kN/m².

20.3 An increment of vertical stress equal to 100 kN/m² is applied to confined specimens of the soils listed below. For each soil list the mechanism(s) that is probably an important cause of strain.

a. Montmorillonite (Fig. 5.16*b*).
b. London clay, depth = 3.05 m (Fig. 7.8).
c. Manglerud clay, depth = 6 m (Fig. 7.7).
d. Ottawa sand, $\bar{\sigma}_v$ = 34.5 MN/m² (Fig. 20.4).
e. Laurentian clay, $\bar{\sigma}_v$ = 500 kN/m² (Fig. 20.9).

20.4 Two meters of fill are placed over a deposit of Cambridge clay (see Fig. 20.5). Will more settlement occur if the overconsolidation ratio is 2 than if it is 1? Explain.

CHAPTER 21

Drained Shear Strength

Chapter 11 presented fundamental concepts of strength theory and strength data for dry cohesionless soil. As long as stresses are expressed in terms of effective stresses, the concepts and data presented in Chapter 11 also hold for saturated cohesionless soils wherein there are no excess pore pressures. Chapter 21 extends Chapter 11 to cover the strength behavior of clays and compares this behavior of clays with that of sands. As in Chapter 20, Chapter 21 is limited to the condition where the pore pressure is either static or zero. Excess pore pressures generated by strains are permitted to dissipate by the free movement of pore water, i.e., *drained* conditions.

Drained tests are usually called *consolidated-drained*, abbreviated CD, in order to have consistent terminology with undrained tests presented in Part V. It is really necessary, however, to use only the word drained, since if the excess pore pressure is equal to zero at all times the clay must be consolidated at the start of the test. Since drained tests on clay are run at a slow rate in order to permit the excess pore pressures to dissipate, drained tests are also called *slow tests.*

To present the principles of drained strength, use will be made of actual test data on remolded Weald clay obtained by Henkel (1956, 1959). Weald clay is an estuarine clay of the Cretaceous period and has the following properties:

$$\begin{aligned} \text{Liquid limit} &= 43\% \\ \text{Plastic limit} &= 18\% \\ \text{Plasticity index} &= 25\% \\ \text{Percent clay minus 0.002 mm} &= 40\% \\ \text{Activity} &= 0.6 \\ \text{Specific gravity} &= 2.74 \end{aligned}$$

The Weald clay is used for illustration both because of the wealth of high-quality test data available and because this clay has typical behavior. Later in this chapter deviations from typical behavior are presented and discussed.

21.1 NORMALLY CONSOLIDATED CLAY

On identical specimens of remolded Weald clay we run the following three tests:

1. Consolidated to 69 kN/m² and then failed in a standard triaxial test.
2. Consolidated to 207 kN/m² and then failed in a standard triaxial test.
3. Consolidated to 690 kN/m² and then failed in a standard triaxial test.

The results of these three tests are shown in Fig. 21.1.

In the standard triaxial test the vertical stress σ_v is the major principal stress and the horizontal stress σ_h is the intermediate and minor principal stress. Figure 21.1*a* presents half the deviator stress, q, versus vertical strain. Figure 21.1*b* presents volumetric strain versus vertical strain. Figure 21.1*c* shows the effective stress path for each of the tests, and a line drawn through the peak value of q for each test. This line, called the K_f-line, is a plot of q_f versus $\bar{p}_f$. Figure 21.1*d* presents void ratio versus $\bar{p}_0$, $\bar{p}_f$, and q_f.

Strength Versus Effective Stress

Based on the data shown in Fig. 21.1*c* plus many other points not shown, the effective stress-strength relation of remolded, normally consolidated Weald clay is found to be a straight line through the origin. Thus there appears to be no essential difference between the form of the effective stress-strength relationship, as found from drained tests, of sand and of normally consolidated clay; neither soil exhibits a q_f intercept at $\bar{p}_f$ equal to zero, meaning neither soil exhibits "cohesion." The slope of the K_f-line is $\bar{\alpha}$:

$$\bar{\alpha} = \tan^{-1}\frac{q_f}{\bar{p}_f}$$

and (from geometry, as noted in Chapter 11)

$$\bar{\phi} = \sin^{-1}\tan\bar{\alpha}$$

For the Weald clay $\bar{\alpha} = 20\frac{1}{2}°$ and $\bar{\phi} = 22°$.

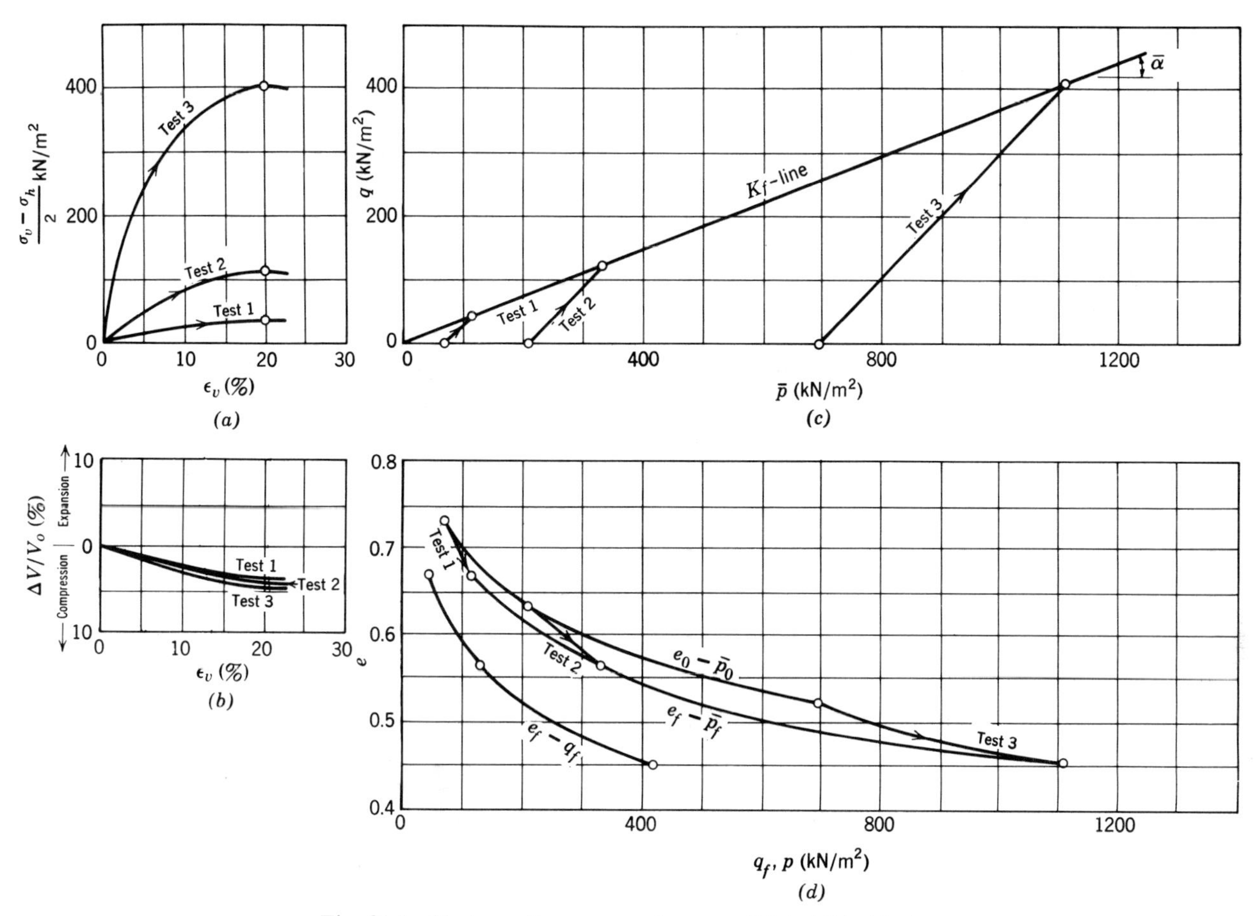

Fig. 21.1 Test results on normally consolidated Weald clay.

Figure 21.2 shows stress-strain and stress path data for normally consolidated Weald clay presented in a normalized fashion. Since the K_f-line is straight, the $\bar{p}$-q plot can be accurately normalized. The stress-strain data can only be approximated by normalized curves.

Stress-Volume Relationship

From Fig. 21.1 we can see that the test data from the drained tests on the normally consolidated Weald clay indicate that $e_0 - \bar{p}_0$, $e_f - \bar{p}_f$, and $e_f - q_f$ are smooth plots. Also shown in this figure are lines indicating how the void ratio changes during the drained tests. The stress-volume data in Fig. 21.1*d* are replotted in Fig. 21.3 in a more convenient form. Water contents are used in place of void ratios since they are more convenient to determine, and for a saturated soil, water content and void ratio are uniquely related. The stress data are plotted on a log scale since it has been found empirically that void ratio or water content versus the various stresses form approximately straight lines which are parallel to one another for normally consolidated clay.

From the data plots in Fig. 21.1*d* or 21.3, we see that there is a three-way relationship among shear strength, effective stress at failure, and water content or void ratio at failure. In Part V we shall see that the same $q_f - \bar{p}_f - w_f$ relation still applies to Weald clay when the clay is sheared without permitting water to escape.

Any factor that alters the void ratio, i.e., the density of the skeleton, should alter the drained strength, which should also be accompanied by a change in the effective stresses existing at failure. In a sense, a change in effective stress can produce a material with a different density and hence a different strength. It is difficult to say which factor—density or effective stress—controls strength: all three of these factors are interrelated. However, it is more useful to use *effective stress* as the primary controlling variable.

Other Loading Conditions

There is an infinite number of possible stress paths for shearing a sample of clay. Figure 9.8 showed four convenient stress paths for the triaxial tests. In general, the strength parameters of normally consolidated soil are approximately independent of the stress path to failure. As a first and a very good general approximation, we can take q_f versus $\bar{p}_f$ for normally consolidated clay as a unique straight line. The value of q_f, of course, depends very much on the stress path of a loading since $\bar{p}_f$ is a

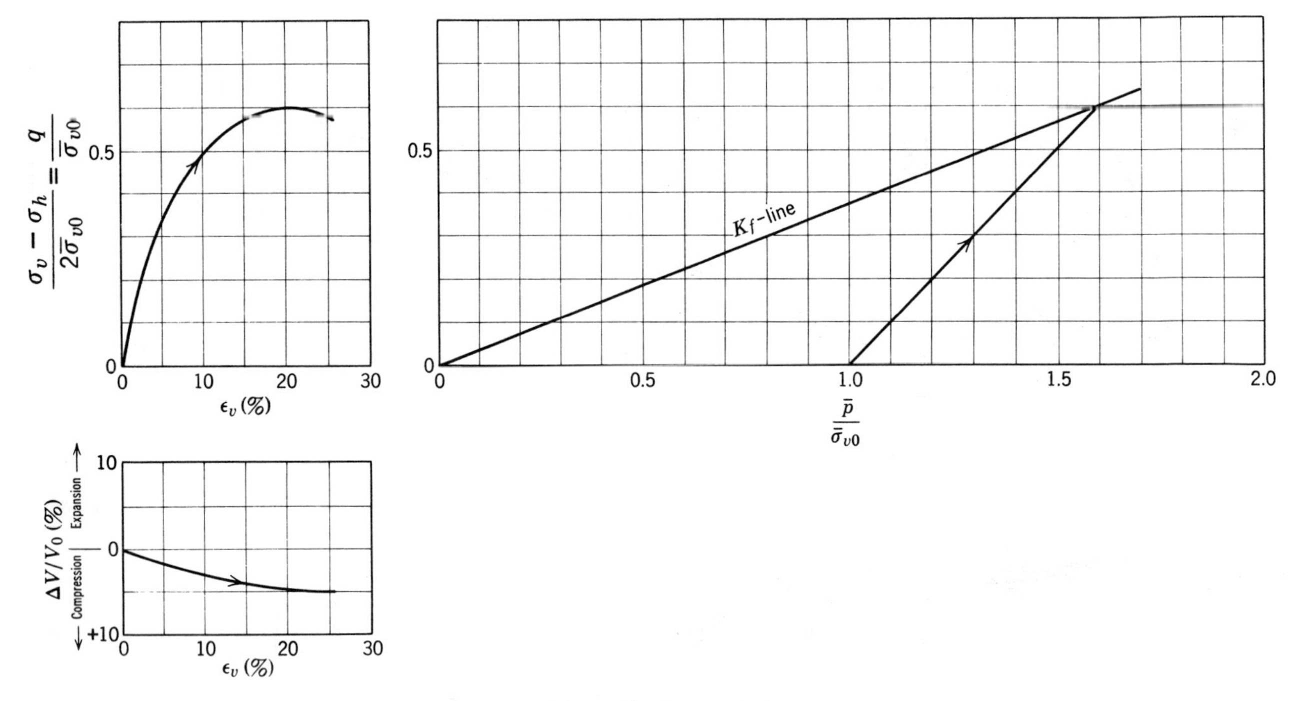

Fig. 21.2 Normalized test results.

function of loading stress path. Examples 21.1 to 21.3 illustrate the estimation of q_f for various loading conditions.

Comparison of Clay and Sand

The preceding discussion illustrates that the behavior of normally consolidated clay during drained shear is essentially similar to that of sand, or more particularly, loose sand. That is, for both classes of soil the failure law is

$$\tau_{ff} = \bar{\sigma}_{ff} \tan \bar{\phi}$$

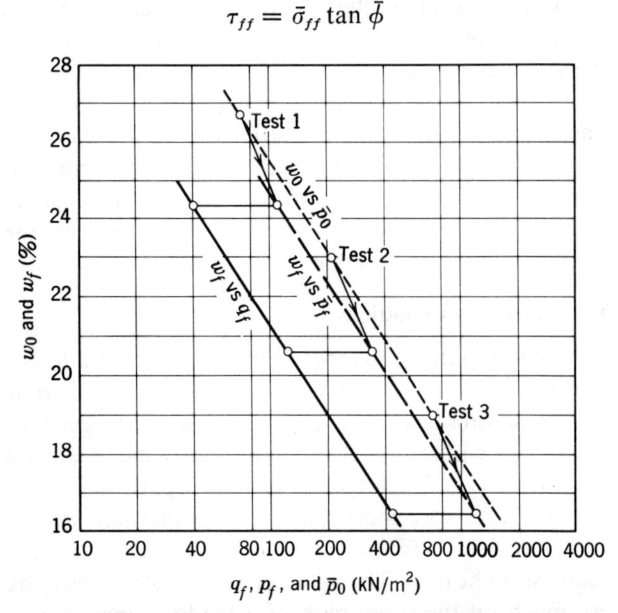

Fig. 21.3 Test results on normally consolidated Weald clay.

Typical Values of $\bar{\phi}$ for Normally Consolidated Clays

Figure 21.4 gives a good indication of typical values of $\bar{\phi}$ for soil. Although there is considerable scatter, there is a definite trend toward decreasing $\bar{\phi}$ with increasing plasticity. It is difficult to determine the effective stress-strength relationship for an extremely plastic clay such as montmorillonite or the Mexico City clay because of the extremely long time required for dissipation of excess pore pressures during the test. Hence the magnitude of $\bar{\phi}$ for very plastic soils is uncertain.

21.2 OVERCONSOLIDATED CLAY

Now let us take several specimens of remolded Weald clay and consolidate them to 827 kN/m². This stress (represented by $\bar{p}_m$ for maximum past $\bar{p}$) is greater than any effective stress to which the specimens have been subjected since remolding. Now the chamber pressure acting on each specimen is reduced to values $\bar{p}_0$ ranging from 34.5 to 483 kN/m², and water is permitted to flow into the specimens so that they come to equilibrium under these reduced effective stresses. The overconsolidation ratio OCR (defined as OCR $= \bar{p}_m/\bar{p}_0$; in one-dimensional compression, OCR is $\bar{\sigma}_{vm}/\bar{\sigma}_{v0}$) is a convenient way to characterize the condition of these specimens. The values of OCR for our specimens range from 1.7 to 24.

Strength Versus Effective Stress

Figure 21.5 shows the stress paths for these tests. As long as $\bar{p}_f > 0.5\bar{p}_m$, the q_f versus $\bar{p}_f$ points fall along the

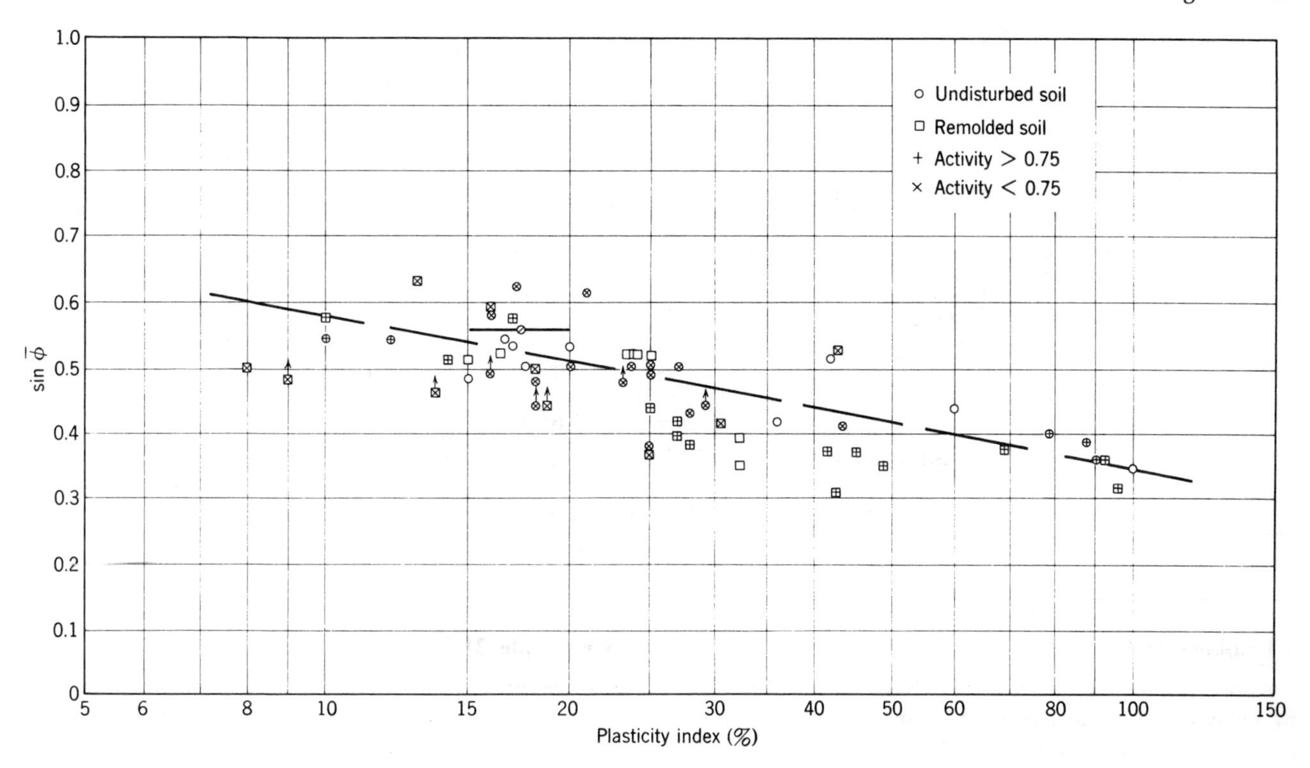

Fig. 21.4 Relationship between sin $\bar{\phi}$ and plasticity index for normally consolidated soils (From Kenney, 1959).

same line obtained from tests on normally consolidated specimens. For lower $\bar{p}_f$ the points fall above the relation from normally consolidated tests. Thus preconsolidation affects the effective stress-strength relation and tends to make the sample stronger at a given $\bar{p}_f$. This preconsolidation effect is difficult to see when results are plotted to the scale of Fig. 21.5*a*, hence the portion of this plot near the origin is magnified in the lower portion of the figure.

▶ Example 21.1

A specimen of Weald clay is consolidated to 689 kN/m², and is then failed by decreasing $\bar{\sigma}_3$ while $\bar{\sigma}_1$ is held constant. Find q_f, $\bar{p}_f$, and w_f.

Solution. On part (*c*) of Fig. 21.1, draw the effective stress path for this loading until it intersects the q_f–$\bar{p}_f$ relation (see Fig. E21.1). The point of intersection gives q_f and $\bar{p}_f$. Then go to Fig. 21.3, enter with *either* q_f or $\bar{p}_f$, and read w_f.

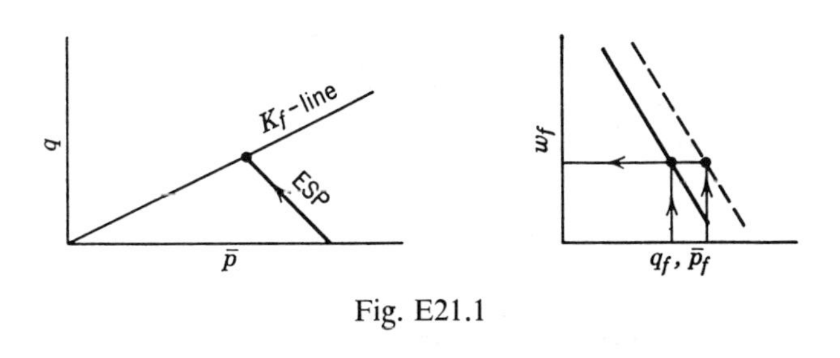

Fig. E21.1

Answers. q_f = 186 kN/m², $\bar{p}_f$ = 503 kN/m², w_f = 19.2%. Note that w increased slightly during shear. ◀

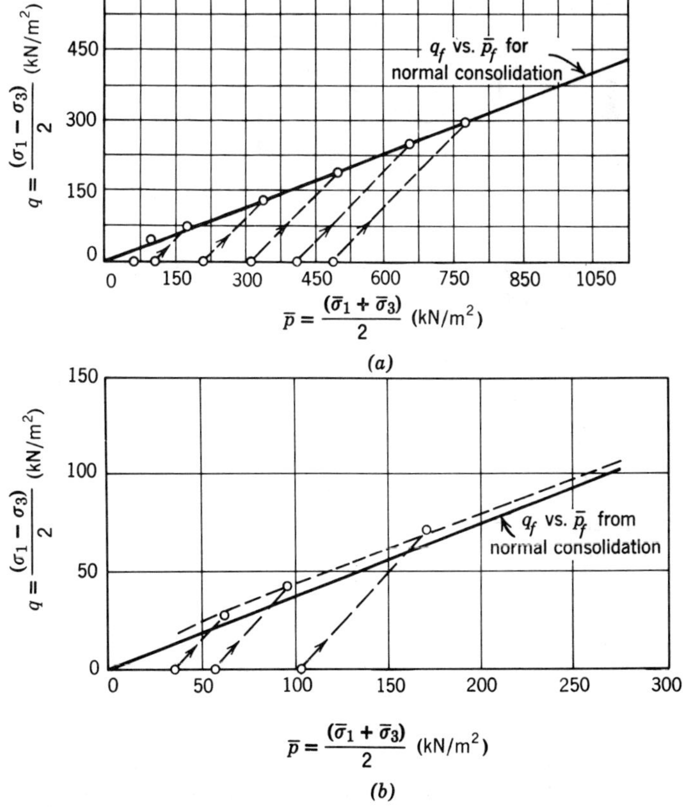

Fig. 21.5 Results of CD tests on overconsolidated Weald clay. $\bar{p}_m$ = 827 kN/m².

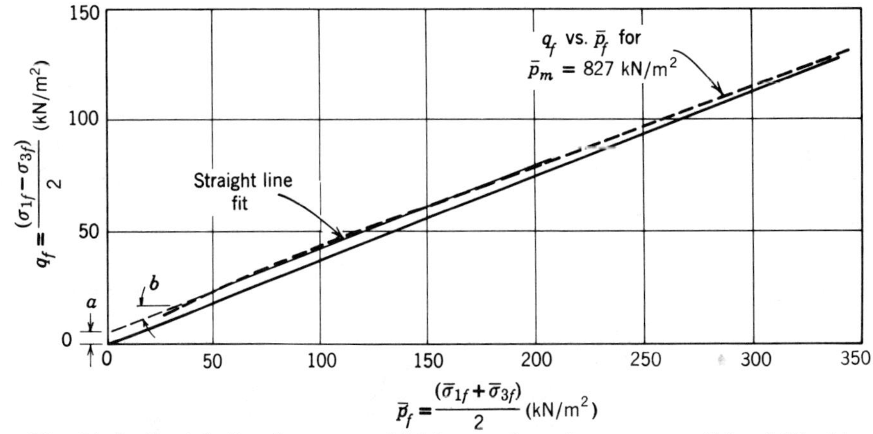

Fig. 21.6 Straight line fit to curved Mohr envelope for overconsolidated Weald clay.

$a = 6.9$ kN/m² $\quad b = 20.3°\quad \sin\bar{\phi} = \tan b \quad \bar{c} = a/\cos\bar{\phi}$

$\bar{\phi} = 21.5° \quad \bar{c} = 7.4$ kN/m²

▶ **Example 21.2**

A specimen is consolidated to 414 kN/m², and is then failed by increasing $\bar{\sigma}_1$ and decreasing $\bar{\sigma}_3$ in such a way that $\bar{p}$ remains constant. Find q_f, $\bar{p}_f$, and w_f.

Solution. For this case the effective stress path is a vertical line (see Fig. 21.2) and $\bar{p}_f = 414$ kN/m². Next turn to Fig. 21.3 to find the answers.

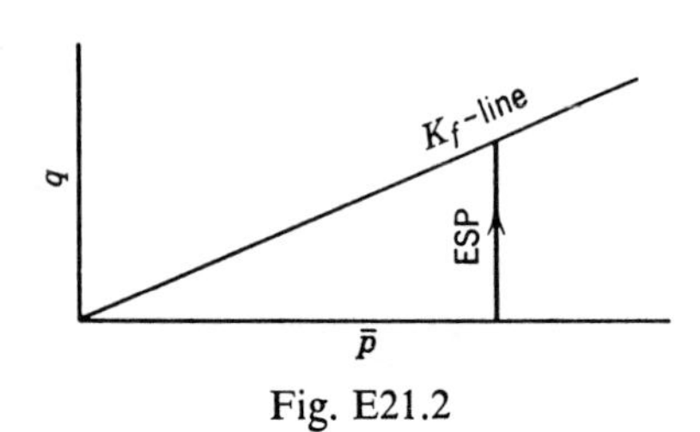

Fig. E21.2

Answers. $q_f = 155$ kN/m², $w_f = 19.9\%$. Note that w decreased during shear. (Incidentally, actual test data for this loading gave $q_f = 155$ kN/m² and $w_f = 19.6\%$.) ◀

The effective stress-strength relation for overconsolidated Weald clay (see Fig. 21.5) has been reproduced in Fig. 21.6. This relation is not a straight line. For calculation purposes, it is usually desirable to replace the curved relation with a straight line, thus expressing strength as

$$\tau_{ff} = \bar{c} + \bar{\sigma}_{ff} \tan\bar{\phi}$$

This straight line is drawn to provide the best fit to the actual curve over the stress range of interest. Such a straight line, when extended to the ordinate, generally produces a *cohesion intercept* even though the effective stress-strength relation itself may curve back through the origin. For the particular straight-line fit shown in Fig. 21.6, the strength parameters for Weald clay are

▶ **Example 21.3**

A specimen is consolidated to 207 kN/m², and is then failed by increasing $\bar{\sigma}_1$ and $\bar{\sigma}_3$, such that $\Delta\bar{\sigma}_3 = \frac{1}{3}\Delta\bar{\sigma}_1$. Find $\bar{p}_f$, q_f, and w_f.

Solution. From the given data:

$$\left.\begin{aligned}\Delta\bar{p} &= \frac{\Delta\bar{\sigma}_1 + \Delta\bar{\sigma}_3}{2} = \tfrac{2}{3}\,\Delta\bar{\sigma}_1\\ \Delta q &= \frac{\Delta\bar{\sigma}_1 - \Delta\bar{\sigma}_3}{2} = \tfrac{1}{3}\,\Delta\bar{\sigma}_1\end{aligned}\right\} \quad \Delta q = \tfrac{1}{2}\Delta\bar{p}$$

The effective stress path rises at a slope of 2 on 1 to the right (see Fig. E21.3). The intersection with q_f versus $\bar{p}_f$ (Fig. 21.1*c*) gives q_f and $\bar{p}_f$. w_f is then read from Fig. 21.3.

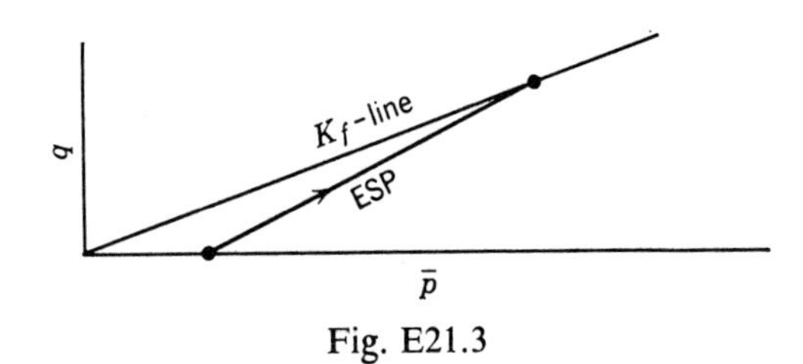

Fig. E21.3

Answers. $q_f = 300$ kN/m², $\bar{p}_f = 807$ kN/m², $w_f = 17.4\%$. Note that there has been a large decrease in water content during shear (from 23% to 17.4%), and a very large axial strain would be necessary in order to achieve failure. ◀

$\bar{\phi} = 21.5°$ and $\bar{c} = 7.4$ kN/m². This situation is similar to that experienced with dense sands.

Stress-Volume Relationships

The relationships between $\bar{p}_0$ and w_0, $\bar{p}_f$ and q_f and w_f are shown in Fig. 21.7. For convenience in separating the relationships, they are given in two diagrams. The straight portions without data points are the relations for normally consolidated specimens.

The effect of preconsolidation is seen in all of these relations. Thus the w_f versus q_f and w_f versus $\bar{p}_f$ relations are different for overconsolidated specimens than for the normally consolidated specimens: for a given $\bar{p}_f$ the water content is smaller (corresponding to the increased strength) in the overconsolidated specimen. However, the relations for the failure condition are affected much less by preconsolidation than is the w_0 versus $\bar{p}_0$ relation. Thus shearing action tends to destroy the effects of preconsolidation but does not quite succeed.

Figure 21.7 is especially useful for illustrating the changes in water content during shear. Some of the specimens increased in water content during shear, whereas others decreased in water content. For example, the specimen that was rebounded to 103 kN/m² increased in water content from 20.5 to 21.3%; i.e., water was sucked into the specimen during the shear part of the test. On the other hand, the specimen that was rebounded to 414 kN/m² decreased in water content during shear from 18.9 to 18.3%. A sample with an overconsolidation ratio of about 4 would not have any net volume change during the shear process.

The q_f-$\bar{p}_f$-w_f Relationship

We have now seen that a soil does not possess a truly unique relationship among shear strength, effective stress at failure, and water content at failure—this relationship is affected by the degree of preconsolidation.[1]

This will be the first of several instances in which we shall see that the q_f-p_f-w_f relation is only approximately unique. Actually, the concept of such a relation is qualitative rather than quantitative. It reminds us that the three quantities are closely interrelated and helps us to envision how changes in either $\bar{p}_f$ or w_f will affect strength and each other.

Having determined this relationship for a given preconsolidation, we can use it to find the strength for any type of loading. Examples corresponding to those for normally consolidated soil will be found among the problems.

Comparison of Clay and Sand

The results presented here again emphasize the similarities in the strength behavior of sand and clay. For both types of soils, the shear strength is strongly dependent on the effective stress existing at the time of failure. Furthermore, in both types of soil, the shear strength is also dependent on the initial void ratio before shear, i.e., the degree to which the soil has been "predensified" by some past action. In a general way, dense sands correspond to heavily overconsolidated clays, whereas loose sands correspond to normally consolidated clays. Note that heavily overconsolidated clays and dense sands both have curved failure envelopes.

Stress by itself is not usually as effective in densifying a sand as it is in densifying a clay. In order to achieve a significant densification of a loose sand, it generally is necessary to apply many cycles of loading and unloading (or vibration). Thus, in detail, we have one difference between the strength behavior of clay and sand: an application and removal of stress (slowly, so that drainage of water can occur) will improve the strength of clay at some given effective stress, but will not significantly alter the strength of the sand.

There is one additional difference between the drained strength behavior of sand and clay: whereas the Mohr envelope for a dense sand does pass through the origin, that for heavily overconsolidated clay generally passes

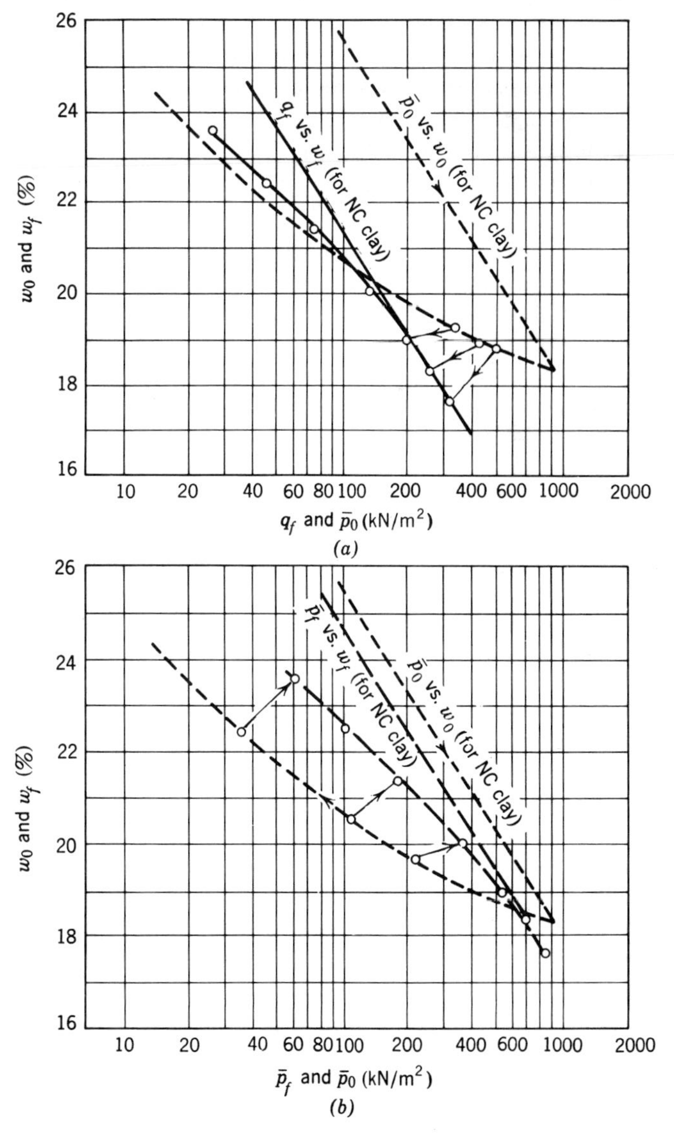

Fig. 21.7 Stress-volume relations for overconsolidated Weald clay. $\bar{p}_m$ = 827 kN/m².

[1] For remolded Weald clay there is such a unique relationship for any given $\bar{p}_m$; i.e., there is a family of such relationships with $\bar{p}_m$ as a parameter and with the relation for normally consolidated clays as an envelope.

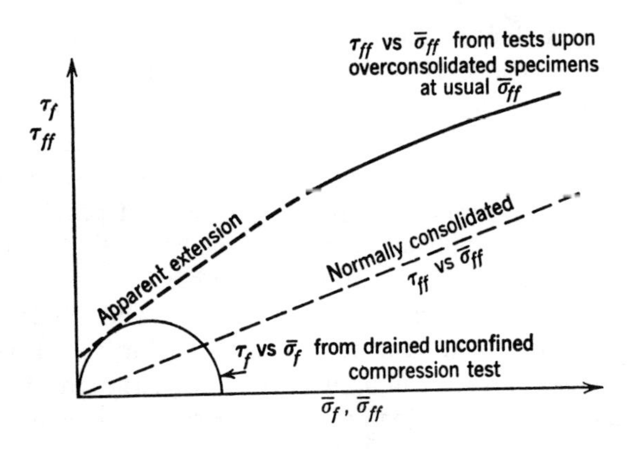

Fig. 21.8 Extension of curved Mohr envelope toward origin.

somewhat above the origin. This actual cohesion intercept presumably results from the formation of bonds among clay particles. An important question is: How close does the envelope pass to the origin; i.e., how large is q_f when $\bar{p}_f = 0$? A preliminary investigation using submerged, unjacketed specimens (so that $\bar{\sigma}_3 = 0$) has given the following results for two remolded clays with $\bar{p}_m = 600$ kN/m² (see Fig. 21.8).

	$\bar{c}$ for straight line fit for $\bar{\sigma}_{ff}$ between 100 and 600 kN/m²	Actual $\bar{c}$
Clay *A*	10 kN/m²	5 kN/m²
Clay *B*	20 kN/m²	8.5 kN/m²

Typical Values of $\bar{c}$ and $\bar{\phi}$ for Overconsolidated Clays

The magnitude of $\bar{c}$ and $\bar{\phi}$ for a given clay depends on how large the preconsolidation stress has been, how long the clay has been under the preconsolidation stress, etc. The effect of preconsolidation can best be illustrated by data for a compacted soil (Fig. 21.9) where the compaction effort supplies the preconsolidation.

The $\bar{c}$ and $\bar{\phi}$ for a given soil also depend on the stress

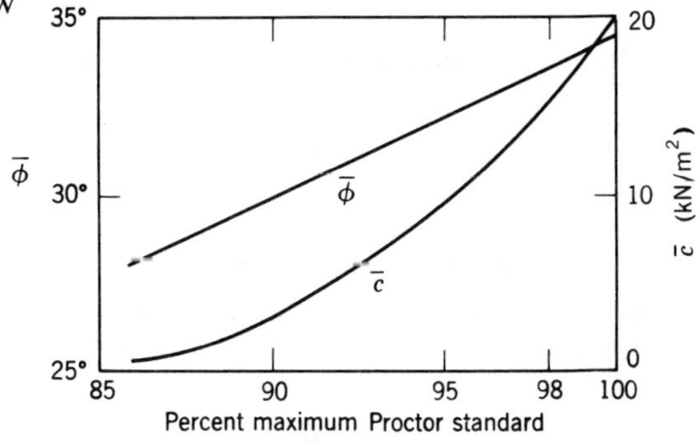

Fig. 21.9 Effect of compactive effort on strength parameters of clayey sand (after Moretto et al., 1963).

range over which a straight line fit is made to the curved Mohr envelope. Thus:

1. When effective stress is a large fraction of preconsolidation stress—when there is low OCR—$\bar{\phi}$ will be slightly less than for normally consolidated clay, while $\bar{c}$ will depend on magnitude of preconsolidation stress (void ratio).
2. When effective stress is very small compared to preconsolidation stress—when there is high OCR—$\bar{c}$ will be relatively small and $\bar{\phi}$ will depend on magnitude of preconsolidation stress (void ratio).

Figure 21.10 illustrates (to an extreme degree) the way in which $\bar{c}$ and $\bar{\phi}$ can vary with the stress range.

The Mohr envelope for an overconsolidated soil seldom is as curved as that in Fig. 21.10 (Fig. 21.5 is more typical). Usually, the $\bar{\phi}$ from a straight-line fit to an overconsolidated clay is about equal to that for normally consolidated specimens. Moreover, the effects of preconsolidation on strength are important numerically only at small effective stresses; at large effective stresses the envelopes for normally and overconsolidated specimens of a clay tend to merge. Thus the major question with regard to overconsolidated clay lies in choosing the appropriate value of $\bar{c}$.

Values of $\bar{c}$ ranging from 4.8 to 24 kN/m² are often used

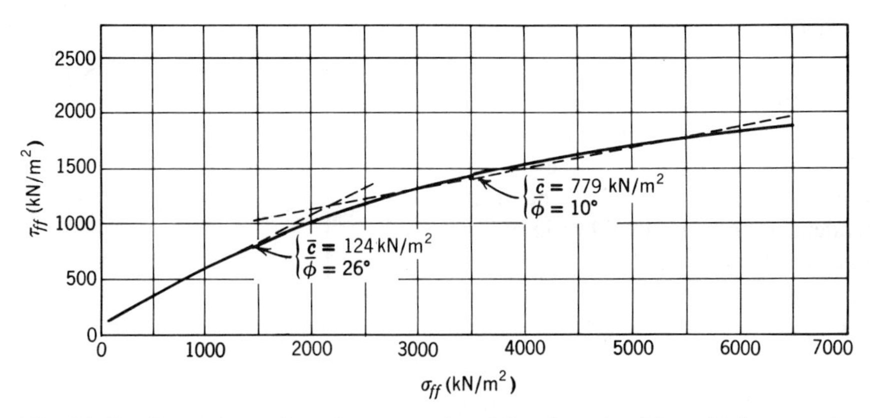

Fig. 21.10 Strength envelope for unweathered London clay (From Bishop, et al., 1965).

for soils. Even larger values undoubtedly are valid for the stiffest of soils, but large "measured" $\bar{c}$ values frequently result from running CD tests too rapidly so that pore pressures develop. Extreme caution must be exercised when choosing a value of $\bar{c}$ for practical calculations.

21.3 HVORSLEV PARAMETERS

It has been seen that neither $\bar{p}_f$ nor w_f alone determines q_f. For a given $\bar{p}_f$, two specimens with different w_f can exist (one normally consolidated, one overconsolidated) and the two specimens will have different strengths. Similarly, for a given w_f, specimens with different $\bar{p}_f$ and consequently different q_f can exist.

Hvorslev (1937) proposed a shear strength theory[2] to take these observations into account. He suggested that strength could be separated into two components, one dependent upon water content alone and the other dependent upon effective stress alone. By using a Coulomb-type expression this would be written

$$\tau_{ff} = f(w_f) + f(\bar{\sigma}_{ff})$$

The effective stress-strength curves for normally consolidated and overconsolidated ($\bar{p}_m = 827$ kN/m²) Weald clay have been replotted in Fig. 21.11. The dashed lines connect points of equal w_f. By including data for other values of $\bar{p}_m$, it can be shown that the lines connecting points of equal w_f are straight lines, or at least approximately so. These lines, each for a different water content, are called *Hvorslev failure lines*.

To reiterate, both w_f and $\bar{p}_f$ (and, of course, q_f) are varying along the usual failure line, such as that shown in Fig. 21.5. Along Hvorslev failure lines only $\bar{p}_f$ varies. Thus along the Hvorslev failure line for $w_f = 23.5\%$, the difference in strength between the normally consolidated specimen with $q_f = 51.7$ kN/m² and the overconsolidated specimen with $q_f = 26.2$ kN/m² is not the result of any volume change and presumably results from different amounts of internal friction mobilized by different $\bar{p}_f$.

[2] See Bjerrum (1954b) for a good description of Hvorslev's theory.

Using the data assembled in Fig. 21.11 and converting back to Coulomb-type parameters we obtain:

Water content	$\bar{c}_e$	$\bar{\phi}_e$
23.5%	9.6 kN/m²	18°
21.3%	17.7 kN/m²	18°

$\bar{\phi}_e$ and $\bar{c}_e$ are often called *true friction* and *true cohesion*, respectively. From many tests, $\bar{\phi}_e$ is found to be independent of water content and $\bar{c}_e$ versus $\log w_f$ is a straight line.

There has been much speculation that $\bar{c}_e$ and $\bar{\phi}_e$ may be descriptive of the actual internal mechanisms of shear resistance; e.g., that $\bar{c}_e$ indicates the magnitude of the bonding between particles. It seems unlikely that the situation is this simple. On the other hand, Gibson (1953) has shown that $\bar{\phi}_e$ of a soil is closely related to the clay content of the soil and the type of clay mineral present and that ϕ_e becomes aery small in those very plastic soils within which the transfer of stress between particles may well be through long-range forces.

In any case, the Hvorslev theory has served an important role by emphasizing that an increase in effective stress has two effects upon the mineral skeleton:

1. It increases the particle-to-particle contact force thus increasing the frictional resistance.
2. It decreases the volume, thus increasing the amount of interlocking.

The difference between $\bar{\phi}$ and $\bar{\phi}_e$ is thus indicative of the role of volume changes with regard to changes in strength. In addition to helping us understand the nature of shear

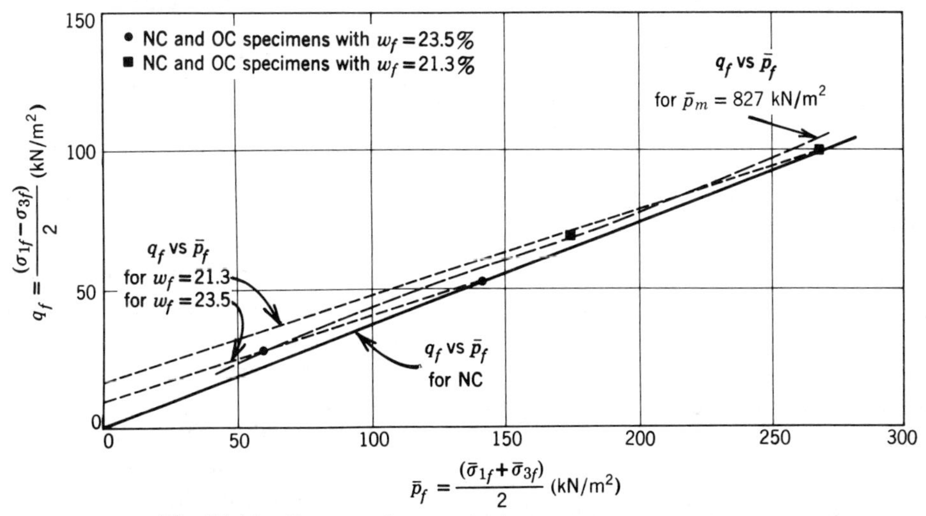

Fig. 21.11 Construction to obtain Hvorslev parameters.

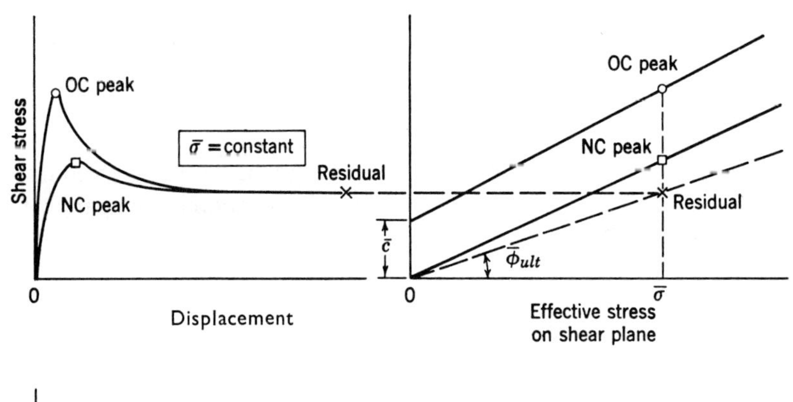

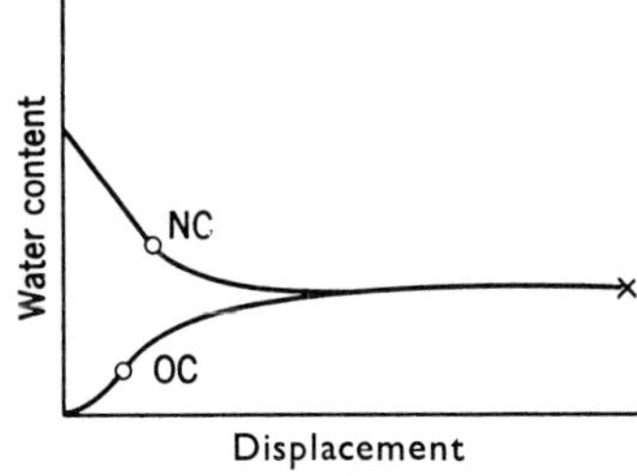

Fig. 21.12 Relationship between peak and ultimate conditions.

resistance, the Hvorslev theory has some direct applications in research, but for practical work the regular strength parameters $\bar{c}$ and $\bar{\phi}$ are needed and must be evaluated by direct measurement in each case.

21.4 ULTIMATE STRENGTH

Thus far we have been concerned only with the peak resistance of clays, which is the strength at the peak point of the stress-strain curve. Even in remolded soils, stress-strain curves drop off somewhat following the peak, and this drop off usually is much more pronounced for natural soils. In the case of sands, a unity of knowledge was achieved by considering the ultimate strength so now we must consider this aspect of the behavior of clays. This question has been considered in detail by Skempton (1964), principally using direct shear tests.[3]

Figure 21.12 indicates the stress-strain behavior of a hypothetical soil when carried well past the peak of the stress-strain curve. This picture has been pieced together from the observed behavior of a number of actual soils. Some of the main features of this picture are:

1. The postpeak drop off in strength becomes more pronounced as the degree of overconsolidation increases, but can be quite noticeable even for normally consolidated soils.
2. In the ultimate condition, the strength at a given effective stress is independent of past stress history. In fact, the ultimate strength of remolded and undisturbed specimens of a given soil have proved to be essentially the same.
3. The strength envelope for the ultimate condition is a straight line through the origin, generally at a position lower than that for the peak strength of the normally consolidated clay.
4. In the ultimate condition, the water content for a given effective stress appears to be independent of past stress history. Actually, it is very difficult to establish this as fact because the failure zone tends to be very thin (perhaps only a few microns in thickness) and the water content of a slice cut from the clay may not be representative of the water content in the actual failure zone.

Thus the overall behavior of clay is essentially the same as that of sand: there is an ultimate condition wherein the strength and void ratio are independent of past history. In this ultimate condition, there really is a unique relationship among strength, effective stress, and density of packing.

However, there is one important difference between the ultimate strength behavior of sands and clays: in clays, the ultimate strength can be significantly less than the peak strength of normally consolidated specimens, whereas the peak and ultimate resistances of loose sand are equal.

Although the progressive breaking of adhesive bonds may play a role in the postpeak drop off of the strength of normally consolidated clays, a second factor that would seem to be even more important is the gradual

[3] Skempton used the phrase *residual strength*; we shall retain the phrase *ultimate strength*. Roscoe et al. (1958) refers to this ultimate condition as the *critical state*.

reorientation of clay particles into parallel, face-to-face arrangements. As discussed in Chapter 6, such reorientation should be accompanied by a decrease in ϕ_μ and also a decrease in the degree of interlocking. This reorientation also seems to play a role with regard to the postpeak loss of strength in overconsolidated clays, inasmuch as some of this loss seems to occur after the clay has reached essentially constant volume.

There is good evidence that this reorientation occurs. Polished, slickensided surfaces have been found in direct shear tests after considerable strain (Fig. 21.13). Examination of failure zones using the electron microscope and X-ray techniques has indicated a highly oriented fabric. Finally, it has been observed that the strength of clays at large strains decreases while the void ratio apparently is decreasing slightly, a phenomenon which can most readily be explained by reorientation.

Magnitude of Ultimate Friction Angle

Figure 21.14 indicates the way in which the ultimate friction angle varies with clay content. For clay contents approaching 100%, the ultimate friction angles are of the same magnitude as ϕ_μ for the sheet minerals.[4] For very low clay contents, $\bar{\phi}_{ult}$ is of the same magnitude as ϕ_μ between quartz particles.

In the general case, where the soil consists of both platelike and granular particles, the granular particles tend to raise $\bar{\phi}_{ult}$ above ϕ_μ for the clay particles by inhibiting to some extent the full orientation of the clay particles and by contributing some measure of their own higher angle of shear resistance. It is significant that the difference between $\bar{\phi}_{ult}$ and $\bar{\phi}$ for normally consolidated soil also increases with increasing clay content. This again indicates that reorientation of clay particles plays a major role in the drop off in strength past the peak of the stress-strain curve.

[4] For additional results, see Herrmann and Wolfskill, 1966. The Norwegian Geotechnical Institute has also obtained excellent results. $\bar{\phi}_{ult}$ depends on the type of clay mineral present, the smallest value having been obtained for sodium montmorillonite. $\bar{\phi}_{ult}$ as small 3° to 4° has been measured.

Fig. 21.13 Slickensided shear surface after large shear displacement.

21.5 A BRIEF LOOK AT CERTAIN COMPLICATIONS

The foregoing picture concerning the strength behavior of clays does not cover the entire story. The strength

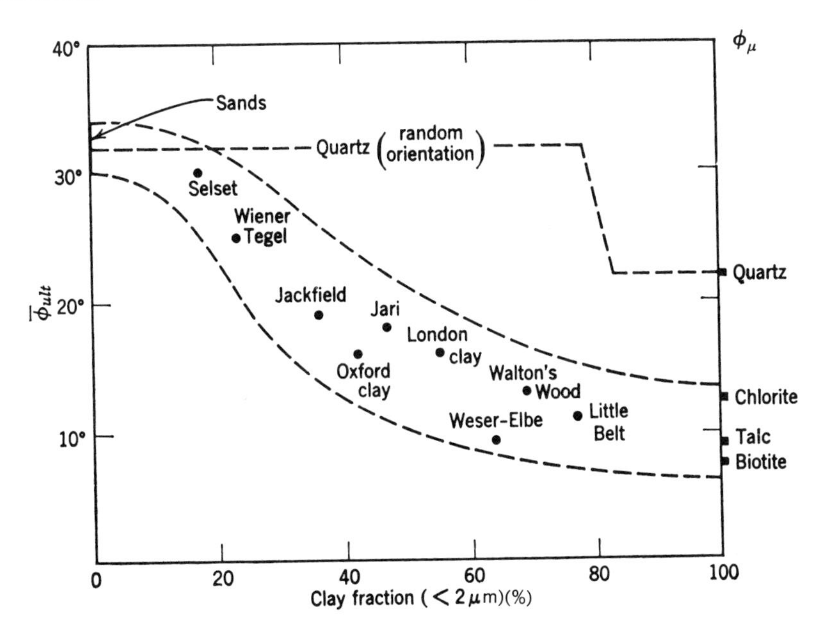

Fig. 21.14 Relations between $\bar{\phi}_{ult}$ and clay content (From Skempton, 1964).

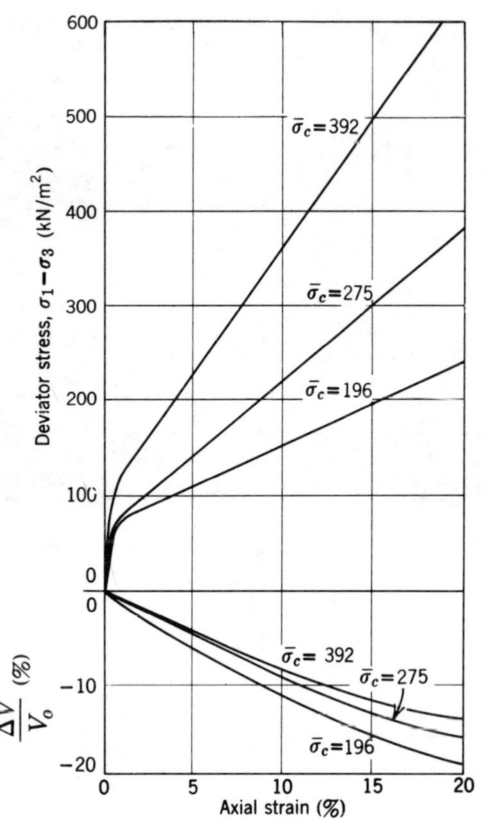

Fig. 21.15 Stress-strain curves for undisturbed Leda clay (from Crawford, 1959).

parameters for a soil in terms of effective stress, $\bar{c}$ and $\bar{\phi}$, are affected by such variables as temperature, the nature of the pore fluid, the rate of straining, and the intermediate principal stress. Whereas a full treatment of these topics is beyond the scope of this text, some of the more important complications will be mentioned.

Rate of Strain

Consideration of the effect of strain rate is conveniently divided into two parts: normally consolidated and overconsolidated clays. In either case, we are speaking only of slow and slower tests, both of which are slow enough so that no excess pore pressures develop.

In the case of normally consolidated clays, there is general agreement that the effect of strain rate upon $\bar{\phi}$ is small (except possibly for the extremely plastic soils which have not been studied adequately), but there is little agreement as to whether $\bar{\phi}$ increases or decreases as the rate of straining decreases. Data by Gibson and Henkel (1954) suggest that the measured $\bar{\phi}$ decreases by perhaps 10% as the strain rate is reduced by a factor of 10. Mitchell (1964) presents evidence of creep under constant load. Bjerrum et al. (1958) indicate that $\bar{\phi}$ remains constant with changing strain rate. Casagrande and Rivard (1959) and Richardson and Whitman (1965) both present evidence that $\bar{\phi}$ increases somewhat with decreasing strain rate. Physically, the question can also be argued two ways, citing viscous shear resistance between particles on one hand and thixotropy and secondary compression on the other hand. Probably the exact nature of the strain rate effect varies from soil to soil.

In the case of overconsolidated clays, most engineers agree that some of the shear strength is lost with time. Gould (1960) has presented some excellent evidence on this point. There is, however, no general agreement as to how much of the strength can be lost. There have been many theories in this regard—all $\bar{c}$ will be lost but $\bar{\phi}$ will remain, etc.—but no real proof exists. The best study of this question has been by Skempton (1964). For fissured clays, such as those found in England, he found that with time strength (for unloading) will decrease until $\bar{c} = 0$ and $\bar{\phi} = \bar{\phi}_{ult}$. Moreover, the loss of strength can occur on a time scale significant to engineering practice (75 years or less). On the other hand, there are a few bits of evidence to indicate that *intact* clays will retain the effects of preconsolidation over a very long period of time.

Intermediate Principal Stress

As in the case of sands, the effect of the intermediate principal stress on the drained strength of clays is still uncertain. The available data from axial compression and axial extension tests (Parry, 1960) indicate that $\bar{c}$ and especially $\bar{\phi}$ for clay are affected little by the magnitude of $\bar{\sigma}_2$. However, these and other similar results possibly were influenced to an unknown degree by experimental error and by inadequate knowledge concerning the magnitude of the circumferential stresses within triaxial specimens.

Sensitive Clays

It has already been noted that large axial strains are required to develop the peak shear resistance during drained loading of normally consolidated clays. This situation is accentuated in quick and sensitive clays (described in Chapter 7) because of the large volume changes that occur in these clays during loading (Fig. 21.15).

Quick clays present special problems in connection with the selection of shear strength parameters for practical calculations. For many practical problems it is necessary to keep the shear stresses to quite small values, usually to values less than the shear stresses at the "knee" or "yield point" shown in the curves in Fig. 21.15. For this reason, it is often reported that the friction angle for quick clays is quite small—even less than the residual friction angle. However, such small angles are not indicative of the full resistance available within such clays at large strains.

21.6 APPARENT COHESION

Thus far in this part we have emphasized the similarities in the strength behavior of sands and clays (and hence of all soils) during drained shear. Now it is time to emphasize one[5] important difference between the behaviors of these materials. This difference arises from the capillary tensions which can develop within the pores of the soil.

These capillary tensions have already been discussed in Chapter 16. It was seen that, owing to the great difference in the particle size within sands and clays, there is a great difference in the capillary tensions that may develop within these two classes of soils. There can be very large capillary tensions within the clays, but only very small capillary tensions can exist within sands.

These capillary tensions must be considered when evaluating the effective stresses that exist within an element of soil located above the water table. For example, consider the situations shown in Fig. 21.16. In both cases the total stresses are the same. However, because of the capillary tensions available within the clay, the *effective stresses* above the water table are larger in the clay than in the sand. Hence, even though $\bar{\phi}$ is much the same for the two materials, the shear resistance of the clay is larger than that of the sand.

[5] Another important difference will be discussed in Part V.

An extreme case of this situation occurs when a sample of saturated soil is removed from the ground. Now the total stresses acting on the sample are zero. If the sample is to have shear resistance there must be either (*a*) a genuine cohesion intercept; i.e., a strength when $\bar{p}_f = 0$; or (*b*) capillary tensions that give $\bar{p}_f > 0$. In sands, neither the genuine cohesion intercept nor the capillary tensions are significant, and an unconfined sample of sand has so little strength that it generally cannot support its weight. A sample of clay may possess a genuine cohesion intercept of small magnitude. More important, however, capillary tensions develop within a clay as soon as it is removed from the ground because of the menisci which form along the outsides of the sample. Hence there will be effective stresses, with a magnitude of 1 atm or even much more, within the clay and as a result the clay will possess significant strength. The clay

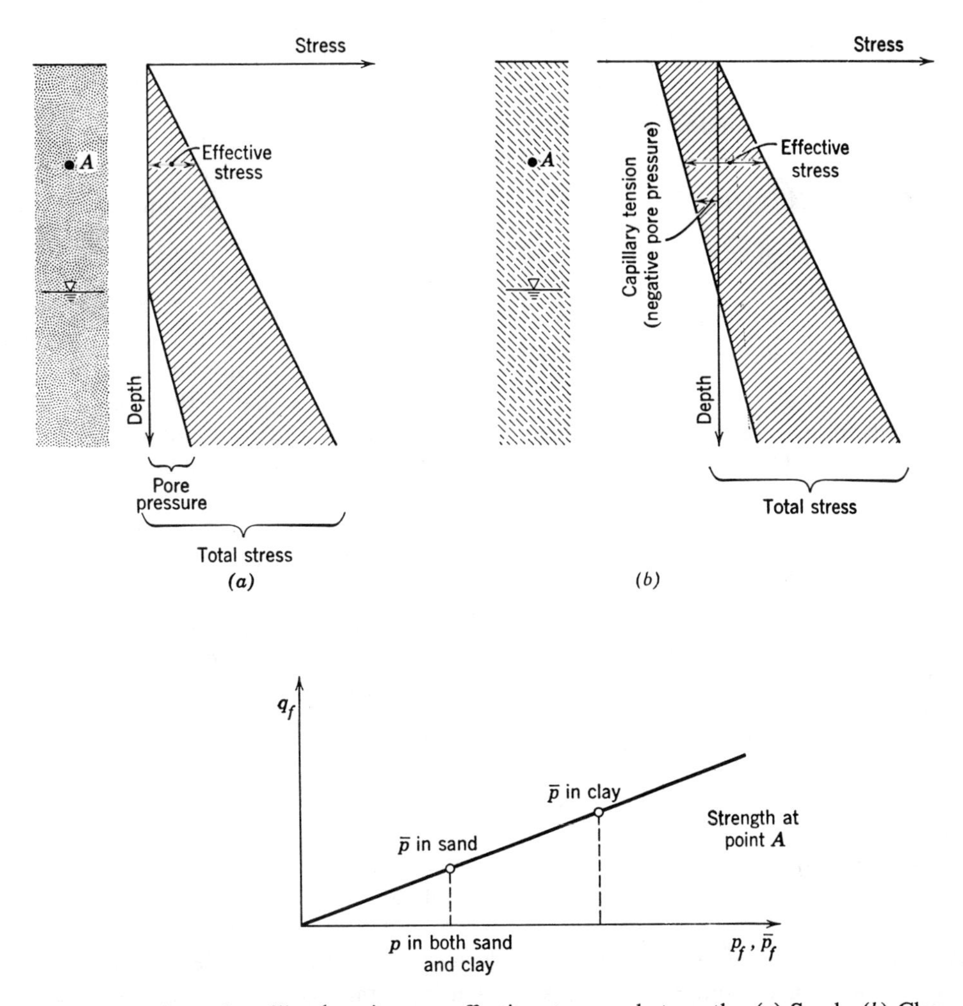

Fig. 21.16 Effect of capillary tensions on effective stress and strength. (*a*) Sand. (*b*) Clay.

appears as a cohesive material and thus this strength imparted by the capillary tensions is called *apparent cohesion*.

Apparent cohesion is potentially present in all fine-grained soils above the water table, becoming more important as the clay content increases. However, apparent cohesion is lost as soon as the soil is submerged below the water table. The role and importance of apparent cohesion will be discussed further in Chapter 23.

21.7 PARTIALLY SATURATED SOILS

The strength of partially saturated soils is controlled by the effective stress within the soil. However, it is more difficult to apply the effective stress principle to partially saturated soils because, as was explained in Chapter 16, the relation between total stress and effective stress involves the pressures in both the liquid and gas phases plus a factor a_w which is related to the degree of saturation. Special techniques are needed to measure these pressures in partially saturated soils, and there is doubt as to just how a_w should be determined. The best procedure to estimate strength is to run tests that duplicate the field conditions as closely as possible: same degree of saturation, same total stress and, if possible, the same pressure in the liquid phase.

21.8 SUMMARY OF MAIN POINTS

At this point, having noted the great similarity between the strength behavior of sand and clay, we are now prepared to summarize the drained strength behavior of all soils. The key, of course, is to express strength as a function of effective stress.

1. In the ultimate condition, achieved after considerable shearing strain, the strength behavior of soil is that of a frictional material. That is, the failure law is

$$\tau_{ff} = \bar{\sigma}_{ff} \tan \bar{\phi}_{\text{ult}}$$

The ultimate friction angle $\bar{\phi}_{\text{ult}}$ is related to the clay content of the soil (Fig. 21.14). This angle is greatest (about 30°) in pure sand and least (as low as 3 or 4°) in pure clay. At the ultimate condition clay platelets are aligned in an oriented, face-to-face configuration.

2. At the point of peak resistance, the strength of a normally consolidated soil is also given by a frictional type of failure law,

$$\tau_{ff} = \bar{\sigma}_{ff} \tan \bar{\phi}$$

This angle $\bar{\phi}$ is related to the clay content of the soil (Fig. 21.4). For loose sands $\bar{\phi}$ and $\bar{\phi}_{\text{ult}}$ are equal. As the clay content increases, $\bar{\phi}$ exceeds $\bar{\phi}_{\text{ult}}$ since at the peak resistance the clay platelets within the failure zone have not yet reached a fully oriented, face-to-face alignment.

3. Densification increases the peak strength of soils. For soils with a significant clay content, large stresses suffice to produce an overconsolidated soil, while stresses alone do not effectively densify predominantly granular soils and cycles of loading and unloading are necessary. The failure envelope for densified soils generally is curved, but for practical calculations the peak strength can be represented by a linear relation,

$$\tau_{ff} = \bar{c} + \bar{\sigma}_{ff} \tan \bar{\phi}$$

For the usual values of $\bar{\sigma}_{ff}$ (0–600 kN/m²) the following are useful guides:

Soil Type	$\bar{c}$	$\bar{\phi}$
Predominantly granular	0	Table 11.3
Predominantly clayey	4.8–24 kN/m²	Roughly equal to $\bar{\phi}$ for NC soil

As $\bar{\sigma}_{ff}$ increases, $\bar{c}$ increases and $\bar{\phi}$ decreases.

4. Any change in effective stress changes the density at failure as well as changing the shear strength. Conversely, any action that changes the density at failure must produce a change in shear strength. For the ultimate condition, there is a unique relationship among effective stress, shear strength, and water content, such that knowledge of any of these three quantities specifies the other two quantities. At the peak resistance, this three-way relation is not quite unique (it is influenced by the degree of preconsolidation, etc.) but still this relation helps to vizualize the effect of changes in effective stress and density upon strength.
5. Capillary tensions must be taken into account when determining the effective stresses within soils located above the water table. Because large capillary tensions are possible in clayey soils, such soils can exhibit a large apparent cohesion even though they possess little or no cohesion intercept $\bar{c}$. This apparent cohesion is fully explained by effective stresses.

It usually is not feasible to obtain undisturbed samples of granular soils and, it is thus necessary to estimate strength by indirect methods, such as a correlation between penetration resistance and friction angle. For clayey soils it is possible to obtain undisturbed samples, and values of $\bar{c}$ and $\bar{\phi}$ should be selected on the basis of strength tests on such samples. Strength tests should also be used to evaluate the strength of compacted fills using samples compacted to the same density that will be used in the fill.

Even though strength tests are performed, there still are uncertainties involved in the selection of $\bar{c}$ and $\bar{\phi}$. For example, should $\bar{c}$ and $\bar{\phi}$ be based upon ultimate or peak resistance? If peak resistance is to be used, should $\bar{c}$ be reduced because shear in the field is slower than that in laboratory tests? Such questions can only be answered by examining actual field situations in which failure has occurred so that the mobilized shear resistance can be computed.

PROBLEMS

21.1 For normally consolidated Weald clay (see Figs. 21.1, 21.2, 21.3, etc.) estimate q_f, w_f, and $\bar{p}_f$ for the following tests:

a. $\bar{\sigma}_c$ = 345 kN/m^2 and failed in drained shear with $\Delta\sigma_h = 0$.
b. $\bar{\sigma}_c$ = 345 kN/m^2 and failed in drained shear with $\Delta\sigma_v = 3\Delta\sigma_h$.

21.2 For the two tests in Problem 21.1, draw the total and effective stress paths if the static pore pressure is 69 kN/m^2.

21.3 For normally consolidated Weald clay prepare a plot of $\bar{p}_0$ versus q_f.

21.4 A specimen of Weald clay is consolidated to $\bar{p}_m$ = 827 kN/m^2 and then allowed to swell to $\bar{p}_0$ = 69 kN/m^2. For a drained test in which $\Delta\sigma_h = 0$, estimate q_f, $\bar{p}_f$, and w_f.

21.5 A specimen of Weald clay is consolidated to $\bar{p}_m$ = 827 kN/m^2 and allowed to swell to a value of $\bar{p}_0$ such that a drained test with $\Delta\sigma_h = 0$ can be run rapidly. What is the proper value of $\bar{p}_0$? Explain your answer.

CHAPTER 22

Stress-Strain Relations for Drained Conditions

In order to solve deformation problems in soil we must extend the material in Chapter 20 by taking a closer look at the stress-strain behavior of soil. Chapter 12 took such a close look at the stress-strain behavior of dry cohesionless soil. As long as the stresses are expressed in terms of effective stresses, all of the stress-strain behavior for dry cohesionless soil presented in Chapter 12 holds for saturated sands under drained conditions. Thus our task in Chapter 22 is to extend Chapter 12 to include the behavior of saturated clays under drained conditions and to compare the behavior of clays with that of sands.

Most of this chapter is devoted to two types of loadings: (*a*) confined compression and (*b*) triaxial loading in which the horizontal stress is kept constant during the test. Toward the end of the chapter we compare the behavior from these two loadings with that from other types.

Because of the relatively high permeability of sands, the volume of the sand can freely change during the loading or unloading in most practical problems. Thus drained loading situations in sand are the rule. On the other hand, the permeability of clay is so low that the general situation is for the volume of a clay to remain essentially constant during the load application in a practical problem. The drainage then occurs after load application, i.e., under constant total stresses.[1] Thus the behavior of a clay during drained conditions at constant total stresses is of considerable importance in engineering problems involving clay. Part V will consider the strains that occur in soil, especially clay, during the loading or unloading process. Chapter 22 examines the strains that occur during drainage at constant load. In a practical problem the two components of deformation must be combined to obtain the total deformation.

[1] The total stresses in the foundation soil can change during consolidation even though the applied surface stresses remain constant (see Fig. 32.10).

22.1 BEHAVIOR DURING CONFINED COMPRESSION

Table 12.2 listed the various stress-strain parameters for confined compression. Example 22.1 shows the determination of the *compression index*, C_c, *coefficient of compressibility*, a_v, and *coefficient of volume change*, m_v, for the Cambridge clay (Fig. 20.5) for the increment of 400 kN/m² to 600 kN/m². For normally consolidated Cambridge clay (i.e., in virgin compression) the e versus log $\bar{\sigma}_v$ curve is essentially a straight line. Hence C_c is a constant. This situation is true of many clays and therefore the e-log $\bar{\sigma}_v$ diagram is a widely used method of presenting the results of confined compression tests.

From Fig. 22.1 we can obtain a good perspective of the virgin compression characteristics of a wide variety of soils subjected to stresses over a very large range. As can be seen in this figure, the more plastic soils exist at higher void ratios and have higher compression indices. Based on the work of Skempton and others, Terzaghi and Peck (1948) suggested the following two expressions applicable for virgin compression:

Remolded soil: $C_c = 0.007\,(w_l - 10\%)$

Undisturbed soil: $C_c = 0.009\,(w_l - 10\%)$

where w_l is the liquid limit in percent.

Table 22.1 presents an accumulation of compression data for a number of clay minerals and natural soils along with the corresponding Atterberg limits. An examination of the data in Table 22.1 leads to the conclusion that any relation between Atterberg limits and compression characteristics is only approximate. Relations between compression characteristics with Atterberg limits should be used as intended—only as an estimate of virgin compression characteristics and never as a substitute for the results of actual tests.

▶ Example 22.1

Given. The following results of an oedometer test on Cambridge clay (see Fig. 20.5):

$$\text{at } \bar{\sigma}_v = 400 \text{ kN/m}^2, \quad e = 1.012$$
$$\text{at } \bar{\sigma}_v = 800 \text{ kN/m}^2, \quad e = 0.870$$

Find. C_c, a_v, and m_v for the stress change of 400 kN/m² to 600 kN/m².

Solution. Noting from Fig. 20.5 that the clay is in virgin compression in the range 4–8 kg/cm², we make the plots shown in Fig. E22.1.

From e versus $\log \bar{\sigma}_v$,

$$C_c = \text{compression index} = -\frac{\Delta e}{\Delta \log \bar{\sigma}_v}$$
$$(C_c)_{400-600} = 0.47$$

From e versus $\bar{\sigma}_v$,

$$a_v = \text{coefficient of compressibility} = -\frac{\Delta e}{\Delta \bar{\sigma}_v}$$
$$(a_v)_{400-600} = 0.043$$

From $\Delta e/(1 + e_0)$ versus $\bar{\sigma}_v$,

$$m_v = \text{coefficient of volume change} = \frac{\epsilon_{vol}}{\Delta \bar{\sigma}_v} = \frac{\epsilon_{vert}}{\Delta \bar{\sigma}_v}$$
$$(m_v)_{400-600} = 0.021$$

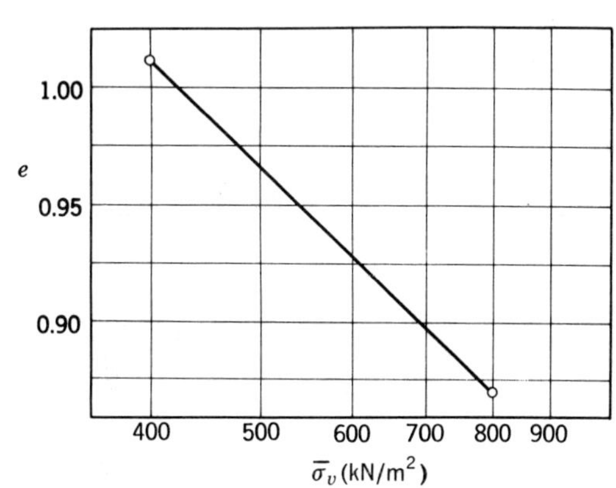

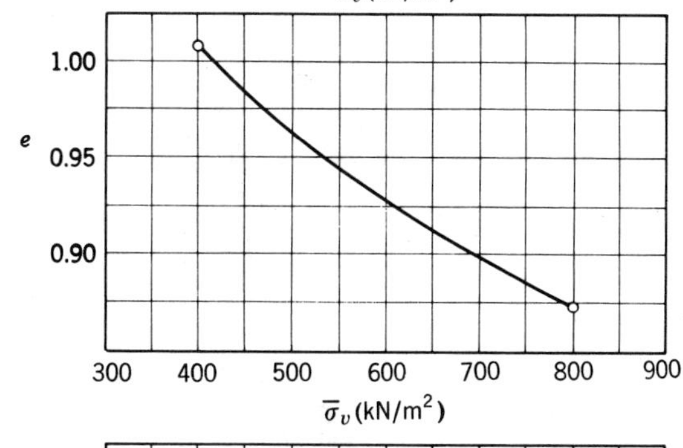

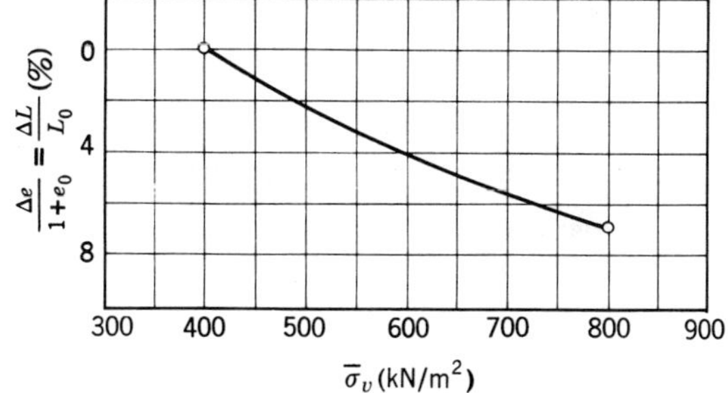

Fig. E22.1 Compression of Cambridge clay. ◀

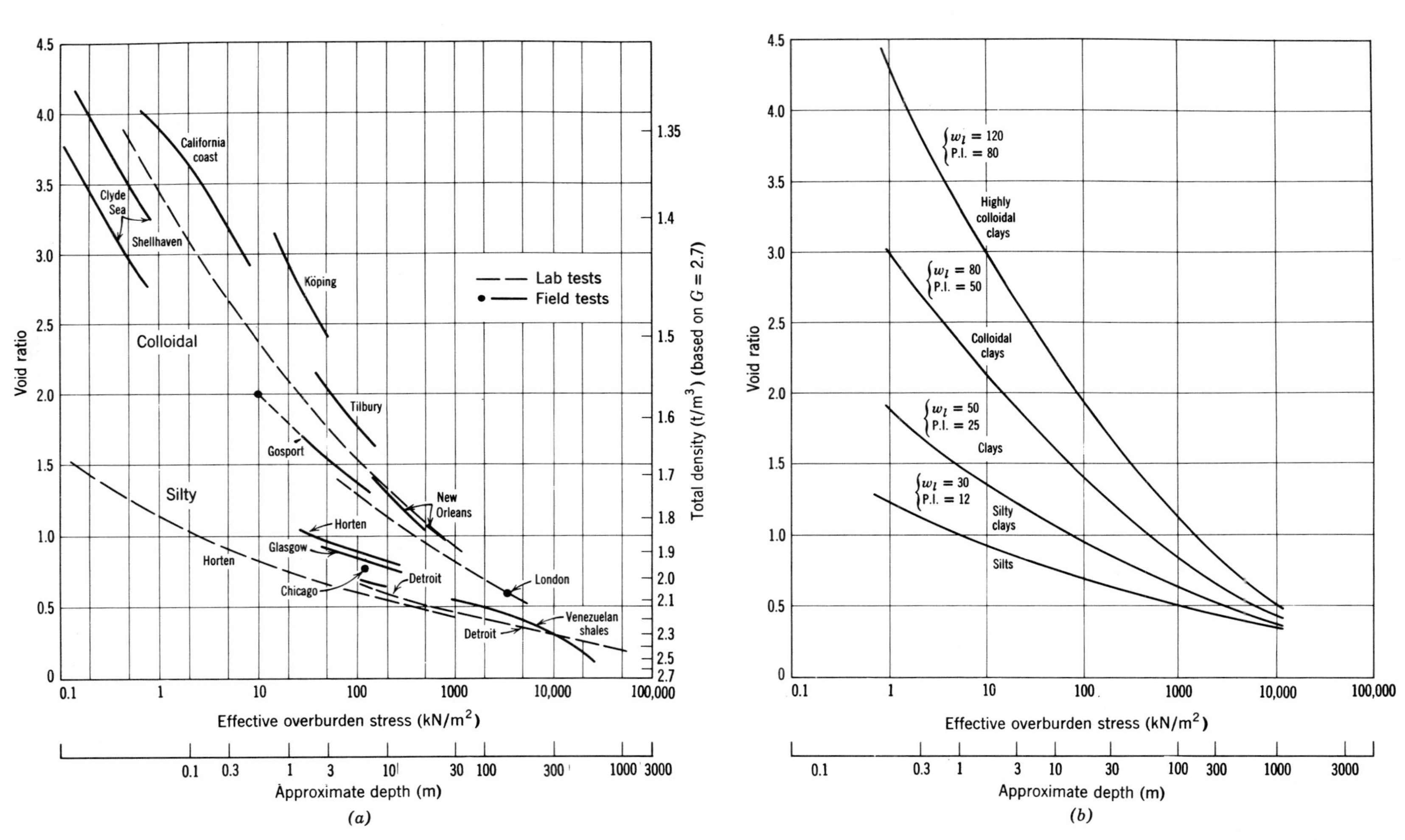

Fig. 22.1 Compression curves. (*a*) Graphs of the relation between void ratio and overburden stress for colloidal and silty clays. (*b*) Graphs of the approximate relation between void ratio and overburden stress for clay sediments, as a function of the Atterberg limits. w_l = liquid limit; P.I. = plasticity index. *Note.* To obtain water content divide void ratio by specific gravity of particles, G $\approx$ 2.7. (*c*) Plot of data by Arango of soils from western United States and Columbia.

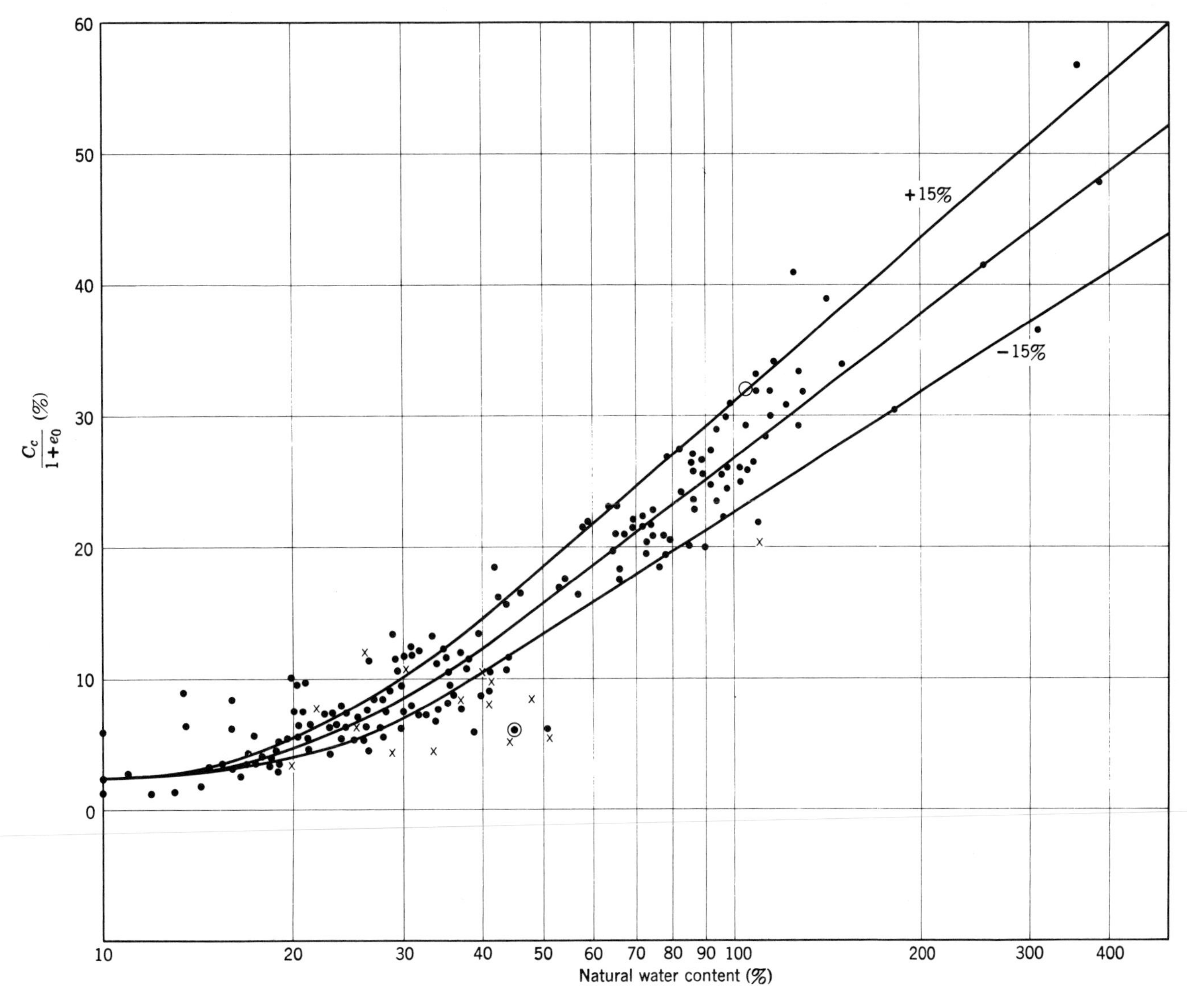

Fig. 22.1(c)

Unloading

Upon unloading a soil sample in confined compression the sample expands, as illustrated in Fig. 20.5. The parameter most commonly used to measure the expansion is

$$C_s = \text{swell index} = \frac{-\Delta e}{\Delta \log \bar{\sigma}_v} \tag{22.1}$$

C_s is always much smaller than C_c for virgin compression. This is illustrated by the data in Table 22.1. By consolidating a series of specimens to different maximum vertical stresses $\bar{\sigma}_{vm}$ before unloading, a series of expansion curves are obtained. Such expansion curves tend to be parallel. Note, for example, in Fig. 20.5 that the unload portion from the first cycle and that from the second cycle are approximately parallel. Thus C_s is more or less the same for all $\bar{\sigma}_{vm}$.

In Fig. 22.2 values of swell index have been plotted against the corresponding liquid limit. C_s increases with increasing liquid limit, but any relation between C_s and w_l will be only approximate.

Reloading

If a clay is subjected to many cycles of load and unload, the compression and recompression curves tend toward each other, i.e., C_c for recompression approximately equals C_s.

The compressibility of a soil depends very much on the stress level in relation to the stress history. For example, we can see from Fig. 20.5 that the compressibility of the Cambridge clay is much greater in the virgin compression range than it is in the recompression range; this means the compression index above $\bar{\sigma}_{vm}$ is much greater than below $\bar{\sigma}_{vm}$. This important fact presents the engineer

Table 22.1 Compression and Swell Indices

COMPRESSION AND SWELL OF CLAY MINERALS[a]

Clay	Exchangeable Ion	w_l	w_p	Compression Index	Swell[b] Index
Montmorillonite	Na^+	710	54	2.6	—
	K^+	660	98	1.0	—
	Ca^{+2}	510	81	2.2	0.51
	H^+	440	55	1.9	0.34
	Mg^{+2}	410	60	1.9	0.44
	Fe^{+3}	290	75	1.6	0.03
	Fe^{+3} dried and rewet	210	63	1.7	0.006
Illite	Na^+	120	53	1.10	0.15
	K^+	120	60	0.62	0.27
	Ca^{+2}	100	45	0.86	0.21
	H^+	100	51	0.61	0.10
	Mg^{+2}	94	46	0.56	0.18
	Fe^{+3}	110	49	—	0.15
	Fe^{+3} dried and rewet	100	46	0.50	0.22
Kaolinite	Na^+	53	32	0.26	—
	K^+	49	29	—	0.06
	Ca^{+2}	38	27	0.21	0.06
	H^+	53	25	0.23	0.05
	Mg^{+2}	54	31	0.24	0.08
	Fe^{+3}	59	37	0.24	0.06
	Fe^{+3} dried and rewet	52	35	0.19	0.15
Attapulgite	Mg^{+2}	270	150	0.77	0.24

COMPRESSION AND SWELL OF CLAY MINERALS[c]

Clay	Exchangeable Ion	w_l	Compression Index	Swell[d] Index
Montmorillonite	Li	576	7.7	2.0
	Na^+	494	6.2	2.5
	K	193	2.0	0.3
	Ca^{2+}	186	2.0	0.8
	Ba^{2+}	168	1.4	0.2
Kaolinite	Na^+	98	0.6	0.2

RECOMPRESSION AND SWELL OF CLAY MINERALS[e]

Mineral	Exchangeable Ion	Pore Fluid	Compression Index		Swell Index	
			1000 to 100 kN/m^2	100 to 10 kN/m^2	1000 to 100 kN/m^2	100 to 10 kN/m^2
Montmorillonite	Na^+	10^{-3} M NaCl	9.0	17.0	9.0	17.0
	Ca^{2+}	10^{-3} M $CaCl_2$	5.0	9.0	—	—
Illite	Na^+	0.001 to 0.1 M NaCl	0.9	1.7	—	—

Table 22.1 ***continued***

COMPRESSION AND SWELL OF NATURAL SOILS

Soil	w_l	w_p	Virgin Compress. Index	Swell Index 1000 to 100 kN/m²	Swell Index 100 to 10 kN/m²	Reference
Expansive Soil A	84	48	—	0.14	0.25	Dawson, 1957
Expansive Soil B	87	42	0.21	0.05	0.15	Dawson, 1957
Extruded Clay Sample	47	26	0.32	0.10	0.10	Dawson, 1957
Boston Blue Clay Undisturbed	41	20	0.35	0.07	0.09	Mitchell, 1956
Boston Blue Clay Remolded	41	20	0.21	0.07	0.07	Mitchell, 1956
Fore River Clay Undisturbed	49	21	0.36	0.09	0.09	Mitchell, 1956
Fore River Clay Remolded	49	21	0.25	0.04	0.04	Mitchell, 1956
Chicago Clay Undisturbed	58	21	0.42	0.07	0.12	Mitchell, 1956
Chicago Clay Remolded	58	21	0.22	0.07	0.09	Mitchell, 1956
Louisiana Clay Undisturbed	74	26	0.33	0.05	0.08	Mitchell, 1956
Louisiana Clay Remolded	74	26	0.29	0.04	0.07	Mitchell, 1956
New Orleans Clay Undisturbed	79	26	0.29	0.04	0.08	Mitchell, 1956
New Orleans Clay Remolded	79	26	0.26	0.04	0.09	Mitchell, 1956
Montana Clay	58	28	0.21	0.04	0.07	Lambe-Martin, 1957
Fort Union Clay	89	20	0.26	0.04	—	Smith-Redlinger, 1953
Beauharnois Clay	56	22	0.55	0.01	0.04	Mitchell, 1956
Cincinnati Clay	30	12	0.17	0.02	0.03	Mitchell, 1956
St. Lawrence Clay	55	22	0.84	0.04	0.08	Mitchell, 1956
Siburua Clay	70	26	0.21	0.08	0.12	Mitchell, 1956

[a] From Cornell, 1951.
[b] Estimated from Cornell data.
[c] From Jimenez Salas and Serratosa, 1953.
[d] Estimated from the Jimenez Salas and Serratosa data.
[e] From Bolt, 1956.

with considerable difficulty in selecting the proper stress-strain parameter for an actual problem. The compressibility, for example, of the Cambridge clay in Fig. 20.5 depends very much on whether the engineer determines it slightly above or slightly below the 304 kN/m² maximum past consolidation stress. A slight error in determining the initial vertical stress or $\bar{\sigma}_{vm}$ can thus result in a large error in the selection of effective compressibility in the actual field problem.

Stress Increment

There is evidence that a clay can develop a structural bond from many years of confinement under a given stress system. This structural bond may enable a clay to carry a small increment of load with very little strain. Common laboratory test practice involves the use of a load increment which is significant relative to the existing effective stress on the sample. A very common technique is to add an increment of stress equal to the existing

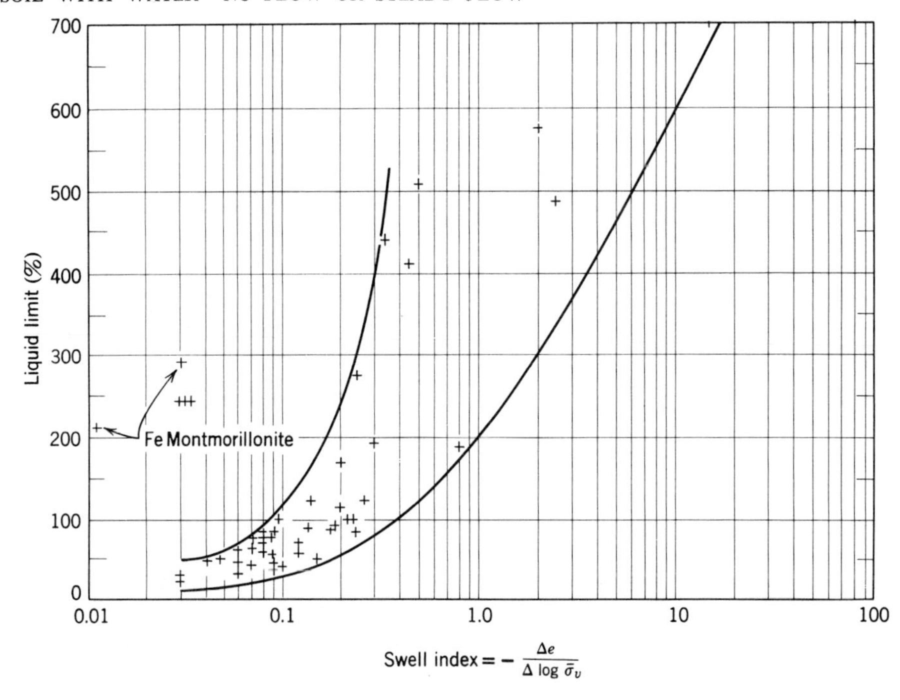

Fig. 22.2 Swell index versus liquid limit.

effective stress in the sample. This procedure may give a very poor indication of the compression which will occur in an actual problem if that problem involves a very small increment. (For further consideration see Taylor, 1942; Leonards, 1962.)

Other Factors Influencing Compressibility

There are a number of other factors that influence the compressibility of a clay. Some of these exist both in the field situation and in the laboratory test, whereas others are peculiar to the test conditions in the laboratory. Among these are temperature of the soil during compression; size of specimen; ratio of diameter to height of specimen; soil disturbance; and friction along the side of the sample during compression in the oedometer. Some of the factors, especially those during laboratory testing, that influence compressibility are discussed by Lambe (1951), Taylor (1942), and Simons (1965).

22.2 BEHAVIOR DURING TRIAXIAL COMPRESSION

As indicated in Chapter 12, the standard triaxial test—constant confining stress and increasing vertical stress—gives a direct measure of Young's modulus and permits a calculation of Poisson's ratio. For example, from the data in Fig. 21.1 we can compute for test 3 that secant E at peak stress is 4140 kN/m² and Poisson's ratio is 0.40.

Stress Level

From the stress-strain curves presented in Chapter 21, (e.g., Fig. 21.1), we see that the stress-strain curve is concave downward and the volumetric strain-axial strain curve is concave upward. In other words, the modulus E is decreasing and Poisson's ratio is increasing with increasing stress level. The general shape of the stress-strain curve has led to the suggestion (Kondner, 1963; Hansen, 1963) of various relationships between $\sigma_v - \sigma_h$ and ϵ, such as

$$\sigma_v - \sigma_h = \sqrt{\frac{\epsilon_v}{a + b\epsilon_v}} \tag{22.2}$$

where a and b are constants. The general form of these empirical expressions indicates that the modulus depends very much on stress level. It is common practice to determine the modulus E for a stress increment up to one-half or one-third of the peak value of stress.

Initial Consolidation Stress

The stress-strain modulus of normally consolidated clay varies directly with the initial consolidation stress. If the stress-strain curve normalizes, as suggested in Fig. 21.2, the stress-strain modulus is proportional to the value of $\bar{\sigma}_{v0}$.

Disturbance

Disturbing the structure of a natural clay usually results in a reduction of the stress-strain modulus.

Overconsolidation

The data in Fig. 20.10 illustrate this general trend: the higher the overconsolidation ratio, the higher the value of the modulus E and the higher the value of Poisson's ratio μ.

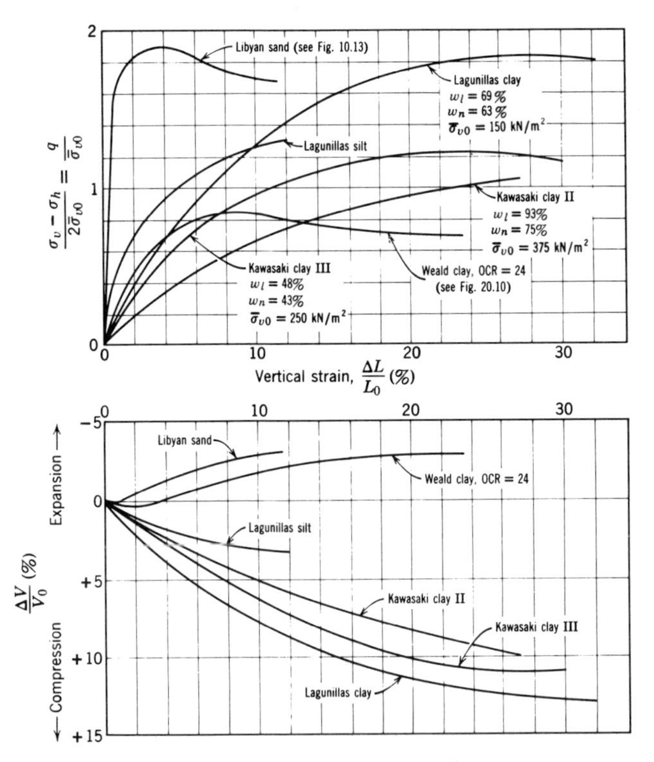

Fig. 22.3 Results of drained triaxial tests (constant horizontal stress).

Soil Type

In Fig. 22.3 the results of drained standard triaxial tests on a variety of soils are plotted. The tests on the Lagunillas and Kawasaki soils were run on undisturbed specimens. The plots in Fig. 22.3 and those in the preceding chapters illustrate several important points:

1. The stress-strain behavior of all soils in drained compression is basically the same; any difference is in magnitude rather than nature. There is a general trend toward lower stress-strain modulus and greater volumetric strain with increased plasticity of soils.
2. Loose sand and normally consolidated clay have the same shaped stress-strain curve in triaxial compression. The curve is concave downward with no pronounced peak value. Whereas both loose sand and normally consolidated clay undergo volume decrease during compression, clay usually shows a greater volume decrease.
3. Dense sand and overconsolidated clay have similar stress-strain curves. Both show a peak followed by a reduced strength. The dense sand has a higher modulus than overconsolidated clay and shows more volumetric expansion than does the overconsolidated clay.

22.3 BEHAVIOR DURING OTHER LOADINGS

The preceding two sections have indicated the nature of stress-strain behavior for confined compression and for standard triaxial compression. As we shall see in the next few chapters, practical problems often involve stress paths quite unlike those of the confined compression and standard triaxial tests. It thus is important to know whether or not the stress-strain behavior of soil depends on the nature of the stress path. Often engineers make calculations in which they assume that the relationship between increment of vertical stress and vertical strain is independent of the increment of horizontal stress.

The data in Figs. 10.22 and 10.23 have already shown that the vertical stress-vertical strain relationship of sands is very sensitive to the stress path of the test. This same situation exists with clay.

Figure 22.4 presents stress-strain data for two triaxial tests, a standard triaxial test along the path AB and an unloading triaxial test along the path AC. The data in Fig. 22.4 show that the *initial stress-strain modulus* for the unload test is *considerably greater* than that for the load test. Also, the sample volume increased, as measured by the change in water content, for the unload

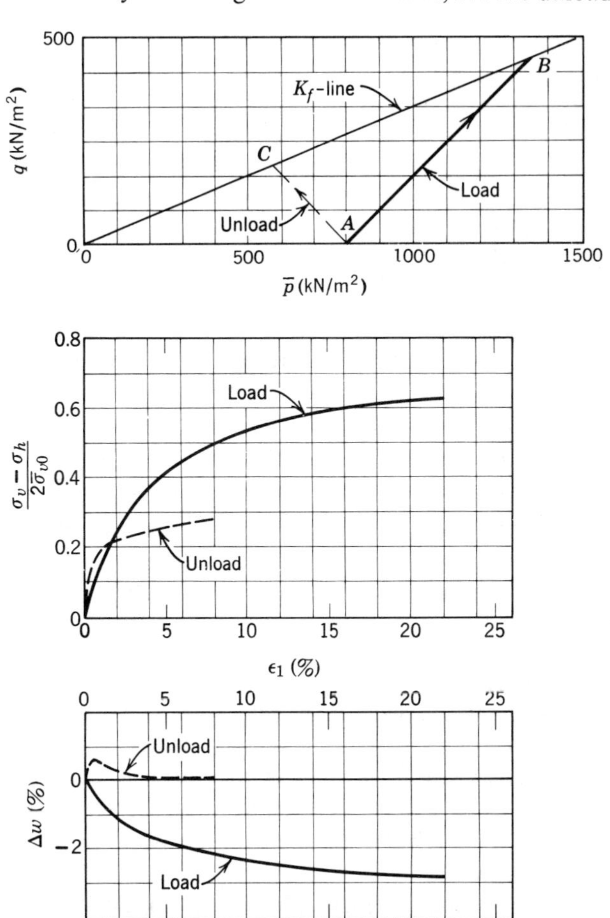

Fig. 22.4 Results of load and unload triaxial tests (Data from Ladd, 1964).

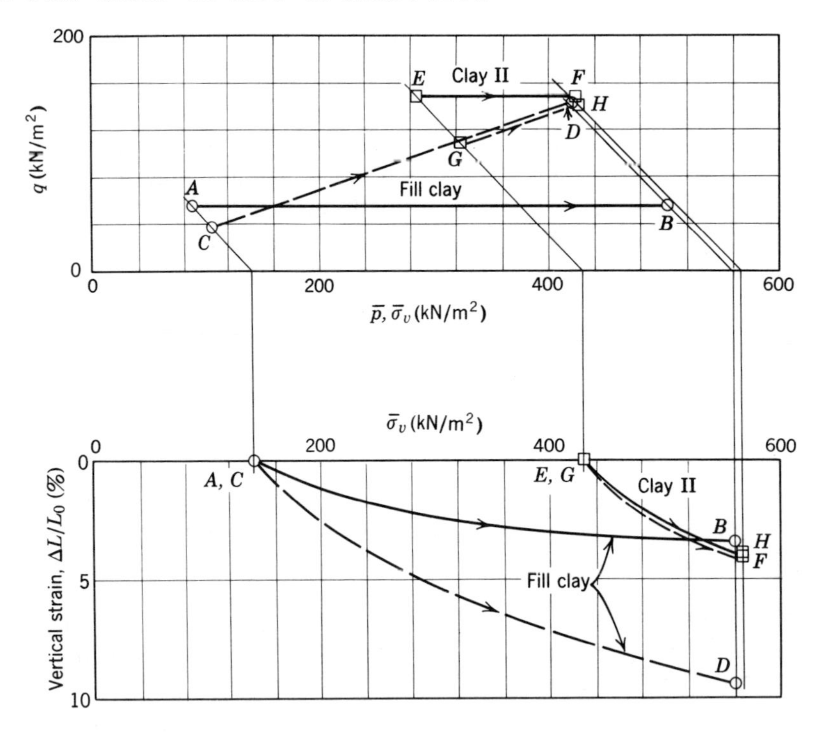

Fig. 22.5 Results of tests on Kawasaki clays.

test, and decreased for the load test. The trends shown in Fig. 22.4 are as we would expect. During the unload test the average effective stress in the sample is decreasing and thus the soil tends to expand. (Note that the strength parameter $\bar{\phi}$ is independent of the stress path, but the shear strength is very much dependent on the stress path because the effective stress on the failure plane at failure is greater in the load test.)

Figure 22.5 presents the results of two pairs of tests on undisturbed specimens of two clays from Kawasaki, Japan (see E16.2–1). *CD* is the effective stress path for a confined compression test on a sample of fill clay. *AB* is the effective stress path for a test between the two same vertical stresses as was used in the confined compression test. *GH* and *EF* are corresponding tests on clay II. In the bottom part of Fig. 22.5 are presented the strain data for the four tests.

For the fill clay much larger vertical strains occurred in the confined compression test than in the test along the horizontal stress path *AB*, even though the initial and final vertical stresses for the two tests are identical. The shear stresses and shear strains in the confined compression test *CD* are greater than in the test *AB*.

On the two tests on clay II the vertical strains are approximately equal. During the test *EF* the shear stress is constant but larger than that during the confined compression test.

The test data in Figs. 22.4 and 22.5 illustrate an important fact: the vertical stress-vertical strain relationships of a clay can be highly dependent on the actual stress path.

22.4 SUMMARY OF MAIN POINTS

The stress-strain behavior of soil is far from that of a linear elastic material. The stress-strain modulus depends very significantly on a number of factors. The most important ones are:

1. *Initial consolidation.* The stress-strain modulus depends on the magnitude and type of initial consolidation. The higher the initial consolidation stress, the higher the stress-strain modulus.
2. *Stress history.* An overconsolidated soil has a higher stress-strain modulus than a normally consolidated soil.
3. *Soil type.* The more plastic a soil, the more compressible it is and the lower its stress-strain modulus.
4. *Disturbance.* Disturbing the structure of a natural soil usually reduces the stress-strain modulus.
5. *Type of loading.* The stress-strain modulus can be highly dependent on the stress path.

PROBLEMS

22.1 Refer to the oedometer test results in Fig. 20.5 and determine C_c, a_v, and m_v for a stress change from 100 kN/m² to 200 kN/m². Determine C_s for the decrement of 200 to 100 kN/m².

22.2 From a freshly remolded batch of Weald clay two specimens are prepared and tested in the following manner:

a. Test 1. Isotropically consolidated in triaxial machine to $\bar{\sigma}_v = \bar{\sigma}_h = 100$ kN/m² and then consolidated to $\bar{\sigma}_v = \bar{\sigma}_h = 200$ kN/m².

b. Test 2. Consolidated in the oedometer to $\sigma_r = 100$ kN/m² and then consolidated to $\sigma_r = 200$ kN/m².

For each test draw the effective stress path and estimate the vertical strain for the loading of 100 to 200 kN/m².

22.3 Using the empirical expression between liquid limit and compression index given in Section 22.1 and the limit data in Fig. 7.9, estimate the compression index for Cambridge clay. Compare with the value given in Example 22.1.

22.4 For the Δq range of zero to half peak value in the load test of Fig. 22.4, determine E.

22.5 For the test on the overconsolidated Weald clay in Fig. 20.10, estimate E and μ for the Δq range of zero to half peak q.

CHAPTER 23

Earth Retaining Structures with Drained Conditions

This chapter considers lateral stresses acting against earth retaining structures. Knowledge of these stresses is needed to evaluate the stability of the structure at hand and to design the components in the structure. Chapter 13 treated lateral stresses exerted by dry sand, and this chapter extends these principles to include the effects of water and to consider soil having a cohesion intercept.

23.1 ACTIVE AND PASSIVE STRESSES

As shown in Fig. 23.1, Figs. 13.4 and 13.5 apply to saturated granular soils as well as to dry granular soils, provided that all stresses are effective stresses. Thus Eqs. 13.1 and 13.2 applying to problems with geostatic stresses involving saturated soils are

$$K_a = \frac{\bar{\sigma}_{ha}}{\bar{\sigma}_v} = \frac{1}{N_\phi} = \frac{1 - \sin \bar{\phi}}{1 + \sin \bar{\phi}} = \tan^2\left(45 - \frac{\bar{\phi}}{2}\right) = \frac{1 - \tan \bar{\alpha}}{1 + \tan \bar{\alpha}} \tag{23.1}$$

$$K_p = \frac{1}{K_a} = \frac{\bar{\sigma}_{hp}}{\bar{\sigma}_v} = N_\phi = \frac{1 + \sin \bar{\phi}}{1 - \sin \bar{\phi}} = \tan^2\left(45 + \frac{\bar{\phi}}{2}\right) = \frac{1 + \tan \bar{\alpha}}{1 - \tan \bar{\alpha}} \tag{23.2}$$

where the bars over $\bar{\phi}$ and $\bar{\alpha}$ remind us that these parameters are based on effective stresses. Note that K_a, the *active stress ratio*, and K_p, the *passive stress ratio*, are both based on *effective stresses* and *not* total stresses.

Figure 23.2 shows both the effective stress path ESP and the total stress path TSP for the drained unloading to the active condition. During the unloading there is a constant pore pressure u_s such as that from static pore pressure conditions in the field or a static backpressure in the laboratory. The total stress path *DE* is parallel to the effective stress path *AC* and is displaced to the right a magnitude equal to u_s. Had the pore pressure u_s been negative, the total stress path would have been to the left of the effective stress path.

Figures 23.1 and 23.2 hold for any soil in which the strength is $\tau_{ff} = \bar{\sigma}_{ff} \tan \bar{\phi}$. If the soil has a cohesion intercept, i.e., $\tau_{ff} = \bar{c} + \bar{\sigma}_{ff} \tan \bar{\phi}$, then the effective stress path to the active state is as shown in Fig. 23.3. For a given strength angle $\bar{\phi}$, the presence of a cohesion intercept $\bar{c}$ means an increased strength, which in turn gives a *lower* active stress and a *higher* passive stress. The decrease in $\bar{\sigma}_{ha}$ may be seen by comparing Figs. 23.2 and 23.3.

Strains Needed to Develop Active and Passive Stresses

Figures 23.4 and 23.5 illustrate the horizontal and vertical strains required to develop active and passive stresses for the normally consolidated clay from Fig. 22.4. The general pattern of behavior is the same as for sand (see Fig. 13.7). Note the general characteristic of more strain being required to develop passive stress than is required to develop active stress. Five times as much horizontal strain is needed to go from the initial condition of $K = 1$ to the passive state K_p as is needed to go from $K = 1$ to the active state K_a. Figure 23.6 presents the stress paths and the strains required to reach active and passive conditions starting from a K_0 condition. We see from this figure that less strain is required to reach the active state starting from K_0 than from $K = 1$, and that more strain is required to go from K_0 to the passive state compared to going from $K = 1$ to the passive state. This situation is logical since the K_0 condition is along the stress path from $K = 1$ to the active state.

The preceding chapters have emphasized that the differences in the stress-strain behavior among different soils are not so much in nature as in magnitude. The soil used to illustrate the strains required to reach active and passive conditions was normally consolidated clay. As we would expect, the strains involved with carrying this soil to failure are greater than those required for a sand. For example, Fig. 23.4 shows that approximately 8% vertical strain and -4% horizontal strain were

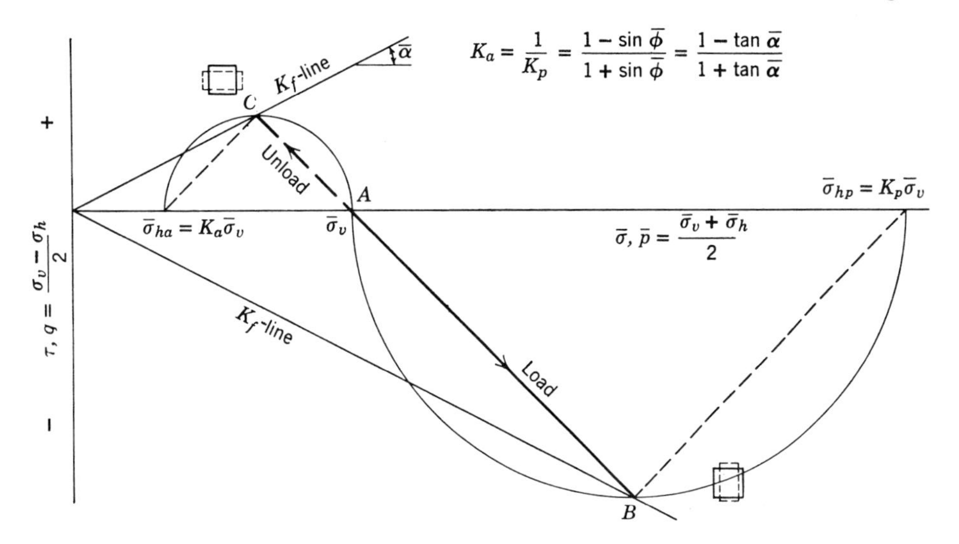

Fig. 23.1 Stress paths and Mohr circles for active and passive states.

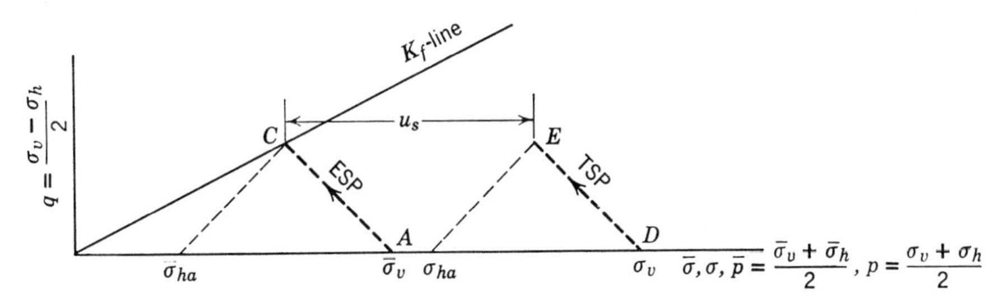

Fig. 23.2 Stress paths for active state.

required to carry the clay from an initial $K = 1$ condition to active stress condition, whereas the data in Fig. 10.22 for the sand from Libya show that less than 1% vertical strain and less than $-\frac{1}{2}$% horizontal strain were required to take the sand to an active condition.

23.2 GRAVITY RETAINING WALLS RETAINING COHESIONLESS SOIL

This section considers problems in which all or a portion of the sand fill behind a retaining wall is saturated with water. The general principles governing the design of gravity retaining walls are still those described in Chapter 13. However, the presence of the water may alter the magnitude of the thrust against the retaining

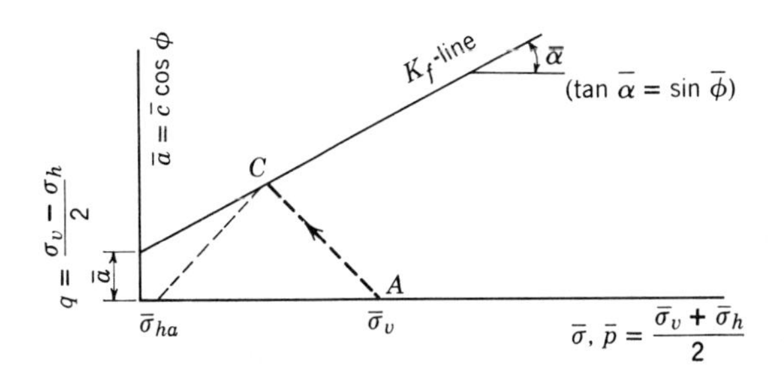

Fig. 23.3 Active state in soil with cohesion intercept.

wall and may also alter both the resistance to sliding along the base of the wall and the bearing capacity of the soil that supports the wall. This section considers only the effect of the water on the active thrust against the wall.

We can use either force system *a* or force system *b* for the analysis of the stability of a soil mass:

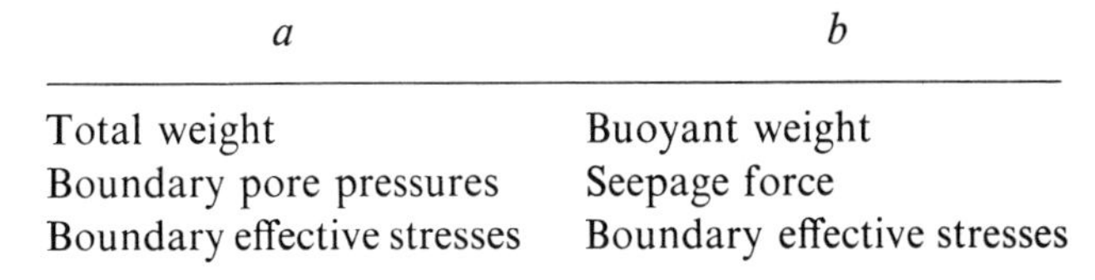

a	*b*
Total weight	Buoyant weight
Boundary pore pressures	Seepage force
Boundary effective stresses	Boundary effective stresses

Submerged Retaining Wall

In the situation shown in Fig. 23.7, water stands to the same elevation against each side of the retaining wall and there is no seepage of water. Such a situation approximates that which might be found where a stream passes through a congested area in cities and towns. In order to find the thrust exerted by the backfill, it is necessary to analyze the equilibrium of a wedge of soil such as *IJM*. Let us assume that there are no shear stresses between the wall and the soil. The analysis for this case (including the effects of a uniform surcharge) is given in Example 23.1. The resulting expression contains exactly

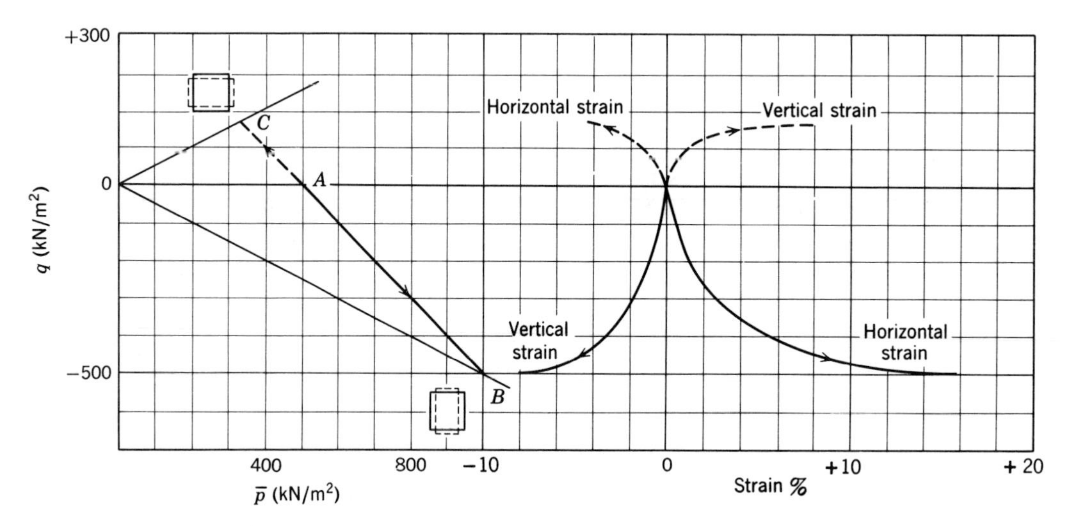

Fig. 23.4 Strains associated with active and passive stresses.

the same trigonometric function as did the corresponding expression for dry soil (Fig. 13.11). Hence the inclination θ of the trial boundary which gives the maximum value of P works out to be identical with that for the case of dry backfill (Example 13.4). The total thrust from the backfill

$$P_a = \tfrac{1}{2}\gamma_w H^2 + \tfrac{1}{2}\gamma_b H^2 K_a + q_s H K_a \qquad (23.3)$$

is made up of three parts:

1. The term $\frac{1}{2}\gamma_w H^2$ is the thrust from the pore water, and has exactly the same magnitude as though the soil were not present.
2. The term $\frac{1}{2}\gamma_b H^2 K_a$ is the thrust exerted by the soil skeleton as the result of its own weight. Note especially that the buoyant unit weight appears in this term.
3. The term $q_s H K_a$ is the thrust exerted by the soil skeleton as the result of the surcharge. This term is exactly the same as for a dry backfill.

Alternatively, this problem might have been analyzed by considering the boundary effective stresses (including shear stresses) and the buoyant unit weight of the soil. Such an analysis gives the resultant of the effective stresses against the wall:

$$\bar{P} = \tfrac{1}{2}\gamma_b H^2 K_a + q_s H K_a$$

The total thrust from the backfill is the sum of the effects

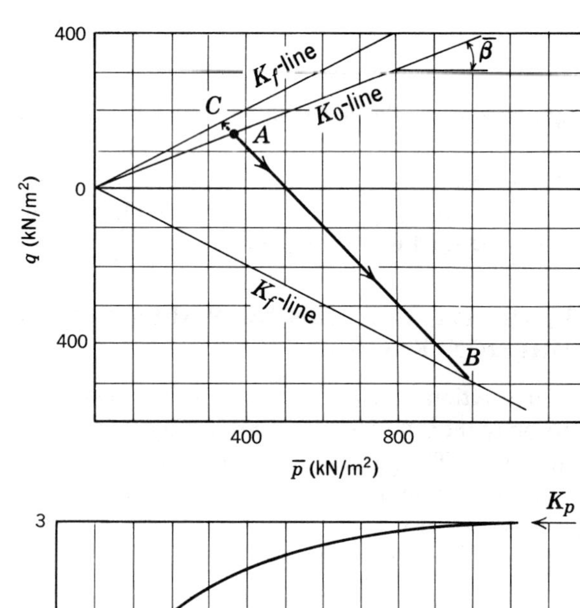

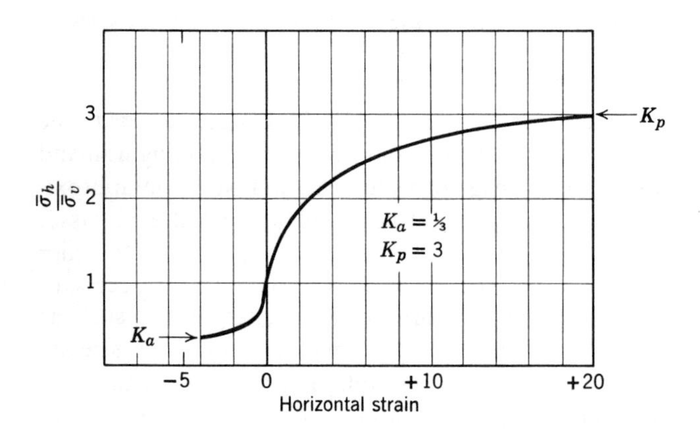

Fig. 23.5 Lateral strains for active and passive stresses.

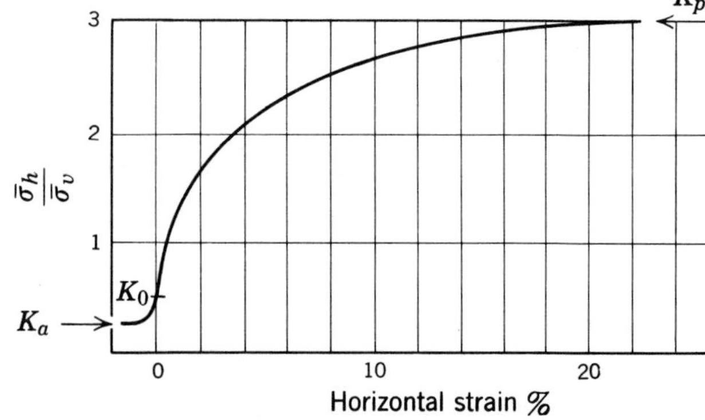

Fig. 23.6 Active and passive stresses from K_0 condition.

of pore water pressure and the effective stress:

$$P_a = \tfrac{1}{2}\gamma_w H^2 + \bar{P}$$

This, of course, is just the result given in Eq. 23.3.

It can further be proved that all of the results presented in Chapter 13 hold for the case where the backfill is submerged and there is no seepage, provided only that the buoyant unit weight is used. Thus the equation in Fig. 13.21 (with γ_b replacing γ) can be used to estimate the thrust exerted by the mineral skeleton in the case of wall friction and sloping wall. The wall friction angle, $\bar{\phi}_w$, must also be expressed in terms of effective stress. The thrust exerted by the pore water must be added in order to obtain the total thrust.

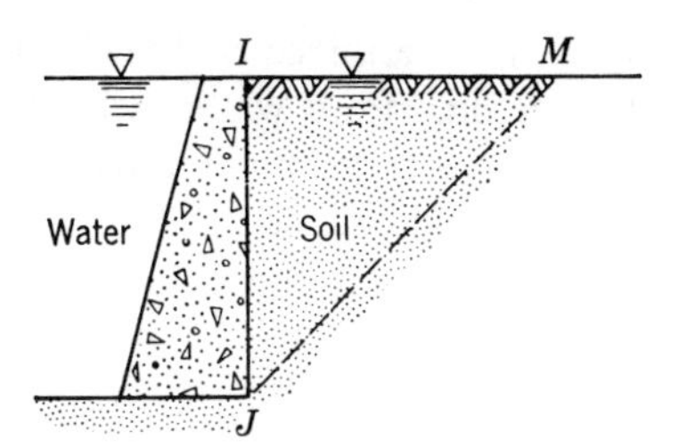

Fig. 23.7 Completely submerged retaining wall.

Note especially that the coefficient of active stress is exactly the same regardless of whether the soil is dry or submerged. However, this coefficient gives the effective stresses and resultant of the effective stresses. The

---

▸ Example 23.1 Analysis of Completely Submerged Backfill

Given: Completely submerged backfill, as in Fig. E.23.1

Find: Expression for total active thrust from backfill.

Solution:

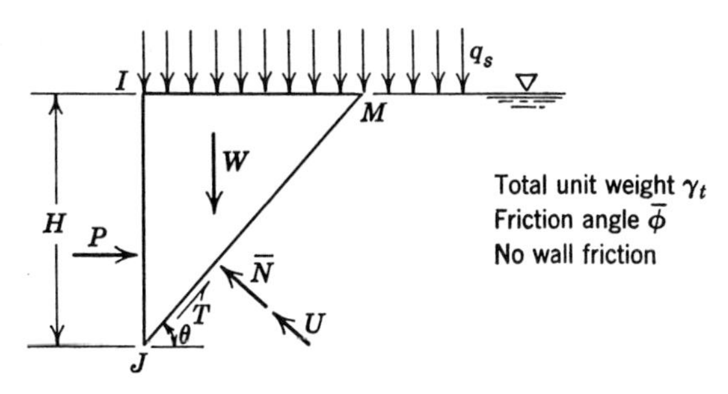

Fig. E23.1

In Fig. E23.1 the pore pressure along the assumed boundary varies linearly from zero at point M to $\gamma_w H$ at point J. The average pore pressure is $\frac{1}{2}\gamma_w H$, and

$$U = \frac{\frac{1}{2}\gamma_w H^2}{\sin\theta}$$

Now

$$T = \bar{N}\tan\bar{\phi}$$

Requirements for equilibrium:

$\Sigma V = 0$:

$$W + q_s H\cot\theta - T\sin\theta - (\bar{N} + U)\cos\theta = 0$$

$$(\tfrac{1}{2}\gamma_t H + q_s)H\cot\theta - \bar{N}(\cos\theta + \sin\theta\tan\bar{\phi}) - \tfrac{1}{2}\gamma_w H^2\cot\theta = 0$$

$$\bar{N} = (\tfrac{1}{2}\gamma_b H + q_s)H\cot\theta\,\frac{1}{\cos\theta + \sin\theta\tan\bar{\phi}}$$

$\Sigma H = 0$:

$$P - (\bar{N} + U)\sin\theta + T\cos\theta = 0$$

$$P - \bar{N}(\sin\theta - \cos\theta\tan\bar{\phi}) - \tfrac{1}{2}\gamma_w H^2 = 0$$

$$P = \tfrac{1}{2}\gamma_w H^2 + (\tfrac{1}{2}\gamma_b H + q_s)H\cot\theta\tan(\theta - \bar{\phi})$$

Note that θ enters exactly as it did in Fig. 13.11.

Maximizing expression for P:

Since θ enters exactly as it did in Fig. 13.11.

$$\theta_{cr} = 45° + \frac{\bar{\phi}}{2}$$

$$P_a = \tfrac{1}{2}\gamma_w H^2 + (\tfrac{1}{2}\gamma_b H^2 + q_s H)\frac{1 - \sin\bar{\phi}}{1 + \sin\bar{\phi}} = \tfrac{1}{2}\gamma_w H^2 + \tfrac{1}{2}\gamma_b H^2 K_a + q_s H K_a$$ ◂

magnitude of K_a is determined by the shear resistance of the mineral skeleton. Pore water, which cannot resist shear, does not affect the magnitude of K_a but does influence the magnitude of the total thrust.

Example 23.2 illustrates the computation of the net horizontal thrust against a completely submerged retaining wall. Only the force exerted by the mineral skeleton is shown in the figure. The force exerted by the pore water of the soil is completely balanced by the water force against the other side of the wall. Thus submerging the wall reduces the net thrust of the soil against the wall.[1] Later, however we shall examine cases in which a saturated backfill is a detriment.

Example 23.2 Example of Completely Submerged Retaining Wall

Given: Retaining wall in Fig. E23.2-1
Find: Net thrust against wall

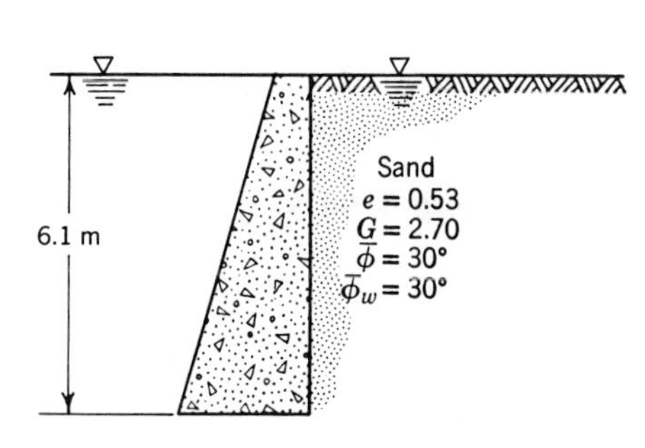

Fig. E23.2-1

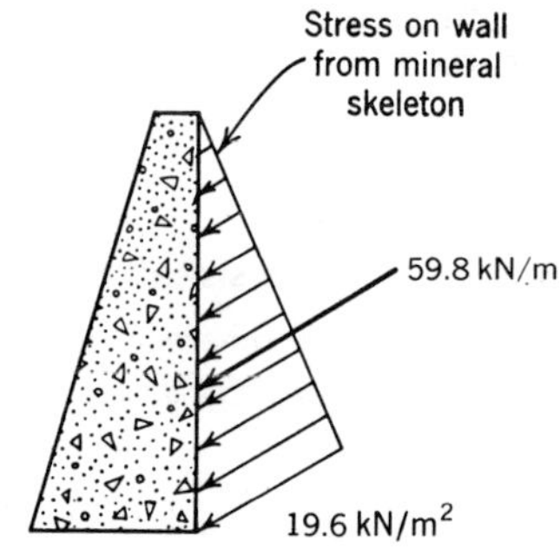

Fig. E23.2-2

Solution:

$$\gamma_d = \left(\frac{1}{1.53}\right)(2.70)(9.81) = 17.31 \text{ kN/m}^3$$

$$\gamma_t = 17.3 + \frac{0.53}{1.53}(9.81) = 20.71 \text{kN/m}^3$$

$$\gamma_b = 20.71 - 9.81 = 10.9 \text{ kN/m}^3$$

From Fig. 13.18,

$$K_a = 0.295$$

$$\bar{P}_a = \tfrac{1}{2}(10.9)(6.1)^2(0.295) = 59.8 \text{ kN/m}$$

Note that this thrust is inclined. Fig. E23.2-2 gives the vector sum of normal and shear stresses per unit area of wall. ◀

Partially Submerged Retaining Wall

A more common situation is to have the phreatic surface below the soil surface. Example 23.3 shows such a situation and analyzes the equilibrium of a wedge behind the wall for the case of zero wall friction. The soil above the water table is presumed to be completely dry.

From the form of Eq. 23.4, it can be seen that the critical value of θ is the same as for the cases of dry and completely submerged backfills. Equation 23.5 gives the resultant thrust exerted against the wall by the mineral skeleton. From the form of Eq. 23.5, it can be deduced that the effective stress against the wall increases γK_a per meter of depth above the water table, and $\gamma_b K_a$ per meter of depth below the water table.

Problems involving a sloping wall and/or wall friction can also be treated using Eq. 23.4 plus the appropriate value of K_a from Fig. 13.18, as illustrated in Example 23.4. Because the stress distribution is not linear with depth, the resultant force of the mineral skeleton on the wall acts higher than at the bottom third point of the wall.

If a more accurate solution is made for a partially submerged backfill by using trial failure surfaces which are curved, it will be found that the location of the critical failure surface does vary somewhat with depth of submergence. However, the predicted thrust will not vary appreciably from that given by Eq. 23.4.

[1] However, the resistance to sliding at the base of the wall and the bearing capacity of the supporting soil are also decreased.

▸ Example 23.3 Analysis of Partially Submerged Backfill

From Fig. E23.3,

$$W = \tfrac{1}{2}\gamma_t(H')^2 \cot\theta + \gamma(H - H')H'\cot\theta + \tfrac{1}{2}\gamma(H - H')^2\cot\theta$$
$$= \tfrac{1}{2}\gamma_t(H')^2\cot\theta + \tfrac{1}{2}\gamma\cot\theta(H^2 - H'^2)$$

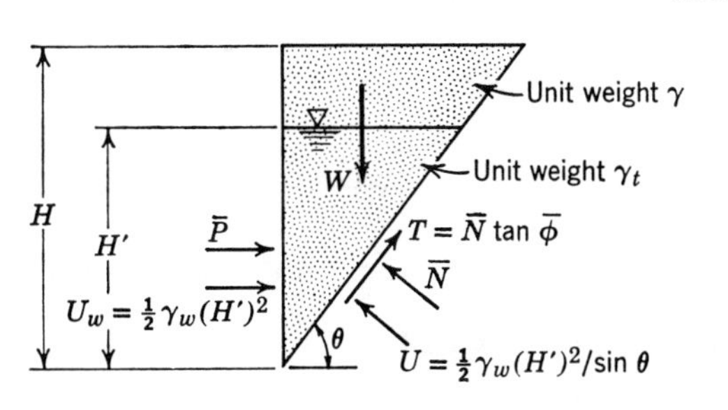

Fig. E23.3

Summation of vertical forces:

$$W - T\sin\theta - \bar{N}\cos\theta - U\cos\theta = 0$$

$$\tfrac{1}{2}\gamma_t(H')^2\cot\theta + \tfrac{1}{2}\gamma\cot\theta\,[H^2 - (H')^2] - \bar{N}(\cos\theta + \sin\theta\tan\bar{\phi}) - \tfrac{1}{2}\gamma_w(H')^2\cot\theta = 0$$

$$\tfrac{1}{2}\gamma_b(H')^2\cot\theta + \tfrac{1}{2}\gamma\cot\theta[H^2 - (H')^2] = \bar{N}(\cos\theta + \sin\theta\tan\bar{\phi}]$$

Summation of horizontal forces:

$$\bar{P} + \tfrac{1}{2}\gamma_w(H')^2 - \bar{N}\sin\theta + T\cos\theta - \tfrac{1}{2}\gamma_w(H')^2 = 0$$

$$\bar{P} = \bar{N}(\sin\theta - \cos\theta\tan\bar{\phi})$$

$$\bar{P} = \{\tfrac{1}{2}\gamma_b(H')^2 + \tfrac{1}{2}\gamma[H^2 - (H')^2]\}\cot\theta\tan(\theta - \bar{\phi}) \qquad (23.4)$$

$$\bar{P}_a = \tfrac{1}{2}\{\gamma_b(H')^2 + \gamma[H^2 - (H')^2]\}K_a$$
$$= \tfrac{1}{2}[\gamma H^2 - (\gamma - \gamma_b)(H')^2]K_a \qquad (23.5)$$ ◀

▸ Example 23.4 Example of Partially Submerged Retaining Wall

Given: Retaining wall in Fig. E23.4

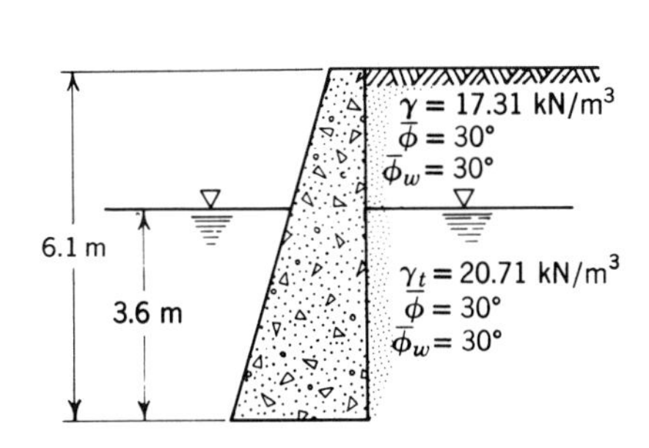

Fig. E23.4

Find: Net horizontal thrust against wall we have

Solution: Using results from Example 23.3

$$\text{Stress at 2.5 m depth} = 17.31(2.5)(0.295) = 12.77\ \text{kN/m}^2$$

$$\text{Additional stress at bottom} = 10.9(3.6)(0.295) = 11.58\ \text{kN/m}^2$$

$$\text{Resultant force} = \tfrac{1}{2}[17.31(6.1)^2 - 6.41(3.6)^2]0.295$$
$$= 95.02 - 12.25 = 82.77\ \text{kN/m of wall}$$

$$\text{Height of resultant} = \frac{95.02(6.1/3) - 12.25(1.2)}{82.77} = 2.16\ \text{m}$$ ◀

Perched Water Table in Backfill

A very serious condition arises when the water table stands high in the backfill, but there is no water standing against the exposed face of the wall. If, for example, the retaining wall rests upon an impermeable stratum, the water table in the backfill might rise to the surface of the backfill during a very heavy and prolonged rainfall.

An illustration of this situation is worked out in Example 23.5. The thrust exerted by the mineral skeleton is just as in Example 23.2. However, there is now a large net thrust from the pore water. The combined thrust is extremely large.

▶ Example 23.5 Example of Perched Water Table in Backfill

Given: Retaining wall in Fig. E23.5-1

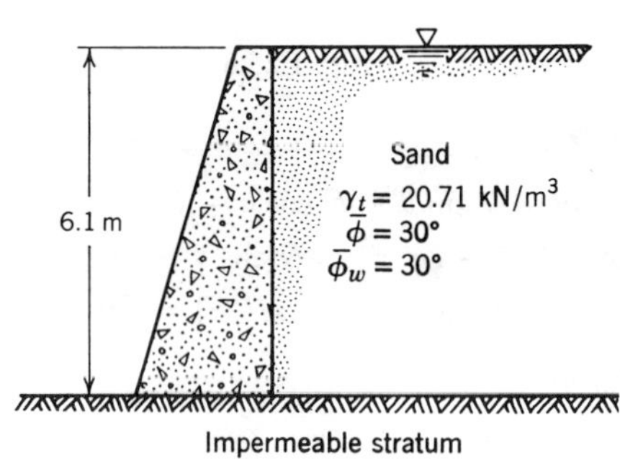

Fig. E23.5-1

Find: Horizontal thrust against wall.
Solution:

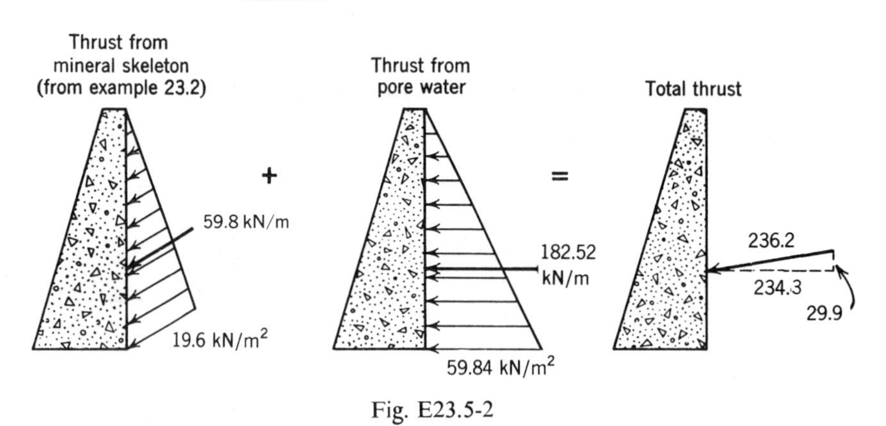

Fig. E23.5-2 ◀

Backfill with Sloping Drain

The foregoing example emphasizes the importance of providing drains to reduce the pore pressures within the backfill. One common form of drain is shown in Fig. 23.8. Such a drain can readily be constructed against natural sloping ground prior to placement of the backfill. The soil used for the drainage layer must be much more permeable than the backfill, and weep holes must be provided through the wall to permit water to escape from the drainage layer. A filter (discussed in Chapter 19) may be needed between the backfill and the drain.

For a steady rainfall on the surface of the backfill, the flow net consists of vertical flow lines and horizontal equipotentials. Hence the pore pressure is zero throughout the backfill. Thus the effective stresses and shear stresses must be in equilibrium with the total (i.e., saturated) weight of the wedge. Hence the analysis takes exactly the same form as the analysis for a dry backfill, except that the saturated unit weight replaces the dry unit weight:

$$\bar{P}_a = \tfrac{1}{2}\gamma_t H^2 K_a$$

Since the pore pressure against the wall is zero, $\bar{P}_a$ equals the total thrust against the wall. Thus a sloping drain effects a large reduction in the thrust caused by a rain-saturated backfill.

Because the flow pattern is so simple, this problem might also have been solved by employing buoyant unit

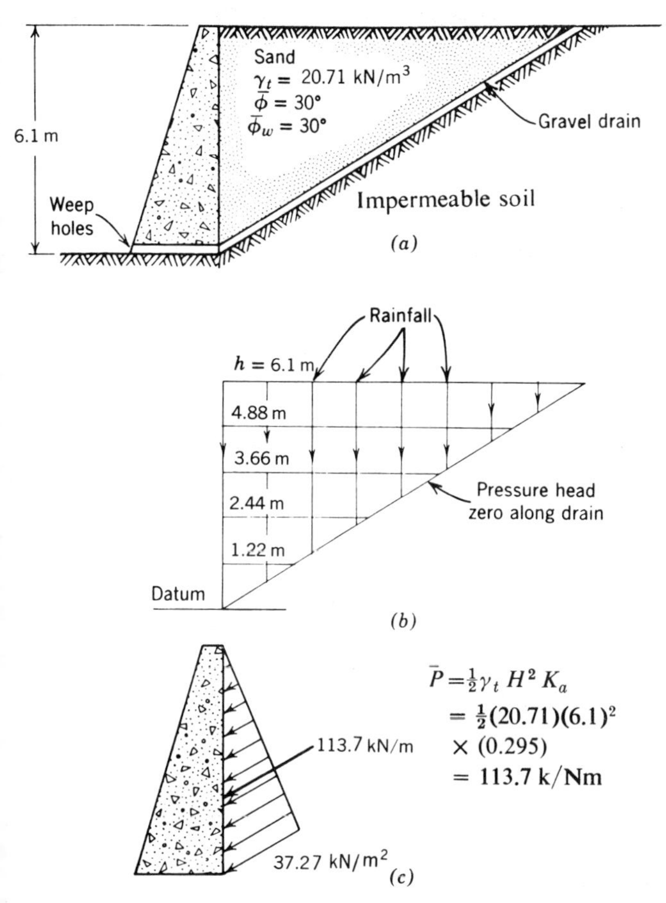

Fig. 23.8 Analysis of backfill with sloping underdrain. (*a*) Arrangement of drain. (*b*) Flow net for rainfall on surface of backfill. (*c*) Stresses on wall from mineral skeleton.

weight, seepage force, and boundary effective stresses. Since the gradient is unity, the seepage force per unit volume j is $1 \cdot \gamma_w = \gamma_w$ acting downwards. The sum of the buoyant unit weight and the seepage force equals the saturated unit weight. Hence we are right back to the system of forces described in the previous paragraph.

Since the gradient is unity, the quantity of seepage per unit of backfill surface area is equal to k. The gravel drain and weep holes must be designed to carry this flow with negligible head loss. The rainfall necessary to cause full saturation of the backfill also equals k. Thus, for a fine sand with a permeability of 10^{-3} cm/sec, an active thrust of 113.7 kN/m will be achieved by a steady rainfall of 3.6 cm/hr—a heavy but not uncommon rainfall. For lesser rains the thrust would be intermediate between 95.02 kN/m (dry backfill) and 113.7 kN/m. Use of a sufficiently permeable backfill can avoid saturation by even the heaviest storm.

Backfill with Vertical Drain

Another common form of drain, together with the flow net for a heavy rainfall, is shown in Fig. 23.9. No simple mathematical solution is available for this case. Various positions must be assumed for the boundary of the trial wedge, the pore pressures along this boundary summed, and the wall thrust required for equilibrium computed. The pore pressure against the wall is zero at all depths. The computation must be repeated for various assumed boundaries until the location giving the largest thrust is found. Since there will be positive pore pressures along all assumed boundaries, the thrust for the case of a vertical drain will exceed that for the case of a sloping drain. This case is illustrated in Example 23.6. Figure E23.6-2 shows the flow net in more detail than Fig. 23.9. Since the pore pressure is zero at all points along the vertical drain, the total head at the drain must equal the elevation head. If there is equal head loss between successive equipotentials, these equipotentials must be spaced uniformly along the vertical drain.

Sample calculations are shown for the assumption that the failure surface is inclined at 45°. Figure E23.6-3 and the accompanying table show the distribution of pore pressure against the assumed failure surface and the calculation of the resultant pore water force U against this surface. Figure E23.6-4 shows the force diagram and a formula, derived from this diagram, for the thrust P. For the failure surface at 45°, the thrust is computed to be 148.8 kN per running meter of wall. Results for various θ are plotted in Fig. E23.6-5. The maximum thrust occurs for θ of approximately 45°.

The result for this case should be compared with the other results in Table 23.1. As expected, the thrust for the case of a vertical drain exceeds that for the case of a sloping drain, but still is much less than that for the case of no drain. Whereas all other cases in Table 23.1 involved a critical failure surface inclined at about 55°, the case with a vertical drain involved a much shallower failure surface. In this latter case, the pore pressures increased with increasing distance from the wall, whereas in all other cases the pore pressure at any depth is independent of distance from the wall.

Figure 23.10 shows the same wall used in Example 23.2. For the simplified condition of no wall friction, this figure shows the force polygon for each of four situations: dry backfill, submerged backfill with water on the outside of the wall, vertical seepage, and vertical

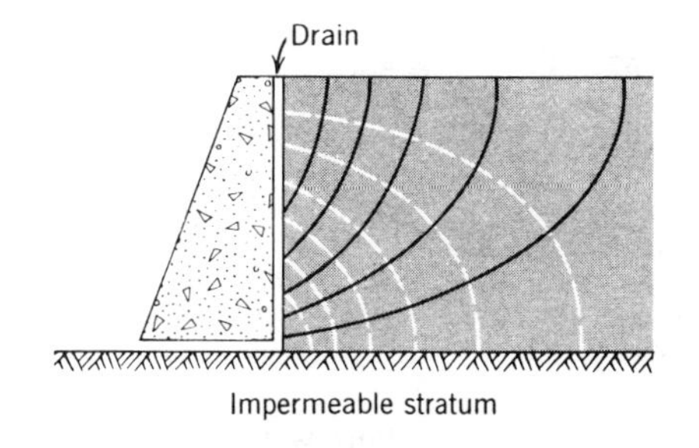

Fig. 23.9 Backfill with vertical drain.

▶ Example 23.6 Analysis of Backfall with Vertical Drain

The wall and drain are shown in Fig E23.6-1. The various steps in the solution appear in Figs. E23.6-2 to E23.6-5.

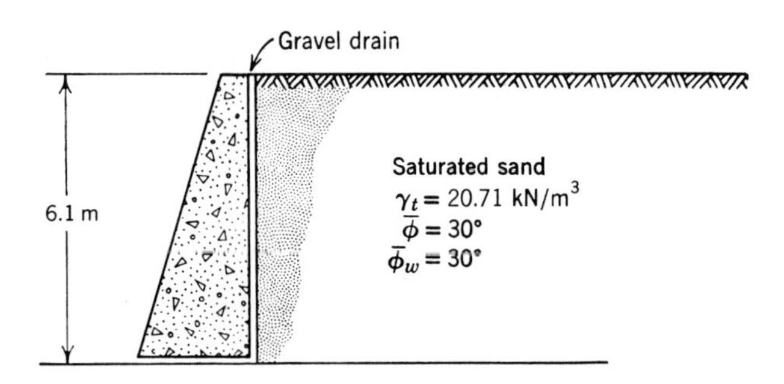

Fig. E23.6-1 The retaining wall and drain.

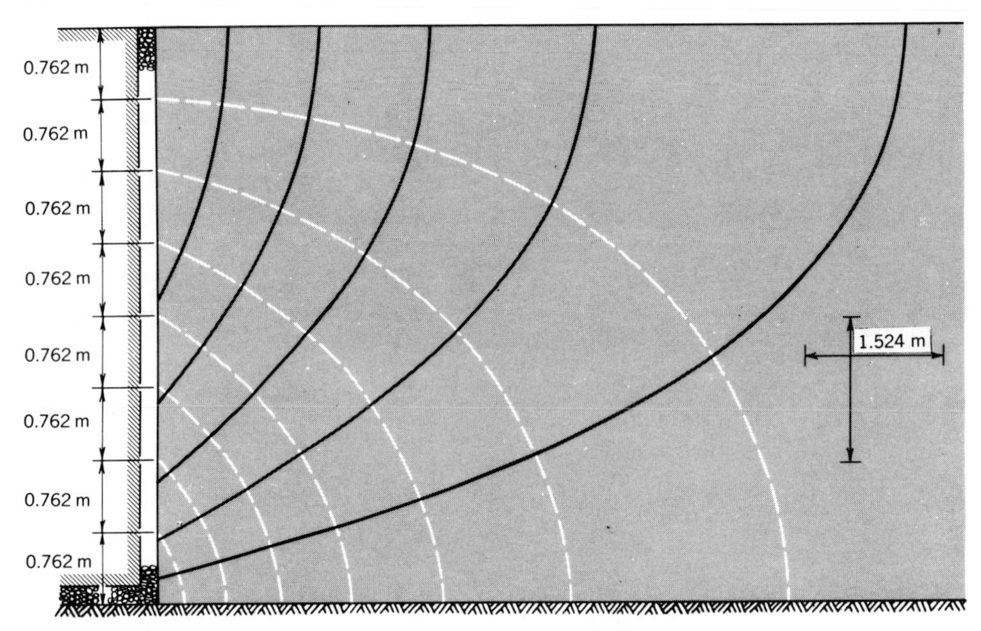

Fig. E23.6-2 Flow net for steady rainfall.

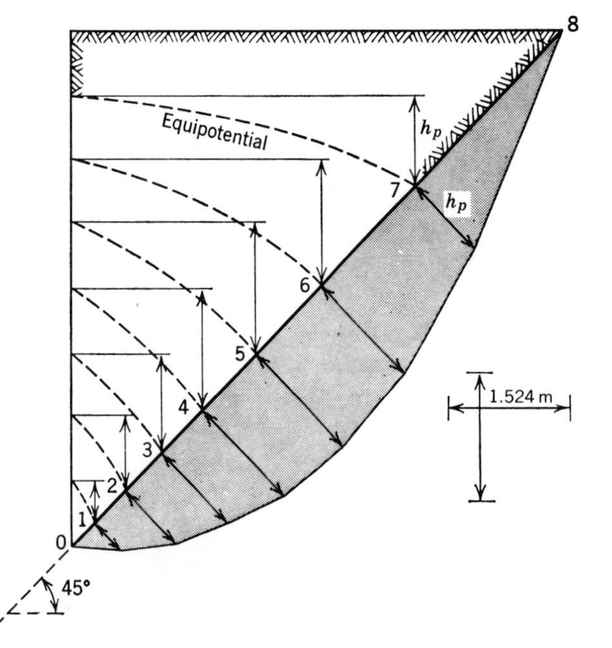

Point	h_p	Interval ΔL	$(h_p)_{ave}\Delta L$
0	0		
		0.366	0.084
1	0.457		
		0.518	0.355
2	0.915		
		0.640	0.673
3	1.189		
		0.701	0.930
4	1.463		
		0.945	1.426
5	1.555		
		1.159	1.767
6	1.494		
		1.677	2.147
7	1.067		
		2.591	1.382
8	0		
			8.764 m²

Fig. E23.6-3 Pore water force for $\theta = 45°$. Pore water force $= 8.764 \times 9.81 = 86.0$ kN/m of wall

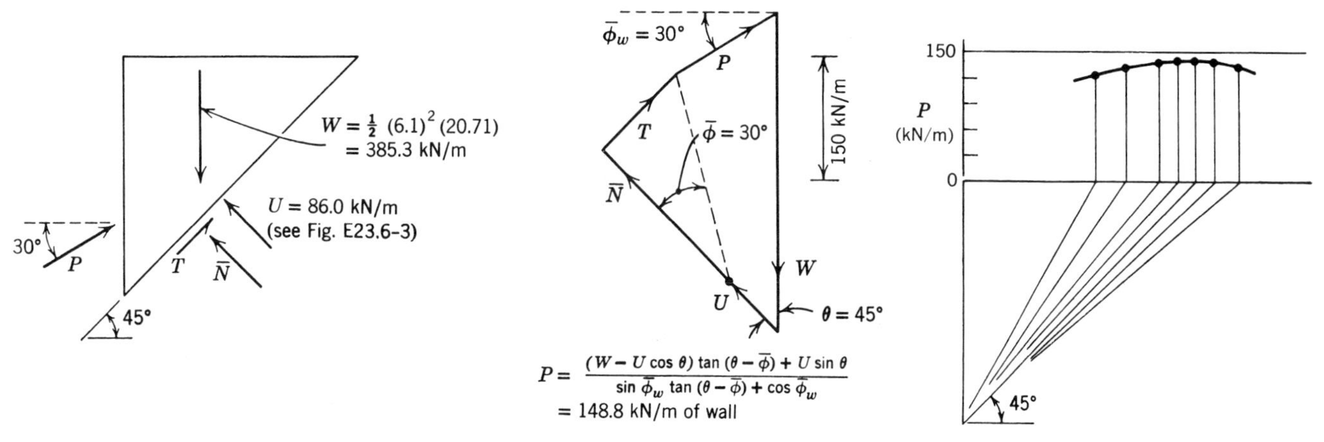

Fig. E23.6-4 Analysis for required horizontal force for case $\theta = 45°$.

Fig. E23.6-5 a convenient way to plot results for various θ. ◀

drain. In each situation the failure surface is assumed to rise at 60° to the horizontal. For the last three situations, the forces are shown for both total forces and for buoyant plus seepage forces. Study of Fig. 23.10 helps one appreciate the influence of water on the lateral forces acting on the wall.

23.3 GRAVITY WALLS RETAINING COHESIVE SOIL

The preceding section on cohesionless backfills actually applies to any soil with no cohesion intercept. Since most normally consolidated clays have no cohesion intercept, the preceding section applies to most normally consolidated clays as well as to cohesionless soils. The present section considers soil that does exhibit a cohesion intercept.

At the outset it should be emphasized that most retaining walls have cohesionless backfills because unless the fill behind a wall is properly drained, very high water pressures can result. Example 23.5 showed that the thrust from the water could far exceed that from the soil. Two things are needed to keep down the water thrust: (*a*) a drain system, and (*b*) a backfill with a high permeability. Thus, where the designer has a choice he uses a permeable (cohesionless) backfill. There are situations, however, where a cohesive soil must be retained.

The following paragraphs discuss a series of highly idealized situations. The examples are intended to introduce and clarify key concepts. The equations and calculation procedures may seldom be of use in practical problems, but understanding of the key concepts will permit rational approaches to practical problems.

Effect of Cohesion upon Passive Thrust

Let us first consider the situation in which the ground water table is at the surface of the backfill, as shown in Fig. 23.11. The passive thrust is applied so slowly no excess pore pressures exist in the backfill. It is assumed that there is no shear stress between the wall and the backfill.

For these simple conditions the stresses within the soil are geostatic, and the magnitude of the horizontal stress at any depth can readily be evaluated using the Mohr circle construction (see Fig. 23.11). The branches of the Mohr envelope are extended backward until they intersect the normal stress axis. The distances A and B are measured from this intersection. We note that the ratio $B/A = N_\phi$ is identical with K_p for a cohesionless soil with zero wall friction. However, when cohesion is present N_ϕ is not equal to the ratio of vertical to horizontal effective stress. Expressing A and B in terms of $\bar{\sigma}_1$, $\bar{\sigma}_3$, and $\bar{c}\cot\bar{\phi}$, we obtain an expression for $\bar{\sigma}_1 = \bar{\sigma}_h$ (Eq. 23.6). This expression can be integrated to give the

Table 23.1 Active Thrusts for Various Conditions

Condition	Source	Horizontal Component (kN/m)	Vertical Component (kN/m)
Fully submerged	Ex. 23.2	51.8	29.9
Dry	Ex. 13.8	82.8	47.8
Saturated with Sloping drain	Fig. 23.8	98.5	56.9
Saturated with vertical drain	Ex. 23.6	128.9	74.4
Saturated without drain	Ex. 23.5	234.3	29.9

Note. For the fully submerged case, only the net thrust, which is the thrust from the mineral skeleton, is listed.

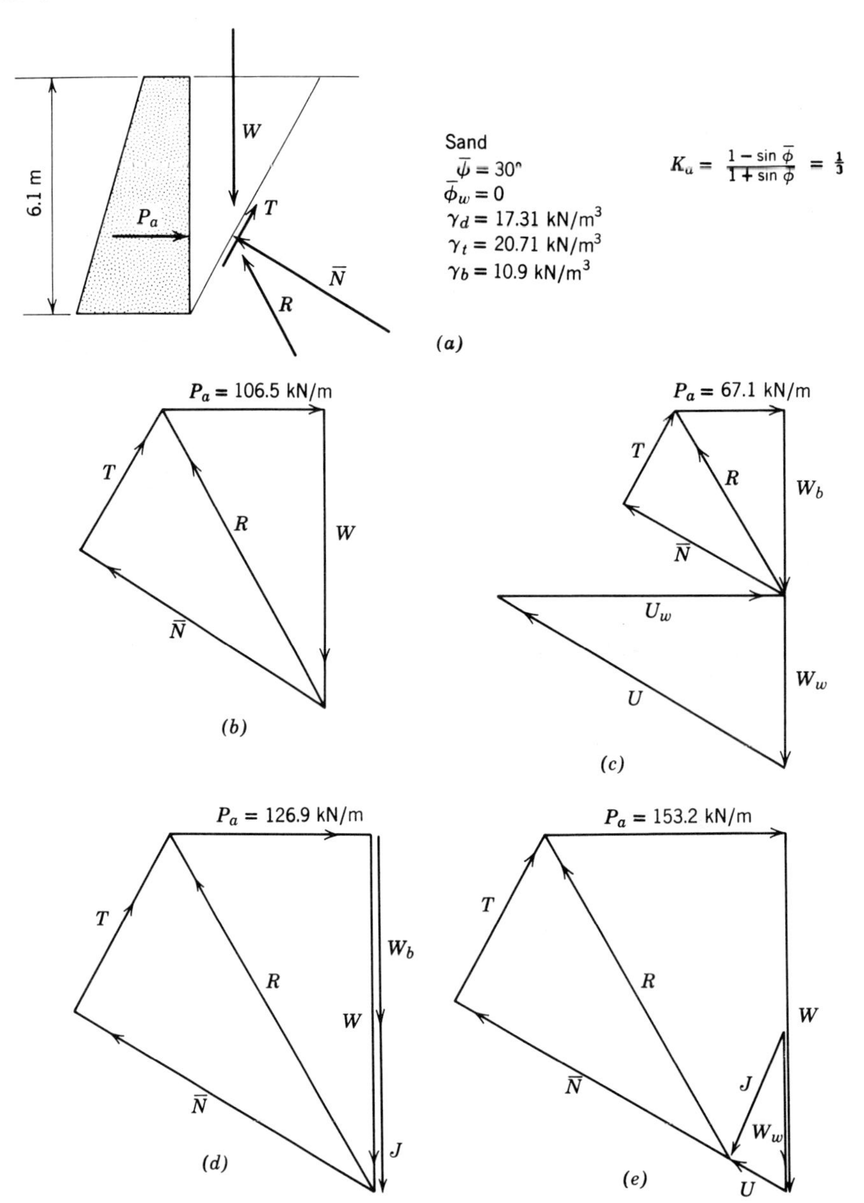

Fig. 23.10 Force polygons for various situations. (*b*) Dry. (*c*) Submerged. (*d*) Vertical seepage. (*e*) Vertical drain.

thrust exerted by the mineral skeleton, and the thrust from the pore water can be added to give the total thrust (Eq. 23.8).

The first term in Eq. 23.6 implies a linear variation of stress with depth, while the second term of this equation implies a constant stress with depth. The resultant thrust will thus be located somewhere between the mid-point and the lower third point, depending on the magnitude of $\bar{c}$.

For a uniform surcharge of magnitude q_s the total thrust will be

$$P_p = \tfrac{1}{2}\gamma_w H^2 + \tfrac{1}{2}\gamma_b H^2 N_\phi + q_s H N_\phi + 2\bar{c}H\sqrt{N_\phi} \tag{23.8a}$$

Solution by Trial Wedge

The problem can also be solved using the Coulomb trial wedge procedure. Although the details of the solution are much more involved, the solution serves to indicate how this powerful tool can be extended to problems involving a cohesion intercept.

We proceed just as in Chapter 13 and in the preceding section. A location for the failure surface is assumed, and the failure wedge is analyzed. Since there is no seepage, this analysis for the effective stress portion of the total thrust can be carried out using the buoyant unit weight of the soil and the boundary effective stresses. The effect of the boundary pore pressures is fully taken into account by using the buoyant unit weight.

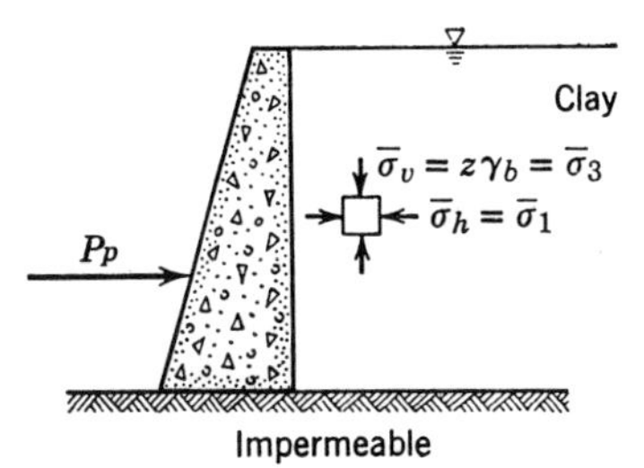

$$\frac{B}{A} = N_\phi = \frac{1 + \sin \bar{\phi}}{1 - \sin \bar{\phi}}$$

$$B = \bar{\sigma}_1 + \bar{c} \cot \bar{\phi}$$
$$A = \bar{\sigma}_3 + \bar{c} \cot \bar{\phi}$$

Combining these equations:

$$\bar{\sigma}_1 = \bar{\sigma}_3 N_\phi + \bar{c} \cot \bar{\phi}\,(N_\phi - 1)$$

It can be shown that

$$\cot \bar{\phi}\,(N_\phi - 1) = 2\sqrt{N_\phi}$$

Hence

$$\bar{\sigma}_1 = \gamma_b z N_\phi + 2\bar{c}\sqrt{N_\phi} \tag{23.6}$$

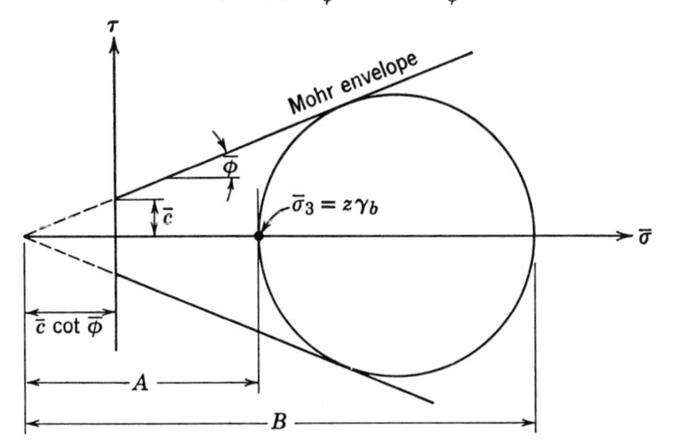

Effective stress thrust against wall:

$$\bar{P}_p = \int_0^H \bar{\sigma}_1\,dz$$

$$\bar{P}_p = \tfrac{1}{2}\gamma_b H^2 N_\phi + 2\bar{c}H\sqrt{N_\phi} \tag{23.7}$$

Total thrust against wall:

$$P_p = \tfrac{1}{2}\gamma_w H^2 + \tfrac{1}{2}\gamma_b H^2 N_\phi + 2\bar{c}H\sqrt{N_\phi} \tag{23.8}$$

Fig. 23.11 Passive thrust for soil with cohesion.

The analysis for the assumption of a plane failure surface is carried through in Example 23.7. The shear stress at any point of the failure surface is given by

$$\tau = \bar{c} + \bar{\sigma} \tan \bar{\phi}$$

where $\bar{\sigma}$ is the normal stress across the failure surface at that point. Consequently, the total shear force acting upon the failure surface can be broken into two parts:

$$C = \bar{c} \times \text{length of the failure surface}$$
$$T = \bar{N} \tan \bar{\phi}$$

The resulting expression for $\bar{P}$ is quite complicated. However, when this expression is minimized with respect to θ, $\theta_{cr} = 45 - \bar{\phi}/2$ is just the same as for the case of no cohesion. Moreover, when $\theta = 45 - \bar{\phi}/2$ is substituted into Eq. 23.10, the equation simplifies to Eq. 23.7.

Example 23.8 illustrates the use of Eqs. 23.6 and 23.7. Note that the term involving $\bar{c}$ contributes a significant portion of the total passive resistance, even though $\bar{c}$ is only 9.58 kN/m². However, the term involving $\bar{c}$ increases only as the first power of the height of the wall, whereas the other terms increase as H^2. Thus for a wall height of 6.1 m the term involving $\bar{c}$ would be relatively unimportant.

Importance of Capillary Tensions

In general, the water table will not be at the surface of the backfill, but rather it will be at some depth below the surface. With a clay backfill the soil immediately above the water table will be saturated through capillary action, and a high degree of saturation will exist to the ground surface. A useful approximation to the actual conditions can be obtained by assuming that the soil is fully saturated to the ground surface.

Figure 23.12 derives an expression for the passive thrust which a wall can sustain for this simplified situation. The analysis follows from that presented in Fig. 23.11. The vertical effective stress is

$$\bar{\sigma}_v = \bar{\sigma}_3 = z\gamma_t - (z - D)\gamma_w$$

Above the water table the pore pressure is negative and the effective stress exceeds the total stress. Consequently, the expression for the horizontal effective stress contains an extra item corresponding to the capillary tension at the point.

The form of the result for the total horizontal stress or for the total passive resistance shows that the effect of the capillary tensions is to introduce an *apparent* cohesion equal to $D\gamma_w \tan \bar{\phi}$. Example 23.9 illustrates the importance of this apparent cohesion.

While Eq. 23.14 is in a form convenient for computation, the significance of each term of this equation may not be readily apparent. The first term is not just the thrust from the pore water, since part of the effect of apparent cohesion is included in this term. Example 23.9 shows the actual variation of effective stress against the wall, reconstructed using Eq. 23.13.

Example 23.9 assumes that the capillary tensions above the water table are able to exert a pull to the right upon the wall. Although there is no doubt that the capillary tensions act within the soil and thus give the soil the strength that has been assumed in this example, there is some doubt whether these tensions can act upon the wall. If these tensions upon the wall are assumed not to exist, then it would be concluded that the applied thrust could be increased slightly.

Backfills of Normally Consolidated Clay

If the backfill is a normally consolidated clay and the

▶ Example 23.7 Alternate Solution for Passive Thrust in Soil with Cohesion

Derive the equation for passive thrust from consideration of the free body in Fig. E23.7.

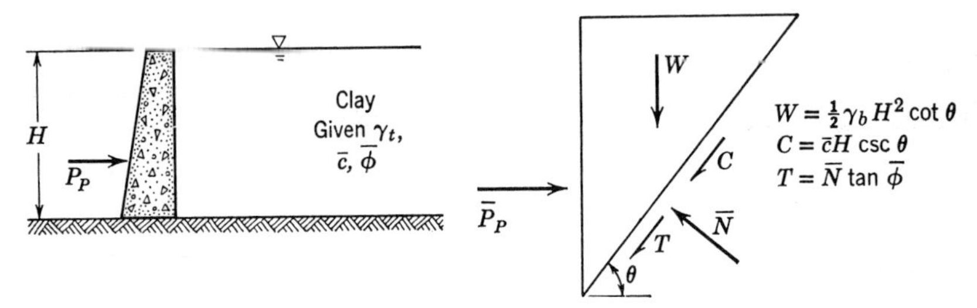

Fig. E23.7

Requirements for equilibrium:
$\Sigma V = 0$:

$$W + C\sin\theta + T\sin\theta - \bar{N}\cos\theta = 0$$

$$\tfrac{1}{2}\gamma_b H^2 \cot\theta + \bar{c}H - \bar{N}(\cos\theta - \sin\theta\tan\bar{\phi}) = 0$$

$$\bar{N} = \frac{\tfrac{1}{2}\gamma_b H^2\cot\theta + \bar{c}H}{\cos\theta - \sin\theta\tan\bar{\phi}} \tag{23.9}$$

$\Sigma H = 0$:

$$\bar{P} - \bar{N}\sin\theta - T\cos\theta - C\cos\theta = 0$$

$$\bar{P} = \bar{N}(\sin\theta + \cos\theta\tan\bar{\phi}) + \bar{c}H\cot\theta$$

$$\bar{P} = \tfrac{1}{2}\gamma_b H^2\cot\theta\tan(\theta + \bar{\phi}) + \bar{c}H[\cot\theta + \tan(\theta + \bar{\phi})] \tag{23.10}$$

Minimizing expression for $\bar{P}$

$$\frac{\partial\bar{P}}{\partial\theta} = \frac{(\tfrac{1}{2}\gamma_b H^2\tan\bar{\phi} - \bar{c}H)(\sin 2\theta\tan\bar{\phi} - \cos 2\theta)}{[\sin\theta(\cos\theta - \sin\theta\tan\bar{\phi})]^2} = 0 \tag{23.11}$$

As in the case of a cohesionless material, Eq. 23.11 is satisfied when

$$(\sin 2\theta\tan\bar{\phi} - \cos 2\theta) = -\frac{1}{\cos\bar{\phi}}[\cos(2\theta + \bar{\phi})] = 0; \qquad 2\theta + \bar{\phi} = 90^\circ$$

$$\theta_{cr} = 45^\circ - \frac{\bar{\phi}}{2} \tag{23.12}$$

Substituting θ_{cr} in Eq. 23.10 and using trigonmetric identities

$$\bar{P}_p = \tfrac{1}{2}\gamma_b H^2\frac{1 + \sin\bar{\phi}}{1 - \sin\bar{\phi}} + 2\bar{c}H\sqrt{\frac{1 + \sin\bar{\phi}}{1 - \sin\bar{\phi}}}$$

$$= \tfrac{1}{2}\gamma_b H^2 N_\phi + 2\bar{c}H\sqrt{N_\phi} \tag{23.7}$$ ◀

---

water table is at the very surface of the backfill, the clay at the surface will have zero shear strength and the thrust against a retaining wall will be exactly the same as for a sand of equal weight and friction angle. In general, however, the water table will lie below the surface, and there will be capillary tensions within the pore water of the backfill—thus giving an apparent cohesion. Because a normally consolidated clay undergoes a large volume decrease as it is compressed and sheared, a large horizontal strain will be required to mobilize full passive resistance from a normally consolidated backfill, as discussed in Section 23.1.

▶ Example 23.8 Passive Thrust for Saturated Backfill

Given the wall and backfill in Fig. E23.8-1, find the passive resistance.

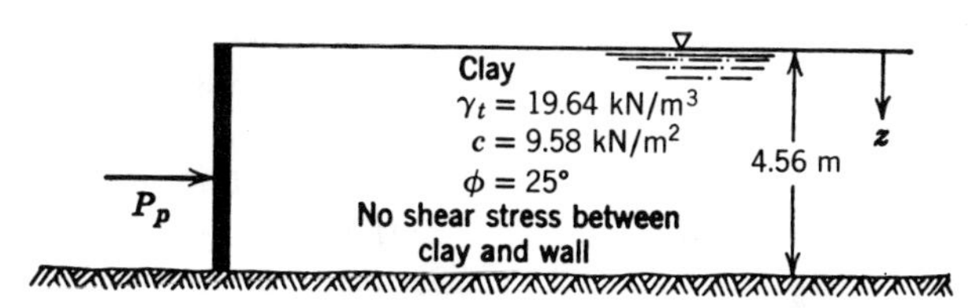

Fig. E23.8-1

Solution:

$$P_p = \tfrac{1}{2}\gamma_w H^2 + \tfrac{1}{2}\gamma_b H^2 \frac{1 + \sin \bar{\phi}}{1 - \sin \bar{\phi}} + 2\bar{c}H\sqrt{\frac{1 + \sin \bar{\phi}}{1 - \sin \bar{\phi}}}$$

$$= 102.4 + 102.6(2.46) + 87.6\sqrt{2.46}$$

$$= 102.4 + 252.4 + 137.4 = 492.2 \text{ kN/m of wall}$$

$$\bar{x} = \frac{102.4(1.52) + 252.4(1.52) + 137.4(2.28)}{492.2}$$

$$= 1.73 \text{ m}$$

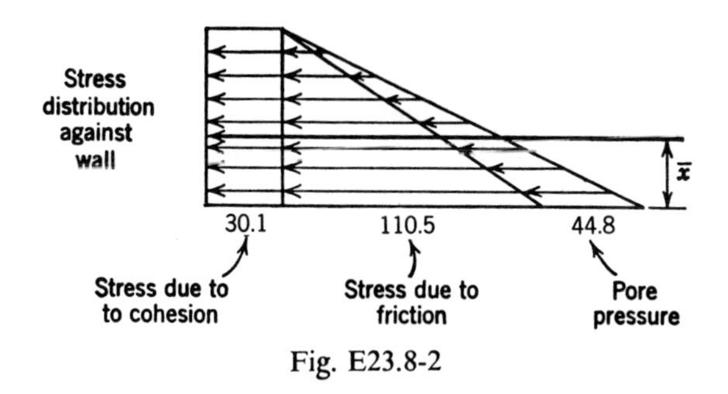

Fig. E23.8-2

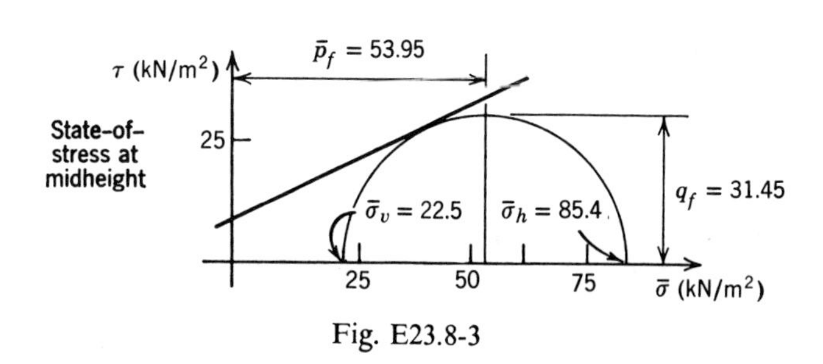

Fig. E23.8-3

The distribution of normal stresses corresponding to the several components of the thrust are shown in Fig. E23.8-2. The Mohr circle for the stresses at mid-depth is given in Fig. E23.8-3. ◀

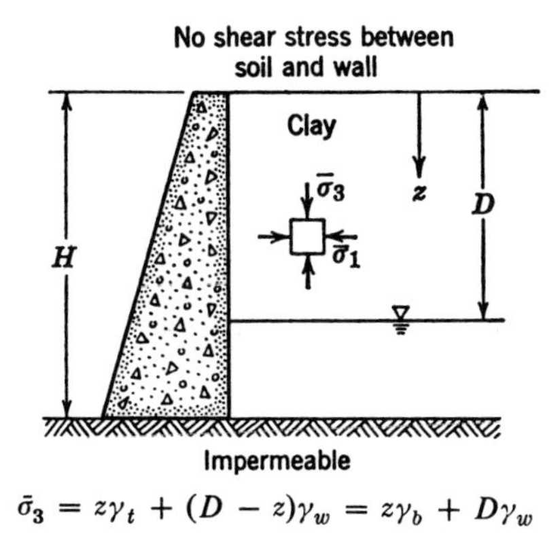

$$\bar{\sigma}_3 = z\gamma_t + (D - z)\gamma_w = z\gamma_b + D\gamma_w$$

From Fig. 23.11:

$$\bar{\sigma}_1 = \bar{\sigma}_3 N_\phi + 2\bar{c}\sqrt{N_\phi}$$

$$= z\gamma_b N_\phi + D\gamma_w N_\phi + 2\bar{c}\sqrt{N_\phi} \qquad (23.13)$$

$$\sigma_1 = \bar{\sigma}_1 - (D - z)\gamma_w$$

$$= z\gamma_w + z\gamma_b N_\phi + D\gamma_w(N_\phi - 1) + 2\bar{c}\sqrt{N_\phi}$$

$$N_\phi - 1 = \frac{1 + \sin \bar{\phi}}{1 - \sin \bar{\phi}} - 1 = \frac{2 \sin \bar{\phi}}{1 - \sin \bar{\phi}} = 2 \tan \bar{\phi} \frac{\cos \bar{\phi}}{1 - \sin \bar{\phi}}$$

$$= 2 \tan \bar{\phi} \sqrt{\frac{1 + \sin \bar{\phi}}{1 - \sin \bar{\phi}}} = 2 \tan \bar{\phi} \sqrt{N_\phi}$$

Hence

$$\sigma_1 = z\gamma_w + z\gamma_b N_\phi + 2(\bar{c} + D\gamma_w \tan \bar{\phi})\sqrt{N_\phi}$$

$$P_p = \tfrac{1}{2}\gamma_w H^2 + \tfrac{1}{2}\gamma_b H^2 N_\phi + 2(\bar{c} + D\gamma_w \tan \bar{\phi})H\sqrt{N_\phi} \qquad (23.14)$$

Fig. 23.12 Passive thrust for case of capillary saturation of backfill.

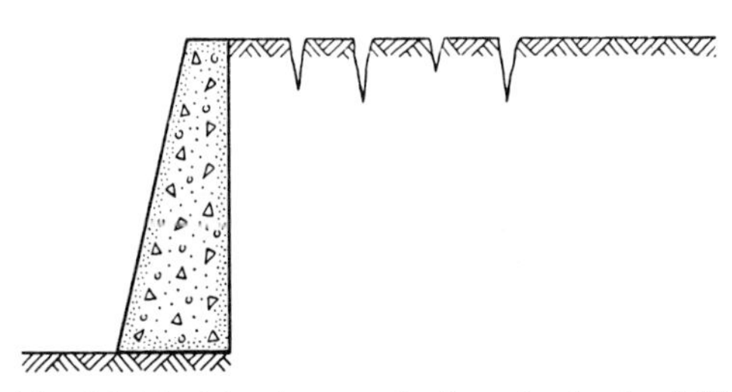

Fig. 23.13 Tension cracks in cohesive backfill.

Effect of Cohesion on Active Thrust

For the active case, assuming no shear stresses between the wall and the backfill, the equations for the horizontal effective and total stresses and for the total thrust against the retaining structure are:

$$\bar{\sigma}_h = \bar{\sigma}_3 = \frac{z\gamma_b}{N_\phi} + \frac{D\gamma_w}{N_\phi} - 2\bar{c}\sqrt{\frac{1}{N_\phi}} + \frac{q_s}{N_\phi} \quad (23.15)$$

$$\sigma_h = \sigma_3 = z\gamma_w + \frac{z\gamma_b}{N_\phi} - 2(\bar{c} + D\gamma_w \tan\bar{\phi})\sqrt{\frac{1}{N_\phi}} + \frac{q_s}{N_\phi} \quad (23.16)$$

$$P_a = \tfrac{1}{2}\gamma_w H^2 + \frac{\tfrac{1}{2}\gamma_b H^2}{N_\phi} - 2(\bar{c} + D\gamma_w \tan\bar{\phi})H\sqrt{\frac{1}{N_\phi}} + \frac{q_s H}{N_\phi} \quad (23.17)$$

Equations 23.15 and 23.17 are obtained from Eqs. 23.13 and 23.14 by replacing N_ϕ by $1/N_\phi$, by reversing the sign of the third term, and by adding the final term (see Chapter 13), which gives the effect of a uniform surcharge. Note that the first term in Eq. 23.17 is not simply the thrust from the pore water but also includes an effect of apparent cohesion. These equations apply to problems in which the retaining wall moves so slowly that there are no excess pore pressures within the backfill.

Example 23.10 presents the computation of active thrust and of the accompanying stress distribution against the wall for a problem in which the water table is assumed to be at the surface of the backfill.

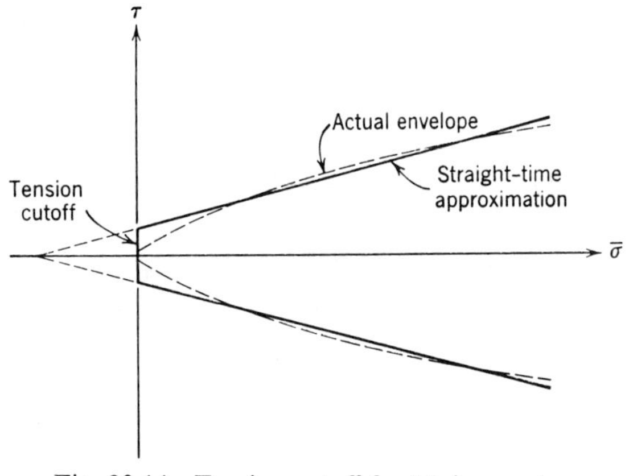

Fig. 23.14 Tension cutoff for Mohr envelope.

The Tension Crack Problem

In the foregoing example, the stress exerted by the mineral skeleton against the wall is tensile over the upper part of the wall. Moveover, the horizontal effective stress is tensile throughout the upper part of the backfill. It seems rather unlikely that the mineral skeleton of a soil can sustain tensile stresses, at least over an extended period of time. Consequently, there is a tendency for *tension cracks* to open at the surface of a backfill behind a retaining wall (see Fig. 23.13). These tension cracks are associated with the horizontal stretching which is inherent in the active lateral stress situation. They are not a problem in the passive stress situation because there is horizontal compression.

The actual failure envelope for an overconsolidated clay is curved and passes through or close to the origin. However, the straight line envelope, used to approximate the actual envelope, may pass appreciably above the origin. To preserve the accuracy of the approximation it becomes necessary to cut off the straight line failure envelope at the origin (see Fig. 23.14). With this modification to the failure envelope the soil skeleton will be unable to withstand tension.

It now becomes necessary to modify the method of analysis used to determine active thrust. To do this, let us first determine the height a vertical bank of soil can stand by itself. This problem is analyzed in Fig. 23.15.

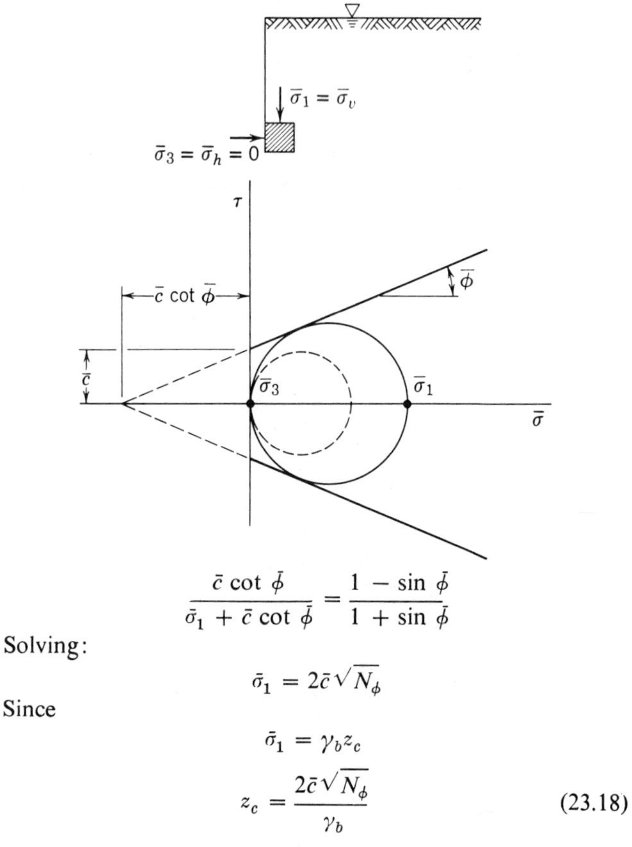

$$\frac{\bar{c}\cot\bar{\phi}}{\bar{\sigma}_1 + \bar{c}\cot\bar{\phi}} = \frac{1 - \sin\bar{\phi}}{1 + \sin\bar{\phi}}$$

Solving:

$$\bar{\sigma}_1 = 2\bar{c}\sqrt{N_\phi}$$

Since

$$\bar{\sigma}_1 = \gamma_b z_c$$

$$z_c = \frac{2\bar{c}\sqrt{N_\phi}}{\gamma_b} \quad (23.18)$$

Fig. 23.15 Depth of tension cracks.

▶ Example 23.9 Passive Thrust with Capillary Saturation of Backfill

Given the backfill and water table in Fig. E23.9-1, find the total passive resistance.

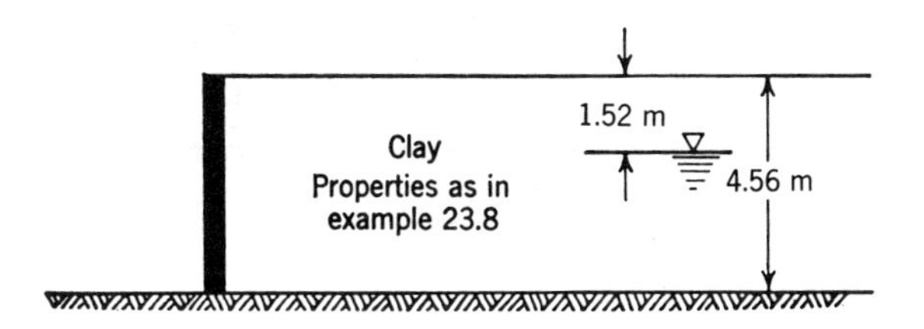

$$\text{Total cohesion} = \text{actual cohesion plus apparent cohesion}$$

$$= 9.58 + (1.52)(9.81)(\tan 25°) = 9.58 + 6.95 = 16.53 \text{ kN/m}^2$$

From Eq. 23.14

$$P_p = 102.4 + 252.4 + 2(16.53)(4.56)\sqrt{2.46}$$

$$= 102.4 + 252.4 + 237.0 = 591.8 \text{ kN/m of wall}$$

Fig. E23.9-1

The stresses are plotted in Fig. E23.9-2

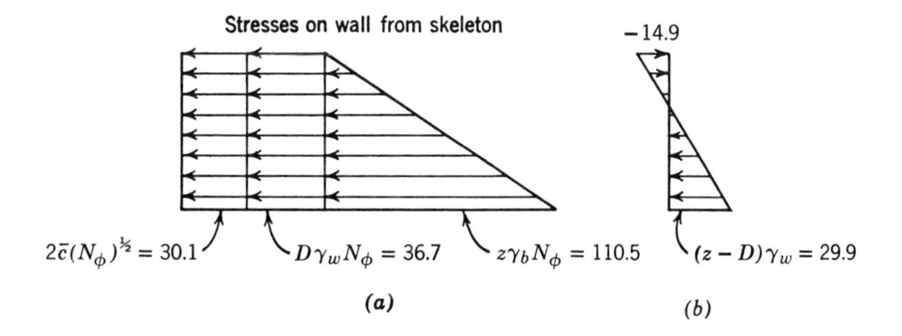

Stresses at midheight (in kN/m²)

$\sigma_v = 44.9$	$\sigma_h = 129.5$
$u = 7.5$	$\bar{\sigma}_h = 122.0$
$\bar{\sigma}_v = 37.4$	$q_f = 42.3$
	$\bar{p}_f = 79.7$

Fig. E23.9-2 (*a*) Stresses on wall from skeleton (kN/m²); see Eq. 23.13. (*b*) Stresses on wall from pore water.

The solid circle in this figure shows the largest circle that can be drawn with $\bar{\sigma}_3 = 0$. Using the equation for the largest circle that can be fitted inside of a sloping Mohr envelope, it is possible to solve for the largest $\bar{\sigma}_1$ that can be supported with $\bar{\sigma}_3 = 0$. By expressing $\bar{\sigma}_1$ in terms of unit weight and depth, it is finally possible to find the maximum possible height of unsupported soil, which also is the maximum possible depth of tension cracks. The dashed circle in the figure indicates the state-of-stress that must exist at shallower depths. Thus only part of the cohesion can be mobilized at shallower depths if there is to be no horizontal tensile stress.

Now let us return to the trial wedge method, breaking the failure wedge up into two parts, as shown in Fig. 23.16. It is assumed that there are no horizontal effective stresses along *KL*; i.e., this line is a tension crack. The

▶ Example 23.10 Active Thrust for Backfill with Cohesion

Given: The backfill of Example 23.8

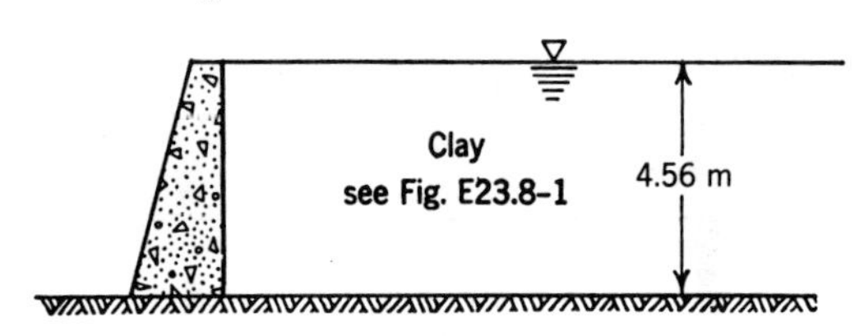

Fig. E23.10-1

Find: The active thrust

Solution: The total thrust is:

$$P_a = 102.4 + \frac{102.6}{2.46} - \frac{87.6}{\sqrt{2.46}}$$

$$= 102.4 + 41.7 - 55.9 = 88.2 \text{ kN/m of wall}$$

The distribution of stresses and the Mohr circle for stresses at mid-depth, appear in Fig. E23.10-2.

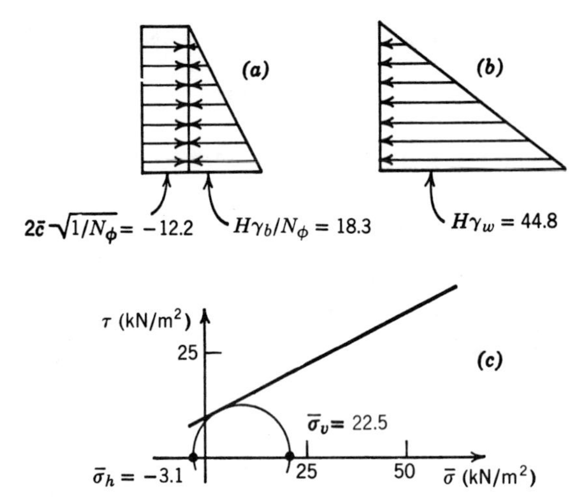

Fig. E23.10-2 (*a*) Stress on wall from skeleton (kN/m²). (*b*) Stress on wall from pore water (kN/m²). (*c*) Stresses at midheight. ◀

cohesive and frictional resistance along JL is just sufficient to permit the wedge JKL to stand by itself. At point L the full cohesive resistance is mobilized, but at points between L and J less than the full cohesive resistance is mobilized. The cohesive and frictional resistance along LM, together with the thrust on IM, must be sufficient to hold the wedge $IKLM$ in equilibrium.

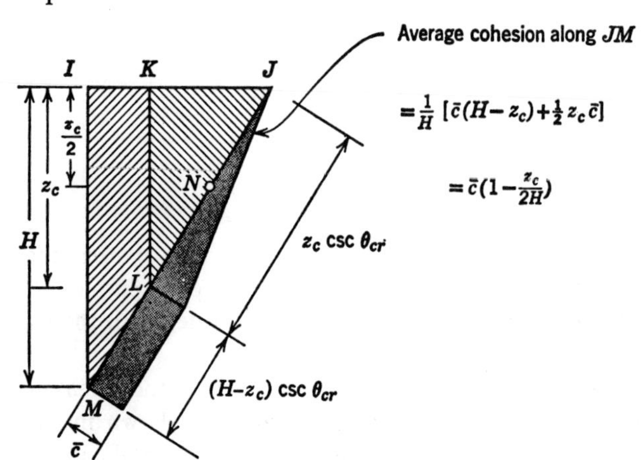

Fig. 23.16 Trial wedge with tension crack.

Of course the wedge IJM must also be in equilibrium. The system of forces acting upon this wedge is just the same as that used to arrive at Eq. 23.16, except that the average cohesion along JM is

$$\bar{c}\left(1 - \frac{z_c}{2H}\right)$$

instead of $\bar{c}$. Thus we can use Eq. 23.17 to get the thrust that is necessary for equilibrium:

$$P_a = \tfrac{1}{2}\gamma_w H^2 + \frac{\tfrac{1}{2}\gamma_b H^2}{N_\phi} - 2\bar{c}\left(H - \frac{z_c}{2}\right)\sqrt{\frac{1}{N_\phi}} \quad (23.19)$$

The thrust and stress distribution against the wall of Example 23.10 have been re-evaluated in Example 23.11

▶ Example 23.11 Analysis of Active Thrust Considering Tension Crack

Using data from Example 23.10:
The depth of tension crack is

$$z_c = \frac{2(9.58)\sqrt{2.46}}{9.83} = 3.06 \text{ m}$$

Then

$$\text{average mobilized cohesion} = (9.58)(1 - \tfrac{3.06}{9.12}) = 6.4 \text{ kN/m}^2$$

$$P_a = 102.4 + 41.7 - 37.2 = 106.9 \text{ kN per meter of wall}$$

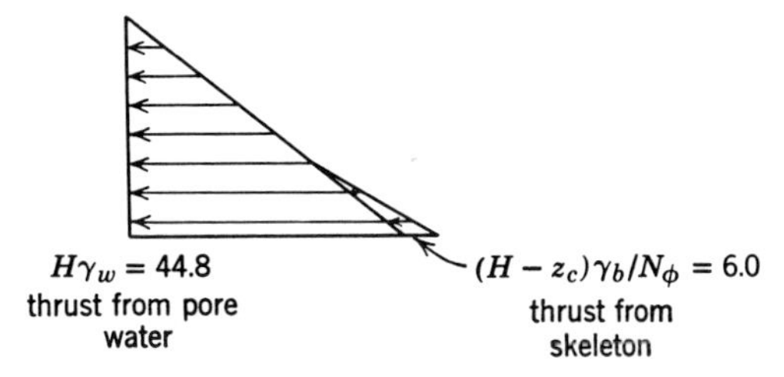

Fig. E23.11-1 Stress distribution against wall.

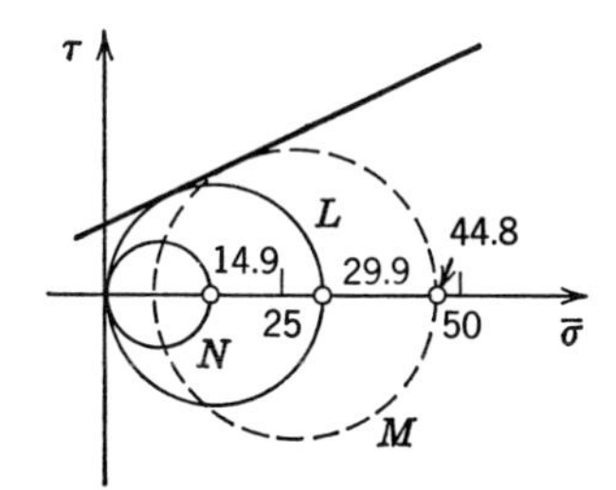

Fig. E23.11-2 Mohr circles for various points. ◀

taking into account the effect of the tension crack. The active thrust has thereby been increased by 20% over the value computed in Example 23.10. The effect of tension cracks is quite important in active thrust problems. If the water table actually is at the surface of a backfill it is unlikely that tension cracks will actually be observed, but the foregoing analysis nonetheless serves to take into account the fact that the backfill cannot withstand horizontal tensions.

Water Table below Ground Surface

Example 23.12 determines the active thrust from a backfill with apparent cohesion resulting from capillary tensions. If the apparent cohesion were neglected in this example (which would be equivalent to having the water table at the ground surface), the active thrust would be 66.4 kN/m—2.5 times the computed value.

There is some doubt that these capillary tensions can exist against the retaining wall. Furthermore, it seems likely that the effect within the soil of these capillary tensions may be partially destroyed because of the development of tension cracks.

Example 23.13 shows one possible analysis that takes these tension cracks into account (but still permits some capillary tensions against the wall). Now the criterion for the depth of the tension crack is that the *total* horizontal stress must be zero; i.e., the crack stands completely open and the horizontal effective stress equals the capillary tension. The first step in the analysis is to find the depth of the tension crack z_c. At this depth, $\sigma_h = 0$ with all frictional resistance mobilized. This depth works out to be 0.92 m. Above this depth, less than full frictional resistance is mobilized. Below this depth, horizontal stress is required even though full friction is mobilized. As a result of this tension crack the active thrust is raised from 26.3 to 32.3 kN/m, an increase of about 23%.

A second possible analysis assumes that there can be no capillary tensions at all against the wall then, in addition to the thrust computed in Example 23.13, a force corresponding to the tensions between 0.92 and 2.44 m depth must be added:

$$P_a = 32.3 + \tfrac{1}{2}(1.52)^2(9.81) = 32.3 + 11.3 = 43.6 \text{ kN/m}$$

Table 23.2 summarizes the results obtained for the various versions of Example 23.12. Note that the capillary tensions enter into the problem in two ways: (*a*) pulling directly upon the wall; and (*b*) influencing the effective stresses within the soil. There is doubt as to

▶ Example 23.12 Active Thrust for Backfill with Apparent Cohesion

Given: The retaining wall and backfill in Fig. E23.12-1

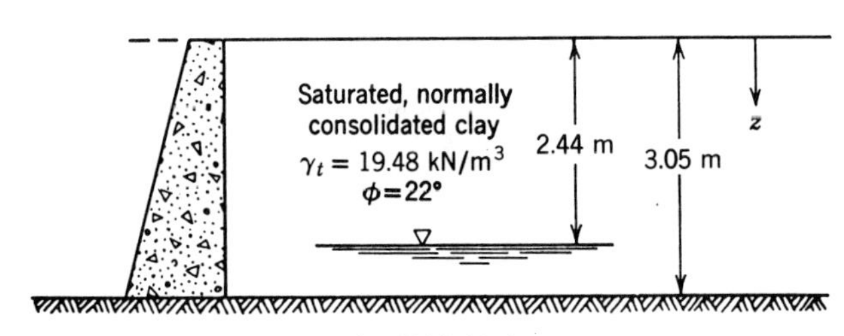

Fig. E23.12-1

Find: The active thrust, assuming the pore water can carry tension

Solution:

$$\text{Apparent cohesion} = (2.44)(9.81)\tan 22° = 9.67\ \text{kN/m}^2$$

Active thrust:

$$\begin{aligned} P_a &= \tfrac{1}{2}(9.81)(3.05)^2 + \tfrac{1}{2}(9.67)(3.05)^2(0.455) - (2)(9.67)(3.05)\sqrt{0.455} \\ &= 45.6 + 20.5 - 39.8 = 26.3\ \text{kN/m of wall} \end{aligned}$$

Fig. E23.12-2 shows the distribution of stresses, while Fig. E23.12-3 shows the Mohr circle for the stresses at midheight.

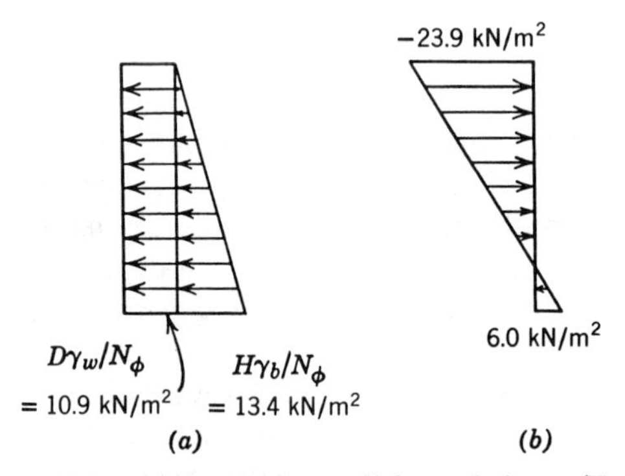

Fig. E23.12-2 (*a*) Stresses on wall from skeleton (Eq. 23.15). (*b*) Pore pressures on wall.

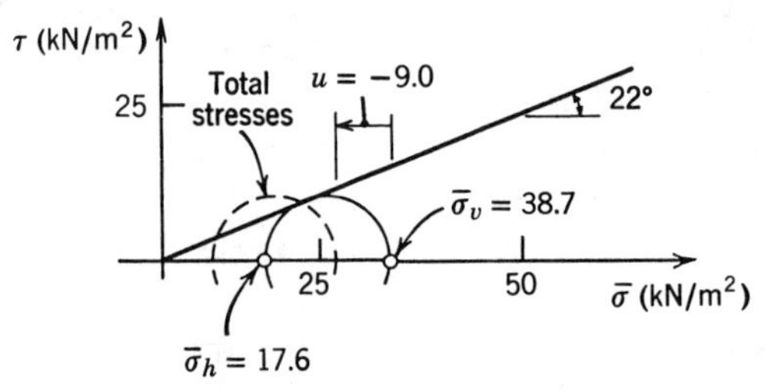

Fig. E23.12-3 State of stress at midheight. ◀

the presence of the first effect, but there can be no question as to the importance of the second effect.

Examples 23.12 and 23.13 make it evident that apparent cohesion plays an important role in the design of retaining walls for backfills with a significant clay content. However, in actual practice, it is quite difficult to estimate the magnitude of the apparent cohesion. Moreover, the magnitude of the apparent cohesion will vary throughout the year as the result of rainfalls, etc. It is little wonder that engineers prefer to avoid cohesive backfills when possible and to be conservative in the wall design when such backfills are unavoidable. Even though the types of computations illustrated in Examples 23.12 and 23.13 may be of limited use in practice, they serve to indicate the considerations necessary to make reasonable estimates of the shearing resistance of cohesive backfills.

Table 23.2 Effect of Capillary Tensions on Active Thrust (for conditions shown in Example 23.12)

Assumption	Thrust (kN/m of wall)
1. No capillary tensions at wall or in soil (water table at ground surface)	66.1
2. Tension cracks in soil; no capillary tensions on wall	43.6
3. Tension cracks in soil; capillary tensions on wall below tension cracks	32.3
4. Capillary tensions over whole height of wall	26.3

23.4 ANCHORED BULKHEADS

The principles of anchored bulkheads retaining dry backfills, presented in Section 13.8, also hold for those retaining wet backfills. Because anchored bulkheads are commonly waterfront structures, the influence of pore water pressures on the wall is usually an important consideration. Just as was true with gravity walls, the water pressure can greatly increase the lateral thrust on a bulkhead.

Figure 23.17 shows the lateral stress distribution for two conditions: (*a*) submerged with the water table at the same elevation inside and outside of the bulkhead, and (*b*) perched water table of 1.52 m. The first situation would exist where there is good drainage from the backfill and through the wall. The second can arise from an impervious backfill and a varying water level outside the wall, such as a tide change.

Figure 23.17 shows that a difference in water levels can result in a significant lateral thrust on the wall. Remember that the K value for water is one, since it has no shear strength, and thus a small difference in water levels on a retaining structure can cause a large water thrust. This example shows again the great importance of backfill drainage.

The water pressure diagram in Fig. 23.17 is based on the condition of no seepage. This is a common assumption in wall design. Seepage will occur, however, unless there is an impermeable layer at the bottom of the wall. Actually, the flow situation should be determined and a

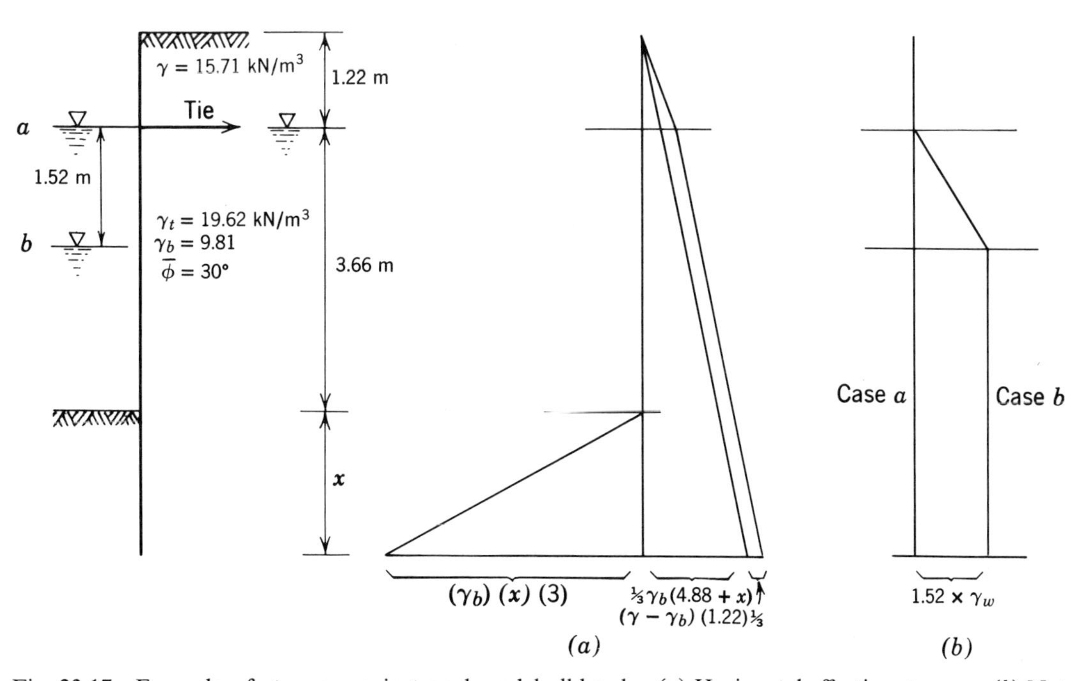

Fig. 23.17 Example of stresses against anchored bulkhead. (*a*) Horizontal effective stresses. (*b*) Net water pressure.

▸ Example 23.13 Active Thrust for Backfill with Apparent Cohesion and Tension Crack

The distribution of stress in general terms is shown in Fig. E23.13-1

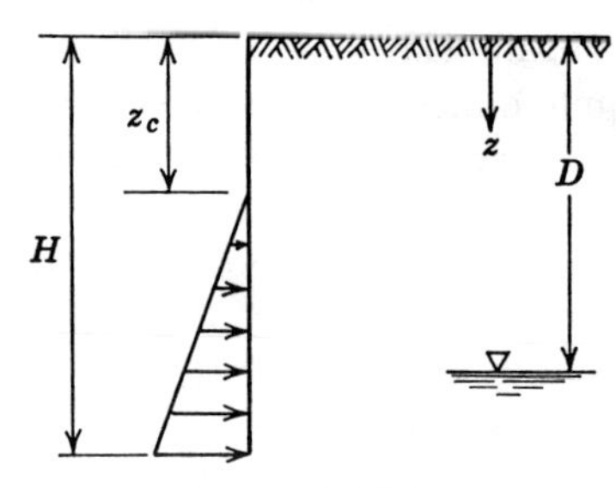

Fig. E23.13-1

At all depths:

$$\sigma_v = \gamma_t z$$
$$u = (z - D)\gamma_w$$
$$\bar{\sigma}_v = \gamma_b z + D\gamma_w$$

For $z > z_c$

$$\bar{\sigma}_h = \frac{\bar{\sigma}_v}{N_\phi} = \frac{\gamma_b z}{N_\phi} + \frac{D\gamma_w}{N_\phi}$$

$$\sigma_h = \frac{\gamma_b z}{N_\phi} + \frac{D\gamma_w}{N_\phi} + (z - D)\gamma_w$$

$$= \frac{\gamma_b z}{N_\phi} + z\gamma_w + D\gamma_w\left(\frac{1}{N_\phi} - 1\right)$$

$$= z\left(\frac{\gamma_b}{N_\phi} + \gamma_w\right) - \frac{2D\gamma_w \tan\bar{\phi}}{\sqrt{N_\phi}}$$

At $z = z_c$

$$\sigma_h = 0$$

$$z_c = \frac{2D\gamma_w \tan\bar{\phi}\,\sqrt{N_\phi}}{\gamma_b + N_\phi\gamma_w}$$

For the problem in Example 23.12:

$$z_c = \frac{2(2.44)(9.81)(0.404)\sqrt{2.2}}{9.67 + (2.2)(9.81)} = 0.92 \text{ m}$$

At bottom of wall:

$$\sigma_h = 3.05(4.40 + 9.81) - \frac{2 \times 9.67}{\sqrt{2.2}} = 43.3 - 13.0 = 30.3 \text{ kN/m}^2$$

$$\text{Total active thrust } P_a = \tfrac{1}{2}(2.13)(30.3) = 32.3 \text{ kN/m of wall}$$

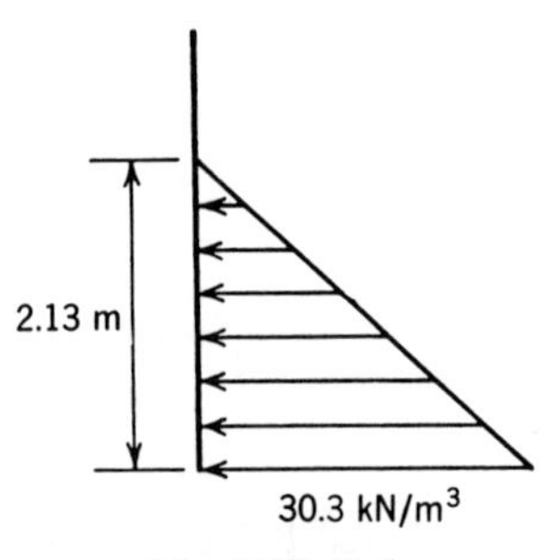

Fig. E23.13-2

Example 23.13. (*continued*)

The distribution of stresses for this case appears in Fig. E23.13-2. The following table gives the stresses at various depths. Mohr circles for 3 of these states of stress appear in

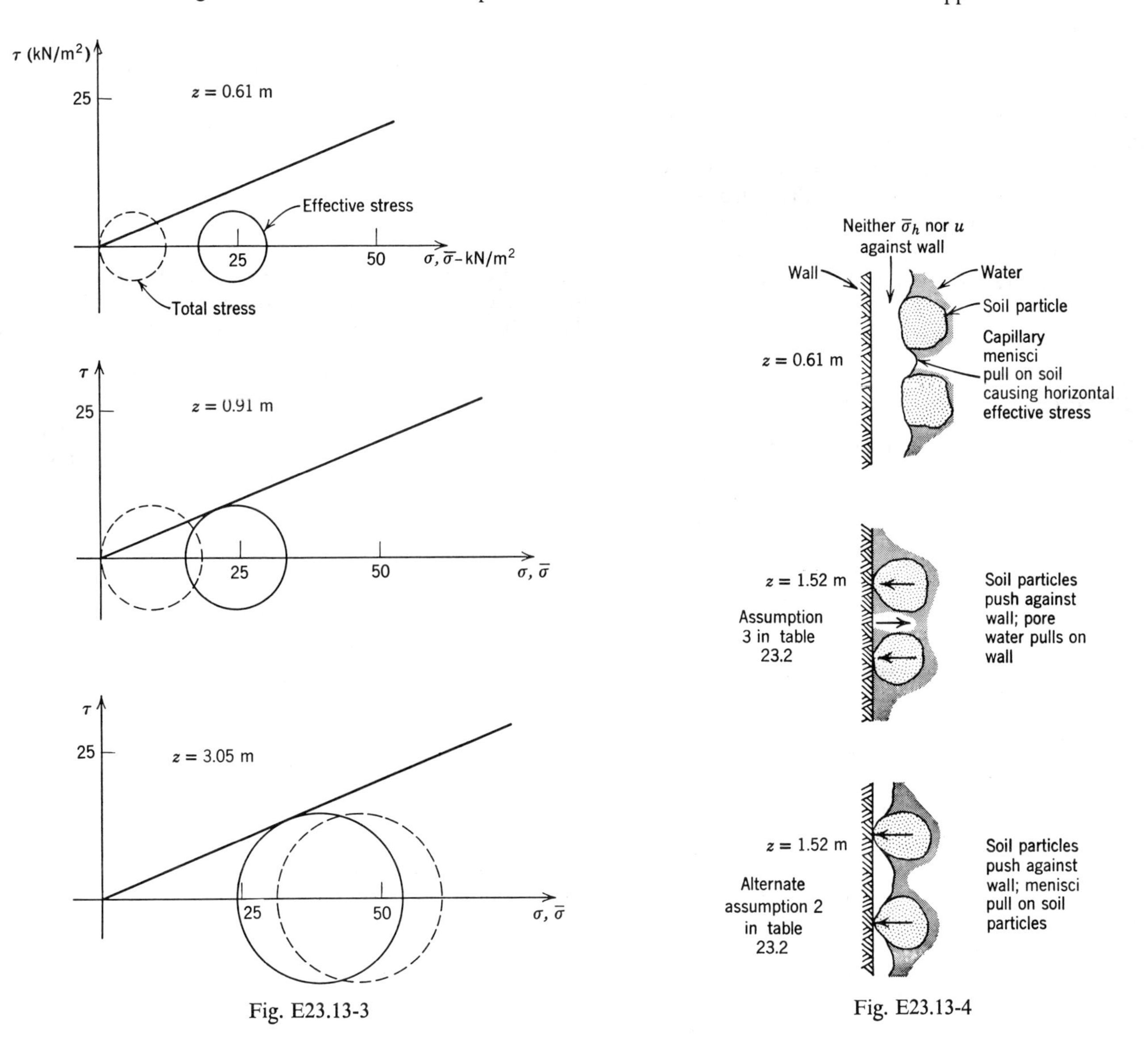

Fig. E23.13-3

Fig. E23.13-4

Fig. E23.13-3. Fig. E23.13-4 shows the ways in which the soil particles, pore water, and wall act upon one another.

z (m)	σ_v (kN/m²)	u (kN/m²)	$\bar{\sigma}_v$ (kN/m²)	$\bar{\sigma}_h$ (kN/m²)	σ_h (kN/m²)
0.61	11.9	−17.9	29.8	17.9	0
0.91	17.7	−15.0	32.7	15.0	0
1.52	29.6	− 9.0	38.6	17.5	8.5
3.05	59.4	6.0	53.4	24.3	30.3

◀

water pressure distribution computed from the flow pattern, just as was done in Example 18.1. Example 18.1 shows that seepage under the wall gives a lower net water pressure on the inside of the wall than does the static case.

The effective stress distribution shown is based on full mobilization of shear strength—the soil on both sides of the wall is at failure.

In addition to designing a wall to take lateral stresses, the engineer must check the stability of the wall against a failure wherein shear develops in the soil below the wall. Also, the engineer must be careful to locate the anchor far enough away from the wall to be outside the critical failure surface.

23.5 SUMMARY OF MAIN POINTS

1. Active stress is the minimum lateral stress and exists when the shear strength of the soil is fully mobilized. In cohesionless soil the active stress equals $K_a\bar{\sigma}_v$. If there is no wall friction,

$$K_a = \frac{1}{N_\phi} = \frac{1 - \sin\bar{\phi}}{1 + \sin\bar{\phi}} = \frac{1 - \tan\bar{\alpha}}{1 + \tan\bar{\alpha}}$$

2. Passive stress is the maximum lateral stress and exists when the shear strength of the soil is fully mobilized. In cohesionless soil the passive stress equals $K_p\bar{\sigma}_v$. If there is no wall friction,

$$K_p = \frac{1}{K_a} = N_\phi = \frac{1 + \sin\bar{\phi}}{1 - \sin\bar{\phi}} = \frac{1 + \tan\bar{\alpha}}{1 - \tan\bar{\alpha}}$$

3. The lateral stress ratios K_a and K_p refer to effective stresses. The principles of lateral stress in dry soils, presented in Chapter 13, hold also for wet soils as long as the stresses are effective stresses.
4. For a given value of $\bar{\phi}$, a cohesion intercept $\bar{c}$ means a higher strength. The higher strength permits a lower active stress and a higher passive stress.
5. In evaluating the stability on a trial wedge, either of the following force systems can be used:
   a. Total weight, boundary pore pressures, and boundary effective stresses.
   b. Buoyant weight, seepage force, and boundary effective stresses.
   The two force systems are exactly equivalent.
6. In general, the total lateral thrust against a retaining structure is composed of three components:
   a. Thrust exerted by pore water.
   b. Thrust exerted by the soil skeleton due to soil weight.
   c. Thrust exerted by the soil skeleton as a result of a surcharge.
7. The water thrust can be very large. In order to minimize the lateral thrust on a wall, the designer must design a drainage system and select a pervious backfill.
8. At the active state in soil with a cohesion intercept the stress in the mineral skeleton near the surface will be in tension. There is therefore a tendency for tension cracks to open at the surface of the backfill.

PROBLEMS

23.1 Refer to the wall in Example 23.2 and consider a wedge of soil cut by a failure plane rising at an angle of 50° to the horizontal. Draw to scale the force diagram for this wedge.

23.2 Repeat Problem 23.1 for the wall in Fig. 23.11 and a failure plane at $32\frac{1}{2}°$ with the horizontal.

23.3 Refer to Fig. 23.16. For wedge *IKLM* and for wedge *JKL* draw the force diagrams. Use the soil properties given in Example 23.8 and $\theta = 57\frac{1}{2}°$.

23.4 For the situation shown in Fig. P23.4, determine the total thrust for both active and passive cases.

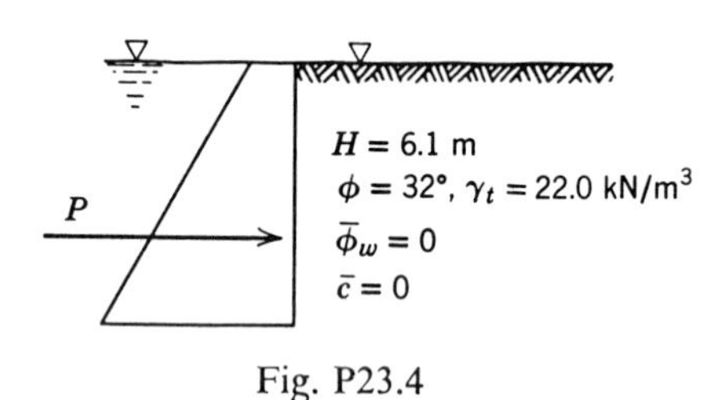

Fig. P23.4

23.5 Find the active thrust on the wall and the moment of thrust about point *A* of the wall in Fig. P23.5.

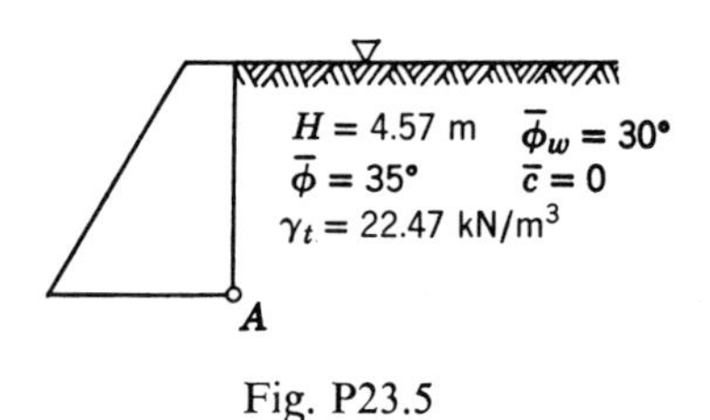

Fig. P23.5

23.6 Repeat Problem 23.5 with a uniform surcharge of 4.8 kN/m².

23.7 Draw the force diagram for a trial wedge formed by $\theta = 60°$ in Example 23.6.

23.8 For the wall in Fig. 23.17 with the 1.52 m difference in water levels, determine x and the tie force for a factor of safety of one relative to rotation about the tie.

23.9 A given soil has the following properties: $\bar{\phi} = 25°$, $\bar{c} = 9.58$ kN/m², $\gamma_t = 19.64$ kN/m³. The vertical effective stress is 95.8 kN/m² and the static pore pressure is 28.7 kN/m². The horizontal effective stress is initially equal to the vertical stress. On a p–q plot show the ESP and TSP for going to active state and for going to passive state. At all times the excess pore pressure is zero. On the plot label the active and passive stresses.

23.10 For Problem 23.9 compute N_ϕ and use it to compute the active stress and passive stress.

23.11 By constructing stress paths, evaluate the active thrust for the wall in Fig. 23.10, for the four water conditions treated in that figure. (*Hint.* Draw stress paths for conditions at midheight, as in Fig. 23.2. Multiply the horizontal stresses found by the wall height to get the force per meter of wall.)

CHAPTER 24

Earth Slopes with Drained Conditions

Several important problems involving slope stability have already been described in Chapter 1. Such problems most often arise in connection with the construction of highways, canals, and basements. Slope stability is an extremely important consideration in the design and construction of earth dams. There are also important problems involving the stability of natural slopes. The result of a slope failure can often be catastrophic, involving the loss of considerable property and many lives. Yet the cost of flattening a slope to achieve greater stability can be tremendous. Hence, although safety must be assured, undue conservatism must be avoided.

The analysis of slopes was touched upon briefly at the end of Chapter 13, but the important features of slope stability problems begin to emerge only when the effects of pore pressure and cohesion are considered.

The trial wedge method, described in Chapters 13 and 23 in connection with the equilibrium of backfill behind retaining structures, is generally used to analyze the stability of slopes. The methods of calculation are more complicated than those presented in the two earlier chapters, but it is still relatively easy to compute the stresses that must exist if the slope is to be stable. The first half of this chapter will be devoted to the mechanics involved in such computations.

However, with clayey soils it is often extremely difficult to decide how much shear strength actually is available within the slope. Such an estimate of available shear strength requires knowledge of the following.

1. The shear strength parameters in terms of effective stress. Whereas the principles and methods of testing described in Chapter 21 may be used to establish the general magnitude of the strength parameters, it is often difficult to be precise about exactly what strength is available in a given problem. Failures of actual slopes provide one of the best methods for learning how the strength of a soil as measured in the laboratory compares with the actual *in situ* strength. Several examples of such failures will be discussed in Section 24.8.
2. The pore pressures acting within the slope. In this chapter these pore pressures are assumed to be known from natural ground water conditions. In some problems this is a reasonable assumption, but in many problems it is not. This subject will be treated further in Chapter 31.

There are many similarities between the analysis of the stability of retaining structures and the analysis of slope stability. However, there is an important difference in the philosophy of design. With retaining structures, it is assumed that the full strength of the backfill is mobilized and the reserve strength is provided in the retaining structure. However, with slopes there must be a reserve of strength within the soil itself. The design of slopes, and the important question of the choice of safety factor, are discussed in Section 24.9.

24.1 INFINITE SLOPES IN SAND

As a first step in understanding the effect of pore pressures and flowing water upon slope stability, it is convenient to consider infinite slopes in sand using the concepts and principles developed in Section 13.9.

Submerged Slopes

Figure 24.1 shows a sand slope submerged below the surface of a body of static water. Such a slope might be found near the shore of a lake. The water both above and within the sand is in a hydrostatic condition; i.e., there is no flow of water within the soil.

In order to ascertain the maximum possible stable slope i_{max} we follow the same type of analysis used in Fig. 13.33. The following forces on the element must be in equilibrium:

1. Total weight of the element—soil plus water.
2. The resultant of the boundary effective stresses. The effective stresses acting on the vertical sides of the unit element must exactly balance each other. Hence the resultant $\bar{N}$ acts normal to the boundary

CD. The bar indicates that $\bar{N}$ is the resultant of the *effective* stresses.

3. The resultant of the shear stresses around the boundary. The shear stresses on the vertical side of the element must cancel each other, leaving only the resultant T of the shear stresses on CD. Since the pore water can carry no shear, T must be carried entirely by the mineral skeleton.
4. The resultant of the boundary pore pressures. Since there is no seepage, this force is buoyancy (as proved in Chapter 17), which is the volume of the element times the unit weight of the water.

Some readers may be puzzled by the fact that the resultant of the boundary pore pressures is vertical even when the top and bottom of the unit element are not horizontal. Figure 24.2 shows the actual boundary pore pressures and proves that the resultant is indeed vertical. Note that the pressures vary along the top and bottom of the element as well as along the sides of the element, and that the resultants against the two vertical sides are not equal. (In other words, buoyancy acts upward regardless of the shape of the submerged mass.)

Because the resultant of the boundary pore pressures is vertical, the final result of the analysis is the same as in Fig. 13.33 except that the buoyant unit weight replaces the total unit weight. The final result is that the slope is stable for $i \leq \bar{\phi}$, i.e., the maximum value of i is the strength angle $\bar{\phi}$ in terms of effective stresses. Thus the maximum stable slope angle is the same for a given sand whether the slope is completely dry or completely submerged under water.

Seepage Parallel to Slope

Figure 24.3 shows the case of a slope with seepage parallel to the slope. The flow net consists of straight lines with the flow lines parallel to the slope and the equipotential lines perpendicular to the slope. Such a condition is often reached in the lower portions of natural slopes, as noted in Fig. 24.4.

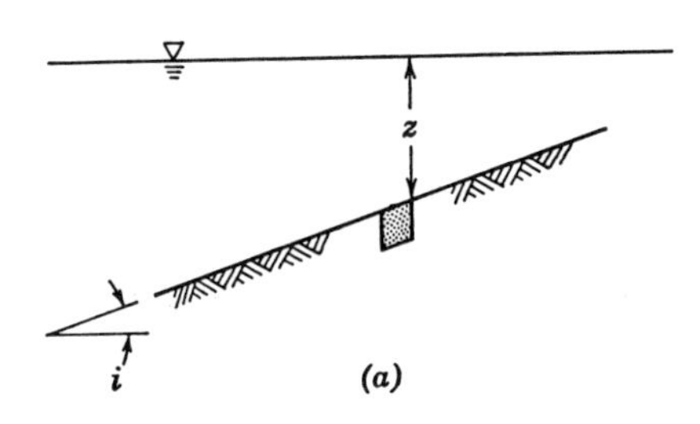

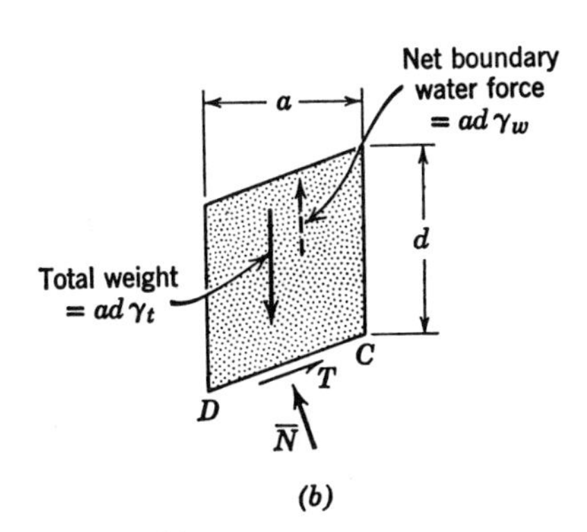

By analysis of Fig. 13.33,

$$\bar{N} = ad\gamma_b \cos i$$
$$T = ad\gamma_b \sin i$$
$$\bar{\sigma} = \gamma_b d \cos^2 i$$
$$\tau = \gamma_b d \cos i \sin i$$

If full resistance is mobilized so that $\tau = \bar{\sigma} \tan \bar{\phi}$, then

$$i = \bar{\phi}$$

Fig. 24.1 Analysis of submerged infinite slope. (*a*) Submerged slope. (*b*) Analysis of equilibrium.

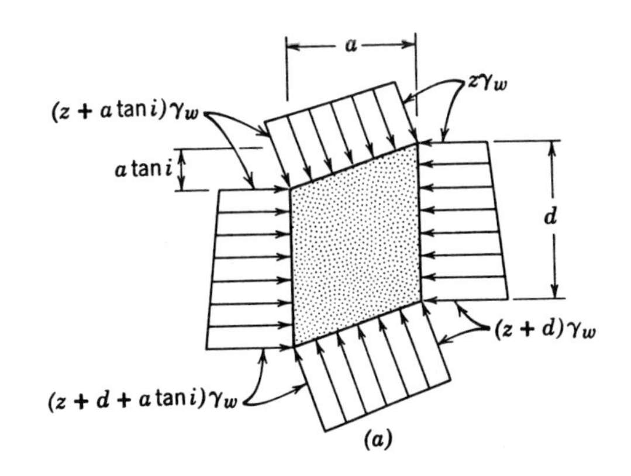

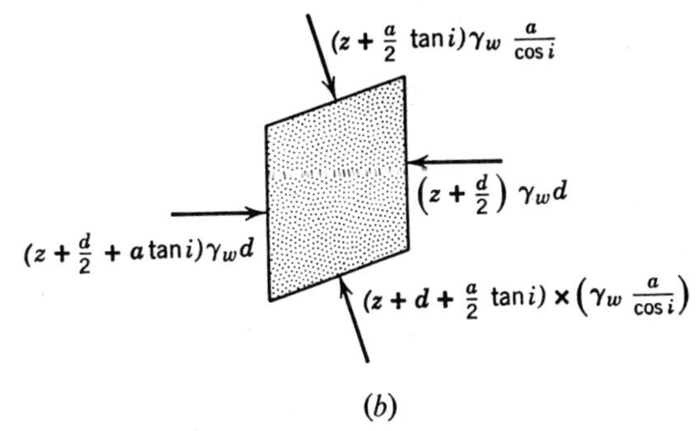

(*b*)

Sum of horizontal forces:

$$\left(z + \frac{a}{2}\tan i\right)\gamma_w \frac{a}{\cos i}\sin i - \left(z + d + \frac{a}{2}\tan i\right)\gamma_w \frac{a}{\cos i}\sin i + \left(z + \frac{d}{2} + a\tan i\right)\gamma_w d - \left(z + \frac{d}{2}\right)\gamma_w d = -d\gamma_w \frac{a}{\cos i}\sin i + a\gamma_w d \tan i = 0$$

Sum of vertical forces:

$$\left(z + \frac{a}{2}\tan i\right)\gamma_w \frac{a}{\cos i}\cos i - \left(z + d + \frac{a}{2}\tan i\right)\gamma_w \frac{a}{\cos i}\cos i = ad\gamma_w$$

Fig. 24.2 Resultant boundary pore pressure for submerged slope. (*a*) Boundary pore pressures. (*b*) Resultants of boundary pore pressures.

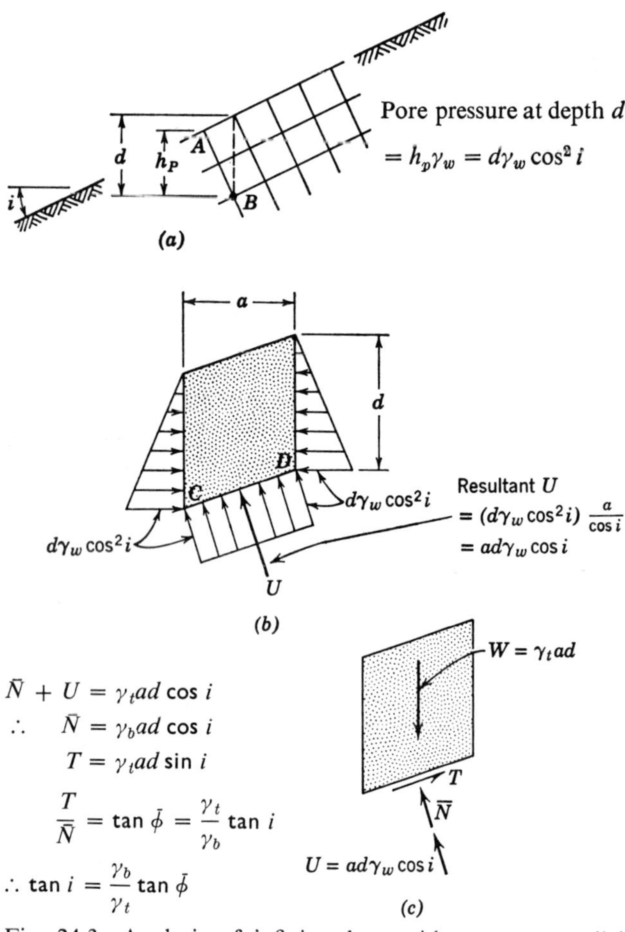

Fig. 24.3 Analysis of infinite slope with seepage parallel to slope. (*a*) Flow net. (*b*) Boundary pore pressures. (*c*) Analysis of force equilibrium (moments balanced by side forces.)

Figure 24.3*b* shows the distribution and magnitude of the boundary pore pressures. The procedure used to determine the pore pressure at any depth is derived in Fig. 24.3 (see Fig. E18.4). The total head is constant along the equipotential *AB*. Since the pressure head is zero at point *A*, the pressure head at point *B* must equal the difference in elevation between *A* and *B*. The resultant boundary water force acts normal to the side *CD* of the unit element because the water forces on the two vertical faces cancel each other. Hence the existence of boundary water pressures does not affect the magnitude of the shear force *T* required for equilibrium, but does affect the value of the resultant $\bar{N}$ of the effective normal stresses. The shear stress acting on *CD* is proportional to the total unit weight, whereas the effective normal stress $\bar{\sigma}$ is proportional to the buoyant unit weight. Since the ratio γ_b/γ_t is typically about one-half for sands, the maximum possible stable slope is about half of $\bar{\phi}$; i.e., seepage reduces the maximum slope to about half that for no flow.

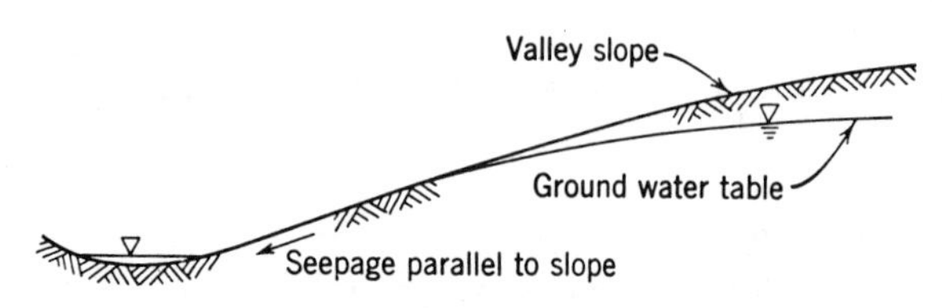

Fig. 24.4 Seepage below a natural slope.

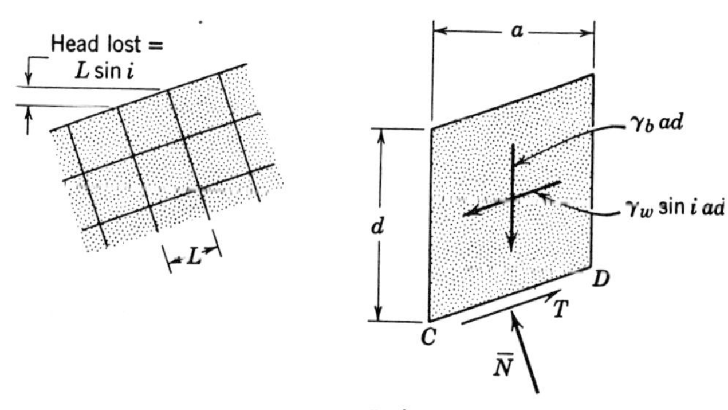

$$\text{Gradient } i = \frac{L \sin i}{L} = \sin i$$

Summing ⊥ to *CD*:

$$\bar{N} = \gamma_b ad \cos i$$

Summing = to *CD*:

$$T = \gamma_b ad \sin i + \gamma_w ad \sin i = \gamma_t ad \sin i$$

Fig. 24.5 Alternative analysis for case of parallel seepage.

Figure 24.5 presents an alternate method for ascertaining the maximum possible stable slope for the case of parallel seepage. Here the following forces are placed in equilibrium:

1. The buoyant weight of the element.
2. The resultant of the boundary normal effective stresses.
3. The resultant of the shear stresses acting over the boundary of the element.
4. The seepage force. According to Chapter 17, the seepage force equals the product of the volume of the element times the unit weight of water times the gradient. As shown in the figure, the gradient equals sin *i*. The seepage forces act parallel to the flow lines, i.e., parallel to the slope.

Solving for the $\bar{N}$ and T required for equilibrium gives the same results obtained in Fig. 24.3 (In Fig. 17.16 a similar comparison was made for vertical flow.)

24.2 INFINITE SLOPES IN CLAY

From the equations in Fig. 24.3 the shear stresses and effective normal stress must satisfy the relation

$$\frac{\tau}{\bar{\sigma}} = \frac{\gamma_t}{\gamma_b} \tan i \tag{24.1}$$

This equation, which is based solely upon statics, must be satisfied regardless of whether or not the soil has a

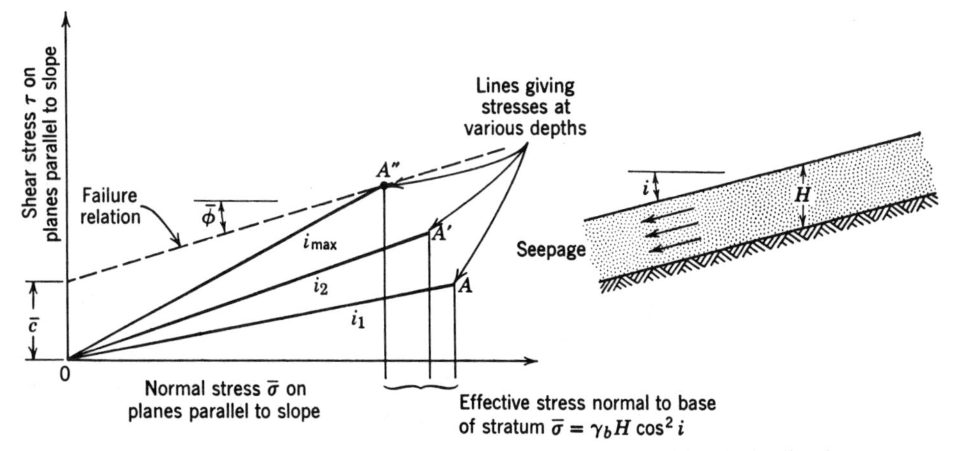

Fig. 24.6 Development of failure as stratum is gradually inclined.

cohesion intercept. If there is no cohesion intercept, the preceding section on sands applies. As discussed in Chapter 21, most normally consolidated clays have a zero cohesion intercept.

Now let us imagine the following hypothetical experiment. A stratum of soil having a cohesion intercept is slowly tipped, always maintaining seepage parallel to the slope. Figure 24.6 compares, for various slope angles, the stress conditions required for equilibrium with those at failure. The values of $\bar{\sigma}$ and τ along each of the lines OA, OA', OA'' represent the stresses upon planes parallel to the slope but at various depths below the surface. These lines extend only as far as the effective normal stress against the base of the stratum. At some angle of tipping, the full strength of the soil is reached at the base of the stratum, whereas the stresses above this level are still less than failure. Any attempt to tip the stratum further will cause the stratum to slide as a rigid body along a failure plane at its base.

The combinations of slope angle, stratum depth, and strength parameters which will just give limiting equilibrium may be found by substituting the values of τ and $\bar{\sigma}$ required for equilibrium (Eq. 24.1) into the failure law $\tau = \bar{c} + \bar{\sigma} \tan \bar{\phi}$. This result can be expressed as

$$\frac{\bar{c}}{\gamma_t H_c} = \cos^2 i \left(\tan i - \frac{\gamma_b}{\gamma_t} \tan \bar{\phi}\right) \tag{24.2}$$

where H_c is the depth (measured vertically) to the failure plane. If there is no seepage, as with a totally submerged slope, then the equation becomes

$$\frac{\bar{c}}{\gamma_b H_c} = \cos^2 i \,(\tan i - \tan \bar{\phi}) \tag{24.3}$$

Examples 24.1 and 24.2 illustrate that various combinations of parameters will satisfy these equations.

The foregoing results lead to an important practical conclusion: the inclination of a submerged slope in a cohesive soil may exceed the friction angle $\bar{\phi}$ of the soil, provided that the depth of the soil is less than a critical value determined by the magnitude of the unit weight and cohesion intercept. (A similar statement can be made for the case where there is seepage.) As already noted in Chapter 23, a vertical bank is possible in a cohesive soil provided that the height of the bank is less than some critical value.

▶ Example 24.1

Given. $\bar{c}$ = 4.8 kN/m², $\bar{\phi}$ = 20°, γ_t = 19.62 kN/m³, seepage parallel to slope.

Find. Combinations of slope angle and depth of stratum which will produce failure.

Solution. From Eq. 24.2:

$$H_c = \frac{\bar{c}}{\gamma_t} \frac{1}{\cos^2 i \,[\tan i - (\gamma_b/\gamma_t) \tan \bar{\phi}]}$$
$$= \frac{4.8}{19.62} \frac{1}{\cos^2 i (\tan i - \tfrac{1}{2} 0.364)}$$
$$\approx 0.245 \frac{1}{\cos^2 i (\tan i - 0.182)}$$

Values of H_c and i satisfying this equation are plotted in Fig. E24.1.

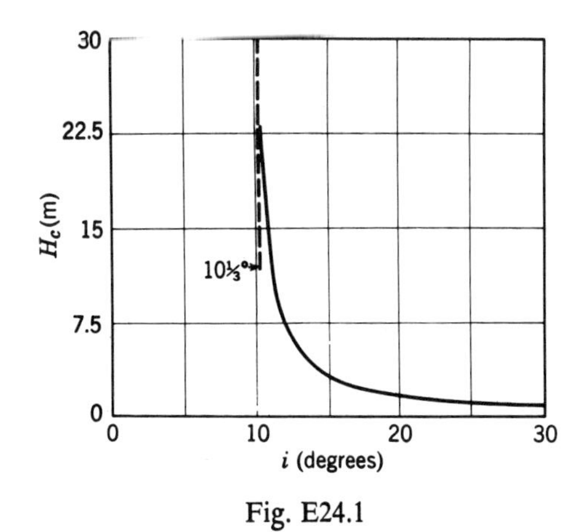

Fig. E24.1 ◀

▶ Example 24.2

Given. $H = 3.05$ m, $\gamma_t = 19.23$ kN/m³, $\bar{\phi} = 12°$, submerged slope without seepage.

Find. Combinations of slope angle and $\bar{c}$ which will give failure.

Solution. From Eq. 24.3:

$\bar{c} = \gamma_b H_c \cos^2 i\,(\tan i - \tan\bar{\phi}) = 28.73 \cos^2 i\,(\tan i - 0.212)$

Values of $\bar{c}$ and i satisfying this equation are plotted in Fig. E24.2.

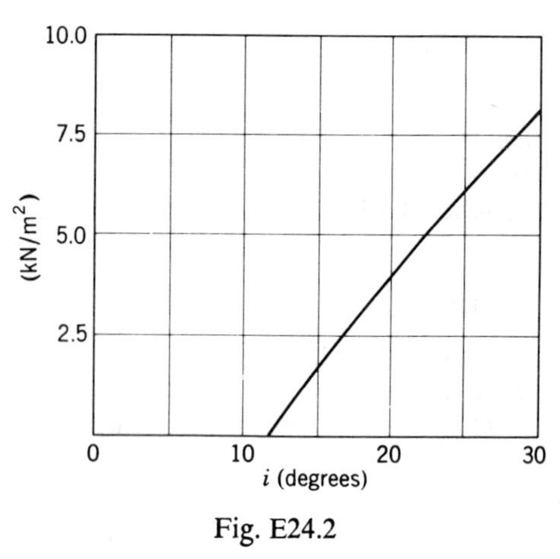

Fig. E24.2 ◀

Equations 24.2 and 24.3 have practical value in cases where soil overlies a rock surface at shallow depth and parallel to the slope.[1] In addition, there have been numerous cases in which a shallow slide, with sliding plane parallel to the slope, develops in a soil mass of great depth. This happens because the strength parameters of an overconsolidated soil are not constant with depth. A typical situation is depicted in Fig. 24.7. Here weathering has weakened the soil near the surface, destroying most of the cohesion intercept. The sliding plane develops on the stronger, less weathered soil which exists at depth. Equations 24.2 and 24.3 may be used to analyze the equilibrium of such slides. If a clay has no cohesion intercept, then of course the maximum slope angle is related directly to the friction angle, as in the case of sands.

The safety factor F for a slope is usually defined as

$$F = \frac{\text{available shear strength}}{\text{shear stress required for equilibrium}} \qquad (24.4)$$

This safety factor must be evaluated for the most critical surface through the slope. For the case of soil having vertical thickness H overlying rock, this critical surface is the soil-rock interface. Thus, for seepage parallel to the slope,

$$F = \frac{\bar{c} + \bar{\sigma}\tan\bar{\phi}}{\tau} = \frac{\bar{c} + \gamma_b H \cos^2 i \tan\bar{\phi}}{\gamma_t H \sin i \cos i} \qquad (24.5)$$

[1] See page 431 of Taylor (1948) for a similar equation which applies when the top flow line is parallel to but at some distance below the sloping ground surface.

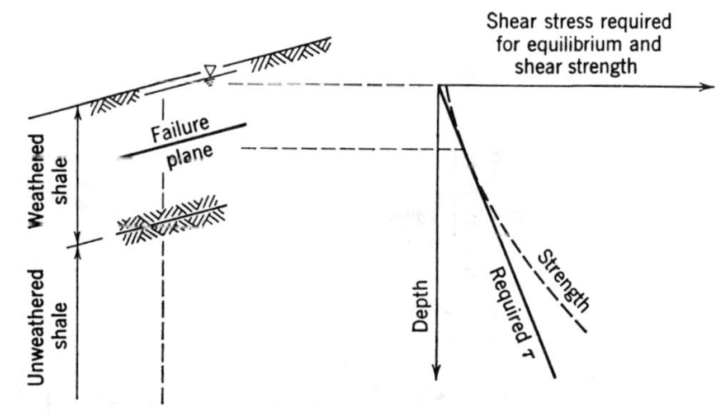

Fig. 24.7 Development of failure plane within weathered soil or rock.

If the strength parameters $\bar{c}$ and $\bar{\phi}$ vary with depth, as in Fig. 24.7, then F must be evaluated for several depths until the minimum value of F is found.

For a slope with a given inclination and a given seepage condition, the shear stress required for equilibrium can be determined with great accuracy. Uncertainty as to the stability hence arises from uncertainty as to the shear strength. The safety factor F for an infinite slope thus expresses the amount by which the shear strength can be in error and still have equilibrium. When an infinite slope in nature fails, it usually means that the shear strength of the soil has decreased through weathering and other geological processes. Such a failure may take the form of a gradual downhill creep or may involve a very sudden and extensive slip. Actually, the depth to the phreatic surface will generally vary somewhat throughout the year, and so too will the shear stress required for equilibrium. The worst condition usually occurs during severe rains, and most failures also occur during such periods. In evaluating F for an infinite slope, the worst condition—the phreatic line at the surface of the slope—is generally assumed.

24.3 GENERAL BEHAVIOR OF SLOPES OF LIMITED HEIGHT

In many problems $\bar{c}$ is large enough so that the critical depth becomes quite large: 7.5 m, 30 m, or even much more. When H_c approaches the height of the slope, the problem must be treated as a slope of limited height.

The first signs of imminent failure of a slope are usually an outward or upward bulging near the toe and the development of cracks near the crest of the slope. Failure involves a downward and outward motion of soil until a new position of equilibrium is achieved. During this movement the sliding mass often breaks up into smaller blocks. Often the surface of sliding is more or less circular, as in Fig. 24.8*a*. In some problems the location of the failure surface and the shape of the sliding mass are influenced by weak strata within the soil, as indicated in Figs. 24.8*b* and 24.8*c*.

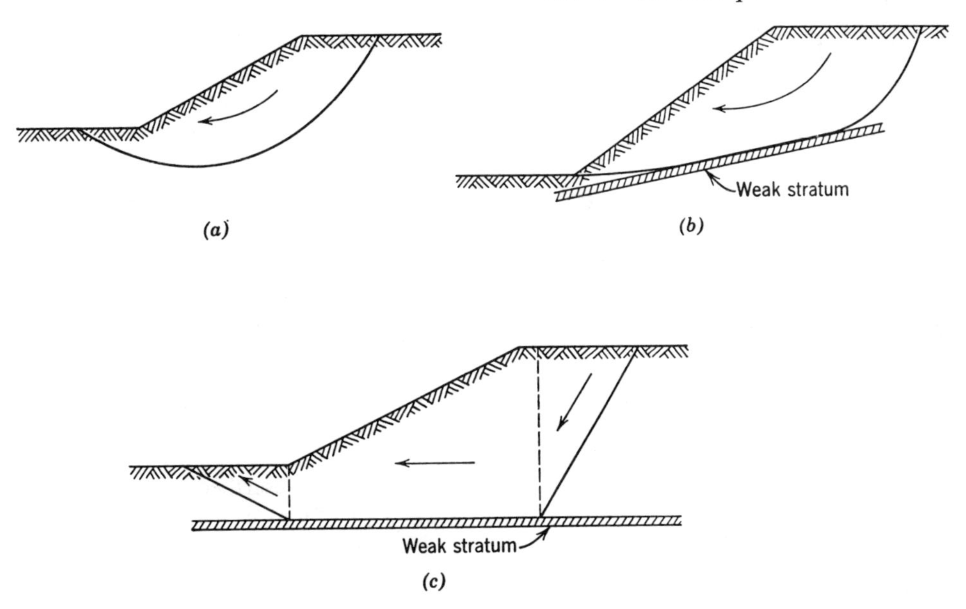

Fig. 24.8 Types of failure surfaces. (*a*) Circular failure surface. (*b*) Noncircular failure surface. (*c*) Sliding block failure.

With slopes of limited height it is necessary to work with curved failure surfaces or compound surfaces made up of several straight lines. The next three sections will present practical methods for handling two special forms of failures: a circular sliding surface (Fig. 24.8*a*) and a wedge-shaped sliding mass (Fig. 24.8*c*). These methods will suffice for most practical problems. References will be given to methods that may be used for more complicated types of failure surfaces. Although a full calculation of safety factor for a given slope requires many trial failure surfaces, the mechanics of the calculation will be demonstrated by considering a single failure surface.

24.4 CIRCULAR FAILURE SURFACES; EQUILIBRIUM OF FREE BODY AS A WHOLE

The free body cut from the slope by an assumed circle of failure is acted upon by: (*a*) the pull of gravity upon the mass of soil in the free body; (*b*) pore water pressure distributed along all or a portion of the boundaries of the free body; (*c*) a normal effective stress distributed along the assumed failure surface; and (*d*) a shear stress distributed along the assumed failure surface. For a homogeneous soil, the weight of the free body W is the area of the body times the unit weight. The boundary water force U is the sum of the boundary water pressures which are obtained from a flow net, as in Fig. E18.4.

Now let us use the definition of a safety factor contained in Eqs. 24.4 and 24.5. Then at each point on the failure surface the mobilized shear resistance is

$$\tau_m = \frac{\bar{c}}{F} + \bar{\sigma}\frac{\tan \bar{\phi}}{F} \tag{24.6}$$

Then the stresses distributed along the failure surface can be replaced by the following three resultant forces, as shown in Fig. 24.9*b*:[2]

1. Resultant of cohesion R_c. The line of action of R_c is completely determined by the variation of $\bar{c}$ along the failure surface, and the magnitude can be expressed in terms of $\bar{c}$ and the unknown F.
2. Resultant of normal effective stress $\bar{N}$. Both the magnitude and line of action of $\bar{N}$ are unknown at this stage, although $\bar{N}$ must by definition be normal to the failure arc.
3. Resultant of friction R_ϕ. R_ϕ must be normal to $\bar{N}$, and $R_\phi = \bar{N} \tan \bar{\phi}/F$. However, the line of action of R_ϕ is unknown. Different distributions of normal stress, all giving the same $\bar{N}$ and same β, will in general give different r_ϕ (Taylor, 1948).

Thus there are four unknown quantities: F, the magnitude of $\bar{N}$, β (an angle describing the line of direction of $\bar{N}$), and r_ϕ (a distance describing the line of action of R_ϕ).

Since there are four unknowns and only three equations of static equilibrium, the problem is statically indeterminate, and a unique solution is impossible without consideration of the deformation characteristics of the soil. A range of solutions, all of which satisfy statics, can be found by assuming the value of one of the four unknowns.

For example, assuming $r_\phi = r$ leaves three unknowns (F, $\bar{N}$, and β), which can be determined from the three equations of equilibrium. The technique of solution is

[2] See Taylor (1948) for a detailed discussion of these resultant forces. This approach is also discussed in Whitman and Moore (1963).

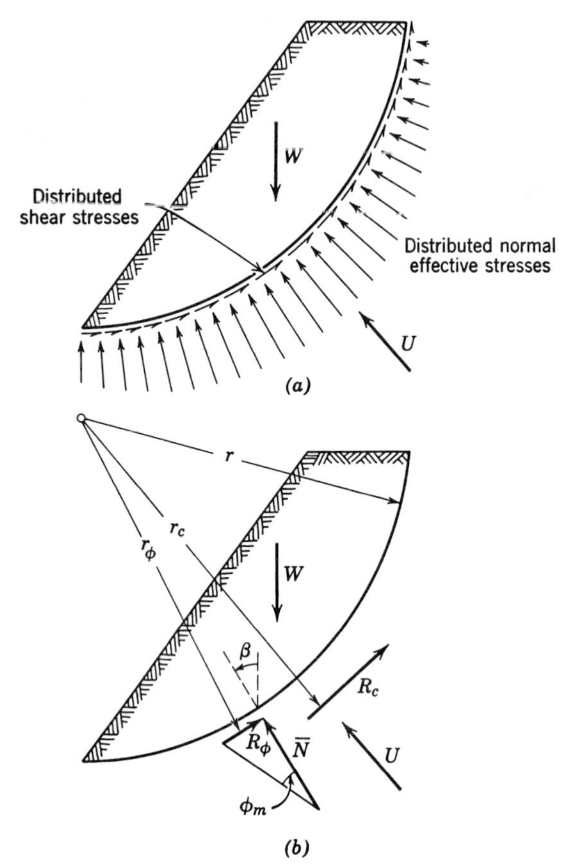

Fig. 24.9 Forces acting on free body with circular failure arc. (*a*) Distributed shear and normal stress. (*b*) Resultant forces.

illustrated in Example 24.3.[3] Assuming $r_\phi = r$ is equivalent to assuming that all of the normal stress is concentrated at a single point along the failure arc. This certainly is not a reasonable assumption, but it can be shown that the value of F computed on the basis of this assumption is a *lower bound* for all safety factors that satisfy statics and Eq. 24.6.

Similarly, an *upper bound* for the safety factor can be obtained by assuming that the effective stresses are concentrated only at the two endpoints of the assumed failure arc (Frohlich, 1955). The calculation of this upper bound is also illustrated in Example 24.3. For the slope of Example 24.3, the upper and lower bounds are $F = 1.62$ and 1.27, respectively.

In the actual slope, the normal stresses will be distributed along the failure arc in some unknown way. By assuming that these stresses are distributed in a pattern similar to a half sine wave, Taylor (1937, 1948) derived a relation between r_ϕ/r and the central angle of the failure arc (Fig. 24.10). Applying to the slope of Example 24.3 gives $F = 1.34$.

Although any value of F between 1.61 and 1.27 may satisfy statics, detailed study has shown that only a very limited range of values of F corresponds to intuitively reasonable stress distributions. For the slope of Example 24.3, this range is from $F = 1.30$ to 1.36. Any value of F within this range must be regarded as being *equally correct*. This is as closely as the safety factor can be evaluated by consideration of statics alone without taking into account the stress-strain properties of the soil. Fortunately, this range of uncertainty often is small enough for practical purposes.

[3] This procedure for analyzing slope stability is generally known as the *friction circle method*.

24.5 CIRCULAR FAILURE SURFACES: METHOD OF SLICES

The method described in the preceding section provides a very satisfactory means for determining the safety factor for slopes in homogeneous soil. However, if the slope is composed of more than one soil, or if unusual patterns of seepage exist, intuition ceases to answer the question: What is a reasonable distribution of stresses along the failure surface? Moreover, many engineers dislike using the graphical and trial and error procedures required by the friction circle method. For these reasons, other methods of analysis have been developed.

The normal stress acting at a point of the failure arc should be influenced mainly by the weight of soil lying above that point. This reasonable statement forms the basis for the *method of slices*. In this method the failure mass is broken up into a series of vertical slices and the equilibrium of each of these slices is considered. Figure 24.11 shows one slice with the unknown forces that act on it. These forces include the resultants X_i and $\bar{E}_i$ of shear and normal effective stresses along the side of the slice, as well as the resultants T_i and $\bar{N}_i$ of the shear and normal effective stresses acting along the failure arc. Also acting on this slice are the resultants U_l and U_r of the pore water pressures against the sides of the slice and U against the failure arc. These pore water pressures are assumed to be known.

Table 24.1 lists the number of unknown forces and unknown locations for these forces for a failure mass

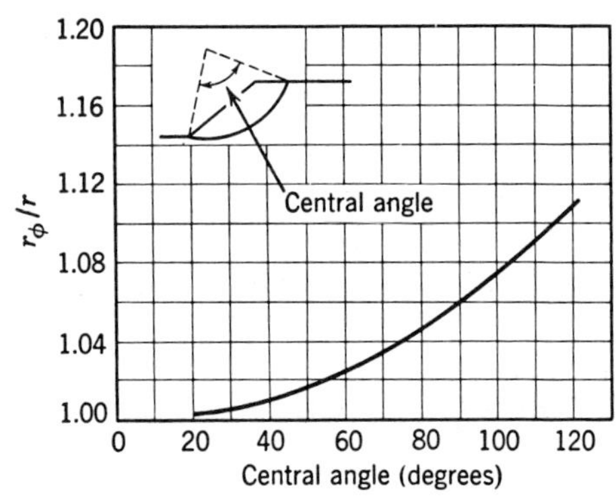

Fig. 24.10 Curve for r_ϕ/r as function of central angle of failure arc (From Taylor, 1948).

▶ Example 24.3

Given. The slope, failure surface, flow net, and strength parameters in Fig. E24.3-1.

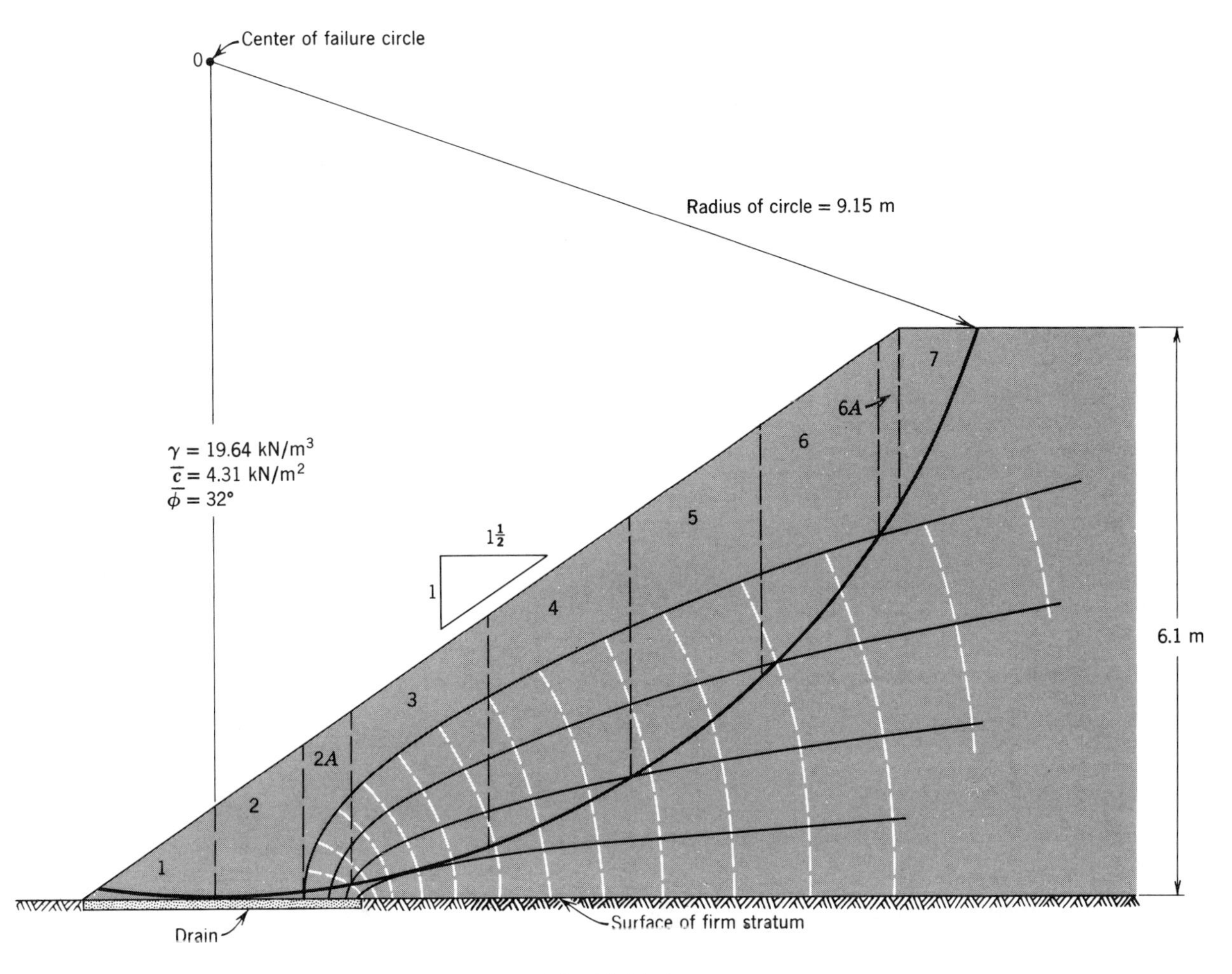

Fig. E24.3-1

Find. The safety factor.

Solution. The first step is to find the weight of the free body above the failure surface. This may be done conveniently by breaking the free body into a series of vertical slices as shown in the figure. Slices 2 to 6*A* are approximately trapezoids, and their weight can be computed by multiplying the unit weight of the soil times the width of the slice times the average height of the slice. Slices 1 and 7 may similarly be treated as triangles. The calculation of the resulting weight is given in Table E24.3.

The next step is to determine the resultant of the pore water pressures along the failure arc. Figure E24.3-2 illustrates the evaluation of the pore water force on the base of one slice: slice 4. The forces on the several slices is summed vectorally, giving the resultant force U. This force must act through the center of the failure circle.

The next step is to construct a force polygon. This is done as follows (refer to Fig. 24.9):

1. Lay off the line of actions of W and U and find their intersection (point A in Fig. E24.3-3).

2. Determine graphically the resultant Q of W and U. Q must act through point A.

3. Determine the line of action of R_c. The moment of the cohesive stresses about O is $\bar{c}L_a r/F$ where L_a is the length of the failure arc. However, the resultant R_c is $\bar{c}L/F$ where L is the length of the chord of the failure arc, because components of $\bar{c}$ normal to the chord

Example 24.3 (*continued*)

Table E24.3

Slice	Width (m)	Average Height (m)	Weight (kN)	Moment about O (kN-m)
1	1.37	0.49	13.2	−9.0
2	0.98	1.28	24.6	12.1
2*A*	0.55	1.77	19.1	24.0
3	1.52	2.26	67.5	154.6
4	1.52	2.74	81.8	311.7
5	1.52	2.84	84.8	452.0
6	1.34	2.56	67.4	455.6
6*A*	0.18	2.04	7.2	54.1
7	0.98	1.16	22.3	180.6
			W = 387.9	1635.7

Resultant lies 1635.7/387.9 = 4.2 m to right of center of circle.

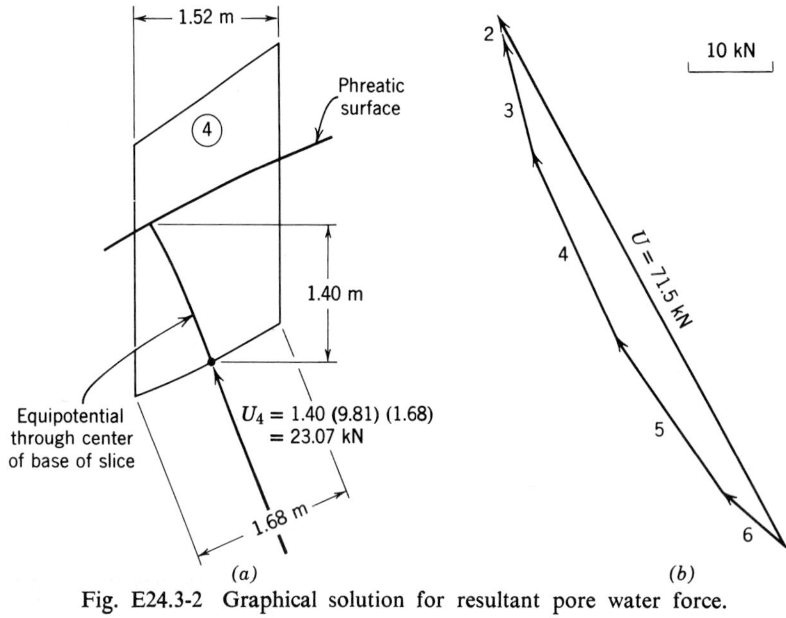

Fig. E24.3-2 Graphical solution for resultant pore water force.

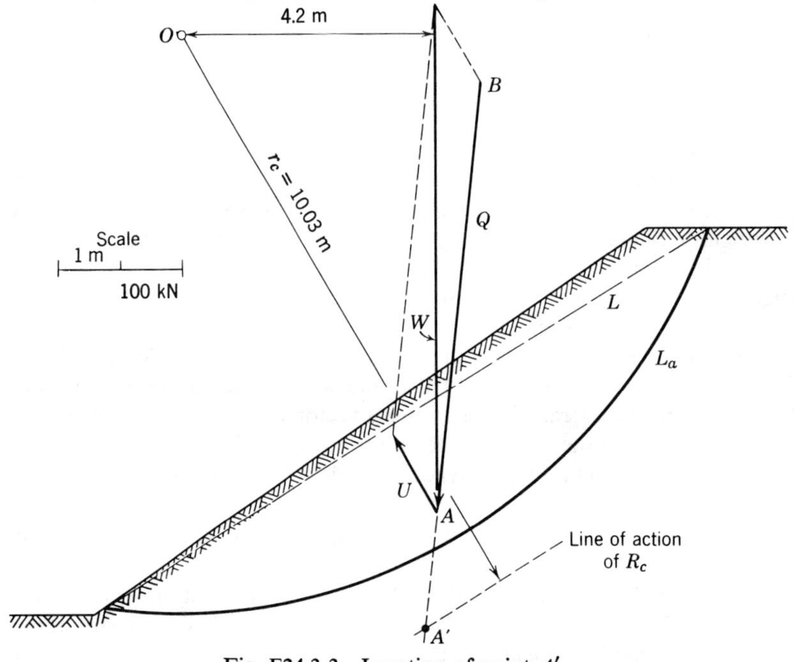

Fig. E24.3-3 Location of point A'.

Example 24.3 (*continued*)

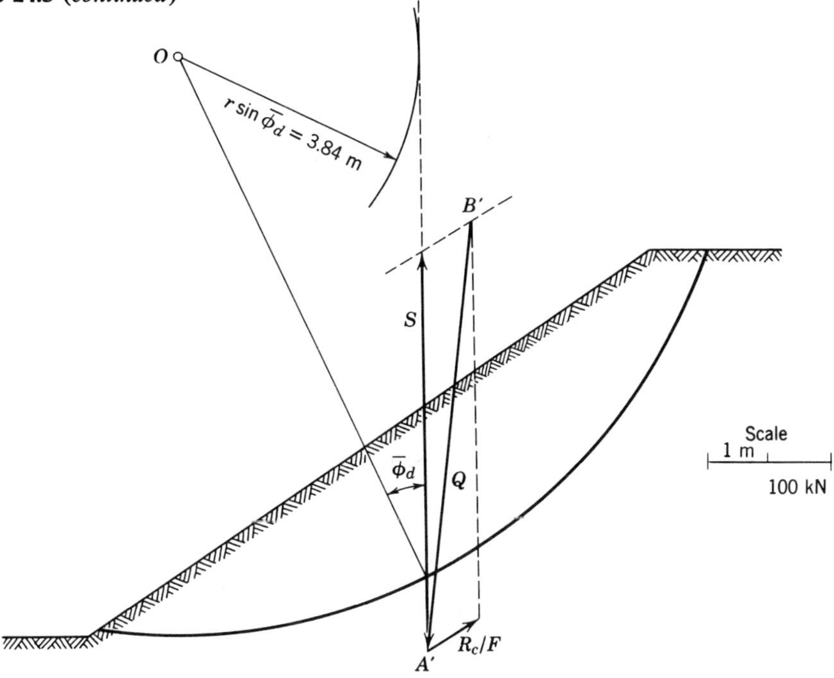

Fig. E24.3-4 Force equilibrium.

cancel and give no net force. Hence

$$r_c R_c = r_c \frac{\bar{c}L}{F} = \frac{\bar{c}L_a r}{F} \qquad \text{or} \qquad r_c = \frac{L_a}{L}\, r$$

r_c is 10.03 m for this case.

4. Determine the location of point A' by the intersection of Q and R_c. The force S, the resultant of $\bar{N}$ and R_ϕ, must act through A' (see Fig. E24.3-4). Assuming $r_\phi = r$, the line of action of S must make an angle $\bar{\phi}_d$ with the radius through the intersection of S with the failure arc, where $\bar{\phi}_d$ is given by

$$\tan \bar{\phi}_d = \frac{\tan \bar{\phi}}{F}$$

Thus S must pass tangent to a circle having a radius $r \sin \bar{\phi}_d$. This is the *friction circle.*

5. Equilibrium is satisfied by a closed force polygon involving Q, S, and R_c/F. A trial and error procedure is necessary to find the solution. Several friction circles are assumed, thus permitting the polygon to be closed. For each assumed circle, two safety factors are obtained:

$$F_\phi = \frac{\tan \bar{\phi}}{\tan \bar{\phi}_d}$$

$$F_c = \frac{\bar{c}L}{R_c}$$

The correct solution is that giving $F_\phi = F_c$. For example, for the solution in Fig. E24.3-4,

$$r \sin \bar{\phi}_d = 3.84 \text{ m} \qquad \bar{\phi}_d = 25°; \qquad F_\phi = 1.34$$

$$\frac{R_c}{F} = 45.96 \text{ kN}, \qquad L\bar{c} = 11.59(4.31) = 49.93 \text{ kN}, \quad F_c = 1.09$$

The correct safety factor satisfying statics is $F = 1.27$ (see Fig. E24.3-5).

This is the answer for the given circle. Now other circles must be analyzed until the circle giving the smallest F is found. The circle given actually is the critical circle.

As discussed in the text, the foregoing solution with $r = r_\phi$ gives a lower bound. Figure E24.3-6 shows a trial solution based on the assumption that normal stress against the failure arc is concentrated at the two ends. Then S does not pass tangent to the friction circle; rather it acts as shown.

Example 24.3 (*continued*)

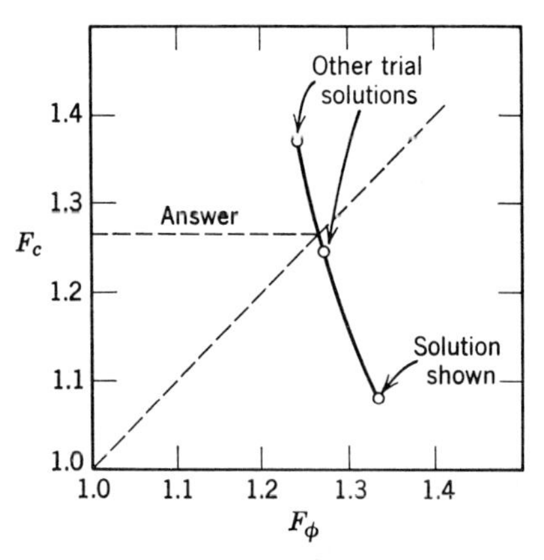

Fig. E24.3-5 Safety factors.

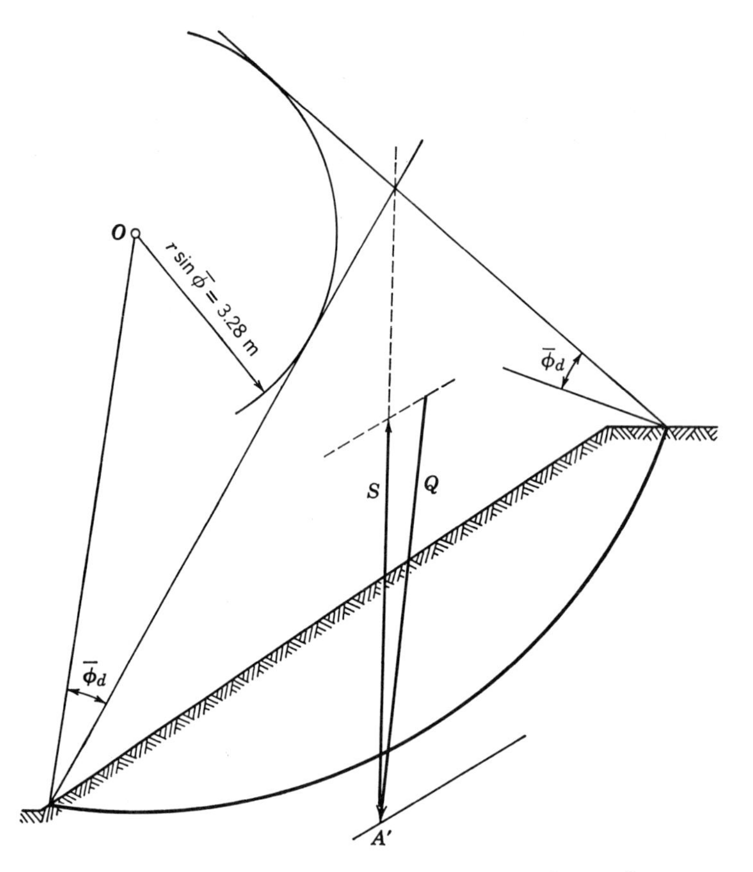

Fig. E24.3-6 Solution for upper bound. Assumed $\bar{\phi}_d = 21°$, $F_\phi = 1.63$, $R_c/F = 31.37$ kN, $F_c = 49.93/31.37 = 1.59$. Further trials give $F = 1.61$. ◀

which is broken into n vertical slices. If the slices are made so thin that the coordinates a_i (which determine the location of the resultants $\bar{N}_i$ along the segments of the failure arc) can be taken as zero, then there are $4n - 2$ unknowns versus $3n$ equations, or $n - 2$ extra unknowns. Breaking the mass up into a series of vertical slices does not remove the problem of statical indeterminacy. Hence in order to obtain values of the safety factor by using the method of slices, it is still necessary to make assumptions to remove the extra unknowns. The value of safety factor computed thereby will, of course, depend on the reasonableness of the assumptions that have been made.

Usually assumptions are made regarding the forces that act against the sides of the slices. If the problem is to become statically determinate, exactly $n - 2$ assumptions must be made. A discussion of the best way in which to make these assumptions, and of the techniques for solving the resulting system of simultaneous equations is beyond the scope of this text (see Morgenstern and Price, 1965; Whitman and Bailey, 1967). Careful analysis shows that there are severe limitations on the way in which these assumptions can be made: the shear forces on the side of the slices cannot exceed the shear resistance of the soil, and the side forces $\bar{E}_i$ should fall at a distance above the failure arc between one-third and one-half of the height of the slice. Hence, although a wide range of safety factors can be computed based on the assumptions, there is only a narrow range of safety factors corresponding to an intuitively reasonable distribution of stress along the failure arc and within the failure mass. For the slope in Example 24.3, this range is again from 1.30 to 1.36.

Use of a method of slices that takes full account of side forces and fully satisfies equilibrium requires use of a computer (see Whitman and Bailey, 1967). Even then there are considerable complexities involved in the use of such a method. This method can and should be used for advanced stages of slope stability studies, and is especially useful for the study of noncircular failure surfaces. For many problems, however, it is sufficient to use approximate methods, which do not fully satisfy the requirements of static equilibrium but have been found to give reasonably correct answers for most problems. Several such methods will now be described.

Features Common to All Approximate Methods

In all of these methods, the safety factor is defined in terms of moments about the center of the failure arc:

$$F = \frac{M_R}{M_D} = \frac{\text{Moment of shear strength along failure arc}}{\text{Moment of weight of failure mass}} \quad (24.7)$$

The denominator is the driving moment and may be evaluated as in Example 24.3. Note that the moment arm for the weight of any slice is equal to $r \sin \theta_i$. Hence we may write

$$M_D = r \sum_{i=1}^{i=n} W_i \sin \theta_i$$

where r is the radius of the failure arc, n is the number of slices, and W_i and θ_i are as defined in Fig. 24.11. Similarly, the resisting moment may be written as[4]

$$M_R = r \sum_{i=1}^{i=n} (\bar{c} + \bar{\sigma}_i \tan \bar{\phi}) \Delta l_i = r\left(\bar{c}L + \tan \bar{\phi} \sum_{i=1}^{i=n} \bar{N}_i\right)$$

where Δl_i is the length of the failure arc cut by the ith slice and L is the length of the entire failure arc. Thus

[4] The following derivations assume that $\bar{c}$ and $\bar{\phi}$ are constant along the failure arc. The equations may be generalized by including $\bar{c}$ and $\bar{\phi}$ inside the summations.

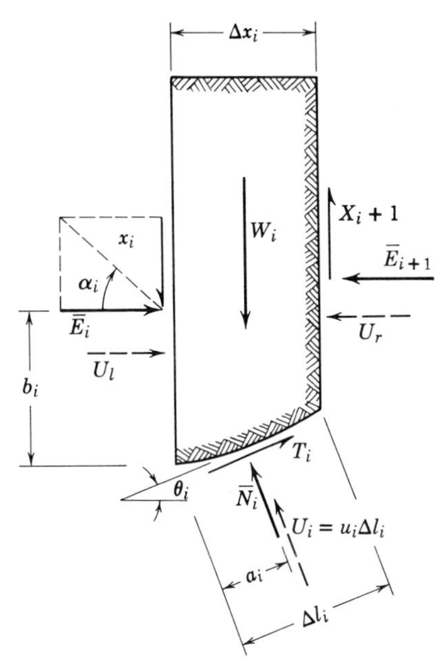

Fig. 24.11 Complete system of forces acting on a slice.

Eq. 24.7 becomes

$$F = \frac{\bar{c}L + \tan \bar{\phi} \sum_{i=1}^{i=n} \bar{N}_i}{\sum_{i=1}^{i=n} W_i \sin \theta_i} \quad (24.8)$$

Equation 24.8 is a perfectly accurate equation. If the $\bar{N}_i$ used in this equation satisfy statics, then an accurate

Table 24.1 Unknowns and Equations for *n* Slices

Unknowns Associated with Force Equilibrium	
n	Resultant normal forces $\bar{N}_i$ on the base of each slice or wedge
1	Safety factor, which permits the shear forces T_i on the base of each slice to be expressed in terms of $\bar{N}_i$
$n - 1$	Resultant normal forces $\bar{E}_i$ on each interface between slices or wedges
$n - 1$	Angles α_i which express the relationships between the shear force X_i and the normal force $\bar{E}_i$ on each interface
$3n - 1$	Unknowns, versus $2n$ equations
Unknowns Associated with Moment Equilibrium	
n	Coordinates a_i locating the resultant $\bar{N}_i$ on the base of each wedge or slice
$n - 1$	Coordinates b_i locating the resultant $\bar{E}_i$ on each interface between wedges or slices
$2n - 1$	Unknowns, versus n equations
Total Unknowns	
$5n - 2$	Unknowns, versus $3n$ equations

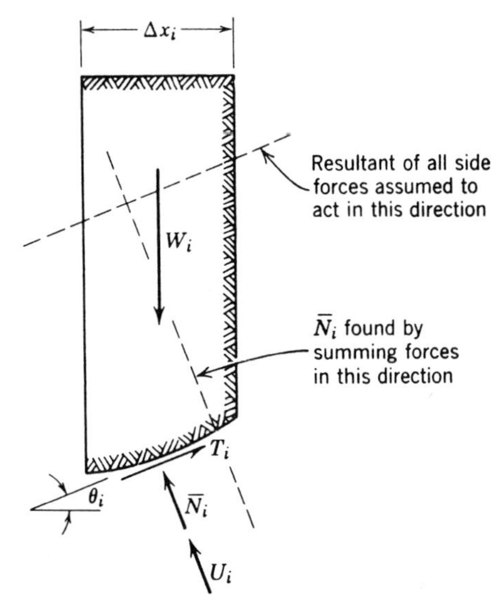

Fig. 24.12 Forces considered in ordinary method of slices.

value of F will result. Moreover, the definition of F in Eq. 24.8 is perfectly consistent with the definition of F in Eqs. 24.4 and 24.5. However, the approximate methods discussed below do not use values of $\bar{N}_i$ that satisfy statics.

If any external forces other than gravity act on the failure mass (such as the weight of a building upon the slope), the moment of these forces is included in M_D. Pore pressures on the failure arc do not contribute to M_D, since their resultant passes through the center of the arc.

Ordinary Method of Slices

In this method,[5] it is assumed that the forces acting upon the sides of any slice have *zero resultant in the direction normal to the failure arc for that slice.* This situation is depicted in Fig. 24.12. With this assumption,

$$\bar{N}_i + U_i = W_i \cos \theta_i$$

or

$$\bar{N}_i = W_i \cos \theta_i - U_i = W_i \cos \theta_i - u_i \Delta l_i \quad (24.9)$$

Combining Eqs. 24.8 and 24.9.

$$F = \frac{\bar{c}L + \tan \bar{\phi} \sum_{i=1}^{i=n} (W_i \cos \theta_i - u_i \Delta l_i)}{\sum_{i=1}^{i=n} W_i \sin \theta_i} \quad (24.10)$$

The use of Eq. 24.10 to compute F is illustrated in Example 24.4.

Here the assumption regarding side forces involves $n - 1$ assumptions, while there are only $n - 2$ unknowns. Hence the system of slices is overdetermined and in general it is not possible to satisfy statics. Thus the safety factor computed by this method will be in error. Numerous examples have shown that the safety factor obtained in this way usually falls *below the lower bound* of solutions that satisfy statics. In some problems, F from this method may be only 10 to 15% below the range of equally correct answers, but in other problems

[5] Also known as Swedish Circle Method or Fellenius Method. Consideration of slices within the trial wedge was first proposed by Fellenius (1936).

Example 24.4

Given. Slope in Example 24.3.
Find. Safety factor by ordinary method of slices.
Solution. See Table E24.4.

Table E24.4

Slice	W_i (kN)	$\sin \theta_i$	$W_i \sin \theta_i$ (kN)	$\cos \theta_i$	$W_i \cos \theta_i$ (kN)	u_i (kN/m)	Δl_i (m)	U_i (kN)	$\bar{N}_i$ (kN)
1	13.2	−0.03	−0.4	1.00	13.2	0	1.34	0	13.2
2	24.6	0.05	1.2	1.00	24.6	0	0.98	0	24.6
2A	19.1	0.14	2.7	0.99	18.9	1.4	0.58	0.8	18.1
3	67.5	0.25	19.6	0.97	65.5	10.0	1.62	16.2	49.3
4	81.8	0.42	34.4	0.91	74.4	13.9	1.71	23.8	50.6
5	84.8	0.58	49.2	0.81	68.7	12.0	1.89	22.7	46.0
6	67.4	0.74	49.9	0.67	45.2	5.3	2.04	10.8	34.4
6A	7.2	0.82	5.9	0.57	4.1	0	0.37	0	4.1
7	22.3	0.87	19.4	0.49	10.9	0	2.23	0	10.9
			181.9				12.76		251.2

$$F = \frac{4.31(12.76) + 251.2 \tan 32°}{181.9} = \frac{55.00 + 156.97}{181.9} = \frac{211.97}{181.9} = 1.17$$

Note. That $r \sum W_i \sin \theta_i = 9.15(181.9) = 1664.4$ kNm should equal the moment in the last column of Table E24.3. The slight difference results from rounding errors. ◀

the error may be as much as 60% (e.g., see Whitman and Bailey, 1967).

Despite the errors, this method is widely used in practice because of its early origins, because of its simplicity, and because it errs on the safe side. Hand calculations are feasible, and the method has been programmed for computers. It seems unfortunate that a method which may involve such large errors should be so widely used, and it is to be expected that more accurate methods will see increasing use.

Simplified Bishop Method of Slices

In this newer method[6] it is assumed that the forces acting on the sides of any slice have *zero resultant in the vertical direction.* The forces $\bar{N}_i$ are found by considering the equilibrium of the forces shown in Fig. 24.13. A value of safety factor must be used to express the shear forces T_i, and it is assumed that this safety factor equals the F defined by Eq. 24.8. Then:

$$\bar{N}_i = \frac{W_i - u_i\,\Delta x_i - (1/F)\bar{c}\,\Delta x_i \tan\theta_i}{\cos\theta_i[1 + (\tan\theta_i \tan\bar{\phi})/F]} \qquad (24.11)$$

Combining Eqs. 24.8 and 24.11 gives

$$F = \frac{\sum_{i=1}^{i=n} [\bar{c}\,\Delta x_i + (W_i - u_i\,\Delta x_i)\tan\bar{\phi}][1/M_i(\theta)]}{\sum_{i=1}^{i=n} W_i \sin\theta_i} \qquad (24.12)$$

[6] The method was first described by Bishop (1955); the simplified version of the method was developed further by Janbu et al. (1956).

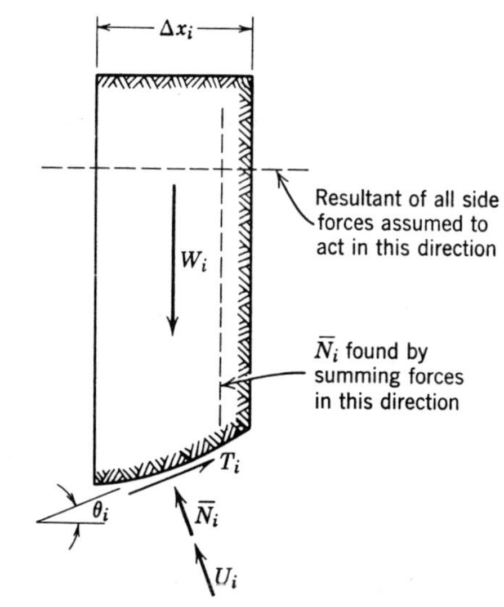

Fig. 24.13 Forces considered in simplified Bishop method of slices.

where

$$M_i(\theta) = \cos\theta_i\left(1 + \frac{\tan\theta_i \tan\bar{\phi}}{F}\right) \qquad (24.13)$$

Equation 24.12 is more cumbersome than Eq. 24.10 from the ordinary method, and requires a trial and error solution since F appears on both sides of the equation. However, convergence of trials is very rapid. Example 24.5 illustrates the tabular procedure which may be used. The chart in Fig. 24.14 can be used to evaluate the function M_i.

▸Example 24.5

Given. Slope in Example 24.3.
Find. Safety factor by simplified Bishop method of slices.
Solution. See Table E24.5.

Table E24.5

(1) Slice	(2) Δx_i (m)	(3) $\bar{c}\Delta x_i$ (kN)	(4) $u_i\Delta x_i$ (kN)	(5) $W_i - u_i\Delta x_i$ (kN)	(6) (5) tan $\bar{\phi}$ (kN)	(7) (3) + (6) (kN)	(8) M_i $F = 1.25$	$F = 1.35$	(9) (7) ÷ (8) $F = 1.25$	$F = 1.35$
1	1.37	5.9	0	13.2	8.3	14.2	0.97	0.97	14.6	14.6
2	0.98	4.2	0	24.6	15.4	19.6	1.02	1.02	19.2	19.2
2*A*	0.55	2.4	0.8	18.3	11.4	13.8	1.06	1.05	13.0	13.1
3	1.52	6.6	15.2	52.3	32.7	39.3	1.09	1.08	36.1	36.4
4	1.52	6.6	21.1	60.7	37.9	44.5	1.12	1.10	39.7	40.5
5	1.52	6.6	18.2	66.6	41.6	48.2	1.10	1.08	43.8	44.6
6	1.34	5.8	7.1	60.3	37.7	43.5	1.05	1.02	41.4	42.7
6*A*	0.18	0.8	0	7.2	4.5	5.3	0.98	0.95	5.4	5.6
7	0.98	4.2	0	22.3	13.9	18.1	0.93	0.92	19.5	19.7
									232.7	236.4

For assumed $F = 1.25$ $\quad F = \dfrac{232.7}{181.9} = 1.28$

$F = 1.35$ $\quad F = \dfrac{236.4}{181.9} = 1.30$

A trial with assumed $F = 1.29$ would give $F = 1.29$. ◂

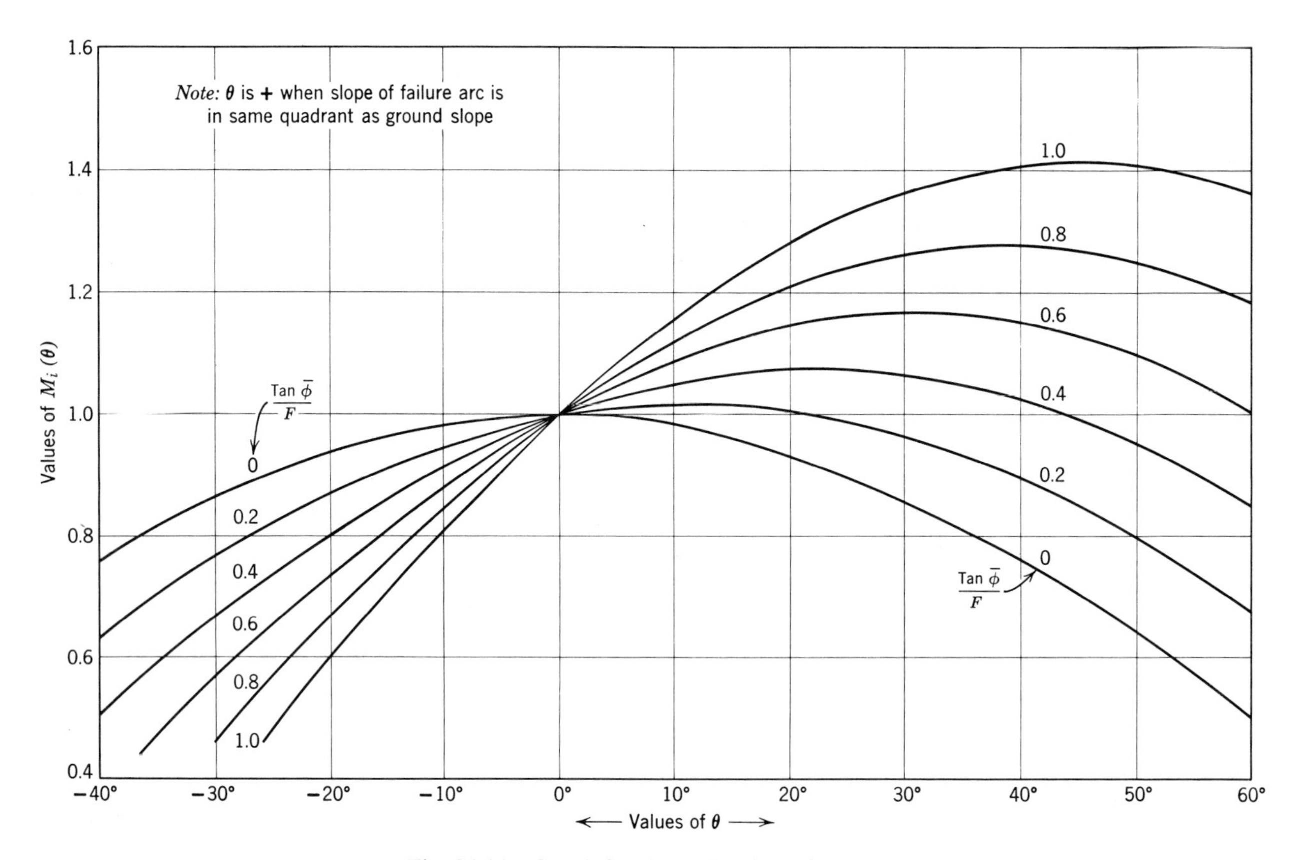

Fig. 24.14 Graph for determination of $M_i(\theta)$.

The simplified Bishop method also makes $n - 1$ assumptions regarding unknown forces and hence overdetermines the problem so that in general the values of $\bar{N}_i$ and F are not exact. However, numerous examples have shown that this method gives values of F which fall within the range of equally correct solutions as determined by exact methods. There are cases where the Bishop method gives misleading results; e.g., with deep failure circles when F is less than unity (see Whitman and Bailey, 1967). Nonetheless, the Bishop method is recommended for general practice. Hand calculations are possible, and computer programs are available.

Other Methods of Slices

There are numerous other versions of this method. In one such method the inclinations α_i of the side forces are assumed (see Lowe and Karafiath, 1960; Sherard et al., 1963). Often all α_i are taken equal to the inclination of the slope. This method also overdetermines the system of slices but gives very satisfactory answers. In its present form, this method requires a trial and error graphical solution.

24.6 WEDGE METHOD

In many problems, the potential or actual failure surface can be approximated closely by two or three straight lines. This situation arises when there are weak strata within or beneath the slope and also when the slope rests upon a very strong stratum. Figures 24.8*c* and 24.15*a* illustrate situations where the failure surfaces are almost exactly composed of straight lines. Figure 24.8*b* shows a situation where use of straight lines gives a very satisfactory approximation. A general version of the method of slices can be used for such problems. However, a satisfactory and usually very accurate estimate of the safety factor can be obtained by the *wedge method.*

In this method, the potential failure mass is broken up into two or three wedges, as shown in Fig. 24.15*b*. The shear resistance along the several segments of the failure surface is expressed in terms of the applicable strength parameters and a safety factor F, which is the same for all segments. In Fig. 24.15*b* there are three unknown forces, (P, $\bar{N}_1$, and $\bar{N}_2$), the unknown inclination α of the force between the wedges, and the unknown safety factor. Thus there are five unknowns but only four equations of force equilibrium (two for each wedge), and the system is statically indeterminate. In order to make the system determinate the value of α is assumed. Then the safety factor can be computed.

The wedge method is illustrated in Example 24.6. The strength of the core of the dam is represented by a c with $\bar{\phi} = 0$. The conditions for which such strength

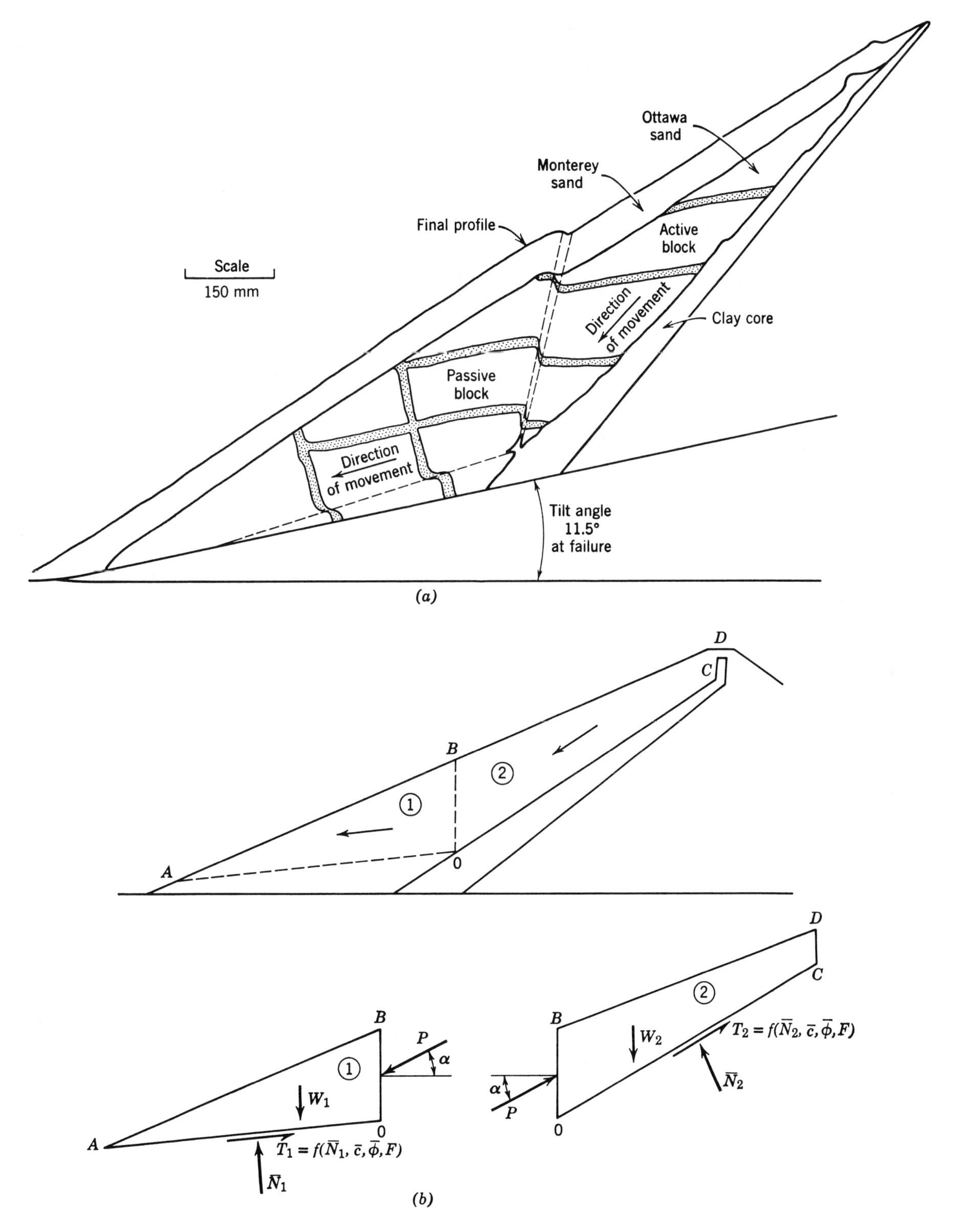

Fig. 24.15 Wedge method of stability analysis. (*a*) Wedge failure in model of sloping core dam (Sultan and Seed, 1967). (*b*) Wedges and forces.

▶ Example 24.6

Given. Sloping core dam shown in Fig. E24.6-1.

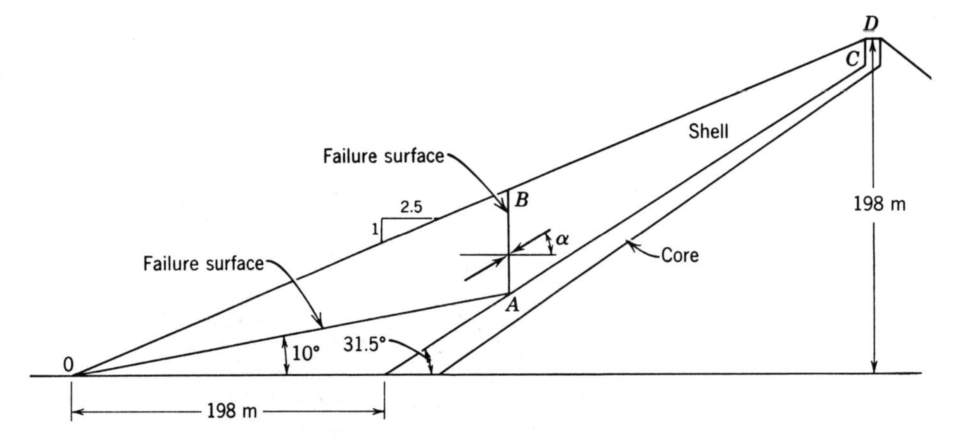

Fig. E24.6-1 Shell: $\bar{\phi} = 40°, \gamma = 17.28 \text{ kN/m}^3$; core: $c = 96 \text{ kN/m}^2$; no pore pressures.

Find. Safety factor by wedge method, assuming $\alpha = \bar{\phi}_m$.

Solution. The weights of the two wedges may be found by scaling the areas from the sketch.

$$W_{OBA} = \tfrac{1}{2}(297)(58)(17.28) = 148\,830 \text{ kN/m}$$

$$W_{ABCD} = \left[\tfrac{1}{2}(33.5)(54) + (233)\frac{54 + 14}{2}\right](17.28) = 152\,520 \text{ kN/m}$$

The available shear resistance along $AC = 96(265) = 25\,440$ kN/m. A trial and error procedure is used, assuming various F until the force polygons close. The diagram (Fig. E24.6-2) shows a trial with $F = 1.65$ which gives

$$\bar{\phi}_m = \tan^{-1}\frac{\tan\bar{\phi}}{1.65} = \tan^{-1}\left(\frac{0.839}{1.65}\right) = 27°$$

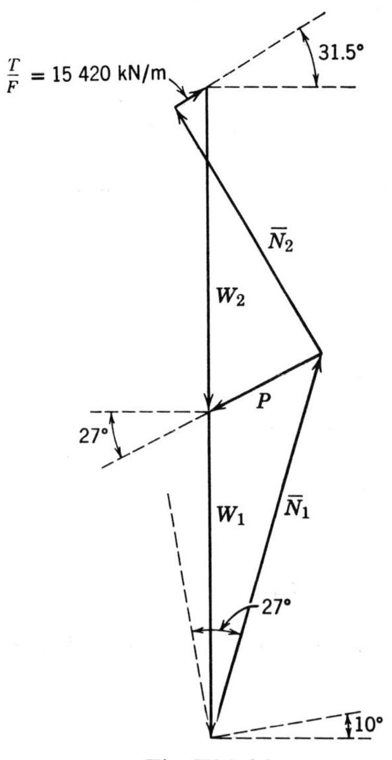

Fig. E24.6-2

Example 24.6 (*continued*)

This trial value establishes the direction of $\bar{N}_1$ and P and thus the polygon for wedge 1 may be closed.

The polygon for wedge 2 may then be constructed. A cohesive force of 15 420 kN is required for closure. Thus

$$F = \frac{25\,440}{15\,420} = 1.65$$

This result checks the trial value and hence the answer is correct. ◀

parameters are applicable are discussed in Chapter 31. No pore pressures are considered in this example. If there are pore pressures, the resultants of these pore pressures on the bottom of the wedges and between the wedges must be included in the equations of force equilibrium, but otherwise the procedure is the same.

In this example α was assumed equal to $\bar{\phi}_m$ where $\tan \bar{\phi}_m = \tan \bar{\phi}/F$. That is, the ratio of mobilized to available strength is the same on the plane between the wedges as on the failure surface. The effect of the assumed value of α has been studied by Seed and Sultan (1967). Another common assumption is α equal to the inclination of the slope. The computed safety factors by these two assumptions are within a few percent of each other.

The safety factor, of course, depends on the location of the assumed failure surface. The safety factor can also be changed by using an inclined surface between the wedges (Seed and Sultan, 1967).

24.7 FINAL COMMENTS ON METHODS OF ANALYSIS

Sections 24.4 to 24.6 have presented in detail methods for computing the safety factor for a given cross section and given failure arc. There are additional considerations involved in applying these methods to practical problems.

It is necessary to make a trial and error search for the failure surface having the smallest factor of safety. When using circular failure surfaces, it is convenient to establish a grid for the centers of circles, to write at each grid point the smallest safety factor for circles centered on the grid point, and then to draw contours of equal safety factor. Figure 24.16 shows an example of contours of equal safety factor. In making this analysis, only circles passing tangent to the underlying firm stratum were considered, but in many problems it would also be necessary to consider shallower circles.

Tension cracks generally occur near the crest of a slope

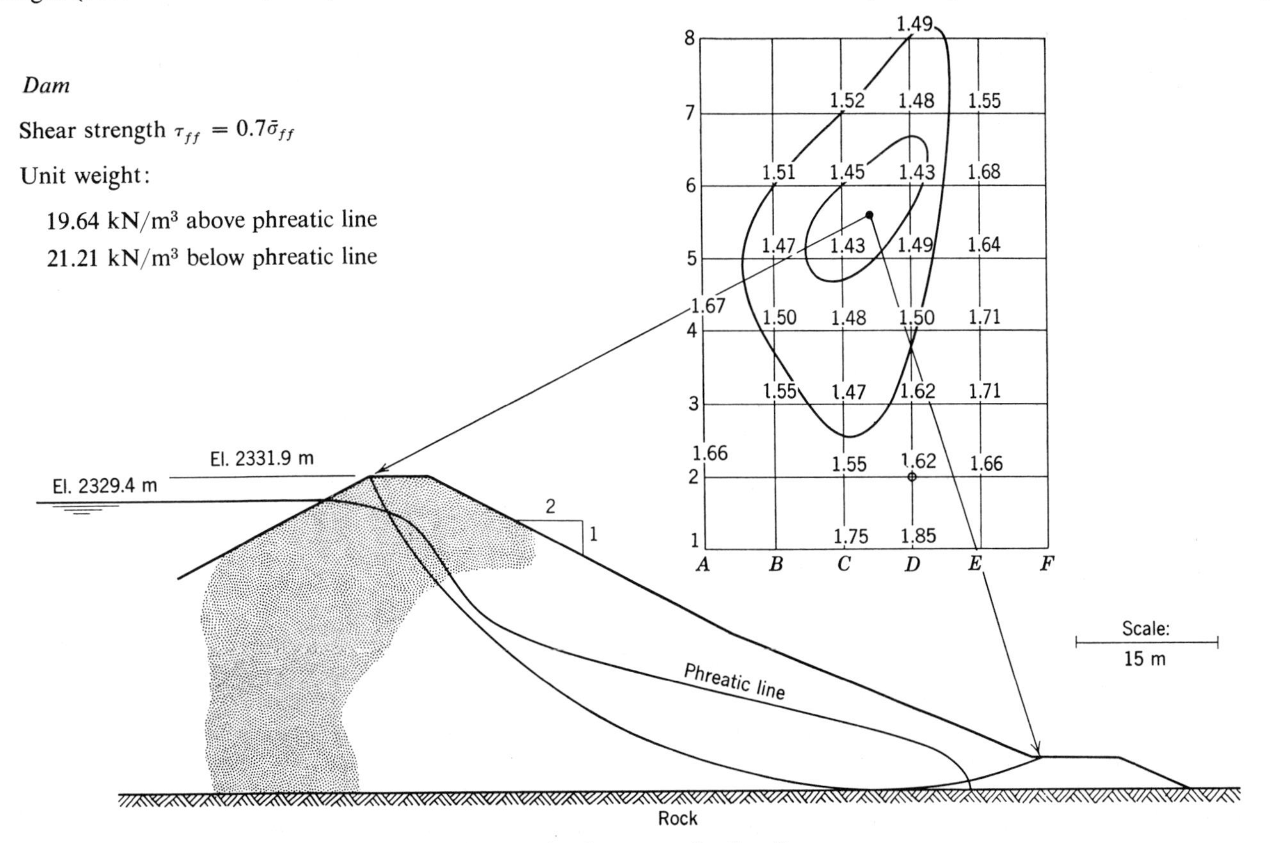

Fig. 24.16 Contours of safety factor.

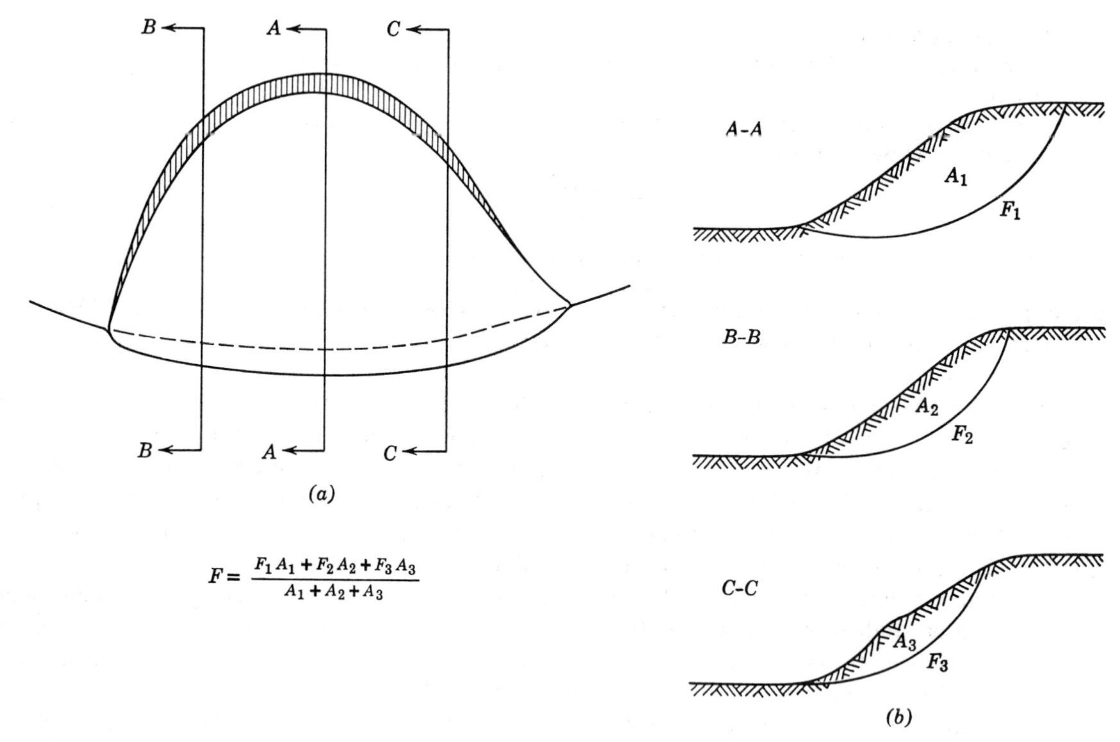

Fig. 24.17 Approximate treatment of three-dimensional effects. (*a*) Plan view of landslide. (*b*) Safety factors for different cross sections.

and reduce the overall stability of a slope by decreasing the cohesion which can be mobilized along the upper part of a potential failure surface. This effect is just the same as that discussed in Section 23.3. The methods of analysis presented in this chapter may readily be modified to include a vertical crack in place of the topmost portion of a sloping failure surface.

The methods of analysis described in the foregoing sections consider only the stresses in a single vertical cross section through the slope. There is no rigorous method for treating three-dimensional effects. If three-dimensional effects appear to be important, the best available approach is to consider three parallel cross sections through the slope, compute the safety factor for each, and then compute a weighted safety factor using the total weight above the failure surface in each cross section as the weighting factor (see Fig. 24.17).

There are other methods of stability analysis in addition to those described in Sections 24.4 to 24.6. Methods of slices that consider fully the forces between slices have already been mentioned. Methods have been developed for failure surfaces which are spirals (Terzaghi, 1943). A graphical version of the ordinary method of slices is often used (May and Brahtz, 1936) and is subject to all the errors and limitations of that method. Methods based upon finite difference and finite element procedures are currently being developed. The main result of studies based on such more sophisticated methods is to learn how to use the simpler methods more effectively. With proper attention to detail, the simpler methods (Bishop's simplified method of slices, the wedge method, and sometimes the ordinary method of slices) will give safety factors within $\pm 10\%$ of being correct for the assumed strength parameters. Thus the main uncertainty in slope stability analysis lies in the proper choice of strength parameters.

In the past, slope stability calculations have involved considerable tedium. This situation has been relieved by the widespread availability of computers (e.g., see Whitman and Bailey, 1967). It is essential, however, that any engineer using such computer programs be fully aware of the limitations of the calculation method which has been used as a basis for the program. Stability charts, giving combinations of $\bar{c}$ and $\bar{\phi}$ required for stability in typical situations, are available and are quite useful for preliminary analysis (Taylor, 1948; Bishop and Morgenstern, 1960).

24.8 ANALYSIS OF ACTUAL LANDSLIDES

This section summarizes results from analyses of several actual landslides for which the pore pressures within the slope prior to failure are known. These are all landslides which have occurred at some time following creation of the slope, at a time when the pore pressures are controlled by natural ground water conditions. All such failures result from long-term changes in either strength parameters or pore water pressures. Other slope failures that occur during or immediately after construction are discussed in Chapter 31.

Unfortunately there are relatively few landslides for which pore pressures are known and for which appropriate shear strength parameters have been determined from laboratory tests of high quality. It is of the greatest importance that future landslides be analyzed to provide engineers with valuable information regarding the nature and magnitude of shear strength.

Landslides in Intact Clays

Figure 24.18 shows stability analyses for a slope in clay at Lodalen near Oslo, Norway (Sevaldson, 1956). The slope was cut about 30 years prior to the failure. Extensive field investigations and laboratory studies were carried out to determine the pore pressure in the slope at the time of failure and the shear parameters of the clay. The shear parameters, as established from triaxial tests using the peak point of the stress-strain curve, were $\bar{c} = 12$ kN/m² and $\bar{\phi} = 32°$. The weighted safety factor based on Bishop's simplified method of slices was 1.05.

Table 24.2 lists several additional case studies. In each case, the computed safety factor was reasonably close to unity and the critical failure surface as determined from the analysis was in reasonable agreement with the observed actual failure surface. These case studies thus provide confirmation for the method of stability analysis and for the choice of strength parameters based on triaxial tests. All of these cases involved intact, nonfissured clays.

Landslides in Fissured Clays

Figure 24.19 shows a landslide involving a stiff-fissured clay-shale (Henkel and Skempton, 1955). Because of the geometry of the failure, stability could be analyzed with great accuracy using equations for an infinite slope. Strength parameters for this clay-shale based on peak resistance were $\bar{c} = 7.2$ kN/m² and $\bar{\phi} = 21°$. These strength parameters lead to a calculated safety factor of $F = 1.45$, which is inconsistent with the fact that failure occurred. However, if $\bar{c}$ is taken as zero, the safety factor drops to a reasonable value of $F = 1.07$. This result indicates that use of strength parameters based on peak resistance may lead to unsafe estimates for safety factor of slopes in fissured clays.

Skempton (1964) has analyzed a number of failures in the stiff London clay and has compared the strength developed in the field with the peak and ultimate strength measured in laboratory tests. The results are shown in Fig. 24.20, where average shear stress along the failure surface is plotted against average normal effective stress.

The three solid circles on this diagram correspond to landslides which developed some years after a slope was cut. In one case where the failed slope came into

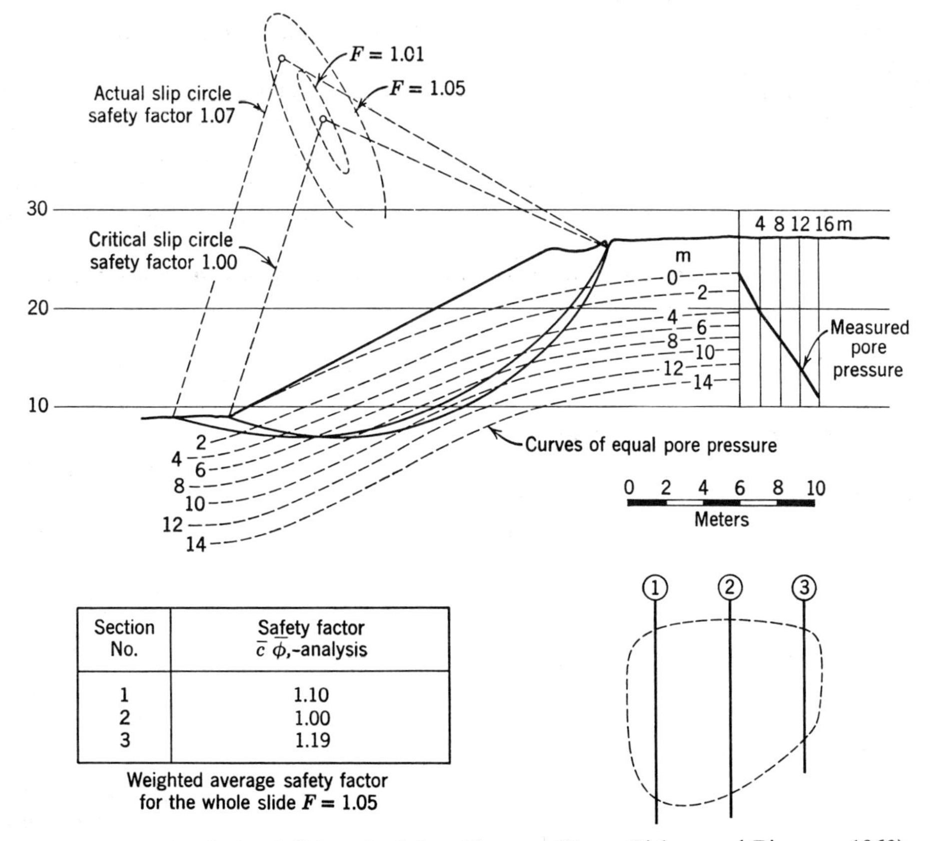

Section No.	Safety factor $\bar{c}\,\bar{\phi}$,-analysis
1	1.10
2	1.00
3	1.19

Fig. 24.18 Analysis of slide at Lodalen, Norway (From Bishop and Bjerrum, 1960).

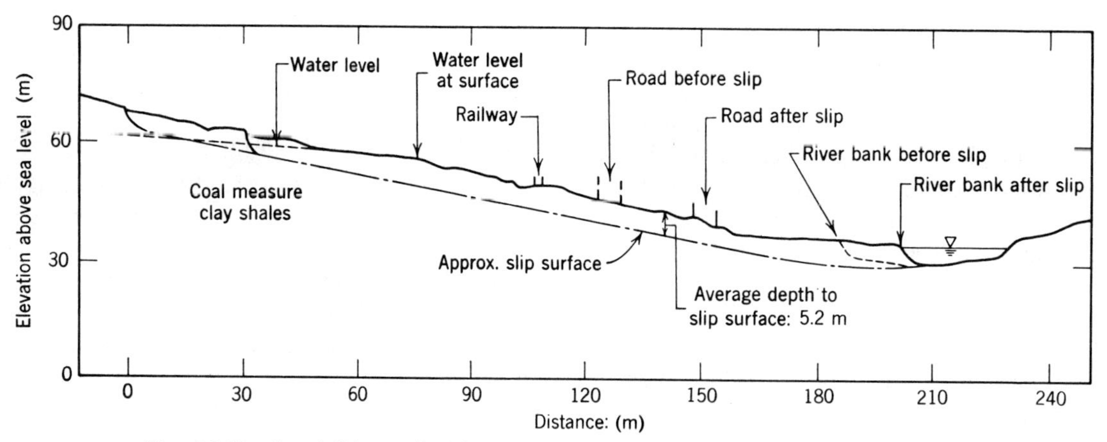

Fig. 24.19 Landslide at Jackfield, England (From Bishop and Bjerrum, 1960).

equilibrium again after only moderate movement, another analysis was made to find the strength available after failure. The box marked "natural slopes" indicates conditions at nature-made slopes in which landslides have recently developed.

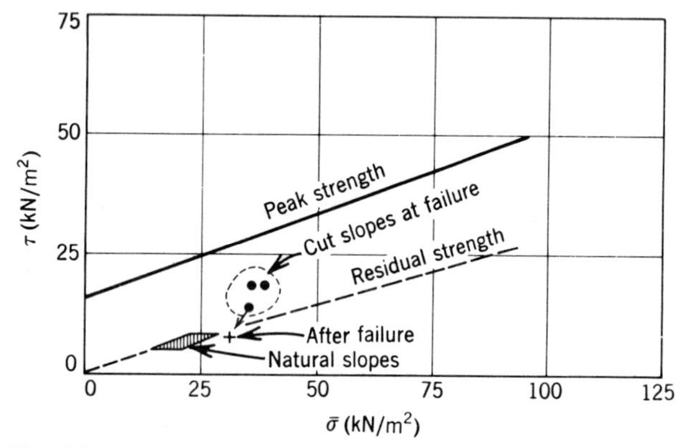

Fig. 24.20 Strength mobilized during landslides in London clay (After Skempton, 1964).

The inference from these results is that only the ultimate or residual strength of the London clay can be relied upon to sustain a slope; the additional strength that exists at the peak of the stress-strain curve is lost with time. Similar conclusions apparently apply for other overconsolidated clays which exhibit fissuring, notably the clay-shales of the Dakotas, Montana, and Saskatchewan. The processes that cause a gradual reduction in strength to the residual value are termed *progressive failure*. Bjerrum (1967) has discussed these processes and the extent to which they may develop in various overconsolidated soils.

24.9 DESIGN OF SLOPES

The inclination of most man-made slopes, such as those along highways, are chosen primarily on the basis of experience. Analyses are made when an unusually high slope must be cut, or when there is some reason to believe that trouble might develop. Analyses are usually

Table 24.2 Safety Factors from Analysis of Landslides in Intact Clays

Location	Soil Type	Computed Safety Factor	Reference
Lodalen, Norway	Lightly overconsolidated	1.05	Sevaldson (1956)
Drammen, Norway	Normally consolidated	1.15	Bjerrum and Kjaernsli (1957)
Selset, England	Overconsolidated	1.03	Skempton and Brown (1961)
Breckenridge, Canada	Lightly overconsolidated	1.12	Crawford and Eden (1967)
Siburua, Venezuela	Compacted plastic clay	1.02	Wolfskill and Lambe (1967)

made whenever a slide does develop, as an aid in the choice of remedial works. In the design of earth dams the slopes are usually first chosen on the basis of experience and then checked by a full analysis. As will be discussed in Chapter 31, the methods presented in this chapter are primarily useful for checking *long-term stability*; e.g., stability of the downstream slope of an earth dam once steady seepage has been established through the dam, or stability of a cut slope some years after cutting.

Although some questions remain regarding the accuracy of the mechanics of slope stability analysis, in practical situations the greatest uncertainties lie in the estimation of the pore pressures and especially in the selection of strength parameters. As defined in Eq. 24.4, a safety factor indicates the degree to which the expected strength parameters can be reduced before failure would occur, and hence essentially is a safety factor against an error in the estimation of these parameters. For intact homogeneous soils, when the strength parameters have been chosen on the basis of good laboratory tests and a careful estimate of pore pressure has been made, a safety factor of at least 1.5 is commonly employed. With fissured clays and for nonhomogeneous soils larger uncertainties will generally exist and more caution is necessary. Peck (1967) has recently documented the difficulties and frustrations in estimating stability in a particularly difficult problem.

24.10 SUMMARY OF MAIN POINTS

In evaluating the forces on a free body element of soil, one can correctly account for the effects of water by considering either:

1. Boundary water forces along with the total soil weight.
2. Seepage forces along with the buoyant soil weight.

These two approaches give identical results since the boundary water forces equal buoyancy plus seepage. In stability problems it is usually more convenient to work with boundary water forces and total soil weight.

The study of infinite slopes is helpful, both because the fundamentals of stability problems can clearly be seen and because the results are useful in certain practical problems. The maximum stable slope of a submerged sand is approximately the same as for the sand in a dry condition. For both cases i_{max} equals the strength angle $\bar{\phi}$. Seepage within a slope generally reduces stability.

A general slope stability problem is statically indeterminate. There are various techniques for solving stability problems, depending on which assumption is used to make the problem determinate. The Bishop method and wedge method give good accuracy and are recommended for practical use, especially where calculations must be made by hand. Where computer facilities are available, the engineer may use the more sophisticated Morgenstern method to check simpler solutions and for cases where neither circular nor wedge-shaped failure surfaces are suitable.

The greatest uncertainties in stability problems arise in the selection of the pore pressure and strength parameters. The error associated with the method of analysis, of the order of 10% difference in computed factor of safety for the better available techniques, is small compared to that arising from the selection of strength parameters. This is the reason why a factor of safety against loss of strength is used for stability problems.

PROBLEMS

24.1 An infinite slope at $i = 28°$ consists of sand with a friction angle $\bar{\phi}$ equal to 30°, a dry unit weight of 17.28 kN/m^3, and a void ratio of 0.52. During a heavy rain the sand becomes saturated and vertical downward seepage under a gradient of unity occurs. Will the slope flatten? What is the maximum stable slope during the rain?

24.2 Compute the maximum stable slope angle for a layer of normally consolidated Weald clay having a vertical thickness of 6.1 m and with seepage parallel to the slope. The slope is infinite.

24.3 With the numerical values in the table prepared in Example 24.4, show on sketches the forces acting on slices 3 and 6. Are these forces in equilibrium? Explain.

24.4 Repeat Problem 24.3 but use the numerical values in the table prepared in Example 24.5.

24.5 Repeat Example 24.3, using a failure arc centered over the boundary between slices 2*A* and 3 with the center 7.62 m above the firm stratum.

24.6 Repeat Example 24.4 using the failure arc described in Problem 24.5.

24.7 Repeat Example 24.5 using the failure arc described in Problem 24.5.

24.8 Repeat Example 24.6 with the following variations:

a. Assume the angle α is equal to the inclination of the slope.

b. Using $\alpha = \bar{\phi}_m$, and keeping the location of point *A* the same, move point *B* *up* the slope so that *AB* is inclined 10° to the vertical.

c. Using $\alpha = \bar{\phi}_m$ and with *AB* vertical, move points *A* and *B* downslope so that *OA* is inclined at 5°.

CHAPTER 25

Shallow Foundations with Drained Conditions

25.1 GENERAL BEHAVIOR OF SHALLOW FOUNDATIONS

Chapter 14 introduced the subject of shallow foundations and treated in detail the behavior of shallow foundations resting on dry soil. Chapter 25 extends Chapter 14 in two respects:

1. It covers the situation where some or all of the subsoil is saturated.
2. It presents certain considerations especially pertinent to foundations on silt and clay.

The general behavior of footings described in Chapter 14 holds for all types of soil. Moreover, the expressions presented in Chapter 14 for the bearing capacity and settlement of foundations on dry sand apply equally well for foundations on saturated sand, as long as the stresses are effective stresses. As we shall see in later parts of the present chapter, raising the water table into a soil reduces the effective stresses at any given depth in the soil and thereby results in a lower bearing capacity and greater settlement.

However, the detailed behavior of a shallow foundation depends very much on the type of soil on which the foundation rests. Table 25.1 lists important general differences between the behavior of sand and the behavior of clay as a foundation material. Silt lies between the sand and the clay. The remaining portion of Chapter 25, as well as Chapter 32, will discuss and illustrate the items in Table 25.1.

Figure 14.9 showed plots of maximum differential settlement against maximum distortion and maximum total settlement for a variety of structures resting on sand. Figure 25.1 presents similar data for structures resting on clay. Whereas the maximum differential settlement for a footing on sand tends to be almost as large as the maximum total settlement, in the case of clay the maximum differential is usually considerably smaller than the maximum settlement. This important point is more dramatically illustrated by the data in Figs. 25.2, 25.3, and 25.4.

Figures 25.2, 25.3, and 25.4 present settlement data for the shells of tanks of approximately the same size (most tanks were large storage tanks of approximately

Table 25.1 Comparison of Sand and Clay as Foundation Material

Item	Sand	Clay (normally consolidated or slightly overconsolidated)
Factor controlling footing design	$\Delta\rho$, especially under cycles of load or dynamic load	ρ_{max} and $\Delta\rho$
Settlement magnitude	Small	Large
Settlement rate	Fast	Slow
Settlement pattern	Irregular; larger ρ at edges of footing	Dished shape
Relation between $\Delta\rho_{max}$ and ρ_{max}	$\Delta\rho_{max}$ often close to ρ_{max}	$\Delta\rho_{max}$ usually much less than ρ_{max}
Effect of given $\Delta\rho$ on structure	Relatively large because ρ is irregular and occurs fast	Relatively small because ρ is regular and occurs slowly

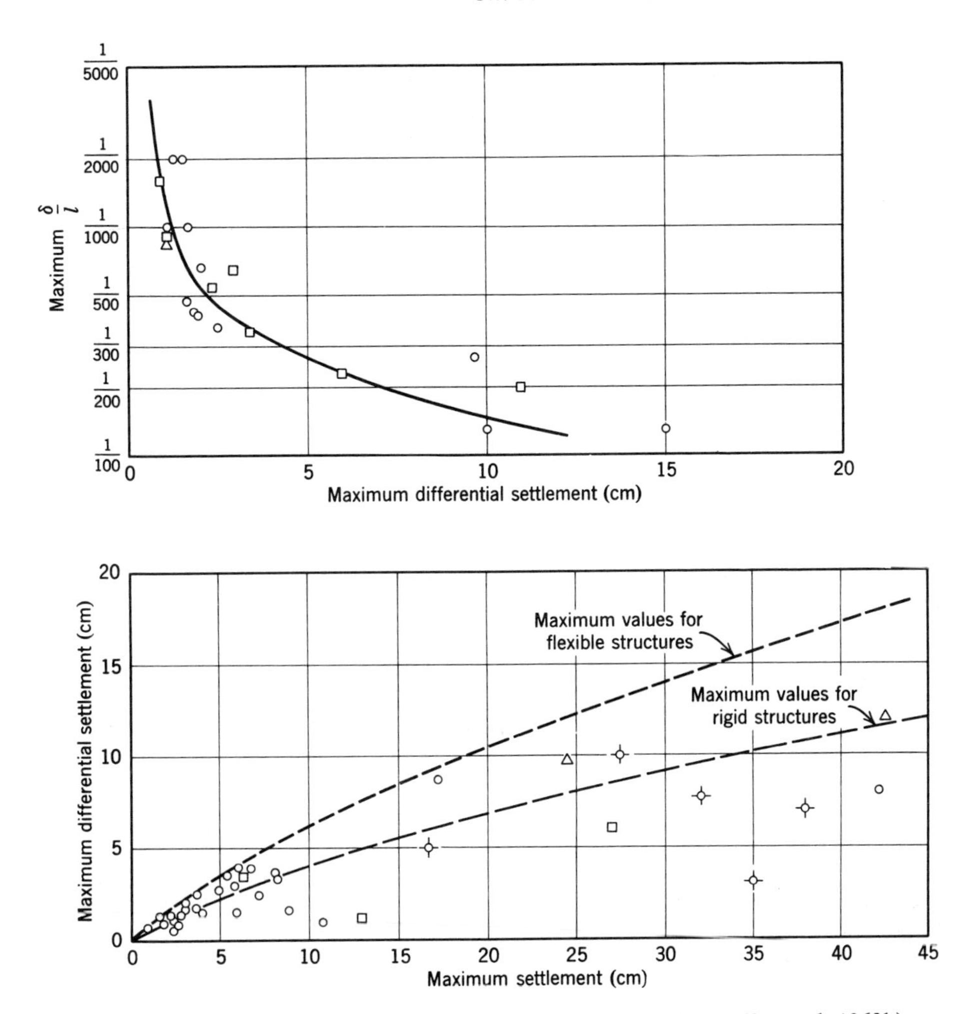

Fig. 25.1 Settlement of structures on clay (From Bjerrum, 1963a and 1963b).

60 to 75 m in diameter and about 15 m high). The data in Fig. 25.2 are for tanks resting on a thick deposit of sand—similar to Example 8.9 and Examples 14.12 to 14.14. The data in Fig. 25.3 were obtained on tanks resting on 2.4 m of hydraulically placed sand, overlying 0.6 to 1.2 m of soft silt, which in turn overlies rock. The data in Fig. 25.4 were obtained from tanks resting on the Kawasaki subsoil shown in Example 16.2, i.e., 15 m of sand-silt overlying 40 m of soft clay.

We see from Fig. 25.2 that the settlements are very small and that the maximum differential settlement tends to be just slightly smaller than the maximum total settlement. Figure 25.3 shows relatively large settlements from a thin compressible subsoil and, as was the case for the tanks on the Libyan sand, the maximum differential settlement tends to be as large as the maximum total settlement. Figure 25.4 shows the settlement of the tanks overlying soft clay can be very large, in excess of a meter. However, the differential settlements are much less than the maximum settlement. Further, the data show that most of the differential settlement occurs during the first loading of the tank, and that the

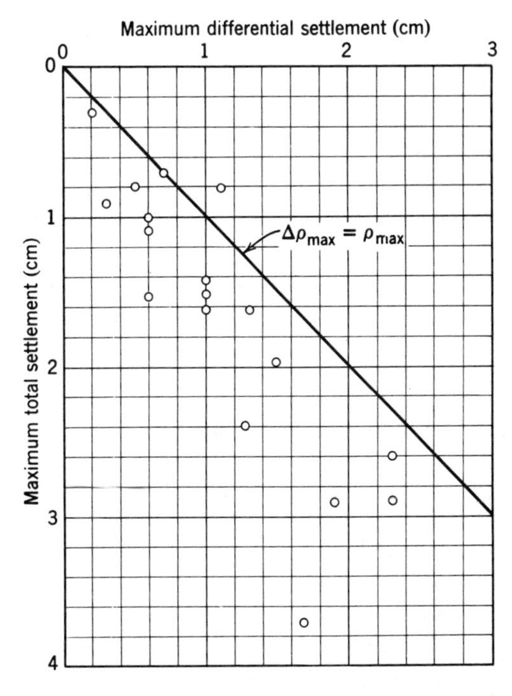

Fig. 25.2 Settlement of tanks on Libyan sand.

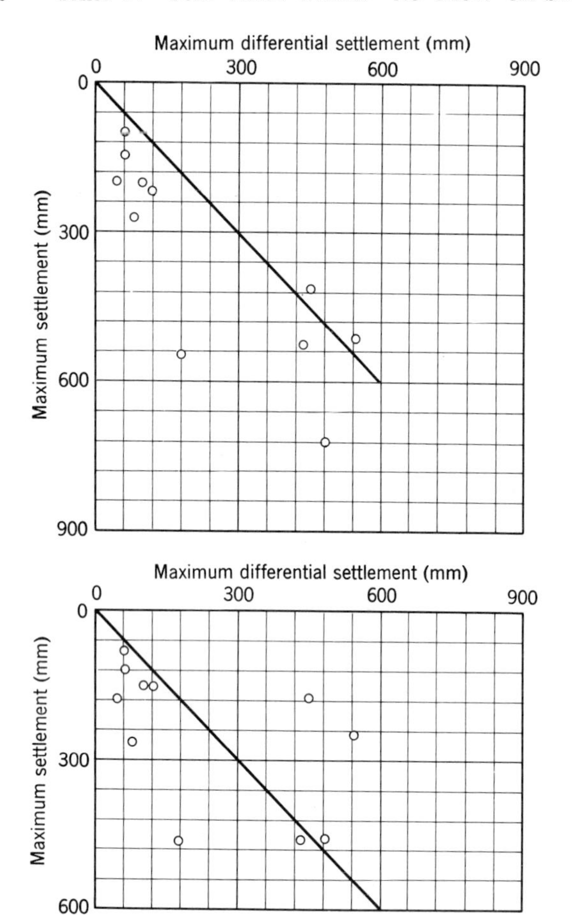

Fig. 25.3 Settlement of tanks.

additional settlement occurring with time contributes little to the differential settlement of the tank.

The settlements of the tanks on the Libyan sand occurred as the load was placed in the tank, each cycle of load and unload causing a little additional settlement. The settlements in Fig. 25.3 took approximately 6 months to occur. The settlement of the tanks overlying the Kawasaki clay occurred very slowly. Settlement predictions indicate that more than 10 years will be required for about three-quarters of the total settlement of the tanks to occur. The reasons for this time lag were pointed out in Chapter 2 and are treated in Chapter 27.

25.2 ALLOWABLE SETTLEMENT

The principles and values of allowable settlement given in Chapter 14 (Fig. 14.8) hold for all types of soil. As noted in Table 25.1, a given magnitude of differential settlement is more detrimental to the structural integrity of a building resting on sand than would be true for the same building resting on clay. Two reasons exist for this difference. First, the differential settlements on sand tend to be more irregular in pattern than those on clay. The settlements on sand are generally caused by scattered zones of loose soil, whereas settlements in clay are due to the higher stresses under the center portion of the structure. Second, settlements due to clay compression occur over a longer period, thus the structure has more time to adjust to the settlement. Plastic flow in the concrete or steel frame of the structure can occur; this would better accommodate settlement which occurs gradually over a period of many years.

It should be emphasized that the maximum allowable settlements suggested in Fig. 14.8 are only a guide, and the engineer must use his judgment to select the actual maximum allowable settlement for each particular case. The fact that exceptions to the sound general principles in Fig. 14.8 can occur is well illustrated by the data in Fig. 25.5. This figure presents the results of a line of levels run down the first floor corridor of a building which presumably was initially level. The three-story steel frame building rests over a thick deposit of soft soil. As the data show, there are very large differential settlements along the length of the building, with a maximum distortion of $\frac{1}{22}$ occurring between points B and C. This maximum distortion is much greater than the angular distortion indicated in Fig. 14.8 as that at which structural damage might occur. The building for which the

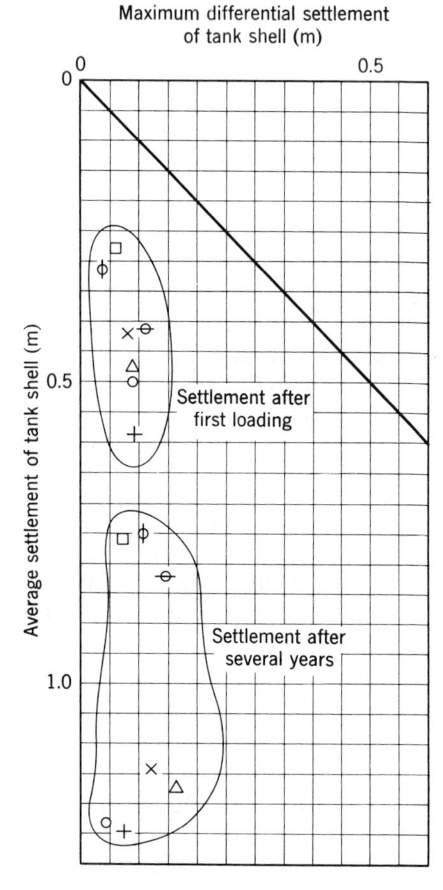

Fig. 25.4 Settlement of Kawasaki tanks.

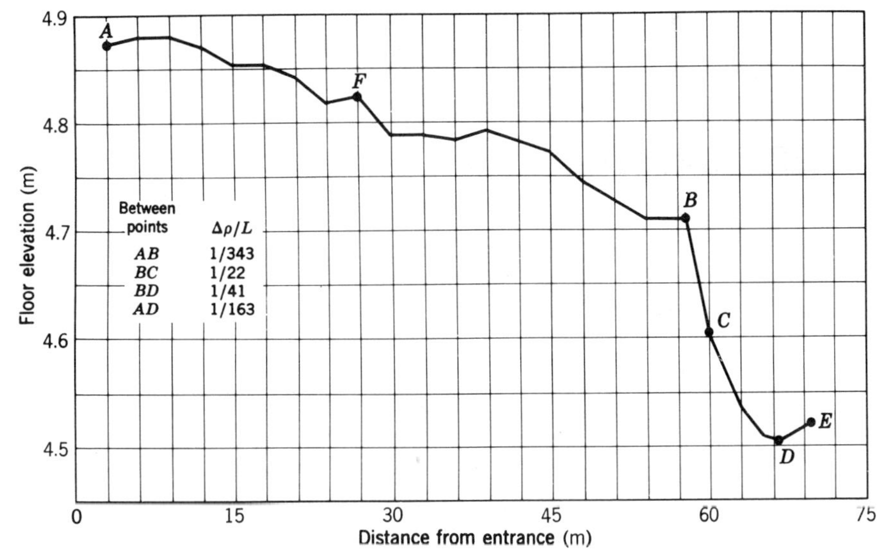

Fig. 25.5 Elevation of first floor corridor of three-story steel frame building.

settlement data are shown in Fig. 25.5 showed no structural distress and was still in full satisfactory use when the floor elevations were measured.

25.3 BEARING CAPACITY

All of the concepts and formulas presented in Chapter 14 for dry sand also hold for saturated sand, as long as the stresses are effective stresses. In Eq. 14.6

$$(\Delta q_s)_u = \frac{\gamma B}{2} N_\gamma + \gamma d N_q \qquad (14.6)$$

the unit weight that should be used is that which contributes to the effective stress in the soil. Raising the phreatic surface in a soil reduces the unit weight which generates effective stress from the total unit weight to the buoyant weight and thereby reduces the ultimate bearing capacity. If a dry subsoil became saturated, the ultimate bearing capacity of a surface footing would be reduced by the ratio of the buoyant unit weight over the dry unit weight. Since this ratio of unit weights is typically 0.5 to 0.7, the bearing capacity of a footing on the surface of a saturated soil would be about 0.5 to 0.7 of that for the dry soils.

Figure 25.6 shows the bearing capacity equation, Eq. 14.6, altered to allow for the phreatic surface being at the bottom of the footing. Examples 25.1 and 25.2 illustrate the use of this equation from Fig. 25.6. As can be seen, saturating the soil below the bottom of the footing reduced the bearing capacity from 798 kN/m² to 664 kN/m².

Formulas cannot be readily used to solve situations where the rupture surface passes partially above the phreatic surface and partially below the phreatic surface,

▶ **Example 25.1**

Given. A strip footing of 2.44 m width resting on dry soil having $\bar{\phi} = 30°$, $\gamma = 15.71$ kN/m³. The footing is 1.2 m below ground surface.

Find. $(\Delta q_s)_u$ for general shear.

Solution. Use Eq. 14.6. From Fig. 14.13b,

$$N_\gamma = 20$$

$$N_q = 22$$

$$(\Delta q_s)_u = \frac{(15.71)(2.44)(20)}{2} + (15.71)(1.2)(22)$$

$$= 383 + 415 = \underline{\underline{798 \text{ kN/m}^2}}$$ ◀

▶ **Example 25.2**

Given. The situation in Example 25.1 except the phreatic surface is at the bottom of the footing. The buoyant unit weight is 10.21 kN/m³.

Find. $(\Delta q_s)_u$ for general shear.

Solution. Again use Eq. 14.6, but modify unit weight. The bearing capacity factors are the same as in Example 25.1.

$$(\Delta q_s)_u = \frac{(10.21)(2.44)(20)}{2} + (15.71)(1.2)(22)$$

$$= 249 + 415 = 664 \text{ kN/m}^2$$ ◀

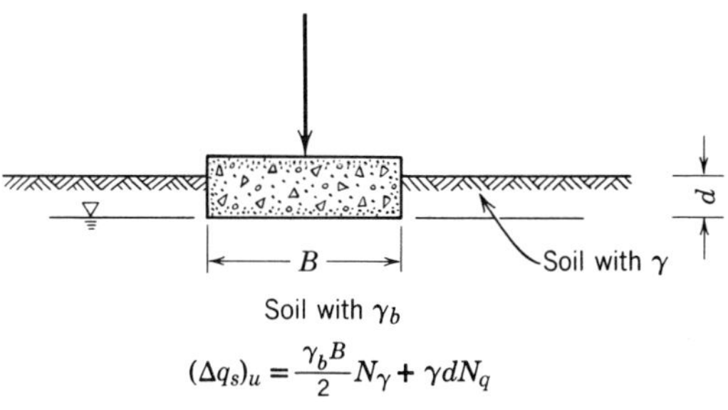

Fig. 25.6 Footing on saturated sand.

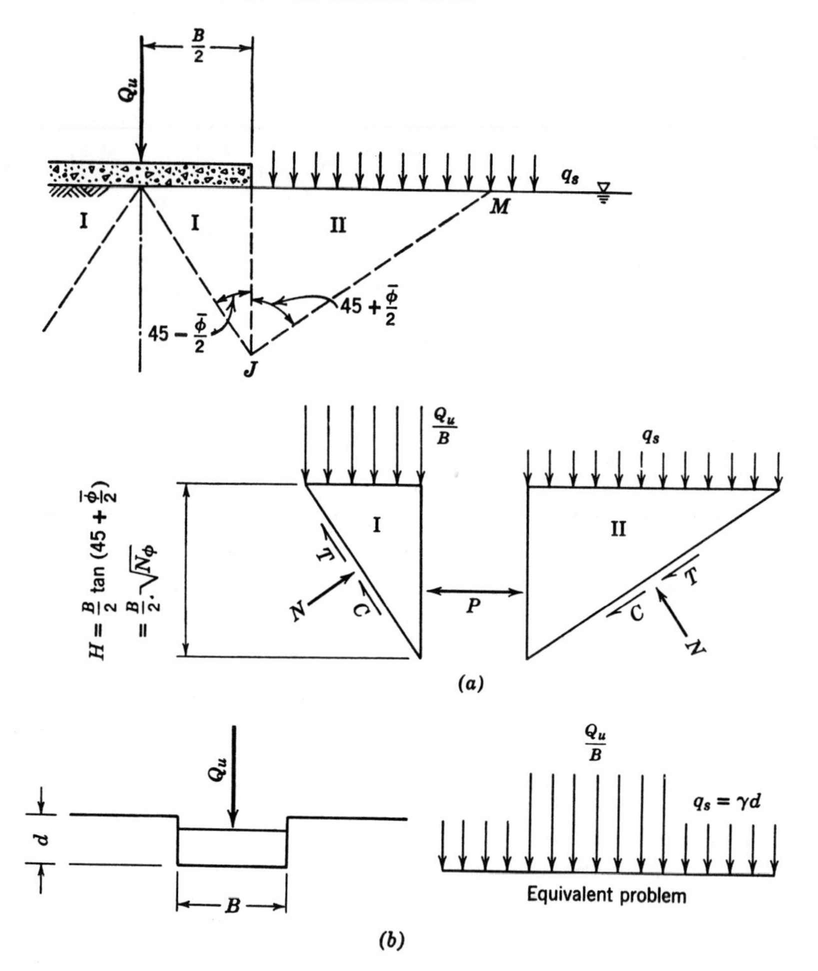

Fig. 25.7 Derivation of bearing capacity equation using Rankine wedges. (*a*) Derivation. (*b*) Equivalent problem.

Maximum force P that can be applied to passive wedge II:

$$P = P_{p\text{II}} = q_s H N_\phi + \tfrac{1}{2}\gamma_b H^2 N_\phi + 2\bar{c}H\sqrt{N_\phi}$$

$$P = q_s \frac{B}{2}(N_\phi)^{3/2} + \tfrac{1}{8}\gamma_b B^2 N_\phi^2 + \bar{c}BN_\phi \qquad (25.1)$$

Maximum Q_u/B that can be applied to active wedge I:

$$P = P_{a\text{I}} = \frac{Q_u}{B}\frac{H}{N_\phi} + \tfrac{1}{2}\gamma_b H^2 \frac{1}{N_\phi} - 2\bar{c}H\sqrt{\frac{1}{N_\phi}}$$

$$\frac{Q_u}{B} = \frac{2P}{B}\sqrt{N_\phi} - \tfrac{1}{4}\gamma_b B\sqrt{N_\phi} + 2\bar{c}\sqrt{N_\phi} \qquad (25.2)$$

Using Eq. 25.1:

$$\frac{Q_u}{B} = \frac{\gamma B}{4}(N_\phi^{5/2} - N_\phi^{1/2}) + 2\bar{c}(N_\phi^{3/2} + N_\phi^{1/2}) + q_s N_\phi^2 \qquad (25.3)$$

From Eq. 25.3

$$\frac{Q_u}{B} = \bar{c}\,N_c + \gamma_b\,B\frac{N\gamma}{2} + \gamma d N q \qquad (25.4)$$

$$\left.\begin{aligned} N_c &= 2[N_\phi^{3/2} + N_\phi^{1/2}] \\ \frac{N\gamma}{2} &= \tfrac{1}{4}[N_\phi^{5/2} - N_\phi^{1/2}] \\ Nq &= N_\phi{}^2 \end{aligned}\right\} \qquad (25.5)$$

or through a zone with capillary pressures, or through nonhomogeneous soil. For such complex situations, the safety against a shear rupture may be determined employing the principles for analyzing slopes, as presented in Chapter 24.

The present chapter, as well as all of Part IV, is limited to those situations in which the magnitude of the pore water pressures is known independently of the loads applied to the surface; i.e., where the pore pressures are controlled at some known level or where they are determined by ground water conditions. The pore water pressure in sands generally meets this condition because the permeability of sand is high enough to permit any excess pore pressures to dissipate during foundation loading. In clays, on the other hand, the permeability is generally so low that the foundation loading generates significant pore pressures. Thus the shear strength of an impermeable soil which is effective in resisting a shear rupture immediately following the placement of a foundation load is not the drained strength. Chapter 32 treats bearing capacity for undrained and partially drained situations.

Except in heavily overconsolidated clays, the bearing capacity for undrained loading is less than that for drained loading, and thus controls the foundation design. With heavily overconsolidated clays, settlement rather than bearing capacity usually controls design. Hence only a crude treatment of bearing capacity under drained conditions will suffice for most problems involving clay.

We can readily extend the bearing capacity formulas to include the effects of a cohesion intercept just as the formulas for the lateral forces on retaining structures were extended in Chapter 23 to include the effects of cohesion. Figure 25.7 presents the derivation of the bearing capacity equation based on the Rankine wedges. As was the case with Fig. 14.11, the derivation in Fig. 25.7 serves to illustrate how the foundation soil behaves to support the footing load. The resulting equation, Eq. 25.3, is too approximate to be of practical value.

The Terzaghi bearing capacity equation, Eq. 14.6, extended to include a cohesion intercept is

$$(\Delta q_s)_u = \bar{c}N_c + \frac{\gamma B}{2}N_\gamma + \gamma d N_q \qquad (25.6)$$

Equation 25.6 consists of Eq. 14.6 with the addition of the cohesion term $\bar{c}N_c$. Figure 14.13*b* gives the values of the bearing capacity factors N_c, N_γ, and N_q as a function of friction angle $\bar{\phi}$. Example 25.3 illustrates the use of Eq. 25.6.

As discussed in Chapter 22, loose sand and normally consolidated clay have the same type of stress-strain behavior, and dense sand and overconsolidated clay also have similar stress-strain behavior. Thus the discussion in Chapter 14 on local shear and general shear for sands can be appropriately used for clay. We would therefore expect that in a normally consolidated clay, just as in a loose sand, local shear would be reached at a much lower value than the ultimate. On the other hand, in overconsolidated clay, just as in dense sand, the difference between the bearing capacity based on local and that based on general shear would be quite similar. The following equation considers the effect of local shear and may be used to find the bearing capacity $(\Delta q_s)_b$:

$$(\Delta q_s)_b = f\bar{c}N_c + \frac{\gamma B}{2}N_\gamma + \gamma d N_q \qquad (25.7)$$

where N_c, N_γ, and N_q are obtained from Fig. 14.13 and f is a factor varying between 1 for a stiff clay to $\frac{2}{3}$ for a soft clay.

▸Example 25.3

Given. The situation in Example 25.2 except the soil has a value of $\bar{c}$ = 14.4 kN/m².

Find. $(\Delta q_s)_u$ for general shear.

Solution.

$$(\Delta q_s)_u = \bar{c}N_c + \frac{\gamma B N_\gamma}{2} + \gamma d N_q \qquad (25.4)$$

Same bearing capacity factors as in Example 25.2:

$$(\Delta q_s)_u = (14.4)(37) + \frac{(10.21)(2.44)(20)}{2} + (15.71)(1.2)(22)$$

$$(\Delta q_s)_u = 533 + 249 + 415 = 1197 \text{ kN/m}^2$$ ◂

25.4 METHODS FOR PREDICTING SETTLEMENT

The general principles and methods for predicting settlement given in Chapter 14 hold for saturated sand as well as for dry sand, as long as the soil stresses used to make the prediction are effective stresses. Since saturating a dry sand reduces the effective stress at any depth, it increases the compressibility of the sand and thus the settlement of any overlying structure.

Example 25.4 illustrates the influence of saturation on the settlement of a structure overlying sand. For the case with dry sand to an infinite depth, the predicted settlement of the footing was 37 mm. Saturating the sand by bringing the phreatic surface to the bottom of the footing resulted in a predicted settlement of 46 mm, an increase of 25%.

The empirical methods for predicting settlement, described in Section 14.10, can be used with saturated sands as well as with dry sands, although the interpretation of load test data and penetration test data from saturated soil can be more complicated.

Example 25.5 illustrates a not uncommon situation. The small load test plate obtains its settlement from strains in soil above the phreatic surface, whereas the

▶Example 25.4

Given. A rigid, round footing, 3 m in diameter, loaded to 240 kN/m², resting on the surface of sand with $\mu = 0.45$ and $E = 9580$ kN/m² at $\bar{\sigma}_{v0} = 24$ kN/m². E varies as $\sqrt{\bar{\sigma}_{v0}}$.

Find. The settlement of the center of the footing top:

a. Dry sand, $\gamma = 17.28$ kN/m³.

b. Saturated sand, $\gamma_t = 20.82$ kN/m³ and water table at the bottom of the footing.

Solution.

$$\rho = \Delta q_s \frac{R}{E}\frac{\pi}{2}(1 - \mu^2) = \frac{240 \times 1.5 \times \pi(0.797)}{E \times 2} = \frac{451}{E}$$

Case *a*:

$$\bar{\sigma}_{v0} \text{ at ave. pt.} = 17.28 \text{ kN/m}^3 \times 2.25 \text{ m} = 38.9 \text{ kN/m}^2$$

$$E = 9580 \frac{\text{kN}}{\text{m}^2}\sqrt{\frac{38.9}{24}} = 12\,196$$

$$\rho = \frac{451}{12\,196} = \underline{\underline{0.037 \text{ m}}}$$

Case *b*:

$$\bar{\sigma}_{v0} \text{ at ave. pt.} = (20.82 - 9.81)2.25 = 24.8 \text{ kN/m}^2$$

$$E = 9580\sqrt{\frac{24.8}{24}} = 9738$$

$$\rho = \frac{451}{9738} = \underline{\underline{0.046 \text{ m}}}$$ ◀

settlement of the actual footing comes primarily from strains in soil below the phreatic surface. As can be seen from Example 25.5, capillary pressures above the phreatic surface can contribute significantly to the effective stresses in the bulb of stress of the load test plate. The capillary stresses are of negligible significance to the actual footing.

More care is needed in running penetration tests in soils below the water table than in dry sands. For example, if in the standard penetration test the water is removed from the casing (as is common), a large gradient and even a quick condition can result in the soil at the location of the penetration test device. The upward gradient reduces the strength of the soil and can give an erroneously low penetration resistance.

There are situations in which saturating a sand can cause settlements far in excess of that suggested by Example 25.4. It is not uncommon for a wind laid soil (loess) or residual soil to have its particles cemented together with soluble materials or with clay. Water coming into such a soil, either from surface runoff or from a rise in the water table, can destroy the interparticle cementation with a resulting collapse of the soil structure. Terzaghi and Peck (1967) discuss large settlements and slope failures in loess due to submergence. Holtz and Hilf (1961) discuss similar problems associated with canals, and Jennings and Knight (1957) and Brink and Kantey (1961) describe large settlements which have occurred in Southern Africa due to collapse of the soil skeleton from wetting.

For certain conditions (as follows), the theoretical procedures of Chapter 14 can be used to predict the settlement of structures founded on clay.

1. The load from the structure is applied slowly compared to the rate at which consolidation can occur. A drained test should be used to obtain the stress-strain data.
2. The degree of saturation is low enough (usually less than 75%) so that no water is squeezed out of the soil by the loading. There is no standard laboratory testing technique for use in this case, and the factors affecting the test results are poorly understood (see Jennings and Burland, 1962).
3. The initial settlement is obtained when the load is applied so quickly that no consolidation occurs during load application. This case is discussed in Chapter 32.
4. The final settlement is obtained in cases where the safety factor for undrained loading is so large (say, greater than 4) that no yielding occurs during undrained loading; i.e., problems where the final settlement is independent of the stress path. There are two such common situations:

   a. Heavily overconsolidated clays. Results from drained tests should be used to estimate the final settlement, although the modulus E from drained and undrained tests are usually similar for such soils.

   b. Where strains are primarily one-dimensional so that strains result only from volume changes. This case is discussed in the next section.

Normally, however, buildings are not placed on a deep deposit of soft clay because the resulting settlements would be too large to be tolerated. Common situations involving soft clay are: (*a*) a structure overlying a relatively thin stratum of clay; (*b*) a structure resting on a stronger soil, which in turn overlies a compressible clay. For these situations the resulting settlements, although still large, are within reason. The next section presents methods that are useful for predicting settlements for these two conditions. These methods can also be used to study the settlement of a soil deposit during deposition under its own weight, and can be used to approximate crudely the settlement of buildings which are founded on a thick deposit of compressible clay.

25.5 SETTLEMENT AND HEAVE FROM ONE-DIMENSIONAL STRAIN

This section considers vertical movement—settlement or heave—that results from one-dimensional strain. As

▶ **Example 25.5**

Given. The settlement from the load test (Fig. E25.5) = 0.0015 m.

Find. Predicted settlement for actual footing in Fig. E25.5.

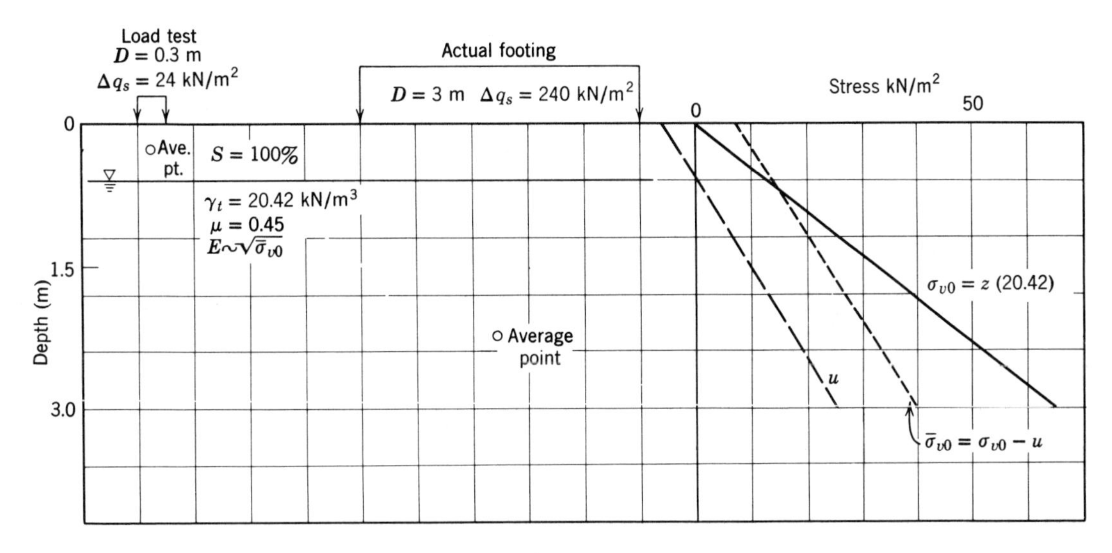

Fig. E25.5

Solution.

$$\rho = \Delta q_s \frac{R}{E}\frac{\pi}{2}(1-\mu^2) = \Delta q_s \frac{R}{E}(1.25)$$

From load test:

$$E = \frac{24 \text{ kN/m}^2 \times 0.15 \text{ m} \times 1.25}{0.0015 \text{ m}} = 3000 \text{ kN/m}^2$$

$$\bar{\sigma}_{v0} \text{ at } d = 0.225 \text{ m} \quad (20.42)(0.225) + (9.81)(0.375) = 8.3$$

For actual footing:

$$\bar{\sigma}_{v0} \text{ at } d = 2.25 \text{ m} \quad (20.42)(2.25) - (9.81)(1.65) = 29.8$$

$$E = 3000\left(\frac{29.8}{8.3}\right)^{1/2} = 5684 \text{ kN/m}^2$$

$$\rho = \Delta q_s \frac{R}{E}(1.25) = \frac{240 \times 1.5 \times 1.25}{5684} = 0.079 \text{ m}$$

we have noted several times in this book, the effective stress path for one-dimensional strain is approximately a straight line, the K_0-line. Figure 25.8*a* shows the stress path *IF* for a soil element undergoing one-dimensional strain. During the compression from *I* to *F*, the element of soil is both decreasing in volume and changing in shape in such a fashion that the lateral expansion from the change in element shape is exactly balanced by the lateral compression from the change in volume; this is one-dimensional strain.

One-dimensional settlement can occur when the boundary conditions of the subsoil and the conditions of applied stress impose a K_0 effective stress path on the average soil element. Figure 25.8*b* shows two situations in which the boundary conditions force a one-dimensional settlement. The case at the left in Fig. 25.8*b*,[1] which is hardly likely to occur in the field, is similar to the laboratory oedometer test. The case shown at the right in Fig. 25.8*b* resembles many field problems. Even though a deposit may be experiencing two- or three-dimensional strain, a particular point may happen to experience only one-dimensional strain. As suggested in Fig. 25.8*c*, a soil element at an initial K_0 condition can be loaded by some nonuniform pattern of surface load in

[1] In this case and in the oedometer test, all of the applied load does not go straight down into the soil; some of it spreads into the rock or the oedometer ring. The case with infinite lateral extent is thus more nearly one-dimensional, since both vertical stress transmission and strain are one-dimensional.

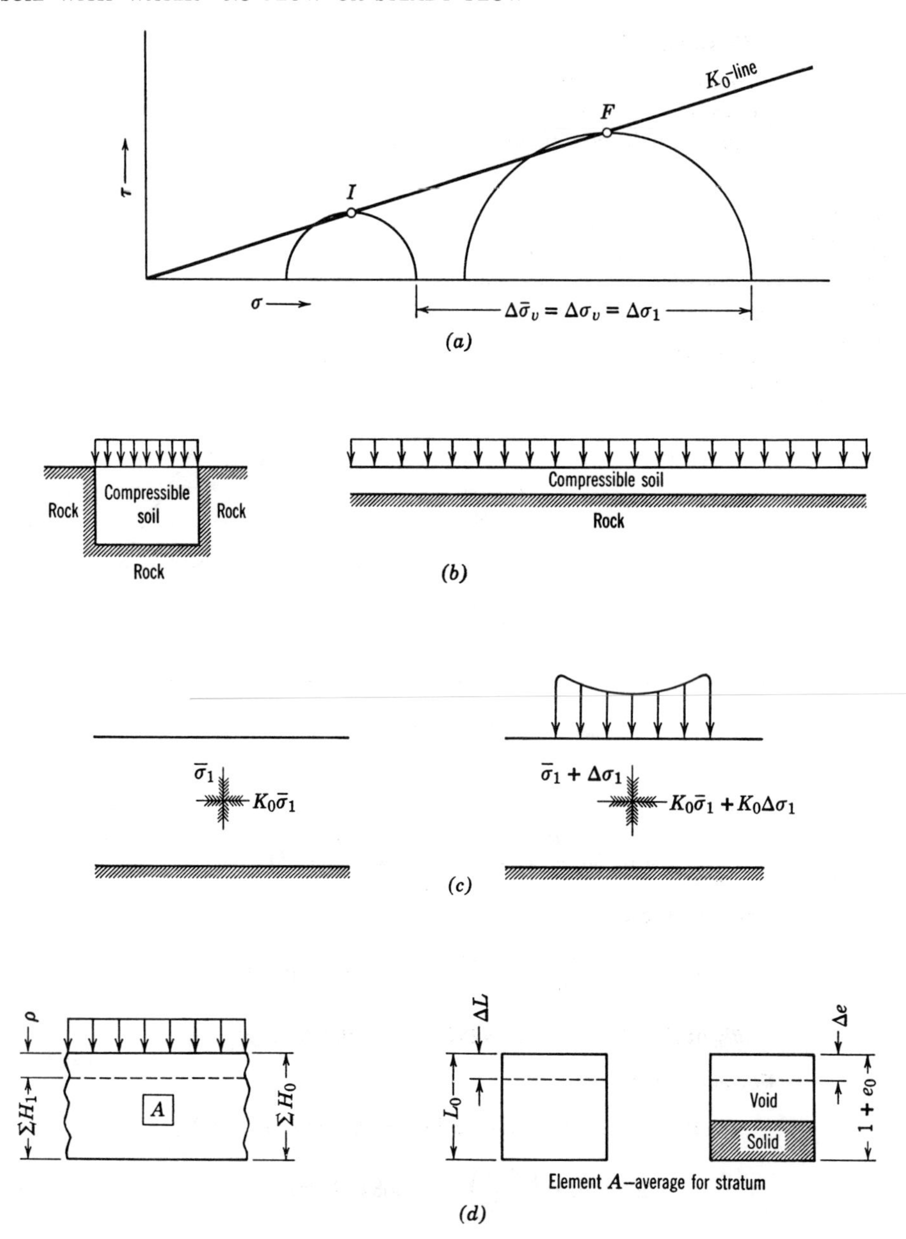

Fig. 25.8 One-dimensional settlement. (*a*) One-dimensional strain. (*b*) One-dimensional settlement imposed by boundary conditions. (*c*) One-dimensional settlement from K_0 stresses. (*d*) Settlement $= \rho = \Sigma H_0 - \Sigma H_1 = (\Delta L/L_0)\Sigma H_0 = [\Delta e/(1 + e_0)]\Sigma H_0$.

such a way that the stresses induced in the average element of the compressing layer keep the soil in a K_0 condition. If this point should happen to be the average or typical point for the entire deposit, then the settlement can be estimated using the procedures outlined in this section.

If the strain at a point is one-dimensional, then these strains must result from volume changes. With a saturated soil such volume changes can occur only if water flows to or from the soil. In practice, this situation may come about in two ways:

1. When the loading is applied so slowly that full drainage occurs.
2. During the consolidation that follows an undrained loading.

Since the final compression under a given increment of load generally is almost the same for both types of loading, the following procedures apply to both loadings.

Figure 25.8*d* shows a layer of compressible soil undergoing one-dimensional compression. The layer of original thickness ΣH_0 is compressed to a final thickness of ΣH_1 with a settlement of ρ which is equal to the original thickness minus the final thickness. From this relationship we can develop the following expressions for settlement (refer back to Chapter 12 for discussion of m_v, a_v, and C_c):

$$\rho = \sum H_0 - \sum H_1$$

$$\rho = \sum H_0 \epsilon_{\text{vertical}} = \sum H_0 \epsilon_{\text{volume}} \tag{25.8}$$

where

$$\epsilon_{\text{vertical}} = \frac{\Delta L}{L_0} = \epsilon_{\text{volume}} = -\frac{\Delta e}{1 + e_0}$$

or

$$\rho = \sum H_0 m_v \Delta\bar{\sigma}_v \tag{25.9}$$

where

$$m_v = \text{coefficient of volume change} = -\frac{\Delta e}{(1 + e_0)} \frac{1}{\Delta\bar{\sigma}_v}$$

$$= -\frac{\epsilon_{\text{volume}}}{\Delta\bar{\sigma}_v}$$

or

$$\rho = \frac{\sum H_0}{1 + e_0} a_v \Delta\bar{\sigma}_v \tag{25.10}$$

where

$$a_v = \text{coefficient of compressibility} = -\frac{\Delta e}{\Delta\bar{\sigma}_v}$$

$$\rho = \frac{\sum H_0}{1 + e_0} C_c \log_{10}\left(\frac{\bar{\sigma}_{v0} + \Delta\bar{\sigma}_v}{\bar{\sigma}_{v0}}\right) \tag{25.11a}$$

where

$$C_c = \text{compression index} = -\frac{\Delta e}{\Delta \log_{10} \bar{\sigma}_v}$$

$$\bar{\sigma}_{v0} = \text{initial vertical effective stress}$$

When $\Delta\bar{\sigma}_v \ll \bar{\sigma}_{v0}$, this may be rewritten approximately as

$$\rho = \frac{\sum H_0 \ \Delta\bar{\sigma}_v}{(1 + e_0) \ \bar{\sigma}_{v0}} 0.435 \ C_c \tag{25.11b}$$

In these equations L refers to length, e to void ratio; the subscript 0 refers to the initial condition and 1 to the final condition as far as the thickness of the clay is concerned.

Example 25.6 is a one-dimensional settlement problem arising from the placement of a 4.5 m thick layer of fill over a large lateral area. As shown in the example, all of the equations for settlement give the same answer except for Eq. 25.11*b*. The expression for settlement in terms of C_c is approximate.

In Example 25.6 both the stress distribution and the strain are one-dimensional because of the large lateral extent of the fill relative to the thickness of the compressible layer. An element at mid-depth of the clay was selected as the average for the clay layer. Actually, the average element occurs slightly above mid-depth since the $e - \log \bar{\sigma}_v$ plot is straight, i.e., more strain occurs in the upper half of the clay layer than in the lower half. The virgin compression portion of the oedometer test was used to compute settlement even though the laboratory test indicates a slight overconsolidation. The virgin compression curve was used since a study of the geology at the site of Example 25.6 indicated that the clay was normally consolidated. Also, the soil sample was obtained from a depth slightly greater than the mid-depth of the clay.

Example 25.7 illustrates the computation of heave for one-dimensional situations. The problem consists of determining the heave that will occur on the removal of the 4.5 m fill placed in Example 25.6. As is generally true, the heave is much less than the settlement. As indicated in the oedometer test results of Example 25.6, recompression results in a much smaller settlement than virgin compression. This fact is the basis of the technique of "preloading." By using the fill in Example 25.6 to preload the soil, most of the settlement occurred before a structure was placed. Consolidation of the clay under the fill resulted in a settlement of 0.65 m; removal of the fill resulted in a heave of 0.076 m; replacement of a structure load equal to that of the 4.5 m of fill would have caused a settlement only slightly larger than the 0.076 m of heave.

Example 25.8 illustrates a very common situation: settlement or heave resulting from the lowering or raising of the ground water table. In the example the phreatic surface in the subsoil of Example 25.6 is lowered 3 m from elevation −2.9 to elevation −5.9. If the dewatering causes no change in degree of saturation and/or void ratio of the overlying soil, the unit weight of the silt does not change and thus the total stresses throughout the clay remain constant. The pore pressure, however, has been reduced by the ground water lowering. Figure E25.8-1 shows the initial (static) pore pressure distribution. In order to determine the equilibrium pore pressures following the ground water lowering, the engineer must know the equilibrium flow conditions. Since the clay in the example is underlain by sand, the pressure head at the bottom of the clay will remain at its static value, 8.7 m. The pore pressure distribution throughout the silt at equilibrium will be essentially hydrostatic because the permeability of the silt is so much greater than the clay. For the condition of constant permeability through the clay, there is a straight line distribution of pore pressure for steady pumping, as noted in the figure. The pressure head has decreased 3 m at the top of the clay, remained unchanged at the bottom of the clay, and decreased 1.5 m at mid-depth. A pore pressure change of 1.5 m equals 14.7 kN/m². For this change in pore pressure there will be an equal change in effective stress causing a settlement of 0.152 m. In Fig. E25.8-2 is shown the stress paths for the water table lowering. The effective stress path for the lowering is

▶ Example 25.6 One-Dimensional Settlement

Problem. A 4.5 m depth of fill is placed over a large lateral area having the profile shown in Fig. E25.6-1. Determine the component of settlement due to compression of the clay. (This example is based on an actual field case in Lagunillas, Venezuela. The case is described in Lambe, 1961.)

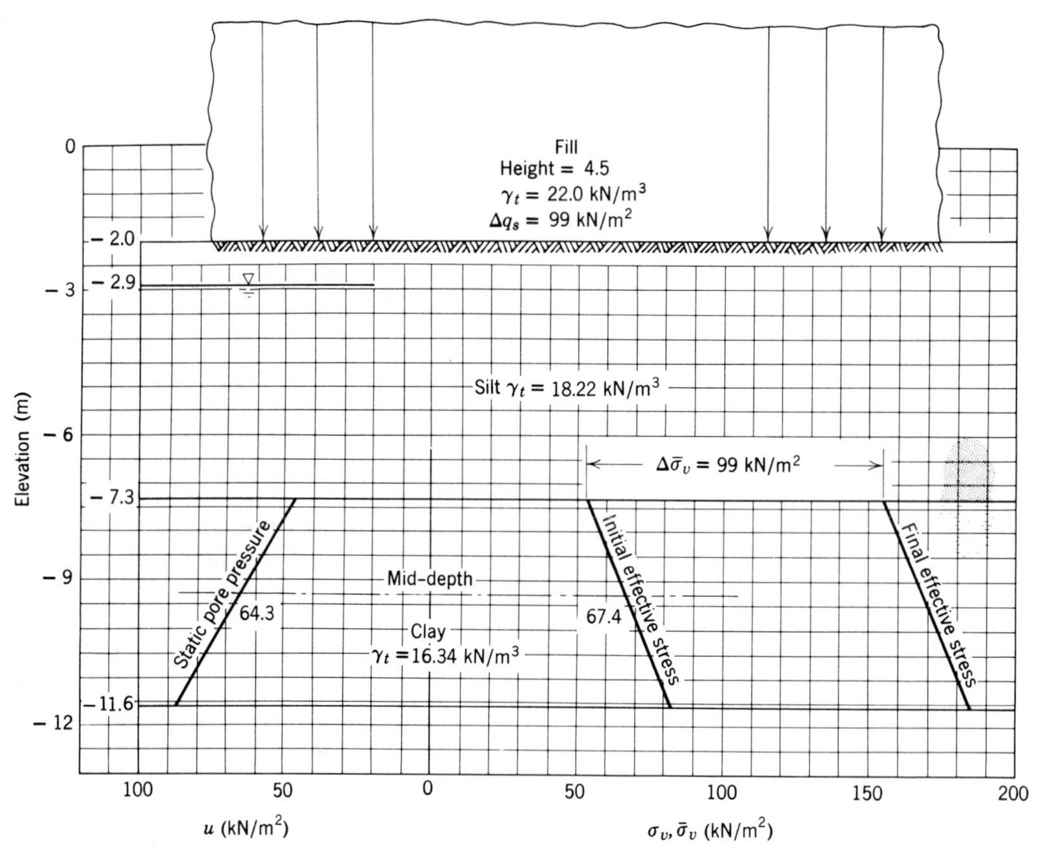

Fig. E25.6-1 Subsoil profile.

Stress Calculations.
Initial stresses:

Elevation (m)	Description	$\Delta\sigma_v$	$\bar{\sigma}_v$ (kN/m²)	u (kN/m²)	$\bar{\sigma}_v$ (kN/m²)
−2.0	Ground surface		0		
		5.3 × 18.22 = 96.6			
−7.3	Silt-clay boundary		96.6	43.2	53.4
		2.15 × 16.34 = 35.1			
−9.45	℄ clay		131.7	64.3	67.4

Increment of stress:

$$\Delta\sigma_v = \Delta q_s = 4.5 \text{ m} \times 22.0 \text{ kN/m}^3 = 99.0 \text{ kN/m}^2$$

Stresses @ ℄ clay:

$$\bar{\sigma}_{v0} = 67.4 \text{ kN/m}^2$$

$$\Delta\bar{\sigma}_v = 99.0 \text{ kN/m}^2$$

$$\bar{\sigma}_{v1} = 166.4 \text{ kN/m}^2$$

Lab Compression Test. Figure E25.6-2 shows results of lab oedometer test. Since clay is normally consolidated the virgin compression curve is used in problem.

Example 25.6 *(continued)*

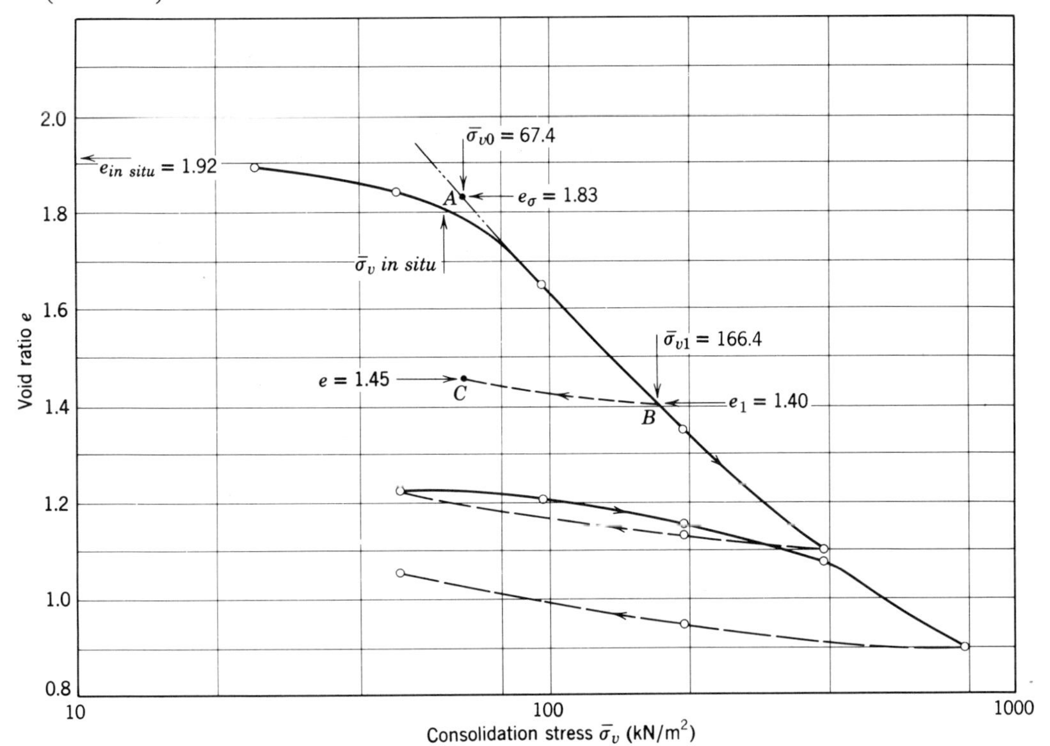

Fig. E25.6-2 Oedometer test results—Lagunillas clay.

Settlement Calculation.

$$\rho = \frac{\sum H_0}{1 + e_0} \Delta e$$

$$= \frac{4.3}{1 + 1.83} (1.83 - 1.40) = 0.653 \text{ m}$$

Alternate settlement calculations:

1. $$\rho = \sum H_0 m_v \Delta\bar{\sigma}_v = 4.3\left(-\frac{\Delta e}{1 + e_0} \cdot \frac{1}{\Delta\bar{\sigma}_v}\right) \Delta\bar{\sigma}_v = 0.653 \text{ m}$$

2. $$\rho = \frac{\sum H_0}{1 + e_0} a_v \Delta\bar{\sigma}_v = \frac{4.3}{2.83}\left(-\frac{\Delta e}{\Delta\bar{\sigma}_v}\right) \Delta\bar{\sigma}_v = 0.653 \text{ m}$$

3. $$\rho = \frac{\sum H_0}{1 + e_0} 0.435 C_c \cdot \frac{\Delta\bar{\sigma}_v}{\bar{\sigma}_{v,\text{ave}}} = \frac{4.3}{2.83} \times \frac{1.025}{1.204} \times 0.435 \times 1.055$$

$$= 0.588 \text{ m}$$

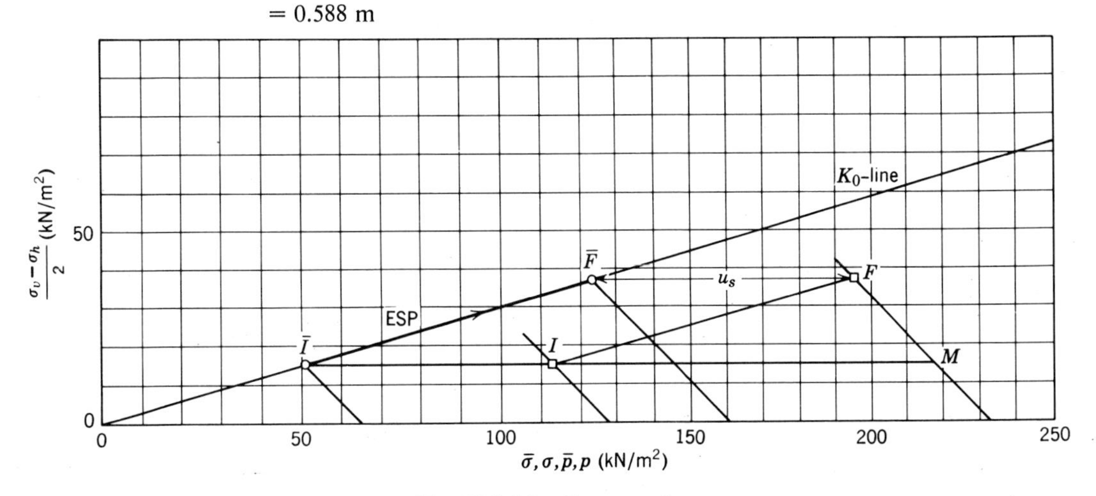

Fig. E25.6-3 Stress paths. ◄

▶ **Example 25.7 One-Dimensional Heave**

Problem. The 4.5 m fill in Example 25.6 is removed. Determine the component of heave due to expansion of the clay.

Solution.
Stress decrement:

$$\Delta\bar{\sigma}_v = 99.0 \text{ kN/m}^2$$

Lab test. On the $e - \log \bar{\sigma}_v$ plot in Example 25.6 a line BC has been drawn parallel to expansion lines from 4 to 2 and from 8 to 2.
Heave calculation:

$$\rho\uparrow = \frac{\Sigma H_0}{1 + e_0}\Delta e = \frac{4.3 - 0.653}{1 + 1.40} \times 0.05$$
$$= 0.076 \text{ m}$$ ◀

$\overline{IF}$. Since the lateral extent of the clay is very large and the dewatering occurred over a large area, the strains are one-dimensional and the effective stress path must proceed along the K_0 line. This imposed field boundary condition means that the vertical total stress remained constant but that the lateral total stress reduced, as shown in Fig. E25.8-2.

Example 25.8 is properly handled as a one-dimensional settlement problem. The engineer should not, however, assume that all dewatering problems are one-dimensional. In the case of a building excavation the dewatering may take place over a small area and field conditions therefore do not impose one-dimensional strain. Dewatering can contribute lateral as well as vertical strains.

Settlement and area subsidence from the reduction of pore fluid pressures from pumping of water or oil is very common. In Mexico City and Tokyo, for example, very large settlements have occurred over extensive areas because of water pumping.

25.6 SUMMARY OF MAIN POINTS

1. The general principles of the behavior of shallow foundations on dry sand and the methods of predicting their settlement, presented in Chapter 14, apply to saturated sand as long as soil stresses are expressed as effective stresses.
2. The bearing capacity equations for sand can be extended, by adding a cohesion term, to handle drained conditions in clay. The bearing capacity of clay is usually controlled by undrained or partially drained conditions (treated in Chapter 32).
3. The theoretical methods for predicting settlement, presented in Chapter 14, can be used to predict the settlement of clay for either a drained loading or an undrained loading. The appropriate type of modulus must be employed. Where the safety factor during loading is very large, the final settlement after an undrained loading followed by consolidation can be predicted using data from drained loadings.
4. Vertical movement, settlement, and heave involve essentially one-dimensional strain for situations where the loaded area is large relative to the thickness of the compressible soil. One-dimensional movement requires volume change within the compressing soil, and can be predicted from the simple expressions given in this chapter.

PROBLEMS

25.1 A strip footing 1.5 m wide rests on dry soil having: $\bar{\phi} = 25°$, $\gamma = 15.4$ kN/m³, $G = 2.75$, and E varies as $\sqrt{\bar{\sigma}_{v0}}$. The footing is 0.9 m below ground surface. Determine the settlement and reduction of bearing capacity (for general shear) resulting from a rise in the water table (and saturation) to the bottom of the footing.

25.2 Compute the settlement for Example 25.8 if the soil below elevation -11.6 m is rock and has a permeability 1/100 times the permeability of the overyling clay.

▸ Example 25.8 Settlement Caused by Water Table Lowering

Problem. The phreatic surface in the Lagunillas subsoil is to be lowered from elevation −2.9 to −5.9 (no fill has been placed). Determine the settlement due to clay compression.

Solution. The initial and final pore pressures appear in Fig. E25.8-1.

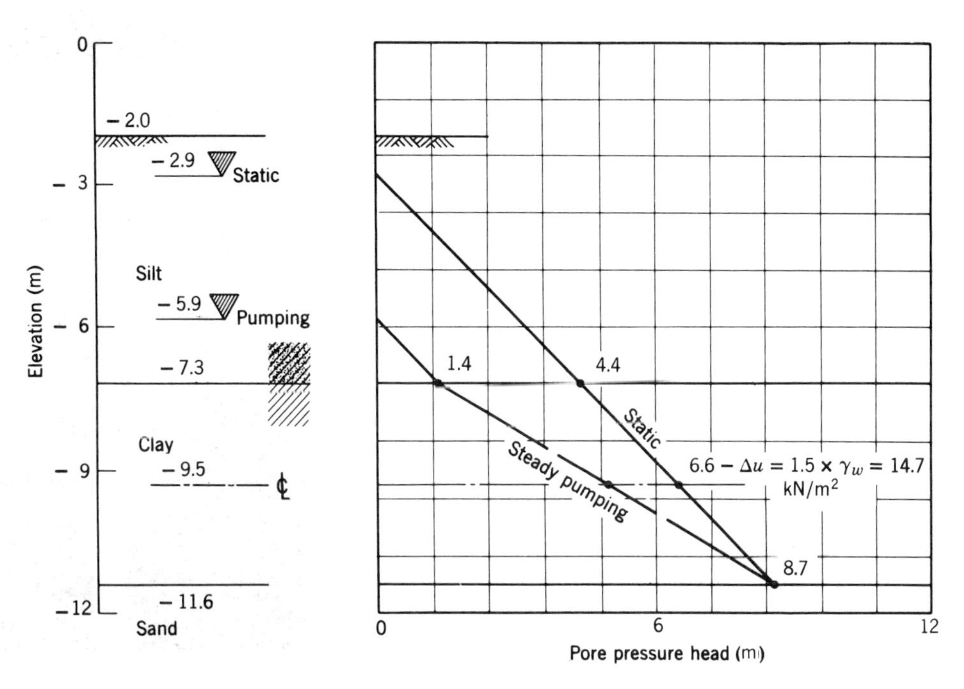

Fig. E25.8-1

Stress increment:

$$\Delta\bar{\sigma}_v = -\Delta u = 1.5 \times 9.81 = 14.7 \text{ kN/m}^2$$

The corresponding stress path is $\bar{I}\bar{F}$ in Fig. E25.8-2.

Settlement calculation:

$$\rho = \frac{\sum H_0}{1 + e_0} \Delta e = \frac{4.3}{2.83}(1.83 - 1.73) = 0.152 \text{ m}$$

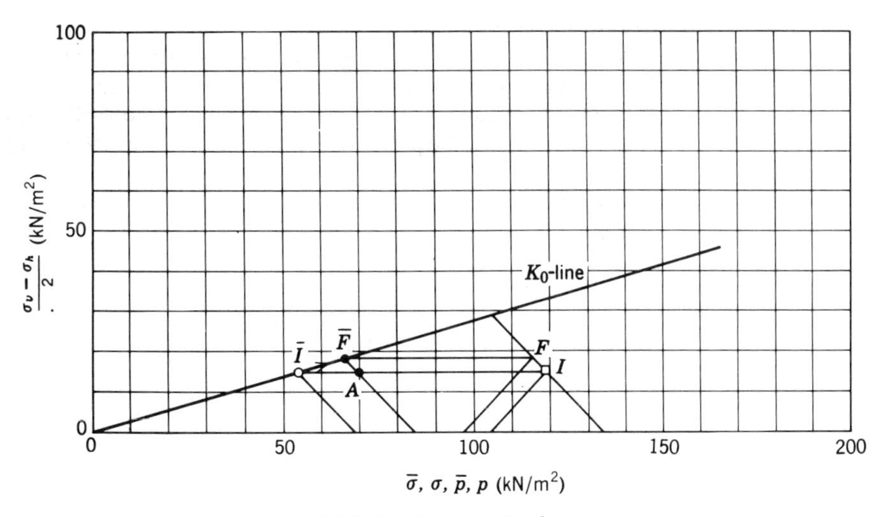

Fig. E25.8-2 Stress paths for water.

LAURITS BJERRUM

Dr. Bjerrum was born August 6, 1918 in Farsö, Denmark. He did his undergraduate work at the Technical University of Denmark and his graduate work at the Federal Institute of Technology, Zürich, Switzerland.

After working in Denmark and Switzerland, Dr. Bjerrum went to Norway in 1951 to become the first Director of the new Norwegian Geotechnical Institute. Dr. Bjerrum successfully built NGI into the best institute of its type in the world. From Dr. Bjerrum and his staff at NGI have come many publications. Their work on the fundamentals of shear strength, especially on sensitive clays, and the stability of natural slopes is classic.

An excellent speaker, Dr. Bjerrum has lectured at a number of universities in the United States and in Europe, especially at M.I.T. and the University of Illinois. Dr. Bjerrum was President of the International Society of Soil Mechanics and Foundation Engineering during the period of 1965 to 1969. He was a Rankine Lecturer and a Terzaghi Lecturer.

Dr. Bjerrum died suddenly on February 27, 1973, while in England to attend a Rankine Lecture.

PART V

Soil with Water-Transient Flow

Part IV considered the interaction between pore fluid and mineral skeleton and demonstrated the importance of *effective stress*. Effective stress determines shear strength. Change in effective stress determines the magnitude and type of volume change.

A key step in the evaluation of effective stress is the evaluation of the pore water pressure. In all of Part IV, pore water pressure was determined solely by the hydraulic boundary conditions applied to the pore phase. Pore pressure was in no way influenced by the weight of the soil or by external loads applied to the soil. Part V treats those situations in which pore pressure *is* influenced by loads applied to the soil mass. As discussed in Chapter 2, when a load change is suddenly applied to a soil, the change is carried either by the pore fluid or jointly by the pore fluid and by the mineral skeleton. The change in pore pressure will cause water to move through the soil, and the effective stresses within the soil will change with time. Hence the properties of the soil will change with time.

Part V treats the most general type of problem and thus ties together many of the concepts presented in previous sections.

CHAPTER 26

Pore Pressure Developed during Undrained Loading

Chapter 2 presented and illustrated an essential concept in soil mechanics: a load applied to an element of soil is carried partly by the pore phase and partly by the soil skeleton. If we confine a saturated sample of soil in an oedometer, as shown in Fig. 26.1, and apply an increment of vertical stress $\Delta\sigma_1$, we find that an increment in pore pressure Δu is built up. The pore pressure within the sample now is no longer in equilibrium with the pore pressures at the boundary of the sample, hence fluid flow commences and the additional pore pressure induced by the loading dissipates. Since pore pressure, and thus total head, is changing with time during the fluid flow, this flow is transient (or unsteady).

26.1 UNDRAINED LOADINGS

In many problems, it is possible to separate the effects of a loading into two distinct phases:

1. *Undrained loading*, during which an increment of pore pressure is developed but there is no flow of pore fluid, i.e., no change in the water content of the soil.
2. *Dissipation*, during which the total load applied to the soil remains constant and fluid flow occurs to dissipate the additional pore pressure.

This idealized situation can easily be created in a laboratory triaxial test by keeping the drainage line from the sample closed as an increment of load is applied and then by opening the drainage line to permit dissipation. This idealized situation also is often realized in the field whenever the interval during which load is applied is very short compared to the time required for dissipation of the pore pressures. As discussed in Chapter 2, this condition frequently occurs with clays. Of course there are also many practical problems, especially those involving silts, where dissipation begins while the total load still is being changed and continues after the final load is reached. In these problems the effects of a loading cannot be separated into two distinct stages. With coarse granular soils, dissipation generally occurs so quickly that no measurable additional pore pressure actually develops.

Figure 26.2 illustrates an undrained loading followed by dissipation during constant load. Prior to the loading, the pore pressures are hydrostatic, as indicated by the line marked u_s. The total head is the same at all depths and there is no flow of water. Within the clay, the loading causes an *increment of pore pressure* Δu and a corresponding increase in the total head. Because of fast rate of dissipation within the fill, sand and gravel, the pore pressure and total head within these strata are not changed by the loading. Hence a gradient exists at the top and bottom surfaces of the clay and flow begins as a result of these gradients and the pore pressures within the clay begin to decrease. Assuming that the location of the water table remains unchanged, the final or *steady-state pore pressures* u_{ss} are in this problem just the same as the *static pore pressures* u_s. Transient flow will continue as long as the *excess pore pressure* u_e, where

$$u_e = u - u_{ss}$$

differs from zero. In this problem, the *initial excess pore pressure* u_0 is just equal to the increment of pore pressure Δu during the undrained loading.

In order to estimate the strength and compressibility at any given point in the subsoil at any given time, we must know the effective stress at the point at that time. To evaluate the effective stress we must in turn know the pore pressure at the point at that time. In general this means that we must know: (*a*) the initial pore pressure immediately following the change in boundary water conditions or change in total stress; (*b*) the final equilibrium pore pressure; and (*c*) the pore pressure during the intervening transient condition. The final equilibrium condition, where pore pressure is determined solely by hydraulic boundary conditions, has already been discussed in Part IV. Chapter 26 treats in detail the

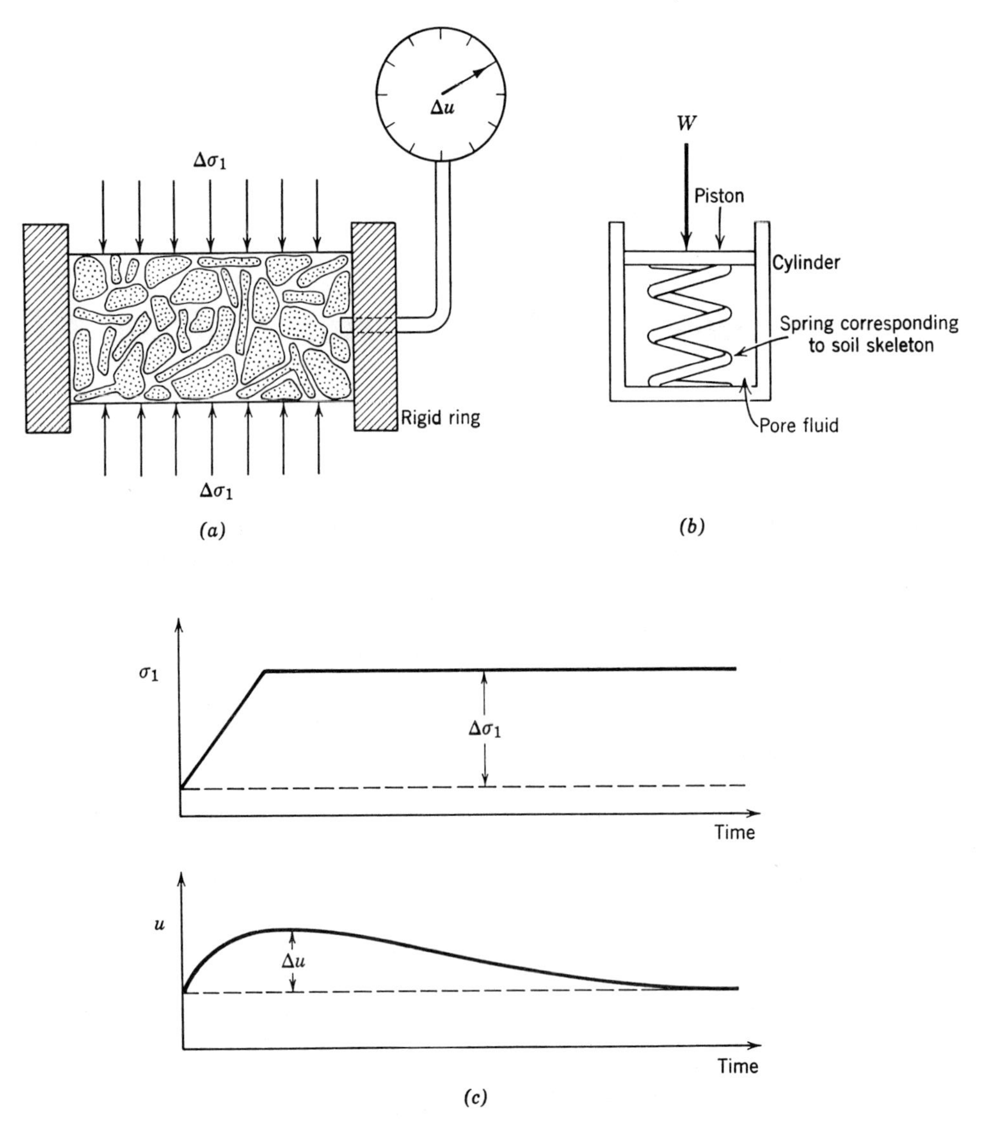

Fig. 26.1 Loading in oedometer. (*a*) Soil-water system. (*b*) Spring analogy.

initial pore pressures caused by a change in total stress. Chapter 27 discusses the transient change in pore pressure from the initial to the equilibrium condition.

Pore Pressure Parameters

Figure 26.1*b* shows a spring analogy of our soil-water system in the oedometer. The spring corresponds to the soil skeleton and the water corresponds to the pore fluid. When the load W is applied to the piston with no escape of water permitted, part of W is carried by the spring and part is carried by the water. Intuitively, we would expect nearly all of W to be carried by the water and very little carried by the spring. Similarly, we would expect that most of the increment of stress $\Delta\sigma_1$ would be carried by the pore pressure Δu. Experimental data show that this intuitive expectation is indeed correct.

It is convenient to express the pore pressure built up by a change in total stress as a ratio, $\Delta u/\Delta\sigma$. Such a ratio of pore pressure increment to total stress increment is termed a *pore pressure parameter*. Figure 26.3 illustrates the type of data that would be obtained if the pore pressure were measured in an oedometer test. The slope of the u versus σ_1 line is the pore pressure parameter C, equal to $\Delta u/\Delta\sigma_1$. For the oedometer test on saturated soil the plot of u versus σ_1 is essentially a straight line at a slope of 45°; thus essentially all of $\Delta\sigma_1$ is carried by the pore pressure.

The direct and practical way to determine a pore pressure parameter is to apply the type of stress system of interest, measure the developed pore pressure, and then divide the pore pressure increment by the total stress increment. Thus C would be obtained from $\Delta u/\Delta\sigma_1$ where Δu and $\Delta\sigma_1$ would be measured from the setup shown in Fig. 26.1*a*. The following four sections derive expressions that can also be used to determine the parameters. The purpose of these derivations is not,

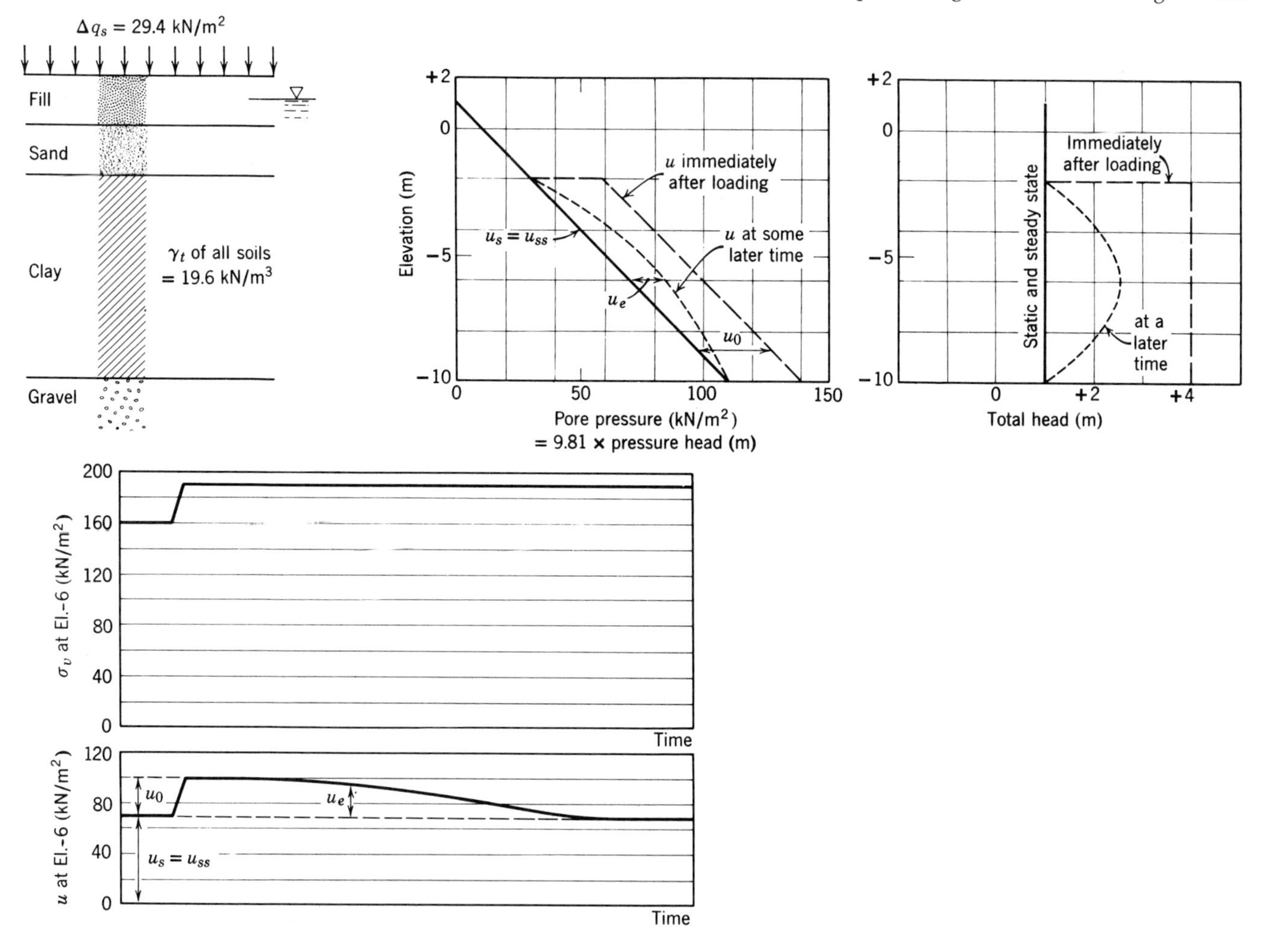

Fig. 26.2 Undrained surface loading.

however, to develop equations for the determination of the parameters, but to provide a deeper understanding of the nature of the parameters.

Other Situations Causing Transient Flow

Figure 26.4 illustrates that excess pore pressures and transient flow may occur even if there is no change in the total load applied to the soil. In this case the transient flow is produced by a lowering of the water table in the strata above the clay while the piezometric level in the underlying gravel remains constant. Prior to the lowering, the pore pressures are hydrostatic; $u = u_s$. Long after the lowering, the pore pressures will be the steady-state pressures u_{ss}. In this final equilibrium condition the total head varies across the clay and there is upward flow of water from the gravel to the sand. Assuming that the unit weight of the soil is unchanged as a result of the lowering, the vertical total stress does not change at any point. Assuming that the lowering occurs very quickly compared to the time required for dissipation of excess pore pressure, the pore pressures within the clay at the end of lowering will still be given by the line marked u_s. At this time, the initial excess pore pressure is

$$u_0 = u_s - u_{ss}$$

After dissipation begins and the pore pressures decrease, the excess pore pressure at any time is

$$u_e = u - u_{ss}$$

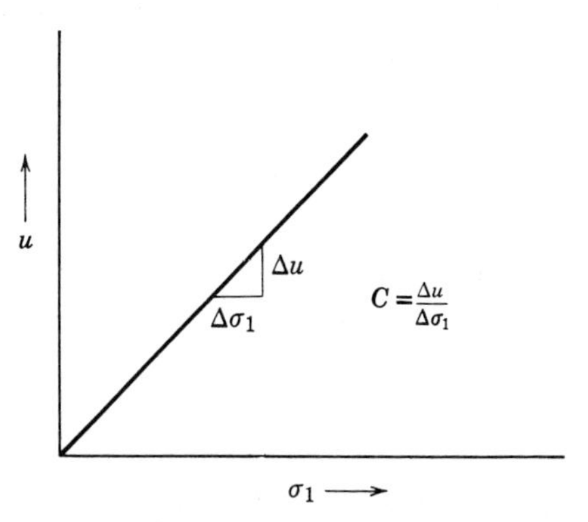

Fig. 26.3 Results of loading in oedometer.

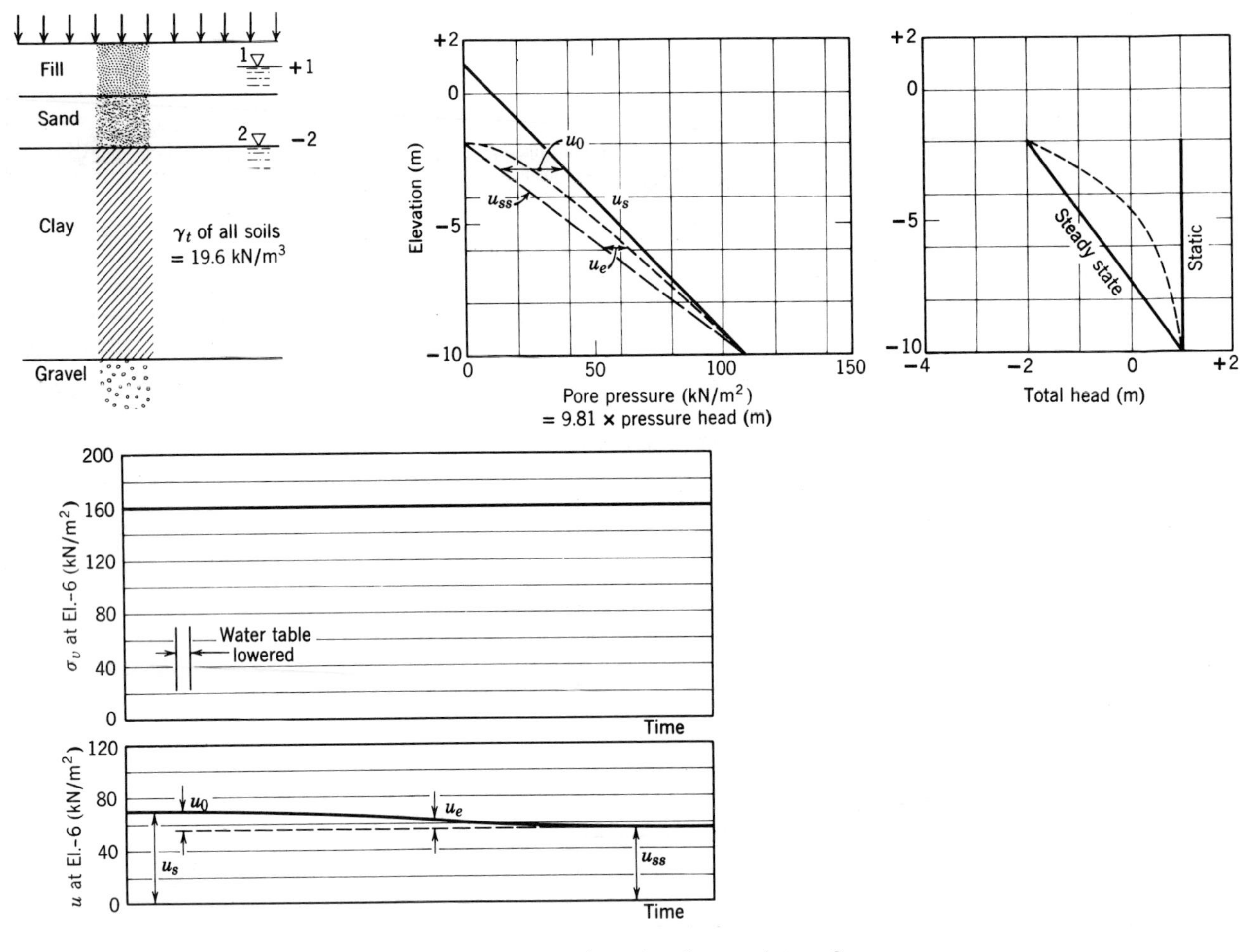

Fig. 26.4 Water lowering from +1 to −2.

The example in Figure 26.4 brings out the important point that excess pore pressures are referenced to the final steady-state pore pressures.

26.2 PORE PRESSURE DEVELOPED IN THE OEDOMETER TEST

Let us look more closely at the apportionment of a total stress increment to a soil specimen in an oedometer. The compressibility of each of the two phases is measured separately and used to apportion a stress applied to the two-phase, saturated soil sample.

Figures 26.5*a* and 26.5*b* present compression plots for the soil skeleton and for the pore fluid. The soil skeleton was tested in one-dimensional compression with free escape of pore fluid permitted, as in Chapter 22. The slope of the plot of volumetric strain-effective stress is[1]

$$C_{c1} = +\frac{\Delta V}{V_0} \cdot \frac{1}{\Delta \bar{\sigma}_1} = -\frac{\Delta e}{1 + e_0} \cdot \frac{1}{\Delta \bar{\sigma}_1}$$

[1] Note that C_{c1} is identical to m_v, the coefficient of volume change, defined by Eq. 12.12. Note that a volume decrease is taken as positive.

From a compression test on the pore fluid,

$$C_w = +\frac{\Delta V}{V_0} \cdot \frac{1}{\Delta u}$$

Under the loading of $\Delta\sigma_1$, the change in volume of the soil skeleton ΔV_{sk} must be equal to the change in volume of the pore fluid ΔV_p:

$$\Delta V_{sk} = \Delta V_p$$

Using the compression coefficients, we can express the volume changes of the soil skeleton ΔV_{sk} and the pore fluid ΔV_p as follows:

$$\Delta V_{sk} = +V_0 C_{c1}\, \Delta \bar{\sigma}_1$$

and

$$\Delta V_p = +n V_0 C_w\, \Delta u$$

in which

V_0 = initial total volume of soil-water system
C_{c1} = compressibility of soil skeleton and is obtained

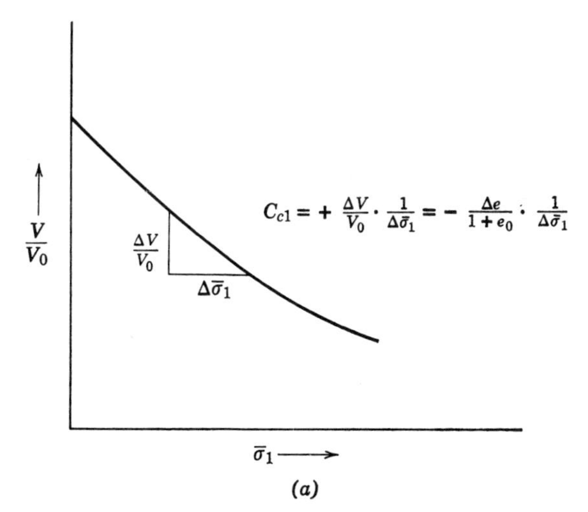

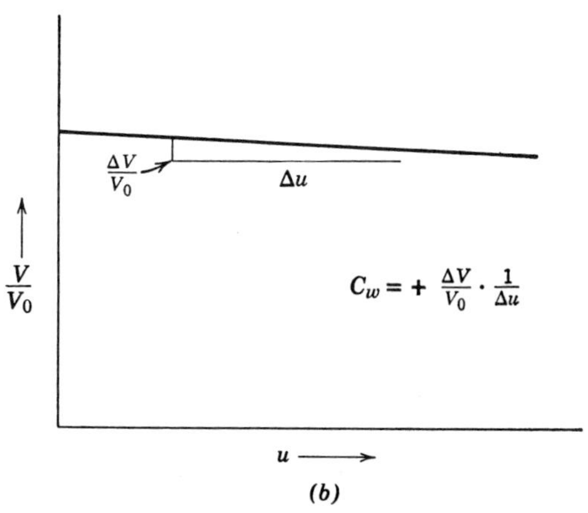

Fig. 26.5 One-dimensional deformation. (*a*) Soil skeleton. (*b*) Water.

from one-dimensional compression test where $\Delta u = 0$

n = porosity

C_w = compressibility of water

Δu = change in pore pressure

Since the change in volume of the soil skeleton is equal to the change in volume of the pore fluid, the following expression can be written:

$$V_0 C_{c1}\, \Delta\bar{\sigma}_1 = n V_0 C_w\, \Delta u$$

By definition

$$\bar{\sigma} = \sigma - u$$

therefore

$$\Delta\bar{\sigma}_1 = \Delta\sigma_1 - \Delta u$$

Substituting this expression for $\Delta\bar{\sigma}_1$ into the volume equilibrium expression and cancelling V_0 gives us

$$C_{c1}(\Delta\sigma_1 - \Delta u) = nC_w\, \Delta u$$

and

$$C = \frac{\Delta u}{\Delta \sigma_1} = \frac{1}{1 + n(C_w/C_{c1})} \tag{26.1}$$

The ratio C given by Eq. 26.1 is the pore pressure parameter for an undrained loading in the oedometer. Table 26.1 lists values of C computed from measured values of C_w and C_{c1}. For all typical, saturated soils C is essentially unity.

This derivation assumes that the soil particles are incompressible. In fact, the compressibility of the minerals composing soil particles is about thirty times less than the compressibility of water, and hence this assumption is justified. The volume change ΔV_{sk} is the change in the volume delineated by the boundaries of the soil (the sides and bottom of the oedometer and the piston) and results from particles slipping past one

Table 26.1

VALUES OF PARAMETER C

Material ($S = 100\%$)	C	Reference
Vicksburg buckshot clay slurry	0.99983	M.I.T. Test
Lagunillas soft clay	0.99957	M.I.T. Test
Lagunillas sandy silt	0.99718	M.I.T. Test

VALUES OF PARAMETER B

Material	$S(\%)$	B	Reference
Sandstone	100	0.286	
Granite	100	0.342	
Marble	100	0.550	Computed from
Concrete	100	0.582	compressibi-
Dense sand	100	0.9921	lities given by
Loose sand	100	0.9984	Skempton (1961)
London clay (OC)	100	0.9981	
Gosport clay (NC)	100	0.9998	
Vicksburg buckshot clay	100	0.9990	M.I.T.
Kawasaki clay	100	0.9988 to 0.9996	M.I.T.
Boulder clay	93	0.69	Measured by
	87	0.33	Skempton
	76	0.10	(1954)

VALUES OF PARAMETER A

Material ($S = 100\%$)	A (at failure)	Reference
Very loose fine sand	2 to 3	Typical
Sensitive clay	1.5 to 2.5	values
Normally consolidated clay	0.7 to 1.3	given by
Lightly overconsolidated clay	0.3 to 0.7	Bjerrum
Heavily overconsolidated clay	−0.5 to 0	

Material ($S = 100\%$)	A (for foundation settlement)	Reference
Very sensitive soft clays	>1	
Normally consolidated clays	$\frac{1}{2}$ to 1	From Skempton
Overconsolidated clays	$\frac{1}{4}$ to $\frac{1}{2}$	and Bjerrum
Heavily overconsolidated sandy clays	0 to $\frac{1}{4}$	(1957)

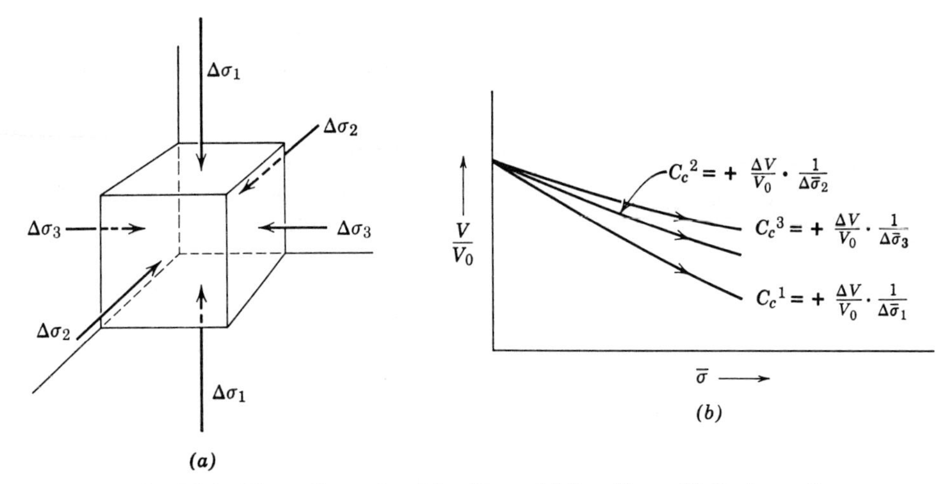

Fig. 26.6 Three-dimensional loading. (*a*) Loading. (*b*) Deformation.

another. Since the pore fluid is the only compressible element within this volume, ΔV_{sk} must equal ΔV_p.

This derivation also assumes that C_{c1} as measured during a drained loading equals the ratio of change in volume to change in *effective* stress during an undrained loading. The validity of this assumption is discussed in Chapter 28.

26.3 PORE PRESSURE DEVELOPED BY AN INCREMENT OF ISOTROPIC STRESS

Figure 26.6 shows a soil element subjected to a three-dimensional loading; also shown are soil skeleton compression data obtained on the soil element under a uniaxial load change with zero pore pressure. In other words, the top compression curve is a plot of unit volume versus $\bar{\sigma}_3$ with $\bar{\sigma}_2$ and $\bar{\sigma}_1$ held constant.

For the three-dimensional loading, the total change in volume of the soil skeleton is

$$\Delta V_{sk} = +V_0 C_c^1 \,\Delta\bar{\sigma}_1 + V_0 C_c^2 \,\Delta\bar{\sigma}_2 + V_0 C_c^3 \,\Delta\bar{\sigma}_3$$

and

$$\Delta V_p = +nV_0 C_w \,\Delta u$$

For the special case where the applied stress in all three principal directions is the same, i.e., uniform stress application or isotropic stress application,

$$\Delta\sigma_1 = \Delta\sigma_2 = \Delta\sigma_3 = \Delta\sigma$$

and

$$\Delta\bar{\sigma}_1 = \Delta\bar{\sigma}_2 = \Delta\bar{\sigma}_3 = \Delta\bar{\sigma} = \Delta\sigma - \Delta u$$

we know

$$\Delta V_{sk} = \Delta V_p$$

therefore

$$nV_0 C_w \,\Delta u = V_0(\Delta\sigma - \Delta u)(C_c^1 + C_c^2 + C_c^3)$$

$$\frac{\Delta u}{\Delta\sigma} = \frac{C_c^1 + C_c^2 + C_c^3}{nC_w + C_c^1 + C_c^2 + C_c^3} \qquad (26.2)$$

If

$$C_c^2 = C_c^3$$

then

$$\frac{\Delta u}{\Delta\sigma} = \frac{C_c^1 + 2C_c^3}{nC_w + C_c^1 + 2C_c^3} \qquad (26.2a)$$

If

$$C_c^1 = C_c^2 = C_c^3$$

i.e., if the soil element is isotropic,

$$B = \frac{\Delta u}{\Delta\sigma} = \frac{1}{1 + n(C_w/C_{c3})} \qquad (26.2b)$$

where

$$C_{c3} = +\frac{\Delta V}{V_0} \cdot \frac{1}{\Delta\bar{\sigma}}$$

and

$$\Delta\bar{\sigma} = \text{uniform all-around pressure change}$$

Equation 26.2 gives the pore pressure parameter B which is (pore pressure change/total stress change) for a three-dimensional loading. For the special case of isotropic soil and uniform stress application, the pore pressure parameter B can be determined from the simple expression in Eq. 26.2*b*. As can be seen, Eq. 26.2*b* for this special case of three-dimensional compression is very similar to Eq. 26.1 for one-dimensional compression. In most soils C_{c1} is close to being equal to C_{c3} and thus the pore pressure parameter C is approximately equal to the pore pressure parameter B. This fact is illustrated by the typical values of pore pressure parameters given in Table 26.1.

26.4 PORE PRESSURE DEVELOPED BY AN INCREMENT OF UNIAXIAL STRESS

A soil element subjected to a uniaxial loading with freedom to strain laterally, as in the triaxial test, is shown in Fig. 26.7. Figure 26.7*b* shows compression curves obtained in the soil skeleton in a drained loading.

With the sample at $V/V_0 = 1$, the major principal effective stress is increased while the immediate and minor principal effective stresses are held constant; the compression curve shown is obtained. The two expansion curves are obtained by holding the major principal effective stress constant and by decreasing in turn $\bar{\sigma}_2$ and $\bar{\sigma}_3$. A volume increase normally occurs under such an unloading.

When we add an increment of stress $\Delta\sigma_1$ perpendicular to the major principal plane we get

$$\Delta\bar{\sigma}_1 = \Delta\sigma_1 - \Delta u$$

and

$$\Delta\bar{\sigma}_2 = \Delta\bar{\sigma}_3 = -\Delta u$$

Since

$$\Delta V_p = \Delta V_{sk}$$

we have

$$nV_0C_w\,\Delta u = V_0C_c^1(\Delta\sigma_1 - \Delta u) + V_0C_s^2(-\Delta u) + V_0C_s^3(-\Delta u)$$

$$\frac{\Delta u}{\Delta\sigma_1} = \frac{C_c^1}{nC_w + C_s^2 + C_s^3 + C_c^1}$$

$$\frac{\Delta u}{\Delta\sigma_1} = \frac{1}{1 + n(C_w/C_c^1) + C_s^2/C_c^1 + C_s^3/C_c^1} \qquad (26.3)$$

If

$$C_s^2 = C_s^3$$

then

$$D = \frac{\Delta u}{\Delta\sigma_1} = \frac{1}{1 + n(C_w/C_c^1) + 2C_s^3/C_c^1} \qquad (26.3a)$$

If

$$C_c^1 = C_s^2 = C_s^3$$

i.e., if the soil element is elastic and isotropic,

$$D = \frac{\Delta u}{\Delta\sigma_1} = \frac{1}{n(C_w/C_c^1) + 3} \qquad (26.3b)$$

The pore pressure parameter D thus is the fraction of total stress increment carried by pore pressure for a one-dimensional loading with no drainage permitted. As Eq. 26.3*b* shows, when the soil is elastic and isotropic,

$$D = \frac{1}{n(C_w/C_c^1) + 3}$$

For a saturated soil under normal pressure conditions, the term $n(C_w/C_c^1)$ is essentially equal to zero and the pore pressure parameter D is therefore equal to one-third. The following table illustrates the changes in stress for this situation:

$\Delta\sigma_1 = 3$	$\Delta u = 1$	$\Delta\bar{\sigma}_1 = 2$
$\Delta\sigma_2 = 0$		$\Delta\bar{\sigma}_2 = -1$
$\Delta\sigma_3 = 0$		$\Delta\bar{\sigma}_3 = -1$

Note that the sum $\Delta\bar{\sigma}_1 + \Delta\bar{\sigma}_2 + \Delta\bar{\sigma}_3 = 0$; i.e., there is no change in the *effective* isotropic stress. This is the correct result if there is to be zero (or very, very little) change in the volume of the soil.

26.5 PORE PRESSURE DEVELOPED BY TRIAXIAL STRESS

Figure 26.8 shows an element of soil subjected to a triaxial loading in which $\Delta\sigma_2$ is equal to $\Delta\sigma_3$. This three-dimensional loading can be considered to be made up of an isotropic stress of $\Delta\sigma_3$ plus a deviator stress equal to $\Delta\sigma_1 - \Delta\sigma_3$ acting on the major principal plane. This superposition results in the loading condition shown in Fig. 26.8 in which the increment of stress on the minor and intermediate principal planes is equal to $\Delta\sigma_3$ and that on the major plane is equal to $\Delta\sigma_1$. This superposition of loading is exactly that which occurs in the standard triaxial test in which *no* drainage is permitted. In this test, the sample is initially subjected to an all-around stress $\Delta\sigma_3$ and then failed under an increasing axial stress of $\Delta\sigma_1 - \Delta\sigma_3$.

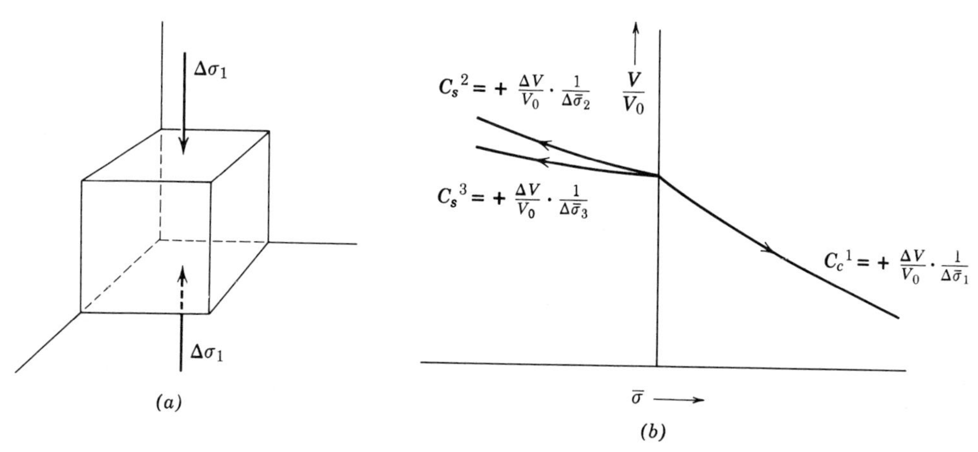

Fig. 26.7 Uniaxial loading. (*a*) Loading. (*b*) Deformations.

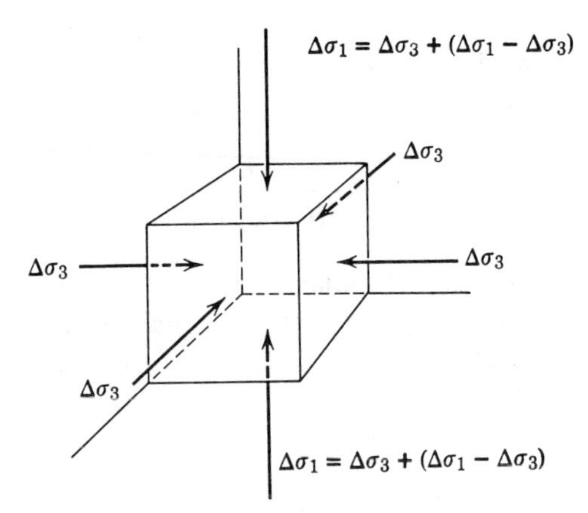

Fig. 26.8 Isotropic loading followed by uniaxial loading.

Equation 26.2*b* shows that for isotropic soil, the increment of pore pressure resulting from the increase in all-around pressure is

$$\Delta u = \frac{\Delta\sigma_3}{1 + nC_w/C_{c3}}$$

Equation 26.3*a* shows that for the condition of equal expansibility in the intermediate and minor principal directions, the increment of pore pressure arising from the deviator stress is

$$\Delta u = \frac{\Delta\sigma_1 - \Delta\sigma_3}{1 + n(C_w/C_c^1) + 2(C_s^3/C_c^1)}$$

Adding the pore pressure increase from the all-around stress increase to that resulting from $\Delta\sigma_1 - \Delta\sigma_3$ gives

$$\Delta u = \frac{\Delta\sigma_3}{1 + n(C_w/C_{c3})} + \frac{\Delta\sigma_1 - \Delta\sigma_3}{1 + n(C_w/C_c^1) + 2(C_s^3/C_c^1)}$$

or

$$\Delta u = B\,\Delta\sigma_3 + D(\Delta\sigma_1 - \Delta\sigma_3) \tag{26.4}$$

For an element of soil saturated with an incompressible pore fluid, and for a stress range normally encountered in soil engineering, Eq. 26.4 reduces to

$$\Delta u = \Delta\sigma_3 + \frac{\Delta\sigma_1 - \Delta\sigma_3}{1 + 2C_s^3/C_c^1} \tag{26.4a}$$

or

$$\Delta u = \Delta\sigma_3 + A(\Delta\sigma_1 - \Delta\sigma_3) \tag{26.4b}$$

where

$$A = \frac{1}{1 + 2(C_s^3/C_c^1)} = \frac{1}{1 + C_{s2}/C_c^1} \tag{26.5}$$

where

$$C_{s2} = C_s^2 + C_s^3$$

For the special case of an isotropic and elastic soil mass saturated with an incompressible pore fluid,

$$\Delta u = \Delta\sigma_3 + \tfrac{1}{3}(\Delta\sigma_1 - \Delta\sigma_3) \tag{26.6}$$

26.6 THE PORE PRESSURE PARAMETER *A*

Sections 26.2 to 26.5 derived expressions for the pore pressure parameters. These derivations are summarized in Fig. 26.9, and typical values for the pore pressure parameters are given in Table 26.1. The tabulated values show that each of the parameters C and B is unity for saturated soil. The parameter A, however, can be far from unity, varying from below zero to greater than one. Thus the soil engineer working with the parameters for saturated soils normally need worry only about the value of A.

The direct way to determine A is from Eq. 26.4*b* rewritten as

$$A = \frac{\Delta u - \Delta\sigma_3}{\Delta\sigma_1 - \Delta\sigma_3} \tag{26.7}$$

For the usual undrained triaxial test in which $\Delta\sigma_3 = 0$, Eq. 26.7 reduces to

$$A = \frac{\Delta u}{\Delta\sigma_1} \tag{26.8}$$

Figure 26.10 illustrates the determination of A from the usual undrained triaxial test. The test was started with zero pore pressure and an isotropic effective stress system as represented by the point S. During the test σ_3 was held constant and σ_1 increased, giving a total stress path having a slope of one to one as indicated by the line ST. The effective stress path SU was determined by subtracting the measured pore pressure Δu from the total stress path. The pore pressure parameter from Eq. 26.8 for the stage of the test represented by the point U is

$$A = \frac{UT}{2TY} = 0.60$$

The procedure, illustrated in Fig. 26.10, is the usual[2] way that parameter A is determined.

Figure 26.11 indicates that the parameter A can be expressed as a tangent of an angle on a plot of stress paths, and, further, that the parameter can be determined from various types of tests. Test 1 consists of a triaxial test in which σ_3 is kept constant, whereas test 2 is a triaxial test in which σ_1 is kept constant. As can be seen from Fig. 26.11, for both tests the value of A is equal to the tangent of the angle at the point V.

Figure 26.11 suggests that A can be determined from the location of the effective stress path and does not

[2] Lambe (1963) presents the results of the determination of the pore pressure parameter A on a particular soil both from an undrained test (as described in Fig. 26.10) and from drained tests using Eq. 26.5. The value of A determined from the undrained test was from 3 to 20% greater than that determined from the drained tests for strains up to 5%. This agreement is considered close in view of the assumptions involved in the derivation of the equations, especially Eq. 26.5, which is based on a superposition of two stress systems.

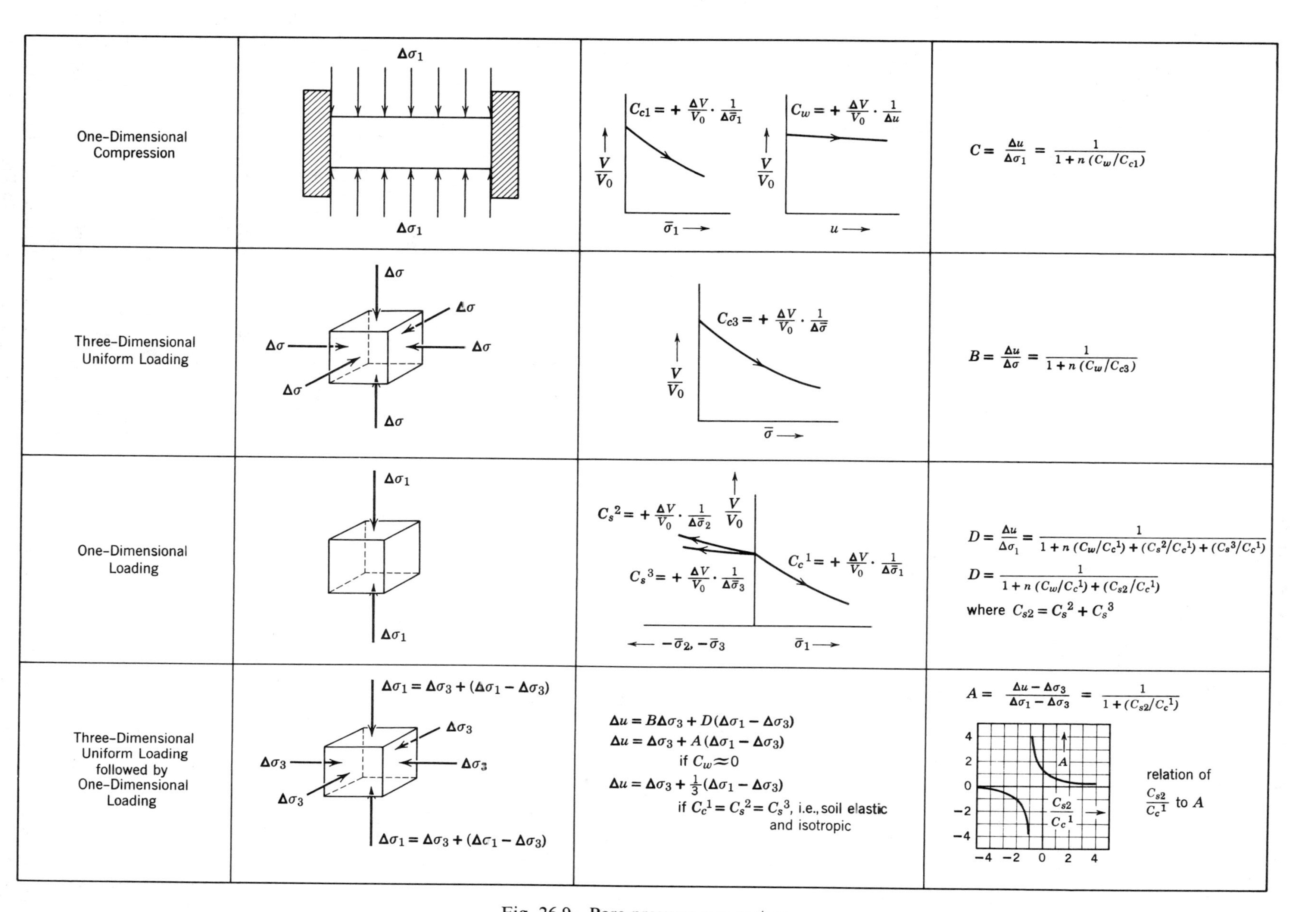

Fig. 26.9 Pore pressure parameters.

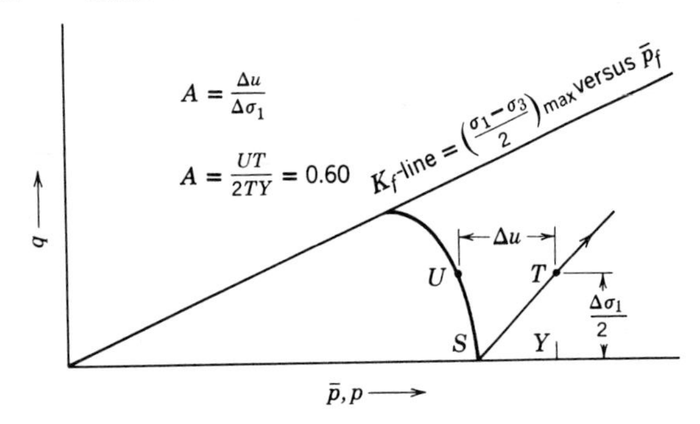

Fig. 26.10 The determination of A from triaxial test.

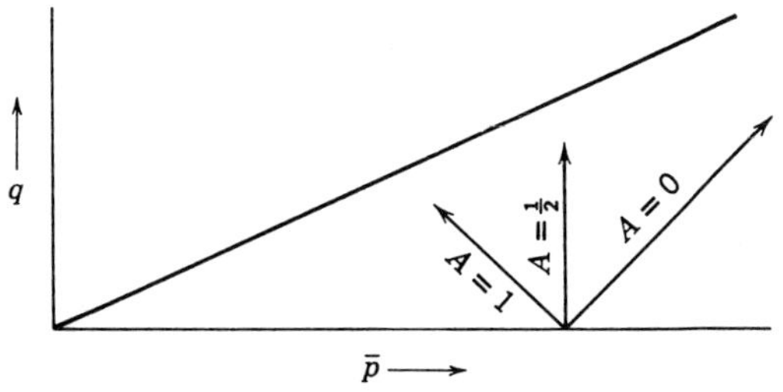

Fig. 26.12 Special values of A.

depend on the total stress path. Using this principle,[3] we can associate certain effective stress paths with certain values of the pore pressure parameter A. Some of these are indicated in Fig. 26.12. An effective stress path with a slope of 1:1 to the right indicates $A = 0$; a vertical effective stress path gives $A = \frac{1}{2}$; and an effective stress path with a 1:1 slope to the left gives $A = 1$. Paths to the right and below that for $A = 0$ indicate values of A that are negative. The values of A for paths to the left and below the stress path for $A = 1$ indicate A values greater than one.

Since the notion of a pore pressure parameter having a value greater than one or less than zero may seem unusual, a look into the types of soil structures that produce such values is worthwhile. A pore pressure parameter greater than one is associated with a loose structure, either in sand or clay, which collapses upon load application. If, for example, to a sample of very loose sand under an all-around effective stress of 70 kN/m² we added $\Delta\sigma_1$ equal to 14 kN/m², we could cause the structure to collapse and $\Delta\sigma_1$ plus some of the effective stress already acting on the sand could be transferred to pore pressure. Thus we could get an increment of pore pressure in excess of $\Delta\sigma_1$, i.e., a value of A greater than one.

[3] This is only approximately true, as indicated by Fig. 26.13d.

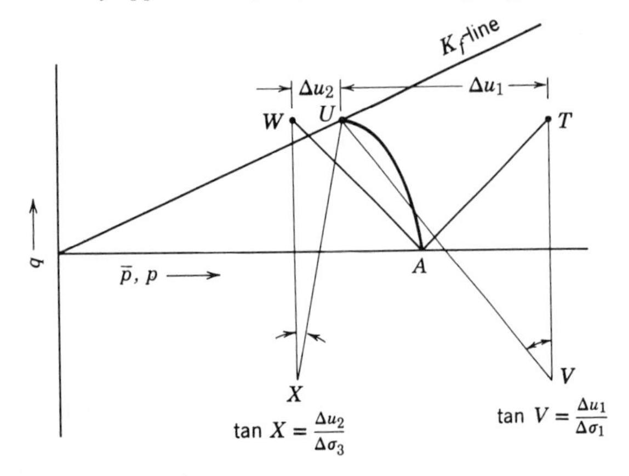

Test 1: $\Delta\sigma_3 = 0$. $A = \Delta u_1/\Delta\sigma_1 = TU/TV = \tan V$
Test: 2: $\Delta\sigma_1 = 0$. $A = (\Delta u_2 - \Delta\sigma_3)/(-\Delta\sigma_3) = 1 - WU/WX = 1 - \tan X = \tan V$

Fig. 26.11 The determination of A from triaxial test.

On the other hand, if we load a sample of sand or clay which tends to expand upon loading, we can develop negative pore pressures. In general, a heavily overconsolidated clay or a very dense sand tends to expand when subjected to a shear stress. Thus such a sample could be subjected to a shear stress that would result in a negative pore pressure, which would give a negative value of A.

The theoretical considerations and numerical information already presented clearly show that the pore pressure parameter A is not a constant soil property. It is a serious, but unfortunately not uncommon error to think of A as a constant. Figure 26.13 illustrates four of the factors that influence the value of A. [Lambe (1963) presents numerical data on the influence of these factors on the value of A for a particular soft clay.]

The parameter A depends very much on the strain to which the soil element under consideration has been subjected. Figure 26.13a illustrates that the parameter A is increasing as the applied shear stress increases. A curved effective stress path indicating a varying parameter A is usual, and a straight effective stress path indicating a constant value of A is unusual.

Figure 26.13b suggests that A depends on the initial stress system of the soil being tested. If, for example, a soil is initially under an isotropic stress system S and sheared to failure U, the parameter at the point W is higher than if the sample were brought to the condition T, all excess pore pressures were allowed to dissipate, and then the sample were sheared. For the anisotropic initial state of stress T, the stress path TV, indicating a much lower value of A than was obtained for the initial isotropic consolidation, would not be unusual.

Figure 26.13c suggests that A depends very much on the stress history of the sample. A soft, normally consolidated clay tends to have a parameter A not too far from unity, whereas an overconsolidated sample has a lower value as indicated by the stress path SV.

The value of A can depend on the total stress path, i.e., the type of stress change. In Fig. 16.13d are shown the stress path ST resulting from a loading from S and a stress path SU resulting from an unloading starting at S.

The parameter A from these two types of loading may be similar but might well be somewhat different.

Since the parameter A can depend so significantly on a number of factors, the engineer should be cautious in using A values stated in the literature. The values given in Table 26.1 are marked to indicate that the first group of A values apply to failure conditions, whereas the second group of values apply to lower strains such as those that exist in the normal foundation settlement problem. One should use the values in Table 26.1 only as a first approximation and should actually measure A for a particular problem, taking into consideration the various factors noted in Fig. 26.13. In other words, the soil under consideration should be subjected to the stress and strain conditions expected to exist in the problem under consideration.

26.7 THE ESTIMATION OF PORE PRESSURE IN THE FIELD

The major reason for determining the values of the pore pressure parameters is to estimate the magnitude of initial excess pore pressure produced at a given point in the subsoil by a change in the total stress system. This section will present two examples of the estimation of initial excess pore pressure—one for a loading condition and one for the unloading condition.

Loading Example

Figure 26.14 presents the prediction of the pore pressure in a layer of foundation clay caused by the application of a heavy soil preload applied to the ground surface. As shown, the preload consisted of a truncated cone of soil applying a maximum vertical stress at the surface equal to 225 kN/m². The piezometer under consideration, *P*21, is directly under the center of the load at an elevation of −9.45 m.

The initial stresses at *P*21, computed using the techniques presented in Chapter 8, are: vertical effective stress = 67.4 kN/m² and static pore pressure = 64.3 kN/m². Using the stress distribution chart presented in Chapter 8, the following increments of stress were computed: $\Delta\sigma_3 = 85.2$ kN/m² and $\Delta\sigma_1 = 195.8$ kN/m². The actual preload shape was replaced by a cylinder having the same weight.

A sample of the clay was obtained and tested in the triaxial machine. The stress-strain curves shown in Fig. 26.14 were obtained. For values of strain greater than approximately 3%, the pore pressure parameter A was 0.85.

Substituting the computed increments of stress and the measured pore pressure parameter A in Eq. 26.4*b* yields the computed value of initial excess pore pressure of 179.2 kN/m² which is a head of 18.3 m of water. The head measured by *P*21 was 17.7 m.

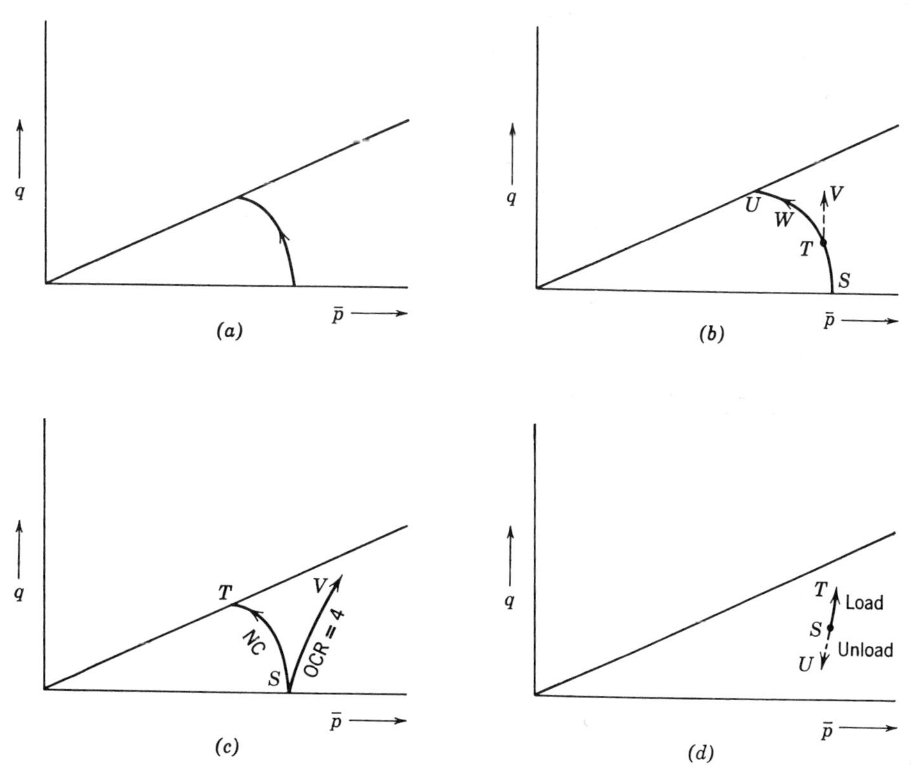

Fig. 26.13 Factors influencing A. (*a*) Strain. (*b*) Initial stress system. (*c*) Stress history. (*d*) Type of stress change.

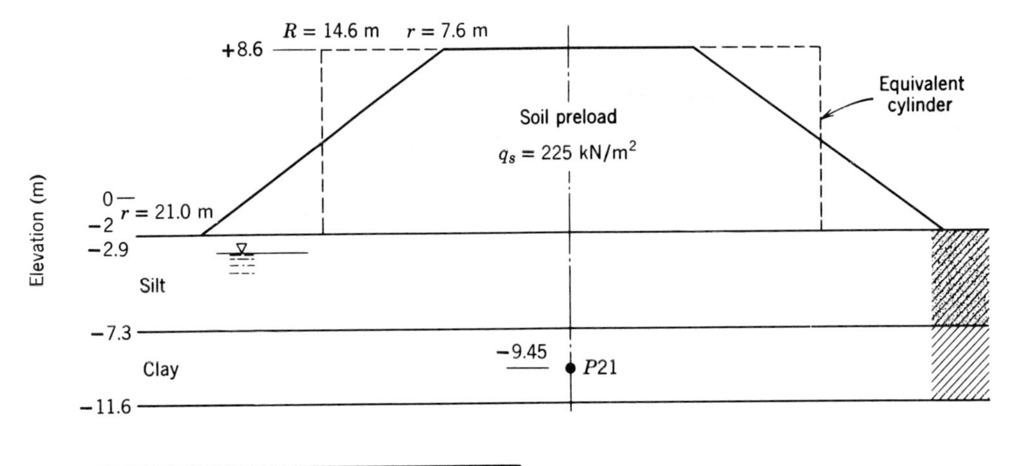

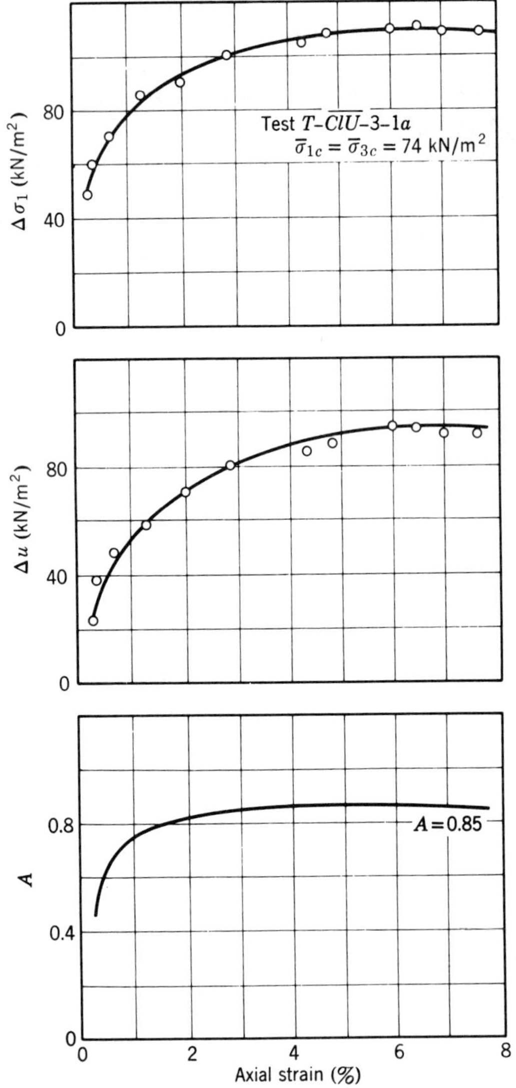

STRESS AT $P21$

Initial:

$$\bar{\sigma}_v = 67.4 \text{ kN/m}^2$$

$$u_s = 64.3 \text{ kN/m}^2$$

Increments from preload:

$$\Delta\sigma_3 = 85.2 \text{ kN/m}^2$$

$$\Delta\sigma_1 = 195.8 \text{ kN/m}^2$$

Pore pressure parameter A from lab test,

$$A = 0.85$$

Excess pore pressure
Calculated:

$$\Delta u = \Delta\sigma_3 + A(\Delta\sigma_1 - \Delta\sigma_3)$$

$$\Delta u = 85.2 + 0.85(110.6) = 179.2 \text{ kN/m}^2$$

$$\Delta u = 18.3 \text{ m water}$$

Measured on Field piezometer $P21$

$$\Delta u = 17.7 \text{ m}$$

Fig. 26.14 Determination of Δu_i.

Unloading Example

Figure 26.15 shows the determination of initial excess pore pressure for an unloading caused by an excavation for a building. Excavation stage 1 consisted of removing soil from elevation +6.86 m to elevation +4.88, an excavation of 1.98 m. Stage 2 took the ground surface at the construction site from elevation +4.88 m to elevation +2.29 m, an excavation of 2.59 m. The excavation was 43.3 by 71.6 m in plan. Of interest are the initial excess pore pressures developed at piezometer *P*3 (elevation −14.51) and *P*4 (elevation −18.81). Both piezometers are approximately under the center of the excavation.

Also shown in Fig. 26.15 are the stress paths for the unloading for piezometer *P*4. Point *A* represents the $p - q$ values for initial total stresses at *P*4 and $\bar{A}$ represents the effective stresses. The horizontal distance between $\bar{A}$ and *A* is the static pore pressure of 218.6 kN/m². The first stage of excavation results in a total stress path of *AB* and the second stage results in a total stress path of *BC*. The locations of points *B* and *C* were found by computing the $p - q$ values, allowing for the decrements in σ_1 and σ_3 that occur at point *P*4 for the first and second stages of excavation.

In the laboratory a sample of clay was subjected to the stress system *A* for total stresses and $\bar{A}$ for effective stresses and then unloaded along the total stress path *ABC*. The pore pressure during the test was measured permitting the location of $\bar{B}$ and $\bar{C}$. $\overline{ABC}$ is thus the

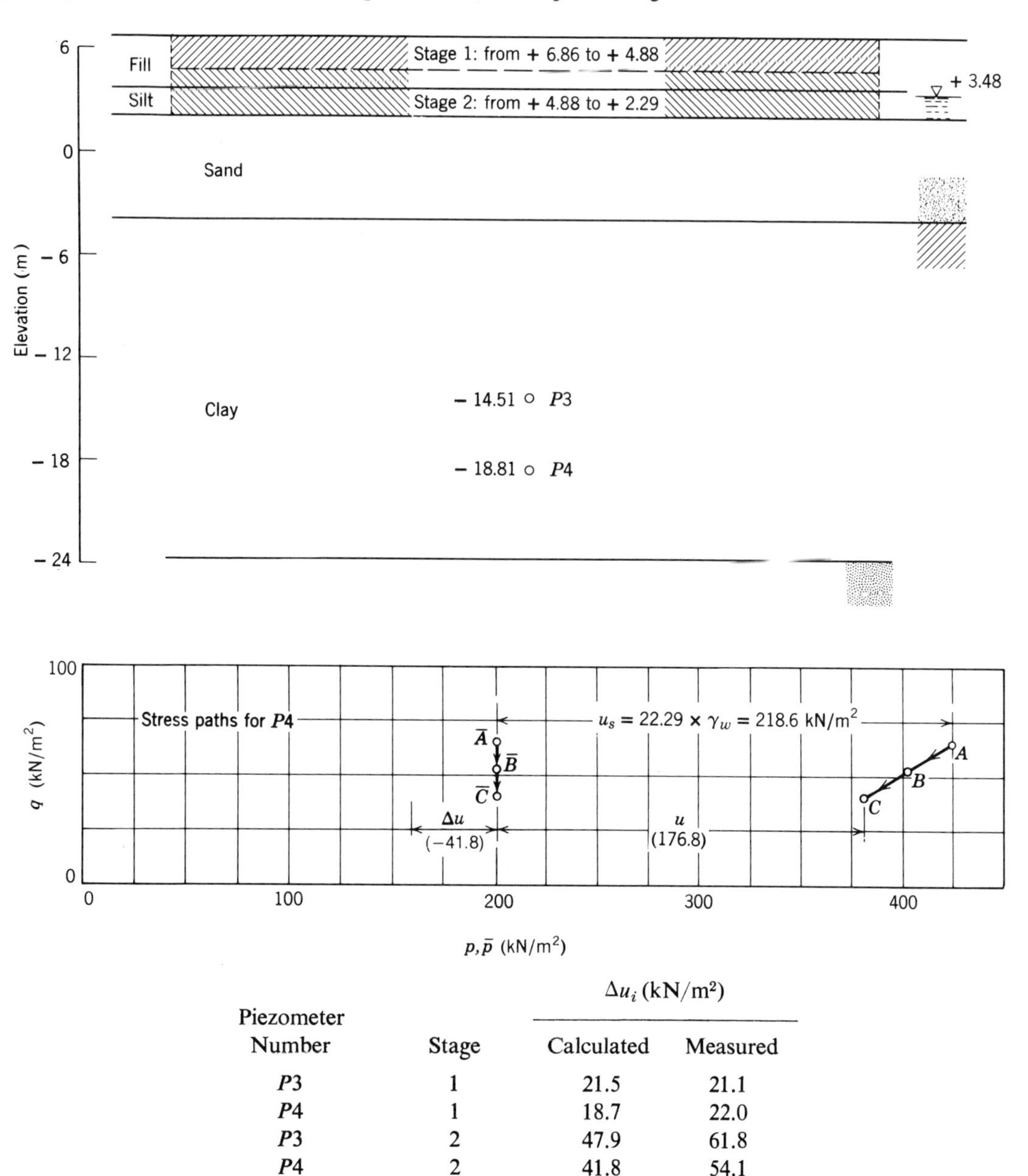

Piezometer Number	Stage	Δu_i (kN/m²) Calculated	Measured
*P*3	1	21.5	21.1
*P*4	1	18.7	22.0
*P*3	2	47.9	61.8
*P*4	2	41.8	54.1

Fig. 26.15 Determination of Δu_i for unloading.

effective stress path for the unloading. As shown in the figure, the pore pressure is 176.8 kN/m² and the excess pore pressure is −41.8 kN/m².

As can be seen from the effective stress path in Fig. 26.15, the laboratory test yielded a pore pressure parameter A which was approximately constant during the unloading and equal to 0.50. By using this pore pressure parameter A, the decrements of stress based on elastic theory, and Eq. 26.4*b*, initial excess pore pressures for piezometers $P3$ and $P4$ for both stages of construction can be computed. In the bottom part of Fig. 26.15 are listed the computed and measured values of excess pore pressures.

In the unloading example, the pore pressure at $P4$ for stage 2 was predicted directly from the laboratory test without resorting to pore pressure parameters. This was possible since the laboratory test duplicated the stress path for soil at $P4$. Pore pressure parameters are used when laboratory tests duplicating the field situation are not available.

The two examples (Figs. 26.14 and 26.15) are real field cases—the loading case (Lambe, 1963) is for a preload at Lagunillas, Venezuela. The excavation is for the Student Center built on the M.I.T. campus during 1963–1965. For the five comparisons between calculated and measured excess pore pressures, excellent agreement was obtained for the loading case and the first stage of the unloading. The difference between the calculated and measured excess pore pressures for the second stage of unloading was 30%.

26.8 PORE PRESSURE IN SOIL NOT SATURATED WITH WATER

Pore pressure parameters can be derived for soil whose pores are filled with air or partly with air and partly with water. For these situations, the pore phase can be highly compressible relative to the compressibility of the skeleton with the result that the pore pressure parameter is very small. Consider, for example, Eq. 26.2*b* for the parameter B written as:

$$B = \frac{1}{1 + n \dfrac{C_{\text{pore phase}}}{C_{\text{skeleton}}}} \tag{26.9}$$

Substituting in Eq. 26.9 numerical data from a triaxial test (Lambe, 1951) on a coarse, well-graded sand gives, for a dry soil,

$$B = \frac{1}{1 + 0.311[(3.33 \times 10^{-2})/(3.96 \times 10^{-5})]}$$

$$= \frac{1}{1 + 262} = 0.00380$$

As illustrated by the preceding example, loading a dry sand at constant mass—an undrained condition—results in essentially all of the applied stress being carried by the soil skeleton and almost none of the applied stress being carried by the pore air. This behavior results from the fact that the pore fluid—air—is much more compressible than the soil skeleton.

When the pore fluid consists of air and water (i.e., partly saturated soil) the parameters have values intermediate between those for dry and those for saturated soil. Because of the very high compressibility of the air relative to the water and soil skeleton, the air has a very significant effect on the pore pressure parameters. If, for example, our soil in the preceding example were 50% saturated we would get

$$B = \frac{1}{1 + 131} = 0.00767$$

Thus the parameter B is essentially zero for our soil even though it is 50% saturated. Obviously, the percent saturation must be very high for a significant increment of pore pressure to be developed during an undrained loading. For example, the values listed in Table 26.1 show that a boulder clay at 93% saturation has a value of B of only 0.69.

Equation 16.15 gives an expression for equivalent pore pressure u^* as

$$u^* \approx u_a + a_w(u_w - u_a) \tag{16.15}$$

In a partly saturated soil it is possible to measure both the water pressure u_w and the air pressure u_a. The simplest situation is that where the air pressure is maintained at atmospheric pressure and the water pressure is measured. This simple situation exists in the setup shown in Fig. 16.5. Figure 26.16 shows the condition at the point where the pore pressure is being measured.

The sensing element of the piezometer consists of a porous material (stone, brass, etc.) with pores small enough to develop menisci. The porous material is initially saturated with water and will thus admit water and exclude air through the development of menisci, as shown in Fig. 26.16. If the difference between the air pressure and water pressure exceeds the "breakthrough" pressure of the sensing element, air enters the pore sensing element and destroys the reliability of the measuring system. For measuring pore pressures in partly saturated sands, elements with only small breakthrough pressures are required. Elements with high breakthrough pressures are required for partly saturated fine-grained soils. Elements with breakthrough pressures of 200 to 400 kN/m² are used with fine-grained soils.

Lambe (1948) described a system similar to that shown in Fig. 26.16 and presents (1950) measured values of water pressure in partly saturated cohesionless soil. Bishop (1961) describes the measurement of both air pressure and water pressure in the triaxial test and presents test data.

There are a number of other direct and indirect techniques for obtaining pore water pressures in partly saturated soils used by experimenters working in agriculture or in highway engineering (e.g., see Croney and Coleman, 1961). As noted in Chapter 16, pore water pressure is not an unique function of percent saturation (or water content) but depends on the history of the soil. One should therefore be cautious of the reliability of pore water pressures inferred from measured percent saturation or water content data, as is necessary in some of the indirect techniques.

26.9 SUMMARY OF MAIN POINTS

1. An increment of stress applied to an element of soil is carried partly by the pore fluid as pore pressure and partly by the soil skeleton as effective stress. The relative portions of the stress increment carried by the pore fluid and by the skeleton depend on their relative compressibilities.
2. A convenient way to express the portion of a stress increment carried by the pore fluid is with *pore pressure parameters*. A pore pressure parameter is the ratio of the pore pressure increment to the total stress increment. Three pore pressure parameters are of importance:

$$C = \frac{\Delta u}{\Delta \sigma_1} \quad \text{for loading in the oedometer}$$

$$B = \frac{\Delta u}{\Delta \sigma} \quad \text{for isotropic loading}$$

$$A = \frac{\Delta u - \Delta \sigma_3}{\Delta \sigma_1 - \Delta \sigma_3} \quad \text{for triaxial loading}$$

3. In saturated soil the compressibility of the soil skeleton is almost infinitely greater than that of the pore water, and thus essentially all of a stress increment applied to a saturated soil is carried by the pore fluid in the oedometer loading and in isotropic loading, i.e., $C = B = 1$. In dry soil the compressibility of the pore air is almost infinitely greater than the compressibility of the soil skeleton, and thus essentially all of the increment in total stress applied to the dry soil element in an oedometer loading and in an isotropic loading is carried by the soil skeleton, i.e., $C = B = 0$. In partly saturated soils the very high compressibility of air relative to water and the soil skeleton results in low values of the parameters C and B until the percent saturation approaches 100%.

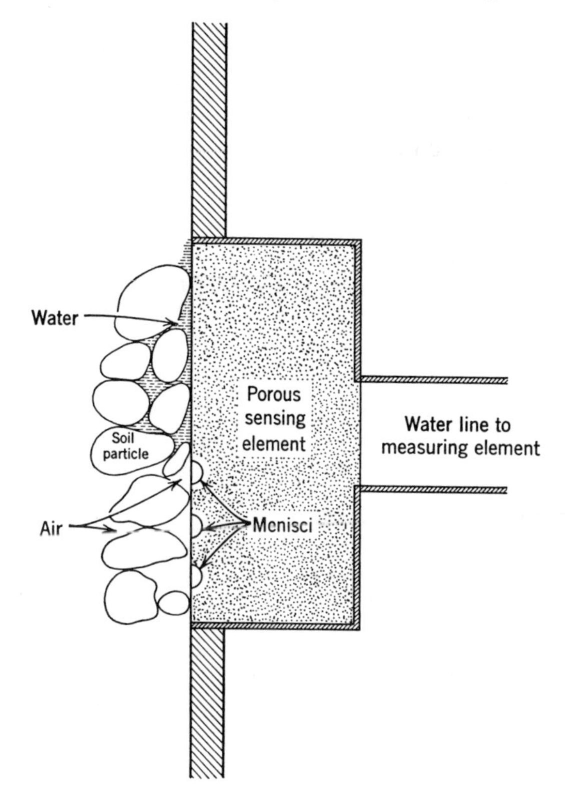

Fig. 26.16 Device for measuring u_a in partly saturated soil.

4. A total stress increment on a soil element which generates shear stress within the soil generally causes both a pore pressure increment and an effective stress increment. The value of the pore pressure increment developed by shear depends on the nature of the soil, type of stress, magnitude of strain, and time. In general, soft, loose soils have high values of A and the higher the shear strain the higher the value of A.

PROBLEMS

26.1 Compute the pore pressure parameter C for the clay in Example 22.1.

26.2 From the data in Fig. 29.6 compute the pore pressure parameter A at 5% strain for each of the four densities.

26.3 For the Amuay clay (Fig. 30.3) plot the pore pressure parameter A at 1% strain versus the initial value of $\bar{p}$.

26.4 Draw the TSP and the ESP for an element at mid-depth of the clay for the loading in Fig. 26.2.

26.5 Draw the TSP and the ESP for an element at mid-depth of the clay for the water table lowering in Fig. 26.4.

CHAPTER 27

Consolidation Theory

In Chapter 26 we studied the excess pore pressures caused by an undrained loading, which is a change in loading or boundary conditions that takes place in a time that is short compared to the dissipation time of excess pore pressures. As soon as the loading is complete, water begins to flow owing to the gradient caused by the excess pore pressures, and the soil changes in volume. If the excess pore pressures are positive so that the soil tends to decrease in volume, the process is called *consolidation*. If the excess pore pressures are negative so that the soil tends to increase in volume, the process is called *swell* or *heave*. The mathematical theory describing the dissipation of excess pore pressures (positive or negative) and associated deformation of the soil is called *consolidation theory*.

An introduction to consolidation theory has already appeared in Chapter 2. Chapter 27 presents a more concise formulation of the basic equations and a number of solutions of practical value.

27.1 CONSOLIDATION EQUATION

The process of consolidation (or swell) is governed by: (*a*) the equations of equilibrium for an element of soil; (*b*) stress-strain relations for the mineral skeleton; and (*c*) a continuity equation for the pore fluid.

Equilibrium equations have been discussed in Section 13.4, and stress-strain relations have been discussed in Chapter 12. The continuity equation was derived in Chapter 18:

$$k_z \frac{\partial^2 h}{\partial z^2} + k_x \frac{\partial^2 h}{\partial x^2} = \frac{1}{1+e}\left(e\frac{\partial S}{\partial t} + S\frac{\partial e}{\partial t}\right) \qquad (18.2)$$

where

z = coordinate in vertical direction
x = coordinate in horizontal direction
k_z, k_x = permeabilities in respective directions
e = void ratio
h = total head
S = degree of saturation
t = time

The left side of Eq. 18.2 is the net flow of water into an element of soil during two-dimensional flow. The right side is the increase in the volume of water within the element. If the soil is fully saturated, the right side also represents the change in the volume of the element. The main assumptions involved in the derivation of Eq. 18.2 are the validity of Darcy's law and a limitation to small strains.[1]

We now develop the governing equations for several special cases. All of these derivations assume full saturation so that $S = 1$ and $\partial S/\partial t = 0$.

One-Dimensional Consolidation with Linear Stress-Strain Relation

Figures 26.2 and 26.4 show two common situations leading to flow only in the vertical direction and also strain only in the vertical direction. In both cases there is a stratum of clay sandwiched between strata which are stiffer than the clay and which consolidate instantaneously compared to the clay. Thus there are excess pore pressures only in the clay and nearly all of the settlement arises because of volume changes within the clay. In both cases the horizontal dimension over which change occurs is very great compared to the thickness of the consolidating stratum. Hence all vertical sections have the same distribution of pore pressures and stress with depth. Flow of water occurs only in the vertical direction and there is no horizontal strain.

For these simple conditions, the equations governing consolidation (or swell) are the following:

Equilibrium:

$$\sigma_v = \gamma_t z + \text{surface stress}$$

Stress-strain:

$$\frac{\partial e}{\partial \bar{\sigma}_v} = -a_v \qquad (12.10)$$

Continuity:

$$k\frac{\partial^2 h}{\partial z^2} = \frac{1}{(1+e)}\frac{\partial e}{\partial t}$$

[1] See Gibson et al. (1967) for a concise formulation of the equations involved in the consolidation process.

The second and third of these equations may be combined to give

$$\frac{k(1+e)}{a_v}\frac{\partial^2 h}{\partial z^2} = -\frac{\partial \bar{\sigma}_v}{\partial t} \tag{27.1}$$

A further useful modification to the equation is made by breaking the total head into its component parts:

$$h = h_e + \frac{u}{\gamma_w} = h_e + \frac{1}{\gamma_w}(u_{ss} + u_e)$$

where

h_e = elevation head
u_{ss} = steady-state pore pressure
u_e = excess pore pressure

By definition, $\partial^2 h_e/\partial z^2 = 0$. Moreover, in the equilibrium condition the pore pressure varies linearly with depth so that $\partial^2 u_{ss}/\partial z^2 = 0$. Hence Eq. 27.1 becomes

$$\frac{k(1+e)}{\gamma_w a_v}\frac{\partial^2 u_e}{\partial z^2} = -\frac{\partial \bar{\sigma}_v}{\partial t} \tag{27.2}$$

The coefficient in this equation is called the *coefficient of consolidation* c_v:

$$c_v = \frac{k(1+e)}{\gamma_w a_v} = \frac{k}{\gamma_w m_v} \tag{27.3}$$

where m_v is the coefficient of volume change as defined in Eq. 12.12.

Finally, Eq. 27.2 may be again modified by expressing effective stress in terms of total stress and pore pressure. By definition, $\partial u_{ss}/\partial t = 0$, and hence

$$c_v \frac{\partial^2 u_e}{\partial z^2} = \frac{\partial u_e}{\partial t} - \frac{\partial \sigma_v}{\partial t} \tag{27.4}$$

Equation 27.4 is Terzaghi's *consolidation equation*, whose derivation marked the birth of modern soil mechanics.

Solutions of Eq. 27.4 for typical initial and boundary conditions will be presented in Sections 27.2 and 27.3.

Two-Dimensional Consolidation of Isotropic Elastic Material

In most actual problems, surface loadings cause excess pore pressures which vary both horizontally and vertically. This may happen when a tank is placed on a stratum of clay having a thickness greater than the diameter of the tank. The resulting consolidation will involve horizontal as well as vertical flow and horizontal as well as vertical strains. Equation 18.2 applies to such problems provided that flow is two-dimensional, such as under a long embankment (plane strain) or under a circular tank (axisymmetrical loading). For more general problems a term in $\partial^2 h/\partial y^2$ would be added to the left side.

The further modification of Eq. 18.2 into a form suitable for solution will be illustrated for the case of plane strain. The volume change may be expressed using Eq. 12.6 (see Fig. 27.1):

$$\frac{\Delta V}{V_0} = \frac{(1-2\bar{\mu})}{\bar{E}}(\bar{\sigma}_x + \bar{\sigma}_y + \bar{\sigma}_z) \tag{27.5}$$

where

$\bar{E}$ = Young's modulus for the mineral skeleton
$\bar{\mu}$ = Poisson's ratio for the mineral skeleton

Using Eq. 12.5b with the lateral strain $\epsilon_y = 0$ gives:

$$\bar{\sigma}_y = \bar{\mu}(\bar{\sigma}_x + \bar{\sigma}_z) = \bar{\mu}(\bar{\sigma}_v + \bar{\sigma}_h) \tag{27.6}$$

Combining Eqs. 18.2, 27.5 and 27.6, and expressing total head in terms of excess pore pressure and effective stress in terms of total stress and pore pressure, we have

$$\frac{k\bar{E}}{2(1-2\bar{\mu})(1+\bar{\mu})}\left(\frac{\partial^2 u_e}{\partial x^2} + \frac{\partial^2 u_e}{\partial z^2}\right) = \frac{\partial u_e}{\partial t} - \frac{1}{2}\frac{\partial}{\partial t}(\sigma_v + \sigma_h) \tag{27.7}$$

Equation 27.7 must be solved in conjunction with equations of equilibrium for the soil (Eqs. 13.10). Solutions to this set of equations, and to the similar set of equations for the axisymmetric case, will be presented in Section 27.6.

Radial Consolidation

The term radial consolidation is used for axisymmetric problems in which there is transient radial flow but zero axial flow. Such a situation may occur during consolidation of a triaxial test specimen or when vertical sand drains are used to speed consolidation of a soil deposit. In this case, the left-hand side of Eq. 27.4 becomes

$$c_v\left(\frac{\partial^2 u_e}{\partial r^2} + \frac{1}{r}\frac{\partial u_e}{\partial r}\right)$$

where r is the radius (Scott, 1963).

Analogy to Other Physical Problems

If $\partial \sigma_v/\partial t = \partial \sigma_h/\partial t = 0$, Eqs. 27.4 and 27.7 are forms of the diffusion equation, which is a common differential

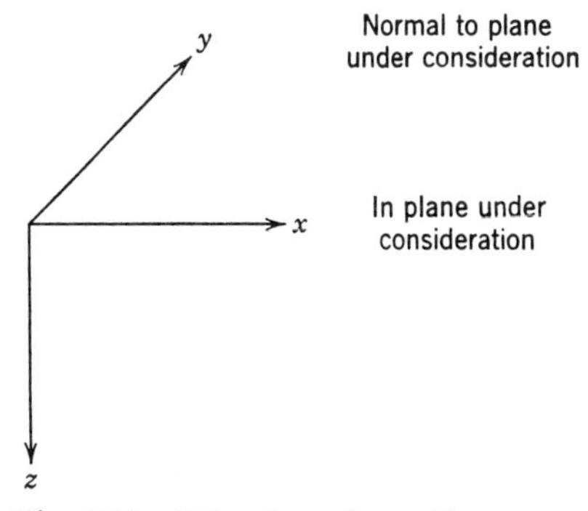

Fig. 27.1 Direction of coordinate axes.

equation applicable to numerous physical problems. In particular, the equations for transient heat flow are basically identical to these equations for consolidation, with temperature replacing excess pore pressure. Solutions have been obtained for many problems in heat flow involving a variety of initial and boundary conditions, and these solutions often may be used to considerable advantage in the study of consolidation.

27.2 SOLUTION FOR UNIFORM INITIAL EXCESS PORE PRESSURE

The simplest case of consolidation is the one-dimensional problem in which: (*a*) the total stress is constant with time, so that $\partial\sigma_v/\partial t = 0$; (*b*) the initial excess pore pressure is uniform with depth; and (*c*) there is drainage at both the top and bottom of the consolidating stratum. These conditions are met by the loading in Fig. 26.2 provided that the loading is applied in a time that is very small compared to the consolidation time so that literally no consolidation occurs before the loading is complete. The total vertical stress at any point will then be constant during the consolidation process.

For this problem, it is convenient to convert Eq. 27.4 by introducing nondimensional variables:

$$Z = \frac{z}{H} \tag{27.8a}$$

$$T = \frac{c_v t}{H^2} \tag{27.8b}$$

where z and Z are measured from the top of the consolidating stratum and H is *one-half* of the thickness of the consolidating stratum. (The reason for this choice of H will be apparent later.) The nondimensional time T is called the *time factor*. With these variables, Eq. 27.4 becomes

$$\frac{\partial^2 u_e}{\partial Z^2} = \frac{\partial u_e}{\partial T} \tag{27.9}$$

We now need a solution to Eq. 27.9 satisfying the following conditions:

Initial condition at $t = 0$:

$$u_e = u_0 \text{ for } 0 \leq Z \leq 2$$

Boundary condition at all t:

$$u_e = 0 \text{ for } Z = 0 \text{ and } Z = 2$$

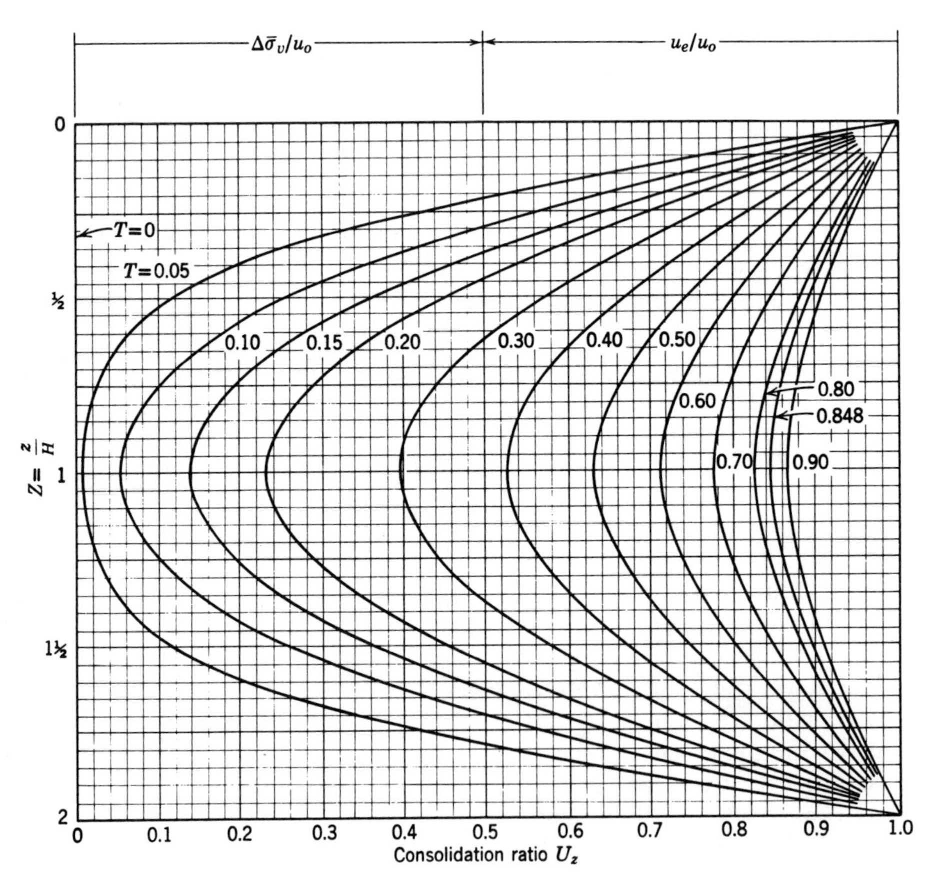

Fig. 27.2 Consolidation ratio as function of depth and time factor: uniform initial excess pore pressure.

where u_0 is the initial excess pore pressure. The solution is (e.g., see Taylor 1948)

$$u_e = \sum_{m=0}^{m=\infty} \frac{2u_0}{M} (\sin MZ) e^{-M^2 T} \quad (27.10)$$

where

$$M = \frac{\pi}{2}(2m + 1) \quad (27.11)$$

and m is a dummy variable taking on values 1, 2, 3, ... This solution may be conveniently portrayed in graph form (Fig. 27.2) where the consolidation ratio

$$U_z = 1 - \frac{u_e}{u_0}$$

is shown as a function of Z and T.

Example 27.1 illustrates the use of Fig. 27.2 to evaluate excess pore pressure, velocity of flow, and effective

▸Example 27.1

Given. The stratum of clay and loading shown in Fig. E27.1-1. This is the same profile and loading as in Example 25.6.

Find. At elevation −8.37 m and 4 months after loading

a. Excess pore pressure.
b. Pore pressure.
c. Vertical effective stress.
d. Velocity of flow.

Solution. Because the overlying and underlying soils are much more permeable than the clay, there is double drainage.

$$H = 2.13 \text{ m}, \quad Z = \frac{(8.37 - 7.3)}{2.13} = 0.5, \quad T = \frac{1.26(0.33)}{(2.13)^2} = 0.092$$

Interpolating in Fig. 27.2, $U_z = 0.24$
Thus:

$$u_e = 99.0(1 - 0.24) = 75.2 \text{ kN/m}^2$$

$$u = u_{ss} + u_e = 53.7 + 75.2 = 128.9 \text{ kN/m}^2$$

$$\bar{\sigma}_v = (\bar{\sigma}_v)_0 + \Delta\bar{\sigma}_v = 60.4 + 99.0(0.24) = 60.4 + 23.8 = 84.2 \text{ kN/m}^2$$

The stresses and pore pressures after 4 months are shown in Fig. E27.1-2. The slope of the tangent at $Z = 0.5$ to the interpolated curve for $T = 0.092$ is shown in Fig. E27.1-3. In terms of gradient this becomes

$$i = \frac{1}{\gamma_w} \frac{U_z}{Z} \frac{u_0}{H} = \frac{(0.95)(99.0)}{(9.81)(2.13)} = 4.50$$

The superficial seepage velocity is thus

$$v = ki = 0.018(4.50) = 0.081 \text{ m/yr upward}$$

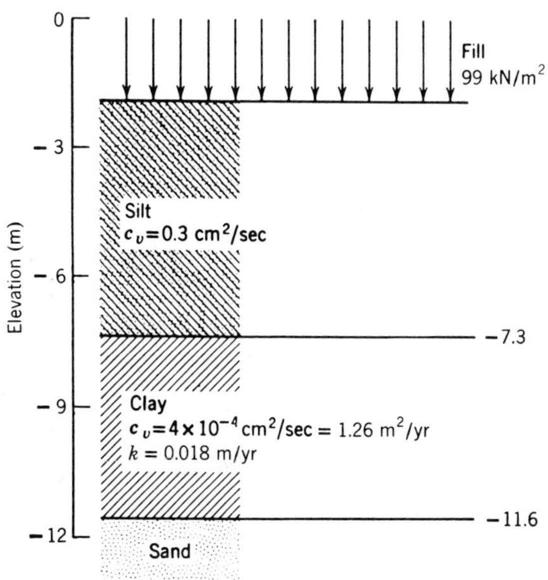

Fig. E27.1-1

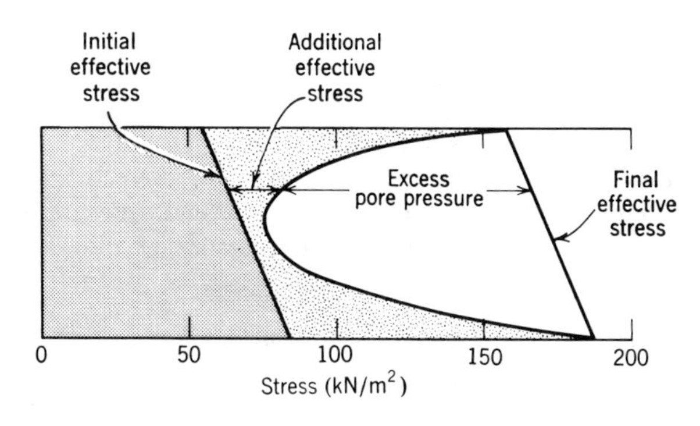

Fig. E27.1-2

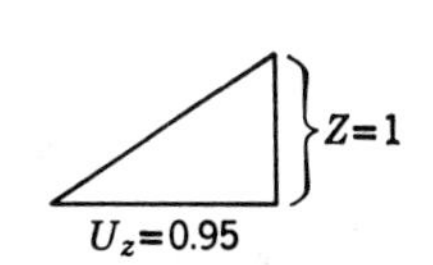

Fig. E27.1-3 ◂

stress at various stages of the consolidation process. Some important features of the process are revealed by a careful examination of Fig. 27.2 and Example 27.1:

1. Immediately following application of the load, there are very large gradients near the top and bottom of the strata but zero gradient in the interior. Thus the top and bottom change in volume quickly, whereas there is no volume change at mid-depth until T exceeds about 0.05.
2. For $T > 0.3$, the curves of U_z versus Z are almost exactly sine curves; i.e., only the first term in the series of Eq. 27.10 is now important.
3. The gradient of excess pore pressure at mid-depth ($Z = 1$) is always zero so that no water flows across the plane at mid-depth.

Average Consolidation Ratio

Of particular interest is the total compression of the stratum at each stage of the consolidation process, which may be found by summing the vertical compressions at the various depths. This compression is conveniently expressed by the *average consolidation ratio* U:

$$U = \frac{\text{compression at time } T}{\text{compression at end of consolidation}}$$

As shown in Fig. 27.3*a*, U may be interpreted as an area on the U_z versus Z diagram. By integrating Eq. 27.10, an expression for U as a function of T can be obtained. This expression is graphed in Fig. 27.3*b*. Example 27.2 illustrates the use of this graph. The first step always is to estimate the final settlement, using methods discussed in Chapter 25. Then Fig. 27.3*b* is used to find the settlement at various times during consolidation. Note that initially U decreases rapidly but that the rate of settlement then slows. Since U approaches 1 asymptotically, theoretically consolidation is never "complete."

▶Example 27.2

Given. Stratum of clay and loading in Fig. E27.1-1.
Find. Settlement as a function of time.
Solution. The final settlement has already been computed in Example 25.6 as 0.653 m. The dimensionless plot in Fig. 27.3 can now be converted into a time-settlement plot, as in Fig. E27.2.

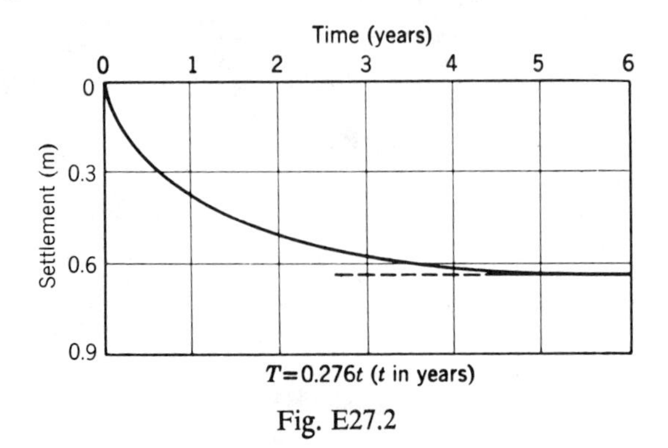

Fig. E27.2 ◀

However, U exceeds 99% at $T = 3$ and at $T = 1$ consolidation is about 92% complete. For engineering purposes, $T = 1$ is often taken as the "end" of consolidation. For $T = 1$,

$$t = \frac{H^2}{c_v} = \frac{\gamma_w m_v H^2}{k} \tag{27.12}$$

Equation 27.12 should be compared with Eq. 2.2, which was derived on a purely intuitive basis.

Single Drainage

All of the results in this section may still be used if there is an impermeable boundary at either the top or the bottom of the consolidating stratum. Now H is the total thickness of the stratum. Hence, for a stratum of given thickness, consolidation will proceed four times more quickly if there is double drainage rather than single drainage. Example 27.3 illustrates the use of Figs. 27.2 and 27.3 for a problem involving single drainage. Thus H is properly interpreted as the thickness per drainage surface.

27.3 EVALUATION OF c_v

The key to estimating the rate of dissipation of pore pressures or rate of settlement lies in selecting the proper

▶Example 27.3

Given. Stratum of clay and loading in Fig. E27.1-1, but with a stiff impervious clay underlying the soft clay so that there is single drainage from the top.
Find. Results requested in Examples 27.1 and 27.2.
Solution.

$$H = 4.3 \text{ m}, \quad Z = 0.25, \quad T = 0.023$$

Interpolating in Fig. 27.2, $U_z = 0.12$,

$$u_e = 99.0(1 - 0.12) = 87.1 \text{ kN/m}^2$$
$$u = 53.7 + 87.1 = 140.8 \text{ kN/m}^2$$
$$\bar{\sigma}_v = 60.4 + 99.0(0.12) = 72.3 \text{ kN/m}^2$$

The stresses and pore pressures at 4 months are shown in Fig. E27.3. The slope of the tangent is about 1.1, leading to

$$v = \frac{(1.1)(99.0)}{(9.81)(4.3)} 0.018 = 0.046 \text{ m/yr upward}$$

The time-settlement curve is the same as in Example 27.2 but with 4 yrs in place of 1 yr, etc. Thus the settlement at 1 yr is ≈0.18 m.

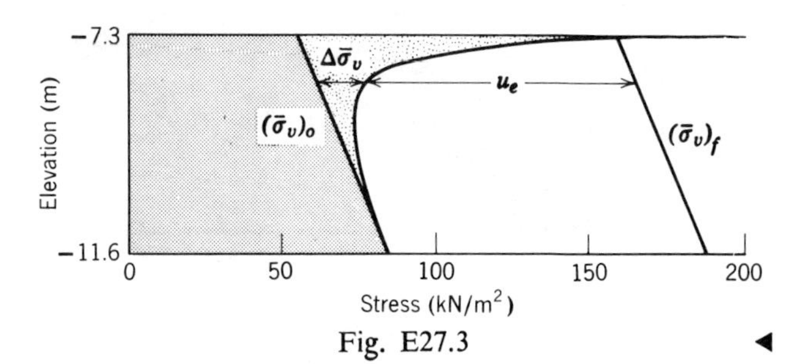

Fig. E27.3 ◀

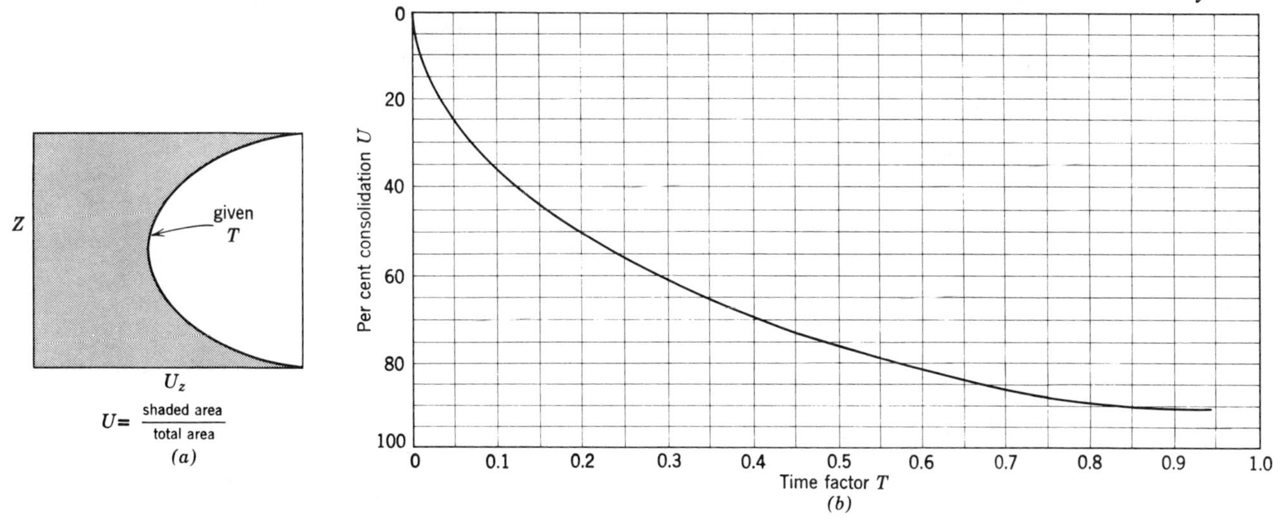

Fig. 27.3 Average consolidation ratio: linear initial excess pore pressure. (a) Graphical interpretation of average consolidation ratio. (b) U versus T.

value for c_v. This is generally done by observing the rate of compression of an undisturbed sample during an oedometer (or consolidation) test (see Sections 9.1 and 20.2).

Figure 27.4 shows a typical set of dial readings, showing change in thickness with time, obtained during one increment of load. The form of such actual time versus compression curves is similar to, but not exactly the same as, the theoretical curves predicted from consolidation theory. The following *fitting methods* are commonly used to determine c_v from such test results (Lambe, 1951).

Square root method. Extend a tangent to the straight-line portion of the observed curve back to intersect zero time and obtain the corrected zero point d_s. Through d_s draw a straight line having an inverse slope 1.15 times the tangent. Theoretically, this straight line should cut the observed compression-time curve at 90% compression. Thus the time to 90% compression is 12.3 minutes. From Fig. 27.3, the dimensionless time T for 90% compression is 0.848. Substituting these results, with H equal to the thickness of the sample per drainage surface (1.31 cm in this case) into Eq. 27.8*b*, c_v is determined to be 26.2×10^{-4} cm²/sec.

Log method. As shown in Fig. 27.4*b*, tangents are drawn to the two straight-line portions of the observed curve. The intersection of these curves defines the d_{100} point. The corrected zero point d_s is located by laying off above a point in the neighborhood of 0.1 minute a distance equal to the vertical distance between this point and one at a time which is four times greater. The 50% compression point is halfway between d_s and d_{100}, or at a time of 3.3 minutes. From the theoretical curve, $T = 0.197$ for 50% compression. Using Eq. 27.8b, c_v is then computed at 22.7×10^{-4} cm²/sec.

Discussion of results. Obviously, these fitting methods contain arbitrary steps that compensate for differences between actual and theoretical behavior. A correction for the initial point is usually required because of apparatus errors or the presence of a small amount of air in the specimen. An arbitrary determination of d_{90} or d_{100} is required because compression continues to occur even after excess pore pressures are dissipated. This *secondary compression* occurs because the mineral skeleton has time-dependent stress-strain properties (Chapter 20); the importance of secondary compression will be discussed in Section 27.7. The fitting methods have been developed to provide the best possible estimates for c_v. It is hardly surprising that the two methods yield somewhat different results. The square root method usually gives a larger value of c_v than does the log method, and this method is usually preferred.

In addition to the problems involved in evaluating c_v from a given increment, c_v varies from increment to increment and is different for loading and unloading. Figure 27.5 shows typical results. Moreover, c_v usually varies considerably among samples of the same soil.

Thus it is quite difficult to select a value of c_v for use in a particular engineering problem and hence it is difficult to predict accurately the rate of settlement or heave. Often the actual observed rate of settlement or heave of a structure is two to four times faster than the rate predicted on the basis of c_v as measured using undisturbed samples (e.g., see Bromwell and Lambe, 1968). Such differences arise partially because of the difficulties in measuring c_v, partially because of shortcomings in the linear theory of consolidation, and partially because of the two- and three-dimensional effects discussed in Section 27.6. Predictions of rate of consolidation are useful only to indicate in advance of construction the approximate time required for consolidation. If the actual rate of consolidation is critical to the design, as in certain stability problems where the excess

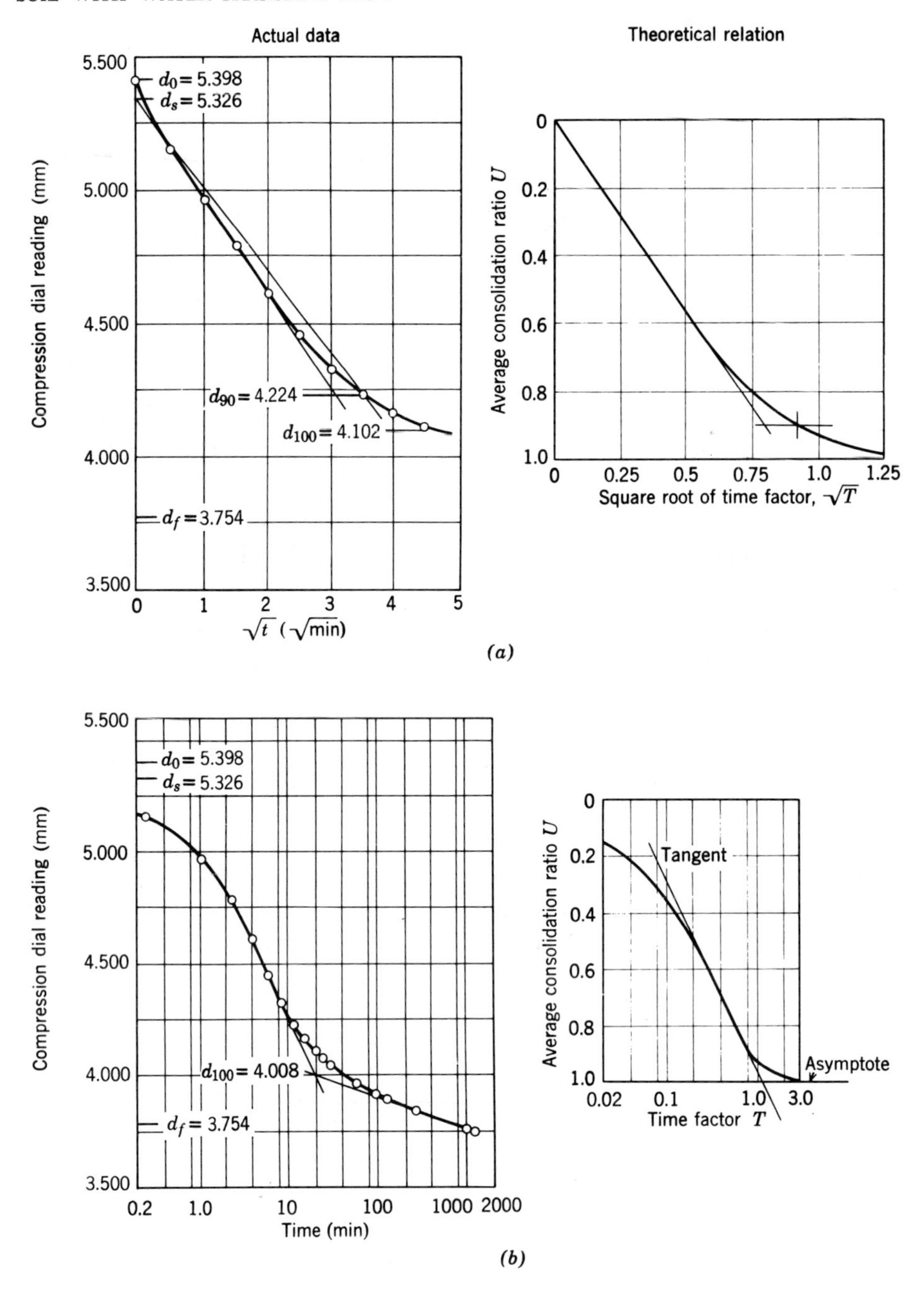

Fig. 27.4 Time-compression compression curves from laboratory tests. Analyzed for c_v by two methods. (*a*) Square root of time fitting method. (*b*) Log time fitting method. (From Taylor, 1948.)

pore pressures must be known accurately, pore pressures must actually be measured in the field as construction proceeds.

Table 27.1 gives typical values of c_v for a range of soils. From the form of Eq. 27.3, c_v should increase with increasing permeability and decreasing compressibility. Permeability is by far the more important parameter.

27.4 OTHER ONE-DIMENSIONAL SOLUTIONS

This section presents solutions for several common problems in which one or more of the conditions stated at the beginning of Section 27.2 are changed.

Table 27.1 Typical Values for Coefficient[a] of Consolidation c_v

Liquid Limit	Lower Limit for Recompression	Undisturbed Virgin Compression	Upper Limit Remolded
30	3.5×10^{-2}	5×10^{-3}	1.2×10^{-3}
60	3.5×10^{-3}	1×10^{-3}	3×10^{-4}
100	4×10^{-4}	2×10^{-4}	1×10^{-4}

Source: U.S. Navy, 1962.

[a] c_v in cm²/sec.

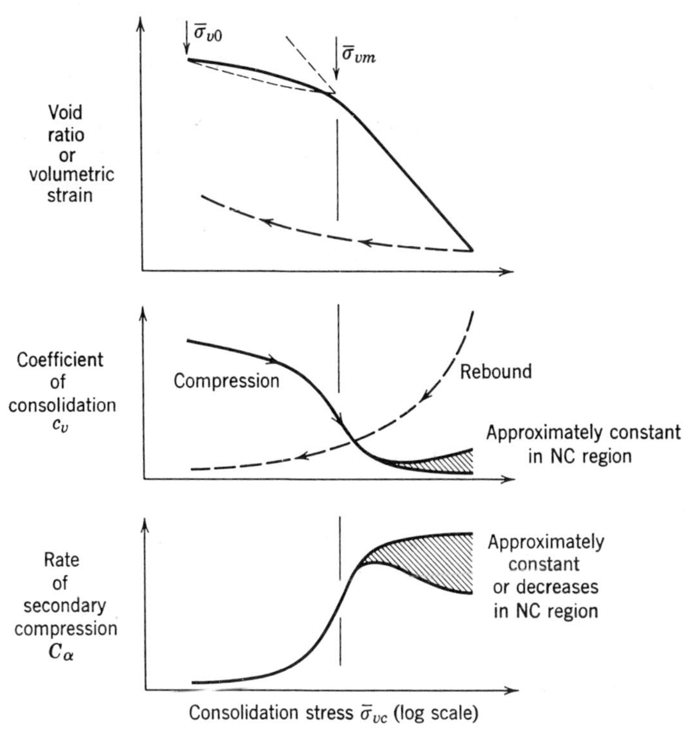

Fig. 27.5 Typical variation in coefficient of consolidation and rate of secondary compression with consolidation stress.

Triangular Initial Excess Pore Pressure

This situation arises when the boundary conditions change at only one boundary of a stratum. Figure 26.4 depicts such a situation. The piezometric level in the sand above the clay is lowered by pumping while the piezometric level in the gravel, which connects to some distant source of water, remains unchanged. This change sets up a gradient through the clay and the eventual equilibrium condition will involve upward seepage. However, before equilibrium is reestablished the pore pressures represented by the triangular area in Fig. 26.4 must dissipate. This consolidation process involves increasing effective stresses and settlement of the clay.

Figure 27.6 gives the consolidation ratio U_z as a function of dimensionless depth and time. In contrast to Fig. 27.2, the curves are no longer symmetrical about mid-depth. Consolidation begins at the boundary where the piezometric level was changed, since gradients are initially very large in this region. Gradually the final equilibrium gradient is established throughout the stratum. The relation between time and *average* consolidation ratio U proves to be exactly the same as for uniform initial excess pore pressure! Thus Fig. 27.3 applies for *any* linear initial distribution of excess pore pressure, provided that there is double drainage. When using Figs. 27.3 and 27.6 for the situation in Fig. 26.4, H is one-half the total thickness of the stratum. At first sight this is surprising since there would appear to be single drainage, but the mathematical solution proves that the system acts as though there is double drainage. Example 27.4 illustrates the solution to such a problem.

Ground water lowering during construction may cause undesirable settlements of adjacent structures, but if properly controlled such dewatering may be used to preconsolidate clay prior to construction of a building. Another common situation giving undesirable settlements is pumping of water from an aquifer lying beneath compressible clay. Such pumping lowers the piezometric level at the bottom of the clay and leads to a gradual compression of the lower part of the clay. This type of pumping has caused the great subsidence in Mexico City (Fig. 1.3) and elsewhere.

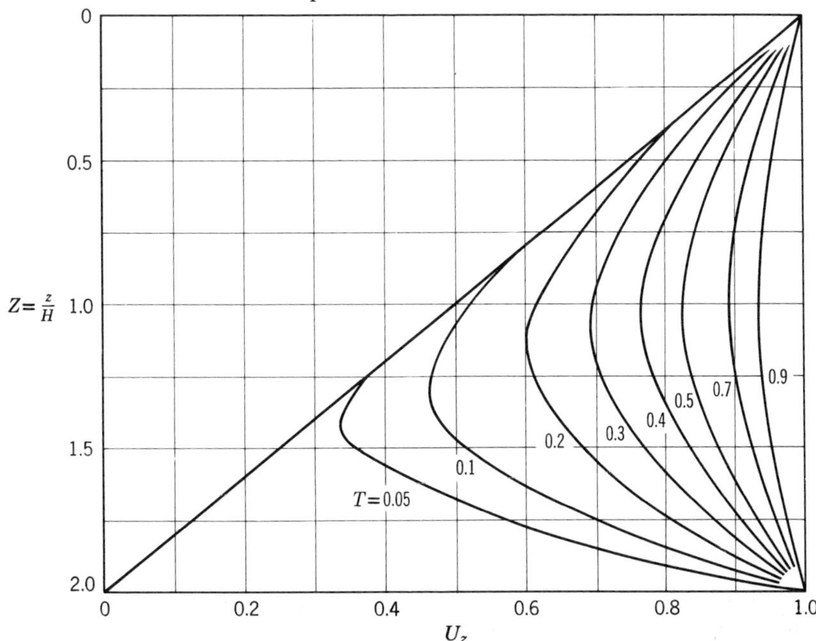

Fig. 27.6 U_z versus Z for triangular initial excess pore pressure distribution.

▶Example 27.4

Given. Soil profile of Fig. E27.1-1 with ground water lowering of 3.05 m as shown in Example 25.8.

Find. Results requested in Examples 27.1 and 27.2.

Solution.

$$H = 2.13 \text{ m}, \quad Z = 1.5, \quad T = 0.092$$

Interpolating in Fig. 27.6, $U_z = 0.5$

$$u_e = 29.9(1 - 0.5) = 15.0 \text{ kN/m}^2$$

$$u = 53.7 - 15.0 = 38.7 \text{ kN/m}^2$$

$$\Delta\bar{\sigma}_v = 60.4 + (0.75 - 0.50)29.9 = 67.9 \text{ kN/m}^2$$

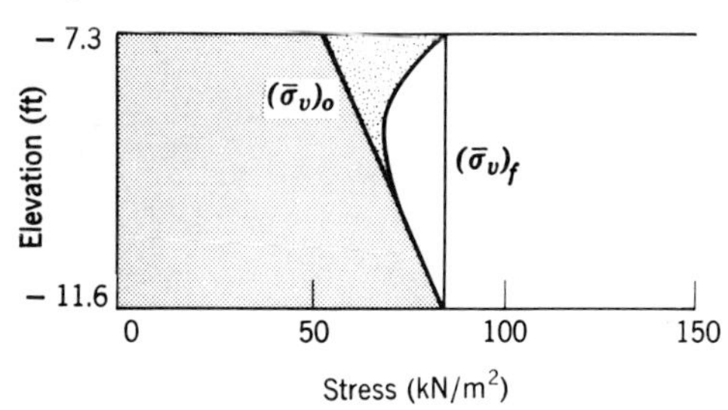

Fig. E27.4-1

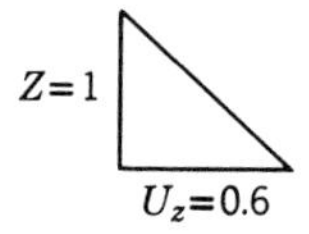

Fig. E27.4-2

The stresses and pore pressures at 4 months are shown in Fig. E27.4-1. The slope of the tangent is about as in Fig. E27.4-2. Hence, the transient portion of the gradient and velocity are:

$$i = \frac{1}{9.81}(0.6)\frac{29.9}{2.13} = 0.86$$

$$v = 0.018(0.86) = 0.015 \text{ m/yr upward}$$

The final settlement is computed in Example 25.8 to be 0.152 m. Using $T = 0.276\, t$ (Example 27.2) the time-settlement curve may be constructed. At 1 yr, the settlement is

$$\rho = \frac{0.152}{0.653}(0.380) = 0.088 \text{ m} = 88 \text{ mm}$$

Time-Varying Load

In many problems, the time required to increase the load to its final value is a significant part of the time required for consolidation. Hence initially there is an interval during which consolidation occurs simultaneously with the increase of the load (Fig. 27.7). During this interval Eq. 27.4 with $\partial\sigma_v/\partial t \neq 0$ applies. This initial interval is followed by an ordinary consolidation process with $\partial\sigma_v/\partial t = 0$.

The consolidation due to every load increment proceeds independently of the consolidation due to the preceding and succeeding load increments. Therefore the rate of consolidation during the period of transition can be computed by a process of superposition. Computer programs are available which will yield U versus time for *any* time history of loading (Jordan and Schiffman, 1967).

An approximate but very satisfactory method may be used for the specific case of a linear increase in load followed by a constant load (Taylor, 1948). The settlement at any time t during the construction period is determined by the following rule:

$$\text{settlement} = \begin{pmatrix}\text{settlement for instantaneous}\\ \text{loading, computed using } 0.5t\end{pmatrix} \times \begin{pmatrix}\text{fraction of final}\\ \text{load in place}\end{pmatrix}$$

Following the construction period of duration t_1, the settlement is computed by

$$\text{settlement} = \begin{pmatrix}\text{settlement for instantaneous}\\ \text{loading, computed using } t - 0.5t_1\end{pmatrix}$$

Example 27.5 illustrates the use of this procedure. Charts applicable to this specific case have been prepared by Schiffman (1958).

More than One Consolidating Layer

Figure 27.8 depicts a common situation involving two compressible layers. The behavior of even such a relatively simple system begins to be quite complicated, and depends upon the relative values of k and m_v for the two strata. Figure 27.9 shows values of U_z as a function of Z (based on total thickness) and T (using c_v of upper strata) for a specific problem in single drainage. Even though the upper strata consolidates much more quickly than the lower strata, pore pressures must still exist within the upper strata to provide a gradient for the flow of water and such pore pressures must persist so long as there are pore pressures within the lower strata.

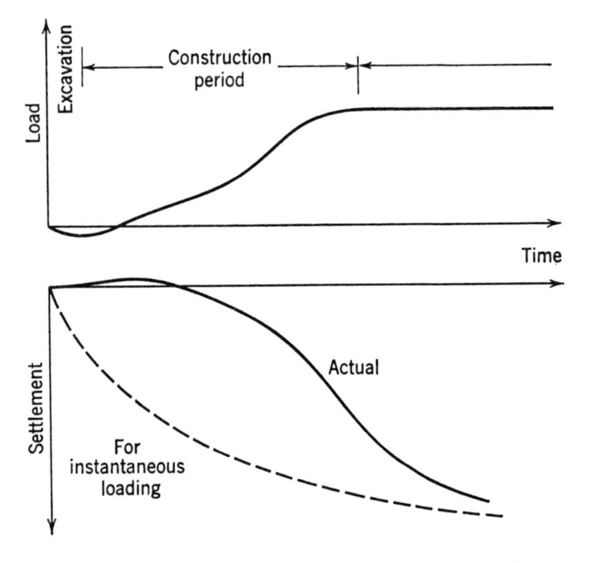

Fig. 27.7 Settlement from time-varying load.

▸Example 27.5

Given. Soil profile and loading of Fig. E27.1-1 with the load increased linearly during 1 yr.

Find. Time-settlement relationship.

Solution. See Fig. E27.5.

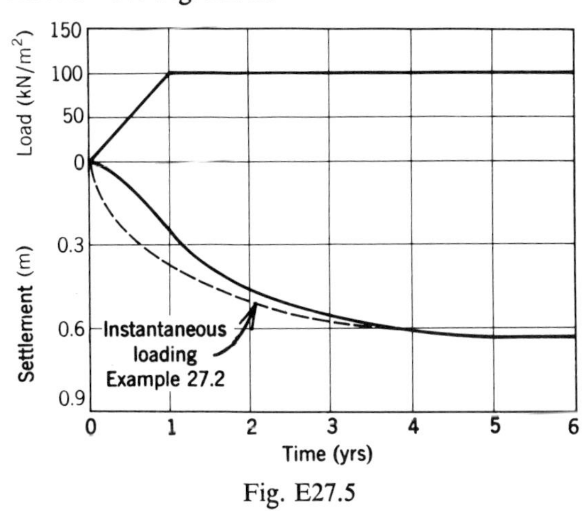

Fig. E27.5

Sample computations:

$t = 3$ mo. From curve for instantaneous loading, settlement at 1.5 mo is 0.12 m. One-quarter of the load is in place.

$$\text{settlement} = 0.12(0.25) = 0.03 \text{ m}$$

$t = 2$ yr. From curve for instantaneous loading, settlement at $t = (2 - 0.5) = 1.5$ yr is:

$$\text{settlement} = 0.46 \text{ m}$$ ◀

Because of the complexity of the problem, no generally applicable charts are possible. Computer programs have been written to handle any possible combination of strata thicknesses and properties (Jordan and Schiffman, 1967). Approximate methods of computation have been developed for two strata with double drainage (U.S. Navy, 1962). If there is double drainage and if one

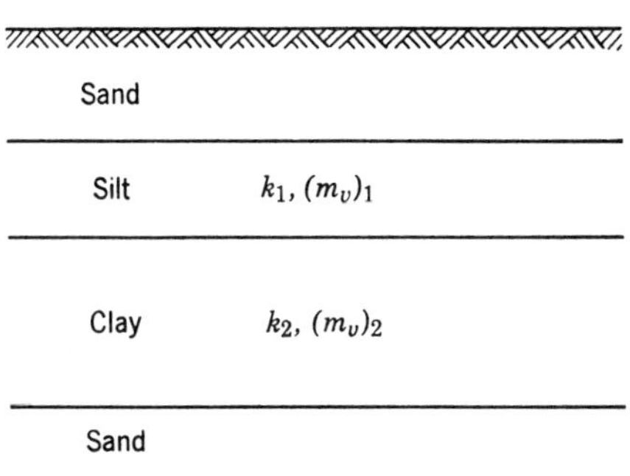

Fig. 27.8 Consolidation problem with two compressible strata.

stratum has a permeability or c_v greater than 20 times the permeability or c_v of the second stratum, the consolidation can reasonably be evaluated in two separate stages: (*a*) first the more permeable stratum consolidating with single drainage and (*b*) then the less permeable stratum consolidating with double drainage.

Other Solutions

Many other solutions are available in the literature. They may be divided roughly into three categories:

1. Solutions to Eq. 27.4 for other boundary or initial conditions. Several useful solutions are presented by Terzaghi (1943). Solutions applicable to the process of sediment formation are of special importance (Gibson, 1958).
2. Solutions that consider continuous variation of k and m_v with depth, as in Schiffman and Gibson (1964).
3. Solutions that consider the variation of k and m_v with stress. There are many versions of this nonlinear problem. Mikasa (1965) has studied the very plausible case where the ratio k/m_v remains constant but the two parameters change. Then the variation

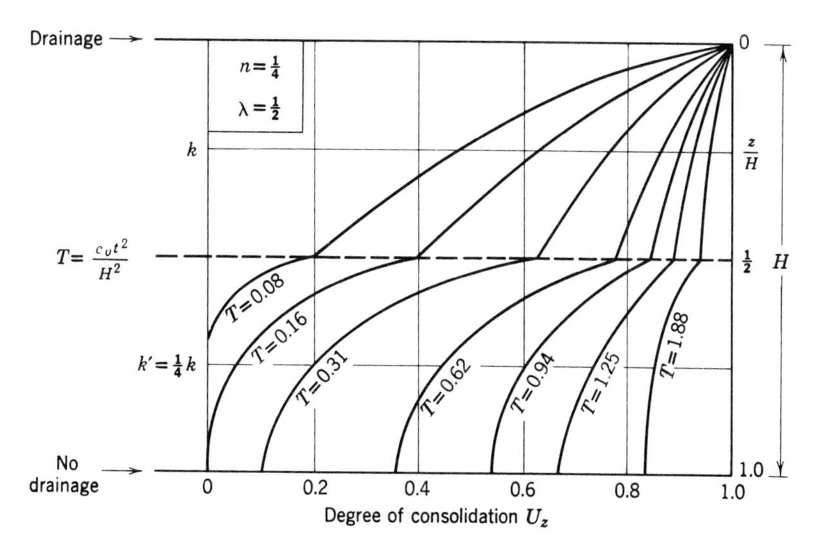

Fig. 27.9 Consolidation of two layers. c_v and k for bottom layer are $\frac{1}{4}$ of values for upper layer. T is based upon c_v of upper layer (From Luscher, 1965).

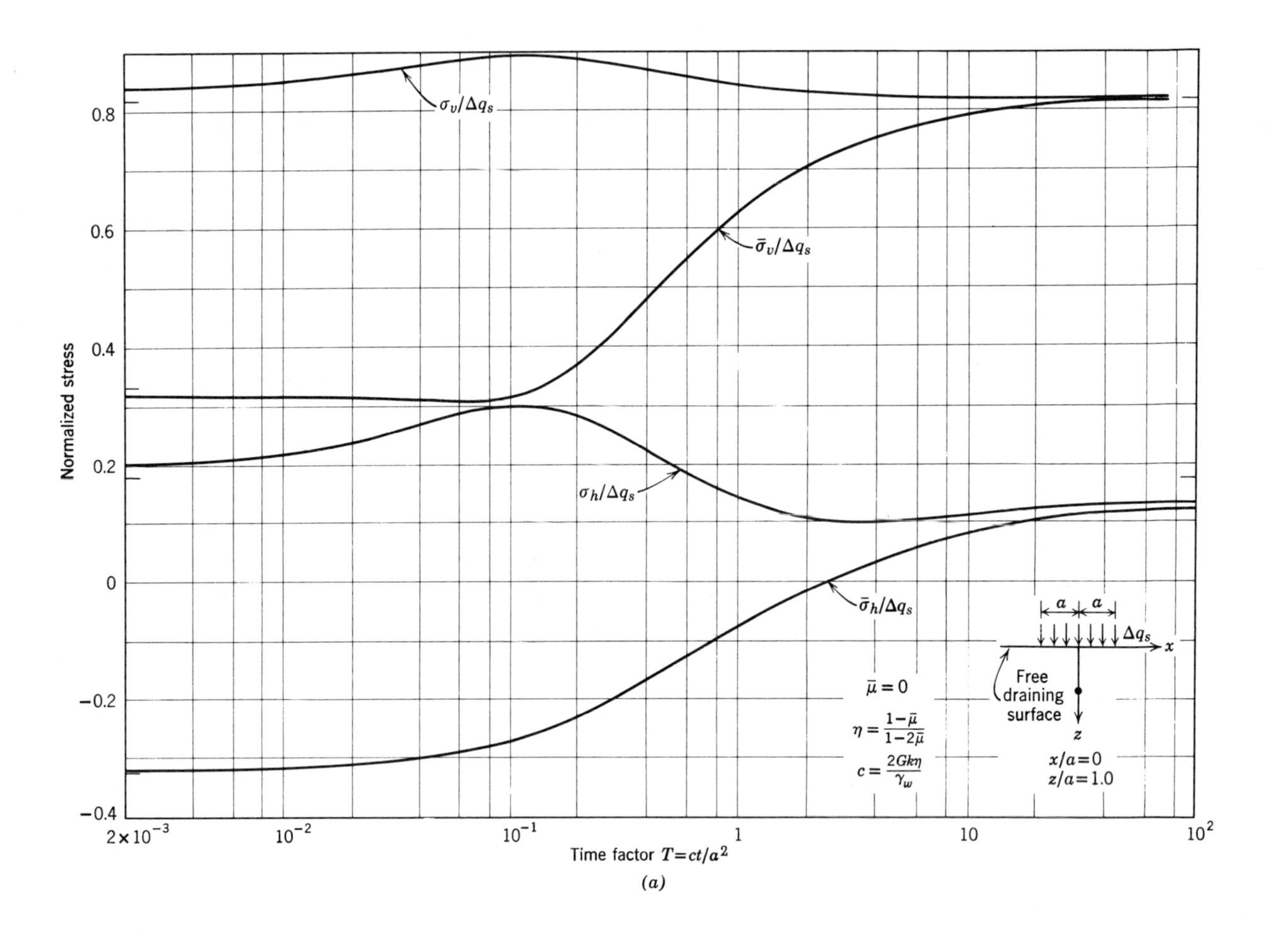

(a)

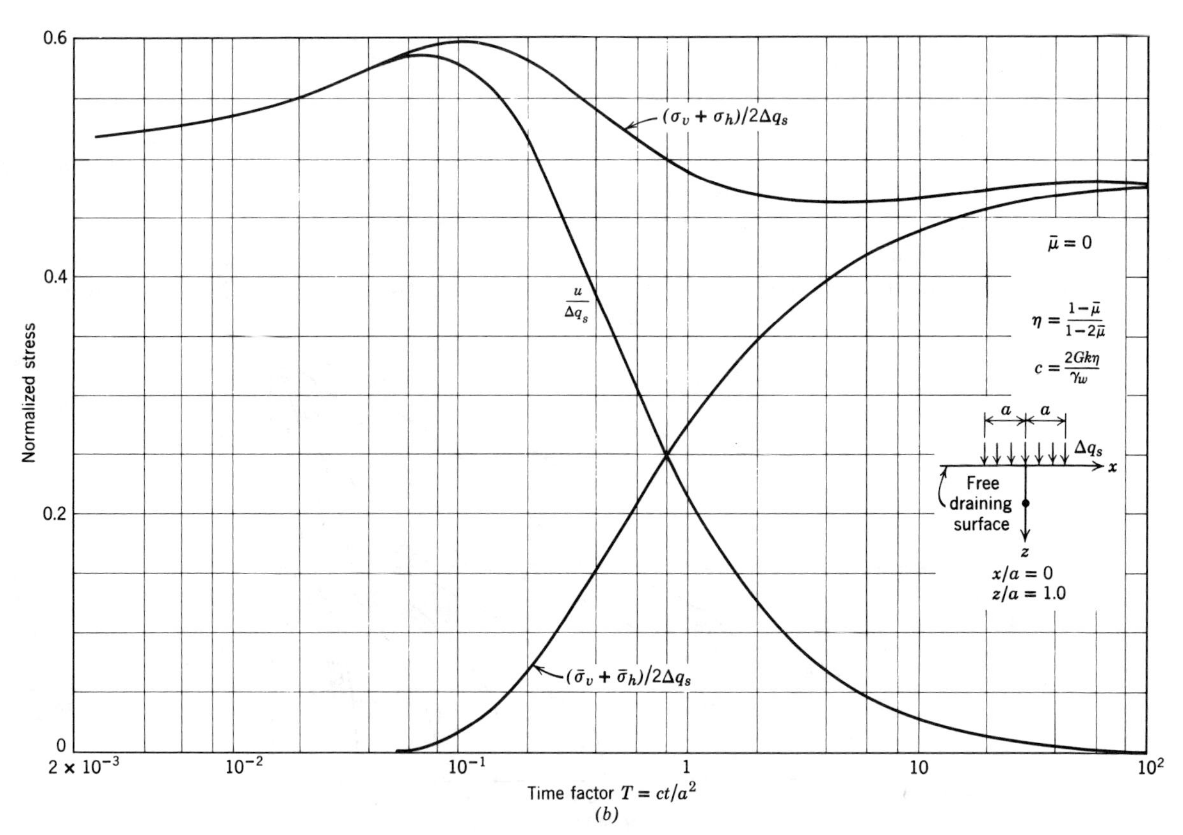

(b)

Fig. 27.10 Two-dimensional consolidation beneath a strip load. (From Schiffman et al., 1967).

of strain (and hence settlement) with depth and time remains as given by the usual consolidation theory, but excess pore pressure follows a different law.

27.5 TWO- AND THREE-DIMENSIONAL CONSOLIDATION

A true three-dimensional theory of consolidation couples the equilibrium of total stresses and the continuity of the soil mass. A pseudo three-dimensional theory uncouples these two phenomena, under the assumption that the total stresses are constant, so that the rate of change of excess pore pressure is equal to the rate of change of volume at all points in the soil (Schiffman, Chen, Jordan, 1967). This condition is only strictly true in special cases. One-dimensional consolidation is one such case. Here the fact of one-dimensional compression provides a situation in which the increment of total stress is uniform and is equal to the applied load. Thus, in using the effective stress equation and the appropriate stress-strain relationship, there is a direct relationship between excess pore pressure and volume change.

The difference between pseudo three-dimensional consolidation and three-dimensional consolidation manifests itself in the variation of total stresses with time. The variation of total stresses is illustrated by the results shown in Fig. 27.10, for the consolidation of a halfspace under a strip load. This is a plane-strain problem. Equation 27.7 applies along with two equilibrium equations which provide the distribution of σ_h and σ_v. Part *a* shows the variation of total and effective vertical and horizontal stresses at a particular point during the consolidation process for these calculations. In plane-strain problems with stress boundary conditions, stresses computed from elastic theory are independent of the elastic constants. The total stresses are then the same at the beginning and at the end of consolidation. They vary, however, with time. Both σ_v and σ_h, and hence their sum (see last term in Eq. 27.7), achieve values during consolidation greater than the initial and final values. As a result (see Fig. 27.10*b*), the excess pore pressure at this point first *increases* before it starts to dissipate.

Such an increase in pore pressure during the early stages of consolidation (called the Mandel-Cryer effect) has been noted in experiments and in theoretical solutions for several different multidimensional problems. Study of the results in Fig. 27.10 further shows that the greatest shear stress at the point is reached at an intermediate stage in the consolidation process. Such increases in pore pressure and shear stress cannot be predicted by any solution that ignores the possible change in total stress. These special effects are greatest when $\bar{\mu}$ of the mineral skeleton is zero and are less important for more realistic values of $\bar{\mu}$.

Circular Load on a Stratum

Figure 27.11 shows the effect of the ratio of loaded radius to strata thickness upon the progress of consolidation, calculated by a theory that fully considers changes in total stress. The curve for $a/H = \infty$ is from the one-dimensional theory for vertical consolidation of a circular slab.[2] Even for $a/H = 1$ (diameter twice the thickness), radial drainage significantly speeds the consolidation process. Such results indicate that use of the ordinary one-dimensional theory can be quite conservative and more results of this type will provide a basis for estimating the actual effect of radial drainage.

27.6 PSEUDO TWO- AND THREE-DIMENSIONAL CONSOLIDATION

Most currently available two- and three-dimensional consolidation solutions are based upon the assumption that the total stresses remain constant. Despite their shortcomings, many of these solutions have proved useful in practice.

Radial Consolidation in Triaxial Test

Figure 27.12 presents results for radial consolidation, neglecting the effect of possible changes in radial total stress. Two boundary conditions on axial strains are possible: (*a*) equal strain at all radii, so that the radial distribution of axial stress changes as consolidation proceeds and (*b*) free strain where the axial stress remains the same at all radii. From the practical standpoint, there is little difference in the results.

If the effect of total stress changes is again ignored, these results for radial drainage may be combined with axial drainage to obtain the average consolidation U_{vh} for a triaxial sample with both types of drainage (see Scott, 1963):

$$U_{vh} = 1 - (1 - U_v)(1 - U_h) \qquad (27.13)$$

where U_v is obtained from Fig. 27.3 and U_h from Fig. 27.12. This theory has been used to study the effect of radial drainage and the influence of side drains in the triaxial test (Bishop and Henkel, 1962; Bishop and Gibson, 1963).

Drain Wells

Vertical sand-filled holes, or other forms of vertical drains, are sometimes used to speed consolidation of a clay stratum (see Sections 34.6 and 34.7). By use of closely spaced wells, the drainage path is changed from

[2] This curve applies for a uniform load on soil underlain by a *smooth* rigid base. This is not the same as the usual one-dimensional case, which effectively assumes a *rough* rigid base.

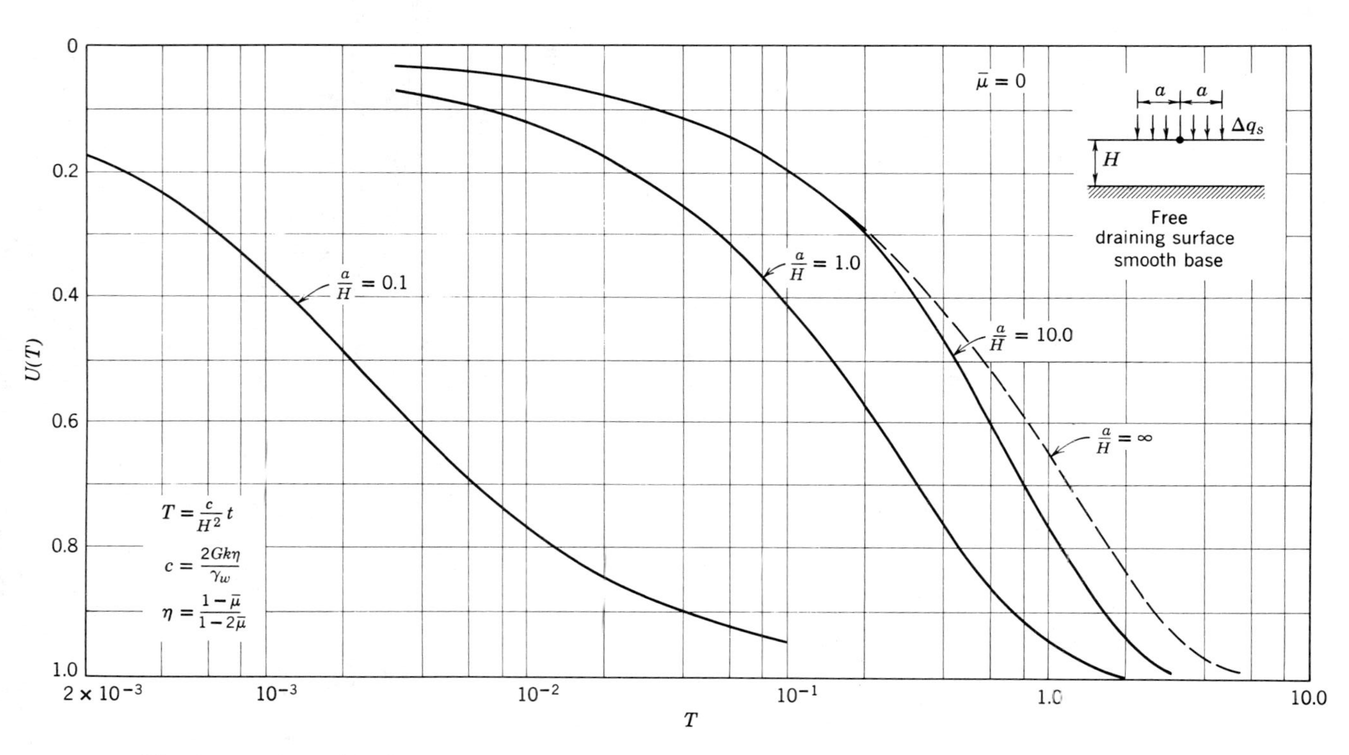

Fig. 27.11 Consolidation of stratum under circular load (From Gibson, Schiffman, and Pu, 1967).

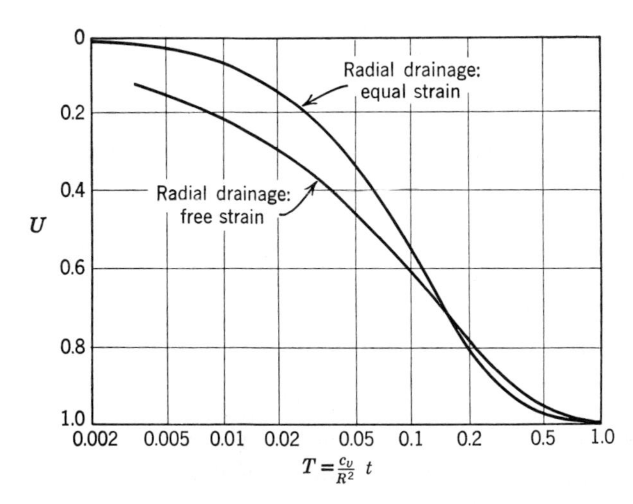

Fig. 27.12 Average consolidation ratio for radial drainage in triaxial test (After Scott, 1963).

the thickness of the stratum to one-half of the well spacing. The fact that the horizontal permeability of soil generally exceeds the vertical permeability adds to the attractiveness of this arrangement.

Consolidation by drain wells involves radial flow. The theory is based on one-dimensional (vertical) strain along with three-dimensional water flow. Figure 27.13 gives the average consolidation ratio for the ideal case. Here $T_r = c_v t / r_e^2$ where r_e is one-half of the well spacing and $r_w = r_e/n$ is the radius of the drain well. There are many practical problems associated with the use of sand drains. Some of these are the shearing of the drains due to slip planes developing in the soil and the development of disturbed "smear" zones of soil on the periphery of the drain. An extensive series of case histories has been compiled (Moran, Proctor, Mueser, and Rutledge, 1958). A review of theories of sand drains and numerical methods of solution has been presented by Richart (1959).

27.7 SECONDARY COMPRESSION

Deviations from the results predicted by the Terzaghi theory of consolidation were first reported by Buisman (1936) and Gray (1936).

A typical set of results from one increment of a consolidation test involving measurement of pore pressures is sketched in Fig. 27.14. These results are based upon careful tests first conducted by Taylor (1942) and since repeated by numerous investigators (see Crawford, 1965 for a recent summary).

The compression during such a test may conveniently be divided into two phases:

1. The compression that occurs while the excess pore pressures dissipate. This compression proceeds with time according to the consolidation theory set forth in this chapter and is generally called *primary compression.*
2. The slow continued compression that continues after the excess pore pressures have substantially dissipated is called *secondary compression.* Actually there must be small excess pore pressures during secondary compression to cause water to flow from the soil. However, secondary compression proceeds very slowly and the velocity of flow is very small. Hence the associated excess pore pressures are immeasurably small.

Secondary compression occurs because the relationship between void ratio and effective stress is usually somewhat time dependent: the longer the clay remains under a constant effective stress, the denser it becomes. Figure 27.15 shows typical curves of stress versus void ratio for a normally consolidated clay, obtained using different durations of continued load following the completion of consolidation.

Figure 27.16 illustrates the effect of specimen thickness on the relative importance of primary and secondary compression for a given soil. As the specimen becomes thinner, the time required to dissipate excess pore pressures becomes shorter. If it were possible to test a very thin specimen, compression could indeed occur in two distinct phases: an *instantaneous compression* and a *delayed compression.* For specimens of finite thickness, the instantaneous and delayed effects are both present during the so-called primary compression. For very thick layers of clay, much of the compression that occurs as excess pore pressures dissipate may actually be delayed compression. Instantaneous compression and the rate of delayed compression are properties of the soil

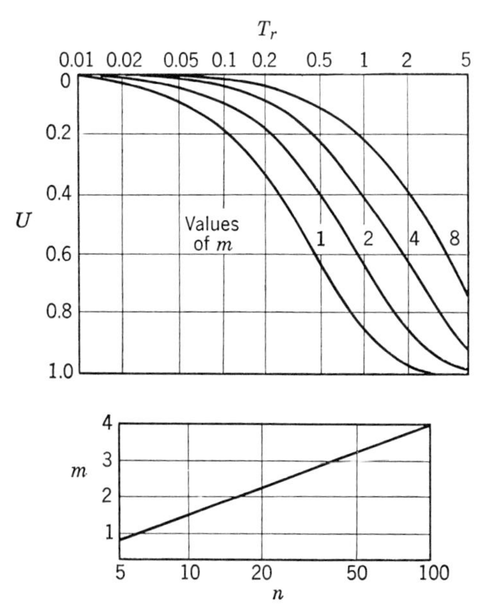

Fig. 27.13 Average consolidation ratio for radial drainage by vertical drains (After Scott, 1963).

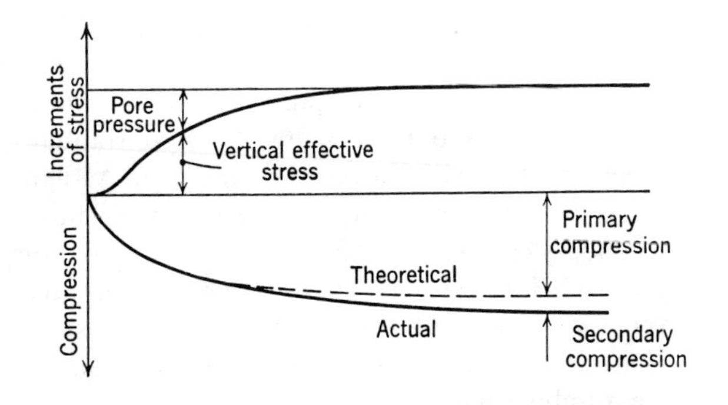

Fig. 27.14 Primary and secondary compression.

skeleton. The relative importance of primary and secondary compression depend on the time required to dissipate pore pressures and hence on the thickness of the soil.

The relative importance of secondary and primary compression varies with the type of soil and also with the ratio of stress increment to initial stress.

The magnitude of secondary compression is often expressed by the slope C_α of the final portion of the time compression curve on semi-log paper (Fig. 27.17). Table 27.2 gives typical values for this slope C_α. The time rate of secondary compression is largest for highly plastic soils and especially for organic soils.

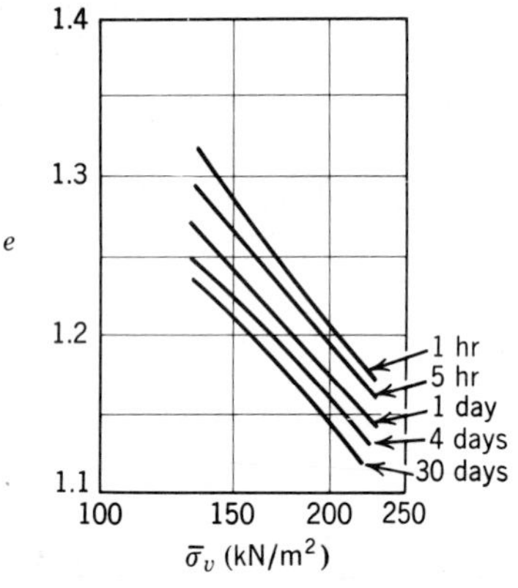

Fig. 27.15 e versus log $\bar{\sigma}_v$ as function of duration of secondary compression (After Bjerrum, 1967).

The ratio of secondary to primary compression is largest when the ratio of stress increment to initial stress is small. This is illustrated in Fig. 27.18, which shows that the usual form of time-compression curve occurs only when the stress increment is large. Fortunately, most problems involving important settlements involve relatively large increments of stress.

Taylor (1942) was the first person to propose a rational theory of secondary compression. This theory modeled the soil skeleton as a viscoelastic material. Recent work in this area is directed at the developing models of behavior and numerical techniques for solving secondary compression problems with complicated rheologic models.

The phenomenon of secondary compression greatly complicates prediction of the time history and final magnitude of settlement. Bjerrum (1967) has discussed this subject. Secondary compression also makes it difficult to determine c_v accurately from laboratory tests.

27.8 SUMMARY OF MAIN POINTS

1. The differential equation of continuity, which is the basis for the study of consolidation, equates the *net* flow to the change in volume of the soil.

Table 27.2 Typical Values for Rate of Secondary Compression C_α

	C_α
Normally consolidated clays	0.005 to 0.02
Very plastic soils; organic soils	0.03 or higher
Precompressed clays with OCR > 2	less than 0.001

From Ladd, 1967.

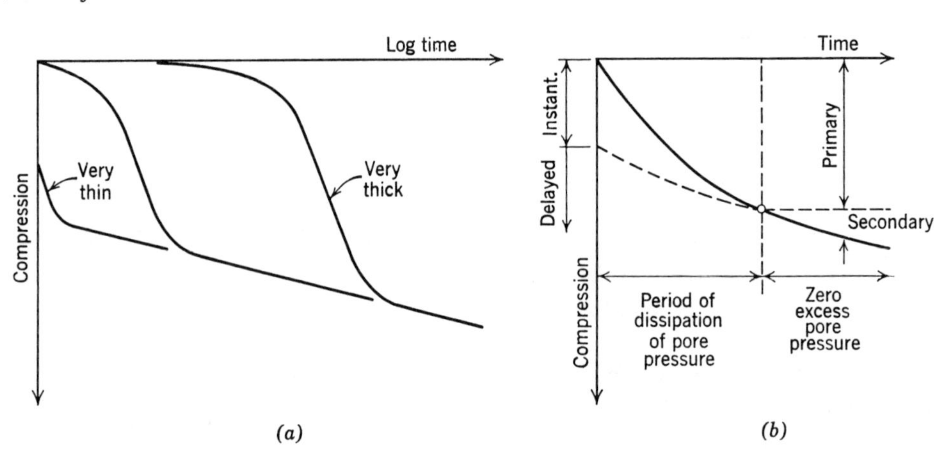

Fig. 27.16 Relation of instantaneous and delayed compression to primary and secondary compression. (*a*) For different thicknesses. (*b*) For a given thickness.

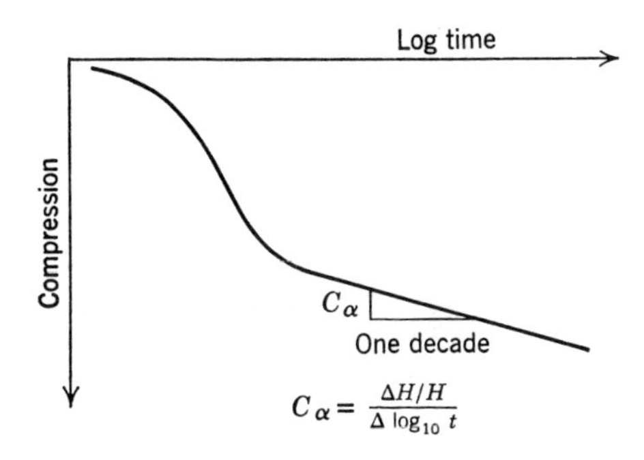

Fig. 27.17 Definition of rate of secondary compression.

2. The time required for essentially complete consolidation is proportional to

$$\frac{H^2 m_v}{k}$$

where

H = length of drainage path
m_v = compressibility of soil
k = permeability of soil

The great difference in the time required to consolidate different soils results primarily from differences in k and H.

3. Solutions in chart form are available for many different one-dimensional problems.
4. Multidimensional problems are complicated by the fact that the total stress at a point generally changes as the soil mass strains during consolidation. Useful solutions for multidimensional problems are being developed.
5. The biggest problem facing the engineer who wishes to apply consolidation theory is the choice of a suitable value for c_v. Because of uncertainty in the evaluation of c_v, the theory usually gives only an order-of-magnitude estimate for the consolidation time.
6. In many problems, secondary compression may be more important than primary consolidation.

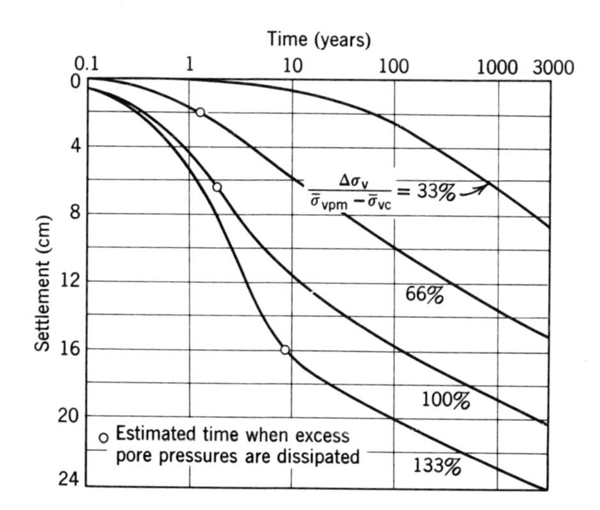

Fig. 27.18 Time-settlement curves for different magnitudes of stress increment. $\bar{\sigma}_{vpm}$ is a pseudo maximum past stress. (From Bjerrum, 1967).

PROBLEMS

27.1 The clay layer shown in Fig. P27.1 has drainage at both its top and bottom surfaces. The coefficient of consolidation c_v is 0.5 m²/yr. A surface loading of 50 kN/m² is applied to this site by placing a fill.

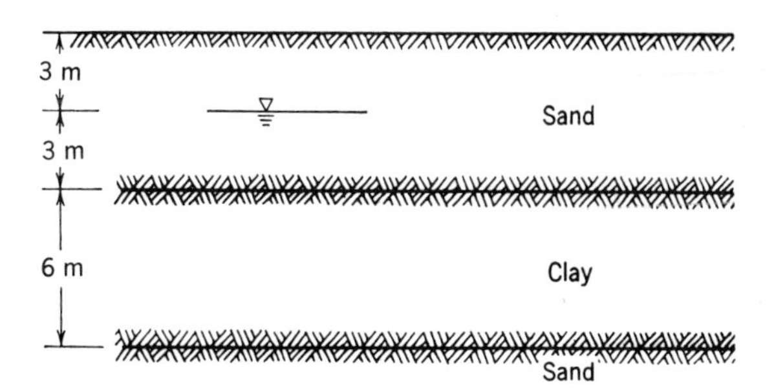

Fig. P27.1

a. Make a neat drawing showing:
(1) The distribution of pore pressure before the fill is placed.
(2) The distribution of pore pressure *immediately after* the fill is placed.
(3) The distribution of pore pressure 6 yrs after the fill is placed.
b. How long will it take for the average consolidation of the layer to exceed 90%?

27.2 After consolidation under the load in Problem 27.1 is complete, the piezometric level in the lower sand stratum is lowered 4.5 m by pumping while the piezometric level in the upper sand strata remains unchanged. Answer all questions asked in Problem 27.1. Does the clay layer compress or expand as a result of this change?

27.3 The soil of Fig. P27.1 is subjected instantaneously to a loading of 50 kN/m². Six years later this load is removed instantaneously. Assuming c_v is the same for both loading and unloading, construct plots of the following quantities versus time:
a. Vertical total stress at mid-depth of the clay.
b. Pore pressure at mid-depth of the clay.
c. Vertical effective stress at mid-depth of the clay.
d. Average excess pore pressure within the clay.

27.4 Considering the difference in the behavior of soil during loading and unloading, answer the following questions relative to Problem 27.3.
a. Will the excess pore pressures after unloading dissipate more or less quickly than dissipation after the initial loading?

b. Will the final thickness of the clay be the same as, or more or less than the initial thickness?

27.5 The soil in Fig. P27.1 is subjected to a load which increases linearly with time for 1 year until the load reaches 50 kN/m^2, and thereafter the load remains constant. Construct the curve of average consolidation ratio versus time.

27.6 Assume that the clay used for the test results shown in Fig. 27.4 came from a stratum 18 m thick with double drainage. How long will it take for the entire stratum to consolidate ($T = 1$)?

27.7 A clay with $c_v = 10^{-3}$ cm^2/sec is formed into a triaxial specimen 5 cm in diameter and 15 cm long. Full drainage is provided at both ends and around the cylindrical surface. Neglecting the effect of changes in total stress during consolidation, how long will be required to achieve 95% consolidation following an increment in chamber pressure?

CHAPTER 28

Drained and Undrained Stress-Strain Behavior

Chapter 26 assumed that stress-strain relations obtained from drained tests could be used to predict the excess pore pressures caused by an undrained loading. This assumption implies that there is an intimate relationship between stress-strain behavior during the two types of loading. This relationship is explored in Chapter 28. The key to understanding the relationship is the principle of effective stress, which tells us that strength is related to *effective* stress and that change in volume is related to change in *effective* stress.

As discussed in Chapter 27, the condition of undrained loading is of great practical importance in the case of clays and silts but only infrequently of importance in the case of sands. Hence this chapter concentrates on the drained and undrained behavior of clays, but the behavior of both sands and clays (and hence of all soils) is basically the same. A more complete treatment of the important concepts developed in this chapter is given by Ladd (1967).

28.1 CONFINED AND ISOTROPIC COMPRESSION

Figure 28.1*a* shows the total stress path and the effective stress path during and subsequent to an increment of undrained loading in an oedometer test. At the beginning of this increment, the total vertical stress is 600 kN/m² and there is a static pore pressure of 200 kN/m². Thus the vertical effective stress is 400 kN/m². The horizontal effective stress is 200 kN/m², corresponding to $K_0 = 0.5$.

Now the vertical stress is quickly increased by an increment of 400 kN/m². Assuming that the time to apply this increment is very small compared to the time required for consolidation, there will be no change in the volume of the soil as the increment is applied. Since the soil is held against lateral strain, zero volume change means zero shear strain. Since there is neither volumetric nor shear strain, it would be expected that there is neither change in effective stress nor change in shear stress. This is exactly what happens. The change in pore pressure is equal to the change in total vertical stress (see Section 26.2) so that the effective vertical stress remains unchanged. The horizontal effective stress also remains unchanged, but the horizontal total stress increases by the amount of the pore pressure increase. The difference between the vertical and horizontal total stresses does not change. Thus the effective stress remains at point *A* during this undrained loading, while the total stress path is *DE*.

The final step is to permit drainage while holding the vertical total stress constant. As drainage occurs, the pore pressure decreases. The result is an increase in vertical effective stress, and the horizontal effective stress must also increase to maintain the K_0-condition. Thus the effective stress path during drainage is *AB*. Since the shear stress is increasing, the horizontal total stress must decrease, and hence the total stress path rises to the left along *EF*.

Figure 28.2 gives the relationship between vertical stress and vertical strain, in terms of both total and effective stress. Note again that strain occurs only when effective stress changes. During drainage there is both volume change and shear strain, corresponding to the increase in both average effective stress and shear stress.

Nearly all oedometer tests are actually carried out in this way. Since pore pressures usually are not measured in these tests, the relationship between effective stress and volume is actually determined only for discreet values of effective stresses: the endpoint for each increment where the excess pore pressures have dissipated and the effective stress is hence known.

It is possible to perform an oedometer test in such a way that no excess pore pressures are developed. It is only necessary to apply stress so slowly that dissipation can occur simultaneously with the loading; i.e., to perform a true drained test. Several such tests have been carried out (e.g., see Crawford 1964) and thus it has been

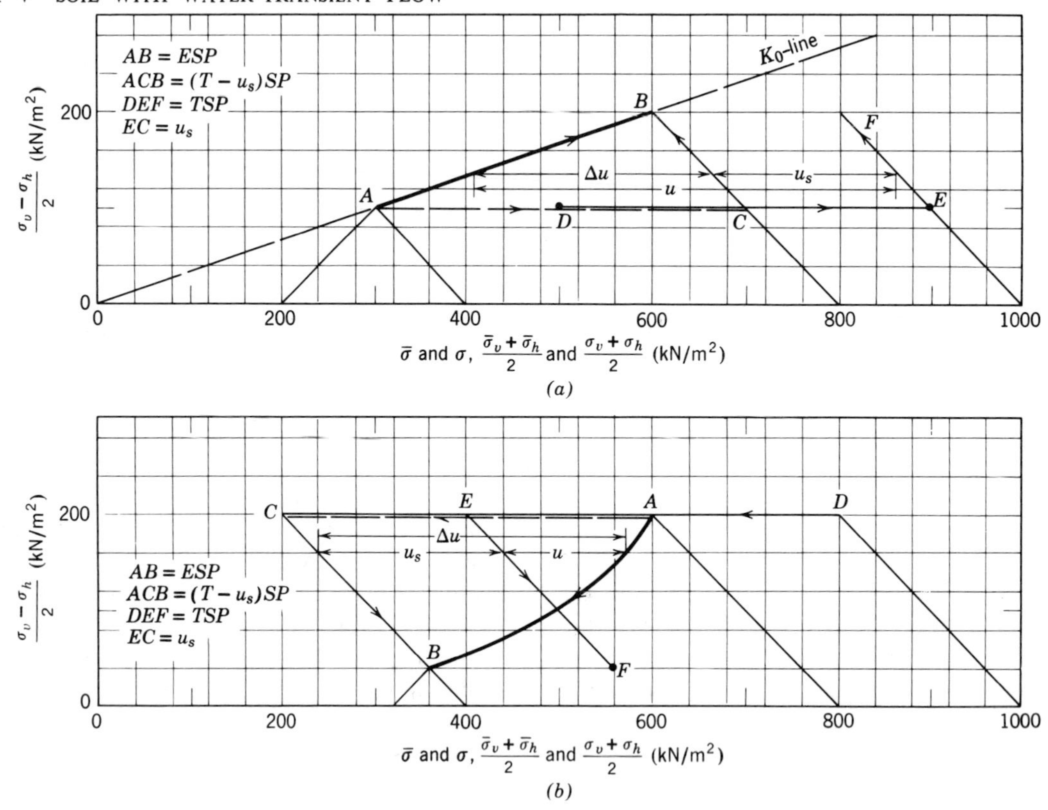

Fig. 28.1 Stress paths for oedometer test. (*a*) Loading. (*b*) Unloading.

possible to compare curves of volume change versus effective stress from drained tests and tests with increments of undrained loading followed by drainage.

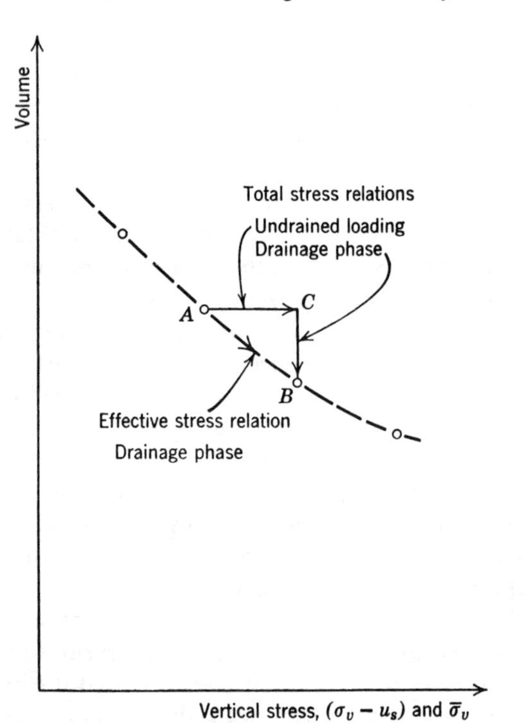

The "drained" stress-strain relation is usually obtained from a series of increments *AB*.

Fig. 28.2 Stress-strain relationships during oedometer test.

Curves from the two types of tests are quite similar and in many cases almost identical. Differences do arise and should be expected because, as we saw in Chapter 22, the volumetric stress-strain relation is not unique but is influenced by such factors as the size of the load increment.

Figure 28.1*b* shows total and effective stress paths for an increment of undrained unloading followed by dissipation of the excess pore pressures. The total stress path is *DE* followed by *EF*. The effective stress path remains at *A* during the undrained unloading, and then follows *AB*. The excess pore pressure is negative, corresponding to the decrease in vertical total stress. The negative excess pressure causes water to be sucked into the soil, and the soil thus swells during the dissipation phase.

Behavior during isotropic compression is quite similar to that during confined compression, except that there never is any shear stress in the soil. It would thus be expected that only volume change occurs during drainage. Because soil seldom is purely isotropic, some shear strain may actually occur during the drainage phase.

28.2 VARIOUS TYPES OF UNDRAINED TESTS

In Chapter 20, drained stress-strain behavior was studied using a triaxial test in which the soil was first consolidated under an all-around confining stress, and

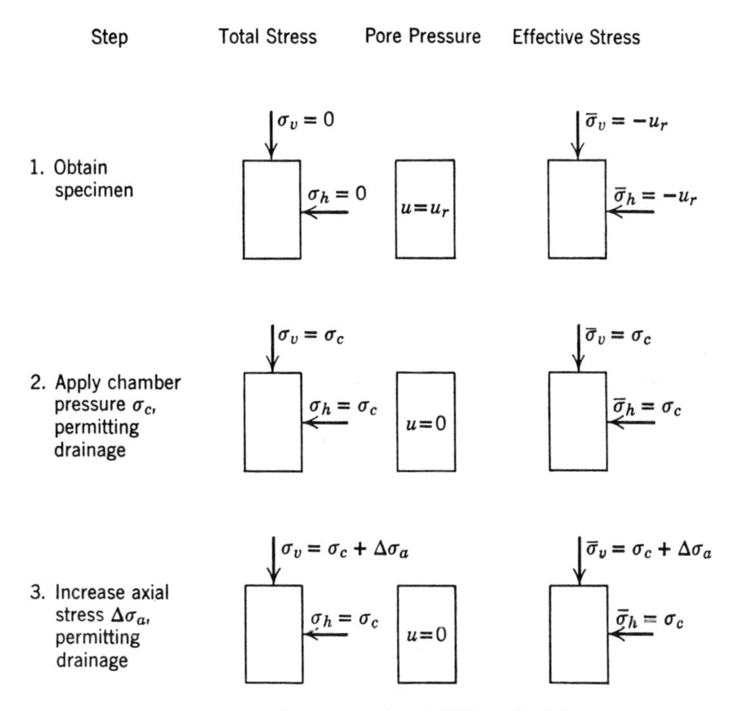

Fig. 28.3 Consolidated-drained (CD) triaxial test.

was then compressed (and hence sheared) by increasing the vertical stress. This is called a *consolidated drained* (CD) test. Figure 28.3 shows the sequence of steps during a series of CD tests on different specimens using different confining stresses. In this figure, u_r denotes the residual pore pressure after sampling of the soil. This is a negative pore pressure associated with capillary tensions. In steps 2 and 3, it is assumed that the drainage line is held at zero pore pressure. If some other static pore pressure u_s is used, u_s must be subtracted from the effective stresses shown for steps 2 and 3.

In the laboratory, the undrained condition is usually studied by means of triaxial tests in which drainage is prevented. There are two types of such tests.

Unconsolidated Undrained Tests

As illustrated by the sequence of events shown in Fig. 28.4, *unconsolidated undrained* (UU) tests represent an opposite extreme from CD tests. Here no consolidation is allowed in step 2, and no drainage occurs in step 3. A UU test can be performed using the type of triaxial testing equipment described in Chapter 9. The drainage line from the specimen must be kept closed. Since there is no concern about the time for water to flow into or out of the specimen, there are no restrictions on the rate at which the additional axial stress may be applied in step 3.

If the soil under test has a very low permeability, it is *in principle* possible to perform a UU test without closing the drainage line from the specimen. That is, if the time lapse between steps 2 and 3 is very short and if the axial stress is applied rapidly, there will be insufficient time for movement of water to or from the specimen. Later in the chapter we shall see that this principle forms the basis for a special type of UU test: the *unconfined compression test*. However, *in actuality* a UU test performed in this way seldom provides high-quality results.

Since the stresses may be applied rapidly, and must be applied rapidly for some forms of the test, a UU test is also called a *Quick* (Q) test.

During a UU test the water content of a soil remains unchanged from its value at the end of the preliminary step. If the soil is saturated, the volume likewise remains unchanged. Excess pore pressures generally develop during both step 2 and step 3. The use of the word "unconsolidated" to describe this test is somewhat misleading, since a truly unconsolidated, saturated clay would be little more than a soup. Use of "unconsolidated" in connection with a UU test really means that there has been no *further consolidation* beyond that which either nature or man provided prior to step 1.

Consolidated Undrained Tests

This type of test (Fig. 28.5) combines step 2 of a CD test with step 3 of a UU test. Consolidation is permitted during step 2 but not during step 3, and hence the name *consolidated undrained* (CU) has emerged.[1] Previous comments concerning step 2 of a CD test and step 3 of a UU test apply here.

[1] Such a test is also called a *consolidated quick* (Q_c or *R*—simply because *R* falls between *Q* and *S* in the alphabet) test.

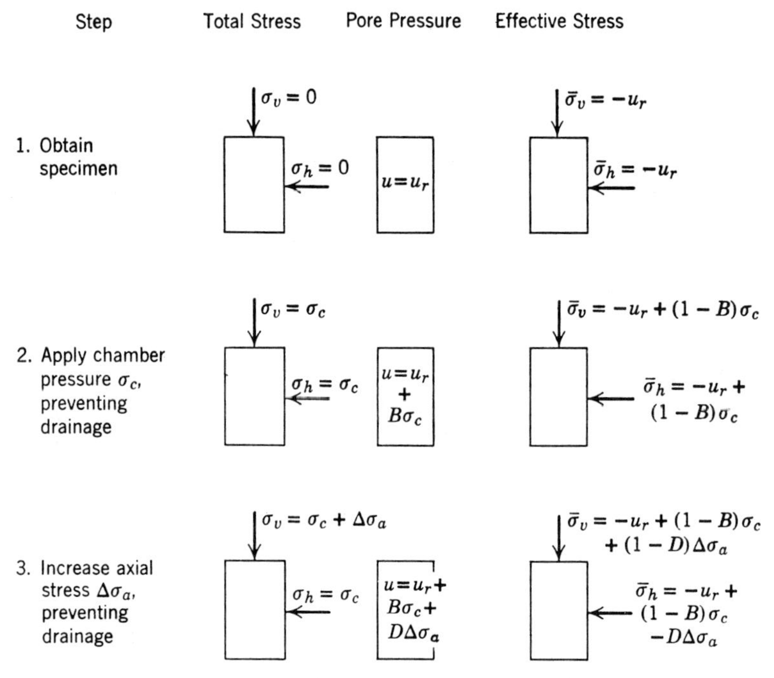

Fig. 28.4 Unconsolidated-undrained (UU) triaxial test.

Measurement of Pore Pressure

If the purpose of a triaxial test is solely to measure the strength during undrained shear, there is no need to measure the pore pressures developed during the test. However, measurement of pore pressures permits a determination of the effective stresses existing during undrained loading and leads to an understanding of the relationship between undrained and drained strength. The symbols $\overline{\text{UU}}$ and $\overline{\text{CU}}$ are used to denote undrained tests in which pore pressures are measured.

Techniques used to measure pore pressures during undrained triaxial tests have already been described in Chapter 17.

Use of Direct Shear Tests

Using a principle first developed by Taylor (1952), O'Neill (1962) has perfected a technique for carrying out CU tests using the direct shear apparatus. The normal load acting on the specimen is varied during shear in such a way that the thickness of the specimen is held constant.

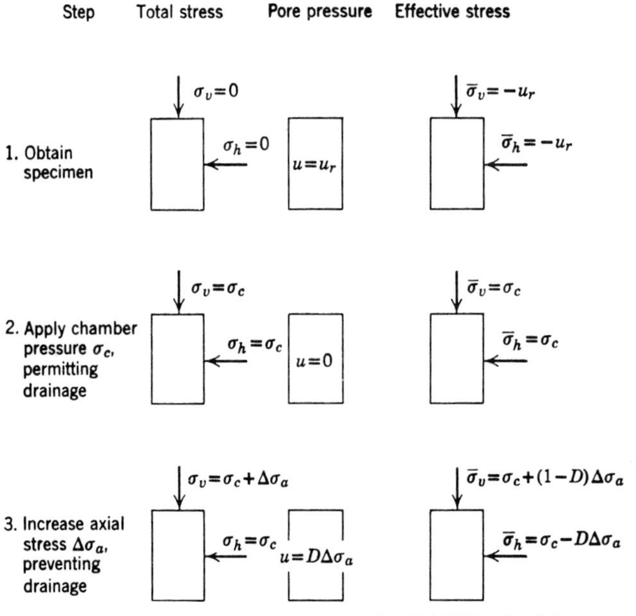

Fig. 28.5 Consolidated-undrained (CU) triaxial test.

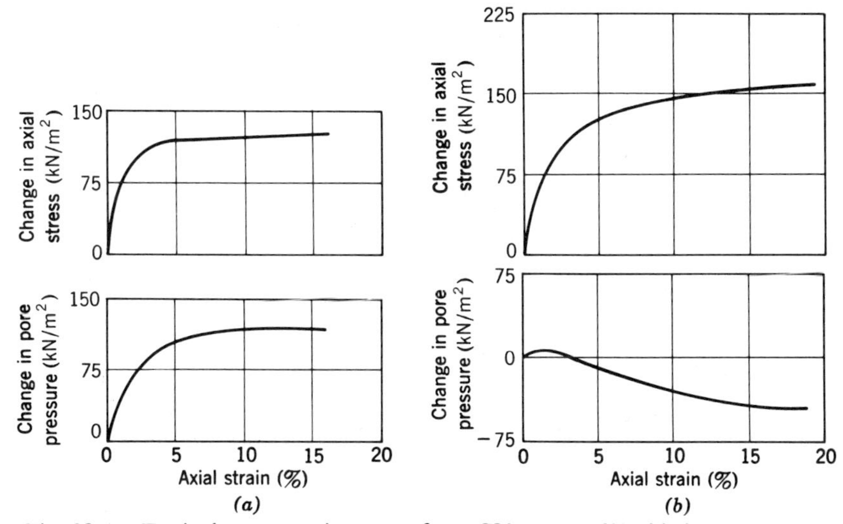

Fig. 28.6. Typical stress-strain curves from CU tests on Weald clay. (*a*) Specimens normally consolidated to 207 kN/m². (*b*) p_m = 827 kN/m²; p = 69 kN/m².

28.3 RELATIONSHIP BETWEEN DRAINED AND UNDRAINED STRENGTH

In earlier chapters dealing with strength (e.g., Chapter 21) we saw that soil would change in volume during shear if such volume changes were permitted. If volume changes cannot occur during shear, we would expect the magnitude of the strength to be different from that found when volume changes do occur. However, as we shall see in this section, drained and undrained strength, which usually differ in magnitude for a given specimen, are related through the effective stress principle. To show this, results of CU tests on Weald clay will be compared to the results of CD tests on this same soil (Henkel, 1956). In this section and in Section 28.4 the soil is assumed to be completely saturated. Section 28.7 considers the strength behavior of partially saturated soils.

Typical Stress-Strain Behavior

Figure 28.6 shows typical stress-strain curves from tests on a normally consolidated specimen and a heavily overconsolidated specimen of the Weald clay. Figures

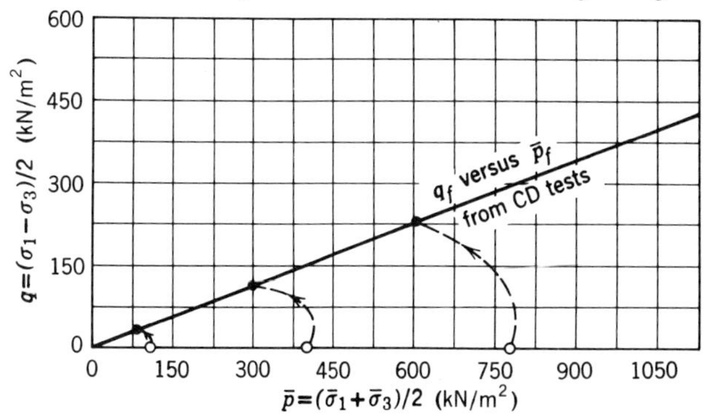

Fig. 28.7 Effective stress paths from CU tests on normally consolidated Weald clay.

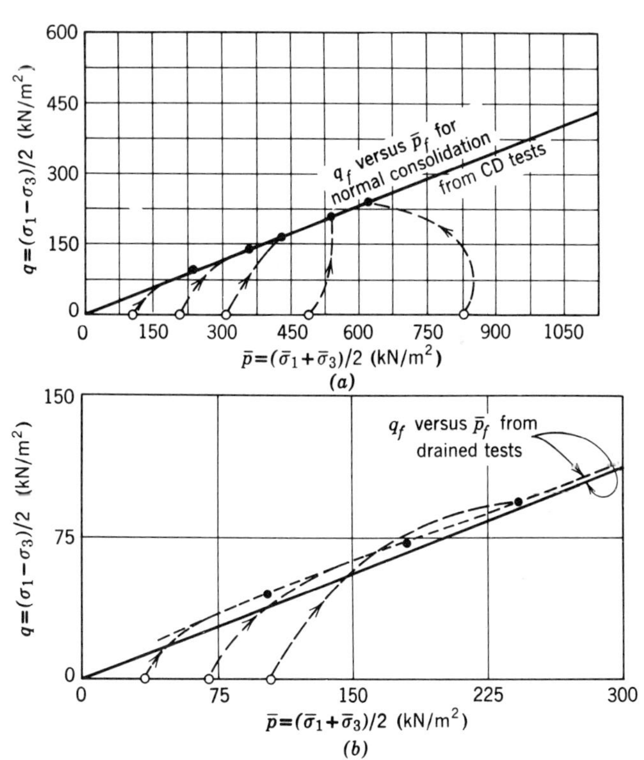

Fig. 28.8 Effective stress paths from CU tests on overconsolidated Weald clay. $\bar{p}_m$ = 827 kN/m².

28.7 and 28.8 show, respectively, effective stress paths for tests upon a family of normally consolidated specimens and upon a family of specimens all previously consolidated to $\bar{p}_m$ = 827 kN/m². The dots at the end of the effective stress paths give the values of $\bar{p}_f$ and q_f at the peaks of the stress-strain curves. The q_f versus $\bar{p}_f$ relations drawn on these figures will be discussed later.

These results follow the general patterns discussed in Chapter 26. Positive excess pore pressures developed during tests on normally consolidated specimens, and, as shown by the effective stress paths in Fig. 28.7, $\bar{p}$

decreased during these tests (after a slight increase at the start of test). The decrease in $\bar{p}$, which usually means an increase in volume, was just enough to compensate for the decrease in volume that usually accompanies the shearing of a normally consolidated clay (Chapter 21). Since there was no net volume change, the strain necessary to fail the specimen was less in a CU test than in a CD test.

In the heavily overconsolidated specimen, negative excess pore pressures developed and $\bar{p}$ increased markedly during this test. Thus the tendency toward volume increase during shear of a heavily overconsolidated clay (Chapter 21) is counteracted by an increase in effective stresses. For heavily overconsolidated specimens, the tendency toward volume expansion exists out to large strains and consequently the excess pore pressure induced by undrained shear continues to increase to large strains. These decreasing pore pressures imply increasing effective stress, and the stress-strain curve continues to rise out to very large strains.

Induced Pore Pressures

From the effective stress paths, it is possible to determine the values of the pore pressure and pore pressure parameter A at any stage of loading, using the procedures developed in Chapter 26. These procedures are illustrated in Examples 28.1 and 28.2. A_f depends on the degree of preconsolidation. Values for Weald clay are given in Fig. 28.9 as a function of overconsolidation ratio ($\bar{p}_m/\bar{p}_0$). For OCR of about 4, there is no excess pore pressure at failure during undrained shear. The range of values given on this chart is typical for many clays, although the location of the crossover point varies.

▶ Example 28.1

Given. Normally consolidated Weald clay with $\bar{p}_0 = 206.8$ kN/m².
Find. σ_1, $\bar{\sigma}_1$, σ_3, $\bar{\sigma}_3$, and u when
a. $q = 34.5$ kN/m².
b. q is at its peak value.
Solution. Figure E28.1 is a blown-up version of the q versus $\bar{p}$ diagram of Fig. 28.7, using an effective stress path interpolated between those for $\bar{p}_0 = 110$ kN/m² and $\bar{p}_0 = 407$ kN/m².

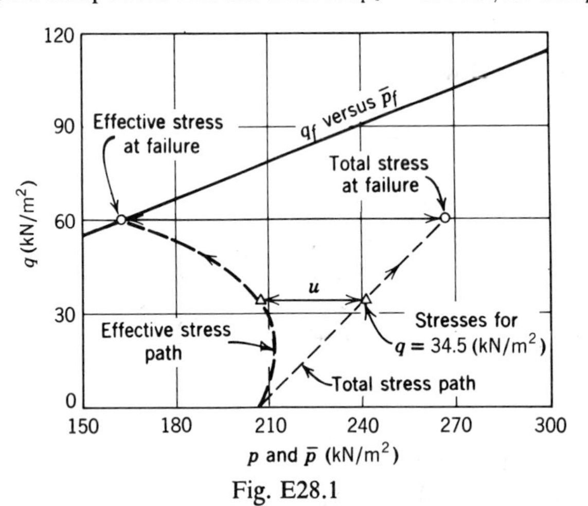

Fig. E28.1

Stress (kN/m²)	(a)	(b)	
q	34.5	60.0	given
p	241.3	266.8	from graph
σ_1	275.8	326.8	$p + q$
σ_3	206.8	206.8	$p - q$
$\bar{p}$	206.8	160.6	from graph
$\bar{\sigma}_1$	241.3	220.6	$\bar{p} + q$
$\bar{\sigma}_3$	172.3	100.6	$\bar{p} - q$
$u = \Delta u$	34.5	106.2	$p - \bar{p}$
A	0.50	0.89	$\Delta u/\Delta(\sigma_1 - \sigma_3)$

◀

▶ **Example 28.2**

Given. A specimen of Weald clay with $\bar{p}_m = 827\ \text{kN/m}^2$ and $\bar{p}_0 = 69$ kN.

Find. σ_1, $\bar{\sigma}_1$, σ_3, $\bar{\sigma}_3$, u, and A at failure.

Solution. Figure E28.2 is copied from Fig. 28.8.

$$
\begin{aligned}
q_f &= 70.3\ \text{kN/m}^2\\
p_f &= 139.3\\
\sigma_{1f} &= 209.6\\
\sigma_{3f} &= 69.0\\
\bar{p}_f &= 179.3\\
\bar{\sigma}_{1f} &= 249.6\\
\bar{\sigma}_{3f} &= 109.0\\
u_f &= -40.0\\
A_f &= -0.28
\end{aligned}
$$

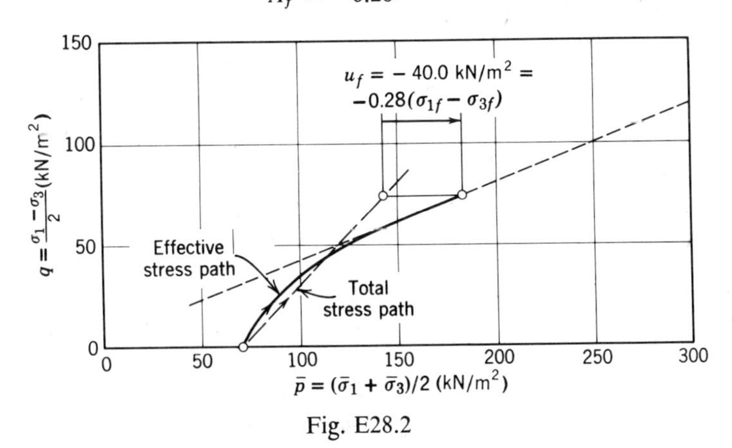

Fig. E28.2 ◀

Effective Stress Path for Various Loadings

Thus far we have considered only undrained triaxial tests in which failure was induced by increasing axial stress, i.e., a compression loading test. Other loadings that increase the axial strain are possible; these include holding axial stress constant while decreasing lateral stress (compression unloading), or increasing axial stress and decreasing lateral stress in such a way that p is a constant. If a group of specimens, all consolidated to the same effective stress and having the same water content, are subjected to these various types of undrained tests, an important fact emerges: *the effective stress path and the undrained strength are the same for each type of loading.*

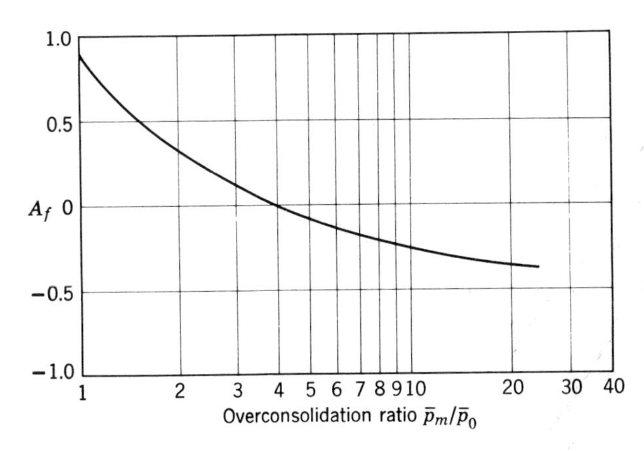

Fig. 28.9 Pore pressure parameter A_f for Weald clay.

Other loadings are also possible; one example is an extension unloading in which the axial stress is decreased while the lateral stress is held constant. When all possible undrained loadings are considered, we find that effective stress path and undrained strength are somewhat dependent on loading (see Section 29.4). However, as a first approximation, it can be said that *undrained strength and undrained effective stress path are dependent only upon the initial conditions existing before shear and are independent of the way in which shear is applied.*

Relations between Drained and Undrained Strength

The effective stress-strength relations found using drained tests (i.e., the curves from Chapter 21) have been superimposed upon the undrained effective stress paths in Figs. 28.7 and 28.8. An important fact is immediately evident. The relationship between q_f and $\bar{p}_f$ is the same regardless of whether the clay is sheared with full drainage or with no drainage.

Figures 28.10 and 28.11 show the relationships between stresses at failure and water content at failure. Here the data points are from undrained tests while the lines and curves are those from Chapter 21. Now we see another important fact: the $q_f - \bar{p}_f - w_f$ relations obtained for drained shear also apply to undrained shear. Now we can understand the concept toward which we

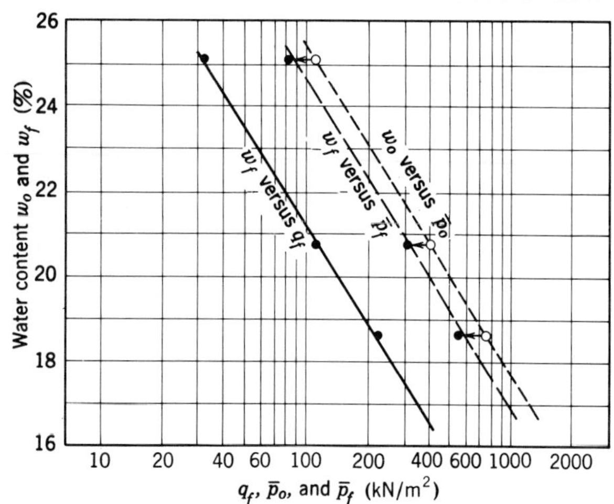

Fig. 28.10 Stress-volume relationships for normally consolidated Weald clay.

started working in Chapter 21: *for a given clay with a given stress history, there is a unique* $q_f - \bar{p}_f - w_f$ *relation which applies independent of the type of loading and the degree of drainage during loading.*

The foregoing result provides a complete unity for the shearing resistance of clay under a variety of loading conditions. That is, regardless of how the soil is sheared, the relationship between strength and effective stress remains the same. However, if two specimens of a given clay are consolidated to the same stress $\bar{p}_0$, and one then is sheared with full drainage and the other without further drainage, different values of strength q_f will result. This difference is explained by the difference in the pore pressures, and hence effective stresses, existing within the two specimens.

By using the pore pressure parameter A_f it is possible to derive an expression connecting undrained shear

▶ **Example 28.3**

Given. Normally consolidated Weald clay with $\bar{p}_0 = 207$ kN/m².

Find. q_f and w_f for both drained and undrained shear with σ_1 increasing while σ_3 remains contants.

Solution.

Drained shear: Construct the effective stress path and find $\bar{p}_f$ and q_f. Then find w_f using either the $\bar{p}_f$ versus w_f or q_f versus w_f relations (see Fig. E28.3).

$$\bar{p}_f = 331 \text{ kN/m}^2, \quad q_f = 124 \text{ kN/m}^2, \quad w_f = 20.6\%$$

Undrained shear: Enter the stress-volume diagram with the given $\bar{p}_0$ and find w_f. From q_f versus w_f, find q_f. $\bar{p}_f$ can be found from q_f versus $\bar{p}_f$.

$$w_f = 23.0\%, \quad q_f = 60 \text{ kN/m}^2, \quad \bar{p}_f = 162 \text{ kN/m}^2$$

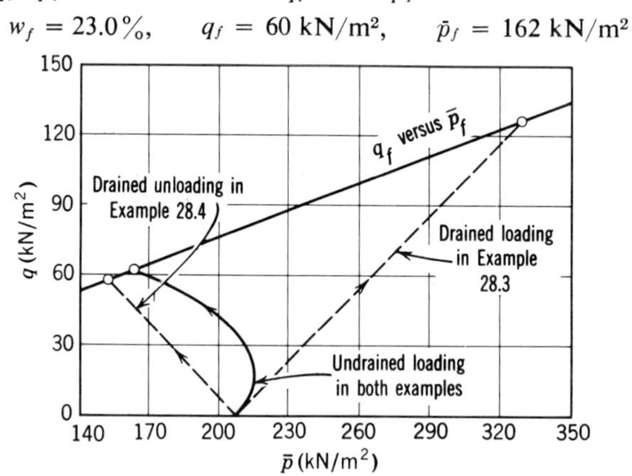

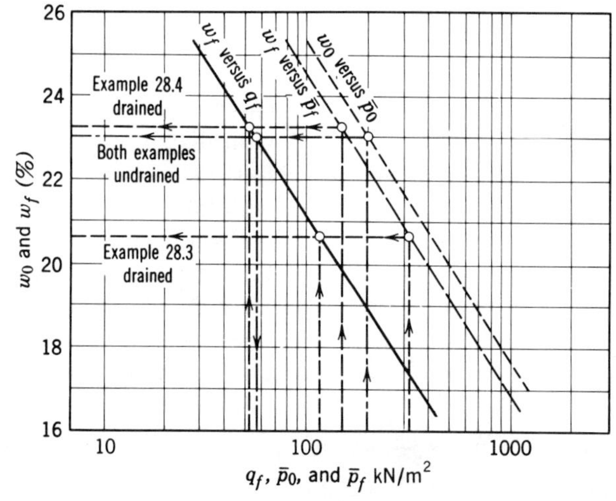

Fig. E28.3 ◀

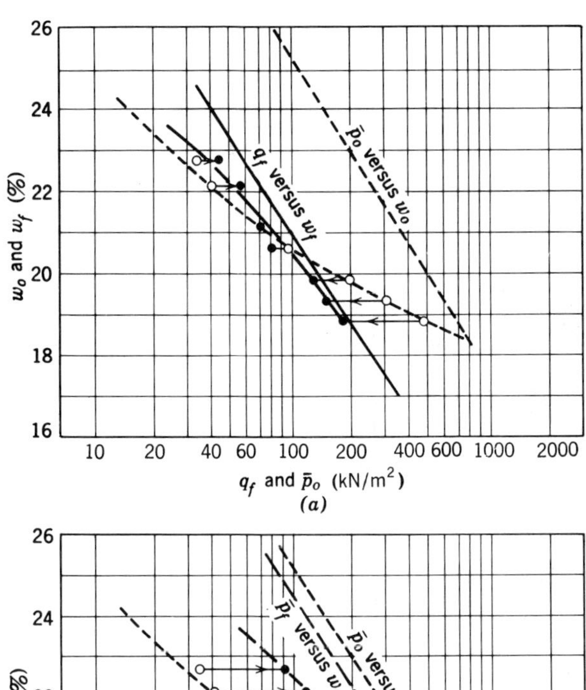

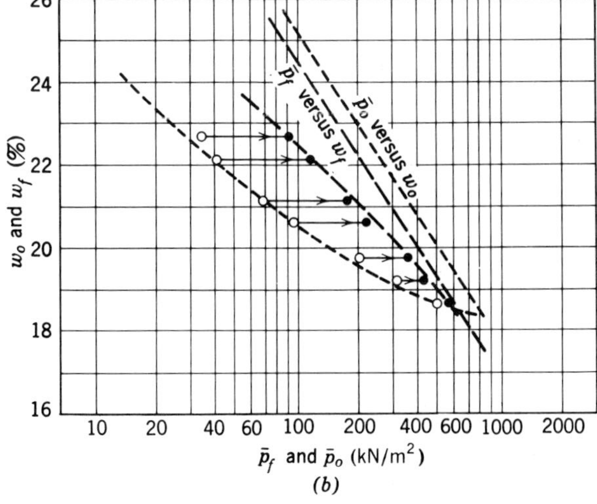

Fig. 28.11 Stress-volume relations for overconsolidated Weald clay. $\bar{p}_m = 827$ kN/m².

strength and initial consolidation stress. This derivation is presented in Fig. 28.12. The result (Eq. 28.1) emphasizes an important point: the undrained shear strength depends upon the conditions existing before shear, i.e., upon $\bar{p}_0$ and also upon A_f, $\bar{\phi}$, and $\bar{c}$, which are functions of stress history. For normally consolidated Weald clay with $A_f = 0.89$, $\bar{c} = 0$, and $\bar{\phi} = 22°$, we find $q_f = 0.29\bar{p}_0$.

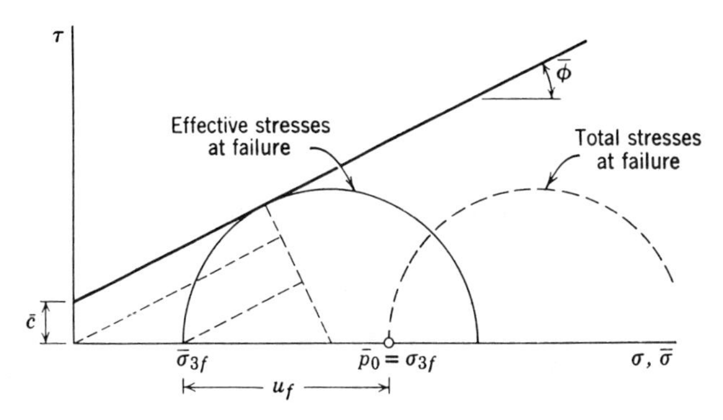

$$\text{At failure: } \tfrac{1}{2}(\sigma_{1f} - \sigma_{3f}) = q_f = \frac{\bar{c}\cos\bar{\phi} + \bar{\sigma}_{3f}\sin\bar{\phi}}{1 - \sin\bar{\phi}}$$

If

$$\Delta u = A_f(\Delta\sigma_1 - \Delta\sigma_3),$$

then

$$u_f = A_f(\sigma_{1f} - \sigma_{3f}) = 2Aq_f$$

$$q_f = \frac{\bar{c}\cos\bar{\phi} + (\bar{p}_0 - 2A_f q_f)\sin\bar{\phi}}{1 - \sin\bar{\phi}} \tag{28.1}$$

If $\bar{c} = 0$

$$\frac{q_f}{\bar{p}_0} = \frac{\sin\bar{\phi}}{1 + (2A_f - 1)\sin\bar{\phi}} \tag{28.2}$$

If $A_f = 1$,

$$\frac{q_f}{\bar{p}_0} = \frac{\sin\bar{\phi}}{1 + \sin\bar{\phi}}$$

If $A_f = \frac{1}{2}$

$$\frac{q_f}{\bar{p}_0} = \sin\bar{\phi}$$

Fig. 28.12 Equation for undrained strength in terms of effective stress-strength parameters and A_f.

Relative Magnitude of Drained and Undrained Strength

The foregoing subsections have established two important principles: (*a*) *the $q_f - \bar{p}_f - w_f$ relation is unique for a soil with a given stress history and* (*b*) *the effective stress path and undrained strength depend only*

▶ Example 28.4

Repeat Example 28.3 with σ_3 decreased and σ_1 constant.

Solution. Follow same steps as in Example 28.3. The undrained strength is the same for both examples. Note also that $q_f = 0.29\bar{p}_0$.

	$\bar{p}_f$ (kN/m²)	w_f (%)	q_f (kN/m²)
Drained loading	331	20.6	124
Undrained loading and unloading	162	23.0	60
Drained unloading	152	23.2	55

◀

▶ Example 28.5

Given. Overconsolidated Weald clay with $\bar{p}_m = 827\ \text{kN/m}^2$ and $\bar{p}_0 = 207\ \text{kN/m}^2$.
Find. q_f and w_f for both drained and undrained shear with σ_1, increasing while σ_3 remains constant.
Solution. Follow same steps as in Example 28.3. The diagrams are given in Fig. E28.5 and the answers appear in the table in Example 28.6.

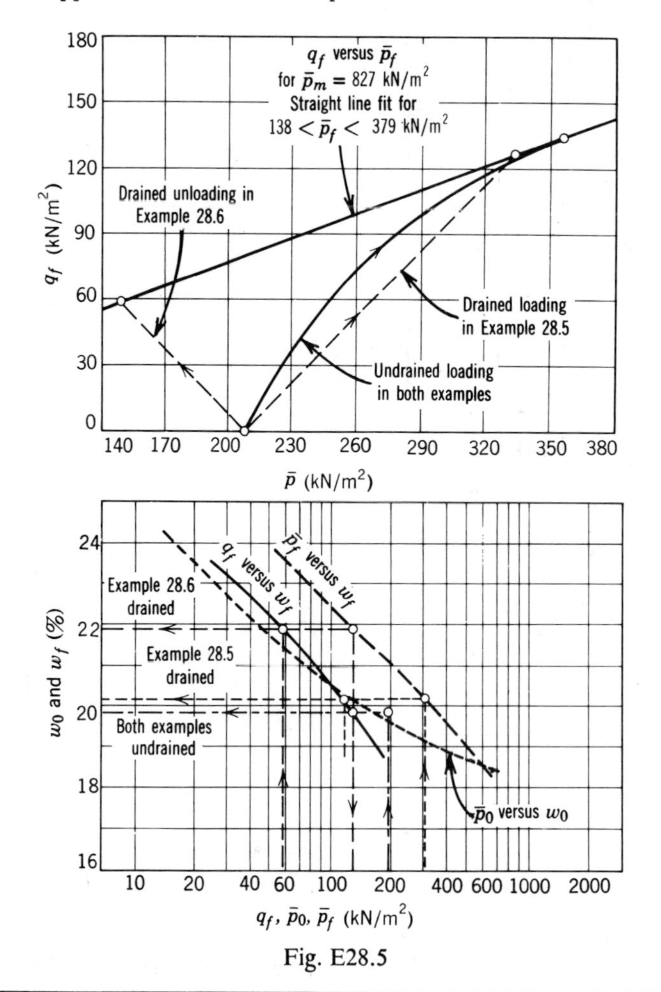

Fig. E28.5

◀

on the conditions just before shear. One useful way to illustrate these principles is to evaluate the relative magnitude of drained and undrained strength for various loading conditions. This evaluation is made in Examples 28.3 to 28.6. These examples remind us of two important points concerning the strength of clay with a given stress history:

1. The specimen with the greater density (as indicated by the smaller water content) has the greater strength.
2. The specimen under the greater effective stress (as measured by $\bar{p}_f$) has the greater strength.

▶ Example 28.6

Repeat Example 28.5 with σ_3 decreased and σ_1 constant.
Solution. Again follow the same steps. See Fig. E28.5.

	$\bar{p}_f$ (kN/m^2)	w_f (%)	q_f (kN/m^2)
Drained loading	334.4	20.1	127.6
Undrained loading and unloading	358.5	19.9	136.5
Drained unloading	146.2	21.9	60.7

◀

Table 28.1 Relative Magnitude of Drained and Undrained Strength

	Normally Consolidated Clay	Heavily Overconsolidated Clay
Triaxial compression loading (σ_1 increasing with σ_3 constant)	CD > CU	CU ≈ CD
Triaxial compression unloading (σ_1 constant with σ_3 decreasing)	CU ≈ CD	CU ≫ CD

Note. These comparisons apply for specimens with the same initial effective stress.

As noted in Chapter 21, strength, effective stress, and density are all interrelated, and it is impossible to say whether the strength is controlled by effective stress or by water content. However, generally it is more useful to consider effective stress the controlling variable.

The examples also indicate that it is difficult to give simple rules concerning the relative magnitude of drained and undrained strength. This relation depends on the type of loading and the degree of preconsolidation. Table 28.1 provides a useful guide. The importance of these relations will be illustrated in Chapter 31.

28.4 THE $\phi = 0$ CONCEPT

Next we consider a series of unconsolidated undrained tests. Three specimens are selected and all are consolidated to 110 kN/m². This brings the specimens to the end of step 1 in the UU test program. Now the confining pressures are changed to say 69, 138, and 690 kN/m², without allowing further consolidation and then sheared undrained. The result, within experimental scatter, is that $q_f = 33.1$ kN/m² for each specimen!

This is what has happened. When the confining pressure was changed in step 2, the pore pressure in the fully saturated specimens changed just as much as did the confining pressure, and the effective stress remained unchanged and equal in each specimen. Thus $\bar{p}_0 = 110$ kN/m² in each specimen, and each specimen behaved during shear just as did the CU specimen in Fig. 28.7 with $\bar{p}_0 = 110$ kN/m². This state of affairs is illustrated in Fig. 28.13. For example, consider the specimen under a chamber pressure of 207 kN/m². The stresses observed during the various stages of the test are shown in Table 28.2. For the other specimens, $\bar{p}$ and q are the same, and p, u, and σ_3 change in step with each other.

Now three more specimens are selected and consolidated to 772 kN/m². Again UU tests are performed, using chamber pressures of 69, 138, and 690 kN/m². These specimens all have a strength of 260.6 kN/m². The stress changes for the test with 207 kN/m² confining stress are given in Table 28.3. Note that the pore pressure is shown as negative at the end of step 2 and becomes less negative (algebraically greater) when the clay is sheared. Although direct measurements of negative pore pressures

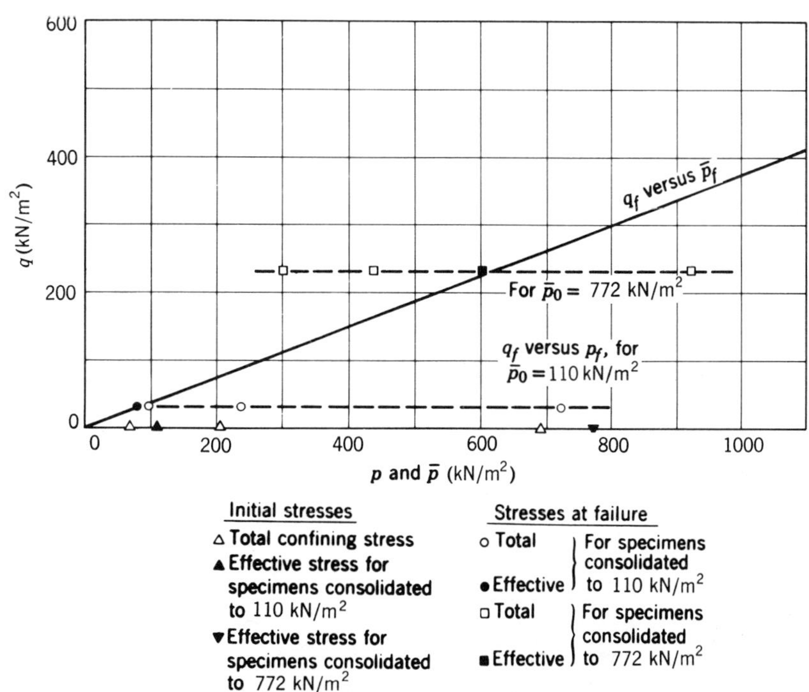

Fig. 28.13 Two series of UU tests on normally consolidated Weald clay.

Table 28.2[a]

End of Step	σ_3	p	u	$\bar{p}$	q
1	110	110	0	110	0
2	207	207	97	110	0
3	207	240.1	153	87.1	34.1

[a] Stresses and pressures in kN/m².

of this magnitude are lacking, the circumstantial evidence regarding the existence of such negative pore pressures is overwhelming.

Example 28.7 presents still another example of behavior during a UU test.

▶ **Example 28.7**

Given. A normally consolidated specimen of Weald clay at 19.5%.

Find. The strength and pore pressure at failure in a UU test using a chamber pressure of 690 kN/m².

Solution. Enter Fig. 28.10 with w_f = 19.5%, and find q_f = 165 kN/m² and $\bar{p}_f$ = 441 kN/m². u_f = 855 − 441 = 414 kN/m². ◀

Such results are further proof of the importance of the effective stress principle. Out of this study emerges another important observation:

> Undrained strength is independent of *changes* in the total stress p.

Thus, when undrained strength is plotted versus total average principal stress p, a horizontal line results as on Fig. 28.13. When such a relation was first observed, it was taken to imply a cohesive material, i.e., $\phi = 0$. Although it is now recognized that this relation implies nothing about the internal mechanism of shear resistance, it is still known as the $\phi = 0$ concept. For each clay, there exists a whole family of horizontal lines "relating" undrained strength and total stress. Each line corresponds to a different consolidation stress $\bar{p}_0$, or we may equally well say that there is a different water content for each line.

Finally, there is one more experiment to perform. A specimen is consolidated to 110 kN/m². Then, working quickly, the specimen is removed from the triaxial cell, stripped of its jacket, and then compressed axially; in other words, an unconfined compression test is carried out on this specimen. It will be found that the axial stress at failure is very nearly 66.2 kN/m² or q_f = 33.1 kN/m². This is the same strength obtained in the UU tests upon specimens consolidated to 110 kN/m². The effective stresses that give this sample its strength result from negative pore pressures as seen in Table 28.4. This test again illustrates the validity of the $\phi = 0$ concept, and also shows the relation of an unconfined compression test to the more general form of UU test.

The $\phi = 0$ concept is of considerable practical importance, as we shall see in Chapter 31.

Table 28.3[a]

End of Step	σ_3	p	u	$\bar{p}$	q
1	772	772	0	772	0
2	207	207	−565	772	0
3	207	434	−173	607	260.6

[a] Stresses and pressures in kN/m².

Table 28.4[a]

End of Step	σ_3	p	u	$\bar{p}$	q
1	110	110	0	110	0
2	0	0	−110	110	0
3	0	0	−54	87	33.1

[a] Stresses and pressures in kN/m².

28.5 RELATION OF DRAINED AND UNDRAINED STRESS-STRAIN CURVES

To investigate this relationship, imagine a triaxial test with axial loading and with the drainage line fully open to atmospheric pressure. The total stress path is AB in Fig. 28.14. Points A and B also represent the initial and final effective stress conditions. However, there may be a variety of effective stress paths, depending upon the rate at which the axial load is applied.

1. If the load is applied very slowly compared to the rate of consolidation, AB is also the effective stress

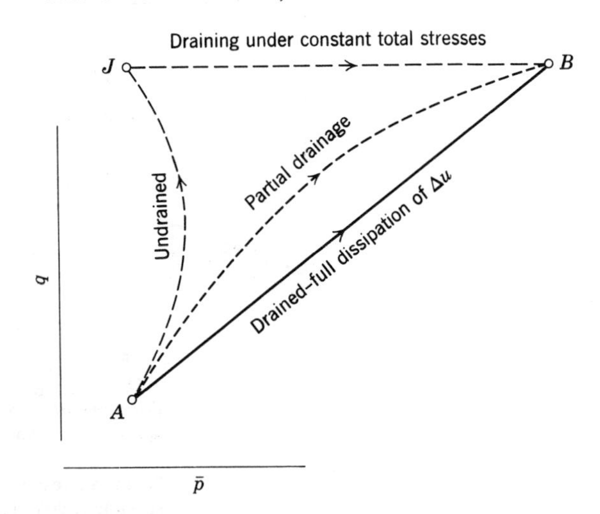

Fig. 28.14 Effective stress paths for varying degrees of drainage during triaxial test.

path. Both volume change and shear strain occur during the loading.

2. If the load is applied very quickly compared to the rate of consolidation, the effective stress path is *AJB*. During the loading there is no drainage, and with a saturated soil only shear strains occur along the path *AJ*. Line *JB* corresponds to the subsequent consolidation. There is no volume change along *AJ*. If soil were isotropic there would be no shear strain along *JB*, but in actuality some shear strain may occur.
3. If the rate of loading and rate of consolidation are similar, partial drainage will occur during the loading, giving an intermediate effective stress path. Both volume change and shear strain occur along such a path.

Several important concepts can be established by comparing in more detail the strains along the path *AJB* with the strains developed along the path *AB*.

Strains during Undrained Loading

The ratio of change of axial stress to change in axial strain during a triaxial test is analogous to Young's modulus (Section 12.1). For a drained loading (Section 22.2), this ratio relates *effective* stress and strain and will henceforth be denoted by $\bar{E}$. The following derivation, applicable to an ideal isotropic material, relates $\bar{E}$ to undrained Young's modulus E: the ratio of *total* stress to strain during undrained loading.

If the mineral skeleton is isotropic, then Eq. 12.5a may be applied to the undrained loading in two forms:

Total stress:

$$\epsilon_v = \frac{1}{E}\,\sigma_v \qquad (28.3)$$

Effective stress:

$$\epsilon_v = \frac{1}{\bar{E}}(\bar{\sigma}_v - 2\bar{\mu}\bar{\sigma}_h) \qquad (28.4)$$

where $\bar{\mu}$ is Poisson's ratio for the mineral skeleton. For the undrained loading, since the pore pressure parameter A equals $\frac{1}{3}$ for an isotropic mineral skeleton,

$$\bar{\sigma}_v = \tfrac{2}{3}\sigma_v$$

$$\bar{\sigma}_h = -\tfrac{1}{3}\sigma_v$$

Hence

$$\epsilon_v = \frac{1}{\bar{E}}(\tfrac{2}{3}\sigma_v + \tfrac{2}{3}\bar{\mu}\sigma_v) = \frac{2}{3}\frac{\sigma_v}{\bar{E}}(1 + \bar{\mu})$$

or

$$E = \frac{3}{2(1 + \bar{\mu})}\,\bar{E} \qquad (28.5)$$

Since $\bar{\mu}$ is typically about 0.3, approximately $E = 1.15\bar{E}$.

With actual soils, the ratio $E/\bar{E}$ is typically much greater than this theoretical value. Values of 3 or 4 are not uncommon for normally consolidated clays. Because no volume change occurs during an undrained loading, the axial strain during an undrained loading (path *AJ*) is less than for the same loading drained (path *AB*).

If the pore pressures caused by an undrained loading are known, the strains caused by the loading can always be computed by applying Eqs. 12.5 to the mineral skeleton; i.e., by using the system

$$\bar{\sigma}, \quad \bar{E}, \quad \bar{\mu}$$

Often it is convenient to evaluate strains directly in terms of total stresses. For a saturated isotropic soil, this can be done by using Eqs. 12.5 with

$$\sigma, \quad E, \quad \mu = 0.5$$

Use of Poisson's ratio equal to 0.5 gives the condition of zero volume change. Since shear stress is unchanged by pore pressure, the shear modulus G is the same whether the loading is drained or undrained. The use of undrained modulus together with $\mu = 0.5$ will be illustrated in Chapter 32.

Final Strains after Consolidation

If the mineral skeleton were elastic, the final strains at point B in Fig. 28.14 would be the same for any stress path. Thus the shear strain along *AJ* plus the volume change along *JB* would just equal the shear and volume strains occurring simultaneously along *AB* (Fig. 28.15*b*). However, since the mineral skeleton usually does not remain elastic, the final strains are *path dependent*. According to the results in Chapters 12 and 22, the greater the value of $\bar{p}$ during straining, the lower will be the strain that occurs. Hence we can logically reason that the least vertical strain will occur for the path *AB*, that the most vertical strain will occur for the path *AJB*, and that an intermediate strain will occur for a partial drainage path (Fig. 28.15*c*). In the extreme case where a failure condition is reached during the undrained loading so that the shear strains during this loading are extremely large (Fig. 28.15*e*), clearly the final strains will be much greater for the path *AJB* than for the path *AB*. This path-dependent nature of strains will be discussed further in Chapter 30.

28.6 COMPRESSION OF PARTIALLY SATURATED SOILS

Figure 28.16 depicts in a general way the events that occur during one-dimensional compression of a partially saturated soil in an oedometer.

Initially the soil is quite compressible, since the pore fluid (air plus water) offers little resistance to compression until the degree of saturation exceeds 85% (Table 26.1). During this initial phase the effective stress increases while the pore pressure changes but little. Very little

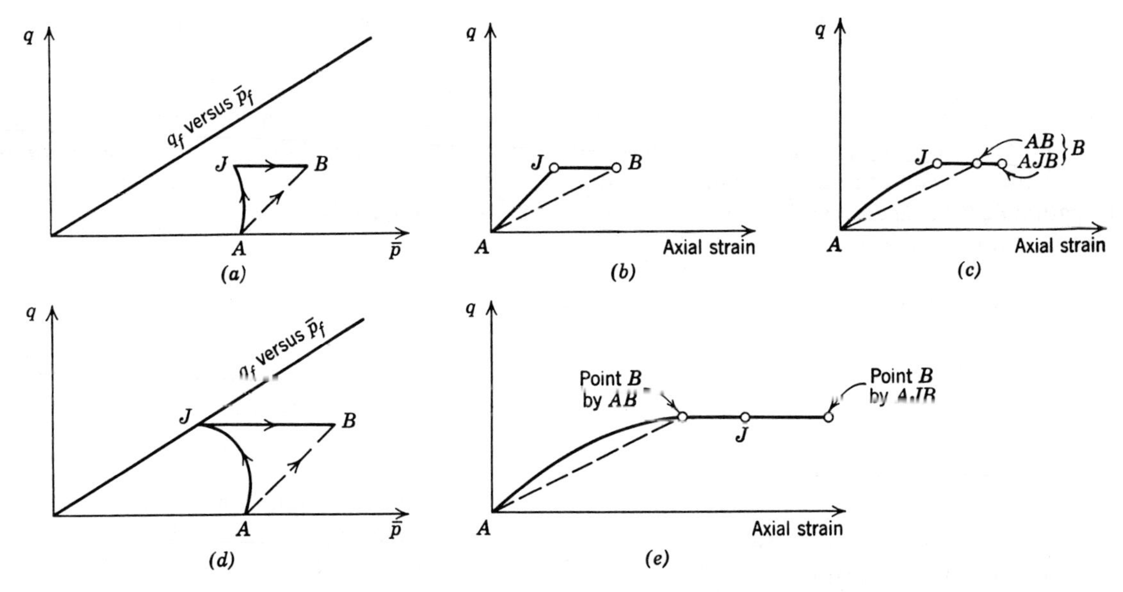

Fig. 28.15 Axial strains as a function of stress path. (*a*) Non-failure loading. (*b*) Elastic response to non-failure loading. (*c*) Nonelastic response to non-failure loading. (*d*) Failure loading. (*e*) Response to failure loading.

water will be squeezed from the soil during this phase of a drained loading, thus it matters little whether drainage is permitted or prevented.

If the load increase is sufficient to compress and dissolve all of the air in the pores, the soil will become fully saturated and now any further increase in load will be carried entirely by the pore fluid. If drainage is permitted, water will flow from the specimen as in a consolidation test upon an initially saturated soil. Once the applied load reaches its maximum value and becomes constant, a drained soil may again become only partially saturated.

While this qualitative picture may easily be understood, detailed quantitative analysis is quite difficult.

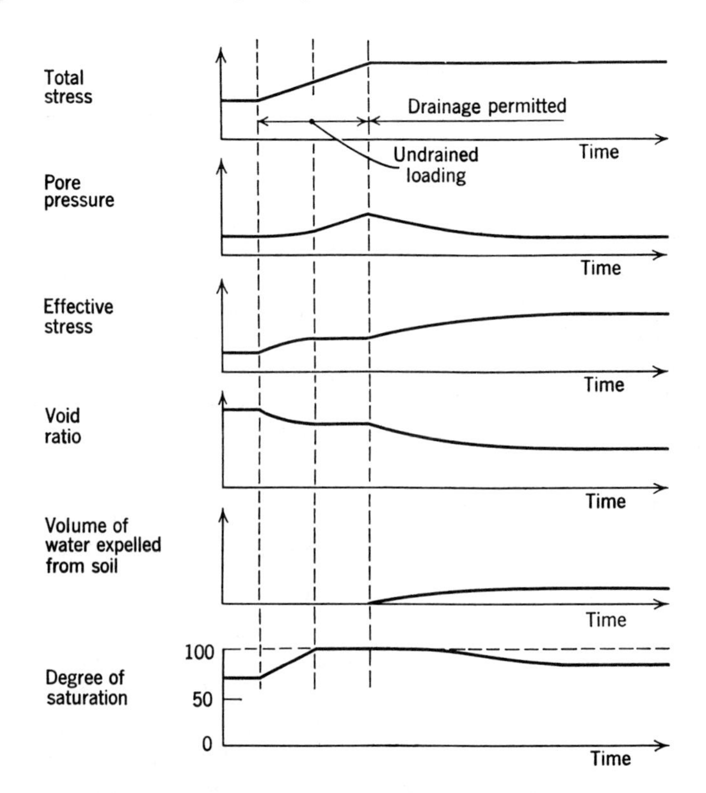

Fig. 28.16 Compression of partially saturated soil in an oedometer.

In a general way, the principle of effective stress must apply; i.e., the volume of the soil at any time is related to the effective stress at that time. However, both the pore air pressure and the pore water pressure must be considered when evaluating effective stress (Eq. 16.15) and there are great difficulties in the evaluation of the parameter a_w (called χ in some references). For discussion of these considerations, see Bishop and Blight (1963) and Blight (1965). The best approach to estimating compression of partially saturated soils is to apply, both initially and finally, the total stress, pore air pressure, and pore water pressure existing or expected *in situ*.

28.7 STRENGTH OF PARTIALLY SATURATED SOILS

As noted in Section 21.7, the shear strength of partially saturated soils is related to effective stress. This is true for undrained as well as drained loading. However, effective stress must be evaluated by Eq. 16.15 with the attendant difficulties in determining the parameter a_w (Bishop and Blight, 1963).

Clearly the $\phi = 0$ concept does not in general apply to partially saturated soils. Figure 28.17 shows the typical relation between confining stress and strength during UU tests. For the lower range of confining stresses, the soil remains partially saturated and effective stress increases as the confining stress is increased. Once the confining stress becomes large enough to cause full saturation, further increases in confining stress do not cause the effective stress to increase and from this point the $\phi = 0$ concept does apply. Sometimes a relationship between undrained strength and total stress is used:

$$\tau_{ff} = c_u + \sigma_{ff} \tan \phi_u \tag{28.6}$$

where c_u and ϕ_u are called *total stress strength* parameters. Values of c_u and ϕ_u depend very much upon the range of σ_{ff} which is of interest.

Great care must be taken to duplicate expected field conditions (total stress, pore water pressure, and pore air pressure) when performing tests to evaluate the undrained strength of partially saturated soils.

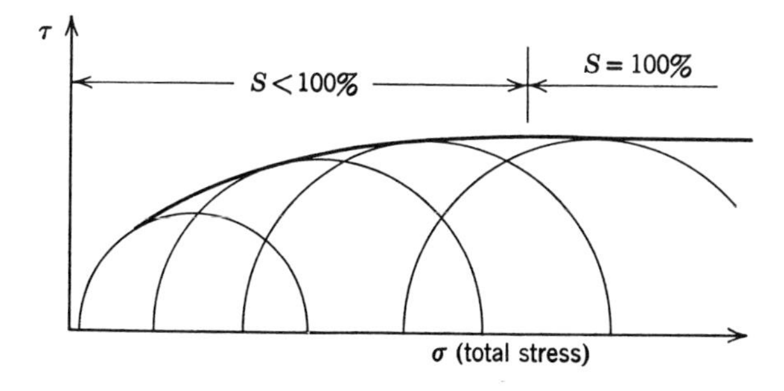

Fig. 28.17 Unconsolidated undrained tests on partially saturated soil.

28.8 SUMMARY OF MAIN POINTS

The most important points in this chapter concern the strength of saturated soils.

1. As a first approximation, there is a unique relationship for each soil among q_f, $\bar{p}_f$ and w_f, which governs the shear strength for all types of loadings and drainage conditions. This relationship is really unique only in the ultimate condition, but at least it is a useful guide to the behavior of all soils.
2. As a first approximation, the effective stress path for undrained shear and the undrained shear strength are dependent only on the initial conditions and are independent of the details of the loading process. While this principle is a useful guide to the solution of many problems, in actuality the pore pressures induced during shear, and hence the stress path and undrained strength, are somewhat sensitive to the details of the loading.
3. Starting from a given initial condition, undrained strength may either exceed or be less than drained strength, depending upon the type of loading and degree of overconsolidation. Evaluation of the water content changes during shear and of the effective stresses at failure provides the key to deciding which strength is greater.
4. As a first approximation, undrained strength is changed only by a change in the effective stress to which the soil has been consolidated and the accompanying change in water content. Undrained strength is independent of changes in total stress unless a change in water content occurs. Thus, with regard to changes in total stress with constant volume, clay behaves as though $\phi = 0$.
5. As a first approximation, it makes no difference whether the undrained strength of a soil is measured by an unconfined compression test, triaxial compression test, or vane shear test, as long as the testing procedure leaves the soil at its natural water content. Actually, there are differences among the values of undrained strength measured in these various procedures, because of disturbance to the mineral skeleton and because soil is really anisotropic.
6. All soils obey the principles stated above. However, in practical terms there are great differences in the strength characteristics of various soils because of (*a*) differences in permeability and hence in the rate of dissipation of excess pore pressures; and (*b*) differences in the ability of the pore water to support capillary tensions.

This chapter makes some other important points regarding strains.

1. As a first approximation, the e versus $\bar{\sigma}$ relation during confined compression is the same for a

continuation drained (slow) loading as for increments of undrained loading followed by interval of consolidation.

2. Whenever the shear stresses during an undrained loading approach the shear strength, the strains resulting from this undrained loading followed by consolidation will exceed the strains caused by the same stresses applied slowly.

PROBLEMS

28.1 Refer to Fig. 28.4. Assuming that $u_r = -\sigma_c$, that the soil is saturated, and that $A = \frac{1}{3}$, write expressions for the pore pressures and effective stresses at steps 2 and 3.

28.2 Refer to Figs. 28.7 and 28.10. Find the shear strength and water content at failure for specimens consolidated to 110 kN/m^2 effective stress:

a. Triaxial compression loading, drained.
b. Triaxial compression loading, undrained.
c. Triaxial compression unloading, drained.
d. Triaxial compression unloading, undrained.

28.3 Refer to Figs. 28.8 and 28.11. Find the shear strength and water content at failure for specimens consolidated to 827 kN/m^2 and then rebounded to 110 kN/m^2. Answer the same four cases as in Problem 28.2.

28.4 Refer to Figs. 28.7 and 28.13. A specimen of clay is first normally consolidated to an effective stress of 110 kN/m^2. Then, without permitting further drainage, the chamber pressure is increased to 455 kN/m^2. Then, still without permitting further drainage, the specimen is loaded in triaxial compression. Find the shear strength and the values of $\bar{\sigma}_1$, $\bar{\sigma}_3$, and pore pressure at failure. (*Hint.* Draw total and effective stress paths.)

CHAPTER 29

Undrained Shear Strength

Undrained shear occurs in practical problems whenever external loads change at a rate much faster than the rate at which the induced pore pressures can dissipate. In Chapter 27 we saw that excess pore pressures dissipate relatively slowly from a clay and relatively rapidly from a sand. The condition of undrained shear is thus of great practical importance in the case of clays. With sands, undrained strength is relatively unimportant for static loadings but may be very important for problems involving dynamic loadings.

In this chapter we study the magnitude of undrained strength in various situations. In Chapter 28 we learned the influence of certain important factors: undrained strength increases with (*a*) decreasing water content, (*b*) increasing consolidation stress, and (*c*) increasing maximum past consolidation stress. We will return to these factors at the end of the chapter. Now we begin by considering the influence of other factors.

29.1 UNDRAINED STRENGTH OF SATURATED SAND

The behavior of saturated sands during undrained shear is basically similar to that which has been described for clays. That is, either positive or negative pore pressures are induced, depending upon whether the sand tends to decrease or increase in volume during drained shear. Figures 29.1 and 29.2 show typical results from CU tests upon a loose and a moderately dense sand. Both specimens were under an effective stress of 69 kN/m² at the end of step 2 before additional axial stress was applied in step 3. Figure 29.2 shows effective stress paths for these tests, together with the stress paths that would apply for drained tests. In both cases the full friction angle $\bar{\phi}$ is reached relatively early in the loading (i.e., the effective stress path reaches the failure line), but the shear resistance continues to increase because the pore pressure changes lead to increased effective stresses.

Going a step further, it can be said that the undrained strength behavior of all soils is basically similar to that which has been described for clays. The effective stress path for undrained shear of a soil depends on the tendency toward expansion or contraction during shear; i.e., it depends on the initial density considered in relation to the initial effective stress. The effective stress paths can take on a variety of forms, as suggested in Figs. 28.8 and 29.2.

Cavitation of Pore Water

In a sand, it is not possible for the pore water pressure to be less than about -1 atm (-101 kN/m²). If the pore pressure falls below this limit, the pore water will cavitate. Thus in order to achieve the results shown in Fig. 29.1 for the moderately dense sand, the initial pore pressure must have been at least 380 kN/m²; say an initial pore pressure of 480 kN/m² together with a chamber pressure of 550 kN/m².

Figure 29.3 shows typical results for a test in which the pore water cavitated at an axial strain of somewhat less than 5%. From this point onward, the specimen no longer remained at constant volume even though it was undrained. The pore pressure, and hence $\bar{\sigma}_3$, remained constant following cavitation, and the behavior was then essentially the same as in drained shear. These results remind us of an important point: it is not the water content of a soil that controls the shear strength, rather it is the density or tightness of packing of the mineral skeleton.

If cavitation occurs during undrained shear, the peak undrained shear resistance will not depend solely on the initial effective stress. Rather, the total confining stress will also influence the strength. For example, let us consider the influence of σ_3 upon the results shown in Fig. 29.3. This effect is shown in Table 29.1. In constructing this tabulation, it is assumed that the pore water cavitates in both tests. The ratio $\bar{\sigma}_{1f}/\bar{\sigma}_{3f}$ must be the same for both tests, and it can be found from the results in Fig. 29.3.

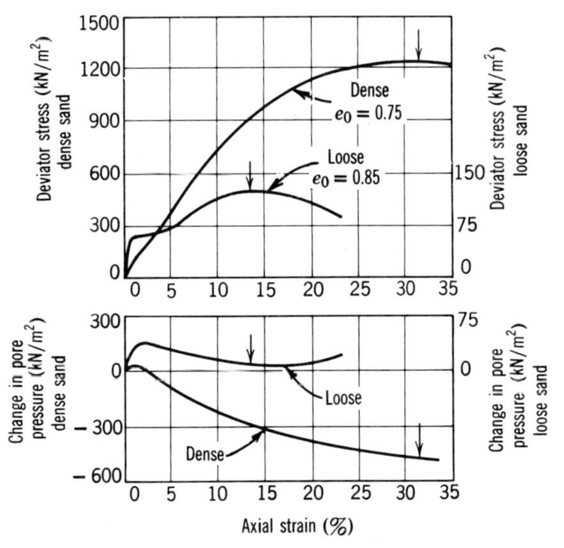

Fig. 29.1 Stress-strain curves for undrained triaxial compression of a saturated sand (From Leonards, 1962).

Cavitation is very likely to occur during undrained shear of a saturated dense sand. With a dense uniform sand of medium coarseness and rounded grains, cavitation will occur unless the initial pore water pressure is about 100 atm (10 MN/m²). With more typical sands, there will be cavitation within dense specimens during shear unless the initial pore water pressure is from 15 to 30 atm (1.5 to 3.0 MN/m²). Shear failure of a saturated sand at constant volume thus generally occurs only in sands in a medium to a loose state.

The $\phi = 0$ Concept

If a series of UU tests is performed upon specimens of a fine sand in a medium to loose state, it will be found that the $\phi = 0$ principle applies except for very low confining stresses. In Fig. 29.4 all specimens have the same initial effective stress ($\bar{p}_0 = 36.5$ kN/m²) but different chamber pressures and different pore pressures. At low chamber pressures (and initial pore pressures) cavitation occurs during shear.

As long as there is no cavitation, undrained strength depends only on $\bar{p}_0$ and the $\phi = 0$ concept applies to sands as well as clays. Clearly, the fact that undrained strength is independent of *changes* in total stress has nothing to do with the internal mechanism of shear resistance.

These results do emphasize one difference between the undrained strength behavior of sand and clay: an unconfined compression test ($\sigma_3 = 0$) can be used to

Table 29.1

	Figure 29.3	Another Case	Notes
σ_3	276 kN/m²	414 kN/m²	given
u_f	−83	−83	cavitation
$\bar{\sigma}_{3f}$	359	497	$\sigma_3 - u_f$
$\bar{\sigma}_{1f}$	1427	1972	
$\sigma_{1f} - \sigma_{3f}$	1068	1475	

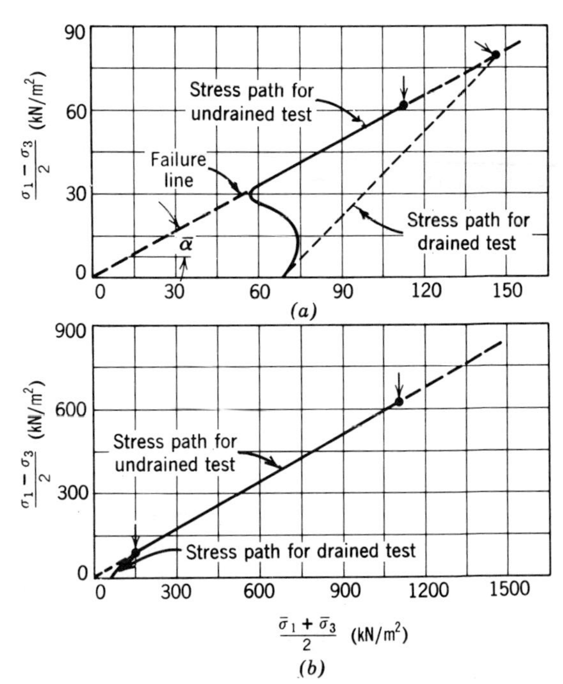

Fig. 29.2 Stress-paths for undrained triaxial compression of a saturated sand. (*a*) Loose, $e_0 = 0.85$. (*b*) Moderately dense, $e_0 = 0.75$.

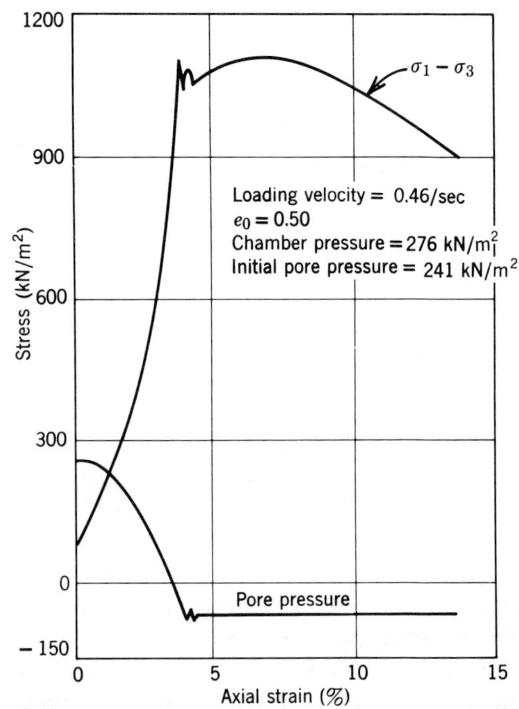

Fig. 29.3 Cavitation during undrained triaxial test of a dense sand (From Whitman and Healy, 1962).

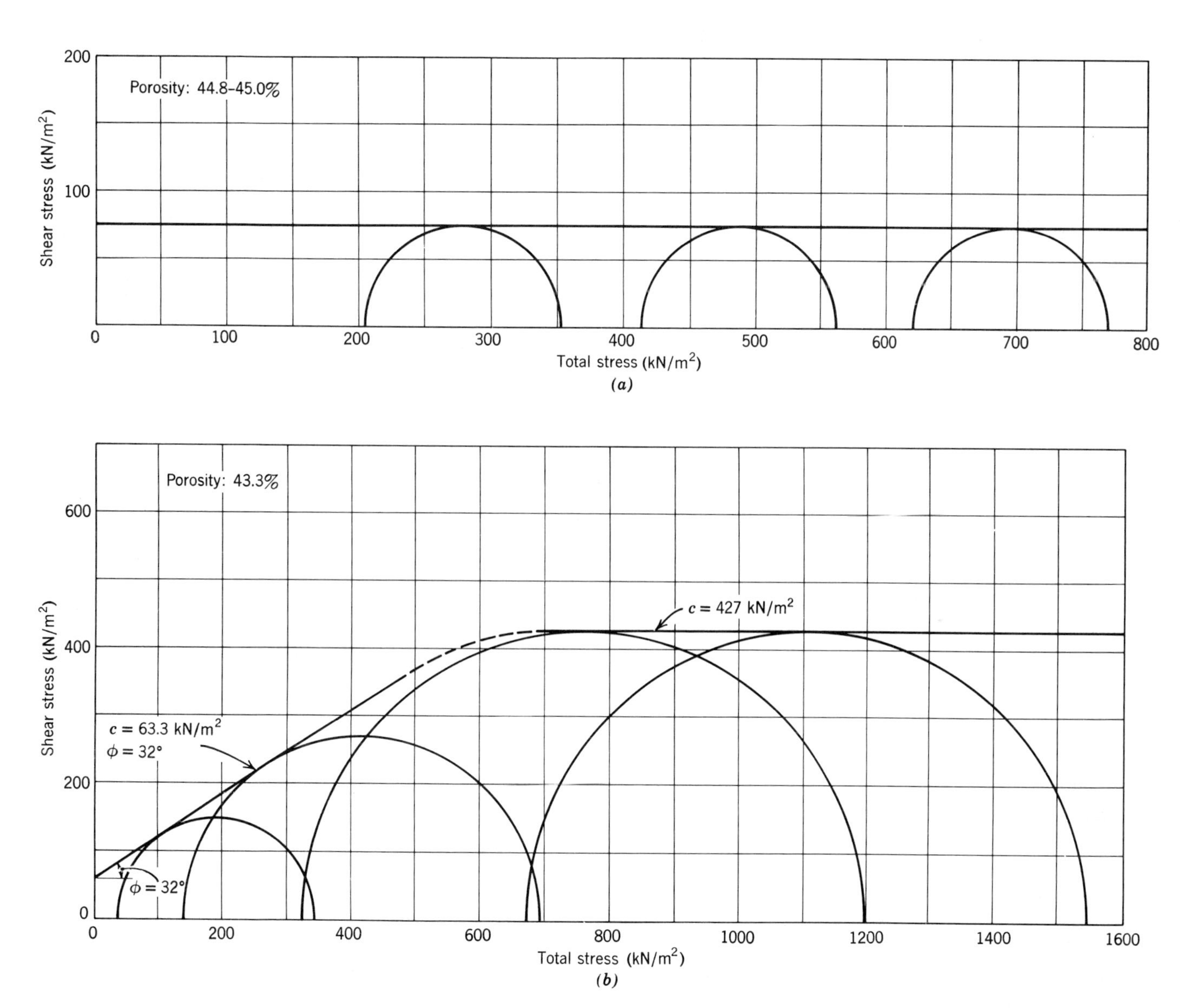

Fig. 29.4 Results of UU tests on saturated sand. Consolidation stress = 36.5 kN/m², initially fully saturated (After Bishop and Eldin, 1950).

give the UU strength of clay but not of sand. This is because large negative pore pressures can exist within the tiny pores among clay particles but not within the larger pores among sand particles. The limit of pore pressures before cavitation is related to the capillary rise of water within a soil (Chapter 16) and to the apparent cohesion which can exist above the water table (Chapter 21).

29.2 SENSITIVE CLAYS AND VERY LOOSE SANDS

In Chapter 28 and in the preceding section we have emphasized the unity between drained and undrained strength and have suggested that the q_f versus $\bar{p}_f$ relation (i.e., the $\bar{c}$ and $\bar{\phi}$) is the same for both tests. Now we must consider some deviations and exceptions to this

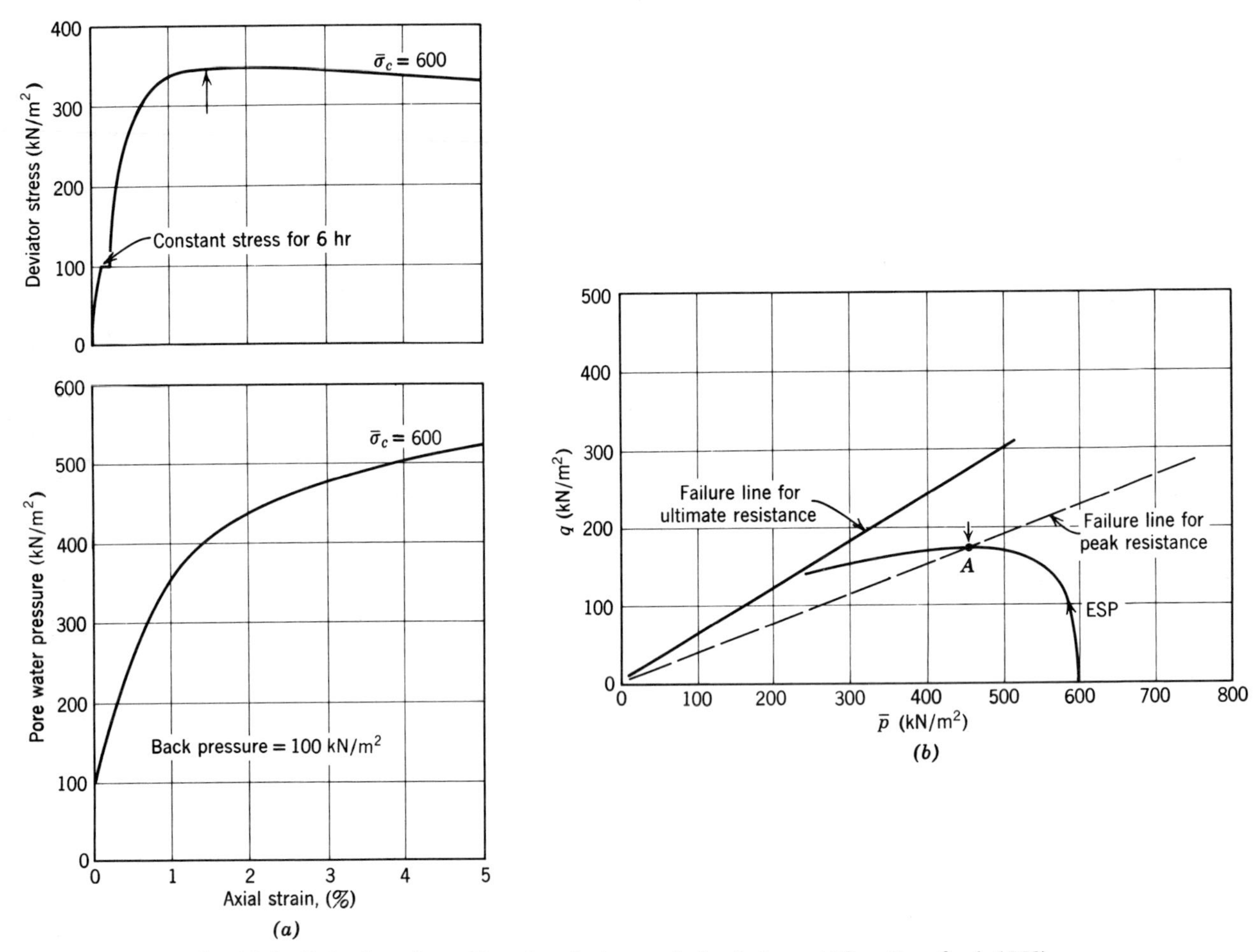

Fig. 29.5 Behavior of sensitive clay during undrained shear (After Crawford, 1959).

simplified picture. The most important of these exceptions occurs in the case of sensitive clays and very loose sands.

Figure 29.5 shows the results of a CU test upon an undisturbed sample of a normally consolidated sensitive clay. The deviator stress reaches a peak at a rather small axial strain, and then decreases with further strain. The pore pressure continues to increase even after the deviator stress has peaked. The effective stress path has a form quite different from that which we encountered with Weald clay, since now the q_f, $\bar{p}_f$ point representing peak strength (point marked with an arrow) lies well below the q_f versus $\bar{p}_f$ relation from drained shear.

This type of behavior results from the very loose metastable skeleton of the sensitive clays. The behavior of this clay during drained shear has already been discussed in Section 21.5: the clay experiences a great decrease in volume. Consequently, large positive pore pressures are induced during undrained shear. As the clay is sheared undrained, two opposing trends develop: (*a*) more and more of the potentially available friction is mobilized; and (*b*) the effective stresses decrease. Thus the overall shear resistance, which is related to the product of the effective stress and the mobilized friction factor, reaches a peak before full frictional resistance is mobilized. At very large strains, when all of the available friction is finally mobilized, the overall shear resistance is small because the effective stress is so small.

This important point is illustrated in Example 29.1. Note especially the decrease in $\bar{\sigma}_3$ from the initial value of 600 kN/m² and the large value of the pore pressure parameter A during the latter stages of the test.

▶ Example 29.1

Given. The data presented in Fig. 29.5.

Find. The principal effective stresses, mobilized friction angle, and pore pressure parameter A (*a*) for an axial strain of 1.5%; (*b*) at the end of the effective stress path (roughly 8% axial strain).

Solution.

Strain (%)	1.5	8	
$\bar{\sigma}_1$ (kN/m²)	625	380	$\bar{p} + q$
$\bar{\sigma}_3$ (kN/m²)	275	100	$\bar{p} - q$
mobilized $\bar{\phi}$	23°	36°	$\sin^{-1} q/\bar{p}$
q (kN/m²)	175	1400	
Δu (kN/m²)	325	500	$600 - \bar{\sigma}_3$
A	0.9	1.8	$\Delta u/2q$

◀

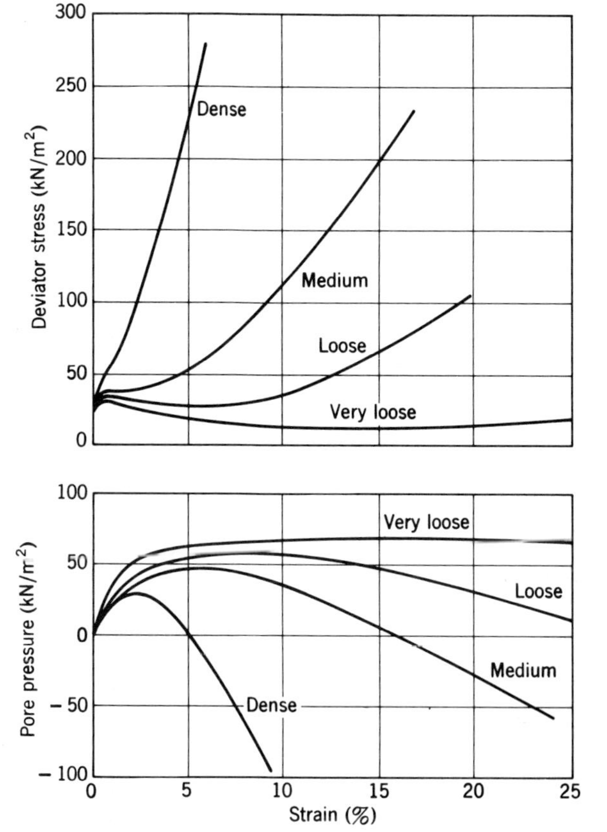

Fig. 29.6 Stress-strain curves for undrained triaxial tests on saturated sand at four densities. Specimens consolidated to 10 MN/m² (From Healy, 1963).

This same type of behavior occurs in very loose sands, as shown by the data presented in Fig. 29.6. The peak deviator stress in the test upon the very loose sand occurred at a very small axial strain of about $\frac{1}{4}\%$. At this point, the mobilized friction angle was about 10°, even though the friction angle of the sand in drained shear at this density was about 30°.

The difference between the friction angles mobilized at peak resistance in drained and undrained tests on undisturbed specimens of marine soils is shown in Fig. 29.7. The discrepancy is greatest in soils with the least plasticity—fine sands and silts. That is to say, the tendency toward a metastable structure is greatest in relatively nonplastic soils.

Once again we have seen that the concept of a unique q_f-p_f-w_f relation really holds only at very large strains. While the concept of such a relation helps us to understand the connection between drained and undrained strength in all soils, the relation does not apply to the peak resistance of soils with a metastable skeleton. For soils such as remolded Weald clay, the peak undrained shear resistance occurs at the same time that full frictional resistance is mobilized, and the q_f-p_f-w_f relation is the same for drained and undrained shear.

The loss of strength with remolding accounts for the phenomenon of *liquefaction* in quick clays and very loose sands. If a hillside of such a material starts to slide, the soil loses its strength and flows away like a liquid (see Fig. 1.13). Liquefaction has been involved in a number of important slope failures, notably in the slide at Fort Peck Dam (Casagrande, 1965). Whether or not a flow slide caused by liquefaction will occur in a sand is related to whether the sand tends to expand or to decrease in volume during shear. Casagrande introduced the concept of a *critical void ratio* (Taylor, 1948). If a sand is at an *in situ* void ratio greater than the critical void ratio for that sand, the sand is highly susceptible to flow slides. The critical void ratio of uniform, fine sand with subrounded grains corresponds to a relative density of 20 to 30% for small initial effective confining stress (up to 10 kN/m²) and a relative density of about 50% for an initial effective confining stress[1] of 1000 kN/m².

[1] Data by Gonzalo Castro at Harvard University, 1968.

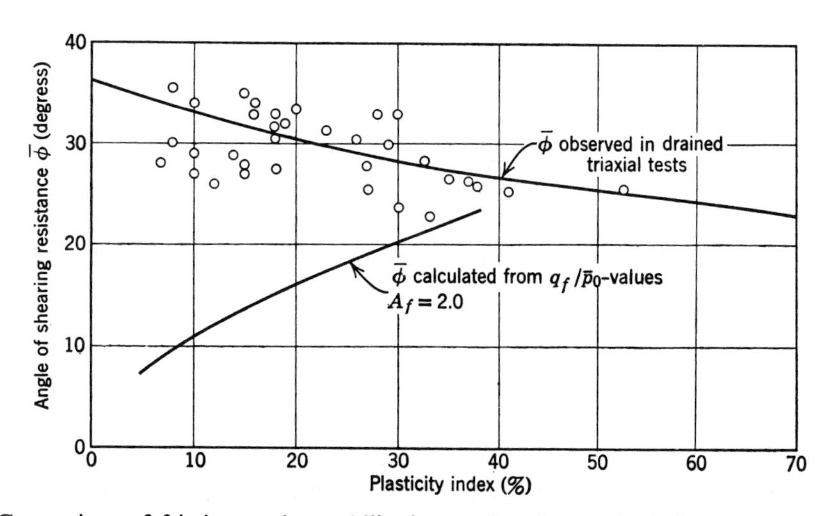

Fig. 29.7 Comparison of friction angles mobilized at peak resistance in drained and undrained tests.

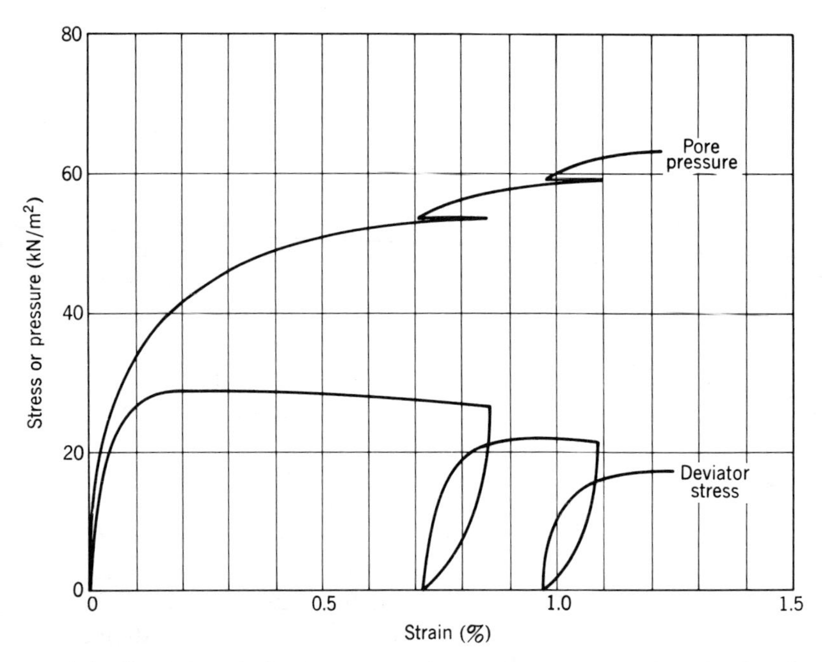

Fig. 29.8 Effect of repeated loading on undrained strength of very loose saturated sand. Specimen consolidated to 69 kN/m²; void ratio = 0.834. (From Healy, 1963.)

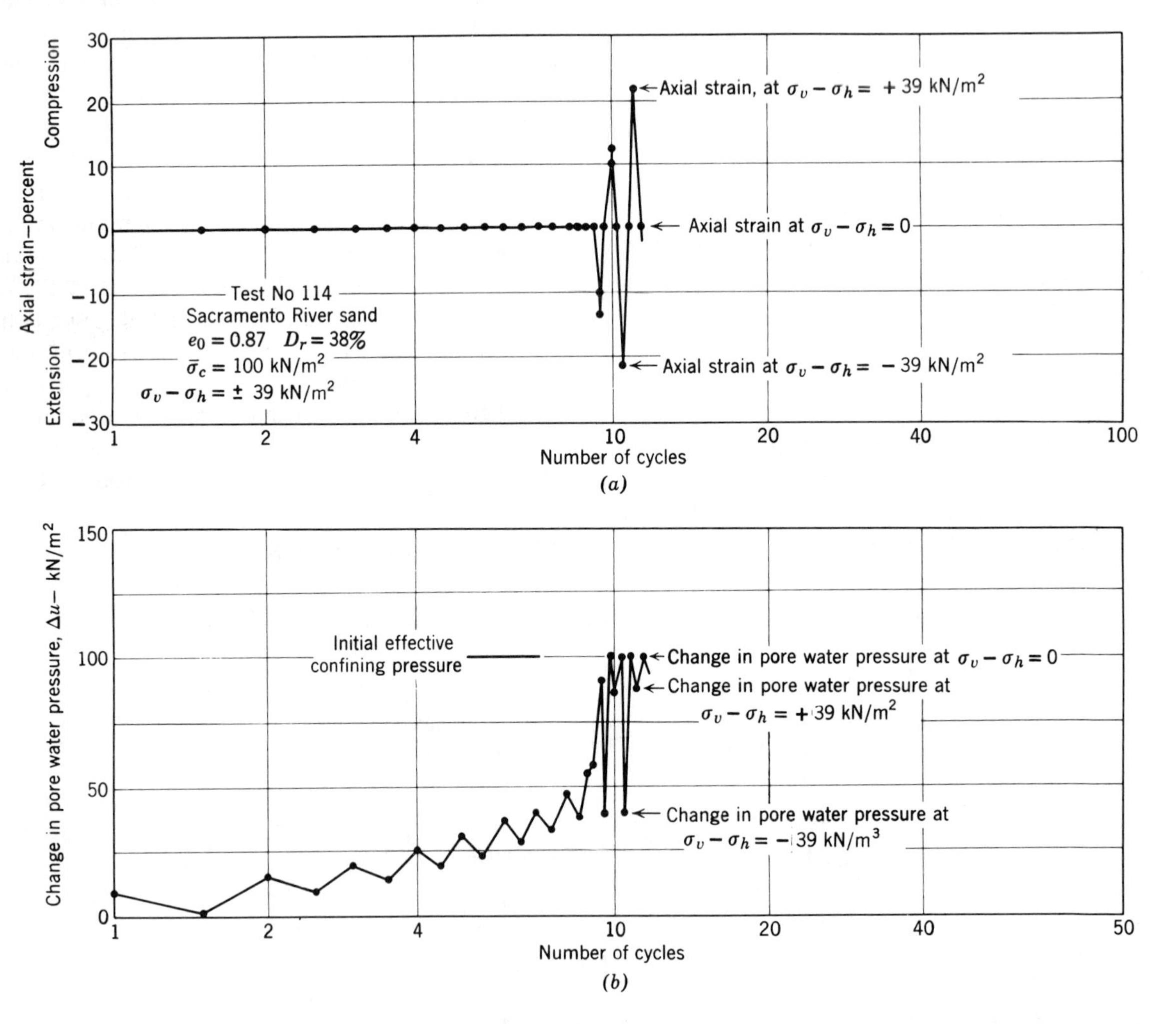

Fig. 29.9 Pore pressure and axial strain versus number of cycles during repeated triaxial loading of loose saturated sand. (*a*) Axial strain versus number of cycles. (*b*) Observed change in pore water pressure versus number of cycles. (From Seed and Lee, 1966.)

29.3 STRENGTH DURING REPETITIVE LOADING

During repeated undrained application of a shear stress, it is possible for a soil to fail at a shear stress less than the shear strength during a single loading. This is especially true when the direction of the shear stress reverses during each cycle of loading. This occurs because the excess pore pressures do not return to zero after each unloading, but rather accumulate as shown in Fig. 29.8. As the pore pressures increase during each cycle of loading, the shear resistance decreases. The increase in pore pressure is caused by a progressive rearrangement of the soil particles during each succesive cycle of loading. In a drained test, these rearrangements would lead to a large decrease in volume, but in an undrained test they permit the soil to be under a much smaller effective stress while at constant volume.

With sandy soils, this behavior during repeated loading may cause nearly total loss of resistance to shear, similar to that during liquefaction. This behavior, which may lead to catastrophic failures during earthquakes (see Chapter 31) has been studied by Seed and Lee (1966). Figure 29.9 shows a typical set of results from their repeated load triaxial tests. In this test, little or no strain was observed until the ninth cycle of loading. In the ninth cycle, large strains suddenly developed and within a few cycles these strains exceeded 20%, implying a total failure. The pore pressures had been building up during the first eight cycles, and in the ninth cycle the pore pressure became equal to the confining stress so that the lateral effective stress dropped to zero. The same effect also develops to a lesser degree (momentary or partial liquefaction) in dense sands (see Fig. 29.10). As shown in Fig. 29.11, the stress-strain relation during repetitive loading can be much lower than during a single loading. Figure 29.12 shows the relation between stress to cause failure (20% strain) and number of pulses of loading; this relation will vary depending on the sand and its void ratio. The susceptibility to liquefaction is greatest in the case of a uniform fine sand.

Loss of strength during cyclic loading also occurs in clays (Fig. 29.13), but total loss of strength does not occur until after very large strains have already developed.

29.4 OTHER TEST CONDITIONS AFFECTING STRENGTH

In Chapter 28 we introduced the concept that the effective stress path and strength for an undrained loading are independent of the way in which the loading is applied. It was mentioned that this rule is only approximately true, and now we must mention a few of the complications.

Intermediate principal stress. The undrained strength of a soil may be decreased by as much as 20% if the clay is sheared with $\sigma_2 = \sigma_1$ (extension test) rather than with $\sigma_2 = \sigma_3$ (compression test). This difference arises because the induced pore pressures are greater in the test with $\sigma_2 = \sigma_1$ (Hirschfeld, 1958).

Strain-rate. Increasing the rate at which a saturated soil is sheared increases the undrained strength. For example, the undrained strength typically increases two-fold between a time to failure of an hour and a time to failure of 5 msec (Whitman, 1957).

There is a general agreement that undrained strength is less in a test of long duration (say several months) than in a test of conventional duration (say several minutes). However, there is little agreement as to the magnitude of this time effect. Housel (1965) has suggested that the strength of normally consolidated clays may drop to as little as 50% of its value during tests of conventional duration. Other results (e.g., Bjerrum et al., 1958; Peck and Raamont, 1965) suggest that the drop is no more than 25% provided that samples of good quality are used.

In tests of long duration upon overconsolidated soils the undrained strength may be quite low (Casagrande and Wilson, 1951), although these results may have been influenced by leakage of water into the specimens during the tests.

In all cases where it has been possible to measure the pore pressures during undrained tests at various rates of loading, it has been found that the change in undrained strength results from a difference in induced pore pressure (Richardson and Whitman, 1964). Increasing the rate-of-strain means smaller induced pore pressures.

Duration of consolidation. The time during which the soil remains under the consolidating stress (step 2 of the CU program) influences undrained strength: the longer the time of consolidation, the greater the undrained strength (Taylor, 1955). Again this happens because the pore pressures induced by shear are different in tests with varying consolidation times. Presumably this effect is associated with secondary consolidation (Chapter 27). The longer a specimen remains under the consolidating stress, the denser it becomes and hence the smaller the pore pressures induced by shear.

Discussion. Changes in temperature, changes in the concentration of ions in the pore fluid, and other environmental changes also can alter the magnitude of the pore pressure induced during undrained shear and hence can alter the undrained shear strength.

Each of the factors described in this section has very little effect on the q_f versus $\bar{p}_f$ relation (either that for peak strength or that for the ultimate condition). However, the magnitude of the pore pressures induced during shear, and hence the undrained shear strength, is moderately sensitive to the details of the loading process.

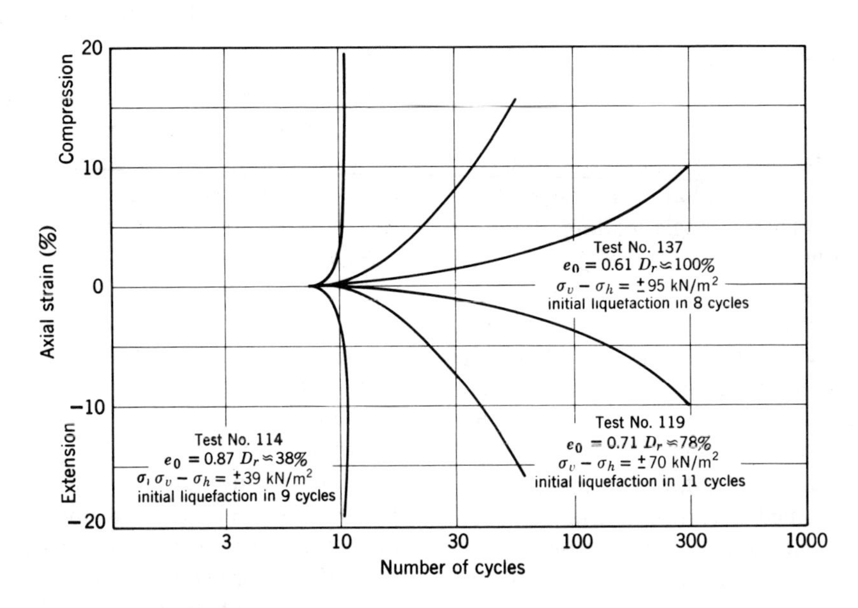

Fig. 29.10 Axial strain versus number of loading cycles for saturated sands at several initial densities. Sacramento River sand; $\bar{\sigma}_0 = 100$ kN/m² (From Seed and Lee, 1966).

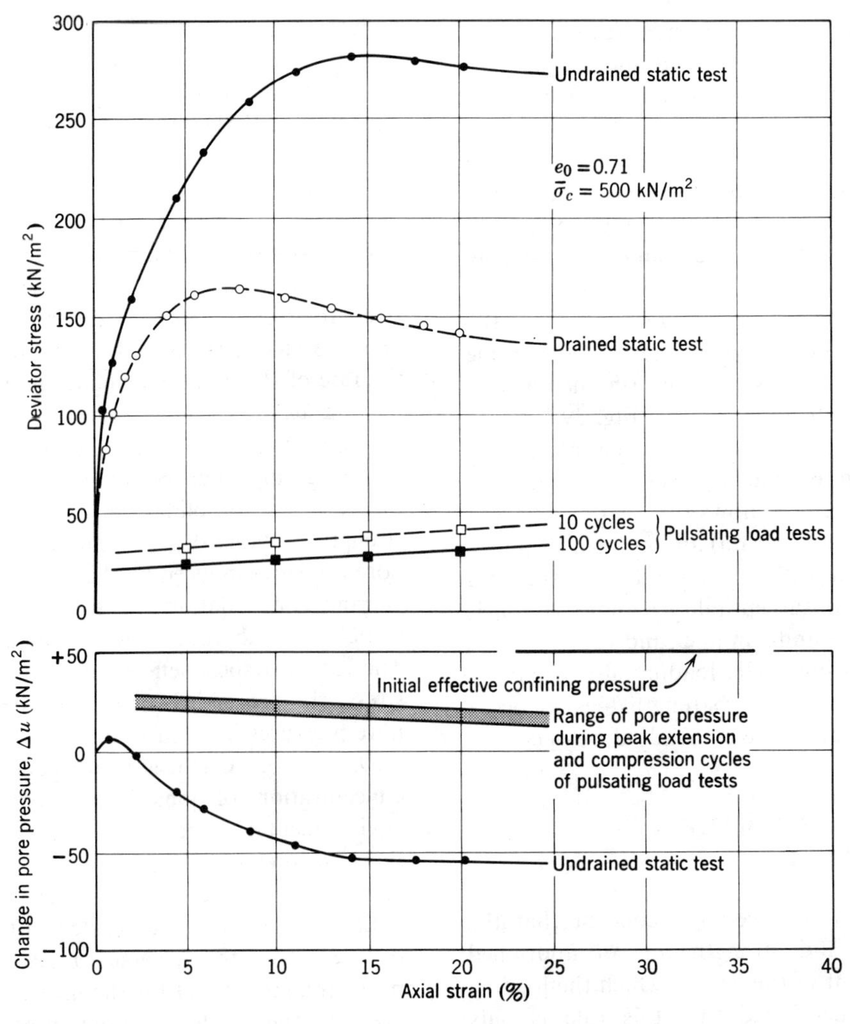

Fig. 29.11 Comparison of strength and pore pressure during single and repeated loading. Sacramento River sand (From Seed and Lee, 1966).

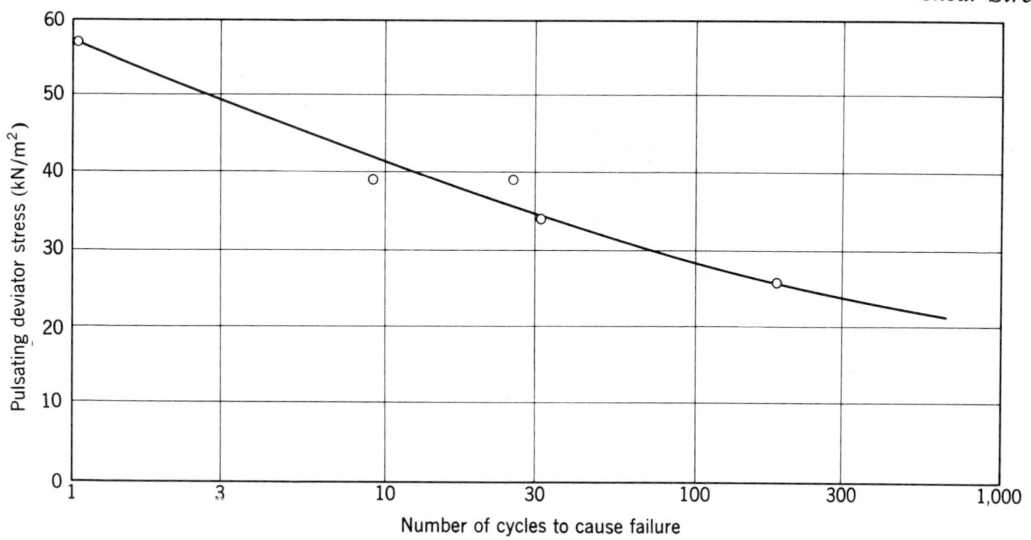

Fig. 29.12 Relationship between pulsating deviator stress and number of cycles required to cause failure Sacremento River sand initial void ratio = 0.87; initial confining stress = 100 kN/m² (From Seed and Lee, 1967).

29.5 CONSOLIDATION TO NONISOTROPIC STRESS

Thus far we have discussed only the case where prior to shear the soil is consolidated under an isotropic stress, i.e., $\bar{\sigma}_{10} = \bar{\sigma}_{30}$. Since the state of stress before shear has proved to have a controlling influence with regard to undrained strength, it is natural to wonder what will happen if $\bar{\sigma}_{30} < \bar{\sigma}_{10}$. For example, natural soils are usually consolidated *in situ* to a K_0-condition:

$$\bar{\sigma}_{30} = \bar{\sigma}_h = K_0\bar{\sigma}_v = K_0\bar{\sigma}_{10}$$

Theory

Figure 29.14 shows two effective stress paths which might be followed to arrive at a given q_0, $\bar{p}_0$ condition. Stress path 1 involves no lateral strain during any stage of the loading, but shear stresses are present throughout the loading. The second stress path involves first consolidation under an isotropic stress (path 2*A*, involving inward lateral strain) followed by an undrained shear until the stress state q_0, $\bar{p}_0$ is reached (path 2*B*, involving outward lateral strain). It has been found that these two stress paths will lead to approximately the same water content for the stress state q_0, $\bar{p}_0$ (Henkel, 1960).

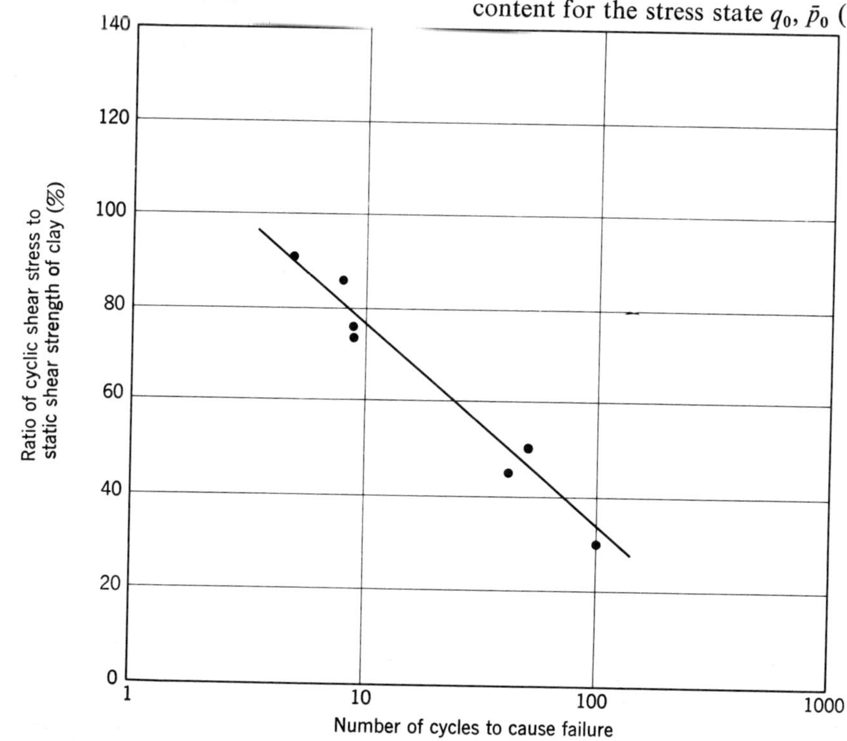

Fig. 29.13 Strength of samples of silty clay under cyclic loading conditions (From Seed and Wilson, 1967).

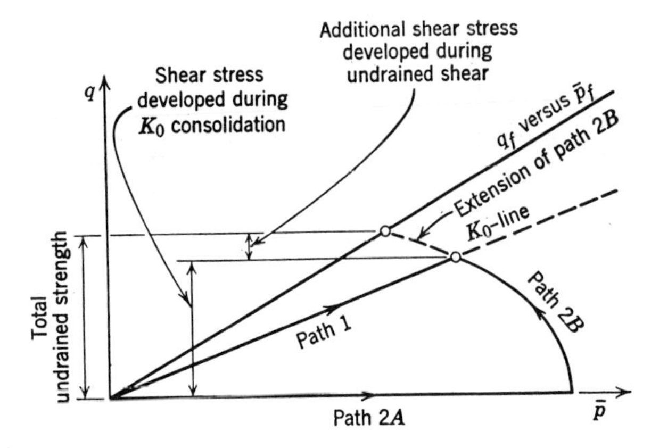

Fig. 29.14 Theory for undrained shear starting from K_0-condition.

The specimen which has been consolidated along stress path 1 is to be sheared undrained. It would seem reasonable that the effective stress path for this undrained shear would simply be the extension of stress path 2*B*. That is, the undrained strength for a specimen normally consolidated at the K_0-condition to water content w_0 is the same as the undrained strength of a specimen normally consolidated under isotropic stress to the same water content w_0. This conclusion is simply an application of the principle that, as a first approximation, undrained strength is uniquely related to water content. Note that the additional shear stress developed during undrained shear is a rather small portion of the total shear strength.

As an aid to understanding this principle, let us answer the following question. Suppose we have two specimens normally consolidated to the same $\bar{\sigma}_{10}$. For specimen *A*, $\bar{\sigma}_{30} = \bar{\sigma}_{10}$, while for specimen *B*, $\bar{\sigma}_{30} = K_0\bar{\sigma}_{10}$. What is the relationship between the undrained strengths of the two specimens? The solution to this question is worked out in Example 29.2. The conclusion is that specimen *B* is weaker than specimen *A*, which might be expected since $\bar{p}_0$ is less for specimen *B* than for specimen *A* and hence specimen *B* has the greater water content. The ratio of the strength of specimen *B* to that of specimen *A* is typically between 0.75 and 1.0.

Thus (assuming the foregoing theory to be correct), if an actual K_0 consolidation condition is simulated by isotropic consolidation to the same $\bar{\sigma}_{10}$, the undrained strength will be overestimated by an error of as much as 33% in case of normally consolidated soils.

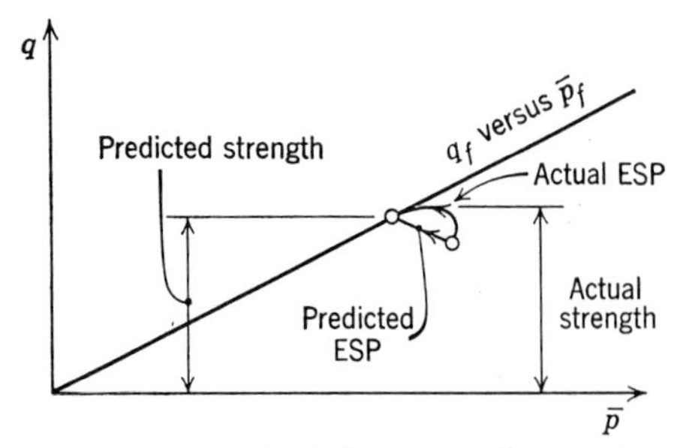

Fig. 29.15 Actual typical effective stress path for undrained shear starting from K_0-condition.

Experimental Results

A typical stress path for undrained shear of a clay consolidated to the K_0-condition is shown in Fig. 29.15. The stress path deviates considerably from that predicted, presumably because the clay had remained at the consolidation condition (the initial point) for some time instead of quickly passing through this stress condition. The magnitude of the peak undrained strength is somewhat greater than predicted by the theory. Ladd (1963) provides experimental data for the relative undrained strength of isotropically and anisotropically consolidated clays.

29.6 REMOLDING AND DISTURBANCE

For many soils there is a great difference between the peak undrained strength of the soil as it exists in the ground and the peak undrained strength of the soil after it has been remolded without change of water content. The ratio of undisturbed to remolded strength has been defined as *sensitivity*.

Figure 29.16 depicts the stress paths for undrained shear of undisturbed and remolded specimens of a sensitive clay. Both specimens are at the same water content but under very different effective stresses. During the remolding, most of the effective stress which had been carried by the mineral skeleton is transferred to the pore water. Figure 29.17 will help to understand what has happened. The physical processes active during remolding have been discussed in Chapter 7.

There is no such thing as a *truly* undisturbed sample. Occasionally the soil of interest in an actual problem can be exposed by excavation and a block sample can be cut by hand. This process results in a relatively high quality

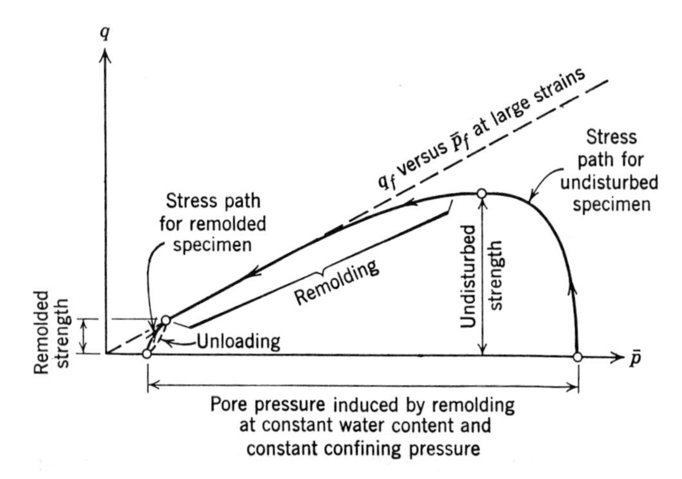

Fig. 29.16 Stress paths for undisturbed and remolded soils.

▶ Example 29.2

Given. q_f versus $\bar{p}_f$ and A for undrained shear starting from isotropic consolidation.
Find. Undrained strength starting from K_0 consolidation.
Solution. According to Eq. 28.2, the undrained strength is proportional to the isotropic stress corresponding to the appropriate stress path:

$$\frac{(q_f)_B}{(q_f)_A} = \frac{(\bar{p}_0')_B}{(\bar{p}_0)_A}$$

Specimen A:

$$(\bar{p}_0)_A = \bar{\sigma}_{10} \qquad \text{(given)}$$

Specimen B:

$$\bar{p}_0' = \bar{p}_0 + (2A_0 - 1)q_0$$

where A_0 is the value of A for loading to the K_0 loading

$$\bar{p}_0 = \frac{1 + K_0}{2}\bar{\sigma}_{10}$$

$$q_0 = \frac{1 - K_0}{2}\bar{\sigma}_{10}$$

$$\bar{p}_0' = \bar{\sigma}_{10}\left[\frac{1 + K_0}{2} + (2A_0 - 1)\frac{1 - K_0}{2}\right]$$
$$= \bar{\sigma}_{10}[K_0 + A_0(1 - K_0)]$$

Hence

$$\frac{(q_f)_B}{(q_f)_A} = K_0 + A_0(1 - K_0) \xleftarrow{\text{Answer}}$$

A_0 typically is somewhat less than A_f; say $0.5 < A_0 < 1$. K_0 typically has values between 0.65 and 0.5.

K_0 \ A_0	1.0	0.5
0.5	1.0	0.75
0.65	1.0	0.82

◀

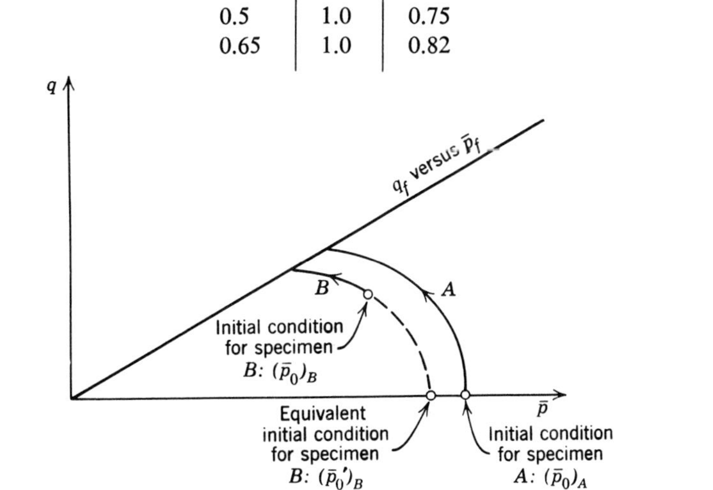

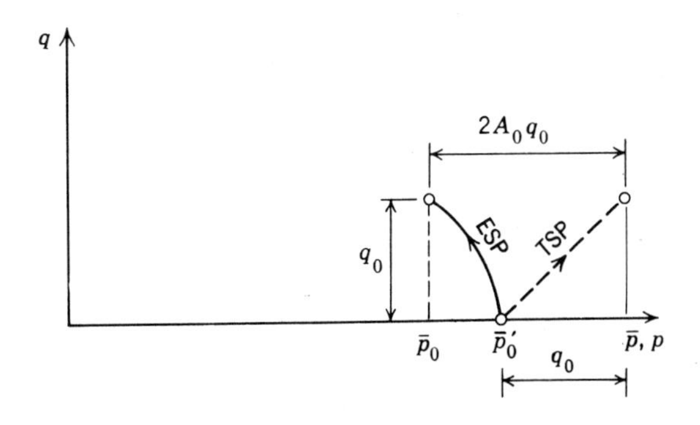

Fig. E29.2

◀

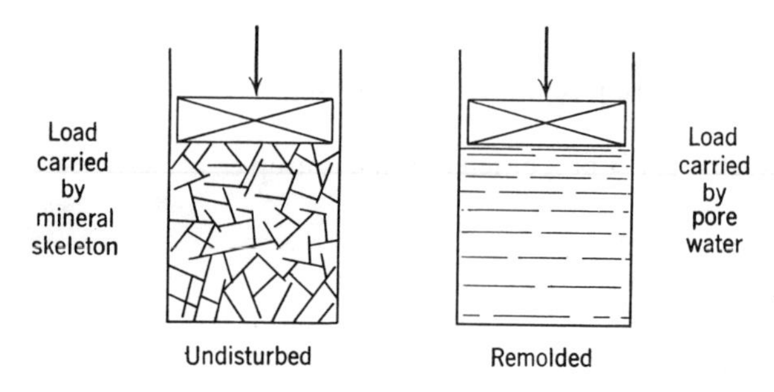

Fig. 29.17 Mechanistic picture of load transfer during remolding.

of soil sample. Unfortunately, the usual situation requires that the sample of soil be extracted by a sampler lowered into the soil through a borehole. The quality of sample obtained by this process tends to be considerably inferior to that obtained by hand cutting.

Even if the process of cutting a chunk of soil from the subsoil, transporting it to the laboratory, trimming a test specimen, and mounting the specimen in the triaxial apparatus were done in a "perfect" fashion, there would have been an inevitable change in stresses acting on the soil. The soil in the ground was subjected to a system of total stresses which have been completely removed by the time the specimen has been mounted in the shear apparatus. Consider, for example, a sample of soil consolidated to a K_0-system of effective stresses, as illustrated by the point C in Fig. 29.18. By the time the element of soil has been removed from the ground and placed in the test apparatus all total stresses have been removed, and the pore water pressures have become negative—resulting in an isotropic effective stress of $\bar{\sigma}_{ps}$ as represented by the point H. In other words, the sample under the effective stresses represented by point C in the ground would exist at the effective stresses shown by point H in the laboratory if a perfect sampling operation had been conducted. (The point H is determined by loading a specimen in the laboratory to point C, removing the total stresses and measuring the negative pore pressure. The effective stress $\bar{\sigma}_{ps}$ is equal to the negative pore pressure.)

Unfortunately, the process of sampling, trimming, and mounting the soil in the test equipment can have a significant influence on the structure of the soil. All of the changes in the soil structure associated with the sampling operation are termed sampling "disturbance." Many experimenters have studied soil disturbance (e.g., Hvorslev, 1949, Schmertmann, 1955, Ladd and Lambe, 1963, Skempton and Sowa, 1963).

An indication of the large effect of disturbance on a clay can be obtained by measuring the negative pore pressure in the soil specimen prior to testing and comparing it with that which would exist had the sampling been "perfect." Test data presented in Fig. 29.18 for the Kawasaki clay show the measured stress $\bar{\sigma}_s$ is approximately one-third of that measured for perfect sampling $\bar{\sigma}_{ps}$. In other words, disturbance during the sampling operation resulted in almost two-thirds of the effective stress in the sample being destroyed. (The actual stress path between the points C and I is not known—only the locations of the two points C and I are known.)

Figure 29.18 also illustrates the effect of sampling disturbance upon undrained strength. When the element of soil in the field is loaded to failure, the stress path CD in Fig. 29.18 is obtained. When a sample is taken of the soil at point C, brought to the laboratory, a test specimen prepared, and an unconfined compression test run, the effective stress path IJ is obtained. Unfortunately, the unconfined compression test gave an undrained strength equal to only 40% of that achieved *in situ*. Further, the unconfined compression test required five times as much strain to reach failure as occurred for the loading C to D. This is the usual effect of disturbance—it increases the strains for a loading.

One possible way to avoid soil disturbance is to use field tests to get stress-strain and strength data. Such a procedure, although obvious, is not easy to carry out. Small-scale field tests load only a small fraction of the soil involved under the actual structure. Frequently the soil of most interest is far below the ground surface. Should a pit be excavated so that the field test can be run on the soil in question, the soil has then undergone a change in stresses similar to that which occurs during the sampling operation. Further, the interpretation of field tests is frequently difficult because of uncertain boundary conditions in the field.

29.7 PRACTICAL METHODS OF MEASURING UNDRAINED STRENGTH

Table 29.2 lists some of the more common methods for measuring undrained shear strength. The vane shear device has been discussed in Chapter 7, as has the so-called standard penetration test. Table 7.4 gives a correlation between unconfined compressive strength (twice the undrained shear strength) and blow count in the standard penetration test. All of the laboratory

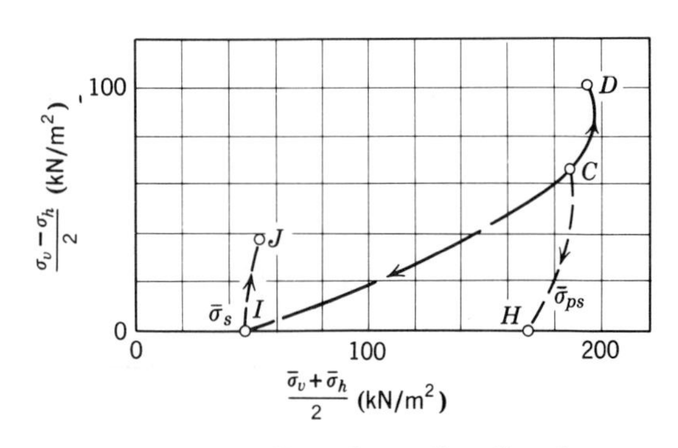

Fig. 29.18 Effect of sampling disturbance.

Table 29.2 Common Methods for Measuring Undrained Strength

Method	Comment
In-situ measurements	
1. Vane test	Usually considered to give best result, but is limited as to strength of soil with which it can be used
2. Penetration test	Gives crude correlation to strength
Measurements upon undisturbed samples	
1. Unconfined compression	Best general purpose test; underestimates strength because disturbance decreases effective stress
2. UU test at *in situ* confining pressure	Most representative of laboratory tests, because of compensating errors
3. CU test at *in situ* confining pressure	Overestimates strength, because disturbance leads to smaller water content upon reconsolidation

procedures depend on obtaining good undisturbed samples.

If there really were a unique relationship among q_f, $\bar{p}_f$, and w_f, all of these procedures which shear the soil at the *in situ* water content would give the same undrained shear strength. In actuality, as we have seen, the $q_f - \bar{p}_f - w_f$ relation is only approximately unique, and undrained strength is sensitive to the details of the applied loading. Since the details of the loading differ for the several methods of Table 29.2, it is natural that each will give somewhat different results. Because of sampling disturbance, unconfined compression tests on even good quality samples usually somewhat underestimate the *in situ* undrained strength, often by a factor of 2 or even more. Use of CU tests compensates for the effects of disturbance; indeed, such tests usually overestimate strength since the density of the soil increases during reconsolidation because disturbance has increased the compressibility of the mineral skeleton.

Whereas the foregoing paragraphs have emphasized the difficulties inherent in sampling, the *in situ* measurements also are not without their difficulties. The standard penetration test provides only a crude estimate of strength. Problems arise with the vane device because of disturbance as the device is inserted into the ground, rate-of-strain, etc. It generally (but not always) has been found that properly conducted vane tests and unconfined compression tests upon *good* undisturbed samples give strengths which agree within 25%. The vane test usually, but not always, gives a larger strength for a given soil than does the unconfined compression test.

In short, because the undrained strength of a soil is somewhat sensitive to test conditions, it is difficult to establish undrained strength within about ±20% at best. In the last analysis, the true test of any of these methods is how well they predict actual failures. We shall return to this question in Chapter 31.

The choice of the method to be used for any particular engineering problem will depend upon a number of factors, especially availability of equipment and economics. The vane device is especially useful when strength varies considerably over a site and with depth, for this device permits, within a reasonable time, many measurements to establish the extent and pattern of the variations. Where soil properties are reasonably uniform, on the other hand, the behavior of the soil will be most clearly established by means of a relatively few carefully conducted laboratory tests on samples of good quality.

For uniform, normally consolidated clays, the best procedure is to consolidate samples to effective stresses greater than twice those existing *in situ*, and then to correct the measured undrained strength by the ratio of the effective stress *in situ* to the consolidation stress used in the laboratory test. This procedure overcomes errors caused by sampling disturbance.

Use of advanced testing techniques, such as plane strain triaxial tests and simple shear tests, permits a better simulation of all components of the *in situ* stresses and changes in stress caused by loading.

29.8 MAGNITUDE OF UNDRAINED STRENGTH IN VARIOUS SOILS

Here we define undrained shear strength as the peak value of q. Henceforth in this book we shall use the symbol s_u to denote shear strength; i.e., $s_u = q_f$ in an undrained test.[2]

Normally Consolidated Soil

According to Eq. 28.2, the undrained strength of a given normally consolidated soil should increase linearly with overburden stress and hence linearly with depth. Strength variations of this type have already been shown in Figs. 7.7, 7.8, and 7.10.

The ratio of undrained strength to effective overburden stress, $s_u/\bar{\sigma}_{vo}$, is a useful way to characterize the undrained strength of normally consolidated soil.[3] Figure 29.19 shows a correlation between this ratio and plasticity index. The "special clays" include those which have thixotropic behavior or which tend to dilate during shear. Many remolded clays have a $s_u/\bar{\sigma}_{vo}$ ratio of about 0.3 ± 0.1.

Relations such as those in Fig. 29.19 are useful for preliminary estimates concerning the undrained strength of normally consolidated soils.

[2] The symbol c is often used in the literature. The literature also often quotes values for undrained *compressive* strength, which is equal to $2s_u$.

[3] In the literature this ratio is often expressed as c/p, where $p = \bar{\sigma}_{vo}$ and is not the same as the p used in this book.

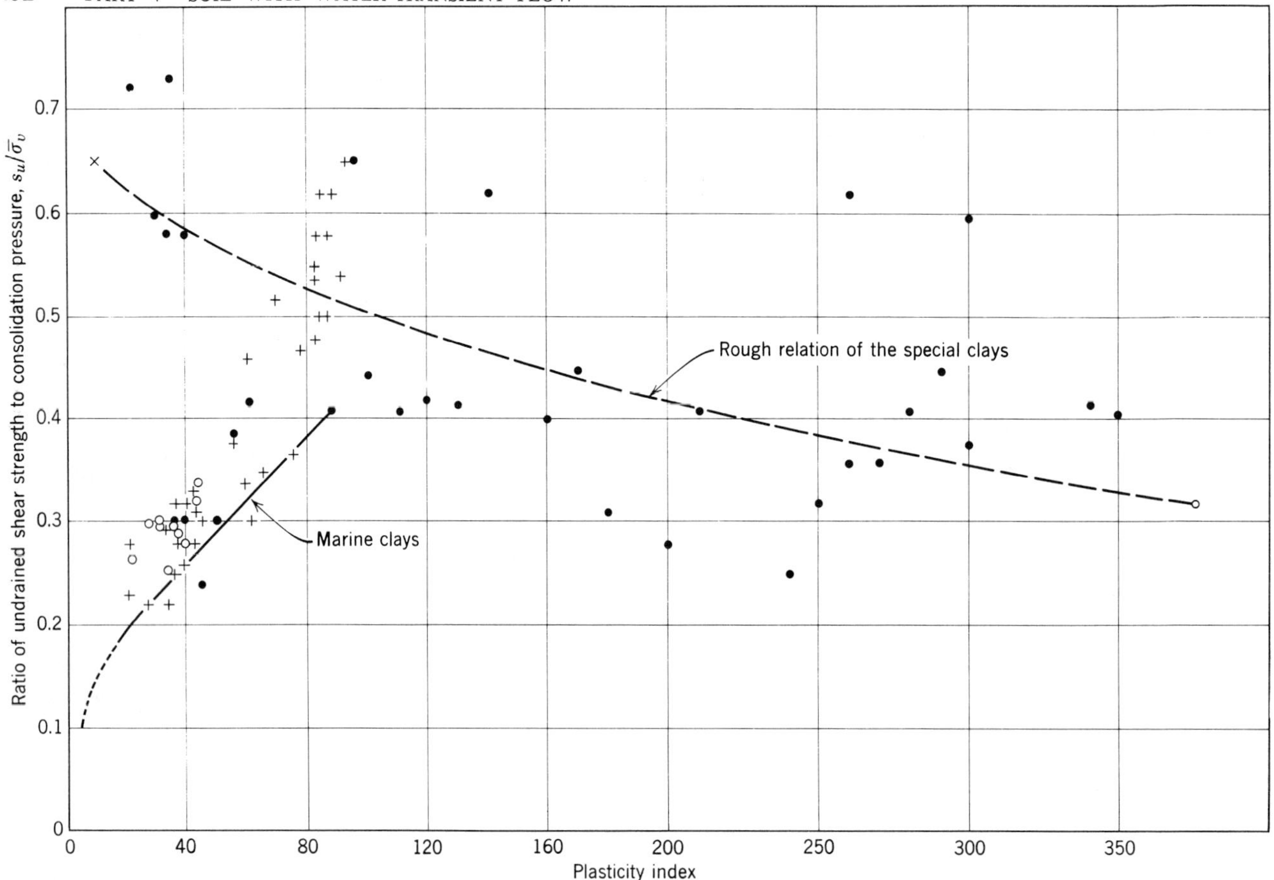

Fig. 29.19 $s_u/\bar{\sigma}_{10}$ ratio as a function of plasticity index (From Osterman, 1959).

Overconsolidated Soils

In overconsolidated soils, undrained strength depends on the maximum past value of $\bar{\sigma}_v$ as well as the present value of this stress. Figure 29.20 shows the relationship of undrained strength of remolded Weald clay, isotropically consolidated, to the ratio $\bar{p}_0/\bar{p}_m$. Example 29.3 illustrates the use of these data to compute the variation of undrained strength with depth in a case where erosion has removed some of the overburden. To simplify the problem, unit weights have been assumed constant with depth and isotropic consolidation has been assumed. In this example, the clay just a short distance below the present ground surface has considerable strength as the result of the preconsolidation. If the depth of overburden removed had been greater, the curve of s_u versus depth would be nearly vertical. There are phenomena other than overburden that can produce a preconsolidation effect: weathering, partial drying, indeed any effect that tends to reduce the void ratio of a soft, normally consolidated clay. Figure 7.7*b* shows a weathered crust over the top of a soft, normally consolidated clay.

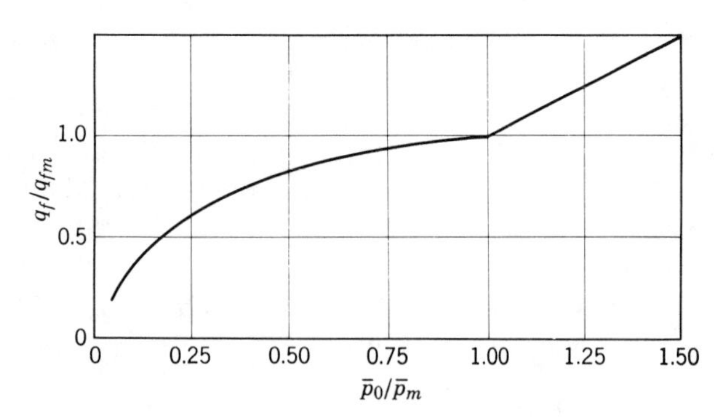

Fig. 29.20 Relationship of undrained strength to overconsolidation ratio.

Figure 7.9 shows the undrained strength versus depth relation in the Boston clay. Past events have conspired to leave the strength more-or-less constant with depth. Many clay deposits have almost a constant undrained strength with depth, at least to the extent it is reasonable to assume a uniform strength for calculation purposes.

It is impossible to correlate undrained strength of overconsolidated soils directly to index properties because these index properties do not adequately reflect the effects of stress history. The natural water content considered in relation to the liquid and plastic limits gives some idea of the degree of overconsolidation, but does not suffice to permit quantitative estimates of undrained strength. Table 7.4, which correlates strength to blow count in the standard penetration test, gives an idea of the possible range of undrained strength.

▶ **Example 29.3**

Given. Past and present soil profiles as shown, in Fig. E29.3, with $q_{fm} = 0.29\bar{p}_m$ and q_f/q_{fm} as given in Fig. 29.20.

Find. q_f versus depth for present profile.

Solution. The pertinent stresses are worked out in Fig. E29.3

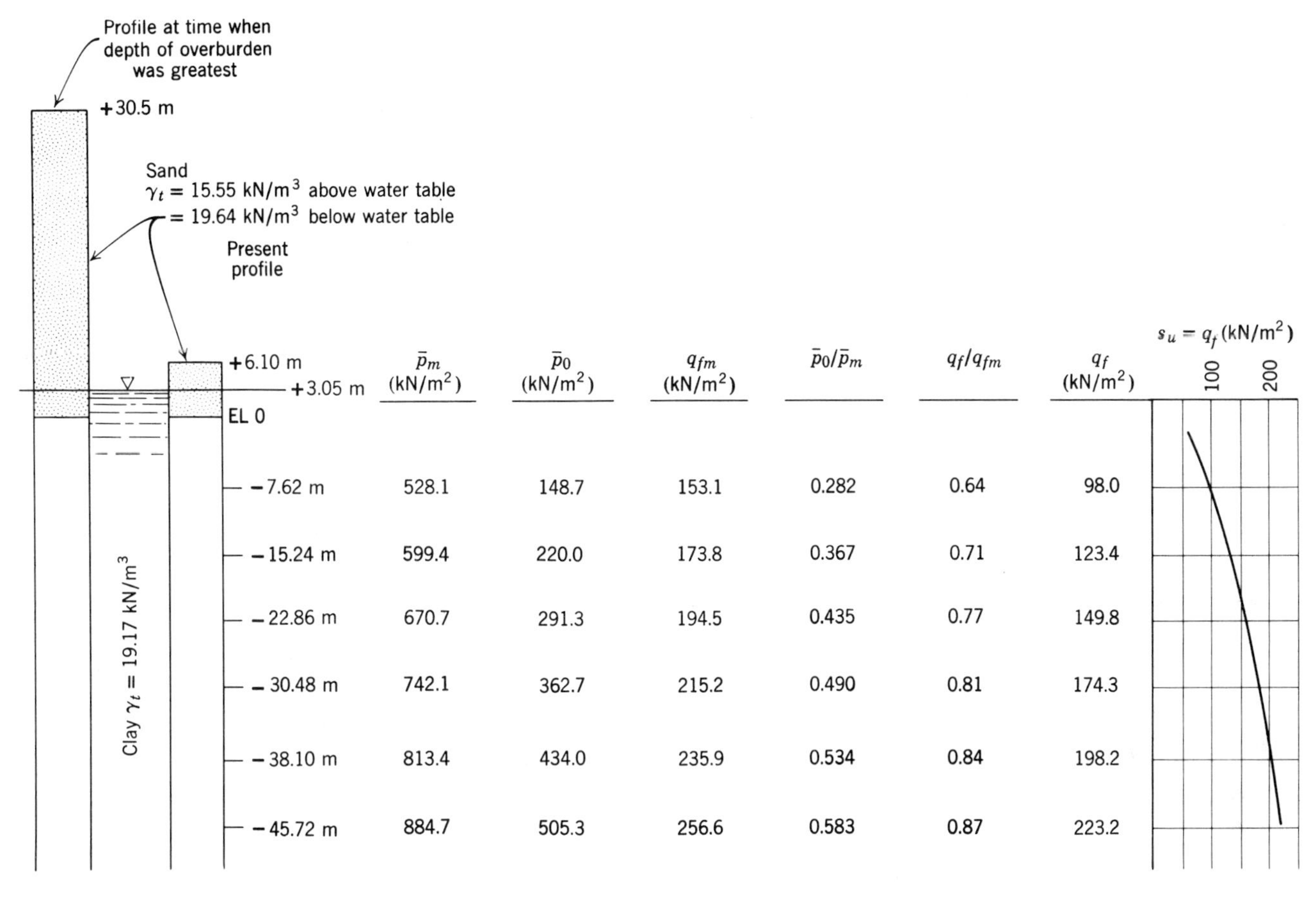

Elevation	$\bar{p}_m$ (kN/m^2)	$\bar{p}_0$ (kN/m^2)	q_{fm} (kN/m^2)	$\bar{p}_0/\bar{p}_m$	q_f/q_{fm}	q_f (kN/m^2)
−7.62 m	528.1	148.7	153.1	0.282	0.64	98.0
−15.24 m	599.4	220.0	173.8	0.367	0.71	123.4
−22.86 m	670.7	291.3	194.5	0.435	0.77	149.8
−30.48 m	742.1	362.7	215.2	0.490	0.81	174.3
−38.10 m	813.4	434.0	235.9	0.534	0.84	198.2
−45.72 m	884.7	505.3	256.6	0.583	0.87	223.2

Fig. E29.3 ◀

29.9 HISTORICAL NOTE

Foundation engineers 50 years ago were taught that sands were cohesionless and that $\phi = 0$ for saturated clays, with intermediate values for intermediate materials. Clays were thought to be cohesive in the same sense that steel is cohesive, and clays and sands were treated as quite different materials. Today it is realized that the main difference between sands and clays rests with their relative permeabilities and relative capillary heads.

Terzaghi's discovery of the effective stress concept in the early 1920s of course marks the starting point for this new understanding. Once it was understood that the phenomenon of consolidation existed, it was a logical step to explain the dependence of the undrained shear strength of clay upon the stress to which the clay had been consolidated. A major breakthrough came with the realization that excess pore pressures are generated by the application of shear stress even though the average normal stress remains unchanged (Casagrande and Albert, 1930). Now it was possible to relate undrained and drained strengths of clay. Rendulic (1936, 1937), working in Terzaghi's laboratory in Vienna, devised the first systems for measuring pore water pressures, and thus gave the first actual confirmation of the hypothesis of the unifying role of effective stress.

The intervening years have seen the improvement of experimental techniques, especially those for the measurement of pore pressure, and the collection of data to confirm and show the limitations of the effective stress principle. Taylor at M.I.T. made especially important contributions to experimental technique. Rutledge (1947), then at Northwestern University, pointed out the relation of water content to strength. Finally, Skempton (1954) and Bjerrum (1954), through their efforts to develop theoretical relations between volume changes in drained tests and excess pore pressures in undrained tests, have provided a clearer and more concise picture of the importance of effective stress.

29.10 SUMMARY OF MAIN POINTS

This chapter has emphasized that it is not a simple matter to obtain accurate measurements of undrained strength. In particular, great care must be taken in sampling and in preparation of test specimens. To obtain very accurate strengths, all aspects of the *in situ* stress conditions should be reproduced in the tests. Stress history has a great effect upon undrained strength. Strength during repetitive undrained shear can be much less than during a single loading.

PROBLEMS

29.1 Refer to Fig. 29.18. What is the value of pore pressure at point J for the unconfined compression test specimen?

29.2 Refer to Fig. 29.18. Derive the following equation for $\bar{\sigma}_{ps}$:

$$\bar{\sigma}_{ps} = \bar{\sigma}_{v0}[K_0 + A_u(1 - K_0)]$$

where

$$A_u = \frac{\Delta u - \Delta\sigma_h}{\Delta\sigma_v - \Delta\sigma_h}$$

is an A parameter for undrained unloading from K_0 stresses to isotropic stresses.

29.3 The concept of a unique relationship between effective stress and undrained strength for a soil is only valid under certain conditions. List the factors discussed in Chapter 29 that can influence this relationship.

29.4 Refer to the lower part of Fig. 29.4. For the four tests shown on this figure:

a. In which tests did cavitation occur?

b. What was the value of pore pressure at which cavitation occurred?

c. Plot the four Mohr circles in terms of effective stresses and show the pore pressure at failure for each test.

Answer

a. Cavitation occurred in the tests with σ_3 = 36.5 kN/m² and 138 kN/m² (i.e., those tests with $\phi > 0$).

b.
$$-u_{max} = \frac{c}{\tan\phi} = \frac{63.3}{\tan 32^\circ} = 101.3\ \text{kN/m}^2$$

c. Draw effective stress envelopes through the origin ($\bar{c} = 0$) and at $\bar{\phi} = 32°$. All circles must be tangent to this envelope.

σ_3	$\bar{\sigma}_3$	u	
36.5	137.8	−101.3	cavitation
138	239.3	−101.3	cavitation
310	393.3	−83.3	no cavitation
676	393.3	+282.7	no cavitation

CHAPTER 30

Stress-Strain Relations for Undrained Conditions

Engineers frequently must estimate the undrained (or initial) settlement of loaded areas. Such estimates are often based on equations from the theory of elasticity. To use these equations, it is necessary to evaluate the modulus of soil for undrained loading conditions. This can be done either by evaluating the shear modulus G or the undrained Young's modulus E. Since Poisson's ratio is $\frac{1}{2}$ for undrained loading, Eq. 12.4 indicates that theoretically $E = 3G$.

Unfortunately, it is extremely difficult to evaluate undrained modulus accurately. As is true for a drained loading (Chapters 12 and 22), the modulus during undrained loading is very sensitive to stress level. Moreover, undrained modulus is affected by the many factors that affect undrained strength, e.g., rate of loading, time of consolidation, intermediate principal stress, and especially sampling disturbance. The influence of these factors on modulus is considerably greater than their influence on strength; i.e., the details of the loading affect the early portion of a stress-strain curve more than they affect the peak of the curve.

This chapter gives general guidance as to the magnitude of undrained modulus for various conditions, and the data presented here may be used for very crude estimates of settlements. Where accurate estimates are required, it is necessary to perform tests on the best possible samples with careful attention to duplication of the loading conditions expected *in situ*.

30.1 RELATIONSHIP TO WAVE VELOCITIES

Values of shear modulus applicable for very small stress changes may be determined by measuring the propagation velocity for shear waves, either *in situ* or in laboratory tests on undisturbed specimens. Such values of modulus are directly useful for a variety of dynamic problems, and are often used to provide an upper bound to the value of modulus applicable to larger stress changes.

Since shear modulus should be the same for both drained and undrained loadings, the shear wave velocity through a saturated soil should differ only slightly from that through a dry soil having the same void ratio and carrying the same effective stress. Indeed, it has been found that the data in Fig. 12.10 are applicable to both dry and saturated sands, with only slight differences arising from the change in mass density as the result of saturation (Hardin and Richart, 1963).

Going a step further, it has been found that the shear modulus of *any* soil is, as a first approximation, related solely to void ratio and effective stress, independently of grain size characteristics. Hardin and Black (1968) find that the following equation applies for sand with angular particles and for several clays:

$$G = 3230 \frac{(2.973 - e)^2}{1 + e} \sqrt{\bar{\sigma}_c} \qquad (30.1)$$

where G and $\bar{\sigma}_c$ (the average principal stress to which the soil has been consolidated) are in kN/m². With only slight modification to the numerical coefficients, the same equation applies to sands with rounded grains.

Undrained Young's modulus for small stress changes can be evaluated by measuring the rod velocity in laboratory specimens. With accurate measurements, it has been found that E is very nearly equal to $3G$, provided that comparisons are made at the same level of strains.

If a saturated soil were truly incompressible, the dilatational modulus D and dilatational velocity C_D would be infinite (see Eq. 12.8). Actually, of course, water is only relatively incompressible. The dilatational velocity C_D through saturated soil is typically about 1500 m/sec (Fig. 30.1) and is much greater than C_D through dry soil. At usual levels of effective stress, C_D is controlled by the compressibility of the pore phase and is affected little by the compressibility of the mineral skeleton; hence C_D is more-or-less independent of effective stress. Dilatational velocity, which can be

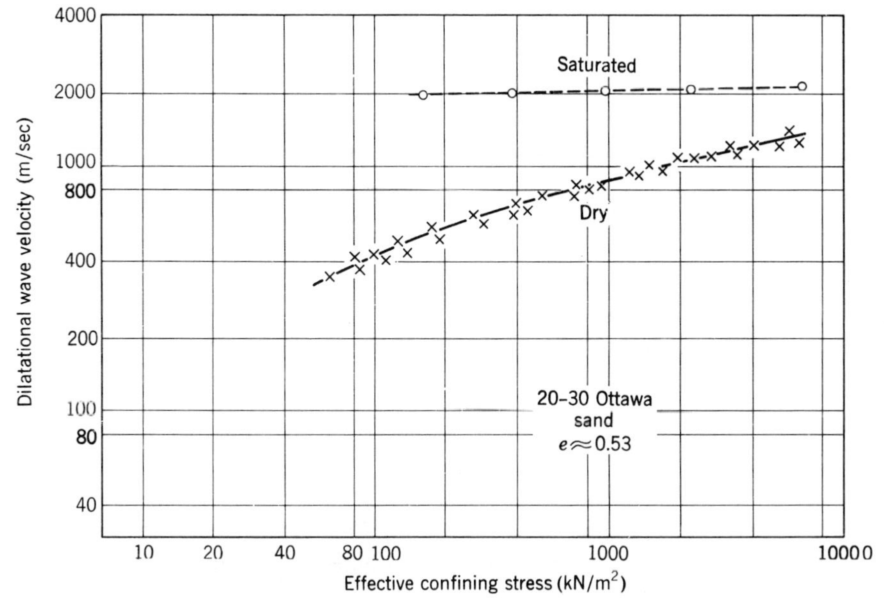

Fig. 30.1 Measured dilatational velocity through dry and saturated sand (Data obtained by Shell Research and Development Laboratory using pulse technique.)

measured easily in the field, unfortunately does not provide useful information regarding the stiffness of the mineral skeleton of a saturated soil.

30.2 YOUNG'S MODULUS FOR LARGER LOADS

Figure 30.2 presents a typical undrained stress-strain curve for normally consolidated clay obtained from a standard triaxial test with increasing axial stress and constant lateral stress. Values of undrained Young's modulus E, as computed from different stages of this test, are the following:

Initial Tangent Modulus, E_i, equals the slope of the $\sigma_1 - \sigma_3$ versus ϵ plot at the start of the test. As shown in Fig. 30.2, $E_i = 8570$ kN/m².

Secant Modulus at Failure, E_f, equals the slope of the line between the origin and the point of failure on the plot of $\sigma_1 - \sigma_3$ versus ϵ. E_f from Fig. 30.2 $= 4000$ kN/m².

Secant Modulus, at any specified stress or strain level. Among a number of specified levels of stress or strain that have been used are $\epsilon = 2\%$, $\epsilon = 5\%$, $\sigma_1 - \sigma_3$ at half of the value of $(\sigma_1 - \sigma_3)_f$ (also called the modulus at a factor of safety equal to two).

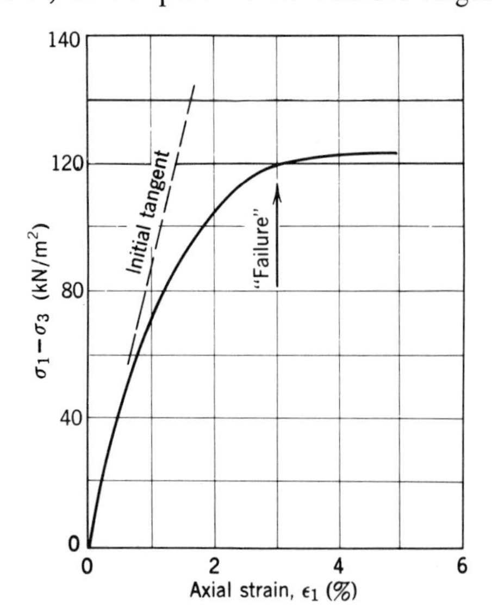

Stress-strain modulus:

$$E_f = \frac{(\sigma_1 - \sigma_3)_f}{\epsilon_f} = \frac{120 \text{ kN/m}^2}{0.03} = 4000 \text{ kN/m}^2$$

$$E_i = 8570 \text{ kN/m}^2$$

Fig. 30.2 Typical stress-strain curve from undrained triaxial test on normally consolidated clay.

It is quite difficult to determine E_i accurately from such tests, since the slope of the stress-strain curve changes rapidly even at small strains. The initial modulus as determined from the first loading of a triaxial test usually is much less than the modulus computed from wave velocity.

Relation to Consolidation Stress

For normally consolidated clay, it is often assumed that modulus is proportional to consolidation stress; i.e., the stress at any strain is proportional to consolidation stress. Figure 30.3 presents for three clays the results of undrained triaxial tests in the form of stress paths through which strain contours have been drawn.[1] This is a particularly informative type of plot. If the effective stress paths are geometrically similar and the strain contours are straight radial lines for a group of tests, then a plot of $q/\bar{\sigma}_c$ versus strain would be unique. If the plot is unique, the modulus is then proportional to the consolidation pressure. The plots indicate there is variation from this unique relationship, although as a

[1] The characteristics of these clays are presented in Table 30.1.

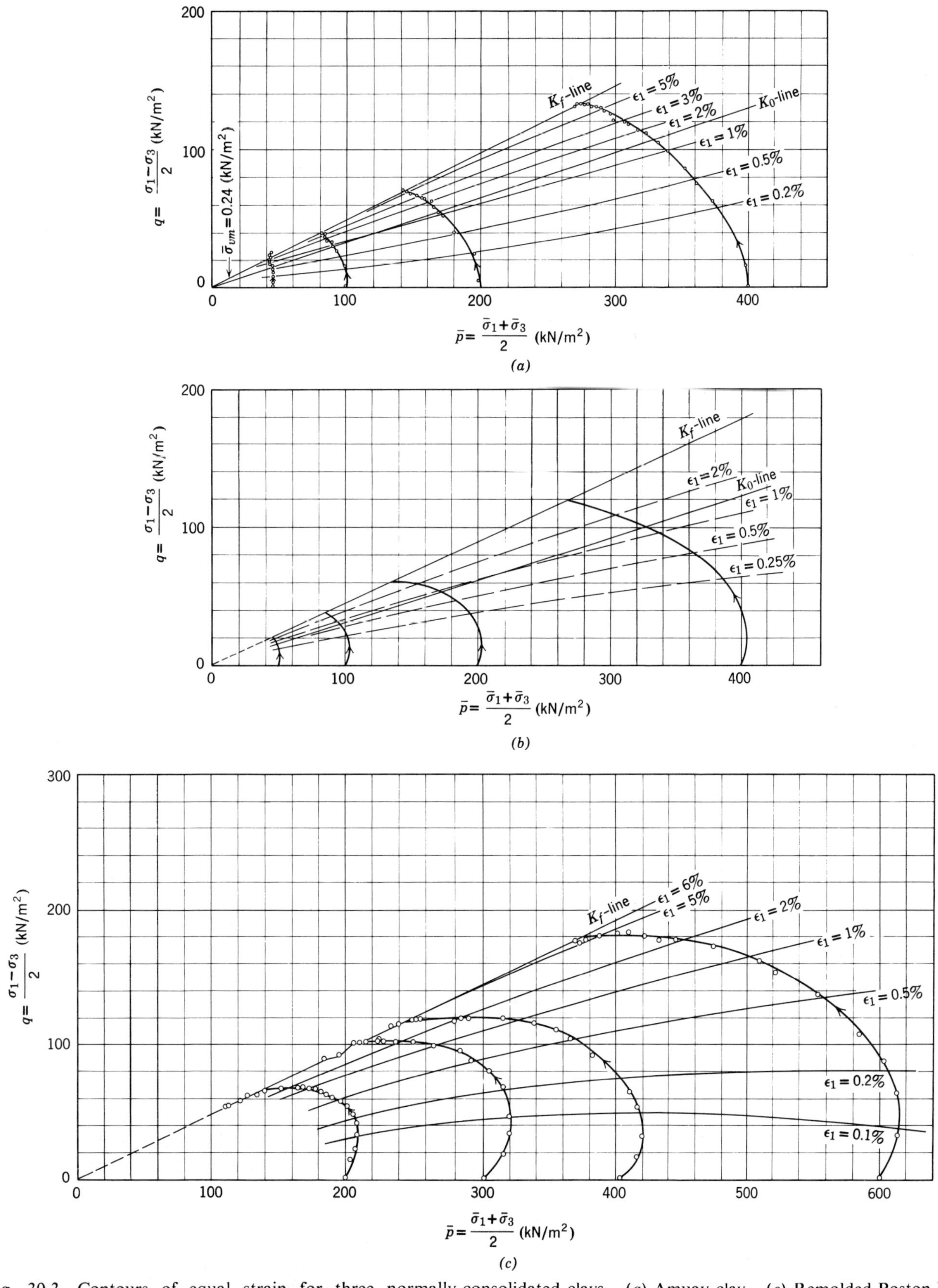

Fig. 30.3 Contours of equal strain for three normally consolidated clays. (*a*) Amuay clay. (*c*) Remolded Boston blue clay. (*b*) Lagunillas clay (From Lambe, 1964).

crude first approximation, we can assume this uniqueness. In general the modulus is less than proportional, i.e., the strain contours curve downward, especially the contours for the smaller strains. In view of Eq. 30.1, such deviation is to be expected for smaller strains. On the other hand, since undrained strength of a normally consolidated clay is proportional to $\bar{\sigma}_c$, it is reasonable that the secant modulus at large strains might be approximately proportional to $\bar{\sigma}_c$.

Table 30.1 Description and Classification Data on Six Clays

Undisturbed clays

1. *Amuay Clay*, Amuay, Venezuela
   (Block sample from 2 m depth).
   Clay is slightly overconsolidated.
   $w_n = 47–55\%$, $w_l = 71\%$, PI $= 42\%$
2. *Boston Blue Clay*, M.I.T. Campus, Cambridge, Mass. (Also called *Cambridge clay*.)
   (76 mm diameter fixed piston samples from depths d of 11 to 30 m).
   Clay is overconsolidated for $d = 11$ to 20 m.
   Clay is slightly overconsolidated for $d = 20$ to 25 m.
   Clay is normally consolidated for $d \geq 25$ m.
   $w_n = 40 \pm 5\%$, $w_l = 42–55\%$, PI $= 25 \pm 25\%$
3. *Kawasaki Clay I*, Kawasaki, Japan[a]
   (76 mm diameter tube samples from depths of 20 to 26 m).
   Clay is normally consolidated and has a sensitivity of 10 ± 5
   $w_n = 67\%$ (46–79), $w_l = 70\%$ (51–83), PI $=$ 34% (20–45)
   Activity = 1.03 (0.74–1.62)
4. *Lagunillas Clay*, Lagunillas, Venezuela
   (76 mm diameter shelby tube samples from a depth of 6 m).
   Clay is normally consolidated and has a sensitivity below 10
   $w_n = 60\%$ (40–73), $w_l = 61\%$ (50–79), PI $=$ 37% (29–49)
   Activity = 0.8 (0.6–0.9)

Remolded clays

Preparation: Slurry of clay with a water content equal to two to four times the liquid limit is placed in a 241 mm diameter oedometer and consolidated to a pressure of 100 to 150 kN/m². The sample is then extruded from the oedometer and cut into 14 specimens for triaxial testing.

1. *Boston Blue Clay* (B.B.C.)
   Consolidation pressure of large sample = 150 kN/m²
   Water content of large sample = $28 \pm 2\%$
   $w_l = 33 \pm 3\%$, PI $= 15 \pm 2\%$
2. *Vicksburg Buckshot Clay* (V.B.C.)
   Consolidation pressure of large sample = 100 kN/m².
   Water content of large sample = $46 \pm 2\%$
   $w_l = 64 \pm 2\%$, PI $= 39 \pm 1.5\%$
   Activity = 0.7

[a] The strength behavior of the Kawasaki clay is similar in many respects to that of much less plastic clays. This unusual behavior is explained by the fact that the clay contains a high percentage of volcanic glass, shell and diatoms.

Figure 30.4 shows stress-strain curves for five normally consolidated clays, normalized by dividing stress by $\bar{\sigma}_c$. The general similarities are evident. Figure 30.5 shows the ratio of secant modulus to $\bar{\sigma}_c$ for these clays, plotted against the safety factor.

Load Cycle

As was true with drained loading, the stress-strain modulus for undrained loading is greater on a subsequent cycle of loading than on the initial one. Part of the "strain" apparent in the initial loading results from seating deformations, closing sample cracks, etc. These test errors are particularly large in unconfined compression testing and lead to lower moduli in this type of test than in the standard triaxial test. Tests on undisturbed London clay (Ward, Samuels, Butler, 1959) showed that the stress-strain modulus from a second loading cycle was about 1.4 to 1.5 times that obtained from the initial loading.

Overconsolidation

Overconsolidation tends to make a soil stiffer and stronger, although the effect of overconsolidation is less on the stiffness than on the strength. Figure 30.6, presenting $E/\bar{\sigma}_c$ versus overconsolidation ratio for four clays, suggests that the modulus increases with overconsolidation ratio. At high values of the overconsolidation ratio the trend is not clear.

Time

The effects of time on the stress-strain modulus may be considered under three headings:

Thixotropic Effects. The term "thixotropy" is used to describe a strength increase with time at constant composition. Mitchell (1960), Skempton and Northey (1952), and Moretto (1948) have presented extensive data on thixotropic effects. The effects are generally most significant at low strains and in remolded soils having a high liquidity index. In a clay exhibiting thixotropic effects, the stress-strain modulus increases with increasing time of soil storage before testing. As an extreme example, unconfined tests on specimens of Laurentian

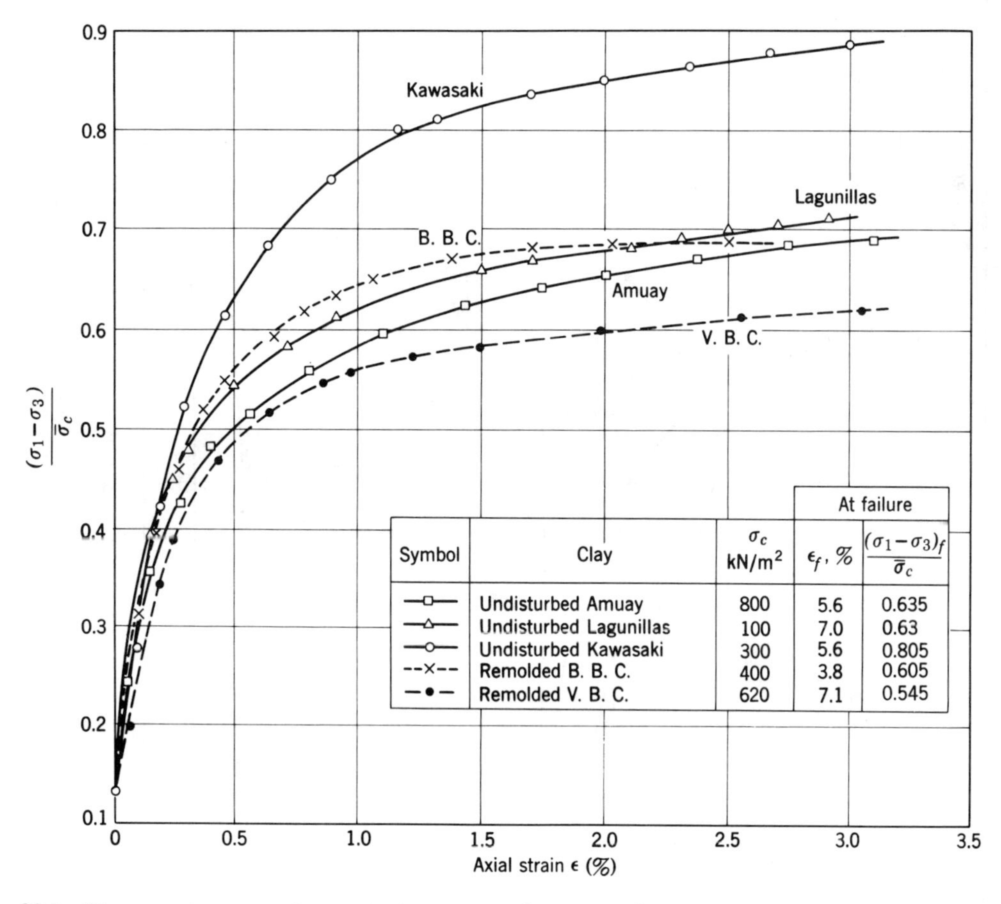

Fig. 30.4 Stress-strain curves from triaxial tests on five normally consolidated clays (From Ladd, 1964).

clay remolded at the liquid limit (Moretto, 1948) showed E values at FS = 1.5 of about 40, 250, 550, and 700 kN/m^2 at test times of 0, 7, 28, and 120 days, respectively.

Aging Effects "Aging" refers to the length of time allowed for consolidation (in excess of primary consolidation) prior to shear testing, i.e., aging increases the time allowed for "secondary" consolidation. The effects of aging on the stress-strain modulus of normally consolidated Vicksburg buckshot clay as measured in CU tests are summarized in Table 30.2 (primary consolidation was over in less than one day). Similar large increases in modulus with aging have been observed with CU tests on normally consolidated samples of remolded Boston blue clay (Bailey, 1961) and kaolinite (Wissa, 1961) and on undisturbed samples of normally consolidated marine clay (Bjerrum and Lo, 1963).

Table 30.2

Time of Aging (days)	$E/\bar{\sigma}_c$	
	FS = 3	FS = 1.5
3	175	110
10	230	135
60	300	210

Strain-Rate Effects. Strain-rate refers to the rate of strain (change in axial strain per unit time) that is applied during undrained shear. The rate of strain can have very large effects on the stress-strain modulus, as has been shown by many investigators with both normally consolidated clays (Bjerrum, Simons and Torblaa, 1958; Crawford, 1959; Casagrande and Wilson, 1951) and overconsolidated clays (Casagrande and Wilson, 1951). Table 30.3 presents data by Richardson and Whitman (1963) from CU tests on normally consolidated and overconsolidated samples of remolded Vicksburg buckshot clay, which illustrate possible effects. The rate of strain in the fast tests corresponds to that conventionally used in unconfined compression tests, whereas the slow rate is several times slower than ordinarily employed in CU tests.

By comparing the available data from dynamic tests (a test in which the sample is failed in 1 msec) with those from static tests (one in which a specimen is failed in 10 min), Whitman (1964) concluded that the dynamic modulus is 1.5 to 2.0 times greater than the static modulus.

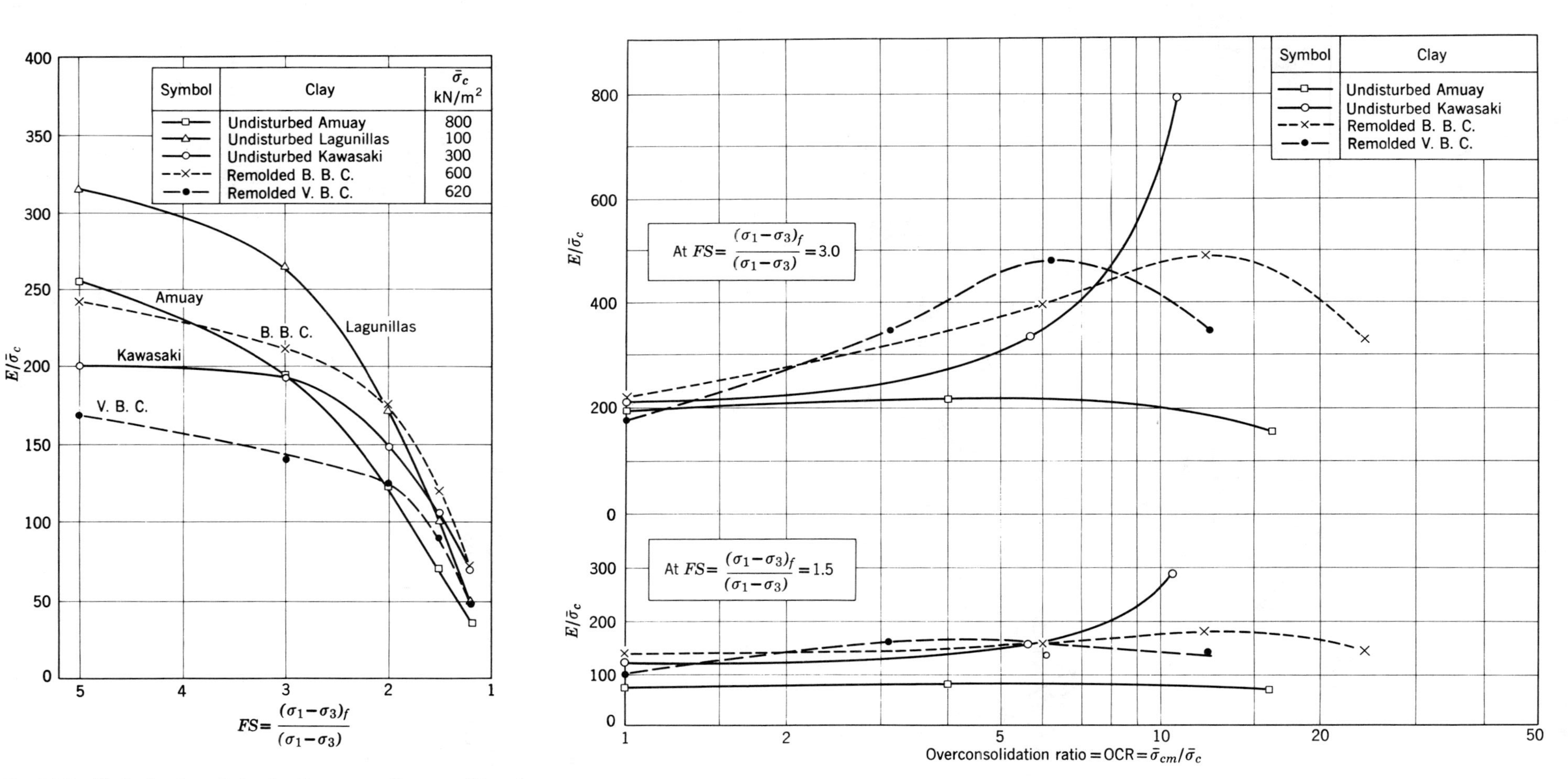

Fig. 30.5 Undrained modulus for five normally consolidated clays (From Ladd, 1964).

Fig. 30.6 Effect of overconsolidation on modulus (From Ladd, 1964).

Table 30.3

	$E/\bar{\sigma}_c$	
	FS = 3	FS = 1.5
Normally consolidated		
Fast tests[a]	250	160
Slow tests[b]	120	60
Overconsolidated (OCR = 16)		
Fast tests[a]	450	200
Slow tests[b]	250	140

[a] Fast tests: 1% strain in 1 min.
[b] Slow tests: 1% strain in 500 min.

30.3 STRAINS ALONG VARIOUS STRESS PATHS

All of the data presented so far in this chapter came from tests on specimens initially consolidated to an isotropic stress system and then loaded by a vertical stress while the horizontal stress was held constant. While such data are very useful in comparative studies on soils, two questions arise in their use in practical problems: (*a*) Does this simple loading condition represent field problems? (*b*) If not, are strain data from standard triaxial tests equal to those obtained by loading along other paths? In short, the answer to each question for most problems is "No".

Stress paths for even a simple foundation problem (see Example 8.9) indicate that only at some considerable depth below the center of the foundation does the total stress path move upward at an angle of 45° like that in the standard triaxial test. Further, in most deformation problems, the initial state of consolidation is not isotropic but anisotropic. In heave problems the soil is unloaded rather than loaded. We can thus see that the total stress path for actual field problems can be far different from that for the standard triaxial test.

Chapter 10 presented stress-strain data for drained tests for a large variety of initial stress conditions and stress paths. As can be seen in Figs. 10.20 to 10.23, the stress-strain behavior of soil is very dependent on the stress path.

We can gain some impression of the importance of effective stress path on strain by examining the results of the three triaxial tests shown in Figs. 30.7 and 30.8. Three specimens of remolded Boston blue clay were all consolidated to identical initial conditions of $\bar{\sigma}_v$ = 400 kN/m² and $\bar{\sigma}_h$ = 216 kN/m² (point A in Fig. 30.7). In

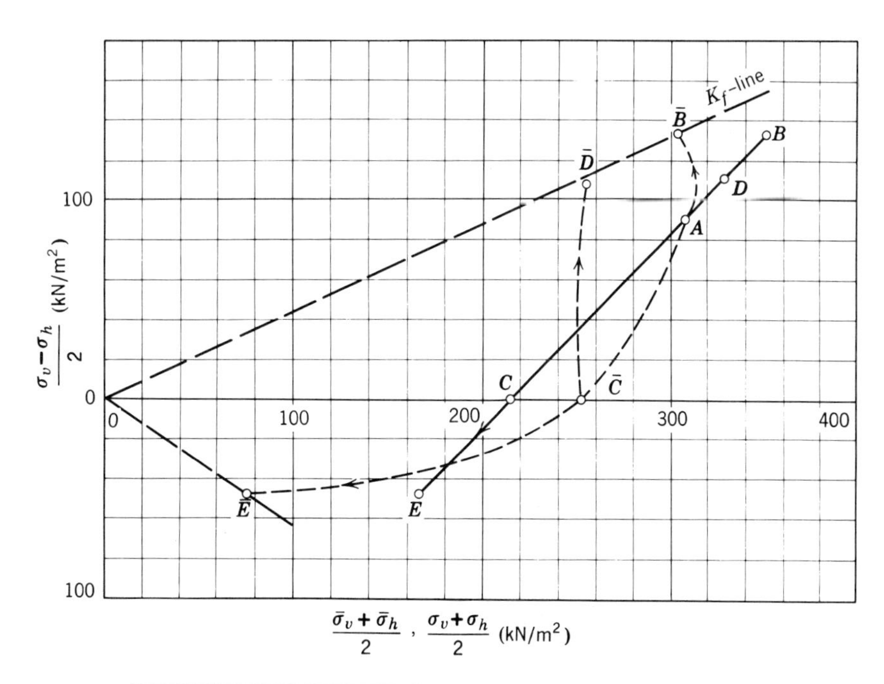

TEST	TSP	ESP	TYPE OF TEST
1	*AB*	$A\bar{B}$	CAU
2	*ACE*	$A\bar{C}\bar{E}$	CAU-RE
3	*ACD*	$A\bar{C}\bar{D}$	CA-UU

Fig. 30.7 Strength tests on Boston blue clay.

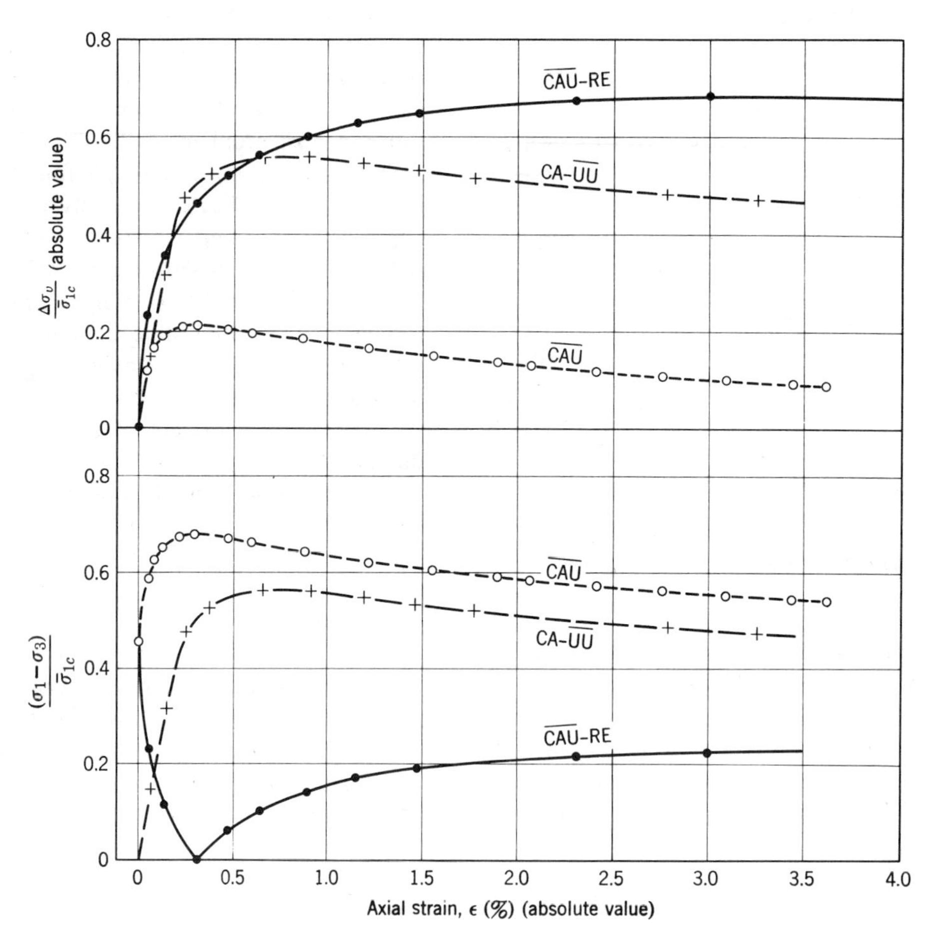

Fig. 30.8 Stress-strain data on Boston blue clay. *Note.* Zero strain on CA–UU test taken at $\sigma_1 - \sigma_3 = 0$. (From Ladd, 1964.)

test 1, the specimen was carried to failure by increasing the vertical stress while holding the horizontal stress constant. This test,[2] termed $\overline{\text{CAU}}$, has a total stress path AB and an effective stress path $A\bar{B}$. Such a test approximates a loading condition far below the center of a foundation. In test 2, the soil specimen was held at constant horizontal stress and the vertical stress reduced until failure was reached. This test, termed $\overline{\text{CAU}}$-RE, approximates conditions in the soil at the bottom of a deep excavation. In test 3, the specimen was held at constant horizontal stress and the vertical stress was reduced until it was equal to the horizontal stress, as might occur in a perfect sampling operation, and the sample was then failed in undrained shear by increasing the vertical stress at constant horizontal stress. This test, termed CA-$\overline{\text{UU}}$, has the stress paths as follows: total stress ACD and effective stress $A\bar{C}\bar{D}$. Strains for test 3 were measured with the initial zero strain taken at the moment the vertical stress became equal to the horizontal stress. Figure 30.8 presents the stress-strain data for the three tests. As can be seen, there is a very great difference in the stress-strain behavior for the three types of stress paths—e.g., the strains at failure are 0.3% for test 1, 14% for test 2, and 0.7% for test 3.

These illustrative stress-strain data emphasize that the actual stress path to failure can have a very great influence on the stress-strain behavior of the soil. That is to say, soil is not an isotropic material.

[2] The letter A in the test designations denotes consolidation under an anisotropic stress, i.e., $\bar{\sigma}_{ch} \neq \bar{\sigma}_{cv}$. Generally $\bar{\sigma}_h = K_o\bar{\sigma}_v$ during consolidation in such a test. Thus a $\overline{\text{CAU}}$ test is a consolidated undrained test with anisotropic consolidation; the bar denotes that pore pressures were measured during the test. The Symbol $\overline{\text{CIU}}$ is sometimes used for the standard triaxial test with isotropic consolidation.

30.4 DISTURBANCE

Estimates of settlement made using modulus measured in laboratory tests generally exceed the observed settlements, often by a factor of 4 or 5. Although there is sometimes failure to consider all of the factors discussed in Sections 30.2 and 30.3, the main reason for these discrepancies undoubtedly is disturbance during sampling. There is some evidence that the modulus during a second cycle of loading on an undisturbed sample gives a reasonable estimate for the *in situ* modulus, the effects of disturbance having been largely eliminated as the result of the initial loading.

There have been attempts to backfigure the stress-strain modulus of a soil from the measured heave or settlement of a field structure. Some experimenters have suggested that stress-strain data, obtained on specimens trimmed from large undisturbed block samples, are not too far different from the behavior of the soil in the field. Other experimenters have found a large discrepancy, leading them to suggest empirical correction factors or relationships. Bjerrum (1964), for example, suggested multiplying the modulus from laboratory unconfined compression tests by 5, or taking the modulus as 200 to 400 times the undrained shear strength. Such relations are too crude for use in important problems, but they are helpful in indicating the importance of sample disturbance and in giving values for approximate analyses.

30.5 SUMMARY OF MAIN POINTS

A convenient way to characterize the stress-strain behavior of a soil is with a modulus that is stress divided by strain. The modulus of soil for undrained loading is not a unique property but varies widely with stress level, stress history, time, type of loading, and soil disturbance. In general the modulus of a soil *decreases* with:

1. An increase in deviator stress.
2. Soil disturbance.

It *increases* with:

1. An increase in consolidation stress.
2. An increase in overconsolidation ratio.
3. An increase in aging.
4. An increase in strain rate.

PROBLEMS

30.1 Determine the initial tangent modulus for the three tests shown in Fig. 30.8. Determine the secant moduli at FS = 1.5 and FS = 3.0 for the three tests. Compare these values with the results for Boston blue clay shown on Figs. 30.4 and 30.5.

30.2 Use Eq. 30.1 to calculate $E/\bar{\sigma}_c$ for *BBC*. Refer to Fig. 20.5 for consolidation data. Compare your results with the values obtained in Problem 30.1.

30.3 List the factors that can influence the value of E for a saturated soil. Indicate whether the effect of each factor is likely to be small or large for most practical situations.

CHAPTER 31

Earth Retaining Structures and Earth Slopes with Undrained Conditions

There are numerous practical problems in which the soil within a slope or within the backfill behind a retaining structure is stressed quickly compared to the consolidation time for the soil:

1. When a slope, with or without a retaining structure, is excavated quickly.
2. During construction of an earth dam.
3. During a *rapid drawdown*, when the level of water standing against a slope or retaining structure drops rapidly.

Generally, it is only with clays or clayey soils that the loading time is small compared to the consolidation time, but upon occasion these conditions will develop within sands as well.

For such problems, it is appropriate to use undrained strength in analyses to determine safety factor or to estimate lateral thrust. Since undrained strength is determined by the initial conditions prior to the loading, it is not necessary to determine the effective stresses that would exist at failure. Stability analyses based on undrained strength are called *total stress analyses* (s_u-analyses) and are generally much simpler than *effective stress analyses* ($\bar{c}$, $\bar{\phi}$-analyses) described in Chapters 23 and 24.

Equations and charts applicable for total stress analyses are given in Section 31.1. Sections 31.2 to 31.4 discuss the very important question of the relationship between analyses based upon total stress versus those based upon effective stress and develop useful rules as to when each method should be used. The remaining sections of the chapter then discuss several important classes of practical problems.

31.1 ANALYSES OF STABILITY

If a soil undergoing undrained loading is saturated, so that the $\phi = 0$ concept applies, analysis for stability is greatly simplified. Procedures and results appearing in Chapters 23 and 24 may be used directly, with $\bar{\phi}$ taken as zero and $\bar{c}$ replaced by the undrained strength s_u. This so-called $\phi = 0$ analysis is a special case of the s_u-analysis. In applying these analyses, pore pressure should be taken as zero along any failure surface where undrained strength is used. This step does not imply that pore pressures actually are zero, but rather is done to be consistent with the assumption that undrained strength can be expressed independently of the effective stress at failure.

Active thrust. For $\bar{\phi} = 0$, Eq. 23.17 reduces to

$$P_a = \tfrac{1}{2}\gamma_t H^2 - 2s_u H + q_s H \qquad (31.1)$$

The corresponding critical failure plane is inclined at 45°. An analysis based on circular failure arcs leads to the same equation but with a coefficient of 1.92 replacing the coefficient 2 in the second term on the right side.

As discussed in Section 23.3, use of Eq. 31.1 implies that there can be tensile stress between the backfill and the retaining wall and within the backfill. Since soil generally will not support such tensile stresses, tension cracks tend to develop. The depth of tension cracks, or the maximum unsupported height of vertical cuts, is (see Fig. 23.15)[1]

$$z_c = \frac{2s_u}{\gamma_t} \qquad (31.2)$$

Eq. 23.18 then may be reduced to

$$P_a = \tfrac{1}{2}\gamma_t H^2 - 2s_u\left(H - \frac{s_u}{\gamma_t}\right) \qquad (31.3)$$

The use of Eqs. 31.1 to 31.3 is illustrated in Example 31.1.

If the backfill carries a surcharge q_s, the resulting horizontal compressive stresses will tend to close the

[1] When undrained shear applies, the condition along a tension crack is that the *total* horizontal stress must be zero, and hence γ_t replaces γ_b in the derivation.

Example 31.1

Given. Retaining wall in Fig. E31.1.

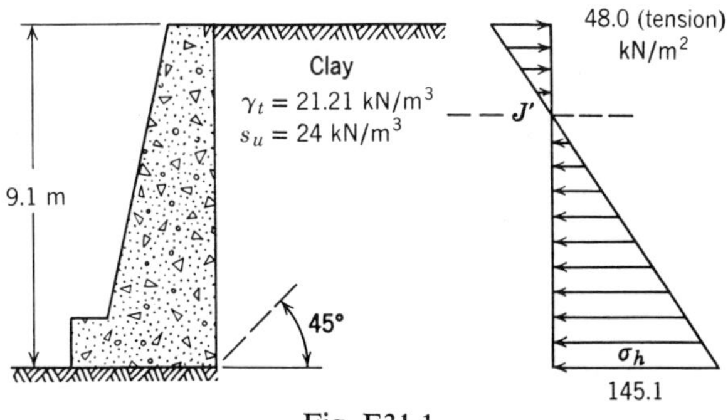

Fig. E31.1

Find. Active thrust with and without tension cracks.
Solution.
a. Without cracks
At base:

$$\sigma_h = 21.21z - 48 \text{ kN/m}^2$$

$$P_a = (0.5)(21.21)(9.1^2) - (2)(24)(9.1) = 878.2 - 436.8 = 441.4 \text{ kN/m}^2$$

b. With tension cracks

$$z_c = \frac{2(24)}{21.21} = 2.26 \text{ m}$$

$$P_a = 878.2 - 2(24)(7.97) = 495.6 \text{ kN/m}^2$$

The horizontal stress at the base of the wall is still

$$\sigma_h = 21.21(6.84) = 145.1 \text{ kN/m}^2$$ ◀

---

tension cracks. If $q_s > 2s_u$, there will be no cracks and Eq. 31.1 may be used. A special solution is needed if $q_s < 2s_u$.

Passive thrust. Similarly, Eq. 23.8 reduces to

$$P_p = \tfrac{1}{2}\gamma_t H^2 + 2s_u H \tag{31.4}$$

Stability of slopes. When an entire slope is in cohesive soil and it is appropriate to use undrained strength throughout the slope, then the equations in Chapter 24 become greatly simplified. For a circular failure surface, using any method of slices, the safety factor is

$$F = \frac{\sum_{i=1}^{i=n} s_u \Delta l_i}{\sum_{i=1}^{i=n} W_i \sin \theta_i} \tag{31.5}$$

If the undrained strength is constant throughout the slope, the numerator is simply $s_u L_a$.

Figure 31.1 gives a chart which may be used for rapid evaluation of the safety of simple slopes having uniform undrained strength. Use of this chart is illustrated in Example 31.2. When the depth of soil beneath a slope becomes very large (D large) then the maximum possible height of the slope is the same for all slope inclinations less than 54°. Gibson and Morgenstern (1962) present a chart for the case where undrained strength is proportional to depth. In this case, the safety factor is not affected by the depth of soil beneath the toe of the slope.

Example 24.6 has presented a case in which a failure surface must pass partly through a free draining soil where strength is appropriately expressed in terms of effective stress and partly through a clay where under some conditions undrained strength should be used. In such cases, the parameters $\bar{c}$ and $\bar{\phi}$, together with appropriate pore pressures, apply along one portion of the surface and the parameters $\phi = 0$ and $c = s_u$ (with zero pore pressure) apply along the other part.

▶ **Example 31.2**

Given. A 1 on 1 slope, 18.3 m high, in a soil with an undrained strength of 144 kN/m² and γ_t = 20.42 kN/m³.

Find. Safety factor against slope failure.

Solution. Assuming that the soil extends to considerable depth beneath the slope, the undrained strength required for equilibrium is

$$\frac{s_u}{(20.42)(18.3)} = 0.181$$

$$s_u = 67.6 \text{ kN/m}^2$$

$$F = \frac{144}{67.6} = 2.13$$ ◀

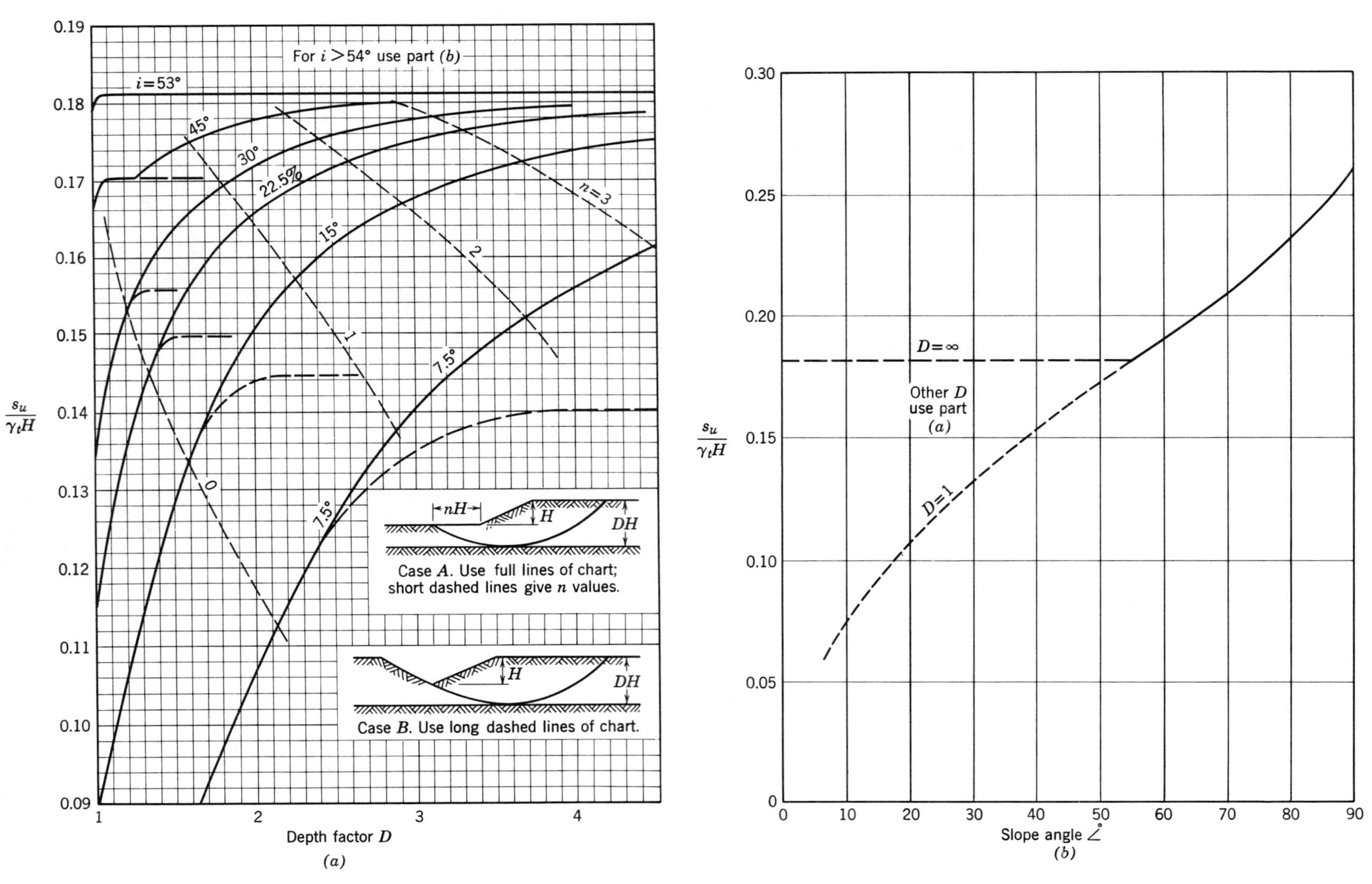

Fig. 31.1 Stability charts for $\phi = 0$ (From Taylor, 1948).

Partially saturated soils. If soil is not fully saturated, the strength during undrained loading is not independent of changes in total stress. As discussed in Section 28.7, undrained strength for such conditions can be expressed approximately in terms of two parameters c_u and ϕ_u, and the procedures and equations in Chapters 23 and 24 with zero pore pressures may then be used to estimate thrusts or safety factors. However, because of the many factors that influence c_u and ϕ_u, results from such analyses must be interpreted with great care (see Section 31.6).

31.2 PROBLEMS INVOLVING LOADING

This section develops the key features of problems in which the total stresses acting within the soil mass are increased. Examples of such problems include foundations for buildings (discussed in more detail in Chapters 32 and 33), embankments constructed upon soft foundations, and construction of earth dams. In order to develop the key features of these problems, a highly idealized problem will be considered: passive thrust against an overconsolidated clay. The problem is shown in Example 31.3.

Initially, before the vertical wall is constructed and the passive thrust is applied, the surface of the soil is level. The water table is 1.52 m below the surface of the ground. The zone above the water table is assumed to be saturated and to sustain capillary tensions which increase linearly with height above the water table. Step 1 in the analysis evaluates the stresses at mid-depth of the soil and the total horizontal force against a vertical plane. It is assumed that $K_0 = 1$, as is appropriate for such an overconsolidated soil. The horizontal thrust P equals the total horizontal stress at mid-depth times the depth of the stratum.

Step 2 analyzes conditions at the "end-of-construction" when the vertical wall is in place and the horizontal thrust has been applied. Since it is envisioned that construction takes place "instantaneously," the undrained strength must be used to determine the permissible thrust at this stage.

This evaluation of the permissible thrust at the end-of-construction has required no knowledge of the pore pressures existing at this stage. However, it is now useful to evaluate what these pore pressures might be. This is done in step 3 for a point at mid-depth. If the maximum possible thrust were to be applied, so that the soil is at a failure condition, then the value of $\bar{p}_f$ that must exist can be determined from the relation (see Fig. 11.6):

$$q_f = s_u = \bar{c} \cos \bar{\phi} + \bar{p}_f \sin \bar{\phi}$$

In other words, even though the undrained shear strength is used, the relationship between strength and effective stress still must apply. Since p_f has been established from the result of step 2, finding $\bar{p}_f$ means that the pore pressure is now known. The important result of this step is that the pore pressure *increases* as a result of the loading. The pore pressure caused by the permissible thrust will be less than that calculated here, but will still be larger than for the initial condition.

Following application of the thrust, the soil will begin to consolidate and the pore pressures will begin to return to the initial values determined by natural ground water conditions. Thus the pore pressures within the soil will *decrease* in this example, and hence the effective stresses will increase. This means that the soil will become stronger with time. The thrust that can be resisted if there are no excess pore pressures has already been calculated in Example 23.9 and indeed it is greater than the thrust which can be resisted by undrained shear. Thus the end-of-construction condition controls the magnitude of the thrust that can be applied with the specified safety factor.

Table E31.3 summarizes the stresses for:

1. The initial condition.
2. With failure at the end-of-construction.
3. With failure in the long term.

In Fig. E31.3–2, path OCD is the effective stress path that would occur if the soil were brought to failure at the end-of-construction and then kept at failure by increasing the thrust as consolidation occurs. Path OAB applies when the permissible thrust of 335 kN/m is applied and held constant as consolidation occurs. It is clear that the margin of safety increases as consolidation occurs and the stress changes from point A to point B.

The clay in Example 31.3 was lightly overconsolidated. If the clay had been normally consolidated, the excess pore pressures in the end-of-construction condition would have been greater and hence the end-of-construction condition would have been even more critical than the long-term situation. With a heavily overconsolidated clay, the excess pore pressure at the end-of-construction would be nearly zero and might even possibly be somewhat negative. Hence with a heavily overconsolidated clay it is conceivable that the long-term condition might be more critical than the end-of-construction. The relationship between undrained (end-of-construction) and drained (long-term) resistance as deduced from this example should be compared with the entries in Table 28.1 for triaxial compression loading.

Example 31.3, although highly idealized, leads to practical conclusions applicable to a great number of problems involving *loading* of soil in a time short compared to the time required for consolidation:

1. Conditions at the end-of-loading generally are critical and control the permissible load.
2. The permissible load hence generally may be

▶ Example 31.3

Given. Stratum of soil, with properties shown in Fig. E31.3-1, in which wall is constructed to resist a passive thrust.

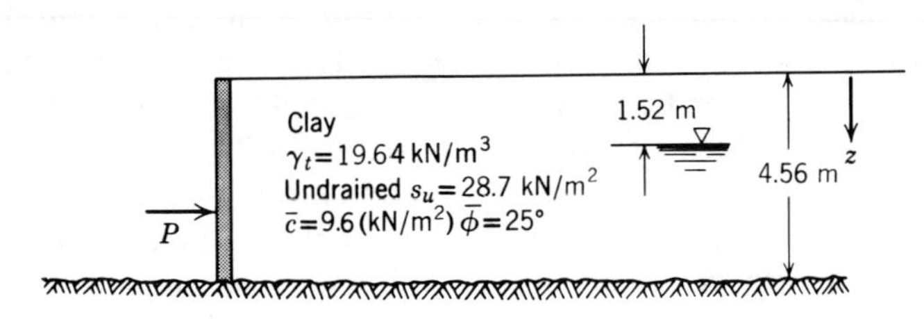

Fig. E31.3-1

Find. The maximum permissible thrust against the wall for a safety factor of 2, assuming the thrust is applied instantaneously and thereafter held constant.

Solution.

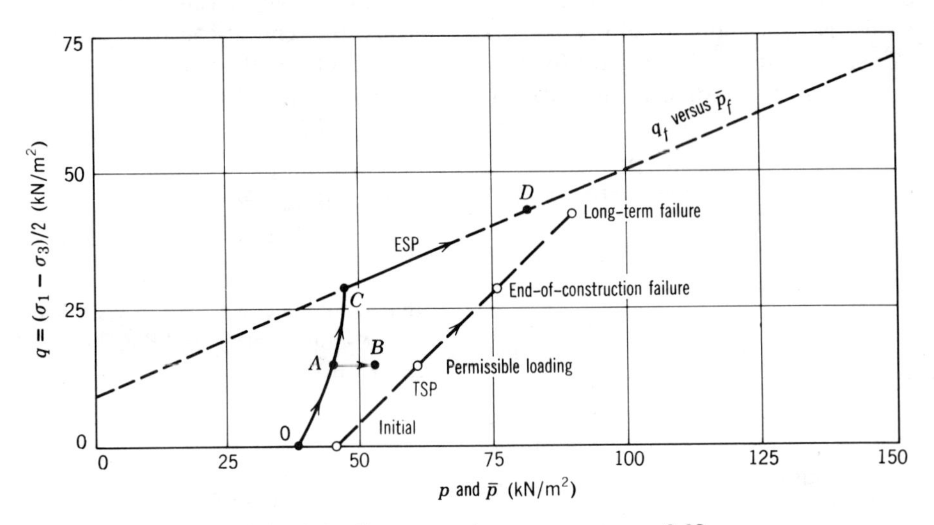

Fig. E31.3-2 Stresses and pressures at z = 2.28 m

Step 1—initial conditions (see Fig. E31.3-2). At z = 2.28 m

$$\sigma_v = (2.28)(19.64) = 44.8 \text{ kN/m}^2$$
$$u = (0.76)(9.81) = 7.5 \text{ kN/m}^2$$
$$\bar{\sigma}_v = 44.8 - 7.5 = 37.3 \text{ kN/m}^2$$
$$\bar{\sigma}_h = K_0\bar{\sigma}_v = 37.3 \text{ kN/m}^2$$
$$\sigma_h = 37.3 + 7.5 = 44.8 \text{ kN/m}^2$$
$$P = (4.56)(44.8) = 204.3 \text{ kN/m}$$

Step 2—"end of construction". From Eq. 31.4, the maximum possible thrust is

$$P_p = \tfrac{1}{2}(19.64)(4.56)^2 + 2(28.7)(4.56) = 465.9 \text{ kN/m}$$

The corresponding total horizontal stress at mid-depth is

$$\sigma_h = \frac{465.9}{4.56} = 102.2 \text{ kN/m}^2$$

The permissible thrust at this stage is

$$204 + \frac{466 - 204}{2} = 335 \text{ kN/m}$$

Example 31.3 (*continued*)

Step 3—pore pressure and effective stress at "end of construction"

$$q_f = s_u = 28.7 \text{ kN/m}^2 = \bar{c} \cos \bar{\phi} + \bar{p}_f \sin \bar{\phi} = (9.6)(0.906) + \bar{p}_f(0.422)$$

$$\bar{p}_f = 47.4 \text{ kN/m}^2$$

$$p_f = \frac{\sigma_v + \sigma_h}{2} = \frac{44.8 + 102.2}{2} = 73.5 \text{ kN/m}^2$$

Hence

$$u_f = 73.5 - 47.4 = 26.1 \text{ kN/m}^2$$

$$\bar{\sigma}_v = 44.8 - 26.1 = 18.7 \text{ kN/m}^2$$

$$\bar{\sigma}_h = 102.2 - 26.1 = 76.1 \text{ kN/m}^2$$

Step 4—Long-term stability

1. Pore pressures determined by water table.
2. This case already analyzed in Example 23.9.

$$P_p = 591.8 \text{ kN/m}$$

This is less critical than the end-of-construction condition. Hence the permissible thrust is

$$P = 335 \text{ kN/m}$$

Table E 31.3 Summary of Stresses for Initial and Failure Conditions

Case	Stresses (in kN/m²) at z = 2.28 m								
	σ_v	u	$\bar{\sigma}_v$	σ_h	$\bar{\sigma}_h$	p	$\bar{p}$	q	$q - 8.7$
Initial	44.8	7.5	37.3	44.8	37.3	44.8	37.3	0	—
End-of-const.	44.8	26.1	18.7	102.2	76.1	73.5	47.4	28.7	20.0
Long term	44.8	7.5	37.3	129.5	122.0	87.2	79.7	42.4	33.7

Failure criteria in terms of q_f and $\bar{p}_f$:

$$q_f = \bar{c} \cos \bar{\phi} + \bar{p}_f \sin \bar{\phi}$$ ◀

evaluated by a total stress analysis based on undrained strength.

3. It is not necessary to determine the pore pressures developed by the loading. However, by determining these pore pressures and comparing them with the pore pressures that will exist at a later time, the engineer can assure himself that the end-of-loading is really the critical condition.
4. It generally is not necessary to determine the load that may be carried after consolidation, although such a calculation may also be useful to ensure that the end-of-loading is really the critical condition.

31.3 PROBLEMS INVOLVING UNLOADING

This section develops the key features of problems in which the total normal stresses acting within the soil mass are decreased. Such decreases occur whenever an excavation is made into horizontal ground or when a flat slope is steepened. In order to develop the key features of such problems, another highly idealized case will be considered: active thrust from a normally consolidated clay. The problem is shown in Example 31.4. The clay is assumed to have the properties of Weald clay. The clay above the water table is assumed to be saturated and to sustain capillary tensions which increase linearly with height above the water table. For simplicity, tension cracks will be ignored, although the overall conclusions from the example would still be the same if tension cracks were considered.

Step 1 analyzes the stresses at mid-depth of the soil prior to excavation and construction of the wall. For simplicity, the coefficient of lateral stress at rest, K_0, has been assumed equal to unity. More typically K_0 would be less than unity for this normally consolidated soil, but the conclusions reached on the basis of this example still remain valid.

The undrained strength applicable to this problem is evaluated in step 2. Since the clay is normally consolidated, the undrained strength varies linearly with depth. Because both the total vertical stress and the shear strength vary linearly with depth, the horizontal thrust against the wall in the undrained condition also varies linearly with depth. Thus the total horizontal thrust may be calculated using an average value for the undrained strength. This calculation is performed in step 3.

▶ Example 31.4

Given. Stratum of normally consolidated Weald clay (see Fig. E31.4-1).

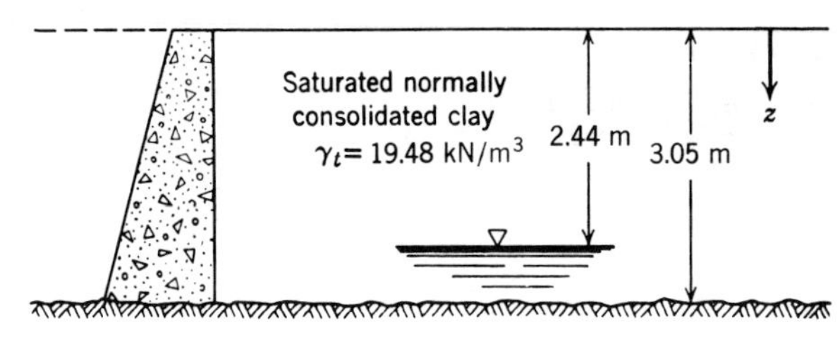

Fig. E31.4-1

Find. Active thrust for design of gravity retaining wall.

Solution.

Step 1—Initial conditions. Stresses assuming $K_0 = 1$ and no tension cracks are given in Table E31.4-1.

Table E31.4-1

z (m)	$\sigma_v = \sigma_h$ (kN/m²)	u_0 (kN/m²)	$\bar{\sigma}_v = \bar{\sigma}_h = \bar{p}_0$ (kN/m²)
0	0	−23.9	23.9
1.525	29.7	−9.0	38.7
3.05	59.4	6.0	53.4

$$P = (3.05)(29.7) = 90.6 \text{ kN/m}$$

Step 2—Evaluation of undrained strength. See Fig. 28.10, part of which is reproduced in Fig. E31.4-2.

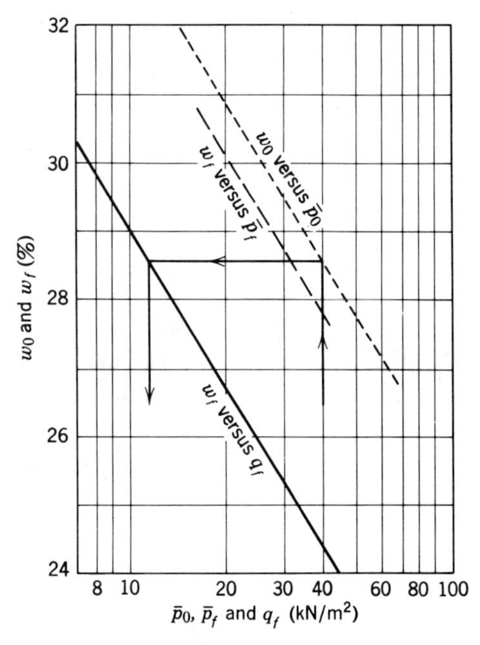

Fig. E31.4-2

Example 31.4 (*continued*)

Table E31.4-2

z (m)	$q_f = s_u$ kN/m²	
0	7.2	ave. s_u = 11.6 kN/m²
1.525	11.6	
3.05	16.0	

Step 3—"End-of-construction"

$$P_a = \tfrac{1}{2}\gamma_t H^2 - 2s_u H = \tfrac{1}{2}(19.48)(3.05)^2 - 2(11.6)(3.05) = 90.6 - 70.8 = 19.8 \text{ kN/m}$$

The corresponding σ_h is 6.5 kN/m²

Step 4—Pore pressure and effective stress at "end-of-construction." $\bar{p}_f$ may be read from Fig. E31.4-3 or calculated from $s_u = q_f = \bar{p}_f \sin \bar{\phi}$. At $z = 1.525$ m

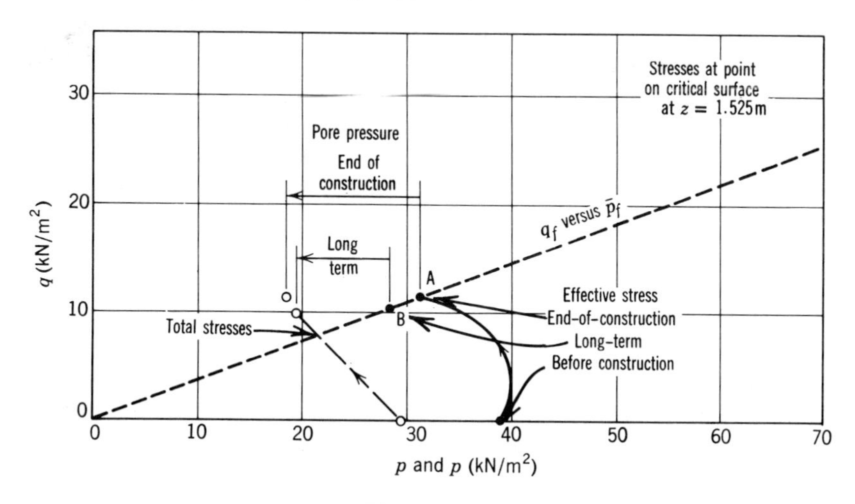

Fig. E31.4 3

$$\bar{p}_f = 30.9 \text{ kN/m}^2$$

$$p_f = \frac{29.7 + 6.5}{2} = 18.1 \text{ kN/m}^2$$

$u_f = -12.8$ kN/m²: a *decrease* from initial condition

$\bar{\sigma}_v = 42.5$ kN/m²; $\bar{\sigma}_h = 19.3$ kN/m²

Step 5—Long-term stability. See Example 23.12.

$$P_a = 26.3 \text{ kN/m}$$

Thus thrust is greatest *after* excess pore pressures have dissipated and the wall must be designed for this thrust of 26.3 kN/m.

Table E31.4-3 Summary of Stresses

Case	Stresses in kN/m² at z = 1.525 m							
	σ_v	u	$\bar{\sigma}_v$	σ_h	$\bar{\sigma}_h$	p	$\bar{p}$	q
Initial	29.7	−9.0	38.7	29.7	38.7	29.7	38.7	0
End-of-const.	29.7	−12.8	42.5	6.5	19.3	18.1	30.9	11.6
Long-term	29.7	−9.0	38.7	8.6	17.6	19.2	28.2	10.5

◄

This evaluation of the active thrust at the end-of-construction required no knowledge of the pore pressures existing at this stage. However, it once again is useful to evaluate what these pore pressures might be. As in Example 31.3, this pore pressure is found by first inserting $q_f = s_u$ into the appropriate equation for shear strength in terms of effective stress and thereby finding $\bar{p}_f$, and then comparing p_f and $\bar{p}_f$ to find the pore pressure. This calculation is shown in step 4. The important result of this step is that the pore pressure *decreases* as a result of the excavation.

Following excavation and construction of the wall, the soil will begin to swell, and the pore pressures will begin to return to the initial values determined by natural ground water conditions. Thus the pore pressures within the soil will *increase*, and the effective stresses will decrease with time. This means that the soil will become weaker with time, and the thrust against the wall will correspondingly increase. The thrust that must be resisted after all excess pore pressures have dissipated has already been calculated in Example 23.12, and indeed it is greater than the thrust existing at the end-of-construction. Thus the long-term stability condition controls the magnitude of the thrust for which the retaining wall must be designed.

Table E31.4-3 summarizes the average stresses for the three conditions. In Fig. E31.4-3, path OA shows the effective stresses developed at mid-depth as a result of construction of the wall. At point A the clay is at failure under undrained conditions. Path AB shows the subsequent change in stresses as the excess pore pressures dissipate with the soil remaining in the failure condition.

The clay in Example 31.4 was normally consolidated. The pore pressure decreased only slightly as the result of excavation, and hence the long-term condition was only slightly more critical than the end-of-construction condition. If the clay had been overconsolidated, the excess pore pressures at end-of-construction would have been much smaller, and hence the long-term would have been much more critical than the end-of-construction condition. The relationship between undrained (end-of-construction) and drained (long-term) resistance as deduced from this example should be compared with the entries in Table 28.1 for triaxial compression unloading.

Although Example 31.4 is highly idealized, it leads to practical conclusions applicable to a great number of problems where soil is *unloaded* by excavation in a time short compared to the time required for consolidation.

1. Conditions existing long after construction generally are critical and control the safety factor of a slope or the thrust for which a retaining structure must be designed.
2. The safety factor or design thrust for the long term situation may be evaluated by an effective stress analysis using pore pressures determined by natural ground water conditions.
3. It generally is not necessary to determine the safety factor or thrust immediately after excavation, although such a calculation may be useful to ensure that the long-term condition really is the critical situation.
4. It also generally is not necessary to determine the pore pressures developed immediately at the end of excavation. However, by determining these pore pressures and comparing them with the pore pressures that will exist at a later time, the engineer can assure himself that the long-term is really the critical condition.
5. In problems where an excavation is temporary and is to remain open for a time that is short compared to the time required for excess pore pressures within the adjacent soil to dissipate, the stability of the excavation may be analyzed using undrained strength. However, excess pore pressures tend to dissipate very rapidly around and beneath excavations, and when this happens use of undrained strength may not be conservative (see Sections 31.5 and 32.3).

31.4 TOTAL VERSUS EFFECTIVE STRESS ANALYSIS

As has been noted, the examples presented in Sections 31.2 and 31.3 are highly idealized. Several important stages involved in actually getting the wall into place have been ignored. Moreover, clay soil immediately behind a retaining wall would generally be replaced in actual problems by a free draining granular backfill. However, while these examples are largely academic, their simplicity has laid the basis for understanding of the proper choice of methods of analysis in more practical—and hence more complex—situations.

Before restating these principles in more general form, it is desirable to consider in some detail the relationship between s_u-analysis and $\bar{c}$, $\bar{\phi}$-analysis.

Critical Failure Planes

There are three questions which very carefully have been ignored until this time: (*a*) Why has the strength s_u in a $\phi = 0$ analysis been taken as one-half the deviator stress, i.e., as q_f? (*b*) On what plane does failure occur in an undrained triaxial test upon a clay? (*c*) Where is the failure surface located in the retaining wall problem? These three points are interrelated, and the answer to these questions involves an extremely important theoretical consideration. This consideration has been left until now to avoid complicating the foregoing sections.

The $\phi = 0$ analysis used for the end-of-construction case is, as was shown in Section 31.1, associated with a critical plane rising at 45°. However, Weald clay also has an angle of shearing resistance $\bar{\phi} = 22°$. Thus, if the Mohr-Coulomb failure theory were correct, the failure plane should rise at an angle of $(45 + \bar{\phi}/2) = 56°$. Thus there appears to be some contradiction to the use of a $\phi = 0$ analysis in connection with a soil whose shear strength is really controlled by effective stress.

This dilemma can be resolved by examining the effective stresses that exist within the clay mass behind a retaining wall, such as that in Example 31.4. These effective stresses are summarized in Example 31.5. For both end-of-construction and long-term stability conditions, stresses are shown both for a plane rising at 56° and a plane at 45°. Examination of the diagrams shows that, in each case, the maximum shear stress occurs on the 45° plane, while the shear stresses on the 56° plane are exactly $\bar{\sigma}$ tan 22°. This, of course, is exactly the state of affairs that would exist in triaxial tests with the $\bar{\sigma}_1$ and $\bar{\sigma}_3$ that exist at $z = 1.525$ m.

When a specimen fails in triaxial compression, it generally is difficult to determine the inclination of the plane on which failure occurs. In fact, it is not even necessary to know on what plane failure occurs. The maximum shear stress that exists at failure can be determined, as can the maximum obliquity, i.e., the maximum value of $\tau/\bar{\sigma}$. Which of these is used in stability analysis depends upon the assumptions used to derive the equations for the limiting equilibrium condition.

In the $\phi = 0$ analysis for a retaining wall, failure is assumed to occur when the *maximum* shear stress at a point reaches a stated value. Thus the value of s_u used in equations derived on this basis should be the *maximum* shear stress achieved in a corresponding triaxial test[2], q_f. On the other hand, in a solution based on effective stress, failure is assumed to occur when the maximum value of $\tau/\bar{\sigma}$ at a point reaches a stated value, i.e., when the maximum obliquity at a point reaches $\bar{\phi}$. Thus when employing an equation based on this form of solution, the values of $\bar{c}$ and $\bar{\phi}$ should be based on the maximum obliquity achieved in a corresponding triaxial test. Thus the rule to apply is:

> Be consistent, and evaluate shear strength in a way consistent with the assumptions made in the limiting analysis theory.

It is thus seen that there is nothing inconsistent with using a theory that requires a 45° failure plane in connection with a material having an angle of shearing resistance of 22°. Both end-of-construction and long-term stability can be analyzed using a failure plane at either 45° or $(45 + \bar{\phi}/2)°$, provided that the shear strength appropriate to this surface is used. The question still remains: At what inclination is the actual failure surface?

The soft clay used in this series of examples would likely bulge when tested in triaxial compression and simply sag behind a retaining wall. Evidence from triaxial tests and bank failures where definite slip planes have developed indicates that the failure surface in a clay is inclined to the horizontal at an angle which is closer to $(45 + \bar{\phi}/2)$ than 45°. This is further evidence that the shear resistance of a clay comes from a frictional mechanism rather than a cohesive mechanism. However observed failure plane angles do not check exactly with the $\bar{\phi}$ from CD tests, and this situation is the departure point for many hypotheses regarding strength theories (Gibson, 1953). Fortunately, as has been seen from the discussion in this subsection, it is possible to carry out a reasonable analysis of a thrust on a retaining wall without knowing the actual inclination of the failure plane.

[2] Skempton (1948) has presented a rigorous analysis for the case of a vertical bank.

Unity between Analysis by Total and Effective Stress

In principle, stability at the end-of-construction can be analyzed in terms of effective stress as well as in terms of undrained strength. For such an analysis, it is necessary to estimate the excess pore pressures caused by the loading or unloading. However, in practice it is more convenient to determine stability at-end-of construction (assuming that undrained conditions apply at this stage) by total stress analysis. Analysis in terms of effective stress requires an extra step: estimation of the excess pore pressures. No increased accuracy results from this extra step, since the very same factors that make it difficult to select a completely accurate value of s_u also make it difficult to estimate accurately the excess pore pressures caused by the construction (see Chapter 29).

In principle it also is possible to analyze long-term stability in terms of total stress as well as in terms of effective stress. This may be done as follows. A series of triaxial tests would be run in such a way that at failure the pore pressure and total normal stress on the failure plane in each test just equal the pore pressure and total normal stress at a point on the expected failure surface within the soil mass. These tests would then give the long-term strength available at corresponding points of the expected failure plane. These strengths are then used as values of s_u in an analysis with $\phi = 0$. However, this approach is much more cumbersome than performing tests to establish $\bar{c}$ and $\bar{\phi}$ and then performing stability analysis in terms of effective stress.

Thus, in any problem, the engineer is completely free to choose between an analysis in terms of effective stress or in terms of total stress. However, for practical reasons one type of analysis will generally be favored over the other for specific types of problems.

▶ **Example 31.5**

Given. Stresses computed in Example 31.4.
Find. Stresses upon planes inclined at 45° and at $(45° + \bar{\phi}/2)$ to the horizontal.
Solution. See Figs. E31.5-1. and E31.5-2.

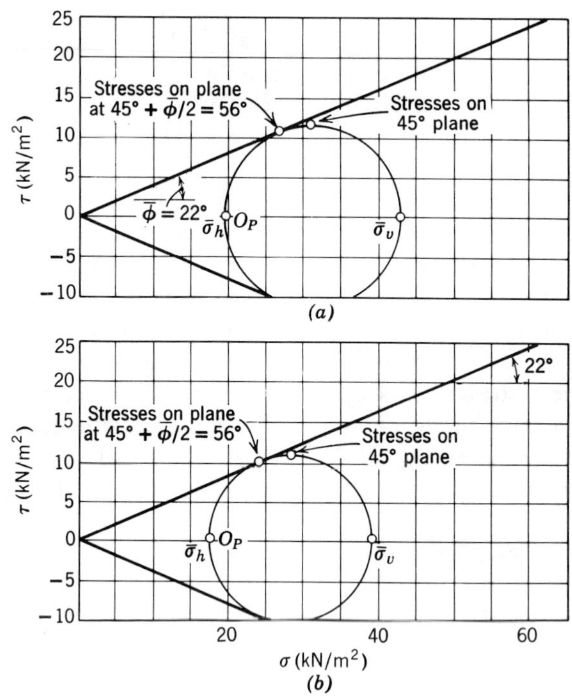

Fig. E31.5-1 (*a*) End-of-construction. (*b*) Long-term.

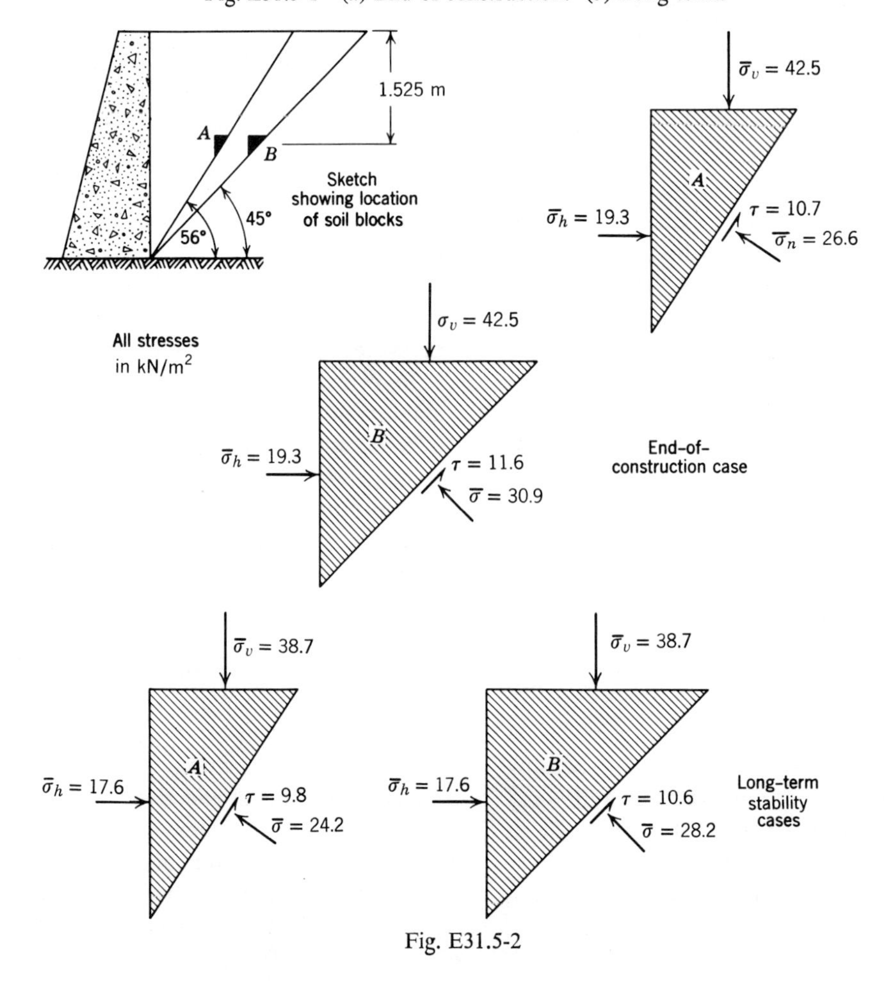

Fig. E31.5-2 ◀

Choice between Total and Effective Stress Analysis

Table 31.1 summarizes the preferred choices for methods of analysis in various situations.

Situations 1 (end-of-construction with saturated soil) and 2 (long-term stability) have already been discussed in detail in Sections 31.2 and 31.3. Analysis of situation 3 (end-of-construction with partially saturated soil) involves some difficult problems and will be discussed in detail in Section 31.6. Situation 4 involves problems where the effective stresses are a minimum at some time other than either end-of-construction or long-term. Examples are stage construction of embankments (Lobdell, 1959) and embankments upon foundations containing thin permeable seams (Ward et al., 1955). It is extremely difficult to make accurate analyses for such problems in advance of construction. The best approach is through an effective stress analysis using estimated pore pressures. It is absolutely necessary in such problems to measure actual pore pressures during construction and to recheck stability using these measured pressures.

31.5 STABILITY OF NATURAL SLOPES AND CUTTINGS

Figure 31.2 depicts the change in safety factor of a slope with time during and after formation of the slope. During excavation, the average shear stress along the potential failure surface increases. Following completion of excavation, the average shear stress remains constant but the effective stress across the failure surface decreases, so that the safety factor continues to decrease. A failure will occur at such time as the safety factor drops below unity.

Failure can occur during excavation. If the excavation has been rapid enough to prevent dissipation of excess pore pressures caused by unloading, such failures can be analyzed by the total stress method (s_u, $\phi = 0$). Table 31.2 refers to four cuttings into intact clay where failures were correctly explained (computed $F \approx 1.0$ for slope which failed) by $\phi = 0$ analysis. The table also gives one example where the $\phi = 0$ analysis did not give the correct result, since the fissures permitted dissipation of the negative excess pore pressures as excavation progressed.

The stability of a slope at the end of excavation is clearly no guarantee that it will remain stable thereafter. Indeed, the long-term situation is generally critical for a slope. The analysis of natural slopes and of cut slopes in the long-term, using a $\bar{c}$, $\bar{\phi}$-analysis with pore water pressures corresponding to natural ground water conditions, has already been discussed in Section 24.8. A $\phi = 0$ analysis definitely is not appropriate for such conditions (Bishop and Bjerrum, 1960). Neither the driving nor resisting stresses are really constant in the long-term situation, as is depicted in Fig. 31.3. In fissured soils, the strength parameter $\bar{c}$ may be gradually decreasing. Seasonal rises in ground water decrease the effective stress and hence the resistance along a potential failure surface. Such rises also increase the weight of the

Table 31.1 Choice of Total Versus Effective Stress Method of Stability Analysis

Situation	Preferred Method	Comment
1. End-of-construction with saturated soil; construction period short compared to consolidation time	s_u-analysis with $\phi = 0$ and $c = s_u$	$\bar{c}$, $\bar{\phi}$-analysis permits check during construction using actual pore pressures
2. Long-term stability	$\bar{c}$, $\bar{\phi}$-analysis with pore pressures given by equilibrium ground water conditions	
3. End-of-construction with partially saturated soil; construction period short compared to consolidation time	Either method: c_u, ϕ_u from UU tests or $\bar{c}$, $\bar{\phi}$ plus estimated pore pressures	$\bar{c}$, $\bar{\phi}$-analysis permits check during construction using actual pore pressures
4. Stability at intermediate times	$\bar{c}$, $\bar{\phi}$-analysis with estimated pore pressures	Actual pore pressures must be checked in field

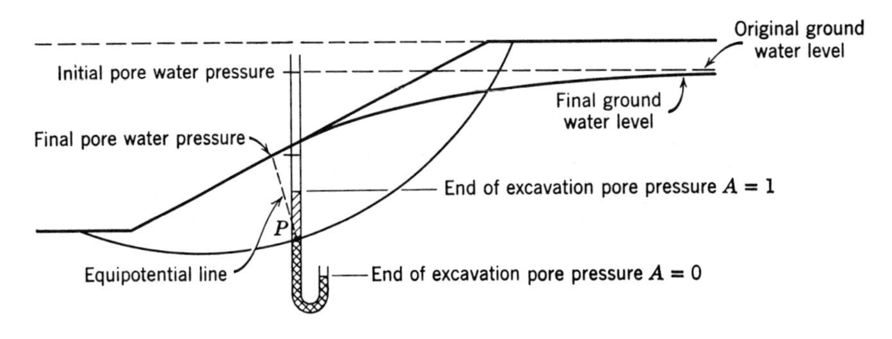

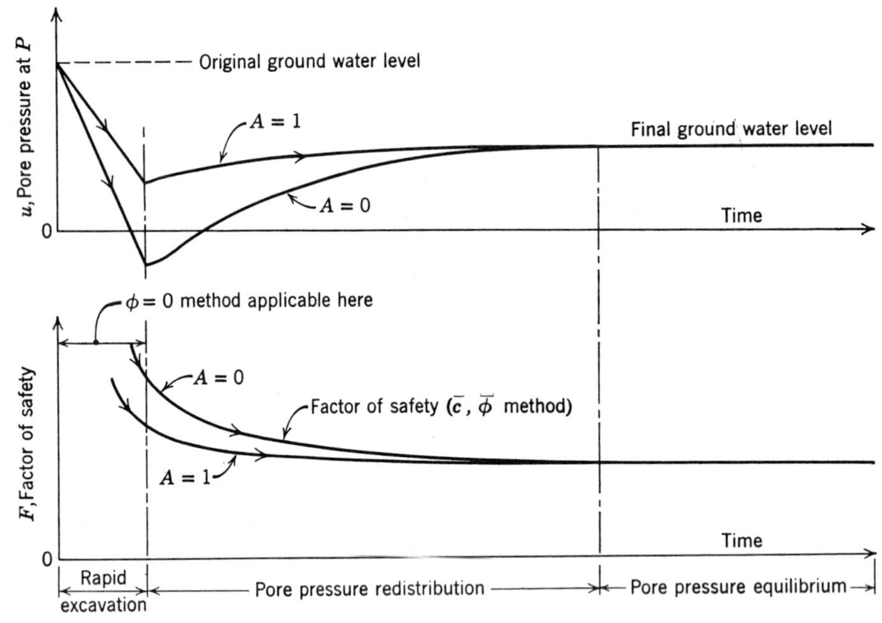

Fig. 31.2 Changes in pore pressure and safety factor during and after excavation of a cut in clay (From Bishop and Bjerrum, 1960).

soil above the failure surface and hence the driving moment. Erosion at the toe of the slope, causing a small slump, may increase the driving force and decrease the resistance. Failure in the long-term usually results from some such small change within a slope already close to instability.

Table 31.2 End-of-Construction Failures in Excavations

Location	Soil	Computed Safety Factor ($\phi = 0$ analysis)
Huntspill, England	Intact clay	0.90
Congress St., Chicago	Intact clay	1.10
Skattsmanso, Sweden	Intact clay	1.06
Skattsmanso, Sweden	Intact clay	1.03
Bradwell, England	Stiff-fissured clay	1.7

From Bishop and Bjerrum, 1960.

Failures in natural slopes of quick clay and very loose sand deserve special mention. Suppose that quick clay within a slope is consolidated to the stress condition at point A in Fig. 29.5*b*. If additional shear strain were to occur slowly enough to prevent development of excess pore pressures, the slope would remain stable because of the distance between point A and the failure line for ultimate resistance. However, if additional shear strain occurs rapidly, so that undrained conditions prevail, the shear resistance of the soil will decrease. Thus, if a sudden additional shear, caused by erosion at the toe of the slope, occurs within a slope in this condition, the entire soil mass within the slope may decrease in strength thus causing a major landslide (see Fig. 1.13). The slides described by Bjerrum (1954), Jakobson (1952), Meyerhof (1957), and Hutchinson (1961) all occurred in this way.

The stability of natural slopes in quick clay and loose sand should be analyzed by a $\bar{c}$, $\bar{\phi}$-analysis using pore pressures associated with natural ground water conditions. However, the values of $\bar{c}$ and $\bar{\phi}$ must be evaluated from special tests which determine effective stress conditions at incipient instability during additional undrained shear (Bjerrum and Landva, 1966). Use of $\bar{c}$ and $\bar{\phi}$ corresponding to peak resistance in undrained tests will lead to a great overestimate of the stability of slopes in quick clay (Bjerrum, 1961).

31.6 STABILITY OF DAMS AND EMBANKMENTS

Figure 31.4 depicts the changes in shear stress, pore pressure, and safety factor within an earth dam, starting with construction of the dam and continuing through operation of the reservoir. During construction shear stress on potential failure surfaces of course increases. Pore pressure also increases, since soil already in place is loaded as subsequent lifts are placed. Following completion the excess pore pressures begin to dissipate, only to increase again as the reservoir is filled. Filling the reservoir causes the shear stresses within the upstream slope to decrease because of the favorable effect of water pressure against the slope, while the average shear stresses under the downstream slope remain unchanged or increase slightly. The upstream slope may be subjected to additional shearing several times during operation of the reservoir as the result of rapid drawdown.

Inspection of Fig. 31.4 reveals that the critical times for the upstream slope are at end-of-construction and during rapid drawdown, whereas the critical times for the downsteam slope are end-of-construction and steady seepage during full reservoir. Analysis of end-of-construction and steady seepage are discussed in this section; rapid drawdown is discussed in Section 31.7.

End-of-Construction

As noted in Table 31.1, either total stress (s_u) analysis or effective stress ($\bar{c}$, $\bar{\phi}$) analysis may be used to analyze stability at end-of-construction, but there are difficulties inherent in either approach. Table 31.3 summarizes the steps involved in the two types of analysis, while Table 31.4 compares the two methods.

For an s_u-analysis, it is necessary to estimate the shear strength that will be available at the end-of-construction, considering both the factors that affect the undrained strength of partially saturated soil and the dissipation of pore pressures that may occur during construction. The estimation of undrained strength has been discussed briefly in Section 28.7. The key to success is duplication in laboratory tests of the initial stresses and pore pressures existing just after the soil is compacted and of the subsequent changes in stress caused by placement of the overlying soil. There are some major uncertainties involved in estimating suitable values for undrained strength, and estimating the effect of partial dissipation of excess pore pressures is even more difficult.

For a $\bar{c}$, $\bar{\phi}$-analysis, it is necessary to estimate the pore pressures developed within the dam during construction. Bishop and Bjerrum (1960) describe how this may be done, using a ratio of induced pore pressure to major principal stress determined from laboratory tests or from past experience with earth dams. Estimation of pore pressures involves many uncertainties, especially if there is partial dissipation during construction.

As noted in Table 31.4, from the standpoint of reliability there is no basic difference between the two methods. The gaps in our knowledge which make it difficult to estimate the pore pressures make it equally difficult to evaluate properly undrained strength (Whitman, 1960). There is one major advantage to the use of $\bar{c}$, $\bar{\phi}$-analysis: the pore pressures assumed during design of a dam can be checked by field measurements and the design can be modified during construction if necessary (Bishop, 1957).

Instability at end-of-construction is most likely when the soil is compacted at a water content near or above the optimum water content; i.e., when the initial degree of saturation is relatively high (see Section 26.8 and Chapter 34). Hence from the standpoint of stability, it is desirable to compact soil well dry of optimum water content. However, there are other practical considerations, such as economy or the desire for a plastic dam that will resist cracking, which may require a high placement water content. Sherard et al. (1963) discuss the many practical questions involved in design and construction for end-of-construction stability. Dams which failed or developed excessive deformations during construction are discussed by Bishop et al. (1960) and Linell and Shea (1960).

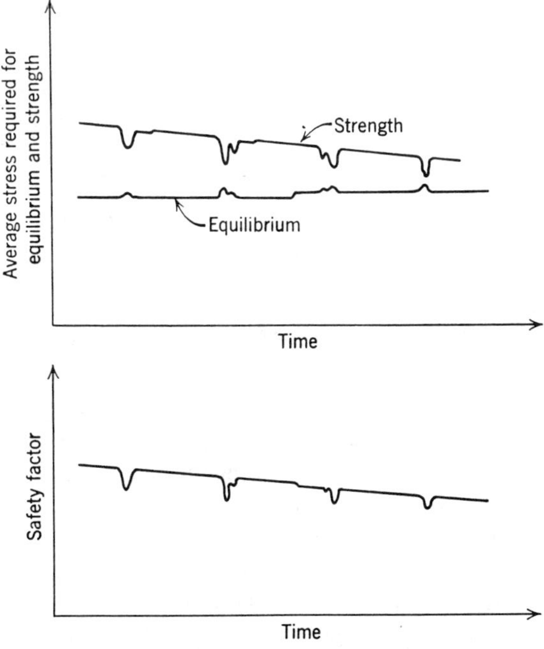

Fig. 31.3 Changes in driving and resisting forces in the long term.

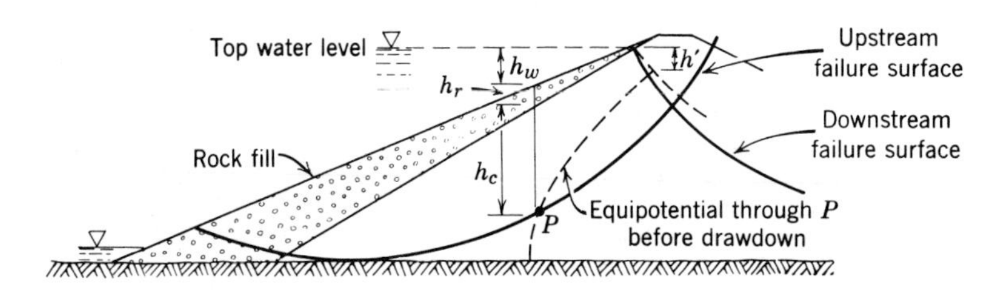

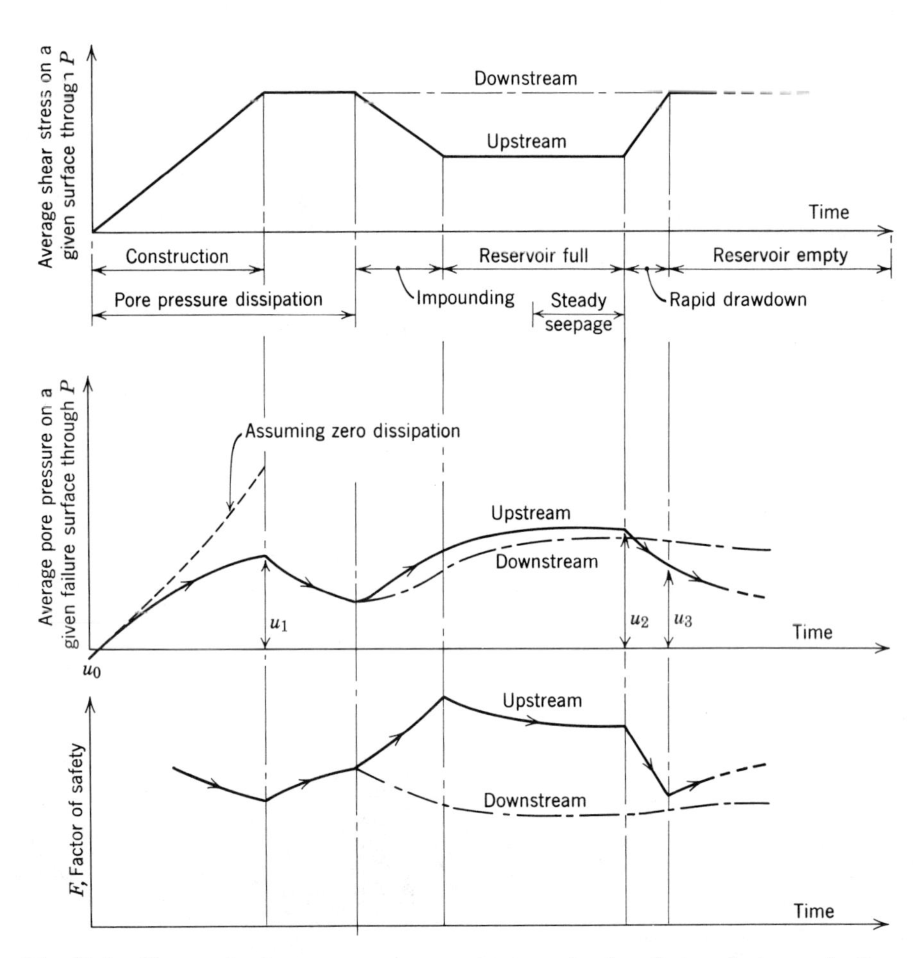

Fig. 31.4 Changes in shear stress, pore pressure, and safety factor during and after construction of earth dam (Based upon Bishop and Bjerrum, 1960).

Stability of Downstream Slopes during Steady Seepage

Analyses for the stability of a downstream slope under conditions of steady seepage are always accomplished using a $\bar{c}$, $\bar{\phi}$-analysis with pore pressures estimated from a flow net. The essential features of such flow nets, and the methods used to determine pore pressure from a flow net, have been discussed in Chapter 18. The design of a downstream slope to minimize the chances of instability during steady seepage involves use of permeable soil in the downstream shell and/or construction of filters to drain away the seepage, so that the phreatic line within the downstream slope remains low (Sherard et al. 1963). If seepage were to break out on the downstream slope, leading to local instability at this point, a process of gradual erosion and undermining of the dam may begin. This type of failure, known as *piping*, has been a common cause for the total or partial failure of earth dams (Middlebrooks, 1953).

Embankments on Soft Foundations

When a low embankment is constructed on a soft weak stratum of clay, there may be a bearing capacity failure within the clay even though the slope of the embankment itself is quite stable. Full understanding of this problem

Table 31.3 Requirements for Total and Effective Stress Analyses

Requirement	Comment
TOTAL STRESS ANALYSIS	
Total stresses in soil mass due to body forces and external loads	Common to both methods
Tests to determine strength of soil when subject to changes in total stress similar to stress changes within soil mass	Accuracy of tests is always doubtful, since strength depends upon induced pore pressures, and induced pore pressures depend upon many details of the test procedures; tests are easy to conduct
EFFECTIVE STRESS ANALYSIS	
Total stresses in soil mass due to body forces and external loads	Common to both methods
Tests to determine relationship between strength and effective stress	Can be done with considerable accuracy, since this relation is not very sensitive to test conditions; tests are somewhat time consuming
Determination of changes in pore pressure resulting from changes in external loads	Accuracy always doubtful because of the many factors which affect the magnitude of pore pressure changes

requires consideration of topics covered in Chapter 32. The stability of embankments upon soft foundations is often analyzed by assuming a failure surface such as that shown in Fig. 31.5. The most critical condition usually occurs at end-of-construction, and an s_u-analysis ($\phi = 0$) is suitable at least for the portion of the failure surface passing through the foundation. Table 31.5 lists several case studies which demonstrate the applicability of the s_u-analysis. Where instability during construction of an embankment is feared, it often is necessary to resort to either sand drains within the foundation to speed dissipation of pore pressures or to staged construction, which permits time for pore pressure dissipation to occur. In such problems, the actual pore pressures developed within the foundation must be observed in the field, and stability must be re-estimated from time to time using a $\bar{c}$, $\bar{\phi}$-analysis together with these measured pore pressures.

31.7 RAPID DRAWDOWN

A *rapid drawdown* is a sudden lowering in the level of water standing against the slope. Upstream slopes of earth dams, as well as the natural slopes adjacent to a reservoir, experience rapid drawdown when the level of the reservoir is dropped suddenly. Rapid drawdown also occurs when the stage of a river falls following flood and when sea level drops following a storm tide. Unless pore pressures within a slope can adjust immediately to the falling water level, high pore pressures may exist within a slope following a rapid drawdown. Morgenstern (1963) lists slope failures caused by rapid drawdown.

Events following a rapid drawdown may usefully, but approximately be divided into two stages (see Fig. 31.6). If the drawdown time is much less than the time in which consolidation adjustments can occur within the slope, the pore pressures immediately following the drawdown will equal the pore pressures before drawdown plus the change in pore pressure due to the change in water load against the slope. In time, consolidation adjustments will occur, but pore pressures will still remain high until the excess water drains from the slope and a new equilibrium is reached corresponding to the low level of water against the slope. With free-draining

Table 31.4 Comparison of Total and Effective Stress Analyses

Criteria	Total Stress	Effective Stress
Simplicity and amount of calculation or testing	Much less effort, since effective stress analysis has the extra step of determining the pore pressure changes	
Reliability	No difference: the same gaps in our knowledge which make it difficult to predict pore pressure changes make it difficult to create the proper test conditions for undrained tests	
Clarity of result		Clearer, because strength is controlled by effective stress. It is possible to check design by pore pressure measurements during construction

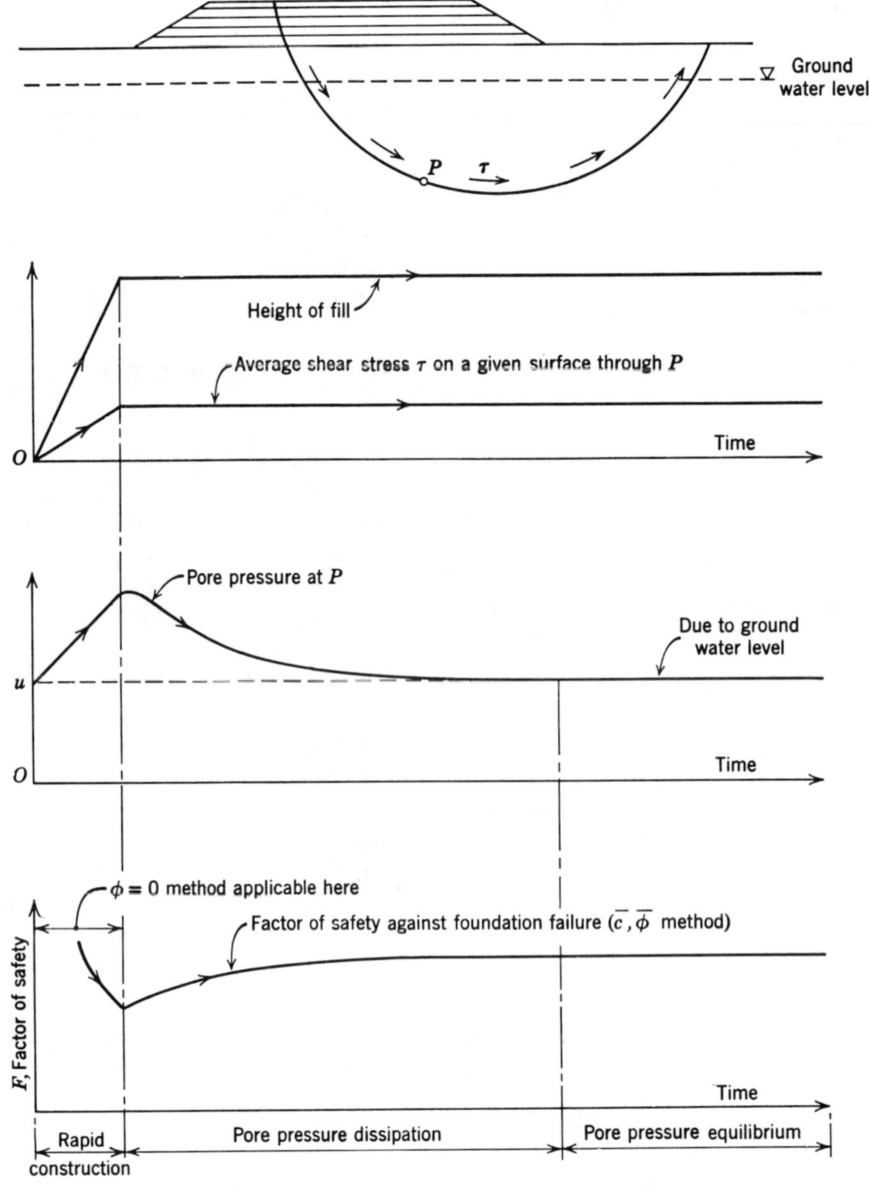

Fig. 31.5 Changes in shear stress, pore pressure, and safety factor during and after construction of embankment (From Bishop and Bjerrum, 1960).

Table 31.5 End-of-Construction Failures of Fills on Saturated Clay Foundation

Location	Computed Safety Factor ($\phi = 0$ analysis)
Chingford	1.05
Gosport	0.93
Panama 2	0.93
Panama 3	0.98
Newport	1.08
Bromma II	1.03
Bocksjön	1.10
Huntington	0.98

From Bishop and Bjerrum (1960).

soils, such as coarse sands and gravels, the consolidation time will generally be less than any actual drawdown time so that the stage depicted in Fig. 31.6*b* never occurs, and stability of slopes in such soils can be analyzed using a transient flow net as shown in Fig. 31.6*c*. With slowly draining soils, the situation depicted in Fig. 31.6*b* is critical with regard to stability of slopes.

Consolidation Time Much Longer than Drawdown Time

Stability in this condition (Fig. 31.6*b*) may be analyzed either using an s_u-analysis or a $\bar{c}$, $\bar{\phi}$-analysis. For a $\bar{c}$, $\bar{\phi}$-analysis, it is necessary to calculate the change in pore pressures caused by the change in water load against the

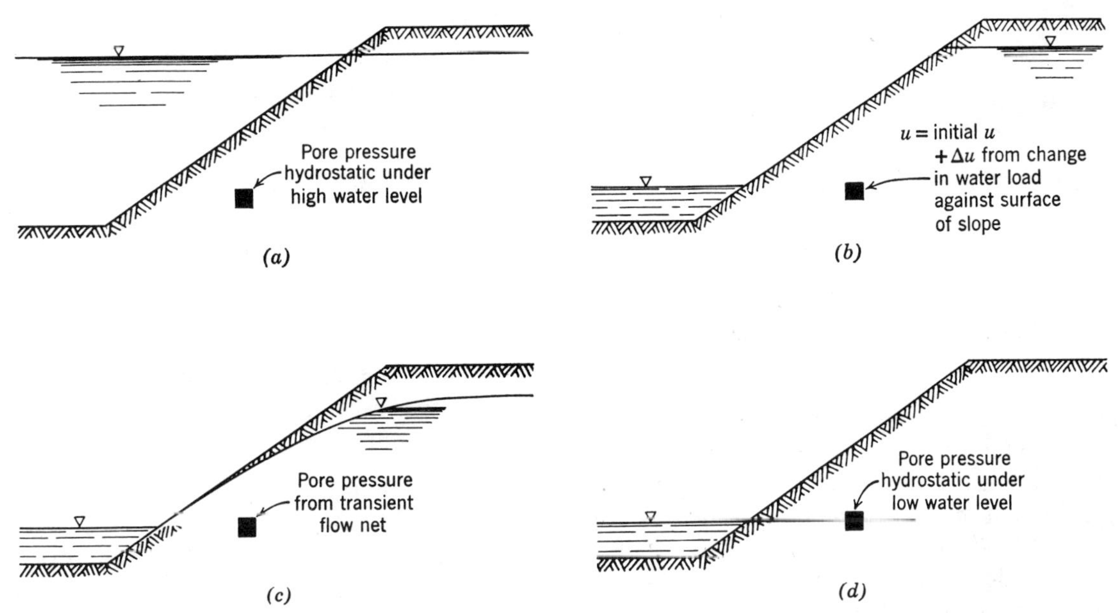

Fig. 31.6 Response of slope to rapid drawdown. (*a*) Initial equilibrium condition. (*b*) After drawdown but before consolidation adjustment. (*c*) After consolidation adjustment. (*d*) Final equilibrium condition.

slope. Bishop (1954) has provided equations for making these calculations, based upon the conservative assumption that the soil within the slope is saturated. Morgenstern (1963) presents stability charts for approximate calculations concerning this case.

For an s_u-analysis, it is necessary to determine the undrained strength available taking into consideration the stresses to which the soil is consolidated immediately prior to drawdown. Lowe and Karafiath (1960) outlined in detail the method for determining this shear strength and for accomplishing the necessary stability calculations.

Consolidation Time Much Less than Drawdown Time

Stability of slopes for this situation (Fig. 31.6*c*) is analyzed using a $\bar{c}$, $\bar{\phi}$-analysis with pore pressures determined from a flow net (Renius, 1955). Such flow nets represent the flow and pore pressure conditions at a particular instant of time. As time progresses the phreatic surface lowers and for a complete analysis it is necessary to construct a series of transient flow nets. Generally, however, the condition immediately after drawdown is critical and it suffices to construct a single flow net corresponding to this situation.

31.8 STABILITY DURING EARTHQUAKES

During large earthquakes there are numerous landslides in natural slopes and slumping of embankments. Most of these failures have been of relatively minor consequence, but there have been some failures of major consequence (Seed, 1966a). The major portion of the damage caused by the Alaskan earthquake of 1964 resulted from landslides (Seed and Wilson, 1967; Shannon, 1966). Figure 31.7 shows a landslide involving approximately 23 000 000 m^3 of earth which occurred during the Chilean earthquake of 1960 (Duke and Leeds, 1963). There has been at least one catastrophic failure of an earth dam as a result of an earthquake (Seed, 1966b). Failures such as these result in part from increased shear stresses caused by seismic loading, but major failures usually have been caused by decrease or loss of strength during cyclic loading (see Section 29.3). Earth dams, and other major cut or natural slopes whose failure might cause loss of life or extensive damage, should be carefully studied from the standpoint of susceptibility to failure during an earthquake.

All currently available methods for analyzing the safety of a slope during an earthquake involve a modification of the conventional limiting equilibrium analysis to

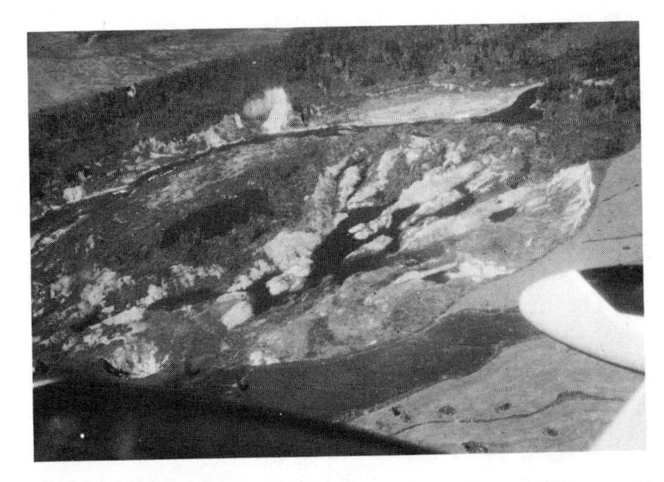

Fig. 31.7 Landslide near Lago Riñihue, Chile, during 1960 earthquake.

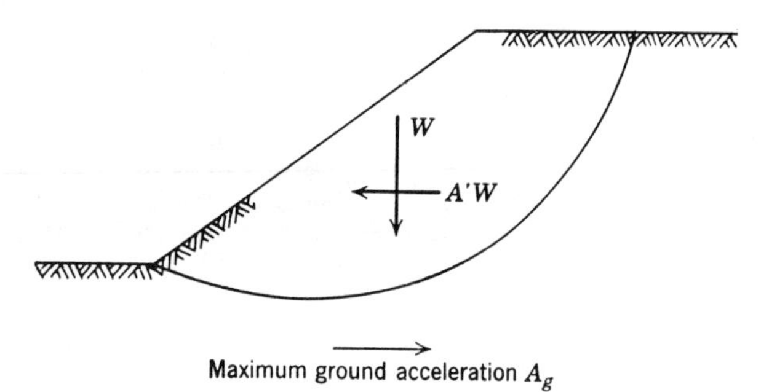

Fig. 31.8 Seismic force upon potential sliding mass.

include a seismic force (see Fig. 31.8.) The seismic force is proportional to the weight of the potential sliding mass times a seismic coefficient A', which is in some way related to the acceleration of the underlying earth. Until recently values for A' were simply assumed (usually 0.1 or 0.2) and the slope was required to have a safety factor of at least unity under the combined effects of the weight W and this seismic force $A'W$. Recently two more realistic approaches to the evaluation of safety have been proposed.

The method of Newmark (1965), which has already been discussed in Section 15.3, treats the soil as a rigid-plastic material. The first step in applying the method is to determine, using the shear strength available along the potential failure surface, the maximum possible acceleration $A'g$ which can be transmitted to the potential sliding mass. If A' is greater or equal to A, where Ag is the maximum acceleration of the underlying earth during an earthquake, there will be no plastic deformation of the soil. However, plastic deformation will occur any time that A exceeds A', and the final step in Newmark's method is estimation of the permanent displacement resulting from each pulse of ground motion with A' greater than A. If the total estimated permanent displacement is small (say several centimeters), the slope is then deemed to be safe for the assumed earthquake. On the other hand, if the estimated permanent displacement is quite large (say 4.5 m or more), then the slope would be deemed unsafe. This method does not consider the deformation that may be caused by shear stresses less than the shear stress at failure, although this deficiency may be overcome through proper choice of a yield shear stress. This method is not satisfactory for problems in which there may be a major loss of shear strength as a result of cyclic loading.

The method proposed by Seed (1966b) involves the following steps:

1. Starting from some assumed motion in the underlying earth and treating the slope as an elastic deformable body with damping, the average seismic coefficient A' is calculated.
2. Using conventional limiting equilibrium analysis (as in Chapter 24), the average shear stress along the potential failure surface is calculated, first without the seismic force and then including the seismic force.
3. One or more laboratory tests using cyclic loading are performed, corresponding to one or more points along the potential failure surface, with initial stresses equal to the calculated stresses without the seismic loading and with additional stresses equal to those calculated including the seismic load. The maximum strain observed during these laboratory tests is taken as an indication of the safety of the slope during an earthquake.

This method does not directly take into account the limitation upon the maximum acceleration within the slope because of limited shear strength of the slope, but it does indicate the magnitude of the deformation that may be caused by shear stresses less than the strength of the soil. The method is especially well suited for analyses of problems where liquefaction may occur.

Sherard (1967) discusses a number of considerations, other than simply safety factors, which must be considered when deciding upon the safety of an earth dam during an earthquake.

31.9 SUMMARY OF MAIN POINTS

1. In problems where the stresses applied to a soil mass are changed in a time short compared to the consolidation time for the mass, stability may be considered in two stages: end-of-construction corresponding to undrained shear, and long-term corresponding to drained shear.
2. If the soil is saturated, stability at the end of construction may best be analyzed by a total stress analysis using $\phi = 0$ and a strength equal to the undrained strength. In such an analysis it is not necessary to determine the pore pressures along a potential failure surface. However, knowledge of the pore pressure at this stage will help in deciding whether end-of-construction or long-term is more critical. Stability at end-of-construction may also be analyzed in terms of effective stress ($\bar{c}$, $\bar{\phi}$-analysis) using pore pressures estimated by the methods discussed in Chapter 26, but this method is difficult to apply because of the uncertainties in such estimates of pore pressure.
3. If the soil is partially saturated, then stability at end-of-construction may be analyzed either using an s_u-analysis or a $\bar{c}$, $\bar{\phi}$-analysis with estimated pore pressures. Either method must be used with great care, and the reliability is the same for both methods. There is an advantage to use of a $\bar{c}$,

$\bar{\phi}$-analysis: stability may be recomputed during actual construction using measured values of the actual pore pressures.

4. Long-term stability should always be analyzed using a $\bar{c}$, $\bar{\phi}$-analysis with pore pressures corresponding to equilibrium ground water conditions. With quick clays and very loose sands, special techniques must be used to evaluate $\bar{c}$ and $\bar{\phi}$.
5. In problems involving loading of a soil mass, the end-of-construction condition usually is critical, whereas long-term stability is generally critical when soil masses are unloaded. There are some special problems for which intermediate conditions may be more critical. The key to establishing the critical condition lies in a study of the variation of pore water pressure with time.

PROBLEMS

31.1 Refer to Example 31.4. Determine the available passive resistance for both the end-of-construction and long-term stability cases. Construct a diagram showing total and effective stress paths for the "typical point" at mid-depth.

31.2 Refer to Example 31.3. Determine the active thrust for both the end-of-construction and long-term stability cases. Construct a diagram showing total and effective stress paths for the "typical point" at mid-depth. Neglect the possibility of tension cracks; i.e., assume that both the water-wall interface and the mineral skeleton can take tension.

31.3 A temporary excavation, with a 1.5 horizontal on 1 vertical slope, is to be made into a stratum of clay 30 m thick with an undrained shear strength of 72 kN/m^3. The total unit weight of the clay is 19.64 kN/m^3. If the safety factor must be at least 1.5, what is the maximum permissible depth of excavation?

CHAPTER 32

Shallow Foundations with Undrained Conditions

Chapter 14 introduced the subject of shallow foundations and treated in detail the behavior of shallow foundations resting on dry soil. Chapter 25 extended Chapter 14 by covering the situation where the subsoil contained water. Chapter 25 also treated the bearing capacity and vertical movement of foundations loaded under drained conditions. Chapter 32 extends further the coverage of shallow foundations given in Chapters 14 and 25 by treating the very common and important situation of foundations loaded under undrained conditions followed by drainage.

Figure 28.14 shows the effective stress paths for a soil element subjected to two extreme drainage conditions: full drainage *AB*, and no drainage *AJ* followed by drainage *JB*. These paths can be considered as representative of average conditions within the soil beneath a footing. The drained condition *AB*, treated in Chapter 25, normally yields the minimum settlement and the maximum factor of safety against a bearing capacity failure. The undrained loading *AJ* followed by consolidation normally yields the maximum settlement and the minimum factor of safety—the most critical situation.

Between these two limiting situations of drained and undrained lies an infinite number of partially drained conditions, such as the one shown in Fig. 28.14. The engineer normally needs to consider only the two limiting cases as far as stability is concerned. To predict deformation, however, he must consider the loading condition that will occur in the actual structure. As we may readily infer, each problem must be examined to determine the extent of the drainage that will occur during loading. Principles presented in Chapter 27 permit an approximation to be made of the time needed for complete consolidation.

By comparing the time in which the load is built to its final value with the time required for complete consolidation, the engineer can place his problem in one of three categories:

1. *Drained loading*, when consolidation time is much less than loading time.
2. *Undrained loading*, when consolidation time is much greater than loading time.
3. *Partially drained loading*, when consolidation time and loading time are of the same order of magnitude.

Generally the permeability of sand is so large that deformation problems in sand should be handled as drained loading problems. On the other hand, the permeability of clay is so low that *undrained loading* frequently approximates the field situation. The permeability of silt, lying between those of clay and sand, is such that general statements, as have been made for sand and clay, are not appropriate; each case must be examined.

32.1 FOUNDATION STABILITY

The factor of safety against a shear failure of a shallow foundation can be determined using the principles of limiting equilibrium analysis given in Chapter 31. Figure 32.1 illustrates this approach. A strip load of intensity 100 kN/m^2 is resting on a deposit of clay having the shear strength versus depth relationship shown in Fig. 32.1. A trial failure surface is drawn and divided into slices of equal arc length Δl. The average undrained shear strength for each slice is obtained from the strength-depth plot. Dividing the resisting moment by the overturning moment gives a factor of safety of 1.84.

The approach illustrated in Fig. 32.1 can be used for situations consisting of nonuniform subsoil (Fig. 32.2) and for irregular cross sections (Fig. 32.3).

When the footing load does not extend in a direction perpendicular to the plane of the paper a distance which

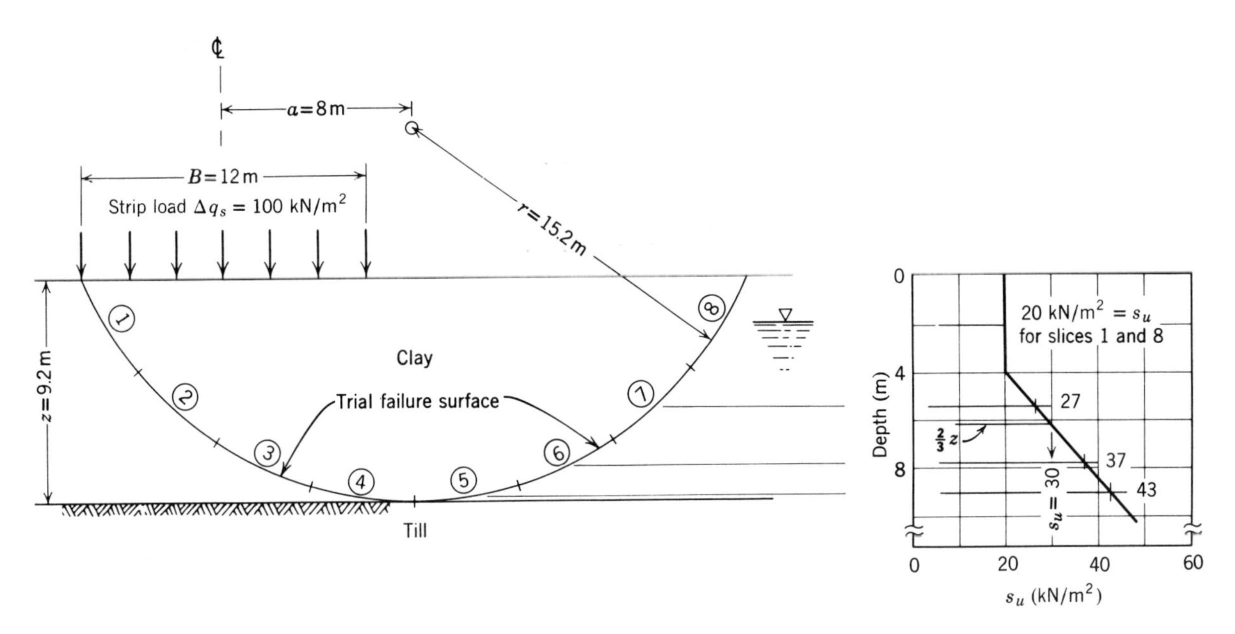

$$\text{Factor of safety} = \frac{\text{resisting moment}}{\text{overturning moment}}$$

Resisting moment: $r\ \Sigma s_u \Delta l = r\mathrm{L}\ \Sigma s_u$

$r = 15.2, \quad \mathrm{L} = 4.6 \quad \Sigma s_u = 2(20 + 27 + 37 + 43) = 2(127) = 254$

Overturning moment: $B\ \Delta q_s a$

$$FS = \frac{(15.2\ \text{m})(4.6\ \text{m})(254\ \text{kN/m}^2)}{(12\ \text{m})(100\ \text{kN/m}^2)(8\ \text{m})} = \frac{17\,760}{9600} = 1.84$$

$$\text{Average } s_u \text{ for trial surface} = \frac{127}{4} = 32\ \text{kN/m}^2$$

s_u at $\frac{2}{3}z = 30\ \text{kN/m}^2$

Bearing capacity: Δq_s for FS of 1 $= 184\ \text{kN/m}^2 = (\Delta q_s)_u$

$$(\Delta q_s)_u = N_c s_u = (5.14)(30\ \text{kN/m}^2) = 154\ \text{kN/m}^2$$

Fig. 32.1 Calculation of bearing capacity by trial wedge method.

is long relative to the width of the load, the analysis for stability becomes more difficult and more approximate. Such a situation arises when the load in Fig. 32.3 is caused by a storage tank rather than a strip footing. Significant resistance to a rotational failure is furnished by the shear strength on the soil surfaces parallel to the page. To neglect this contribution to resisting moment results in a factor of safety which is lower than the actual one.

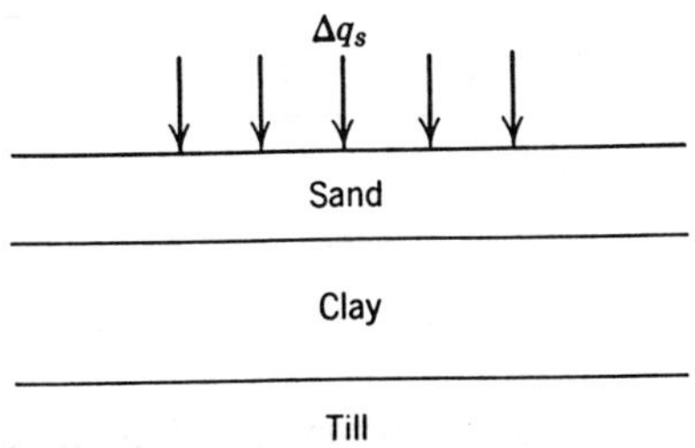

Δq_s

Sand

Clay

Till

Fig. 32.2 Shallow footing resting on nonuniform subsoil.

32.2 BEARING CAPACITY EQUATIONS

The derivation for the bearing capacity equation (Fig. 25.7) showed that for drained loading the applied footing

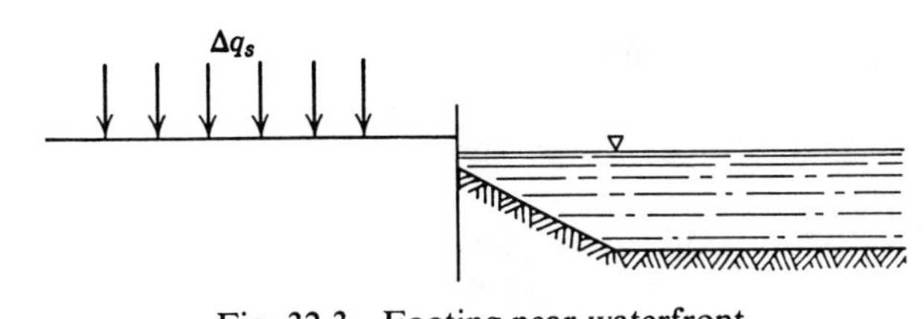

Fig. 32.3 Footing near waterfront.

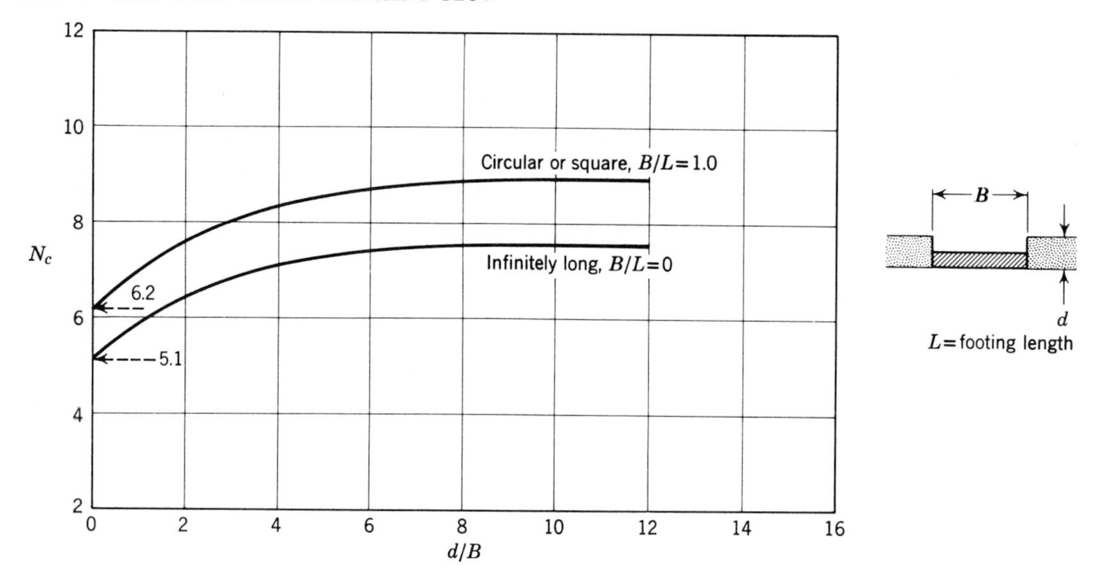

Fig. 32.4 Bearing capacity factors for footings on clay. (From Skempton, 1951.)

load generated shear strength within the soil, which helped resist failure. In undrained loading, the soil strength depends primarily on conditions prior to foundation loading, and therefore the applied footing load does not contribute to bearing capacity. The bearing capacity equation for undrained loading can thus be simplified to

$$\text{Bearing capacity} = (\Delta q_s)_u = N_c s_u + \gamma_t d \qquad (32.1)$$

where

N_c = the bearing capacity factor
s_u = the undrained shear strength
γ_t = total unit weight of soil
d = depth of footing base

Terzaghi and Peck (1967) gave the following values of N_c:

$N_c = 5.14$ for continuous footing
$N_c = 6.2$ for round or square footing

$N_c = 5\left(1 + 0.2\frac{B}{L}\right)$ for footing of width B and length L

Table 32.1 Values of N_c from Field Cases

N_c	Source
5.8 to 8.6	Six cases summarized by Skempton (1951)
5	Transcona elevator—Peck and Bryant (1953)
5.3	Oil storage tank—Brown and Patterson (1964)
5.6	Oil storage tank—Bjerrum and Overland (1957)

Skempton (1951) gave values of N_c as a function of the foundation and subsoil boundary geometry, as shown in Fig. 32.4. Bjerrum and Overland (1957) suggested reduced values for N_c for a localized or edge failure. Table 32.1 lists values of N_c backfigured from field cases in which failure was involved. As can be seen, these values vary from 5 to almost 9.

The use of Eq. 32.1 is illustrated in Fig. 32.1. Since the footing is continuous, N_c is taken as 5.14. The average undrained shear strength for the potential failure surface is approximately the strength at two-thirds the depth of the failure surface (this approximation is reasonable only when the strength variation with depth is not too irregular or large). As shown in Fig. 32.1, the bearing capacity from Eq. 32.1 is 154 kN/m², which agrees reasonably well with the 184 kN/m² computed by the stability analysis.

32.3 STABILITY OF EXCAVATIONS

The determination of the factor of safety of an excavation against a shear failure can be handled as a foundation loading problem. The foundation load acts upward and consists of the weight of soil excavated. Figure 32.5

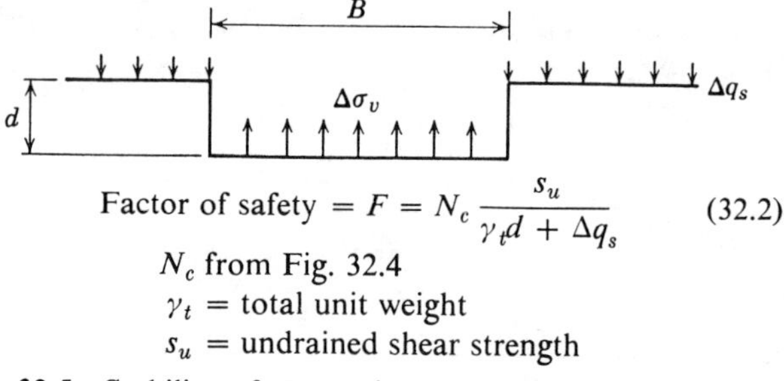

$$\text{Factor of safety} = F = N_c \frac{s_u}{\gamma_t d + \Delta q_s} \qquad (32.2)$$

N_c from Fig. 32.4
γ_t = total unit weight
s_u = undrained shear strength

Fig. 32.5 Stability of excavation (From Bjerrum and Eide, 1956).

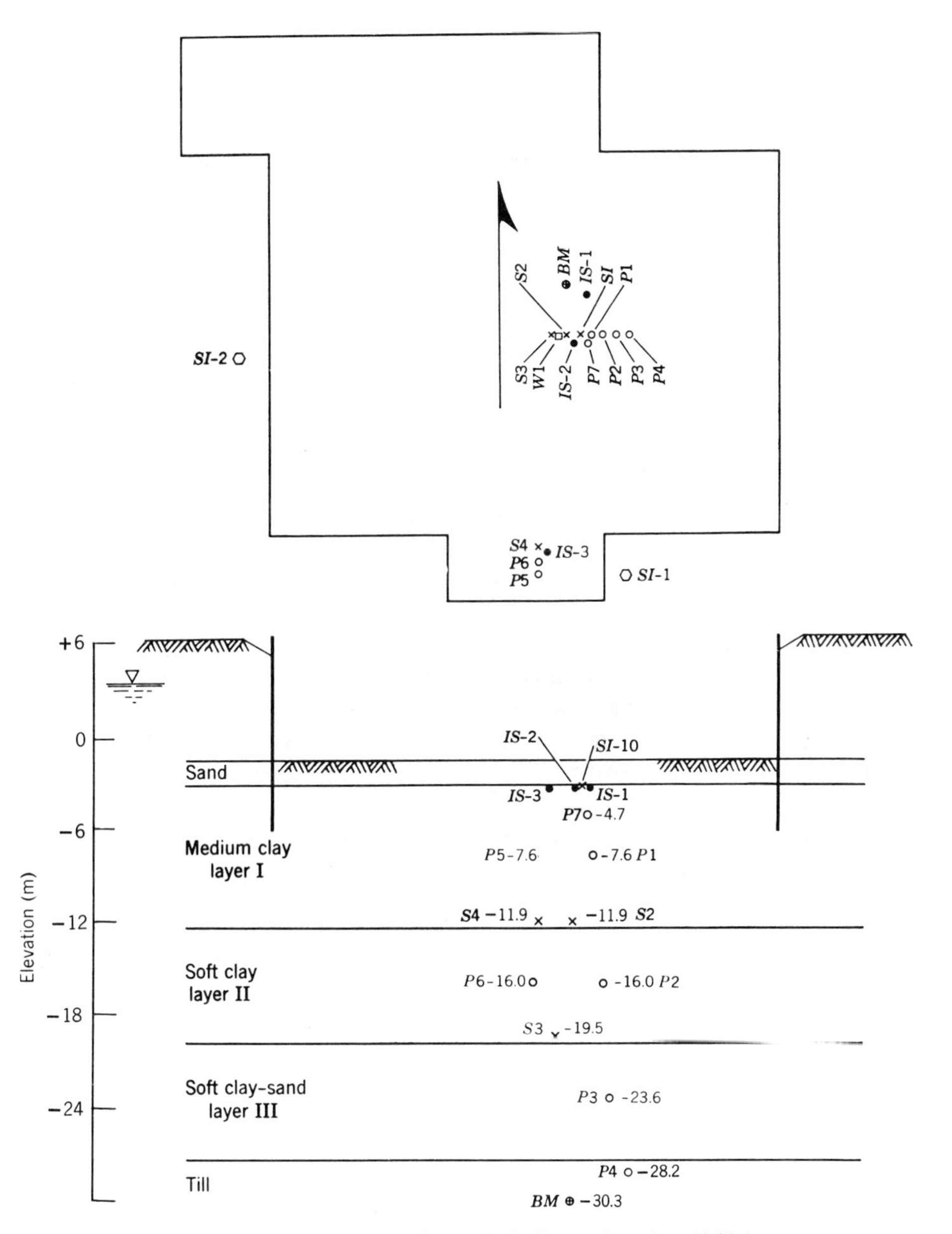

Fig. 32.6 Excavation for CAES (From Lambe, 1967a).

illustrates this approach and presents an equation (Eq. 32.2) for computing the factor of safety of an excavation.

As an illustration of the determination of the factor of safety of an excavation using the bearing capacity equation, consider the actual excavation shown in Fig. 32.6. The upper part of this figure shows the plan of the excavation for the M.I.T. Center for Advanced Engineering Study (Lambe, 1967a) and the lower part of the figure shows a cross section through the excavation. Located in Fig. 32.6 are various devices used to measure pore water pressure and movement within the subsoil. Figure 32.7 presents the calculations for the determination of the factor of safety of the CAES excavation.

To base a stability analysis of an excavation on undrained shear strength may not be a safe procedure.

Excavation $\approx$ 32.6 m $\times$ 33.5 m in plan
(consider square with B = 32.9 m)
Depth of excavation = +6.4 to −1.5 = 7.9 m
Stress release at bottom of excavation = $\Sigma\gamma\,\Delta z$

$$\Delta\sigma_v = 135.5\ \text{kN/m}^2$$

Average undrained shear strength in clay = 37.3 kN/m²

$$N_c \text{ from Fig. 32.4}\left(\text{for } \frac{d}{B} = \frac{7.9}{32.9} = 0.24\right)$$

$$N_c = 6.6$$

$$F = N_c \frac{s_u}{\Delta\sigma_v} = \frac{(6.6)(37.3\ \text{kN/m}^2)}{135.5\ \text{kN/m}^2} = 1.8$$

Fig. 32.7 Stability of CAES excavation.

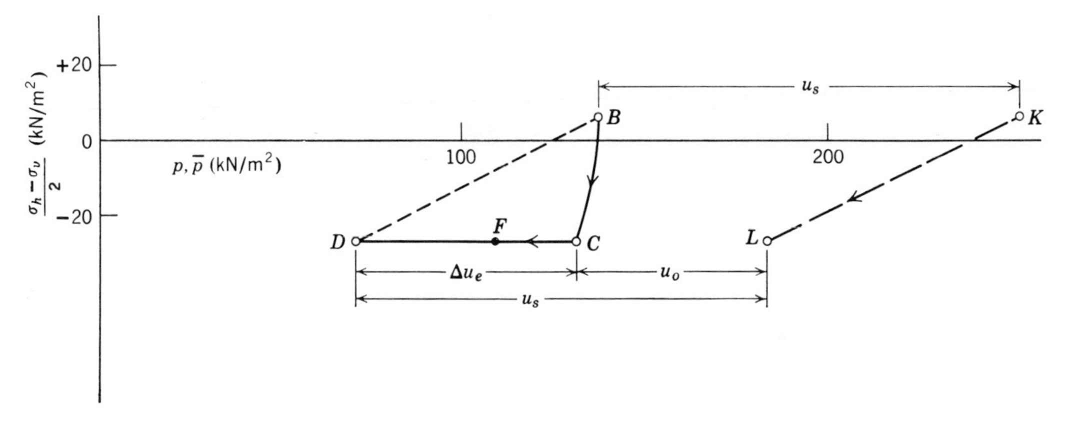

Fig. 32.8 Stress paths for $P1$ at CAES.

Consider, for example, the stress paths for piezometer $P1$ (in the middle of clay layer I, as shown in Fig. 32.6), which are presented in Fig. 32.8. KL is the total stress path for the excavation and BC is the effective stress path. Immediately following excavation the pore pressure at piezometer $P1$ is CL, which is less than the static pore pressure KB and LD; i.e., an excess pore pressure of DC exists at $P1$ immediately after excavation. Note that the excess pore pressure u_e, equal to DC, is negative. Even though the CAES excavation was open for only 24 days, a significant increase occurred in the pore pressure at $P1$.

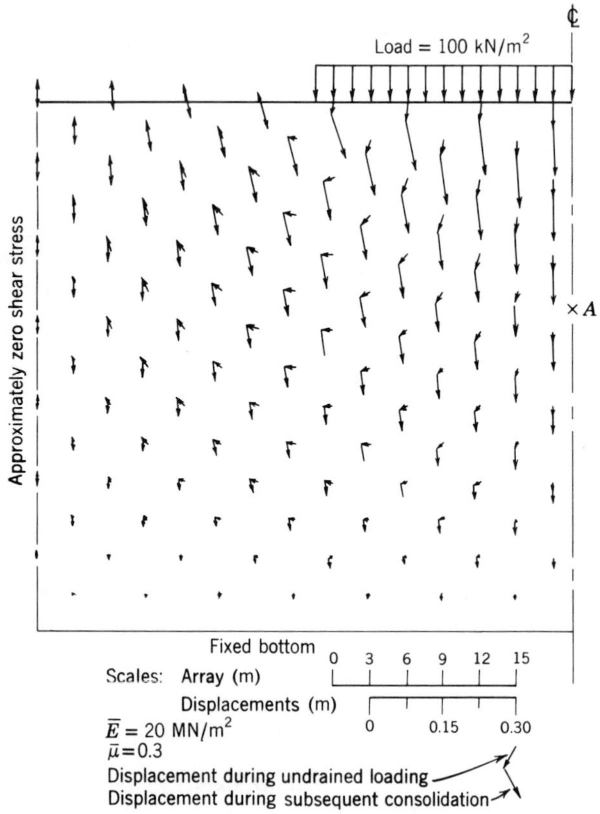

The measured pore pressure changed from LC to LF, reflecting a decrease in the negative excess pore pressure equal to FC. During the dissipation of the negative excess pore pressure the strength of the soil at $P1$ decreased. Thus the factor of safety of an excavation decreased with time. This tendency is opposite that which occurs with a foundation loading. In general, the critical time for an excavation is at the end of the unloading period—just before load is replaced in the excavation.

Load = 100 kN/m²

Approximately zero shear stress

×A

Fixed bottom

Scales: Array (m) 0 3 6 9 12 15

Displacements (m) 0 0.15 0.30

$\bar{E}$ = 20 MN/m²

$\bar{\mu}$=0.3

Displacement during undrained loading

Displacement during subsequent consolidation

Fig. 32.9 Computed displacements in elastic body subjected to strip load.

32.4 MOVEMENT DURING UNDRAINED LOADING FOLLOWED BY CONSOLIDATION

As a vehicle for studying the fundamentals of settlement during and following an undrained loading, let us consider the problem of a strip load supported by an elastic stratum (see Fig. 32.9). The stratum is underlain by a rigid base such that there is neither horizontal nor vertical movement at the bottom of the elastic stratum. This is a plane strain problem; hence there are no movements perpendicular to the plane. The movements and stresses at the end of consolidation were computed using the values of $\bar{E}$ and $\bar{\mu}$ shown in Fig. 32.9. The movements and stresses at the end of the undrained loading were computed using $E = 3\bar{E}/2(1 + \bar{\mu})$ and $\mu = 0.5$ (see Section 28.5).

The vectors in Fig. 32.9 show the movement in two stages: first the movement during undrained application of the load, and then subsequent movement as the stratum consolidates with constant surface load. During the undrained loading, the surface of the body immediately beneath the load moves downward, but the portion of the surface which is not loaded moves upward. Points within the body move outward. These types of motions are necessary in order to maintain the constant volume condition. During consolidation all points on the surface move downward. Thus the settlement of the surface under the loaded area increases, and the surface outside the loaded area ends with a net downward movement.

During consolidation the stresses acting on a typical element change. For example, the results in Table 32.2 apply at point *A* of Fig. 32.9. Changes in stress as the result of consolidation occur in plane strain problems whenever there is a boundary condition upon displacements; they also occur in axisymmetric and three-dimensional problems. The stress paths for these stress increments are shown in Fig. 32.10; the stresses existing prior to the addition of the load are not shown. Actually, the stress paths during consolidation are generally not straight lines but are shown as such because calculations were not made for intermediate times during consolidation.

Table 32.3 gives the corresponding strains. During the undrained loading there are shear strains ($\epsilon_v - \epsilon_h$) but zero volume change. During consolidation there are volume decreases and also additional shear strains; both are necessary to maintain continuity of the body.

32.5 MOVEMENT PREDICTION BY THE STRESS PATH METHOD

The principles of the stress path method have been explained in Section 14.9. When applied to a problem involving undrained loading (or unloading) followed by consolidation, prediction of settlement (or heave) by the stress path method involves the following steps:

1. Establish the subsoil conditions and select one or more average points for the soil layers that will contribute to the vertical movement.
2. Determine the initial stresses and pore pressures at the average points.
3. Determine the total stress changes at the average points, both as a result of the undrained loading and as a result of the subsequent consolidation.
4. Duplicate the initial stresses and total stress changes in laboratory tests, and measure the resulting strains—which are the expected strains at the average points. The laboratory tests involve first an undrained loading and then consolidation to the final total stresses minus the steady-state pore pressures.
5. Use the strains measured in step 4 to predict the magnitude of the initial and final settlements.
6. Determine the time required for the excess pore pressures to dissipate, and from this determine the rate of settlement during consolidation.

Table 32.2

Stress Increments (kN/m²)	At End of Undrained Loading	At End of Consolidation
σ_v	92	92
σ_h	42	15
$(\sigma_v + \sigma_h)/2$	67	53
$(\sigma_v - \sigma_h)/2$	25	39
u	67	0
$\bar{\sigma}_v$	25	92
$\bar{\sigma}_h$	−25 (tensile)	15
$(\bar{\sigma}_v + \bar{\sigma}_h)/2$	0	53

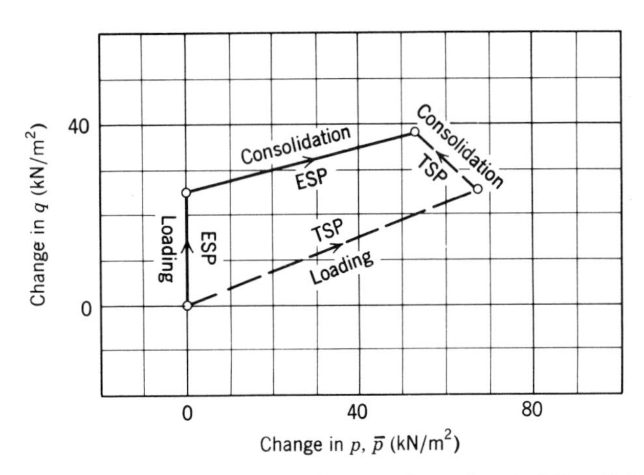

Fig. 32.10 Paths for changes in stress for point *A* of Fig. 32.9.

To illustrate the application of this approach, we will use the stress path method to estimate the settlement at the center of a storage tank caused by filling and the heave caused by emptying. Figure 32.11 shows the tank, approximate stress paths for the average point in the soil, and stress-strain data for the soil. Elastic theory has been used to determine the increments of total stress caused by the loading and unloading. Changes in total stress as the result of consolidation have been neglected. Settlement (or heave) has been determined by multiplying the vertical strain at the average point times the thickness of the straining soil layer. The strains and settlements during the various stages of loading and unloading are summarized in Fig. 32.12.

The foundation soil in our illustrative problem is clay—low permeability; the loading is relatively fast—one month to load and one week to unload; the boundary conditions consist of a thick clay layer with drainage at the top of the layer and at the bottom of the layer. These factors combine to make our case essentially

Table 32.3

Strains (%)		
ϵ_v	0.16	0.38
ϵ_h	−0.16 (tensile)	−0.11 (tensile)
$\epsilon_v + \epsilon_h$	0	0.27
$\epsilon_v - \epsilon_h$	0.32	0.49

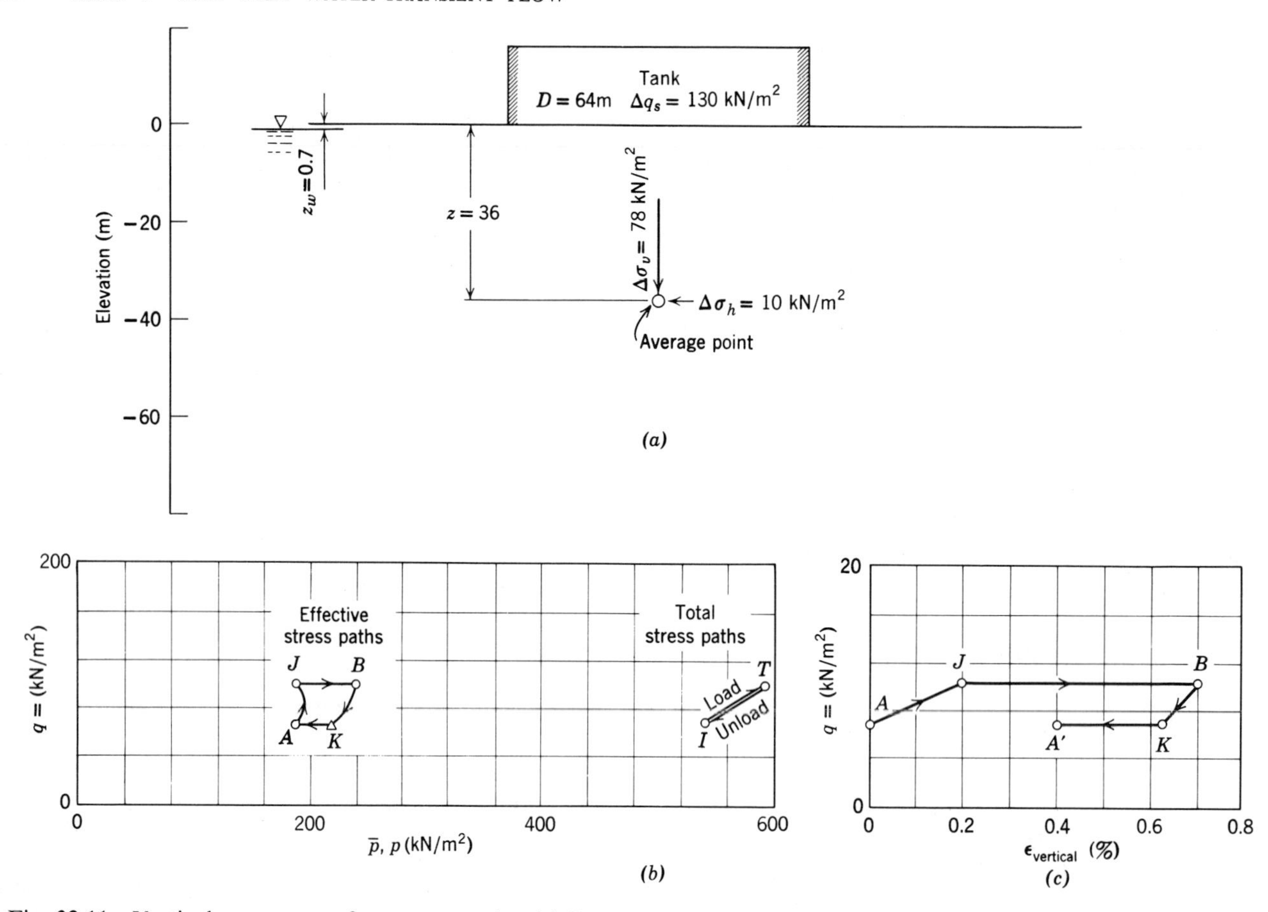

Fig. 32.11 Vertical movement of a storage tank. (*a*) Stress increments at average point. (*b*) Stress paths for average point. (c) Strains at average point.

an undrained loading followed by consolidation under constant load. There may be some consolidation at the very top and at the very bottom of the clay during loading, but the amount is negligible.

Symbols and terms used for the several stages of movements are:

1. ρ_i—*initial settlement* (settlement during undrained loading).
2. ρ_c—*consolidation settlement* (settlement during consolidation under constant foundation load).
3. $\rho_t = \rho_i + \rho_c$ = *total settlement*.

When the movement is downward it is *settlement* and when it is upward it is *heave*. Usually there is no confusion with using the symbol ρ for both settlement and heave; where such confusion is possible an arrow upward (↑) will indicate heave and an arrow downward (↓) will indicate settlement.

Initial settlement sometimes is referred to as "elastic" or "shear" settlement, but use of these terms is confusing. Use of "elastic" has arisen because elastic theory often has been employed to predict initial settlement. However, Section 32.6 shows that elastic theory can be used

	Conditions at Average Point				Vertical Movement of
Stage	Type of Stress	Change in Δu	Type of Strain	Vertical Strain	Tank Center
1	*AJ* (undrained loading)	from 0 to +44 kN/m²		0.2%↓	72 m × 0.002 = 0.144 m↓
2	*JB* (consolidation at constant load)	from +44 to 0		0.5%↓	72 m × 0.005 = 0.360 m↓
3	*BK* (undrained unloading)	from 0 to −29		0.07%↑	72 m × 0.0007 = 0.050 m↑
4	*KA* (expansion at constant load)	from −29 to 0		0.23%↑	72 m × 0.0023 = 0.166 m↑

Fig. 32.12 Vertical movement of storage tank.

to predict all components of settlement. Use of "shear" has arisen because in a saturated soil all of the initial settlement arises from shear distortions of the soil. Section 32.4 indicated that shear distortions also occur during consolidation.

32.6 MOVEMENT PREDICTION BY ELASTIC THEORY

Elastic theory usually is used in step 3 of the stress path method to estimate the stress increments caused by loading or unloading. Elastic theory may also be used in step 5 by deducing modulus from the measured stress-strain curves and then inserting these values of modulus in equations derived from elastic theory.

As an illustration of the use of elastic theory to estimate settlements during an undrained loading followed by consolidation, let us apply Eq. 14.14 to the tank problem in Fig. 32.11:

$$\rho = \frac{R(\Delta q_s)}{E} I_\rho \qquad (14.14)$$

The radius R is 32 m and the bearing stress Δq_s is 130 kN/m². The modulus E and the influence coefficient I_ρ, a function of Poisson's ratio μ, must be chosen as appropriate for the loading condition.

To estimate the initial settlement, we must use a value of E appropriate for undrained loading. A suitable value of E is obtained from Fig. 32.11c for the part of the stress-strain curve from A to J. This value[1] is $E = 39\ 000$ KN/m². For undrained loading, $\mu = 0.5$, and hence the value of I_ρ is 1.5 (see Fig. 14.20). Inserting these numbers in Eq. 14.14 gives $\rho_i = 0.16$ m. This result is similar to that (0.14 m) obtained using the stress path method.

To estimate the total settlement, we must use $\bar{E}$ for drained loading. A suitable value may be obtained from Fig. 32.11c as the slope of a straight line from A to B. This value is $\bar{E} = 11\ 200$ kN/m². For a drained loading, we must use the $\bar{\mu}$ of the mineral skeleton. A typical value is $\bar{\mu} = 0.3$, and hence from Fig. 14.20 we find $I_\rho = 1.8$. Inserting these numbers into Eq. 14.14 gives $\rho_t = 0.67$ m. This estimate may be compared with the estimate of 0.50 m obtained using the stress path method.

Note that elastic theory may be used to estimate initial settlement and total settlement. The settlement during consolidation ρ_c may be obtained by subtracting initial settlement from total settlement.

As has been discussed in Section 14.8, elastic theory can be used to provide useful estimates for settlements even though soil certainly is not a linearly elastic material.

[1] E has been computed as $\Delta\sigma_v/\epsilon_v$. Since $\Delta\sigma_h \neq 0$, this is not exactly correct, but is a reasonable approximation. The evaluation of I_ρ assumes that the stress increments are unaffected by the presence of sand starting at a depth of 72 m. The same assumption was made in the computation for $\Delta\sigma_h$ and $\Delta\sigma_v$ shown in Fig. 32.11.

The key lies in the exercise of proper judgment when choosing values for the "elastic" constants E and μ. The modulus E is the most critical parameter, and its value must be selected with an eye toward the magnitude of both the initial stress and the stress change. Of course, elastic theory may be used most readily when there is at all stages of the loading a large safety factor against ultimate failure.

Elastic theory is *not* a substitute for the stress path method. Rather, use of equations such as Eq. 14.14 is one way of accomplishing step 5 in the stress path method.. Steps 1 to 4 of the stress path method are essential for the selection of suitable values for the modulus to be used in such equations.

32.7 OTHER METHODS FOR PREDICTING MOVEMENTS

This section describes other methods that are commonly used to predict settlement or heave. In Fig. 32.13 the stress paths implied by these methods are compared with the "correct" stress path (path AJB) as deduced in Fig. 32.11. It must be emphasized that the accuracy of any of the common methods depends on how close the model upon which the method is based agrees with the actual problem at hand. (Lambe, 1967b, presents a numerical example in which pore pressure and movement are predicted by the techniques described here.)

One–dimensional. The one-dimensional strain method (described in Chapter 25) is very widely used for all types of settlement and heave problems. For those actual problems that involve essentially one-dimensional strain, this approach is obviously sound. For a problem such as our tank, an analysis based on one-dimensional strain should not be expected to yield a good approximation of the true settlement. (Note the significant lateral deformations in the foundation to the tank shown in Fig. 32.15.) In the one-dimensional settlement technique the effective stress path AM (Fig. 32.13) is implied. The vertical strain resulting from a one-dimensional compression of a sample from the stress $\bar{\sigma}_{vA}$ to $\bar{\sigma}_{vB}$ is obtained and multiplied by the thickness of the soil stratum in the field. A comparison of this stress path (AM) with the actual one (AJB) shows that they are far from similar and strains along the two paths should be quite different.

Initial plus one–dimensional. This technique consists of employing an elastic analysis for the initial settlement and a one-dimensional analysis for the consolidation settlement; the stress path AJ plus AM is implied. In most problems this addition will result in a considerable overestimation of movement. To approximate the actual stress path AJB with a path AJ plus AM is not theoretically sound and should not be employed.

Skempton–Bjerrum. The Skempton-Bjerrum technique (1957) consists of using the elastic method for the

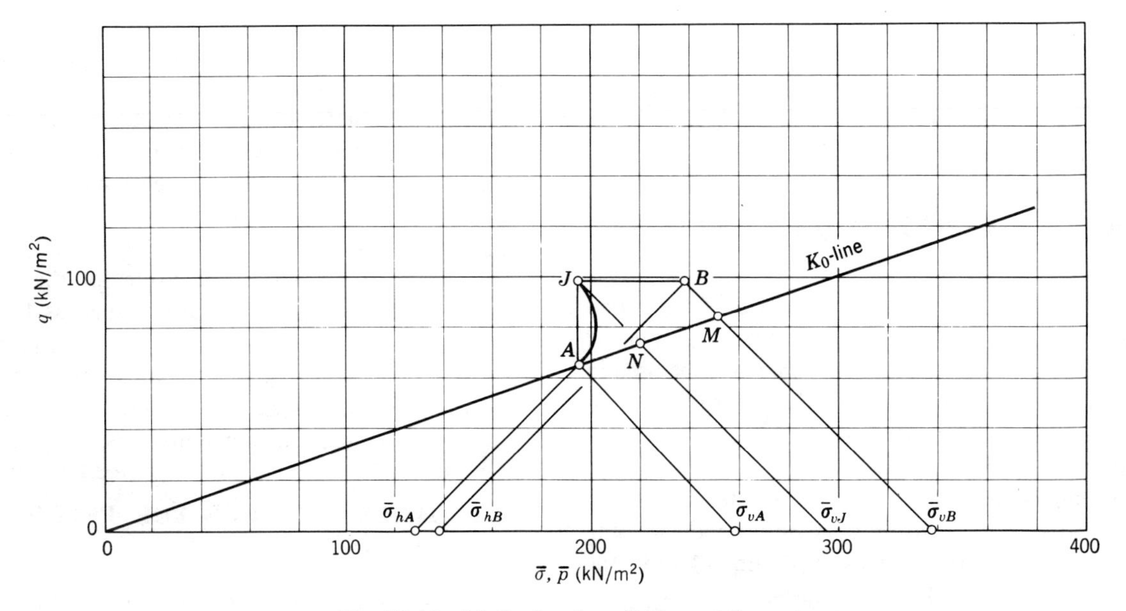

Fig. 32.13 Methods of predicting settlement.

initial settlement (*AJ*), estimating the developed excess pore pressure (*JB*) by the expressions in Chapter 26, and obtaining the consolidation settlement for one-dimensional strain for an increment of vertical stress equal to the excess pore pressure, i.e., for the stress path *NM*. In other words, the Skempton-Bjerrum method assumes a stress path *AJ* plus *NM*—a discontinuous and therefore incorrect path. It is superior theoretically to the preceding method and for a problem such as ours will give a smaller settlement.

On an important movement problem the actual effective stress path should be estimated. If one of the common techniques is based on a path not too unlike the actual, the method can probably be used with good accuracy. This comparison may indicate the desirability of actually stressing an element of soil along the estimated field stress path, i.e., the desirability of using the stress path method.

Chapter 14 presented a widely used empirical method of estimating settlement on sand: the use of the results of the standard penetration test. We can also use the the results of the penetration test in clay to approximate settlement involving undrained loading, although the reliability of this approach is not high.

32.8 MEASURED FOUNDATION MOVEMENT

Figures 32.14, 32.15, and 32.16 present measured pore pressure and movement data during construction. These actual field data illustrate some of the principles that have been discussed in the preceeding pages and also show some of the complexities that can arise.

Figure 32.14 presents data from a site of a warehouse. The subsoils consisted of hard yellow clay overlain by a 1.8 m layer of organic silt which in turn was overlain by 3.3 m of firm sand and gravel fill. In order to precompress the 1.8 m of soft silt and to raise the surface of the site to the desired elevation, a depth of 4.1 m of sandy gravel was placed over the site which was approximately 90 by 76 m. The relatively large lateral extent of the building site as compared to the thickness and depth of soft soil ensures that the strains in the soft soil will be essentially vertical.

The field data in Fig. 32.14 show the following interesting points:

1. More than three-quarters of the total settlement resulted from strain within the soft soil.
2. The maximum measured excess pore pressure was only 70% of the stress applied by the fill.
3. A large portion of the total settlement and excess pore pressure dissipation occurred during the loading process.
4. Upon preload removal the pore pressures became negative and the site heaved slightly.

Figure 32.15 presents not only settlement and pore pressure data but also lateral displacement data in a soft soil loaded by a storage tank. The most significant fact illustrated by the field data is the occurrence of very large lateral movements relative to the settlements. As can be seen, a horizontal displacement of 190 mm occurred at the top of the soft clay even though the settlement of the tank was less than 300 mm.

Figure 32.16 presents pore pressure and movement data obtained during the excavation for the CAES shown in Fig. 32.6. (This case is described in detail by Lambe, 1967a.) Interesting observations to be drawn from these data include:

1. Most of the heave and settlement resulted from the soil immediately under the excavation.
2. Most heave occurred while the foundation load was constant.
3. A significant increase in pore pressure occurred before loading started.

32.9 COMPLICATIONS IN THE PREDICTION OF FOUNDATION MOVEMENT

Complications to the prediction of foundation movement can be so numerous and significant that the engineer should consider a movement prediction as an approximate analysis.

Boundary Conditions

Establishing the boundary conditions in a soil movement or stability problem can be very difficult. It is usually easier to locate the limits of the soft and weak soils than it is to determine the drainage conditions of the layer or layers under consideration. The prediction of vertical movement in the subsoil shown in Fig. 32.14 assumed that essentially no settlement would occur from compression of the soil below the soft layer. A check on this assumption was made by installing and observing a settlement rod with its sensing element in the soil below the soft layer. The field data in Fig. 32.14 show the assumption of no settlement below the soft layer was not too far wrong.

The subsoil profile shown in Fig. 32.16 indicates glacial till below the soft clay. In predicting the rate of compression of the clay, a determination had to be made as to whether or not the glacial till was pervious and therefore acted as a drainage surface. Field pore pressure measurements in the glacial till indicated that it did serve as a drainage surface. A study of the rate of movement and pore pressure dissipation indicate that the clay had drainage layers in addition to that at the top and that at the bottom.

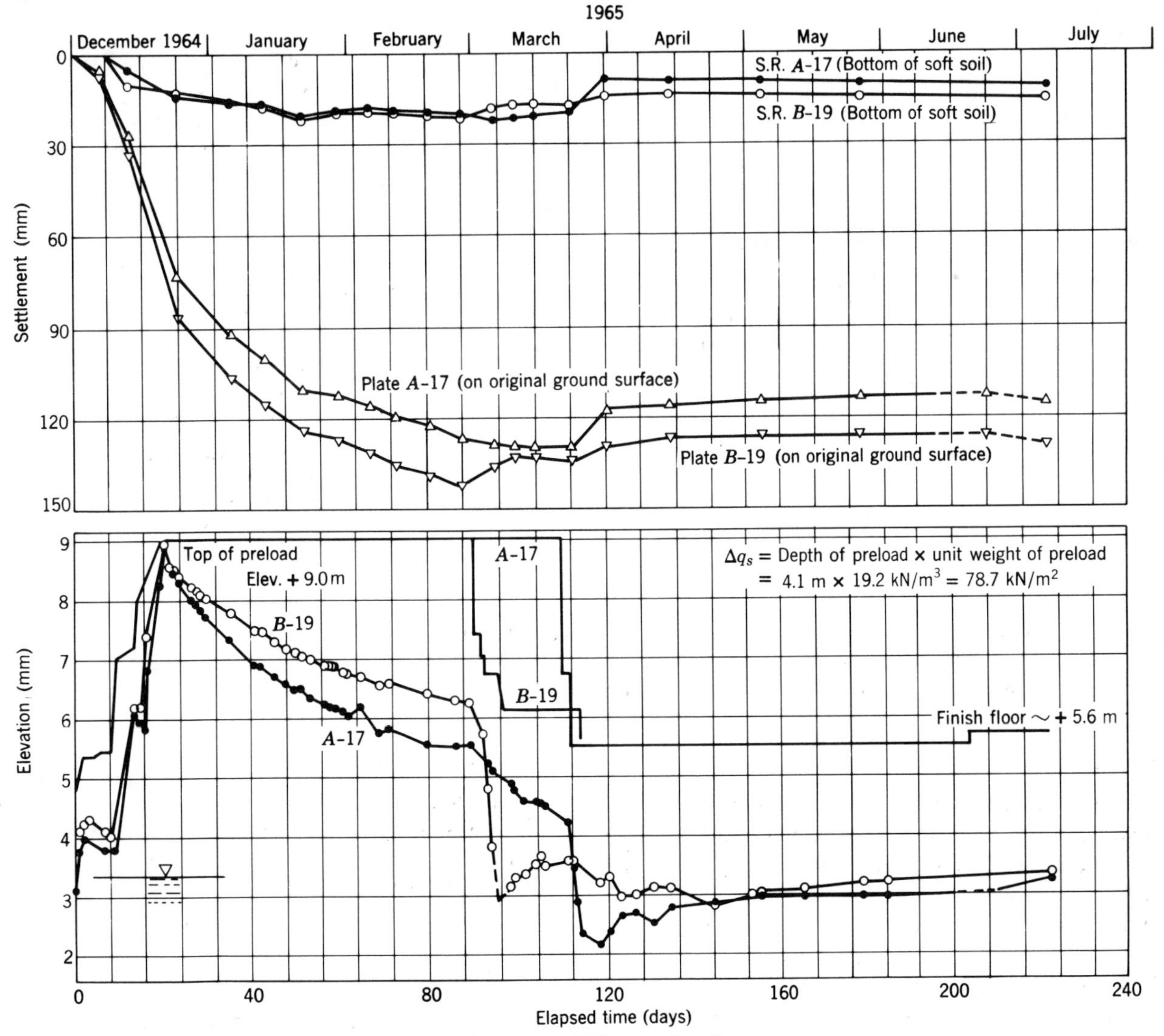

Fig. 32.14 Fill on soft soil.

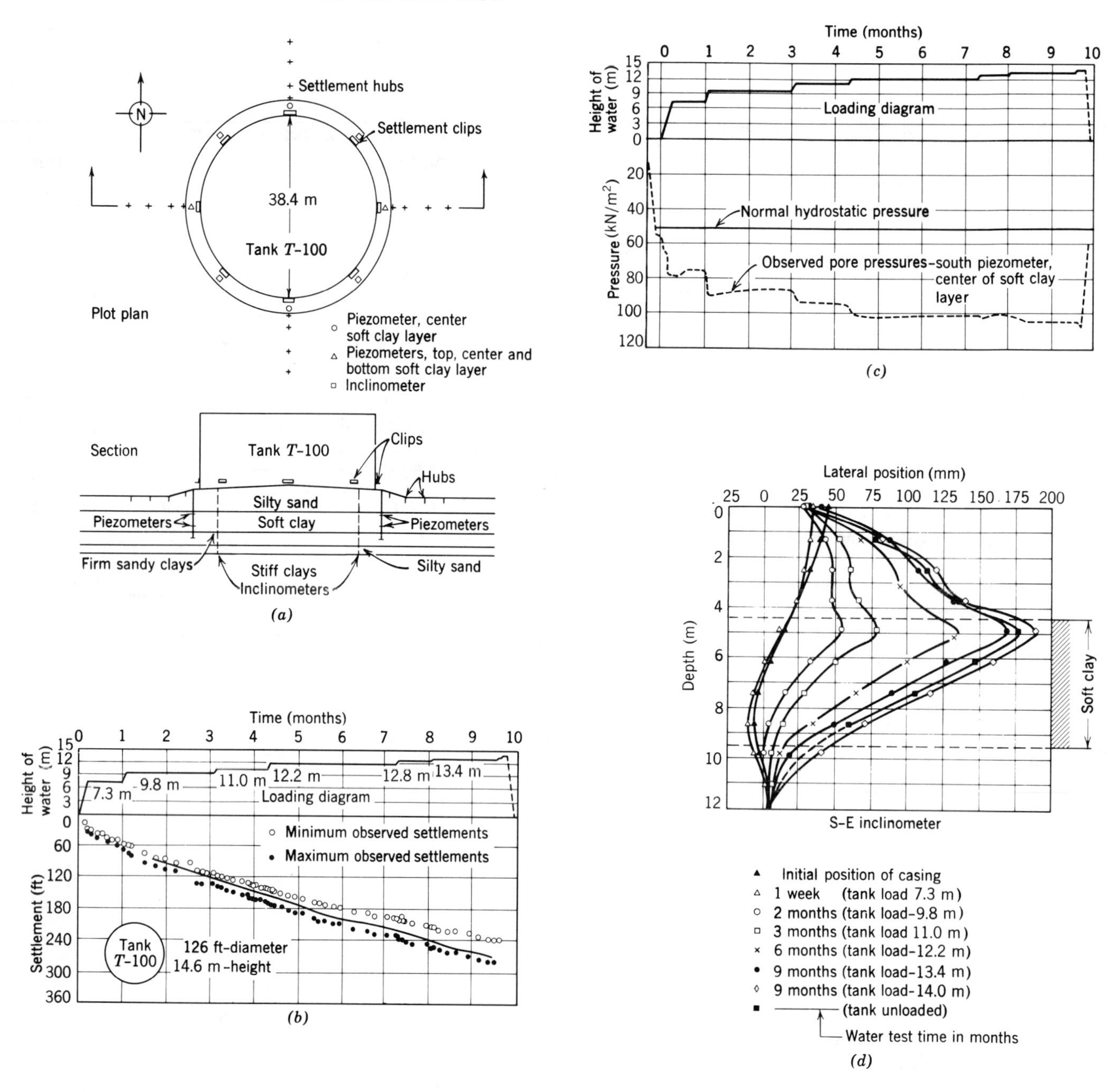

Fig. 32.15 Tank on soft soil. (*a*) Preloading instrumentation, tank *T*-100, Pascagoula Refinery. (*b*) Observed and predicted shell settlements, tank *T*-100, Pascagoula Refinery, (*c*) Observed pore water pressures in soft clays, tank *T*-100, Pascagoula Refinery. (*d*) Observed lateral soil displacements, tank *T*-100, Pascagoula Refinery. (From Darragh, 1964.)

Soil Properties

The selection of the proper values of soil properties for a prediction of foundation movement can tax the judgment of the experienced soil engineer. Soil properties tend to scatter widely, disturbance of soil during the sampling and specimen preparation alters the properties, and certain of the laboratory stress-strain tests are difficult to run. The scatter in natural soils is illustrated in the profiles shown in Chapter 7. As described in Chapter 29, sample disturbance tends to decrease undrained strength, increase compressibility, and alter the pore pressure characteristics of fine-grained soils. The stress path that should be used for the laboratory test in a given problem might well involve stress conditions that are not easily obtained with standard laboratory equipment.

Soil Stresses

The initial vertical stresses in subsoil can usually be obtained with good accuracy; lateral stresses can only

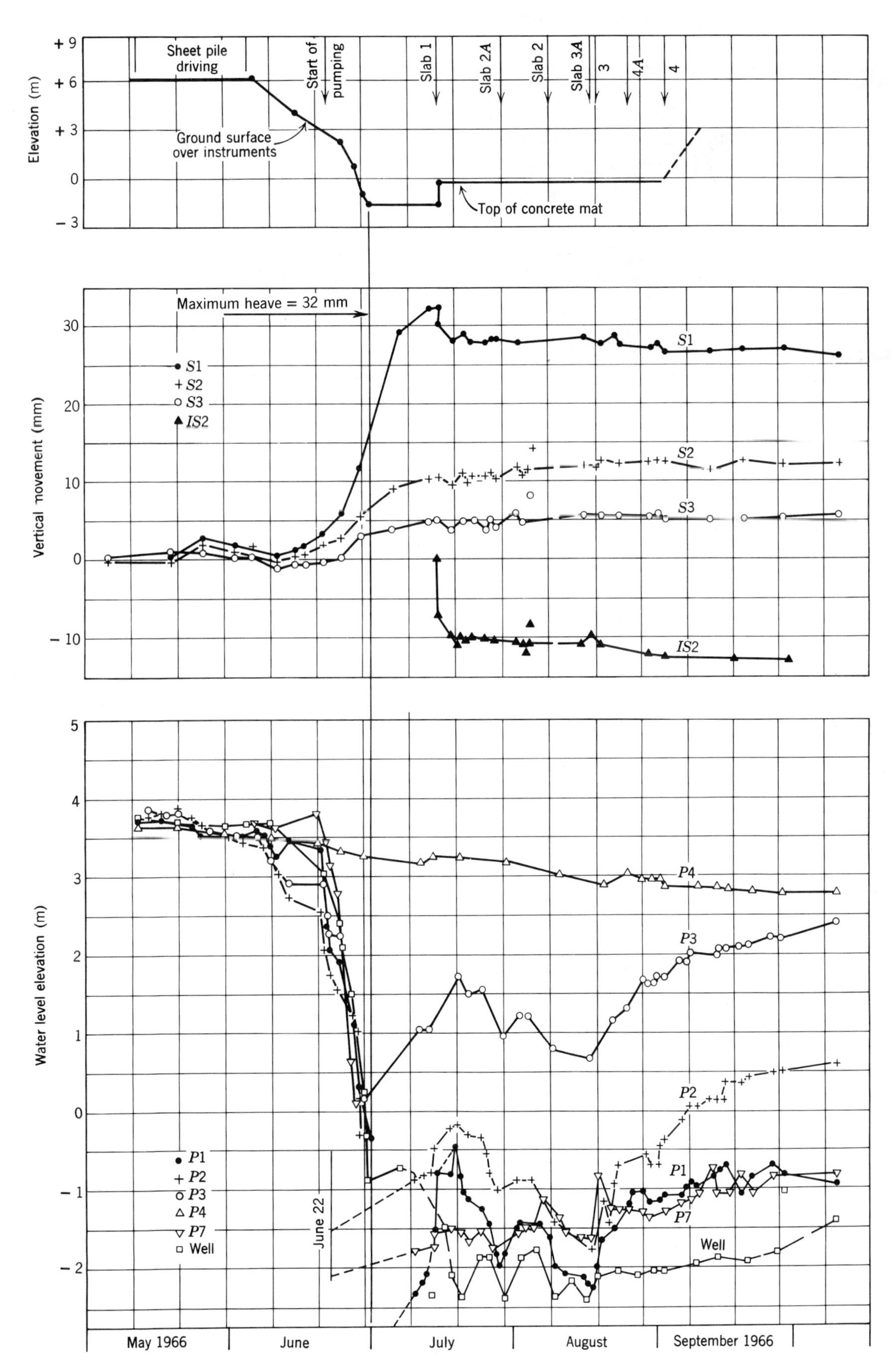

Fig. 32.16 Foundation behavior at center of site. (From Lambe, 1967a.)

Fig. 32.17 Apartment building in Niigata, Japan after 1964 earthquake (courtesy of H. Kishida).

be estimated. The main uncertainty with soil stresses arises, however, when we attempt to compute the increments of stress induced at various locations within the subsoil. The magnitude and distribution of surface stress can seldom be obtained with any degree of accuracy. Even having these stresses, we are forced to use a stress distribution theory involving simplifying assumptions of the soil properties. The stress increments induced in the subsoil tend to change during consolidation.

32.10 DYNAMIC PROBLEMS

The methods described in Section 15.1 may be used for analysis of dynamically loaded foundations resting on soils containing water. Obviously, values of shear modulus G and Poisson's ratio μ appropriate for undrained loading should be used. The evaluation of G has been discussed in Section 30.1. For a saturated soil, μ is 0.5. Whitman (1966) presents field evidence of the applicability of these methods to foundations resting on clayey soil.

These same methods may be used to analyze the effect of foundation flexibility on the response of buildings to earthquakes (Parmalee, 1967). One very important problem in earthquake zones is the possibility that sands will loose bearing capacity as the result of shaking by an earthquake. Figure 32.17 shows an apartment building in Niigata, Japan which tilted severely during an earthquake in 1964 because of liquefaction of the underlying sand. There was no structural damage to this building. Liquefaction, with resulting loss of bearing capacity and subsidence, was widespread in Niigata during this earthquake. The fundamentals of liquefaction have been discussed in Section 29.3. The application of these fundamentals to analysis of the Niigata liquefaction is given by Seed and Idriss (1967).

32.11 SUMMARY OF MAIN POINTS

1. This chapter considers the stability and vertical movement of shallow foundations subjected to undrained loading.
2. The factor of safety against a shear failure for a loading or for an unloading can be determined either by a stability analysis or by a bearing capacity equation.
3. For most soils the critical condition in a foundation loading occurs immediately following the load application, and the undrained shear stength of the soil prior to loading is the appropriate type of strength to use in the analysis.
4. In an unloading, the critical situation occurs at the end of the unloading period, i.e., at the start of loading, and the undrained shear strength existing in the subsoil at this moment is the appropriate value to use in an analysis. The undrained shear strength at the start of loading will often be less than that prior to construction.
5. A foundation undergoes an *initial settlement* during undrained loading and a *consolidation settlement* during the period after the loading is completed. The *total settlement* is the sum of the two.
6. There are numerous techniques for estimating the movement of a foundation. The engineer should estimate the stress paths for the average point in the actual structure and then use for his prediction a technique based on stress paths that approximate the field situation.
7. Data from field cases suggest that the usual condition during foundation loading or unloading is "partially drained" rather than either drained or undrained.
8. There are many complications to predicting movement—especially rate of movement—of actual foundations. The most important difficulties are:
   a. Determination of initial horizontal stress.
   b. Determination of field drainage conditions.
   c. Determination of stress increments caused by the foundation loading or unloading.
   d. Selection of soil parameters.

PROBLEMS

32.1 The bearing capacity equations (such as Eq. 32.1) for clay usually employ the undrained shear strength. List and discuss situations where the use of undrained strength would not be logical.

32.2 Determine the factor of safety against bearing capacity failure of the situation in Fig. 32.1 if the soil is normally consolidated Weald clay. (See Chapters 28, 29, and 30 for the properties of Weald clay.)

32.3 For the following values of pore pressure parameter A, compare the consolidation component of settlement as computed by the Skempton-Bjerrum method with that computed by the one-dimensional method:

$$A = 0.5, \quad A = 1.0, \quad A = 1.5$$

32.4 Why is such widespread use made of the one-dimensional method to predict the settlement of buildings on clay?

CHAPTER 33

Deep Foundations

Chapters 14, 25, and 32 have presented the fundamentals of *shallow foundations*—foundations where the soil support is applied near the usable portion of the structure. Chapter 33 considers *deep foundations*—those where the soil support is applied at some depth below the usable portion of the structure. Chapter 1 illustrated these two different types of foundations.

The basic situation for a deep foundation is where soft soil exists near the ground surface, as illustrated in Fig. 33.1. A deep foundation is employed to carry the building loads through the soft soil to the firmer soil below. Even though the deep foundation is an obvious solution to the problem of soft soils, it may not be the most satisfactory solution or the most economical solution. A partial or full flotation—as used for the Center for Advanced Engineering Study described in Chapter 32—may be more satisfactory than the deep foundation. Further, for certain situations, the improvement of the soft soil through a technique such as preloading (described in Chapter 34) can be a superior scheme to a deep foundation.

The ideal foundation for a given situation depends on a number of factors, including (*a*) type of soft soil, (*b*) extent of soft soil, (*c*) type of structure, (*d*) value to the owner of basement space gained in a flotation scheme, (*e*) length of construction time available, and (*f*) availability of soil for preloading. Deep foundations have been selected for the majority of the situations requiring the construction of a building over soft ground. For these situations, deep foundations have been used more frequently than justified—apparently because of the widespread (and mistaken) feeling that deep foundations present no construction problems and yield no foundation settlement.

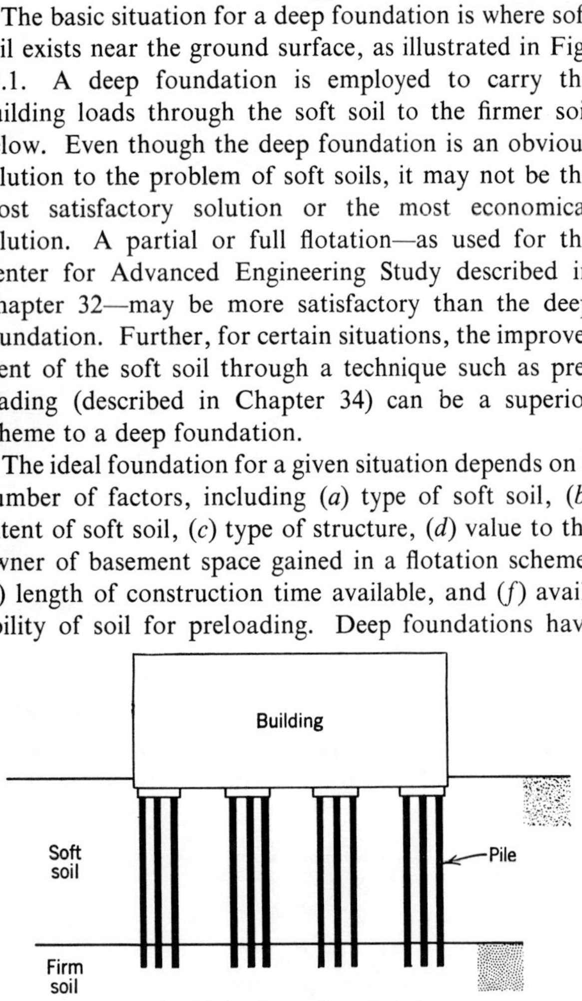

Fig. 33.1 Deep foundation.

Many types of deep foundations have been used. The most common is the *pile foundation*. This chapter considers only pile foundations, although most of the fundamentals presented also apply to other deep foundations, such as caisson foundations. A pile can be installed by: (*a*) inserting it into a bored hole; (*b*) pushing it into place by a static load; or, most commonly, (*c*) thumping it into place by a *pile hammer*. A pile that receives the majority of its support by friction or adhesion from the soil along its shaft is a *friction pile*. A pile that receives the majority of its support from soil near its tip is a *point-bearing pile*. Wooden piles, concrete piles, steel piles, steel shells filled with concrete are all common. Figure 33.2 gives values of usual maximum length and maximum design load for various types of piles.

A pile foundation—even a single pile—is statically indeterminate to a very high degree. The chance of a precise analysis of a pile foundation is thus even more remote than is true for most problems in soil mechanics. Empirical knowledge and the results of pile tests at the actual site are usually necessary for the proper solution to a given pile foundation problem. This chapter can only hope to identify the fundamental soil mechanics phenomena involved with deep foundations and to direct the reader to more detailed treatments of this important, highly complex subject. There are a number of good "state-of-art" treatments of deep foundations, such as Kerisel (1967), Vesic (1967b), Chellis (1962), and especially Horn (1966). The Proceedings of the International Conferences on Soil Mechanics and Foundation Engineering contain many papers on deep foundations.

33.1 SUPPORT OF A SINGLE PILE

The load applied to a single pile is carried jointly by the soil beneath the tip of the pile and by the soil around the shaft, and the maximum load that the pile can support—the *pile capacity*—is (Fig. 33.3)

$$Q = Q_p + Q_s \tag{33.1}$$

where

$$Q_p = \text{point resistance}$$

$$Q_p = A_p(\Delta q_s)_u$$

$$= A_p\left(cN_c + \frac{\gamma B}{2} N_\gamma + \gamma d N_q\right) \tag{33.2}$$

and

$$Q_s = \text{shaft resistance}$$

$$Q_s = \sum (\Delta L)(a_s)(s_s) \tag{33.3}$$

In Eqs. 33.2 and 33.3

A_p = area of pile tip
$(\Delta q_s)_u$ = ultimate bearing capacity as given in Eq. 25.6
ΔL = increment of pile length
a_s = area of pile surface in length ΔL in contact with soil
s_s = unit shaft resistance

The strength parameters c and ϕ may be either in terms of effective stresses or total stresses, depending upon the nature of the problem. Similarly, the unit weight γ may be interpreted in different ways in different problems.

Point Resistance

The principles of bearing capacity given in Chapters 14, 25, and 32 for shallow foundations hold also for deep foundations. The location of the failure surface for a deep foundation is less well known than for shallow foundations, and depending on the location of the failure surface assumed, investigators have calculated various values for the bearing capacity factors. There is a general feeling that the factors, especially N_q, are higher, and probably much higher for deep foundations than for shallow foundations. Figure 33.4 gives values of N_q as a function of ϕ as proposed by various investigators. Figure 33.5 shows some of the patterns of failure that have been assumed in the theoretical analyses.

With free-draining soils, the excess pore pressures caused by the loading of a deep foundation can dissipate readily; therefore *drained* conditions exist. In nonfree-draining soils, the excess pore pressures generated by loading a deep foundation may or may not dissipate depending upon the situation, especially the type of loading. Under a long-duration steady load—such as that arising from the weight of the structure—the excess pore pressures can dissipate. Under a short-duration

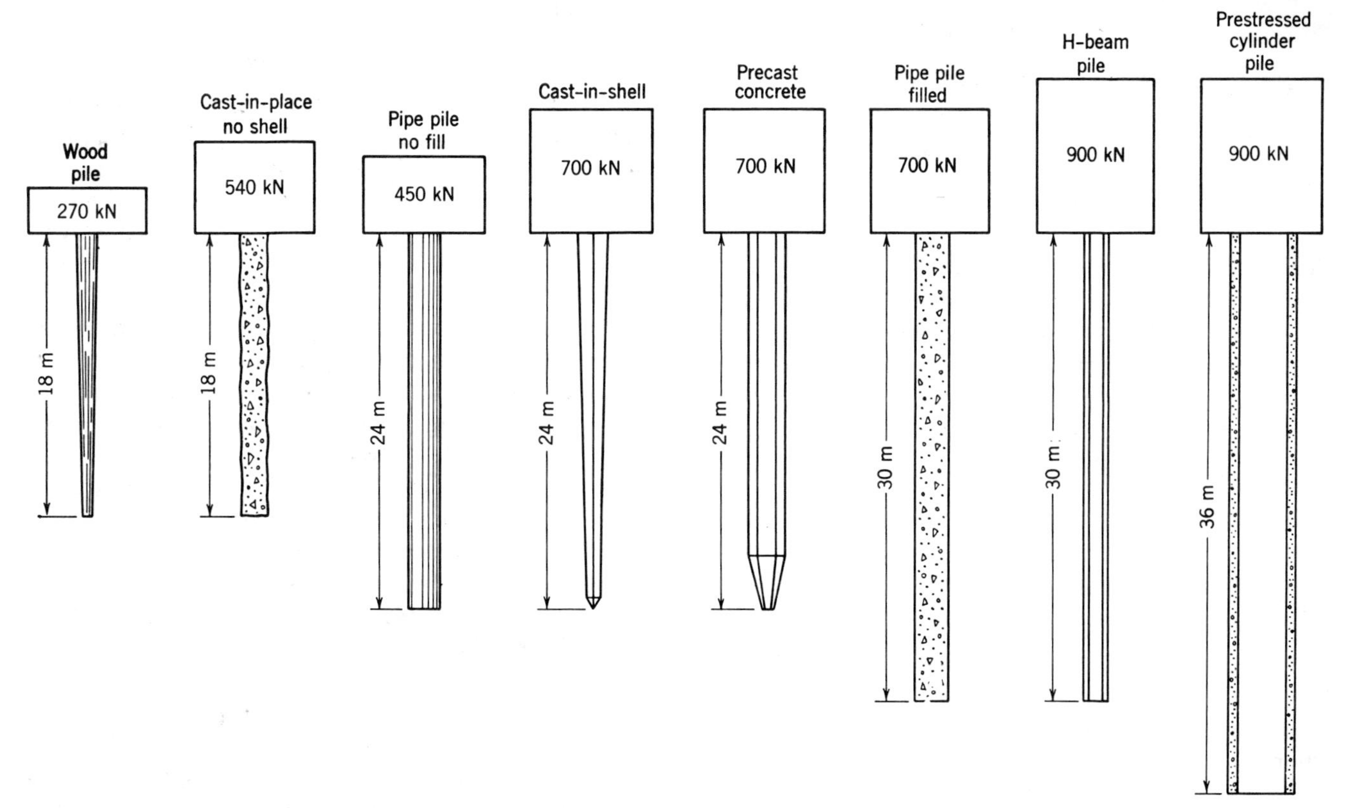

Fig. 33.2 Usual maximum length and maximum load for various piles (design values). Greater lengths and higher loads are common (From Carson, 1965).

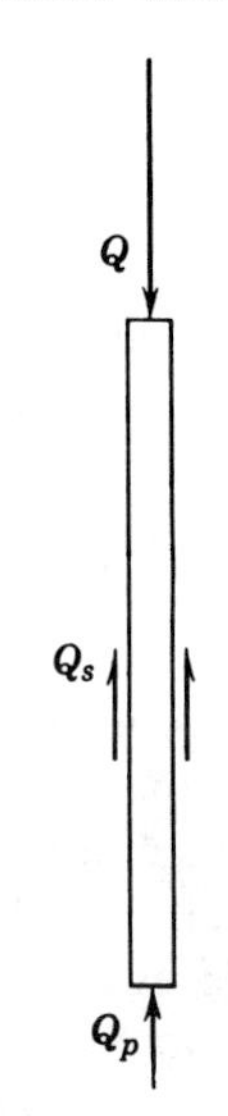

Q = pile capacity

Q_p = point resistance

Q_s = shaft resistance (skin resistance)

$$Q = Q_p + Q_s$$

Point Resistance:

$$Q_p = A_p(\Delta q_s)_u$$

A_p = area of pile point

$$(\Delta q_s)_u = \bar{c}N_c + \frac{\gamma BN\gamma}{2} + \gamma\, dNq \qquad (25.6)$$

Shaft Resistance:

$$Q_s = \Sigma\, (\Delta L)(a_s)(s_s)$$

ΔL = increment of pile length

a_s = area of pile surface in ΔL

s_s = unit shaft resistance

Fig. 33.3 Pile capacity.

load—such as that caused by a wind force on the structure—the excess pore pressures cannot readily dissipate. Since undrained conditions result in the minimum bearing capacity in soft, cohesive soils, the reasonable procedure in computing point resistance of a pile in clay is to assume undrained bearing capacity. This procedure is obviously somewhat conservative.

Employing these principles and the fact that cohesion in sands is zero, we can simplify Eq. 33.2 as follows:

For free-draining soils (sands), $c = \bar{c} = 0$ and $\phi = \bar{\phi}$. Hence

$$Q_p = A_p\left(\frac{\gamma B}{2} N_\gamma + \gamma d N_q\right) \qquad (33.2a)$$

and, since $(\gamma B/2)N_\gamma$ is small compared to $\gamma d N_q$, we can simplify Eq. 33.2a to

$$Q_p = A_p(\gamma d N_q) \qquad (33.2b)$$

For this case γd is $\bar{\sigma}_{v0}$.

For nonfree-draining soils (clays), where the $\phi = 0$ concept applies,

$$Q_p = A_p(cN_c + \gamma d) \qquad (33.2c)$$

Here c is average undrained strength s_u and $\gamma d = \sigma_{v0}$:

$$Q_p = A_p(s_u N_c + \sigma_{v0}) \qquad (33.2d)$$

Shaft Resistance

In computing the shaft resistance (Fig. 33.6), we must consider not only the type of soil but also the method of installing the pile. The method of installation may have a significant effect on the degree of soil disturbance, the lateral stress acting on the pile, the friction angle, and even the area of contact. In stiff clays, there is, for example, evidence that the shafts of bored piles do not always fully contact the soil. For bored piles in stiff clays the value of a_s may thus be less than the area of the pile shaft. The real difficulty in evaluating shaft stress lies, however, in the selection of the proper value of unit shaft resistance.

In free draining soil the unit shaft resistance equals

$$s_s = \tau_{ff} = \bar{\sigma}_{ff} \tan \bar{\phi}$$

where

$$\bar{\phi} = \phi_\mu \text{ for a steel-soil contact}^1$$

and

$$\bar{\phi} = \bar{\phi}_{cv} \text{ for concrete-soil and wood-soil contact}$$

(see Table 11.1), and

$$\bar{\sigma}_{ff} = \bar{\sigma}_{h0} = K\bar{\sigma}_{v0}$$

A pile in sand is usually pushed or driven (since a bored hole would not remain open without horizontal support) and because of the high load required to push a pile in sand, nearly all piles in sand are driven. The vibrations from driving a pile in sand do two things: (*a*) densify the sand, as discussed in Chapter 15 and (*b*) increase the value of *K*.

Penetration tests run in a sand prior to pile driving and after pile driving indicate significant densification of the sand for distances as large as eight diameters away from the pile. Increasing the density results in an increase in the friction angle. Driving of a pile displaces soil laterally and thus increases the horizontal stress acting on the pile. Horn (1966) summarized the results of studies of the horizontal stress acting on piles in sand. His summary, Table 33.1, shows a wide divergence of opinion as to the value of the horizontal effective stress. It would seem logical that *K* must exceed one and a value of two would seem to be reasonable.

It seems logical that the unit resistance (adhesion) of clay on the shaft of a wood or concrete pile should be approximately equal to the shear strength of the soil. Because of the smoothness of a steel pile the clay

[1]By definition, ϕ_μ is an effective stress parameter.

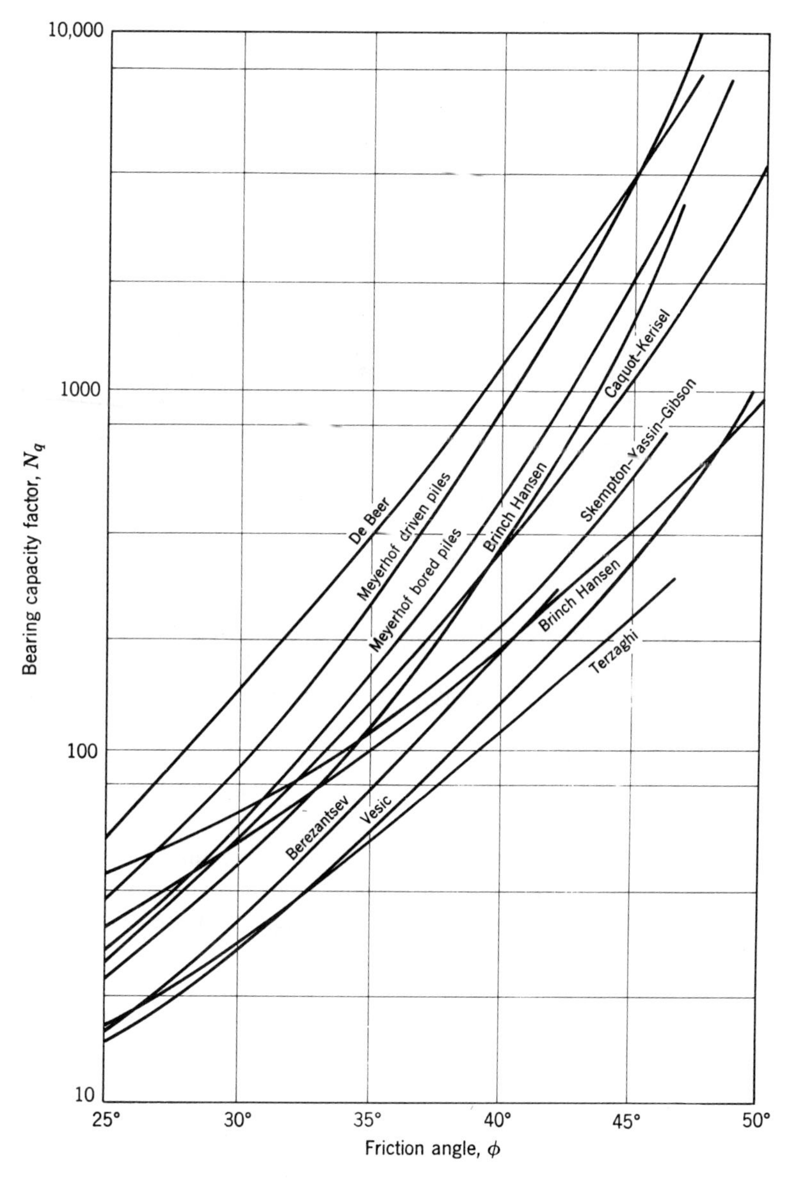

Fig. 33.4 Bearing capacity factors for deep circular foundations. (From Vesic, 1967).

Table 33.1 Horizontal Stress on Pile Driven in Sand

Reference	Relationship	Basis of Relationship
Brinch Hansen and Lundgren (1960)	(*a*) $\bar{\sigma}_h = \cos^2 \bar{\phi} \cdot \bar{\sigma}_v = 0.43\bar{\sigma}_v$ if $\bar{\phi} = 30°$ (*b*) $\bar{\sigma}_h = 0.8\sigma_v$	(*a*) Theory (*b*) Pile test
Henry (1956)	$\bar{\sigma}_h = K_p \cdot \bar{\sigma}_v = 3\bar{\sigma}_v$	Theory
Ireland (1957)	$\bar{\sigma}_h = K \cdot \bar{\sigma}_v = (1.75 \text{ to } 3) \cdot \bar{\sigma}_v$	Pulling tests
Meyerhof (1951)	$\bar{\sigma}_h = 0.5\bar{\sigma}_v$; loose sand $\bar{\sigma}_h = 1.0\bar{\sigma}_v$; dense sand	Analysis of field data
Mansur and Kaufman (1958)	$\bar{\sigma}_h = K\bar{\sigma}_v$; $K = 0.3$ (Compression) $K = 0.6$ (Tension)	Analysis of field data

From Horn, 1966.

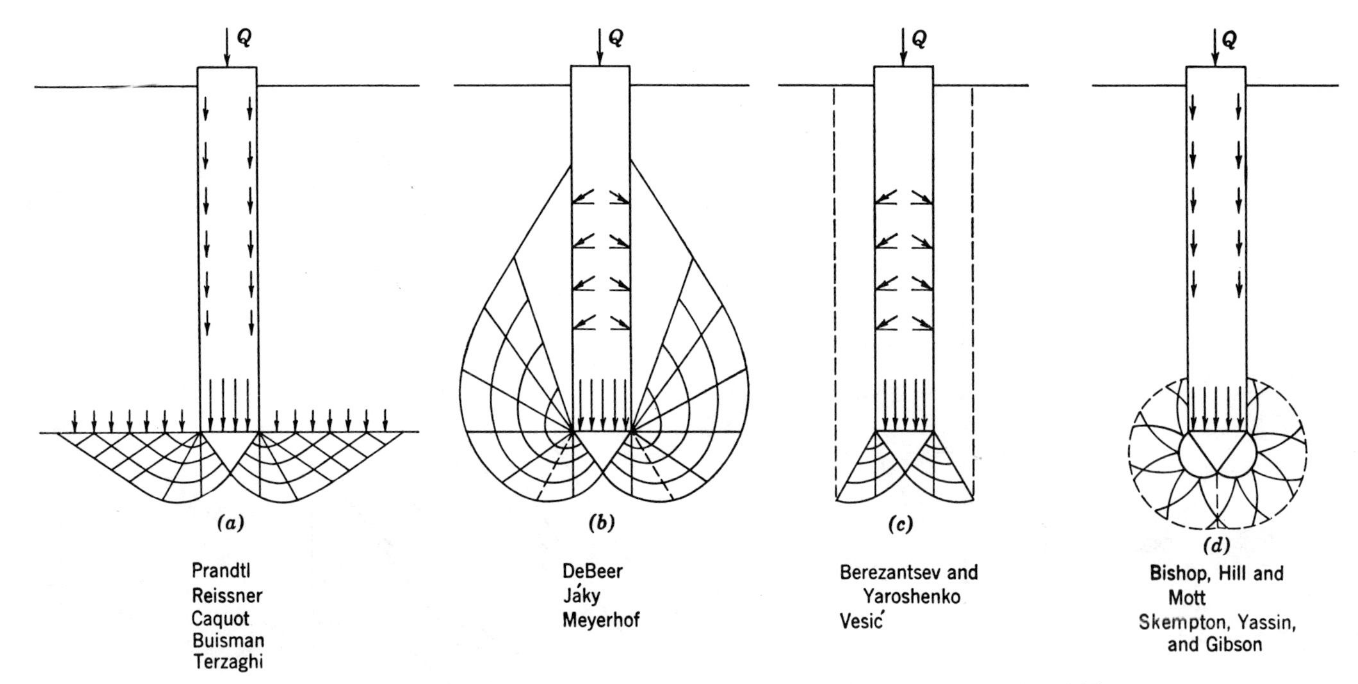

Fig. 33.5 Assumed failure patterns under deep foundations. (From Vesic, 1967).

adhesion may be slightly less than the shear strength. To use the shear strength of the clay as the adhesion along the shaft of the pile is considered to be a good approximation.

As with bearing capacity, the maximum load on a pile would normally occur for a duration so short that a clay would not fully drain, and thus it seems reasonable to use the undrained shear strength of the clay to approximate the adhesion of the clay on the pile shaft. However, several factors must be considered in selecting the appropriate value of undrained strength.

With his classic paper Casagrande (1932) alerted the profession to the possibility of disturbance by driving piles into clay. Casagrande pointed out that disturbance of a natural clay by pile driving could result in a great increase in compressibility and loss of strength. Cummings, Kerkhoff, and Peck (1950) described the results of an investigation in which changes in shear strength from pile driving were measured. Their results showed that the shear strength next to the pile was reduced by pile driving, but that one month after the piles had been driven the strength had returned to its initial value, and that eleven months after driving the strength was considerably greater than its initial value. This seems logical for most situations in which piles are driven into clay. The pile driving results in a decrease of strength due to disturbance and an increase in pore pressure, but some or all of the strength will recover following dissipation of the excess pore pressure and consolidation of the soil. Since the horizontal stress after pile driving is greater than prior to pile driving, and since reconsolidation leads to a decreased void ratio, the strength may well be greater after consolidation than before pile driving. Seed and Reese (1957) present field measurements showing the magnitude and dissipation of excess pore pressure near a pile and the accompanying regain of strength.

Since the piles in a foundation are not subjected to their full load until after the building has been completed, it is logical to use the reconsolidated strength for design purposes. Peck (1961) has compared, for a large number of piles, adhesion as deduced from load tests with undrained strength as determined by unconfined compression tests upon undisturbed samples. For normally consolidated clays, undrained strength provided a conservative estimate for adhesion, but with overconsolidated soils the observed adhesion was generally less than the undrained strength. With bored piles the loss of strength from disturbance would be less, but at the same time the horizontal effective stress following consolidation would also be less.

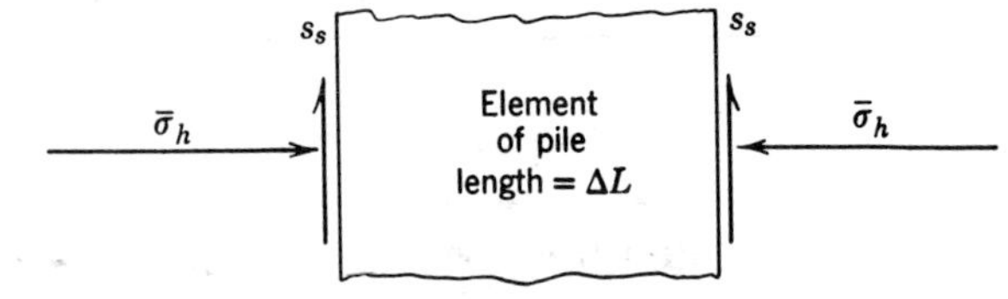

Cohesionless soil:

$$s_s = s_d = \bar{\sigma}_{hf} \tan \bar{\phi} \approx \bar{\sigma}_{hd} \tan \bar{\phi}$$

Cohesive soil:

$$s_s \approx s = \bar{c} + \bar{\sigma}_{hf} \tan \bar{\phi} \approx s_u \text{ for } \bar{\sigma}_c = \bar{\sigma}_{hd}$$

$\bar{\sigma}_{hd}$ = horizontal effective stress at time pile loaded.

Fig. 33.6 Shaft resistance on pile element.

In summary, the capacity[2] of a single pile equals the point resistance Q_p plus the shaft resistance Q_s. For free-draining soil:

$$Q = A_p\bar{\sigma}_{v0}N_q + \sum (\Delta L)(a_s)(K\bar{\sigma}_{v0} \tan \bar{\phi}) \qquad (33.4)$$

where

$K = 1$ to 3

$\bar{\phi} = \phi_\mu =$ for steel piles and $\bar{\phi}_{cv}$ for wood or concrete piles

Nonfree-draining soil:

$$Q = A_p(s_uN_c + \sigma_{v0}) + \sum (\Delta L)(a_s)(s_u) \qquad (33.5)$$

where

$s_u =$ CU strength of remolded soil, reconsolidated to $\bar{\sigma}_c = \bar{\sigma}_{v0}$

N_c from Fig. 32.4

Example 33.1 illustrates the computation for the capacity of a concrete-filled pipe pile driven into sand; Example 33.2 illustrates the computation of the capacity for the same pile driven into clay. Dividing these computed pile capacities by two, a common value used for the factor of safety on piles, yields a design capacity of 470 kN for the pile in sand and 122 kN for the pile in clay. In designing a pile foundation, we must consider not only the capacity of the pile as far as the support from the soil is concerned, but also the strength of the pile itself. The strength of the pile in design is determined by a procedure specified by the building code applicable in the area involved. If the pile in Examples 33.1 and 33.2 consisted of a steel shell 6.4 mm in thickness, a design capacity as a structural member is 800 kN (based on an allowable 58.6 MN/m² compressive load in the steel and 6.2 MN/m² compressive load in the concrete). The computed capacity of 800 kN has a large factor of safety in it.

▶ Example 33.1

Given. Concrete-filled pipe pile:

diameter $= 0.3$ m

penetration $= 12$ m

Soil: sand with

$\gamma_t = 18.9$ kN/m³

$\phi_\mu = 30°$

Find. Pile capacity Q.

Solution.

$$Q = A_p\bar{\sigma}_{v0}N_q + \sum (\Delta L)(a_s)(K\bar{\sigma}_{v0} \tan \bar{\phi}) \qquad (33.4)$$

$A_p = 0.0707$ m²

$\bar{\sigma}_{v0}$ at 12 m $= 12(18.9 - 9.81) = 109$ kN/m²

N_q from Fig. 33.4 $= 30$

Since strength varies linearly, we can work with mid-depth as average for entire pile:

$\sum \Delta L = 12$ m, $\quad a_s = 0.943$ m²/m, $\quad K$ taken as 2

$\bar{\sigma}_{v0}$ at mid-depth $= 54.5$ kN/m²

$\therefore$

$$Q = (0.0707)(109)(30) + (12)(0.943)(2)(54.5)(0.577)$$
$$= 231 + 712 = \underline{\underline{943 \text{ kN}}}$$ ◀

[2] There are many variations of the basic static equation (33.4). Bulletin No. 25 (1968) of the Danish Geotechnical Institute presents these variations. McClelland et al. (1967) describe the use of the static equation to heavily loaded pipe piles.

▶ Example 33.2

Given. Same pile as in Example 33.1. The soil is clay with:

$\gamma_t = 18.9$ kN/m³

$s_u = \frac{1}{3}\bar{\sigma}_{v0}$

Find. Pile capacity Q.

Solution.

$$Q = A_p(s_uN_c + \sigma_{v0}) + \sum (\Delta L)(a_s)(s_u) \qquad (33.5)$$

s_u at tip $= \frac{1}{3}(109) = 36.3$ kN/m²

s_u at mid-depth $= 18.2$ kN/m²

N_c from Fig. 32.4 $= 9$

$$Q = (0.0707)[(36.3)(9) + 227] + (12)(0.943)(18.2)$$
$$= 39 + 206 = \underline{\underline{245 \text{ kN}}}$$ ◀

33.2 PILE DRIVING AND FORMULAS FOR DRIVING RESISTANCE

Piles are usually forced into the ground by a *pile driver* or *pile hammer*. In medieval times piles were driven by men manually swinging hammers. Next came the *drop hammer*, which consists of a weight raised by ropes or cables and allowed to drop freely striking the top of the pile. Typically, a 1 to 1.5 t mass is dropped freely from a height of 6 to 9 m. After the drop hammer came the following:

1. *Single-acting hammer*, which uses steam or compressed air acting against a piston to raise a ram, which drops freely to strike the pile.
2. *Double-acting hammer* in which the ram is not only raised but is forced down by the steam or compressed air.
3. *Differential-acting hammer* in which steam or compressed air acts to raise the ram and force it down as in the double-acting hammer, but unlike the double-acting hammer, the air or steam pressure remains constant.
4. *Diesel pile hammer* in which an explosion of atomized diesel oil raises a ram, which is allowed to fall freely.

Table 33.2 Pile Hammer Characteristics

Rated energy (kJ)	Make of Hammer	Type	Size	Blows per min	Mass Striking Parts (kg)	Total Mass (kg)	Length of Hammer (m)	Air flow (l/sec)	ASME Boiler (kW)	Steam or Air (kN/m²)	Size of Hose (mm)	$\sqrt{E \times W}$ Rating[a]
ENERGY OVER 150 kJ												
153.88	Super-Vulcan	Differential	400C	100	18 140	37 650	5.11	2199	522	1034	127	52.83
ENERGY 60 to 150 kJ												
81.36	Vulcan	Single-act.	020	60	9070	17 690	4.57	829	207	827	76	27.16
81.36	McKiernan-Terry	Single-act.	S20	60	9070	17 530	5.61	812	209	1034	76	27.16
68.07	Super-Vulcan	Differential	200C	98	9070	17 710	4.01	824	194	979	76	24.85
ENERGY 40 to 60 kJ												
56.95	Vulcan	Single-act.	014	60	6 350	12 470	4.42	605	149	758	76	19.02
50.85	McKiernan-Terry	Single-act.	S14	60	6 350	14 330	4.52	595	142	689	76	17.97
48.82	Super-Vulcan	Differential	140C	103	6 350	12 694	3.73	673	157	965	76	17.61
44.07	McKiernan-Terry	Single-act.	S10	55	4 540	10 070	4.29	472	104	552	63	14.14
44.07	Vulcan	Single-act.	010	50	4 540	8 505	4.57	473	117	724	63	14.14
ENERGY 30 to 40 kJ												
35.26	Vulcan	Single-act.	08	50	3630	7600	4.57	415	95	572	63	11.31
35.26	McKiernan-Terry	Single-act.	S8	55	3630	8210	4.37	401	89	552	63	11.31
33.15	Super-Vulcan	Differential	80C	111	3630	8110	3.45	588	134	827	63	10.97
33.15	Vulcan	Differential	8M	111	3630	8350	3.20	588	134	827	63	10.97
ENERGY 15 to 30 kJ												
26.95	Union	Double-act.	0	110	1360	6577	3.07	378	—	862	51	6.05
26.92	McKiernan-Terry	Double-act.	11B3	95	2270	6577	3.38	425	94	689	63	7.82
26.44	Vulcan	Single-act.	06	60	2950	5080	3.96	295	70	689	51	8.83
26.04	Super-Vulcan	Differential	65C	117	2950	6752	3.68	468	113	1034	51	8.76
22.04	McKiernan-Terry	Single-act.	S5	60	2270	5613	4.04	283	63	552	51	7.07
21.70	McKiernan-Terry	Compound	C5	110	2270	5390	2.67	276	42	689	63	7.02
20.48	Super-Vulcan	Differential	50C	120	2270	5344	3.10	415	93	827	51	6.82
20.48	Vulcan	Differential	5M	120	2270	5850	2.84	415	93	827	51	6.82
20.34	Vulcan	Single-act.	1	60	2270	4580	3.96	267	60	559	51	6.79
17.76	McKiernan-Terry	Double-act.	10B3	105	1360	4922	2.84	354	78	689	63	4.91
17.26	Union	Double-act.	1	125	725	4540	2.49	283	—	689	38	3.54
ENERGY 8 to 15 kJ												
12.20	McKiernan-Terry	Single-act.	S3	65	2270	[illegible]990	3.76	189	43	552	38	5.26
11.87	McKiernan-Terry	Double-act.	9B3	145	725	3175	2.49	283	63	689	51	2.93
11.23	Union	Double-act.	1½A	135	680	4173	2.54	212	—	689	38	2.76
9.84	Vulcan	Single-act.	2	70	1360	3220	3.66	159	37	552	38	3.66
9.84	Super-Vulcan	Differential	30C	133	1360	3192	2.72	230	52	827	38	3.66
9.84	Vulcan	Differential	3M	133	1360	3850	2.41	230	52	827	38	3.66
ENERGY under 8 kJ												
6.64	Vulcan	Differential	DGH900	238	410	2268	2.11	274	56	538	38	1.65
4.96	Union	Double-act.	3	160	320	2132	1.93	142	—	689	32	1.26
4.88	McKiernan-Terry	Double-act.	7	225	360	2268	1.85	212	47	689	38	1.33
0.60	Union	Double-act.	6	340	45	413	1.17	35	—	689	19	0.16
0.52	Vulcan	Differential	DGH100A	303	45	357	1.27	35	6	414	25	0.15
0.48	McKiernan-Terry	Double-act.	3	400	31	306	1.49	52	—	689	25	0.12
0.43	Union	Double-act.	7A	400	36	245	1.09	33	—	689	19	0.12

DIESEL HAMMERS

McKiernan-Terry Corp.	Delmag	Link-Belt Speeder Corp.
Model No. DE-20 = 21.70 kJ	Model No. D-5 = 12.34 kJ	Model No. 105 = 10.17 kJ
Model No. DE-30 = 30.37 kJ	Model No. D-12 = 30.51 kJ	Model No. 312 = 24.40 kJ
	Model No. D-22 = 53.83 kJ	Model No. 520 = 40.67 kJ

From Carson, 1965.

[a] E = rated striking energy in (kilojoules); W = mass of striking parts in tonnes.

5. *Vibratory driver*, which employs pair(s) of counter-rotating eccentric weights phased such that the lateral force components cancel and the vertical components add.

Carson (1965) describes pile driving and pile-driving equipment. Table 33.2 (from Carson, 1965) gives pertinent characteristics of a number of commercial pile-driving hammers. Table 33.3 (from Davisson, 1966) compares many of the vibratory drivers.

The energy from the pile driver is expended both in useful work—forcing the pile into the ground—and in nonuseful work—compressing the pile cushion, compressing the pile, etc. Because of the energy lost in nonuseful work, a high-energy hammer is usually more effective in pile driving.

Dynamic pile formulas are widely used to determine the static capacity of a pile. These formulas are derived starting with the relation

Energy input = energy used + energy lost

The energy used equals the driving resistance times the pile movement. Thus by knowing the energy input, estimating the energy lost on the basis of experience, and observing the pile movement during a blow, we can compute the driving resistance.

The most commonly used dynamic pile formula, known as the Engineering News formula, is

$$R = \frac{166.64E}{s + 2.54} \tag{33.6}$$

A superior formula (Boston Building Code, 1964) is

$$R = \frac{141.64E}{s + 2.54\sqrt{w_p/w_r}} \tag{33.7}$$

In these two equations:

R = the allowable pile load in kilonewtons
E = the energy per blow in kilonewton-meters (kJ)
s = the average penetration in millimeters per blow for the final 150 mm of driving (minimum permissible value of s = 1.25 mm)
w_p = the weight of the pile and other driven parts
w_r = the weight of the striking part of the hammer (minimum permissible value of $w_p/w_r = 1.0$)

Table 33.2 gives values of E and w_r for a number of commercial hammers. These two formulas are illustrated by Example 33.3.

The Hiley dynamic pile formula is superior to both the Engineering News formula and the Boston Building Code formula in that it better accounts for the energy lost in pile driving. Carson (1965) gives this formula along with tables of values of the various coefficients needed to approximate the temporary compression in the pile, pile cap, and soil.

Table 33.3 Vibratory Pile Drivers

Make and Model		Total Weight (kN)	Available Power kW	Frequency (Hz)	Force (kN)[a] Frequency (Hz)
Foster	2–17	27.6	25.4	18–21	
(France)	2–35	40.5	52.2	14–19	276/19
	2–50	49.8	74.6	11–17	449/17
Menck	MVB22–30	21.4	37.3		214/
(Germany)	MVB6.5–30	8.9	5.6		62/
	MVB44–30	38.3	74.6		431/
Muller	MS–26	42.7	53.7		
(Germany)	MS–26D	71.6	108.1		
Uraga	VHD–1	37.4	29.8	16.3–19.7	191/19.7
(Japan)	VHD–2	52.9	59.7	16.3–19.7	383/19.7
	VHD–3	68.5	89.5	16.3–19.7	574/19.7
Bodine (U.S.A.)	B	97.9	745.7	0–150	280/100–778/100
Russian	BT–5	12.9	27.6	42	214/42
	VPP–2	21.8	40.3	25	218/25
	100	17.8	27.6	13	196/13
	VP	48.9	59.7	6.7	156/6.7
	VP–4	115.2	155.1		881/

From Davisson, 1966.

[a] Forces given are present maximums. These usually can be raised or lowered by changing weights in the oscillator.

▶ **Example 33.3**

Given. The pile, subsoil, and driving record shown in Fig. 33.7.

Find. The allowable pile load by:

(*a*) Engineering News formula.

(*b*) Boston Code formula.

Solution. From Table 33.2

$$E = 35.26 \text{ kJ}$$

From Fig. 33.7

$$s = \tfrac{25}{16} = 1.563 \text{ mm/blow}$$

From Table 33.2

$$w_r = 3630 \times 9.81 = 35\,610 \text{ N}$$

$$\text{Pile weight (steel shell)} = 24\,020 \text{ N}$$

(*a*) Engineering News formula:

$$R = \frac{166.64E}{s + 2.54} \qquad (33.6)$$

$$R = \frac{166.64 \times 35.26}{1.563 + 2.54} = \underline{\underline{1432 \text{ kN}}}$$

(*b*) Boston Code formula:

$$R = \frac{141.64E}{s + 2.54\sqrt{w_p/w_r}} \qquad (33.7)$$

$$R = \frac{141.64 \times 35.26}{1.563 + 2.54\underline{\sqrt{24\,020/35\,610}}^{*}} = \underline{\underline{1217 \text{ kN}}}$$

* Must use the minimum permissible value of w_p/w_r i.e., 1.0

◀

Note that the dynamic pile-driving formulas, Eqs. 33.6 and 33.7, yield an "allowable" pile load and not the pile capacity. Presumably the formulas have incorporated in them a factor of safety in equating the dynamic resistance to the static resistance. The factor of safety in the Engineering News formula is six times the efficiency of the impact—the factor of safety is thought to be somewhere between two and five.

Because of the difficulty of evaluating the many energy losses involved with pile driving, it is doubtful that a dynamic pile driving formula can do much better than approximate the pile driving resistance. The discussion in Section 33.1 clearly shows that the capacity of a pile during driving or immediately after driving may be far different from the static capacity. This inequality is especially true for friction piles in clay. The static capacity of a friction pile in clay may be several times that computed from dynamic driving formulas.

In spite of their serious limitations, the dynamic pile driving formulas have considerable utility for the soil engineer. On an important pile foundation job, one or more static pile tests, as described in the next section, are usually performed. Having the measured static pile capacity and the computed dynamic resistance, the soil engineer can establish a driving specification based on blows per final 25 mm of penetration to be used in driving the production piles on the job.

A pile can be forced into the ground by the application of a static load to the top of the pile. Based on 35 years of experience with driving piles in Lake Maracaibo, the Creole Petroleum Corporation (Trinkunas, 1967) developed correlations between the long-term static capacity of piles and the capacity of a pile forced into the ground with a static load. Creole frequently used loads as large as 1800 kN and occasionally 2700 kN to force piles into the ground.

In driving piles for a pile foundation the soil engineer may be faced with many practical considerations such as selection of pile driving equipment, sequence of pile driving, the need to employ some technique like pre-augering or jetting to assist pile driving, and the difficulty of inspecting piles to be sure they are in the right location and at a proper alignment.

33.3 PILE LOAD TEST

A *pile load test* consists of applying increments of static load to a test pile and measuring the deflection of the pile. The load is usually jacked onto the pile using either a large dead weight or a beam connected to two uplift anchor piles to supply the reaction for the jack. Figure 33.7 presents the results of a typical static load test on one of the piles for the M.I.T. Center for Space Research. As noted in Fig. 33.7, the test pile consisted of a 324 mm diameter steel shell driven from ground surface elevation +6.1 m to elevation −34.1 m, and then filled with concrete. The pile was then loaded in increments of about 180 kN to a maximum load of 1246 kN, twice the design load of the pile. The pile was then unloaded in decrements of about 445 kN. As can be seen in Fig. 33.7, measurements were taken of the movement of not only the top of the pile but also the bottom of the pile (by a rod attached to the bottom of the pile and protected by a pipe).

A static pile load test can be conducted for any or all of three reasons:

1. To indicate for the contractor the type of driving conditions that will be encountered on the job at hand.
2. To furnish information to the soil engineer to develop driving criteria, as described in the preceding section.
3. To obtain test data needed to convince the building authorities that the pile is adequate to support the design load.

Because of the time effects discussed in Section 33.1 and because of the group action discussed in the following section, the results of a static load test are not always easy to interpret. Note in particular that a load test on a pile into clay should not be made until the clay has had a chance to reconsolidate. The details for carrying out a

Table 33.4 Reduction Factors for Pile Groups in Clay

Spacing between Pile Centers (pile diameters)	Reduction Factor
10	1
8	0.95
6	0.90
5	0.85
4	0.75
3	0.65
$2\frac{1}{2}$	0.55

From Kerisel, 1967.

static pile load test in a given area are usually specified by the building code for that area. For example, the Boston Building Code (1964) specifies in detail the type of equipment, physical setup, loading procedures, etc., that must be used. This code also contains the commendable requirement that the results of the pile load test must be analyzed by a competent engineer. The Boston Code specifies that the settlement under the designed pile load shall not exceed 9.5 mm, and the settlement under twice the design load shall not exceed 25.4 mm. The settlement of the test pile in Fig. 33.7 under the design load of 623 kN was 6.6 mm and under 1246 kN, twice the design load, it was 16.3 mm. The test pile thus met the requirements of the Boston Code.

33.4 CAPACITY OF A PILE GROUP

In general, the capacity of a cluster of piles is not equal to the sum of the capacity for each pile in the cluster acting as a single pile. The ratio of cluster capacity to the sum of the individual capacities is termed *group efficiency* or *reduction factor*. The group efficiency of friction piles in clay is normally less than one, whereas the group efficiency for friction piles in sand is greater than one. The group efficiency of point-bearing piles is normally less than one.

There is a considerable body of theoretical and empirical knowledge on the supporting capacity of a single pile since this subject has been studied extensively. On the other hand, there is relatively little information on the supporting capacity of a pile group because of the considerable difficulty in conducting full-scale tests on pile groups. On most building jobs it would be awkward to make available a space large enough to test an entire pile cluster. Further, to obtain a reaction potential large enough to load to failure a pile cluster could be a major problem.

For friction piles in clay, Kerisel (1967) proposed the reduction factors given in Table 33.4.

Vesic (1967a) attributes the increase in bearing capacity of a pile group in homogeneous sand to increased skin resistance of the piles in the group. He reports from his tests skin efficiencies as high as three, whereas the point efficiencies were all approximately one. His tests indicated that the efficiency of a full pile group increased with pile spacing to a maximum at spacings of three pile diameters and then dropped slightly with a further increase in pile spacing.

The reason that the efficiency of a pile group in sand is greater than one is that the driving of adjacent piles increases the horizontal effective stress and thus the shaft resistance of the piles in place. In addition, the driving of adjacent piles tends to increase the relative density of the sand, thereby causing an increase in the friction angle of the sand.

33.5 NEGATIVE SKIN FRICTION

Under the applied load Q the pile in Fig. 33.3 moved downward relative to the soil at the tip of the pile and relative to the soil around the shaft of the pile. Thus both Q_p and Q_s acted upward—i.e., they combined to support the load Q acting downward on the pile. In several situations some or all of the soil along the shaft of a pile may move downward relative to the pile, thereby reversing the direction of Q_s. Q_s is thus no longer a supporting force, but it becomes a force to be carried by the pile and must be so considered in the design of the pile. Shaft resistance acting downward on the pile is known as *negative skin friction*.

Figure 33.8 shows two classic situations wherein negative skin friction can develop. In Fig. 33.8*a* fill overlies soft soil and a pile goes through both the fill and soft soil to firm soil below. This situation can arise either from the placement of fill around a pile already driven through the soft soil into the firm soil, or, as is the more common situation, where piles are driven through the profile consisting of fill, soft soil, and firm soil. The fill causes the soft soil to compress and thus

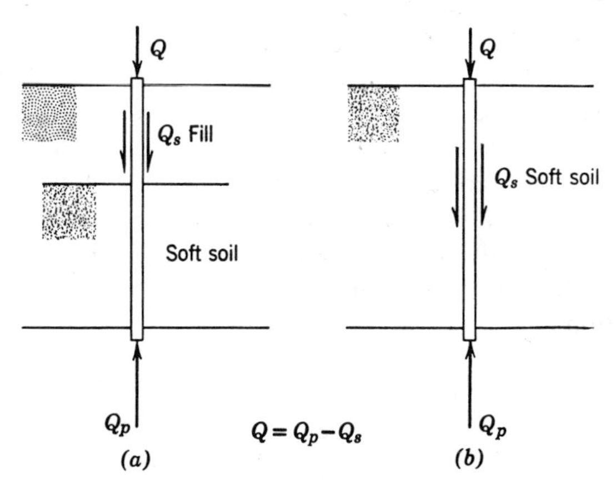

Fig. 33.8 Negative skin friction.

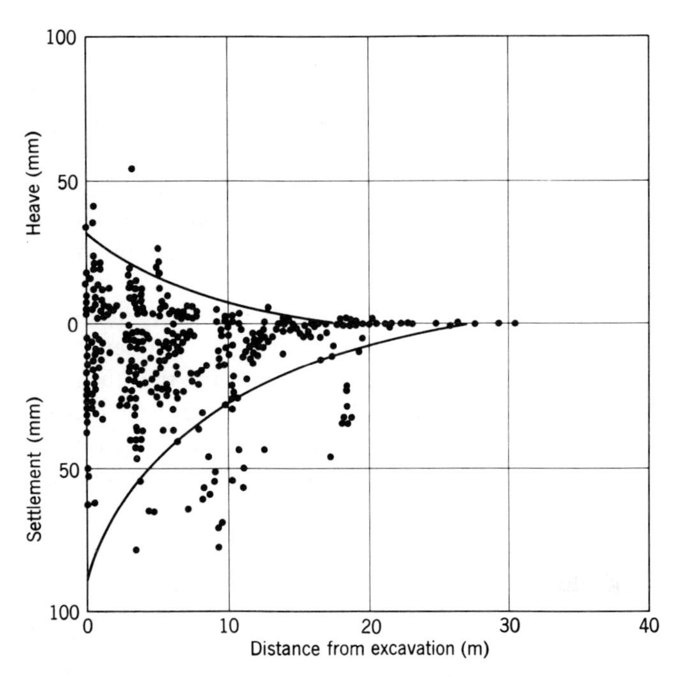

Fig. 33.9 Movements of buildings due to pile driving, Chicago, Ill. (From Horn, 1966). Reference: Ireland (1955).

both the fill and most of the soft soil move downward and drag on the pile. Whereas the placement of fill over soft soil can result in settlements of many centimeters, only about 25 mm of relative displacement between the pile shaft and the surrounding soil is needed to mobilize fully the skin friction of the soil on the pile.

In Fig. 33.8*b* a pile has been driven through soft soil. The disturbance of the soil by the pile can cause a great increase in the compressibility of the soil and develop high excess pore pressures in the soil around the pile shaft. Settlement of the soft soil can thus result.

Settlement of soft soil around a pile can also be caused by construction on nearby sites. Pile driving from an adjacent site can cause an increase in excess pore pressure; dewatering at even considerable distances from the pile can cause an increase in effective stress in the soft soil resulting in settlement.

While negative skin friction from the situation shown in Fig. 33.8*b* is more common than that in Fig. 33.8*a*, it is nowhere near as serious. The situation in Fig. 33.8*a* is particularly serious because the placement of fill over soft soil usually involves large settlements and because fill is frequently granular material with very high strength properties, and thus high capacity for negative skin friction.

There are many cases where piles in situations like those shown in Fig. 33.8 have been actually pulled away from the structure the piles were intended to support. Negative skin friction large enough to force piles into firm bearing soil at the tip can easily be mobilized. The soil engineer should be very reluctant to drive piles through a fill freshly placed over soft soil.

In designing a pile foundation the engineer must give consideration to possible negative skin friction and for those situations where it is expected, it should be allowed for in the design of the pile foundation. Van Weele (1964) reports that negative skin friction is a very common and widespread problem in Holland where there are many areas in which piles have been driven through fill placed on soft ground. Tests and actual experiences in Holland have emphasized the importance of including negative skin friction as part of the design load.

Johannessen and Bjerrum (1965) describe in detail an interesting field test involving negative skin friction. Two hollow steel piles approximately 47 cm in diameter and 55 m long were driven through a thick deposit of marine clay, and then 10 m of fill was placed on top of the clay. One of the piles was instrumented in such a way that the movement at five locations along the length of the pile could be measured. The surface of the fill settled about 1.2 m due to consolidation of the clay, and the top of the pile gradually moved down to a total shortening of 14.3 mm. It was concluded that the stresses in the pile near the tip were of the order of 200 MN/m² and that the total negative skin friction was about 2230 kN, a value high enough to force the pile point into the bedrock. The distribution of pile compression indicated that the developed adhesion between the pile and the clay was distributed approximately the same as the effective vertical stresses in the clay. Johannessen and Bjerrum postulated that the soil adhesion at any point along the pile was

$$s_s = \bar{\sigma}_h \tan \bar{\phi} = \bar{\sigma}_v K \tan \bar{\phi}$$

and, backfigured $K \tan \bar{\phi}$ as 0.20.

33.6 INFLUENCE OF DEEP FOUNDATION CONSTRUCTION ON ADJACENT STRUCTURES

Although practicing soil engineers well know that deep foundation construction frequently has detrimental effects on adjacent structures, very few cases are described in the literature. The lack of written case histories is due, at least in part, to possible legal action seeking damages if such cause and effect were acknowledged. Another contributing factor to the scarcity of written case histories is the unfortunate lack of responsibility of the designing soil engineer for the construction of the project he has designed. In deep foundation work the designing soil engineer usually severs his connection with the project following the design, leaving the contractor with the problem of construction. The sharp division between designing engineer and contractor in the United States fosters this unfortunate isolation of the

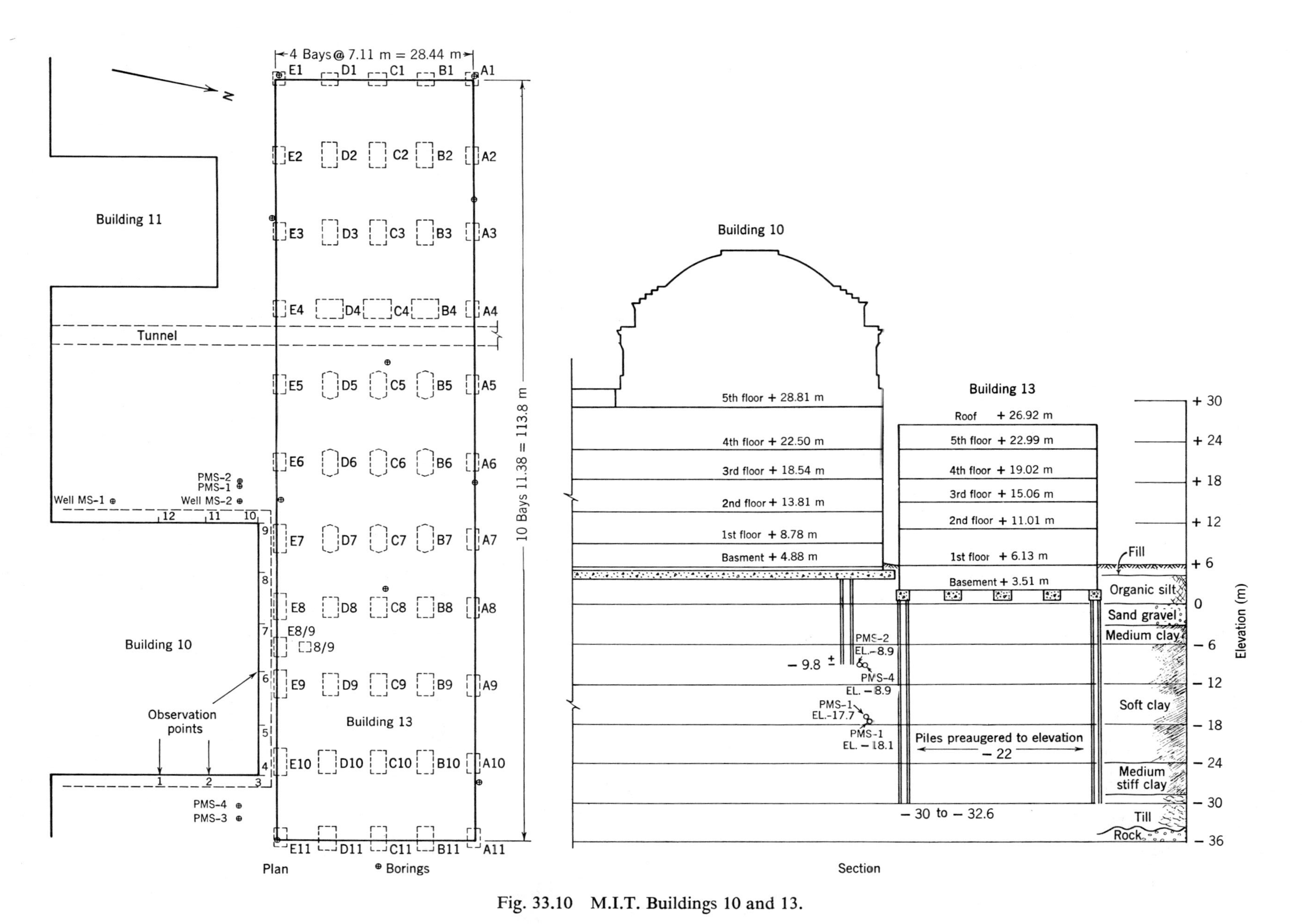

Fig. 33.10 M.I.T. Buildings 10 and 13.

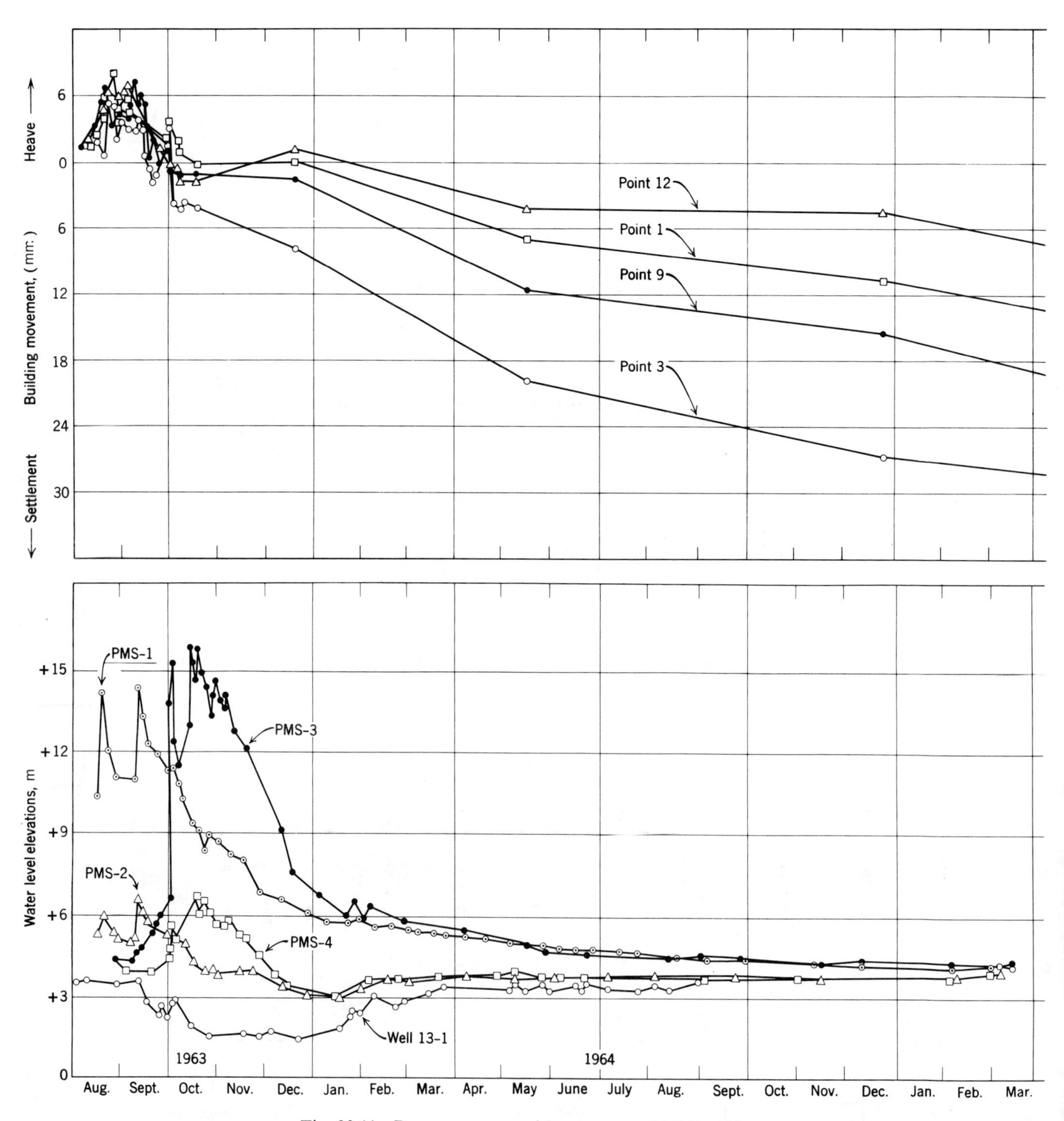

Fig. 33.11 Pore pressures and movements—Building 10.

designing engineer. The designer should give much more thought to the construction of a deep foundation—both from the view of obtaining a good and economical foundation for his client and from the view of not having undesirable effects on adjacent structures.

Pile driving can cause significant movements in nearby structures because of the displacement of soil and because of the high pore pressures developed in clay subsoils. This is particularly true where a large number of long displacement piles are driven in a clay foundation. Horn (1966) describes several case histories including one where piles driven in cohesionless soil caused settlements as large as 150 mm within the pile-driving area and ground settlements as far as 25 m away from the

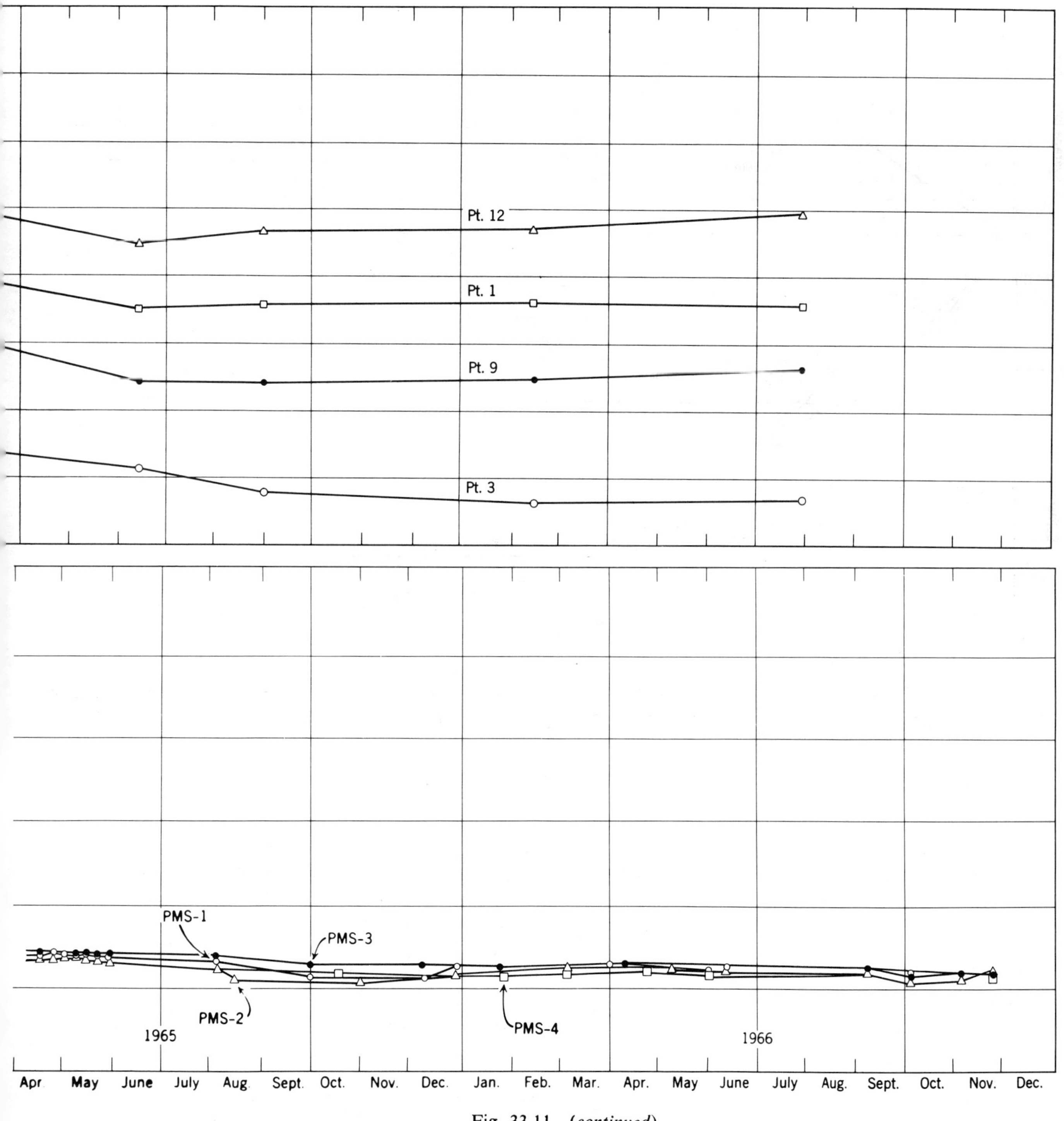

Fig. 33.11 *(continued)*

site. Horn also reports a study by Ireland which suggests that driving piles into clay can cause structure movements for a distance approximately equal to the length of the piles driven. Figure 33.9 shows Ireland's data on a number of buildings in the Chicago area.

A thorough program of foundation evaluation carried out on the M.I.T. campus has emphasized the extent and importance of the influence of deep foundation construction on adjacent structures. For example, periodic measurements of the phreatic surface in 45 observation wells on the campus showed that dewatering for foundation construction lowered the water table over a very wide area. In fact, the Student Center dewatering lowered the phreatic surface over an area extending

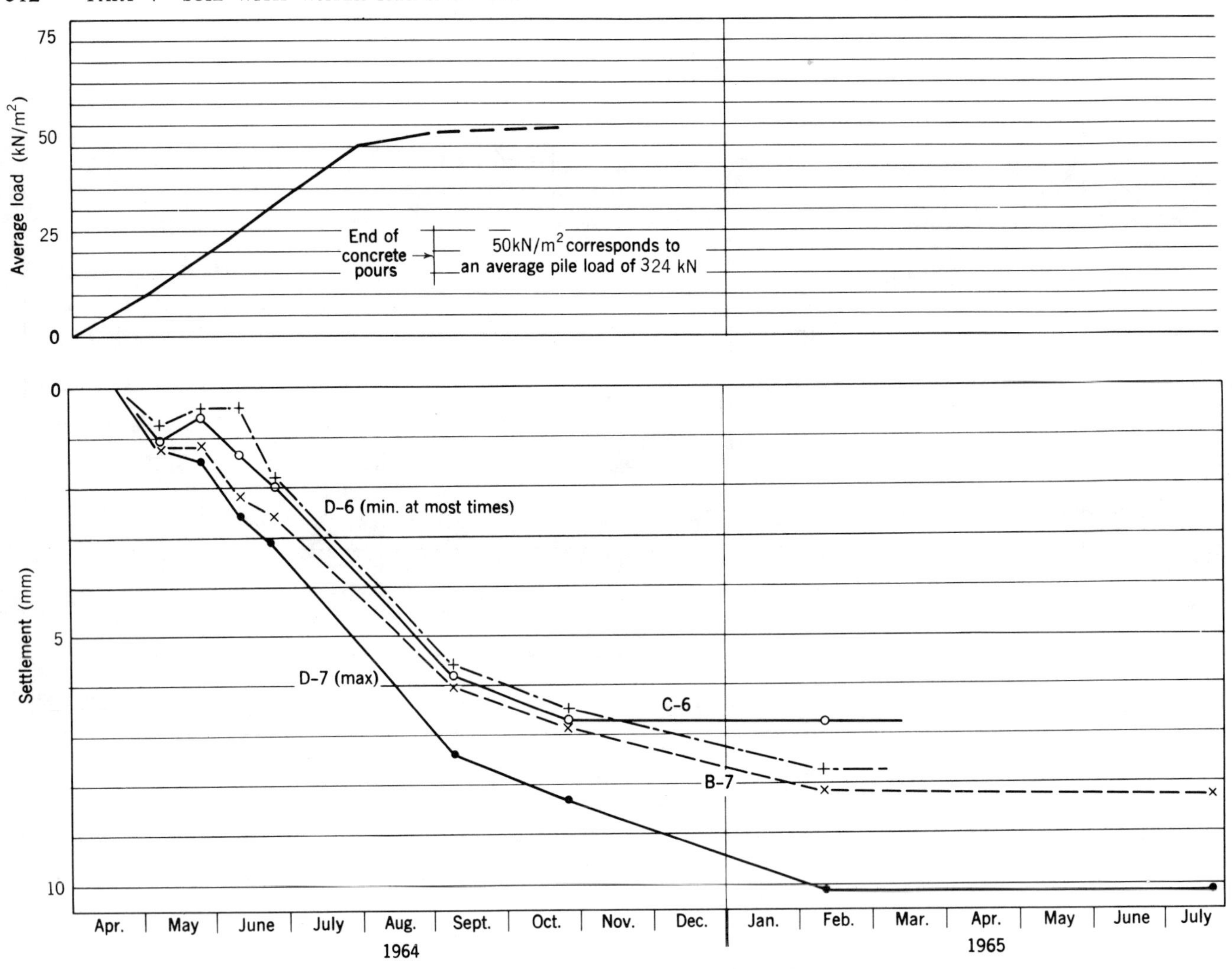

Fig. 33.12 Column settlement—Building 13.

550 m away from the construction site. Piezometer readings indicated that pile driving could cause an increase in pore pressure in clay subsoils as far as 30 m away from the location where the pile was being driven. It was only within 6 m of the pile-driving operation, however, that the pore pressure increases were significant.

Lambe and Horn (1965) describe a study which showed the influence on Building 10 of construction of the nearby Building 13 on the campus. Figure 33.10 shows the two buildings in both plan and cross section. Building 13 rests on 619 piles of the type shown in Fig. 33.7. The design load of each pile was 623 kN. Figure 33.11 shows the pore pressures developed under Building 10 and the vertical movements that occurred at points on Building 10. As can be seen, excess pore pressures of about 12 m head were developed by the pile driving. These excess pore pressures dissipated rapidly. During the driving, Building 10 rose about 6 mm and then settled as the excess pore pressures in the foundation clay dissipated. As can be seen, settlements continued to occur even after the excess pore pressures in the clay were essentially zero. The maximum settlement occurred at point 8 and was slightly in excess of 30 mm.

The foundation study on the M.I.T. campus has shown that generally foundations consisting of long point-bearing piles encounter more construction difficulties than those involving partial or full flotation. Further, the pile foundation construction has more influence on adjacent structures than does shallow foundation construction. On the other hand, smaller settlements are encountered with buildings resting on point-bearing piles than those having floating foundations. The difference in performance between deep foundations and floating foundations, however, is not significant. For example, Fig. 33.12 indicates that the settlement of the pile foundation for Building 13 is between 7 and 10 mm. The maximum settlement measured on the M.I.T. Student Center, a partial flotation foundation, was 15 mm during construction and 5 mm further settlement during the first 2 years after completion of the building.

33.7 SUMMARY OF MAIN POINTS

1. *Deep foundations* are used to carry structure loads through soft, weak soil to firm support. A foundation of *piles* is the most common type of deep foundation.
2. The pile capacity Q is normally carried by *point resistance* Q_p plus *shaft resistance* (also termed *skin resistance* Q_s). Equations 33.3 and 33.4 give pile capacity for cohesionless and cohesive soils, respectively.
3. In a situation where the soil surrounding the pile shaft moves downward with respect to the pile, the shaft resistance in that region acts downward. The downward shaft resistance is called *negative skin friction* and must be considered as a pile load in design.
4. The soil strength which is effective in generating support to a pile is the strength at the time the support is desired. Since soil strength depends on the effective stress and on the strength parameters, pile support, especially in clay, can be highly time dependent.
5. A pile is usually forced into the ground by a *pile driver*. Tables 33.2 and 33.3 list many pile drivers along with their pertinent characteristics.
6. *Dynamic pile formulas*, such as Eqs. 33.6 and 33.7, are often used to estimate the allowable load on a single pile. Dynamic pile formulas employ the rated energy of the pile driver, characteristics of the pile, and the measured penetration for the final part of pile driving to compute the pile-driving resistance, which is then used to estimate the pile capacity under a static load.
7. A dynamic formula is generally a highly unreliable method to determine the pile capacity under a static load because of: (*a*) the difficulty in evaluating correctly the energy lost during pile driving; and (*b*) the difficulty in relating pile resistance during driving to pile capacity under a static load.
8. Pile foundations enjoy a reputation among many engineers as the ideal solution to construction on soft ground. Facts do not support this reputation. There are difficulties in constructing pile foundations and the construction of a pile foundation can have detrimental effects on nearby structures.
9. This chapter identifies and treats the underlying soil mechanics fundamentals of deep foundations. It does not consider many practical aspects of deep foundations—such as uplift and lateral loads on piles and the influence of pile type on capacity. The references noted in this chapter cover these topics.

CHAPTER 34

The Improvement of Soil

Usually the soil at a site to be developed is not ideal from the viewpoint of soil engineering. In some cases, the engineer can *avoid* potential soil problems by choosing another site or by removing the undesirable soil and replacing it with desirable soil. In the early days of highway construction this procedure was widely employed; e.g., highways were routed around swamps. As time went on, the decision to avoid bad soils was made less frequently. The increase in speed of vehicles forced stricter alignment standards on tracks, highways, and runways. With the growth of cities and industrial areas the supply of sites with good foundation conditions became depleted. Increasingly the soil engineer has been forced to construct at sites selected for reasons other than soil conditions.

A second approach to the problem of bad soils is to *adapt* the design for the conditions at hand. For example, floating foundations and deep foundations can be designed to avoid many of the settlement and stability problems associated with soft foundation soils.

A third approach available to the soil engineer is to *improve* the soils. This approach is becoming more feasible and more attractive. Soil improvement is frequently termed *soil stabilization*, which in its broadest sense is the alteration of any property of a soil to improve its engineering performance. Examples of soil improvement are: increased strength (as for a pavement subgrade), reduced compressibility (as for the foundation of a structure), and reduced permeability (as for the foundation of a dam). Soil improvement may be a temporary measure to permit the construction of a facility, or it may be a permanent measure to improve the performance of the completed facility.

Soil improvement techniques can be classified in various ways: according to the nature of the process involved, the material added, the desired result, etc. For example, on the basis of process, we have mechanical stabilization, chemical stabilization, thermal stabilization, and electrical stabilization. The many techniques of soil improvement are discussed by Lambe (1962). A great many empirical data on soil improvement have been obtained from extensive field experience. Sherard et al. (1963) treat soil improvement for dams; the Road Research Laboratory (1952) treats soil improvement for roads and airfields; Fruco and Associates (1966) treat soil improvement for deep excavations; Leonards (1962) and the ASCE (1964) treat compaction, dewatering, and preloading. The ASCE Specialty Conference (1968) was limited to the placement and improvement of soil for foundations.

The most common and important method of soil improvement is *densification*. Three methods of densification are considered in this chapter: (*a*) *compaction* (densification with mechanical equipment, usually a roller); (*b*) *preloading* (densification by placing a temporary load); and (*c*) *dewatering* (removal of pore water and/or reduction of pore pressure). These techniques (as well as others) may be used alone or in combination with each other.

This chapter presents the soil mechanics features of densification. The wealth of available information concerning equipment and techniques for soil improvement and the mass of empirical data are beyond the scope of this text. Any engineer designing a soil improvement scheme would do well to study the available information, however, in order to appreciate the many practical problems involved in this type of work.

34.1 FIELD COMPACTION

An existing deposit of soil can be rolled with compactors to densify it. The compaction of in-place soils is usually limited to the top 0.3 m or so of a subgrade prior to placement of fill or to the compaction of sand. Sands can sometimes be densified with rollers for a depth of 1 to 2 m. Most compaction, however, is done on soil freshly placed in layers.

The field compaction process can include any or all of

the following steps:

1. Select borrow soil.
2. Load soil from borrow pit, haul it to the site, and dump it.
3. Spread the dumped soil into layers; the thickness of the layers may vary from a few centimeters to perhaps 0.6 m depending on the soil type and the compaction equipment.
4. Alter the moisture content of the placed soil: lower it by partial drying or raise it by the addition of water.
5. Mix the dumped soil to make it more nearly uniform and to break up lumps.
6. Roll the soil either according to a specified procedure or until specified properties are obtained.

The details of the compaction process and the equipment used for each step are tailored for the particular job at hand.

During the first half of the twentieth century spectacular developments were made in the size and variety of field compaction equipment. The weight of available compaction equipment grew from approximately 22 to 1800 kN.

The smooth-wheel roller, the rubber-tired roller, the sheeps-foot roller, and the vibratory roller are the principal types of compaction equipment. In cohesive soils, high densities can be obtained with most types of roller; however, the vibratory rollers are the least effective and the rubber-tired rollers with high tire pressures (up to 1 MN/m^2) are usually the most effective. In cohesionless soils, both the vibrating rollers and the rubber-tired rollers are effective in obtaining high densities. (See Foster, 1962, for a treatment of field compaction.)

The control of field compaction by soil technicians is very important in order to obtain the desired soil properties and especially in order to obtain a reasonably uniform material. Depending on the situation, the technician may make measurements of density, water content, and classification characteristics at some given rate, usually based on so many tests per volume of fill placed, e.g., one set of field tests per 4000 m^3 of fill placed. Field control may also be based on in-place strength or some other engineering property.

34.2 COMPACTION TESTS

The soil engineer must select the details of the compaction process to give the optimum combination of engineering properties desired for the problem at hand at the least cost. In order to make this selection, he needs to know the relationships between soil behavior and placement details for his particular soil. This information comes from compaction principles (such as presented in this chapter), laboratory tests, and field tests.

Laboratory compaction tests are run primarily because they are very much cheaper and quicker to perform than field compaction tests. There are many types of laboratory tests, each selected with the intention of duplicating some type of field compaction. The earliest and most common type of compaction test consists of placing soil in a mold and then dropping a hammer on the soil a specified number of times. This type of test is frequently termed a *dynamic* compaction test. In the *kneading* compaction test the soil in the mold is pushed a specified number of times with a tamper at a specified stress. In the *static* compaction test the soil is subjected to a static stress of a given magnitude. The details of various laboratory compaction tests are given by Lambe (1951).

If a cohesive soil is compacted with a given type and amount of compactive effort at various water contents, a compaction curve such as the one shown in Fig. 34.1 is obtained. This compaction curve shows that as the molding water content is increased, the dry density[1] increases to a peak and then decreases. The density and moisture content at peak density are, respectively, *maximum dry density* and *optimum moisture content* for that particular type of compaction and compactive effort. For the Standard Proctor compaction test shown in Fig. 34.1, the maximum dry density (unit weight) is 18.7 kN/m^3 and the optimum moisture content is 11%.

The computed relationship between water content and dry density at a constant degree of saturation may also be plotted on the same scale as the compaction curve. As can be seen in Fig. 34.1, the degree of saturation increases with increasing water content to a value somewhat above that at optimum moisture content and then tends to remain approximately constant.

The moisture-density relationship for a particular soil depends on the amount and type of compaction, as illustrated by Figs. 34.2 and 34.3. Figure 34.2 shows the results of four laboratory tests using dynamic compaction. The compactive effort was decreased from test 1 through test 4. As the data illustrate, for a given type of compaction the higher the compactive effort the higher the maximum density and the lower the optimum water content. Further, as the molding water content increases, the influence of compactive effort on density tends to decrease. The points of maximum dry density and optimum water content for the various compactive efforts tend to fall along a line that goes in the same general direction as the lines of constant degree of saturation.

Figure 34.3 shows the results of static compaction in

[1] The term "dry density" is commonly used in compaction literature for dry unit weight.

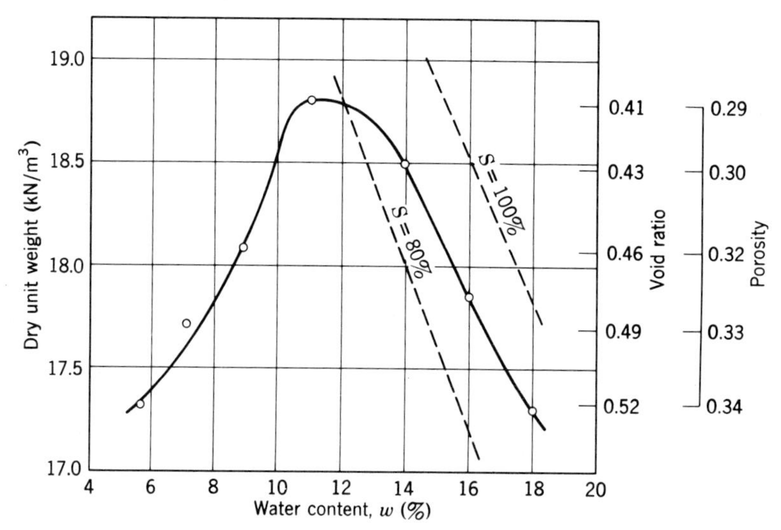

Fig. 34.1 Standard Proctor compaction test. (From Lambe, 1951).

which the compacting stress is decreased going from test 1 toward test 4. As illustrated in this figure, the higher the compactive stress the higher is the maximum density.

Figure 34.4 presents a comparison of field and laboratory compaction on the same soil. The figure illustrates the difficulty of choosing a laboratory test that reproduces a given field compaction procedure.

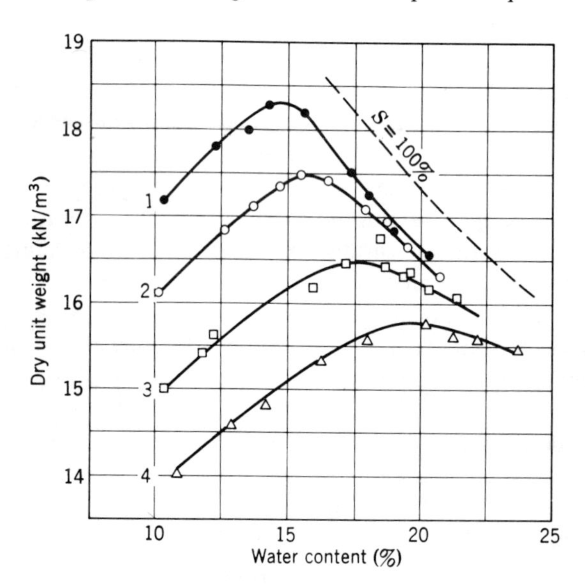

No.	Layers	Blows per Layer	Hammer Mass	Hammer Drop	
1	5	55	4.54 kg	457 mm	(mod. AASHO)
2	5	26	4.54	457	
3	5	12	4.54	457	(std. AASHO)
4	3	25	2.50	305	

Note. 150 mm diameter mold used for all tests.

Fig. 34.2 Dynamic compaction curves for a silty clay. (From Turnbull, 1950).

The laboratory curves generally yield a somewhat lower optimum water content than the actual field optimum.

By varying the laboratory procedure, the moisture-density relation can be shifted to achieve a better correlation with a particular field compaction procedure. There is some evidence that specific types of laboratory compaction correlate better with certain types of field compaction; e.g., kneading compaction and sheepsfoot rollers. Nevertheless, the majority of field compaction is controlled by dynamic laboratory tests.

As discussed in Chapter 15, vibrations can be very effective in compacting cohesionless soils. On the other

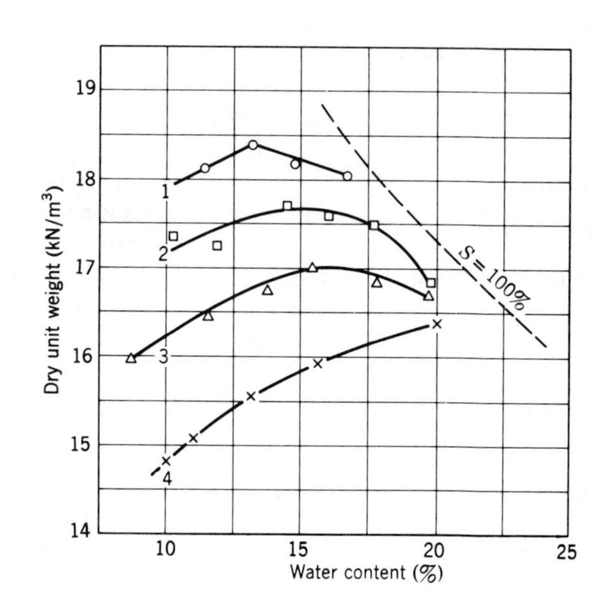

Fig. 34.3 Static compaction curves for a silty clay. (1) 13.8 MN/m² static load. (2) 6.9 MN/m² static load. (3) 3.45 MN/m² static load. (4) 1.38 MN/m² static load. *Note.* Compaction on top of soil sample. (From Turnbull, 1950.)

hand, cohesionless soils do not respond to variations in compacting moisture content and compactive effort in the manner characteristic of fine-grained soils. Figure 34.5 shows the typical compaction curve for cohesionless soils. The low density that is obtained at low water contents is due to capillary forces resisting rearrangements of the sand grains. This phenomenon is known as *bulking*. It is general practice to measure the density of a compacted cohesionless soil in terms of *relative density* (defined in Chapter 3) as is done with natural cohesionless soils.

34.3 EFFECT OF COMPACTION ON SOIL STRUCTURE

Figure 34.6 suggests the effects of compaction on soil structure. For a given compactive effort and dry density, the soil tends to be more flocculated for compaction on the dry side as compared to compaction on the wet side. In other words, the soil at point A is more flocculated than the soil at point C. For a given molding water content, increasing the compactive effort tends to disperse the soil, especially on the dry side of optimum—point A versus point E—and to some extent on the wet side of optimum—point C versus point D.

The soil structures illustrated in Fig. 34.6 follow from the principles given in Chapter 5. Increasing the moisture content tends to increase the interparticle repulsions, thereby permitting a more orderly arrangement of the soil particles to be obtained with a given amount of effort. Increasing the compactive effort at a given moisture content tends to work the particles into a more nearly parallel arrangement.

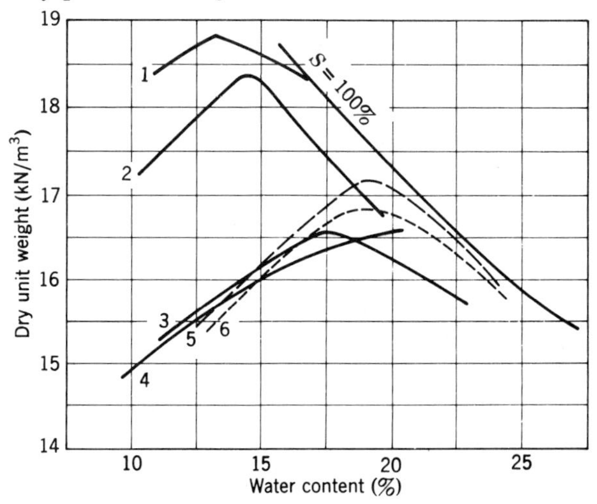

Fig. 34.4 Comparison of field and laboratory compaction. (1) Laboratory static compaction, 13.8 MN/m². (2) Modified AASHO. (3) Standard AASHO. (4) Laboratory static compaction 1.38 MN/m². (5) Field compaction, rubber-tired load, 6 coverages. (6) Field compaction, sheepsfoot roller, 6 passes. *Note.* Static compaction, from top and bottom of soil sample. (From Turnbull, 1950.)

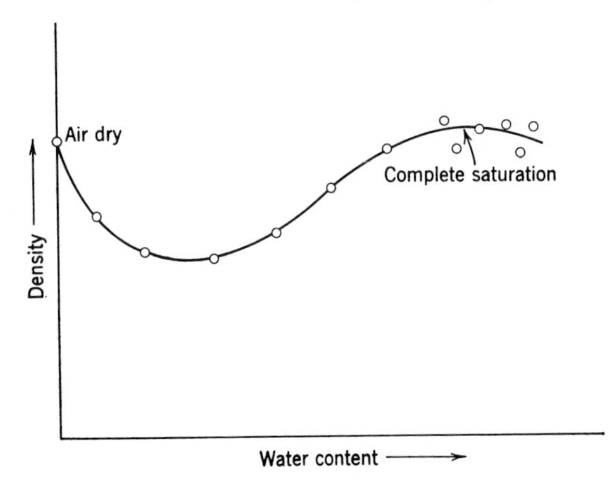

Fig. 34.5 Typical compaction curve for cohesionless sands and sandy gravels. (From Foster, 1962.)

34.4 EFFECT OF COMPACTION ON SOIL STRESSES

To initiate our consideration of the changes in stress within a soil caused by compaction, let us review what a static load does to soil. Figure 28.1 shows the effective stress paths, the total stress paths, and the paths for total stress minus static pore pressure for a load-unload cycle in the oedometer. We see from Fig. 28.1 that the application of a static load to a confined soil sample causes a positive excess pore pressure and that the removal of a static load causes a negative excess pore pressure. Chapter 27 considered the rate at which such excess pore pressures dissipate. Following the load-unload cycle shown in Fig. 28.1 (increase $\bar{\sigma}_v$ from 400 to 800 kN/m², then reduce back to 400 kN/m²), we see that the horizontal effective stress has been increased from 200 kN/m² to 320 kN/m² and that the ratio of horizontal to vertical effective stress, K, has increased from 0.5 to 0.8. The application and removal of static load to a confined soil sample increases the lateral effective stress.

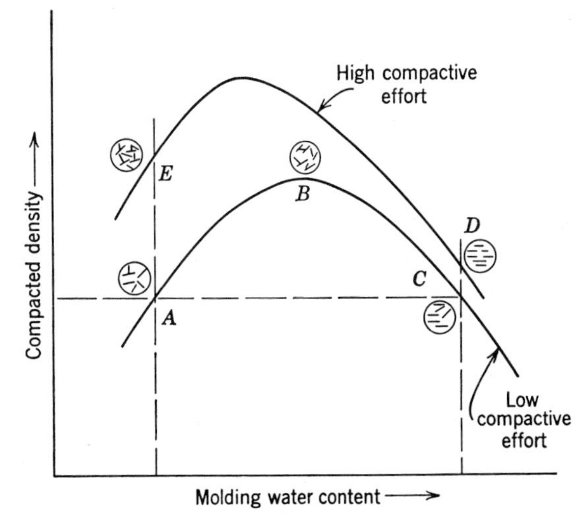

Fig. 34.6 Effects of compaction on structure. (From Lambe, 1962).

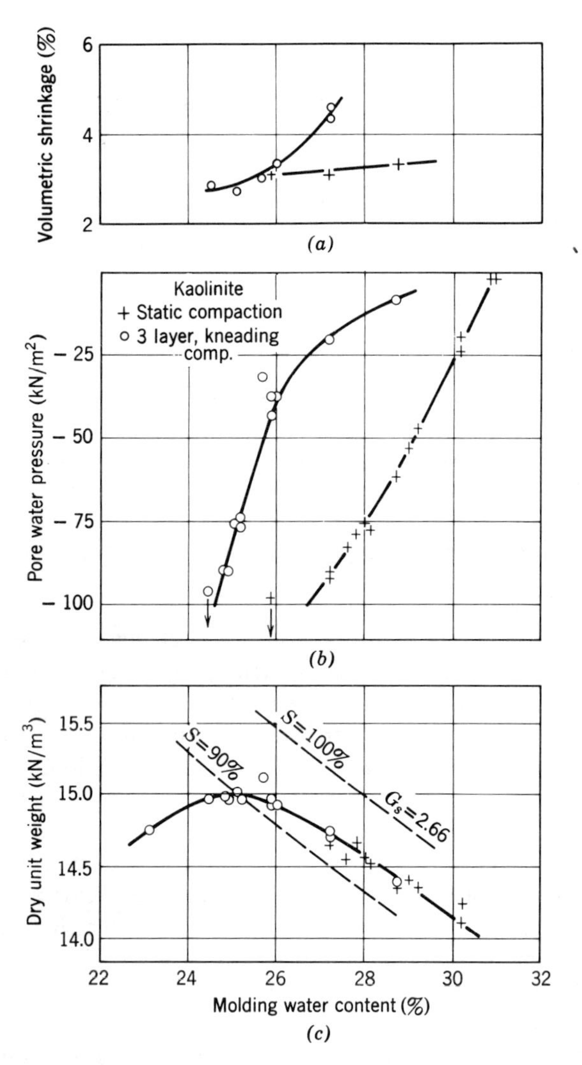

Fig. 34.7 Pore pressures in compacted kaolinite. (From Lambe, 1961).

Any fine-grained soil with a given structure has an equilibrium water content, meaning that the soil at this structure would pick up this amount of water if it were freely available. In general, the molding water content of a fine-grained soil at optimum water content or dryer is less than the equilibrium water content; therefore a water deficiency exists. If water is not available to satisfy this deficiency, capillary menisci and pore water tensions develop. A soil can thus be compacted with a dynamic load or with a static load, and if free water is not available, negative pore water pressures will develop.

Figure 34.7 presents experimental compaction and pore water pressure data on kaolinite. The static compaction and kneading compaction tests were chosen to give approximately the same water content-dry density curve, shown in Fig. 34.7*c*. Plotted above the density-water content curve are the measured pore water pressures obtained at each of the points on the curve of density versus moisture content. From the data presented in this figure two observations can be made: (*a*) at a given density and molding water content the pore water tensions in the sample compacted by a static effort are greater than those in the sample compacted with a kneading effort and (*b*) the pore water tensions get smaller as the molding water content is increased.

Plotted in Fig. 34.7*a* are values of the shrinkage of the compacted samples upon drying. The magnitude of shrinkage upon drying is generally greater the more nearly parallel are the soil particles. These shrinkage data indicate that kneading compaction gives a more dispersed structure than does static compaction.

In summary, compaction, either static or dynamic, can cause a significant change in the total stresses and pore pressures within the compacted soil. The nature and magnitude of these stresses depend on the soil and the amount of compaction imparted to the soil. In general, compaction increases the lateral effective stress.

34.5 EFFECT OF COMPACTION ON ENGINEERING BEHAVIOR

The nature and magnitude of compaction in a fine-grained soil has a significant influence on the engineering behavior of the compacted soil. This important influence is illustrated by Figs. 34.8 to 34.10 and Table 34.1.

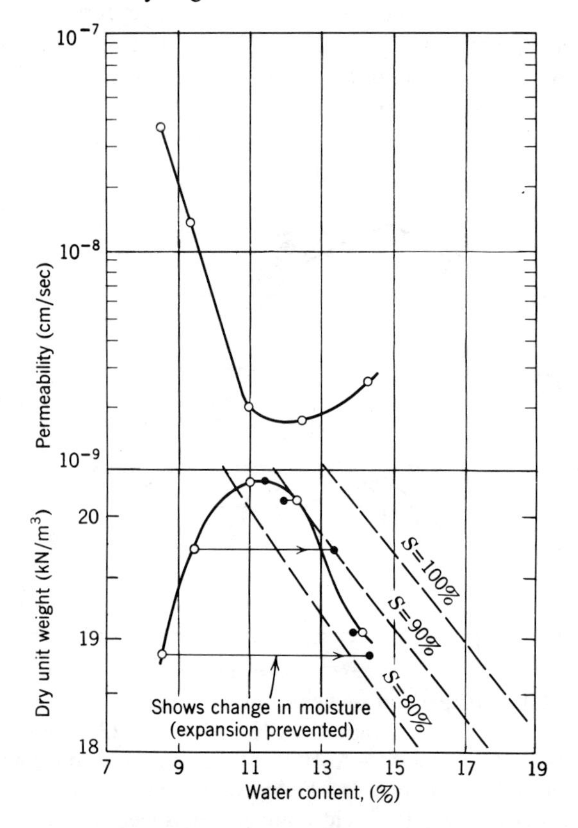

Fig. 34.8 Compaction-permeability tests on Siburua clay (From Lambe, 1962).

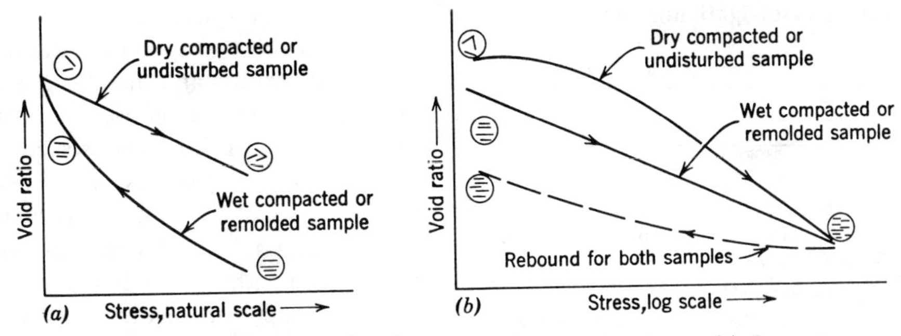

Fig. 34.9 Effect of one-dimensional compression on structure. (*a*) Low-stress consolidation. (*b*) High-stress consolidation. (From Lambe, 1958.)

As suggested by Fig. 34.8, increasing the molding water content results in a decrease in permeability on the dry side of optimum moisture content and a slight increase in permeability on the wet side of optimum. Increasing the compactive effort reduces the permeability since it both increases the dry density, thereby reducing the voids available for flow, and increases the orientation of particles.

Figure 34.9 illustrates the difference in compaction characteristics between two saturated clay samples at the same density, one compacted on the dry side of optimum and one compacted on the wet side. At low stresses the sample compacted on the wet side is more compressible than the one compacted on the dry side. On the other hand, at high applied stresses the sample compacted on the dry side is more compressible than the sample compacted on the wet side.

The test data obtained by Seed and Chan (1959) plotted in Fig. 34.10 show the influence of molding water content on both structure and stress-strain relationships for compacted samples of kaolinite. Samples compacted dry of optimum tend to be more rigid and stronger than samples compacted wet of optimum. Shear strains, by aligning soil particles, tend to destroy some of the differences in structure built up during compaction.

The engineer must consider not only the behavior of the soil as compacted but the behavior of the soil in the completed structure, especially at the time when the

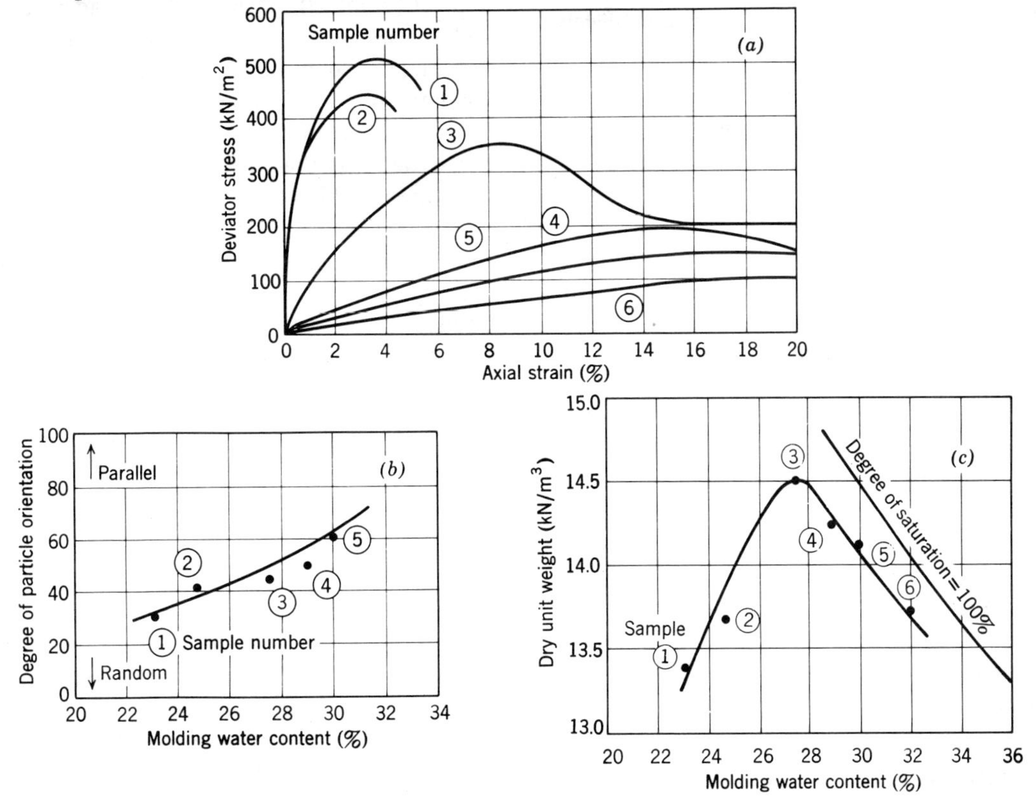

Fig. 34.10 Influence of molding water content on structure and stress-strain relationship for compacted samples of kaolinite. (*a*) Stress versus strain relationships for compacted samples. (*b*) Degree of particle orientation versus water content. (*c*) Dry density versus water content. (From Seed and Chan, 1959.)

Table 34.1 Comparison Dry-of-Optimum with Wet-of Optimum Compaction

Property	Comparison
Structure	
Particle arrangement	Dry side more random
Water deficiency	Dry side more deficiency, therefore more water imbibed, more swell, lower pore pressure
Permanence	Dry-side structure more sensitive to change
Permeability	
Magnitude	Dry side more permeable
Permanence	Dry side permeability reduced much more by permeation
Compressibility	
Magnitude	Wet side more compressible in low-stress range, dry side in high-stress range
Rate	Dry side consolidates more rapidly
Strength	
As molded	
Undrained	Dry side much higher
Drained	Dry side somewhat higher
After saturation	
Undrained	Dry side somewhat higher if swelling prevented; wet side can be higher if swelling permitted
Drained	Dry side about the same or slightly greater
Pore-water pressure at failure	Wet side higher
Stress-strain modulus	Dry side much greater
Sensitivity	Dry side more apt to be sensitive

stability or deformation of the structure is most critical. Chapter 7 noted some of the many changes that may occur during the life of a natural soil. Similarly, there are many changes which can occur in a compacted soil. For example, consider an element of compacted soil in a dam core. As the height of the dam increases, the total stresses on the soil element increase. When the dam is performing its intended function of retaining water, the percent saturation of the compacted soil element is increased by the permeating water. Thus the engineer designing the earth dam must consider not only the strength and compressibility of the soil element as compacted, but also its properties after it has been subjected to increased total stresses and saturated by permeating water.

34.6 PRELOADING

Preloading involves the placement of a surface load prior to construction in order to *precompress* the foundation soil. Consider, for example, the situation shown in Fig. 34.11. A warehouse is to be constructed over a stratum of soil that is so compressible and so weak that large settlements and maybe a shear failure can be expected if the warehouse is constructed on the unimproved soil. Prior to construction of the warehouse, a pile of soil (preload) is placed over the building site. Since the lateral extent of the preload is large relative to the thickness of the soft soil, one-dimensional strains may be assumed within the soft soil.

In the lower part of Fig. 34.11 are shown stresses and strains for the point P in the soft soil. Prior to preload placement the total stresses at point P are represented by J, the effective stresses by A, and the distance AJ is the static pore pressure at point P. The rate of preload placement relative to the rate of consolidation of the soft soil is such that no measurable excess pore pressures are developed in the soft soil—i.e., a drained loading occurs. JK is the total stress path for the loading; and KL is the total stress path for the unloading. The corresponding effective stress paths are AB and BD. Figure 34.11c shows the vertical strain plotted against q for the loading RS and unloading ST. Figure 34.11d shows vertical strain plotted against vertical effective stress.

The placement and removal of the preload has transformed the soft foundation soil from a normally consolidated deposit (point A) to an overconsolidated deposit (point D). Following preloading the foundation soil has all of the desirable characteristics of an overconsolidated deposit as compared to a normally consolidated one—it is less compressible and stronger. The bearing capacity of the soft soil has been increased and the settlements that will result from the construction of the warehouse have been significantly decreased.

Preloading is a powerful technique for the soil engineer. As can be inferred from the preceding discussion and from principles presented in this text, there are circumstances where preloading is most attractive. These include the following cases:

1. Soil (or other material) is readily available for use as preload.
2. The foundation soil drains rapidly so that the time required for preloading is relatively short. This requires a short drainage path and/or a high coefficient of consolidation.

On some occasions a preload greater than the structure load is used. This situation is termed *surcharging* or *overloading*. The excess of the preload over the actual structure load is termed *surcharge* or *overload*. Use of a surcharge reduces the time required for the foundation soil to consolidate to the actual structure load. In addition, if the soil is consolidated to a higher effective stress than will be imposed by the structure, the amount of secondary compression that will occur under the structure load can be greatly reduced.

34.7 DEWATERING

Dewatering is a technique of soil improvement whereby the amount and/or pressure of pore water is reduced. Dewatering usually causes densification. Many parts of this book have discussed the deleterious effects that water can have on soil and earth retaining structures. Upward-flowing water can cause a quick condition; an increase in pore water pressure for a given total stress will cause a reduction in the effective stress and thus soil strength; water can add a very significant lateral thrust to earth retaining structures like retaining walls.

In soil engineering it is frequently highly desirable and sometimes essential to remove pore water from the soil or at least to reduce the pressure of the pore water. Sometimes dewatering is done as a temporary measure to permit construction (such as for the basement of a building below the phreatic surface) and sometimes as a permanent measure to protect a structure (such as a drain under a dam). Mansur and Kaufman (1962) present many of the theoretical and practical aspects of dewatering.

There are many techniques of dewatering, including: (*a*) vertical drains (as used in embankments); (*b*) horizontal drains (as used to dewater natural slopes); (*c*) ditches (along a highway); and (*d*) well points (for an excavation). Dewatering can be assisted by the application of a direct electric current. This process is called *electro-osmosis*. Casagrande (1953) describes the successful use of electro-osmosis on several field projects. An illustration of the use of electro-osmosis for improving a dam foundation is given by Fetzer (1967).

Vertical sand drains are frequently used in conjunction with preloading to accelerate the consolidation of fine-grained soils. The mechanics of radial flow to drain wells was described in Chapter 27. One of the following soil conditions must generally exist in order to obtain significant acceleration of consolidation by use of vertical drains:

1. The soft soil layer is thick, giving a long vertical drainage path.
2. The horizontal permeability is many times larger than the vertical permeability.

The design of sand drain installations is beyond the scope of this book. For a review of sand drain theories,

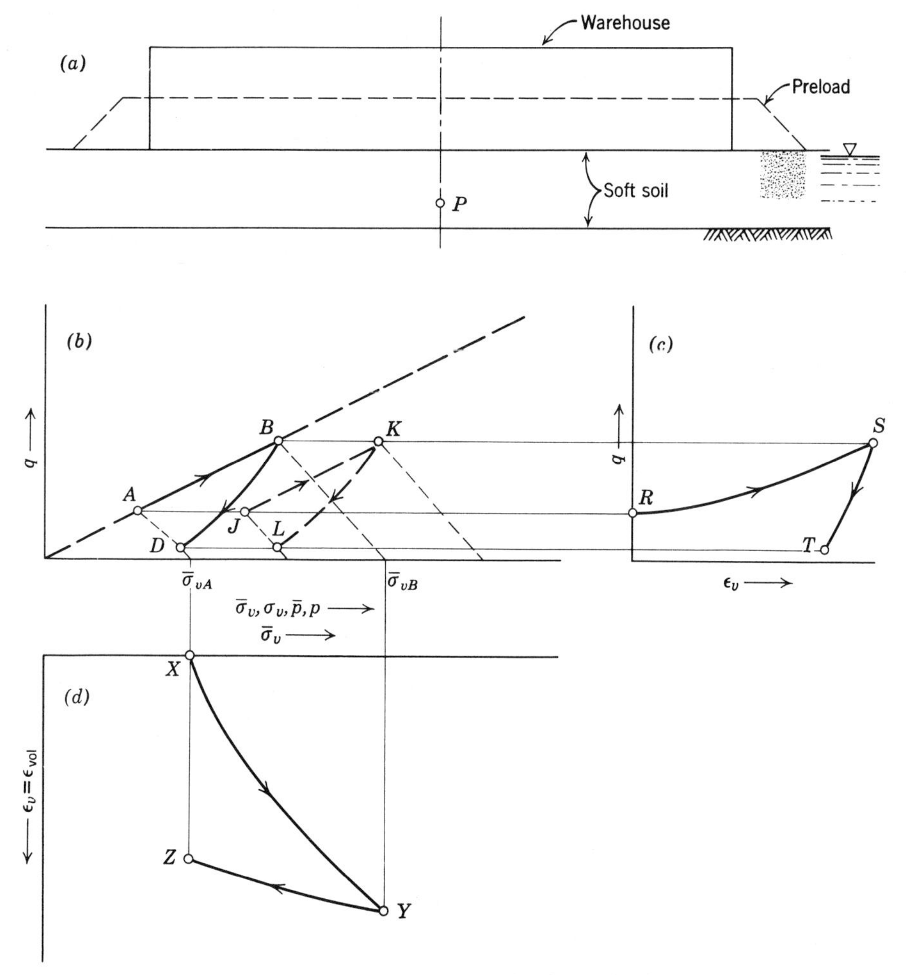

Fig. 34.11 Stresses and strains in preloaded soil.

see Richart (1959). There are many practical problems involved with vertical sand drains that the designer must consider. Moran, Proctor, Mueser, and Rutledge (1958) discuss these problems, describe existing design procedures, and analyze numerous field installations.

As illustrated in Example 25.8, lowering the phreatic surface can cause an increase in the effective stress within the soil and thus compress the soil. Frequently such dewatering is used in conjunction with preloading to improve the soil at a given site.

Dewatering can be a very useful and economical technique for improving soil; however, the soil engineer must examine the situation at hand, giving consideration to such factors as: (*a*) the probable effectiveness of the dewatering; (*b*) the amount of water that must be removed; (*c*) the time required for the dewatering; and (*d*) possible damage to nearby structures.

34.8 FIELD OBSERVATIONS

Even if a preloading or dewatering scheme is designed using the best procedures available, its success is not guaranteed. Actual field conditions may vary markedly from those assumed in the design. The soil engineer frequently can compensate for these differences by altering the design, changing the time required, etc., if the actual field performance of the soil improvement scheme is measured. In order to measure the performance, field devices must generally be installed. These may include observation wells, piezometers, vertical movement indicators, and horizontal movement indicators.

These instruments could be used to answer such questions as:

1. What changes in effective stress resulted from well-point pumping (observation wells and piezometers)?
2. How long must a preload or surcharge be in place (piezometers and vertical movement indicators)?
3. How rapidly can preload be placed on a soft foundation without causing failure (piezometers, horizontal and vertical movement indicators)?

Field measurements also make it possible for the engineer to evaluate the effectiveness of a design, and thereby they lead to improved designs and design procedures for future jobs. Many types of instruments for making field measurements are described by Shannon et al. (1962) and the Bureau of Reclamation (1963).

34.9 SUMMARY OF MAIN POINTS

1. The situation frequently arises where the most economical solution to a soil problem is to *improve* the soil.
2. The most common and generally useful techniques for soil improvement are *densification* and *dewatering.*
3. Densification can be done as soil is placed or on soil *in situ.* An effective means of densifying soil is to roll it with compaction equipment, possibly aided by the addition of water. Vibrations are particularly effective in densifying cohesionless soils.
4. Compaction rearranges soil particles and moves them closer together, and it generally increases the ratio of horizontal effective stress to vertical effective stress.
5. Compaction normally increases soil strength and reduces its permeability and compressibility.
6. *Preloading*, a sort of static compaction, precompresses soil by means of an applied load.
7. Dewatering, the removal of water from a soil and the decrease in pore water pressure, is frequently an economical technique for improving soil or for reducing the loads acting on earth retaining structures.

PART VI *Appendices*

APPENDIX A

Symbols

ENGLISH

Symbol*	Represents	Reference
A		
A	Activity of a clay	Ch. 3
A	Area	Fig. 5.16
A	Attractive stress between particles	Ch. 5
A	Pore pressure parameter	Ch. 26
A'	Attractive force between particles	Ch. 5
A'	Seismic coefficient	Ch. 15
A_c	Area of contact	Ch. 6
A_f	Pore pressure parameter A at failure	Ch. 28
A_v	Area of voids	Ch. 17
A_0	Value of pore pressure parameter A for loading to K_0 loading	Ex. 29.2
Å	Angstrom units	Ch. 4, 5
a	Area	Ch. 19
a_m, a_w, a_a	Area ratios	Ch. 16
a_v	Coefficient of compressibility	Ch. 12, Eq. 12.10, Table 12.2
a_w	Coefficient in partly saturated soils	Ch. 28
B		
B	Bulk modulus	Eq. 12.6
B	Pore pressure parameter	Eq. 26.26
B	Water-plasticity ratio	Fig. 3.4
B	Width of footing	Ch. 14
C		
CAU	Consolidated anisotropic undrained	Ch. 30
CAU, $\overline{\text{CAU}}$	Consolidated undrained test with anistropic consolidation	
CD	Consolidated drained	Ch. 28
CIU, $\overline{\text{CIU}}$	Consolidated undrained test with isotropic consolidation	Ch. 28
CU	Consolidated undrained	Ch. 28
cm	Centimeters	
C	Compressibility	Ch. 26
C	Concentration	Fig. 5.13
C	Pore pressure parameter	Eq. 26.1
C_D	Dilatational wave velocity	Eq. 12.9, Ch. 30
C_L	Rod wave velocity	Eq. 12.9
C_S	Shear wave velocity	Eq. 12.9
C_c	Compression index	Ch. 12, Eq. 12.11, Table 12.2
C_α	Coefficient of secondary compression	Ch. 27
C_s	Swell index	Eq. 22.1
c	Cohesion	Fig. 7.7
$\bar{c}$	Cohesion intercept based on effective stresses	Fig. 7.10, Ch. 21
c_e	"True" cohesion, Hvorslev parameter	Ch. 21
c_u	Total stress strength parameter	Eq. 28.6
c_v	Coefficient of consolidation	Eq. 27.3
D		
D	Constrained modulus	Ch. 12 Table 12.2
D	Depth	Ch. 23
D	Diameter	Ch. 3
D	Pore pressure parameter	Eq. 26.3a
D_r	Relative density	Ch. 3
D_0	Smallest dimension of test footing	Ch. 14
D_{10}	Diameter at which 10% of the soil is finer	Ch. 3
D_{15}	Diameter at which 15% of the soil is finer	Ch. 19
D_{50}	Diameter at which 50% of the soil is finer	Ch. 19

* A bar over a stress denotes effective stress. A bar over a test indicates that pore pressures were measured during the test. A bar over E and μ, i.e., $\bar{E}$ and $\bar{\mu}$, indicates that the modulus and ratio are for the mineral skeleton.

Symbol	*Represents*	*Reference*
D_{60}	Diameter at which 60% of the soil is finer	Ch. 3
D_{85}	Diameter at which 85% of the soil is finer	Ch. 19
d	Depth of base of footing below ground surface	Ch. 14
d	Diameter	Ch. 6
d	½ distance between particles	Fig. 5.16
d_{90}, d_{100}	Dial reading at 90%, 100% consolidation	Ch. 27

E

Symbol	*Represents*	*Reference*
E	Energy if kilojoules imparted by pile driver	Ch. 33
E	Young's modulus	Ch. 12
$\bar{E}$	Effective normal force or slice	Ch. 24
$\bar{E}$	Young's modulus for soil skeleton	Chs. 27, 28
E_f	Secant modulus at failure	Ch. 30
E_i	Initial tangent modulus	Ch. 30
ESP	Effective Stress Path	Ch. 20
e	Eccentricity	Ch. 14
e	Void ratio	Fig. 3.1
e_{cv}	Void ratio at constant volume	Ch. 11
e_f	Void ratio at failure	Ch. 21
$e_{\max}$	Void ratio of soil in loosest condition	Ch. 3, p. 29
$e_{\min}$	Void ratio of soil in densest condition	Ch. 3, p. 29
e_0, e_i	Initial void ratio	Ch. 11

F

Symbol	*Represents*	*Reference*
F	Force	Ch. 13
F	Friction force	Fig. 5.16
F, FS	Factor of safety	Chs. 13, 24, 31
F_a	Air-mineral, or air-air contact force	Ch. 5
F_m	Mineral-mineral contact force	Ch. 5
F_w	Water-mineral or water-water contact force	Ch. 5
f	Coefficient of friction	Ch. 6
f	Factor coefficient for local shear	Eq. 25.6a
f	Frequency	Ch. 15
f_n	Undamped natural frequency	Ch. 15, Eq. 15.3

G

Symbol	*Represents*	*Reference*
G	Shear modulus	Ch. 12 Eqs. 12.4, 30.1
G_m	Specific gravity of total mass	Fig. 3.1
G, G_s	Specific gravity of solids	Fig. 3.1
G_w	Specific gravity of water	Fig. 3.1
g	grams	Ch. 5
g	Acceleration of gravity	Ch. 12

H

Symbol	*Represents*	*Reference*
H	Head	Ch. 18
H	Height	Ch. 13
H	Thickness of soil mass per drainage surface	Eq. 2.1
HP	Horsepower	Ch. 33
H_c	Vertical depth to failure plane	Eq. 24.2
h, h_t	Total head	Chs. 17, 18
h_c	Capillary head	Ch. 16
h_{cn}	Minimum capillary head	Ch. 16
h_{cr}	Capillary rise	Ch. 16
h_{cs}	Saturated capillary head	Ch. 16
h_{cx}	Maximum capillary head	Ch. 16
h_e	Elevation head	Ch. 17
h_p	Pressure head	Ch. 17

I

Symbol	*Represents*	*Reference*
I_f	Flow index	Fig. 3.4
I_l	Liquidity index	Fig. 3.4
I_p	Plasticity index	Fig. 3.4
I_t	Toughness index	Fig. 3.4
I_ρ	Influence coefficient	Eq. 14.14
I_0	Mass moment of inertia	Eq. 15.6b
i	Angle of slope with horizontal	Chs. 11, 13
i	Gradient	Ch. 17
i_c	Critical gradient	Eq. 17.5

J

Symbol	*Represents*	*Reference*
J	Seepage force	Ch. 17
j	Seepage force per unit volume	Ch. 17

K

Symbol	*Represents*	*Reference*
K	Absolute permeability	Eq. 19.6
K	Lateral stress ratio	Ch. 8, Eq. 16.6
K_a	Active stress ratio	Ch. 13
K_f-line	Line through $\bar{p}_f$ versus q_f	Fig. 9.8, Ch. 11
K_p	Passive stress ratio	Ch. 13
K_0	Lateral stress ratio for one-dimensional strain	Ch. 8, Eq. 8.12, Eq. 10.1
K_0-line	Line through $\bar{P}$ versus q for soil subjected to one-dimensional strain	Fig. 20.7
kg	Kilograms	Ch. 8
kW	kilowatts	Ch. 33
kJ	kilojoules	Ch. 33
kN	kilonewtons	Ch. 8
k	Permeability	Eq. 2.1, Ch. 17
k	Spring constant	Ch. 15
k_e	Effective permeability	Ch. 18

L

Symbol	*Represents*	*Reference*
LI	Liquidity Index	Fig. 3.4
L	Length	Chs. 5, 17
l	Length	Ch. 18

M

Symbol	*Represents*	*Reference*
M	Mass	Eq. 15.6a

Symbol	Represents	Reference
M	Moment	Ch. 24
m	Compressibility of mineral skeleton	Eq. 2.1
m_v	Coefficient of volume change	Eq. 12.12, Table 12.2
m	Meters	Chap. 5
me	Milliequivalents	Ch. 5
mm	Millimeters	
mμ	Millimicrons	Fig. 4.1
N		
NC	Normally Consolidated	Ch. 21
N	Normal force	Ch. 2
N	Standard penetration resistance	Ch. 7
$\bar{N}$	Effective normal force	Ch. 24
N_c	Bearing capacity coefficient	Chs. 22, 32
N_ϕ	$\frac{1 + \sin \bar{\phi}}{1 - \sin \bar{\phi}}$ = flow factor	Ch. 14, Eq. 11.4
N_ϕ, N_γ, N_q	Bearing capacity coefficients	Ch. 14
n	Porosity	Fig. 3.1
n_d	Number of head drops	Ch. 18
n_f	Number of flow channels	Ch. 18
n_{max}	Porosity of soil in loosest condition	Ch. 3
n_{min}	Porosity of soil in densest condition	Ch. 3
O		
OC	Overconsolidated	Ch. 21
OCR	Overconsolidation Ratio	Ch. 21
O_P	Origin of planes	Ch. 8
P		
PI	Plasticity Index	Fig. 3.4
P	Force	Ch. 13
P_a	Active force	Chs. 13, 31
P_p	Passive force	Chs. 13, 31
p	Pressure	Figs. 7.7, 7.10
p	$\frac{\sigma_1 + \sigma_3}{2}, \frac{\sigma_v + \sigma_h}{2}$	Ch. 8
$\bar{p}$	$\frac{\bar{\sigma}_1 + \bar{\sigma}_3}{2}, \frac{\bar{\sigma}_v + \bar{\sigma}_h}{2}$	Ch. 21
p_f	p at failure	Ch. 21
$\bar{p}_f$	$\bar{p}$ at failure	Ch. 21
p_i, p_0	Initial value of p	Ch. 21
$\bar{p}_i, \bar{p}_0$	Initial value of $\bar{p}$	Ch. 21
$\bar{p}_m$	Maximum value of $\bar{p}$	Ch. 28
Q		
Q	Line load surcharge	Ch. 13
Q	Periodic dynamic force	Eq. 15.1
Q	Pile capacity	Ch. 33
Q	Quick, i.e., undrained	Ch. 28
Q	Rate of flow	Ch. 17
Q_p	Pile point resistance	Ch. 33
Q_s	Pile shaft resistance	Ch. 33
Q_{ult}	Ultimate load	Ch. 14
q	Rate of flow per unit area	Ch. 18

Symbol	Represents	Reference
q	$\frac{\sigma_1 - \sigma_3}{2}, \frac{\sigma_v - \sigma_h}{2}$	Ch. 8
q_f	q at failure	Ch. 21
q_s	Surface stress	Ch. 8
q_0, q_i	Initial q	Ch. 21
R		
R	Reynolds number	Ch. 17
R	Pile allowable load	Ch. 33
R	Radius	Ch. 8
R	Repulsive stress between particles	Ch. 16
R	Resultant force	Ch. 24
R'	Repulsive force between particles	Ch. 5
r	Radius	Ch. 24
r_e	Radius of well spacing	Ch. 27
r_w	Radius of well	Ch. 27
S		
S	Percentage saturation	Fig. 3.1
SL	Shrinkage limit	Fig. 7.15
S_T	Sensitivity	Fig. 7.7
s	Average pile penetration per blow for final 150 mm of driving	Ch. 33
s	Shear strength of adhered junctions	Eq. 6.2
s_m	Shear strength of material composing particles	Ch. 6
s_s	Unit shaft resistance on pile	Ch. 33
s_u	Undrained shear strength	Ch. 29
T		
TSP	Total stress path	Ch. 20
T	Shear force	Chs. 6, 13
T	Tangential force	Ch. 2
T	Time factor	Ch. 27
T_s	Surface tension	Ch. 16
$(T - u_s)SP$	Path for total stress minus static pore pressure	
t	time	Eq. 2.1
U		
U	Average consolidation ratio	Ch. 27
U	Force from water	Ch. 24
U	Unconfined compression	Ch. 28
UU, $\overline{\text{UU}}$	Unconsolidated undrained test	Chs. 28, 29
U_z	Consolidation ratio	Ch. 27
u_a	Pore air pressure	Ch. 16
u_e	Excess pore pressure	Ch. 27
u_s	Static pore pressure	Ch. 17
u_f	Pore pressure at failure	Ch. 28
u_{ss}	Pore pressure based on steady flow	Chs. 26, 27

Symbol	Represents	Reference
u, u_w	Pore water pressure	Ch. 16
u^*	Equivalent pore pressure	Ch. 16
V		
V	Total volume	Fig. 3.1
V_g	Volume of gas	Fig. 3.1
V_s	Volume of solids	Fig. 3.1
V_v	Volume of voids	Fig. 3.1
V_w	Volume of water	Fig. 3.1
V_0	Initial volume	
v	Velocity	Ch. 17, Eq. 17.2
v_c	Critical velocity	Ch. 17
v_s	Seepage velocity	Ch. 17, Eq. 17.3
W		
W	Total weight	Fig. 3.1
W_g	Weight of gas	Fig. 3.1
W_s	Weight of solids	Fig. 3.1
W_w	Weight of water	Fig. 3.1
w	Water content	Fig. 3.1
w_f	Final water content	Ch. 21
w_l	Liquid limit	Fig. 3.4
w_n	Natural water content	Fig. 3.4
w_p	Plastic limit	Fig. 3.4
w_p	Weight of pile	Ch. 33
w_r	Weight of striking part of pile hammer	Ch. 33
w_s	Shrinkage limit	Fig. 3.4
w_0, w_i	Initial water content	Ch. 21
X		
X	Coordinate	Ch. 27
X	Shear force on slice	Ch. 24
x	Distance along X-axis	Fig. 5.13, Ch. 8
$\bar{x}$	Centroidal distance	Ch. 13
Y		
Y	Coordinate	Ch. 27
y	Distance along Y-axis	Ch. 8
Z		
Z	Coordinate	Ch. 27
z	Depth	
z	Distance along Z-axis	Ch. 8
z	Valence	Fig. 5.13
z_c	Depth of tension cracking	Ch. 23
z_w	Depth to phreatic surface	

GREEK

Symbol	Represents	Reference
A α *alpha*		
α	inclination angle of force	Ch. 24
α	slope of q_f versus p_f	Fig. 9.8, Ch. 11
$\bar{\alpha}$	slope of q_f versus $\bar{p}_f$	Ch. 21
B β *beta*		
β	slope of K_0-line for NC soil	Chs. 8, 10
$\bar{\beta}$	slope of K_0-line for NC soil	Fig. 20.7
Γ γ *gamma*		
γ	unit shear strain	Chs. 10, 12, Eq. 12.3
γ, γ_t	total unit weight	Fig. 3.1
γ_b	buoyant unit weight	Fig. 3.1
γ_d	dry unit weight	Fig. 3.1
$\gamma_{d\max}$	dry unit weight of soil in densest condition	Ch. 3
$\gamma_{d\min}$	dry unit weight of soil in loosest condition	Ch. 3
γ_s	unit weight of soil particles	Fig. 3.1
γ_w	unit weight of water	Fig. 3.1
γ_0	unit weight of water at 4°C	Fig. 3.1
Δ δ *delta*		
Δ	change, e.g., $\Delta\sigma$	Eq. 2.1
Δq_s	increment of surface stress	Ch. 14
$(\Delta q_s)_a$	allowable bearing stress	Ch. 14
$(\Delta q_s)_f$	bearing capacity	Ch. 14
$(\Delta q_s)_l$	ultimate bearing capacity for local shear	Ch. 14
$(\Delta q_s)_u$	ultimate bearing capacity	Ch. 14
δ	difference in orientation between the theoretical and observed failure planes	
$\delta = \Delta\rho$	differential settlement	Ch. 14
$\frac{\delta}{l}$	angular distortion	Ch. 14
E ϵ *epsilon*		
ϵ	dielectric constant	Fig. 5.13
ϵ	strain	Chs. 10, 12
ϵ_{vol}	volumetric strain	Ch. 10
ϵ_h, ϵ_v	horizontal ϵ, vertical strain ϵ	Ch. 10
ϵ_x, ϵ_y, ϵ_z	strain in x-direction, y-direction. z-direction	Ch. 12
ϵ_1, ϵ_2, ϵ_3	principal strains	Ch. 12
Z ζ *zeta*		
H η *eta*		
Θ θ *theta*		
θ	angle between failure surface and horizontal	Ch. 24
θ	angle between normal stress and major principal stress	Ch. 8
θ	average slope angle of asperities	Ch. 6
θ_{cr}	slope of failure plane	Ch. 11
I ι *iota*		
K κ *kappa*		

Symbol	*Represents*	*Reference*
Λ λ *lambda*		
M μ *mu*		
μ	friction angle	Fig. 6.2
μm	micrometer	Ch. 3
μ	Poisson's ratio	Chs. 8, 12, 14
μ	viscosity	Ch. 17
$\bar{\mu}$	Poisson's ratio for mineral skeleton	Chs. 27, 28
N ν *nu*		
ν	ratio of horizontal to vertical strains after failure in a plane strain triaxial compression test	Ch. 14
Ξ ξ *xi*		
O o *omicron*		
Π π *pi*		
P ρ *rho*		
ρ	mass density	Ch. 12
ρ	vertical movement	Ch. 14
ρ_c	consolidation settlement	Ch. 14
ρ_d	dynamic motion	Ch. 15
ρ_i	initial settlement	Ch. 14
ρ_{max}	maximum settlement	
ρ_{min}	minimum settlement	
ρ_t	total settlement	Ch. 14
ρ_0	settlement of test footing	
Σ σ *sigma*		
Σ	sum	Fig. 5.14
σ	normal stress	Eq. 8.1
$\bar{\sigma}$	effective normal stress	Ch. 16
$\bar{\bar{\sigma}}$	mineral-mineral contact stress	Ch. 16
$\bar{\sigma}_c$	consolidation stress in isotropic stress system	Ch. 20
$\bar{\sigma}_{cm}$	maximum past isotropic consolidation stress	Ch. 20
σ_{ff}, $\bar{\sigma}_{ff}$	normal stress on failure plane at failure	Ch. 11
σ_h, $\bar{\sigma}_h$	horizontal normal stress	Chs. 8, 16
σ_{h0}	initial total horizontal stress	
$\bar{\sigma}_{h0}$	initial effective horizontal stress	
$\bar{\sigma}_{ps}$	isotropic effective stress for perfect sampling	Ch. 26
σ_v, $\bar{\sigma}_v$	vertical normal stress	Chs. 8, 16
$\bar{\sigma}_{vm}$	maximum past vertical consolidation stress	Ch. 20, Fig. 7.9
σ_{v0}	initial total vertical stress	
$\bar{\sigma}_{v0}$	initial effective vertical stress	
σ_1, σ_2, σ_3, $\bar{\sigma}_1$, $\bar{\sigma}_2$, $\bar{\sigma}_3$	principal stresses	Chs. 8, 16
T τ *tau*		
τ	shear stress	Ch. 8
τ_{ff}	shear stress on failure surface at failure	Eqs. 11.2, 11.3
τ_h	horizontal shear stress	Eq. 8.1
τ_m	mobilized shear stress	Eq. 24.6
τ_{max}	Maximum shear stress	Ch. 8
τ_v	vertical shear stress	Eq. 8.1
τ_θ	Shear stress on plane oriented at angle θ from major principal plane	Ch. 8
Υ υ *upsilon*		
Φ ϕ *phi*		
ϕ	friction angle	Ch. 11
$\bar{\phi}$	friction angle based on effective stresses	Chs. 20, 21
ϕ_{cv}, $\bar{\phi}_{cv}$	ϕ at constant volume	Ch. 11
$\bar{\phi}_e$	true friction angle-Hvorslev parameter	Ch. 21
ϕ_u	Total stress strength parameter	Eq. 28.6
ϕ_{ult}, $\bar{\phi}_{ult}$	ϕ for ultimate strength	Ch. 21
ϕ_w	friction angle between retaining wall and soil	Ch. 13
ϕ_w, $\bar{\phi}_w$	friction angle between wall and soil	Ch. 23
ϕ_μ	particle-to-particle friction angle	Ch. 6
X χ *chi*		
χ	coefficient in partly saturated soil	Ch. 21
Ψ ψ *psi*		
ψ	electric potential	Ch. 5
Ω ω *omega*		

SUBSCRIPTS

a	axial allowable air active
b	bearing capacity buoyant
c	capillary chamber or cell consolidation contact critical
cm	maximum isotropic
cr	critical
cv	constant volume

Symbol	*Represents*	*Reference*
cn	minimum capillary	associated with capillary heads, Ch. 16
cr	capillary rise	
cs	saturation capillary	
cx	maximum capillary	
D	driving, dilitational	
d	drained	
	dry	
e	effective	
	elevation	
	excess	
f	failure	
ff	failure surface at moment of failure	
g	gas	
h	horizontal	
i	initial	
l	local shear	
m	mass	
	maximum past	
	mineral	
	mobilized	
max	maximum	
min	minimum	
N, n	natural	
p	passive, pressure	
p	pore fluid	
R	resisting	
s	shear, surface, solids, static	
sk	skeleton	
T	temperature	
	transformed	
t	time	
	total	
u	ultimate	
	undrained	
v	vertical, void	
vm	maximum vertical	
$v0$	initial vertical	
vol	volumetric	
w	water, wall, well	
x, y	directions	
z	vertical direction	
0	initial	
	no lateral strain	
	value from model	
1, 2, 3	principal stresses	
I, II, III	sequence of tests	
θ	angular direction	
μ	as in ϕ_u	
℄	center line	

SPECIAL

Symbol	*Represents*	*Reference*
$\mathfrak{S}$	Shape factor	Eq. 18.1
$^{\circ}$	degrees	Ch. 6
$\int$	integral	Ch. 8
/	per	Ch. 5
℄	center line	Ch. 14
$\underline{\nabla}$	water level	Fig. 1.8

APPENDIX B

Conversion Factors

LENGTH

To Convert From	To	Multiply By
1. Inches	feet	0.083333
	angstrom units	2.54×10^{8}
	microns	25400
	millimeters	25.4
	centimeters	2.54
	meters	0.0254
2. Feet	inches	12.0
	angstrom units	3.048×10^{9}
	microns	304800
	millimeters	304.80
	centimeters	30.48
	meters	0.3048
3. Angstrom units	inches	3.9370079×10^{-9}
	feet	3.28084×10^{-10}
	microns	0.0001
	millimeters	1×10^{-7}
	centimeters	1×10^{-8}
	meters	1×10^{-10}
4. Microns	inches	3.9370079×10^{-5}
	feet	3.2808399×10^{-6}
	angstrom units	1×10^{4}
	millimeters	1×10^{-3}
	centimeters	1×10^{-4}
	meters	1×10^{-6}
5. Millimeters	inches	3.9370079×10^{-2}
	feet	3.2808399×10^{-3}
	angstrom units	1×10^{7}
	microns	1×10^{3}
	centimeters	1×10^{-1}
	meters	1×10^{-3}
6. Centimeters	inches	0.39370079
	feet	0.032808399
	angstrom units	1×10^{8}
	microns	1×10^{4}

To Convert From	To	Multiply By
6. Centimeters (*contd.*)	millimeters	10
	meters	1×10^{-2}
7. Meters	inches	39.370079
	feet	3.2808399
	angstrom units	1×10^{10}
	microns	1×10^{6}
	millimeters	1×10^{3}
	centimeters	1×10^{2}

AREA

To Convert From	To	Multiply By
1. Square meters	square feet	10.76387
	square centimeters	1×10^{4}
	square inches	1550.0031
2. Square feet	square meters	9.290304×10^{-2}
	square centimeters	929.0304
	square inches	144
3. Square centimeters	square meters	1×10^{-4}
	square feet	1.076387×10^{-3}
	square inches	0.1550031
4. Square inches	square meters	6.4516×10^{-4}
	square feet	6.9444×10^{-3}
	square centimeters	6.4516

VOLUME

To Convert From	To	Multiply By
1. Cubic centimeters	cubic meters	1×10^{-6}
	cubic feet	3.5314667×10^{-5}
	cubic inches	0.061023744
2. Cubic meters	cubic feet	35.314667
	cubic centimeters	1×10^{6}
	cubic inches	61023.74
3. Cubic inches	cubic meters	1.6387064×10^{-5}
	cubic feet	5.7870370×10^{-4}
	cubic centimeters	16.387064

To Convert From	To	Multiply By
4. Cubic feet	cubic meters	0.028316847
	cubic centimeters	28316.847
	cubic inches	1728

FORCE

To Convert From	To	Multiply By
1. Pounds (avdp)	dynes	4.44822×10^5
	grams	453.59243
	kilograms	0.45359243
	tons (long)	4.464286×10^{-4}
	tons (short)	5×10^{-4}
	kips	1×10^{-3}
	tons (metric)	4.5359243×10^{-4}
	newtons	4.44822
2. Kips	pounds	1000
	tons (short)	0.500
	kilograms	453.59243
	tons (metric)	0.45359243
3. Tons (short)	kilograms	907.18474
	pounds	2000
	kips	2
	tons (metric)	0.907185
4. Kilograms	dynes	980665
	grams	1000
	pounds	2.2046223
	tons (long)	9.8420653×10^{-4}
	tons (short)	11.023113×10^{-4}
	kips	2.2046223×10^{-3}
	tons (metric)	0.001
	newtons	9.806650
5. Tons (metric)	grams	1×10^6
	kilograms	1000
	pounds	2204.6223
	kips	2.2046223
	tons (short)	1.1023112
	kilonewtons	9.806650
6. Kilonewtons	pounds	224.81
	tons (short)	0.1124
	kips	0.22481
	tons (metric)	0.102
	kilograms	101.97

STRESS

To Convert From	To	Multiply By
1. Pounds/square foot	pounds/square inch	0.0069445
	feet of water	0.016018
	kips/square foot	1×10^{-3}
	kilograms/square centimeter	0.000488243
	tons/square meter	0.004882
	atmospheres	4.72541×10^{-4}
	kilonewtons/square meter	0.04788
2. Pounds/square inch	pound/square foot	144
	feet of water	2.3066
	kips/square foot	0.144
	kilograms/square centimeter	0.070307
	tons/square meter	0.70307
	atmospheres	0.068046
	kilonewtons/square meter	6.895
3. Tons (short)/ square foot	atmospheres	0.945082
	kilograms/sqiare meter	9764.86
	tons (metric)/ square meter	9.76487
	pounds/square inch	13.8888
	pounds/square foot	2000
	kips/square foot	2.0
	kilonewtons/square meter	95.76
4. Feet of water (at 4°C)	pounds/square inch	0.43352
	pounds/square foot	62.427
	kilograms/square centimeter	0.0304791
	tons/square meter	0.304791
	atmospheres	0.029499
	inches of Hg	0.88265
	kilonewtons/square meter	2.989
5. Kips/square foot	pounds/square inch	6.94445
	pounds/square foot	1000
	tons (short)/ square foot	0.5000
	kilograms/square centimeter	0.488244
	tons (metric)/square meter	4.88244
	kilonewtons/square meter	47.88
6. Kilograms/ square centimeter	pounds/square inch	14.223
	pounds/square foot	2048.1614
	feet of water (4°C)	32.8093
	kips/square foot	2.0481614
	tons/square meter	10
	atmospheres	0.96784
	kilonewtons/square meter	98.067
7. Tons (metric)/ square meter	kilograms/square centimeter	0.10
	pounds/square foot	204.81614
	kips/square foot	0.20481614

To Convert From	To	Multiply By
7. Tons (metric)/square meter (*contd.*)	tons (short)/square foot	0.102408
	kilonewtons/square meter	9.806650
8. Atmospheres	bars	1.0133
	centimeters of mercury at 0°C	76
	millimeters of mercury at 0°C	760
	feet of water at 4°C	33.899
	kilograms/square centimeter	1.03323
	grams/square centimeter	1033.23
	kilograms/square meter	10332.3
	tons (metric)/suqare meter	10.3323
	pounds/square foot	2116.22
	pounds/square inch	14.696
	tons (short)/square foot	1.0581
	kilonewtons/square meter	101.325
9. Kilonewtons/square meter	pounds/square foot	20.886
	pounds/square inch	0.145
	tons (short)/square foot	0.01044
	feet of water (at 4°C)	0.3346
	meters of water	0.1020
	kips/square foot	0.02089
	kilograms/square centimeter	0.01020
	tons (metric)/square meter	0.1020
	atmospheres	0.00987

UNIT WEIGHT

To Convert From	To	Multiply By
1. Grams/cubic centimeter	tons (metric)/cubic meter	1.00
	kilograms/cubic meter	1000.00
	pounds/cubic inch	0.036127292
	pounds/cubic foot	62.427961
	kilonewtons/cubic meter	
2. Tons (metric)/cubic meter	grams/cubic centimeter	1.00
	kilograms/cubic meter	1000.00
	pounds/cubic inch	0.036127292
	pounds/cubic foot	62.427961
	kilonewtons/cubic meter	
3. Kilograms/cubic meter	grams/cubic centimeter	0.001
	tons (metric)/cubic meter	0.001
	pounds/cubic inch	3.6127292×10^{-5}
	pounds/cubic foot	0.062427961
	kilonewtons/cubic meter	
4. Pounds/cubic inch	grams/cubic centimeter	27.679905
	tons (metric)/cubic meter	27.679905
	kilograms/cubic meter	27679.905
	pounds/cubic foot	1728
	kilonewtons/cubic meter	
5. Pounds/cubic foot	grams/cubic centimeter	0.016018463
	tons (metric)/cubic meter	0.016018463
	kilograms/cubic meter	16.018463
	pounds/cubic inch	$5.78703704 \times 10^{-4}$
	kilonewtons/cubic meter	
6. Kilonewtons/cubic meter	grams/cubic centimeter	0.1020
	tons (metric)/cubic meter	0.1020
	kilograms/cubic meter	101.98
	pounds/cubic inch	0.003685
	pounds/cubic foot	6.3654

TIME

To Convert From	To	Multiply By
1. Milliseconds	seconds	10^{-3}
	minutes	1.66666×10^{-5}
	hours	2.777777×10^{-7}
	days	1.1574074×10^{-8}
	months	3.8057×10^{-10}
	years	3.171416×10^{-11}
2. Seconds	milliseconds	1000
	minutes	1.66666×10^{-2}
	hours	2.777777×10^{-4}
	days	1.1574074×10^{-5}
	months	3.8057×10^{-7}
	years	3.171416×10^{-8}
3. Minutes	milliseconds	60000
	seconds	60
	hours	0.0166666
	days	6.944444×10^{-4}

To Convert From	To	Multiply By
3. Minutes (*contd.*)	months	2.283104×10^{-5}
	years	1.902586×10^{-6}
4. Hours	milliseconds	3600000
	seconds	3600
	minutes	60
	days	0.0416666
	months	1.369860×10^{-3}
	years	1.14155×10^{-4}
5. Days	milliseconds	86400000
	seconds	86400
	minutes	1440
	hours	24
	months	3.28767×10^{-2}
	years	0.0027397260
6. Months	milliseconds	2.6283×10^{9}
	seconds	2.6283×10^{6}
	minutes	43800
	hours	730
	days	30.416666
	years	0.08333333
7. Years	milliseconds	3.1536×10^{10}
	seconds	3.1536×10^{7}
	minutes	525600
	hours (mean solar)	8760
	days (mean solar)	365
	months	12

VELOCITY

To Convert From	To	Multiply By
1. Centimeters/second	microns/second	10,000
	meters/minute	0.600
	feet/minute	1.9685
	miles/hour	0.022369
	feet/year	1034643.6
2. Microns/second	centimeters/second	0.0001
	meters/minute	0.000060
	feet/minute	0.00019685
	miles/hour	0.0000022369
	feet/year	103.46436
3. Feet/minute	centimeters/second	0.508001
	microns/second	5080.01
	meters/minute	0.3048
	miles/hour	0.01136363
	feet/year	525600
4. Feet/year	microns/second	0.009665164
	centimeters/second	0.0000009665164
	meters/minute	5.79882×10^{-7}
	feet/minute	1.9025×10^{-6}
	miles/hour	2.16203×10^{-8}

COEFFICIENT OF CONSOLIDATION

To Convert From	To	Multiply By
1. Square centimeters/second	square centimeters/month	2.6280×10^{6}
	square centimeters/year	3.1536×10^{7}
	square meters/month	2.6280×10^{2}
	square meters/year	3.1536×10^{3}
	square inches/second	0.155
	square inches/month	4.1516×10^{5}
	square inches/year	4.8881×10^{6}
	square feet/month	2.882998×10^{3}
	square feet/year	3.39447×10^{4}
2. Square inches/second	square inches/month	2.6280×10^{6}
	square inches/year	3.1536×10^{7}
	square feet/month	1.8250×10^{4}
	square feet/year	2.1900×10^{5}
	square centimeters/second	6.4516
	square centimeters/month	1.6955×10^{7}
	square centimeters/year	2.0346×10^{8}
	square meters/month	1.6955×10^{3}
	square meters/year	2.0346×10^{4}

APPENDIX C

References

IMPORTANT PUBLICATIONS

This section lists important English-language periodicals and proceedings containing important publications.

Journals

Journal of the Soil Mechanics and Foundations Division, American Society of Civil Engineers, 345 E. 47th St., New York. This Journal is one part of the Proceedings of ASCE. Until 1963, more important papers were reprinted in the *Transactions* of ASCE. *Transactions* of ASCE starting in 1964 contain short abstracts of all papers.

Geotechnique, The Institution of Civil Engineers, Great George Street, London. Volume 1 appeared in 1948.

Canadian Geotechnical Journal, University of Toronto Press, Toronto. Volume 1 appeared in 1963.

Journal of the Boston Society of Civil Engineers, 47 Winter Street, Boston, Mass. Papers are collected in three volumes entitled *Contributions to Soil Mechanics*:

Volume 1 1925–1940
Volume 2 1941–1953
Volume 3 1954–1962

Many important papers have also been published by the following societies, but appear in special bulletins or publications rather than in regular journals:

American Society for Testing and Materials
1916 Race Street, Philadelphia, Pa.
Highway Research Board
2101 Constitution Avenue, Washington, D.c.

Proceedings of International Conferences

Proceedings of International Conferences on Soil Mechanics and Foundation Engineering (ICSMFE) are published by the host country.

1st ICSMFE, 1936, Cambridge, Mass., 3 volumes
2nd ICSMFE, 1948, Rotterdam, 7 volumes
3rd ICSMFE, 1953, Zurich, 3 volumes
4th ICSMFE, 1957, London, 3 volumes
5th ICSMFE, 1961, Paris, 3 volumes
6th ICSMFE, 1965, Montreal, 3 volumes
7th ICSMFE, 1969, Mexico City

Proceedings of Regional Conferences

The following list includes conferences held within the several geographic regions of the International Society for Soil Mechanics and Foundation Engineering. Proceedings generally are published by the host country.

Pan-American Conferences (PACSMFE)
1st, 1959, Mexico City, 3 volumes (published in 1960)
2nd, 1963, Brazil, 2 volumes
3rd, 1967, Caracas, 2 volumes

European Conferences (ECSMFE)
Stability of Earth Slopes, 1954, Stockholm. Proceedings are published in the March and June issues of *Geotechnique*, Vol. 5, 1955.
Earth Pressures, 1958, Brussels, 3 volumes.
Pore Pressure and Suction in Soils, 1960, London, 1 volume published by Butterworths in 1961.
Problems of Settlements and Compressibility of Soils, 1963, Wiesbaden, 2 volumes
Shear Strength Properties of Natural Soils and Rocks, 1967, Oslo, 2 volumes

Asian
1st, 1961, New Delhi
2nd, 1964, Tokyo
3rd, 1967, Haifa, 2 volumes

African
1st, 1955, Pretoria
2nd, 1959, Lourenco Marques
3rd, 1963, Rhodesia
4th, 1967, Johannesburg

Australia–New Zealand
There have been five conferences (as of 1968). Proceedings of the first four were published by The Institute of Engineers, Sydney, Australia. Proceedings of the fifth were published by the New Zealand Institution of Engineers, Wellington, New Zealand.

Southeast Asian
1st, 1967, Bangkok

Proceedings of ASCE Soil Mechanics Conferences

Research Conference on the Shear Strength of Cohesive Soils, Boulder, Col., 1960. Proceedings appear in a special volume.

Conference on Design of Foundations to Reduce Settlements, Evanston, Ill., 1964. Papers appear in a special volume and in *Proc. ASCE*, Vol. 90, No. SM5 and Vol. 91, No. SM2.

Specialty Conference on the Stability of Slopes and Embankments, Berkeley, Cal., 1966. Papers are printed in *Proc. ASCE*, Vol. 93, No. SM4, July 1967.

Specialty Conference on Placement and Improvement of Soil to Support Structures, M.I.T., Cambridge, Mass., 1968.

PUBLICATIONS CITED IN TEXT

(listed alphabetically by author or organization).

Aas, G., 1965. "A Study of the Effect of Vane Shape and Rate of Strain on the Measured Values of In-Situ Shear Strength of Clays," *Proc. 6th Inter. Conf. Soil Mech. Found. Eng.* (Montreal), p. 141.

Acum, W. E. A., and L. Fox, 1951. "Computation of Load Stresses in a Three-Layer Elastic System," *Geotechnique*, Vol. 2., 293–300.

Ahlvin, R. G., and H. J. Ulery, 1962. "Tabulated Values for Determining the Complete Pattern of Stresses, Strains and Deflections Beneath a Uniform Circular Load on a Homogeneous Half Space," *Highway Research Board Bulletin*, No. 342, pp. 1–13.

Aldrich, H. P., 1964. "Precompression for Support of Shallow Foundations," *Proc. ASCE Conf. on Design of Foundations for Control of Settlement* (Evanston, Ill.), p. 471, June 1964.

Alpan, I., 1967. "The Empirical Evaluation of the Coefficient K_0 and K_{0r}," *Soil and Foundation* (Jap. Soc. Soil Mech. Found. Eng.), Vol. VII, No. 1, p. 31 (January).

Archard, J. F., 1957. "Elastic Deformations and the Laws of Friction," *Proc. Royal Soc.*, A243, pp. 190–205.

ASTM Standards, 1967. *Bituminous Materials; Soils; Skid Resistance*, Part 11, Standard D-2049-64T, pp. 610–618.

Bailey, W. A., 1961. Effects of Salt on the Shear Strength of Boston Blue Clay, S. B. thesis, Dept. of Civil Engineering, M.I.T., Cambridge, Mass.

Baracos, A., 1957. "The Foundation Failure of the Transcona Grain Elevator," *Engineering J.*, Vol. 40, No. 7 (July).

Barkan, D. D., 1962. *Dynamics of Bases and Foundations* (translated from the Russian by L. Drashevska), McGraw-Hill, New York.

Begemann, H. K. S., 1953. "Improved Method of Determining Resistance to Adhesion by Sounding Through a Loose Sleeve," *Proc. 3rd Inter. Conf. Soil Mech. Found. Eng.* (Zurich), Vol. 1, p. 213.

Bishop, A. W., 1954. "The Use of Pore Pressure Coefficient in Practice," *Geotechnique*, Vol. 4, pp. 148–152.

Bishop, A. W., 1955. "The Use of the Slip Circle in the Stability Analysis of Earth Slopes," *Geotechnique*, Vol. 5, pp. 7–17.

Bishop, A. W., 1957. "Some Factors Controlling the Pore Pressures Set up During the Construction of Earth Dams," *Proc. 4th Inter. Conf. Soil Mech. Found. Eng.* (London), Vol. 2, pp. 294–300.

Bishop, A. W., 1961. "The Measurement of Pore Pressure in the Triaxial Test," *Pore Pressure and Suction in Soils*, Butterworths, London, p. 38.

Bishop, A. W., 1966. "The Strength of Soils as Engineering Materials," Sixth Rankine Lecture, *Geotechnique*, Vol. 16, No. 2, pp. 91–130.

Bishop, A. W., I. Alpan, E. E. Blight, and I. B. Donald, 1960. "Factors Controlling the Strength of Partly Saturated Cohesive Soils", *Proc. ASCE Research Conf. on Shear Strength of Cohesive Soils*, Boulder, Col., pp. 503–532.

Bishop, A. W., and L. Bjerrum, 1960. "The Relevance of the Triaxial Test to the Solution of Stability Problems," *Proc. ASCE Research Conf. on Shear Strength of Cohesive Soils*, Boulder, Col., pp. 437–501.

Bishop, A. W., and G. E. Blight, 1963. "Some Aspects of Effective Stress in Saturated and Partly Saturated Soils," *Geotechnique*, Vol. 13, pp. 177–197.

Bishop, A. W., and G. Eldin, 1950. "Undrained Triaxial Tests on Saturated Sands and Their Significance in the General Theory of Shear Strength," *Geotechnique*, Vol. 2, p. 13.

Bishop, A. W., and R. E. Gibson, 1963. "The Influence of the Provisions for Boundary Drainage on Strength and Consolidation Characteristics of Soils Measured in the Triaxial Apparatus," *ASTM* STP 361 (Laboratory Shear Testing of Soils), pp. 435–451.

Bishop, A. W., and D. J. Henkel, 1962. "The Measurement of Soil Properties" in *The Triaxial Test*, Edward Arnold Ltd., London, second edition.

Bishop, A. W., M. F. Kennard, and A. Penman, 1960. "Pore Pressure Observations at Selset Dam," *Proc. Conf. on Pore Pressure and Suction in Soil*, London, Butterworths, pp. 36–47.

Bishop, A. W., and N. Morgenstern, 1960. "Stability Coefficients for Earth Slopes," *Geotechnique*, Vol. 10, pp. 129–150.

Bishop, A. W., D. L. Webb, and P. I. Lewin, 1965. "Undisturbed Samples of London Clay from the Ashford Common Shaft: Strength-Effective Stress Relationships," *Geotechnique*, Vol. 15, pp. 1–31.

Bjerrum, L., 1954a. "Geotechnical Properties of Norwegian Marine Clays," *Geotechnique*, Vol. 4, p. 49.

Bjerrum, L., 1954b. "Theoretical and Experimental Investigations on the Shear Strength of Soils," Norwegian Geotechnical Institute Publication No. 5, Oslo, 113 pp.

Bjerrum, L., 1955. "Stability of Natural Slopes in Quick Clay," *Geotechnique*, Vol. 5, pp. 101–119.

Bjerrum, L., 1961. "The Effective Shear Strength Parameters of Sensitive Clays," *Proc. 5th Inter. Conf. Soil Mech. Found. Eng.* (Paris), Vol. 1, pp. 23–28.

Bjerrum, L., 1963a. Discussion to European Conference on Soil Mech. Found. Eng. (Wiesbaden), Vol. II, p. 135.

Bjerrum, L., 1963b. "Generelle krav til fundamentering av forskjellige byggverk; tillatte setninger." Den Norske Ingeniørforening. Kurs i fundamentering. Oslo.

Bjerrum, L., 1964. Unpublished lectures on Observed Versus Computed Settlements of Structures on Clay and Sand given at M.I.T.

Bjerrum, L., 1967. "Engineering Geology of Normally Consolidated Marine Clays as Related to the Settlement of Buildings," *Geotechnique*, Vol. 17, pp. 83–118.

Bjerrum, L., 1967: "Progressive Failure in Slopes of Overconsolidated Plastic Clay and Clay Shales," *Proc. ASCE*, Vol. 93, No. SM5 (Part 1), pp. 1–49.

Bjerrum, L., and A. Eggestad, 1963. "Interpretation of Loading Tests on Sand," *Proc. Eur. Conf. Soil Mech. Found. Eng.* (Wiesbaden), Vol. 1, p. 199.

Bjerrum, L., and O. Eide, 1956. "Stability of Strutted Excavations in Clay," *Geotechnique*, Vol. 6, p. 32.

Bjerrum L., and B. Kjaernsli, 1957. "Analysis of the Stability of Some Norwegian Natural Clay Slopes," *Geotechnique*, Vol. 7, pp. 1–16.

Bjerrum, L., S. Kringstad, and O. Kummeneje, 1961. "The Shear Strength of Fine Sand," *Proc. 5th Inter. Conf. Soil Mech. Found. Eng.* (London), Vol. 1, pp. 29–37.

Bjerrum, L., and O. Kummeneje, 1961. "Shearing Resistance of Sand Samples with Circular and Rectangular Cross Sections," Norwegian Geotechnical Institute Publication No. 44, Oslo.

Bjerrum, L., and A. Landva, 1966. "Direct Simple-Shear Tests on a Norwegian Quick Clay," *Geotechnique*, Vol. 16, No. 1, pp. 1–20.

Bjerrum, L., and K. Y. Lo, 1963. "Effect of Aging on the Shear-Strength Properties of a Normally Consolidated Clay," *Geotechnique*, Vol. 13, No. 2, pp. 147–157.

Bjerrum L., and A. Overland, 1957. "Foundation Failure of an Oil Tank in Fredrikstad, Norway," *Proc. 4th Inter. Conf. Soil Mech. Found. Eng.*, Vol. 1, pp. 285–290.

Bjerrum, L., N. Simons, and I. Torblaa, 1958. "The Effect of Time on the Shear Strength of a Soft Marine Clay," *Proc. Brussels Conference on Earth Pressure Problems*, Vol. 1, pp. 148–158.

Blight, G. E., 1965. "A Study of Effective Stresses for Volume Change," *Moisture Equilibria and Moisture Changes in Soils Beneath Covered Areas*, Butterworths, Australia.

Bolt, G. H., 1956. "Physico-Chemical Analysis of the Compressibility of Pure Clays," *Geotechnique*, Vol. 6, p. 86.

Borowicka, H., 1936. "Influence of Rigidity of a Circular Foundation Slab on the Distribution of Pressures over the Contact Surface," *Proc. 1st Inter. Conf. Soil Mech.* (Cambridge), Vol. 2, pp. 144–149.

Borowicka, H., 1938. "The Distribution of Pressure under a Uniformly Loaded Elastic Strip Resting on Elastic-Isotropic Ground," *2nd Cong. Int. Assoc. Bridge and Struct. Eng.* (Berlin), Final Report.

Boston, 1964. "Building Code of the City of Boston."

Boussinesq, J., 1885. *Application des Potentials a L'Etude de L'Equilibre et du Mouvement des Solides Elastiques*, Gauthier-Villars, Paris.

Bowden, F. P., and D. Tabor, 1950. *The Friction and Lubrication of Solids*, Part 1, Oxford University Press, London.

Bowden, F. P., and D. Tabor, 1964. *The Friction and Lubrication of Solids*, Part II, Oxford University Press, London.

Brace, W. F., 1963. "Behavior of Quartz During Indentation," *Journal of Geology*, Vol. 71, No. 5, pp. 581–595.

Brace, W. F., 1966. "Elasticity and Rigidity of Rock" in *Encyclopedia of Earth Sciences*, R. Fairbridge (ed.), Reinholt.

Brink, A. B. A., and B. A. Kantey, 1961. "Collapsible Grain Structure in Residual Granite Soils in Southern Africa," *Proc. Inter. Soc. Soil Mech. Found. Eng.* (Paris), Vol. 1, p. 611.

Bromwell, L. G., 1966. The Friction of Quartz in High Vacuum. Sc.D. thesis, M.I.T., Cambridge, Mass.

Bromwell, L. G., and T. Lambe, 1968. "A Comparison of Laboratory and Field Values of c_v for Boston Blue Clay," Paper presented to 47th Annual Meeting of Highway Research Board.

Brooker, Elmer W., and H. O. Ireland, 1965. "Earth Pressures at Rest Related to Stress History," *Canadian Geotechnical Journal*, Vol. 11, No. 1 (Feb.).

Brown, J. D., and W. G. Paterson, 1964. "Failure of an Oil Storage Tank Founded on a Sensitive Marine Clay," *Canadian Geotechnical Journal*, Vol. 1, p. 205.

Buisman, A. S. K., 1936. "Results of Long Duration Settlement Tests," *Proc. 1st Inter. Conf. Soil Mech. Found. Eng.* (Cambridge), Vol. 1, pp. 103–105.

Bureau of Reclamation, 1963. *Earth Manual*, U.S. Government Printing Office, Washington D.C.

Burmister, D. M., 1956. "Stress and Displacement Characteristics of a Two-Layer Rigid Base Soil System: Influence Diagrams and Practical Applications," *Proc. Highway Research Board*, Vol. 35, pp. 773–814.

Caquot, A., and J. Kérisel, 1949. *Traite de Mechanique des Sols*, Gauthier-Villars, Paris.

Carson, A. B., 1965. *Foundation Construction*, McGraw-Hill, New York.

Casagrande, A., 1932. "The Structure of Clay and Its Importance in Foundation Engineering." *Contributions to Soil Mechanics*, BSCE, 1925–1940, pp. 72–112 (paper published in *J. BSCE*, April 1932).

Casagrande, A., 1936. "The Determination of the Preconsolidation Load and Its Practical Significance," *Proc. 1st Int. Conf. Soil Mech. Found. Eng.* (Cambridge, Mass.), p. 60.

Casagrande, A., 1948. "Classification and Identification of Soils," *Trans. ASCE*, Vol. 113, p. 901.

Casagrande, A., 1965. "Role of the 'Calculated Risk' in Earthwork and Foundation Engineering," *Proc. ASCE*, Vol. 91, No. SM4, pp. 1–40.

Casagrande, A., 1937. "Seepage Through Dams," *Contributions to Soil Mechanics*, BSCE, 1925–1940 (paper first published in *J. New England Water Works Assoc.*, June 1937).

Casagrande, A., and S. G. Albert, 1930. "Research on the Shearing Resistance of Soils," Report by the Massachusetts Institute of Technology.

Casagrande, A., and R. E. Fadum, 1944. "Application of Soil Mechanics in Designing Building Foundations," *Trans. ASCE*, Vol. 109, p. 383.

Casagrande, A., and P. J. Rivard, 1959. "Strength of Highly Plastic Clays," Norwegian Geotechnical Institute Pub. No. 31, Harvard Soil Mechanics System No. 60.

Casagrande, A., and S. D. Wilson, 1951. "Effect of Rate of

Loading on the Strength of Clays and Shales at Constant Water Content," *Geotechnique*, Vol. 2, pp. 251–263.

Casagrande, L., 1953. "Review of Past and Current Work on Electro-Osmotic Stabilization of Soils," Harvard Soil Mechanics Series No. 45, Harvard Univ., Cambridge, Mass. (reprinted Nov. 1959 with supplement).

Cedegren, H. R., 1960. "Seepage Requirements of Filters and Previous Bases, " *J. Soil Mech. Found. Eng. Div. ASCE* (October). Vol. 86, No. SM5.

Cedergren, H. R., 1967. *Seepage, Drainage, and Flow Nets*, John Wiley and Sons, New York.

Chellis, R. D., 1962. "Pile Foundations," Chapter 7 of *Foundation Engineering*, G. A. Leonards (ed.), McGraw-Hill, New York.

Chen, L-S, 1948. "An Investigation of Stress-Strain and Strength Characteristics of Cohesionless Soils by Triaxial Compression Tests," *Proc. 2nd Inter. Conf. Soil Mech. Found. Eng.*, Vol. 5, p. 35.

Christian, J. T., 1966. "Plane-Strain Deformation Analysis of Soil," Report by M.I.T. Dept. of Civil Eng. to U.S. Army Eng. Waterways Experiment Station.

Cooling, L. F., 1948. "Settlement Analysis of Waterloo Bridge," *Proc. 2nd Inter. Conf. Soil Mech. Found. Eng.* (Rotterdam), Vol. II, p. 130.

Cornell University, 1951. "Final Report on Soil Solidification Research," Ithaca, N.Y.

Cornforth, D. H., 1961. "Plane Strain Failure Characterstics of a Saturated Sand," Ph.D., thesis, U. of London. See also *Geotechnique*, Vol. 16, p. 95.

Cornforth, D. H., 1964. "Some Experiments on the Influence of Strain Conditions on the Strength of Sand," *Geotechnique*, Vol. 16, p. 193.

Corps of Engineers, Dept. of Army, 1952. "Seepage Control, Soil Mechanics Design," Washington D.C.

Crandall, S. H., and N. C. Dahl, 1959. *An Introduction to the Mechanics of Solids*, McGraw-Hill, New York.

Crawford, C. B., 1959. "The Influence of Rate of Strain on Effective Stresses in Sensitive Clay," ASTM Spec. Tech. Pub. 254, pp. 36–48.

Crawford, C. B., 1964. "Interpretation of the Consolidation Test," *Proc. ASCE*, Vol. 90, No. SM5, pp. 87–102.

Crawford, C. B., 1965. "Resistance of Soil Structure to Consolidation," *Canadian Geotechnical Journal*, Vol. 2, pp. 90–115.

Crawford, C. B., and K. N. Burn, 1962. "Settlement Studies on the Mt. Sinai Hospital, Toronto," *The Engineering Journal*, Vol. 45, Nalz (December).

Crawford, C. B., and W. J. Eden, 1967. "Stability of Natural Slopes in Sensitive Clay," *Proc. ASCE*, Vol. 93, No. SM4, pp. 419–436.

Croney, D., and J. D. Coleman, 1961. *Pore Pressure and Suction in Soils*, Butterworths, London, p. 31.

Cryer, C. W., 1963. "A Comparison of the Three-Dimensional Consolidation Theories of Biot and Terzaghi," *Quarterly Journal of Mechanics and Applied Mathematics*, Vol. 16 pp. 401–412.

Cummings, A. E., G. O. Kerkhoff, and R. B. Peck, 1950. "Effect of Driving Piles into Soft Clay," *Trans. ASCE*, Vol. 115, pp. 275–285.

Dana, James Dwight: *Manual of Mineralogy*, published first in 1848 and revised many times, latest being by Cornelius S. Hurlbut, Jr. and published by Wiley in 1949.

Danish Geotechnical Institute, 1968. Bulletin No. 25, Copenhagen.

D'Appolonia, D. J., and E. D'Appolonia, 1967. "Determination of the Maximum Density of Cohesionless Soils," *Proc. 3rd Asian Conf. Soil Mech. Found. Eng.*, Vol. 1.

D'Appolonia, E., 1953. "Loose Sands—Their Compaction by Vibroflotation," ASTM Special Technical Publication No. 156, p. 138.

D'Appolonia, D. J., R. V. Whitman, and E. D'Appolonia, 1968: "Sand Compaction with Vibratory Rollers," *ASCE Specialty Conference on Placement and Improvement of Soil to Support Structures*.

Darragh, R. D., 1964. "Controlled Water Tests to Preload Tank Foundations," *Proc. ASCE Conf. on Design of Foundations for Control of Settlement* (Evanston, Ill.)

Davisson, M. T., 1966. "Pile Hammers, Pile Driving and Driving Formulas." Notes for Lecture to New York Metropolitan Section ASCE, Soil Mechanics and Foundations Group.

DeLory, F. A., 1960. Discussion to NRC Tech., Memo. No. 69, *Proc. 14th Canadian Soil Mech. Conf.*

Deresciewicz, H., 1958. *Mechanics of Granular Matter. Advances in Applied Mechanics*, Vol. 5. Academic Press, New York, pp. 233–306.

Dickey, J. W., 1966. Frictional Characteristics of Quartz, S.B. thesis, M.I.T., Cambridge, Mass.

Duke, C. M., and D. J. Leeds, 1963. "Response of Soils, Foundations, and Earth Structures to the Chilean Earthquakes of 1960," *Bull. Semismological Society of America*, Vol. 53, No. 2.

Dunbar, C. O., 1960. *Historical Geology*, John Wiley and Sons, New York.

Durante, V. A., J. L. Kogan, V. I. Ferronsky, and S. I. Nosal, 1957. "Field Investigations of Soil Densities and Moisture Contents," *Proc. Inter. Conf. Soil Mech. Found. Eng.* (London), Vol. 1, p. 216.

Eden, W. J., and M. Bozozuk, 1962. "Foundation Failure of a Silo on Varved Clay," *Engineering Journal*, Vol. 45, No. 9, pp. 54–57 (Sept.).

Eggstad, A., 1963. "Deformation Measurements Below a Model Footing on the Surface of Dry Sand," Wiesbaden Settlement Conf. Vol. 1 p. 233.

Feld, J., 1965. "Tolerance of Structures to Settlement," *ASCE, J. Soil Mech. Found. Eng.*, Vol. 91, No. SM3, pp. 63–77.

Fellenius, W., 1936. "Calculation of the Stability of Earth Dams," *Trans. 2nd Congress on Large Dams* (Washington), Vol. 4, p. 445.

Fetzer, C. A., 1967. "Electro-Osmotis Stabilization of West Branch Dam," *J. Soil Mech. Found. Div. ASCE*, p. 85 (July).

Flint, R. F., 1947. *Glacial Geology and the Pleistocene Epoch*, John Wiley and Sons, New York.

Forssblad, L., 1965. "Investigation of Soil Compaction by Vibroflotation," *Acta Polytechnical Scandinavica*, No. Ci34 (Stockholm).

Foster, C. R., 1962. "Field Problems: Compaction," *Foundation Engineering*, G. A. Leonards (ed.), McGraw-Hill, New York p. 1000–1024.

Frohlich, O. K., 1955. "General Theory of Stability of Slopes," *Geotechnique*, Vol. 5, pp. 37–47. Also see discussion on pp. 48–49.

Fruco and Associates, 1966. "Dewatering and Ground Water Control for Deep Excavations," report prepared for U.S. Army Corps of Engineers, Waterways Experiment Station, Vicksburg, Miss. (January).

Gibbs, H. J., and W. G. Holtz, 1957. "Research on Determining the Density of Sands by Spoon Penetration Testing," *Proc. 4th Inter. Conf. Soil Mech. Found. Eng.* (London), Vol. I, p. 35.

Gibson, R. E., 1953. "Experimental Determination of the True Cohesion and True Angle of Internal Friction in Clays," *Proc. 3rd Inter. Conf. Soil Mech. Found. Eng.* (Zurich), Vol. 1, p. 126.

Gibson, R. E., 1958. "The Progress of Consolidation in a Clay Layer Increasing in Thickness with Time," *Geotechnique*, Vol. 8, pp. 171–182.

Gibson, R. E., G. L. England, and M. J. L. Husey, 1967. "The Theory of One-Dimensional Consolidation of Saturated Clays. I. Finite Non-Linear Consolidation of Thin Homogeneous Layers," *Geotechnique*, Vol. 17, pp. 261–273.

Gibson, R. E., and D. J. Henkel, 1954: "The Influence of Duration of Tests at Constant Rate of Strain on Measured 'Drained' Strength," *Geotechnique*, Vol. 4, pp. 6–15.

Gibson, R. E. and N. Morgenstern, 1962. "A Note on the Stability of Cuttings in Normally Consolidated Clays," *Geotechnique*, Vol. 12, pp. 212–216.

Gibson, R. E., J. K. L. Schiffman, and S. L. Pu, 1967. "Plain Strain and Axially Symmetric Consolidation of a Clay Layer of Limited Thickness," U. of Illinois (Chicago Circle) MATE Report 67-4.

Goodman, R. E., and H. B. Seed, 1966. "Earthquake-Induced Displacements in Sand Embankments," *Proc. ASCE*, Vol. 92, No. SM2, pp. 125–146.

Gorbunov-Possadov, M. I., and V. Serebrajanyi, 1961. "Design of Structures upon Elastic Foundations," *Proc. 5th Inter. Conf. Soil Mech. Found. Eng.* (Paris), Vol. 1, pp. 643–648.

Gould, J. P., 1960: "A Study of Shear Failure in Certain Tertiary Marine Sediments," *Proc. Research Conf. on Shear Strength of Cohesive Soils, ASCE*, pp. 615–641.

Gray, H., 1936a. "Progress Report on Research on the Consolidation of Fine Grained Soils," *Proc. 1st Inter. Conf. Soil Mech. Found. Eng.* (Cambridge) Vol. II, pp. 138–141.

Gray, H., 1936b. "Stress Distribution in Elastic Solids," *Proc. 1st Inter. Conf. Soil Mech. Found. Eng.* (Cambridge) Vol. II, pp. 157–168.

Grim, Ralph E., 1962: *Applied Clay Mineralogy*, McGraw-Hill, New York.

Hall, H. P., 1954. "A Historical Review of Investigations of Seepage Toward Wells," *J. BSCE*, Vol. 41, pp. 251–311.

Hamilton, J. J., 1960. "Earth Pressure Cells; Design Calibration and Performance," Tech. Paper No. 109, Division of Building Research, National Research Council, Ottawa, Canada.

Hansen, J. B., 1957. "General Report—Foundations of Structures," *Proc. 4th Inter. Conf. Soil Mech.* (London), Vol. II, pp. 441–447.

Hansen, J. B., 1953. *Earth Pressure Calculation*, The Danish Technical Press, Copenhagen.

Hansen, J. B., 1963. "Discussion to Kondner," Vol. 89, No. SM4, p. 241 (July).

Hansen, J. B., 1966. Bulletin No. 20, The Danish Geotechnical Institute.

Hansen, J. B., and H. Lundgren, 1960: "*Hauptprobleme der Bodenmechanik*," Springer-Verlag, Berlin.

Hardin, B. O., and W. L. Black, 1968. "Vibration Modulus of Normally Consolidated Clay," *Proc. ASCE*, Vol. 94, No. SM2, pp. 353–369.

Hardin, B. O., and F. E. Richart Jr., 1963. "Elastic Wave Velocities in Granular Soils," *Proc. ASCE*, Vol. 89, No. SM1, pp. 33–65.

Harr, E., 1962. *Groundwater and Seepage*, McGraw-Hill, New York.

Harr, M. E., 1966. *Foundations of Theoretical Soil Mechanics*, McGraw-Hill, New York.

Hassib, M. H., 1951. "Consolidation Characteristics of Granular Soils," Columbia University, New York.

Haythornthwaite, R. M., 1960. "Mechanics of the Triaxial Test for Soils," *Proc. ASCE*, Vol. 86, No. SM5, pp. 35–62.

Haythornthwaite, R. M., 1961. "Methods of Plasticity in Land Location Studies," *Proc. 1st Inter. Conf. Mechanics of Soil-Vehicle Systems* (Italy).

Healy, K. A., 1963. "Preliminary Investigations into the Liquefaction of Sand," Research Report R63-29, Department of Civil Engineering, M.I.T., Cambridge, Mass.

Hendron, A. J., Jr., 1963. The Behavior of Sand in One-Dimensional Compression, Ph.D. thesis, Department of Civil Engineering, University of Illinois (Urbana).

Henkel, D. J., 1956. "The Effect of Overconsolidation on the Behavior of Clays During Shear," *Geotechnique*, Vol. 6, p. 139.

Henkel, D. J., 1959: "The Relationships Between the Strength, Pore-Water Pressure, and Volume-Change Characteristics of Saturated Clays," *Geotechnique*, Vol. IX, p. 119.

Henkel, D. J., 1960. "The Relationship Between the Effective Stresses and Water Content in Saturated Clays," *Geotechnique*, Vol. 10, p. 41.

Henkel, D. J., and A. W. Skempton, 1955. "A Landslide at Jackfield, Shropshire, in a Heavily Overconsolidated Clay," *Geotechnique* Vol. 5, pp. 131–137.

Henry, T. D. C., 1956: *The Design and Construction of Engineering Foundations*, pp. 374–375.

Herrmann, H. G., and L. A. Wolfskill, 1966: "Residual Shear Strength of Weak Clay," Technical Report 3–699, U.S. Army Engineer Waterways Experiment Station.

Hirschfeld, R. C., 1958. Factors Influencing the Constant Volume Strength of Clays, Ph.D. thesis, Harvard University, Cambridge, Mass.

Holtz, W. G., and H. J. Gibbs, 1956. "Shear Strength of Pervious Gravelly Soils," *Proc. ASCE*, Paper No. 867.

Holtz, W. G., and J. W. Hilf, 1961. "Settlement of Soil Foundations Due to Saturation," *Proc. 5th Inter. Soc. Soil Mech. Found. Eng.* (Paris), Vol. I, p. 673.

Horn, H. M., 1966. "Influence of Pile Driving and Pile Characteristics on Pile Foundation Performance." Notes for Lectures to New York Metropolitan Section ASCE, Soil Mechanics and Foundations Group.

Horn, H. M., and D. U. Deere, 1962. "Frictional Characteristics of Minerals," *Geotechnique*, Vol. 12, pp. 319–335.

Hough, B. K., 1957. *Basic Soils Engineering*, Ronald Press, New York.

Housel, W. S., 1965. Discussion, *Proc. ASCE*, Vol. 91, No. SM1, pp. 196–219.

Huntington, W. C., 1957. *Earth Pressures and Retaining Walls*, John Wiley and Sons, New York.

Hutchinson, J. N., 1961. "A Landslide on a Thin Layer of Quick Clay at Furre, Central Norway," *Geotechnique*, Vol. 11, pp. 69–94.

Hvorslev, M. J., 1937: "Über die Festigkeitseigenschafften gestörter bindiger Böden," kbn. (Gad), 159 p.

Hvorslev, J., 1948. "Subsurface Exploration and Sampling of Soils for Civil Engineering Purposes," Report of Committee on Sampling and Testing, Soil Mech. and Foundations Division, ASCE.

Hvorslev, M. J., 1949. "Time Lag in the Observation of Ground-Water Levels and Pressures," U.S. Army Waterways Experiment Station, Vicksburg, Miss.

Hvorslev, M. J., 1960. "Physical Components of the Shear Strength of Saturated Clays," ASCE Research Conf. on Shear Strength of Cohesive Soils, Boulder, Colorado pp. 169–273.

Iller, R. K., 1955. *The Colloid Chemistry of Silica and Silicates*, Cornell University Press, Ithaca, N.Y.

Ireland, H. O., 1955: "Settlement Due to Building Construction in Chicago," Ph.D. Thesis, Univ. of Illinois.

Jakobson, B., 1952. "The Landslide at Surte on the Gota River," *Proc. Royal Swedish Geotechnical Institute*, Vol. 5, p. 87.

Jaky, J., 1944. "The Coefficient of Earth Pressure at Rest," *Journal of the Society of Hungarian Architects and Engineers*, pp. 355–358.

Janbu, N., L. Bjerrum, and B. Kjaernsli, 1956. "Veiledning ved løsing av fundamenteringsoppgaver" (Soil mechanics applied to some engineering problems), in Norwegian with English summary, Norwegian Geotechnical Institute Publ. 16, Oslo.

Jennings, J. E., 1953: "The Heaving of Buildings on Desiccated Clay," *Proc. 3rd ICSMFE (Zurich)*, Vol. I, pp. 390–396.

Jennings, J. E. B., and J. B. Burland, 1962: "Limitations to the Use of Effective Stress in Partly Saturated Soils," *Geotechnique*, Vol. 12, No. 2, pp. 125–144.

Jennings, J. E., and K. Knight, 1957. "The Additional Settlement of Foundations Due to a Collapse of Structure of Sandy Subsoils on Wetting," *Proc. 4th Inter. Soc. Soil Mech. Found. Eng.* (London), Vol. 1, p. 316.

Johannessen, I. J., and L. Bjerrum, 1965. "Measurement of the Compression of a Steel Pile to Rock Due to Settlement of the Surrounding Clay," *Proc. 6th Inter. Conf. Soil Mech. Found. Eng.* (Montreal), Vol. 2, pp. 261–264.

Jones, A., 1962. "Tables of Stresses in Three-Layer Elastic Systems," *Highway Research Board Bulletin*, No. 342, pp. 176–214.

Jordan, J. C., and R. L. Schiffman, 1967: *User's Manual* for ICES SEPOL-I, Pub. R67-61, Dept. of Civil Engineering, M.I.T., Cambridge, Mass.

Jurgenson, L., 1934. "The Application of Theories of Elasticity and Plasticity to Foundation Problems," *J. BSCE.* p. 184 of "Contributions to Soil Mechanics," *BSCE*, 1940.

Kenney, T. C., 1959: Discussion, *Proc. ASCE*, Vol. 85, No. SM3, pp. 67–79.

Kérisel, J., 1964. "Deep Foundations Basic Experimental Facts," *Proc. Deep Foundations Conf.* (Mexico) (Dec.).

Kérisel, J. L., 1967. "Vertical and Horizontal Bearing Capacity of Deep Foundations in Clay," *Proc. Symposium on Bearing Capacity and Settlement of Foundations*, Duke Univ., Durham, N.C., p. 45.

Kirkpatrick, W. M., 1957. "The Condition of Failure for Sands," *Proc. 4th Inter. Conf. Soil Mech. Found. Eng.* (London), Vol. 1, pp. 172–178.

Kolbuszewski, J. J., 1948. "An Experimental Study of the Maximum and Minimum Porosities of Sands," *Proc. 2nd Inter. Conf. Soil Mech. Found. Eng.* (Rotterdam), Vol. I, p. 158.

Kondner, R. L., 1963. "Hyperbolic Stress-Strain Response: Cohesive Soils," *Proc. ASCE, J. Soil Mech. Found. Eng.*, Vol. 89, No. SM1, p. 115 (Feb.).

Kondner, R. L., and J. S. Zelasko, 1963. "A Hyperbolic Stress-Strain Formulation for Sands," *Proc. 2nd Pan Am. Conf. Soil Mech. Found. Eng.* (Brazil), Vol. 1 pp. 289–324.

Ladd, C. C., 1963. "Stress-Strain Behavior of Anisotropically Consolidated Clays during Undrained Loading," *Proc. 6th Inter. Conf. Soil Mech. Found. Eng.* Vol. 1, pp. 282–286.

Ladd, C. C., 1964a. "Stress-Strain Behavior of Saturated Clay and Basic Strength Principles," Research Report R64-17, Dept. Civil Eng., M.I.T. (April).

Ladd, C. C., 1964b. "Stress-Strain Modulus of Clay From Undrained Triaxial Tests," *Proc. ASCE*, Vol. 90, No. SM5 (Sept.).

Ladd, C. C., 1967. "Strength and Compressibility of Saturated Clays," Pan American Soils Course, Universidad Catolica Andres Bello, Caracas, Venezuela.

Ladd, C. C., and T. W. Lambe, 1963. "The Strength of 'Undisturbed' Clay Determined from Undrained Tests," ASTM-NRC Symposium, Ottawa.

Lambe, T. W., 1948. "The Measurement of Pore Water Pressures in Cohesionless Soils," *Proc. 2nd Inter. Conf. Soil Mech. Fluid. Eng.* (Rotterdam), Vol. VII, p. 38.

Lambe, T. W., 1950. "Capillary Phenomena in Cohesionless Soil," *ASCE*, Separate No. 4. (January).

Lambe, T. W., 1951. *Soil Testing for Engineers*, John Wiley and Sons, New York.

Lambe, T. W., 1953. "The Structure of Inorganic Soil," *Proc. ASCE*, Vol. 79, Separate No. 315 (Oct.).

Lambe, T. W., 1955. "The Permeability of Compacted Fine Grained Soils," ASTM, Special Tech. Pub. No. 163.

Lambe, T. W., 1958. "The Engineering Behavior of Compacted Clay," *J. Soil Mech. Found. Eng., ASCE* (May). [Also in *Trans. ASCE*, Vol. 125, (Part 1), p. 718 (1960).]

Lambe, T. W., 1960. "A Mechanistic Picture of Shear Strength in Clay," *Proc. ASCE Research Conf. Shear Strength of Cohesive Soils*, p. 437.

Lambe, T. W., 1961. "Residual Pore Pressures in Compacted Clay," *Proc. 5th Inter. Conf. Soil Mech. Found. Eng.* (Paris). Vol. I, p. 207.

Lambe, T. W., 1962. "Soil Stabilization," Chapter 4 of *Foundation Engineering*, G. A. Leonards (ed.), McGraw-Hill, New York.

Lambe, T. W., 1964. "Methods of Estimating Settlement," Proc. of the ASCE Settlement Conference, Northwestern University (June).

Lambe, T. W., 1963. "Pore Pressures in a Foundation Clay," *Trans of ASCE*, Vol. 128, Part I, p. 865.

Lambe, T. W., 1967. "Shallow Foundations on Clay," *Proc. of a Symposium on Bearing Capacity and Settlement of Foundations*, Duke University, Durham, N.C.

Lambe, T. W., 1968. "The Behavior of Foundations During Construction," Jour. of the SM and FE Div. Vol. 94, No. SM1, p. 93.

Lambe, T. W., and H. M. Horn, 1965. "The Influence on an Adjacent Building on Pile Driving for the M.I.T. Materials Center," *Proc. 6th Inter. Conf., Soil Mech. Found. Eng.* (Montreal), Vol. II p. 280.

Lambe, T. W., and R. Torrence Martin, 1953–1957. "Composition and Engineering Properties of Soil," *Proc. Highway Research Board* (5 papers).

Lambe, T. W., and R. V. Whitman, 1959. "The Role of Effective Stress in the Behavior of Expansive Soils," *Quarterly of the Colorado School of Mines*, Vol. 54 No. 4 (October).

Lane, K. S., and D. E. Washburn, 1946. "Capillarity Tests by Capillarimenter and by Soil Filled Tubes," *Proc. Highway Research Board.*

Lange, N. A., 1956. *Handbook of Chemistry*, Handbook Publishers, Sandusky, Ohio.

Lee, K. L., and I. Farhoomand, 1967. "Compressibility and Crushing of Granular Soil in Anisotropic Triaxial Compression," *Canadian Geotechnical Journal*, Vol. IX, No. 1, p. 68 (Feb.).

Leonards, G. A. (ed.), 1962. *Foundation Engineering*, McGraw-Hill, New York.

Leslie, D. D., 1963. "Large-Scale Triaxial Tests on Gravelly Soils," *Proc. 2nd Pan Am. Conf. Soil Mech. Found. Eng.* (Brazil), Vol. 1, pp. 181–202.

Linell, K. A., and H. F. Shea, 1960. "Strength and Deformation of Various Glacial Tills in New England," *Proc. ASCE Res. Conf. Shear Strength of Cohesive Soils* (Boulder, Col.), pp. 275–314.

Lobdell, H. L., 1959. "Rate of Constructing Embankments on Soft Ground," *Proc. ASCE*, Vol. 85, No. SM5, pp. 61–78.

Lowe, J., III, and L. Karafiath, 1960. "Stability of Earth Dams upon Drawdown," *Proc. 1st Pan-Am. Conf. Soil Mech. Found. Eng.* (Mexico), Vol. II, pp. 537–560.

Luscher, U., 1965. Discussion, *Proc. ASCE*, Vol. 91, No. SM1, pp. 190–195.

Mansur, C. I., and R. I. Kaufman, 1958: "Pile Tests, Low-Sill Structure, Old River, Louisiana," *Trans. ASCE*, Vol. 123, pp. 715–743.

Mansur, C. I., and R. I. Kaufman, 1962. "Dewatering," Chapter 3 in *Foundation Engineering*, G. A. Leonards (ed.), McGraw-Hill, New York.

Marsal, Rául J., 1959: "Unconfined Compression and Vane Shear Tests in Volcanic Lacustrine Clays," Proc. ASTM Conf. on Soils for Engineering Purposes, Mexico City.

Marsal, R. J., 1963. "Contact Forces in Soils and Rockfill Materials," *Proc. 2nd Pan Am. Conf. Soil Mech. Found. Eng.*, Vol. II, pp. 67–98.

Marsal, R. J., 1963. "Triaxial Apparatus for Testing Rockfill Materials," *Proc. 2nd Pan Am. Conf. Soil Mech. Found. Eng.* (Brazil), Vol. II, p. 99.

May, D. R., and J. H. A. Brahtz, 1936. "Proposed Methods of Calculating the Stability of Earth Dams," *Trans. 2nd Congress on Large Dams* (Washington), Vol. 4, p. 539.

McClelland, Bramlette, J. A. Focht Jr. and W. J. Emrich, 1967. "Problems in Design and Installation of Heavily Loaded Pipe Piles," Proc. ASCE Specialty Conf. on Civil Engineering in the Oceans.

McDonald, D. H., and A. W. Skempton, 1955. "A Survey of Comparisons Between Calculated and Observed Settlements of Structures on Clay," Inst. of Civil Engrs., London.

Mehta, M. R., and A. S. Veletsos, 1959. "Stresses and Displacements in Layered Systems," *Civil Engineering Studies*, Structural Research Series No. 178, Univ. of Illinois.

Meigh, A. C., and I. K. Nixon, 1961. "Comparison of *In Situ* Tests for Granular Soils," *Proc. 5th Inter. Conf. Soil Mech.* (Paris), Vol. 1, p. 499.

Meyerhof, G. G., 1951a. "The Ultimate Bearing Capacity of Foundations," *Geotechnique*, Vol. 2, pp. 301–332.

Meyerhof, G. G., 1951b. "The Tilting of a large Tank on Soft Clay," *Proc. S. Wales Inst. Civil Engrs.* Vol. 67, p. 53.

Meyerhof, G. G., 1953. "The Bearing Capacity of Foundations under Eccentric and Inclined Loads," *Proc. 3rd Inter. Conf. Soil Mech.* (Zurich), Vol. 1, pp. 440–445.

Meyerhof, G. G., 1961. "The Mechanism of Flow Slides in Cohesive Soils," *Geotechnique*, Vol. 7, pp. 41–49.

Meyerhof, G. G., 1965. "Shallow Foundations," *Proc. ASCE*, Vol. 91, No. SM2, pp. 21–31.

Meyerhof, G. G., and T. K. Chaplin, 1953. "The Compression and Bearing Capacity of Cohesive Layers," *British Journal of Applied Physics*, Vol. 4, No. 1, pp. 20–26.

Michaels, A. S., and C. S. Lin, 1954. "The Permeability of Kaolinite," *Industrial and Eng. Chem.*, Vol. 46, pp. 1239–1246 (June).

Middlebrooks, T. A., 1953. "Earth-Dam Practice in the

United States," *Trans. ASCE* (Centennial Volume), p. 697.

Mikasa, M., 1965. Discussion, *Proc. 6th Inter. Conf. Soil Mech. Found. Eng.*, Vol. III, pp. 459–460. Also see "The Consolidation of Soft Clay," *Civil Engineering in Japan*, pp. 21–26 (Japan Society of Civil Engineers, 1965).

Miller, E. T., 1963. "Stresses and Strains in an Array of Elastic Spheres," Report R53-39 by M.I.T. Department of Civil Engineering to U.S. Army Engineer Waterways Experiment Station.

Milligan, V., L. G. Soderman, and A. Rutka, 1962: "Experience with Canadian Varved Clays," Paper 3224, *J. Soil Mechanics and Foundations Division, ASCE*, Aug. 1962.

Mitchell, J. K., 1960. "Fundamental Aspects of Thixotropy in Soils," *J. Soil Mech. Found. Div.*, Vol. 86, No. SM3.

Mitchell, J. K., 1964. "Shearing Resistance of Soils as a Rate Process," *J. Soil Mech. Found. Div.*, Vol. 90, No. SM1, p. 29 (January).

Mohr, H. A., 1962. "Exploration of Soil Conditions and Sampling Operations," *Harvard Bull.*, No. 208 revised).

Moos, A. von, 1962. "Ergebnisse einiger Strassenversuchsdämme auf schlechtem Grund in der Schweiz," *Strasse und Verkehr*, No. 9.

Moran, Proctor, Mueser, and Rutledge, 1958. "Study of Deep Soil Stabilization by Vertical Sand Drains," report prepared for Bureau of Yards and Docks, Department of the Navy (June).

Moretto, O., 1948. "Effect of Natural Hardening on the Unconfined Compressive Strength of Remolded Clays," *Proc. 2nd Inter. Conf. Soil Mech. Found. Eng.* (Rotterdam) Vol. 1, p. 137.

Moretto, O., A. J. L. Bolognesi, A. O. Lopez and E. Nunez, 1963: "Propiedades Y· Comportamiento De Un Suelo Limoso De Baja Plasticidad," Proc. Second Panamerican Conf. on SM and FE. Brazil, Vol. II, p. 131.

Morgenstern, N. R., and V. E. Price, 1965. "The Analysis of the Stability of General Slip Surfaces," *Geotechnique*, Vol. 15, pp. 79–93.

Morgenstern, N., 1963. "Stability Charts for Earth Slopes During Rapid Drawdown, *Geotechnique*, Vol. 13, pp. 121–131.

Muskat, M., 1946. *The Flow of Homogeneous Fluids Through Porous Media*, J. W. Edwards, Ann Arbor, Mich.

Newmark, N. M., 1942. "Influence Charts for Computation of Stresses in Elastic Foundations," Univ. of Illinois Bulletin No. 338.

Newmark, N. M., 1964. "Effects of Earthquakes on Dams and Embankments," *Geotechnique*, Vol. 15, pp. 139–160.

Nixon, I. K., 1949. "$\phi = 0$ Analyses," *Geotechnique*, Vol. 1, pp. 208–209 and 274–276.

O'Neill, H. M., 1962. "Direct-Shear Test for Effective-Strength Parameters," *Proc. ASCE*, Vol. 88 No. SM4, pp. 109–137.

Ortigosa, P., 1968. "Densification of Sands by Vibration—2nd Progress Report," Department of Civil Engineering, M.I.T.

Osterman, J., 1959. "Notes on the Shearing Resistance of Soft Clays," *Acta Polytechnica Scandinavica*, No. 263.

Osterman, J., and G. Lindskog, 1964. "Influence of Lateral Movement in Clay Upon Settlements in Some Test Areas," Swedish Geotechnical Institute, Stockholm.

Osterman, J., and G. Lindskog, 1964. "Settlement Studies of Clay," Swedish Geotechnical Institute, Stockholm, Pub. No. 7.

Overbeek, J. Th., and E. J. W. Verwey, 1948. *Theory of the Stability of the Lyophobic Colloids*, Elsevier Pub. Co., New York.

Panama Canal, 1947. "Report of the Governor of the Panama Canal, Isthmian Canal Studies, Appendix 12: Slopes and Foundations."

Parmalee, R. A., 1967. "Building-Foundation Interaction Effects," *Proc. ASCE*, Vol. 93, No. EM2, pp. 131–152.

Parry, R. H. G., 1960: "Triaxial Compression and Extension Tests on Remolded Saturated Clay," *Geotechnique*, Vol. 10, pp. 166–180.

Peck, R. B., 1961. "Records of Load Tests on Friction Piles," Special Report No. 67, Highway Research Board.

Peck, R. B., 1967: "Stability of Natural Slopes," *Proc. ASCE*, Vol. 93, No. SM4, pp. 403–417.

Peck, R. B., and S. Berman, 1961. "Recent Practice for Foundations of High Buildings in Chicago," *Proc. Symposium of "The Design of High Buildings*," Univ. of Hong Kong (Sept.)

Peck, R. B., and F. G. Bryant, 1953. "The Bearing Capacity Failure of the Transcona Elevator," *Geotechnique*, Vol. 3, pp. 201–208.

Peck, R. B., W. E. Hanson, and T. H. Thornburn, 1953. *Foundation Engineering*, John Wiley and Sons, New York.

Peck, R. B., and T. Raamont, 1965. "Foundation Behavior of Iron Ore Storage Yards," closing discussion, *Proc. ASCE*, Vol. 91, No. SM4, pp. 193–195.

Peck, Ralph B., and Reed, William C., 1954: "Engineering Properties of Chicago Subsoils," Univ. of Illinois Experimental Station, Bull. No. 423.

Penman, A. D. M., 1961. "A Study of the Response Time of Various Types of Piezometer," *Proc. of the Conf. on Pore Pressure and Suction in Soils*, Butterworths, London, p. 53.

Pettijohn, F. J., 1949: *Sedimentary Rocks*, Harper and Brothers, New York.

Polubarinova-Kochina, 1952. *Theory of Ground Water Movement*, State Press, Moscow. Translated by R. de Wiest, Princeton University Press, Princeton, N.J., 1962

Rendulic, L., 1936. Discussion on "Relation Between Void Ratio and Effective Principal Stresses for a Remoulded Silty Clay," *Proc. 1st Inter. Conf. Soil Mech. Found. Eng.* (Cambridge), Vol. 3, pp. 48–51.

Rendulic, L., 1937. "A Fundamental Principle of Soil Mechanics and its Experimental Verification," (in German) *Bauingenieur*, Vol. 18, p. 459.

Renius, E., 1955. "The Stability of Slopes of Earth Dams," *Geotechnique*, Vol. 5, pp. 181–189.

Richardson, A. M., Jr., and R. V. Whitman, 1964. "Effect of Strain-Rate upon Undrained Shear Resistance of Saturated Remolded Fat Clay," *Geotechnique*, Vol. 13, No. 4, pp. 310–346.

Richart, F. E., Jr., 1959. "Review of the Theories for Sand Drains," *Trans. ASCE*, Vol. 124, p. 709.

Richart, F. E., Jr., 1960. "Foundation Vibrations," *Trans. ASCE*, Vol. 127, Part 1, pp. 863–898.

Richart, F. E., Jr., and R. V. Whitman, 1967. "Comparison of Footing Tests with Theory," *Proc. ASCE*, Vol. 93, No. SM6, pp. 143–168.

Road Research Laboratory, 1952. *Soil Mechanics for Road Engineers*, Her Majesty's Stationery Office, London.

Roberts, J. E., 1961. "Small-Scale Footing Studies: A Review of the Literature," Dept. of Civil Engr. Pub., M.I.T. (July).

Roberts, J. E., 1964. Sand Compression as a Factor in Oil Field Subsidence, thesis submitted in partial fulfillment of Sc.D., Dept. of Civil Eng., M.I.T. (Feb.).

Rodin, S., 1961. "Experiences with Penetrometers, with Particular Reference to the Standard Penetration Test," *Proc. 5th Inter. Conf. Soil Mech. Found. Eng.* (Paris), Vol. I, p. 517.

Roscoe, K. H., 1961. Discussion in the *Proc. 5th Inter. Conf. Soil Mech. Found. Eng.*, Vol. 3, pp. 105–107.

Roscoe, K. H., J. R. F. Arthur, and R. G. Jones, 1963. "The Determination of Strains in Soils by an X-Ray Method," *Civil Eng. Pub. Works Rev.*, Vol. 58, pp. 873–876 and 1009–1012.

Roscoe, K. H., A. N. Schofield, and A. Thurairajah, 1963. "An Evaluation of Test Data for Selecting a Yield Criterion for Soils," *Proc. Symposium on Laboratory Shear Testing of Soils*, ASTM Special Technical Publication No. 361, pp. 111–133

Roscoe, K. H., A. N. Schofield and C. P. Wroth, 1958: "On the Yielding of Soils," *Geotechnique*, Vol. 8, pp. 22–53.

Rowe, Peter W., 1952: "Anchored Sheet-Pile Walls," *Proc. Inst. Civil Engrs.*, London, Vol. I, pp. 27–70.

Rowe, P. W., 1963. "Stress-Dilatancy, Earth Pressures and Slopes " *Proc. ASCE* Vol. 89, No. SM3, pp. 37–61.

Rowe, P. W., 1962a. "Anchored Sheet-Pile Walls," *Proc. Inst. Civil Eng.*

Rowe, P. W., 1962b. "The Stress-Dilatancy Relation for Static Equilibrium of an Assembly of Particles in contact," *Proc. Roy. Soc.*, A269, pp. 500–527.

Rowe, P. W., and L. Barden, 1964. "The Importance of Free Ends in Triaxial Testing," *Proc. ASCE*, Vol. 90, No. SM1, pp. 1–27.

Rutledge, P., 1947. "Review of the Cooperative Triaxial Research Program," U.S. Army Corps of Engineers, Waterways Experiment Station.

Salas, J. A. Jimenez, and J. M. Serratosa, 1953: "Compressibility of Clays," *Proc. 3rd ICSMFE*, Vol. I, p. 192.

Saurin, B. F., 1948. "Discussion of $\phi = 0$ Analyses," *Geotechnique*, Vol. 1, pp. 272–274.

Scheidegger, A. E., 1957. *The Physics of Flow Through Porous Media*, Macmillan, New York.

Schiffman, R. L., 1958. "Consolidation of Soil under Time-Dependent Loading and Variable Permeability," *Proc. Highway Research Board*, Vol. 37, p. 584.

Schiffman, R. L., A. Chen, and J. C. Jordan, 1967. "The Consolidation of a Half Plane," U. of Illinois (Chicago Circle) MATE Report 67-3.

Schiffman, R. L., and R. E. Gibson, 1964. "Consolidation of Non-Homogeneous Clay Layers," *Proc. ASCE*, Vol. 90, No. SM5, pp. 1–30.

Schmertmann, J. M., 1955. "The Undisturbed Consolidation of Clay," *Trans. ASCE*, Vol. 120, p. 1201.

Schultz, E. and H. Knausenberger, 1957. "Experiences with Penetrometers," *Proc. 4th Inter. Conf. Soil Mech. Found Eng.* (London), Vol. I, p. 249.

Schultz, E. and E. Menzenbach, 1961. "Standard Penetration Test and Compressibility of Soils," *Proc. 5th Inter. Conf. Soil Mech. Found. Eng.* (Paris), Vol. I, p. 527.

Scott, R. F., 1963. *Principles of Soil Mechanics*, Addison-Wesley Publishing Co., Reading, Mass.

Seaman, L., G. N. Bycroft, and H. W. Kriebel, 1963. "Stress Propagation in Soils—Part III," Report by Stanford Research Institute to Defense Atomic Support Agency, DASA-1266-3.

Seed, H. B., 1966a. "A Method for the Earthquake-Resistant Design of Earth Dams," *Proc. ASCE*, Vol. 92, No. SM1, pp. 13–41.

Seed, H. B., 1966b. "Soil Stability Problems Caused by Earthquakes," Report by Soil Mechanics and Bituminous Materials Laboratory, University of California, Berkeley.

Seed, H. B., and C. K. Chan, 1959. "Structure and Strength Characteristics of Compacted Clays," *J. Soil. Mech. Found. Div. ASCE*, Vol. 85, No. SM5 (October).

Seed, H. B., and C. K. Chan, 1961. "Effect of Duration of Stress Application on Soil Deformation under Repeated Loading," *Proc. 5th Inter. Conf. Soil Mech. Found. Eng.* (London), Vol. 1, pp. 341–345.

Seed, H. B., and R. E. Goodman, 1964. "Earthquake Stability of Slopes of Cohesionless Soils," *Proc. ASCE*, Vol. 90, No. SM6, pp. 43–73.

Seed, H. B., and I. M. Idriss, 1967. "Analysis of Liquefaction: Niigata Earthquake," *Proc. ASCE*, Vol. 93, No. SM3, pp. 83–108.

Seed, H. B., and K. L. Lee, 1966. "Liquefaction of Saturated Sands During Cyclic Loading," *Proc. ASCE*, Vol. 92, No. SM6, pp. 105–134.

Seed, H. B., and L. C. Reese, 1957. "The Action of Soft Clay Along Friction Piles," *Trans. ASCE*, Vol. 22, p. 731.

Seed, H. B., and H. A. Sultan, 1967. "Stability Analysis for a Sloping Core Embankment," *Proc. ASCE*, Vol. 93, No. SM4, pp. 69–84.

Seed, H. B., and S. D. Wilson, 1967. "The Turnagain Heights Landslide, Anchorage, Alaska," *Proc. ASCE*, Vol. 93, No. SM4, pp. 325–353.

Seed, H. B., J. R. Woodward, and R. Lundgren, 1964. "Clay Mineralogical Aspects of the Atterberg Limits," *J. Soil Mech. Found. Div., ASCE*, Vol. 90, No. SM4.

Sevaldson, R. A., 1956. "The Slide in Lodalen, October 6th, 1954," *Geotechnique*, Vol. 6, pp. 1–16.

Shannon, W. L., 1966. "Slope Failures at Seward, Alaska," paper presented to ASCE Soil Mechanics and Foundations Division Conference on Stability and Performance of Slopes and Embankments, Berkeley, Cal.

Shannon, W. L., S. D. Wilson and R. M. Meese, 1962: "Field Problems: Field Measurements," *Foundation Engineering* McGraw-Hill Book Co., pp. 1025–1080.

Shannon, W. L., G. Yamane, and R. J. Dietrich, 1959. "Dynamic Triaxial Tests on Sands," *Proc. 1st Pan Am. Conf. Soil Mech. Found. Eng.*, Vol. 1, pp. 473–489.

Sherard, J. L., 1967: "Earthquake Considerations in Earth Dam Design," *Proc. ASCE*, Vol. 93, No. SM4, pp. 377–401.

Sherard, J. L., R. J. Woodward, S. G. Gizienski, and W. A. Clevenger, 1963. *Earth and Earth Rock Dams*, John Wiley and Sons, New York.

Simons, N. E., 1965. "Consolidation Investigation on Undisturbed Fornebu Clay" Norwegian Geotechnical Institute Pub. No. 62, Oslo.

Skempton, A. W., 1942. "An Investigation of the Bearing Capacity of a Soft Clay Soil," *J. Inst. Civil Engrs.*, Vol. 18, p. 307.

Skempton, A. W., 1948. "The $\phi = 0$ Analysis for Stability and its Theoretical Basis," *Proc. 2nd Inter. Conf. Soil Mech. Found. Eng.* (Rotterdam), Vol. 1, p. 72.

Skempton, A. W., 1948. "A Study of the Geotechnical Properties of Some Post-Glacial Clays," *Geotechnique*, Vol. 1,. p, 7.

Skempton, A. W., 1951. "The Bearing Capacity of Clays," Bldg. Research Congress, England.

Skempton, A. W., 1953. "The Colloidal Activity of Clays," *Proc. 3rd Inter. Conf. Soil Mech. Found. Eng.* (Switzerland), Vol. I, p. 57.

Skempton, A. W., 1953. "Soil Mechanics in Relation to Geology," *Proc. of Yorkshire Geological Soc.*, Vol. 29, p. 33.

Skempton, A. W., 1954. "The Pore-Pressure Coefficient *A* and *B*," *Geotechnique*, Vol. 4, pp. 143–147.

Skempton, A. W., 1961. "Effective Stress in Soils, Concrete and Rocks," *Pore Pressure and Suction in Soils*, Butterworths, London, p. 4.

Skempton, A. W., 1964. "Long Term Stability of Clay Slopes," *Geotechnique*, Vol. 14, p. 77.

Skempton, A. W., 1966. "Large Bored Piles—Summing Up," Symposium on Large Bored Piles, Inst. of Civil Engrs. and Reinf. Conc. Assoc., London (Feb.).

Skempton, A. W., and L. Bjerrum, 1957. "A Contribution to the Settlement Analysis of Foundations on Clay," *Geotechnique*, Vol. 7, p. 168.

Skempton, A. W., and J. D. Brown, 1961. "A Landslip in Boulder Clay at Selset, Yorkshire," *Geotechnique*, Vol. 11, pp. 280–293.

Skempton, A. W., and H. Q. Golder, 1948. "Practical Examples of the $\phi = 0$ Analysis of Stability of Clays," *Proc. 2nd Inter. Conf. Soil Mech. Found. Eng.* (Rotterdam), Vol. 2, p. 63.

Skempton, A. W., and D. J. Henkel. 1953. "The Post-Glacial Clays of the Thames Estuary at Tilbury and Shellhaven," *Proc. 3rd Inter. Conf. Soil Mech. Found. Eng.* (Switzerland), Vol. I, p. 302.

Skempton, A. W., and D. J. Henkel, 1957. "Tests on London Clay from Deep Borings at Paddington, Victoria and the South Bank," *Proc. 4th Inter. Conf. Soil Mech. Found. Eng.* (London), p. 100.

Skempton, A. W., and R. D. Northey, 1952. "The Sensitivity of Clays," *Geotechnique* Vol. III, pp. 30–53.

Skempton, A. W., R. B. Peck, and D. H. McDonald, 1955. "Settlement Analyses of Six Structures in Chicago and London," *Proc. Inst. Civil. Engrs.* (July).

Skempton, A. W., and V. A. Sowa, 1963. "The Behavior of Saturated Clays During Sampling and Testing," *Geotechnique*, Vol. 13, No. 4, pp. 269–290.

Soderman, L. G., and R. M. Quigley, 1965. "Geotechnical Properties of Three Ontario Clays," *Canadian Geotechnical Journal*, Vol. II, No. 2 (May).

Sokolovski, V. V., 1965. *Statics of Granular Media.* Translated from the Russian by J. K. Luscher, Pergamon Press, London.

Sowers, G. F., 1962. "Shallow Foundations," *Foundation Engineering*, G. A. Leonards (ed.), McGraw-Hill, New York, p. 525.

Sowers, G. F., 1963. "Engineering Properties of Residual Soils Derived from Igneous and Metamorphic Rocks," *Proc. 2nd Pan-Am Conf. Soil Mech. Found Engr.* (Brazil), Vol. 1 p. 39.

Sowers, G. B., and B. F. Sowers, 1951. *Introductory Soil Mechanics and Foundations*, Macmillan, New York.

Steinbrenner, W., 1934. "Tafeln zur Setzungsberechnung," *Die Strasse*, Vol. 1, pp. 121–124. See also *Proc. 1st Intern. Conf. Found. Eng.* (Cambridge), Vol. 2, pp. 142–143 1936.

Sultan, H. A., and H. B. Seed, 1967. "Stability of Sloping Core Earth Dams," *Proc. ASCE*, Vol. 93, No. SM4, pp. 45–68.

Szechy, C., 1961. *Foundation Failures*, Concrete Publications Ltd., London.

Taylor, D. W., 1937. "Stability of Earth Slopes," *J. Boston Soc. Civil Engrs.*, Vol. 24, p. 197.

Taylor, D. W., 1939. "A Comparison of Results of Direct Shear and Cylindrical Compression Tests," *Proc. ASTM.*

Taylor, D. W., 1942. "Research on Consolidation of Clays," M.I.T. (August).

Taylor, D. W., 1945. "Review of Pressure Distribution Theories, Earth Pressure Cell Investigations and Pressure Distribution Data," Report to U.S. Army Waterways Experiment Station.

Taylor, D. W., 1948. *Fundamentals of Soil Mechanics*, John Wiley and Sons, New York.

Taylor, D. W., 1952. "A Direct Shear Test with Drainage Control," Symposium on Direct-Shear Testing of Soils, ASTM Spec. Tech. Pub. No. 131, pp. 63–74.

Taylor, D. W., 1955. "Review of Research on Shearing Strength of Clay at M.I.T.: 1948–1953," Report to U.S. Army Corps of Engineers Waterways Experiment Station.

Teng, W. C., 1962. *Foundation Design*, Prentice-Hall, Englewood Cliffs, N.J.

Terracina, F., 1962. "Foundations of the Tower of Pisa," *Geotechnique*, Vol. 12, p. 336.

Terzaghi, K., 1925. *Erdbaumechanik*, Franz Deuticke, Vienna.

Terzaghi, K., 1934. "Large Retaining Wall Test," *Engineering News Record* (Feb. 1, 22; March 8, 20; April 19).

Terzaghi, K., 1943. *Theoretical Soil Mechanics*, John Wiley and Sons, New York.

Terzaghi, K., 1960. *From Theory to Practice in Soil Mechanics*, John Wiley and Sons, New York.

Terzaghi, K., and R. B. Peck, 1967. *Soil Mechanics in Engineering Practice*, 2nd ed. John Wiley and Sons, New York. The first edition was published in 1948.

Timoshenko, S., 1934. *Theory of Elasticity*, McGraw-Hill, New York.

Trinkunas, J., 1967. "Pilotes de Gran Peso y Longitud en Suelos Cohesivos Saturados," *Proc. 3rd Pan-Am. Conf. Soil Mech. Found. Eng.* (Venezuela), Vol. 1, p. 633.

Trollope, D. H., McD. Freeman, and G. M. Peck, 1966. "Tasman Bridge Foundations," *J. Inst. Engrs.* (Australia) (June).

Tschebotarioff, G., 1951. *Soil Mechanics, Foundations, and Earth Structures*, McGraw-Hill, New York, pp. 226–231.

Turnbull, W. J., 1950. "Compaction and Strength Tests on Soil," presented at Annual Meeting ASCE (January).

Turnbull, W. J., A. A. Maxwell, and R. G. Ahlvin, 1961. "Stresses and Reflections on Homogeneous Soil Masses," *Proc. 5th Inter. Conf. Soil Mech. Found. Eng.*, Vol. 2, p. 337.

Twenhofel, W. H., 1939. *Principles of Sedimentation*, McGraw-Hill, New York.

U.S. Navy, 1962. Design Manual—Soil Mechanics, Foundations and Earth Structures, NAVDOCKS DM-7.

Van Olphen H., 1963. *An Introduction To Clay Colloid Chemistry*, John Wiley and Sons, New York.

Van Weele, A. F., 1964. "Negative Skin-Friction on Pile Foundations in Holland," *Proc. Symposium on Bearing Capacity of Piles*, Roorkee, India, pp. 1–10.

Vargas, Milton, 1953. "Some Properties of Residual Clay Soils Occuring in Southern Brazil," *Proc. Inter. Conf. Soil Mech. Found. Eng.* (Switzerland), Vol. I, p. 67.

Vesic, A. S., 1963. "Bearing Capacity of Deep Foundations in Sand," *Highway Research Board Record*, No. 39.

Vesic, A. S., 1967a. "A Study of Bearing Capacity of Deep Foundations," Georgia Inst. of Technology, Atlanta (March).

Vesic, A. S., 1967b. "Ultimate Loads and Settlements of Deep Foundations in Sand," *Proc. Symposium on Bearing Capacity and Settlement of Foundations*, Duke Univ., Durham, N.C., p. 53.

Wagner, A. A., 1957. "The Use of the Unified Soil Classification System by the Bureau of Reclamation," *Proc. 4th Inter. Conf. Soil Mech. Found. Eng.* (London), Vol. I, p. 125.

Wallace, M. I., 1948. Experimental Investigation of the Effect of Degree of Saturation on the Permeability of Sand S.M. thesis, Dept. of Civil Engr., M.I.T., Cambridge, Mass.

Ward, W. H., A. Penman, and R. E. Gibson, 1955. "Stability of a Bank on a Thin Peat Layer," *Geotechnique*, Vol. 5, pp. 154–163.

Ward, W. H., S. G. Samuels, and Muriel E. Butler, 1959. "Further Studies of the Properties of London Clay," *Geotechnique*, Vol. 9, p. 33.

Westergaard, H. M., 1938. "A Problem of Elasticity Suggested by a Problem in Soil Mechanics: A Soft Material Reinforced by Numerous Strong Horizontal Sheets," *Mechanics of Solids*, S. Timoshenko Sixtieth Anniversary Volume, Macmillan, New York.

Whitaker, T., and R. W. Cooke, 1966. "An Investigation of the Shaft and Base Resistances of Large Bored Piles in London Clay," Symposium on Large Bored Piles, Inst. of Civil Engrs. and Reinf. Conc. Assoc., London (Feb.).

Whitman, R. V., 1957. "The Behavior of Soils Under Transient Loadings," *4th Inter. Conf. Soil Mech. Found. Eng.* (London) Vol. 1, p. 207.

Whitman, R. V., 1960. "Some Considerations and Data Regarding the Shear Strength of Clays," *Proc. ASCE Res. Conf. Shear Strength of Cohesive Soils* (Boulder, Co.), pp. 581–614.

Whitman, R. V., 1963. "Stress-Strain-Time Behavior of Soil in One-Dimensional Compression," Report R63-25 by M.I.T. Department of Civil Engineering to U.S. Army Engineer Waterways Experiment Station, 1963.

Whitman, R. V., 1966. "Analysis of Foundation Vibrations," *Vibrations in Civil Engineering*, B. O. Skipp (ed.), Butterworths, London, pp. 159–179.

Whitman, R. V., and W. A. Bailey, 1967. "Use of Computers for Slope Stability Analysis," *Proc. ASCE*, Vol. 93, No. SM4, pp. 475–498.

Whitman, R. V., Z. Getzler, and K. Hoeg, 1963. "Texts upon Thin Domes Buried in Sand," *J. Boston Soc. Civil Eng.*, Vol. 50, pp. 1–22 (January).

Whitman, R. V., and K. E. Healy, 1962. "Shear Strength of Sands During Rapid Loading," *Proc. ASCE*, Vol. 88, No. SM2, pp. 99–132.

Whitman, R. V., and K. A. Healy, 1963. "Shear Strength of Sands During Rapid Loadings," *Trans. ASCE*, Vol. 128, pp. 1553–1594.

Whitman, R. V., and K. Hoeg, 1966. "Development of Plastic Zone Beneath a Footing," Report by M.I.T. Dept. of Civil Eng. to U.S. Army Eng. Waterways Experiment Station.

Whitman, R. V., and F. V. Lawrence, Jr., 1963. Discussion to paper by Hardin and Richart, *Proc. ASCE*, Vol. 89, No. SM5, pp. 112–118.

Whitman, R. V., E. T. Miller, and P. J. Moore, 1964. "Yielding and Locking of Confined Sand," *J. ASCE*, Vol. 90, No. SM4, pp. 57–84.

Whitman, R. V., and P. J. Moore, 1963. "Thoughts Concerning the Mechanics of Slope Stability Analysis," *Proc. 2nd Pan-Am. Conf. Soil Mech. Found. Eng.* (Brazil), Vol. I, pp. 391–411.

Whitman, R. V., A. M. Richardson, and K. A. Healy, 1961. "Time-Lags in Pore Pressure Measurements," *5th Intr. Conf. Soil Mech. Found. Eng.* (Paris). Vol. 1 p. 407.

Whitman, R. V., and F. E. Richart, Jr., 1967. "Design Procedures for Dynamically Loaded Foundations," *Proc. ASCE*, Vol. 93, No. SM6, pp. 169–193.

Wissa, A. E. Z., 1961. A Study of the Effects of Environmental Changes on the Stress-Strain Properties of Kaolinite, S.M. thesis, Department of Civil Engineering, M.I.T., Cambridge, Mass.

Wolfskill, L. A., and T. W. Lambe, 1967. "Slide in the Siburua Dam," *Proc. ASCE*, Vol. 93, No. SM4, pp. 107–133.

Zeevaert, L., 1953. "Pore Pressure Measurements to Investigate the Main Source of Subsidence in Mexico City," *Proc. 3rd Inter. Conf. Soil Mech. Found. Eng.* (Switzerland) Vol. II., p. 299.

INDEX

Numbers in boldface type indicate pages on which entries are defined.